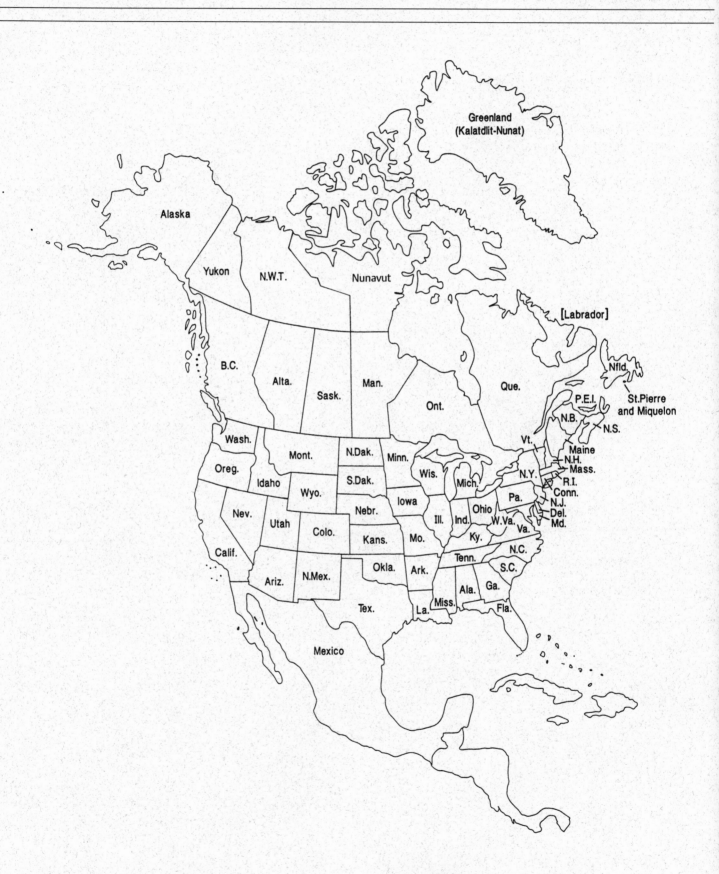

Flora of North America

Contributors to Volumes 19–21

Geraldine A. Allen
Justin W. Allison
Loran C. Anderson
Susan J. Bainbridge
Gary I. Baird
Bruce G. Baldwin
Theodore M. Barkley†
Randall J. Bayer
Mark W. Bierner
A. Linn Bogle
David J. Bogler
Kristin R. Brodeur
Luc Brouillet
Gregory K. Brown
Judith M. Canne-Hilliker
Gerald D. Carr
Robert L. Carr
Kenton L. Chambers
Raymund Chan
Donna M. Cherniawsky
Jerry G. Chmielewski
Curtis Clark
W. Dennis Clark
Jennifer A. Clevinger
Rachel E. Cook
Patricia B. Cox
W. S. Davis
Robert D. Dorn
A. Michele Funston
L. D. Gottlieb

Arthur Haines
Neil A. Harriman
Elizabeth M. Harris
Ronald L. Hartman
Walter C. Holmes
Phillip E. Hyatt
John P. Janovec
Dale E. Johnson
Vesna Karaman-Castro
Brian R. Keener
David J. Keil
Robert W. Kiger
John C. La Duke
Eric E. Lamont
Michelle R. Leonard
Timothy K. Lowrey
Alison McKenzie Mahoney
Staci Markos
John S. Mooring
James D. Morefield
David R. Morgan
Caleb A. Morse
David F. Murray
Amy Trauth Nare
Guy L. Nesom
Kurt M. Neubig
Jörg Ochsmann
Robert J. O'Kennon
Robert Ornduff†
James C. Parks†

Thomas F. Patterson
Donald J. Pinkava
A. Michael Powell
Robert E. Preston
Roland P. Roberts
Edward E. Schilling
Randall W. Scott
John C. Semple
Leila M. Shultz
Beryl B. Simpson
Kunsiri Chaw Siripun
Alan R. Smith
Pamela S. Soltis
David M. Spooner
Phyllis Spurr
John L. Strother
Scott D. Sundberg†
Debra K. Trock
Billie L. Turner
Matt W. Turner
Lowell E. Urbatsch
Linda E. Watson
William A. Weber
Ronald R. Weedon
Mark A. Wetter
Molly A. Whalen
R. David Whetstone
Dieter H. Wilken
Steven J. Wolf
Sharon C. Yarborough

Editors for Volumes 19–21

Theodore M. Barkley†,
Lead and Taxon Editor

Kanchi Gandhi,
Nomenclatural Editor

Luc Brouillet,
Taxon Editor

Robert W. Kiger,
Bibliographic Editor

Helen Jeude,
Technical Editor

Kay Yatskievych,
Managing Editor

John L. Strother,
Taxon Editor

James L. Zarucchi,
Editorial Director

Volume Composition

Pat Harris

Martha J. Hill

Ruth T. King

Kristin Pierce

Liatris cylindracea

Flora of North America

North of Mexico

Edited by FLORA OF NORTH AMERICA EDITORIAL COMMITTEE

VOLUME 21

Magnoliophyta: Asteridae, part 8: Asteraceae, part 3

ASTERALES, part 3 (Aster order)

NEW YORK OXFORD · OXFORD UNIVERSITY PRESS · 2006

Oxford University Press, Inc., publishes works that further
Oxford University's objective of excellence
in research, scholarship, and education.

Oxford New York
Auckland Cape Town Dar es Salaam Hong Kong Karachi
Kuala Lumpur Madrid Melbourne Mexico City Nairobi
New Delhi Shanghai Taipei Toronto

With offices in
Argentina Austria Brazil Chile Czech Republic France Greece
Guatemala Hungary Italy Japan Poland Portugal Singapore
South Korea Switzerland Thailand Turkey Ukraine Vietnam

Copyright © 2006 by Flora of North America Association

Published by Oxford University Press, Inc.
198 Madison Avenue, New York, New York 10016
http://www.oup.com

Oxford is a registered trademark of Oxford University Press

Library of Congress Cataloging-in-Publication Data
(Revised for volumes 19–21)
Flora of North America north of Mexico
edited by Flora of North America Editorial Committee.
Includes bibliographical references and indexes.
Contents: v. 1. Introduction—v. 2. Pteridophytes and gymnosperms—
v. 3. Magnoliophyta: Magnoliidae and Hamamelidae—
v. 22. Magnoliophyta: Alismatidae, Arecidae, Commelinidae (in part), and Zingiberidae—
v. 26. Magnoliophyta: Liliidae: Liliales and Orchidales—
v. 23. Magnoliophyta: Commelinidae (in part): Cyperaceae—
v. 25. Magnoliophyta: Commelinidae (in part): Poaceae, part 2—
v. 4. Magnoliophyta: Caryophyllidae (in part): part 1—
v. 5. Magnoliophyta: Caryophyllidae (in part): part 2—
v. 19. Magnoliophyta: Asteridae (in part): Asteraceae, part 1
v. 20. Magnoliophyta: Asteridae (in part): Asteraceae, part 2
v. 21. Magnoliophyta: Asteridae (in part): Asteraceae, part 3

ISBN-13: 978-0-19-530565-4 (v. 21)
ISBN-10: 0-19-530565-4 (v. 21)
1. Botany—North America.
2. Botany—United States.
3. Botany—Canada.
I. Flora of North America Editorial Committee.
QK110.F55 2002   581.97   92-30459

Printing number: 9 8 7 6 5 4 3 2 1
Printed in the United States of America
on acid-free paper

Contents

FOUNDING MEMBER INSTITUTIONS

Flora of North America Association

Arnold Arboretum
Jamaica Plain, Massachusetts

Agriculture and Agri-Food Canada
Ottawa, Ontario

Canadian Museum of Nature
Ottawa, Ontario

Carnegie Museum of Natural
 History
Pittsburgh, Pennsylvania

Field Museum of Natural History
Chicago, Illinois

Fish and Wildlife Service
United States Department of the
 Interior
Washington, D.C.

Harvard University Herbaria
Cambridge, Massachusetts

Hunt Institute for Botanical
 Documentation
Carnegie Mellon University
Pittsburgh, Pennsylvania

Jacksonville State University
Jacksonville, Alabama

Jardin Botanique de Montréal
Montréal, Québec

Kansas State University
Manhattan, Kansas

Missouri Botanical Garden
St. Louis, Missouri

New Mexico State University
Las Cruces, New Mexico

New York State Museum
Albany, New York

Northern Kentucky University
Highland Heights, Kentucky

The New York Botanical Garden
Bronx, New York

The University of British Columbia
Vancouver, British Columbia

The University of Texas
Austin, Texas

Université de Montréal
Montréal, Québec

University of Alaska Fairbanks
Fairbanks, Alaska

University of Alberta
Edmonton, Alberta

University of California
Berkeley, California

University of California
Davis, California

University of Idaho
Moscow, Idaho

University of Illinois
Urbana-Champaign, Illinois

University of Iowa
Iowa City, Iowa

The University of Kansas
Lawrence, Kansas

University of Michigan
Ann Arbor, Michigan

University of Oklahoma
Norman, Oklahoma

University of Ottawa
Ottawa, Ontario

University of Southwestern
 Louisiana
Lafayette, Louisiana

University of Western Ontario
London, Ontario

University of Wyoming
Laramie, Wyoming

Utah State University
Logan, Utah

*For their support of the preparation of this volume,
we gratefully acknowledge and thank:*

Chanticleer Foundation

National Science Foundation
(Award DEB-0206645)

The Fairweather Foundation

ChevronTexaco

ESRI

*Project Staff — past and present
involved with the preparation of Volumes 19–21*

Justin W. Allison, *Administrative Assistant*
Barbara Alongi, *Illustrator*
Michael Blomberg, *Scanning Specialist*
Trisha K. Consiglio, *GIS Analyst*
Shellie Davis Eldredge, *Editorial Assistant*
Sheila Flinchpaugh, *Illustrator*
Bee F. Gunn, *Illustrator*
Pat Harris, *Editorial Assistant and Compositor*
Linny Heagy, *Illustrator*
Claire A. Hemingway, *Technical Editor*
Martha J. Hill, *Technical Editor and Compositor*
Helen Jeude, *Senior Technical Editor*
Fred Keusenkothen, *Scanning Supervisor*
Ruth T. King, *Editorial Assistant and Compositor*
Marjorie C. Leggitt, *Illustrator*
Asha McElfish, *Editorial Volunteer*
John Myers, *Illustrator and Illustration Compositor*
Amy Trauth Nare, *NSF Summer Intern*
Guy L. Nesom, *Staff Botanist*
Kristin Pierce, *Editorial Assistant and Compositor*
Heidi H. Schmidt, *Editorial Assistant*
Hong Song, *Programmer*
Linda Ann Vorobik, *Illustrator*
Yevonn Wilson-Ramsey, *Illustration Coordinator and Illustrator*
George Yatskievych, *Technical Adviser*

Contributors to Volumes 19–21

Geraldine A. Allen
University of Victoria
Victoria, British Columbia

Justin W. Allison
Botanical Research Institute of
  Texas
Fort Worth, Texas

Loran C. Anderson
Florida State University
Tallahassee, Florida

Susan J. Bainbridge
University of California
Berkeley, California

Gary I. Baird
Brigham Young University-Idaho
Rexburg, Idaho

Bruce G. Baldwin
University of California
Berkeley, California

Theodore M. Barkley†
Botanical Research Institute of
  Texas
Fort Worth, Texas

Randall J. Bayer
Australian National Herbarium
Centre for Plant Biodiversity
  Research
Canberra, Australia

Mark W. Bierner
Boyce Thompson Arboretum
Superior, Arizona

A. Linn Bogle
University of New Hampshire
Durham, New Hampshire

David J. Bogler
Missouri Botanical Garden
St. Louis, Missouri

Kristin R. Brodeur
Oxford, Alabama

Luc Brouillet
Université de Montréal
Montréal, Québec

Gregory K. Brown
University of Wyoming
Laramie, Wyoming

Judith M. Canne-Hilliker
University of Guelph
Guelph, Ontario

Gerald D. Carr
Oregon State University
Corvallis, Oregon

Robert L. Carr
Eastern Washington University
Cheney, Washington

Kenton L. Chambers
Oregon State University
Corvallis, Oregon

Raymund Chan
University of Hawaii at Manoa
Honolulu, Hawaii

Donna M. Cherniawsky
The Provincial Museum of Alberta
Edmonton, Alberta

Jerry G. Chmielewski
Slippery Rock University
Slippery Rock, Pennsylvania

Curtis Clark
California State Polytechnic
  University
Pomona, California

W. Dennis Clark
Arizona State University
Tempe, Arizona

Jennifer A. Clevinger
Walsh University
North Canton, Ohio

Rachel E. Cook
Chicago Botanic Garden
Glencoe, Illinois

Patricia B. Cox
TVA Heritage Program
Knoxville, Tennessee

W. S. Davis
Louisville, Kentucky

Robert D. Dorn
Lingle, Wyoming

A. Michele Funston
Missouri Botanical Garden
St. Louis, Missouri

L. D. Gottlieb
Ashland, Oregon

Arthur Haines
*New England Wild Flower Society
Garden in the Woods
Framingham, Massachusetts*

Neil A. Harriman
*University of Wisconsin-Oshkosh
Oshkosh, Wisconsin*

Elizabeth M. Harris
*University of Central Florida
Orlando, Florida*

Ronald L. Hartman
*University of Wyoming
Laramie, Wyoming*

Walter C. Holmes
*Baylor University
Waco, Texas*

Philip E. Hyatt
*USDA Forest Service
Atlanta, Georgia*

John P. Janovec
*Botanical Research Institute of
Texas
Fort Worth, Texas*

Dale E. Johnson
*Timber Press
Portland, Oregon*

Vesna Karaman-Castro
*Louisiana State University
Baton Rouge, Louisiana*

Brian R. Keener
*The University of Alabama
Tuscaloosa, Alabama
The University of West Alabama
Livingston, Alabama*

David J. Keil
*California Polytechnic State
University
San Luis Obispo, California*

Robert W. Kiger
*Carnegie Mellon University
Pittsburgh, Pennsylvania*

John C. La Duke
*University of North Dakota
Grand Forks, North Dakota*

Eric E. Lamont
*The New York Botanical Garden
Bronx, New York*

Michelle R. Leonard
Sudbury, Ontario

Timothy K. Lowrey
*University of New Mexico
Albuquerque, New Mexico*

Alison McKenzie Mahoney
*Minnesota State University-
Mankato
Mankato, Minnesota*

Staci Markos
*University of California
Berkeley, California*

John S. Mooring
*Santa Clara University
Santa Clara, California*

James D. Morefield
*Department of Conservation
& Natural Resources
Nevada Natural Heritage Program
Carson City, Nevada*

David R. Morgan
*University of West Georgia
Carrollton, Georgia*

Caleb A. Morse
*The University of Kansas
Lawrence, Kansas*

David F. Murray
*University of Alaska,
Museum of the North
Fairbanks, Alaska*

Amy Trauth Nare
Fort Worth, Texas

Guy L. Nesom
*Botanical Research Institute of
Texas
Fort Worth, Texas*

Kurt M. Neubig
*Florida Museum of Natural
History
Gainesville, Florida*

Jörg Ochsmann
Göttingen, Germany

Robert J. O'Kennon
*Botanical Research Institute of
Texas
Fort Worth, Texas*

Robert Ornduff†
*University of California
Berkeley, California*

James C. Parks†
*Millersville University
Millersville, Pennsylvania*

Thomas F. Patterson
*South Texas Community College
McAllen, Texas*

Donald J. Pinkava
*Arizona State University
Tempe, Arizona*

A. Michael Powell
*Sul Ross State University
Alpine, Texas*

Robert E. Preston
*Jones & Stokes
Sacramento, California*

Roland P. Roberts
*Towson University
Towson, Maryland*

Edward E. Schilling
*University of Tennessee
Knoxville, Tennessee*

Randall W. Scott
*Northern Arizona University
Flagstaff, Arizona*

John C. Semple
*University of Waterloo
Waterloo, Ontario*

Leila M. Shultz
Utah State University
Logan, Utah

Beryl B. Simpson
University of Texas
Austin, Texas

Kunsiri Chaw Siripun
Kasetsart University
Kamphaengsaen Campus, Nakhon
Pathom, Thailand

Alan R. Smith
University of California
Berkeley, California

Pamela S. Soltis
University of Florida
Gainesville, Florida

David M. Spooner
University of Wisconsin-Madison
Madison, Wisconsin

Phyllis L. Spurr
University of Northern Kentucky
Highland Heights, Kentucky

John L. Strother
University of California, Berkeley
Berkeley, California

Scott D. Sundberg†
Oregon State University
Corvallis, Oregon

Debra K. Trock
California Academy of Sciences
San Francisco, California

Billie L. Turner
The University of Texas at Austin
Austin, Texas

Matt W. Turner
Austin, Texas

Lowell E. Urbatsch
Louisiana State University
Baton Rouge, Louisiana

Linda E. Watson
Miami University
Oxford, Ohio

William A. Weber
University of Colorado Museum
Boulder, Colorado

Ronald R. Weedon
Chadron State College
Chadron, Nebraska

Mark A. Wetter
University of Wisconsin-Madison
Madison, Wisconsin

Molly A. Whalen
Flinders University
Adelaide, Australia

R. David Whetstone
Jacksonville State University
Jacksonville, Alabama

Dieter H. Wilken
Santa Barbara Botanic Garden
Santa Barbara, California

Steven J. Wolf
California State University
Stanislaus
Turlock, California

Sharon C. Yarborough
Yerington, Nevada

Reviewers

Ray Angelo
New England Botanical Club
Cambridge, Massachusetts

Susan J. Bainbridge
University of California
Berkeley, California

Bruce Bennett
NatureServe Yukon
Whitehorse, Yukon

David E. Boufford
Harvard University Herbaria
Cambridge, Massachusetts

Larry E. Brown
Spring Branch Science Center
  Herbarium
Houston, Texas

Kenton L. Chambers
Oregon State University
Corvallis, Oregon

Tom S. Cooperrider
Kent State University
Kent, Ohio

Craig C. Freeman
The University of Kansas
Lawrence, Kansas

Arthur Haines
New England Wild Flower Society
Garden in the Woods
Framingham, Massachusetts

Robert Kral
Cairo, Georgia

David F. Murray
University of Alaska,
  Museum of the North
Fairbanks, Alaska

Richard W. Spellenberg
New Mexico State University
Las Cruces, New Mexico

Edward G. Voss
University of Michigan Herbarium
Ann Arbor, Michigan

Dieter H. Wilken
Santa Barbara Botanic Garden
Santa Barbara, California

Richard P. Wunderlin
University of South Florida
Tampa, Florida

Acknowledgments

Members of the Flora of North America Association (FNAA), especially those involved in the preparation and production of the three Asteraceae volumes, extend special and heartfelt gratitude to:

Sy Sohmer, Director of the Botanical Research Institute of Texas (BRIT), for inviting Dr. Barkley and the Asteraceae project to locate there;

The Botanical Institute of Texas and staff for warmly and graciously welcoming and housing the Flora of North America (FNA) Editorial Center at BRIT;

Barney Lipscomb, editor of *Sida*, who went to incredible lengths to make sure all new species and combinations were published promptly, ensuring that these taxa could be included in the Asteraceae volumes;

Guy Nesom, BRIT staff botanist and part-time botanist for FNA, who provided many insights concerning numerous treatments, helped with many questions, and provided additional botanical support for the technical staff at BRIT and elsewhere;

Claire A. Hemingway, Mary Ann Schmidt, and Kay Yatskievych for spirited and rewarding discussions regarding finer points of editing and style; and

Justin W. Allison, BRIT Administrative Assistant, for his steadfast competence in communications, statistics, and tracking.

In addition, we appreciate the efforts of many individuals who provided advice, corrections, and support for the Asteraceae team. Some are members of FNAA; they went well beyond their normal jobs in providing extra aid and encouragement. They include: Bruce G. Baldwin, Richard G. Beidleman, Frédéric Coursol, Kanchi Gandhi, Werner Greuter, Vernon L. Harms, Larry Hufford, Philip Jenkins, Robert W. Kiger, Leslie R. Landrum, John McNeill, Sue Meades, David F. Murray, Dan H. Nicolson, Mike Oldham, Jackie M. Poole, John F. Pruski, Alan R. Smith, Richard W. Spellenberg, Robert Vogt, Dieter H. Wilken, George Yatskievych, and James L. Zarucchi.

To the authors who contributed treatments to the Asteraceae volumes, we extend our appreciation and thanks for their hard work, their scholarly contributions, and their enduring patience with relentless rounds of editings and questions.

Many botanists are fortunate to have a significant other who is supportive beyond the call of duty. We wish to acknowledge the crucial support provided the late Theodore M. Barkley and Scott D. Sundberg by their respective spouses, Mary Barkley and Linda K. Hardison, as both were putting finishing touches to their contributions during their last months.

Taxon editors Brouillet and Strother extend special recognition and thanks to Helen Jeude. Throughout the project, she was not only our stalwart "tech" editor, she was also our advisor, comforter, and friend.

Preface for Volume 21

Since the publication of *Flora of North America* volume 5, the ninth volume in the *Flora* series, in early 2005, Guy Baillargeon, Barney L. Lipscomb, Jay A. Raveill (Taxon Editor), Michael A. Vincent (Taxon Editor), and Kay Yatskievych (Managing Editor and Production Coordinator) have become members of the Flora of North America Association Board of Directors. The Board succeeded the former Editorial Committee as the result of a reorganization finalized in 2003, but for the sake of continuity of citation, authorship of *Flora* volumes is still to be cited as "Flora of North America Editorial Committee, eds."

Sadly, in late 2005, two long-standing members of the FNAA Board, John W. Thieret and Grady L. Webster, passed away. They will be greatly missed.

The vast majority of editorial processing for the three, simultaneously-published Asteraceae volumes of the *Flora* was undertaken at the Botanical Research Institute of Texas in Fort Worth. Considerable editorial processing of treatments and writing of suprageneric descriptions and keys were done at the Universit´de Montréal and at the University of California, Berkeley. The maps were prepared at the editorial center at the Missouri Botanical Garden in St. Louis based on taxon distribution statements found in the treatments along with additional data for the indicators showing occurrence in Alaska, Greenland, and the larger Canadian provinces and territories. Pre-press production for the Asteraceae volumes, including typesetting and layout, plus coordination for all aspects of planning, executing, scanning, and labeling the illustrations, was done at the the the St. Louis center.

In addition to her duties as the project's Illustrations Coordinator, Yevonn Wilson-Ramsey prepared the illustrations for numerous taxa of Heliantheae (except those credited here to others). Other illustrations were prepared by Barbara Alongi, for taxa of Heliantheae (*Acanthospermum, Achyrachaena, Acmella, Adenophyllum, Amauriopsis, Amblyopappus, Ambrosia, Bahia, Bartlettia, Bebbia, Flaveria, Florestina, Flourensia, Gaillardia, Jensia, Kyhosia, Lagascea, Lagophylla, Leuciva, Lindheimera, Madia,* and *Picradeniopsis*); by Sheila Flinchpaugh, for *Balduina* (Heliantheae); by Bee F. Gunn, for taxa of Heliantheae (*Pericome, Perityle, Raillardella, Sanvitalia, Sartwellia, Schkuhria, Smallanthus, Sphagneticola,* and *Synedrella*); by Linny Heagy, for all genera in Eupatorieae, including the color frontispiece depicting *Liatris cylindracea,* and taxa of Heliantheae (*Chaenactis, Chamaechaenactis, Chrysogonum, Dicranocarpus, Heterosperma,* and *Sclerocarpus*); by Marjorie C. Leggitt, for taxa of Heliantheae (*Bahiopsis, Blepharipappus, Blepharizonia, Borrichia, Calycadenia, Calyptocarpus, Carlquistia, Centromadia, Cyclachaena, Deinandra, Dicoria, Dimeresia, Dysodiopsis, Dyssodia, Eatonella, Eclipta, Guardiola, Haploësthes, Harmonia, Helianthus, Heliomeris, Heliopsis, Hemizonia, Holocarpha, Hymenopappus, Phoebanthus,* and *Viguiera*); by John Myers, for taxa of

Heliantheae (*Agnorhiza, Balsamorhiza, Berlandiera, Cosmos, Echinacea, Engelmannia, Orochaenactis, Pectis, Rudbeckia, Scabrethia,* and *Wyethia*); and by Linda Ann Vorobik, for taxa of Heliantheae (*Oxytenia, Palafoxia, Parthenice, Parthenium, Pascalia, Peucephyllum, Plateilema, Platyschkuhria, Polymnia, Porophyllum, Psathyrotes, Psathyrotopsis,* and *Pseudoclappia*). Illustrations were scanned by Michael Blomberg and Fred Keusenkothen. Composition and labeling of all artwork was completed by John Myers assisted by Heidi H. Schmidt.

The Flora of North America Association remains deeply grateful to the many people who continue to help create and sustain the *Flora.*

Introduction

Scope of the Work

Flora of North America North of Mexico is a synoptic account of the plants of North America north of Mexico: the continental United States of America (including the Florida Keys and Aleutian Islands), Canada, Greenland (Kalâtdlit-Nunât), and St. Pierre and Miquelon. The *Flora* is intended to serve both as a means of identifying plants within the region and as a systematic conspectus of the North American flora.

The *Flora* will be published in 30 volumes. Volume 1 contains background information that is useful for understanding patterns in the flora. Volume 2 contains treatments of ferns and gymnosperms. Families in volumes 3–26, the angiosperms, are arranged according to the classification system of A. Cronquist (1981). Bryophytes will be covered in volumes 27–29. Volume 30 will contain the cumulative bibliography and index.

The first two volumes were published in 1993, Volume 3 in 1997, and Volumes 22, 23, and 26, the first three of five covering the monocotyledons, appeared in 2000, 2002, and 2002, respectively. Volume 4, the first part of the Caryophyllales, was published in late 2003. Volume 25, the second part of the Poaceae, was published in mid-2003. Volume 5, completing the Caryophyllales plus Polygonales and Plumbaginales, was published in early 2005. The correct bibliographic citation for the *Flora* is: Flora of North America Editorial Committee, eds. 1993+. *Flora of North America North of Mexico.* 12+ vols. New York and Oxford.

Volumes 19–21 treat 2413 species in 418 genera contained in 14 tribes of Asteraceae. For additional statistics, please refer to Table 1.

Contents · General

The *Flora* includes accepted names, selected synonyms, literature citations, identification keys, descriptions, chromosome numbers, phenological information, summaries of habitats and geographic ranges, and other biological observations. Economic uses, weed status, and conservation status are provided from specified sources. Each volume contains a bibliography and an index to the taxa included in that volume. The treatments, written and reviewed by experts from throughout the systematic botanical community, are based on original observations of herbarium specimens and, whenever possible, on living plants. These observations are supplemented by critical reviews of the literature.

Table 1. *Statistics for Volumes 19–21 of Flora of North America.*

Tribe	Total Genera	Total Species	Endemic Genera	Endemic Species	Introduced Genera	Introduced Species	Conservation Taxa
Volume 19							
Mutisieae	7	14	1	4	0	0	0
Cynareae	17	116	0	48	14	50	30
Arctotideae	3	4	0	0	3	4	0
Vernonieae	6	25	1	18	2	2	0
Cichorieae	49	229	7	112	21	64	22
Calenduleae	4	7	0	0	4	7	0
Gnaphalieae	19	111	1	50	5	17	5
Inuleae	3	5	0	0	3	5	0
Plucheeae	3	12	0	3	0	3	0
Anthemideae	26	99	1	37	17	38	4
Volume 20							
Astereae	77	719	33	561	1	5	175
Senecioneae	29	167	8	117	6	20	43
Volume 21							
Heliantheae	148	746	40	470	6	22	131
Eupatorieae	27	159	5	98	0	3	16
Total for Asteraceae	418	2413	97	1518	82	240	426

Italic = introduced

Basic Concepts

Our goal is to make the *Flora* as clear, concise, and informative as practicable so that it can be an important resource for both botanists and nonbotanists. To this end, we are attempting to be consistent in style and content from the first volume to the last. Readers may assume that a term has the same meaning each time it appears and that, within groups, descriptions may be compared directly with one another. Any departures from consistent usage will be explicitly noted in the treatments (see also References).

Treatments are intended to reflect current knowledge of taxa throughout their ranges worldwide, and classifications are therefore based on all available evidence. Where notable differences of opinion about the classification of a group occur, appropriate references are mentioned in the discussion of the group.

Documentation and arguments supporting significantly revised classifications are published separately in botanical journals before publication of the pertinent volume of the *Flora*. Similarly, all new names and new combinations are published elsewhere prior to their use in the *Flora*. No nomenclatural innovations will be published intentionally in the *Flora*.

Taxa treated in full include extant and recently extinct native species, hybrids that are well established (or frequent), and waifs or cultivated plants that are found frequently outside cultivation and give the appearance of being naturalized. Taxa mentioned only in discussions include waifs or naturalized plants now known only from isolated old records and some nonnative, economically important or extensively cultivated plants, particularly when they are relatives of native species. Excluded names and taxa are listed at the ends of appropriate sections, e.g., species at the end of genus, genera at the end of family.

Treatments are intended to be succinct and diagnostic but adequately descriptive. Characters and character states used in the keys are repeated in the descriptions. Descriptions of related taxa at the same rank are directly comparable.

With few exceptions, taxa are presented in taxonomic sequence. If an author is unable to produce a classification, the taxa are arranged alphabetically, and the reasons are given in the discussion.

Treatments of hybrids follow that of one of the putative parents. Hybrid complexes are treated at the ends of their genera, after the descriptions of species.

We have attempted to keep terminology as simple as accuracy permits. Common English equivalents usually have been used in place of Latin or Latinized terms or other specialized terminology, whenever the correct meaning could be conveyed in approximately the same space, e.g., "pitted" rather than "foveolate," but "striate" rather than "with fine longitudinal lines." See *Categorical Glossary for the Flora of North America Project* (R. W. Kiger and D. M. Porter 2001; also available online at http://huntbot.andrew.cmu.edu) for standard definitions of generally used terms. Very specialized terms are defined, and sometimes illustrated, in the relevant family or generic treatments.

References

Authoritative general reference works used for style are *The Chicago Manual of Style,* ed. 14 (University of Chicago Press 1993); *Webster's New Geographical Dictionary* (Merriam-Webster 1988); and *The Random House Dictionary of the English Language,* ed. 2, unabridged (S. B. Flexner and L. C. Hauck 1987). *B-P-H/S. Botanico-Periodicum-Huntianum/Supplementum* (G. D. R. Bridson and E. R. Smith 1991) has been used for abbreviations of serial titles, and *Taxonomic Literature*, ed. 2 (F. A. Stafleu and R. S. Cowan 1976–1988) and its supplements by F. A. Stafleu and E. A. Mennega (1992+) have been used for abbreviations of book titles.

Graphic Elements

All genera and approximately 30 percent of the species in this volume are illustrated. Illustration panels have been enlarged for this and subsequent volumes in the series. The illustrations may show diagnostic traits or complex structures. Most illustrations have been drawn from herbarium specimens selected by the authors. In some cases living material or photographs have been used. Data on specimens that were used and parts that were illustrated have been recorded. This information, together with the archivally preserved original drawings, is deposited in the Missouri Botanical Garden Library and is available for scholarly study.

Specific Information in Treatments

Keys

Dichotomous keys are included for all ranks below family if two or more taxa are treated. For dioecious species, keys are designed for use with either staminate or pistillate plants. Keys are designed also to facilitate identification of taxa that flower before leaves appear. More than one key may be given, and for some groups tabular comparisons may be presented in addition to keys.

Nomenclatural Information

Basionyms of accepted names, with author and bibliographic citations, are listed first in synonymy, followed by any other synonyms in common recent use, listed in alphabetical order, without bibliographic citations.

Vernacular names in common use are given in the appropriate language. In general, such names have not been coined for use in the *Flora*. Those preferred by governmental or conservation agencies are listed if known.

The last names of authors of taxonomic names have been spelled out. The conventions of *Authors of Plant Names* (R. K. Brummitt and C. E. Powell 1992) have been used as a guide for including first initials to discriminate individuals who share surnames. Exceptions include "Alph. Wood" instead of "A. W. Wood" and "K. F. Parker" instead of "K. L. Parker" (Brummitt, pers. comm.)

If only one infraspecific taxon within a species occurs in the flora area, nomenclatural information (literature citation, basionym with literature citation, relevant other synonyms) is given for the species, as is information on the number of infraspecific taxa in the species and their distribution worldwide, if known. A description and detailed distributional information are given only for the infraspecific taxon.

Descriptions

Character states common to all taxa are noted in the description of the taxon at the next higher rank. For example, if flowers are unisexual for all species treated within a genus, that character state is given in the generic description. Characters used in keys are repeated in the descriptions. Characteristics are given as they occur in plants from the flora area. Characteristics that occur only in plants from outside the flora area may be given within square brackets, or instead may be noted in the discussion following the description. In families with one genus and one or more species, the family description is given as usual, the genus description is condensed, and the species are described as usual. Any special terms that may be used when describing members of a genus are presented and explained in the genus description.

In reading descriptions, the reader may assume, unless otherwise noted, that: the plants are green, photosynthetic, and reproductively mature; woody plants are perennial; stems are erect; roots are fibrous; leaves are simple and petiolate; flowers are bisexual, radially symmetric, and pediceled; perianth parts are hypogynous, distinct, and free; and ovaries are superior. Because measurements and elevations are almost always approximate, modifiers such as "about," "circa," or "±" are usually omitted.

Unless otherwise noted, dimensions are length × width. If only one dimension is given, it is length or height. All measurements are given in metric units. Measurements usually are based on dried specimens but these should not differ significantly from the measurements found in fresh or living material.

Chromosome numbers generally are given only if published, documented counts are available from North American material or from an adjacent region. No new counts are published intentionally in the *Flora*. Chromosome counts from nonsporophyte tissue have been converted to the **2n** form. The base number ($x = $) is given for each genus. This represents the lowest known haploid count for the genus unless evidence is available that the base number differs.

Flowering time and often fruiting time are given by season, sometimes qualified by early, mid, or late, or by months. Elevations over 50 m generally are rounded to the nearest 100 m; those 50 m and under are rounded to the nearest 10 m. Mean sea level is shown as 0 m, with the understanding that this is approximate. Elevation often is omitted from herbarium specimen labels, particularly for collections made where the topography is not remarkable, and therefore precise elevation is sometimes not known for a given taxon.

The term "introduced" is defined broadly to refer to plants that were released deliberately or accidentally into the flora and that now exist as wild plants in areas in which they were not recorded as native in the past. The distribution of non-native plants is often poorly documented and presence of the plants in the flora may be ephemeral.

If a taxon is globally rare or if its continued existence is threatened in some way, the words "of conservation concern" appear before the statements of elevation and geographic range.

Criteria for taxa of conservation concern are based on NatureServe's (formerly The Nature Conservancy)—see http://www.natureserve.org—designations of global rank (G-rank) G1 and G2:

G1   Critically imperiled globally because of extreme rarity (5 or fewer occurrences or fewer than 1000 individuals or acres) or because of some factor(s) making it especially vulnerable to extinction.

G2   Imperiled globally because of rarity (5–20 occurrences or fewer than 3000 individuals or acres) or because of some factor(s) making it very vulnerable to extinction throughout its range.

The occurrence of species and infraspecific taxa within political subunits of the *Flora* area is depicted by dots placed on the outline map to indicate occurrence in a state or province. For the 48 contiguous states of the United States and the smaller Canadian provinces, a single dot is used in those units where a taxon is known to occur. In the case of Greenland, the larger Canadian provinces and territories, and the main area of Alaska, a dot's position can vary to indicate more northern, southern, or central/scattered distributions (also western or eastern only for Alaska). In the case of Alaska, the occurrence of a taxon in the Aleutian Islands and/or the panhandle area adjacent to British Columbia may also be indicated. However, the dots for these areas may not be readily seen due to the small map size. The Nunavut boundary on the maps has been provided by the GeoAccess Division, Canada Centre for Remote Sensing, Earth Science. Authors are expected to have seen at least one specimen documenting each geographic unit record and have been urged to examine as many specimens as possible from throughout the range of each taxon. Additional information about taxon distribution may be presented in the discussion.

Distributions are stated in the following order: Greenland; St. Pierre and Miquelon; Canada (provinces and territories in alphabetic order); United States (states in alphabetic order); Mexico (11 northern states may be listed specifically, in alphabetic order); West Indies; Bermuda; Central America (Belize, Costa Rica, El Salvador, Guatemala, Honduras, Nicaragua, Panama); South America; Europe, or Eurasia; Asia (including Indonesia); Africa; Pacific Islands; Australia; Antarctica.

Discussion

The discussion section may include information on taxonomic problems, distributional and ecological details, interesting biological phenomena, economic uses, and toxicity (see "Caution," below).

Selected References

Major references used in preparation of a treatment or containing critical information about a taxon are cited following the discussion. These, and other works that are referred to in discussion or elsewhere, are included in Literature Cited at the end of this volume.

CAUTION

The Flora of North America Editorial Committee **does not encourage, recommend, promote, or endorse** any of the folk remedies, culinary practices, or various utilizations of any plant described within these volumes. Information about medicinal practices and/or ingestion of plants, or of any part or preparation thereof, has been included only for historical background and as a matter of interest. Under no circumstances should the information contained in these volumes be used in connection with medical treatment. Readers are strongly cautioned to remember that many plants in the flora are toxic or can cause unpleasant or adverse reactions if used or encountered carelessly.

Key to boxed codes following accepted names:

- C   of conservation concern
- E   endemic to the flora area
- F   illustrated
- I   introduced to the flora area
- W   weedy, based mostly on R. H. Callihan et al. (1995) and/or D. T. Patterson et al. (1989)

Flora of North America

187m. ASTERACEAE Martinov tribe HELIANTHEAE Cassini, J. Phys. Chim. Hist. Nat. Arts 88: 189. 1819

Annuals, biennials, perennials, subshrubs, shrubs, or trees. **Leaves** usually cauline, sometimes mostly basal or basal and cauline; usually wholly or partly opposite, sometimes mostly whorled or alternate; usually petiolate, sometimes sessile; margins entire or dentate to pinnatifid or palmatifid (faces often gland-dotted). **Heads** usually heterogamous (disciform or radiate), sometimes homogamous (usually discoid, sometimes ± radiant; unisexual in some genera), usually in corymbiform, paniculiform, racemiform, or spiciform arrays, sometimes borne singly or in glomerules, rarely aggregated in second-order heads. **Calyculi** 0 or of 1–15+ bractlets. **Phyllaries** persistent or falling, usually in 3–5+ series, usually distinct (sometimes connate and forming hardened perigynia or winged or tuberculate to spiny burs, e.g., *Ambrosia* spp.), unequal (usually lanceolate to ovate or broader), and herbaceous to chartaceous with margins and/or apices sometimes notably scarious, sometimes in 1–2 series, distinct or connate, subequal (usually linear to lanceolate), and herbaceous with margins and/or apices sometimes notably scarious. **Receptacles** flat to conic or cylindric, paleate or epaleate (sometimes with enations from receptacles among florets, e.g., some *Gaillardia* spp.; sometimes pitted, pit borders often fimbrillate to lacerate). **Ray florets** (rarely 0) usually in 1(–2+) series, usually pistillate, rarely neuter or styliferous and sterile; corollas usually yellow to orange, sometimes cyanic to red, dark brown, or purplish, or white (laminae sometimes marcescent, e.g., *Baileya*, *Zinnia*). **Peripheral (pistillate) florets** 0 or (in disciform heads) in 1–3+ series; corollas (usually present) usually yellow to orange, sometimes ochroleucous or cyanic to reddish, purplish, or brown. **Disc (inner) florets** usually bisexual and fertile, rarely functionally staminate; corollas usually yellow to orange, sometimes ochroleucous or cyanic to reddish, purplish, or brown, sometimes ± zygomorphic (± 2-lipped), lobes (3–)5, usually ± deltate to lance-ovate, sometimes lanceolate to lance-linear; anther bases obtuse or rounded, not tailed (sometimes sagittate), apical appendages usually ovate to lanceolate, rarely 0 (in Ambrosiinae, filaments usually connate, anthers distinct, and plants wind-pollinated); styles abaxially glabrous or papillate to hirsutulous (distally), branches linear, adaxially stigmatic, usually in 2 lines, sometimes continuously, from bases to appendages, appendages usually deltate to lanceolate (abaxially and adaxially papillate to hispidulous). **Cypselae** usually monomorphic, sometimes dimorphic within heads, usually ± columnar to prismatic, sometimes compressed, obcompressed, or flattened, sometimes ± beaked, bodies smooth, rugose, tuberculate, ribbed, or winged (glabrous or hairy; cypselae enclosed within and shed with a hardened perigynium or bur in some Ambrosiinae; enfolded within and shed with subtending phyllaries or paleae in some Madiinae and genera in other subtribes); **pappi** (rarely 0) usually persistent, sometimes readily falling, usually of scales (scales sometimes aristate, sometimes plumose), sometimes of awns or smooth, or barbellulate to barbellate, or plumose bristles, sometimes combinations of scales and/or awns and/or bristles.

Genera ca. 300, species ca. 3300+ (148 genera, 746 species, including 5 hybrids, in the flora): almost wholly New World, mostly subtropical, tropical, and warm-temperate.

Nearly 95% of the species of Heliantheae in the broad sense (both paleate and epaleate genera) are native in the New World. O. Hoffmann (1890–1894) listed 10 subtribes for Heliantheae in a restricted sense (only paleate genera). T. F. Stuessy (1977[1978]) listed 15 subtribes for his Heliantheae, which included transfer of much (but not all) of traditional Helenieae (epaleate genera), as Bahiinae and Gaillardiinae. H. Robinson (1981) listed 35 subtribes for Heliantheae, including transfers of all of traditional Helenieae plus some genera formerly included in Senecioneae (e.g., *Arnica*). Because Robinson's classification of Heliantheae was current and available as plans were being laid for treating the composites for this flora (ca. 1988), it was chosen as the organizational basis for the tribe.

P. O. Karis and O. Ryding (1994, 1994b) tentatively segregated the paleate and epaleate genera of Heliantheae s.l. as Heliantheae s.str. and Helenieae, respectively—a traditional split but with a novel circumscription of Helenieae. Other authors have suggested separating genera of Heliantheae in a broad sense into a dozen or so tribes (e.g., B. G. Baldwin and B. L. Wessa 2000; Baldwin et al. 2002; J. L. Panero and V. A. Funk 2002).

Much of the partitioning and reorganizing of Heliantheae in the past 20 or so years has stemmed from botanists comparing differences and similarities in base sequences in DNA molecules and seeking to have taxa correspond to monophyletic clades. Some botanists see a phylogenetic classification comprising monophyletic taxa as the ultimate goal of systematics (e.g., P. C. van Welzen 1997); other botanists see monophyletic taxa as logically, philosophically, and/or theoretically untenable (e.g., R. K. Brummitt 1997).

SELECTED REFERENCES   Baldwin, B. G. 1996. Phylogenetics of the California tarweeds and the Hawaiian silversword alliance (Madiinae; Heliantheae sensu lato). In: D. J. N. Hind et al., eds. 1996. Proceedings of the International Compositae Conference, Kew, 1994. 2 vols. Kew. Vol. 1, pp. 377–391.   Baldwin, B. G. and B. L. Wessa. 2000. Origin and relationships of the tarweed–silversword lineage (Compositae–Madiinae). Amer. J. Bot. 87: 1890–1908.   Baldwin, B. G., B. L. Wessa, and J. L. Panero. 2002. Nuclear rDNA evidence for major lineages of helenioid Heliantheae (Compositae). Syst. Bot. 27: 161–198.   Barrier, M. et al. 1999. Interspecific hybrid ancestry of a plant adaptive radiation: Allopolyploidy of the Hawaiian silversword alliance (Asteraceae) inferred from floral homeotic gene duplications. Molec. Biol. Evol. 16: 1105–1113.   Bolick, M. R. 1983. A cladistic analysis of the Ambrosiinae Lessing and Engelmanniinae Stuessy. Advances Cladist. 2: 125–141.   Carlquist, S. 1956. On the generic limits of *Eriophyllum* (Compositae) and related genera. Madroño 13: 226–239.   Carlquist, S. 1959. Studies on Madiinae: Anatomy, cytology, and evolutionary relationships. Aliso 4: 171–236.   Carlquist, S., B. G. Baldwin, and G. D. Carr, eds. 2003. Tarweeds and Silverswords: Evolution of the Madiinae (Asteraceae). St. Louis.   Clevinger, J. A. and J. L. Panero. 2000. Phylogenetic analysis of *Silphium* and subtribe Engelmanniinae (Asteraceae: Heliantheae) based on ITS and ETS sequence data. Amer. J. Bot. 87: 565–572.   Johnson, D. E. 1991. Nomenclatural conspectus of annual Eriophyllinae (Asteraceae). Novon 1: 119–124.   Karis, P. O. and O. Ryding. 1994. Tribe Heliantheae. In: K. Bremer. 1994. Asteraceae: Cladistics & Classification. Portland. Pp. 559–624.   Karis, P. O. and O. Ryding. 1994b. Tribe Helenieae. In: K. Bremer. 1994. Asteraceae: Cladistics & Classification. Portland. Pp. 521–558.   Panero, J. L., R. K. Jansen, and J. A. Clevinger. 1999. Phylogenetic relationships of subtribe Ecliptinae (Asteraceae: Heliantheae) based on chloroplast DNA restriction site data. Amer. J. Bot. 413–427.   Robinson, H. 1981. A revision of the tribal and subtribal limits of the Heliantheae (Asteraceae). Smithsonian Contr. Bot. 51.   Schilling, E. E. 1997. Phylogenetic analysis of *Helianthus* (Asteraceae) based on chloroplast DNA restriction site data. Theor. Appl. Genet. 94: 925–933.   Schilling, E. E., C. R. Lander, R. D. Noyes, and L. H. Rieseberg. 1998. Phylogenetic relationships in *Helianthus* (Asteraceae) based on nuclear ribosomal DNA internal transcribed spacer region sequence data. Syst. Bot. 23: 177–187.   Schilling, E. E. and J. L. Panero. 1996. Relationships in Heliantheae subtribe Helianthinae based on chloroplast DNA restriction site analysis. In: D. J. N. Hind et al., eds. 1996. Proceedings of the International Compositae Conference, Kew, 1994. 2 vols. Kew. Vol. 1, pp. 361–376.   Schilling, E. E. and J. L. Panero. 2002. A revised classification of subtribe Helianthinae (Asteraceae: Heliantheae). I. Basal lineages. Bot. J. Linn. Soc. 140: 65–76.   Sharp, W. M. 1935. A critical study of certain epappose genera of the Heliantheae–Verbesininae of the natural family Compositae. Ann. Missouri Bot. Gard. 22: 51–152.   Stuessy, T. F. 1977[1978]. Heliantheae—systematic review. In: V. H. Heywood et al., eds. 1977[1978]. The Biology and Chemistry of the Compositae. 2 vols. London, New York, and San Francisco. Vol. 2, pp. 621–671.   Turner, B. L. and M. C. Johnston. 1957. Chromosome numbers and geographic distribution of *Lindheimera*, *Engelmannia*, and *Berlandiera* (Compositae–Heliantheae–Melampodinae). SouthW. Naturalist 1: 125–132.   Turner, B. L. and A. M. Powell. 1977[1978]. Helenieae—systematic review. In: V. H. Heywood et al., eds. 1977[1978]. The Biology and Chemistry of the Compositae. 2 vols. London, New York, and San Francisco. Vol. 2, pp. 699–737.   Urbatsch, L. E., B. G. Baldwin, and M. J. Donoghue. 2000. Phylogeny of the coneflowers and relatives (Heliantheae: Asteraceae) based on nuclear rDNA internal transcribed spacer (ITS) sequences and chloroplast DNA restriction site data. Syst. Bot. 25: 539–565.   Urbatsch, L. E. and R. K. Jansen. 1995. Phylogenetic affinities among and within the coneflower genera (Asteraceae: Heliantheae), a chloroplast DNA analysis. Syst. Bot. 20: 28–39.

1. Receptacles usually wholly epaleate (rarely bearing conic to setiform enations, e.g., *Gaillardia* in Gaillardiinae, or ± membranous paleae, e.g., *Amblyolepis setigera* in Gaillardiinae, *Chaenactis carphoclinia* in Chaenactidinae, and *Eriophyllum ambiguum* in Baeriinae; see also, first lead of couplet 11, Madiinae; receptacles deeply pitted in *Balduina* in Gaillardiinae, the pit borders sometimes interpreted as coalesced paleae).

2. Leaves and/or phyllaries dotted or streaked with pellucid (schizogenous) glands containing strong-scented oils . . . . 187m.14. Pectidinae, p. 221

2. Leaves and/or phyllaries rarely dotted or streaked (never with pellucid, schizogenous glands containing strong-scented oils, plants sometimes with sessile or stipitate, surface glands and sometimes strong-scented).

3. Annuals 1–5 cm; phyllaries 2–3; ray florets 0; disc florets 2–3; pappi of ca. 20 basally connate, subulate, plumose scales . . . . 187m.11. Dimeresiinae, p. 182

3. Annuals, perennials, subshrubs, or shrubs, (1–)5–200(–300) cm; phyllaries 2–50+; ray florets 0 or (1–)4–21(–60+); disc florets (1–)5–60(–300); pappi 0 or of smooth to barbellate (rarely, if ever, plumose) awns, bristles, and/or scales.

4. Leaves (often somewhat succulent) opposite, sessile or nearly so, blades usually oblong to linear or filiform (not lobed); cypselae clavate to cylindric and 8–15-ribbed.

5. Phyllaries 2–5 in 1 series, subequal . . . . 187m.15. Flaveriinae, p. 245

5. Phyllaries 12–16+ in 2–3+ series, unequal.

6. Cypselae hairy; pappi of 12–25, or ca. 50 bristles or setiform to subulate scales . . . . 187m.16. Clappiinae, p. 251

6. Cypselae glabrous; pappi usually 0, sometimes 1–5 (minute) subulate scales . . . . 187m.17. Jaumeinae, p. 253

4. Leaves (seldom succulent) opposite or alternate, petiolate or sessile, blades often lobed; cypselae usually obpyramidal to obconic, sometimes columnar or flattened, seldom clavate or cylindric, often 4–5-angled (not both clavate to cylindric and 8–15-ribbed; sometimes cylindric and 5–10-nerved, e.g., Chaenactidinae, *Arnica* spp.).

7. Phyllaries often ± conduplicate and navicular; disc corollas usually 4-lobed; cypselae strongly flattened or weakly 3–4-angled, usually callous-margined, often ciliate . . . . 187m.20. Peritylinae, p. 316

7. Phyllaries usually flat to weakly navicular; disc corollas (4–)5-lobed; cypselae usually obpyramidal to obconic, sometimes columnar, seldom clavate or cylindric, often 4–5-angled (seldom strongly compressed or flattened, callous-margined, and ciliate).

8. Cypselae stoutly obconic to obpyramidal (lengths usually 1–2, rarely to 3.5 times diams.).

9. Phyllaries: margins usually notably membranous to scarious; disc corollas usually whitish, sometimes purplish or yellowish (tubes, throats, and lobes glabrous or hairy, hairs not moniliform); cypselae usually 4-angled and 12–16-ribbed . . . . 187m.19. Hymenopappinae, p. 309

9. Phyllaries: margins seldom scarious; disc corollas orange to yellow or partly or wholly purple-brown or reddish (tubes, throats, and lobes often hairy, hairs often moniliform); cypselae usually obpyramidal, sometimes clavate, columnar, or obconic (not both 4-angled and 12–16-ribbed) . . . . 187m.23. Gaillardiinae, p. 415

8. Cypselae narrowly clavate or columnar to obconic or obpyramidal (lengths usually 3+ times diams., if stouter, usually ± compressed).

10. Leaves usually sessile, sometimes obscurely petiolate (rarely truly petiolate); pappi 0, or of scales (scales not medially thickened) . . . . 187m.21. Baeriinae, p. 335

10. Leaves usually petiolate (± sessile in some spp. of *Arnica*, *Chaenactis*, *Hulsea*); pappi 0, or of scales (scales usually notably medially thickened) . . . . 187m.22. Chaenactidinae, p. 364

1. Receptacles wholly or partly paleate.

   11. Heads disciform or discoid; pistillate florets: corollas sometimes none (cypselae shed with accessory structures or within burs); staminate florets: anthers usually distinct (staminal filaments coherent or connate in some spp.) . . . . 187m.1. Ambrosiinae (in part), p. 8

   11. Heads usually disciform, discoid, or radiate; pistillate florets: corollas usually present (cypselae seldom shed with accessory structures, sometimes shed within perigynia, each formed from single phyllary, not shed within burs); bisexual florets: anthers usually connate (staminal filaments usually distinct).

      12. Plants often with tack-glands or pit-glands on stems, leaves, and/or phyllaries; phyllaries (or paleae functioning as phyllaries) usually in 1+ series (each often wholly or partly investing ovary of subtended floret); paleae often in 1 series between ray and disc florets, often connate in a ring, sometimes each disc floret subtended by palea; ray corolla laminae often flabellate (lobes often $^1/_2$+ lengths of laminae); pappi usually of coarse, plumose and/or woolly bristles or subulate, plumose and/or woolly scales, sometimes none . . . . 187m.18. Madiinae, p. 254

      12. Plants without tack-glands or pit glands; phyllaries in (1–)2–7+ series (seldom each inner phyllary wholly or partly investing ovary of subtended floret); paleae seldom restricted to 1 series between ray and disc florets, all or nearly all disc florets subtended by paleae; laminae of ray corollas seldom flabellate (lobes mostly 0–$^1/_{10}$ lengths of laminae); pappi usually of awns, bristles, and/or scales (seldom plumose), sometimes 0.

         13. Calyculi usually of 3–8(–21+) bractlets or bracts, sometimes 0; phyllaries usually in ± 2 series, usually ± equal; disc cypselae obcompressed to obflattened (often winged), or ± equally 4-angled and fusiform to linear . . . . 187m.12. Coreopsidinae, p. 183

         13. Calyculi usually 0; phyllaries in 1–7+ series; disc cypselae seldom obcompressed or 4-angled and fusiform to linear.

            14. Phyllaries (at least inner) usually falling with cypselae; ray florets 0 or pistillate and fertile.

               15. Disc florets bisexual and fertile; anther thecae pale; pappi of subulate, often ± plumose scales or bristles . . . 187m.10. Galinsoginae (in part, *Galinsoga*), p. 176

               15. Disc florets usually functionally staminate (bisexual and fertile in Milleriinae, *Guizotia*); anther thecae usually dark (blackish to purplish); pappi 0.

                  16. Ray florets 6–18 (corollas hairy at bases of tubes); disc florets bisexual, fertile . . . . 187m.4. Milleriinae, p. 40

                  16. Ray florets 3–20+ (corollas seldom hairy at bases of tubes); disc florets functionally staminate.

                     17. Cypselae each shed with associated paleae and/or florets . . . . 187m.1. Ambrosiinae (in part), p. 8

                     17. Cypselae each shed with (and enclosed within) a phyllary or shed without accessory structures . . . . 187m.2. Melampodiinae (in part), p. 32

            14. Phyllaries persistent (in fruit); ray florets 0 or pistillate and fertile, or styliferous and sterile, or neuter.

               18. Receptacles spheric to high-conic or columnar (mostly 8–20+ mm).

                  19. Phyllaries subequal or unequal (outer usually longer than inner); ray florets 0 or 3–21+, neuter; disc florets 100–200+, bisexual and fertile; stigmatic papillae usually in 2 lines; pappi usually 0 or coroniform (vestigial), rarely of 2–4 scales . . . . 187m.6. Rudbeckiinae, p. 42

                  19. Phyllaries subequal or unequal (outer usually shorter, rarely longer, than inner); ray florets 0 or 3–40+, usually pistillate and fertile, sometimes styliferous and sterile, or neuter; disc florets 4–200+, usually bisexual and fertile, sometimes functionally staminate; stigmatic papillae usually continuous, rarely in 2 lines; pappi usually of scales, sometimes of bristles, rarely of awns or 0 . . . . 187m.8. Ecliptinae (in part), p. 64

[18. Shifted to left margin.—Ed.]
18. Receptacles mostly flat to convex or conic (mostly 0–5 mm).
  20. Leaves mostly cauline and alternate (proximal sometimes opposite), or mostly opposite (distal sometimes alternate); ray florets usually neuter, or styliferous and sterile, sometimes 0; pappi 0 or of (fragile or caducous) scales or awns.
    21. Phyllaries 45–60 in 4–5 series (resin-nerved); anther thecae pale; stigmatic papillae in 2 lines; cypselae clavate to columnar and 8–15-ribbed . . . . 187m.13. Varillinae, p. 220
    21. Phyllaries 4–35+ in 1–6+ series; anther thecae dark; stigmatic papillae continuous; cypselae clavate to columnar or prismatic (3–4-angled), or compressed to flattened.
      22. Disc corollas yellow (bases often dilated, clasping tops of ovaries) . . . . 187m.7. Zaluzaniinae, p. 63
      22. Disc corollas yellow to orange or brown-purple (bases not clasping tops of ovaries) . . . . 187m.9. Helianthinae, p. 135
  20. Leaves usually cauline and opposite, sometimes mostly basal and/or mostly alternate; ray florets usually pistillate and fertile (if neuter, leaves mostly basal or alternate), sometimes 0; pappi usually persistent, usually of awns, bristles, and/or scales, sometimes 0.
    23. Disc florets functionally staminate.
      24. Anther thecae green (staminal filaments hairy); Arizona . . . . 187m.5. Guardiolinae, p. 41
      24. Anther thecae dark or pale (not green, staminal filaments not hairy); e, se United States.
        25. Ray florets 7–13, corollas yellow; disc florets 40–80; cypselae (obliquely inserted on receptacles) 30–40-ribbed or -nerved (not beaked) . . . . 187m.2. Melampodiinae (in part), p. 32
        25. Ray florets 2–6, corollas pale yellow to whitish; disc florets 12–30+; cypselae (patently inserted on receptacles) 3–6-ribbed or -nerved (finely striate between ribs, apices often minutely beaked) . . . . 187m.3. Polymniinae, p. 38
    23. Disc florets bisexual and fertile.
      26. Disc corollas lavender, pink, purple, or white; anther thecae cream to purple . . . . 187m.24. Marshalliinae, p. 456
      26. Disc corollas usually orange to yellow, sometimes brown, pink, purple, red, or white; anther thecae pale or dark (not violet).
        27. Heads discoid; pappi of 15–30, ± plumose bristles or subulate scales (desert shrubs, leaf blades mostly linear-filiform) . . . . 187m.10. Galinsoginae (in part), p. 176
        27. Heads discoid or radiate; pappi 0 or of awns, bristles, or (seldom plumose) scales . . . . 187m.8. Ecliptinae (in part), p. 64

187m.1. ASTERACEAE Martinov (tribe HELIANTHEAE) subtribe AMBROSIINAE Lessing, Linnaea 5: 151. 1830 (as Ambrosieae)

Ambrosieae Cassini; *Iveae* Rydberg

Annuals, biennials, perennials, subshrubs, or shrubs (usually taprooted, perennating bases sometimes woody, sometimes rhizomes), mostly 3–150(–400+) cm. **Leaves** usually cauline, sometimes basal or basal and cauline; mostly opposite (distal sometimes alternate); petiolate or sessile; blades (palmately or pinnately nerved) deltate to linear or filiform (and intermediate shapes) sometimes 1–3+-pinnately or -ternately lobed, ultimate margins entire or toothed, faces usually hairy (scabrellous, hispidulous to sericeous, strigillose, or tomentose) and/or stipitate-glandular, or glabrous, often gland-dotted. **Heads** usually disciform, sometimes discoid (then usually unisexual) or radiate (*Parthenium*, laminae minute in *Parthenice*), borne singly or in (usually ebracteate, sometimes bracteate) corymbiform, glomerate, paniculiform, racemiform, or spiciform arrays. **Calyculi** 0. **Involucres** campanulate, cup-shaped, cylindric, hemispheric, obconic, rotate, saucer-shaped, turbinate, or urceolate (in pistillate heads, sometimes forming ± ovoid to fusiform, often ± spiny or winged burs). **Phyllaries** persistent or falling, mostly (3–)5–80+ in 1–8(–12+) series, distinct or connate, ovate or lance-ovate to linear, unequal to subequal, outer mostly herbaceous, inner membranous to scarious (sometimes indurate in fruit, accrescent and ultimately ovate to elliptic in *Dicoria*). **Receptacles** usually flat to convex, sometimes conic to columnar or hemispheric, usually paleate (at least staminate heads; paleae usually spatulate to filiform, sometimes setiform, often gland-dotted; tips sometimes dilated and strigillose or ciliate; sometimes paleae 0). **Ray florets** 0 or 5(–8), pistillate, fertile; corollas whitish to yellowish, laminae seldom conspicuous. **Peripheral (pistillate) florets** 0 or 1–10+; corollas (sometimes 0) usually ochroleucous or whitish, sometimes yellowish or pinkish, tubular to ± filiform (styles: stigmatic papillae in 2 lines, appendages essentially none). **Disc (functionally staminate) florets** 5–60+; corollas mostly ochroleucous or whitish, sometimes yellowish or pinkish (usually hairy and/or gland-dotted or stipitate-glandular), tubes shorter than to longer than cylindric or campanulate to funnelform throats, lobes 5, ± deltate; (staminal filaments usually connate, anthers usually distinct, ± connate in *Parthenice* and *Parthenium*) anther thecae usually pale; stigmatic papillae 0. **Cypselae** (sometimes shed within burlike or nutlike, spiny or winged perygynia, or shed with accessory structures) fusiform, obovoid, prismatic, or pyriform, sometimes obcompressed and plumply biconvex, sometimes flattened and cucullate (lengths usually 1–2+ times diams., sometimes finely striate or 1–5-nerved, corky-winged in *Dicoria* and winged margins ± toothed), usually smooth, sometimes ± tuberculate or warty, glabrous or hispidulous, scabrellous, strigillose, or villous and/or gland-dotted; **pappi** 0 (cypselae sometimes with apical tufts of hairs, or seeming pappi actually enations from ovary walls).

Genera 12, species 76 (11 genera, 45 species in the flora): mostly subtropical and warm-temperate New World.

In most genera of Ambrosiinae, proximal heads are erect and pistillate and distal heads are pendulous and functionally staminate. In functionally staminate florets, the staminal filaments are usually connate and the anthers are distinct or weakly coherent. Unlike nearly all other composites, such plants are wind-pollinated. Wind-borne pollen from some members of Ambrosiinae (especially ragweeds) are sources of "hay-fever" for some people.

H. Robinson (1981) affirmed inclusion of *Parthenice* and *Parthenium* in Ambrosiinae and suggested connections between Ambrosiinae and Clibadiinae and/or Ecliptinae.

SELECTED REFERENCES Jackson, R. C. 1960. A revision of the genus *Iva*. Univ. Kansas Sci. Bull. 41: 793–876. Karis, P. O. 1995. Cladistics of the subtribe Ambrosiinae (Asteraceae: Heliantheae). Syst. Bot. 20: 40–54. Miao, B. M., B. L. Turner, and T. J. Mabry. 1995. Chloroplast DNA variations in sect. *Cyclachaena* of *Iva* (Asteraceae). Amer. J. Bot. 82: 919–923. Miao, B. M., B. L. Turner, and T. J. Mabry. 1995b. Systematic implications of chloroplast DNA variation in the subtribe Ambrosiinae (Asteraceae: Heliantheae). Amer. J. Bot. 82: 924–932. Miao, B. M., B. L. Turner, and T. J. Mabry. 1995c. Molecular phylogeny of *Iva* (Asteraceae, Heliantheae) based on chloroplast DNA restriction site variation. Pl. Syst. Evol. 195: 1–12.

1. Pistillate and functionally staminate florets in separate heads (cypselae shed within hardened, often prickly, spiny, tuberculate, or winged perigynia, forming "burs" or nutlike structures).
   2. Staminate heads: phyllaries partially or wholly connate; receptacles ± flat or convex; pistillate heads: phyllaries 12–30(–80+) in 1–8+ series, outer (1–)5–8 distinct or ± connate, the rest ± connate (becoming indurate, their distinct tips forming straight or hooked spines, tubercles, or wings) . . . . 244. *Ambrosia*, p. 10
   2. Staminate heads: phyllaries distinct to bases; receptacles conic to columnar; pistillate heads: phyllaries 30–75+ in 6–12+ series, outer 5–8 distinct, the rest proximally connate (becoming indurate, their distinct tips usually forming hooked spines) . . . . 245. *Xanthium*, p. 19
1. Pistillate and functionally staminate florets usually together in same heads (sometimes some heads staminate; cypselae not enclosed within perigynia).
   3. Cypselae shed with accessory structures (at least 2 paleae, sometimes florets and/or a phyllary as well); anthers ± connate.
      4. Annuals, biennials, perennials, subshrubs, or shrubs; heads usually radiate (corolla laminae often inconspicuous), sometimes ± disciform (*P. alpinum*, a perennial); cypselae shed with subtending phyllary plus 2 contiguous disc florets and their investing paleae . . . . 246. *Parthenium*, p. 20
      4. Annuals; heads obscurely radiate (corolla laminae minute) or disciform; cypselae shed with 2 adjacent, ± fleshy paleae . . . . 247. *Parthenice*, p. 23
   3. Cypselae usually shed free of accessory structures (associated with subtending phyllaries in *Dicoria*); anthers weakly coherent or distinct.
      5. Cypselae strongly obcompressed to obflattened (subtended by accrescent phyllaries, margins corky-winged and ± irregularly toothed) . . . . 248. *Dicoria*, p. 24
      5. Cypselae sometimes ± obcompressed (not subtended by accrescent phyllaries, margins not corky-winged and toothed).
         6. Heads in (bracteate) racemiform or spiciform arrays (heads 1–2 per bract) . . . . . 249. *Iva*, p. 25
         6. Heads in (± ebracteate) paniculiform arrays, or heads 3–6+ per bract, or heads borne singly (± scattered).
            7. Leaves mostly opposite (distal sometimes alternate, blades 3–5-nerved, ± deltate, ovate, or rhombic, proximal 1–3 pairs often 3–5-lobed, margins usually toothed) . . . . 250. *Cyclachaena*, p. 28
            7. Leaves all or mostly alternate (blades, some or all, laciniately pinnately lobed or 1–3-pinnately lobed).
               8. Subshrubs or shrubs (phyllaries, paleae, and cypselae ± villous) . . . . 251. *Oxytenia*, p. 29
               8. Annuals or perennials (rarely woody at bases; phyllaries, paleae, and cypselae glabrous or strigillose and/or hispidulous).
                  9. Leaf blades laciniately pinnately lobed, abaxial faces ± lanate, adaxial faces ± tomentose . . . . 252. *Leuciva*, p. 29
                  9. Leaf blades 1–3-pinnately lobed, abaxial and adaxial faces hispid, hispidulous, and/or ± scabrellous.
                     10. Heads usually in loose, (± bracteate or ebracteate) paniculiform arrays (sometimes 3–6+ distal to axil of each bract); herbaceous phyllaries usually 5; lobes of functionally staminate corollas soon reflexed . . . . 253. *Hedosyne*, p. 30
                     10. Heads mostly borne singly (in leaf axils or remote from axils, ± scattered); herbaceous phyllaries usually 3+; lobes of functionally staminate corollas usually erect at flowering . . . . 254. *Chorisiva*, p. 31

244. AMBROSIA Linnaeus, Sp. Pl. 2: 987. 1753; Gen. Pl. ed. 5, 425. 1754 • Ragweed [Greek *ambrosia*, "food of the gods," allusion unclear]

John L. Strother

Franseria Cavanilles; *Hymenoclea* Torrey & A. Gray

Annuals, perennials, or shrubs, 10–400+ cm (usually rhizomatous). **Stems** erect, decumbent, or prostrate, branched. **Leaves** usually cauline; opposite ± throughout or opposite (proximal) and alternate or mostly alternate; sessile or petiolate; blades (or lobes) deltate, elliptic, filiform, lanceolate, linear, obovate, ovate, or rhombic (and most intermediate shapes), usually pinnately, sometimes palmately lobed, ultimate margins entire or toothed, faces hairy or glabrate, usually gland-dotted or stipitate-glandular. **Heads** discoid (unisexual, pistillate proximal to or intermixed with staminates, staminates usually in racemiform to spiciform arrays; rarely, single plants all or mostly staminate or pistillate). **Pistillate heads:** phyllaries 12–30(–80+) in 1–8+ series, outer (1–)5–8 distinct or ± connate, herbaceous, the rest (sometimes interpreted as paleae) ± connate, usually with free tips forming tubercles, spines, or wings (the whole becoming a hard perigynium or "bur"); florets 1(–5+), corollas 0. **Staminate heads:** involucres cup-shaped to saucer-shaped, 1.5–6+ mm diam.; phyllaries 5–16+ in ± 1 series, ± connate; receptacles ± flat or convex; paleae spatulate to linear, membranous, sometimes villous, hirtellous, and/or gland-dotted or stipitate-glandular, sometimes none; florets 5–60+; corollas whitish or purplish, ± funnelform, lobes 5, erect or incurved; staminal filaments connate, anthers distinct or weakly coherent. **Cypselae** (black) ± ovoid or fusiform, enclosed within globose to obovoid, pyramidal, pyriform, obconic, or fusiform, hard, smooth, tuberculate, spiny, or winged "burs"; **pappi** 0. $x = 18$.

Species 40+ (22 in the flora): tropical to subtropical and temperate New World, mostly North America, some established in Old World.

SELECTED REFERENCES Payne, W. W. 1964. A re-evaluation of the genus *Ambrosia* (Compositae). J. Arnold Arbor. 45: 401–430.   Peterson, K. M. and W. W. Payne. 1973. The genus *Hymenoclea* (Compositae: Ambrosieae). Brittonia 25: 243–256.   Strother, J. L. and B. G. Baldwin. 2002. Hymenocleas are ambrosias (Compositae). Madroño 49: 143–144.

1. Shrubs.
    2. Leaf blades and/or lobes filiform (0.5–1.5 mm wide); burs winged, wings 5–20+, oblanceolate to cuneiform or flabellate to orbiculate.
        3. Wings on burs mostly around middles, 2–3 × 1–2 mm; flowering (May–)Aug–Nov . . . . 1. *Ambrosia monogyra*
        3. Wings on burs ± scattered, 3–4(–6) × 2–4(–8) mm; flowering Mar–May(–Jun) . . . . 2. *Ambrosia salsola*
    2. Leaf blades and/or lobes mostly deltate, elliptic, lanceolate, ovate, rhombic, or triangular (broader than filiform); burs spiny, spines 8–30(–80+), ± lanceolate (sometimes navicular at bases) to subulate or acerose (tips sometimes uncinate).
        4. Leaf blades (green) hirsutulous to hirtellous abaxially.
            5. Leaves: petioles 0–2 mm, blades elliptic to ovate, margins spiny-toothed . . . . 3. *Ambrosia ilicifolia*
            5. Leaves: petioles 10–35 mm, blades lanceolate to narrowly triangular, margins coarsely toothed (not spiny) . . . . 4. *Ambrosia ambrosioides*
        4. Leaf blades (often gray, silvery, or white) ± densely pilosulous, puberulent, scabrellous, strigillose, or tomentulose abaxially.
            6. Leaf blades (1–)2–3-pinnately lobed; pistillate heads often intermixed with staminates . . . . 5. *Ambrosia dumosa*
            6. Leaf blades not pinnately lobed (irregularly toothed to ± laciniate in *A. eriocentra*); pistillate heads proximal to staminates.

7. Leaf blades elliptic, lance-linear, or rhombic (margins irregularly toothed to ± laciniate); pistillate florets 1; burs densely villous and stipitate-glandular .... 6. *Ambrosia eriocentra*
7. Leaf blades deltate, lance-deltate, ovate, or rounded-deltate; pistillate florets (1–)2(–3); burs puberulent to tomentulose and/or ± stipitate-glandular.
   8. Leaf blades: bases cordate to truncate, abaxial faces densely puberulent (including veins) .... 7. *Ambrosia cordifolia*
   8. Leaf blades: bases cuneate to truncate, abaxial faces tomentulose (mostly between veins).
      9. Leaf blades ovate to rounded-deltate; burs usually tomentulose (little, if at all, stipitate-glandular) .... 8. *Ambrosia chenopodiifolia*
      9. Leaf blades deltate to lance-deltate; burs usually stipitate-glandular (little, if at all, tomentulose) .... 9. *Ambrosia deltoidea*
1. Annuals, perennials, or subshrubs (usually rhizomatous).
   10. Annuals.
      11. Leaves mostly opposite (usually some blades palmately 3–5-lobed) .... 10. *Ambrosia trifida*
      11. Leaves mostly opposite (in *A. bidentata*, opposite and alternate, or mostly alternate, usually some or all blades 1–2-pinnately lobed, except in *A. bidentata* with 0–4 basal lobes).
         12. Leaf blades with (1–)2(–4) basal lobes or not lobed; peduncles of staminate heads 0–0.5 mm; burs 5–8 mm .... 11. *Ambrosia bidentata*
         12. Leaf blades mostly 1–2-pinnately lobed; peduncles of staminate heads 0.5–2 mm; burs 2–5 mm.
            13. Involucres of staminate heads (2–)3–5(–7) mm diam. (usually each with 1–5+ black nerves); burs ± fusiform to obpyramidal, 3–5 mm, spines 8–18+, 2–4(–5) mm .... 12. *Ambrosia acanthicarpa*
            13. Involucres of staminate heads 2–3+ mm diam. (usually without black nerves); burs ± globose to pyriform, 2–3 mm, spines or tubercles 3–5+, 0.1–0.5+ mm .... 13. *Ambrosia artemisiifolia*
   10. Perennials or subshrubs.
      14. Stems ± prostrate or decumbent (usually beaches or strand).
         15. Staminate involucres 2–3+ mm diam.; burs ± pyriform, 1–2 mm, spines or tubercles 0–5+, 0.1–0.5+ mm .... 14. *Ambrosia hispida*
         15. Staminate involucres 4–6+ mm diam.; burs fusiform to ± pyriform, 4–7+ mm, spines 8–16+, 0.5–1.5+ mm .... 15. *Ambrosia chamissonis*
      14. Stems ± erect (rarely beaches or strand).
         16. Leaf blades sometimes 1–2(–3)-pinnately lobed (if lobed, lobes mostly deltate to lanceolate); involucres of staminate heads usually ± saucer-shaped (cup-shaped in *A. cheiranthifolia* from s Texas).
            17. Leaf blades elliptic to lance-elliptic, 50–80(–180) × 12–20(–50+) mm, 1–2(–3)-pinnately lobed (lobes ± deltate), abaxial faces densely finely scabrellous .... 16. *Ambrosia tomentosa*
            17. Leaf blades elliptic, lanceolate, oblanceolate, oblong, or ovate, 20–60(–100+) × 8–45(–75+) mm, 1(–2)-pinnately lobed or not lobed, abaxial faces densely sericeous, strigillose, or strigose.
               18. Leaves: petioles none, blades lanceolate or lance-elliptic to lance-oblong or oblanceolate, 20–50(–70+) mm, rarely ± pinnately lobed (lobes ± deltate) .... 17. *Ambrosia cheiranthifolia*
               18. Leaves: petioles 10–45+ mm, blades elliptic to ovate, 45–60(–100+) mm, usually 1–2-pinnately lobed (lobes ± lanceolate) .... 18. *Ambrosia grayi*
         16. Leaf blades usually 1–4-pinnately lobed (lobes or non-lobed blades mostly lanceolate to linear); involucres of staminate heads cup-shaped.
            19. Burs: spines (1–)8–13+, subulate, tips usually uncinate.
               20. Leaf blades mostly linear (some 1-pinnate, lobes linear); involucres of staminate heads 4–6+ mm diam .... 19. *Ambrosia linearis*
               20. Leaf blades lanceolate to ovate, laciniately 2–4-pinnately lobed (lobes ± lanceolate); involucres of staminate heads 1.5–3+ mm diam .... 20. *Ambrosia confertiflora*

[19. Shifted to left margin.—Ed.]
19. Burs: spines or tubercles 0 or 1–6, stoutly conic to ± acerose, tips straight.
   21. Leaf blades deltate to elliptic, 15–35(–75) × 12–25(–45) mm, laciniately (1–)2(–3)-pinnately lobed; burs ± fusiform, 2–2.5 mm, strigillose . . . . 21. *Ambrosia pumila*
   21. Leaf blades deltate to lanceolate, 20–60(–140) × 8–35(–50+) mm, pinnately toothed or 1(–2)-pinnately lobed; burs ± obpyramidal to globose, 2–3 mm, hirsutulous . . . 22. *Ambrosia psilostachya*

1. **Ambrosia monogyra** (Torrey & A. Gray) Strother & B. G. Baldwin, Madroño 49: 143. 2002

Hymenoclea monogyra Torrey & A. Gray, Mem. Amer. Acad. Arts, n. s. 4: 79. 1849

Shrubs, 30–150(–400) cm. **Stems** erect. **Leaves** mostly alternate; petioles 0; blades mostly filiform, 5–30(–65+) × 0.5–1.5 mm, sometimes with 3(–5+) filiform lobes, abaxial faces glabrous or glabrate, often vernicose, adaxial faces densely scabrellous (white). **Pistillate heads** ± intermixed with staminates; florets 1. **Staminate heads:** peduncles 0–0.5 mm; involucres ± cup-shaped, 2–4 mm diam., ± glabrate, ± vernicose; florets 5–12+. **Burs:** bodies fusiform to pyriform, 4–5 mm, stipitate-glandular, wings 7–12+, mostly around middles, oblanceolate to cuneiform, 2–3 × 1–2 mm. 2*n* = 36.

Flowering (May–)Aug–Nov. Desert washes, ravines; 10–500 m; Ariz., Calif., N.Mex., Tex.; Mexico (Baja California, Baja California Sur, Chihuahua, Sinaloa, Sonora).

Ambrosia monogyra may occur in western Nevada.

2. **Ambrosia salsola** (Torrey & A. Gray) Strother & B. G. Baldwin, Madroño 49: 143. 2002 • Burrobush

F

Hymenoclea salsola Torrey & A. Gray, Mem. Amer. Acad. Arts, n. s. 4: 79. 1849

Shrubs, 20–80(–150+) cm. **Stems** erect. **Leaves** mostly alternate; petioles 0; blades mostly filiform, 15–35(–65+) × 0.5–1.5 mm, sometimes with 3(–5+) filiform lobes, abaxial faces sparsely scabrellous, glabrescent, often vernicose, adaxial faces densely scabrellous (white). **Pistillate heads** ± intermixed with staminates; florets 1. **Staminate heads:** peduncles 0–0.5 mm; involucres ± cup-shaped, 1–3(–4) mm diam., ± glabrate, ± vernicose or gland-dotted; florets 5–15+. **Burs:** bodies plumply fusiform, 3–4(–6) mm, stipitate-glandular, wings 5–20+, ± scattered, broadly cuneiform, flabellate, or orbiculate, 3–4(–6) × 2–4(–8) mm. 2*n* = 36.

Flowering Mar–May(–Jun). Desert washes, sandy and gravelly benches, lava talus; 200–1800 m; Ariz., Calif., Nev., Utah; Mexico (Baja California, Sonora).

Variety *salsola* refers to plants bearing burs with 6–14 wings in 2–3+ cycles, their tips ± spreading to patent; var. *pentalepis* (Rydberg) Strother & B. G. Baldwin refers to plants bearing burs with 5–9(–13) wings in 1(–2) cycles, their tips ± spreading to patent; var. *fasciculata* (A. Nelson) Strother & B. G. Baldwin [= *Hymenoclea salsola* var. *patula* (A. Nelson) K. M. Peterson & W. W. Payne, an illegitimate name] refers to plants bearing burs with 10–18 wings in 2–3+ cycles, their tips antrorsely ± appressed (see K. M. Peterson and W. W. Payne 1973).

Hybrids between *Ambrosia salsola* and *A. dumosa* have been called *A.* ×*platyspina* (Seaman) Strother & B. G. Baldwin.

3. **Ambrosia ilicifolia** (A. Gray) W. W. Payne, J. Arnold Arbor. 45: 425. 1964

Franseria ilicifolia A. Gray, Proc. Amer. Acad. Arts 11: 77. 1876

Shrubs, 30–50(–120+) cm. **Stems** erect. **Leaves** mostly alternate; petioles 0–2 mm; blades (green) elliptic to ovate, 25–60+ × 12–35+ mm, bases rounded to truncate, margins spiny-toothed, abaxial and adaxial faces ± hirtellous (on veins) and stipitate-glandular. **Pistillate heads** clustered, proximal to staminates; florets 2. **Staminate heads:** peduncles 2–8(–12) mm; involucres ± saucer-shaped, 9–15+ mm diam., ± hirtellous and stipitate-glandular; florets 20–40+. **Burs:** bodies ± globose, 6–8 mm, stipitate-glandular, spines 40–50+, scattered, subulate, 4–6 mm, tips straight or ± uncinate. 2*n* = 36.

Flowering Jan–Apr. Sandy washes, benches; 30–500 m; Ariz., Calif.; Mexico (Baja California, Sonora).

4. **Ambrosia ambrosioides** (Cavanilles) W. W. Payne, J. Arnold Arbor. 45: 410. 1964

Franseria ambrosioides Cavanilles, Icon. 2: 79, plate 200. 1793

Shrubs, 60–150+ cm. **Stems** erect. **Leaves** mostly alternate; petioles 10–35 mm; blades lanceolate to narrowly triangular, 50–150+ × 18–55+ mm, bases truncate to cordate, margins coarsely toothed (not spiny), abaxial and adaxial faces ± hirsutulous and gland-dotted. **Pistillate heads** clustered on lateral axes, proximal to staminates; florets

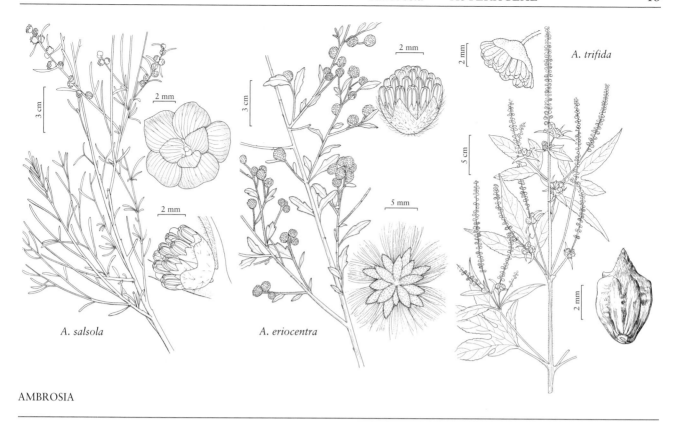

A. salsola

A. eriocentra

A. trifida

AMBROSIA

4–5. **Staminate heads:** peduncles 2–4(–12) mm; involucres ± saucer-shaped, 4–6+ mm diam., ± hirsutulous; florets 40–60+. **Burs:** bodies ± fusiform, 6–8+ mm, stipitate-glandular, spines 60–80+, scattered, subulate, 4–6 mm, tips uncinate. $2n = 36$.

Flowering Mar–May. Sandy soils, washes, banks; (100–)500–1000+ m; Ariz., Calif.; Mexico (Baja California, Baja California Sur, Chihuahua, Durango, Sonora).

Ambrosia ambrosioides has been reported from California; it may occur there.

5. **Ambrosia dumosa** (A. Gray) W. W. Payne, J. Arnold Arbor. 45: 422. 1964 • Bursage

Franseria dumosa A. Gray in J. C. Frémont, Rep. Exped. Rocky Mts., 316. 1845

Shrubs, 10–40(–60+) cm. **Stems** erect. **Leaves** mostly alternate; petioles 2–8(–12+) mm; blades (white) ovate to elliptic, 10–25 (–45) × 8–15(–30) mm, (1–)2–3-pinnately lobed, abaxial and adaxial faces densely strigillose. **Pistillate heads** intermixed with staminates (sometimes wanting or staminates sometimes wanting, plants unisexual); florets (1–)2. **Staminate heads:** peduncles 0–1(–2) mm; involucres shallowly cup-shaped, 3–5 mm diam., ± strigillose; florets 8–15+. **Burs:** bodies ± globose, 3–5+ mm, pilosulous and/or gland-dotted, spines 12–25+, scattered, ± subulate (± navicular at bases), 2–4 mm, tips straight. $2n = 36, 72, 108, 126$.

Flowering Mar–May(–Dec). Rocky or sandy washes, benches; (-100–)100–1200(–1500) m; Ariz., Calif., Nev., Utah; Mexico (Baja California, Sonora).

6. **Ambrosia eriocentra** (A. Gray) W. W. Payne, J. Arnold Arbor. 45: 423. 1964 E F

Franseria eriocentra A. Gray, Proc. Amer. Acad. Arts 7: 355. 1868

Shrubs, 30–80(–120+) cm. **Stems** erect. **Leaves** mostly alternate; petioles 0–2 mm; blades (± white abaxially) elliptic, lance-linear, or rhombic, 10–40(–60) × 5–15(–40) mm, bases ± cuneate, margins irregularly toothed to ± laciniate, abaxial faces densely strigillose, adaxial faces tomentulose and pilosulous. **Pistillate heads** clustered, proximal to staminates; florets 1. **Staminate heads:** peduncles 0–0.5 mm; involucres saucer-shaped to shallowly cup-shaped, 4–8+ mm diam., tomentulose and/or pilose; florets 12–25+. **Burs:** bodies obconic to fusiform, 3–5+ mm, densely villous and stipitate-glandular, spines 12–20+, scattered or mostly distal, ± subulate, 3–4 mm, tips straight. $2n = 36$.

Flowering Apr–May(–Jul). Sandy and gravelly washes; 700–1700 m; Ariz., Calif., Nev., Utah.

7. **Ambrosia cordifolia** (A. Gray) W. W. Payne,
   J. Arnold Arbor. 45: 421. 1964

Franseria cordifolia A. Gray in
A. Gray et al., Syn. Fl. N. Amer.
1(2): 445. 1884

Shrubs, 30–50+ cm. **Stems** erect.
Leaves mostly alternate; petioles
3–35 mm; blades deltate to
rounded-deltate, 10–35(–80) ×
10–30(–70) mm, bases cordate to
truncate, margins coarsely
toothed, abaxial faces densely puberulent (including
veins), adaxial faces strigillose to scabrellous and gland-
dotted. **Pistillate heads** clustered, proximal to staminates;
florets (1–)2. **Staminate heads:** peduncles 0–1 mm; involucres cup-shaped, 2–3(–5) mm diam., strigillose; florets 8–30+. **Burs:** bodies ± globose to fusiform, 3–4+
mm, tomentulose and stipitate-glandular, spines 8–20+,
scattered, ± subulate (basally navicular), 1–2 mm, tips
straight or uncinate.

Flowering Mar–Apr. Sandy washes, benches; 700–
1200 m; Ariz.; Mexico (Sinaloa, Sonora).

8. **Ambrosia chenopodiifolia** (Bentham) W. W. Payne,
   J. Arnold Arbor. 45: 419. 1964

Franseria chenopodiifolia Bentham,
Bot. Voy. Sulphur, 26. 1844

Shrubs, 30–80+ cm. **Stems** erect.
Leaves mostly alternate; petioles 5–
20 mm; blades ovate to rounded-
deltate, 20–35(–50) × 18–30(–45)
mm, bases broadly cuneate to
± truncate, margins toothed,
abaxial faces ± densely tomentulose
(between veins), adaxial faces sparsely tomentulose, glabrescent. **Pistillate heads** clustered, proximal to staminates;
florets 2(–3). **Staminate heads:** peduncles 0–3 mm; involucres ± cup-shaped, 3–5 mm diam., tomentulose; florets
12–20+. **Burs:** bodies obconic to fusiform, 4–6 mm,
tomentulose (little, if at all, stipitate-glandular), spines 15–
20+, scattered or on distal 1/2, ± subulate (the proximal
basally flattened), 2–3 mm, tips usually uncinate, sometimes straight. 2*n* = 72.

Flowering Mar–Apr. Clays or sandy soils, among lava
rocks; 50–200 m; Calif.; Mexico (Baja California, Baja
California Sur).

9. **Ambrosia deltoidea** (Torrey) W. W. Payne, J. Arnold
   Arbor. 45: 421. 1964

Franseria deltoidea Torrey,
Smithsonian Contr. Knowl. 6(2):
15. 1853

Shrubs, 30–80+ cm. **Stems** erect.
Leaves mostly alternate; petioles
5–12+ mm; blades deltate to
lance-deltate, 12–25(–35) × 5–
12(–18) mm, bases cuneate to
truncate, margins toothed,
abaxial faces ± densely tomentulose (between veins),
adaxial faces sparsely tomentulose, glabrescent. **Pistillate
heads** clustered, proximal to staminates; florets 2–3.
Staminate heads: peduncles 0.5–3(–5) mm; involucres
± cup-shaped, 4–8 mm diam., tomentulose; florets 12–
30+. **Burs:** bodies ± fusiform to globose, 3–4 mm, usually
stipitate-glandular (little, if at all, tomentulose), spines
20–30+, scattered or on distal 1/2, ± subulate (the
proximal basally flattened), 1–3 mm, tips usually
straight, sometimes uncinate. 2*n* = 36.

Flowering Feb–May. Sandy washes, benches; 200–
1000 m; Ariz.; Mexico (Baja California, Sonora).

Some specimens are intermediate for the characteristics that distinguish *Ambrosia deltoidea* and *A.
chenopodiifolia*. The type of *A. deltoidea* may be better
treated as conspecific with that of *A. chenopodiifolia*.

10. **Ambrosia trifida** Linnaeus, Sp. Pl. 2: 987. 1753
    · Giant ragweed, grande herbe à poux   F

Ambrosia aptera de Candolle; *A.
trifida* var. *integrifolia* (Muhlenberg
ex Willdenow) Torrey & A. Gray;
A. trifida var. *texana* Scheele

Annuals, 30–150(–400+) cm.
Stems erect. **Leaves** mostly opposite; petioles 10–30(–70+) mm;
blades rounded-deltate to ovate or
elliptic, 40–150(–250+) × 30–70
(–200+) mm, usually some blades palmately 3(–5)-lobed,
bases truncate to cuneate (sometimes decurrent onto petioles), margins usually toothed, rarely entire, abaxial and
adaxial faces ± scabrellous and gland-dotted. **Pistillate
heads** clustered, proximal to staminates; florets 1. **Staminate heads:** peduncles 1–3+ mm; involucres ± saucer-
shaped, 2–4 mm diam., scabrellous (often with 1–3 black
nerves); florets 3–25+. **Burs:** bodies ± pyramidal, 3–5
(–7+) mm, glabrous or glabrate, spines 4–5, ± distal,
± acerose, 0.5–1 mm, tips straight (bases ± decurrent as
ribs). 2*n* = 24, 48.

Flowering Jul–Nov. Disturbed sites, waste places,
damp soils; 0–1600 m; Alta., B.C., Man., N.B., N.S.,

Ont., P.E.I., Que., Sask.; Ala., Alaska, Ariz., Ark., Calif., Colo., Conn., Del., D.C., Fla., Ga., Idaho, Ill., Ind., Iowa, Kans., Ky., La., Maine, Md., Mass., Mich., Minn., Miss., Mo., Mont., Nebr., N.H., N.J., N.Mex., N.Y., N.C., N.Dak., Ohio, Okla., Oreg., Pa., R.I., S.C., S.Dak., Tenn., Tex., Utah, Vt., Va., Wash., W.Va., Wis., Wyo.; Mexico (Chihuahua).

The name *Ambrosia* ×*helenae* Rouleau applies to hybrids between *A. artemisiifolia* and *A. trifida*. Hybrids between *A. bidentata* and *A. trifida* have been recorded. *Ambrosia trifida* may be no longer extant in British Columbia.

11. **Ambrosia bidentata** Michaux, Fl. Bor.-Amer. 2: 182. 1803 E

Annuals, 10–30(–100+) cm. **Stems** erect. **Leaves** mostly opposite; petioles 0–0.5 mm; blades ± lanceolate to lance-linear, 15–40+ × 3–6(–10+) mm, bases rounded to cordate, margins entire or with (1–)2(–4) basal lobes, abaxial and adaxial faces ± piloso-hispid and gland-dotted. **Pistillate heads** clustered, proximal to staminates; florets 1. **Staminate heads:** peduncles 0–0.5 mm; involucres obliquely cup-shaped (lateral lobe longer than others), 2.5–4 mm diam., piloso-hispid; florets 6–8+. **Burs:** bodies pyramidal, 5–8 mm, piloso-hispid, spines 4(–5), ± distal, ± acerose, 0.5–1 mm, tips straight.

Flowering Jul–Oct. Dry, disturbed sites; 200–500 m; Ala., Ark., Conn., D.C., Ga., Ill., Ind., Iowa, Kans., Ky., La., Md., Minn., Miss., Mo., N.C., Ohio, Okla., Tenn., Tex., Va., W.Va.

Hybrids between *Ambrosia bidentata* and *A. trifida* have been recorded.

12. **Ambrosia acanthicarpa** Hooker, Fl. Bor.-Amer. 1: 309. 1833 E

Franseria acanthicarpa (Hooker) Coville

Annuals, 10–80+ cm. **Stems** erect. **Leaves** opposite (proximal) and alternate; petioles 5–15(–25+) mm; blades ± deltate, 15–40(–85+) × 12–35(–80+) mm, 1–2-pinnately lobed, bases cuneate, ultimate margins entire or toothed, abaxial faces hispid and strigillose to sericeous, adaxial faces strigillose and gland-dotted. **Pistillate heads** clustered, proximal to staminates; florets 1. **Staminate heads:** peduncles 0.5–2 mm; involucres shallowly cup-shaped (usually each with 1–5+ black nerves), 3–5(–7) mm diam., sparsely hirsute or glabrous; florets 6–12(–20+). **Burs:** bodies fusiform to obpyramidal, 3–5 mm, ± hirsutulous or glabrate, spines

8–18+, scattered, stoutly subulate (sometimes basally flattened or navicular), 2–4(–5) mm, tips straight or uncinate. $2n = 36$.

Flowering (Apr–)Jul–Nov. Canyons, sandy flats, dunes; 10–3000 m; Alta., Man., Sask.; Ariz., Calif., Colo., Idaho, Kans., Minn., Mo., Mont., Nebr., Nev., N.Mex., N.Dak., Okla., Oreg., S.Dak., Tex., Utah, Wash., Wyo.

13. **Ambrosia artemisiifolia** Linnaeus, Sp. Pl. 2: 988. 1753 • Petite herbe à poux E F

Ambrosia artemisiifolia var. *elatior* (Linnaeus) Descourtilz; *A. artemisiifolia* var. *paniculata* (Michaux) Blankinship; *A. elatior* Linnaeus; *A. glandulosa* Scheele; *A. monophylla* (Walter) Rydberg

Annuals, 10–60(–150+) cm. **Stems** erect. **Leaves** opposite (proximal) and alternate; petioles 25–35(–60+) mm; blades deltate to lanceolate or elliptic, 25–55(–90+) × 20–30(–50+) mm, 1–2-pinnately lobed, bases cuneate, ultimate margins entire or toothed, abaxial faces sparsely pilosulous to strigillose, adaxial faces strigillose, both gland-dotted. **Pistillate heads** clustered, proximal to staminates; florets 1. **Staminate heads:** peduncles 0.5–1.5 mm; involucres shallowly cup-shaped (usually without black nerves), 2–3+ mm diam., glabrous or hispid to pilosulous; florets 12–20+. **Burs:** bodies ± globose to pyriform, 2–3 mm, ± pilosulous, spines or tubercles 3–5+, near middles or distal, ± conic to acerose, 0.1–0.5+ mm, tips straight. $2n = 34, 36$.

Flowering Jul–Oct. Wet to dry soils, disturbed sites; 0–1000+ m; Alta., B.C., Man., N.B., Nfld. and Labr. (Nfld.), N.W.T., N.S., Ont., P.E.I., Que., Sask.; Ala., Alaska, Ariz., Ark., Calif., Colo., Conn., Del., D.C., Fla., Ga., Idaho, Ill., Ind., Iowa, Kans., Ky., La., Maine, Md., Mass., Mich., Minn., Miss., Mo., Mont., Nebr., Nev., N.H., N.J., N.Mex., N.Y., N.C., N.Dak., Ohio, Okla., Oreg., Pa., R.I., S.C., S.Dak., Tenn., Tex., Utah, Vt., Va., Wash., W.Va., Wis., Wyo.

Hybrids between *Ambrosia artemisiifolia* and *A. psilostachya* have been called *A.* ×*intergradiens* W. H. Wagner. The name *Ambrosia* ×*helenae* Rouleau applies to hybrids between *A. artemisiifolia* and *A. trifida*.

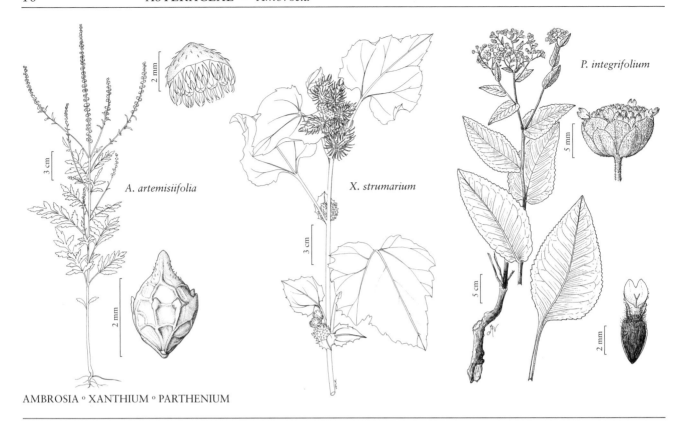

A. artemisiifolia

X. strumarium

P. integrifolium

AMBROSIA ° XANTHIUM ° PARTHENIUM

14. Ambrosia hispida Pursh, Fl. Amer. Sept., 2: 743. 1813

Perennials, 10–50(–80+) cm. **Stems** prostrate or decumbent. **Leaves** mostly opposite; petioles 5–25 mm; blades rounded-deltate to ovate or elliptic, 20–35+ × 10–30+ mm, 2–3-pinnately lobed, bases cuneate, ultimate margins entire or toothed, abaxial and adaxial faces strigillose. **Pistillate heads** clustered, proximal to staminates; florets 1. **Staminate heads:** peduncles 1–2 mm; involucres obliquely cup-shaped, 2–3+ mm diam., strigillose; florets 5–20+. **Burs:** bodies ± pyriform, 1–2 mm, ± strigillose, spines or tubercles 0–5+, near middles or distal, stoutly conic, 0.1–0.5+ mm, tips straight. *2n* = 104.

Flowering Apr. Sandy beaches or strand; 0–10 m; Fla.; West Indies (Bahamas); Central America.

15. Ambrosia chamissonis (Lessing) Greene, Man. Bot. San Francisco, 188. 1894

Franseria chamissonis Lessing, Linnaea 6: 507. 1831;
F. chamissonis subsp. *bipinnatisecta* (Lessing) Wiggins & Stockwell;
A. chamissonis var. *bipinnatisecta* (Lessing) J. T. Howell

Perennials, 50–150(–350+) cm. **Stems** prostrate or decumbent. **Leaves** mostly alternate; petioles 15–30(–45+) mm; blades rounded-deltate to ovate or obovate, 20–45(–70+) × 10–25(–50+) mm overall, not lobed or 1–2(–3)-pinnately lobed, bases cuneate, ultimate margins entire or toothed, abaxial and adaxial faces strigillose to sericeous. **Pistillate heads** clustered, proximal to staminates; florets 1. **Staminate heads:** peduncles 0.5–2 mm; involucres shallowly cup-shaped, 4–6+ mm diam., strigillose; florets 8–25(–50). **Burs:** bodies fusiform to ± pyriform, 4–7+ mm, ± strigillose to pilosulous, spines 8–16+, scattered, stoutly conic or flattened- to triquetrous-subulate, 0.5–1.5+ mm, tips straight. *2n* = 36.

Flowering Apr–Nov. Sandy beaches and strand; 0–10 m; B.C.; Alaska, Calif., Oreg., Wash.; Mexico (Baja California).

16. Ambrosia tomentosa Nuttall, Gen. N. Amer. Pl. 2: 186. 1818 [E]

Franseria discolor Nuttall; *F. tomentosa* (Nuttall) A. Nelson

Perennials, 10–30+ cm. **Stems** erect. **Leaves** mostly alternate; petioles 8–25(–35+) mm; blades elliptic to lance-elliptic, 50–80(–180+) × 12–20(–50+) mm overall, (1–) 2(–3)-pinnately lobed (lobes ± deltate), bases cuneate, ultimate margins with ± deltate teeth, abaxial faces densely finely scabrellous (silvery gray), adaxial faces strigillose. **Pistillate heads** clustered, proximal to staminates; florets (1–)2. **Staminate heads:** peduncles 2–9 mm; involucres saucer-shaped (with black nerves), 3–6+ mm diam., strigillose; florets 25–40+. **Burs:** bodies ± pyriform, somewhat flattened, 1.5–2.5(–4) mm, finely scabrellous, spines (1–) 8–12+, scattered, stoutly conic, 0.5–1+ mm, tips straight.

Flowering Jun–Sep. Seasonally wet clays or sands, disturbed sites; (200–)1200–2100 m; Ariz., Colo., Idaho, Ill., Iowa, Kans., Mo., Mont., Nebr., N.Mex., N.Dak., S.Dak., Tex., Utah, Wis., Wyo.

17. Ambrosia cheiranthifolia A. Gray in W. H. Emory, Rep. U.S. Mex. Bound. 2(1): 87. 1859

Perennials or subshrubs, 10–30+ cm (rhizomatous and/or soboliferous, colonial). **Stems** erect. **Leaves** opposite (distal sometimes alternate); petioles 0; blades lanceolate or lance-elliptic to lance-oblong or oblanceolate, 20–50 (–70+) × 8–12(–25+) mm, rarely ± pinnately lobed (lobes ± deltate), bases cuneate (then rounded), margins entire or toothed, abaxial faces densely strigose (silvery gray), adaxial faces strigillose. **Pistillate heads** clustered (in axils), proximal to staminates; florets 1(–2). **Staminate heads:** peduncles 1–4+ mm; involucres cup-shaped (sometimes with black nerve in each lobe), 3–4+ mm diam., strigillose; florets 8–18+. **Burs:** bodies ± pyriform to ± globose, 2–3+ mm, glabrous or strigose, spines (1–)3–5+, scattered, stoutly conic, 0.4–0.8+ mm, tips straight (often each with black nerve). **2n = 72.**

Flowering Jun–Nov. Seasonally wet clays or sands, scrublands; 10–30+ m; Tex.; Mexico (Tamaulipas).

18. Ambrosia grayi (A. Nelson) Shinners, Field & Lab. 17: 174. 1949 [E]

Gaertneria grayi A. Nelson, Bot. Gaz. 34: 35. 1902, based on *Franseria tomentosa* A. Gray, Mem. Amer. Acad. Arts, n. s. 4: 80. 1849, not *Ambrosia tomentosa* Nuttall 1818

Perennials, 10–30+ cm. **Stems** erect. **Leaves** mostly alternate; petioles 10–45+ mm; blades elliptic to ovate, 45–60(–100+) × 30–45(–75+) mm, usually 1–2-pinnately lobed (lobes ± lanceolate), bases cuneate, ultimate margins serrate, abaxial faces densely sericeous or strigillose (silver or grayish), adaxial faces strigillose. **Pistillate heads** clustered, proximal to staminates; florets (1–)2. **Staminate heads:** peduncles 1–2 mm; involucres saucer-shaped, 3.5–5+ mm diam., strigillose; florets 8–25(–40+). **Burs:** bodies ± pyriform to globose, 3–5+ mm, stipitate-glandular, spines 8–12+, scattered, stoutly conic, 0.5–1+ mm, tips straight. **2n = 36.**

Flowering Aug–Sep. Wet or seasonally wet flats, depressions, alkaline, clay soils, prairies; 600–1200 m; Colo., Kans., Nebr., Okla., Tex.

19. Ambrosia linearis (Rydberg) W. W. Payne, J. Arnold Arbor. 45: 425. 1964 [E]

Gaertneria linearis Rydberg, Bull. Torrey Bot. Club 32: 133. 1905

Perennials or subshrubs, 20–40+ cm. **Stems** erect. **Leaves** mostly alternate; petioles 0; blades mostly linear (some 1-pinnate, lobes linear), 12–20(–35+) × 1(–2+) mm, bases cuneate, margins entire, abaxial faces strigillose (whitish), adaxial faces scabrellous. **Pistillate heads** clustered, proximal to staminates; florets 1. **Staminate heads:** peduncles 0.5–1 mm; involucres cup-shaped (with black nerves), 4–6+ mm diam., strigillose; florets 12–20+. **Burs:** bodies pyriform to obconic, 2–4(–5) mm, strigillose, spines (4–)5–9(–13+), scattered or mostly distal, subulate, (0.8–)1–1.5 mm, tips usually uncinate.

Flowering Jul. Seasonally wet, sandy soils; 2000–2100 m; Colo.

Ambrosia linearis is in the Center for Plant Conservation's National Collection of Endangered Plants.

20. **Ambrosia confertiflora**   de Candolle in A. P. de Candolle and A. L. P. P. de Candolle, Prodr. 5: 526. 1836

Franseria confertiflora (de Candolle) Rydberg

Perennials, 20–80(–150+) cm. **Stems** erect. **Leaves** mostly alternate; petioles 10–35 mm; blades lanceolate to ovate, 40–85(–150) × 20–35(–55) mm overall, laciniately 2–4-pinnately lobed (lobes ± lanceolate); bases cuneate to truncate, ultimate margins entire, abaxial and adaxial faces strigillose to sericeous (often grayish) and gland-dotted. **Pistillate heads** clustered, proximal to staminates; florets 1(–2). **Staminate heads:** peduncles 0.5–2 mm; involucres cup-shaped, 1.5–3+ mm diam., strigillose; florets 5–20+. **Burs:** bodies pyramidal to pyriform, 1–2 mm, strigillose to pilosulous, spines (1–)5–12+, mostly distal, stoutly conic to subulate, 0.5–1 mm, tips uncinate. $2n$ = 72, 108.

Flowering (May–)Sep–Dec. Waste places, disturbed sites; 10–2000 m; Ariz., Calif., Colo., Kans., N.Mex., Okla., Tenn., Tex., Utah; Mexico.

21. **Ambrosia pumila**   (Nuttall) A. Gray, Proc. Amer. Acad. Arts 17: 217. 1882

Franseria pumila Nuttall, Trans. Amer. Philos. Soc., n. s. 7: 344. 1840

Perennials, 10–20(–40+) cm. **Stems** erect. **Leaves** mostly alternate (sometimes crowded at bases of stems); petioles 5–35(–55) mm; blades deltate to elliptic, 15–35 (–75) × 12–25(–45) mm overall, laciniately (1–)2(–3)-pinnately lobed, bases cuneate to truncate, ultimate margins entire, abaxial and adaxial faces ± strigillose to sericeous (often grayish) and gland-dotted. **Pistillate heads** clustered, proximal to staminates; florets 1. **Staminate heads:** peduncles 0.5–6 mm; involucres obliquely cup-shaped, 3–5 mm diam., ± strigillose; florets 8–25+. **Burs:** bodies ± fusiform, 2–2.5 mm, strigillose, spines or tubercles 0 or 1–5, mostly distal, stoutly conic, 0.1–0.5 mm, tips straight. $2n$ = 72.

Flowering May–Jun(–Oct). Disturbed sites, damp to wet, alkaline soils; 100–200 m; Calif.; Mexico (Baja California).

22. **Ambrosia psilostachya**   de Candolle in A. P. de Candolle and A. L. P. P. de Candolle, Prodr. 5: 526. 1836 • Herbe à poux vivace

Ambrosia psilostachya var. *californica* (Rydberg) S. F. Blake; *A. psilostachya* var. *coronopifolia* (Torrey & A. Gray) Farwell; *A. psilostachya* var. *lindheimeriana* (Scheele) Blankinship; *A. rugelii* Rydberg

Perennials, 10–60(–100+) cm. **Stems** erect. **Leaves** proximally opposite, distally alternate; petioles 0–25 mm (often ± winged); blades deltate to lanceolate, 20–60(–140) × 8–35(–50+) mm, pinnately toothed to 1-pinnately lobed, bases cuneate to truncate, ultimate margins entire or toothed, abaxial and adaxial faces hirsutulous to strigose and gland-dotted. **Pistillate heads** clustered, proximal to staminates; florets 1. **Staminate heads:** peduncles 0.5–2 mm; involucres obliquely cup-shaped, 2–4(–5) mm diam., hirsutulous; florets 5–15(–30+). **Burs:** bodies ± obpyramidal to globose, 2–3 mm, hirsutulous, spines or tubercles 0 or 1–6, mostly distal, stoutly conic to acerose, (0.1–)0.5–1 mm, tips straight. $2n$ = 18, 27, 36, 45, 54, 63, 72, 100–104, 108, 144.

Flowering Jul–Oct(–Dec). Disturbed sites, often wet, alkaline, clay soils; 0–2200 m; Alta., B.C., Man., N.S., Ont., P.E.I., Que., Sask.; Ala., Ariz., Ark., Calif., Colo., Fla., Ga., Idaho, Ill., Ind., Iowa, Kans., La., Maine, Mass., Mich., Minn., Miss., Mo., Mont., Nebr., Nev., N.H., N.Mex., N.Y., N.C., N.Dak., Ohio, Okla., Oreg., Pa., S.C., S.Dak., Tenn., Tex., Utah, Vt., Wash., Wis., Wyo.; Mexico (Baja California, Chihuahua, Coahuila).

Hybrids between *Ambrosia psilostachya* and *A. artemisiifolia* have been called *A.* ×*intergradiens* W. H. Wagner. Some botanists consider the type of *A. cumanensis* Kunth (1818) to be conspecific with that of *A. psilostachya.*

245. XANTHIUM Linnaeus, Sp. Pl. 2: 987. 1753; Gen. Pl. ed. 5, 424. 1754 • Lampourde [Greek *xanthos*, yellow, evidently alluding to an ancient name for a plant that produced a yellow dye]

John L. Strother

Annuals (coarse), 10–200+ cm. **Stems** erect, branched. **Leaves** cauline; mostly alternate (proximal 2–6 sometimes opposite); petiolate; blades lanceolate, linear, ovate, rounded-deltate, or suborbiculate, often ± palmately or pinnately lobed, ultimate margins entire or ± toothed, faces hirtellous or ± strigose, usually gland-dotted as well. **Heads** discoid, either pistillate (proximal) or functionally staminate (distal), in racemiform to spiciform arrays or borne singly (in axils). **Pistillate heads:** involucres ± ellipsoid, 2–5+ mm diam. at anthesis (6–20+ mm diam. at maturity); phyllaries 30–75+ in 6–12+ series, outer 5–8 distinct, the rest (sometimes interpreted as paleae) proximally connate, their distinct tips mostly ± hooked (the distal 1–3 usually longer, stouter, and not hooked), the whole becoming a hard, prickly perigynium (a bur); florets 2, corollas 0. **Staminate heads:** involucres ± saucer-shaped, 3–5 mm diam.; phyllaries 6–16+ in 1–2+ series, distinct to bases; receptacles conic to columnar; paleae spatulate to cuneiform or linear, membranous, distally ± villous or hirtellous; florets 20–50+, corollas whitish, ± funnelform, lobes 5, erect or reflexed (filaments connate, anthers distinct or weakly coherent). **Cypselae** (black) ± fusiform, enclosed in obovoid to ellipsoid, hard, prickly, 2-chambered burs; **pappi** 0. $x = 18$.

Species 2–3 (2 in the flora): New World, introduced nearly worldwide.

1. Nodal spines 0; leaf blades suborbiculate to ± pentagonal or deltate . . . . 1. *Xanthium strumarium*
1. Nodal spines (1–)3-lobed, 15–30+ mm; leaf blades ± lanceolate to ovate or lance-linear . . . . 2. *Xanthium spinosum*

1. Xanthium strumarium Linnaeus, Sp. Pl. 2: 987. 1753 • Cocklebur, lampourde glouteron [F]

Xanthium americanum Walter; *X. chasei* Fernald; *X. chinense* Miller; *X. curvescens* Millspaugh & Sherff; *X. cylindricum* Millspaugh & Fernald; *X. echinatum* Murray; *X. echinellum* Greene ex Rydberg; *X. globosum* C. Schull; *X. inflexum* Mackenzie & Bush; *X. italicum* Moretti; *X. orientale* Linnaeus; *X. oviforme* Wallroth; *X. pensylvanicum* Wallroth; *X. speciosum* Kearney, *X. strumarium* var. *canadense* (Miller) Torrey & A. Gray, *X. strumarium* var. *glabratum* (de Candolle) Cronquist; *X. varians* Greene; *X. wootonii* Cockerell

Plants 10–80(–200) cm; nodal spines 0. **Leaves:** petioles 20–100(–140+) mm; blades suborbiculate to ± pentagonal or deltate, 4–12(–18+) × 3–10(–18+) cm, sometimes palmately 3–5-lobed, abaxial faces green, hirtellous. **Burs** 10–30+ mm. $2n = 36$.

Flowering Jul–Oct. Damp or seasonally wet, often alkaline, soils, waste places, margins of agriculture; 10–2000 m; Alta., B.C., Man., N.B., N.S., Ont., P.E.I., Que., Sask.; Ala., Alaska, Ariz., Ark., Calif., Colo., Conn., Del., D.C., Fla., Ga., Idaho, Ill., Ind., Iowa, Kans., Ky., La., Maine, Md., Mass., Mich., Minn., Miss., Mo., Mont., Nebr., Nev., N.H., N.J., N.Mex., N.Y., N.C., N.Dak., Ohio, Okla., Oreg., Pa., R.I., S.C., S.Dak., Tenn., Tex., Utah, Vt., Va., Wash., W.Va., Wis., Wyo.; Mexico; Central America; South America; widely introduced in Old World.

Recognition of a dozen or more taxa (treated as species, subspecies, varieties, and/or forms) has been proposed for plants treated together here as *Xanthium strumarium*. Bases for the various taxa mostly involved subtle differences in the burs.

2. Xanthium spinosum Linnaeus, Sp. Pl. 2: 987. 1753 • Spiny cocklebur, clotbur, lampourde épineuse

Xanthium ambrosioides Hooker & Arnott; *X. spinosum* var. *inerme* Bel

Plants 10–60(–120+) cm; nodal spines usually in pairs, simple or 2–3-partite, 15–30+ mm. **Leaves:** petioles 1–15(–25+) mm; blades ± ovate to lanceolate or lance-linear, 4–8(–12+) × 1–3(–5+) cm, often pinnately 3(–7+)-lobed, abaxial faces gray to white, densely strigose. **Burs** 10–12(–15+) mm. $2n = 36$.

Flowering Jul–Oct. Damp or seasonally wet, alkaline soils, waste places, margins of agriculture; 10–1000 m; N.B., Ont., Que., Sask.; Ala., Ariz., Calif., Colo., Conn., Del., Fla., Ga., Idaho, Ill., Ind., Iowa, Kans., Ky., Maine, Md., Mass., Mich., Miss., Mo., Mont., Nebr., Nev., N.H., N.J., N.Mex., N.Y., N.C., Ohio, Oreg., Pa., R.I., S.C., Tenn., Tex., Utah, Va., Wash., W.Va.; Mexico; Central America; South America; widely established in Old World.

Some authors have contended that *Xanthium spinosum* originated in South America and is introduced and/or naturalized everywhere else that it is found.

246. PARTHENIUM Linnaeus, Sp. Pl. 2: 988. 1753; Gen. Pl. ed. 5, 426. 1754

• Feverfew [Greek *parthenos*, virgin, or *parthenion*, ancient name of a plant; allusion unclear]

John L. Strother

Annuals, biennials, perennials, subshrubs, or shrubs [treelets], 1–120[–400] cm. **Stems** ± erect, usually branched. **Leaves** usually cauline, sometimes in rosettes; alternate; petiolate or sessile; blades elliptic, lanceolate, linear, lyrate, oblanceolate, obovate, ovate, rounded-deltate, spatulate, sometimes (1–)2-pinnately lobed, ultimate margins entire or toothed, faces usually hairy and gland-dotted (at least the abaxial). **Heads** usually radiate, sometimes ± disciform (*P. alpinum*), usually borne in corymbiform or paniculiform arrays (in glomerules in *P. argentatum*, borne singly in *P. alpinum* and *P. ligulatum*). **Involucres** ± hemispheric [rotate], 3–8(–12+) mm diam. **Phyllaries** falling, 10(–16) in 2 series, usually distinct (partially connate in *P. alpinum*), outer 5(–8) herbaceous to scarious, inner 5–8 ± scarious to membranous. **Receptacles** flat to conic; paleae cuneate to flabelliform, scarious or membranous, distally papillate and/or fimbrillate, all or the peripheral each ± enfolding a disc floret. **Ray (pistillate) florets** 5(–8), fertile; corollas ochroleucous, tubes stout, glandular, laminae oblong to reniform or orbiculate, or ± coroniform (none in *P. alpinum*). **Disc florets** 12–60+, functionally staminate; corollas ochroleucous, funnelform, lobes 5 (filaments distinct, anthers connate). **Cypselae** (black) oblanceoloid, obovoid, or pyriform, often ± obcompressed (shed together with subtending phyllary and 2 contiguous disc florets and their investing paleae); **pappi** 0 (shoulders of cypselae may bear 1–3 pappus-like, triangular to ovate, or ± subulate enations). $x = 9$.

Species ca. 16 (7 in the flora): New World, mostly warm-temperate and tropical, some temperate; introduced in Old World.

SELECTED REFERENCES Mears, J. A. 1975. The taxonomy of *Parthenium* section *Partheniastrum* DC. (Asteraceae–Ambrosiinae). Phytologia 31: 463–482. Rollins, R. C. 1950. The guayule rubber plant and its relatives. Contr. Gray Herb. 172: 1–73.

1. Plants 1–2 cm (plants cespitose or forming mats); heads borne singly.
    2. Corollas of pistillate florets bearing laminae 1–2 mm . . . . 1. *Parthenium ligulatum*
    2. Corollas of pistillate florets lacking laminae . . . . 2. *Parthenium alpinum*
1. Plants (5–)15–120 cm (plants not cespitose or forming mats); heads in glomerules, or in simple or compound, corymbiform or paniculiform arrays.
    3. Shrubs.
        4. Leaf blades oval-elliptic to obovate (sometimes pinnately 3–7-lobed or round-toothed, ultimate margins entire) . . . . 3. *Parthenium incanum*
        4. Leaf blades lanceolate to oblanceolate (margins mostly entire, some with 1–5 sharp teeth) . . . . 4. *Parthenium argentatum*
    3. Annuals, biennials, or perennials.
        5. Perennials; leaf blades usually crenate to serrate, sometimes coarsely toothed or somewhat lobed (then mostly toward bases); pappus-like enations from shoulders of cypselae 0 or 2(–4), erect to spreading, ± subulate or threadlike . . . . . 5. *Parthenium integrifolium*
        5. Annuals or biennials (sometimes persisting); leaf blades mostly 1–2-pinnately lobed; pappus-like enations from shoulders of cypselae 2, erect, ± deltate to ovate.

[6. Shifted to left margin.—Ed.]

6. Biennials (sometimes flowering first year or persisting); leaf blades ± pinnately lobed, abaxial faces strigillose (usually with erect hairs 1–2 mm as well) . . . . 6. *Parthenium confertum*
6. Annuals (rarely persisting); leaf blades (1–)2-pinnately lobed, abaxial faces scabrellous (seldom with erect hairs 1–2 mm) . . . . 7. *Parthenium hysterophorus*

1. **Parthenium ligulatum** (M. E. Jones) Barneby, Leafl. W. Bot. 5: 20. 1947 E

Parthenium alpinum (Nuttall) Torrey & A. Gray var. *ligulatum* M. E. Jones, Contr. W. Bot. 13: 16. 1910

Perennials, 1–2 cm (underground caudices 2–5+ cm, branched; plants cespitose or forming mats). **Leaf blades** oblanceolate to spatulate, 5–18+ × 1–2(–4+) mm, margins entire, faces strigilloso-sericeous (gray) and obscurely gland-dotted. **Heads** radiate, borne singly. **Peduncles** 0–4 mm. **Phyllaries:** outer 5 ± ovate, 2.5–3 mm, inner 5 ± orbiculate, 3–4+ mm. **Pistillate florets** 5; corolla laminae oval to elliptic, 1–2 mm, retuse. **Disc florets** 15–25+. **Cypselae** oblanceoloid to pyriform, 3.5–4 mm; pappus-like enations 2, erect to spreading, ± subulate, 0.5–1 mm (a third, ± subulate element sometimes at apex of adaxial face). $2n = 36, 72$.

Flowering May–Jun. Pale shale or sandstone outcrops with pinyon-juniper; 1800–2000; Colo., Utah.

2. **Parthenium alpinum** (Nuttall) Torrey & A. Gray, Fl. N. Amer. 2: 285. 1842 E

Bolophyta alpina Nuttall, Trans. Amer. Philos. Soc., n. s. 7: 348. 1840; *Parthenium tetraneuris* Barneby; *P. alpinum* var. *tetraneuris* (Barneby) Rollins

Perennials, 1–2 cm (underground caudices 2–5+ cm, branched; plants cespitose or forming mats). **Leaf blades** oblanceolate to spatulate, 6–18(–35) × 1–3(–4+) mm, margins entire, faces strigilloso-sericeous (gray) and obscurely gland-dotted. **Heads** ± disciform, borne singly. **Peduncles** 0–5(–30) mm. **Phyllaries:** outer 5–8 ± linear, 4 mm, inner 5–8 ± orbiculate, 4–5 mm. **Pistillate florets** 5–8; corolla laminae 0 (tubes ± compressed, obscurely 2–4-lobed). **Disc florets** 18–28+. **Cypselae** oblanceoloid, 4 mm (narrowly winged); pappus-like enations 2, erect to spreading, ± subulate, 0.5–1 mm (a third, ± subulate element sometimes at apex of adaxial face). $2n = 36$.

Flowering May–Jul. Shale and calcareous outcrops, red clays; 1300–1700 m; Colo., N.Mex., Wyo.

Parthenium alpinum is in the Center for Plant Conservation's National Collection of Endangered Plants.

3. **Parthenium incanum** Kunth in A. von Humboldt et al., Nov. Gen. Sp. 4(fol.): 204. 1818; 4(qto.): 260. plate 391. 1820 • Mariola

Shrubs, 30–100+ cm. **Leaf blades** oval-elliptic to obovate, 15–25 (–40+) × 6–15(–25+) mm, sometimes pinnately (3–)5–7-lobed or round-toothed, ultimate margins entire, faces tomentose (gray to white) and gland-dotted. **Heads** radiate, in glomerules of 3–5+ on branched stalks 1–5(–12+) cm, forming compound, corymbiform arrays. **Peduncles** 1–3+ mm. **Phyllaries:** outer 5 oblong, 1.5–2 mm, inner 5 orbiculate, 2–2.5 mm. **Pistillate florets** 5; corolla laminae ovate, 1–1.5 mm. **Disc florets** 8–20(–30+). **Cypselae** obovoid, 1.5–2 mm; pappus-like enations 2(–3), erect to spreading, ± subulate, 0.5–1 mm. $2n = 54$.

Flowering (May–)Jul–Nov. Openings in desert scrub, often on limestone soils; 1000–1500 m; Ariz., Nev., N.Mex., Tex., Utah; Mexico.

4. **Parthenium argentatum** A. Gray in W. H. Emory, Rep. U.S. Mex. Bound. 2(1): 86. 1859 • Guayule

Shrubs, 30–100+ cm. **Leaf blades** lanceolate to oblanceolate, 15–25 (–40+) × 6–15(–25+) mm, margins mostly entire, some with 1–2(–5) sharp teeth, faces densely strigillose (gray to white) and obscurely or not at all gland-dotted. **Heads** radiate, in glomerules of 3–5+ at ends of ± ebracteate stalks 8–15(–20+) cm. **Peduncles** 1–2(–6+) mm. **Phyllaries:** outer 5 oval-elliptic, 2.5–3+ mm, inner 5 ± orbiculate, 3.5–4 mm. **Pistillate florets** 5; corolla laminae ovate, 1.2–1.5 mm. **Disc florets** 20–30+. **Cypselae** ± obovoid, 2.5–3 mm; pappus-like enations 2(–4), erect to spreading, ± subulate, 0.3–0.8 mm. $2n = 36$.

Flowering Apr–Jun. Calcareous soils; 1000–1500 m; Tex.; Mexico.

Guayule has sometimes been used for commercial production of natural rubber. Hybrids between *Parthenium argentatum* and *P. incanum* have been noted.

5. Parthenium integrifolium Linnaeus, Sp. Pl. 2: 988. 1753 [E] [F]

Parthenium auriculatum Britton; *P. hispidum* Rafinesque; *P. hispidum* var. *auriculatum* (Britton) Rollins; *P. integrifolium* var. *auriculatum* (Britton) Cornelius ex Cronquist; *P. integrifolium* var. *henryanum* Mears; *P. integrifolium* var. *hispidum* (Rafinesque) Mears; *P. integrifolium* var. *mabryanum* Mears; *P. radfordii* Mears

Perennials, 30–60(–100+) cm. **Leaf blades** ovate to lanceolate, 30–350+ × 20–120+ mm, margins usually crenate to serrate, sometimes coarsely toothed or somewhat lobed (then mostly toward bases), faces hispid to hirtellous or ± scabrous, gland-dotted. **Heads** radiate, borne in corymbiform to paniculiform arrays. **Peduncles** 1–8(–12+) mm. **Phyllaries:** outer 5(–6) lanceolate to broadly ovate, 3–5 mm, inner 5(–6) ± orbiculate, 4–6 mm. **Pistillate florets** 5(–6); corolla laminae ovate to oblong or orbiculate, 1–2+ mm. **Disc florets** 15–35+. **Cypselae** ± obovoid, 3–4+ mm; pappus-like enations 0 or 2(–4), erect to spreading, ± subulate or threadlike, fragile, 0.3–0.6+ mm. **2***n* = 72.

Flowering May–Sep. Glades and barrens, prairies, disturbed sites; 10–500 m; Ala., Ark., Conn., Ga., Ill., Ind., Iowa., Kans., Ky., La., Mass., Mich., Minn., Miss., Mo., N.Y., N.C., Ohio, Okla., Pa., S.C., Tenn., Tex., Va., W.Va., Wis.

As evidenced by the synonymy, *Parthenium integrifolium* as here circumscribed has been variously partitioned by other taxonomists. For the present, I see no justification for segregating species from, or for formal recognition of infraspecific taxa within *P. integrifolium*.

6. Parthenium confertum A. Gray, Proc. Amer. Acad. Arts 17: 216. 1882

Parthenium confertum var. *divaricatum* Rollins; *P. confertum* var. *lyratum* (A. Gray) Rollins; *P. confertum* var. *microcephalum* Rollins

Biennials (sometimes flowering first year or persisting), 10–30(–60+) cm. **Leaf blades** ovate or rounded-deltate to elliptic, 30–80(–120+) × 10–20(–40+) mm, usually ± pinnately or sub-bipinnately lobed, ultimate margins entire, faces strigillose, usually also with erect hairs 1–2 mm and gland-dotted. **Heads** disci-form or obscurely radiate, borne in open, ± paniculiform arrays. **Peduncles** 2–8(–12+) mm. **Phyllaries:** outer 5 lance-ovate to elliptic, 2.5–3+ mm, inner 5 ± orbiculate, 3–3.5 mm. **Pistillate florets** 5; corolla laminae 0 or ± coroniform, 0.1–0.5 mm. **Disc florets** 20–30+. **Cypselae** ± obovoid, 2–3 mm; pappus-like enations 2, ± erect, deltate to ovate, 0.5–1 mm (sometimes a third, subulate spur near apex adaxially). **2***n* = 36, 68, 72.

Flowering Mar–Oct. Sandy plains, openings in mesquite grasslands; 20–2000 m; Ariz., N.Mex., Tex.; Mexico.

7. Parthenium hysterophorus Linnaeus, Sp. Pl. 2: 988. 1753 [F]

Annuals, (10–)30–120+ cm. **Leaf blades** ovate to elliptic, 30–180+ × 10–50(–90+) mm, (1–)2-pinnately lobed (ultimate lobes lanceolate to linear, 3–50 × 2–15 mm), faces sparsely to densely scabrellous and gland-dotted (seldom with additional erect hairs 1–2 mm). **Heads** obscurely radiate, borne in open, paniculiform arrays. **Peduncles** 1–8(–15+) mm. **Phyllaries:** outer 5(–6), lance-elliptic, 2–4 mm, inner 5(–6) ovate to ± orbiculate, 2.5–4 mm. **Pistillate florets** 5(–6); corolla laminae reniform or orbiculate to oblong, 0.3–1 mm. **Disc florets** 12–30[–60]. **Cypselae** obovoid, 1.5–2(–3.5) mm; pappus-like enations erect, deltate to ovate, 0.5–1 mm (sometimes a third, subulate spur near apex adaxially). **2***n* = 34.

Flowering Mar–Nov. Disturbed sites; 0–300 m; Ala., Ark., Conn., Del., D.C., Fla., Ill., Kans., La., Md., Mass., Mich., Miss., Mo., N.J., N.Mex., N.Y., Ohio, Okla., Tex., Va.; Mexico; introduced in subtropical, tropical, and warm-temperate New World and Old World.

In some places, especially in Old World, *Parthenium hysterophorus* causes serious contact dermatitis in humans. The plants should be handled with caution.

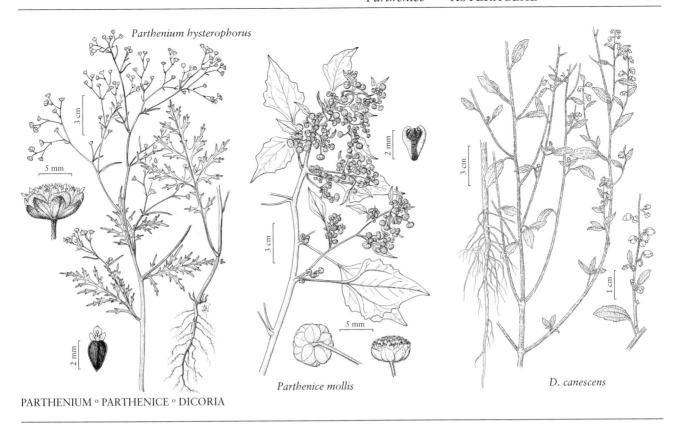

Parthenium hysterophorus

Parthenice mollis

D. canescens

PARTHENIUM ° PARTHENICE ° DICORIA

247. **PARTHENICE** A. Gray, Smithsonian Contr. Knowl. 5(6): 85. 1853 • [No etymology in protologue; evidently alluding to similarities to members of genus *Parthenium*]

John L. Strother

Annuals, 20–200 cm. **Stems** erect. **Leaves** cauline; alternate; petiolate; blades (3-nerved) deltate to ovate, margins entire or toothed, faces usually softly pubescent, sometimes rough-hairy. **Heads** obscurely radiate or disciform, borne in loose, paniculiform arrays. **Involucres** hemispheric, 3–5 mm. **Phyllaries** persistent (outer) or falling, 13, distinct, outer 5 herbaceous (becoming reflexed), inner 6–8 membranous (each becoming cupped around a cypsela). **Receptacles** convex; paleae of 2 kinds: peripheral purplish-mottled, becoming somewhat fleshy and arcuate-clavate, shed together in pairs with each cypsela; inner 0 or narrowly conic, persistent. **Ray florets** 6–8; corollas yellowish (minute), glandular. **Disc florets** 10–25+, functionally staminate; corollas yellowish, funnelform, distally glandular and hairy (anthers ± connate). **Cypselae** (ray) strongly obcompressed to obflattened, obovate, tuberculate, shed with 2 adjacent, ± fleshy paleae; **pappi** 0 or rudimentary. $x = 18$.

  Species 1: Arizona, nw Mexico.

SELECTED REFERENCE Sauck, J. R. 1975. Distribution, chromosomes, and taxonomy of *Parthenice mollis* (Compositae). Madroño 23: 227–234.

1. Parthenice mollis A. Gray, Smithsonian Contr. Knowl. 5(6): 85. 1853 ☐F

Leaf blades mostly 10–30, 8–25 cm, distal smaller. **Heads** 10–100+ together. **Cypselae** obovoid, somewhat cucullate, 1.8–2 mm, often tuberculate, glabrous. $2n = 36$.

Flowering mostly summer. Desert slopes, washes, southern exposures; 900–1400 m; Ariz.; Mexico (Baja California Sur, Chihuahua, Sinaloa, Sonora).

248. DICORIA   Torrey & A. Gray in W. H. Emory, Rep. U.S. Mex. Bound. 2(1): 86, plate 30. 1859 • [Greek *di*, two, and *koris*, bug, alluding to the two, "buglike" cypselae of the original species]

John L. Strother

Annuals, perennials, or subshrubs [shrubs], 10–90+ cm. **Stems** erect, virgately to divaricately branched. **Leaves** cauline; proximally opposite (first 2–10+ pairs in early growth, Feb–Apr), otherwise alternate; petiolate; blades (3-nerved) lance-linear to lanceolate (proximal) or ± deltate or ovate to elliptic or lanceolate (distal), margins entire or toothed, faces sericeous to strigillose, sometimes with coarser, erect hairs, usually gland-dotted as well. **Heads** disciform or discoid (then functionally staminate), in (mostly ebracteate) racemiform to paniculiform arrays [borne singly or loosely aggregated in 2s or 3s]. **Involucres** ± cup-shaped to saucer-shaped, 3–5+ mm diam. **Phyllaries** persistent (outer) or tardily falling, distinct, outer (4–)5(–7) in 1 series, ± herbaceous, inner 0–4 (each subtending a pistillate floret), scarious to membranous (accrescent, ultimately ovate to elliptic in fruit). **Receptacles** convex; paleae cuneiform to linear, membranous, ± villous to hispid distally, sometimes wanting. **Pistillate florets** (0–)1–4; corollas 0. **Functionally staminate florets** 5–15+; corollas whitish, funnelform, lobes 5, erect, reflexed, or incurved (anthers distinct or weakly coherent). **Cypselae** strongly obcompressed, often slightly cucullate, ± obovate to elliptic, margins corky-winged, irregularly toothed, faces smooth or warty, sometimes gland-dotted; **pappi** 0 (cypselae often each with apical tuft of white hairs). $x = 18$.

Species 2 (1 in the flora): w North America, nw Mexico.

Traits that have been used to distinguish species and/or infraspecific taxa of dicorias in the flora (presence/absence of erect hairs on stems, shape, size, and toothing of distal leaves, size of cypselae, and relative development of inner phyllaries; A. Cronquist 1994) vary and combine so capriciously that I find no justification for recognition of more than a single, polymorphic species in the flora area.

1. Dicoria canescens A. Gray in W. H. Emory, Rep. U.S. Mex. Bound. 2(1): 87, plate 30. 1859 F

Dicoria brandegeei A. Gray; *D. canescens* subsp. *brandegeei* (A. Gray) Cronquist; *D. canescens* var. *brandegeei* (A. Gray) Cronquist; *D. canescens* subsp. *clarkiae* (P. B. Kennedy) D. D. Keck; *D. canescens* subsp. *hispidula* (Rydberg) D. D. Keck; *D. canescens* var. *hispidula* (Rydberg) Cronquist; *D. canescens* var. *witherillii* (Eastwood) Cronquist; *D. clarkiae* P. B. Kennedy; *D. paniculata* Eastwood

Leaves: petioles 5–20+ mm; blades mostly 10–30(–120) × 3–20(–30+) mm. **Peduncles** 1–3+ mm. **Involucres** 2–3+ mm. **Phyllaries:** outer ± lanceolate, ± herbaceous, sericeous to strigillose, inner ultimately obovate to ± orbiculate, usually ± cucullate, scarious to membranous, usually glandular-pubescent, each ± investing a cypsela. **Paleae** 1.8–2+ mm. **Functionally staminate florets:** corollas 2.5–3 mm. **Cypselae** 3–8+ mm. **2n** = 36.

Flowering year round, mostly Oct–Feb. Alkaline soils, desert washes, flats; -20–1700 m; Ariz., Calif., Colo., Nev., N.Mex., Utah; Mexico (Baja California, Sonora).

249. IVA Linnaeus, Sp. Pl. 2: 988. 1753; Gen. Pl. ed. 5, 426. 1754 • Marsh elder

[Etymology uncertain; perhaps for *Ajuga iva*, a mint with similar odor]

John L. Strother

Annuals, perennials, subshrubs, or shrubs, 10–350 cm. **Stems** usually erect, sometimes decumbent to sprawling, often freely branched. **Leaves** cauline; mostly opposite (distal sometimes alternate); petiolate or sessile; blades (1- or 3-nerved) deltate, elliptic, filiform, lanceolate, linear, obovate, ovate, spatulate, or trullate, margins entire or toothed, faces glabrous or ± scabrellous, often gland-dotted. **Heads** discoid or ± disciform, in (± bracteate) racemiform or spiciform arrays (heads 1–2 in axil of each bract, bracts sometimes leaflike). **Involucres** ± hemispheric or turbinate to campanulate or urceolate, 2–10+ mm diam. **Phyllaries** persistent, 3–15+ in 1–3+ series, distinct or ± connate, all ± herbaceous or inner scarious to membranous. **Receptacles** flat or convex to hemispheric; paleae cuneiform or spatulate to linear or setiform, ± membranous, sometimes all or partially wanting. **Pistillate florets** usually 1–8+, rarely 0; corollas whitish (inconspicuous), ± tubular. **Functionally staminate florets** 3–20+; corollas whitish to pinkish, funnelform, lobes 5, soon reflexed (filaments ± connate, anthers coherent or distinct). **Cypselae** plumply obovoid to pyriform, often ± obcompressed, sometimes scabrellous or hispidulous distally, usually ± gland-dotted; **pappi** 0. *x* = 18.

Species ca. 9 (7 in the flora): mostly temperate North America, some subtropical, some introduced in Old World.

M. Bolick (1983), P. O. Karis (1995), and B. Miao et al. (1995, 1995b, 1995c) considered *Iva* in the sense of R. C. Jackson (1960) to include species that have closer relatives outside *Iva* in the broad sense than within. I agree and treat the extradited species in *Chorisiva*, *Cyclachaena*, *Hedosyne*, *Leuciva*, and *Oxytenia*.

Ivas are wind pollinated, are related to the ragweeds (*Ambrosia* spp.), and may be similarly associated with pollen allergies.

Records of *Iva asperifolia* Lessing from Florida are evidently based on specimens that are treated as members of *I. angustifolia*.

1. Annuals (seldom persisting); leaf blades deltate, filiform, lanceolate, lance-linear, or ovate (rarely succulent or fleshy); heads in ± spiciform arrays; peduncles mostly 0.5–1 mm.
    2. Leaves petiolate, blades mostly deltate to lanceolate, margins toothed . . . . 5. *Iva annua*
    2. Leaves ± sessile, blades mostly lance-linear to filiform, margins entire.
        3. Involucres ± prismatically turbinate, 2.5–3 mm; outer phyllaries ± connate . . . . . 6. *Iva angustifolia*
        3. Involucres ± hemispheric, 1.5–2 mm; outer phyllaries distinct . . . . 7. *Iva microcephala*

1. Perennials, subshrubs, or shrubs; leaf blades elliptic, lanceolate, oblanceolate, obovate, ovate, or spatulate (usually ± succulent or fleshy); heads in ± racemiform arrays; peduncles mostly 1–7 mm.

    4. Subshrubs or shrubs, 50–350 cm; leaf blades ovate or elliptic to lanceolate, margins usually toothed . . . . 1. *Iva frutescens*

    4. Perennials or subshrubs, 10–50(–100) cm; leaf blades mostly obovate or elliptic to lanceolate, oblanceolate, or spatulate, margins usually entire, rarely toothed.

        5. Leaf blades glabrous; involucres campanulate to urceolate, 5–7+ mm; cypselae 4–5 mm . . . . 2. *Iva imbricata*

        5. Leaf blades usually strigose to scabrellous; involucres ± hemispheric, 2–3.5 mm; cypselae 2–3 mm.

            6. Aerial stems usually woody at bases; outer phyllaries distinct . . . . 3. *Iva hayesiana*

            6. Aerial stems usually herbaceous; outer phyllaries ± connate . . . . 4. *Iva axillaris*

1. Iva frutescens Linnaeus, Sp. Pl. 2: 989. 1753   E F

Iva frutescens subsp. *oraria* (Bartlett) R. C. Jackson; *I. frutescens* var. *oraria* (Bartlett) Fernald & Griscom

Subshrubs or shrubs, 50–350 cm. Stems erect. **Leaves:** petioles 5–10+ mm; blades ovate or elliptic to lanceolate, 30–60(–120+) × 5–30(–40) mm, margins usually toothed, faces closely scabrellous, gland-dotted. **Heads** in ± racemiform arrays. **Peduncles** 1–3+ mm. **Involucres** ± hemispheric, 2–4 mm. **Phyllaries:** outer 5 distinct, ± herbaceous. **Paleae** linear to subulate, 2.5–3 mm. **Pistillate florets** (2–)5; corollas 0.7–1 mm. **Functionally staminate florets** 3–8 (–15+); corollas 2–3 mm. **Cypselae** 2–3 mm. **2n** = 34.

Flowering Jul–Oct. Mostly tidelands, brackish to saline marshes, beaches; 0–10 m; N.S.; Ala., Conn., Del., Fla., Ga., La., Maine, Md., Mass., Miss., N.H., N.J., N.Y., N.C., Pa., R.I., S.C., Tex., Va.

2. Iva imbricata Walter, Fl. Carol., 232. 1788

Perennials or subshrubs, 30–50 (–100) cm. Stems ± decumbent to erect. **Leaves:** petioles 0–2 mm; blades narrowly lance-elliptic to lanceolate or spatulate, 1–3(–6+) cm × 4–10(–20) mm, margins usually entire, rarely toothed, faces glabrous, rarely gland-dotted. **Heads** in racemiform arrays. **Peduncles** 2–7+ mm. **Involucres** ± campanulate to urceolate, 5–7+ mm. **Phyllaries:** outer 6–10 distinct, ± herbaceous. **Paleae** linear to spatulate (distally dilated), 5–6 mm. **Pistillate florets** 2–5; corollas 0.3–1.5 mm. **Functionally staminate florets** 8–12+; corollas 4–5 mm. **Cypselae** 4–5 mm. **2n** = 34.

Flowering Jul–Oct. Dunes, beaches, at or near tidelands; 0–10 m; Ala., Fla., Ga., La., Miss., N.C., S.C., Tex., Va.; West Indies (Bahamas, Cuba).

3. Iva hayesiana A. Gray, Proc. Amer. Acad. Arts 11: 78. 1876

Perennials or subshrubs, 10–100 cm. Stems sprawling to erect (usually woody at bases). **Leaves:** petioles 0–3(–10) mm; blades lance-elliptic to oblanceolate, 2–5(–10) cm × 5–10(–18) mm, margins usually entire, rarely toothed, faces sparsely strigose to scabrellous, gland-dotted. **Heads** in ± racemiform arrays. **Peduncles** 2–3+ mm. **Involucres** ± hemispheric, 2.5–3.5 mm. **Phyllaries:** outer 5 distinct, ± herbaceous. **Paleae** linear to cuneiform, 1.5–2.5 mm. **Pistillate florets** 5; corollas 0.5–1 mm. **Functionally staminate florets** 5–12(–20); corollas 1.5–2.5 mm. **Cypselae** 2–2.5 mm.

Flowering year round. Alkaline flats, depressions; 50–500 m; Calif.; Mexico (Baja California).

4. Iva axillaris Pursh, Fl. Amer. Sept. 2: 743. 1813   E F

Iva axillaris var. *robustior* Hooker

Perennials, 10–40(–60) cm (rhizomatous). Stems erect (usually herbaceous). **Leaves:** petioles 0–3 mm; blades narrowly ovate or obovate to elliptic or spatulate, 15–25(–45) × 3–8(–15) mm, margins rarely toothed, faces usually ± strigose to scabrellous, gland-dotted. **Heads** in ± racemiform arrays. **Peduncles** 1–2 mm. **Involucres** ± hemispheric, 2–3 mm. **Phyllaries:** outer 3–5 ± connate, ± herbaceous. **Paleae** linear, 1.5–2 mm. **Pistillate florets** 3–5(–8); corollas 0.5–1.5 mm. **Functionally staminate florets** 4–8(–20+); corollas 2–2.5 mm. **Cypselae** 2.5–3 mm. **2n** = 36, 54.

Flowering May–Oct. Seasonally wet, saline habitats; 10–2500 m; Alta., B.C., Man., Sask.; Ariz., Calif., Colo., Idaho, Kans., Minn., Mont., Nebr., Nev., N.Mex., N.Dak., Okla., Oreg., S.Dak., Tex., Utah., Wash., Wis., Wyo.

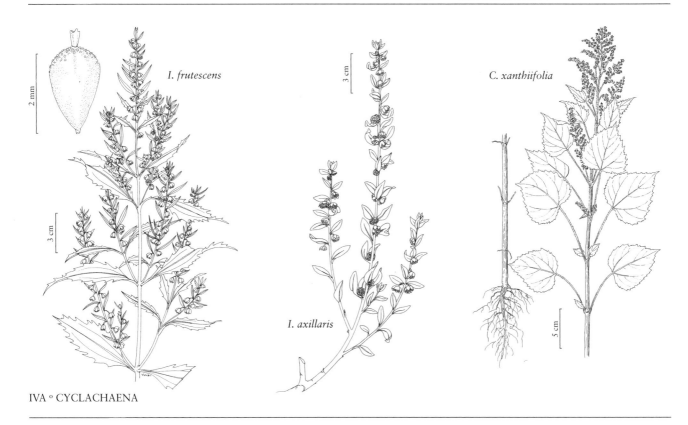

I. frutescens

I. axillaris

C. xanthiifolia

IVA ° CYCLACHAENA

5. Iva annua Linnaeus, Sp. Pl. 2: 988. 1753

Iva annua var. *caudata* (Small) R. C. Jackson; *I. annua* var. *macrocarpa* (S. F. Blake) R. C. Jackson

Annuals, (10–)50–100(–150+) cm. **Stems** erect. **Leaves:** petioles 5–20 (–30) mm; blades deltate or ovate to elliptic, trullate, or lanceolate, 30–100(–150+ × 8–45(–80) mm, margins ± toothed, faces ± scabrellous, gland-dotted. **Heads** in ± spiciform arrays. **Peduncles** 0–1 mm. **Involucres** ± hemispheric, 3–4(–5) mm. **Phyllaries:** outer 3–5 distinct, ± herbaceous. **Paleae** linear, 2–2.5 mm. **Pistillate florets** 3–5; corollas 0.5–1 mm. **Functionally staminate florets** 8–12+; corollas 2–2.5 mm. **Cypselae** 2–3 mm. $2n = 34$.

Flowering Jul–Nov. Disturbed sites, moist soils; 10–500 m; Ala., Ark., Colo., Fla., Ga., Ill., Ind., Iowa, Kans., Ky., La., Maine, Mass., Mich., Miss., Mo., Nebr., N.Mex., N.C., N.Dak., Ohio, Okla., Pa., S.C., S.Dak., Tenn., Tex., Va., W.Va., Wis.; Mexico (Tamaulipas).

6. Iva angustifolia Nuttall ex de Candolle in A. P. de Candolle and A. L. P. P. de Candolle, Prodr. 5: 529. 1836 E

Iva angustifolia var. *latior* Shinners; *I. texensis* R. C. Jackson

Annuals (rarely persisting), 20–60 (–100+) cm. **Stems** usually erect. **Leaves:** petioles 0–2 mm; blades lance-linear to linear, 12–25(–45) × 1–4(–7+) mm, margins entire, faces scabrellous to hispidulous, gland-dotted. **Heads** in ± spiciform arrays. **Peduncles** 0–1 mm. **Involucres** ± prismatically turbinate, 2.5–3 mm. **Phyllaries:** outer 3–5 ± connate, ± herbaceous. **Paleae** linear, 1.5–2 mm. **Pistillate florets** 0–2+; corollas 0.3–0.8+ mm. **Functionally staminate florets** 2–5+; corollas 1.5–2 mm. **Cypselae** 2–3 mm. $2n = 32$.

Flowering Sep–Nov. Disturbed sites, moist soils; 0–300 m; Ark, Fla., Kans., La., Okla., Tex.

7. Iva microcephala Nuttall, Trans. Amer. Philos. Soc., n. s. 7: 346. 1840 E

Annuals, 10–70(–100) cm. **Stems** erect. **Leaves:** petioles 0–1 mm; blades linear to filiform, 10–35 (–60) × 0.6–1.5(–3) mm, margins entire, faces ± scabrellous, gland-dotted. **Heads** in ± spiciform arrays. **Peduncles** 0.5–1 mm. **Involucres** hemispheric, 1.5–2 mm. **Phyllaries:** outer 5 distinct, ± herbaceous. **Paleae** cuneiform to setiform, 1–2 mm.

Pistillate florets 2–3+; corollas 0.5–0.8 mm. **Functionally staminate florets** 3–5; corollas 1.8–2 mm. **Cypselae** 1–1.5 mm. $2n = 32$.

Flowering Sep–Nov. Disturbed sites, moist or wet soils; 1–100 m; Ala., Fla., Ga., N.C., S.C.

250. CYCLACHAENA Fresenius, Index Seminum (Frankfurt) 1836: 4. 1838 • [Greek *cyclo-*, circular, and Latin *achenium*, achene; allusion uncertain, perhaps to the ring of cypselae in each fruiting head] E

John L. Strother

Annuals, 30–120(–200) cm. **Stems** erect, strictly, if at all, branched. **Leaves** cauline; mostly opposite (distal sometimes alternate); petiolate; blades (3–5-nerved) ± deltate, ovate, or rhombic (proximal 1–3 pairs, often 3–5-lobed), margins usually toothed, faces densely to sparsely scabrellous to strigillose (abaxial often canescent), both usually gland-dotted. **Heads** ± disciform, in (± ebracteate) paniculiform arrays. **Involucres** turbinate to hemispheric, 3–5+ mm diam. **Phyllaries** persistent, 10–12+ in 2+ series, distinct, outer 5 herbaceous, inner scarious to membranous. **Receptacles** convex to conic; paleae spatulate to linear, ± membranous, sometimes 0. **Pistillate florets** 5; corollas whitish, tubular, or 0. **Functionally staminate florets** 5–10 (–20+); corollas whitish, funnelform, lobes 5, erect (filaments connate, anthers distinct). **Cypselae** plumply obovoid to obcompressed, finely striate, sparsely strigillose or hispidulous distally, little, if at all, gland-dotted; **pappi** 0. $x = 18$.

Species 1: North America; introduced in Europe.

1. Cyclachaena xanthiifolia (Nuttall) Fresenius, Index Seminum (Frankfurt) 1836: 4. 1836 (as xanthifolia) • Ive à feuilles de lampourde E F

Iva xanthiifolia Nuttall, Gen. N. Amer. Pl. 2: 185. 1818 (as xanthifolia)

Leaves: petioles 1–7(–12+) cm; blades 6–12(–20+) × 5–12(–18+) cm. **Peduncles** 1–6(–12+) mm. **Involucres** 2–3 mm. **Phyllaries:** outer 5 ± herbaceous, sericeous to strigillose, inner scarious to membranous, glabrous or glabrate. **Paleae** 2–2.5 mm. **Pistillate florets:** corollas 0.1–0.5 mm, or 0. **Functionally staminate florets:** corollas 2–2.5 mm. **Cypselae** 2–2.5 (–3) mm. $2n = 36$.

Flowering Jul–Oct. Disturbed sites: abandoned fields, bottomlands, flood plains, stream banks; 1–1300 m; Alta., B.C., Man., N.B., N.S., Ont., P.E.I., Que., Sask.; Ariz., Calif., Conn., Colo., D.C., Idaho, Ill., Ind., Iowa, Kans., Ky., Maine, Mass., Mich., Minn., Mo., Mont., Nebr., Nev., N.H., N.J., N.Mex., N.Y., N.Dak., Ohio, Okla., Oreg., Pa., R.I., S.Dak., Tex., Utah, Vt., Va., Wash., Wis., Wyo.

Cyclachaena xanthiifolia is thought to be native to North American prairies and is evidently adventive east of the Mississippi River and in western states. It was recorded once from California as a weed in commercially grown carrots (specimen in CAS).

251. **OXYTENIA** Nuttall, Proc. Acad. Nat. Sci. Philadelphia 4: 20. 1848 • Copper-weed [Greek *oxytenes*, acuminate, "in allusion to the rigid narrow foliage"] E

John L. Strother

Subshrubs or shrubs, 50–200 cm. **Stems** erect, virgately branched. **Leaves** cauline; alternate; petiolate or ± sessile; blades mostly pinnately lobed (lobes 3–7+, linear to filiform), distal (not lobed) linear to filiform, faces usually sericeous to strigillose, sometimes glabrate or glabrous, usually gland-dotted. **Heads** disciform, in (ebracteate or nearly so) paniculiform arrays, or borne singly or in glomerules of 1–5+. **Involucres** ± hemispheric, 4–5+ mm diam. **Phyllaries** persistent, 10–15+ in 2–3 series, distinct, outer 5–7 herbaceous, inner scarious to membranous (± villous). **Receptacles** convex; paleae spatulate to cuneiform, membranous, distally ± villous. **Pistillate florets** 5; corollas 0. **Functionally staminate florets** 10–25+; corollas whitish, funnelform, lobes 5, erect (filaments connate, anthers distinct or weakly coherent). **Cypselae** plumply obovoid, obcompressed or weakly 3–4-angled, smooth, ± villous, little, if at all, gland-dotted; **pappi** 0. *x* = 18.

Species 1: w United States.

1. **Oxytenia acerosa** Nuttall, Proc. Acad. Nat. Sci. Philadelphia 4: 20. 1848 E F

Euphrosyne acerosa (Nuttall) Panero; *Iva acerosa* (Nuttall) R. C. Jackson

Leaves: petioles 0–20+ mm; blades or lobes mostly 20–150 × 1–1.5 (–3) mm. **Peduncles** 1–3+ mm. **Involucres** 2–3+ mm. **Phyllaries:** outer 5–7 ± herbaceous, sericeous to strigillose, inner scarious to membranous, usually villous. **Functionally staminate florets:** corollas 2.5–3 mm. **Paleae** 1.8–2+ mm. **Cypselae** 1.5–2.5 mm. *2n* = 36.

Flowering Jun–Sep. Wet or seasonally wet, alkaline soils; 100–2200 m; Ariz., Calif., Colo., Nev., N.Mex., Utah.

252. **LEUCIVA** Rydberg in N. L. Britton et al., N. Amer. Fl. 33: 8. 1922 • Woolly sumpweed [Greek *leuc-*, white, and *Iva*, a related genus, perhaps alluding to white indument of leaves]

John L. Strother

Perennials (sometimes flowering in first year), 20–60+ cm. **Stems** erect, branching from bases and/or distally. **Leaves** cauline; mostly alternate; petiolate; blades lanceolate to oblanceolate, laciniately pinnately lobed, abaxial faces ± lanate, adaxial ± tomentose, both usually gland-dotted. **Heads** discoid or disciform, in (ebracteate) paniculiform arrays. **Involucres** ± hemispheric, 2–4+ mm diam. **Phyllaries** persistent, 5+ in 1(–2) series, distinct, outer 5 herbaceous, inner 0 or scarious to membranous. **Receptacles** convex, epaleate. **Pistillate florets** 0 or 3–5; corollas whitish, tubular, or 0. **Functionally staminate florets** 5–12+; corollas whitish, funnelform, lobes 5, erect (filaments connate, anthers coherent or distinct). **Receptacles** convex, epaleate. **Cypselae** pyriform, ± obcompressed, densely gland-dotted; **pappi** 0. *x* = 18?

Species 1: sw United States, n Mexico.

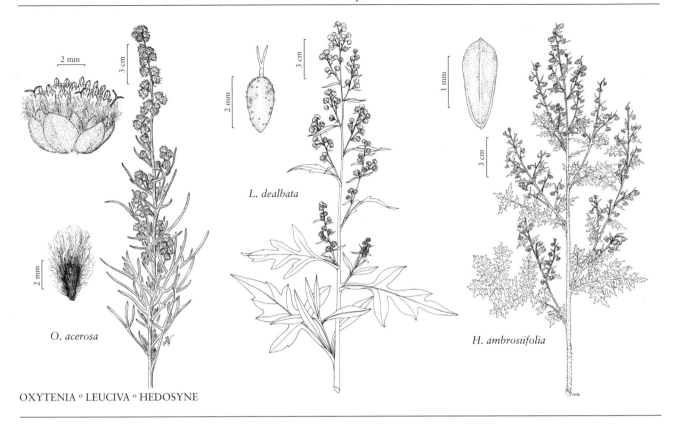

O. acerosa

L. dealbata

H. ambrosiifolia

OXYTENIA ∘ LEUCIVA ∘ HEDOSYNE

1. Leuciva dealbata (A. Gray) Rydberg in N. L. Britton et al., N. Amer. Fl. 33: 8. 1922 F

Iva dealbata A. Gray, Smithsonian Contr. Knowl. 3(5): 104. 1852; *Euphrosyne dealbata* (A. Gray) Panero

Leaves: petioles 1–5+ mm; blades 7–10(–18+) × 1–6(–12) cm overall, lobes 3–5+ mm wide. **Peduncles** 1–10+ mm. **Involucres** 2–3+ mm. **Phyllaries:** outer 5 sparsely strigose or glabrous abaxially, margins ciliolate, inner ciliolate. **Pistillate florets:** corollas 0.1–0.7 mm, or 0. **Functionally staminate florets:** corollas 1.5–2 mm. **Cypselae** 1.5–2 mm. $2n = 72$.

Flowering Jul–Nov. Calcareous desert soils; 1000–2000 m; Ariz., N.Mex., Tex.; Mexico (Chihuahua, Coahuila, Durango, Nuevo León, San Luis Potosí, Zacatecas).

253. HEDOSYNE Strother, Madroño 47: 204. 2001 • [Greek *hedosyne*, delight]

John L. Strother

Annuals, 10–50(–100) cm. **Stems** erect, strictly branched. **Leaves** cauline; mostly alternate; petiolate; blades deltate or ovate to lanceolate, 1–3-pinnately lobed (lobes oblong to lance-linear), ultimate margins entire or toothed, faces ± scabrellous and/or hispid, usually gland-dotted. **Heads** ± disciform, usually in loose, (± bracteate or ebracteate) paniculiform arrays (sometimes 3–6+ distal to axil of each bract). **Involucres** ± hemispheric, 4–5 mm diam. **Phyllaries** persistent, 10–12+ in 2+ series, distinct, outer 5 herbaceous, inner scarious to membranous. **Receptacles** hemispheric; paleae spatulate to linear, membranous, sparsely hairy or glabrate, usually gland-dotted. **Pistillate florets** 5–10; corollas 0. **Functionally staminate**

florets 5–10+; corollas whitish, funnelform, lobes 5, soon reflexed (filaments ± connate, anthers weakly coherent or distinct). **Cypselae** pyriform, ± obcompressed, finely striate, glabrous, little, if at all, gland-dotted; **pappi** 0. *x* = 18.

Species 1: sw United States, n Mexico.

1. Hedosyne ambrosiifolia (A. Gray) Strother, Madroño 47: 204. 2001 [F]

Euphrosyne ambrosiifolia A. Gray, Smithsonian Contr. Knowl. 3(5): 102. 1852 (as ambrosiaefolia); *Cyclachaena ambrosiifolia* (A. Gray) Rydberg; *Iva ambrosiifolia* (A. Gray) A. Gray

Leaves: petioles 5–12(–45) mm; blades 3–5(–9) × 4–5(–8) cm overall, lobes 1–3 mm wide. **Peduncles** 3–12+ mm. **Involucres** 2–3+ mm. **Phyllaries:** outer 5 sparsely strigose or glabrous. **Paleae** 1–1.5 mm.

Functionally staminate florets: corollas 1.5–2 mm. **Cypselae** 1.4–1.7 mm. *2n* = 36.

Flowering Jul–Sep. Disturbed sites (roadsides, washes, etc.), in sandy, gypseous, or calcareous soils; 1000–2000 m; Ariz., N.Mex., Tex.; Mexico (Chihuahua, Coahuila, Durango, Nuevo León, San Luis Potosí, Sonora, Zacatecas).

254. CHORISIVA (A. Gray) Rydberg in N. L. Britton et al., N. Amer. Fl. 33: 8. 1922

• [Greek *choris-*, separate, and *Iva*, a related genus; allusion recondite, perhaps "separate from *Iva*" or to "scattered" arrangement of heads] [E]

John L. Strother

Iva Linnaeus sect. *Chorisiva* A. Gray in A. Gray et al., Syn. Fl. N. Amer. 1(2): 247. 1884

Annuals, 5–25(–40) cm. **Stems** erect, diffusely branched. **Leaves** cauline; mostly alternate; petiolate; blades ovate or deltate to lanceolate, usually 1–2-pinnately lobed (lobes lance-ovate to linear), ultimate margins entire or toothed, faces ± scabrellous, usually gland-dotted. **Heads** ± discoid, mostly borne singly (in leaf axils or remote from axils, ± scattered). **Involucres** ± turbinate to campanulate, 2–3+ mm diam. **Phyllaries** persistent, 3–8+ in 1–2+ series, distinct, outer 3+ herbaceous, inner scarious to membranous, glabrous. **Receptacles** convex; paleae linear to setiform, membranous, usually with dilated, strigillose or ciliate tips. **Pistillate florets** 2–3 (–5); corollas yellowish, tubular. **Functionally staminate florets** 5–10+; corollas yellowish, funnelform, strigillose and gland-dotted distally, lobes 5, erect (filaments ± connate, anthers weakly coherent or distinct). **Cypselae** obovoid, obcompressed, smooth or corky-tuberculate, glabrous (not corky-winged, not gland-dotted); **pappi** 0.

Species 1: sw United States.

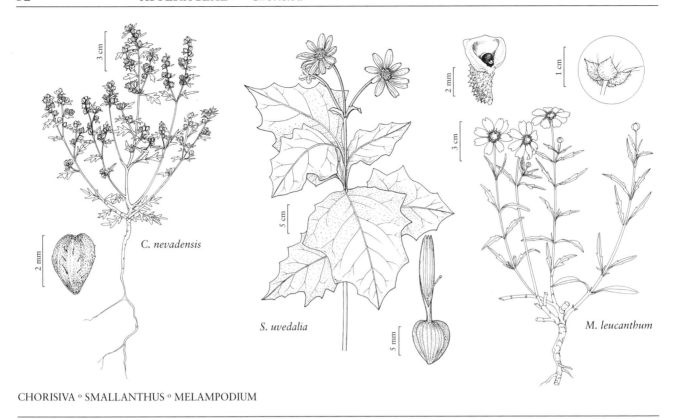

CHORISIVA ° SMALLANTHUS ° MELAMPODIUM

1. **Chorisiva nevadensis** (M. E. Jones) Rydberg in
N. L. Britton et al., N. Amer. Fl. 33: 9. 1922 [E] [F]

Iva nevadensis M. E. Jones, Amer.
Naturalist 17: 973. 1883;
Euphrosyne nevadensis
(M. E. Jones) Panero

Leaves: petioles 2–20 mm; blades
5–20 × 5–15 mm overall, lobes 1–
1.5 mm wide. **Peduncles** 0.5–5+
mm. **Involucres** 2–5+ mm high.
Phyllaries: outer 3 ± scabrellous,
inner scarious to membranous. **Paleae** 1.5–2 mm.
Pistillate florets: corollas 0.5–1+ mm, ± strigillose.

Functionally staminate florets: corollas 1.5–2 mm.
Cypselae 1.5–2 mm.

Flowering May–Oct. Sandy or gravelly flats, washes,
usually alkaline soils; 1000–2200 m; Calif., Nev.

187m.2. ASTERACEAE Martinov (tribe HELIANTHEAE) subtribe MELAMPODIINAE Lessing, Linnaea
5: 149. 1830 (as Melampodieae)

Annuals, perennials, or subshrubs [treelets], 10–120(–300+)[–1200+] cm. **Leaves** cauline;
opposite; petiolate or sessile; blades mostly deltate or ovate to lanceolate or linear (and most
intermediate shapes), sometimes palmately or pinnately lobed, ultimate margins dentate, denticulate, entire, or toothed, faces glabrous or hairy, often gland-dotted. **Heads** radiate, borne
singly or in open, corymbiform arrays. **Calyculi** 0. **Involucres** cylindric or ovoid to hemispheric
(sometimes cupulate). **Phyllaries** persistent and/or falling, 8–25+ in 2 series (outer 2–6 distinct
or connate, herbaceous, contrasting with distinct, more membranous to scarious, inner ones,
each of the inner often enveloping its subtended ray ovary, maturing with the cypsela, forming

a perigynium). **Receptacles** flat to convex, conic, or columnar (mostly 1–2 mm), paleate (paleae lanceolate to spatulate or obovate, scarious to membranous, often conduplicate, often falling together as a unit with spent disc florets). **Ray florets** 3–20+, pistillate, fertile; corollas yellow, orange, or white. **Disc florets** 3–150+, functionally staminate; corollas yellow, orange, or whitish, tubes shorter than funnelform or campanulate throats, lobes 5, deltate; anther thecae dark; stigmatic papillae 0. **Cypselae** weakly to strongly compressed, obovoid to ± prismatic (often 4-angled; sometimes each enclosed within a smooth, bullate, tuberculate, or prickly perigynium derived from its subtending phyllary); **pappi** 0 (or rudimentary).

Genera 5, species 80 (3 genera, 11 species in the flora): subtropical, tropical, and warm-temperate New World.

Circumscription of Melampodiinae in the sense of H. Robinson (1981) is narrower than the traditional one (e.g., T. F. Stuessy 1973), which included most non wind-pollinated genera of Heliantheae with functionally staminate disc florets.

SELECTED REFERENCE Stuessy, T. F. 1973. A systematic review of the subtribe Melampodiinae (Compositae, Heliantheae). Contr. Gray Herb. 203: 65–80.

1. Cypselae: each shed separate from its subtending phyllary . . . . 255. *Smallanthus*, p. 33
1. Cypselae: each enclosed within and shed with perigynium formed from inner, subtending phyllary (the units called fruits).
   2. Heads borne singly, usually pedunculate; fruits smooth or bullate to tuberculate (1–4 mm) . . . . 256. *Melampodium*, p. 34
   2. Heads often in clusters of 2–3, mostly sessile; fruits ± prickly (4–8 mm) . . . 257. *Acanthospermum*, p. 36

255. SMALLANTHUS Mackenzie ex Small, Man. S.E. Fl., 1406, 1509. 1933 • [For John Kunkel Small, 1869–1938, American botanist]

John L. Strother

Perennials [annuals, shrubs], 100–300[1200+] cm. **Stems** erect. **Leaves** cauline; opposite; petiolate (petioles usually winged) [sessile]; blades mostly deltate to ovate, usually ± palmately lobed, ultimate margins dentate to denticulate, faces hirtellous, pilosulous, or puberulent, gland-dotted (at least abaxially). **Heads** radiate, borne singly or (2–5) in crowded, corymbiform arrays. **Involucres** hemispheric, 8–15 mm diam. **Phyllaries** persistent, 12–13[–25+] in 2 series (outer 4–6 ovate to lanceolate [orbiculate], herbaceous, inner as many as rays, more membranous to scarious, narrower and shorter). **Receptacles** flat to convex, paleate (paleae obovate to spatulate, scarious). **Ray florets** 7–13[–25+], pistillate, fertile; corollas yellow [white or orange] (tubes hairy, laminae linear to elliptic [ovate]). **Disc florets** [20–]40–80[–150+], functionally staminate; corollas yellow [orange], tubes shorter than abruptly campanulate [funnelform] throats, lobes 5, deltate. **Cypselae** (obliquely inserted on receptacles, each shed separate from subtending phyllary) obovoid [quadrangular], somewhat compressed, finely 30–40-ribbed or -striate (not narrowed at bases, not apically beaked); **pappi** 0 (cypselae sometimes hairy at apices). $x = 16$.

Species ca. 20 (1 in the flora): United States, Mexico, South America.

SELECTED REFERENCES Robinson, H. 1978. Studies in the Heliantheae (Asteraceae). XII. Re-establishment of the genus *Smallanthus*. Phytologia 39: 47–53. Turner, B. L. 1988. A new species of, and observations on, the genus *Smallanthus* (Asteraceae–Heliantheae). Phytologia 64: 405–409.

1. Smallanthus uvedalia (Linnaeus) Mackenzie ex Small, Man. S.E. Fl., 1509. 1933 ⊡ ⊡

Osteospermum uvedalia Linnaeus, Sp. Pl. 2: 923. 1753; *Polymnia uvedalia* (Linnaeus) Linnaeus; *P. uvedalia* var. *densipilis* S. F. Blake; *P. uvedalia* var. *floridana* S. F. Blake

Leaves: petioles 3–12+ cm, blades 10–35(–60+) × 10–35+ cm, larger usually 3–5-lobed. **Ray laminae** 12–30+ mm. **Cypselae** 5–6 mm. **2n** = 32.

Flowering Jun–Aug(–Oct). Thickets, forest margins, often wet sites; 10–300+ m; Ala., Ark., Del., D.C., Fla., Ga., Ill., Ind., Kans., Ky., La., Md., Mich., Miss., Mo., N.J., N.Y., N.C., Ohio, Okla., Pa., S.C., Tenn., Tex., Va., W.Va.; introduced in Bermuda.

B. L. Turner (1988) included types of *Smallanthus uvedalia* and *S. maculatus* (Cavanilles) H. Robinson within a single species circumscription. If that circumscription is accepted, the range of *S. uvedalia* extends through eastern Mexico and Central America to Panama.

256. MELAMPODIUM Linnaeus, Sp. Pl. 2: 921. 1753; Gen. Pl. ed. 5, 392. 1754

• [Often said (erroneously) to be from Greek *melampodion*, blackfoot; evidently traceable to *Melampus*, a soothsayer of renown in Greek mythology]

John L. Strother

Annuals, perennials, or subshrubs, 5–50(–150+) cm. **Stems** erect or prostrate. **Leaves** cauline; opposite; petiolate (petioles usually winged) or sessile; blades deltate, lance-elliptic, lanceolate, lance-linear, linear, linear-oblong, ovate, or rhombic, sometimes pinnately lobed, ultimate margins entire or toothed, faces usually hairy, usually gland-dotted (at least abaxial). **Heads** radiate, borne singly (from forks of branches, peduncles often uncinate). **Involucres** mostly hemispheric (sometimes ± cupulate), [4–]5–20(–30+) mm diam. **Phyllaries** persistent (outer) or falling with cypselae, 8–20+ in 2 series (outer [2–]5 often ± connate, herbaceous, inner each investing a ray ovary, forming a perigynium, shed with enclosed cypsela). **Receptacles** flat or convex to conic (their tips, paleae, and spent florets often shed as a unit), paleate (paleae lanceolate to linear, scarious, conduplicate, often with dilated, erose tips). **Ray florets** [3–]5–13+, pistillate, fertile; corollas yellow, orange, or cream-white, sometimes purplish abaxially (tubes often wanting). **Disc florets** [3–]45–70[–100+], functionally staminate; corollas ochroleucous, yellow, or orange, tubes shorter than funnelform throats, lobes 5, ± deltate. **Cypselae** each enclosed within and shed with a smooth, sculpted, or tuberculate, scarious, coriaceous, or hardened perigynium (the ultimate "fruits" ± compressed, ± D-shaped); **pappi** 0. *x* = 12.

Species ca. 36 (7 in the flora): United States, Mexico, West Indies, Central America, South America; introduced in Old World.

SELECTED REFERENCE Stuessy, T. F. 1972. Revision of the genus *Melampodium* (Compositae: Heliantheae). Rhodora 74: 1–70, 161–219.

1. Perennials; ray corollas cream-white.
    2. Outer phyllaries connate ¹/₂–³/₅ their lengths; ray laminae 7–13 × 2.5–8 mm . . . .
       . . . . 1. *Melampodium leucanthum*
    2. Outer phyllaries connate ¹/₆–¹/₃+ their lengths; ray laminae 2–8+ × 1–3+ mm . . . . . .
       . . . . 2. *Melampodium cinereum*
1. Annuals; ray corollas yellow to yellow-orange.
    3. Leaf blades deltate or rhombic to lanceolate, lengths 1–2(–3) times widths.
        4. Outer phyllaries ovate, 3.5–6 mm, connate ¹/₄–¹/₃ their lengths; disc florets 40–70+;
           fruits 2.8–4 mm . . . . 3. *Melampodium divaricatum*

4. Outer phyllaries oblong-elliptic, (6–)8–15+ mm, connate 0–¹/₆ their lengths; disc
florets 30–45; fruits 4–7 mm .... 4. *Melampodium perfoliatum*
[3. Shifted to left margin.—Ed.]
3. Leaf blades lance-elliptic or oblanceolate to linear, lengths (3–)4–8+ times widths.
   5. Outer phyllaries connate ³/₅–²/₃ their lengths; disc florets 25–40 .... 5. *Melampodium appendiculatum*
   5. Outer phyllaries connate 0–¹/₆ their lengths; disc florets 4–10.
      6. Peduncles 4–30+ mm; rays 7–12, laminae 1.2–1.5 mm; disc florets 8–10 .... 6. *Melampodium longicorne*
      6. Peduncles 0–3(–12+) mm; rays 5–8, laminae 0.6–1.1 mm; disc florets 4–6 .....
      .... 7. *Melampodium strigosum*

1. Melampodium leucanthum Torrey & A. Gray, Fl. N. Amer. 2: 271. 1842 F

Perennials or subshrubs, 12–40 (–60) cm. **Leaf blades** lanceolate, linear-oblong, or linear, 20–35 (–45) × 1–10(–12) mm, lengths 3–8+ times widths, pinnately lobed, lobes 1–6, ultimate margins entire. **Peduncles** 3–7 cm. **Outer phyllaries** 5, connate ¹/₂–³/₅ their lengths, ovate, 5–7 mm. **Ray florets** 8–13; corollas cream-white (sometimes purplish abaxially), laminae oblong-elliptic, 7–13 × 2.5–8 mm. **Disc florets** 25–50. **Fruits** 1.5–2.6 mm. $2n = 20, 40$.

Flowering Mar–Oct. Open sites, grasslands, roadcuts, arid or desert scrublands; (200–)1000–2000(–2500) m; Ariz., Colo., Kans., N.Mex., Okla., Tex.; Mexico (Chihuahua, Coahuila, Sonora).

2. Melampodium cinereum de Candolle in A. P. de Candolle and A. L. P. P. de Candolle, Prodr. 5: 518. 1836

Melampodium cinereum var. *hirtellum* Stuessy; *M. cinereum* var. *ramosissimum* (de Candolle) A. Gray

Perennials or subshrubs, 10–20+ cm. **Leaf blades** linear-oblong to linear, 10–55 × 1–14 mm, lengths 3–10+ times widths, pinnately lobed, lobes 1–10, ultimate margins entire. **Peduncles** 1–7 cm. **Outer phyllaries** 5, connate ¹/₆–¹/₃+ their lengths, ovate, 3–7+ mm. **Ray florets** 7–13; corollas cream-white (sometimes purplish abaxially), laminae oblong-elliptic, 2–8+ × 1–3+ mm. **Disc florets** 25–50. **Fruits** 1.3–2.2 mm. $2n = 20, 40$.

Flowering Jan–Jul. Open sites, sand, gravel, loam, or clay soils; 0–300 m; Tex.; Mexico (Coahuila, Nuevo León, Tamaulipas).

Along the interface of the distributions of *Melampodium cinereum* and *M. leucanthum*, some plants are intermediate for traits used to characterize the two species; perhaps the two should be treated as one.

Plants of *Melampodium cinereum* with leaf blades linear-oblong and basally strigose with hairs 0.3–0.6 mm have been treated as var. *cinereum*, plants with leaf blades

linear-oblong and basally strigose with hairs 0.6–1.5 mm as var. *hirtellum*, and plants with leaf blades linear and basally strigose with hairs 0.3–0.6 mm as var. *ramosissimum* (T. F. Stuessy 1972).

3. Melampodium divaricatum (Richard) de Candolle in A. P. de Candolle and A. L. P. P. de Candolle, Prodr. 5: 520. 1836 I

Dysodium divaricatum Richard in C. H. Persoon, Syn. Pl. 2: 489. 1807

Annuals, 15–100+ cm. **Leaf blades** ± rhombic to lanceolate, 40–150+ × 25–95+ mm, lengths 1–2(–3+) times widths, margins coarsely toothed or entire. **Peduncles** 1–12+ cm. **Outer phyllaries** 5, connate ¹/₄–¹/₃ their lengths, ovate to orbiculate, 3.5–6 mm. **Ray florets** 8–13; corollas yellow-orange, laminae oblong-elliptic, 3.5–7+ × 1.6–3+ mm. **Disc florets** 40–70+. **Fruits** 2.8–4+ mm. $2n = 24$.

Flowering year round. Disturbed sites; 0–10+ m; introduced; Fla.; Mexico; West Indies (Antilles); Central America; South America; introduced in Old World.

4. Melampodium perfoliatum (Cavanilles) Kunth in A. von Humboldt et al., Nov. Gen. Sp. 4(fol.): 215. 1818; 4(qto.): 474. 1820 I

Alcina perfoliata Cavanilles, Icon. 1: 11, plate 15. 1791

Annuals, 20–150+ cm. **Leaf blades** ± deltate to rhombic-deltate, 40–200+ × 25–150+ mm, lengths 1–2(–3+) times widths, margins ± toothed or entire. **Peduncles** 1–10+ cm. **Outer phyllaries** 5, connate 0–¹/₆ their lengths, oblong to elliptic, (6–)12–15+ mm. **Ray florets** 8–13; corollas yellow-orange, laminae oblong-elliptic, 2.5–5+ × 1.6–3+ mm. **Disc florets** 30–45+. **Fruits** 4–7+ mm. $2n = 22, 24$.

Flowering year round. Disturbed, usually moist sites; 10–50+ m; introduced; Calif.; Mexico; West Indies (Antilles); Central America.

Melampodium perfoliatum may be no longer present in California.

5. Melampodium appendiculatum B. L. Robinson, Proc. Amer. Acad. Arts 36: 457. 1901

Annuals, 10–40+ cm. **Leaf blades** lanceolate to lance-linear or linear, 25–80+ × 4–18+ mm, lengths 4–8+ times widths, margins ± crenate or entire. **Peduncles** (1–)2–8+ cm. **Outer phyllaries** 5, connate ³/₅–²/₃ their lengths, ovate, 3.5–4.5 mm. **Ray florets** 8–13; corollas yellow-orange, laminae oblong-elliptic, 2–8+ × 1–2.5+ mm. **Disc florets** 25–40+. **Fruits** 2.3–2.8 mm.

Flowering Sep–Nov. Openings in pine forests; 800–2000+ m; Ariz.; Mexico (Chihuahua, Durango, Sonora).

6. Melampodium longicorne A. Gray, Pl. Nov. Thurb. 321. 1854

Annuals, 10–60+ cm. **Leaf blades** lance-elliptic or oblanceolate to lance-linear, 25–50+ × 3–12+ mm, lengths 3–4(–6+) times widths, margins ± dentate or entire. **Peduncles** 4–30+ mm. **Outer phyllaries** 5, connate 0–¹/₆ their lengths, ovate to elliptic, 4–6+ mm. **Ray florets** 7–12; corollas yellow, laminae oblong-elliptic, 1.2–1.5+ × 0.7–1+ mm. **Disc florets** 8–10+. **Fruits** 3–3.5+ mm. **2n = 60**.

Flowering Aug–Oct. Sandy, gravelly, or loamy soils; 1200–1500 m; Ariz., N.Mex.; Mexico (Chihuahua, Sonora).

7. Melampodium strigosum Stuessy, Rhodora 74: 51. 1972

Annuals, 5–35+ cm. **Leaf blades** oblanceolate to oblong-linear, 15–55+ × 3–12+ mm, lengths 4–6+ times widths, sometimes pinnately 2–4-lobed, ultimate margins ± dentate or entire. **Peduncles** 0–3(–11+) mm. **Outer phyllaries** 5, connate 0–¹/₆ their lengths, lance-elliptic, 3–6 mm. **Ray florets** 5–8; corollas yellow, laminae oblong-ovate, 0.6–1.1+ × 0.5–10.8+ mm. **Disc florets** 5–8+. **Fruits** 2.2–3 mm. **2n = 40**.

Flowering Aug–Sep. Openings in pine forests; 1200–1700 m; Ariz., Colo., Tex.; Mexico.

Specimens of *Melampodium strigosum* have been misidentified as *M. hispidum* Kunth, a synonym of *M. sericeum* (see T. F. Stuessy 1972). The station for *M. strigosum* in Colorado is ca. 1400 km distant from stations in Arizona and Texas.

257. ACANTHOSPERMUM Schrank, Pl. Rar. Hort. Monac. 2: plate 53. 1820, name conserved • [Greek *acantha*, prickle, and *sperma*, seed, alluding to prickly "fruits"] [1]

John L. Strother

Annuals (sometimes persisting), 10–60(–120) cm. **Stems** erect to ± prostrate (repeatedly "forked"). **Leaves** cauline; opposite; petiolate or ± sessile; blades mostly elliptic to deltate, rhombic, or ovate, sometimes lyrate, ultimate margins entire or toothed, faces usually pilosulous to sericeous or scabrellous, sometimes glabrate or glabrescent, usually gland-dotted. **Heads** radiate, 1(–3) in "forks" of branches (terminal, appearing axillary by sympodial growth). **Involucres** ± hemispheric, 3–5 mm diam. (becoming ± rotate in fruit). **Phyllaries** persistent (outer) or falling, 10–13 in 2 series (outer 4–6 herbaceous, inner 5–8 each investing a ray ovary, enlarging in fruit to form a perigynium, shed with enclosed cypsela). **Receptacles** convex, paleate (paleae cuneate to spatulate, ± conduplicate or flattish, membranous). **Ray florets** 5–8, pistillate, fertile; corollas yellowish (tubes shorter than to equaling laminae, laminae ovate to elliptic or linear). **Disc florets** 3–8(–12+), functionally staminate; corollas yellowish, tubes shorter than funnelform or campanulate throats, lobes 5, deltate. **Cypselae** each enclosed within and shed with an often hardened, ± prickly perigynium (the ultimate "fruits" plumply ellipsoid to fusiform, or ± compressed); **pappi** 0 or rudimentary. *x* = 11.

Species 6 (3 in the flora): introduced; mostly tropical to warm-temperate New World; also introduced in Old World.

SELECTED REFERENCE   Blake, S. F. 1921. Revision of the genus *Acanthospermum*. Contr. U.S. Natl. Herb. 20: 383–392.

1. Fruits 7–9+ mm, 5–7-ribbed, terminal spines 0 . . . . 1. *Acanthospermum australe*
1. Fruits 2–6 mm, 3-ribbed or not notably ribbed, terminal spines 2.
    2. Leaf blades rhombic-ovate to obovate, (20–)40–120(–150+) mm; fruits not notably ribbed, prickles ± scattered . . . . 2. *Acanthospermum hispidum*
    2. Leaf blades ovate to lyrate, 10–30(–45) mm; fruits usually 3-ribbed, prickles mostly along 2 ribs and around apices . . . . 3. *Acanthospermum humile*

1. **Acanthospermum australe** (Loefling) Kuntze, Revis. Gen. Pl. 1: 303. 1891   F   I

Melampodium australe Loefling, Iter Hispan., 268. 1758;

Plants 10–60(–120+) cm. **Stems** ± procumbent. **Leaf blades** deltate to ± rhombic or ovate, 13–37 × 7–32 mm, faces sparsely scabrellous to glabrate or glabrescent, gland-dotted. **Fruits** plumply ellipsoid to fusiform, weakly compressed, 7–9+ mm, 5–7-ribbed, lacking terminal spines, prickles ± uncinate, mostly along ribs. *2n* = 22.

Flowering year round, mostly Jul–Aug. Disturbed, often sandy sites; 0–300 m; introduced; Ala., Ark., D.C., Fla., Ga., La., Mass., Miss., N.C., Oreg., Pa., S.C., Tex., Va.; South America; also introduced in Mexico, West Indies, Central America, Europe, Asia, Africa, Pacific Islands (Hawaii).

Reports of *Acanthospermum xanthioides* (Kunth) de Candolle from New York are evidently based on specimens of *A. australe*.

2. **Acanthospermum hispidum** de Candolle in A. P. de Candolle and A. L. P. P. de Candolle, Prodr. 5: 522. 1836   I

Plants 10–60+ cm. **Stems** erect. **Leaf blades** rhombic-ovate to obovate, (20–)40–120(–150+) mm, faces finely pilosulous, gland-dotted. **Fruits** ± compressed, ± cuneate to obovate, 4–6+ mm, not notably ribbed, terminal spines 2, divergent, 3–4 mm, often 1 ± uncinate, prickles seldom notably uncinate, ± scattered. *2n* = 22.

Flowering year round, mostly Aug–Oct. Disturbed, often sandy sites; 0–100+ m; introduced; Ont.; Ala., Fla., Ga., N.J., Oreg., S.C., Va.; South America; also introduced in Mexico, West Indies, Central America, Europe, Asia, Africa, Pacific Islands (Hawaii).

3. **Acanthospermum humile** (Swartz) de Candolle in A. P. de Candolle and A. L. P. P. de Candolle, Prodr. 5: 522. 1836   I

Melampodium humile Swartz, Prodr., 114. 1788

Plants 10–20(–30+) cm. **Stems** decumbent to erect. **Leaf blades** oval to lyrate, 10–30(–45) mm, faces ± pilosulous to sparsely sericeous, gland-dotted. **Fruits** strongly compressed, ± cuneate, 2–3(–4) mm, usually 3-ribbed, terminal spines 2, divergent, 2–3(–4) mm, often 1 ± uncinate, prickles ± uncinate, mostly along 2 ribs and around apices.

Flowering all year. Disturbed, often sandy sites; 0–10+ m; introduced; Ala., Fla., N.Y., S.C.; West Indies; also introduced in Central America, South America.

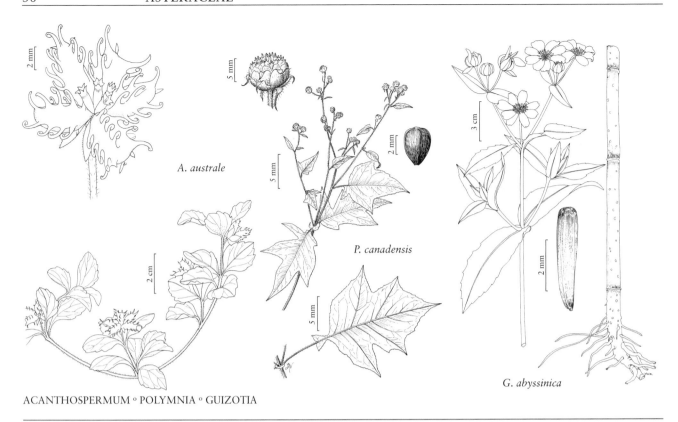

ACANTHOSPERMUM ∘ POLYMNIA ∘ GUIZOTIA

187m.3. ASTERACEAE Martinov (tribe HELIANTHEAE) subtribe POLYMNIINAE H. Robinson, Phytologia 41: 43. 1978 [E]

Annuals or perennials, mostly 50–150+ cm (often rhizomatous). **Leaves** all or mostly cauline; opposite; petiolate; blades (pinnately nerved) deltate or rounded-rhombic to cordate or ovate, often raggedly pinnately lobed, ultimate margins coarsely toothed to denticulate or entire, faces pilose or pilosulous to hirtellous, or glabrate, or glabrous, usually gland-dotted and/or stipitate-glandular. **Heads** usually radiate (rarely, laminae reduced, heads technically ± disciform), (2–5+) in loose to congested, corymbiform arrays. **Calyculi** 0. **Involucres** hemispheric. **Phyllaries** persistent, 6–21+ in 2 series (distinct, outer 2–6 ovate to linear, ± herbaceous, inner ovate to lanceolate, subequal to or shorter than outer, more scarious, similar to paleae). **Receptacles** flat to convex, paleate (paleae obovate to oblanceolate or spatulate, herbaceous to scarious). **Ray florets** 2–6, pistillate, fertile; corollas pale yellowish to whitish. **Disc florets** 12–30+, functionally staminate; corollas pale yellow, tubes shorter than abruptly dilated throats, lobes 5, deltate to lance-ovate; anther thecae pale; stigmatic papillae 0. **Cypselae** plumply pyriform, ± obcompressed, 3–6-angled or -ribbed, finely striate between ribs, sparsely hirtellous or glabrate; **pappi** 0.

Genus 1, species 3 (3 species in the flora): c, e North America.

Traditionally, *Polymnia* has been included in Melampodiinae. It was treated at tribal rank, as Polymnieae (H. Robinson) Panero, by J. L. Panero and V. A. Funk (2002).

258. POLYMNIA Linnaeus, Sp. Pl. 2: 926. 1753; Gen. Pl. ed. 5, 396. 1754 • [Greek *Polymnia*, muse of sacred music and dance] E

John L. Strother

Annuals or perennials, 50–150+ cm (often rhizomatous). **Stems** erect, branched distally. **Leaves** all or mostly cauline; opposite; petiolate (petioles often ± winged, often with connate-perfoliate basal appendages); blades (pinnately nerved) deltate or rounded-rhombic to cordate or ovate, often raggedly pinnately lobed (lobes 3–11), ultimate margins coarsely toothed to denticulate or entire, faces pilose or pilosulous to hirtellous, or glabrate, or glabrous, gland-dotted and/or stipitate-glandular. **Heads** usually radiate, rarely ± disciform, (2–5+) in loose to congested, corymbiform clusters. **Calyculi** 0. **Involucres** hemispheric, 4–15+ mm diam. **Phyllaries** persistent, 6–21+ in 2 series (distinct, outer 2–6 ovate to linear, ± herbaceous, inner ovate to lanceolate, subequal to or shorter than outer, more scarious, similar to paleae). **Receptacles** flat to convex, paleate (paleae obovate to oblanceolate or spatulate, herbaceous to scarious). **Ray florets** 2–6, pistillate, fertile; corollas pale yellowish to whitish (tubes pilosulous, laminae ± cuneate to linear, 3-lobed, sometimes absent or nearly so). **Disc florets** 12–30+, functionally staminate; corollas pale yellow, tubes shorter than abruptly dilated throats, lobes 5, deltate to lance-ovate. **Cypselae** plumply pyriform, ± obcompressed, 3–6-angled or -ribbed, finely striate between ribs (narrowed at bases, patently inserted on receptacles, often minutely beaked), sparsely hirtellous or glabrate; **pappi** 0. $x = 15$.

Species 3 (3 in the flora): c, e North America.

SELECTED REFERENCE Wells, J. R. 1965. A taxonomic study of *Polymnia* (Compositae). Brittonia 17: 144–159.

1. Annuals; blades of larger leaves shallowly, if at all, pinnately lobed . . . . 3. *Polymnia cossatotensis*
1. Perennials; blades of larger leaves usually deeply pinnately lobed.
    2. Stems pilosulous to villous and/or stipitate-glandular; cypselae 3-angled or -ribbed
    . . . . 1. *Polymnia canadensis*
    2. Stems glabrous or glabrate; cypselae 4–6-angled or -ribbed . . . . 2. *Polymnia laevigata*

1. **Polymnia canadensis** Linnaeus, Sp. Pl. 2: 926. 1753 E F

Perennials. Stems pilosulous to villous and/or stipitate-glandular. **Leaves:** petioles 3–8(–12+) cm, sometimes raggedly winged; blades 4–20(–40+) × 2–12(–25+) cm, larger usually deeply pinnately lobed, lobes 5–7. **Cypselae** 3–4 mm, 3-angled or -ribbed. $2n = 30$.

Flowering May–Aug(–Oct). Damp, shaded sites, calcareous soils; 100–1000 m; Ont.; Ala., Ark., Conn., D.C., Ga., Ill., Ind., Iowa, Kans., Ky., Mich., Minn., Mo., N.Y., N.C., Ohio, Okla., Pa., Tenn., Vt., Va., W.Va., Wis.

2. **Polymnia laevigata** Beadle, Bot. Gaz. 25: 278. 1898 E

Perennials. Stems glabrous or glabrate. **Leaves:** petioles 3–10+ cm, seldom, if ever, winged; blades 4–20(–40+) × 2–12(25+) cm, larger usually deeply pinnately lobed, lobes 5–11. **Cypselae** ca. 3 mm, 4–6-angled or -ribbed. $2n = 30$.

Flowering Jul. Damp, shaded sites, calcareous soils; 10–300 m; Ala., Fla., Ga., Ky., Mo., Tenn.

3. Polymnia cossatotensis Pittman & V. M. Bates, Sida
13: 481, figs. 1, 2. 1989 ☐E☐

Annuals. Stems pilosulous to villous and/or stipitate-glandular.
Leaves: petioles 2–10 cm, not winged; blades 4–12(–18+) × 2–8(–12+) cm, shallowly, if at all, pinnately lobed. **Cypselae** 4.5–6 mm, 5-angled or -ribbed.

Flowering Sep–Oct. Cherty outcrops; 300–500 m; Ark.

187m.4. ASTERACEAE Martinov (tribe HELIANTHEAE) subtribe MILLERIINAE Bentham & Hooker f., Gen. Pl. 2: 190. 1873 (as Millerieae) ☐I☐

Annuals [perennials, subshrubs, or shrubs], [3–]20–200[–500] cm [rooting at nodes]. **Leaves** cauline; opposite (all or proximal) or alternate (distal); sessile [petiolate]; blades rhombic to lanceolate or oblanceolate [elliptic, spatulate, trullate], margins entire or toothed, faces glabrous or puberulent to pilose [scabrellous, strigose], gland-dotted (at least abaxial). **Heads** radiate [discoid], in corymbiform arrays [borne singly, in paniculiform arrays]. **Calyculi** 0. **Involucres** campanulate [hemispheric, sometimes gibbous]. **Phyllaries** persistent, [3–]10–13 in 2 series (distinct [connate], subequal to unequal, outer 5–6 oblong [lanceolate, linear, suborbiculate], herbaceous [accrescent], inner narrower, more scarious). **Receptacles** conic to hemispheric, paleate (paleae oblong to lanceolate, membranous to scarious, plane to cucullate). **Ray florets** [0 or 1–]6–18[–20], pistillate, fertile; corollas yellow [purplish or whitish] (tubes hairy at bases [glabrous], apices of laminae usually notably 3-lobed). **Disc florets** [3–]100 [–200+], bisexual, fertile [functionally staminate]; corollas yellow [green, orange], tubes (cylindric, hairy) shorter [longer] than campanulate [cylindric, funnelform] throats, lobes 5, deltate to lance-ovate; anther thecae dark; stigmatic papillae in 2 lines [0]. **Cypselae** ± prismatic to terete [ellipsoid, obovoid], weakly compressed, 3–4-angled, glabrous (shining); **pappi** 0.

Genera 6, species 32 (1 species in the flora): introduced; mostly New World, some Old World.

H. Robinson (1981) removed genera from traditional Milleriinae and placed them in other tribes and in other subtribes of Heliantheae. Members of Milleriinae (in the sense of Robinson) are mostly Mexican and Central American. *Guizotia* is unusual in being centered in Africa.

259. GUIZOTIA Cassini in F. Cuvier, Dict. Sci. Nat. ed. 2, 59: 237, 247, 248. 1829, name conserved • [For Pierre Guizot, 1787–1874, French historian, politician] ☐I☐

John L. Strother

Annuals [perennials, subshrubs, or shrubs], [3–]20–200 cm. **Stems** erect [creeping], branched. **Leaves** mostly cauline; opposite (distal sometimes alternate); sessile; blades rhombic to lanceolate or oblanceolate [spatulate], margins entire or serrate, faces glabrous or puberulent to pilose, gland-dotted (at least abaxial). **Heads** radiate, in corymbiform arrays [borne singly]. **Involucres** campanulate [hemispheric], [8–]10–15[–20] mm diam. **Phyllaries** persistent, 10–13 in 2 series (distinct, outer 5–6+ oblong, herbaceous, inner narrower, more scarious). **Receptacles**

conic to hemispheric, paleate (paleae oblong to lanceolate, plane [cucullate], membranous to scarious). **Ray florets** 6–18, pistillate, fertile; corollas yellow (hairy at bases of tubes). **Disc florets** 100+, bisexual, fertile; corollas yellow, tubes cylindric (hairy), shorter than campanulate throats, lobes 5, deltate. **Cypselae** weakly compressed, 3–4-angled, glabrous (shining); **pappi** 0. $x = 15$.

Species 6 (1 in the flora): introduced; Africa.

Taxonomic affinities of *Guizotia* generally have been acknowledged as obscure; the genus has been placed in Coreopsidinae, Melampodiinae, Milleriinae, and Verbesininae.

SELECTED REFERENCE Baagøe, J. 1974. The genus *Guizotia* (Compositae). A taxonomic revision. Bot. Tidsskr. 69: 1–39.

1. Guizotia abyssinica (Linnaeus f.) Cassini in F. Cuvier, Dict. Sci. Nat. ed. 2, 59: 248. 1829 [F] [I]

Polymnia abyssinica Linnaeus f., Suppl. Pl., 383. 1782

Receptacular paleae 5–7+ mm. **Ray corolla laminae** 8–14+ mm. **Disc corollas** 4–5 mm. **Cypselae** 4–5 mm. $2n = 30$ (from India).

Flowering summer–fall. Disturbed sites; 10–200+ m; introduced; Ont.; Ark., Calif., Conn., Fla., Ill., Kans., Md., Mass., Mich., N.Y., Ohio, Pa., Tenn., Va.; Africa; also introduced in Asia.

In the flora area, *Guizotia abyssinica* has been recorded sporadically at widely scattered stations (evidently often from birdseed wastes); it may be persistently established at relatively few stations.

187m.5. ASTERACEAE Martinov (tribe HELIANTHEAE) subtribe GUARDIOLINAE H. Robinson, Phytologia 41: 41. 1978

Perennials or subshrubs, (10–)30–100[–200] cm. **Leaves** cauline; opposite; ± petiolate; blades deltate or rounded-deltate to ovate [lanceolate to linear] (bases often hastate), margins usually dentate, faces glabrous or glabrate. **Heads** radiate, in corymbiform arrays [borne singly]. **Calyculi** 0. **Involucres** narrowly cylindric [campanulate]. **Phyllaries** persistent, 3–5 in 1 series (distinct, oblong or lanceolate to linear, subequal, herbaceous, conduplicate, streaked with 8–12+ translucent nerves, margins often membranous). **Receptacles** convex, paleate (paleae similar to phyllaries, membranous, conduplicate, each enfolding a floret). **Ray florets** 1–5, pistillate, fertile; corollas whitish (tubes often longer than laminae). **Disc florets** 3–20+, functionally staminate; corollas whitish, tubes longer than abruptly dilated, campanulate to funnelform throats, lobes 5, lanceolate to lance-linear; anther thecae green (filaments hairy); stigmatic papillae 0 or vestigial. **Cypselae** ± ellipsoid [fusiform to clavate], ± obcompressed, smooth or fine-ribbed, glabrous (bases carunculate); **pappi** 0 [5–6, erose scales].

Genus 1, species 10 (1 species in the flora): w United States, Mexico.

Guardiola has been included in Melampodiinae (G. Bentham 1873) and in Coreopsidinae (T. F. Stuessy 1973). According to H. Robinson (1981), the "glands on the anther appendages" in *Guardiola* "are distinct and massive, unlike...other Heliantheae."

260. GUARDIOLA   Cervantes ex Bonpland in A. von Humboldt and A. J. Bonpland, Pl. Aequinoct. 1: 143, plate 41. 1807   •   [For "M. le marquis de Guardiola"]

John L. Strother

Perennials or subshrubs, (10–)30–100 cm. **Stems** mostly erect, branched. **Leaves** cauline; opposite; ± petiolate; blades deltate or rounded-deltate to ovate [lanceolate to linear], margins usually dentate, faces glabrous or glabrate. **Heads** radiate, in corymbiform arrays [borne singly]. **Involucres** narrowly cylindric [campanulate], 2–3[–6] mm diam. **Phyllaries** persistent, 3–5 in 1 series (distinct, oblong or lanceolate to linear, herbaceous, conduplicate, streaked with 8–12+ translucent nerves, margins often membranous). **Receptacles** convex, paleate (paleae similar to phyllaries, membranous, each enfolding a floret). **Ray florets** 1–5, pistillate, fertile; corollas whitish (tubes often longer than laminae). **Disc florets** 3–20+, functionally staminate; corollas whitish, tubes longer than abruptly ampliate, campanulate to funnelform throats, lobes 5, lanceolate to lance-linear (filaments hairy; anthers green). **Cypselae** ± ellipsoid [fusiform to clavate], ± obcompressed, smooth or fine-ribbed, glabrous (bases carunculate); **pappi** 0 [5–6, erose scales]. *x* = 12.

Species ca. 10 (1 in the flora): sw United States, Mexico.

1. **Guardiola platyphylla**   A. Gray, Smithsonian Contr. Knowl. 5(6): 91. 1853   [F]

Leaf blades coriaceous, 20–70 × 15–50 mm. **Phyllaries** 7–10 mm. **Ray laminae** 3–5(–7) mm. **Disc corollas** 9–11 mm. **Cypselae** 5–6 mm. *2n* = 24.

Flowering spring–late summer. Hillsides, canyons, among rocks; 800–1600 m; Ariz.; Mexico (Chihuahua, Sonora).

187m.6. ASTERACEAE   Martinov (tribe HELIANTHEAE) subtribe RUDBECKIINAE H. Robinson, Phytologia 41: 43. 1978

Annuals, biennials, or perennials, 20–300 cm. **Leaves** basal and cauline or mostly cauline; alternate; petiolate or sessile; blades mostly deltate, elliptic, lanceolate, linear, lyrate, oblanceolate, oblong, ovate, pandurate, or spatulate, often 1–2-pinnately lobed or -pinnatifid, ultimate margins entire, dentate, serrate, or toothed, faces glabrous or hairy, often gland-dotted. **Heads** usually radiate, sometimes discoid, borne singly or in loose, ± corymbiform or paniculiform arrays. **Calyculi** 0. **Involucres** hemispheric to rotate. **Phyllaries** persistent, (5–)15–30+ in 1–3 series (distinct, elliptic, lanceolate, linear, ovate, or triangular, usually subequal, sometimes unequal, outer longer, all herbaceous, distally or throughout, or inner scarious, at least outer soon reflexed). **Receptacles** subhemispheric, ovoid, conic, or columnar (8–70+ mm high), paleate (paleae strongly conduplicate, each partly investing its subtended floret). **Ray florets** 0 or 3–25+, neuter; corollas yellow or orange, brown-purple, maroon, or reddish, sometimes bicolor (orange, brown-purple, maroon, or reddish plus yellow, laminae often drooping or reflexed). **Disc florets** 50–800+, bisexual, fertile; corollas yellowish green to yellow or brown-purple, tubes shorter than or equaling funnelform or cylindric throats, lobes 5, deltate to triangular or

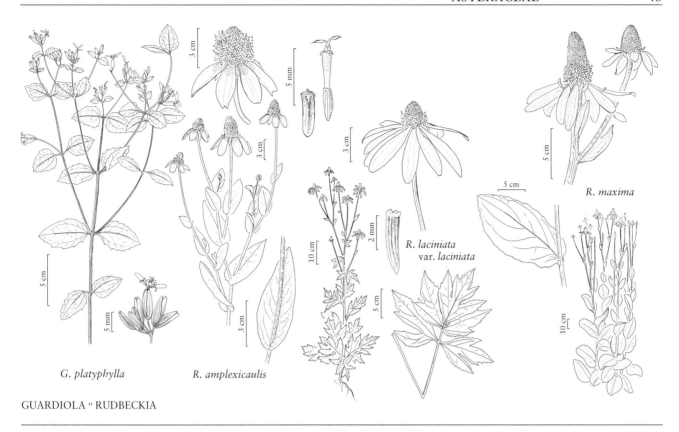

R. maxima

R. laciniata
var. laciniata

G. platyphylla   R. amplexicaulis

GUARDIOLA ∘ RUDBECKIA

obovate; anther thecae dark; stigmatic papillae in 2 lines. **Cypselae** subterete, or obpyramidal and 4-angled, or strongly compressed and linear-oblanceolate to oblong-oblique, glabrous or hairy; **pappi** 0, or coroniform, or of 2–8+ scales or teeth.

Genera 2, species 30 (2 genera, 27 species in the flora): North America, Mexico.

The genera of Rudbeckiinae have traditionally been included in Verbesininae (e.g., G. Bentham 1873) or in Helianthinae (e.g., T. F. Stuessy 1977[1978]). Sometimes, Rudbeckiinae has included *Echinacea* (e.g., P. O. Karis and O. Ryding 1994). Here, we follow H. Robinson (1981) in placing *Echinacea* in Ecliptinae.

In keys and descriptions for *Ratibida* and *Rudbeckia*, "discs" refers to receptacles at late flowering with paleae and florets in place and included in assessing shapes and in measurements of lengths and diameters.

1. Involucres (early flowering) hemispheric to rotate, 15–30+ mm diam.; phyllaries 15–30+ in 2–3 series, subequal; cypselae ± 4-angled, not strongly compressed, margins not pectinate or ciliate . . . . 261. *Rudbeckia*, p. 44
1. Involucres rotate, 8–12+ mm diam.; phyllaries 14–28+ in 2 series, unequal (outer notably longer than inner); cypselae strongly compressed, abaxial margin of each usually pectinate or ciliate . . . . 262. *Ratibida*, p. 60

261.   RUDBECKIA Linnaeus, Sp. Pl. 2: 906. 1753; Gen. Pl. ed. 5, 387. 1754
    • Coneflower [For Olaus (Olof) Johannes Rudbeck, 1630–1702, and Olaus (Olof) Olai Rudbeck, 1660–1740, father and son, professors at Uppsala University, predecessors of Linnaeus]   E

Lowell E. Urbatsch

Patricia B. Cox

Annuals, biennials, or perennials, mostly 50–300 cm (mostly fibrous rooted or rhizomatous, sometimes taprooted). **Stems** 1–15+, erect, branched distally, glabrous or hairy, sometimes glaucous. **Leaves** basal and cauline; alternate; petiolate or sessile; blades elliptic, lanceolate, linear, oblanceolate, ovate, or spatulate, often pinnately lobed to 1–2-pinnatifid, ultimate margins entire, dentate, serrate, or coarsely toothed, faces glabrous or hairy, sometimes glaucous, sometimes gland-dotted. **Heads** radiate or discoid, borne singly or in ± corymbiform or paniculiform arrays. **Involucres** (early flowering) hemispheric to rotate, 15–30+ mm diam. **Phyllaries** persistent, 5–20 in 1–2(–3) series (narrowly triangular to lanceolate or narrowly elliptic, subequal, herbaceous, distally or throughout). **Receptacles** subspheric to ovoid, or conic to columnar, paleate (paleae mostly tan proximally, green to maroon distally, obovate, concave, each ± clasping a floret, apices acute to cuspidate or truncate to rounded, abaxial tips glabrous or hairy, sometimes gland-dotted, resin ducts 2–3, maroon, 1 medial and 1 near each margin; receptacles plus paleae and florets = discs, 8–80 × 5–30 mm). **Ray florets** 0 or 5–25+, neuter; corollas (spreading to drooping or reflexed) usually yellow to yellow-orange or bicolor (laminae often proximally maroon or each with a maroon splotch, distally yellow), sometimes wholly maroon (orangish red to maroon in *R. graminifolia*. **Disc florets** 50–800+, bisexual, fertile; corollas yellow, yellowish green, or brown-purple (often bicolor), tubes shorter than cylindric to funnelform throats, lobes 5, triangular. **Cypselae** (black) ± obpyramidal and 4-angled (often minutely cross rugose), faces glabrous, angles sometimes hairy; **pappi** 0, ± coroniform, or of 2–8+ unequal scales. x = 16, 18, 19.

Species 23 (23 in the flora): North America; introduced in Europe.

The species of *Rudbeckia* are distributed among three major clades or lineages. Although relationships among the lineages are not robustly resolved, the lineages are treated here as sections (as they have been traditionally). *Rudbeckia hirta* and sometimes other species of the genus are used in experimental studies relating to initiation of flowering and hairy root culture. Most species are rich sources of phytochemicals that may offer potential for pharmaceutical or other uses.

SELECTED REFERENCES Cox, P. B. and L. E. Urbatsch. 1994. A taxonomic revision of *Rudbeckia* subg. *Macrocline* (Asteraceae: Heliantheae: Rudbeckiinae). Castanea 59: 300–318. Perdue, R. E. Jr. 1957. Synopsis of *Rudbeckia* subgenus *Rudbeckia*. Rhodora 59: 293–299.

1. Annuals (stems glabrous); leaves sessile, blades elliptic, lanceolate, oblanceolate, oblong, or ovate (not lobed, bases auriculate and clasping, faces glabrous, glaucous); phyllaries in 2 series (lengths of outer 2–4+ times inner, inner sometimes interpreted as paleae); paleae surpassing cypselae, margins ciliate, faces usually glabrous; pappi 0 . . . . 261a. *Rudbeckia* sect. *Dracopis*, p. 45
1. Annuals, biennials, or perennials (if epappose annuals, stems hairy); leaves petiolate or sessile, blades elliptic, lanceolate, linear, oblanceolate, ovate, or spatulate (often lobed, faces glabrous or hairy, sometimes glaucous; if leaves sessile and not lobed and bases auriculate and clasping, then robust perennials); phyllaries in 2(–3) series; paleae sometimes surpassing cypselae, margins usually ciliate, rarely eciliate, faces glabrous or hairy; pappi usually coroniform or of 2–6+ scales, sometimes 0 (sometimes cypselae each with glandular hairs around apices in *R. heliopsidis*).
    2. Leaves often bluish green, glaucous; receptacles usually conic to columnar (paleae not surpassing cypselae, except in *R. laciniata*: paleae surpassing cypselae, receptacles sometimes hemispheric to ovoid, disc corolla lobes yellow); ray florets 0 or 8–15+, corollas bright yellow; disc corollas proximally yellow to yellowish green, distally yellow or greenish to brown-purple; cypselae (3–)3.5–7.5 mm; pappi coroniform, or of 2–6 scales 0.1–2.5 mm. . . . . 261b. *Rudbeckia* sect. *Macrocline*, p. 46
    2. Leaves green, not glaucous; receptacles usually conic to hemispheric, rarely columnar (paleae surpassing cypselae); ray florets 6–25+, corollas usually yellow-orange proximally, yellow distally, sometimes with basal maroon splotch (orangish red to maroon in *R. graminifolia*); disc corollas proximally yellow to yellowish green, distally usually brown-purple, lobes sometimes yellowish or greenish; cypselae 1.5–3.5(–4) mm; pappi coroniform, or of 8+ unequal scales 0.1–2 mm, or 0 (sometimes cypselae each with glandular hairs around apices in *R. heliopsidis*) . . . . 261c. *Rudbeckia* sect. *Rudbeckia*, p. 52

261a. **RUDBECKIA** Linnaeus sect. **DRACOPIS** (Cassini) A. Gray in A. Gray et al., Syn. Fl. N. Amer. 1(2): 263. 1884 [E]

Obeliscaria Cassini subg. *Dracopis* Cassini in F. Cuvier, Dict. Sci. Nat. ed. 2, 35: 273. 1825; *Dracopis* (Cassini) Cassini

Annuals, 25–60(–120) cm (taprooted). **Stems** green (glaucous, glabrous). **Leaves** bluish green, glaucous; basal (seldom persisting to flowering) petiolate or sessile, blades elliptic, lanceolate, oblanceolate, oblong, or ovate, not lobed, bases (cauline) auriculate and clasping, margins crenate, entire or serrate, apices acute or acuminate, faces glabrous; cauline sessile, blades elliptic, lanceolate, oblanceolate, oblong, or ovate, bases auriculate and clasping, margins crenate, entire or serrate, apices acute or acuminate, faces glabrous. **Heads** in loose, corymbiform arrays or borne singly. **Phyllaries** in 2 series (lengths of outer 2–5+ times inner, inner similar to and sometimes interpreted as paleae). **Receptacles** ovoid to conic; paleae surpassing cypselae (margins ciliate), apices obtuse to acute, often mucronate, faces hairy subapically, glabrous near apices. **Ray florets** 6–10+; corollas yellow, sometimes partly orange or maroon (laminae often with proximal maroon splotch). **Discs** 15–30 × 8–15 mm. **Disc florets** to 400+; corollas proximally greenish yellow, distally purplish; styles ca. 5 mm, branches ca. 1.7 mm, proximal ½ stigmatic, apices subulate. **Cypselae** 1.8–2.5 mm; **pappi** 0. $x = 16$.

Species 1: c, e United States.

Rudbeckia sect. *Dracopis* is sometimes recognized at generic rank as *Dracopis*. DNA-based phylogenetic analyses support its placement in *Rudbeckia*, where it was treated by Gray as a section. The single species resembles those in sect. *Macrocline*; it is often placed there based on DNA evidence; the relationship is not uniformly conclusive.

1. **Rudbeckia amplexicaulis** Vahl, Skr. Naturhist. Selsk. 2(2): 29, plate 4. 1793 • Clasping or clasping-leaf coneflower E F

Dracopis amplexicaulis (Vahl) Cassini

Leaf blades 3–15 × 0.5–4 cm. **Involucres** 1–4 cm diam. **Phyllaries** spreading to reflexed, green, linear to lanceolate, herbaceous. **Ray laminae** spreading, eventually reflexed, elliptic to obovate, 12–30 × 7–15 mm, abaxially hirsute. **Disc corollas** 2.8–3.5 mm. **Cypselae:** each face 4–5-striate and minutely cross-rugose, glabrous; **pappi** 0 (cypselae each with ring of tan tissue at apex, ca. 0.1 mm). $2n = 32$.

Flowering late spring–summer. Open sites, moist soils; 0–400 m; Ala., Ark., Fla., Ga., Ill., Kans., La., Miss., Mo., N.Mex., N.Dak., Okla., S.C., Tex.

Rudbeckia amplexicaulis grows mainly in the Gulf coastal plain and Mississippi Embayment. It may be adventive elsewhere. It is used in "native" meadow and roadside plantings and has become a problem in some agricultural crops.

261b. RUDBECKIA Linnaeus sect. MACROCLINE Torrey & A. Gray, Fl. N. Amer. 2: 312. 1842   E

Rudbeckia subg. *Macrocline* (Torrey & A. Gray) P. B. Cox & Urbatsch

Perennials, 50–300 cm (robust, rhizomatous or fibrous rooted). **Stems** green, bluish green, or purplish (usually glaucous). **Leaves** often bluish green, glaucous; basal (usually persistent to flowering) petiolate or sessile; blades linear, lanceolate to ovate, or oblong to elliptic, usually lobed, bases attenuate, cuneate, or rounded, ultimate margins entire, crenate, dentate, or serrate, apices acute to obtuse, faces glabrous or hairy, sometimes glaucous; cauline petiolate or sessile, blades linear or ovate to pandurate, sometimes lobed, bases auriculate, rounded, or attenuate, ultimate margins entire or dentate to serrate, apices acute to obtuse, faces glabrous or hairy, sometimes glaucous. **Heads** borne singly or in loose, corymbiform to paniculiform arrays. **Phyllaries** in 2(–3) series. **Receptacles** usually conic to columnar (hemispheric to ovoid in *R. laciniata*); paleae not surpassing cypselae (except in *R. laciniata*), margins ciliate, apices acute to obtuse or rounded to truncate, attenuate to apiculate, usually hairy. **Ray florets** 0 or 8–15+; corollas bright yellow. **Discs** 12–60(–80) × 10–30 mm. **Disc florets** (100–)150–300 (–600+); corollas proximally yellow to yellowish green, distally yellow or greenish to brown-purple; anther appendages sometimes abaxially gland-dotted; styles 3–7 mm, branches 1–2.2 mm, proximal $^2/_3$–$^4/_5$ stigmatic, apices acute to rounded. **Cypselae** (3–)3.5–7.5 mm; **pappi** usually coroniform or of 2–6, unequal scales, 0.1–2.5 mm, sometimes 0. $x = 18$.

Species 13 (13 in the flora): North America.

1. Basal and cauline leaves 0.2–1.5 cm wide . . . . 9. *Rudbeckia mohrii*
1. Basal and cauline leaves 2–15(–45) cm wide.
  2. Blades of basal and proximal cauline leaves elliptic, lanceolate, or ovate, usually 1–2-pinnatifid or -pinnately compound (leaflets/lobes 3–11, distal cauline leaves sometimes lobed as well).
    3. Blades of all but distalmost leaves usually 1–2-pinnatifid or pinnately lobed (basal leaves not lobed and proximal cauline usually 3-lobed in var. *heterophylla*, known only from Florida); receptacles hemispheric or globose to ovoid; disc corolla lobes yellow (rhizomes elongate, slender, plants usually colonial); Rocky Mountains and e United States . . . . 7. *Rudbeckia laciniata*
    3. Blades of basal leaves ± pinnatifid to pinnately lobed (distal leaves sometimes not lobed); receptacles columnar, conic, cylindric, or ovoid; disc corolla lobes yellowish green, greenish, or purplish (rhizomes stout, plants usually not colonial); w United States (including Rocky Mountains).

4. Leaves sparsely to densely hairy on abaxial or both faces.
   5. Leaves moderately to densely hairy (both faces); rays 0; pappi of 4 scales, to 1 mm; Washington . . . . 2. *Rudbeckia alpicola*
   5. Leaves sparsely hairy (abaxial faces); rays 8–21; pappi coroniform or of scales, 0.5–1.5 mm; California . . . . 4. *Rudbeckia californica*
4. Leaves usually glabrous, sometimes sparsely hairy on veins abaxially.
   6. Rays 7–15; phyllaries to 1.5 cm (apices acute to rounded); discs 15–35 mm; California . . . . 6. *Rudbeckia klamathensis*
   6. Rays 0; phyllaries to 4 cm (apices attenuate); discs 20–60 mm; Colorado Rockies, Utah . . . . 10. *Rudbeckia montana*
[2. Shifted to left margin.—Ed.]
2. Blades of basal, and proximal and mid, cauline leaves elliptic or lanceolate (not lobed, sometimes coarsely toothed, margins of distal cauline leaves crenate, dentate, entire, or toothed, not lobed).
   7. Mid (and often proximal) cauline leaves sessile, bases of blades usually auriculate (clasping) to truncate, sometimes rounded, faces (one or both) glabrate or hairy, or glabrous.
      8. Leaves bluish green (fresh, heavily glaucous; dried blades white under reflected UV light, at least abaxially); discs 40–80 mm; adjoining areas of Arkansas, Louisiana, Oklahoma, Texas (introduced South Carolina) . . . . 8. *Rudbeckia maxima*
      8. Leaves green (fresh; dried blades dark under reflected UV light, at least abaxially); discs 12–25 mm; Alabama, Florida, Louisiana, Texas.
         9. Cauline leaf bases auriculate, clasping; heads (usually 10+) in paniculiform arrays; discs 12–16 mm; paleae 4–6 mm; Alabama, Florida . . . . 3. *Rudbeckia auriculata*
         9. Cauline leaf bases auriculate or rounded, not clasping; heads (usually to 15) in ± corymbiform arrays; discs 12–25 mm; paleae 6–8 mm (stems often purplish); w Louisiana, e Texas . . . . 13. *Rudbeckia scabrifolia*
   7. Mid cauline leaves petiolate, bases of blades attenuate to cuneate (not clasping), faces (one or both) glabrous or hairy.
      10. Leaf margins serrate or entire, faces usually sparsely to densely hairy (mostly abaxially), rarely glabrous; rays 0; n Rocky Mountains and Pacific states . . . 12. *Rudbeckia occidentalis*
      10. Leaf margins entire, serrate, serrulate, or toothed, faces glabrous or sparsely hairy; rays 7–16; Pacific states or se United States.
         11. Leaves bluish green (heavily glaucous); California, s Oregon . . . . 5. *Rudbeckia glaucescens*
         11. Leaves green (sometimes lightly glaucous); se United States.
            12. Basal leaf blades 15–50 × 3–9 cm (lengths ± 5 times widths); discs 20–45 mm; paleae 6–8 mm, apices acute (appressed in young heads); cypselae 5–7.5 mm; w Louisiana, e Texas . . . . 14. *Rudbeckia texana*
            12. Basal leaf blades 15–60 × 2–8 cm (lengths ± 7 times widths); discs 10–30 mm; paleae 5–6 mm, apices acute to acuminate (erect to spreading in young heads); cypselae 3–5.5 mm; n Florida, s Georgia . . . . 11. *Rudbeckia nitida*

2. Rudbeckia alpicola Piper, Erythea 7: 173. 1899
• Washington, Wenatchee Mountain, or showy coneflower [C] [E]

Rudbeckia occidentalis Nuttall var. *alpicola* (Piper) Cronquist

Perennials, to 150 cm (rhizomes stout, plants not colonial, roots fibrous). **Leaves** green, ovate to elliptic or deltate, pinnate to pinnatifid or lyrate-pinnatifid, herbaceous, faces moderately to densely hairy (hairs 1-seriate); basal petiolate, 25–70 × 8–45 cm, lobes 3–9, bases rounded to acute, apices acute; cauline petiolate or sessile, 12–50 × 5–40 cm, bases attenuate to cuneate, ultimate margins dentate to lobed, apices acute. **Heads** borne singly or (2–10) in ± corymbiform arrays. **Phyllaries** to 6 cm (foliaceous, faces scabrous). **Receptacles** columnar; paleae 5–7 mm, apices obtuse to acute, often apiculate, abaxial tips hairy. **Ray florets** 0. **Discs** 30–80 × 18–30 mm. **Disc florets** 300–500; corollas brown-purple, 4.2–5.8 mm; style branches ca. 2 mm, apices acute. **Cypselae** 3.5–5 mm; **pappi** of 4 scales, to 1 mm.

Flowering mid summer–fall. Thickets, bogs, along streams; of conservation concern; 200–1500 m; Wash.

Rudbeckia alpicola is known only from Chelan and Kittitas counties.

3. Rudbeckia auriculata (Perdue) Kral, Rhodora 77: 49. 1975 • Alabama or eared coneflower C E

Rudbeckia fulgida Aiton var. *auriculata* Perdue, Rhodora 63: 119. 1961

Perennials, to 300 cm (rhizomatous, roots fibrous). **Leaves** green, oblong to elliptic or ovate to pandurate, herbaceous, margins crenate, dentate, entire or serrate, apices acute, faces scabrous, abaxial often glabrescent; basal petiolate, 20–65 × 8–20 cm, bases attenuate; cauline sessile, 15–40 × 4–15 cm, bases auriculate to truncate. **Heads** (usually 10+) in paniculiform arrays. **Phyllaries** to 1 cm. **Receptacles** conic to ovoid; paleae 4–6 mm, apices obtuse to acute, abaxial tips hairy. **Ray florets** 8–14; laminae oblanceolate, 18–30 × 4–8 mm, abaxially sparsely hairy. **Discs** 12–16 × 10–18 mm. **Disc florets** 150–200; corollas brown-purple, 3.5–4.2 mm; anther appendages glabrous or gland-dotted; style branches to ca. 1.8 mm, apices acute to obtuse. **Cypselae** 3.5–5 mm; **pappi** of 4–6, unequal scales, to 2 mm. $2n = 36$.

Flowering summer–fall. Sandy stream banks, wet roadsides; of conservation concern; 10–90 m; Ala., Fla., Ga.

Rudbeckia auriculata is known from eight sites in Alabama, one in Florida, and one in Georgia.

4. Rudbeckia californica A. Gray, Proc. Amer. Acad. Arts 7: 357. 1868 • California coneflower E

Perennials, 50–180 cm (rhizomes stout, plants not colonial, roots fibrous). **Leaves** green, blades lanceolate to ovate or elliptic, herbaceous, sometimes pinnately lobed, bases attenuate, ultimate margins entire or coarsely toothed, apices acute, faces (abaxial) sparsely hairy or glabrous (adaxial); basal petiolate, 20–60 × 5–15 cm; cauline petiolate or sessile, 10–30 × 3–15 cm. **Heads** borne singly or (3–12) in ± corymbiform arrays. **Phyllaries** to 2 cm. **Receptacles** conic to columnar; paleae 5–6 mm, apices obtuse to acute, abaxial tips densely hairy. **Ray florets** 8–21; laminae linear to elliptic or oblong, 3–6 × 1–1.6 cm, abaxially hairy. **Discs** 15–60 × 14–25 mm. **Disc florets** 200–300+; corollas greenish yellow, 4–5 mm; style branches ca. 1.5 mm, apices acute to rounded. **Cypselae** 4–5 mm; **pappi** coroniform or of scales, to 1.5 mm. $2n = 36$.

Flowering summer–fall. Meadows, seeps, streamsides; 1400–2400 m; Calif.

Rudbeckia californica grows in the central Sierra Nevada.

5. Rudbeckia glaucescens Eastwood, Leafl. W. Bot. 2: 55. 1937 • Waxy coneflower E

Rudbeckia californica A. Gray var. *glauca* S. F. Blake

Perennials, to 150 cm (roots fibrous). **Leaves** bluish green (heavily glaucous), blades lanceolate to elliptic (not lobed), leathery, bases attenuate, margins entire or remotely serrulate, apices acute, faces glabrous; basal petiolate, 20–50 × 4–10 cm; cauline petiolate or sessile, 10–25 × 2–8 cm. **Heads** borne singly or (2–10) in ± corymbiform arrays. **Phyllaries** to 1.5 cm. **Receptacles** conic to columnar; paleae 4–6.5 mm, apices acute, often attenuate, abaxial tips hairy. **Ray florets** 7–15; laminae elliptic to oblong, 25–40 × 8–14 mm, abaxially hairy. **Discs** 15–35 × 14–22 mm. **Disc florets** 250–400+; corollas yellowish green, 3–4 mm; style branches ca. 1 mm, apices acute. **Cypselae** 4–5.5 mm; **pappi** coroniform or of ± connate scales, to 1.2 mm. $2n = 36$.

Flowering summer–fall. Meadows, seeps, streamsides, 60–1300 m; Calif., Oreg.

Rudbeckia glaucescens often grows on serpentine and often with *Darlingtonia*.

6. Rudbeckia klamathensis P. B. Cox & Urbatsch, Castanea 59: 309. 1994 • Klamath coneflower C E

Rudbeckia californica A. Gray var. *intermedia* Perdue, Rhodora 59: 289. 1957

Perennials, 50–100 cm (rhizomatous, roots fibrous). **Leaves** greenish blue (± glaucous), blades ovate to elliptic, often lobed, herbaceous, bases attenuate to rounded, margins entire, coarsely dentate, or crenate, apices acute, faces glabrous or sparsely hairy (at least on abaxial veins); basal petiolate, 15–65 × 4–15 cm; cauline petiolate or sessile, 7–35 × 4–14 cm. **Heads** borne singly or (2–10) in ± corymbiform arrays. **Phyllaries** to 1.5 cm (margins ciliate, hairs appressed, apices acute to rounded). **Receptacles** ovoid to conic; paleae 5–6 mm, apices acute, abaxial tips hairy. **Ray florets** 7–15; laminae elliptic or oblong, 25–35 × 7–14 mm, abaxially hairy. **Discs** 15–35 × 12–22 mm. **Disc florets** 180–300+; corollas yellowish green, 3.5–4 mm; style branches ca. 1.5 mm, apices acute to obtuse. **Cypselae** 4.5–6 mm; **pappi** coroniform or of 5–8 scales, to 1.5 mm. $2n = 36$.

Flowering late summer–fall. Along streams; of conservation consern; 1000–1600 m; Calif.

7. Rudbeckia laciniata Linnaeus, Sp. Pl. 2: 906. 1753
  • Cutleaf coneflower E F

Perennials, 50–300 cm (rhizomes often elongate, slender, plants colonial, roots fibrous). **Leaves** green, blades broadly ovate to lanceolate, all but distalmost 1–2-pinnatifid or pinnately compound, leaflets/lobes 3–11, bases cuneate to attenuate or cordate, margins entire or dentate, apices acute to acuminate, faces glabrous or hairy (sometimes with translucent patches); basal (often withering before flowering) petiolate, 15–50 × 10–25 cm; cauline petiolate or sessile, mostly lobed to pinnatifid, sometimes not lobed, 8–40 × 3–20 cm. **Heads** (2–25) in loose, corymbiform arrays. **Phyllaries** to 2 cm (8–15, ovate to lanceolate, margins mostly ciliate, glabrous or hairy). **Receptacles** hemispheric or ovoid to globose; paleae 3–7 mm, apices (at least of proximal) truncate or rounded, abaxial tips densely hairy. **Ray florets** 8–12; laminae elliptic to oblanceolate, 15–50 × 4–14 mm, abaxially hairy. **Discs** 9–30 × 10–23 mm. **Disc florets** 150–300+; corollas yellow to yellowish green (lobes yellow), 3.5–5 mm; style branches 1–1.5 mm, apices acute to rounded. **Cypselae** 3–4.5 mm; **pappi** coroniform or of 4 scales, to 1.5 mm.

Varieties 5 (5 in the flora): North America.

Cultivars of *Rudbeckia laciniata* are grown as ornamentals. The cultivar 'golden-glow' is widely planted and occasionally escapes cultivation. Among the varieties traditionally recognized in floristic treatments, vars. *ampla* and *heterophylla* are the most distinctive. Detailed investigation may show that the other varieties, from eastern North America, represent broadly intergrading forms that should be subsumed under var. *laciniata*.

1. Receptacles ovoid; discs (17–)20–30 mm; w of Great Plains . . . . . . 7a. *Rudbeckia laciniata* var. *ampla*
1. Receptacles globose or hemispheric; discs 10–20 mm; Great Plains and e United States.
    2. Basal and proximal cauline leaves not lobed, adaxial faces moderately to densely hairy; Levy County, Florida . . . .
        . . . . 7d. *Rudbeckia laciniata* var. *heterophylla*
    2. Basal and proximal cauline leaves lobed, adaxial leaf faces sparsely hairy or glabrous; e North America (not Levy County, Florida).
        3. Proximal leaves usually with 0, 3, or 5 lobes; se United States . . . .
            . . . . 7c. *Rudbeckia laciniata* var. *digitata*
        3. Proximal leaves usually 1–2-pinnatifid or with 5–11 lobes; e North America (not se United States).

[4. Shifted to left margin.—Ed.]
4. Proximal cauline leaves 2-pinnatifid, mid cauline leaves 5–11-lobed; paleae 3.1–4.1 mm; cypselae 3.5–4 mm; pappi 0.7–1.5 mm . . . .
    . . . . 7b. *Rudbeckia laciniata* var. *bipinnata*
4. Proximal cauline leaves pinnatifid, mid cauline leaves 5–9-lobed; paleae 4.4–6.1 mm; cypselae 4.2–6 mm; pappi 0.1–0.7 mm . . . .
    . . . . 7e. *Rudbeckia laciniata* var. *laciniata*

7a. Rudbeckia laciniata Linnaeus var. **ampla** (A. Nelson) Cronquist in C. L. Hitchcock et al., Vasc. Pl. Pacif. N.W. 5: 280. 1955 • Rocky Mountain cutleaf coneflower E

Rudbeckia ampla A. Nelson, Bull. Torrey Bot. Club 28: 234. 1901

Leaves: basal 20–50 × 10–25 cm, blades pinnately compound to pinnatifid (basal leaflets or lobes often pinnatifid); proximal and mid cauline blades 5–9-lobed; adaxial faces glabrous or sparsely hairy (hairs arching, 1-seriate). **Receptacles** ovoid; paleae 4.5–6.5 mm. **Ray laminae** 25–50 × 7–12 mm. **Discs** (17–)20–30 × 12–20 mm. **Cypselae** 4–5.5 mm; **pappi** mostly longer than 0.7 mm. **2n** = 38.

Flowering summer–fall. Wet sites, along streams, open meadows; 1700–2700 m; B.C.; Ariz., Colo., Mont., N.Mex., S.Dak., Wyo.

Variety *ampla* grows west of the Great Plains. It is introduced in British Columbia.

7b. Rudbeckia laciniata Linnaeus var. **bipinnata** Perdue, Rhodora 64: 328. 1962 • Northeastern cutleaf coneflower E

Leaves: basal 15–40 × 10–25 cm, blades pinnately compound to 2-pinnatifid; proximal cauline 2-pinnatifid, mid cauline blades 5–11-lobed (distal usually 3-lobed); adaxial faces sparsely to moderately hairy. **Receptacles** globose to ovoid; paleae 3.1–4.1. **Ray laminae** 20–45 × 7–12 mm. **Discs** 15–19 × 10–20 mm. **Cypselae** 3.5–4 mm; **pappi** mostly 0.7+ mm. **2n** = 72, ca. 72, 102+.

Flowering late summer–fall. Wet habitats, along streams and edges of woods; 10–300 m; Conn., Del., Md., Mass., N.H., N.Y., Pa.

7c. Rudbeckia laciniata Linnaeus var. **humilis** A. Gray in A. Gray et al., Syn. Fl. N. Amer. 1(2): 262. 1884 • Southeastern cutleaf coneflower [E]

Rudbeckia digitata Miller; *R. laciniata* var. *digitata* (Miller) Fiori

Leaves: basal 12–45 × 10–25 cm, blades 0, 3, or 5-lobed; proximal and mid cauline blades similar (distal usually 3-lobed or not lobed); adaxial faces sparsely to moderately hairy. **Receptacles** hemispheric to globose; paleae 4–5.5 mm. **Ray laminae** 20–45 × 5–11 mm. **Discs** 10–20 × 10–17 mm. **Cypselae** 3.5–5.2 mm; **pappi** to 0.7 mm. **2***n* = 38.

Flowering late summer–fall. Wet sites along streams and edges of woods; 70–1500 m; Ala., Ga., Md., N.C., S.C., Tenn., Va.

7d. Rudbeckia laciniata Linnaeus var. **heterophylla** (Torrey & A. Gray) Fernald & B. G. Schubert, Rhodora 50: 172. 1948 • Florida coneflower [E]

Rudbeckia heterophylla Torrey & A. Gray, Fl. N. Amer. 2: 312. 1842

Leaves: basal 15–30 × 5–10 cm, blades usually not lobed, rarely 3-lobed; cauline blades not lobed; adaxial faces moderately to densely hairy. **Receptacles** globose to ovoid; paleae 4–5 mm. **Ray laminae** 15–45 × 8–17 mm. **Discs** 10–15 × 8–20 mm. **Cypselae** 4–5.5 mm; **pappi** to 0.7 mm.

Flowering spring–fall. Deep sandy soils, along streams, edges of woods; 0–20 m; Fla.

Variety *heterophylla* is known only from Levy County.

7e. Rudbeckia laciniata Linnaeus var. **laciniata** [E] [F]

Leaves: basal 15–40 × 10–25 cm, blades pinnately compound to pinnatifid; proximal and mid cauline blades 5–9-lobed (distal cauline 3-lobed or not lobed); adaxial leaf faces glabrous or sparsely hairy. **Receptacles** globose to ovoid; paleae 4.4–6.1 mm. **Ray laminae** 20–45 × 7–18 mm. **Discs** 15–20 × 10–20 mm. **Cypselae** 4.2–6 mm; **pappi** to 0.7 mm. **2***n* = 36, 54.

Flowering summer–fall. Wet sites, along streams, edges of woods; 10–600 m; Man., N.B., N.S., Ont., P.E.I., Que.; Ala., Ark., Conn., Fla., Ga., Ill., Ind., Iowa, Kans., Ky., La., Maine, Md., Mich., Minn., Miss., Mo., Nebr., N.H., N.Y., N.C., N.Dak., Ohio, Okla., Pa., S.C., S.Dak., Tenn., Tex., Vt., Va., W.Va., Wis.

8. Rudbeckia maxima Nuttall, Trans. Amer. Philos. Soc., n. s. 7: 354. 1840 • Great or cabbage coneflower [E] [F]

Perennials, to 250 cm (rhizomatous, roots fibrous). **Leaves** bluish green (heavily glaucous, dried blades white under UV light, at least abaxially), blades elliptic, ovate to obovate or pandurate (not lobed), leathery, margins crenate, dentate, or entire, apices acute to rounded, faces glabrous; basal petiolate, elliptic to ovate, 15–65 × 4–15 cm, bases attenuate to cuneate; cauline petiolate or sessile, 7–50 × 4–14 cm, blades ovate to pandurate, bases cuneate to auriculate (and clasping). **Heads** borne singly or (3–15) in ± corymbiform arrays. **Phyllaries** to 1.5 cm (ovate to lanceolate, margins ciliate). **Receptacles** ovoid to conic; paleae 6–8 mm, apices mostly rounded, abaxial tips glabrous (hairy subapically). **Ray florets** 10–20; laminae elliptic to oblanceolate, 30–80 × 8–15 mm, abaxially sparsely hairy. **Discs** 40–80 × 15–35 mm. **Disc florets** 300–600+; corollas maroon (at least distally), 4–6 mm; style branches ca. 2.2 mm, apices acute. **Cypselae** 6.5–7 mm; **pappi** of 4–6 scales to 1.5 mm. **2***n* = 36.

Flowering spring–summer. Mesic pastures, roadsides; 10–70 m; Ark., La., Okla., S.C., Tex.

Rudbeckia maxima was introduced in South Carolina and possibly elsewhere through horticultural and agricultural activities. It is a cultivated ornamental and is sometimes a problematic pasture weed.

9. Rudbeckia mohrii A. Gray, Proc. Amer. Acad. Arts 17: 217. 1882 • Mohr's or grassy coneflower [E]

Perennials, to 110 cm (rhizomatous, roots fibrous). **Leaves** green (glaucous), blades (3–5-nerved) narrowly lanceolate to linear (not lobed), leathery, bases attenuate, margins entire (sometimes ciliate), apices acute, faces glabrous; basal petiolate, 10–40 × 0.5–1.5 cm; cauline petiolate or sessile, 2–25 × 0.2–0.8 cm. **Heads** (3–15) in ± corymbiform arrays. **Phyllaries** to 1 cm. **Receptacles** ovoid or ellipsoid to hemispheric; paleae 4–5 mm, glabrous or sparsely hairy, apices obtuse, acuminate. **Ray florets** 7–14; laminae elliptic to oblanceolate, 12–30 × 5–10 mm, abaxially glabrous. **Discs** 9–15 × 7–17 mm. **Disc florets** 100–200+; corollas yellowish green proximally, maroon distally, 3–4 mm; style branches ca. 1 mm, apices acute to blunt. **Cypselae** 3.4–4.2 mm; **pappi** ± coroniform, of connate or distinct scales to 1.2 mm. **2***n* = 36.

Flowering summer–fall. Wet pine savannas, along ditches and bayous, shallow water; 0–50 m; Fla., Ga.

10. **Rudbeckia montana** A. Gray, Proc. Amer. Acad. Arts 17: 217. 1882 • Montane coneflower E

Rudbeckia occidentalis Nuttall var. *montana* (A. Gray) Perdue

Perennials, to 150 cm (rhizomes stout, plants not colonial, roots fibrous). **Leaves** greenish blue (± glaucous), blades elliptic to ovate, usually pinnatifid to pinnately lobed (lobes mostly opposite, ovate to elliptic), ± leathery, bases attenuate to cuneate, ultimate margins entire or coarsely dentate, apices acute, faces usually glabrous, sometimes sparsely hairy (at least abaxially on veins); basal petiolate, 17–60 × 10–25 cm; cauline petiolate or sessile, 8–30 × 5–20 cm (blades among heads not lobed). **Heads** borne singly or in ± corymbiform arrays. **Phyllaries** to 4 cm (margins sometimes ciliate, apices attenuate). **Receptacles** ovoid to conic; paleae (proximally transparent to light brown, distally greenish) 5–8 mm, apices acute to ± rounded, abaxial tips hairy. **Ray florets** 0. **Discs** 20–60 × 12–30 mm. **Disc florets** 200–500+; corollas maroon proximally, greenish distally, 4–5 mm; style branches ca. 1.5 mm, apices acute to rounded. **Cypselae** 5.2–7 mm; **pappi** coroniform, to 1.8 mm. *2n* = 36.

Flowering summer. Hillside seeps, streams; 2400–2800 m; Colo., Utah.

11. **Rudbeckia nitida** Nuttall, J. Acad. Nat. Sci. Philadelphia 7: 78. 1834 • Shiny coneflower, St. John's or black-eyed Susan C E

Rudbeckia glabra de Candolle

Perennials, to 140 cm (rhizomatous, roots fibrous). **Leaves** green (lightly glaucous), blades elliptic to lanceolate (not lobed), leathery, bases attenuate to cuneate, margins crenate, entire, or toothed, apices acute, faces glabrous or sparsely hairy; basal petiolate, 15–60 × 2–8 cm; cauline petiolate or sessile, 5–50 × 2–9 cm. **Heads** borne singly or in ± corymbiform arrays. **Phyllaries** to 2 cm (margins sometimes ciliate, sparsely hairy). **Receptacles** ovate to columnar; paleae (recurved prior to flowering, erect to spreading in young heads) 5–6 mm, apices acute to acuminate, abaxial tips hairy. **Ray florets** 8–15; laminae oblong to oblanceolate, 12–60 × 5–15 mm, abaxially sparsely hairy. **Discs** 10–30 × 12–20 mm. **Disc florets** 200–300+; corollas yellowish green proximally, maroon distally, 3.5–4.5 mm; style branches ca. 1.2 mm, apices acute to acuminate. **Cypselae** 3–5.5 mm; **pappi** ± coroniform, to 2 mm. *2n* = 36.

Flowering spring–summer. Wet pinelands, swales, ditches, bayous; of conservation concern; 0–40 m; Fla., Ga.

Rudbeckia nitida grows in northern Florida and southern Georgia (a report for Alabama has not been confirmed). It is cultivated as an ornamental. According to R. Kral (1983), it is threatened or endangered and is associated with savanna or bog dicots and monocots, particularly *Eriocaulon, Sarracenia, Lachnocaulon, Rhexia, Xyris,* and the composites *Coreopsis, Helianthus,* and *Liatris.*

12. **Rudbeckia occidentalis** Nuttall, Trans. Amer. Philos. Soc., n. s. 7: 355. 1840 • Western coneflower E

Perennials, to 200 cm (rhizomatous, roots fibrous). **Leaves:** green, blades broadly ovate to lanceolate (rarely lobed), herbaceous, bases attenuate to cuneate or broadly rounded, ultimate margins entire or serrate, apices acute, faces sparsely to densely hairy (mostly adaxially), rarely glabrous; basal petiolate, 12–30 × 3–9 cm; cauline petiolate or sessile, 5–25 × 2–10 cm. **Heads** in ± corymbiform arrays. **Phyllaries** to 3 cm (margins mostly ciliate, hairy, especially abaxially). **Receptacles** ovoid to columnar; paleae (proximally light brown, distally green, becoming maroon with age) 5–7 mm, apices acute to acuminate, abaxial tips densely hairy. **Ray florets** 0. **Discs** 17–45 × 12–20 mm. **Disc florets** 200–500+; corollas yellowish green proximally, blackish maroon distally, 4–6 mm; style branches ca. 1.2 mm, apices acute to rounded. **Cypselae** 3.5–5 mm; **pappi** coroniform, to 1.2 mm. *2n* = 36.

Flowering summer–fall. Open meadows, streamsides, seeps; 1000–2800 m; Calif., Idaho, Mont., Nev., Oreg., Utah, Wash., Wyo.

Rudbeckia occidentalis is sometimes grown as an ornamental.

13. **Rudbeckia scabrifolia** L. E. Brown, Phytologia 61: 367, fig. 1. 1986 • Roughleaf coneflower E

Perennials, to 200 cm (rhizomatous, roots fibrous). **Leaves** green (dried, dark under reflected UV light), blades elliptic to lanceolate (not lobed), ± leathery, bases attenuate to cuneate or rounded, margins entire or coarsely toothed, apices acute, faces hairy; basal petiolate, 15–65 × 2–15 cm; cauline sessile (proximalmost often petiolate), 5–50 × 2–12 cm (mids ± pandurate, bases auriculate). **Heads** (to 15) in ± corymbiform arrays. **Phyllaries** to 2 cm (margins ciliate). **Receptacles** ovoid to ellipsoid; paleae 6–8 mm, (apical margins densely ciliate) apices acute, abaxial tips hairy. **Ray florets** 10–15; corollas yellow, laminae oblong to oblanceolate, 20–40 × 5–12 mm, abaxially hairy. **Discs** 12–25 × 12–20 mm. **Disc florets** 150–400+; corollas greenish yellow proximally, maroon distally, 3.5–5 mm; style branches ca. 2 mm, apices acute. **Cypselae** 3.5–5 mm; **pappi** coroniform, to 2.5 mm. $2n = 36$.

Flowering late spring–fall. Hillside seeps; 10–60 m; La., Tex.

Rudbeckia scabrifolia occurs in western Louisiana and eastern Texas. It is in the Center for Plant Conservation's National Collection of Endangered Plants.

14. Rudbeckia texana (Perdue) P. B. Cox & Urbatsch, Phytologia 67: 366. 1989 • Texas coneflower E

Rudbeckia nitida Nuttall var. *texana* Perdue, Rhodora 64: 328. 1962

Perennials, to 150 cm (rhizomatous, roots fibrous). **Leaves** green, blades elliptic to lanceolate (not lobed), ± leathery, bases attenuate to cuneate, margins entire, serrate, or toothed, apices acute to acuminate, faces glabrous or sparsely hairy; basal 15–50 × 3–9 cm; cauline petiolate or sessile, 5–50 × 2–12 cm. **Heads** borne singly or (2–5) in ± corymbiform arrays. **Phyllaries** to 2.5 cm (margins sometimes ciliate). **Receptacles** ovoid to ellipsoid; paleae 6–8 mm, apices acute (appressed in young heads), abaxial tips hairy. **Ray florets** 10–16; laminae oblong to oblanceolate, 20–50 × 8–12 mm, abaxially hairy. **Discs** 20–45 × 10–20 mm. **Disc florets** 200–500+; corollas proximally greenish yellow, distally maroon, 3.5–5 mm; style branches ca. 2 mm, apices acute. **Cypselae** 5–7.5 mm; **pappi** coroniform, to 1.5 mm. $2n = 36$.

Flowering late spring–fall. Swales, prairies, ditches, bayous; 0–50 m; La., Tex.

Rudbeckia texana grows in western Louisiana and eastern Texas.

261c. RUDBECKIA Linnaeus sect. **RUDBECKIA** E

Annuals, biennials, or perennials, to 100(–200) cm (rhizomatous, fibrous rooted, or taprooted). **Stems** green. **Leaves** green, not glaucous; basal (usually withering before flowering) usually petiolate, blades linear, lanceolate to ovate, or elliptic to oblanceolate, sometimes lobed, bases attenuate, cuneate, or rounded, margins entire, dentate, or serrate, apices acute to obtuse, faces glabrous or hairy, sometimes gland-dotted; cauline usually petiolate (at least proximal), distal usually sessile, blades linear or ovate to pandurate, sometimes lobed, lobes 3(–5), bases auriculate, attenuate, or cordate, margins entire or serrate, apices acute to obtuse, faces glabrous or hairy, sometimes gland-dotted. **Heads** borne singly or in loose, corymbiform to paniculiform arrays. **Phyllaries** in 2(–3) series. **Receptacles** usually conic to hemispheric, rarely columnar; paleae surpassing cypselae, margins sometimes ciliate, apices obtuse or rounded to acute or attenuate to apiculate, faces glabrous or hairy abaxially. **Ray florets** 6–25+; corollas usually yellow-orange proximally, yellow distally, sometimes with basal maroon splotch (orangish red to maroon in *R. graminifolia*). **Discs** 8–30 × 5–25 mm. **Disc florets** 50–800+; corollas proximally yellow to yellowish green, distally usually brown-purple, lobes sometimes yellowish or greenish; anther appendages sometimes gland-dotted; styles 3–5+ mm, branches 1–1.8 mm, proximal $^1/_2$–$^4/_5$ stigmatic, apices acute to rounded. **Cypselae** 1.5–3.5(–4) mm; **pappi** usually coroniform (sometimes cypselae each with glandular hairs around apices in *R. heliopsidis*). $x = 19$.

Species 9 (9 in the flora): e North America.

1. Leaves 0.2–1 cm wide (grasslike, lengths 10+ times widths); rays orangish red to maroon; Florida panhandle . . . . 16. *Rudbeckia graminifolia*
1. Leaves (0.4–)1–15 cm wide (if less than 1 cm wide, lengths less than 10 times widths); rays yellow to yellow-orange (at least distally); e North America.
   2. Annuals, biennials, or perennials (taprooted or roots fibrous); pappi 0 or coroniform (to 0.1 mm).
      3. Stems and leaves coarsely hispid to hirsute; style branch apices subulate; e North America . . . . 19. *Rudbeckia hirta*
      3. Stems and leaves softly pilose to woolly; style branch apices acute to obtuse; Alabama, Florida, Georgia, South Carolina . . . . 21. *Rudbeckia mollis*
   2. Perennials (usually rhizomatous and/or stoloniferous, roots fibrous, caudices sometimes woody); pappi 0 or coroniform (0.1–0.5 mm; sometimes cypselae each with glandular hairs around apices in *R. heliopsidis*).
      4. Proximal cauline leaves elliptic or ovate, usually 3(–5)-lobed; paleae cuspidate (tips awnlike, 1.5+ mm) . . . . 23. *Rudbeckia triloba*
      4. Proximal cauline leaves elliptic, linear, spatulate, or ovate, rarely lobed (3–5–lobed in *R. subtomentosa*); paleae acute, obtuse, or rounded.
         5. Stems and leaves densely hirsute; proximal cauline leaves 3(–5)–lobed; heads (usually 8–25) in loose, corymbiform to paniculiform arrays . . . . . 22. *Rudbeckia subtomentosa*
         5. Stems and leaves glabrous or sparsely to moderately hairy; proximal cauline leaves not lobed; heads borne singly or (2–12) in corymbiform arrays.
            6. Leaves: abaxial faces glabrous or hirsute to strigose (not gland-dotted); paleae: abaxial tips usually glabrous, rarely pilose.
               7. Plants stoloniferous (rosettes forming at stolon apices; stem branches spreading); basal leaves lanceolate to broadly ovate or elliptic . . . 15. *Rudbeckia fulgida*
               7. Plants not stoloniferous (rosettes at bases of aerial stems; stem branches ascending); basal leaves linear to narrowly spatulate . . . . . 20. *Rudbeckia missouriensis*
            6. Leaves: abaxial faces glabrous or strigose and gland-dotted; paleae: abaxial tips canescent or strigose, sometimes gland-dotted as well.
               8. Leaf blades elliptic, lanceolate, or ovate (flat); heads usually (4–8) in ± corymbiform arrays, sometimes borne singly; rays 6–12 (± spreading); pappi coroniform, to 1.5 mm (sometimes cypselae each with glandular hairs around apices); Alabama, Georgia, North Carolina, South Carolina, Virginia . . . . 18. *Rudbeckia heliopsidis*
               8. Leaf blades elliptic, lanceolate, or ovate (± conduplicate); heads usually borne singly; rays 12–25 (reflexed); pappi coroniform, to 0.5 mm; midwestern and sc United States . . . . 17. *Rudbeckia grandiflora*

15. Rudbeckia fulgida Aiton, Hort. Kew. 3: 251. 1789

• Orange coneflower   E

Perennials, to 120 cm (stoloniferous, rosettes forming at stolon apices). **Stems** glabrous or moderately hirsute (branches spreading). **Leaves:** blades lanceolate to broadly ovate or elliptic (not lobed), herbaceous, bases attenuate to cordate, margins usually entire or serrate, sometimes lacerate, apices acute, faces glabrous or hirsute to strigose; basal petiolate, 5–30 × 1–8 cm; cauline petiolate, 2–25 × 0.5–7 cm, bases attenuate to cordate or auriculate. **Heads** borne singly or (2–7) in corymbiform arrays. **Phyllaries** to 2 cm. **Receptacles** hemispheric to ovoid; paleae 2.5–4 mm, (apical margins usually ciliate) apices obtuse to acute, abaxial tips usually glabrous. **Ray florets** 10–15; laminae elliptic to oblanceolate, 15–25 × 3–6 mm, abaxially strigose. **Discs** 12–16 × 10–18 mm. **Disc florets** 50–500+; corollas proximally yellowish green, brown-purple distally, 3–4.2 mm; style branches ca. 1.3 mm, apices rounded. **Cypselae** 2.2–4 mm; **pappi** coroniform, to 0.2 mm.

Varieties 7 (7 in the flora): e North America.

1. Basal leaf blade lengths ± 3 times widths . . . . 15b. *Rudbeckia fulgida* var. *fulgida*
1. Basal leaf blade lengths to 2 times widths.
   2. Cauline leaves not notably smaller distally.

3. Stems densely villous-hirsute; basal leaf margins mostly coarsely crenate; cauline leaf margins sharply serrate (teeth remote); Illinois, Indiana, Ohio . . . .
. . . . 15a. *Rudbeckia fulgida* var. *deamii*
3. Stems glabrous or sparsely villous-hirsute; basal leaf margins entire or crenate; cauline leaf margins coarsely serrate to lacerate; ne United States . . . . 15e. *Rudbeckia fulgida* var. *speciosa*

[2. Shifted to left margin.—Ed.]
2. Cauline leaves notably smaller distally (except var. *umbrosa*).
    4. Ray laminae 25–40 mm; palea margins eciliate (Indiana, Michigan, Ohio) . . . .
. . . . 15f. *Rudbeckia fulgida* var. *sullivantii*
    4. Ray laminae 10–30 mm; palea margins ciliate.
        5. Leaf bases (basal and proximal cauline) broadly rounded to cordate (Alabama, Arkansas, Georgia, Indiana, Kentucky, Mississippi, Missouri, North Carolina, Ohio, South Carolina, Tennessee, Virginia)
. . . . 15g. *Rudbeckia fulgida* var. *umbrosa*
        5. Leaf bases (basal and proximal cauline) usually acute, attenuate, cuneate, or rounded.
            6. Cauline leaf blades lanceolate to ovate; ray laminae 15–25+ mm; Arkansas, Missouri, Oklahoma, Texas . . . .
. . . . 15c. *Rudbeckia fulgida* var. *palustris*
            6. Cauline leaf blades oblanceolate to broadly spatulate or pandurate; ray laminae 10–15 mm; Alabama, Florida, Georgia, North Carolina, South Carolina, Tennessee, Virginia, West Virginia . . .
. . . . . 15d. *Rudbeckia fulgida* var. *spathulata*

15a. Rudbeckia fulgida Aiton var. **deamii** (S. F. Blake) Perdue, Rhodora 59: 297. 1958 • Deam's coneflower E

Rudbeckia deamii S. F. Blake, Rhodora 19: 113. 1917

Stems densely villous-hirsute (hairs retrorse). **Leaves:** basal blades ovate, 1.5–6 cm wide, lengths to 2 times widths, bases rounded to attenuate, margins mostly coarsely crenate, faces hairy; cauline sessile distally, elliptic to lanceolate (distalmost ovate), not notably smaller distally, bases ± auriculate (clasping), margins sharply serrate (teeth remote), faces densely villous-hirsute. **Phyllaries** reflexed, 1–2 × 0.2–0.6 cm, densely hairy. **Receptacles** 11–17 mm diam.; palea margins ciliate, faces mostly glabrous. **Ray florets** 12–20; laminae 15–25 mm.

Flowering late summer–fall. Stream banks, woodland ridges; 100–300 m; Ill., Ind., Ohio.

15b. Rudbeckia fulgida Aiton var. **fulgida** E

Rudbeckia acuminata C. L. Boynton & Beadle; *R. foliosa* C. L. Boynton & Beadle; *R. tenax* C. L. Boynton & Beadle; *R. truncata* Small

Stems villous-hirsute (hairs antrorse to spreading). **Leaves:** basal blades lanceolate to narrowly elliptic to ovate or oblanceolate, 2–4.5 cm wide, lengths ± 3 times widths, bases cuneate, margins entire or serrate, faces glabrous or moderately hairy; cauline sessile or petiolate, blades lanceolate to narrowly ovate, somewhat smaller distally, bases attenuate to cuneate, margins entire or denticulate, faces glabrous or moderately hairy. **Phyllaries** reflexed, 1–2.2 × 0.2 cm, mostly strigoso-hispid. **Receptacles** 11–18 mm diam.; palea margins ciliate, faces glabrous. **Ray florets** 8–14; laminae 8–20(–30) mm. $2n$ = 38, ca. 76.

Flowering late summer–fall. Dry to mesic, sunny to shady sites; 0–700 m; Ala., Del., Fla., Ga., Ill., Ind., Ky., Md., N.J., N.Y., N.C., Ohio, Pa., Tenn., Va., W.Va.

15c. Rudbeckia fulgida Aiton var. **palustris** (Eggert ex C. L. Boynton & Beadle) Perdue, Rhodora 59: 297. 1958 • Prairie or marsh coneflower E

Rudbeckia palustris Eggert ex C. L. Boynton & Beadle, Biltmore Bot. Stud. 1: 16. 1901

Stems glabrous or sparsely hairy (hairs antrorse to spreading). **Leaves:** basal blades lanceolate to narrowly elliptic to ovate, 2–4 cm wide, lengths to 2 times widths, bases cuneate, margins entire or serrate, faces glabrous or sparsely hairy; cauline petiolate (proximal) or sessile (distal), blades lanceolate to ovate, notably smaller distally, bases attenuate to subauriculate (distal), margins entire or serrate, faces sparsely to moderately hairy. **Phyllaries** reflexed, 0.8–2.2 × 0.2–0.5 cm, pilose (abaxially) or glabrous (adaxially). **Receptacles** 11–18 mm diam.; palea margins ciliate, faces glabrous. **Ray florets** 8–14; laminae 15–25 mm. $2n$ = 38.

Flowering summer–fall. Mesic to wet sites; 50–600 m; Ark., Mo., Okla., Tex.

15d. Rudbeckia fulgida Aiton var. **spathulata**
(Michaux) Perdue, Rhodora 59: 298. 1958

• Orange coneflower   [E]

Rudbeckia spathulata Michaux, Fl. Bor.-Amer. 2: 144. 1803

Stems glabrous or sparsely strigose (hairs spreading). **Leaves:** basal blades mostly narrowly oblanceolate, 0.5–2 cm wide, lengths to 2 times widths, bases attenuate, margins entire, serrate or dentate, faces sparsely strigose; cauline usually petiolate, rarely sessile, blades oblanceolate to broadly spatulate or pandurate, notably smaller distally, bases attenuate, margins entire or serrate, faces moderately hairy. **Phyllaries** spreading to reflexed, 0.5–1 × 0.2–0.5 cm, pilose abaxially, glabrous adaxially. **Receptacles** 10–14 mm diam.; palea margins ciliate, faces glabrous. **Ray florets** 10–16; laminae 15–25 mm.

Flowering late summer–fall. Wet woodlands, meadows, and clearings; 30–200 m; Ala., Fla., Ga., N.C., S.C., Tenn., Va., W.Va.

15e. Rudbeckia fulgida Aiton var. **speciosa**
(Wenderoth) Perdue, Rhodora 59: 297. 1958

• Showy coneflower   [E]

Rudbeckia speciosa Wenderoth, Index Sem. (Marburg). 1828

Stems glabrous or sparsely villous-hirsute (hairs spreading). **Leaves:** basal blades broadly lanceolate to ovate, 2–6.5 cm wide, lengths to 2 times widths, bases mostly rounded, margins entire or crenate, faces sparsely to moderately hairy; cauline petiolate (proximal) or ± sessile (distal), elliptic to lanceolate (becoming somewhat panduriform distally), not notably smaller distally, bases cuneate, attenuate, or auriculate, margins coarsely serrate to lacerate, faces scabrous to strigose. **Phyllaries** reflexed, 1–2.2 × 0.2–0.7 cm, glabrous or sparsely hairy. **Receptacles** 11–18 mm diam.; palea margins ciliate, faces glabrous. **Ray florets** 12–21; laminae 20–40 mm. **2n** = ca. 76.

Flowering summer–fall. Mesic open woodlands; 30–200 m; Ont., Que.; Ala., Ark., Conn., Del., Ga., Ill., Ky., Mass., Mich., Mo., N.J., N.Y., Ohio, Pa., Va., W.Va., Wis.

Variety *speciosa* has rarely been introduced in Canada.

15f. Rudbeckia fulgida Aiton var. **sullivantii**
(C. L. Boynton & Beadle) Cronquist, Rhodora 47: 400. 1945 • Sullivant's coneflower   [E]

Rudbeckia sullivantii C. L. Boynton & Beadle, Biltmore Bot. Stud. 1: 15. 1901 (as sullivanti); *R. speciosa* Wenderoth var. *sullivantii* (C. L. Boynton & Beadle) B. L. Robinson

Stems glabrous or sparsely hirsute (hairs spreading). **Leaves:** basal blades mostly broadly elliptic to ovate, 1.5–3 cm wide, lengths to 2 times widths, bases broadly rounded, margins sharply toothed (not incised), faces glabrous or sparsely hairy; cauline petiolate (proximal) to nearly sessile (distal), shapes ovate to lanceolate, notably smaller distally, bases rounded to attenuate, margins entire or serrate, faces glabrous or sparsely hairy. **Phyllaries** spreading to reflexed, 1–2 × 0.2–0.5 cm, glabrous or sparsely hairy. **Receptacles** 11–18 mm diam.; palea margins eciliate, faces glabrous. **Ray florets** 12–21; laminae 25–40 mm. **2n** = 76.

Flowering summer–fall. Swamps, shorelines, fens, sedge meadows; 30–200 m; Ark., Ill., Ind., Mich., Mo., N.Y., Ohio, Pa., W.Va.

15g. Rudbeckia fulgida Aiton var. **umbrosa**
(C. L. Boynton & Beadle) Cronquist, Rhodora 47: 400. 1945 • Shady coneflower   [E]

Rudbeckia umbrosa C. L. Boynton & Beadle, Biltmore Bot. Stud. 1: 16. 1901; *R. chapmanii* C. L. Boynton & Beadle

Stems glabrous or sparsely hirsute (hairs antrorse to spreading). **Leaves:** basal blades ovate, 1.5–3.5 cm wide, lengths to 2 times widths, bases broadly rounded to cordate, margins coarsely dentate, faces glabrous or sparsely hairy; cauline petiolate (proximal) to nearly sessile (distal), ovate to lanceolate, not notably smaller distally, bases rounded to attenuate, margins usually coarsely serrate, sometimes entire, faces sparsely to moderately hairy. **Phyllaries** reflexed, 1–2.2 × 0.25–0.5 cm, glabrous or sparsely hairy. **Receptacles** 10–15 mm diam.; palea margins ciliate, faces glabrous. **Ray florets** 8–12; laminae 10–30 mm. **2n** = 76.

Flowering late summer–fall. Wet woodlands, bottoms; 30–400 m; Ala., Ark., Ga., Ind., Ky., Miss., Mo., N.C., Ohio, S.C., Tenn., Va.

16. Rudbeckia graminifolia (Torrey & A. Gray) C. L. Boynton & Beadle, Biltmore Bot. Stud. 1: 12. 1901 · Grassleaf coneflower E

Echinacea atrorubens (Nuttall) Nuttall var. *graminifolia* Torrey & A. Gray, Fl. N. Amer. 2: 306. 1842

Perennials, to 80 cm (roots fibrous). **Stems** moderately strigose (hairs ascending). **Leaves:** blades narrowly lanceolate to elliptic (grasslike, lengths 10+ times widths, not lobed), herbaceous, bases attenuate, margins entire, apices acute, faces glabrous or sparsely hirsute; basal petiolate or sessile, 10–25 × 0.5–1 cm; cauline sessile, 1–25 × 0.2–1 cm (distal smaller), bases attenuate. **Heads** mostly borne singly. **Phyllaries** to 1 cm (spreading to reflexed). **Receptacles** hemispheric to ovoid; paleae 4–5 mm, (apical margins glabrous) apices acuminate-cuspidate, awn-tipped, abaxial tips sparsely strigose. **Ray florets** 8–16; corollas orangish red to maroon, laminae elliptic to obovate, 10–25 × 3–6 mm, abaxially sparsely strigose. **Discs** 10–15 × 8–15 mm. **Disc florets** 50–300+; corollas proximally maroon, distally brown-purple, 3.9–4.4 mm; style branches ca. 1.8 mm, apices obtuse to rounded. **Cypselae** 2–3 mm; **pappi** coroniform, to 0.5 mm. *2n* = 38.

Flowering spring–fall. Wet, sandy flatwoods sites; 0–30 m; Fla.

Rudbeckia graminifolia grows in the Apalachicola region.

17. Rudbeckia grandiflora (Sweet) C. C. Gmelin ex de Candolle in A. P. de Candolle and A. L. P. P. de Candolle, Prodr. 5: 556. 1836 · Largeflower or rough coneflower E

Centrocarpha grandiflora Sweet, Brit. Fl. Gard., ser. 2, 1: plate 87. 1831

Perennials, to 120 cm (roots fibrous, caudices often woody). **Stems** proximally glabrous or sparsely hairy (hairs spreading), distally strigose (hairs ascending). **Leaves:** blades elliptic, lanceolate, or ovate (± conduplicate, not lobed), bases cuneate to rounded, margins entire or remotely serrate, apices acute, faces strigose, abaxially gland-dotted; basal petiolate, 10–35 × 2–11 cm; cauline petiolate (proximal) to nearly sessile (distal), 4–30 × 1.5–9 cm. **Heads** mostly borne singly. **Phyllaries** to 15 mm (strigose and gland-dotted). **Receptacles** hemispheric to ovoid; paleae 5–6.5 mm, (apical margins glabrous) acuminate-cuspidate, awn-tipped, abaxial tips sparsely strigose. **Ray florets** 12–25; laminae elliptic to obovate (reflexed), 20–50 × 5–10 mm, abaxially hairy and gland-dotted. **Discs** 10–30 × 15–25

mm. **Disc florets** 200–800+; corollas greenish yellow basally and in lobes, otherwise maroon, 3.5–5 mm; style branches ca. 1.8 mm, apices obtuse. **Cypselae** 2–3 mm; **pappi** coroniform, to 0.5 mm.

Varieties 2 (2 in the flora): mostly c, e, and s United States.

1. Stems glabrous or sparsely hairy proximally and hairy distally (hairs ascending, mostly shorter than 0.5 mm) ... 17a. *Rudbeckia grandiflora* var. *alismifolia*
1. Stems hairy (hairs spreading proximally, ascending distally, ca. 1 mm).................. .... 17b. *Rudbeckia grandiflora* var. *grandiflora*

17a. Rudbeckia grandiflora (Sweet) C. C. Gmelin ex de Candolle var. **alismifolia** (Torrey & A. Gray) Cronquist, Rhodora 47: 401. 1945 (as alismaefolia) E

Rudbeckia alismifolia Torrey & A. Gray, Fl. N. Amer. 2: 310. 1842 (as alismaefolia)

Stems glabrous or sparsely hairy proximally and hairy distally (hairs ascending, mostly shorter than 0.5 mm). **Leaves** scabrous to puberulent. *2n* = 38.

Flowering late spring–summer. Prairies, open sites; 30–200 m; Ark., Ky., La., Miss., Tex.

17b. Rudbeckia grandiflora (Sweet) C. C. Gmelin ex de Candolle var. **grandiflora** E

Stems hairy (hairs spreading proximally, ascending distally, ca. 1 mm). **Leaves** hirsute (hairs longer than 1 mm). *2n* = 38.

Flowering late spring–summer. Mostly mesic to dry prairies, woodlands; 30–300 m; Ark., Ga., Ill., Kans., Ky., Miss., Mo., Ohio, Okla., Tex.

Variety *grandiflora* has been collected in Ontario; it evidently is no longer found there.

18. Rudbeckia heliopsidis Torrey & A. Gray, Fl. N. Amer. 2: 310. 1842 · Sunfacing coneflower, Little River black-eyed Susan C E

Perennials, to 120 cm (rhizomatous). **Stems** villous to glabrate. **Leaves:** blades elliptic, lanceolate, or ovate (flat, not lobed), bases acute to rounded or cuneate, margins entire or serrate, apices acute, faces glabrous or sparsely strigose and gland-dotted; basal petiolate, blades 10–20 × 2–5 cm; cauline petiolate (proximal) or nearly sessile (distal), blades 2–20

× 1–4 cm. **Heads** usually (4–8) in ± corymbiform arrays, sometimes borne singly. **Phyllaries** to 1.5 cm (faces sparsely to moderately hairy and gland-dotted). **Receptacles** hemispheric to ovoid; paleae 3.2–4.2 mm, apices rounded to acute, abaxial tips canescent and gland-dotted. **Ray florets** 6–12; laminae elliptic to oblanceolate (spreading), 15–30 × 3–6 mm, abaxially strigose and gland-dotted. **Discs** 10–15 × 7–15 mm. **Disc florets** 100–200+; corollas proximally yellowish green, distally brown-purple, 3–4 mm; style branches ca. 1.4 mm, apices obtuse. **Cypselae** 2–3 mm; **pappi** coroniform, to 1.5 mm (sometimes cypselae each with glandular hairs around apices).

Flowering summer–fall. Mesic to wet woodlands, meadows; of conservation concern; 70–200 m; Ala., Ga., N.C., S.C., Va.

19. Rudbeckia hirta Linnaeus, Sp. Pl. 2: 907. 1753
  • Black-eyed Susan [E]

Annuals, biennials, or perennials, to 100 cm (taprooted or roots fibrous). **Stems** hispid to hirsute (hairs spreading, 1+ mm). **Leaves:** blades elliptic, lanceolate, or ovate (not lobed), bases attenuate to cuneate, margins entire or serrate, apices acute, faces hispid to hirsute; basal petiolate, blades 8–30 × 0.5–7 cm; cauline petiolate or sessile, blades (sometimes pandurate) 3–20 × 0.4–4 cm. **Heads** borne singly or (2–5) in loose, corymbiform arrays. **Phyllaries** to 3 cm (faces hispid to hirsute). **Receptacles** hemispheric to ovoid; paleae 4–6 mm, apices acute, often attenuate, abaxial tips hirsute to hispid. **Ray florets** 8–16; laminae (usually uniformly yellow to yellow-orange or with a basal maroon splotch, sometimes mostly maroon) elliptic to oblong or oblanceolate, 15–45 × 5–10 mm, abaxially hispid to hirsute. **Discs** 12–22 × 10–20 mm. **Disc florets** 250–500+; corollas proximally yellowish green, distally brown-purple, 3–4.2 mm; style branches ca. 1.5 mm, apices subulate. **Cypselae** 1.5–2.7 mm; **pappi** 0.

Varieties 4 (4 in the flora): e North America.

Some strains of *Rudbeckia hirta* are cultivated and/or used in seed mixes for "re-naturalization" and erosion control.

1. Annuals, biennials, or perennials; stems branched at bases or proximal to or at or near mid heights, leafy mostly toward bases (leaves smaller distally); peduncles usually at least ¹/₂ plant heights (Gulf Coastal Plain, Florida to Texas).
    2. Stems branched mostly at or near mid heights; basal leaves oblanceolate, faces hispid to ± sericeous; Georgia to Texas . . . .
        . . . . 19a. *Rudbeckia hirta* var. *angustifolia*
    2. Stems branched at or near bases (plants often scapiform); basal leaves obovate to nearly orbiculate, faces scabrous to hirsute; c, s Florida
        . . . . 19b. *Rudbeckia hirta* var. *floridana*
1. Biennials or perennials; stems branched mostly beyond mid heights, leafy ± throughout; peduncles to ¹/₃ plant heights.
    3. Leaves: basal blades broadly ovate to broadly elliptic, 2.5–7 cm wide (lengths mostly 2 times widths), margins coarsely toothed; cauline (sometimes sessile) lanceolate, ovate, or pandurate (mostly Appalachian Highlands to Illinois) . . . . 19c. *Rudbeckia hirta* var. *hirta*
    3. Leaves: basal blades lanceolate to oblanceolate, 1–2.5(–5) cm wide (lengths 3–5 times widths), margins entire or serrulate; cauline blades spatulate, oblanceolate, or broadly linear . .
        . . . . 19d. *Rudbeckia hirta* var. *pulcherrima*

19a. Rudbeckia hirta Linnaeus var. **angustifolia** (T. V. Moore) Perdue, Rhodora 59: 296. 1958 [E]

Rudbeckia floridana T. V. Moore var. *angustifolia* T. V. Moore, Pittonia 4: 176. 1900; *R. divergens* T. V. Moore

Annuals, biennials, or perennials. **Stems** branched mostly at or near mid heights, leafy toward bases. **Leaves:** basal blades oblanceolate, 0.5–2.5 cm wide; cauline narrowly elliptic (smaller distally); margins entire or serrate; faces hispid to ± sericeous. **Peduncles** at least ¹/₂ plant heights. $2n = 38$.

Flowering spring–summer. Meadows, pastures, old fields, roadsides; 20–80 m; Ala., Fla., Ga., La., Miss., S.C., Tex.

Variety *angustifolia* grows along the Gulf coastal plain.

19b. Rudbeckia hirta Linnaeus var. **floridana** (T. V. Moore) Perdue, Rhodora 59: 296. 1958 [E]

Rudbeckia floridana T. V. Moore, Pittonia 4: 176. 1900

Annuals, biennials, or perennials. **Stems** branched at or near bases, leafy toward bases (plants often scapiform). **Leaves:** basal blades obovate to nearly orbiculate, 1–4 cm wide; cauline blades ovate, elliptic, or oblanceolate (smaller distally); margins entire or serrate; faces scabrous to hirsute. **Peduncles** at least ¹/₂ plant heights.

Flowering spring–fall. Pastures, old fields, roadsides; 0–20 m; Fla.

Variety *floridana* grows in central to southern Florida.

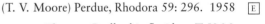

19c. Rudbeckia hirta Linnaeus var. **hirta** [E]

Rudbeckia amplectens T. V. Moore; *R. brittonii* Small; *R. hirta* var. *brittonii* (Small) Fernald; *R. hirta* var. *monticola* (Small) Fernald; *R. monticola* Small

Biennials or perennials. Stems branched mostly beyond mid heights, leafy ± throughout. **Leaves:** basal blades broadly ovate to broadly elliptic, 2.5–7 cm wide (lengths mostly 2 times widths); cauline blades lanceolate, ovate, or pandurate; margins coarsely toothed; faces hirsute to hispid. **Peduncles** less than ⅓ plant heights. $2n = 38$.

Flowering spring–fall. Open woodlands, fields, roadsides; 30–300 m; Ala., Ark., Conn., Del., Ga., Ill., Ind., Ky., Maine, Md., Mass., Mich., Miss., N.H., N.J., N.Y., N.C., Ohio, Pa., S.C., Tenn., Vt., Va., W.Va.

Variety *hirta* grows mostly from the Appalachian highlands to Illinois. A report of var. *hirta* from Ontario has not been confirmed.

19d. Rudbeckia hirta Linnaeus var. **pulcherrima** Farwell, Rep. (Annual) Michigan Acad. Sci. 6: 209. 1904 [E]

Rudbeckia bicolor Nuttall; *R. longipes* T. V. Moore; *R. sericea* T. V. Moore; *R. serotina* Nuttall; *R. serotina* var. *corymbifera* (Fernald) Fernald & B. G. Schubert; *R. serotina* var. *lanceolata* (Bischoff) Fernald & B. G. Schubert; *R. serotina* var. *sericea* (T. V. Moore) Fernald & B. G. Schubert

Annuals or perennials. Stems branched mostly beyond mid heights, leafy ± throughout. **Leaves:** basal blades lanceolate to oblanceolate 1–2.5(–5) cm wide (lengths 3–5 times widths); cauline blades spatulate, oblanceolate, or broadly linear; margins entire or serrulate; faces scabrous to hirsute. **Peduncles** to ⅓ plant heights. $2n = 38$.

Flowering spring–fall. Prairies, meadows, roadsides; 10–500 m; Alta., B.C., Man., N.B., Nfld. and Labr. (Nfld.), N.S., Ont., P.E.I., Que., Sask.; Ala., Ark., Calif., Colo., Conn., Del., D.C., Fla., Ga., Idaho, Ill., Ind., Iowa, Kans., Ky., La., Maine, Md., Mass., Mich., Minn., Miss., Mo., Nebr., N.H., N.J., N.Mex., N.Y., N.C., N.Dak., Ohio, Okla., Oreg., Pa., R.I., S.C., S.Dak., Tenn., Tex., Utah, Vt., Va., Wash., W.Va., Wis., Wyo.

20. Rudbeckia missouriensis Engelmann ex C. L. Boynton & Beadle, Biltmore Bot. Stud. 1: 17. 1901 · Missouri or Missouri orange coneflower [E] [F]

Rudbeckia fulgida Aiton var. *missouriensis* (Engelmann ex C. L. Boynton & Beadle) Cronquist

Perennials, to 80 cm (not rhizomatous, not stoloniferous, rosettes at bases of aerial stems). **Stems** (branches ascending) moderately hirsute (hairs spreading, 1+ mm). **Leaves:** blades linear to spatulate (not lobed), bases attenuate to cuneate, margins entire or remotely serrulate, apices acute to rounded, faces hirsute; basal petiolate, 5–20 × 0.5–2 cm; cauline petiolate (proximal) or sessile (distal), 2–15 × 0.4–1.5 cm. **Heads** borne singly or (2–12) in loose, corymbiform arrays. **Phyllaries** to 1.5 cm (faces hairy, more densely abaxially). **Receptacles** mostly hemispheric; paleae 5–6.5 mm, apices rounded to acute, abaxial tips glabrous. **Ray florets** 9–15; laminae elliptic to oblanceolate, 10–25 × 5–8 mm, abaxially sparsely strigose. **Discs** 8–15 × 10–17 mm. **Disc florets** 150–250+; corollas proximally greenish yellow, distally purple brown, 4–5.5 mm; style branches ca. 1.5 mm, apices obtuse. **Cypselae** 1.5–2.7 mm; **pappi** coroniform, ca. 0.1 mm. $2n = 38$.

Flowering late spring–fall. Dry, rocky prairies, limestone glades; 10–80 m; Ark., Ill., Ky., La., Mo., Okla., Tex.

21. Rudbeckia mollis Elliott, Sketch Bot. S. Carolina 2: 453. 1823 · Softhair coneflower [E]

Annuals, biennials, or perennials, to 100 cm (taprooted). **Stems** softly pilose to woolly (hairs spreading, 1–4 mm, longer ones toward bases). **Leaves:** blades oblong to oblong-lanceolate (not lobed), margins entire or serrate, apices acute to obtuse, faces softly pilose to woolly, gland-dotted; basal petiolate, 5–12 × 0.5–2 cm, bases attenuate; cauline sessile, (elliptic to pandurate) 1.5–10 × 1–4 cm (mid largest, bases cuneate to auriculate). **Heads** borne singly or (2–12) in loose, corymbiform arrays. **Phyllaries** to 1.5 cm (faces hairy and gland-dotted). **Receptacles** conic to hemispheric; paleae 5–6.5 mm, apices acute, abaxial tips hirsute and gland-dotted. **Ray florets** 10–16; laminae linear to oblanceolate, 20–40 × 4–7 mm, abaxially hairy and gland-dotted. **Discs** 10–20 × 12–18 mm. **Disc florets** 200–400+; corolla bases yellowish green, otherwise brown-purple, 3–4.2 mm; style branches ca. 1.5 mm, apices acute to obtuse. **Cypselae** 2.5–3.5 mm; **pappi** 0 or coroniform, to 0.1 mm. $2n = 38$.

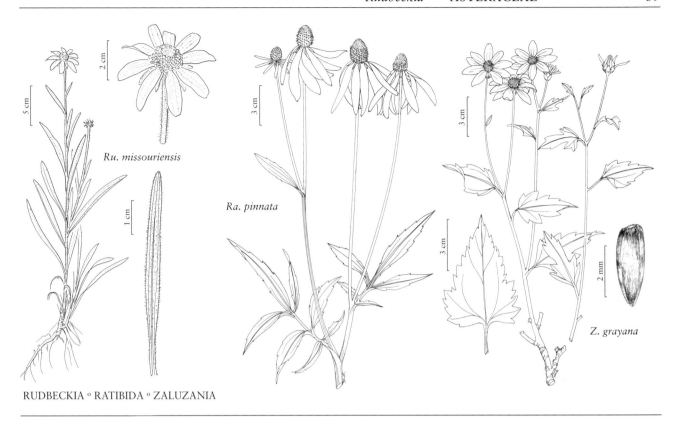

Ru. missouriensis

Ra. pinnata

Z. grayana

RUDBECKIA ∘ RATIBIDA ∘ ZALUZANIA

Flowering late spring–summer. Dry, sandy soils; 30–90 m; Ala., Fla., Ga., S.C.

22. Rudbeckia subtomentosa Pursh, Fl. Amer. Sept. 2: 575. 1813 • Sweet coneflower E

Perennials, to 200 cm (rhizomatous, rhizomes stout). Stems densely hirsute (hairs mostly antrorse, to 0.5 mm). Leaves: blades ovate to elliptic (not lobed), margins denticulate to serrate, apices acute to obtuse or acuminate, faces densely hirsute and gland-dotted (glands fewer adaxially); basal 15–30 × 3–10 cm, bases attenuate; cauline petiolate, ovate to elliptic, proximal 3–25 × 1–15 cm, usually 3–5-lobed, bases truncate to cuneate or rounded. Heads (8–25) in loose, corymbiform to paniculiform arrays. Phyllaries to 1.5 cm (faces hairy and ± gland-dotted). Receptacles conic to hemispheric; paleae 4–6 mm, apices acute, abaxial tips hirsute and gland-dotted. Ray florets 10–16; laminae (yellow to yellow-orange) linear to oblanceolate, 20–40 × 5–8 mm, abaxially sparsely hairy, abundantly gland-dotted. Discs 10–17 × 5–15 mm. Disc florets 200–400+; corollas yellowish green on basal $^1/_2$, otherwise brown-purple, 3–4.2 mm; style branches ca. 1 mm, apices acute. Cypselae 2–3.5 mm; pappi coroniform, to ca. 0.2 mm. $2n = 38$.

Flowering late summer–fall. Mesic to wet prairies, stream banks, and woodland openings; 20–300 m; Ark., Conn., Ill., Ind., Iowa, Kans., Ky., La., Mass., Mich., Miss., Mo., N.Y., N.C., Okla., Tenn., Tex., Wis.

Rudbeckia subtomentosa is often cultivated as an ornamental.

23. Rudbeckia triloba Linnaeus, Sp. Pl. 2: 907. 1753 • Brown-eyed Susan E

Perennials, to 150 cm (rhizomatous). Stems glabrate to hirsute or strigose (hairs 1–2 mm, basal retrorse, others spreading). Leaves: blades ovate to subcordate or elliptic (not lobed), margins serrate, apices acute to acuminate, faces hirsute to strigose; basal petiolate, 10–30 × 2–8 cm, bases truncate or rounded to cordate; cauline petiolate or sessile, ovate to elliptic, proximal usually 3–5-lobed, 2–20 × 1.5–8 cm (smaller, fewer lobed distally), bases rounded to attenuate, sometimes clasping. Heads (10–30) in paniculiform arrays. Phyllaries to 1.5 cm (faces moderately hirsute). Receptacles conic to subhemispheric; paleae 5–6.5 mm, apices cuspidate (tips awnlike, 1.5+ mm), glabrous. Ray florets 8–15; laminae (corollas yellow to yellow-orange with basal maroon splotches) linear to oblanceolate, 8–30 × 3–8 mm, abaxially sparsely strigose. Discs 8–15 × 10–20 mm. Disc florets 150–300+; corollas yellowish green basally, otherwise brown-purple, 3–4 mm; style branches ca. 1.2 mm, apices obtuse to rounded. **Cypselae** 1.9–2.8 mm; **pappi** coroniform, to 0.2 mm.

Varieties 3 (3 in the flora): e North America.

1. Cauline leaves (at least some) 5–7-lobed (Florida, Kentucky, North Carolina, Virginia) . . . . 23a. *Rudbeckia triloba* var. *pinnatiloba*
1. Cauline leaves (at least some) 3-lobed.
    2. Ray laminae 8–17 mm; discs 10–15 mm diam. (20–100 m; relatively widespread) . . . . 23c. *Rudbeckia triloba* var. *triloba*
    2. Ray laminae 18–30 mm; discs 15–20 mm diam. (100–1200 m; Iowa, Kentucky, North Carolina, Tennessee) . . . . 23b. *Rudbeckia triloba* var. *rupestris*

23a. Rudbeckia triloba Linnaeus var. **pinnatiloba** Torrey & A. Gray, Fl. N. Amer. 2: 309. 1842 [E]

Rudbeckia pinnatiloba (Torrey & A. Gray) Beadle

Cauline leaves (at least some) 5–7-lobed. **Discs** 10–15 mm diam. **Ray laminae** 10–20(–25) mm.

Flowering summer–fall. Mesic woodlands; 70–1000 m; Ala., Fla., Ky., N.C., Tenn., Va.

23b. Rudbeckia triloba Linnaeus var. **rupestris** (Chickering) A. Gray in A. Gray et al., Syn. Fl. N. Amer. 1(2): 260. 1884 [E]

Rudbeckia rupestris Chickering, Bot. Gaz. 6: 188. 1881

Cauline leaves (at least some) 3-lobed. **Discs** 15–20 mm diam. **Ray laminae** 20–30 mm.

Flowering mid summer–fall. Mesic woodlands; 100–1200 m; Iowa, Ky., N.C., Tenn.

23c. Rudbeckia triloba Linnaeus var. **triloba** [E]

Rudbeckia beadlei Small; *R. triloba* var. *beadlei* (Small) Fernald

Cauline leaves (at least some) 3-lobed. **Discs** 10–15 mm diam. **Ray laminae** 8–17 mm. $2n = 38$, ca. 57, 57.

Flowering Jul–Oct. Mesic to wet woodlands, thickets, pastures, roadsides, meadows; 20–300 m; Ont., Que.; Ala., Ark., Colo., Conn., Del., Ga., Ill., Ind., Iowa, Kans., Ky., La., Md., Mass., Mich., Minn., Miss., Mo., Nebr., N.J., N.Y., N.C., Ohio, Okla., Pa., Tenn., Tex., Utah, Vt., Va., W.Va., Wis.

Variety *triloba* is introduced in Canada.

262. **RATIBIDA** Rafinesque, Fl. Ludov., 73. 1817 • Prairie coneflower, Mexican-hat [Derivation unknown]

Lowell E. Urbatsch

Patricia B. Cox

Perennials, 15–120 cm. **Stems** 1–12+, erect, branched from bases, throughout, or only among heads, strigoso-hirsute, gland-dotted. **Leaves** basal and cauline; alternate; petiolate (petioles relatively shorter distally); blades lanceolate to ovate or oblanceolate to obovate or oblong, pinnately lobed to 1–2-pinnatifid (lobes usually rounded), ultimate margins entire or serrate, faces usually strigoso-hirsute, usually gland-dotted. **Heads** radiate, borne singly or in loose, corymbiform arrays. **Involucres** ± rotate, 8–16 mm diam. **Phyllaries** persistent, 5–15 in 2 series (usually reflexed in fruit, unequal, outer longer). **Receptacles** subspheric to columnar, paleate (paleae conduplicate, linear to oblong-oblique, chartaceous proximally, herbaceous distally, apices truncate, usually densely strigose and gland-dotted, each margin usually with an adaxial resin gland; receptacles plus paleae and florets = discs, 5–70 × 5–18 mm). **Ray florets** 3–15+, neuter; corollas yellow, bicolor (maroon/yellow), or wholly maroon. **Disc florets** 50–400+, bisexual, fertile; corollas yellowish green, often purplish distally, tubes shorter than cylindric throats, lobes 5, obovate to triangular (gland-dotted and, sometimes, hairy abaxially). **Cypselae** (black) strongly compressed, linear-oblanceolate to oblong-oblique, margins (at least abaxial) usually ciliate or pectinate-fimbriate, apices truncate and, often, ciliate, faces glabrous or sparsely hairy; **pappi** of 1–2 toothlike projections, or coroniform, or 0. *x* = 16.

Species 7 (4 in flora): North America, Mexico.

SELECTED REFERENCES Jackson, S. W. 1963. Hybridization among three species of *Ratibida*. Univ. Kansas Sci. Bull. 44: 3–27. Richards, E. L. 1968. A monograph of the genus *Ratibida*. Rhodora 70: 348–393.

1. Plants taprooted; leaves 2-pinnately lobed; peduncles to 6.5 cm (ribs not prominent; heads held just beyond leaves); discs subspheric to ovoid, 5–15 mm; pappi coroniform (of connate scales 0.2–1.5 mm) . . . . 4. *Ratibida tagetes*
1. Plants fibrous rooted or taprooted; leaves 1–2-pinnatifid or -pinnately lobed; peduncles (1.5–)6.5–48+ cm (ribs tan; heads held well beyond leaves); discs columnar, ellipsoid-globular, or ovoid, 10–70 mm; pappi 0, or of 1–2 toothlike projections.
  2. Plants fibrous rooted; discs ellipsoid to globular or ovoid, 10–25 mm; pappi 0, or of 1–2 toothlike projections . . . . 3. *Ratibida pinnata*
  2. Plants taprooted; discs columnar, 10–70 mm; pappi usually of 1–2 toothlike projections, sometimes 0.
    3. Style branches ca. 1.8 mm, proximal $^{1}/_{2}$ stigmatic, apices subulate; cypselae: abaxial margins narrowly winged to fimbriate or pectinate-fimbriate, adaxial margins pectinate-fimbriate . . . . 2. *Ratibida peduncularis*
    3. Style branches 1–2.5 mm, proximal $^{2}/_{3}$–$^{3}/_{4}$ stigmatic, apices rounded; cypselae: abaxial margins glabrous, adaxial margins glabrous or ciliate . . . . 1. *Ratibida columnifera*

1. Ratibida columnifera (Nuttall) Wooton & Standley, Contr. U.S. Natl. Herb. 19: 706. 1915 • Upright prairie coneflower, redspike Mexican-hat.

Rudbeckia columnifera Nuttall, Cat. Pl. Upper Louisiana, no. 75. 1813; *Ratibida columnaris* (Sims) D. Don

Perennials, to 105+ cm; taprooted. **Leaves** 2–15+ × 0.8–6 cm, 1–2-pinnatifid, lobes 3–14, narrowly linear-lanceolate to narrowly oblong-ovate, 1–16 mm wide, faces hirsute, gland-dotted. **Heads** 1–15, held well beyond leaves. **Peduncles** (1.5–)6.5–48+ cm (ribs tan, prominent). **Phyllaries** 5–14, outer linear, 4–14 × 0.5–2 mm, inner lanceolate-ovate, 0.8–3 × 0.8–2 mm. **Paleae** 2.3–3.5 × 0.5–3 mm, resin glands oval-oblanceolate, 0.6–1.5 mm. **Ray florets** 4–12; corollas yellow, purplish yellow, or maroon, sometimes bicolor (maroon/yellow), laminae lanceolate-ovate to elliptic-oblanceolate, 7–35 × 4–17 mm. **Discs** columnar, 10–50 × 7–12 mm. **Disc florets** 200–400+; corollas greenish yellow, often purplish distally, 1–2.5 mm; style branches 0.5–1.4 mm, proximal $^{2}/_{3}$–$^{3}/_{4}$ stigmatic, apices rounded. **Cypselae** oblong, 1.2–3 × 1.2–2 mm, abaxial margins glabrous, adaxial margins and apices glabrous or ciliate; **pappi** (tan) usually of 1–2 toothlike projections, sometimes 0. **2***n* = 28.

Flowering Mar–Nov. Prairies; 10–2200 m; Alta., B.C., Man., Ont., Sask.; Ariz., Ark., Colo., Ill., Ind., Iowa, Kans., La., Mass., Minn., Mo., Mont., Nebr., N.J., N.Mex., N.C., N.Dak., Okla., S.Dak., Tenn., Tex., Utah, Wis., Wyo.; Mexico (Chihuahua, Coahuila, Hidalgo, Nuevo León, San Luis Potosí, Tamaulipas, Veracruz).

Ratibida columnifera is grown as an ornamental and is often included in wild flower plantings. Such activities may extend the geographic range of the species to roadsides and prairie-like habitats.

Some variants of *Ratibida columnifera* have been treated as varieties or forms. The most prominent of these is forma *pulcherrima* (de Candolle) Fernald, which is characterized by its showy, purplish yellow to purple rays; it is more frequent in the southwestern part of the range of the species. In the typical form, rays are yellow (E. L. Richards 1968).

Some authors have argued that *Ratibida columnaris* (Sims) D. Don is the correct name for this species; J. L. Reveal (1968) and E. L. Richards (1968) provided synoptic discussions of the issue.

2. Ratibida peduncularis (Torrey & A. Gray) Barnhart, Bull. Torrey Bot. Club 24: 411. 1897 • Naked prairie coneflower or Mexican-hat

Lepachys peduncularis Torrey & A. Gray, Fl. N. Amer. 2: 315. 1842

Perennials, to 115+ cm; taprooted. **Leaves** 1.6–14 × 1–6.5 cm, 2-pinnatifid to lyrate-pinnate, lobes 3–10, narrowly linear-lanceolate to narrowly ovate, 0.8–4.5 mm wide, faces hirsute and gland-dotted. **Heads** 1–12+, held well beyond leaves. **Peduncles** 14–45+ cm (ribs tan, prominent). **Phyllaries** 7–14, outer linear, 6–12 × 0.8–1.5 mm, inner lance-ovate, 1–5 × 0.8–1.5 mm. **Paleae** 3–4.3 × 1.5–3 mm, resin glands oval, 1–1.8 mm. **Ray florets** 4–14; corollas yellow, purplish yellow, or maroon, often bicolor (maroon/yellow), laminae mostly ovate to elliptic, 5–15 × 4–12 mm. **Discs** columnar, 15–70 × 8–13 mm. **Disc florets** 200–400+; corollas greenish yellow, often purplish distally, 2.5–3 mm; style branches ca. 1.8 mm, proximal ¹/₂ stigmatic, apices subulate. **Cypselae** oblong, 2.3–5 × 1.2–3 mm, abaxial margins narrowly winged to fimbriate or pectinate-fimbriate, adaxial margins pectinate-fimbriate; **pappi** (tan) of 1–2 toothlike projections.

Varieties 2 (2 in the flora): sc United States, ne Mexico.

1. Leaves 2-pinnatifid; cypselae pectinate-fimbriate (both margins) . . . .
. . . . 2a. *Ratibida peduncularis* var. *peduncularis*
1. Leaves lyrate-pinnate; cypselae pectinate-fimbriate (adaxial margins) . . . .
. . . . 2b. *Ratibida peduncularis* var. *picta*

2a. Ratibida peduncularis (Torrey & A. Gray) Barnhart var. **peduncularis**

Leaves 2-pinnatifid, 2–12 × 1–5 cm, lobes 4–10 (linear-lanceolate to ovate-obovate, mostly 1–10 mm wide). **Ray corollas** mostly yellow to purplish yellow. **Cypselae** pectinate-fimbriate (both margins). **2***n* = 28.

Flowering Mar–Nov. Sandy sites; 0–300 m; La., Tex.; Mexico (Tamaulipas).

Variety *peduncularis* has been reported to hybridize with *Ratibida columnifera* (S. W. Jackson 1963).

2b. Ratibida peduncularis (Torrey & A. Gray) Barnhart var. **picta** (A. Gray) W. M. Sharp, Ann. Missouri Bot. Gard. 22: 74. 1935 E

Lepachys peduncularis Torrey & A. Gray var. *picta* A. Gray, Smithsonian Contr. Knowl. 3(5): 107. 1852

Leaves lyrate-pinnate, 2–14 × 1–6.5 cm, lobes 3–7 (terminal lobes lanceolate-ovate, toothed to parted, 3–50 mm wide). **Ray corollas** mostly purple, sometimes with some yellow. **Cypselae** pectinate-fimbriate (adaxial margins). **2***n* = 28.

Flowering Apr–Oct. Sandy seacoasts; 0–50 m; Tex.

Variety *picta* has been reported to hybridize with var. *peduncularis* (E. L. Richards 1968) and *Ratibida columnifera* (S. W. Jackson 1963).

3. Ratibida pinnata (Ventenat) Barnhart, Bull. Torrey Bot. Club 24: 410. 1897 • Grayhead or pinnate prairie coneflower E F

Rudbeckia pinnata Ventenat, Descr. Pl. Nouv., plate 71. 1802

Perennials, to 125+ cm; fibrous rooted (arising from stout rhizomes or woody caudices). **Leaves** 5–40 × 3–15+ cm, pinnatifid to pinnate, lobes 3–9, narrowly lanceolate to ovate, 1–15 × 0.2–3.5 cm, faces strigose, gland-dotted. **Heads** mostly 1–12, held well beyond leaves. **Peduncles** 3–27+ cm (ribs tan, prominent). **Phyllaries** 10–15, outer linear, 3–15 × 1–3 mm, inner lanceolate-ovate, 3–6 × 0.7–3 mm. **Paleae** 1.2–5 × 1–1.8 mm, resin glands linear to oblanceolate, 2–3.3 mm. **Ray florets** 6–15; corollas yellow, 2.5–3.8 mm, tubes ca. 1–3 mm, hirsute, laminae linear-elliptic to oblong-oblanceolate, 16–60 × 4–15 mm. **Discs** ellipsoid to globular or ovoid, 10–25 × 10–18 mm. **Disc florets** 100–200+; corollas greenish yellow, often purplish distally, 2.5–3.8 mm; style branches ca. 1.8 mm, proximal ¹/₂ stigmatic, apices subulate. **Cypselae** linear-oblanceoloid, 2–4 × 1–2.3 mm, margins usually glabrous, sometimes adaxial ciliate; **pappi** 0 or of 1–2 toothlike projections. **2***n* = 28.

Flowering May–Oct. Prairies, woodland openings and borders, limestone outcrops; 10–300 m; Ont.; Ala., Ark., Fla., Ga., Ill., Ind., Iowa, Kans., Ky., La., Mass., Mich., Minn., Miss., Mo., Nebr., N.Y., Ohio, Okla., S.Dak., Tenn., Vt., Wis.

4. Ratibida tagetes (E. James) Barnhart, Bull. Torrey Bot. Club 24: 410. 1897 • Green Mexican-hat or prairie coneflower, shortray prairie coneflower

Rudbeckia tagetes E. James, Account Exped. Pittsburgh 2: 68. 1823

Perennials, to 55 cm; taprooted. **Leaves** (proximal sometimes lanceolate, usually pinnately lobed; cauline mostly 1–2-pinnately lobed) 0.5–9 × 0.2–6.7 cm, lobes 2–7, linear to narrowly lanceolate-obovate, 0.5–4 mm wide, faces strigose and gland-dotted. **Heads** 1–8+, held just beyond leaves. **Peduncles** 0.3–6.5 cm (± uniformly green, not ribbed). **Phyllaries** 10–12, outer lanceolate, 1.2–6 × 0.5–2.5 mm, inner lanceolate-ovate, 1.2–4 × 0.3–1.5 mm. **Paleae** oblong, 2–4 × 1.5–2.5 mm, resin glands linear to oblanceolate, 1–2 mm. **Ray florets** 5–10; corollas yellow or purplish yellow to purple, laminae linear to elliptic-oblanceolate, 3–10 × 2–6 mm. **Discs** subspheric to ovoid or ellipsoid, 5–15 × 5–14 mm. **Disc florets** 50–200+; corolla greenish yellow, sometimes purplish distally, 1.2–2.5 mm; style branches 0.8–1.2 mm, proximal ¹/₂ stigmatic, apices subulate. **Cypselae** oblong-oblique, 1.9–2.8 × 1.2–2 mm, adaxial margins usually winged and distally ciliate, apices ciliate; **pappi** coroniform (of connate scales, 0.2–1.5 mm). $2n = 32$.

Flowering May–Oct. Prairies, high plains, rocky hillsides; 300–2500 m; Ariz., Colo., Kans., N.Mex., Okla., Tex.; Mexico (Chihuahua).

187m.7. ASTERACEAE Martinov (tribe HELIANTHEAE) subtribe ZALUZANIINAE H. Robinson, Phytologia 41: 44. 1978

Perennials or subshrubs [shrubs], 30–80[–250+] cm. **Leaves** mostly cauline [basal]; mostly alternate (proximal sometimes opposite); petiolate; blades deltate to cordate [ovate to lanceolate], sometimes ± palmately 3-lobed [dissected], ultimate margins [entire] toothed, faces ± strigillose [often whitened with close-set hairs] and/or gland-dotted. **Heads** radiate [discoid], in loose, corymbiform [paniculiform] arrays [borne singly]. **Calyculi** 0. **Involucres** hemispheric. **Phyllaries** persistent, 10–25 in 2–3+ series (distinct, lanceolate to linear, subequal, ± herbaceous). **Receptacles** convex to conic, paleate (paleae often greenish, ± conduplicate, herbaceous to scarious, entire or 3-toothed). **Ray florets** [0, or 4–]8–10, pistillate, fertile; corollas yellow (laminae often with 9–12+ brownish nerves). **Disc florets** 30–100+, bisexual, fertile; corollas yellow, tubes shorter than or about equaling cylindric to urceolate throats (bases of tubes often dilated over tops of ovaries and/or cypselae, tubes and throats often glandular-puberulent), lobes 5, deltate; anther thecae dark; stigmatic papillae continuous. **Cypselae** (blackish) ± compressed [clavate to prismatic, ± 3–4-angled], often arcuate, glabrous [hairy]; **pappi** 0 (ray and disc) [of 3–8, subulate to setiform scales or bristles (ray)].

Genera 3, species 13 (1 species in the flora): sw United States, Mexico.

Other members of Zaluzaniinae in the sense of H. Robinson (1981) have been placed in Neurolaeninae and in Helianthinae.

263. ZALUZANIA Persoon, Syn. Pl. 2: 473. 1807 • [No etymology in protologue; perhaps for Adam Zaluziansky von Zaluzian, 1558–1613, Polish physician/botanist]

John L. Strother

Perennials or subshrubs [shrubs], 30–80[–250+] cm. **Stems** erect, branched. **Leaves** cauline; mostly alternate (proximal sometimes opposite); petiolate; blades deltate to cordate [ovate to lanceolate], sometimes ± palmately 3-lobed [dissected], ultimate margins [entire] toothed, faces ± strigillose [often whitened with close-set hairs] and/or gland-dotted. **Heads** radiate [discoid], in loose, corymbiform [paniculiform] arrays. **Involucres** hemispheric, 4–8 mm diam. **Phyllaries** persistent, 10–25 in 2–3+ series (distinct, lanceolate to linear, subequal, ± herbaceous). **Receptacles** convex to conic, paleate (paleae often greenish, ± conduplicate, herbaceous to scarious, entire or 3-toothed). **Ray florets** [0 or 4–]8–10, pistillate, fertile; corollas yellow (laminae often with 9–12+ brownish nerves). **Disc florets** 30–100+, bisexual, fertile; corollas yellow, tubes shorter than or about equaling cylindric to urceolate throats (bases of tubes often dilated over tops of ovaries and/or cypselae), lobes 5, ± deltate. **Cypselae** (blackish) ± compressed [clavate to prismatic, ± 3–4-angled], often arcuate, glabrous [hairy]; **pappi** 0 (ray and disc) [of 3–8, subulate to setiform scales or bristles (ray)]. $x = 17, 18$.

Species 10 (1 in the flora): sw United States, Mexico.

1. Zaluzania grayana B. L. Robinson & Greenman, Proc. Amer. Acad. Arts 34: 531. 1899 〔F〕

Gymnolomia triloba A. Gray, Proc. Amer. Acad. Arts 17: 217. 1882, not *Zaluzania triloba* Persoon; *Viguiera triloba* (A. Gray) J. Olsen

Leaf blades 2–8 × 1–6 cm, ultimate margins coarsely toothed, faces sparsely strigillose and gland-dotted. **Ray laminae** 15–20 mm. **Cypselae** 2–3 mm. *2n* = 34.

Flowering summer–fall. Rocky sites, canyon walls; 1500–2700 m; Ariz., N.Mex.; Mexico (Chihuahua).

187m.8. ASTERACEAE Martinov (tribe HELIANTHEAE) subtribe ECLIPTINAE Lessing, Linnaea 6: 153. 1831 (as Ecliptae)

Enceliinae Panero; Engelmanniinae Stuessy; Spilanthinae Panero; Verbesininae Bentham & Hooker f.; Zinniinae Bentham & Hooker f.

Annuals, perennials, subshrubs, or shrubs, (3–)10–200(–400+) [2500+] cm. **Leaves** usually cauline, sometimes mostly basal, or basal and cauline; opposite, alternate, or both; petiolate or sessile; blades cordate, deltate, elliptic, lanceolate, linear, rhombic, orbiculate, or ovate (and intermediate shapes), often 1(–2+)-palmately or -pinnately lobed, ultimate margins entire or toothed, faces glabrous or hairy (often hispid or scabrous), often gland-dotted or stipitate-glandular. **Heads** usually radiate, sometimes discoid, borne singly or in cymiform, corymbiform, paniculiform, racemiform, or spiciform arrays. **Calyculi** 0. **Involucres** cylindric, or campanulate to hemispheric or broader, or rotate. **Phyllaries** persistent or falling, usually 8–40+ in 2–4+ series, usually distinct, usually lanceolate, linear, orbiculate, or ovate, subequal or unequal (outer usually shorter, rarely longer, than inner). **Receptacles** convex, conic, or flat, paleate (paleae usually falling, oblong to linear, herbaceous to chartaceous or scarious, usually conduplicate). **Ray florets** usually (1–)3–40+, sometimes 0, usually pistillate and fertile, sometimes styliferous and sterile, or neuter; corollas usually yellow to orange, sometimes pink to purple, red, brown, or white (sometimes sessile, persistent, and becoming papery, e.g., in *Heliopsis*, *Sanvitalia*, and *Zinnia*). **Disc florets** 4–200+, bisexual and fertile, or functionally staminate; corollas usually yellow to orange, sometimes brown, greenish, maroon, pink, purple, or red, rarely white or whitish, tubes shorter to longer than campanulate, cylindric, or funnelform throats, lobes (4–)5, deltate to lanceolate (usually equal); anther thecae usually dark (collars not continuous around filaments); stigmatic papillae usually continuous, sometimes none, rarely in 2 lines. **Cypselae** obpyramidal or prismatic (3–4-angled), or terete to ovoid, or strongly compressed or flattened and ± orbiculate to obovate or cuneate (lengths seldom more than 2 times diams.), sometimes ribbed, sometimes winged, glabrous or hairy; **pappi** usually persistent, sometimes fragile or readily falling, usually of scales, sometimes of bristles or bristlelike awns or coroniform, rarely of awns or 0.

Genera 75, species 600 (31 genera, 124 species, including 4 hybrids, in the flora): mostly subtropical and warm-temperate North America, especially arid areas, also in the Old World.

The circumscription of Ecliptinae adopted by H. Robinson (1981) and followed here differs from the treatment by P. O. Karis and O. Ryding (1994), who placed *Eclipta* as unassigned to a subtribe and placed *Echinacea* in Rudbeckiinae, *Acmella*, *Heliopsis*, *Sanvitalia*, and *Zinnia* in

Zinniinae, *Berlandiera*, *Chrysogonum*, *Engelmannia*, *Lindheimera*, and *Silphium* in Engelmanniinae, and *Balsamorhiza* (including *Agnorhiza*), *Borrichia*, *Calyptocarpus*, *Encelia*, *Enceliopsis*, *Flourensia*, *Geraea*, *Helianthella*, *Jefea*, *Lasianthaea*, *Melanthera*, *Pascalia*, *Phoebanthus*, *Sphagneticola*, *Synedrella*, *Verbesina*, *Wedelia*, and *Wyethia* (including *Scabrethia*) in Verbesininae. Based on studies of chloroplast DNA sequences, J. L. Panero (2005) placed *Encelia*, *Enceliopsis*, *Flourensia*, *Geraea*, and *Helianthella* together in Enceliinae and *Acmella* in Spilanthinae, indicating a sister relationship between Spilanthinae and traditionally circumscribed Zinniinae.

1. Ray florets usually 5–21 (more in "double" cultivars), sometimes 0 (in *Zinnia anomala*, subshrubs or shrublets to 12 cm, leaves linear, Texas), corollas usually yellow to orange, sometimes purple, red, or whitish (usually persistent, sessile, becoming papery).
    2. Leaf margins serrate to coarsely toothed . . . . 264. *Heliopsis*, p. 67
    2. Leaf margins entire.
        3. Leaves petiolate or sessile; cypselae ± terete or obscurely 3–4-angled to compressed or flattened (all usually tuberculate and usually bearing uncinate hairs, none, some, or all in each head winged) . . . . 265. *Sanvitalia*, p. 70
        3. Leaves sessile; cypselae 3-angled or flattened (none with uncinate hairs, none winged) . . . . 266. *Zinnia*, p. 71
1. Ray florets usually (2–)5–35, sometimes 0, corollas usually yellow to orange, sometimes white (seldom sessile, laminae usually borne on tubes, never persistent and becoming papery).
    4. Disc florets functionally staminate (only ray florets produce cypselae).
        5. Phyllaries 8–10 in 2 series.
            6. Perennials; leaves opposite, petiolate, margins crenate . . . . 267. *Chrysogonum*, p. 74
            6. Annuals; leaves mostly alternate (distal sometimes opposite), ± sessile, margins coarsely toothed (mostly distal 1/2) . . . . 268. *Lindheimera*, p. 75
        5. Phyllaries 12–45+ in (2–)3–4 series.
            7. Ray florets 8–35 (in 1–3 series); cypselae (shed alone without accessory structures) . . . . 269. *Silphium*, p. 77
            7. Ray florets usually (2–)8(–13); cypselae (each shed together with subtending phyllary and 2–4 adjacent paleae and disc florets).
                8. Phyllaries usually 14–22 in 2–3 series (broadly obovate to orbiculate, without linear appendages); ray corollas pale yellow to orange-yellow (abaxially red to maroon or with greenish or red to maroon veins) . . . . 270. *Berlandiera*, p. 83
                8. Phyllaries 18–24+ in ± 3 series (at least outer with relatively short, expanded, indurate bases and longer, linear, herbaceous tips); ray corollas yellow (without notably colored veins) . . . . 271. *Engelmannia*, p. 87
    4. Disc florets bisexual, fertile.
        9. Leaves mostly basal, or basal and cauline, or cauline, mostly alternate.
            10. Cypselae prismatic, or nearly so, 3–4-angled.
                11. Ray florets usually 8–21 (corollas mostly pink, red, purple, or white, yellow in *E. paradoxa* var. *paradoxa*, laminae often drooping or reflexed); receptacles mostly conic (heights often 2+ times diams.) . . . . 272. *Echinacea*, p. 88
                11. Ray florets (1–)5–25+ (corollas yellow or orange, laminae not drooping or reflexed), sometimes 0; receptacles flat to convex.
                    12. Leaves mostly basal (cauline usually notably smaller than basal); pappi 0 . . . . 273. *Balsamorhiza*, p. 93
                    12. Leaves basal and cauline, or mostly cauline; pappi usually coroniform, sometimes of 1–4+ scales or 0.
                        13. Leaves mostly cauline (blades narrowly oblong to linear, 5–25 mm wide) . . . . 274. *Scabrethia*, p. 99
                        13. Leaves basal and cauline, or mostly cauline (blades mostly 30–120 mm wide).

14. Leaves mostly elliptic, lanceolate, or oblong (basal and cauline, basal usually notably larger than cauline, cauline mostly sessile) .... 275.   *Wyethia*, p. 100

14. Leaves mostly orbiculate, ovate, or rounded-deltate (mostly cauline, mostly petiolate, proximal and distal usually ± similar) .... 276.   *Agnorhiza*, p. 104

10. Cypselae compressed to flattened.

15. Cypselae winged; pappi persistent, of 2(–3) scales (scales often aristate or subulate, without additional scales) .... 277.   *Verbesina* (in part), p. 106

15. Cypselae sometimes thin-edged (margins sometimes ciliate or corky-thickened, never truly winged); pappi usually of (1–)2, subulate scales or bristlelike awns plus 2–4+ shorter scales, sometimes 0 (rarely of 2–3 aristate scales without additional scales).

16. Perennials (scapiform); leaves all or mostly basal; involucres 20–30+ mm diam. .... 278.   *Enceliopsis*, p. 112

16. Perennials (rarely scapiform), subshrubs, or shrubs; leaves usually cauline, sometimes basal and cauline; involucres 4–30 mm diam.

17. Perennials (rhizomatous); leaves linear to filiform ...... 279.   *Phoebanthus*, p. 113

17. Perennials or shrubs (not rhizomatous); leaves mostly deltate, elliptic, lanceolate, or ovate (and most intermediate shapes, not linear to filiform).

18. Ray florets 8–21, pistillate and fertile ...... 280.   *Helianthella* (in part), p. 114

18. Ray florets 0 or 8–40, neuter, or styliferous and sterile.

19. Subshrubs or shrubs (glabrous or ± scabrellous, usually vernicose); phyllaries 12–40 in 2–4+ series (subequal or unequal, outer longer) .... 281.   *Flourensia*, p. 117

19. Annuals, perennials, or shrubs (glabrous or canescent, hirtellous, scabrellous, strigose, or tomentose, often gland-dotted or glandular-puberulent to stipitate-glandular, seldom vernicose); phyllaries 18–30(–50+) in 2–3+ series (subequal or unequal, outer shorter).

20. Perennials (*E. nutans*), subshrubs, or shrubs; pappi usually 0, sometimes fragile, of 2 weak, villous scales .... 282.   *Encelia*, p. 118

20. Annuals or perennials; pappi usually persistent, of 2 subulate scales .... 283.   *Geraea*, p. 122

[9. Shifted to left margin.—Ed.]

9. Leaves mostly cauline, mostly opposite.

21. Heads discoid (corollas white or whitish); pappi 0, or readily falling, of 2–12 barbellate bristles or awns .... 284.   *Melanthera*, p. 123

21. Heads radiate or discoid (corollas seldom white or whitish); pappi 0, or usually persistent and coroniform or cyathiform (each an erose, fimbriate, or lacerate cup, with or without additional awns or bristles, borne on rostrums), or of 2–4+ awns, bristles, and/or scales.

22. Pappi usually coroniform or cyathiform (cypselae often rostrate, each with apical boss or neck), sometimes 0 (prostrate perennials).

23. Subshrubs or shrubs (erect); cypselae (some or all) strongly compressed, notably winged .... 285.   *Wedelia*, p. 125

23. Perennials (prostrate); cypselae strongly biconvex to plumply 3–4-angled (not compressed, not winged, epidermes usually corky, often tuberculate) .... 286.   *Sphagneticola*, p. 126

22. Pappi usually of 2–4+ awns, bristles, and/or scales (not cyathiform, cypselae not rostrate), sometimes coroniform or 0 (not prostrate perennials).

24. Some or all cypselae winged (each bordered by wing of membranous or corky tissue different from that of body of cypsela).

25. Heads in glomerules or borne singly (sessile or subsessile in axils); cypselae winged (rays, not discs, wings lacerate) . . . . 287. *Synedrella*, p. 127
25. Heads borne singly or in corymbiform, dichasiiform, or paniculiform arrays (not sessile); cypselae winged (rays and discs, wings not lacerate).
   26. Phyllaries 9–30 in 1–4 series (outer 2–5 similar to others, unlike foliage); pappi of 2(–3) persistent (often aristate or subulate) scales without additional scales . . . . 277. *Verbesina* (in part), p. 106
   26. Phyllaries 26–38+ in 3–4+ series (outer 2–6+ similar to foliage in shape, texture, and indument); pappi of 2–3 fragile or persistent awns or subulate scales plus 2–8+, distinct or basally connate, erose or lacerate scales (often each cypsela with additional seta on inner shoulder) . . . . 288. *Jefea*, p. 127
      [24. Shifted to left margin.—Ed.]
24. Cypselae sometimes sharp-edged (not winged).
   27. Cypselae 3–4-angled (weakly or not at all compressed or obcompressed, epidermes usually thick, corky).
      28. Corollas white or whitish (paleae linear-filiform, not conduplicate) . . . . 289. *Eclipta*, p. 128
      28. Corollas yellow to orange (paleae lanceolate to ovate, conduplicate).
         29. Leaves elliptic, linear, oblanceolate, obovate, or ovate, glabrous or puberulent to villous and/or sericeous (outer phyllaries elliptic, oblanceolate, or ovate) . . . . 290. *Borrichia*, p. 129
         29. Leaves lanceolate to lance-linear, sparsely scabrous (outer phyllaries lance-linear to linear) . . . . 291. *Pascalia*, p. 131
   27. Cypselae (all or at least disc) strongly compressed or obcompressed or flattened (epidermes seldom thick and corky).
      30. Annuals or perennials (mostly 5–30+ cm; larger leaves mostly 1–5+ cm); involucres 3–8 mm diam.; phyllaries 5 in 1(–2) series, or 8–15+ in 1–3 series.
         31. Disc florets 25–100(–200+); receptacles conic; cypselae ellipsoid to obovoid; pappi 0, or fragile, of 1–3 bristlelike awns . . . . 292. *Acmella*, p. 132
         31. Disc florets 10–20; receptacles convex; cypselae cuneate; pappi of 2(–5+) stout awns . . . . 293. *Calyptocarpus*, p. 133
      30. Perennials (coarse, 10–150 cm; larger leaves mostly 5–25 cm); involucres 10–30 mm diam.; phyllaries 12–35 in 2–5 series.
         32. Leaf blades oblanceolate to lanceolate or lance-linear (longer usually 8–25 cm), margins entire . . . . 280. *Helianthella* (in part), p. 114
         32. Leaf blades rounded-deltate to ovate or lance-ovate (longer usually 5–8 cm), margins coarsely serrate . . . . 294. *Lasianthaea*, p. 133

264. HELIOPSIS Persoon, Syn. Pl. 2: 473. 1807, name conserved • Oxeye, sunflower everlasting, héliopside [Greek *helios*, sun, and *-opsis*, likeness]

Alan R. Smith

Perennials [annuals], 30–150 cm. **Stems** erect or trailing, branched from bases or ± throughout. **Leaves** cauline; opposite; petiolate; blades ± 3-nerved from bases, deltate or ovate to lanceolate, bases ± cuneate to subtruncate, margins serrate or coarsely toothed, faces glabrous or hairy. **Heads** radiate [discoid], borne singly. **Involucres** turbinate to hemispheric, 8–14 mm diam. **Phyllaries** persistent, 12–20 in 2–3 series (ovate or oblong, ± equal, outer more foliaceous than inner). **Receptacles** convex to conic, paleate (paleae tardily falling, yellowish, conduplicate, becoming chartaceous). **Ray florets** [0] 5–20, pistillate, fertile; corollas yellow to orange (laminae persistent, sessile, becoming papery). **Disc florets** 30–150+, bisexual, fertile; corollas yellow or brown to purple, tubes much shorter than narrowly cylindric throats, lobes 5, deltate. **Cypselae** (brown to black-brown) subterete or obscurely 3(ray)- or 4(disc)-angled (not winged);

pappi 0, or persistent, coroniform (each a laciniate crown plus 1–3 toothlike scales). *x* = 14.

Species ca. 18 (3 in the flora): North America, Mexico, Central America, South America (to Bolivia).

Most species of *Heliopsis* are known only from Mexico.

SELECTED REFERENCE   Fisher, T. R. 1957. Taxonomy of the genus *Heliopsis* (Compositae). Ohio J. Sci. 57: 171–191.

1. Aerial stems 1–2+ (from creeping rhizomes 4–10+ cm × 1–2 mm, rhizome internodes ca. 20+ mm); heads 1–3(–5); rays 6–8(–13) . . . . 3. *Heliopsis gracilis*
1. Aerial stems 1–10+ (from ± erect caudices, or from creeping rhizomes to 4 cm × 2–6 mm, rhizome internodes 2–10 mm); heads 1–15+; rays 9–18.
    2. Leaf blades (1.5–)3–6 × 0.8–3 cm, faces sparsely pubescent to glabrescent; peduncles (6–)15–30 cm; cypselae rugulose to subtuberculate . . . . 1. *Heliopsis parvifolia*
    2. Leaf blades 6–12(–15) × 2.5–6(–12) cm, faces glabrous, pubescent, or scabrous; peduncles 9–25 cm; cypselae smooth. . . . . 2. *Heliopsis helianthoides*

1. Heliopsis parvifolia A. Gray, Smithsonian Contr. Knowl. 5(6): 86. 1853 • Mountain oxeye

Perennials, 30–50(–80) cm. **Aerial stems** (from ± erect caudices) 1–10+, stramineous to reddish, glabrous or sparingly hairy. **Leaf blades** deltate-lanceolate, (1.5–)3–6 × 0.8–3 cm, margins irregularly dentate to subentire (proximal teeth larger), apices acuminate to obtuse, faces sparsely pubescent to glabrescent. **Heads** 1–10+. **Peduncles** (6–)15–30 cm. **Involucres** 12–20 mm diam. **Phyllaries** densely pubescent on margins and apices, abaxial faces glabrescent. **Paleae** lanceolate to oblong, apices acuminate, faces glabrous. **Ray florets** 9–11; corollas golden yellow (young) to pale yellow (old), laminae (1–)2–3 cm × 5–14 mm. **Disc florets** 8–50+; corollas greenish yellow to yellow-brown (lobes brighter than tubes), 3–4.5 mm, glabrous. **Cypselae** 4.5–5 mm, glabrous, rugulose to subtuberculate; **pappi** 0. *2n* = 28.

Flowering late spring–early fall. Open, rocky mountain slopes, canyons; 1200–2500 m; Ariz., N.Mex., Tex; Mexico (Chihuahua, Coahuila, Durango, Sonora, Tamaulipas).

2. Heliopsis helianthoides (Linnaeus) Sweet, Hort. Brit., 487. 1826 • False sunflower, oxeye, smooth oxeye, héliopside faux-hélianthe E F

Buphthalmum helianthoides Linnaeus, Sp. Pl. 2: 904. 1753

Perennials, (40–)80–150 cm. **Aerial stems** (from creeping rhizomes to 4 cm × 2–6 mm, rhizome internodes mostly 2–10 mm) 1–10+, stramineous to reddish brown, glabrous or hairy. **Leaf blades** ovate to deltate-lanceolate, 6–12(–15) × 2–6(–12) cm, margins regularly to irregu-larly and coarsely dentate, apices acute to acuminate, faces glabrous, sparsely pubescent, moderately to densely scabrellous, or scabrous. **Heads** 1–15+. **Peduncles** 9–25 cm. **Involucres** 12–25 mm diam. **Phyllaries** glabrescent to densely pubescent on margins, apices, and abaxial faces. **Paleae** lanceolate to oblong, apices obtuse, faces glabrous. **Ray florets** 10–18; corollas golden yellow, laminae mostly 2–4 cm × 6–13 mm. **Disc florets** 10–75+; corollas yellowish to brownish yellow (lobes brighter than tubes), 4–5 mm, glabrous. **Cypselae** 4–5 mm, glabrous or pubescent on angles, smooth; **pappi** 0 or of 2–4 minute, toothlike scales. *2n* = 28 (variety unknown).

Varieties 2 (2 in the flora): e North America.

As T. R. Fisher (1957) noted, intermediates occur between var. *helianthoides* and var. *scabra*, and artificial hybrids show only slightly decreased pollen stainabili-ties and normal meiotic pairing. Intermediates are espe-cially common in Missouri, Illinois, and New England, and may also be encountered elsewhere in areas of sym-patry. In addition to the key characters, var. *helianthoides* generally has longer petioles, has smaller heads on shorter peduncles, and occupies less-open habitats than var. *scabra*.

1. Leaf blades ovate, 8–12(–15) × 4–8(–12) cm, abaxial faces glabrous or sparsely pubescent, adaxial faces glabrous or minutely scabrellous . . . . 2a. *Heliopsis helianthoides* var. *helianthoides*
1. Leaf blades deltate to narrowly ovate-lanceolate, 6–12 × 2–5 cm, faces moderately to densely scabrellous to scabrous . . . . 2a. *Heliopsis helianthoides* var. *scabra*

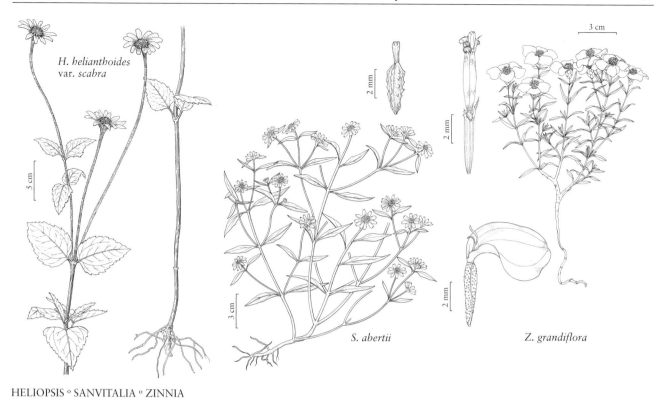

HELIOPSIS ° SANVITALIA ° ZINNIA

2a. Heliopsis helianthoides (Linnaeus) Sweet var. helianthoides [E]

Heliopsis helianthoides var. *solidaginoides* (Linnaeus) Fernald

Leaf blades ovate, 8–12(–15) × 4–8(–12) cm, abaxial faces glabrous or sparsely pubescent, adaxial faces glabrous or minutely scabrellous.

Flowering late spring–summer. Open woods, thickets, marshes, meadows, road banks, and waste places; 0–800 m; Ont.; Ala., Ark., Conn., Del., Ga., Ill., Ind., Ky., La., Md., Mass., Mich., Miss., Mo., N.J., N.Y., N.C., Ohio, Pa., S.C., Tenn., Vt., Va., W.Va., Wis.

2b. Heliopsis helianthoides (Linnaeus) Sweet var. scabra (Dunal) Fernald, Rhodora 44: 340. 1942

• Héliopside scabre [E] [F]

Heliopsis scabra Dunal, Mém. Mus. Hist. Nat. 5: 56, plate 8. 1819; *H. helianthoides* subsp. *occidentalis* T. R. Fisher; *H. helianthoides* var. *occidentalis* (T. R. Fisher) Steyermark

Leaf blades deltate to narrowly ovate-lanceolate, 6–12 × 2–5 cm, faces moderately to densely scabrellous to scabrous.

Flowering late spring–early fall. Open woods, prairies, old pastures, edges of fields, meadows, road banks, ditches; 0–2300 m; Man., N.B., Nfld. and Labr. (Nfld.), Ont., Que., Sask.; Ark., Colo., Conn., Ill., Ind., Iowa, Kans., Ky., La., Maine, Mass., Mich., Minn., Mo., Nebr., N.H., N.Mex., N.Dak., Ohio, Okla., R.I., S.Dak., Tex., Vt., Wis.

Variety *scabra* is reported as introduced in British Columbia, Newfoundland, Prince Edward Island, and perhaps Quebec (H. J. Scoggan 1978–1979, part 4).

Varieties (subspp.) *scabra* and *occidentalis* intergrade so completely and intermediate plants are so numerous that it seems futile to draw taxonomic lines. In general, some plants traditionally treated as var. *scabra* (in the strict sense) have somewhat narrower leaf blades and longer petioles than most plants treated as var. *occidentalis*.

3. Heliopsis gracilis Nuttall, Trans. Amer. Philos. Soc., n. s. 7: 353. 1840 • Smooth or pinewoods oxeye [E]

Heliopsis helianthoides (Linnaeus) Sweet var. *gracilis* (Nuttall) Gandhi & R. D. Thomas; *H. laevis* Persoon var. *minor* Hooker; *H. minor* (Hooker) C. Mohr

Perennials, 30–40(–80) cm. **Aerial stems** (from creeping rhizomes 4–10+ cm × 1–2 mm, rhizome internodes ca. 20+ mm) 1–2+, stramineous to red-brown, glabrous. **Leaf blades** ovate-lanceolate to lanceolate, 4–8 × 1.5–3 cm, margins regularly dentate to subentire, apices acuminate, abaxial faces glabrous or inconspicuously short-hairy, adaxial faces sparingly scabrellous. **Heads** 1–3(–5). **Peduncles** mostly 10–15 cm. **Involucres** 7–15 mm diam. **Phyllaries** minutely pubescent on margins and apices, less so on abaxial faces. **Paleae** lanceolate, apices obtuse, faces glabrous. **Ray florets** 6–8(–13); corollas golden yellow, laminae 10–22 × 4–8 mm. **Disc florets** 10–20+; corollas pale brown-yellow (lobes dull yellow), 3.5–4 mm, glabrous. **Cypselae** dark brown, 4–5 mm, glabrous or minutely pubescent on margins, smooth or nearly so; **pappi** coroniform (each a laciniate crown plus 1–3 toothlike scales).

Flowering spring–summer. Open wooded slopes, especially with pines; 0–100 m; Ala., Ark., Fla., Ga., La., Tex.

A. Cronquist (1980) and others have alluded to intermediates between *Heliopsis gracilis* and *H. helianthoides* var. *scabra* in Louisiana; I have seen none. *Heliopsis gracilis* is distinguishable from var. *scabra* by the abaxially and adaxially glabrous or nearly glabrous leaf blades, smaller plant size, single or relatively few heads, and long-creeping, narrow rhizomes. Presence of *H. gracilis* in southeast Texas (D. S. Correll and M. C. Johnston 1970) has not been confirmed.

I am uncertain whether the type of *Heliopsis laevis* var. *minor*, collected from near New Orleans, is referable to *H. gracilis*, as A. Cronquist (1980) thought, or to *H. helianthoides* var. *scabra*, where it was placed by T. R. Fisher (1957) and others. If the latter, then var. *minor* has priority over var. *scabra*.

265. SANVITALIA Lamarck, J. Hist. Nat. 2: 176, plate 33. 1792 • [Etymology unclear; possibly for some member of the Italian family Sanvitali or for a Spanish botanist named Sanvital]

John L. Strother

Annuals or perennials, (3–)10–30 cm. **Stems** prostrate to erect, branched from bases or ± throughout. **Leaves** cauline; opposite; petiolate or sessile; blades obovate or spatulate to linear, bases rounded to ± cuneate, margins entire [toothed or lobed], faces hairy. **Heads** radiate, borne singly. **Involucres** hemispheric to ± rotate, 4–12+ mm diam. **Phyllaries** persistent, 8–21 in 2–3 series (lanceolate to linear, outer distally herbaceous, others each with stiff, subulate appendage). **Receptacles** convex to conic, paleate (paleae conduplicate, scarious). **Ray florets** 5–20, pistillate, fertile; corollas white or yellow (laminae sessile, persistent, becoming papery). **Disc florets** 15–60, bisexual, fertile; corollas distally yellow to orange (sometimes drying white), tubes much shorter than funnelform throats, lobes 5, deltate. **Cypselae** ± terete or obscurely 3–4-angled to compressed or flattened (all usually tuberculate and usually bearing uncinate hairs; none, some, or all in each head winged); **pappi** persistent, of 3–4 awns. $x = 8, 11$.

Species 5 (3 in the flora): sw United States, Mexico, Central America, South America.

SELECTED REFERENCES Strother, J. L. 1979. Extradition of *Sanvitalia tenuis* to *Zinnia* (Compositae–Heliantheae). Madroño 26: 173–179. Torres, A. M. 1964. Revision of *Sanvitalia* (Compositae–Heliantheae). Brittonia 16: 417–433.

1. Phyllaries 8–11; ray cypselae subterete, obscurely 3-angled, sulcate on each angle and on abaxial faces . . . . 3. *Sanvitalia abertii*
1. Phyllaries 12–21; ray cypselae 3-angled (clearly 3-faced, adaxial faces often 2–3-nerved).
　　2. Disc cypselae strongly dimorphic within heads: outer wingless, inner 1–2-winged
　　　. . . . . 1. *Sanvitalia procumbens*
　　2. Disc cypselae ± monomorphic within heads, ± 4-angled to laterally compressed, none winged . . . . 2. *Sanvitalia ocymoides*

1. **Sanvitalia procumbens** Lamarck, J. Hist. Nat. 2: 176, plate 33. 1792   [I]

Stems procumbent to erect, 3–15 cm. **Leaf blades** ovate to lance-linear, 10–60 × 4–31 mm. **Phyllaries** 13–21, unequal. **Ray corollas** 2–9 mm. **Cypselae:** rays 2.5–3.5 mm with awns 1–3 mm, clearly 3-faced, adaxial faces often 2–3-nerved; discs strongly dimorphic within single heads: outer 4-angled, wingless, inner ± flattened, 1–2-winged. $2n = 16, 32$.

Flowering summer–fall. Ruderal; 10–1000 m; introduced; Calif., Md., Tex.; Mexico; Central America.

Sanvitalia procumbens has been seen only sporadically in the flora; it probably is not a resident.

2. **Sanvitalia ocymoides** de Candolle in A. P. de Candolle and A. L. P. P. de Candolle, Prodr. 5: 628. 1836

Stems procumbent to erect, 5–10 cm. **Leaf blades** ovate to oblanceolate or obovate, 1–35 × 7–16 mm. **Phyllaries** 12–20, subequal. **Ray corollas** 1.5–2.5 mm. **Cypselae:** rays 4 mm (with awns 2–3 mm), clearly 3-faced, adaxial faces often 2–3-nerved; discs ± monomorphic within heads, ± 4-angled to compressed, none winged. $2n = 32$.

Flowering summer–late fall. Open places, compact, clay soils; 0–30 m; Tex.; Mexico (Coahuila, Nuevo León, Tamaulipas).

3. **Sanvitalia abertii** A. Gray, Mem. Amer. Acad. Arts, n. s. 4: 87. 1849 (as aberti)   [F]

Stems erect, 5–29 cm. **Leaf blades** lance-linear, 20–60 × 2–12 mm. **Phyllaries** 8–11, subequal. **Ray corollas** 1.5–2.5+ mm. **Cypselae:** rays 4 mm (with awns 2–3 mm), subterete, obscurely 3-angled, each sulcate on angles and on abaxial face; discs 4-angled, innermost ± compressed, not winged. $2n = 22$.

Flowering late summer–early fall. Open places, scrublands, pinyon-juniper woodlands; 1400–2400 m; Ariz., Calif., N.Mex., Tex.; Mexico (Baja California, Chihuahua, Coahuila, Sonora).

266. ZINNIA Linnaeus, Syst. Nat. ed. 10, 2: 1189, 1221, 1377. 1759, name conserved • Zinnia [For Johann Gottfried Zinn, 1727–1759, professor of botany, Göttingen, known for botanical studies in Mexico]

Alan R. Smith

Annuals or subshrubs [perennials], 10–100(–200+) cm. **Stems** prostrate or erect. **Leaves** cauline; opposite or subopposite; sessile [petiolate]; blades (1-, 3-, or 5-nerved from bases) acerose, elliptic, lance-linear, lanceolate, linear, oblong, or ovate, bases rounded to cuneate, margins entire, faces hairy (often scabrous or scabrellous), usually gland-dotted. **Heads** usually radiate (rarely ± discoid in *Z. anomala*), borne singly. **Involucres** campanulate, cylindric, to hemispheric or broader, 5–25 mm diam. **Phyllaries** persistent, 12–30+ in 3–4+ series (orbiculate to obovate or oblong, unequal, often colored or dark-banded distally, outer shorter). **Receptacles** conic, paleate (paleae yellowish, often reddish to purplish distally, chartaceous to scarious, conduplicate, apices rounded to acute, sometimes fimbriate). **Ray florets** usually 5–21 (more in "double" cultivars, sometimes 0 in *Z. anomala*), pistillate, fertile; corollas yellow, orange, red, maroon, purple, or white (laminae persistent, sessile or nearly so, becoming papery, sometimes much reduced). **Disc florets** 20–150+, bisexual, fertile; corollas usually yellow to reddish, sometimes purple-tinged, tubes much shorter than cylindric throats, lobes 5, lance-ovate (usually unequal, usually villous or velutinous adaxially). **Cypselae** 3-angled (ray) or flattened (disc; not winged); **pappi** 0, or persistent, of 1–3(–4) awns or toothlike scales. $x = 12$ (11, 10).

Species ca. 17 (5 in the flora): United States, Mexico, Central America, South America (one species to Argentina, Bolivia).

A. M. Torres (1963) recognized subg. *Diplothrix*, comprising six species, including the three perennial species treated here, and subg. *Zinnia*, comprising 11 species, mostly annuals. This division is reflected in the first couplet of the key.

Zinnia angustifolia Kunth (= *Z. linearis* Bentham), native to northern and western Mexico, is commonly grown as an ornamental in the United States and has been reported from Utah (S. L. Welsh et al. 1993); the record was likely from a cultivated source. The species also persists in gardens in California; it is not known outside of cultivation. It can be distinguished from other zinnias by the combination of annual habit, plants to 50 cm, leaf blades linear to narrowly elliptic (mostly 2–7 cm × 4–8 mm), involucres mostly hemispheric, usually much less than 1 cm high or wide, bright orange ray corollas (white-rayed and other color variants known in cultivation), and lobes of disc flowers glabrous or nearly so. Hybrids between *Z. angustifolia* and *Z. violacea* are known in the horticultural trade.

The lack of articulation of the corolla tubes in the ray florets of *Zinnia verticillata* Andrews (= *Z. peruviana*) and the bilateral disposition of vascular bundles (continuous with vasculature of the ovary walls) in the ray florets led D. Don (1830) to conclude that true ray "corollas" in *Zinnia* are lacking, being replaced instead by de novo petaloid structures that mimic ray corollas of other Compositae.

SELECTED REFERENCES Don, D. 1830. On the origin and nature of the ligulate rays in *Zinnia*; and on a remarkable multiplication observed in the parts of fructification of that genus. Trans. Linn. Soc. London 16: 155–158. Torres, A. M. 1963. Taxonomy of *Zinnia*. Brittonia 15: 1–25.

1. Annuals, 30–50(–200) cm; leaf blades elliptic, lanceolate, oblong, or ovate; involucres campanulate, or hemispheric or broader, 10–25 mm diam.
  2. Involucres narrowly to broadly campanulate, 9–18 × 10–20 mm; paleae apically obtuse, erose or subentire; pappi usually of 1 awn (disc cypselae) . . . . 4. *Zinnia peruviana*
  2. Involucres hemispheric or broader, 10–15 × 15–25 mm; paleae apically fimbriate; pappi 0 . . . . 5. *Zinnia violacea*
1. Subshrubs, to 22 cm; leaf blades linear to acerose; involucres narrowly campanulate, cylindric, or subhemispheric, 5–10mm diam.
  3. Leaf blades 1-nerved, linear to acerose; ray florets 4–7, corollas usually white, sometimes pale yellow . . . . 1. *Zinnia acerosa*
  3. Leaf blades 1- or 3-nerved (some larger leaves), linear; ray florets usually 5–8, corollas usually yellow (laminae sometimes 0).
    4. Involucres 8–10 mm; ray laminae 0–6(–9) mm . . . . 2. *Zinnia anomala*
    4. Involucres 5–8 mm; ray laminae mostly 10–18 mm . . . . 3. *Zinnia grandiflora*

1. Zinnia acerosa (de Candolle) A. Gray, Smithsonian Contr. Knowl. 3(5): 105. 1852 • Desert or shrubby or southern zinnia

Diplothrix acerosa de Candolle in A. P. de Candolle and A. L. P. P. de Candolle, Prodr. 5: 611. 1836; *Zinnia pumila* A. Gray

Subshrubs, to 16 cm (rounded or flat-topped). **Stems** greenish to gray, much branched, pilose. **Leaf blades** 1-nerved, linear to acerose, 8–20 × 1–2 mm, scabrous to glabrescent. **Peduncles** 5–35 mm. **Involucres** campanulate, 3–5 × 5–7 mm. **Phyllaries** suborbiculate to oblong, becoming scarious proximally, appressed-hairy distally, apices obtuse, ciliate. **Paleae** uniformly yellow, apices obtuse, erose. **Ray florets** 4–7; corollas usually white, sometimes pale yellow, laminae oblong to suborbiculate, 7–10 mm. **Disc florets** 8–13; corollas yellow or tinged with purple (drying reddish), 3–6 mm, lobes 1 mm. **Cypselae** 2.4–4 mm, 3-angled (ray) or compressed (disc), ribbed, strigose or distally ciliate; **pappi** usually of 1–3 unequal awns, sometimes reduced to teeth. $2n = 20, 40,$ or 22.

Flowering spring–fall. Rocky open slopes, flats, calcareous soils; 700–1900 m; Ariz., N.Mex., Tex.; Mexico

(Chihuahua, Coahuila, Durango, Nuevo León, San Luis Potosí, Sonora, Zacatecas).

Attribution of *Zinnia acerosa* to Utah (S. L. Welsh et al. 1993) was based on *Atwood et al. 9704* (BRY), from Moab, Grand County; the specimen was indicated as "possibly cultivated" by the collector, and it is well outside the known range of the species.

2. Zinnia anomala A. Gray, Smithsonian Contr. Knowl. 3(5): 106. plate 10. 1852 • Shortray zinnia

Subshrubs, to 12 cm (rounded). **Stems** yellowish, much branched, strigose. **Leaf blades** 1- or 3-nerved, linear, 15–30 × 1–4 mm, strigose. **Peduncles** to 25 mm. **Involucres** broadly cylindric, 8–10 × 8–10 mm. **Phyllaries** round to oblong, becoming scarious, appressed-hairy distally, apices rounded, ciliate. **Paleae** yellowish (often red-tipped), apices obtuse, erose. **Ray florets** 0 or 5–8; corollas yellow, laminae suborbiculate to ovate, 0–6(–9) mm. **Disc florets** ca. 20; corollas reddish, 5.5–6.2 mm, lobes 1 mm. **Cypselae** 7–8 (ray) or 3–6 mm (disc), 3-angled (ray) or compressed (disc), ribbed, distally ciliate or scabrellous; **pappi** usually of 2 or 3 equal or unequal awns. $2n$ = ca. 48, ca. 84.

Flowering spring–fall. Open, rocky roadsides, disturbed sites; 200–1500 m; Tex.; Mexico (Coahuila, Nuevo León, Zacatecas).

A. M. Torres (1963) speculated that *Zinnia anomala*, which often lacks rays or has inconspicuous rays, might be an octoploid derivative of the rather similar *Z. grandiflora*, which has showy yellow rays.

3. Zinnia grandiflora Nuttall, Trans. Amer. Philos. Soc., n. s. 7: 348. 1840 • Plains or Rocky Mountain zinnia
F

Subshrubs, 8–22 cm (rounded or flat-topped). **Stems** greenish, much branched, strigillose. **Leaf blades** 1- or 3-nerved (some larger leaves), linear, 10–30 × 2–3 mm, strigose to scabrous. **Peduncles** to 11 mm. **Involucres** narrowly campanulate to cylindric, 5–8 × 5–8 mm. **Phyllaries** oblong, often becoming scarious, glabrous or appressed-hairy distally, apices obtuse, erose-ciliate (red-tipped). **Paleae** yellowish (often red-tipped), apices obtuse, erose. **Ray florets** 3–6; corollas bright yellow, laminae ovate to orbiculate, mostly 10–18 mm. **Disc florets** 18–24; corollas red or green, to 10 mm, lobes 1 mm. **Cypselae** 4–5 mm, 3-angled (ray) or angular or compressed (disc), ribbed, scabrellous; **pappi** 0 or of (1–)2(–4) unequal awns. $2n$ = 42.

Flowering spring–fall. Dry, often slopes, mesas, shortgrass prairies, calcareous soils; 600–2200 m; Ariz., Colo., Kans., N.Mex., Okla., Tex.; Mexico (Chihuahua, Coahuila, Sonora, Zacatecas).

4. Zinnia peruviana (Linnaeus) Linnaeus, Syst. Nat. ed. 10, 2: 1221. 1759 • Peruvian zinnia

Chrysogonum peruvianum Linnaeus, Sp. Pl. 2: 920. 1753; *Zinnia multiflora* Linnaeus

Annuals, mostly 30–50(–100) cm. **Stems** greenish, becoming purplish or yellowish, unbranched or sparingly branched distal to bases, strigose. **Leaf blades** 3–5-nerved, ovate to elliptic or broadly lanceolate, 25–70 × 8–35 mm, scabrellous. **Peduncles** 10–50(–70) mm. **Involucres** narrowly to broadly campanulate, 9–18 × 10–20 mm. **Phyllaries** obovate to oblong, becoming scarious, glabrous, apices rounded, usually entire or erose, sometimes ciliate. **Paleae** red to purple or yellow, apices obtuse, erose or subentire. **Ray florets** 6–15(–21); corollas usually scarlet red or maroon, sometimes yellow, laminae linear to spatulate, 8–25 mm. **Disc florets** 12–50; corollas yellow, 5–6 mm, lobes ca. 1 mm. **Cypselae** 7–10 mm, 3-angled (ray) or compressed (disc), ribbed, ciliate; **pappi** usually of 1 stout awn 4–6 mm (from shoulders of cypselae). $2n$ = 24.

Flowering summer–fall. Rocky roadsides, ravines, calcareous soils; 1200–1600 m; Ariz., Fla., Ga., N.C., S.C.; Mexico; West Indies (Hispaniola); Central America; South America; introduced in Asia (China), South Africa, Australia.

Zinnia peruviana is presumably native in southern Arizona and reported as naturalized in southeastern United States.

5. Zinnia violacea Cavanilles, Icon. 1: 57, plate 81. 1791 • Elegant or garden zinnia

Zinnia elegans Jacquin

Annuals, to 100(–200) cm. **Stems** greenish, becoming yellowish to purplish, unbranched or sparingly branched distal to bases, hirsute to strigose or scabrous. **Leaf blades** 3–5-nerved, ovate to oblong, mostly 60–100 × 20–60 mm, scabrellous to glabrate. **Peduncles** to 85 mm. **Involucres** ± hemispheric or broader, 10–15 × 5–25 mm. **Phyllaries** obovate, becoming scarious, glabrous or sparsely hairy, apices rounded, erose or fimbriate. **Paleae** red to purple, apices rounded to acute, fimbriate. **Ray florets** 8–21 (more in "double" cultivars); corollas usually red (white, yellow, or purple in cultivars), laminae spatulate to obovate, 10–35 mm. **Disc florets** 100–150+; corollas yellow, 7–9 mm, lobes 1–2.5 mm. **Cypselae** 6–10 mm, 3-angled (ray) or ± compressed (disc), not or faintly ribbed, ciliolate; **pappi** 0. $2n = 24$.

Flowering summer–fall. Disturbed sites; to 500? m; introduced; Conn., Fla., Ga., Ky., La., N.C., Ohio, Pa., S.C., Tex.; Mexico; West Indies (Cuba); Central America; South America (Bolivia); also introduced in Asia.

Zinnia violacea is perhaps adventive in Costa Rica, Panama, Cuba, Bolivia, China, and Malesia. The most widely cultivated *Zinnia*, it is reported to have escaped from cultivation and apparently naturalized in ten eastern and southern states but is nowhere common in the flora area. It is not as weedy as *Z. peruviana*, possibly because it lacks awns and thus is not as easily dispersed by animals.

267. CHRYSOGONUM Linnaeus, Sp. Pl. 2: 920. 1753; Gen. Pl. ed. 5, 391. 1754
• [Greek *chrysos*, gold, and *gonos*, seed, apparently alluding to the bright yellow, hemispheric capitula or to the fertile cypselae from the cypsela-complexes of the ray florets] E

Guy L. Nesom

Perennials, 2–30(–50) cm (aerial stems from fibrous-rooted rhizomes to 5 cm). **Stems** erect to erect-ascending (flowering) or prostrate (vegetative stolons), branched from bases or ± throughout (villous). **Leaves** basal and cauline; opposite; petiolate; blades elliptic-ovate to deltate-ovate, bases cordate or truncate to cuneate, margins crenate, faces hairy. **Heads** radiate, borne singly or in pairs. **Involucres** cupulate-hemispheric, 7–10 mm diam. **Phyllaries** persistent (outer), 8–10 in 2 series (outer spreading, oblong-oblanceolate, foliaceous, inner erect, greenish, smaller, ± scarious). **Receptacles** flat to shallowly convex, paleate (paleae oblanceolate, scarious). **Ray florets** 5(–6), pistillate, fertile; corollas yellow. **Disc florets** 25–50, functionally staminate; corollas yellow, tubes shorter than narrowly funnelform throats, lobes 5, deltate. **Cypselae** (blackish brown) strongly obflattened, obovoid to obovate (each basally adnate to 1 inner phyllary, 3 paleae, and 3 disc florets, the "cypsela-complexes" falling as units); **pappi** persistent, coroniform (asymmetric). $x = 16$.

Species 1: e United States.

Each cypsela-complex includes an elaiosome (a fleshy, oil-bearing structure at the base of the phyllary) and two associated paleae. Ants carry cypsela-complexes into their underground nests, chew off the elaiosomes, and return the undamaged cypselae or complexes to the surface.

Chrysogonum virginianum has been divided into two species (e.g., J. K. Small 1933) or treated as a single species with two varieties (e.g., A. E. Radford et al. 1968; T. F. Stuessy 1977). Stuessy mapped many intermediates between two varieties, not only in the area of North Carolina and South Carolina where their ranges meet, but also practically throughout both their ranges. G. L. Nesom (2001) found three allopatric taxa, separated primarily by habit (production or not of stolons) and morphology of flowering stems. Plants treated here as *C. virginianum* var. *australe* might justifiably be treated at specific rank.

All three taxa of *Chrysogonum* are now used as garden plants and are sold in nurseries under horticultural names.

SELECTED REFERENCES Nesom, G. L. 2001. Taxonomic review of *Chrysogonum* (Asteraceae: Heliantheae). Sida 19: 811–820. Stuessy, T. F. 1977. Revision of *Chrysogonum* (Compositae, Heliantheae). Rhodora 79: 190–202.

1. Chrysogonum virginianum Linnaeus, Sp. Pl. 2: 920. 1753 • Green and gold   E F

Leaf blades 2.5–10 cm, faces minutely strigoso-hirsutulous to villoso-hirsute and stipitate-glandular (hairs ca. 0.1 mm). **Peduncles** 2–22 cm. **Paleae** ca. 4 mm. **Ray corollas** 6–17 mm, laminae broadly elliptic, apices 3-toothed. **Disc corollas** 2.5–2.7 mm. **Cypselae** 3–4.5 × 2–2.5 mm.

Varieties 3 (3 in the flora): e United States.

1. Plants not stoloniferous; earliest flowering stems leafless, later ones leafy and mostly 15–35(–50) cm ... 1a. *Chrysogonum virginianum* var. *virginianum*
1. Plants stoloniferous (colonial, mat-forming); flowering stems leafless and/or leafy, leafy stems (2–)15–25 cm.
  2. Longest stolon internodes 2–6 cm; earliest flowering stems leafless, mostly 2–25 cm (later flowering stems leafy, 15–25 cm) . . . .
    . . . . 1b. *Chrysogonum virginianum* var. *brevistolon*
  2. Longest stolon internodes 12–60 cm; earliest (and all other) flowering stems leafless, 2–10 cm . . . . . 1c. *Chrysogonum virginianum* var. *australe*

1a. Chrysogonum virginianum Linnaeus var. **virginianum**   E

Chrysogonum virginianum var. *dentatum* A. Gray

Plants not stoloniferous (not colonial or mat-forming). **Stems** (flowering): earliest leafless, later stems leafy, 15–35(–50) cm. **Leaf blades** deltate-ovate, bases abruptly cuneate to truncate or subcordate, faces minutely strigoso-hirsutulous (hairs ascending-appressed, 0.1–0.3 mm). **Peduncles** 2–22 cm. **Outer phyllaries** 6–11(–15) × 3–7 mm. **2n** = 32.

Flowering late Mar–Jun(–Jul, later with moisture). Moist to dry woodlands, forests, partially sunny sites, clearings, edges; 10–700 m; D.C., Ky., Md., N.Y., N.C., Ohio, Pa., S.C., Va., W.Va.

1b. Chrysogonum virginianum Linnaeus var. **brevistolon** G. L. Nesom, Sida 19: 817, figs. 1, 3. 2001   E F

Plants stoloniferous (colonial and mat-forming; longest stolon internodes 2–6 cm). **Stems** (flowering): earliest leafless, later stems leafy, mostly 2–25 cm. **Leaf blades** deltate-ovate, bases abruptly cuneate to truncate or subcordate, faces strigoso-hirsutulous (hairs 0.5–0.3 mm). **Peduncles** 2–22 cm. **Outer phyllaries** 6–10 × 3–4 mm. **2n** = 32.

Flowering Mar–May(–early Jun). Moist to dry woodlands, partially sunny sites; 10–400 m; Ala., Ga., Ky., N.C., S.C., Tenn.

1c. Chrysogonum virginianum Linnaeus var. **australe** (Alexander ex Small) H. E. Ahles, J. Elisha Mitchell Sci. Soc. 80: 173. 1964   E

Chrysogonum australe Alexander ex Small, Man. S.E. Fl., 1415, 1509. 1933

Plants stoloniferous (colonial, mat-forming; longest stolon internodes 12–60 cm). **Stems** (flowering): leafless, 2–10 cm. **Leaf blades** elliptic-ovate, bases gradually cuneate, faces villoso-hirsute (hairs spreading, 1–3 mm). **Peduncles** 2–10 cm. **Outer phyllaries** 7–10 × 2.5–4 mm.

Flowering mid Mar–Apr(–May). Pine-oak, longleaf pine, beech-oak-magnolia woods, ravine slopes, limestone outcrops, flood plains and terraces, sand or sandy loam; 10–50 m; Ala., Fla., Ga., La., Miss.

268. **LINDHEIMERA** A. Gray & Engelmann, Proc. Amer. Acad. Arts 1: 47. 1847 • [For Ferdinand Jacob Lindheimer, 1801–1879, German expatriate, botanist/intellect, settled in Texas]

John L. Strother

Annuals, 10–120(–300+) cm. **Stems** erect, distally branched. **Leaves** cauline; mostly alternate (distal sometimes opposite); petiolate (proximal) or sessile; blades mostly lanceolate, oblanceolate, or ovate, bases mostly cuneate, margins coarsely toothed to crenate or ± entire, faces ± scabrellous and/or ± hispid. **Heads** radiate, borne singly or in (leafy to bracteate) loose, corymbiform arrays. **Involucres** ± turbinate to hemispheric, 5–12 mm diam. **Phyllaries** persistent (outer), (8–)10

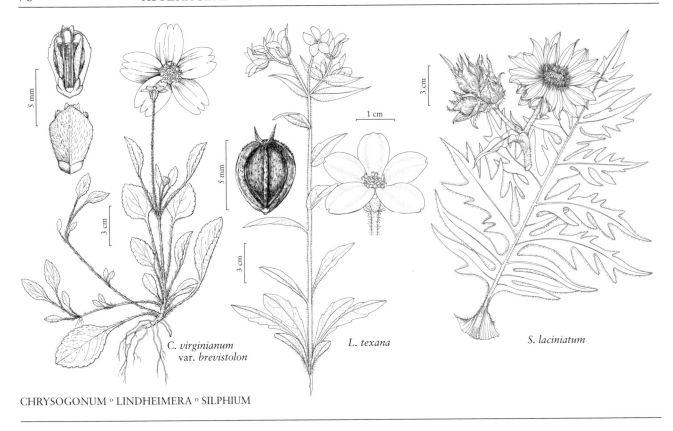

C. *virginianum*
var. *brevistolon*

L. *texana*

S. *laciniatum*

CHRYSOGONUM ∘ LINDHEIMERA ∘ SILPHIUM

in 2 series (outer 4–5 becoming reflexed, usually lance-linear, narrower and shorter than inner). **Receptacles** an outer ring supporting ray florets plus an inner column supporting disc florets, paleate (paleae linear, conduplicate). **Ray florets** (4–)5, pistillate, fertile; corollas lemon to bright yellow. **Disc florets** 12–25+, functionally staminate; corollas dull yellow to brownish, tubes shorter than funnelform throats, lobes 5, deltate. **Cypselae** obcompressed, oblanceolate to obovate (winged, faces ± scabrous to scabrellous; each shed with its subtending phyllary and 2 contiguous paleae with their florets); **pappi** 0 (shoulders of cypsela wings often projecting as deltate to cornute processes, sometimes interpreted as pappus). *x* = 8.

Species 1: sc United States, n Mexico.

1. **Lindheimera texana** A. Gray & Engelmann, Proc. Amer. Acad. Arts 1: 47. 1847 • Star daisy F

Leaf blades 4–12(–18+) × 1–3(–5+) cm. **Peduncles** 1–3(–6) cm. **Phyllaries:** outer ± lance-linear to linear, 8–12 mm, scabrellous and/or hispid; inner lance-ovate to elliptic, 9–15 mm, scabrellous to glabrate. **Ray laminae** 10–14 × 4–9 mm. **Disc corollas** 3–4 mm. **Cypselae** 4.5–6 mm, ± scabrellous to glabrate (pappus-like processes 0.5–1 mm). *2n* = 16.

Flowering Mar–Jun. Sandstones, clays, alkaline soils; 10–500 m; Okla., Tex.; Mexico (Coahuila).

Leaves, peduncles, and phyllaries of *Lindheimera texana* sometimes bear stipitate glands reminiscent of glands in some members of Madiinae.

269. SILPHIUM Linnaeus, Sp. Pl. 2: 919. 1753; Gen. Pl. ed. 5, 391. 1754 • [Greek *silphion*, an unknown plant appearing on ancient coins of the city of Cyrene] [E]

Jennifer A. Clevinger

Perennials, 20–250+ cm (fibrous rooted, rhizomatous, or taprooted). **Stems** usually erect, usually branched (terete or square, often vernicose with resinous exudates). **Leaves** basal and cauline (basal persistent or withering before flowering); whorled, opposite, subopposite, or alternate (sometimes all conditions on one plant); petiolate or sessile; blades (1- or 3-nerved) deltate, elliptic, linear, ovate, or rhombic, sometimes 1–2-pinnately lobed or -pinnatifid, bases cordate or truncate to cuneate, margins entire or toothed, faces glabrous or hairy (sometimes stipitate-glandular). **Heads** radiate, in paniculiform or racemiform arrays. **Involucres** campanulate to hemispheric, 10–30 mm diam. **Phyllaries** persistent, 11–45 in 2–4 series (outer broader, foliaceous, inner smaller, thinner, each subtending a ray floret). **Receptacles** flat to slightly convex, paleate (paleae oblong, lanceolate, or linear). **Ray florets** 8–35+ in 1–4 series, pistillate, fertile; corollas yellow or white. **Disc florets** 20–200+, functionally staminate; corollas yellow or white, tubes much shorter than narrow, cylindric throats, lobes 5, deltate. **Cypselae** (black to brown) obflattened (± winged, shed alone without accessory structures); **pappi** 0, or persistent, of 2 awns (± confluent with cypsela shoulders and wings). $x = 7$.

Species 12 (12 in the flora): North America.

In keys and descriptions here, "plants scapiform" indicates plants with distal cauline leaves notably smaller than basal and proximal cauline leaves, and "plants caulescent" indicates plants with leaves ± uniform in size throughout; "leaves persistent" indicates leaves persisting to flowering time, and "leaves caducous" indicates leaves withering before flowering time.

SELECTED REFERENCE Perry, L. M. 1937. Notes on *Silphium*. Rhodora 39: 281–297.

1. Stems square (distal leaf or petiole bases connate-perfoliate) . . . . 5. *Silphium perfoliatum*
1. Stems terete or slightly 4-angled (distal leaves not perfoliate).
    2. Leaves (at least proximal) usually laciniate, pinnatifid, or 1–2-pinnately lobed (lobes 3–15).
        3. Ray florets 6–12 . . . . 4. *Silphium compositum* (in part)
        3. Ray florets 14–40.
            4. Corollas white . . . . 1. *Silphium albiflorum*
            4. Corollas yellow.
                5. Phyllaries appressed, apices acute to obtuse . . . . 3. *Silphium terebinthinaceum* (in part)
                5. Phyllaries reflexed or appressed, apices acuminate to caudate . . . . . . 2. *Silphium laciniatum*
    2. Leaves (proximal and distal) seldom lobed (sometimes with 2 basal lobes, then hastate, sagittate, or auriculate).
        6. Plants taprooted (scapiform; cauline leaves notably smaller distally).
            7. Ray flowers 6–12 . . . . 4. *Silphium compositum* (in part)
            7. Ray flowers 17–29 . . . . 3. *Silphium terebinthinaceum* (in part)
        6. Plants fibrous rooted (caulescent; cauline leaves not notably smaller distally).
            8. Leaves (at least proximal leaves) petiolate (never sessile), bases truncate, cordate, auriculate, hastate, or sagittate.
                9. Stems glabrous; basal leaves caducous . . . . 9. *Silphium brachiatum*
                9. Stems hispid; basal leaves persistent . . . . 6. *Silphium wasiotense*
        8. Leaves (proximal and distal) petiolate (blade bases round to tapered) or sessile (blade bases round to cordate).
            10. Phyllaries: abaxial faces usually glabrous (rarely scabrous in outer series).
                11. Ray florets 20–36+ . . . . 7. *Silphium integrifolium* (in part)
                11. Ray florets 8–17 . . . . 12. *Silphium asteriscus* (in part)

[10. Shifted to left margin.—Ed.]

10. Phyllaries: abaxial faces glandular, hirsute, hispid, or scabrous.
    12. Leaf faces stipitate-glandular . . . . 11. *Silphium glutinosum*
    12. Leaf faces eglandular.
        13. Basal leaves persistent.
            14. Leaf faces shaggy-hispid; phyllaries hispid; corollas pale yellow; cypselae 6–10 mm; Alabama, Georgia, Tennessee . . . . 10. *Silphium mohrii*
            14. Leaf faces hirsute, hispid, or scabrous; phyllaries scabrous; corollas bright yellow; cypselae 10–17 mm; Louisiana, Texas . . . . 8. *Silphium radula* (in part)
        13. Basal leaves caducous.
            15. Cauline leaves opposite (clasping stems) . . . . 7. *Silphium integrifolium* (in part)
            15. Cauline leaves opposite or alternate (not clasping stems).
                16. Ray florets 20–30+ . . . . 8. *Silphium radula* (in part)
                16. Ray florets 12–20 . . . . 12. *Silphium asteriscus* (in part)

1. **Silphium albiflorum** A. Gray, Proc. Amer. Acad. Arts 19: 4. 1883 • White rosinweed E

Plants scapiform, 20–75 cm; taprooted. Stems terete, glabrous, hirsute, hispid, or scabrous. Leaves: basal persistent, petiolate; cauline usually alternate, rarely subopposite, petiolate or sessile; blades deltate, elliptic, linear, ovate, or rhombic, 2–40 × 1.5–26 cm, sometimes (proximal) 1–2-pinnately lobed, bases attenuate, cuneate, or truncate, ultimate margins unevenly toothed or entire (distal blades), apices acuminate to apiculate, faces scabrous to hirsute. Phyllaries 23–36 in 2–3 series, outer reflexed, apices acuminate, abaxial faces hispid. Ray florets 14–30; corollas white. Disc florets 70–130; corollas white. Cypselae 12–26 × 7–12 mm; pappi 2–5 mm.

Flowering late spring–summer. Prairies, open fields, calcareous soils; 100–500 m; Tex.

2. **Silphium laciniatum** Linnaeus, Sp. Pl. 2: 919. 1753 • Compass plant E F

Silphium laciniatum var. *robinsonii* L. M. Perry

Plants scapiform, (40–)100–300 cm; taprooted. Stems terete, hirsute, hispid, or scabrous. Leaves: basal persistent, petiolate or sessile; cauline petiolate or sessile; blades lanceolate, linear, ovate, or rhombic, 4–60 × 1–30 cm, usually (proximal) 1–2-pinnately lobed, bases attenuate to truncate, ultimate margins unevenly toothed or entire, apices acute, faces hirsute, hispid, or scabrous. Phyllaries 25–45 in 2–3 series, outer reflexed or appressed, apices acuminate to caudate, abaxial faces hispid to scabrous, ± stipitate-glandular. Ray florets 27–38; corollas yellow. Disc florets 100–275; corollas yellow. Cypselae 10–18 × 6–12 mm; pappi 1–3 mm. 2*n* = 14.

Flowering summer–early fall. Prairies, open, disturbed sites; 50–600 m; Ont.; Ala., Ark., Ill., Ind., Iowa, Kans., Ky., La., Mich., Minn., Miss., Mo., Nebr., N.Mex., N.Y., Ohio, Okla., Pa., S.Dak., Tenn., Tex., Wis.

3. **Silphium terebinthinaceum** Jacquin, Hort. Bot. Vindob. 1: 16, plate 43. 1770 • Prairie-dock E

Plants scapiform, (40–)100–250+ cm; taprooted. Stems terete, glabrous. Leaves: basal persistent, petiolate; cauline alternate, petiolate or sessile; blades cordate, deltate, lanceolate, ovate, or sagittate, 3–40 × 1–40 cm, sometimes (proximal) pinnately lobed, bases attenuate, cordate, hastate, round, or truncate, ultimate margins coarsely toothed or entire, apices acuminate, acute, or obtuse, faces glabrous or scabrous. Phyllaries 23–33 in 2–3 series, outer appressed, apices obtuse to acute, abaxial faces glabrous. Ray florets 17–29; corollas yellow. Disc florets 120–140; corollas yellow. Cypselae 7–13 × 4–10 mm; pappi 0–1 mm.

Varieties 2 (2 in the flora): e North America.

1. Leaves pinnately lobed . . . . 3a. *Silphium terebinthinaceum* var. *pinnatifidum*
1. Leaves not lobed . . . . . . 3b. *Silphium terebinthinaceum* var. *terebinthinaceum*

3a. **Silphium terebinthinaceum** Jacquin var. **pinnatifidum** (Elliott) A. Gray, Manual, 220. 1848 E

Silphium pinnatifidum Elliott, Sketch Bot. S. Carolina 2: 462. 1823

Leaves pinnately lobed (lobes to 13, sinuses relatively shallow to deep, sometimes almost to midrib; ultimate margins entire or toothed). 2*n* = 14.

Flowering mid summer–early fall. Wet or dry prairies, fens, open, disturbed sites; 100–500 m; Ala., Ga., Ill., Ky., Mich., Ohio, Tenn.

3b. Silphium terebinthinaceum Jacquin var. **terebinthinaceum** E

Silphium rumicifolium Small; *S. terebinthinaceum* var. *lucy-brauniae* Steyermark

Leaf blades not lobed (margins entire or toothed). $2n = 14$.

Flowering mid summer–early fall. Wet or dry prairies, fens, disturbed sites free of underbrush; 100–500 m; Ont.; Ala., Ark., Ga., Ill., Ind., Iowa, Ky., Mich., Miss., Mo., Nebr., N.Y., N.C., Ohio, S.C., Tenn., Va., Wis.

4. Silphium compositum Michaux, Fl. Bor.-Amer. 2: 145. 1803 • Kidney-leaf rosinweed E

Silphium compositum subsp. *ovatifolium* (Torrey & A. Gray) C. R. Sweeney & T. R. Fisher; *S. compositum* subsp. *reniforme* (Rafinesque ex Nuttall) C. R. Sweeney & T. R. Fisher; *S. compositum* var. *reniforme* (Rafinesque ex Nuttall) Torrey & A. Gray; *S. compositum* subsp. *venosum* (Small) C. R. Sweeney & T. R. Fisher; *S. compositum* var. *venosum* (Small) Kartesz & Gandhi; *S. lapsuum* Small; *S. orae* Small; *S. ovatifolium* (Torrey & A. Gray) Small; *S. reniforme* Rafinesque ex Nuttall; *S. venosum* Small

Plants scapiform, 50–250 cm; taprooted. **Stems** terete, glabrous (sometimes glaucous). **Leaves:** basal persistent, petiolate; cauline alternate, petiolate or sessile; blades cordate, deltate, elliptic, hastate, ovate, reniform, or sagittate, 2–35 × 0.5–52 cm, sometimes (proximal) pinnately or palmately lobed, bases oblique, truncate, attenuate, cordate, sagittate, or hastate, ultimate margins toothed, apices acute, faces glabrous, hispid, or scabrous. **Phyllaries** 11–18 in 2–3 series, outer appressed to reflexed, apices obtuse to cuspidate, abaxial faces sparsely scabrous. **Ray florets** 6–12; corollas yellow. **Disc florets** 20–85; corollas yellow. **Cypselae** 6–12 × 4–10 mm; **pappi** 1–4 mm. $2n = 14$.

Flowering summer–early fall. Open, pine and oak forests, sandy soils, fields, roadsides, meadows; 0–1600 m; Ala., Fla., Ga., N.C., S.C., Tenn., Va., W.Va.

5. Silphium perfoliatum Linnaeus, Syst. Nat. ed. 10, 2: 1232. 1759 • Cup plant E

Plants caulescent, 75–300 cm; fibrous rooted. **Stems** square, glabrous, hispid, or scabrous. **Leaves:** basal caducous; cauline usually opposite, rarely whorled (in 3s), petiolate or sessile; blades deltate, lanceolate, or ovate, 2–41 × 0.5–24 cm, bases attenuate or truncate (distal connate-perfoliate), margins entire, dentate, or bidentate, apices acuminate to acute, faces scabrous to hispid. **Phyllaries** 25–37 in 2–3 series, outer appressed, apices acute to acuminate, abaxial faces scabrous or hispid. **Ray florets** 17–35; corollas yellow. **Disc florets** 85–150 (–200); corollas yellow. **Cypselae** 8–12 × 5–9 mm; **pappi** 0.5–1.5 mm.

Varieties 2 (2 in flora): e North America.

1. Stems glabrous or sparsely scabrous; peduncles glabrous . . . . . 5a. *Silphium perfoliatum* var. *connatum*
1. Stems glabrous, hispid, or scabrous; peduncles usually scabrous to hispid . . . .
. . . . 5b. *Silphium perfoliatum* var. *perfoliatum*

5a. Silphium perfoliatum Linnaeus var. **connatum** (Linnaeus) Cronquist, Brittonia 29: 223. 1977 E

Silphium connatum Linnaeus, Mant. Pl., 574. 1771; *S. perfoliatum* subsp. *connatum* (Linnaeus) Cruden

Stems glabrous or sparsely scabrous. **Peduncles** glabrous. $2n = 14$.

Flowering summer–early fall. River bottoms; 50–500 m; N.C., Va., W.Va.

Variety *connatum* is known only from New River Valley.

5b. Silphium perfoliatum Linnaeus var. **perfoliatum** E

Stems glabrous, hispid, or scabrous. **Peduncles** usually scabrous to hispid. $2n = 14$.

Flowering summer–early fall. Wet prairies, open forests, and river bottoms; 50–500 m; Ont.; Ala., Ark., Conn., Ill., Ind., Iowa, Kans., Ky., La., Mass., Mich., Minn., Miss., Mo., Nebr., N.J., N.Y., N.C., N.Dak., Ohio, Okla., Pa., S.Dak., Tenn., Tex., Vt., Va., W.Va., Wis.

6. **Silphium wasiotense** Medley, Sida 13: 285, figs. 1, 2e–h, 3. 1989 (as wasiotensis) • Appalachian rosinweed E

Plants caulescent, 60–150 cm; fibrous rooted. **Stems** terete, hispid (never glaucous). **Leaves:** basal persistent, petiolate; cauline opposite, petiolate (proximal) or sessile (distal); blades cordate, deltate, lanceolate, or ovate, 1–28 × 1.5–20 cm, sometimes (proximal) lobed, bases auriculate, cordate, or truncate (proximal and distal), ultimate margins toothed or entire, apices acute to acuminate, faces hispid. **Phyllaries** 12–14 in 2–3 series, outer reflexed, apices acute to obtuse, abaxial faces hispid. **Ray florets** 6–14; corollas yellow. **Disc florets** 30–45; corollas yellow. **Cypselae** 4–8 × 4–6 mm; **pappi** 0.5–1.5 mm.

Flowering late summer–fall. Dry, open sites in mesic forests, along roadcuts; 300–600 m; Ky., Tenn.

7 **Silphium integrifolium** Michaux, Fl. Bor.-Amer. 2: 146. 1803 • Wholeleaf rosinweed E

Plants caulescent, 40–200 cm; fibrous rooted. **Stems** terete to slightly square, sparsely scabrous or glabrous (sometimes glaucous). **Leaves:** basal caducous; cauline opposite, sessile; blades lanceolate to ovate, 2–23 × 0.1–11 cm, bases round to caudate, margins finely serrate or entire, apices acute to acuminate, faces hispid, scabrous, or glabrous. **Phyllaries** 17–37 in 2–3 series, outer appressed, apices acute to acuminate, abaxial faces glabrous, hispid, pilose, or scabrous, sometimes stipitate-glandular. **Ray florets** 12–36; corollas yellow. **Disc florets** 70–225; corollas yellow. **Cypselae** 9–14 × 6–10 mm; **pappi** 1–4 mm.

Varieties 2 (2 in flora): e North America.

1. Phyllaries: abaxial faces usually hispid, pilose, or scabrous, rarely glabrous; ray florets 12–22 . . .
. . . . 7a. *Silphium integrifolium* var. *integrifolium*
1. Phyllaries: abaxial faces glabrous; ray florets 20–36+ . . . . 7b. *Silphium integrifolium* var. *laeve*

7a. **Silphium integrifolium** Michaux var. **integrifolium** E

Silphium integrifolium var. *deamii* L. M. Perry; *S. integrifolium* var. *neglectum* Settle & Fisher

Plants 40–200 cm. **Leaves:** blades scabrous, pilose, hispid, or glabrous. **Phyllaries:** abaxial faces usually hispid, pilose, or scabrous, rarely glabrous. **Ray florets** 12–22. **Disc florets** 70–140. **2***n* = 14.

Flowering mid summer–early fall. Prairies, along roads and railroads; 50–600 m; Ala., Ark., Ill., Ind., Iowa, Kans., Ky., La., Mich., Miss., Mo., Nebr., Okla., S.Dak., Tenn., Wis.

7b. **Silphium integrifolium** Michaux var. **laeve** Torrey & A. Gray, Fl. N. Amer. 2: 279. 1842 E

Silphium speciosum Nuttall, Trans. Amer. Philos. Soc., n.s. 7: 341. 1840

Plants 100–150 cm. **Leaves:** blades glabrous. **Phyllaries:** abaxial faces glabrous. **Ray florets** 20–36+. **Disc florets** 130–225. **2***n* = 14.

Flowering mid summer–early fall. Prairies, along roads and railroads; 100–600 m; Ont.; Ark., Iowa, Kans., Mo., Nebr., N.Mex., Okla., Tex., Wyo.

8. **Silphium radula** Nuttall, Trans. Amer. Philos. Soc., n. s. 7: 341. 1840 • Roughleaf rosinweed E

Plants caulescent, 30–250 cm; fibrous rooted. **Stems** terete, scabrous, hispid, or hirsute. **Leaves:** basal persistent or caducous, petiolate; cauline alternate or opposite, petiolate or sessile; blades lanceolate to ovate, 2–22 × 0.5–8 cm, bases attenuate to round, margins coarsely dentate or entire, apices acuminate to acute, faces scabrous, hispid, or hirsute. **Phyllaries** 18–37 in 2–4 series, outer appressed or reflexed, apices acute, abaxial faces scabrous. **Ray florets** 12–30+; corollas yellow. **Disc florets** 100–175; corollas yellow. **Cypselae** 10–17 × 7–15 mm; **pappi** 2–5 mm.

Varieties 2 (2 in the flora): sc United States.

1. Basal leaves caducous; ray florets 20–30+ . . . .
. . . . 8a. *Silphium radula* var. *radula*
1. Basal leaves persistent; ray florets 12–18 . . . . .
. . . . 8b. *Silphium radula* var. *gracile*

8a. **Silphium radula** Nuttall var. **radula** E

Silphium reverchonii Bush; *S. simpsonii* Greene var. *wrightii* L. M. Perry

Plants 50–250 cm. **Basal leaves** caducous. **Heads** to 30. **Ray florets** 20–30+. **Disc florets** 125–175. **2***n* = 14.

Flowering summer. Prairies, fencerows, roadsides; 0–400 m; Ark., Kans., La., Mo., Okla., Tex.

6–8.5 × 4–7 mm; **pappi** 0.5–2 mm. **2***n* = 14.

8b. Silphium radula Nuttall var. **gracile** (A. Gray) J. A. Clevinger, Novon 14: 277. 2004 • Slender rosinweed [E]

Silphium gracile A. Gray, Proc. Amer. Acad. Arts 8: 653. 1873

Plants 30–100 cm. **Basal leaves** persistent. **Heads** to 15. **Ray florets** 12–18. **Disc florets** 100–150. $2n = 14$.

Flowering summer. Prairies, fencerows, roadsides; 0–300 m; La., Tex.

9. Silphium brachiatum Gattinger, Bot. Gaz. 9: 192. 1884 • Cumberland rosinweed [E]

Plants caulescent, 65–200 cm; fibrous rooted. **Stems** terete, glabrous. **Leaves:** basal caducous; cauline opposite, petiolate; blades lanceolate, hastate, ovate, or sagittate, 1–24 × 0.5–18 cm, bases cuneate, or round (distal), or cordate, truncate, hastate, or sagittate (proximal), margins coarsely dentate or entire, apices acuminate to acute, faces sparsely hirsute, hispid, or scabrous. **Phyllaries** 11–15 in 2–3 series, outer reflexed or appressed, apices acute, abaxial faces scabrous. **Ray florets** 7–9; corollas yellow. **Disc florets** 20–40; corollas yellow. **Cypselae** 7–8 × 5–6 mm; **pappi** 0.5–1 mm. $2n = 14$.

Flowering late summer–fall. Open forests, calcareous soils, roadcuts; 200–600 m; Ala., Ga., Tenn.

10. Silphium mohrii Small, Bull. Torrey Bot. Club 24: 493, plate 315. 1897 • Shaggy rosinweed [E]

Plants caulescent, 40–160 cm; fibrous rooted. **Stems** terete, shaggy-hispid. **Leaves:** basal persistent, petiolate; cauline usually alternate, sometimes opposite, petiolate or sessile; blades lanceolate to narrowly ovate, 1.5–26 × 0.5–14 cm, bases attenuate to acute, margins coarsely toothed or entire, apices acuminate, faces shaggy-hispid. **Phyllaries** 14–19 in 2–3 series, outer appressed, apices acuminate to acute, abaxial faces shaggy-hispid. **Ray florets** 10–20; corollas pale yellow. **Disc florets** 40–85; corollas pale yellow. **Cypselae** 6–10 × 4–7 mm; **pappi** 1–2 mm.

Flowering late summer–early fall. Prairies, clearings, fence rows, sandy soils; 100–400 m; Ala., Ga., Tenn.

11. Silphium glutinosum J. R. Allison, Castanea 66: 183, figs. 11, 12A–C. 2001 • Sticky rosinweed [E]

Plants caulescent 80–150(–180) cm; fibrous rooted. **Stems** terete, scabrous, stipitate-glandular. **Leaves:** basal caducous; cauline usually opposite, rarely alternate, petiolate or sessile; blades lanceolate to ovate, 3–22 × 1–12 cm, bases attenuate to rounded, margins toothed or entire, apices acute to acuminate, faces scabrous, stipitate-glandular. **Phyllaries** 19–23 in 3–4 series, outer appressed to reflexed, apices acute to obtuse, abaxial faces scabrous, stipitate-glandular. **Ray florets** 8–16; corollas pale yellow. **Disc florets** 50–70; corollas pale yellow. **Cypselae** 6–8.5 × 4–7 mm; **pappi** 0.5–2 mm. $2n = 14$.

Flowering summer–early fall. Limestone glades, along rocky streams; 100–300 m; Ala.

James Allison (in protologue) suggested that *Silphium perplexum* is a hybrid between *S. glutinosum* and *S. integrifolium*. It has glandular hairs on its stems, leaves, and phyllaries as in *S. glutinosum*; it is also found on chalk, has taller stems, larger leaf blades, and deeper yellow corollas. Further study is needed to determine its status.

12. Silphium asteriscus Linnaeus, Sp. Pl. 2: 920. 1753 • Starry rosinweed [E][F]

Plants caulescent, 30–200 cm; fibrous rooted. **Stems** terete, glabrous, hirsute, hispid, or scabrous. **Leaves:** basal persistent or caducous, petiolate or sessile; cauline alternate, opposite, or whorled, petiolate or sessile; blades simple, lanceolate, falcate, or ovate, 1.5–25 × 0.5–5 cm, bases attenuate to round, margins dentate, serrate, or entire, apices acuminate, acute, or obtuse, faces glabrous, hirsute, hispid, or scabrous. **Phyllaries** 12–26 in 2–3 series, outer appressed or reflexed, apices acute to acuminate, abaxial faces hispid or scabrous. **Ray florets** 8–21; corollas yellow. **Disc florets** 35–150; corollas yellow. **Cypselae** 6–15 × 4–11 mm; **pappi** 0–5 mm.

Varieties 5 (5 in the flora): e United States.

1. Leaf faces glabrous or sparsely hairy.
   2. Cauline leaves mostly whorled (in 3s) . . . . 12a. *Silphium asteriscus* var. *trifoliatum*
   2. Cauline leaves mostly opposite . . . . 12b. *Silphium asteriscus* var. *latifolium*
1. Leaf faces hirsute, hispid, or scabrous.
   3. Basal leaves persistent . . . . 12c. *Silphium asteriscus* var. *simpsonii*
   3. Basal leaves caducous.

[4. Shifted to left margin.—Ed.]
4. Paleae puberulent and stipitate-glandular . . . . .
. . . . 12d. *Silphium asteriscus* var. *dentatum*
4. Paleae puberulent to scabrous (not glandular)
. . . . 12e. *Silphium asteriscus* var. *asteriscus*

12a. Silphium asteriscus Linnaeus var. **trifoliatum**
(Linnaeus) J. A. Clevinger, Novon 14: 276. 2004
• Whorled rosinweed   E

Silphium trifoliatum Linnaeus, Sp. Pl. 2: 920. 1753; *S. atropurpureum* Retzius ex Willldenow; *S. laevigatum* Pursh

Plants 75–140 cm. **Leaves:** basal persistent; cauline mostly whorled (in 3s); faces glabrous or sparsely hairy. **Paleae** puberulent. **Ray florets** 11–16. **Disc florets** 35–90. $2n = 14$.

Flowering mid summer–early fall. Prairies, meadows, roadsides, along railways; 100–900 m; Ala., Ga., Ill., Ind., Ky., Md., N.Y., N.C., Ohio, Pa., S.C., Tenn., Va., W.Va.

12b. Silphium asteriscus Linnaeus var. **latifolium**
(A. Gray) J. A. Clevinger, Novon 14: 276. 2004   E

Silphium trifoliatum Linnaeus var. *latifolium* A. Gray in A. Gray et al, Syn. Fl. N. Amer. 1(2): 241. 1884; *S. confertifolium* Small; *S. glabrum* Eggert

Plants 50–150 cm. **Leaves:** basal persistent or caducous; cauline mostly opposite; faces glabrous or sparsely hairy. **Paleae** puberulent. **Ray florets** 12–17. **Disc florets** 35–130.

Flowering mid summer–early fall. Fields, roadsides, disturbed areas; 100–900 m; Ala., Ga., Ky., La., Miss., N.C., S.C., Tenn., Va., W.Va.

Silphium confertifolium has persistent basal leaves and a diminutive stature that is due to environmental variation; it is best included in *S. asteriscus* var. *latifolium.*

12c. Silphium asteriscus Linnaeus var. **simpsonii**
(Greene) J. A. Clevinger, Novon 14: 275. 2004   E

Silphium simpsonii Greene, Pittonia 4: 44. 1899

Plants 25–110 cm. **Leaves:** basal persistent; cauline opposite or alternate; faces hirsute, hispid, or scabrous. **Paleae** puberulent and stipitate-glandular. **Ray florets** 10–15(–20). **Disc florets** 85–110.
Flowering summer. Roadsides, meadows, open fields; 0–200 m; Ala., Fla., Ga., Miss., S.C.

12d. Silphium asteriscus Linnaeus var. **dentatum**
(Elliott) Chapman, Fl. South. U.S., 221. 1860   E F

Silphium dentatum Elliott, Sketch Bot. S. Carolina 2: 468. 1823; *S. asteriscus* var. *angustatum* A. Gray; *S. elliottii* Small; *S. incisum* Greene; *S. nodum* Small

Plants 50–150 cm. **Leaves:** basal caducous; cauline opposite or alternate; faces hirsute, hispid, or scabrous. **Paleae** puberulent. **Ray florets** 8–13. **Disc florets** 35–120. $2n = 14$.

Flowering summer. Meadows, open forests, along roadsides; 0–1000 m; Ala., Fla., Ga., N.C., S.C., Tenn.

Varieties *dentatum* and *angustatum* are variable in the character of stem vestiture, have overlapping geographic ranges, and therefore are best treated as one variety.

12e. Silphium asteriscus Linnaeus var. **asteriscus**   E

Silphium asperrimum Hooker; *S. asteriscus* var. *scabrum* Nuttall; *S. dentatum* Elliott var. *gatesii* (C. Mohr) H. E. Ahles; *S. gatesii* C. Mohr; *S. integrifolium* Michaux var. *gattingeri* L. M. Perry; *S. laevigatum* Pursh; *S. scaberrimum* Elliott

Plants 40–150 cm. **Leaves:** basal caducous; cauline opposite or alternate; faces hirsute, hispid, or scabrous. **Paleae** puberulent to scabrous. **Ray florets** 8–20. **Disc florets** 35–110. $2n = 14$.

Flowering summer–early fall. Prairies, meadows, open forests, roadsides; 50–1000 m; Ala., Ark., Fla., Ga., Ky., La., Miss., Mo., N.C., Okla., S.C. Tenn., Tex., Va.

The name *Silphium asperrimum* has been misapplied to *S. radula.* Based upon examination of a photograph of the type and of the original description, *S. asperrimum* should be synonymized with *S. asteriscus.*

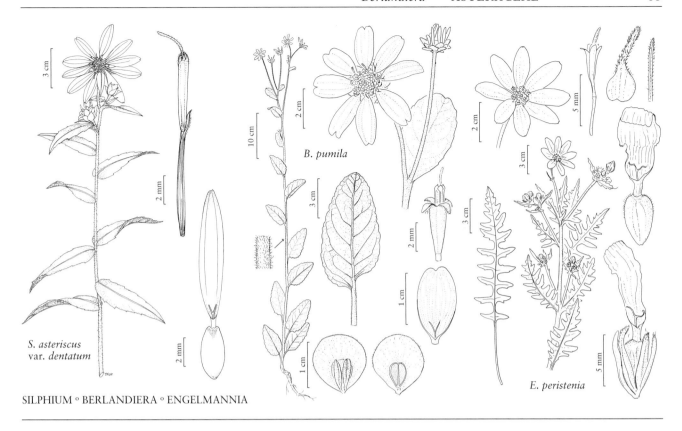

B. pumila

S. asteriscus var. *dentatum*

E. peristenia

SILPHIUM ∘ BERLANDIERA ∘ ENGELMANNIA

270. BERLANDIERA   de Candolle in A. P. de Candolle and A. L. P. P. de Candolle, Prodr. 5: 517. 1836 • [For Jean Louis Berlandier, 1805–1851, Belgian explorer in North America]

Donald J. Pinkava

Perennials or subshrubs, 8–120 cm (stems often arising annually from taproots or woody caudices; herbage usually hirsute, hispid, scabrous, or velvety). **Stems** decumbent to erect, usually branched (scapiform or ± leafy throughout). **Leaves** basal and/or cauline; alternate; petiolate or sessile; blades (± pinnately nerved) elongate-deltate, lanceolate, lyrate, oblanceolate, oblong, obovate, ovate, or spatulate, sometimes pinnately lobed to pinnatifid (usually each with deeper sinuses proximal), bases cordate or truncate to cuneate, ultimate margins usually crenate, dentate, or sinuate, faces hirsute, hispid, scabrous, or velvety. **Heads** radiate, borne singly or in paniculiform to corymbiform arrays. **Involucres** ± campanulate to ± hemispheric or broader, 12–30 mm diam. (broader in age). **Phyllaries** persistent (outer) or falling, usually 14–22 in 2–3+ series (inner broadly obovate to orbiculate, hairy, shed with cypselae). **Receptacles** ± turbinate (distally dilated, apices flattened to slightly depressed), paleate (paleae ± linear-conduplicate, each wrapped round a disc floret, distally dilated, hirtellous). **Ray florets** (2–)8(–13), pistillate, fertile; corollas pale yellow to orange-yellow (abaxially green or red to maroon or with 9–12, green or red to maroon, anastomosing, veins). **Disc florets** 80–200+, functionally staminate; corollas yellow or red to maroon, tubes shorter than throats, lobes 5, ± deltate. **Cypselae** (black) obcompressed or obflattened, mostly obovate (thin edged, not truly winged, hairy, at least adaxially; each adhering to and shed with 2 adjacent paleae, 2 disc florets, and a subtending phyllary); **pappi** 0, or persistent, ± coroniform (inconspicuous ridges or minute crowns of aristate teeth). $x = 15$.

Species 8 (8, including 3 hybrids, in the flora): United States, Mexico.

All *Berlandiera* species have $2n = 30$ and all but perhaps *B. monocephala* (which was unavailable for study) are intercrossable, resulting in all possible artificial interspecific hybrid combinations. Some combinations are not found in nature, e.g., *B. lyrata* × *B. subacaulis*, because distributions of potential parentals do not overlap (D. J. Pinkava 1964, 1967).

SELECTED REFERENCES Nesom, G. L. and B. L. Turner. 1998. Variation in the *Berlandiera pumila* (Asteraceae) complex. Sida 18: 493–502. Pinkava, D. J. 1964. Biosystematic Study of Genus *Berlandiera* DC. (Compositae). Ph.D. dissertation. Ohio State University. Pinkava, D. J. 1967. Biosystematic study of *Berlandiera* (Compositae). Brittonia 19: 285–298.

1. Leaf blades lyrate (pinnate to pinnatifid) or oblanceolate (seldom pinnate, ultimate margins mostly crenate), faces velvety; peduncles hairy (some hairs reddish, bulbous-based, wartlike); ray corollas: abaxial veins red to maroon; disc corollas usually red to maroon, rarely yellow.
    2. Leaves usually lyrate-pinnatifid, rarely not lobed; peduncles hairy (relatively many red-dish wartlike hairs surpassing appressed, white hairs) . . . . 1. *Berlandiera lyrata*
    2. Leaves mostly oblanceolate, usually some pinnatifid near bases; peduncles hairy (rela-tively few reddish wartlike hairs enmeshed in matted whitish hairs; plants intergrades between *B. lyrata* and *B. monocephala*) . . . . 2. *Berlandiera* ×*macrophylla*
1. Leaf blades deltate, lanceolate, lyrate, oblong, oval, or ovate, sometimes pinnately lobed to pinnatifid, faces hirsute, hispid, ± scabrous, or velvety (leaves not both oblanceolate or lyrate-pinnatifid and velvety); peduncles hairy (velvety to scabrous, and with wartlike hairs in *B. subacaulis*); ray corollas: abaxial veins green; disc corollas yellow or red to maroon.
    3. Leaves crowded near bases of stems; heads borne singly or 2–3 together; disc corollas yellow.
        4. Leaf blades oblanceolate (usually not lobed, sometimes lobed at bases, ultimate margins crenate to remotely dentate), faces velvety; peduncles hairy (hairs whitish, matted, none wartlike); Arizona (Sonora) . . . . 3. *Berlandiera monocephala*
        4. Leaf blades mostly oblong (usually sinuate-pinnatifid), faces ± scabrous; peduncles hairy (hispid and with wartlike hairs); Florida . . . . 4. *Berlandiera subacaulis*
    3. Leaves crowded near bases of stems or ± evenly distributed on stems; heads borne singly or 2–20 in paniculiform to corymbiform arrays; disc corollas red to maroon.
        5. Leaves crowded near bases of stems (usually some leaves lobed basally, faces finely hirsute); peduncles hairy (hairs relatively long and fine mixed with shorter, stouter hairs; plants intergrade between *B. pumila* and *B. subacaulis*) . . . . 5. *Berlandiera* ×*humilis*
        5. Leaves evenly distributed along stems (not lobed, faces sometimes finely hirsute); peduncles hairy (hairs not mixed long and short).
            6. Leaf blades usually ovate, faces velvety; peduncles hairy (hairs whitish, rela-tively long and fine, matted); South Carolina to Mississippi, e Texas . . . . 6. *Berlandiera pumila*
            6. Leaf blades elongate-deltate, lanceolate, or ovate, faces finely ± hirsute to ± scabrous; peduncles hairy (hirsute or hairs grayish to reddish, erect or spread-ing, sometimes curling).
                7. Stems lax (sometimes suffrutescent); leaf blades (at least mid-stem) peti-olate, narrowly to broadly ovate (widths 1/2–1 1/4 times lengths), membra-nous, margins, usually crenate, faces finely to coarsely hirsute; peduncles hairy (hairs grayish to reddish, erect or spreading, sometimes curling); intergrades to *B. pumila* and *B. texana*) . . . . 7. *Berlandiera* ×*betonicifolia*
                7. Stems (stiff) erect; leaf blades (at least mid-stem) petiolate to sessile, elon-gate-deltate to lanceolate (widths to 2/3 lengths), chartaceous, margins serrate to dentate, faces hirsute to scabrous; peduncles hairy (densely hirsute) . . . . 8. *Berlandiera texana*

1. **Berlandiera lyrata** Bentham, Pl. Hartw., 17. 1839
   • Chocolate flower

Berlandiera incisa Torrey & A. Gray

Plants 10–60(–120) cm. **Stems** (erect to decumbent) usually branched. **Leaves** evenly distributed along stems; petiolate; blades oblanceolate or obovate to spatulate, often lyrate, sometimes ± pinnatifid (terminal lobes usually shorter than pinnatifid portions, crenate to irregularly incised), membranous to slightly chartaceous, ultimate margins crenate or entire, faces ± velvety. **Heads** in corymbiform arrays. **Peduncles** hairy (some hairs reddish, bulbous-based, wartlike, surpassing white, appressed hairs). **Involucres** 13–17 mm diam. **Ray corollas** deep yellow to orange-yellow, abaxial veins (sometimes whole surfaces) red to maroon, laminae 10–14 × 5.5–8 mm. **Disc corollas** red to maroon (rarely yellow). **Cypselae** 4.5–6 × 2.7–3.7 mm. *2n* = 30.

Flowering nearly year round. Dry, sandy loams, rocky, limestone soils, roadsides, grasslands with mesquite, oak, and juniper; 700–2200 m; Ariz., Colo., Kans., N.Mex., Okla., Tex.; Mexico.

Berlandiera lyrata is cultivated in Arizona. Exceptional specimens that are scapiform (sometimes monocephalic) with mostly undivided leaves *and* with wartlike hairs on peduncles occur at higher elevations (south-central New Mexico, trans-Pecos Texas, and Nuevo León). They have yellow disc corollas, as do most collections from Chihuahua, Durango, Nuevo León, and Tamaulipas.

2. **Berlandiera ×macrophylla** (A. Gray) M. E. Jones, Contr. W. Bot. 12: 48. 1908 (as species)

Berlandiera lyrata Bentham var. *macrophylla* A. Gray in A. Gray et al., Syn. Fl. N. Amer. 1(2): 243. 1884

Plants 15–120 cm. **Stems** (1–8) unbranched or branched near bases. **Leaves** crowded near stem bases; petiolate; blades oblanceolate, sometimes pinnatifid basally (terminal lobes longer than pinnatifid portions), membranous, ultimate margins usually crenate, faces velvety. **Heads** usually borne singly, sometimes 2–3 together. **Peduncles** hairy (hairs reddish, wartlike, enmeshed in whitish, matted hairs). **Involucres** 15–30 cm diam. **Ray corollas** yellow, abaxial veins green, laminae 10–13 × 3–5 mm. **Disc corollas** yellow. **Cypselae** (obovate) 5–6 × 3.5–4 mm. *2n* = 30.

Flowering late spring–fall. Calcareous outcrops and derived soils; 600–2100 m; Ariz.; Mexico (Sonora).

Berlandiera ×macrophylla is a hybrid of *B. lyrata* and *B. monocephala*, with which it occurs.

3. **Berlandiera monocephala** (B. L. Turner) Pinkava, J. Arizona-Nevada Acad. Sci. 36: 8. 2003

Berlandiera lyrata A. Gray var. *monocephala* B. L. Turner, Phytologia 64: 205, fig. 1. 1988

Plants 10–100 cm (taproots much thickened). **Stems** unbranched or branched near bases. **Leaves** crowded near stem bases; petiolate; blades oblanceolate, rarely pinnatifid, membranous, margins usually crenate, faces velvety. **Heads** usually borne singly, rarely 2–3 together. **Peduncles** hairy (hairs whitish, matted). **Involucres** 15–35 mm diam. **Ray corollas** yellow, abaxial veins green, laminae 9–12 × 4.5–6 mm. **Disc corollas** yellow. **Cypselae** (obovate) 5–6 × 3–4 mm. *2n* = 30.

Flowering May–Oct. Grasslands on rolling hills, often with oak, pinyon, and juniper; 1300–2100 m; Ariz.; Mexico (Chihuahua, Sonora).

Roots of *Berlandiera monocephala* are sold in local Mexican drugstores for stomach and lung troubles (B. L. Turner 1988d) and for insect or snake bites.

4. **Berlandiera subacaulis** (Nuttall) Nuttall, Trans. Amer. Philos. Soc., n. s. 7: 343. 1840 (as subacaule) • Green eyes, Florida-dandelion ☒

Silphium subacaule Nuttall, Amer. J. Sci. Arts 5: 301. 1821; *S. nuttallianum* Torrey

Plants 20–50 cm. **Stems** unbranched or branched near bases. **Leaves** crowded near stem bases; petiolate or sessile; blades mostly oblanceolate to spatulate, usually alternately sinuate-pinnatifid, chartaceous, ultimate margins ± crenate, faces ± scabrous. **Heads** usually borne singly. **Peduncles** hairy (hispid and with wartlike hairs). **Involucres** 16–19+ mm diam. **Ray corollas** lemon yellow, abaxial veins green, laminae 10–12.6 × 5–7.8 mm. **Disc corollas** red to maroon. **Cypselae** (oval) 5.2–7 × 3–3.8 mm. *2n* = 30.

Flowering year round. Dry, sandy oak and pine flats, lawns, along roadsides; 10–90 m; Fla.

5. Berlandiera ×humilis Small, Fl. S.E. U.S., 1246, 1340. 1903 (as species) [E]

Plants 10–30 cm. **Stems** (erect) usually branched. **Leaves** crowded near stem bases; petiolate; blades usually lyrate, sometimes pinnatifid, slightly chartaceous, ultimate margins entire, faces sparsely hirsute (at least adaxial). **Heads** in corymbiform arrays. **Peduncles** hairy (hairs relatively fine and usually mixed with shorter, stouter hairs). **Involucres** 16–22 mm diam. **Ray corollas** yellow to orange-yellow, abaxial veins green, laminae 12–17 × 5.5–7.8 mm. **Disc corollas** red to maroon. **Cypselae** (obovate) 5–6 × 3–4.5 mm. **2*n*** = 30.

Flowering Mar–Jun(–Dec). Sandy pine flats, openings; 20–100 m; Fla.

Berlandiera ×humilis is a hybrid of *B. pumila* and *B. subacaulis*. Specimens from Alabama, Georgia, and South Carolina have some leaf-blade bases with 1+ lobes, evidence of introgression with *B. pumila*.

6. Berlandiera pumila (Michaux) Nuttall, Trans. Amer. Philos. Soc., n. s. 7: 342. 1840 [E] [F]

Silphium pumilum Michaux, Fl. Bor.-Amer. 2: 146. 1803; *Berlandiera dealbata* (Torrey & A. Gray) Small; *B. pumila* var. *dealbata* (Torrey & A. Gray) Trelease; *B. tomentosa* (Pursh) Nuttall; *B. tomentosa* var. *dealbata* Torrey & A. Gray; *Polymnia caroliniana* Poiret; *S. tomentosum* Pursh

Plants to 100 cm. **Stems** (erect) usually branched (sometimes suffrutescent). **Leaves** evenly distributed along stems; usually petiolate (at least at mid stem); blades oblong, oval, or ovate (widths ³⁄₄–1¹⁄₄ lengths, not lobed, except intergrades to *B. ×humilis*, bases truncate to subcordate), membranous, margins crenate, faces grayish velvety. **Heads** in paniculiform to corymbiform arrays. **Peduncles** hairy (hairs relatively fine, densely matted). **Involucres** 14–18 mm diam. **Ray corollas** deep yellow to orange-yellow, abaxial veins green, laminae 12–20 × 6–9 mm. **Disc corollas** red to maroon. **Cypselae** (obovate) 4.5–6 × 3–4 mm. **2*n*** = 30.

Flowering Apr–Aug. Sandy soils, roadsides, fields, woodland borders, open woodlots; 30–100 m; Ala., Fla., Ga., S.C., Tex.

7. Berlandiera ×betonicifolia (Hooker) Small, Fl. S.E. U.S., 1246, 1340. 1903 (as species) [E]

Silphium betonicifolium Hooker, Compan. Bot. Mag. 1: 99. 1835; *Berlandiera pumila* (Michaux) Nuttall var. *scabrella* G. L. Nesom & B. L. Turner; *B. texana* de Candolle var. *betonicifolia* (Hooker) Torrey & A. Gray

Plants to 110 cm. **Stems** (erect, flexible, sometimes suffrutescent) usually branched. **Leaves** evenly distributed along stems; petiolate; blades (at least at mid stem) narrowly to broadly ovate (widths ¹⁄₂–1¹⁄₄ lengths), membranous, margins serrate to crenate or doubly crenate, faces finely hirsute to hispid (at least adaxial). **Heads** in paniculiform to corymbiform arrays. **Peduncles** hairy (hairs spreading to erect, relatively fine, curled). **Involucres** 12–18 mm diam. **Ray corollas** deep yellow to orange-yellow, abaxial veins green, laminae 9–18 × 5.4–9 mm. **Disc corollas** red to maroon. **Cypselae** (obovate) 2.5–4 × 4–5.5 mm. **2*n*** = 30.

Flowering Apr–Sep. Sandy-loamy soils, edges of or in woodlands of oak, pine, and/or hickory; 50–200 m; La., Tex.

In most characters, *Berlandiera ×betonicifolia* is intermediate between its parents (*B. pumila* and *B. texana*) in varying degrees and combinations. Drummond's type (K) of *Silphium betonicifolium* resembles *B. texana*, to which it was assigned as a variety by J. Torrey and A. Gray (1838–1843), who commented that other specimens fitted neither varietal description completely. Artificially produced F₁ offspring between extreme forms of the parents resemble closely Drummond's type collection. Character states of *S. betonicifolium* that Hooker described in contradistinction to those of *B. texana* include longer petioles, less woody stems, and more ovate, hirsute leaves of membranous texture; and from *B. pumila* by the coarser and more deeply crenate margins, and peduncles with "beautiful jointed purplish hairs." Artificially produced crosses between F₁ hybrids and *B. pumila* resemble most of the field-collected *B. ×betonicifolia* specimens [called *B. pumila* (green form) by G. L. Nesom and B. L. Turner (1998)] because gene flow of *B. ×betonicifolia* is over a much greater geographic region of overlap with *B. pumila* than with *B. texana*.

8. Berlandiera texana de Candolle in A. P. de Candolle and A. L. P. P. de Candolle, Prodr. 5: 517. 1836 [E]

Berlandiera longifolia Nuttall

Plants to 120 cm. **Stems** (erect, usually suffrutescent) much branched distally. **Leaves** evenly distributed along stems or distally crowded; sessile or petiolate (at least mid stem); blades elongate-deltate to lanceolate (widths to ²/₃ lengths), chartaceous, margins dentate, serrate, or doubly serrate, faces hirsute to ± scabrous. **Heads** in paniculiform or corymbiform arrays. **Peduncles** densely hirsute. **Involucres** 18–27 mm diam. **Ray corollas** deep yellow to orange-yellow, abaxial veins green, laminae 10.5–17 mm. **Disc corollas** red to maroon. **Cypselae** (ovate) 4.5–6 × 3–4.8 mm. $2n = 30$.

Flowering Mar–Nov. Dry, rocky, calcareous and sandy soils, open woodlands, glades, thickets; 100–300 m; Ark., Kans., Mo., Okla., Tex.

---

271. ENGELMANNIA A. Gray ex Nuttall, Trans. Amer. Philos. Soc., n. s. 7: 343. 1840
    • [For George Engelmann, 1809–1884, German-American physician and botanist]

David J. Keil

Perennials, 20–50(–100) cm (taproots or caudices becoming woody). **Stems** erect (coarsely strigose, hispid, or hirsute), usually branched (at least distally, sometimes branched from bases, aerial stems multiple). **Leaves** basal and cauline; alternate; petiolate (basal and proximal cauline) or sessile (distal); blades (pinnately nerved) mostly oblong to lanceolate, usually 1(–2)-pinnately lobed, bases ± cuneate, ultimate margins entire, faces coarsely strigose, hispid, or hirsute. **Heads** radiate, in open, corymbiform arrays. **Involucres** hemispheric, 6–10 mm diam. **Phyllaries** persistent (outer) or falling (inner, with cypselae), mostly 18–24+ in ± 3 series (outer with relatively short, expanded, indurate bases and longer, linear, herbaceous tips, inner broadly ovate, mostly indurate, scarious-margined, herbaceous tips relatively broader and shorter, tending to split along midveins in age). **Receptacles** flat, paleate (paleae linear to narrowly oblong, hirsute-ciliate at tips). **Ray florets** 8–9 (each subtended by an inner phyllary), pistillate, fertile; corollas yellow (laminae oblong-elliptic, entire or minutely 2–3-toothed). **Disc florets** 25–50, functionally staminate; corollas yellow, tubes shorter than campanulate throats, lobes 5, ± deltate (anthers black, appendages deltate, obtuse; styles not branched). **Cypselae** obcompressed or obflattened, obovate (each falling with subtending phyllary, 2–4, indurate paleae, plus sterile ovaries of 2–4 disc florets, margins ± ciliate, faces strigose to pilose); **pappi** persistent or tardily falling, of 2–4, ciliate scales. $x = 9$.

Species 1: sw United States, n Mexico.

1. Engelmannia peristenia (Rafinesque) Goodman & C. A. Lawson, Rhodora 94: 381. 1992 • Engelmann daisy, cutleaf daisy [F]

Silphium peristenium Rafinesque, Atlantic J. 1: 146. 1832; *Engelmannia pinnatifida* Nuttall

Basal leaves (5–)10–30 cm. **Cauline leaves** gradually smaller distally, primary lobes ovate to narrowly linear, their margins sometimes again lobed or toothed. **Involucres** 6–10 mm. **Phyllary** tips hirsute-ciliate. **Ray corollas** 10–16 mm. **Disc corollas** 3.5–5 mm. **Cypselae** 3–4 mm; **pappi** 1–2 mm. $2n = 18$.

Flowering Apr–Jul. Grasslands, pinyon-juniper woodlands, desert scrub, roadsides; 100–2000 m; Ariz., Ark., Colo., Kans., Mo., Nebr., N.Mex., Okla., S.Dak., Tex.; Mexico (Chihuahua, Coahuila, Nuevo León, Tamaulipas).

272. ECHINACEA Moench, Methodus, 591. 1794 · Purple coneflower [Latin, *echinus*, sea urchin, alluding to spiny tips of receptacular paleae] E

Lowell E. Urbatsch

Kurt M. Neubig

Patricia B. Cox

Perennials, to 140 cm (caudices vertical or horizontal; plants usually taprooted, fibrous rooted in *E. purpurea*). **Stems** erect, unbranched or ± branched (glabrous or hairy, hairs appressed, ascending, or spreading, uniseriate). **Leaves** basal and cauline; alternate; petiolate (at least basal and proximal cauline, petioles progressively shorter distally); blades (1-, 3-, or 5-nerved) linear to lanceolate or elliptic to ovate (distal smaller), bases mostly attenuate (decurrent on petioles) to cuneate, sometimes rounded or cordate, margins usually entire, sometimes dentate or serrate, faces usually hairy (hairs uniseriate, usually with 1–4 rings of cells surrounding bases), sometimes glabrate or glabrous. **Heads** radiate, borne singly (on relatively long peduncles). **Involucres** crateriform to hemispheric, 12–40 mm diam. **Phyllaries** persistent, 15–50 in 2–4 series (spreading, recurved, or reflexed, linear or lanceolate to ovate, subequal to unequal, mostly herbaceous, apices mostly attenuate, abaxial faces usually hairy, sometimes glabrate or glabrous). **Receptacles** hemispheric to conic, paleate (paleae orange to reddish purple distally, surpassing disc corollas, bases partially surrounding cypselae, bodies keeled, apices abruptly constricted to awnlike tips; discs 10–45 × 15–40 mm). **Ray florets** 8–21, neuter; corollas dark purple to pale pink, white, or yellow (tubes glabrous or sparsely hairy, laminae spreading, reflexed, or drooping, linear to elliptic or obovate, abaxial faces glabrous or moderately hairy). **Disc florets** 200–300+, bisexual, fertile; corollas pinkish, greenish, reddish purple, or yellow, tubes shorter than throats (often sparsely hairy), lobes 5 (erect or spreading to recurved), triangular (pollen usually yellow, usually white in *E. pallida*). **Cypselae** (tan or bicolored with dark brown band distally) 3- or 4-angled (faces smooth to finely tuberculate, glabrous or sparsely hairy); **pappi** persistent, ± coroniform (with 0–4 ± prominent teeth). *x* = 11.

Species 9 (9 in the flora): e, c North America.

In keys and descriptions, "discs" refers to receptacles at late flowering with paleae and florets included in assessing shapes and in measurements of lengths and diameters.

SELECTED REFERENCES Baskin, J. M., K. M. Snyder, and C. C. Baskin. 1993. Nomenclatural history and taxonomic status of *Echinacea angustifolia*, *E. pallida*, and *E. tennesseensis* (Asteraceae). Sida 15: 597–604. Binns, S. E., B. R. Baum, and J. T. Arnason. 2002. A taxonomic revision of *Echinacea* (Asteraceae: Heliantheae). Syst. Bot. 27: 610–632. Foster, S. 1991. *Echinacea*: The Purple Coneflowers. Austin. Schulthess, B. H., G. E. Giger, and T. W. Baumann. 1991. *Echinacea*—Anatomy, phytochemical pattern, and germination of the achene. Pl. Med. (Stuttgart) 57: 384–388.

1. Basal leaves: blades (1–)5–12 cm wide, bases rounded to cordate (plants fibrous rooted)
   .... 6. *Echinacea purpurea*
1. Basal leaves: blades 0.5–4(–6.5) cm wide, bases usually cuneate to attenuate, sometimes rounded (*E. laevigata*) (plants taprooted).
   2. Herbage (stems and leaves) glabrate (and glaucous); basal leaf blades 3–6.5 cm wide, margins usually serrate or dentate .... 3. *Echinacea laevigata*
   2. Herbage usually sparsely to densely hairy, rarely glabrous; basal leaf blades 0.5–4 cm wide, margins usually entire.
      3. Hairs (stems, leaves, and peduncles) appressed to ascending.
         4. Rays yellow; s, c Missouri, n Arkansas (Ozark region) .... 5a. *Echinacea paradoxa* (in part)
         4. Rays pale pink, dark pink, purple, or white; Kansas, Oklahoma, and Texas.

5. Ray laminae 19–35 mm; ray cypselae glabrous distally; e Kansas, Oklahoma, and Texas .... 2. *Echinacea atrorubens*
5. Ray laminae 30–70 mm; ray cypselae hairy distally (at least on angles); s, c Oklahoma (Arbuckle Mountain region) .... 5b. *Echinacea paradoxa* (in part)
[3. Shifted to left margin.—Ed.]
3. Hairs (stems, leaves, and peduncles) spreading.
  6. Ray laminae 15–40 mm.
    7. Discs 20–35 mm diam. (herbage induments relatively harsh); widespread, plains, w North America .... 1. *Echinacea angustifolia*
    7. Discs 15–25 mm diam. (herbage induments relatively soft); c Tennessee ... 9. *Echinacea tennesseensis*
  6. Ray laminae 40–90 mm.
    8. Discs subspheric, 15–30 × 20–30 mm; paleae 8–11 mm, tips purple, rounded (sw Arkansas, w Louisiana, se Oklahoma, e Texas) .... 7. *Echinacea sanguinea*
    8. Discs conic to hemispheric, 20–40 × 20–37 mm; paleae 9–14 mm, tips pinkish to purple, sharp-pointed.
      9. Ray corollas pink to reddish purple, 3–4 mm wide; pollen usually white, sometimes lemon yellow; cypselae glabrous .... 4. *Echinacea pallida*
      9. Ray corollas rose to pink or white, 4–7 mm wide; pollen yellow; cypselae (rays) hairy (n Arkansas, se Missouri, eastward) .... 8. *Echinacea simulata*

1. Echinacea angustifolia de Candolle in A. P. de Candolle and A. L. P. P. de Candolle, Prodr. 5: 554. 1836 • Narrow-leaved purple coneflower, blacksamson echinacea E

Echinacea angustifolia var. *strigosa* McGregor

Plants to 70 cm (roots usually fusiform, ± branched). **Herbage** moderately to densely hairy (induments relatively harsh, hairs spreading, ca. 1–2 mm). **Stems** mostly green to purplish. **Basal leaves:** petioles 2–12 cm; blades (1-), 3-, or (5-)nerved, elliptic to lanceolate, 7–30 × 0.5–2.5(–4) cm, bases cuneate to attenuate, margins entire (usually ciliate). **Peduncles** 10–30 cm. **Phyllaries** lanceolate to ovate, 6–12 × 1–2.5 mm. **Receptacles:** paleae 9–14 mm, tips purple, straight, sharp-pointed. **Ray corollas** pink to purplish, laminae reflexed, 15–40 × 5–8 mm, sparsely hairy abaxially. **Discs** conic to hemispheric, 15–30 × 20–35 mm. **Disc corollas** 5–7+ mm, lobes usually purple. **Cypselae** often bicolored, tan proximally, dark brown banded distally, 4–5 mm, faces ± smooth, usually glabrous; **pappi** to ca. 1 mm (major teeth 0–4). **2*n*** = 22, 44.

Flowering late spring–mid summer, dry prairies, barrens, rocky to sandy-clay soils; 100–1600 m; Man., Sask.; Colo., Iowa, Kans., Minn., Mo., Mont., Nebr., N.Mex., N.Dak., Okla., S.Dak., Tex., Wyo.

R. L. McGregor (1967, 1968) recognized *Echinacea angustifolia* var. *strigosa* as a complex of diploid and tetraploid populations ranging geographically from southeastern Kansas and central Oklahoma to north-central Texas. He noted that var. *strigosa* is distinguishable by its shorter stature, stems frequently branched, somewhat flexuous, distally strigose to strigose-hirsute that retain, in part, a green color upon drying. A hybrid origin for var. *strigosa* was suggested based on the morphologic intermediacy of natural populations compared to synthesized hybrids between typical *E. angustifolia* and *E. atrorubens* (McGregor 1968). Populations along the southern boundary of the range of var. *strigosa* are tetraploid. In their morphometric analyses, S. E. Binns et al. (2002) did not recognize var. *strigosa*; they found it indistinguishable from typical *E. angustifolia*.

2. Echinacea atrorubens (Nuttall) Nuttall, Trans. Amer. Philos. Soc., n. s. 7: 354. 1840 • Topeka purple coneflower E

Rudbeckia atrorubens Nuttall, J. Acad. Nat. Sci. Philadelphia 7: 80. 1834

Plants to 90 cm (roots elongate-turbinate, ± branched). **Herbage** usually hairy (hairs appressed to ascending, spreading on adaxial leaf faces, to 1.2 mm), rarely glabrous. **Stems** light green to tan. **Basal leaves:** petioles 0–12(–20) cm; blades (1-), 3-, or 5-nerved, usually linear or lanceolate, rarely ovate, 5–30 × 0.5–3 cm, bases attenuate, margins usually entire. **Peduncles** 20–50 cm. **Phyllaries** linear to lanceolate, 6–15 × 1–3 mm. **Receptacles:** paleae 9–15 mm, tips red to orange-tipped, usually straight, sharp-pointed. **Ray corollas** purple, rarely pink or white, laminae reflexed, 19–35 × 2–7 mm, glabrous or sparsely hairy abaxially. **Discs** ovoid to conic, 25–35 × 20–40 mm. **Disc corollas** 4.5–5.5 mm, lobes greenish to pink or purple. **Cypselae** tan, 4–5 mm, faces finely tuberculate, glabrous; **pappi** to 1.2 mm (major teeth 3–4). **2*n*** = 11.

Flowering mostly late spring. Dry, limestone or sand-stone outcrops, prairies; 50–500 m; Kans., Okla., Tex.

3. **Echinacea laevigata** (C. L. Boynton & Beadle) S. F. Blake, J. Wash. Acad. Sci. 19: 273. 1929 • Smooth purple coneflower [C][E]

Brauneria laevigata C. L. Boynton & Beadle in J. K. Small, Fl. S.E. U.S., 1261, 1340. 1903; *Echinacea purpurea* (Linnaeus) Moench var. *laevigata* (C. L. Boynton & Beadle) Cronquist

Plants 35–110 cm (roots fusiform, ± branched). **Herbage** (glaucous) mostly glabrous (leaves sometimes sparsely hairy abaxially). **Stems** green. **Basal leaves:** petioles 4–26 cm; blades 3- or 5-nerved, elliptic to lanceolate-ovate, 10–50 × 3–6.5 cm, bases broadly cuneate to rounded, margins usually serrate or dentate. **Peduncles** 10–40 cm. **Phyllaries** lanceolate, 3–15 × 2–3(–5) mm. **Receptacles:** paleae 7–12 mm, tips orange to brownish purple-tipped, often incurved, sharp-pointed. **Ray corollas** pink to purple, laminae spreading to reflexed, 35–80 × 3–7 mm, sparsely hairy abaxially. **Discs** conic to spheric, 15–30 × 15–40 mm. **Disc corollas** 8–9 mm, lobes purple to greenish (usually erect). **Cypselae** tan, disc cypselae tan, banded, 4–5 mm, usually glabrous (ray cypselae sometimes hairy on angles); **pappi** to 1.2 mm (teeth unequal). $2n = 22$.

Flowering late spring–early summer. Well-drained soils, open wooded hillsides, fields; of conservation concern; 100–1100 m; Ga., N.C., Pa., S.C., Va.

Echinacea laevigata was historically present in Pennsylvania but is now thought to be extirpated; its occurrence in Maryland has not been confirmed; reports of its presence in Alabama and Arkansas are most likely based on misidentifications. It is in the Center for Plant Conservation's National Collection of Endangered Plants.

4. **Echinacea pallida** (Nuttall) Nuttall, Trans. Amer. Philos. Soc., n. s. 7: 354. 1840 • Pale purple coneflower [E][F]

Rudbeckia pallida Nuttall, J. Acad. Nat. Sci. Philadelphia 7: 77. 1834

Plants to 140 cm (roots fusiform to narrowly turbinate, usually branched). **Herbage** sparsely to densely hairy (hairs spreading, ca. 1.5–1.7 mm). **Stems** green to purplish (rarely branched). **Basal leaves:** petioles 5–20+ cm; blades (1-), 3-, or 5-nerved, elliptic to lanceolate, 12–40 × 1–4 cm, bases cuneate to attenuate, margins entire (usually ciliate). **Peduncles** 15–50 cm. **Phyllaries** lanceolate to

ovate, 7–15 × 1–3 mm. **Receptacles:** paleae 9–14 mm, tips purple, usually incurved, sharp-pointed. **Ray corollas** pink to reddish purple, laminae reflexed, 40–90 × 3–4 mm, sparsely hairy abaxially. **Discs** conic to hemispheric, 20–40 × 25–37 mm. **Disc corollas** 5.5–6.7 mm, lobes usually pink to purple (pollen usually white, rarely lemon yellow). **Cypselae** tan or bicolored, 2.5–5 mm, faces ± smooth, usually glabrous; **pappi** to ca. 1 mm (major teeth 0–4). $2n = 22$.

Flowering late spring–summer. Rocky prairies, open wooded hillsides, and glades; 50–1500 m; Ont.; Ala., Ark., Conn., Ill., Ind., Iowa, Kans., La., Maine, Mass., Mich., Mo., Nebr., N.Y., N.C., Okla., Tenn., Tex., Va., Wis.

Echinacea pallida is generally regarded as introduced in Connecticut, Georgia, Maine, Massachusetts, New York, North Carolina, and Virginia.

5. **Echinacea paradoxa** (Norton) Britton in N. L. Britton and A. Brown, Ill. Fl. N. U.S. ed. 2, 3: 476. 1913 [C][E]

Brauneria paradoxa Norton, Trans. Acad. Sci. St. Louis 12: 40, plate 8. 1902

Plants to 90 cm (roots fusiform to elongate-turbinate, branched). **Herbage** sparsely to densely hairy (hairs appressed to ascending). **Stems** usually yellowish green (usually not branched). **Basal leaves:** petioles 0–15 cm; blades (3-) or 5-nerved, usually linear or lanceolate, rarely ovate, 5–35 × 0.5–2(–2.5) cm, bases usually attenuate, margins entire (usually ciliate). **Peduncles** (2–)10–30 cm. **Phyllaries** lanceolate to ovate, 7–12 × 1–4 mm. **Receptacles:** paleae 10–14 mm, tips red to orange, often incurved, sharp-pointed. **Ray corollas** yellow (var. *paradoxa*) or pinkish to white (var. *neglecta*), laminae reflexed, 30–70 × 3–8 mm, sparsely hairy abaxially. **Discs** ovoid to conic, 2–3.5 × 2–3.5 cm. **Disc corollas** 4.5–6.2 mm, lobes pinkish to yellowish. **Cypselae** tan or bicolored (with distal dark brown band), 4–5.5 mm, faces ± tuberculate, usually glabrous (angles of ray cypselae hairy distally in var. *neglecta*); **pappi** to ca. 1.2 mm (major teeth 0–4).

Varieties 2 (2 in the flora): sc United States.

1. Ray corollas yellow; n Arkansas, s Missouri . . . . 5a. *Echinacea paradoxa* var. *paradoxa*
1. Ray corollas light purple, pink, or white; s Oklahoma, n Texas . . . . 5b. *Echinacea paradoxa* var. *neglecta*

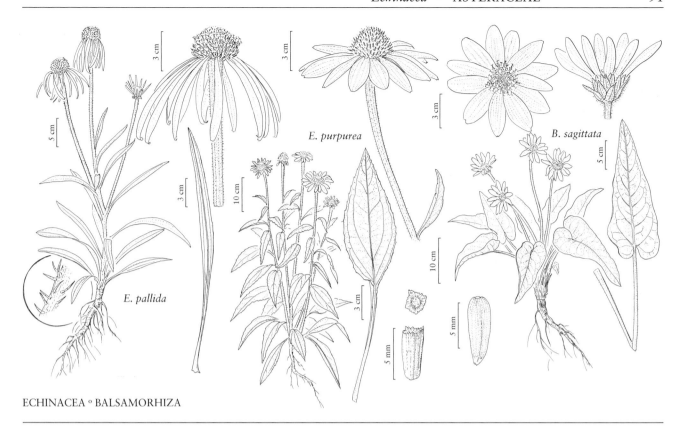

ECHINACEA ° BALSAMORHIZA

5a. Echinacea paradoxa (Norton) Britton var. **paradoxa**
• Yellow coneflower [C] [E]

Ray corollas yellow. **Cypselae** (all) glabrous. $2n$ = 22.

Flowering late spring–early summer. Rocky prairies, glades, bald knobs; of conservation concern; 50–400 m; Ark., Mo.

5b. Echinacea paradoxa (Norton) Britton var. **neglecta** McGregor, Trans. Kansas Acad. Sci. 70: 370. 1968
• Bush's purple coneflower [C] [E]

Echinacea atrorubens (Nuttall) Nuttall var. *neglecta* (McGregor) Binns, B. R. Baum & Arnason

Ray corollas light purple, pink, or white. **Cypselae** (rays) ± hairy (at least on distal margins); disc cypselae glabrous. $2n$ = 22.

Flowering late spring. Rocky prairies, open, wooded hillsides; of conservation concern; 100–400 m; Okla., Tex.

6. Echinacea purpurea (Linnaeus) Moench, Methodus, 591. 1794 • Eastern purple coneflower [E]

Rudbeckia purpurea Linnaeus, Sp. Pl. 2: 907. 1753

Plants 50–120 cm (roots fibrous). **Herbage** usually hairy (hairs spreading to ascending, to 2 mm), sometimes glabrous. **Stems** usually brownish green. **Basal leaves:** petioles 0–17(–25) cm; blades 3- or 5-nerved, ovate to narrowly lanceolate, 5–30 × (1–)5–12 cm, bases usually rounded to cordate, margins usually serrate to dentate, rarely entire. **Peduncles** 8–25 cm. **Phyllaries** linear to lanceolate, 8–17 × 1–8 mm. **Receptacles:** paleae 9–15 mm, tips red-orange, straight or slightly curved, sharp-pointed. **Ray corollas** pink to purple, laminae spreading to recurved, 30–80 × 7–19 mm, sparsely hairy abaxially. **Discs** conic, 14–45 × 20–40 mm. **Disc corollas** 4.5–5.7 mm, lobes greenish to pink or purple. **Cypselae** off-white, 3.5–5 mm, usually glabrous (ray cypselae sometimes hairy on angles); **pappi** ca. 1.2 mm (teeth equal). $2n$ = 22.

Flowering late spring–summer. Rocky, open woods, thickets, prairies, especially near waterways; 10–400+ m; Ont.; Ala., Ark., Fla., Ga., Ill., Ind., Iowa, Kans., Ky., La., Miss., Mo., N.C., Ohio, Okla., Tenn., Tex., Wis.

Echinacea purpurea is introduced in Ontario. It and cultivars derived from it are extensively grown ornamentals in gardens, wildflower roadside plantings, and prairie restoration sites. Because of its popularity as an herbal remedy, it is also grown commercially. As a result of such activities, naturalized and persisting populations may extend the natural range of *E. purpurea*. Selections used for such plantings may differ from native forms.

7. **Echinacea sanguinea** Nuttall, Trans. Amer. Philos. Soc., n. s. 7: 354. 1840 • Sanguine purple coneflower Ⓔ

Echinacea pallida (Nuttall) Nuttall var. *sanguinea* (Nuttall) Gandhi & R. D. Thomas

Plants to 120 cm (roots fusiform to elongate-turbinate, branched). **Herbage** sparsely to densely hairy (hairs spreading, to 1.5 mm) or glabrate. **Stems** mostly green to purplish. **Basal leaves:** petioles 4–12 cm; blades (1-) or 3-nerved, elliptic to lanceolate, 8–30 × 1–3 cm, bases attenuate, margins entire (usually ciliate). **Peduncles** 20–50+ cm. **Phyllaries** lanceolate to ovate, 7–12 × 1–4 mm. **Receptacles:** paleae 8–11 mm, tips purple, slightly curved, usually rounded. **Ray corollas** pink to reddish purple, laminae reflexed, 40–70 × 3–4 mm, glabrous or sparsely hairy abaxially. **Discs** subspheric, 15–30 × 20–30 mm. **Disc corollas** 5.5–6.7 mm, lobes usually purple. **Cypselae** tan to bicolored (with distal dark brown band), 2.5–5 mm, faces ± tuberculate, glabrous; **pappi** to ca. 1 mm (major teeth 0–4). $2n = 22$.

Flowering late spring–early summer. Sandy, open, pine woods, prairies; 0–200 m; Ark., La., Okla., Tex.

8. **Echinacea simulata** McGregor, Sida 3: 282. 1968 • Wavy-leaf purple coneflower Ⓔ

Echinacea speciosa McGregor, Trans. Kansas Acad. Sci. 70: 366. 1967, not (Wenderoth) Paxton 1849; *E. pallida* (Nuttall) Nuttall var. *simulata* (McGregor) Binns, B. R. Baum & Arnason

Plants to 100 cm (roots fusiform, branched). **Herbage** sparsely to densely hairy (hairs spreading). **Stems** mostly green to purplish. **Basal leaves:** petioles 4–20 cm; blades (1-), 3-, or 5-nerved, linear to lanceolate, 5–40 × 0.5–4 cm, bases attenuate, margins entire (usually ciliate). **Peduncles** 20–40+ cm. **Phyllaries** lanceolate to ovate, 7–15 × 1.5–3.5 mm. **Receptacles:** paleae 10–14 mm, tips pinkish to purple, incurved, sharp-pointed. **Ray corollas** rose to pink or white, laminae drooping to reflexed, 40–90 × 4–7 mm, glabrous or sparsely hairy abaxially. **Discs** conic to hemispheric, 20–30 × 20–30 mm. **Disc corollas** 5–6.5 mm, lobes pink to purplish. **Cypselae** tan, 3–4.5 mm, faces smooth, usually glabrous, sometimes (rays) hairy; **pappi** to ca. 1 mm (usually without major teeth). $2n = 22$.

Flowering late spring–summer. Rocky, open, wooded hillsides, prairies; 100–400 m; Ark., Ga., Ill., Ky., Mo., Tenn.

Echinacea simulata has been reported as introduced in Illinois (http://www.natureserve.org).

9. **Echinacea tennesseensis** (Beadle) Small, Man. S.E. Fl., 1421, 1509. 1933 • Tennessee purple coneflower Ⓒ Ⓔ

Brauneria tennesseensis Beadle, Bot. Gaz. 25: 359. 1898; *Echinacea pallida* (Nuttall) Nuttall var. *tennesseensis* (Beadle) Binns, B. R. Baum & Arnason

Plants to 50 cm (roots elongate-turbinate, branched). **Herbage** sparsely to densely hairy (indument relatively soft, hairs spreading, to 2+ mm). **Stems** yellowish green becoming tan. **Basal leaves:** petioles 2–10 cm; blades 1- or 3-nerved, linear to lanceolate, 6–12 × 0.7–1.5 cm, bases attenuate, margins entire (usually ciliate). **Peduncles** 8–25+ cm. **Phyllaries** lanceolate to ovate, 5–10 × 1.5–2.5 mm. **Receptacles:** paleae 9–12 mm, tips purple, 2–3 mm, often incurved, rounded to acute. **Ray corollas** pink to purplish, laminae spreading to reflexed, 20–40 × 3–4 mm, moderately hairy abaxially. **Discs** conic, 10–25 × 15–25 mm. **Disc corollas** 5.5–6.5 mm, lobes usually purple. **Cypselae** tan, 4–5 mm, faces smooth, glabrous; **pappi** to ca. 1.2 mm (major teeth 0–4). $2n = 22$.

Flowering in summer. Dry, rocky hills, barrens; of conservation concern; 100–200 m; Tenn.

Echinacea tennesseensis is in the Center for Plant Conservation's National Collection of Endangered Plants.

SELECTED REFERENCE Drew, M. B. and E. C. Clebsch. 1995. Studies on the endangered *Echinacea tennesseensis* (Asteraceae): Plant community and demographic analysis. Castanea 60: 60–69.

273. BALSAMORHIZA Hooker ex Nuttall, Trans. Amer. Philos. Soc., n. s. 7: 349. 1840

• Balsamroot [Greek *balsamon*, a fragrant gum, and *rhiza*, root; alluding to resiniferous rootstocks] E

William A. Weber

Perennials, 10–45(–100) cm (taproots slender or massive, thick- or thin-barked; caudices unbranched or multibranched). **Stems** erect, branched mostly from bases. **Leaves** mostly basal; opposite or alternate; petiolate (bases persisting as fibrils); blades (mostly pinnately nerved, sometimes 3- or 5-nerved) either rounded-deltate to triangular-deltate with bases sagittate or cordate to truncate and margins entire or crenate (*B.* subg. *Artorhiza*), or blades mostly elliptic, ovate, or lanceolate to lance-ovate or oblong and often 1–2-pinnatifid or -pinnately lobed with bases mostly truncate to cuneate and (if not lobed) margins usually crenate, dentate, or serrate, seldom entire (*B.* subg. *Balsamorhiza*), faces usually hirsute, hispid, pilose, puberulent, scabrous, sericeous, strigose, tomentose, or velutinous and gland-dotted or stipitate-glandular, seldom glabrous. **Heads** radiate, usually borne singly, rarely (2–3+) in ± corymbiform to racemiform arrays (peduncles ± scapiform, usually bearing 2+ leaves or bracts proximally or at mid length). **Involucres** mostly campanulate or turbinate to hemispheric, 11–30+ mm diam. **Phyllaries** persistent, 8–20+ in 2–3+ series (subequal to unequal, outer equaling or surpassing inner). **Receptacles** flat to convex, paleate (paleae persistent, conduplicate, at least at bases, chartaceous). **Ray florets** 5–21+, pistillate, fertile; corollas usually yellow to orange, rarely becoming brick red (*B. rosea*). **Disc florets** (15–)50–150+, bisexual, fertile; corollas yellow to orange, tubes much shorter than cylindric throats, lobes 5, ± deltate (style branches stigmatic in 2 barely distinct lines, appendages filiform). **Cypselae** obscurely prismatic, weakly 3–4-angled (faces usually glabrous, strigose in some *B. careyana* and in *B. rosea*); **pappi** 0. $x = 19$.

Species 12 (12 in the flora): w North America.

Balsamorhiza ×*bonseri* H. St. John refers to a hybrid derivative involving *B. sagittata* and *B. rosea*. The plants have the habit of *B. sagittata* and the reddish ray corollas of *B. rosea*. The cypselae are hairy.

Balsamorhiza ×*terebinthacea* (Hooker) Nuttall and *B. macrophylla* var. *terebinthacea* (Hooker) A. Nelson refer to hybrids derived from *B. hookeri* × *B. deltoidea*.

In the key and descriptions here, "leaves" refers to basal leaves and "leaf blades" refers to blades of basal leaves, unless otherwise indicated.

SELECTED REFERENCES Moore, A. J. and L. Bohs. 2003. An ITS phylogeny of *Balsamorhiza* and *Wyethia* (Asteraceae: Heliantheae). Amer. J. Bot. 90: 1653–1660. Ownbey, M. and W. A. Weber. 1943. Natural hybridization in the genus *Balsamorhiza*. Amer. J. Bot. 30: 179–187. Weber, W. A. 1953. *Balsamorhiza terebinthacea* and other hybrid balsam-roots. Madroño 12: 47–49.

1. Leaves (petioles mostly longer than blades): blades rounded-deltate or deltate to triangular-deltate, bases sagittate or cordate to truncate, margins usually entire, sometimes crenate to dentate (subg. *Artorhiza*).
    2. Heads usually borne singly, sometimes 2–3+; leaves ± silvery to white, faces (at least the abaxial) sericeous, tomentose, tomentulose, or velutinous . . . . 1. *Balsamorhiza sagittata*
    2. Heads 2–3+ or borne singly; leaves green, faces glabrous or finely hispidulous to hirtellous.
        3. Heads usually (2–)3+, sometimes borne singly; leaf margins usually entire, sometimes crenate (to dentate near bases), faces usually finely hispidulous to hirtellous (cypselae strigose or glabrous) . . . . 2. *Balsamorhiza careyana*
        3. Heads usually borne singly, sometimes 2+; leaf margins usually crenate to dentate (at least near bases), sometimes entire, faces usually sparsely hirtellous to hispidulous, sometimes glabrous (usually gland-dotted, sometimes vernicose; cypselae glabrous) . . . . 3. *Balsamorhiza deltoidea*

1. Leaves (petioles mostly shorter than blades): blades mostly lanceolate, lance-elliptic, lance-ovate, linear-oblong, oblong, or ovate, often 1–2-pinnatifid or -pinnately lobed, bases usually truncate to cuneate, sometimes cordate, ultimate margins usually crenate, dentate, or serrate, seldom entire (subg. *Balsamorhiza*).
    4. Leaf blades sometimes pinnately lobed (ultimate margins usually crenate-serrate, dentate, or serrate).
        5. Leaf blades (earliest sometimes, later seldom, pinnately lobed): margins usually dentate to serrate, faces hirsutulous to scabrous (ray corollas yellow, not becoming red or chartaceous; cypselae glabrous) . . . . 4. *Balsamorhiza serrata*
        5. Leaf blades (rarely pinnately lobed): margins crenate-serrate, faces finely strigose to moderately scabrous (ray corollas yellow at anthesis, becoming red, drying pink and chartaceous; cypselae strigose) . . . . 5. *Balsamorhiza rosea*
    4. Leaf blades usually 1–2-pinnatifid.
        6. Leaves grayish, silvery, or white, faces lanate-tomentose, sericeous, tomentose, or villous.
            7. Leaves: faces densely sericeous; n California, s Oregon . . . . 6. *Balsamorhiza sericea*
            7. Leaves: faces lanate-tomentose, tomentose, or villous; n California, Oregon, Washington.
                8. Leaf blades ovate to lanceolate, 10–45 × 3–7(–10) cm (1-pinnatifid, lobes ovate to lanceolate, 20–50 × 5–25 mm), margins plane; Oregon, Washington . . . . 7. *Balsamorhiza incana*
                8. Leaf blades lanceolate to linear-oblong, 10–20 × 3–6(–8) cm (1–2-pinnatifid, primary lobes lance-linear to oblong, 5–40 × 1–10 mm, margins revolute); n California . . . . 8. *Balsamorhiza lanata*
        6. Leaves bright green to gray-green, faces glabrous or hirsute, hirtellous, hispid, hispidulous, pilose, piloso-hirtellous, scabrous, sericeous, strigillose, strigose, subvelutinous, or tomentose (sometimes gland-dotted as well).
            9. Leaf blades 30–60 cm, 1-pinnatifid or nearly so; outer phyllaries usually much surpassing inner.
              10. Leaves: faces strigillose to subvelutinous or tomentose (margins not ciliate); ray laminae 20–30+ mm . . . . 9. *Balsamorhiza macrolepis*
              10. Leaves: faces scabrous or piloso-hirtellous to pilose (at least adaxial, margins ciliate); ray laminae 35–50+ mm . . . . 10. *Balsamorhiza macrophylla*
            9. Leaf blades 6–40 cm, usually 1–2-pinnatifid; outer phyllaries seldom surpassing inner.
              11. Leaf blades bright green, lance-elliptic to lanceolate, (6–)15–25(–40) × (3–)5–9+ cm (primary lobes lanceolate to oblanceolate, 10–45 × 2–15 mm), faces hispidulous to hirtellous . . . . 11. *Balsamorhiza hispidula*
              11.Leaf blades usually gray-green, narrowly to broadly lanceolate or ovate, (8–)20–30(–40) × 2–15 cm (primary lobes oblong or lanceolate to linear, 5–100 × 0.5–15 mm), faces hirsute, sericeous, or strigose . . . . 12. *Balsamorhiza hookeri*

273a. BALSAMORHIZA Hooker ex Nuttall subg. ARTORHIZA (Nuttall) W. M. Sharp, Ann. Missouri Bot. Gard. 22: 115. 1935 [E]

Balsamorhiza [unranked] *Artorhiza* Nuttall, Trans. Amer. Philos. Soc., n. s. 7: 350. 1840; *Balsamorhiza* sect. *Artorhiza* (Nuttall) Jepson

Taproots relatively massive; bark relatively thick, corrugated; rhizomes none; caudices usually multi-branched. **Petioles** mostly longer than leaf blades. **Basal leaves:** blades mostly rounded-deltate or deltate to triangular-deltate, bases sagittate or cordate to truncate, margins usually entire, sometimes crenate. **Cauline leaves** usually at mid stems. **Ray corollas** yellow.

Species 3 (3 in the flora): w North America.

Members of subg. *Artorhiza* inhabit relatively deep, well-developed soils, flower only after at least 4–5 years, and attain ages of possibly 50 years.

1. Balsamorhiza sagittata (Pursh) Nuttall, Trans. Amer. Philos. Soc., n. s. 7: 350. 1840 E F

Buphthalmum sagittatum Pursh, Fl. Amer. Sept. 2: 564. 1813; *Balsamorhiza helianthoides* (Nuttall) Nuttall; *Espeletia helianthoides* Nuttall; *E. sagittata* (Pursh) Nuttall

Plants (15–)20–40(–65) cm. **Basal leaves:** blades ± silvery to white or gray-green, rounded-deltate or deltate to triangular-deltate, 5–25 × 3–15 cm, bases ± cordate, margins entire, apices acute to attenuate, faces sericeous, tomentose, tomentulose, or velutinous (at least abaxially, usually gland-dotted as well), sometimes glabrescent. **Heads** usually borne singly, sometimes 2–3+. **Involucres** hemispheric to turbinate, 12–25 mm diam. **Outer phyllaries** lanceolate to oblanceolate or linear, (15–)20–25(–30+) mm, equaling or surpassing inner, apices acute to acuminate. **Ray laminae** 20–40 mm. **2n** = 38.

Flowering (Apr–)May–Jun(–Jul). Openings, banks, flats, meadows, ridges, sagebrush scrub, conifer forests; (100–)900–2500(–3000) m; Alta., B.C.; Ariz., Calif., Colo., Idaho, Mont., Nev., Oreg., S.Dak., Utah, Wash., Wyo.

Balsamorhiza sagittata grows east of the Cascade-Sierra axis to the Rocky Mountains and Black Hills. It is one of the more spectacular of all spring-flowering plants in the northwestern United States. Hybrids occur along lines of contact between *B. sagittata* and almost all species of sect. *Balsamorhiza* except *B. macrophylla* (a high polyploid).

2. Balsamorhiza careyana A. Gray, Mem. Amer. Acad. Arts, n. s. 4: 81. 1849 (as Balsamorrhiza) E

Balsamorhiza careyana var. *intermedia* Cronquist

Plants (15–)20–60 cm. **Basal leaves:** blades green, rounded-deltate or deltate to triangular-deltate, 15–25 × 6–15 cm, bases cordate or hastate to truncate, margins usually entire, sometimes crenate (to dentate near bases), apices acute to attenuate, faces finely hispidulous to hirtellous (gland-dotted as well). **Heads** usually (2–)3+, sometimes borne singly. **Involucres** hemispheric to turbinate or campanulate, 12–20 mm diam. **Outer phyllaries** oblong to lanceolate or linear, 15–25 mm, usually surpassing inner, apices acute to attenuate. **Ray laminae** 20–30(–40) mm (cypselae strigose or glabrous). **2n** = 38.

Flowering (Mar–)May–Jun(–Jul). Dry scablands, semi-desert soils, openings in pine forests; 500–1000 m; Oreg., Wash.

Hybrids involving *Balsamorhiza careyana* and *B. deltoidea* occur near the Columbia River Gorge; intermediates are found to the east and south. In northern Oregon, plants in some populations have glabrous cypselae and some populations are mixed. The hairiness may come from *B. rosea*; *B. rosea* and *B. careyana* may hybridize profusely, producing mostly plants with the stature of *B. careyana* and with the relatively short, brick-red ray corollas of *B. rosea*. Hybridization also occurs, occasionally, between *B. careyana* and *B. sagittata*; *B. careyana* also hybridizes with any species of sect. *Balsamorhiza* with which it comes in contact. Plants called *Balsamorhiza careyana* var. *intermedia* usually have crenate leaf margins and glabrous cypselae.

3. Balsamorhiza deltoidea Nuttall, Trans. Amer. Philos. Soc., n. s. 7: 351. 1840 E

Balsamorhiza glabrescens Bentham

Plants 20–90 cm. **Basal leaves:** blades green, deltate or rounded-deltate to triangular-deltate, 10–25+ × 7–15(–20) cm, bases cordate to sagittate, margins usually crenate to dentate (at least near bases), sometimes entire, apices acute, faces usually sparsely hirtellous to hispidulous, sometimes glabrous (usually gland-dotted, sometimes vernicose). **Heads** usually borne singly, sometimes 2+. **Involucres** campanulate to hemispheric 15–30+ mm diam. **Outer phyllaries** oblong to lanceolate or lance-linear, (12–)30–40(–60) mm, usually much surpassing inner, apices acute. **Ray laminae** 20–50 mm. **2n** = 38.

Flowering (Mar–)Apr–Jun(–Jul). Open hillsides, openings in chaparral or forests; 60–1800(–2400) m; B.C.; Calif., Oreg., Wash.

Balsamorhiza deltoidea varies in the crenation of the leaf margins. The variability may stem from an old hybridization with *B. hookeri* or other species of *Balsamorhiza*. Otherwise, *B. deltoidea* shows little significant variability except where it hybridizes with *B. careyana* in the narrows of the Columbia River Gorge.

273b. BALSAMORHIZA Nuttall subg. BALSAMORHIZA E

Taproots relatively slender, bark relatively thin, not corrugated; rhizomes sometimes present (relatively short, deep-seated); caudices seldom branched. **Petioles** mostly shorter than leaf blades. **Basal leaves:** blades mostly lanceolate, lance-elliptic, lance-ovate, linear-oblong, oblong, or ovate, often 1–2-pinnatifid or -pinnately lobed, sometimes cordate, ultimate margins usually crenate, dentate, or serrate, seldom entire. **Cauline leaves** mostly proximal to mid stems. **Ray corollas** mostly yellow (becoming brick red and chartaceous in *B. rosea*).

Species 9 (9 in the flora): w North America.

Members of subg. *Balsamorhiza* usually inhabit shallow, rocky soils.

4. Balsamorhiza serrata A. Nelson & J. F. Macbride, Bot. Gaz. 56: 479. 1913 E

Plants 10–20(–30) cm. **Basal leaves:** blades pale green to gray-green, lanceolate or lance-ovate, 5–15(–20+) × 2–5(–10) cm (earliest sometimes, later seldom, pinnately lobed, mostly toward tips), bases usually cuneate to subtruncate, sometimes cordate, margins usually dentate to serrate, apices acute, faces hirsutulous to scabrous. **Heads** usually borne singly. **Involucres** campanulate, 15–25 mm diam. **Outer phyllaries** lanceolate to lance-triangular, 10–22 mm, shorter than inner, apices acute to attenuate (margins ciliate). **Ray laminae** (20–)30–40. $2n = 38$.

Flowering Apr–Jun. Basaltic scablands, sagebrush scrub, openings in forests, meadow borders; (1000–)1400–1500 m; Calif., Nev., Oreg., Wash.

Balsamorhiza serrata hybridizes with *B. careyana*. Leaves of the hybrids tend to be prostrate rather than erect. Most of the hybrids display the larger size of *B. careyana* and lack the pubescence of the heads; *serrata*-like plants never develop more than a single head nor do they have the pubescence of the cypselae.

5. Balsamorhiza rosea A. Nelson & J. F. Macbride, Bot. Gaz. 56: 478. 1913 E F

Balsamorhiza hookeri Nuttall var. *rosea* (A. Nelson & J. F. Macbride) W. M. Sharp

Plants 6–10(–30) cm. **Basal leaves:** blades gray-green, oblong to lance-ovate, 3–10(–20) × 2–5 cm (rarely pinnately lobed), bases weakly cordate or truncate, margins crenate to serrate, apices rounded to acute, faces finely strigose to moderately scabrous (usually gland-dotted as well). **Heads** usually borne singly. **Involucres** hemispheric, 18–20 mm diam. **Outer phyllaries** deltate or ovate to lanceolate, 8–12 mm, not surpassing inner. **Ray laminae** (becoming brick-red, often drying to pink or rose, and chartaceous) (8–)15(–25)

mm (hispidulous abaxially; cypselae strigose). $2n = 38$.

Flowering Apr–May. Dry hills; 300–400 m; Oreg., Wash.

In a hybrid swarm involving *Balsamorhiza rosea* and *B. careyana*, *B. rosea* remains relatively uncontaminated; the dominance among the hybrids appears to lie with *B. careyana*. A record of a hybrid between *B. rosea* and *B. careyana* from the Spokane area is doubtful.

6. Balsamorhiza sericea W. A. Weber, Phytologia 50: 358. 1982 E

Plants 10–30(–40) cm. **Basal leaves:** blades silvery, lanceolate or lance-ovate, 9–30 × 2–7 cm (1-pinnatifid, lobes oblong to oblanceolate, 5–35 × 3–17 mm), bases cuneate, ultimate margins usually entire (plane or weakly revolute, obscurely, if at all, ciliate), apices rounded to acute, faces densely sericeous. **Heads** borne singly. **Involucres** campanulate to hemispheric, 11–20+ mm diam. **Outer phyllaries** broadly ovate to triangular-ovate, 12–20 mm, slightly surpassing inner, apices acuminate to attenuate (margins not ciliate). **Ray laminae** 15–20 mm (adaxially puberulous on veins).

Flowering Apr–May. Serpentine outcrops, among surface rocks, in crevices, hillsides, dry streamsides among cobbles; 400–1800 m; Calif., Oreg.

Balsamorhiza sericea hybridizes with *B. deltoidea*.

7. Balsamorhiza incana Nuttall, Trans. Amer. Philos. Soc., n. s. 7: 350. 1840 E

Plants 20–35(–70) cm. **Basal leaves:** blades white to grayish, lance-ovate to oblong, 10–45 × 3–7(–10) cm (1-pinnatifid, lobes ovate to lanceolate, 20–50 × 5–25 mm), bases broadly cuneate to truncate, margins entire or serrate (plane, not distinctly ciliate), apices obtuse, faces usually densely, sometimes sparsely, tomentose. **Heads** usually borne

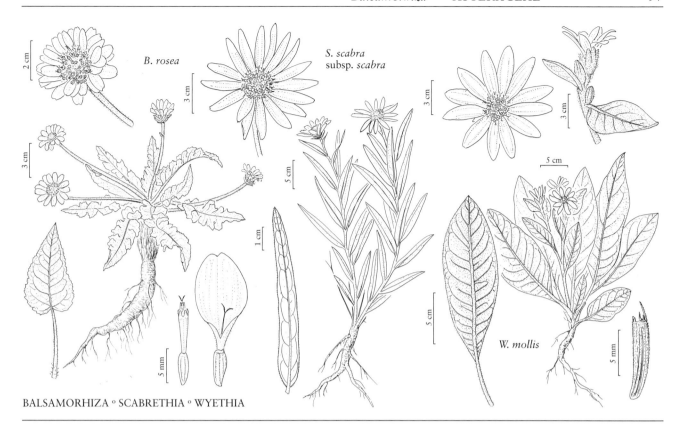

BALSAMORHIZA ○ SCABRETHIA ○ WYETHIA

singly. **Involucres** hemispheric, 20–25(–30) mm. **Outer phyllaries** ovate to lanceolate or linear, 15–20 mm, shorter than inner, apices acute to attenuate (margins ± ciliate). **Ray laminae** 20–40(–50+) mm. $2n$ = 38.

Flowering May–Jun(–Jul). Grassy meadows, dry, rocky sites, openings in conifer forests; 1200–2800 m; Idaho, Mont., Oreg., Wash., Wyo.

Balsamorhiza incana hybridizes with *B. sagittata*.

8. **Balsamorhiza lanata** (W. M. Sharp) W. A. Weber, Phytologia 85: 20. 1999 C E

Balsamorhiza hookeri Nuttall var. *lanata* W. M. Sharp, Ann. Missouri Bot. Gard. 22: 130. 1935

Plants 10–20(–30) cm. **Basal leaves:** blades white to grayish, lanceolate to linear-oblong, 10–20 × 3–6(–8) cm (1–2-pinnatifid, primary lobes lance-linear to oblong, mostly 5–40 × 1–10 mm, secondary lobes or teeth antrorse, divergent), bases cuneate to truncate, ultimate margins mostly entire (± revolute), apices rounded to acute, faces densely lanate-tomentose to villous. **Heads** borne singly. **Involucres** ± hemispheric, 12–20 mm diam. **Outer phyllaries** lanceolate to linear, 10–20 mm, sometimes surpassing inner, apices acute to attenuate. **Ray laminae** (10–)15–20 mm (abaxially puberulent or glabrous).

Flowering Apr–May(–Jun). Roadsides, grassy slopes; of conservation concern; 700–1500 m; Calif., Oreg.

Balsamorhiza lanata is known from a relatively restricted area in northern California and southern Oregon. No hybrids between it and other species have been noted.

9. **Balsamorhiza macrolepis** W. M. Sharp, Ann. Missouri Bot. Gard. 22: 132. 1935 C E

Plants 20–40(–60+) cm. **Basal leaves:** blades green to gray-green, elliptic to lanceolate, 15–40+ × 4–14 cm (1–2-pinnatifid, primary lobes ovate to linear, 15–70 × 3–15 mm, usually lobed or toothed), bases cuneate to truncate, ultimate margins usually entire (slightly revolute, not ciliate), apices rounded to acute, faces strigillose to subvelutinous or tomentose (sometimes gland-dotted as well). **Heads** usually borne singly. **Involucres** hemispheric, (15–)20–30 mm diam. **Outer phyllaries** ovate or oblong to lanceolate, 12–30 (–40) mm, usually surpassing inner, apices obtuse to acute or attenuate. **Ray laminae** 20–30+ mm.

Flowering Apr–Jun. Open, dry or moist, grassy or rocky slopes, valleys; of conservation concern; 90–1400 m; Calif.

Balsamorhiza macrolepis grows in the western foothills of central Sierra Nevada and in the eastern San Francisco Bay area (there mostly extirpated). The tall habit, exhibiting gigas characteristics, suggests that, like

B. macrophylla, this taxon may be a polyploid. No hybrids with other species have been noted.

10. Balsamorhiza macrophylla Nuttall, Trans. Amer. Philos. Soc., n. s. 7: 350. 1840 [E]

Balsamorhiza hookeri Nuttall var. *idahoensis* (W. M. Sharp) Cronquist; *B. macrophylla* var. *idahoensis* W. M. Sharp

Plants 30–45(–100) cm. **Basal leaves** green, ovate to lanceolate, 20–50+ × 8–15 cm (pinnatifid, lobes lanceolate, 20–80+ × 10–40 mm, entire or ± dentate), bases ± cuneate, ultimate margins usually entire (plane or weakly revolute, ciliate), apices obtuse to acute, faces scabrous or piloso-hirtellous to pilose (at least abaxial usually gland-dotted as well). **Heads** usually borne singly. **Involucres** ± hemispheric, 20–30 mm diam. **Outer phyllaries** lance-ovate or lanceolate to lance-linear, 12–30 (–40) mm, equaling or surpassing inner (margins ciliate), apices acute to attenuate. **Ray laminae** 35–50+ mm. $2n = 100 \pm 2$.

Flowering May–Jul. Deep soils, rocky meadows, sagebrush scrublands, conifer forests; 1000–2400 m; Idaho, Utah, Wyo.

Balsamorhiza macrophylla is a high polyploid; it occurs sympatrically with *B. sagittata*. It evidently arose from hybridization between *B. sagittata* and *B. hispidula*. *Balsamorhiza macrophylla* has the multi-branched caudices and massive taproots of the former, and the leaf dissection of the latter. No hybrids with other species are known. Presumably, the high-polyploid chromosome complement precludes interbreeding. Plants of var. *idahoensis* are smaller, are known only from southwestern Idaho and northeastern Utah, and differ from var. *macrophylla* by being pilose, with strongly shaggy-pilose involucres. More study may determine that var. *idahoensis* merits specific rank. The Utah populations are not well understood and deserve attention.

SELECTED REFERENCE Helton, N., D. Wiens, and B. A. Barlow. 1972. High polyploidy and the origin of *Balsamorhiza macrophylla*. Madroño 21: 526–535.

11. Balsamorhiza hispidula W. M. Sharp, Ann. Missouri Bot. Gard. 22: 137, fig. 3, plate 5. 1935 [E]

Balsamorhiza hookeri Nuttall var. *hispidula* (W. M. Sharp) Cronquist

Plants 6–40 cm. **Leaves:** blades bright green, narrowly lance-elliptic to lanceolate, (6–)15–25 (–40) × (3–)5–9+ cm (1–2-pinnatifid, primary lobes lanceolate to oblanceolate, 10–45 × 2–15 mm, secondary lobes usually divergent, antrorse), bases ± cuneate, ultimate margins usually en-

tire (± revolute and/or thickened, often ciliate), apices acute to attenuate, faces ± hispidulous to hirtellous (gland-dotted as well; veins ± scabrous). **Heads** borne singly. **Involucres** campanulate to hemispheric 10–25 mm diam. **Outer phyllaries** lanceolate to linear, 12–20 mm, not surpassing inner, apices acute to attenuate (margins often ciliate). **Ray laminae** 15–40 mm (abaxially glabrous).

Flowering Apr–May. Basalt scablands and desert-steppe areas, juniper and sagebrush scrublands; 1800–2500 m; Ariz., Colo., Idaho, Mont., Nev., Oreg., Utah, Wyo.

Distribution of *Balsamorhiza hispidula* is highly disrupted in the southern part of its range.

12. Balsamorhiza hookeri Nuttall, Trans. Amer. Philos. Soc., n. s. 7: 349. 1840 (as hookerii) [E]

Heliopsis balsamorhiza Hooker, Fl. Bor.-Amer. 1: 310. 1833; *Balsamorhiza hirsuta* Nuttall; *B. hirsuta* var. *lagocephala* W. M. Sharp; *B. hirsuta* var. *neglecta* W. M. Sharp; *B. hookeri* var. *hirsuta* (Nuttall) A. Nelson; *B. hookeri* var. *lagocephala* (W. M. Sharp) Cronquist; *B. hookeri* var. *neglecta* (W. M. Sharp) Cronquist; *B. macrolepis* W. M. Sharp var. *platylepis* (W. M. Sharp) Ferris; *B. platylepis* W. M. Sharp

Plants 10–30 cm. **Basal leaves:** blades usually gray-green, narrowly to broadly lanceolate or ovate, (8–)20–30(–40) × 2–15 cm (1–2-pinnatifid, primary lobes oblong or lanceolate to linear, 5–100 × 0.5–15 mm, secondary lobes usually ± linear, spreading), bases ± truncate to broadly cuneate, ultimate margins usually entire (plane or revolute, ciliate or not), apices obtuse to acute, faces hirsute, sericeous, or strigose (and gland-dotted or finely stipitate-glandular). **Heads** borne singly. **Involucres** campanulate to hemispheric, 15–30 mm diam. **Outer phyllaries** ovate-lanceolate to lanceolate, 10–24+ mm, seldom surpassing inner, apices acuminate to attenuate (margins usually ciliate). **Ray laminae** 15–30(–45) mm. $2n = 38$.

Flowering Apr–May(–Jul). Rocky outcrops, dry meadows, sagebrush scrublands, basalt scablands (north), dry, open forests (south); (300–)1000–1500(–2900) m; Calif., Nev., Oreg., Wash.

At one time or another, most species of subg. *Balsamorhiza* have been synonymized under *B. hookeri*. Nevertheless, a number of taxa are justifiably segregated as species by their morphologic differences and geographic restrictions. One might logically choose either of two taxonomies: recognizing only two species in the entire genus, one representing subg. *Artorhiza* and the other subg. *Balsamorhiza*, or recognizing each slightly differing population as a species. Either course results in an unsatisfactory classification. The present classification is a compromise. A knotty problem persists. A central cluster of populations from eastern Washington to southeastern California display a number of minor and locally discrete morphologies. They tend to be less isolated from each other than are the peripheral populations, although some tend to mimic the latter ones in one or more characteristics. Their evolutionary history may be involved with past hybridizations with each other or with species of subg. *Artorhiza*, gene drift, and polyploidy. At present, it appears impossible to reach a satisfactory classification.

274. SCABRETHIA W. A. Weber, Phytologia 85: 20. 1999 • [Latin *scabra*, rough, and related genus *Wyethia*] E

William A. Weber

Perennials, 20–60 cm (taproots often relatively massive; caudices unbranched or multibranched). **Stems** erect, seldom branched (usually white in age). **Leaves** mostly cauline (proximal scale-like); mostly alternate, sometimes opposite; sessile or nearly so; blades (mid and distal, usually 1 or ± 3-nerved, laterals weaker than midribs) narrowly oblong to linear, bases ± cuneate, margins entire, faces hispid to scabrous, glabrescent (then usually roughened by persistent hair bases). **Heads** radiate, borne singly (peduncles sometimes bracteate). **Involucres** hemispheric to turbinate, (13–)20–30(–40) mm diam. **Phyllaries** persistent, 30–45+ in 3–4+ series (subequal to unequal, outer often surpassing inner). **Receptacles** flat to convex, paleate (paleae conduplicate, papery). **Ray florets** (8–)10–18(–21), pistillate, fertile; corollas yellow. **Disc florets** 40–120+, bisexual, fertile; corollas yellow, tubes much shorter than cylindric throats, lobes 5, ± deltate (style branches stigmatic in 2 barely distinct lines, appendages filiform). **Cypselae** ± prismatic, (3–)4-angled (faces glabrous, smooth or finely honeycombed); **pappi** persistent, coroniform, ± lacerate. *x* = 19.

Species 1: w United States.

1. Scabrethia scabra (Hooker) W. A. Weber, Phytologia 85: 21. 1999 E F

Wyethia scabra Hooker, London J. Bot. 6: 245. 1847

Leaf blades mostly 8–20 cm × 5–20 (–25) mm. **Phyllaries:** bases ± rounded, apices acuminate or attenuate to linear or filiform. **Ray laminae** 15–50 mm. **Cypselae** 7–9 mm; **pappi** 0.01–0.5 mm. *2n* = 38 (subspp. unknown).

Subspecies 3 (3 in the flora): w United States.

The subspecies of *Scabrethia scabra* evidently have evolved along with the dissection of the Colorado Plateau.

1. Outer phyllaries: chartaceous bases ovate to oblong, 2–4(–6) mm, herbaceous tips attenuate to linear or filiform, 15–30+ mm, margins ± hispid (ciliate), faces glabrous or minutely gland-dotted . . . . 1c. *Scabrethia scabra* subsp. *attenuata*
1. Outer phyllaries: chartaceous bases deltate to ovate, 2–5+ mm, herbaceous tips narrowly triangular to attenuate, 10–20 mm, margins (ciliolate or not) and faces minutely hispidulous to hirtellous or ± coarsely hispid.

2. Herbaceous tips of outer phyllaries usually spreading to recurved, margins (not ciliate) and faces minutely hispidulous to hirtellous . . . . 1b. *Scabrethia scabra* subsp. *canescens*
2. Herbaceous tips of outer phyllaries usually ± erect to spreading, margins (± ciliate) and faces ± coarsely hispid . . . 1a. *Scabrethia scabra* subsp. *scabra*

1a. Scabrethia scabra (Hooker) W. A. Weber subsp. **scabra** E F

Outer phyllaries: chartaceous bases deltate to ovate, 2–5+ mm, herbaceous tips usually ± erect to spreading, narrowly triangular to attenuate, 10–20 mm, margins (± ciliate) and faces ± coarsely hispid.

Flowering May–Aug. Desert sands, sandstones, ridges, washes, juniper woodlands; 1100–1600(–2600) m; Colo., Utah, Wyo.

1b. Scabrethia scabra (Hooker) W. A. Weber subsp. **canescens** (W. A. Weber) W. A. Weber, Phytologia 85: 21. 1999 (as cinerea) [E]

Wyethia scabra Hooker var. *canescens* W. A. Weber, Amer. Midl. Naturalist 35: 425. 1946

Outer phyllaries: chartaceous bases deltate to ovate, 2–5+ mm, herbaceous tips usually spreading to recurved, narrowly triangular to attenuate, 10–20 mm, margins and faces minutely hispidulous to hirtellous.

Flowering May–Jul. Desert sandy and rocky sites, dunes, washes; 1000–1700 m; Ariz., Colo., N.Mex., Utah.

1c. Scabrethia scabra (Hooker) W. A. Weber subsp. **attenuata** (W. A. Weber) W. A. Weber, Phytologia 85: 21. 1999 [E]

Wyethia scabra Hooker var. *attenuata* W. A. Weber, Amer. Midl. Naturalist 35: 425. 1946

Outer phyllaries: chartaceous bases ovate to oblong, 2–4(–6) mm, herbaceous tips ± spreading to erect, attenuate to linear or filiform, 15–30+ mm, margins ± hispid (ciliate), faces glabrous or minutely gland-dotted.

Flowering late May–Aug. Desert sands, including dunes, washes, juniper woodlands; 1500–1800 m; Ariz., N.Mex., Utah.

275. **WYETHIA** Nuttall, J. Acad. Nat. Sci. Philadelphia 7: 39, plate 5. 1834 • Mules-ears [For Nathaniel Jarvis Wyeth, 1802–1856, early western American explorer] [E]

William A. Weber

Perennials, 15–60(–100) cm (taproots relatively massive; caudices seldom branched). **Stems** erect (or bending erect from bases), branched mostly from bases. **Leaves** basal and cauline; alternate; usually petiolate (basal), sometimes sessile; blades (usually pinnately nerved, basal larger than cauline) ± deltate, elliptic-ovate, lanceolate, lance-elliptic, lance-linear, or oblong-ovate, bases truncate to cuneate, margins usually entire, rarely dentate to serrate (sometimes ciliate), faces glabrous or hairy (sometimes gland-dotted or finely stipitate-glandular). **Heads** radiate, borne singly or 2–5(–8+) in ± corymbiform to racemiform arrays. **Involucres** hemispheric to campanulate or turbinate, 12–60+ mm diam. **Phyllaries** persistent, 12–36(–48) in 2–3+ series (subequal to unequal, outer sometimes foliaceous, much larger than inner). **Receptacles** flat to convex, paleate (paleae conduplicate, at least bases, papery). **Ray florets** 5–25+, pistillate, fertile; corollas mostly yellow (cream to white in *W. helianthoides*). **Disc florets** 35–150+, bisexual, fertile; corollas yellow, tubes much shorter than cylindric throats; lobes 5, ± deltate to lanceolate (style branches stigmatic in 2 barely distinct lines, appendages ± filiform). **Cypselae** ± prismatic, weakly 3–4-angled (faces glabrous or hairy); **pappi** 0, or persistent, coroniform (usually lacerate) or of 1–4+ ovate to subulate, erose to lacerate (often basally connate) scales. $x = 19$.

Species 8 (8 in the flora): w North America.

Some species formerly included in *Wyethia* are here treated in *Agnorhiza* and *Scabrethia*. *Balsamorhiza* is closely related to *Wyethia*.

SELECTED REFERENCE Weber, W. A. 1946. A taxonomic and cytological study of the genus *Wyethia*, family Compositae, with notes on the related genus *Balsamorhiza*. Amer. Midl. Naturalist 35: 400–452.

1. Involucres 35–60+ mm diam.; outer phyllaries (30–)40–80+ mm (± foliaceous, much surpassing discs).
    2. Leaf blades: faces glabrous or finely stipitate-glandular, sometimes sparsely pilosulous as well (usually shining); cypselae 10–13 mm .... 1. *Wyethia glabra*
    2. Leaf blades: faces densely tomentose to tomentulose (usually gland-dotted as well), glabrescent; cypselae 12–15 mm .... 2. *Wyethia helenioides*

1. Involucres 10–30(–40) mm diam.; outer phyllaries 15–30(–40) mm (not foliaceous, seldom surpassing discs).
    3. Leaf blades: faces tomentose to tomentulose (at least when young, glabrescent, usually gland-dotted) . . . . 3. *Wyethia mollis*
    3. Leaf blades: faces glabrous, glabrate, hirsute, hirsutulous, hirtellous, pilose, pilosulous, scabrellous, or strigillose (not tomentose to tomentulose, often gland-dotted or stipitate-glandular).
        4. Leaf blades: faces glabrous or glabrate (gland-dotted); phyllary margins not ciliate; cypselae glabrous.
            5. Basal leaves: blades lance-elliptic or oblong-lanceolate to lanceolate; rays 8–21(–25), laminae 25–60 mm; cypselae 8–9 mm . . . . 4. *Wyethia amplexicaulis*
            5. Basal leaves: blades narrowly oblong-lanceolate to lance-linear; rays 5–8+, laminae 18–30 mm; cypselae 6–8 mm . . . . 5. *Wyethia longicaulis*
        4. Leaf blades: faces usually hirsute, hirsutulous, hirtellous, pilose, pilosulous, scabrellous, or strigillose (sometimes vernicose, rarely finely stipitate-glandular or gland-dotted), sometimes glabrate; phyllary margins ± ciliate; cypselae strigillose or glabrous
            6. Basal leaves: blades lanceolate to lance-linear; cypselae 7–8 mm, ± strigillose . . . . 6. *Wyethia angustifolia*
            6. Basal leaves: blades elliptic-ovate, lanceolate, or lance-elliptic; cypselae 9–11 mm, strigillose or glabrous.
                7. Phyllaries 36–48; rays 13–25, laminae (cream to white) 25–45 mm; cypselae strigillose distally and on margins . . . . 7. *Wyethia helianthoides*
                7. Phyllaries 16–34; rays 11–12, laminae (yellow) 25–50 mm; cypselae glabrous. . . . . 8. *Wyethia arizonica*

1. Wyethia glabra A. Gray, Proc. Amer. Acad. Arts 6: 543. 1865 E

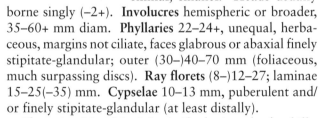

Plants 15–40(–60) cm. **Basal leaves:** blades oblong-lanceolate to elliptic-ovate, 20–30(–40) cm, margins entire or ± serrate-dentate (often undulate), faces glabrous or finely stipitate-glandular, sometimes sparsely pilosulous as well (usually shining); cauline leaves similar, smaller. **Heads** usually borne singly (–2+). **Involucres** hemispheric or broader, 35–60+ mm diam. **Phyllaries** 22–24+, unequal, herbaceous, margins not ciliate, faces glabrous or abaxial finely stipitate-glandular; outer (30–)40–70 mm (foliaceous, much surpassing discs). **Ray florets** (8–)12–27; laminae 15–25(–35) mm. **Cypselae** 10–13 mm, puberulent and/or finely stipitate-glandular (at least distally).

Flowering Mar–May(–Jun). Shady sites, dry foothills; 10–800 m; Calif.

Wyethia glabra grows in the Coast Ranges, often in the fog belt.

2. Wyethia helenioides (de Candolle) Nuttall, Trans. Amer. Philos. Soc., n. s. 7: 353. 1840 E

Alarconia helenioides de Candolle in A. P. de Candolle and A. L. P. P. de Candolle, Prodr. 5: 537. 1836

Plants (15–)25–40(–60) cm. **Basal leaves:** blades (whitish to grayish) oblong-ovate to lance-elliptic, 18–38 cm, margins usually entire, sometimes serrulate or dentate (often undulate), faces densely tomentose to tomentulose (usually gland-dotted as well), glabrescent; cauline leaves (3–6) similar, smaller. **Heads** usually borne singly (–2+). **Involucres** hemispheric or broader, 35–60+ mm diam. **Phyllaries** 16–24, unequal, herbaceous, margins not ciliolate, faces tomentose; outer (30–)40–80+ mm (foliaceous, much surpassing the discs). **Ray florets** 13–21; laminae 20–35(–50) mm. **Cypselae** 12–15 mm, distally strigillose. $2n = 38$.

Flowering (Feb–)Mar–May. Grassy slopes, openings in woodlands; 10–1600(–2000) m; Calif.

Wyethia helenioides grows in the Sierra Nevada foothills and west of the Central Valley, avoiding the fog belt.

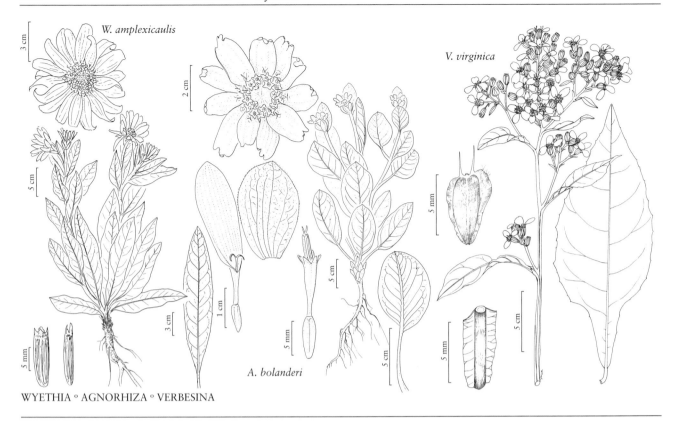

W. *amplexicaulis*

V. *virginica*

A. *bolanderi*

WYETHIA ° AGNORHIZA ° VERBESINA

3. Wyethia mollis A. Gray, Proc. Amer. Acad. Arts 6: 544. 1865 E F

Plants 30–40(–100) cm. **Basal leaves:** blades lanceolate to oblong-ovate (whitish to gray or green), 20–30(–40) cm, margins entire, not ciliate, faces sparsely to densely tomentose to tomentulose (usually gland-dotted as well), glabrescent (nearly bald in age); cauline similar, smaller distally. **Heads** 2–3 in racemiform to corymbiform arrays or borne singly. **Involucres** campanulate to hemispheric, 20–25 (–30) mm diam. **Phyllaries** 12–22, subequal to unequal, herbaceous, margins not ciliate, faces tomentose to tomentulose; outer 20–40 mm (equaling or surpassing inner). **Ray florets** 6–15; laminae 15–45 mm. **Cypselae** 9–10 mm, distally strigillose. $2n = 38$.

Flowering Jun–Jul. Meadows, dry to wet, open sites, openings in conifer forests; 900–2200(–3000) m; Calif., Nev., Oreg.

Wyethia mollis grows in the Sierra Nevada and the Cascade Range.

4. Wyethia amplexicaulis (Nuttall) Nuttall, Trans. Amer. Philos. Soc., n. s. 7: 352. 1840 E F

Espeletia amplexicaulis Nuttall, J. Acad. Nat. Sci. Philadelphia 7: 38. 1834; *Wyethia amplexicaulis* subsp. *major* Piper; *W. amplexicaulis* subsp. *subresinosa* Piper; *W. lanceolata* Howell

Plants 25–50(–100) cm. **Basal leaves:** blades (green, shining) lance-elliptic or oblong-lanceolate to lanceolate, 15–30(–40+) cm, margins usually entire, sometimes denticulate or dentate, not ciliate, faces glabrous (gland-dotted, vernicose); cauline leaves similar, smaller distally (distalmost sessile, usually ± clasping). **Heads** usually 2–8 in ± corymbiform to racemiform arrays, sometimes borne singly. **Involucres** hemispheric to turbinate, 15–30 mm diam. **Phyllaries** 18–36+, subequal or unequal, herbaceous (at least outer), margins not ciliate, faces glabrous; outer 18–32 mm (surpassing inner). **Ray florets** 8–21(–25); laminae 25–60 mm. **Cypselae** 8–9 mm, glabrous. $2n = 38$.

Flowering May–Jul. Wet to dry, open sites, meadows, sagebrush scrub, yellow-pine forests; 400–3000 m; Colo., Idaho, Mont., Nev., Oreg., Utah, Wash., Wyo.

Wyethia amplexicaulis introgressively hybridizes with *W. arizonica* (producing plants called *W.* ×*magna* A. Nelson ex W. A. Weber) across a broad area of central and southern Colorado. Habit of the hybrid is typical of *W. amplexicaulis*; the indument resembles that of *W. arizonica* in varying degrees. Occasionally, *W. amplexicaulis* hybridizes with but does not introgress with *W. helianthoides*.

5. **Wyethia longicaulis** A. Gray, Proc. Amer. Acad. Arts 19: 4. 1883 E

Plants 20–40(–60) cm. **Basal leaves:** blades narrowly oblong-lanceolate to lance-linear, 15–25 (–40) cm, margins entire or denticulate, faces glabrous (usually gland-dotted and ± vernicose, the varnish, in age, becoming whitish and opaque); cauline leaves similar, smaller distally. **Heads** 2–4+ in racemiform to corymbiform arrays or borne singly. **Involucres** broadly hemispheric or campanulate, 10–20 (–30) mm diam. **Phyllaries** 14–18+, unequal to subequal, herbaceous, margins not ciliate, faces glabrous (gland-dotted); outer 18–25+ mm (equaling or surpassing discs). **Ray florets** 5–8+; laminae 18–30 mm. **Cypselae** 6–8 mm, glabrous.

Flowering late May–early Jul. Grassy slopes, openings in forests; 700–1500; Calif.

Wyethia longicaulis is known only from the Coast Ranges of northern California.

6. **Wyethia angustifolia** (de Candolle) Nuttall, Trans. Amer. Philos. Soc., n. s. 7: 352. 1840 E

Alarconia angustifolia de Candolle in A. P. de Candolle and A. L. P. P. de Candolle, Prodr. 5: 537. 1836; *Wyethia angustifolia* var. *foliosa* (Congdon) H. M. Hall

Plants 15–60 cm. **Basal leaves:** blades (dark green) nearly deltate or broadly lanceolate to lance-linear, 15–35 cm, margins entire or irregularly serrate (sometimes undulate), sometimes strigillose, not ciliate, faces hirsute, hirtellous, scabrous, or strigillose (sometimes vernicose); cauline leaves smaller distally (petiolate or sessile). **Heads** usually borne singly, sometimes 2–3+ in corymbiform arrays. **Involucres** ± hemispheric, (15–)20–30+ mm diam. **Phyllaries** 26–40, subequal, ± herbaceous, margins ciliate, faces (abaxial) hispid, hispidulous, or strigillose; outer 15–25 mm (not or little surpassing discs). **Ray florets** (8–)12–14; laminae 15–25 mm. **Cypselae** 7–8 mm, ± strigillose. *2n* = 38.

Flowering May–Jul. Meadows, grassy slopes, chaparral, wet to dry openings in pine and pine-oak forests; 20–2100 m; Calif., Oreg., Wash.

Wyethia angustifolia ranges from southwestern Washington, in and west of the Columbia Gorge, through Oregon into California west of the Cascade-Sierra axis.

7. **Wyethia helianthoides** Nuttall, J. Acad. Nat. Sci. Philadelphia 7: 40, plate 5. 1834 E

Plants 25–40(–80) cm. **Basal leaves:** blades (light green) lance-elliptic to elliptic-ovate, (10–)20–30+ cm, margins usually entire, sometimes denticulate, ± ciliolate, faces sparsely pilose to pilosulous (usually minutely gland-dotted or stipitate-glandular as well); cauline leaves similar, smaller distally (petiolate or sessile). **Heads** borne singly. **Involucres** ± hemispheric, (12–)20–30(–40) mm diam. **Phyllaries** 36–48, subequal, herbaceous, margins villous-ciliate, faces glabrous or glabrate; outer 18–25 mm (not surpassing discs). **Ray florets** 13–25; laminae (cream to white) 25–45 mm. **Cypselae** 9–11 mm, strigillose distally and on margins.

Flowering May–Jul. Meadows, other damp to wet sites, openings in pine forests; 40–2600 m; Idaho, Mont., Nev., Oreg., Wyo.

Hybrids between *Wyethia helianthoides* and *W. amplexicaulis* are infrequent where the species are in contact; no introgression has been noted. The hybrids have been called *Wyethia* ×*cusickii* Piper; they usually bloom later than *W. helianthoides* and earlier than *W. amplexicaulis*.

8. **Wyethia arizonica** A. Gray, Proc. Amer. Acad. Arts 8: 655. 1873 E

Plants 20–30(–100) cm. **Basal leaves:** blades elliptic or lance-elliptic to lanceolate, 12–30 cm, margins entire, not ciliate, faces usually ± hirsutulous or scabrellous, sometimes glabrate; cauline leaves mostly smaller, narrower, distal-most usually petiolate, rarely sessile and clasping. **Heads** borne singly or 2–4+ in ± corymbiform arrays. **Involucres** turbinate to hemispheric, 18–25+ mm diam. **Phyllaries** 16–34, subequal, ± herbaceous, margins ciliate, faces glabrous or strigillose to hispidulous; outer 18–22(–30) mm (not or seldom surpassing discs). **Ray florets** 11–12; laminae (25–)35–50 mm (abaxially hirsutulous). **Cypselae** 9–10 mm, glabrous. *2n* = 38.

Flowering Apr–Jun. Meadows, openings in pine, oak, or spruce-fir forests; 600–2200(–3000) m; Ariz., Colo., N.Mex., Utah.

At the northern margin of the range, glabrate plants with the general habit of *Wyethia arizonica* (introgressants with *W. amplexicaulis*) exhibit greater frost-hardiness (a characteristic of *W. amplexicaulis*) than do other members of the species.

276. AGNORHIZA (Jepson) W. A. Weber, Phytologia 85: 19. 1999 • [Possibly Greek *agnostos*, unknown, and *rhiza*, root, alluding to the initially unknown roots; in protologue of basionym of type species, Greene stated, "Root unknown."]

William A. Weber

Balsamorhiza Nuttall sect. *Agnorhiza* Jepson, Man. Fl. Pl. Calif., 1077. 1925

Perennials, 5–60(–100) cm (taproots relatively massive; caudices sometimes branched). **Stems** erect and branched (mostly distally), or decumbent (and seldom branched distally). **Leaves** basal and cauline (basal usually scalelike on flowering stems, well-formed on non-flowering shoots); alternate; petiolate; blades (3-or 5-nerved or pinnately nerved), cordate-ovate, deltate, deltate-triangular, ovate, ovate-lanceolate, or suborbiculate, bases broadly cuneate, subcordate, or truncate, margins entire or dentate to crenate, faces glabrous or hairy (often gland-dotted). **Heads** radiate or discoid, borne singly (terminal) or in ± racemiform arrays. **Involucres** campanulate to ± hemispheric, 12–40 mm diam. **Phyllaries** persistent, (8–)12–26+ in 2–3+ series (subequal to unequal, outer usually surpassing inner). **Receptacles** flat to convex, paleate (paleae persistent, conduplicate, chartaceous). **Ray florets** 0 or (1–)5–21+, pistillate, fertile; corollas yellow. **Disc florets** 50–150+, bisexual, fertile; corollas yellow, tubes shorter than cylindric throats, lobes 5, lance-deltate (style branches stigmatic in 2, barely distinct lines, appendages filiform). **Cypselae** ± prismatic, 3–4-angled (faces glabrous or strigillose); **pappi** usually persistent, coroniform (± lacerate, sometimes with 1–4 teeth or scales), sometimes 0. $x = 19$.

Species 5 (5 in the flora): California, nw Mexico.

SELECTED REFERENCE Weber, W. A. 1946. A taxonomic and cytological study of the genus *Wyethia*, family Compositae, with notes on the related genus *Balsamorhiza*. Amer. Midl. Naturalist 35: 400–452.

1. Plants (5–)15–30(–45+) cm; stems decumbent, seldom distally branched; cauline leaves: blades ovate, suborbiculate, elliptic, rounded-deltate, or oblong; involucres ± campanulate, 12–25(–30+) mm diam.
    2. Leaves glabrous (shining, finely gland-dotted, glutinous); heads held among or beyond leaves; ray florets (7–)8–12, laminae (12–)20–30(–35) mm; cypselae 7–9 mm; pappi 0 or coroniform, 0.1–0.3 mm . . . . 1. *Agnorhiza bolanderi*
    2. Leaves silky-villous to strigillose, glabrescent; heads in axils of (and overtopped by) leaves; ray florets 5–8(–9), laminae 8–18 mm; cypselae 9–10 mm; pappi coroniform, 1–1.5+ mm . . . . 2. *Agnorhiza ovata*
1. Plants (20–)40–100 cm; stems erect, usually distally branched; cauline leaves: blades deltate, ovate, or ovate-lanceolate; involucres ± hemispheric, (15–)20–40 mm diam.
    3. Ray florets 0 or 2–3 . . . . 3. *Agnorhiza invenusta*
    3. Ray florets 10–23.
        4. Leaf blades: faces tomentulose to pilosulous (and gland-dotted); cypselae 8–12 mm; pappi (0.5–)1–2(–3) mm . . . . 4. *Agnorhiza elata*
        4. Leaf blades: faces sparsely hispid or scabrous (and finely gland-dotted, green, shining, often vernicose); cypselae ca. 6 mm; pappi 0.1–1 mm . . . . 5. *Agnorhiza reticulata*

1. **Agnorhiza bolanderi**   (A. Gray) W. A. Weber,
   Phytologia 85: 19. 1999   [E] [F]

Balsamorhiza bolanderi A. Gray,
Proc. Amer. Acad. Arts 7: 356.
1868; *Wyethia bolanderi* (A. Gray)
W. A. Weber

Plants (10–)15–30 cm. **Stems** decumbent, seldom distally branched.
Cauline leaves: blades oblong,
ovate, rounded-deltate, or suborbiculate, 4–12 cm, bases ± truncate
or cordate, margins entire, faces glabrous (shining, finely
gland-dotted, ± glutinous). **Heads** held among or
beyond leaves. **Involucres** ± campanulate, 15–25(–30+)
mm diam. **Outer phyllaries** ovate to ovate-lanceolate, 10–
40+ mm (seldom surpassing ray corollas). **Ray florets**
(7–)8–12, laminae (12–)20–30(–35) mm. **Cypselae** 7–9
mm, glabrous; **pappi** 0 or coroniform, 0.1–0.3 mm.
2*n* = 38.

Flowering Mar–Apr(–May). Openings in chaparral;
600–800 m; Calif.

Agnorhiza bolanderi is known only from the foothills
of the Sierra Nevada, where it evidently grows only on
serpentine soils.

2. **Agnorhiza ovata**   (Torrey & Gray) W. A. Weber,
   Phytologia 85: 20. 1999

Wyethia ovata Torrey & A. Gray in
W. H. Emory, Not. Milit. Reconn.,
143. 1848; *W. coriacea* A. Gray

Plants (5–)10–30(–45+) cm. **Stems**
decumbent, seldom distally
branched. **Cauline leaves:** blades
broadly ovate or elliptic to
suborbiculate, 7–20 cm, bases
broadly cuneate to truncate or
cordate, faces silky-villous to strigillose, glabrescent.
Heads in axils of (and overtopped by) leaves. **Involucres**
narrowly campanulate, 12–20(–25) mm diam. **Outer
phyllaries** broadly linear to oblong or obovate, 12–35
mm (equaling or surpassing ray corollas). **Ray florets**
5–8(–9), laminae 8–18 mm. **Cypselae** 9–10 mm, glabrous; **pappi** coroniform (lacerate, projected into scales
on angles), 1–1.5 mm. **2***n* = 38.

Flowering May–Jul(–Aug). Grassy slopes, openings
in pine forests; (900–)1300–1900 m; Calif.; Mexico (Baja
California).

3. **Agnorhiza invenusta**   (Greene) W. A. Weber,
   Phytologia 85: 20. 1999   [E]

Helianthus invenustus Greene,
Pittonia 1: 284. 1889; *Balsamorhiza
invenusta* (Greene) Coville; *Wyethia
invenusta* (Greene) W. A. Weber

Plants 20–60(–100) cm. **Stems**
erect, usually distally branched
(densely glandular distally).
Cauline leaves: blades ± deltate to
ovate or ovate-lanceolate, 7–20
cm, bases broadly cuneate to truncate or shallowly cordate, margins usually entire, rarely irregularly crenate,
faces piloso-hispid or hirtellous (usually gland-dotted or
stipitate-glandular as well). **Heads** usually held beyond
leaves. **Involucres** ± hemispheric, 20–30 mm diam. **Outer
phyllaries** linear-triangular to lanceolate, 18–25(–30+)
mm (equaling or slightly surpassing discs). **Ray florets** 0
or 2–3, laminae 6–10 mm. **Cypselae** 7–8 mm, glabrous
(apices developing knoblike projections on angles);
pappi 0.

Flowering Jun–Jul. Openings in chaparral and in pine
and oak forests; (600–)1100–1900(–2300) m; Calif.

Agnorhiza invenusta is known only from the western
foothills of the Sierra Nevada.

4. **Agnorhiza elata**   (H. M. Hall) W. A. Weber, Phytologia
   85: 19. 1999   [E]

Wyethia elata H. M. Hall, Univ.
Calif. Publ. Bot. 4: 208. 1912,
based on *W. ovata* A. Gray, Proc.
Amer. Acad. Arts 7: 357. 1868, not
Torrey & A. Gray 1848

Plants 50–100 cm. **Stems** erect,
usually branched distally. **Cauline
leaves:** blades ovate-lanceolate to
deltate, (8–)10–15(–20) cm, bases
truncate to subcordate, margins entire or finely dentate
or serrulate, faces tomentulose to pilosulous (and gland-
dotted). **Heads** held beyond leaves. **Involucres** hemispheric, 25–40 mm diam. **Outer phyllaries** lanceolate,
20–30 mm (equaling or surpassing discs, tips often
spreading to squarrose). **Ray florets** 10–14(–23), laminae
(30–)50–60 mm. **Cypselae** 8–12 mm, usually glabrous,
sometimes distally strigillose; **pappi** coroniform (lacerate, projected into scales on angles), (0.5–)1–2(–3) mm.
2*n* = 38.

Flowering Jun–Jul(–Aug). Open pine forests; 900–
1200 m; Calif.

Agnorhiza elata is known only from foothills of the
Sierra Nevada.

5. Agnorhiza reticulata (Greene) W. A. Weber,
      Phytologia 85: 20. 1999   [C] [E]

Wyethia reticulata Greene, Bull.
Calif. Acad. Sci. 1: 9. 1884

Plants mostly 40–70 cm. **Stems** erect, usually branched distally. **Cauline leaves:** blades (green) ovate-lanceolate to deltate, 6–15 cm, bases truncate or subcordate, margins entire or serrulate to crenate, faces sparsely hispid or scabrous (and finely gland-dotted, often vernicose). **Heads** held beyond or among the leaves. **Involucres** hemi-spheric, 15–30 mm diam. **Outer phyllaries** oblong to lanceolate or oblanceolate, 12–20+ mm (± equaling discs, tips usually spreading to squarrose). **Ray florets** 10–16 (–21), laminae 20–25 mm. **Cypselae** ca. 6 mm, glabrous; **pappi** coroniform, 0.1–1 mm. *2n* = 38.

Flowering May–Jun. Chaparral, oak woodlands; of conservation concern; 300–500 m; Calif.

Agnorhiza reticulata is known only from foothills of the Sierra Nevada and is considered to be globally endangered.

277. VERBESINA Linnaeus, Sp. Pl. 2: 901. 1753; Gen. Pl. ed. 5, 384. 1754, name conserved • [No etymology in protologue; perhaps from genus name *Verbena* and Latin -*ina*, resemblance]

John L. Strother

Annuals or **perennials** [shrubs, trees], 7–15+ cm (*Verbesina nana*) or 30–200(–400)[–2500+] cm. **Stems** usually erect, usually branched (internodes sometimes winged). **Leaves** basal and/or cauline; opposite (sometimes whorled) or alternate; petiolate or sessile; blades (pinnately nerved or 3- or 5-nerved from at or near bases) mostly rhombic, deltate, ovate, or elliptic to lanceolate or lance-linear (sometimes intermediate shapes), sometimes pinnately or palmately lobed, bases cuneate to rounded or cordate, ultimate margins subentire or toothed, faces glabrous or hairy. **Heads** radiate or discoid, borne singly or in corymbiform, dichasiiform, or paniculiform arrays. **Involucres** hemispheric, turbinate, or campanulate to saucerlike, 5–20+[–30+] mm diam. **Phyllaries** persistent, 9–30[–50+] in 1–4[–5+] series (orbiculate, ovate, or oblong to spatulate, lanceolate, or linear, subequal or unequal, outer shorter or longer than inner, herbaceous to chartaceous). **Receptacles** flat to convex or ± conic, paleate (paleae usually navicular, ± conduplicate, herbaceous to scarious, linear to filiform in *V. encelioides* and *V. nana*). **Ray florets** 0 or (1–)5–30, either pistillate and fertile, or styliferous and sterile, or neuter; corollas yellow to orange or ochroleucous [reddish]. **Disc florets** 8–150[–300+], bisexual, fertile; corollas usually concolorous with rays, tubes much shorter than or ± equaling funnelform or campanulate throats, lobes 5, ± deltate to lance-deltate. **Cypselae** ± flattened, orbiculate, obovate, or oblanceolate to ± elliptic (usually winged); **pappi** persistent [falling], usually of 2, ± subulate scales or awns, sometimes 0. *x* = 17, 18?

Species 200 or more (16 in the flora): mostly subtropical, tropical, and warm-temperate North America.

SELECTED REFERENCES Coleman, J. R. 1966. A taxonomic revision of section *Ximenesia* of the genus *Verbesina* (Compositae). Amer. Midl. Naturalist 76: 475–481. Coleman, J. R. 1966b. A taxonomic revision of section *Sonoricola* of the genus *Verbesina* (Compositae). Madroño 18: 129–137. Olsen, J. S. 1979. Taxonomy of the *Verbesina virginica* complex (Asteraceae). Sida 8: 128–134.

1. Leaves all or mostly alternate (proximalmost sometimes opposite).
  2. Corollas usually ochroleucous, sometimes white.
    3. Ray florets 0 . . . . 1. *Verbesina walteri*
    3. Ray florets (1–)2–12+.
      4. Ray florets (1–)2–3(–7); disc florets 8–12(–15) . . . . 2. *Verbesina virginica*
      4. Ray florets (9–)10–12+; disc florets 20–25+ . . . . 3. *Verbesina microptera*

2. Corollas yellow.
    5. Plants 10–50(–120+)cm (annuals); internodes not winged; disc florets 80–150+
        .... 4. *Verbesina encelioides*
    5. Plants 7–15 or 30–200+ cm (perennating bases ± erect or horizontal); internodes
        (at least proximal) winged; disc florets 40–80+.
        6. Heads (3–)8–25(–50+) in corymbiform to paniculiform arrays; phyllaries 8–12
           in 1(–2) series, ± spreading to reflexed; ray florets (2–)6–8+ .... 5. *Verbesina alternifolia*
        6. Heads 2–5(–10+) in ± corymbiform arrays; phyllaries 16–21+ in 2–3 series,
           ± erect; ray florets 8–13+ .... 6. *Verbesina helianthoides*
1. Leaves all or mostly opposite (distal sometimes alternate).
    7. Internodes winged.
        8. Phyllaries 8–12+ in 2 series; ray florets (0–)1–3(–5); disc florets 8–15+; pappi 3–4
           mm .... 7. *Verbesina occidentalis*
        8. Phyllaries 18–20+ in 2–3 series; ray florets (5–)8; disc florets 20–60+; pappi 0–0.3
           mm .... 8. *Verbesina heterophylla*
    7. Internodes not winged.
        9. Plants mostly 7–15+ cm; leaves: abaxial faces mostly strigoso-sericeous; cypselae
           ± strigillose .... 9. *Verbesina nana*
        9. Plants mostly 30–100(–150+) cm; leaves: faces (at least abaxial) scabrellous to hir-
           tellous, hirsutulous, or hispidulous (not strigoso-sericeous); cypselae glabrous or
           sparsely hirtellous.
           10. Perennating bases horizontal; disc florets 40–60(–80+); cypselae purplish black;
              Florida.
              11. Phyllaries 2–3+ mm; ray florets (5–)11–13; cypselae 4–5 mm; pappi 0.5–1
                 mm .... 10. *Verbesina aristata*
              11. Phyllaries 5–9+ mm; ray florets 0; cypselae 5–7 mm; pappi 0–0.3 mm
                 .... 11. *Verbesina chapmanii*
           10. Perennating bases ± erect; disc florets 60–120+; cypselae dark brown; Arizona,
              California, New Mexico, Texas.
              12. Leaf blades ± lance-linear (lengths 10–15 times widths) .... 12. *Verbesina longifolia*
              12. Leaf blades lance-elliptic, lance-ovate, ovate, ovate-deltate, or rhombic
                 (lengths mostly 1.5–2.5 times widths).
                 13. Pappi 3–4 mm .... 13. *Verbesina dissita*
                 13. Pappi 0–0.5 mm.
                    14. Phyllaries 12–16 in ± 2 series, lance-linear, lance-ovate, or linear,
                       4–7 mm .... 14. *Verbesina oreophila*
                    14. Phyllaries 18–30+ in 3–4 series, elliptic, oblong, orbiculate, or ovate,
                       5–12+ mm.
                       15. Involucres 10–15+ mm diam.; phyllaries elliptic to oblong;
                           cypselae 10 mm .... 15. *Verbesina rothrockii*
                       15. Involucres 15–22+ mm diam.; phyllaries oblong, orbiculate,
                           or ovate; cypselae 7 mm .... 16. *Verbesina lindheimeri*

1. Verbesina walteri Shinners, Sida 1: 253. 1964   E

Athanasia paniculata Walter, Fl. Carol., 201. 1788, not *Verbesina paniculata* Poiret 1808

Plants 100–400 cm (perennating bases ± erect, internodes winged, at least proximal). **Leaves** all or mostly alternate (proximal usually opposite); blades lance-elliptic to lanceolate or lance-linear, 10–20 × 2–7 cm, bases ± cuneate, margins coarsely toothed to subentire, apices acute, faces ± hirtellous. **Heads** 8–35+ in corymbiform to paniculiform arrays. **Involucres** ± saucerlike, 8–10 mm diam. **Phyllaries** 12–15+ in 1–2 series, ± spreading, oblanceolate to linear, 2–5+ mm. **Ray florets** 0. **Disc florets** 40–60+; corollas ochroleucous to white. **Cypselae** dark brown to blackish, obovate, 3–4.5+ mm, faces ± strigillose; **pappi** 1–2 mm (sometimes shorter scales between awns). $2n = 34$.

Flowering Aug–Sep. Bottomlands, flood plains, borders of woodlands; 10–300 m; Ala., Ark., Ga., La., N.C., Okla., S.C., Tex.

Verbesina virginica may be no longer present in Georgia and North Carolina.

2. Verbesina virginica Linnaeus, Sp. Pl. 2: 901. 1753 E F

Verbesina laciniata Walter, *V. virginica* var. *laciniata* A. Gray (based on *V. laciniata* Nuttall, not Walter)

Plants (50–)100–250+ cm (perennating bases ± erect, internodes winged, at least proximal). **Leaves** all or mostly alternate (proximal sometimes opposite); blades lance-ovate or lance-elliptic to lance-linear, 5–12(–22+) × 1–6 (–12+) cm, sometimes pinnately lobed, bases ± cuneate, ultimate margins usually coarsely toothed to subentire, sometimes sinuate, apices acute, faces ± scabrellous to strigillose. **Heads** (20–)60–100+ in corymbiform to paniculiform arrays. **Involucres** ± obconic to turbinate, 3–5 mm diam. **Phyllaries** 8–12+ in 1–2 series, ± erect, spatulate to oblanceolate, 2.5–5(–7) mm. **Ray florets** (1–)2–3(–7); laminae 3–4(–7+) mm. **Disc florets** 8–12 (–15); corollas ochroleucous or white. **Cypselae** dark brown to blackish, oblanceolate, 3.5–5+ mm, faces ± scabrellous; pappi 1.5–3+ mm. $2n = 32, 34$.

Flowering (Jul–)Sep–Oct(–Dec). Bottomlands, flood plains, thickets, borders of woodlands, disturbed sites; 10–300 m; Ala., Ark., D.C., Fla., Ga., Ill., Kans., Ky., La., Md., Miss., Mo., N.C., Ohio, Okla., S.C., Tenn., Tex., Va., W.Va.

Plants of *Verbesina virginica* from near the Atlantic Coast with margins of some or most leaf blades sinuate to pinnately 3–5(–7+)-lobed have been called *V. laciniata* or *V. virginica* var. *laciniata*.

3. Verbesina microptera de Candolle in A. P. de Candolle and A. L. P. P. de Candolle, Prodr. 5: 616. 1836

Plants 50–250 cm (perennating bases ± erect, internodes winged). **Leaves** all or mostly alternate (proximal usually opposite); blades deltate-ovate or deltate to lanceolate, 4–10(–25+) × 2–6(–15) cm, bases broadly cuneate to ± truncate, margins coarsely toothed to subentire, apices obtuse to acute, faces ± hirtellous. **Heads** 30–80+ in ± paniculiform arrays. **Involucres** ± hemispheric, 5–7 mm diam. **Phyllaries** 12–15+ in 2 series, ± erect, spatulate to linear, 2.5–4 mm. **Ray florets** (9–)10–12+; laminae 4–6 mm. **Disc florets** 20–25+; corollas ochroleucous or white. **Cypselae** dark brown to blackish, oblanceolate, 4–5 mm, faces ± hirtellous or glabrous; pappi 1.5–2+ mm. $2n = 34$.

Flowering (May–)Jul–Sep(–Dec). Swales, disturbed sites; 10–100 m; Tex.; Mexico (Nuevo León, Tamaulipas).

4. Verbesina encelioides (Cavanilles) Bentham & Hooker f. ex A. Gray in W. H. Brewer et al., Bot. California 1: 350. 1876 F

Ximenesia encelioides Cavanilles, Icon. 2: 60, plate 178. 1793 (as enceliodes); *Verbesina encelioides* subsp. *exauriculata* (B. L. Robinson & Greenman) J. R. Coleman; *V. encelioides* var. *exauriculata* B. L. Robinson & Greenman

Plants 10–50(–120+) cm (annuals, internodes not winged). **Leaves** all or mostly alternate (proximal usually opposite); blades deltate-ovate or rhombic to lanceolate, 3–8(–12+) × 2–4 (–6+) cm, bases broadly cuneate to ± truncate, margins coarsely toothed to subentire, apices acute to attenuate, faces strigoso-scabrellous to sericeous. **Heads** usually borne singly, sometimes 2–3+ in loose, cymiform or corymbiform arrays. **Involucres** ± hemispheric to saucerlike, 10–20+ mm diam. **Phyllaries** 12–18+ in 1–2 series, ± erect to spreading, lance-ovate or lance-linear to linear, 6–8+ mm. **Ray florets** (8–)12–15+; laminae 8–10 (–20+) mm. **Disc florets** 80–150+; corollas yellow. **Cypselae** dark brown to blackish, narrowly obovate, 3.5–5+ mm, faces ± strigillose; pappi 0.5–1(–2) mm (0 on ray cypselae). $2n = 34$.

Flowering Aug–Oct. Swales, disturbed sites; 10–2500 m; Ala., Ariz., Ark., Calif., Colo., Fla., Ga., Ill., Iowa, Kans., La., Mo., Mont., Nebr., Nev., N.Mex., N.Y., N.C., Okla., Pa., R.I., S.C., S.Dak., Tenn., Tex., Utah, Wyo.; Mexico; West Indies; South America; introduced in Asia, Pacific Islands (Hawaii), Australia.

Native distribution of *Verbesina encelioides* in the flora area is uncertain. Plants of *Verbesina encelioides* from ca. 100°W (e.g., c Texas) and eastward usually have auriculate petiole bases and have been called var. *encelioides*; plants from the west usually lack auricles and have been called var. *exauriculata*.

5. Verbesina alternifolia (Linnaeus) Britton ex Kearney, Bull. Torrey Bot. Club 20: 485. 1893 E

Coreopsis alternifolia Linnaeus, Sp. Pl. 2: 909. 1753

Plants (30–)100–200+ cm (perennating bases ± erect or horizontal rhizomes, internodes winged, at least proximal). **Leaves** all or mostly alternate (proximal sometimes opposite); blades lance-elliptic or lanceolate to lance-linear, 10–25+ × 2–8+ cm, bases narrowly cuneate, margins coarsely toothed to subentire, apices attenuate, faces scabrellous. **Heads** (3–)8–25(–50+) in corymbiform to paniculiform arrays. **Involucres** ± saucerlike, 10–12+ mm

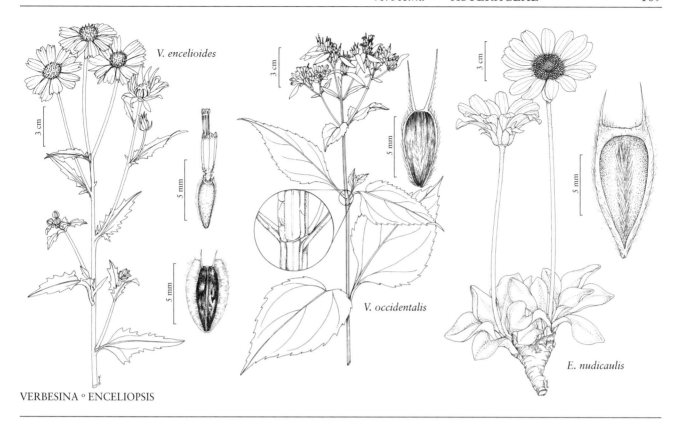

V. encelioides

V. occidentalis

E. nudicaulis

VERBESINA ° ENCELIOPSIS

diam. **Phyllaries** 8–12+ in 1(–2) series, ± spreading to reflexed, spatulate or lance-linear to linear, 3–8+ mm. **Ray florets** (2–)6–8+; laminae 15–25+ mm. **Disc florets** 40–60+; corollas yellow. **Cypselae** dark brown to black, oblanceolate to ± orbiculate, 4.5–5 mm, faces sparsely hirtellous to glabrate; **pappi** 1.5–2 mm. $2n = 68$.

Flowering Aug–Oct. Alluvial flats, along streams, woodlands; 10–600 m; Ont.; Ala., Ark., Del., D.C., Fla., Ga., Ill., Ind., Iowa, Kans., Ky., La., Md., Mich., Miss., Mo., Nebr., N.J., N.Y., N.C., Ohio, Okla., Pa., R.I., S.C., Tenn., Tex., Va., W.Va., Wis.

Verbesina alternifolia may be no longer present in Delaware.

6. Verbesina helianthoides Michaux, Fl. Bor.-Amer. 2: 135. 1803 [E]

Plants 60–120+ cm (perennating bases ± erect or horizontal rhizomes, internodes winged). **Leaves** all or mostly alternate (proximal sometimes opposite); blades ± lance-ovate to lanceolate, 5–12+ × 2–5+ cm, bases ± cuneate, margins ± toothed, apices acute to attenuate, faces strigose to sericeous. **Heads** 2–5(–10+) in ± corymbiform arrays. **Involucres** ± hemispheric, 10–15+ mm diam. **Phyllaries** 16–21+ in 2–3 series, ± erect, lanceolate, 6–9+ mm. **Ray florets**

8–13+; laminae 20–25(–30+) mm. **Disc florets** 40–80+; corollas yellow. **Cypselae** dark brown to black, oblanceolate to elliptic, 5 mm, faces strigillose to glabrate; **pappi** 0.5–1.5 mm. $2n = 34$.

Flowering May–Jul. Moist places in sandy, pine woodlands, post-oak woodlands, disturbed places; 70–500 m; Ala., Ark., Ga., Ill., Ind., Iowa, Kans., Ky., La., Miss., Mo., N.C., Ohio, Okla., Tenn., Tex.

Verbesina helianthoides may be no longer present in Georgia.

7. Verbesina occidentalis (Linnaeus) Walter, Fl. Carol., 213. 1788 [E] [F]

Siegesbeckia occidentalis Linnaeus, Sp. Pl. 2: 900. 1753

Plants 80–200+ cm (perennating bases ± erect, internodes winged). **Leaves** all or mostly opposite (distal sometimes alternate); blades ± ovate or lance-deltate to lanceolate, 6–12(–16+) × 3–6(–10+) cm, bases ± cuneate, margins coarsely toothed to subentire, apices usually acute to attenuate, sometimes rounded, faces scabrellous. **Heads** 20–100+ in corymbiform-paniculiform arrays. **Involucres** campanulate to turbinate, 3–5+ mm diam. **Phyllaries** 8–12+ in 2 series, ± erect, spatulate to oblanceolate, 3–7+ mm. **Ray florets** (0–)1–3(–5); laminae 10–15(–20+) mm.

Disc florets 8–15+; corollas yellow. **Cypselae** dark brown to black, oblanceolate, 5 mm, faces strigose; **pappi** 3–4 mm. $2n = 34$.

Flowering Aug–Oct. Bottomlands, margins of thickets, waste places; 10–300 m; Ala., Del., D.C., Fla., Ga., Ill., Ky., Md., Miss., Mo., N.C., Ohio, Okla., S.C., Tenn., Tex., Va., W.Va.

8. **Verbesina heterophylla** (Chapman) A. Gray, Proc. Amer. Acad. Arts 19: 12. 1883 [E]

Actinomeris heterophylla Chapman, Bot. Gaz. 3: 6. 1878; *Verbesina warei* A. Gray

Plants 40–50(–80+) cm (perennating bases ± horizontal rhizomes, internodes winged). **Leaves** all or mostly opposite (distal sometimes alternate); blades ± ovate to elliptic or lanceolate, 3–7+ × 0.8–2.5+ cm, bases ± cuneate, margins toothed, apices obtuse, faces scabrellous. **Heads** borne singly or 3–9+ in loose, corymbiform arrays. **Involucres** hemispheric or broader, 12–15+ mm diam. **Phyllaries** 18–20+ in 2–3 series, ± erect, narrowly oblong to lanceolate, 4–6+ mm. **Ray florets** (5–)8; laminae 12–15+ mm. **Disc florets** 20–60+; corollas yellow. **Cypselae** purplish black, narrowly obovate to ± elliptic, 5 mm, faces glabrous; **pappi** 0–0.3 mm. $2n = 34$.

Flowering (Apr–)Jun. Pine barrens; 0–20+ m; Fla.

9. **Verbesina nana** (A. Gray) B. L. Robinson & Greenman, Proc. Amer. Acad. Arts 34: 543. 1899

Ximenesia encelioides Cavanilles var. *nana* A. Gray, Smithsonian Contr. Knowl. 5(6): 92. 1853

Plants mostly 7–15+ cm (perennating bases ± erect, internodes not winged). **Leaves** all or mostly opposite (distal sometimes alternate); blades ± deltate to rhombic, 3–8+ × 2–5+ cm, bases ± cuneate, margins raggedly toothed, apices obtuse to acute, faces strigoso-sericeous. **Heads** borne singly or 2–3 together. **Involucres** shallowly bowl-shaped, 15–25+ mm diam. **Phyllaries** 28–35+ in 2–3 series, ± erect, linear to lanceolate, 8–12+ mm. **Ray florets** 12–16+; laminae 10–12+ mm. **Disc florets** 80–150+; corollas yellow to orange. **Cypselae** blackish, ± elliptic, 6.5–8 mm, faces ± strigillose; **pappi** 0.5–1+ mm. $2n = 68$.

Flowering May(–Oct). Silty flats, roadsides, desert scrub; 1000–2000 m; N.Mex., Tex.; Mexico (Coahuila).

10. **Verbesina aristata** (Elliott) A. Heller, Muhlenbergia 1: 8. 1900 [E]

Helianthus aristatus Elliott, Sketch Bot. S. Carolina 2: 428. 1823

Plants 70–150+ cm (perennating bases ± horizontal rhizomes, internodes not winged). **Leaves** all or mostly opposite (distal sometimes alternate); blades ± elliptic, 3–6(–10) × 1.2–2.8(–3.5) cm, bases rounded, margins toothed, apices ± rounded to acute, faces hirsutulous to hispido-scabrellous or scabrellous. **Heads** (3–)12–25+ in ± paniculiform arrays. **Involucres** ± hemispheric, 8–10 mm diam. **Phyllaries** 16–24+ in 2–3 series, ± erect, linear, 2–3+ mm. **Ray florets** (5–)11–13; laminae (15–)25–30+ mm. **Disc florets** 40–60(–80+); corollas yellow. **Cypselae** purplish black, ± obovate, 4–5 mm, faces glabrous; **pappi** 0.5–1 mm. $2n = 34$.

Flowering Jun. Pine woodlands; 0–10+ m; Ala., Fla., Ga.

11. **Verbesina chapmanii** J. R. Coleman, Rhodora 74: 100. 1972 [E]

Plants 50–50(–80+) cm (perennating bases ± horizontal rhizomes, internodes not winged). **Leaves** all or mostly opposite (distal sometimes alternate); blades ± elliptic, 3–6(–10) × 0.8–2.2(–3) cm, bases ± cuneate, margins toothed to subentire, apices ± rounded or obtuse, faces ± scabrellous. **Heads** borne singly or (2–)3+ in dichasial arrays. **Involucres** ± hemispheric to turbinate, 8–16 mm diam. **Phyllaries** 18–24+ in 2–3 series, ± erect, spatulate to lance-linear, 5–9+ mm. **Ray florets** 0. **Disc florets** 40–60(–80+); corollas yellow. **Cypselae** purplish black, ± elliptic, 5–7 mm, faces glabrous or sparsely hirtellous; **pappi** 0–0.3 mm. $2n = 34$.

Flowering Jun–Aug. Pine barrens, bogs, flatwoods; 10–30 m; Fla.

12. **Verbesina longifolia** (A. Gray) A. Gray, Proc. Amer. Acad. Arts 19: 12. 1883

Actinomeris longifolia A. Gray, Smithsonian Contr. Knowl. 5(6): 89. 1853

Plants 60–150+ cm (perennating bases ± erect, internodes not winged). **Leaves** all or mostly opposite (distal sometimes alternate); blades ± lance-linear (lengths 10–15 times widths), 9–12(–20+) × 0.6–1.2(–1.5+) cm, bases rounded, margins obscurely toothed to subentire, apices ± attenuate, faces ± scabrellous. **Heads** borne singly or 3–5+ in loose, corymbiform arrays. **Involucres** ± hemispheric to saucerlike, 11–15+ mm diam. **Phyllaries** 18–24+ in 2–3 series, ± erect to spreading, linear to lance-linear, 5–15+ mm. **Ray florets** 12–13; laminae 15–25+ mm. **Disc florets** 60–100+; corollas yellow. **Cypselae** dark brown, oblanceolate to elliptic, 7 mm, faces glabrous; **pappi** 0–0.5 mm. $2n = 34$.

Flowering (Jul–)Sep–Oct. Rocky slopes; 1500–2400 m; Ariz., N.Mex.; Mexico (Chihuahua, Durango).

13. **Verbesina dissita** A. Gray, Proc. Amer. Acad. Arts 20: 299. 1885

Plants 50–150+ cm (perennating bases ± erect, internodes not winged). **Leaves** all or mostly opposite (distal sometimes alternate); blades ± rhombic to ovate, 5–12+ × 3–6+ cm, bases rounded to cuneate, margins coarsely toothed to subentire, apices ± acute, faces ± scabrellous. **Heads** borne singly or (2–)5–15+ in loose, corymbiform arrays. **Involucres** ± hemispheric, 10–20+ mm diam. **Phyllaries** 24–30+ in 2–3 series, ± erect, spatulate or oblanceolate to lanceolate, 5–12+ mm. **Ray florets** 10–13; laminae 12–25+ mm. **Disc florets** 60–100+; corollas yellow. **Cypselae** dark brown, narrowly lance-elliptic, 8–10 mm, faces glabrous or sparsely hirtellous; **pappi** 3–4 mm.

Flowering May. Chaparral or coastal scrub; 30–200 m; Calif.; Mexico (Baja California).

14. **Verbesina oreophila** Wooton & Standley, Contr. U.S. Natl. Herb. 16: 190. 1913

Plants 50–100+ cm (perennating bases ± erect, internodes not winged). **Leaves** all or mostly opposite (distal sometimes alternate); blades ovate-deltate to lance-elliptic, 3–8(–11) × 1–4(–6) cm, bases subtruncate to cuneate, margins coarsely toothed to subentire, apices acute to attenuate, faces ± hirsutulous to scabrellous. **Heads** borne singly or 2–6 in loose, corymbiform arrays. **Involucres** ± hemispheric, 8–10 mm diam. **Phyllaries** 12–16 in 2 series, ± erect, linear to lance-ovate or lance-linear, 4–7 mm. **Ray florets** 13; laminae 9–12+ mm. **Disc florets** 60–100+; corollas yellow. **Cypselae** dark brown, obovate, 4–5 mm, faces glabrous; **pappi** 0–0.5 mm.

Flowering (Aug–)Sep(–Nov). Rocky slopes, limestone; 1500–2100 m; N.Mex., Tex.; Mexico (Coahuila).

15. **Verbesina rothrockii** B. L. Robinson & Greenman, Proc. Amer. Acad. Arts 34: 541. 1899

Plants 30–60+ cm (perennating bases ± erect, internodes not winged). **Leaves** all or mostly opposite (distal sometimes alternate); blades ovate-deltate or rhombic to lance-elliptic, 3–5 × 1–3 cm, bases subtruncate to cuneate, margins coarsely toothed to subentire, apices obtuse to acute, faces ± hirtellous to scabrellous. **Heads** borne singly. **Involucres** ± hemispheric, 10–15+ mm diam. **Phyllaries** 18–30+ in 3–4 series, ± erect, oblong to elliptic, 6–10+ mm. **Ray florets** 8–13; laminae 15–25+ mm. **Disc florets** 60–100+; corollas yellow. **Cypselae** dark brown, ± elliptic, 10 mm, faces glabrous; **pappi** 0–0.5 mm. $2n = 34$.

Flowering (May–)Aug–Sep. Rocky slopes, igneous or limestone substrates; 1200–1800 m; Ariz., N.Mex.; Mexico (Coahuila, Durango, Zacatecas).

16. **Verbesina lindheimeri** B. L. Robinson & Greenman, Proc. Amer. Acad. Arts 34: 541. 1899   E

Plants 30–70+ cm (perennating bases ± erect, internodes not winged). **Leaves** all or mostly opposite (distal sometimes alternate); blades ovate-deltate or ovate to lance-ovate, 2–6 × 1–5 cm, bases rounded to cuneate, margins toothed to entire, apices obtuse to acute, faces ± scabrellous. **Heads** borne singly. **Involucres** ± hemispheric, 15–22+ mm diam. **Phyllaries** 24–30+ in 3–4 series, ± erect, oblong, orbiculate, or ovate, 5–12+ mm. **Ray florets** ± 13; laminae 25 mm. **Disc florets** 80–120+; corollas yellow. **Cypselae** dark brown, narrowly elliptic, 7 mm, faces glabrous; **pappi** 0–0.5 mm. $2n = 34$.

Flowering (May–)Sep–Nov. Shaded slopes, limestone, juniper scrublands; 100–500 m; Tex.

278. ENCELIOPSIS A. Nelson, Bot. Gaz. 47: 432. 1909 • [Generic name *Encelia* and Greek *-opsis*, resembling] E

Curtis Clark

Perennials 15–100+ cm (caudices or taproots woody). **Stems** erect, branched from bases. **Leaves** basal; alternate; petiolate or sessile; blades 3-nerved, elliptic, ovate, rhombic, or suborbiculate, bases ± cuneate to nearly truncate, margins entire (sometimes corrugate or ruffled), faces densely puberulent or silky-velutinous. **Heads** radiate, borne singly (peduncles much longer than involucres). **Involucres** ± hemispheric or broader, 10–30+ mm diam. **Phyllaries** persistent, 30–65+ in 3–6 series. **Receptacles** convex, paleate (paleae ± conduplicate, folded around and falling with cypselae). **Ray florets** (11–)20–35+, neuter; corollas yellow. **Disc florets** (50–)200–500+, bisexual, fertile; corollas yellow, tubes shorter than to equaling abruptly expanded throats, lobes 5, triangular. **Cypselae** strongly compressed or flattened, ± cuneate (margins ± white, corky, usually ciliate, faces black, glabrous or ± silky-villous); **pappi** 0, or persistent, of 2 awns (often with 2–10+, often connate, minute scales or teeth as well). *x* = 18.

Species 3 (3 in the flora): w North America.

SELECTED REFERENCE Sanders, D. L. and C. Clark. 1987. Comparative morphology of the capitulum of *Enceliopsis*. Amer. J. Bot. 74: 1072–1086.

1. Herbage dull gray; petioles not or barely winged; leaf blades ovate to suborbiculate . . . . . . 1. *Enceliopsis nudicaulis*
1. Herbage silvery; petioles winged (wings merging with blades); leaf blades rhombic or widely elliptic.
    2. Ray corolla laminae deep yellow, 12–30 mm (lengths 1–1.2 times disc diams.) . . . 2. *Enceliopsis argophylla*
    2. Ray corolla laminae light yellow, 30–50 mm (lengths 1.5–2 times disc diams.) . . . . 3. *Enceliopsis covillei*

1. Enceliopsis nudicaulis (A. Gray) A. Nelson, Bot. Gaz. 47: 433. 1909 • Naked-stemmed daisy E F

Encelia nudicaulis A. Gray, Proc. Amer Acad. Arts 8: 656. 1873; *Enceliopsis nudicaulis* var. *bairdii* S. L. Welsh; *E. nudicaulis* var. *corrugata* Cronquist

Perennials, 10–40 cm; herbage dull gray, hairs ± spreading. **Leaves:** petioles not or barely winged; blades suborbiculate to ovate, 2–6 × 2–6 cm. **Peduncles** 15–45 cm. **Involucres** 10–20 mm. **Phyllaries** in 3–5 series, narrowly lanceolate (bases ovate), apices acute. **Ray florets** ± 21; corollas yellow, laminae 20–40 mm. **Cypselae** ± 9 × 3.5 mm, silky-hairy; **pappi** of 2 awns 1–1.5 mm (plus minute scales or teeth). **2*n* = 32.**

Flowering Apr–Aug. Stony hillsides, canyons; 900–2000 m; Ariz., Calif., Idaho, Nev., Utah.

Plants with strongly corrugate leaf margins from Ash Meadows, Nevada, are var. *corrugata*. Cronquist believed that corrugate-leaved plants from other areas in southern Nevada were a result of drying in press; D. L. Sanders (unpubl.) showed a gradual drop-off in corrugation of living leaves in the region to the east of Ash Meadows.

2. Enceliopsis argophylla (D. C. Eaton) A. Nelson, Bot. Gaz. 47: 433. 1909 E

Tithonia argophylla D. C. Eaton in S. Watson, Botany (Fortieth Parallel), 423. 1871

Perennials, 15–80 cm; herbage silvery, hairs fine, ± appressed. **Leaves:** petioles winged, wings merging with blades; blades rhombic or widely elliptic, 4–10 × 2–8 cm. **Peduncles** 30–80 cm. **Involucres** 15–25 mm. **Phyllaries** in 4–6 series, lanceolate to ovate, apices acuminate. **Ray florets** 20–35; corollas deep yellow, laminae 12–30 mm (1–1.2 times disc diams.). **Cypselae** 9–13 × 4–5 mm, hirsute; **pappi** usually of 2 awns 1–2 mm (plus minute scales or teeth), sometimes 0.

Flowering Apr–May. Stony plains, hillsides; 1000–1500 m; Ariz., Nev.

3. Enceliopsis covillei (A. Nelson) S. F. Blake, J. Wash. Acad. Sci. 21: 334. 1931 • Panamint daisy E

Helianthella covillei A. Nelson, Bot. Gaz. 37: 273. 1904; *Enceliopsis argophylla* (D. C. Eaton) A. Nelson var. *grandiflora* Jepson

Perennials, 15–80(–100+) cm; herbage silvery, hairs fine, ± appressed. **Leaves:** petioles winged, wings merging with blades; blades rhombic or widely elliptic, 4–10 × 2–8 cm. **Peduncles** 30–100 cm. **Involucres** 18–30 mm.

Phyllaries in 4–6 series, lanceolate to ovate, apices acuminate. **Ray florets** 20–35; corollas light yellow, laminae 30–50 mm (1.5–2 times disc diams.). **Cypselae** ± 10 × 6.5 mm, glabrous or puberulent; **pappi** of 2 awns ± 1 mm (plus minute scales). $2n = 36$.

Flowering Apr–Jun. Stony hillsides, canyons; 400–1300 m; Calif.

---

279. PHOEBANTHUS S. F. Blake, Proc. Amer. Acad. Arts 51: 520. 1916 • False sunflower [Greek *phoebus*, the sun, and *anthos*, flower] E

Edward E. Schilling

Perennials, 40–100+ cm (aerial stems from relatively slender, horizontal tubers). **Stems** erect, branched mostly distally (among heads, leafy). **Leaves** cauline; all opposite, or proximal opposite and distal alternate, or all alternate; sessile; blades (usually 1-nerved) linear-filiform or linear to linear-lanceolate, bases ± cuneate or rounded, margins entire (revolute), faces scabrous (adaxial) and sparsely hispid-scabrous, gland-dotted as well. **Heads** radiate, borne singly or (2–6) in open, ± corymbiform arrays. **Involucres** hemispheric to campanulate, 10–20 mm diam. **Phyllaries** persistent, 20–40+ in 2–3+ series (green, subequal, herbaceous, strigillose to hispid). **Receptacles** convex, paleate (paleae tan, conduplicate, trilobed or tricuspidate). **Ray florets** 10–20, neuter; corollas yellow (apices 2–4-lobed). **Disc florets** 40–100+, bisexual, fertile; corollas yellow, tubes shorter than narrowly campanulate throats, lobes 5, triangular. **Cypselae** (purplish black) obpyramidal, slightly compressed; **pappi** 0, or persistent or fragile, of 1–2 lacerate scales (over cypsela shoulders) plus 0–6 lacerate or bristlelike shorter scales. $x = 17$.

Species 2 (2 in the flora): se United States.

1. Leaves linear to linear-lanceolate, the larger 3–7 mm wide; phyllaries appressed
.... 1. *Phoebanthus grandiflorus*
1. Leaves linear-filiform, the larger 0.5–2 mm wide; phyllaries spreading. .... 2. *Phoebanthus tenuifolius*

1. Phoebanthus grandiflorus (Torrey & A. Gray) S. F. Blake, Proc. Amer. Acad. Arts 51: 520. 1916 • Florida false sunflower E F

Helianthella grandiflora Torrey & A. Gray, Fl. N. Amer. 2: 333. 1842

Perennials, 50–100+ cm (tubers 2–10 cm). **Leaves** mostly alternate, sometimes proximal opposite; blades linear to linear-lanceolate, the larger 2–7 cm × 3–7 mm. **Involucres** hemispheric, 10–12 × 12–20 mm. **Paleae** 10 × 2 mm (lateral lobes nearly equaling central lobe). **Phyllaries** appressed. **Ray laminae** 20–40 mm. **Disc florets** 50+;

corollas 5–7 mm. **Cypselae** 3.5–4.5 mm; **pappi** 0 or of 1 (–2) lacerate scales 0.4–2 mm plus 0–2 scales 0.2–0.5 mm. $2n = 68$.

Flowering Mar–Nov. Sandhills, pine flatwoods; 10–30+ m; Fla.

Phoebanthus grandiflorus is known only from peninsular Florida.

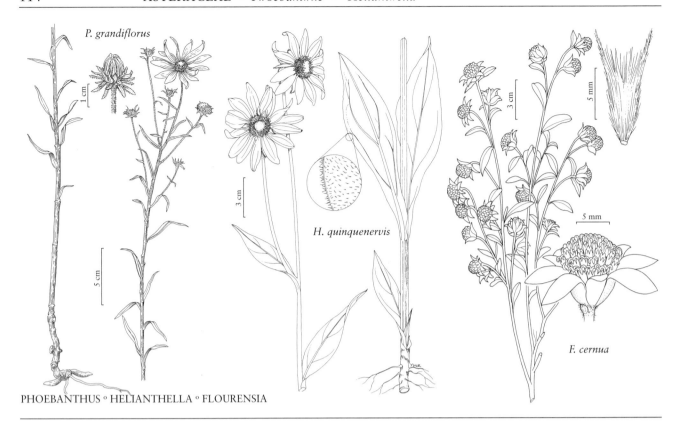

P. grandiflorus

H. quinquenervis

F. cernua

PHOEBANTHUS ○ HELIANTHELLA ○ FLOURENSIA

2. **Phoebanthus tenuifolius** (Torrey & A. Gray) S. F. Blake, Proc. Amer. Acad. Arts 51: 520. 1916

• Pineland false sunflower [E]

Helianthella tenuifolia Torrey & A. Gray, Fl. N. Amer. 2: 333. 1842

Perennials, 40–100+ cm (tubers 2–4 cm). **Leaves** all opposite, or opposite and alternate, or all alternate; blades linear-filiform, the larger 3–7 cm × 0.5–2 mm. **Involucres** campanulate to hemispheric, 10–12 × 10–15 mm. **Paleae** 7–8 × 1.5–2 mm (central lobes usually longer than lateral lobes). **Phyllaries** spreading. **Ray laminae** 20–40 mm. **Disc florets** 40+; corollas 4–5 mm. **Cypselae** 3.5–4 mm; **pappi** of 1–2, lacerate scales 0.2–2 mm plus 0–4 scales 0.2–0.3 mm. $2n = 34$.

Flowering May–Sep. Sandhills, flatwoods; 10–30+ m; Ala., Fla.

Phoebanthus tenuifolius in known from southeastern Alabama and the panhandle of Florida.

280. **HELIANTHELLA** Torrey & A. Gray, Fl. N. Amer. 2: 333. 1842 • Little sunflower [Generic name *Helianthus* and Latin *-ella*, diminutive]

William A. Weber

Perennials, 20–150 cm. **Stems** erect, usually branched. **Leaves** basal and cauline; opposite (proximal) and alternate; usually petiolate (at least proximal); blades (usually 3- or 5-nerved from at or near bases) elliptic, lanceolate, lance-linear, linear, oblanceolate, ovate, ovate-lanceolate, or spatulate, bases mostly cuneate, margins entire, faces glabrous or hairy. **Heads** radiate, borne singly or (2–15+) in corymbiform arrays. **Involucres** campanulate to hemispheric or broader, 10–50 mm diam. **Phyllaries** persistent, 22–32+ in ± 3 series (mostly deltate-ovate or ovate to lanceolate, linear, or oblong, subequal or unequal, outer sometimes foliaceous). **Receptacles** ± convex, paleate (paleae falling with cypselae, conduplicate, scarious or chartaceous). **Ray florets** 8–21, neuter; corollas ± yellow. **Disc florets** 30–200+, bisexual, fertile; corollas yellow or purple to brown, tubes shorter than or equaling throats, lobes 5, deltate (abaxially hairy). **Cypselae** (brownish) ± compressed, obovate to obcordate (margins usually ± winged, usually ± ciliate, faces glabrous or strigose); **pappi** 0, or persistent, of 2 ± subulate scales plus 0–4+ shorter, lacerate scales (sometimes all ± connate). x = 15.

Species 9 (6 in the flora): w North America, n Mexico.

SELECTED REFERENCES Weber, W. A. 1948. The genus *Helianthella* in Oregon. Madroño 9: 186–189. Weber, W. A. 1952. The genus *Helianthella* (Compositae). Amer. Midl. Naturalist 48: 1–35.

1. Heads 5–15+; phyllaries mostly deltate-ovate to oblong (unequal to subequal); ray laminae 8–14+ mm (usually little surpassing discs); disc corollas purple to brown . . . . . 1. *Helianthella microcephala*
1. Heads 1(–3+); phyllaries mostly lanceolate, lance-linear, lance-oblong, or ovate (subequal, outer sometimes foliaceous); ray laminae (11–)15–30(–45+) mm; disc corollas yellow.
  2. Heads ± nodding; paleae scarious, relatively soft.
    3. Plants 20–50 cm; leaf blades oblanceolate or spatulate (broadest distal to middles, not leathery), faces rough-pubescent; involucres 15–20 mm diam.; ray florets 8–14 (corollas pale yellow) . . . . 2. *Helianthella parryi*
    3. Plants usually 50–150 cm; leaf blades elliptic to ovate-lanceolate (broadest near middles, leathery), faces sparsely hirsute or glabrous; involucres 40–50 diam.; ray florets (13–)21 (corollas bright yellow) . . . . 3. *Helianthella quinquenervis*
  2. Heads erect; paleae chartaceous, relatively firm.
    4. Cauline leaves: largest at mid stems; faces of cypselae strigose . . . . 4. *Helianthella uniflora*
    4. Cauline leaves: largest proximal to mid stems; faces of cypselae usually glabrous.
      5. Involucres 15–20(–25) mm diam. (sometimes 2–3 outer phyllaries foliaceous); cypselae strongly compressed . . . . 5. *Helianthella californica*
      5. Involucres 25–40 mm diam. (usually 3–4+ outer phyllaries foliaceous, 3–10+ cm); cypselae weakly compressed (plump in middles, thin at edges) . . . . 6. *Helianthella castanea*

1. Helianthella microcephala (A. Gray) A. Gray, Proc. Amer. Acad. Arts 19: 10. 1883 [E]

Encelia microcephala A. Gray, Proc. Amer. Acad. Arts 8: 657. 1873

Plants 20–60(–80) cm. **Cauline leaves:** largest proximal to mid stems; blades usually 3-nerved, linear-spatulate to oblong-spatulate, 6–25 cm, faces hispidulous to scabrellous or glabrate. **Heads** (3–) 5–15+ in corymbiform arrays, erect. **Involucres** turbinate, campanulate, or hemispheric, (8–)12–15(–20+) mm diam. **Phyllaries** deltate-ovate to oblong (unequal to subequal, margins ciliolate, apices obtuse). **Paleae** chartaceous, relatively firm. **Ray florets** (5–)13; laminae 8–14 mm (usually little surpassing discs). **Disc corollas** purple to brown. **Cypselae** strongly compressed, ± oblanceolate (scarcely winged, faces densely strigose); **pappi** of 2 ± subulate scales plus 0–4 shorter scales. $2n$ = 30.

Flowering (Jul–)Aug–Sep. Gulches, ravines, hillsides, semidesert sites, pinyon-juniper associations; 1700–2800; Ariz., Colo., N.Mex., Utah.

Helianthella microcephala grows in the Colorado Plateau province.

2. Helianthella parryi A. Gray, Proc. Acad. Nat. Sci. Philadelphia 15: 65. 1864 [E]

Plants 20–50 cm. **Cauline leaves:** largest proximal to mid stems; blades usually 5-nerved, oblanceolate or spatulate, faces rough-pubescent. **Heads** usually borne singly, sometimes 2–3+, ± nodding. **Involucres** hemispheric or broader, 15–20 mm diam. **Phyllaries** lanceolate to lance-linear (subequal, margins ciliate). **Paleae** scarious, relatively soft. **Ray florets** 8–14; (corollas pale yellow) laminae 25–30 mm. **Disc corollas** yellow. **Cypselae** strongly compressed, narrowly obovate (margins ciliate, faces strigose); **pappi** of 2 ± subulate scales plus 0–4+ shorter scales.

Flowering Jul–Aug. Upper montane forests; 2400–3600 m; Ariz., Colo., N.Mex.

3. **Helianthella quinquenervis** (Hooker) A. Gray, Proc. Amer. Acad. Arts 19: 10. 1883

Helianthus quinquenervis Hooker, London J. Bot. 6: 247. 1847

Plants (30–)50–150 cm. **Cauline leaves:** largest proximal to mid stems; blades usually 3- or 5-nerved, elliptic or ovate-lanceolate to lanceolate, 10–50 cm, faces sparsely hirsute or glabrous. **Heads** usually borne singly, ± nodding. **Involucres** hemispheric or broader, (25–)40–50 mm diam. **Phyllaries** (outer sometimes ± foliaceous) ovate to lanceolate (subequal to unequal, margins ciliate). **Paleae** scarious, relatively soft. **Ray florets** (13–)21; (corollas bright yellow) laminae 25–30(–40) mm. **Disc corollas** yellow. **Cypselae** strongly compressed, narrowly obovate (margins ciliate, faces strigose); **pappi** of 2 ± subulate scales plus 2–4+ shorter scales.

Flowering Jun–Aug(–Oct). Moist meadows, aspen glades; 1400–3600 m; Ariz., Colo., Idaho, Mont., Nev., N.Mex., Oreg., S.Dak., Wyo.; Mexico (Chihuahua, Nuevo León).

Helianthella quinquenervis grows primarily in the Rocky Mountains, Great Basin ranges, and Black Hills.

4. **Helianthella uniflora** (Nuttall) Torrey & A. Gray, Fl. N. Amer. 2: 334. 1842 [E]

Helianthus uniflorus Nuttall, J. Acad. Nat. Sci. Philadelphia 7: 37. 1834; *Helianthella uniflora* var. *douglasii* (Torrey & A. Gray) W. A. Weber

Plants 40–120 cm. **Cauline leaves:** largest at mid stems; blades usually 3-nerved, usually lanceolate to elliptic, rarely ovate, 12–25 cm, faces puberulent, hirsute, or scabrous. **Heads** usually borne singly, sometimes 2–3+, erect. **Involucres** turbinate to hemispheric or broader, 15–30 mm diam. **Phyllaries** lanceolate to lance-linear (subequal or outer larger, margins sometimes ciliate, faces ± cinereous-pubescent to sparsely puberulent). **Paleae** chartaceous, relatively firm. **Ray florets** 11–13(–21); laminae 15–30(–45) mm. **Disc corollas** yellow. **Cypselae** strongly compressed, narrowly obovate (margins ciliate, faces strigose); **pappi** of 2 ± subulate scales plus 2–4 shorter scales. *2n* = 30.

Flowering May–Jul(–Aug). Grasslands, meadows, sagebrush scrublands, swales, aspen forests, rocky slopes, and spruce-fir forests; 300–3400 m; B.C.; Ariz., Colo., Idaho, Mont., Nev., Oreg., Utah, Wash., Wyo.

Two infraspecific taxa within *Helianthella uniflora* may be distinguished; they have been named at varietal rank.

Variety *douglasii* has stems hirsute; involucres (15–)20–25(–30) mm diam.; outer phyllaries rarely elongated, margins ciliate, abaxial faces sparsely puberulent; ray laminae 30–40 mm; and *2n* = 30. It grows in grasslands in the northern Rocky Mountains and on the east side of the Cascade Range (B.C.; Idaho, Oreg., Wash.) at 300–2500 m where it flowers May–Jul.

Variety *uniflora* has stems sparingly puberulent to scabrous; involucres 15–20(–25) mm diam.; outer phyllaries frequently elongated, margins not ciliate, abaxial faces uniformly cinereous-pubescent; ray laminae 20–30 mm; and *2n* = 30. It grows in meadows, sagebrush scrublands, swales, aspen forests, rocky slopes, and spruce-fir forests of the Rocky Mountains, intermountain plateaus, and Great Basin ranges (Ariz., Colo., Idaho, Mont., Nev., Oreg., Utah, Wyo.) at 1500–3400 m, where it flowers (May–)Jun–Jul(–Aug).

5. **Helianthella californica** A. Gray in War Department [U.S.], Pacif. Railr. Rep. 4(5): 103. 1857 [E]

Plants 15–60 cm. **Cauline leaves:** (nearly all opposite or mostly alternate) largest proximal to mid stems; blades ± 3-nerved, lance-ovate or oblong-lanceolate to lanceolate or linear, 5–26 cm, faces ± hirsutulous to scaberulous or glabrescent. **Heads** usually borne singly, sometimes 2–3+, erect. **Involucres** hemispheric, 15–25 mm diam. **Phyllaries** lanceolate (sometimes outer 2–3 foliaceous; margins usually ciliate, abaxial faces densely hairy or glabrate). **Paleae** chartaceous, relatively firm. **Ray florets** (10–)12–16; laminae 15–20+ mm. **Disc corollas** yellow. **Cypselae** strongly compressed, obovate-cuneate to oblong (winged or almost wingless, margins ciliate, apices deeply notched, faces glabrous); **pappi** 0 or 2 ± subulate scales plus 0–4 shorter scales.

Subspecies 3 (3 in the flora): w United States.

1. Stems branched; cauline leaves nearly all opposite; cypselae almost wingless; pappi 0 . . . . 5a. *Helianthella californica* subsp. *californica*
1. Stems not branched; cauline leaves mostly alternate; cypselae winged; pappi of 2 ± subulate scales plus 0–4 shorter scales.
  2. Leaf blades oblong-lanceolate or lance-linear, (10–)20–45 mm wide; heads usually 2–3+ . . . . . 5b. *Helianthella californica* subsp. *nevadensis*
  2. Leaf blades usually linear to lance-linear, sometimes lance-ovate, 3–6(–18) mm wide; heads borne singly . . . . 5c. *Helianthella californica* subsp. *shastensis*

5a. Helianthella californica A. Gray subsp. **californica**
E

Stems branched. Cauline leaves nearly all opposite; blades oblanceolate to narrowly lanceolate, (8–)12–40 mm wide. Heads borne singly or 2–3+. Cypselae almost wingless; pappi 0. $2n = 30$.

Flowering May–Jun. Open sites, ridges, grasslands; 30–1000 m; Calif.

Subspecies *californica* grows in the San Francisco Bay area and north to Mendocino and Glenn counties.

5b. Helianthella californica A. Gray subsp. **nevadensis** (Greene) W. A. Weber, Phytologia 85: 20. 1999   E

Helianthella nevadensis Greene, Bull. Calif. Acad. Sci. 1: 89. 1885; *H. californica* var. *nevadensis* (Greene) Jepson

Stems not branched. Cauline leaves mostly alternate; blades oblong-lanceolate to lance-linear, (10–)20–45 mm wide. Heads usually 2–3+, seldom borne singly. Cypselae winged; pappi of 2 ± subulate scales plus 0–4 shorter scales.

Flowering Apr–Jul. Meadows, open forest, chaparral; (200–)600–2300+ m; Calif., Nev., Oreg.

Subspecies *nevadensis* grows in the Sierra Nevada and in the Cascade Range.

5c. Helianthella californica A. Gray subsp. **shastensis** (W. A. Weber) W. A. Weber, Phytologia 85: 20. 1999
E

Helianthella californica var. *shastensis* W. A. Weber, Amer. Midl. Naturalist 48: 30. 1952

Stems not branched. Cauline leaves mostly alternate; blades usually linear to lance-linear, sometimes lance-ovate, 3–6 mm (–18+) wide. Heads borne singly. Cypselae winged; pappi of 2 ± subulate scales plus 0–4 shorter scales.

Flowering May–Jun. Pine forests, scrublands; 1300–2500 m; Calif.

Subspecies *shastensis* grows in Shasta, Siskiyou, and Trinity counties.

6. Helianthella castanea Greene, Erythea 1: 127. 1893   E

Plants 15–45 cm. Cauline leaves: largest proximal to mid stems; blades sometimes ± 3-nerved, oblong or elliptic to broadly-oblanceolate, 10–15 cm, faces sparsely hirsute (mostly abaxially and near margins). Heads borne singly, erect. Involucres hemispheric, 25–40 mm diam. Phyllaries (usually outer 3–4+ foliaceous, 3–10 cm, curling over heads) lanceolate to lance-oblong (margins usually ciliate, abaxial faces densely pubescent or glabrate). Paleae chartaceous, relatively firm. Ray florets 12–21; laminae (10–)20–30 mm. Disc corollas yellow. Cypselae weakly compressed (plump in middles, edges thin, faces, glabrous); pappi 0 or of 2 ± subulate scales. $2n = 30$.

Flowering Apr–May. Grassy hillsides; 0–1200 m; Calif.

Helianthella castanea is known only from the San Francisco Bay area.

281. FLOURENSIA   de Candolle in A. P. de Candolle and A. L. P. P. de Candolle, Prodr. 5: 592. 1836 • [For Marie-Jean-Pierre Flourens, 1794–1867, physiologist, perpetual secretary, Académie des Sciences, Paris]

John L. Strother

Subshrubs or shrubs [trees], to 100(–200)[–500+] cm. Stems erect, branched from bases or ± throughout. Leaves cauline; alternate; petiolate [nearly sessile]; blades pinnately nerved, mostly elliptic to lance-oblong or ovate, bases rounded to cuneate, margins entire [toothed], faces glabrous or ± scabrellous, usually gland-dotted and vernicose. Heads discoid or radiate, borne ± singly or in ± spiciform arrays. Involucres campanulate to hemispheric, 4–20 mm diam.

Phyllaries persistent, 12–40 in 2–4+ series (subequal or unequal, outer longer). **Receptacles** flat to conic-ovoid, paleate (paleae conduplicate, cartilaginous to scarious). **Ray florets** 0 or [5–]13–21, either neuter, or styliferous and sterile; corollas yellow. **Disc florets** 10–50[–150], bisexual, fertile; corollas yellow, tubes much shorter than cylindric-funnelform throats, lobes 5, ± deltate. **Cypselae** ± compressed or flattened [subterete], oblong to oblanceolate (not winged, ± sericeous); **pappi** persistent or tardily falling, of 2 subulate scales. $x = 9$.

Species ca. 30 (2 in the flora): sw United States, Mexico, Central America, South America.

SELECTED REFERENCE Dillon, M. O. 1984. A systematic study of *Flourensia* (Asteraceae, Heliantheae). Fieldiana, Bot., n. s. 16: 1–66.

1. Shrubs; leaf blades elliptic to ovate, 10–25(–40+) × 4–15(–20) mm; ray florets 0 . . . . . . 1. *Flourensia cernua*
1. Subshrubs; leaf blades elliptic to lance-oblong, (20–)50–100 × 10–40+ mm; ray florets 13–21 . . . . 2. *Flourensia pringlei*

1. Flourensia cernua de Candolle in A. P. de Candolle and A. L. P. P. de Candolle, Prodr. 5: 593. 1836

• Tarbush F

Shrubs to 100(–200) cm. **Leaf blades** elliptic to ovate, 10–25(–40+) × 4–15(–20) mm (margins sometimes undulate). **Ray florets** 0. **Disc florets** 10–25(–40); corollas 3–4 mm. **Cypselae** 4–6.5 mm; **pappi** 2.5–3.5 mm. $2n = 36$.

Flowering mostly Sep–Nov. Limestone or alkaline or clay soils, gravelly sites, desert scrub; 800–2000+ m; Ariz., N.Mex., Tex.; Mexico.

Plants of *Flourensia cernua* usually have a tarry odor and are often locally co-dominant with *Larrea tridentata* throughout much of the Chihuahuan Desert.

2. Flourensia pringlei (A. Gray) S. F. Blake, Proc. Amer. Acad. Arts 49: 375. 1913

Helianthella pringlei A. Gray, Proc. Amer. Acad. Arts 21: 389. 1886

Subshrubs to 100 cm. **Leaf blades** elliptic to lance-oblong, (20–)50–100 × 10–40+ mm. **Ray florets** 13–21; laminae 10–15+ mm. **Disc florets** 40–50; corollas 5–6 mm. **Cypselae** 8–12 mm, sericeous; **pappi** to 4 mm.

Flowering mostly Aug–Oct. Rocky slopes, disturbed sites; 1500–2100 m (Mexico); N.Mex.; Mexico (Chihuahua, Durango).

282. ENCELIA Adanson, Fam. Pl. 2: 128. 1763 • [For Christoph Entzelt (Christophorus Enzelius), 1517–1583, German naturalist]

Curtis Clark

Perennials, subshrubs, or shrubs (10–)30–150 cm. **Stems** erect, usually branched from bases, often throughout (scapiform in *E. nutans* and *E. scaposa*). **Leaves** usually cauline, sometimes basal (*E. nutans* and *E. scaposa*); alternate (usually drought-deciduous); petiolate (obscurely in *E. scaposa*); blades (1- or 3-nerved) mostly deltate, lanceolate, rhombic, or ovate (narrowly oblanceolate to linear in *E. scaposa*), bases broadly to narrowly cuneate, margins usually entire, rarely toothed, face glabrous or canescent, hirtellous, scabrellous, strigose, or tomentose, often gland-dotted as well. **Heads** radiate or discoid, borne singly or in ± paniculiform arrays (peduncles usually longer than involucres). **Involucres** ± hemispheric or broader, 4–22 mm diam. **Phyllaries** persistent, 18–30(–50+) in 2–3+ series (subequal to unequal, outer shorter). **Receptacles** flat or convex, paleate (paleae ± conduplicate, folded around and falling with cypselae). **Ray florets** 0 or 8–25(–40), neuter; corollas yellow. **Disc florets** 80–100(–200+), bisexual, fertile; corollas yellow or brown-purple, tubes shorter than to equaling abruptly expanded throats, lobes

5, triangular. **Cypselae** strongly compressed, obovate to cuneate (margins ciliate, apices usually ± notched except in *E. scaposa*, faces usually glabrous except in *E. scaposa*); **pappi** usually 0, sometimes readily falling or persistent, of 2 bristlelike awns. *x* = 18.

Species 13 or 14 (8 in the flora): sw United States, Mexico, South America.

Encelias commonly hybridize, especially in disturbed areas: *Encelia farinosa* × *E. frutescens* is common; *E. farinosa* × *E. californica*, *E. farinosa* × *E. actoni*, *E. actoni* × *E. frutescens*, *E. frutescens* × *E. virginensis*, and *E. farinosa* × *Geraea canescens* have been reported.

SELECTED REFERENCE   Clark, C. 1998. Phylogeny and adaptation in the *Encelia* alliance (Asteraceae: Heliantheae). Aliso 17: 89–98.

1. Perennials; leaves all or mostly basal.
  2. Ray florets 20–40 . . . . 1. *Encelia scaposa*
  2. Ray florets 0 (heads nodding in fruit) . . . . 2. *Encelia nutans*
1. Subshrubs or shrubs; leaves cauline.
  3. Ray florets 0 . . . . 3. *Encelia frutescens*
  3. Ray florets 8–25.
    4. Heads in paniculiform arrays; leaves tomentose to strigose.
      5. Leaves tomentose (branching among heads mainly distal; ray florets 11–21, corolla laminae 8–12 mm) . . . . 8. *Encelia farinosa*
      5. Leaves tomentose to strigose (branching among heads mainly proximal; ray florets fewer and smaller) . . . . *Encelia farinosa* × *E. frutescens* (see 8. *E. farinosa*)
    4. Heads borne singly; leaves glabrous or canescent, scabrous, and/or strigose (not tomentose).
      6. Leaves glabrous or glabrate; disc corollas brown (ray laminae lengths 1.5–2 times disc diams.) . . . . 7. *Encelia californica*
      6. Leaves scabrous, strigose, and/or canescent; disc corollas yellow (ray laminae lengths 1–1.2 times disc diams.).
        7. Leaves scabrous to strigose (not canescent) . . . . 4. *Encelia resinifera*
        7. Leaves canescent, sometimes strigose as well.
          8. Leaves sparsely canescent and strigose; rays 11–21, laminae 8–15 mm (relatively deeply toothed) . . . . 5. *Encelia virginensis*
          8. Leaves ± silvery-canescent (not strigose); rays 14–25, laminae 10–25 mm (relatively shallowly toothed) . . . . 6. *Encelia actoni*

1. Encelia scaposa (A. Gray) A. Gray, Proc. Amer. Acad. Arts 8: 657. 1873   E F

Simsia scaposa A. Gray, Smithsonian Contr. Knowl. 5(6): 88. 1853; *Encelia scaposa* var. *stenophylla* Shinners

Perennials, 10–30(–60) cm (caudices 1–2 cm diam.). **Stems** contracted, at soil surface or ± subterranean (except peduncles). **Leaves** mostly basal; petioles ± wanting or merging with blades; blades greenish to cinereous, (proximalmost scalelike) mostly narrowly oblanceolate to linear, 30–100 mm (including attenuate bases, mostly 1–8 mm wide), faces ± hirtellous to scabrellous. **Heads** borne singly. **Peduncles** ± scabrellous. **Involucres** 12–22 mm. **Phyllaries** linear. **Ray florets** 20–40. **Disc corollas** yellow, ca. 5 mm. **Cypselae** (cuneate to obovate) ca. 5 mm (faces ± villous); **pappi** (readily falling) of 2 (± villous) bristlelike awns.

Flowering Mar–Apr. Rocky, desert slopes; 1300–1700 m; N.Mex., Tex.

Although *Encelia scaposa* traditionally has been included within *Encelia*, it seems misplaced here. Its cypselae differ from those of other encelias in lacking a narrowed apical notch and in having more or less villous (rather than mostly glabrous) faces. Molecular evidence suggests that it may be closer to *Flourensia* than to members of the alliance comprising *Encelia*, *Enceliopsis*, and *Geraea*.

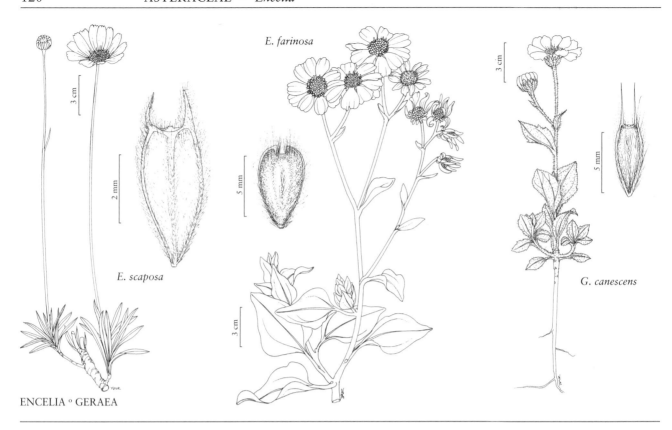

E. farinosa

E. scaposa

G. canescens

ENCELIA ° GERAEA

2. Encelia nutans Eastwood, Zoë 2: 230. 1891 [E]

Enceliopsis nutans (Eastwood)
A. Nelson

Perennials, 10–25 cm (roots swollen, 6–10 cm). **Stems** ± subterranean (except peduncles). **Leaves** basal; petioles 20–35 mm; blades green, broadly ovate, 30–40 mm, faces glabrous or substrigose (sometimes rugose). **Heads** borne singly (nodding in fruit). **Peduncles** hairy. **Involucres** 12–22 mm. **Phyllaries** lanceolate. **Ray florets** 0. **Disc corollas** yellow, 6–8 mm. **Cypselae** 7–12 mm; **pappi** usually 0, rarely of 2 bristlelike awns. **2n** = 36.

Flowering Apr. Flat areas with heavy soils; 1300–1900 m; Colo., Utah.

Encelia nutans shares the apically notched and glabrous-faced cypselae of the genus, but its geophyte habit, especially the ephemeral nature of its aboveground parts, is unique in either *Encelia* or *Enceliopsis*.

3. Encelia frutescens (A. Gray) A. Gray, Proc. Amer. Acad. Arts 8: 657. 1873

Simsia frutescens A. Gray in W. H. Emory, Rep. U.S. Mex. Bound. 2(1): 89. 1859

Subshrubs or shrubs, 50–150 cm. **Stems** with slender branches from trunks, glabrous, developing fissured barks. **Leaves** cauline; petioles 2–7 mm; blades green, elliptic or narrowly ovate, 10–25 mm, apices obtuse, faces strigose. **Heads** borne singly. **Peduncles** strigose. **Involucres** 6–12 mm. **Phyllaries** lanceolate. **Ray florets** 0. **Disc corollas** yellow, 5–6 mm. **Cypselae** 6–9 mm; **pappi** 0 or of 2 bristlelike awns. **2n** = 36.

Flowering Feb–May, Aug–Sep. Desert washes, flats, slopes, roadsides; 0–800 m; Ariz., Calif., Nev.; Mexico (Baja California, Sonora).

Plants of *Encelia frutescens* in the flora area are var. *frutescens*; var. *glandulosa* C. Clark, with glandular, strigose leaves, is found in northeastern Baja California and is not expected here.

4. Encelia resinifera C. Clark, Aliso 17: 201. 1998 [E]

Encelia frutescens (A. Gray) A. Gray var. *resinosa* M. E. Jones ex S. F. Blake, Proc. Amer. Acad. Arts 49: 364. 1913, not *E. resinosa* Brandegee 1907; *E. resinifera* subsp. *tenuifolia* C. Clark

Shrubs, 40–150 cm. **Stems** with slender branches from trunks, glabrate, developing fissured barks. **Leaves** cauline; petioles 2–6 mm; blades green, lanceolate to ovate, 10–25 mm, apices usually acute, sometimes obtuse, faces strigose, gland-dotted. **Heads** borne singly. **Peduncles** strigose to glabrate. **Involucres** 5–11 mm. **Phyllaries** lanceolate. **Ray florets** 8–13; corolla laminae 8–12 mm. **Disc corollas** yellow, 5–6 mm. **Cypselae** 5–8 mm; **pappi** 0. **2*n*** = 36.

Flowering May–Jul, Sep. Sandstone-derived soils; 1100–1700 m; Ariz., Utah.

Plants of *Encelia resinifera* from the Grand Canyon of the Colorado with lengths of leaf blades and ray laminae at least three times their widths are subsp. *tenuifolia*.

5. Encelia virginensis A. Nelson, Bot. Gaz. 37: 272. 1904 [E]

Encelia frutescens (A. Gray) A. Gray var. *virginensis* (A. Nelson) S. F. Blake

Shrubs, 50–150 cm. **Stems** with slender branches from bases, hairy, developing fissured barks. **Leaves** cauline; petioles 2–7 mm; blades gray-green, narrowly ovate to deltate, 12–25 mm, apices acute or obtuse, faces sparsely canescent and strigose. **Heads** borne singly. **Peduncles** canescent. **Involucres** 9–13 mm. **Phyllaries** narrowly ovate. **Ray florets** 11–21; corolla laminae 8–15 mm. **Disc corollas** yellow, 5–6 mm. **Cypselae** 5–8 mm; **pappi** usually 0, rarely of 1–2 bristlelike awns. **2*n*** = 36.

Flowering Apr–Jun, Dec. Desert flats, rocky slopes, roadsides; 500–1500 m; Ariz., Calif., Nev., N.Mex., Utah.

In the mountains of the eastern Mojave Desert in California, *Encelia virginensis* may intergrade with *E. actoni* at higher elevations, probably as a result of hybridization. Plants of *E. virginensis* in New Mexico may be adventive.

6. Encelia actoni Elmer, Bot. Gaz. 39: 47. 1905

Encelia frutescens (A. Gray) A. Gray var. *actoni* (Elmer) S. F. Blake; *E. virginensis* A. Nelson subsp. *actoni* (Elmer) D. D. Keck

Shrubs, 50–150 cm. **Stems** with slender branches from bases, hairy, developing fissured barks. **Leaves** cauline; petioles 6–12 mm; blades silvery green, ovate to deltate, 25–40 mm, apices acute, faces ± silvery-canescent. **Heads** borne singly. **Peduncles** canescent. **Involucres** 8–14 mm. **Phyllaries** ovate. **Ray florets** 14–25; corolla laminae 10–25 mm. **Disc corollas** yellow, 5–6 mm. **Cypselae** 5–7 mm; **pappi** usually 0, rarely of 1–2 bristlelike awns. **2*n*** = 36.

Flowering Feb–Jul. Open areas, rocky slopes, roadsides; 800–1500 m; Calif., Nev.; Mexico (Baja California).

The epithet *actoni* (sometimes spelled *actonii*) is from a place name (Acton, Los Angeles County, California); *actoni* is correct.

7. Encelia californica Nuttall, Trans. Amer Philos. Soc., n. s. 7: 357. 1841

Shrubs, 50–150 cm. **Stems** with slender branches from bases, glabrous, developing smooth or roughened barks. **Leaves** cauline; petioles 5–25 mm; blades green, rhombic or narrowly ovate, 30–60 mm, apices acute, faces glabrous. **Heads** borne singly. **Peduncles** pubescent. **Involucres** 10–12 mm. **Phyllaries** lanceolate. **Ray florets** 15–25; corolla laminae 15–35 mm. **Disc corollas** brown-purple, 5–6 mm. **Cypselae** 5–7 mm; **pappi** 0. **2*n*** = 36.

Flowering Feb–Oct. Coastal scrub; 0–600 m; Calif.; Mexico (Baja California).

8. Encelia farinosa A. Gray ex Torrey in W. H. Emory, Not. Milit. Reconn., 143. 1848 • Brittlebush, incienso [F]

Encelia farinosa var. *phenicodonta* S. F. Blake) I. M. Johnston

Shrubs, 30–150 cm (sap fragrant). **Stems** branched distally, tomentose, developing smooth barks. **Leaves** cauline (clustered near stem tips); petioles 10–20 mm; blades silver or gray, ovate to lanceolate, 20–70 mm, apices obtuse or acute, faces tomentose. **Heads** in paniculiform arrays (branching among heads mainly distal). **Peduncles** glabrous except near heads (± yellow). **Involucres** 4–10 mm.

Phyllaries lanceolate. **Ray florets** 11–21; corolla laminae 8–12 mm. **Disc corollas** yellow or brown-purple, 5–6 mm. **Cypselae** 3–6 mm; **pappi** 0. $2n = 36$.

Flowering Feb–May, Aug–Sep. Coastal scrub, stony desert hillsides; 0–1000 m; Ariz., Calif., Nev.; Mexico (Baja California, Baja California Sur, Sinaloa, Sonora).

Plants of *Encelia farinosa* with brown-purple disc corollas, found along the Colorado and Salt rivers, and common in Baja California, are var. *phenicodonta*. Plants with substrigose leaves, capitulescences branched toward bases rather than distally, and ray florets reduced in both size and number are most often hybrids and backcrosses between *E. farinosa* and *E. frutescens*. P. A. Munz (1959) indicated that I. L. Wiggins had reported var. *radians* Brandegee ex S. F. Blake as occurring in southeastern California; that variety is known only from Baja California.

283. GERAEA Torrey & A. Gray, Proc. Amer. Acad. Arts 1: 48. 1847 • Desert-sunflower [Greek *geraios*, old, alluding to white-haired involucre of *Geraea canescens*]

Curtis Clark

Annuals or perennials, 10–100 cm. **Stems** erect, branched (branches ascending). **Leaves** basal and cauline; alternate; sessile or petiolate; blades (3-nerved) elliptic, lanceolate, oblanceolate, oblong, or ovate, bases auriculate or tapering, margins entire or toothed, faces ± canescent or glandular-puberulent to stipitate-glandular. **Heads** radiate or discoid, borne singly or in open, paniculiform arrays (peduncles usually longer than involucres). **Involucres** hemispheric, 7–15 mm diam. **Phyllaries** persistent, 20–30+ in 2–3+ series (distinct, narrowly lanceolate to linear or narrowly lance-oblong to lance-ovate, subequal to unequal, outer shorter). **Receptacles** convex, paleate (paleae ± conduplicate, folded around and falling with cypselae). **Ray florets** 0 or 10–21, neuter; corollas yellow. **Disc florets** 60–200+, bisexual, fertile; corollas yellow, tubes shorter than gradually to abruptly expanded throats, lobes 5, triangular. **Cypselae** (black) strongly compressed, narrowly cuneate (margins ± white, long-ciliate, faces ± hairy); **pappi** persistent, of 2 awns or subulate scales. $x = 18$.

Species 2 (2 in the flora): sw United States, nw Mexico.

1. Annuals; leaves sessile or petiolate, bases of blades tapering to wings, faces ± canescent; phyllaries narrowly lanceolate to linear, apices acute (margins shaggy-ciliate); ray florets 10–21 . . . . 1. *Geraea canescens*
1. Perennials; leaves sessile, bases of blades ± auriculate, faces glandular-puberulent; phyllaries narrowly lance-oblong to lance-ovate, apices obtuse (faces densely glandular-puberulent); ray florets 0 . . . . 2. *Geraea viscida*

1. Geraea canescens Torrey & A. Gray, Proc. Amer. Acad. Arts 1: 49. 1847 • Hairy desert-sunflower, desert-gold F

Geraea canescens var. *paniculata* (A. Gray) S. F. Blake

Annuals, 10–80 cm; herbage bristly and/or soft-hairy. **Leaves** petiolate (proximal) or sessile; blades green or gray, lanceolate or ovate to elliptic or oblanceolate, 1–10 cm, bases tapering to wings, margins entire or dentate, apices acute, faces ± canescent. **Heads** borne singly or in paniculiform arrays. **Involucres** 7–12 mm. **Phyllaries** green, narrowly lanceolate to linear (margins shaggy-ciliate), apices acute. **Ray florets** 10–21; laminae 1–2 cm. **Disc corollas** 4–5 mm. **Cypselae** 6–7 mm; **pappi** 3–4 mm. $2n = 36$.

Flowering Oct–May. Sandy desert soils; 0–1300 m; Ariz., Calif., Nev., Utah; Mexico (Baja California, Sonora).

2. Geraea viscida (A. Gray) S. F. Blake, Proc. Amer. Acad. Arts 49: 357. 1913 • Sticky desert-sunflower

Encelia viscida A. Gray, Proc. Amer. Acad. Arts 11: 78. 1876

Perennials, 30–100 cm (caudices underground); herbage densely glandular-puberulent and ± bristly. **Leaves** sessile; blades green, ovate to oblong, 3–9 cm, bases ± auriculate, margins entire or dentate, apices obtuse, faces glandular-puberulent to stipitate-glandular. **Heads** usually borne singly. **Involucres** 10–15 mm. **Phyllaries** green, narrowly lance-oblong, apices obtuse (abaxial faces densely glandular-puberulent). **Ray florets** 0. **Disc corollas** 6–8 mm. **Cypselae** 7–10 mm; **pappi** 3–5 mm. $2n = 36$.

Flowering May–Jul. Openings in chaparral; 400–1700 m; Calif.; Mexico (Baja California).

284. MELANTHERA Rohr, Skr. Naturhist.-Selsk. 2: 213. 1792 • [Greek *melan*, black, and Latin, *anthera*, anther]

James C. Parks†

Perennials or subshrubs, 30–220 cm. **Stems** usually erect or spreading, sometimes sprawling or scandent, branched throughout (± 4-angled and sulcate, usually scabro-hispid, caudices usually ± woody, often ± spheric). **Leaves** cauline; opposite; petiolate or nearly sessile; blades (usually 3-nerved) usually lance-elliptic, linear, oblanceolate, ovate-elliptic, or ovate, sometimes deltate to 3-lobed, bases cuneate to truncate (sometimes hastate or with antrorse basal lobes), margins crenate or serrate (often irregularly), faces usually scabro-hispid, sometimes glabrescent. **Heads** discoid [radiate], borne singly or in loose, corymbiform arrays. **Involucres** ± hemispheric or broader, 6–20 mm diam. **Phyllaries** persistent, mostly 8–16 in 2–3 series (mostly ovate to lanceolate, subequal to unequal, bases pallid, with or without green veins, apices herbaceous, tips usually ± mucronate). **Receptacles** flat to convex or convex-hemispheric, paleate (paleae oblanceolate, conduplicate, apices mucronate, pungent, erect or spreading to recurved). **Ray florets** 0 [8–15]. **Disc florets** 20–100+, bisexual; corollas white [pale to bright yellow], tubes shorter than narrowly funnelform to cylindric throats, lobes 5, ± deltate (anther sacs black; style branches stigmatic in 2 lines, appendages lance-triangular, papillate). **Cypselae** (± brown) obpyramidal, (3–)4-angled, not or slightly compressed (striate, rarely verrucose-tuberculate, glabrous); **pappi** readily falling, of 2–12 barbellulate bristles or awns. $x = 15$.

Species ca. 20 (3 in the flora): North America, Mexico, West Indies, Central America, South America, Africa.

Melanthera is common as a strand and weedy plant in Mexico, the West Indies, Central America, South America, and sub-Saharan Africa. In the flora, *Melanthera* is distinguished by its lack of ray florets, white disc corollas, distinctive cypselae, and delicate, caducous pappus awns. *Tilesia* G. F. W. Meyer (= *Wulffia* Necker ex Cassini), *Wollastonia* de Candolle ex Decaisne, and *Echinocephalum* Gardner are closely related to *Melanthera*. The type of *Echinocephalum* (South America; rays neuter, corollas pale yellow) may be congeneric with that of *Melanthera*. Inclusion of African species (all radiate) in *Melanthera* further broadens and complicates its delimitation. H. Wild (1965) broadened the circumscription even more by including the unispecific *Wollastonia* among African *Melanthera* species with styliferous ray florets.

SELECTED REFERENCE Parks, J. C. 1973. A revision of North American and Caribbean *Melanthera* (Compositae). Rhodora 75: 169–210.

1. Leaf blades usually linear to narrowly lanceolate-oblanceolate, rarely narrowly ovate-elliptic; peduncles glabrescent; phyllaries ovate, 4–5 × 2–3 mm; mucros on paleae 0.5–1 mm, straight to slightly recurved . . . . 2. *Melanthera angustifolia*
1. Leaf blades usually ovate, sometimes deltate or 3-lobed, rarely lanceolate-elliptic; peduncles hispid to strigose; phyllaries broadly ovate to lanceolate, 5–7 × 1.5–3 mm, or 3–10 × 3–4 mm; mucros on paleae 0.5–1.5 mm, usually notably recurved.
   2. Plants 30–60 cm; stems sprawling to weakly erect; leaf blades 1.5–4 × 1–1.5 cm; florets 25–40 . . . . 3. *Melanthera parvifolia*
   2. Plants 50–220 cm; stems usually erect, sometimes sprawling or scandent; leaf blades 5–12 × 3–8 cm; florets 30–100+ . . . . 1. *Melanthera nivea*

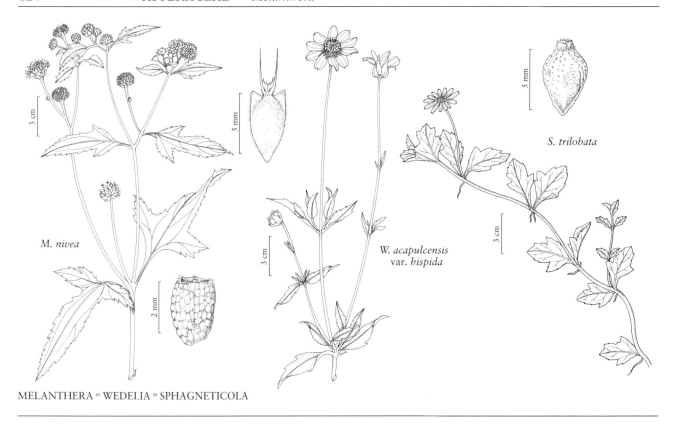

MELANTHERA ∘ WEDELIA ∘ SPHAGNETICOLA

1. Melanthera nivea (Linnaeus) Small, Fl. S.E. U.S.,
1251, 1340. 1903 F

Bidens nivea Linnaeus, Sp. Pl. 2:
833. 1753; *Calea aspera* Jacquin;
Melanthera aspera (Jacquin) Small;
M. carpenteri Small; *M. deltoidea*
Michaux; *M. hastata* Michaux;
M. ligulata Small

Plants 50–220 cm. **Stems** usually
erect, sometimes sprawling or scandent, usually scabro-hispid, some-
times glabrescent. **Leaf blades** usually ovate, sometimes
deltate or 3-lobed, rarely lance-elliptic, 5–12 × 3–8 cm
(often triplinerved), bases truncate to cuneate, margins
crenate to serrate (often irregularly), faces usually strigose
to hispid, sometimes glabrescent. **Heads** borne in axils of
reduced distal leaves (sometimes forming loose,
corymbiform arrays). **Peduncles** 5–14 cm, hispid to strigose. **Involucres** 12–20 mm diam. **Phyllaries** broadly ovate
to lanceolate, 3–10 × 3–4 mm. **Paleae** 4–7 × 1.5–2.5 mm,
apical mucros 0.5–1.4+ mm, usually recurved. **Florets** 30–
100+; corollas 5–10 mm; anther sacs 2+ mm. **Cypselae**
2–3 × 1–2 mm. $2n$ = 30.

Flowering year-round. Moist forests, forest borders,
strands; 0–10(–200) m; Ala., Fla., Ga., Ill., Ky., La., Miss.,
S.C.; Mexico; West Indies; Central America; South
America.

In the flora area, most populations of *Melanthera nivea*
occur along the Atlantic and Gulf coastal plains;
populations are also known from southern Illinois and
Kentucky.

Melanthera nivea ranges widely geographically and
varies ecologically and morphologically. Apparent
ecogeographic races or ecotypes are often distinct; I have
delimited them taxonomically (J. C. Parks 1973). I took
the northern, erect, forest-dwelling plants with sulcate,
purple-mottled stems, hastate leaves, and lanceolate, long-
acuminate, recurved phyllaries and paleae to be *M. nivea*
in the restricted sense. I referred sprawling, bushlike,
highly scabrous plants with moderately long phyllaries
and paleae from open, weedy habitats to *M. aspera* var.
aspera. A strand form from the West Indies and southern Florida, called *M. aspera* var. *glabriuscula* (Kuntze)
J. C. Parks, was distinguished by its less hispid, deltate
leaves, ovate, obtuse phyllaries, paleae with apical mucros only slightly longer than unopened disc florets, and
cypselae frequently obovate with relatively numerous
short hairs on the apex rather than obpyramidal with
relatively few hairs on a truncate-concave apex. Field
observations by floristic botanists have yielded plants of
intermediate morphology. Healthy F_1 offspring were
recovered from 88 genetic crosses among elements of *M.
nivea* in the broad sense (i.e., including *M. aspera*;
unpubl.). On the basis of results from crosses and field
observations, I concluded that it is better taxonomically
to treat the ecotypes as variants of one wide-ranging,
morphologically variable, possibly diverging species.

2. Melanthera angustifolia A. Richard in R. de la Sagra, Hist. Fis. Cuba, 11: 54. 1850

Plants 30–90 cm. **Stems** weakly erect, strigose. **Leaf blades** usually linear to narrowly oblanceolate, rarely narrowly ovate-elliptic, 3–8 × 0.5–1.5 cm, bases cuneate, margins irregularly serrate, faces sparsely strigose. **Heads** borne singly. **Peduncles** 9–20 cm, glabrescent. **Involucres** 6–12 mm diam. **Phyllaries** ovate, 4–5 × 2–3 mm. **Paleae** 4–5 × 1 mm, apical mucros 0.5–1 mm, straight to slightly recurved. **Florets** 30–40; corollas 5.5 mm; anther sacs 2 mm. **Cypselae** 2.5 × 1.5 mm. $2n = 30$.

Flowering Jan–Mar (perhaps year-round). Moist to dry, lightly shaded pine flatwoods; 0–10+ m; Fla.; Mexico (Yucatan); West Indies (Cuba); Central America.

3. Melanthera parvifolia Small, Fl. S.E. U.S., 1251, 1340. 1903 E

Melanthera radiata Small

Plants 30–60 cm. **Stems** sprawling to weakly erect, strigose to hirsute. **Leaf blades** ovate or 3-lobed, 1.5–4 × 1–1.5 cm, bases broadly ± cuneate, hastate, margins coarsely serrate (sometimes undulate), faces strigose to hispid. **Heads** borne singly. **Peduncles** 2.5–10 cm, hispid to strigose. **Involucres** 10–14 mm diam. **Phyllaries** lanceolate, 5–7 × 1.5–3 mm. **Paleae** 6–7 × 2 mm, apical mucros 1–1.5 mm, usually recurved. **Florets** 25–40; corollas 5 mm; anther sacs 1.5 mm. **Cypselae** 2.5–3 × 1.5–2 mm. $2n = 30$.

Flowering Feb–Apr (perhaps year-round). Old coral reefs or porous oölitic rocks in open pine forests; 0–10+ m; Fla.

Melanthera parvifolia is known only from southern Florida, including Big Pine Key.

285. WEDELIA Jacquin, Enum. Syst. Pl., 8, 28. 1760, name conserved • [For George Wolfgang Wedel, 1645–1721, botanist/professor at Jena]

John L. Strother

Subshrubs or shrubs [annuals, perennials], 10–50(–100)[–250] cm. **Stems** mostly erect, branched from bases and/or throughout. **Leaves** cauline; opposite; petiolate or sessile; blades (3-nerved [pinnately nerved]) trullate to lanceolate or lance-linear [deltate, elliptic, filiform, linear, ovate], sometimes ± 3-lobed, bases cuneate to truncate, margins coarsely toothed to subentire, faces hispid [scabrous to scabrellous or strigose to strigillose, ± sericeous, often with finer, uncinate hairs as well], usually gland-dotted. **Heads** radiate (discoid), borne singly [in corymbiform arrays]. **Involucres** obconic to hemispheric, 4–8[–15] mm diam. **Phyllaries** persistent, 8–16+ in 2–3+ series (outer usually larger and/or more herbaceous than inner). **Receptacles** convex, paleate (paleae conduplicate, chartaceous to scarious). **Ray florets** 0 or 4–18, pistillate, fertile [neuter]; corollas yellow to orange [purplish or white]. **Disc florets** 8–150+, bisexual, fertile; corollas yellow to orange [purplish], sometimes marked with purple, tubes shorter than or equaling funnelform or cylindric throats, lobes 5, deltate. **Cypselae** ± dimorphic; peripheral sometimes obcompressed and weakly 3-angled, inner compressed and biconvex or flattened [somewhat 4-angled] (some or all winged, all ± rostrate, bearing central neck or boss apically, some or all each bearing wartlike elaiosome at base); **pappi** persistent, cyathiform (fimbriate cups plus 0–3 coarse bristles or awns borne together on rostra). $x = 13$?

Species 25+ (1 in the flora): tropical and subtropical New World.

African aspilias may belong within *Wedelia*.

Wedelia glauca is treated herein as *Pascalia glauca*, *W. trilobata* as *Sphagneticola trilobata*. A report of *Stemmodontia asperrima* (Sprengel) C. Mohr from ballast may be source of reports of *W. calycina* Richard or *W. frutescens* Jacquin from Alabama. I have seen no specimens to support those reports.

1. **Wedelia acapulcensis** Kunth in A. von Humboldt et al., Nov. Gen. Sp. 4(fol.): 168. 1818; 4(qto.): 215. 1820 [F]

Varieties 7 (1 in the flora): Texas; Mexico; Central America (to Costa Rica).

1a. **Wedelia acapulcensis** Kunth var. **hispida** (Kunth) Strother, Syst. Bot. Monogr. 33: 62. 1991 [F]

Wedelia hispida Kunth in A. von Humboldt et al., Nov. Gen. Sp. 4(fol.): 169. 1818; 4(qto.): 215, plate 371. 1820; *Zexmenia hispida* (Kunth) A. Gray

Leaves ± sessile; blades 20–50(–80+) × 4–12(–35) mm, sometimes obscurely to clearly 3-lobed. **Peduncles** 5–30+ cm. **Ray florets** 0 or 5–8(–13+); laminae 6–15 mm. **Disc florets** 15–80+; corollas 4.5–7.5 mm. **Cypselae** 4–5 mm; **pappi** of lacerate crowns 0.5–1 mm plus 0–3 bristles or awns 4.5–6 mm. $2n$ = ca. 52.

Flowering mostly late spring–early fall. Damp places in otherwise dry sites, igneous or limestone derived soils, thorn woodlands, desert scrublands; 0–1500+ m; Tex.; Mexico.

The name *Wedelia texana* has been widely applied to this taxon; *W. texana* has not been validly published [cf., *Lipochaeta texana* Torrey & A. Gray and *Zexmenia texana* (Torrey & A. Gray) A. Gray].

286. SPHAGNETICOLA O. Hoffmann, Notizbl. Königl. Bot. Gart. Berlin 3: 36. 1900
    • [Evidently from Latin *sphagnum*, a moss, and *cola*, dwelling in, perhaps alluding to usually wet habitats] [I]

John L. Strother

Perennials, mostly 30–200 cm. **Stems** prostrate, branched ± throughout (rooting at nodes, usually ± succulent). **Leaves** cauline; opposite; sessile or petiolate; blades (usually 3-nerved) mostly trullate to lanceolate, often obscurely 3-lobed, bases ± cuneate, margins entire or toothed, faces usually scabrous. **Heads** radiate, borne singly. **Involucres** obconic, 6–12 mm diam. **Phyllaries** persistent, 12–15 in 2–3 series (outer larger, more foliaceous than inner). **Receptacles** convex to conic, paleate (paleae conduplicate, scarious, acute). **Ray florets** 4–10+, pistillate, fertile; corollas yellow to orange. **Disc florets** 30–60+, bisexual, fertile; corollas yellow to orange, tubes shorter than campanulate throats, lobes 5, lance-triangular. **Cypselae** strongly biconvex to plumply 3–4-angled (epidermes usually thick, corky, smooth or tuberculate, apices rostrate, each bearing central neck or boss, rostra obscure in mature fruits); **pappi** 0, or persistent, coroniform (erose or fimbrillate to lacerate, 0.1–0.6 mm). x = 15.

Species 4 (1 in the flora): introduced; Mexico, West Indies, Bermuda, Central America, South America, Asia, Indian Ocean Islands, Pacific Islands, Australia.

1. **Sphagneticola trilobata** (Linnaeus) Pruski, Mem. New York Bot. Gard. 87: 114. 1996 [F][I]

Silphium trilobatum Linnaeus, Syst. Nat. ed. 10, 2: 1233. 1759 (as trilobatu); *Wedelia trilobata* (Linnaeus) Hitchcock

Leaf blades 5–18 × 1–5+ cm. **Ray laminae** 6–15 mm. **Disc corollas** 3.5–6 mm. **Cypselae** 3–5 mm. $2n$ = ca. 40, ca. 50–54, 50–58, 56, ca. 56, 57, and 60.

Flowering nearly year round. Ruderal, wet places; 0–10 m; introduced; Fla.; Mexico; West Indies; Bermuda; Central America; South America; probably introduced in Asia; Indian Ocean Islands; Pacific Islands; Australia.

Sphagneticola trilobata is probably native to subtropics and tropics of the New World and probably introduced in Old World. It is increasingly cultivated and is to be expected as escaped or persisting in warm to hot, sandy spots anywhere in the flora area. Report of it from Louisiana has not been confirmed.

287. **SYNEDRELLA** Gaertner, Fruct. Sem. Pl. 2: 456, plate 171, fig. 7. 1791, name conserved • [Etymology not clear; perhaps alluding to clustering of heads] [I]

John L. Strother

Annuals, 10–80+ cm. **Stems** erect or ascending, branched from bases or ± throughout. **Leaves** cauline; opposite; petiolate; blades (usually 3-nerved) ovate to elliptic, bases cuneate to rounded, margins toothed, faces ± scabrous. **Heads** radiate, (sessile or subsessile) in axillary glomerules or borne singly. **Involucres** cylindric to campanulate, 3–6 mm diam. **Phyllaries** persistent, 2–5+ in 1(–2) series (lanceolate, herbaceous to chartaceous). **Receptacles** convex, paleate (paleae lance-linear, scarious, flat or weakly cupped at bases). **Ray florets** 2–9+ (in 1–2 series), pistillate, fertile; corollas yellowish. **Disc florets** 4–12+, bisexual, fertile; corollas yellowish, tubes about equaling slightly ampliate, cylindric throats, lobes 4, rounded-deltate. **Cypselae** dimorphic, strongly obcompressed or flattened, narrowly oval and winged (ray, wings laciniate) or linear-cuneate and not winged (disc); **pappi** persistent, of 2 triangular scales (ray) or 2–3 subulate scales or awns (disc). *x* = 20?

Species 1: introduced; Mexico, West Indies, Central America, South America; also introduced in Asia, Africa, Pacific Islands, Australia.

1. **Synedrella nodiflora** (Linnaeus) Gaertner, Fruct. Sem. Pl. 2: 456. 1791 [F] [I]

Verbesina nodiflora Linnaeus, Cent. Pl. I, 28. 1755

Leaf blades 3–10 × 2–5 cm. **Ray corollas:** tubes ca. 2 mm, laminae narrowly ovate to linear, 2–4 mm. **Disc corollas** ca. 3 mm. **Cypselae** ca. 4 mm. *2n* = 36, 38, 40.

Flowering (sporadically) year round. Ruderal in ± wet places; 0–10 m; introduced; Fla.; Mexico; West Indies; Central America; South America; also introduced in Asia, Africa, Pacific Islands, Australia.

Synedrella nodiflora probably is a recurrent introduction and probably does not persist in the flora area.

288. **JEFEA** Strother, Syst. Bot. Monogr. 33: 22, fig. 2. 1991 • [Spanish *jefe*, chief; for Billie Lee Turner, b. 1925, Texan, botanist]

John L. Strother

Subshrubs or shrubs, 10–100[–200+] cm. **Stems** erect, branched ± throughout. **Leaves** cauline; opposite (proximal) or alternate; petiolate; blades (usually 3-nerved) deltate, lanceolate, oblanceolate, or suborbiculate, bases cuneate to truncate, margins entire or lobed [toothed], faces scabrellous to hispid and gland-dotted [densely woolly abaxially]. **Heads** radiate, borne singly. **Involucres** campanulate to hemispheric, 6–10+ mm diam. **Phyllaries** persistent, [22–]26–38+ in 3–4+ series (outer 2–6+ spreading, similar to foliage in shape, texture, and indument, inner appressed, more papery to scarious or membranous). **Receptacles** convex to conic, paleate (paleae persistent, lance-linear, conduplicate). **Ray florets** 5–13[–20], pistillate, fertile; corollas yellow to orange. **Disc florets** 30–60[–100+], bisexual, fertile; corollas yellow to orange, tubes shorter than or about equaling funnelform throats, lobes 5, deltate. **Cypselae** weakly obcompressed, 3-angled (peripheral) or strongly compressed (inner) and oblanceolate to rounded-cuneate, some or all winged; **pappi** fragile or persistent, of 2–3 subulate scales or awns plus 2–8+ shorter, distinct or basally connate, erose or lacerate scales (often with an additional seta on inner shoulder of each cypsela). *x* = 14.

Species 5 (1 in the flora): sw United States, Mexico, Central America (Guatemala).

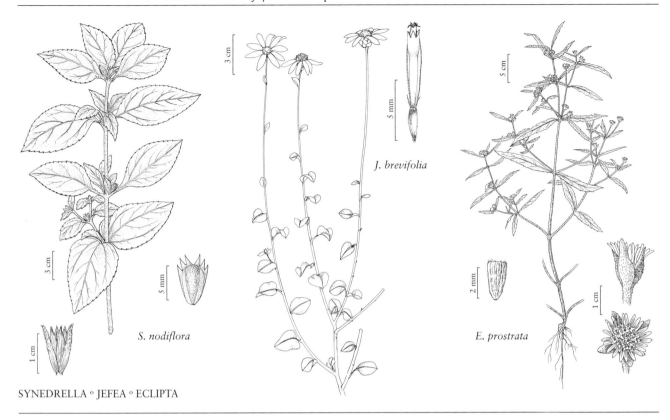

SYNEDRELLA ∘ JEFEA ∘ ECLIPTA

J. brevifolia

S. nodiflora

E. prostrata

1. **Jefea brevifolia** (A. Gray) Strother, Syst. Bot. Monogr. 33: 26. 1991 [F]

Zexmenia brevifolia A. Gray, Smithsonian Contr. Knowl. 3(5): 112. 1852

Leaf blades 8–25+ × 4–20 mm, sometimes obscurely 3(–5)-lobed. **Cypselae** 4–5 × 1.6–2.8 mm, including wings. $2n$ = 28.

Flowering summer–fall. Limestone, desert scrub; 600–1200 [–2400] m; N.Mex., Tex.; Mexico (Chihuahua, Coahuila, Nuevo León, San Luis Potosí, Tamaulipas, Zacatecas).

289. **ECLIPTA** Linnaeus, Mant. Pl., 157, 286. 1771, name conserved   •   [Greek *ekleipsis*, a failing, perhaps alluding to minute or wanting pappus]

John L. Strother

Annuals or perennials, 10–50(–70+) cm. **Stems** erect or decumbent, branched from bases and/ or distally (sometimes rooting at proximal nodes). **Leaves** cauline; opposite; petiolate or sessile; blades (1- or 3-nerved) lanceolate to lance-linear, bases cuneate, margins serrate to subentire, faces sparsely scabrellous. **Heads** radiate, in loose, corymbiform arrays or borne singly. **Involucres** hemispheric, 3–5 mm diam. **Phyllaries** persistent, 8–12+ in 2–3 series (lanceolate to linear, subequal, thin-herbaceous, spreading in fruit). **Receptacles** flat to convex, paleate (paleae linear to filiform, not conduplicate, falling with fruits). **Ray florets** 20–40 (in 2–3+ series), pistillate, fertile; corollas white or whitish. **Disc florets** 15–30+, bisexual, fertile; corollas white or whitish, tubes much shorter than ampliate, cylindric throats, lobes 4–5, ± deltate. **Cypselae** obcompressed, weakly 3–4-angled (not winged, epidermes usually corky and rugose to tuberculate); **pappi** persistent, coroniform (sometimes with 2 teeth). $x = 11$.

Species 1–4 (1 in the flora): mostly warm-temperate to tropical New World; introduced in Old World.

1. **Eclipta prostrata** (Linnaeus) Linnaeus, Mant. Pl., 286. 1771 [F]

Verbesina prostrata Linnaeus, Sp. Pl. 2: 902. 1753; *Eclipta alba* (Linnaeus) Hasskarl

Leaf blades 2–10 cm × 4–30+ mm. **Ray laminae** ca. 2 mm. **Disc corollas** ca. 1.5 mm. **Cypselae** ca. 2.5 mm. $2n = 22$ (as *E. alba*).

Flowering year round, mostly summer–fall. Wet places, often ruderal; 60–1600 m; Ont.; Ala., Ariz., Ark., Calif., Del., D.C., Fla., Ga., Ill., Ind., Iowa, Kans., Ky., La., Md., Mass., Mich., Minn., Miss., Mo., Nebr., N.J., N.Mex., N.Y., N.C., Ohio, Okla., Pa., S.C., S.Dak., Tenn., Tex., Va., W.Va., Wis.; Mexico; West Indies; Central America; South America; introduced in Europe, Asia, Africa, Pacific Islands, Australia.

290. **BORRICHIA** Adanson, Fam. Pl. 2: 130. 1763 • Sea oxeye daisy, seaside tansy or oxeye [For Ole Borch (Olaus Borrichius), 1626–1690, Danish botanist]

John C. Semple

Perennials, subshrubs, or shrubs, to 150 cm (rhizomatous, forming clonal colonies, roots fibrous, sometimes adventitious). **Stems** decumbent to erect, branched ± throughout (sap sticky-resinous). **Leaves** cauline; opposite; ± petiolate or sessile; blades (usually 1-nerved, sometimes obscurely 3- or 5-nerved) elliptic, linear, oblanceolate, obovate, or ovate (usually coriaceous or succulent), bases ± cuneate, margins entire or toothed (teeth often spine-tipped), faces glabrous or puberulent to villous and/or sericeous. **Heads** radiate, borne singly or (3–10) in cymiform arrays. **Involucres** globose or ovoid to hemispheric or broader, (5–)8–13(–18+) mm diam. **Phyllaries** persistent, 10–45 in 2–3 series (outer larger, elliptic to oblanceolate or ovate, apices acute, cuspidate, obtuse, rounded, or spine-tipped). **Receptacles** convex, paleate (paleae lanceolate to ovate, ± conduplicate, partially enclosing cypselae, apices often ± pungent). **Ray florets** 7–30, pistillate, fertile; corollas yellow. **Disc florets** 20–75, bisexual, fertile; corollas yellow, tubes shorter than funnelform throats, lobes 5, lance-triangular (anthers black with orange glands on connectives). **Cypselae** (gray to black) ± obcompressed, or obpyramidal and 3- or 4-angled (faces faintly finely reticulate); **pappi** persistent or tardily falling, (stramineous) coroniform or cupular (3–4-angled). $x = 14$.

Species 3 (3 species, including 1 hybrid, in the flora): United States, Mexico, West Indies.

Leaves of *Borrichia* species are usually heteroblastic: leaves on primary stems are usually ± petiolate (sometimes with spine-tipped teeth on margins of petioles or near bases of blades) and larger and relatively broader than the usually sessile leaves on secondary stems.

SELECTED REFERENCES Semple, J. C. 1978. A revision of the genus *Borrichia* Adans. (Compositae). Ann. Missouri Bot. Gard. 65: 681–693. Semple, J. C. and K. S. Semple. 1977. *Borrichia* ×*cubana* (*B. frutescens* × *arborescens*): Interspecific hybridization in the Florida Keys. Syst. Bot. 2: 292–301.

1. Leaves (at least mid cauline): margins entire or dentate to serrate (teeth remote, mostly toward apices), faces glabrous or sericeous; phyllaries (chartaceous in fruiting heads): apices acute, rounded, or obtuse; paleae: apices obtuse to acute. 1. *Borrichia arborescens*

1. Leaves (at least mid cauline): margins usually dentate to serrate, rarely entire, faces usually villous (longer hairs) and/or sericeous (shorter hairs) to glabrate, rarely glabrous; phyllaries: apices cuspidate or spine-tipped; paleae: apices cuspidate or spine-tipped.

    2. Mid-cauline leaves (petioles or blade bases with spine-tipped teeth): margins dentate to serrate (usually only proximal ¹/₃–¹/₂, rarely most, of each margin); phyllaries (often reflexed, swollen in fruit): apices spine-tipped; paleae: apices spine-tipped . . . . . . 2. *Borrichia frutescens*

    2. Mid-cauline leaves: margins dentate or serrate (at least mid portions, often undulate as well); phyllaries: apices cuspidate; paleae: apices cuspidate . . . . 3. *Borrichia* ×*cubana*

1. **Borrichia arborescens** (Linnaeus) de Candolle in A. P. de Candolle and A. L. P. P. de Candolle, Prodr. 5: 489. 1836

Buphthalmum arborescens Linnaeus, Syst. Nat. ed. 10, 2: 1227. 1759

Stems decumbent to ascending. **Leaves** (at least mid cauline) oblanceolate to linear, 5–10(–15) × 0.4–2 cm, margins entire or dentate to serrate (teeth remote mostly toward apices), faces glabrous or sericeous (sometimes on same leaf). **Heads** borne singly or (3–6) in cymiform arrays (peduncles 0.5–5 cm). **Involucres** ovoid to hemispheric, (8–)10–14(–17) mm. **Phyllaries** 10–16 in 3–4 series (chartaceous in fruiting heads, outer ovate-elliptic, apices obtuse or acute, faces glabrate or sericeous; inner oblanceolate, apices rounded, obtuse, or acute, faces glabrous). **Paleae:** apices obtuse to acute. **Ray florets** 7–15; laminae 5–8 × 2.5–4 mm. **Disc florets** 20–50; corollas 5–6 mm, lobes 1–1.5 mm. **Cypselae** 3.5–5 mm, 4-angled. $2n = 28$.

Flowering late spring–summer. Rocky coasts, sandy marshes, beaches; 0–10 m; Fla.; West Indies; Bermuda.

Borrichia arborescens grows in the Florida Keys, Bermuda, and the West Indies, including Jamaica and Guadeloupe. Plants of *B. arborescens* are usually glabrous; both glabrous and hairy plants occur in the northern West Indies. Common names that include "tree" are misleading because the plants are shrubs.

2. **Borrichia frutescens** (Linnaeus) de Candolle in A. P. de Candolle and A. L. P. P. de Candolle, Prodr. 5: 489. 1836

F

Buphthalmum frutescens Linnaeus, Sp. Pl. 2: 903. 1753

Stems usually erect, sometimes decumbent or arching. **Leaves** (at least mid cauline) obovate or elliptic to oblanceolate, 4.5–8 (–11) × 1–3 cm, (petioles or blade bases usually with 1–2+ spine-tipped teeth) margins dentate to serrate (usually only proximal ¹/₃–¹/₂, rarely most of each margin), faces usually villous (longer hairs) and/or sericeous (shorter hairs), rarely glabrous. **Heads** borne singly (peduncles 2–6 cm). **Involucres** globose, 5–8 mm. **Phyllaries** 20–40 in 3–4 series (reflexed in fruit, outer shorter, apices spine-tipped, inner similar, spines spreading to erect, faces sericeous). **Paleae:** apices spine-tipped. **Ray florets** 15–30; laminae 6–8 × 2.5–6 mm. **Disc florets** 20–75; corollas 5–6.5 mm, lobes 1.75–2.5 mm. **Cypselae** 2.5–4(–5) mm, 3–4-angled. $2n = $ usually 28, rarely 42 ($2n = 3x$).

Flowering late spring–summer. Salt marsh communities, brackish backwaters, limestone rocky shores; 0–10 m; Ala., Fla., Ga., La., Miss., N.C., S.C., Tex., Va.; Mexico; introduced in West Indies, Bermuda.

Borrichia frutescens grows in salt marsh and brackish backwaters along the Atlantic and Gulf coasts from Virginia to the Yucatan Peninsula in Mexico, and inland along the Rio Grande Valley in Texas. It is found on limestone rocky shores in the Florida Keys. It is introduced in Bermuda, the Bahamas, and Cuba.

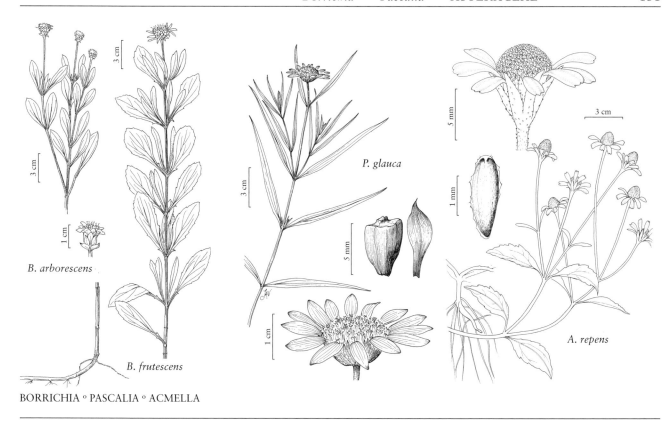

B. arborescens

B. frutescens

BORRICHIA ∘ PASCALIA ∘ ACMELLA

P. glauca

A. repens

3. Borrichia ×cubana Britton & S. F. Blake, Mem. Torrey Bot. Club 16: 116. 1920 (as species).

Stems erect. **Leaves** (at least mid cauline) elliptic, 4–10 × 1–3 cm, margins dentate to serrate (at least mid portions, often undulate as well), faces sericeous to puberulent (hairs unequal) or glabrate. **Heads** borne singly. **Involucres** hemispheric, 7–11 mm. **Phyllaries** 16–25 in 3–4 series (obovate, apices cuspidate, indument similar to leaves). **Paleae:** apices cuspidate (not spine-tipped). **Ray florets** 14–25; laminae 5–7.5 × 2.5–3.5 mm. **Disc florets** 25–60; corollas 5–6 mm, lobes 1.5–2 mm. **Cypselae** 3–5 mm, 3–4-angled. $2n = 28$.

Flowering late spring–summer. Sandy and rocky coasts; 0–10 m; Fla.; West Indies (Cuba).

Borrichia ×*cubana* is a fertile F₁ hybrid between *B. frutescens* and *B. arborescens*. It is found where the parental species are sympatric; backcrosses of the hybrid with either parent appear to be infertile.

291. PASCALIA   Ortega, Nov. Pl. Descr. Dec., 39, plate 4. 1797 • [For D. B. Pascal, French/Italian physician/botanist, once director of royal garden at Parma]   [I]

John L. Strother

Perennials, 30–100 cm (rhizomatous). **Stems** erect, branched ± throughout. **Leaves** cauline; opposite; petiolate or sessile; blades (3-nerved) lanceolate to lance-linear (widest at or proximal to middles, not notably fleshy), bases cuneate, margins entire or toothed, faces sparsely scabrous. **Heads** radiate, borne singly. **Involucres** campanulate to hemispheric, 10–15 mm diam. **Phyllaries** persistent, 20–28 in 2–3 series (linear to lance-linear, herbaceous, subequal or outer longer than inner, spreading in fruit). **Receptacles** hemispheric, paleate (paleae lanceolate to ovate, conduplicate, becoming papery). **Ray florets** 13–21, pistillate, fertile; corollas yellow. **Disc florets** 45–90+, bisexual, fertile; corollas yellow, tubes shorter than campanulate throats, lobes 5, ± deltate. **Cypselae** 3-angled or ± compressed and weakly 4-angled (epidermes usually thick, corky, bases without elaiosomes, apices not rostrate); **pappi** 0, or persistent, coroniform (of connate scales plus 0–2 awns). $x = 33$.

Species 2 (1 in the flora): introduced; South America.

1. Pascalia glauca Ortega, Nov. Pl. Descr. Dec., 39, plate 4. 1797 Ⓕ Ⓘ

Wedelia glauca (Ortega) Hoffmann ex Hicken [sometimes attributed to S. F. Blake]

Leaf blades mostly 5–10 cm. **Ray laminae** 10–15 mm. **Cypselae** 5–6 mm. $2n = 66$ (South America).

Flowering sporadically. Low, wet places; 0–10 m; introduced; Ala., Fla., Ga.; South America.

Pascalia glauca evidently has been introduced into southeastern United States repeatedly. It seems not to have persisted in the flora.

292. ACMELLA Richard in C. H. Persoon, Syn. Pl. 2: 472. 1807 • [From a Singhalese name for a plant now known as *Blainvillea acmella* (Linnaeus) Philipson]

John L. Strother

Annuals or perennials, 10–20(–30+) cm. **Stems** prostrate to erect, usually branched ± throughout. **Leaves** cauline; opposite; petiolate [± sessile]; blades (usually 3-nerved) ovate to rhombic or lanceolate [linear to filiform], bases ± cuneate, margins entire or toothed, faces sparsely pilose to strigillose, glabrescent. **Heads** radiate or discoid [disciform], borne singly at tips of branches [corymbiform arrays]. **Involucres** ± hemispheric to ovoid, 3–6+ mm diam. **Phyllaries** persistent, 8–15+ in 1–3 series (distinct, ovate to linear, subequal or outer longer). **Receptacles** conic, paleate (paleae falling with fruit, ± navicular, membranous to scarious, each about equaling subtended floret). **Ray florets** 0 or 5–20+, pistillate, fertile; corollas yellow to orange [white or purplish] (laminae ovate to linear) [wanting]. **Disc florets** 25–100(–200+) bisexual, fertile; corollas yellow [orange], tubes shorter than campanulate throats, lobes 4–5, deltate. **Cypselae** 2–3-angled (peripheral) or strongly compressed, ellipsoid to obovoid (glabrous or ciliate on the 2–3 angles or ribs); **pappi** 0, or fragile, of 1–3 awnlike bristles. $x = 13$.

Species 30 (2 in the flora): s United States, Mexico, West Indies, Central America, South America; introduced in Asia, Africa, Pacific Islands, Australia.

Acmella pilosa R. K. Jansen has been reported as introduced in Florida (http://www.plantatlas.usf.edu); it differs from *A. repens* mainly by its more densely pilose stems and leaves and more truncate to cordate (versus cuneate) leaf bases.

SELECTED REFERENCE Jansen, R. K. 1985. The systematics of *Acmella* (Asteraceae–Heliantheae). Syst. Bot. Monogr. 8: 1–115.

1.   Petioles 3–40+ mm; leaf blades ovate to lance-ovate, mostly 20–40(–100) × 10–35 mm, margins dentate to denticulate or entire . . . . 1. *Acmella repens*
1.   Petioles 2–4 mm; leaf blades lanceolate, mostly 12–40 × 3–10 mm, margins usually sinuate-dentate, rarely entire . . . . 2. *Acmella pusilla*

1. **Acmella repens** (Walter) Richard in C. H. Persoon, Syn. Pl. 2: 473. 1807 F

Anthemis repens Walter, Fl. Carol., 211. 1788; *Acmella oppositifolia* (Lamarck) R. K. Jansen var. *repens* (Walter) R. K. Jansen; *Spilanthes americana* (Mutis) Hieronymus var. *repens* (Walter) A. H. Moore

Leaves: petioles 3–40+ mm; blades ovate to lance-ovate, mostly 20–40(–100) × 10–35 mm, margins dentate to denticulate or entire. **Cypselae** 1–2.5 mm, ciliate or not, tips of cilia straight; **pappi** 0 or of 1–2(–3) bristles. **2***n* = 52.

Flowering Jan–Dec. Disturbed, wet or drying sites; 0–200 m; Ala., Ark., Fla., Ga., La., Miss., Mo., N.C., S.C., Tenn., Tex.; Mexico (Coahuila).

2. **Acmella pusilla** (Hooker & Arnott) R. K. Jansen, Syst. Bot. Monogr. 8: 46. 1985 I

Spilanthes pusilla Hooker & Arnott, J. Bot. (Hooker) 3: 317. 1841

Leaves: petioles 2–4 mm; blades lanceolate, mostly 12–40 × 3–10 mm, margins usually sinuate-dentate, rarely entire. **Cypselae** 1.2–1.6 mm, ciliate or not, tips of cilia recurved; **pappi** 0 or of 1–2 bristles. **2***n* = 26, 52.

Flowering May–Sep. Disturbed sites; introduced; 0–10 m; Fla., Ga., N.C., S.C.; South America.

293. **CALYPTOCARPUS** Lessing, Syn. Gen. Compos., 221. 1832 • [Greek *kalypto*, covered or hidden, and *karpos*, fruit]

John L. Strother

Perennials, to 30 cm. **Stems** prostrate to decumbent, branched throughout. **Leaves** cauline; opposite; petiolate; blades (± 3-nerved) deltate to ovate or lanceolate, bases cuneate to truncate, margins toothed, faces ± scabrellous. **Heads** radiate, borne singly (in axils). **Involucres** obconic, 3–8 mm diam. **Phyllaries** persistent, 5 in 1(–2) series (linear to lanceolate, herbaceous). **Receptacles** convex, paleate (paleae persistent, scarious). **Ray florets** 3–8, pistillate, fertile; corollas pale yellow. **Disc florets** 10–20, bisexual, fertile; corollas yellow, tubes shorter than funnelform throats, lobes 4–5, ± deltate. **Cypselae** compressed or flattened, cuneate (faces often tuberculate); **pappi** persistent, of 2(–5+) stout awns. *x* = 12.

Species 2 (1 in the flora): s United States, Mexico, Central America.

1. **Calyptocarpus vialis** Lessing, Syn. Gen. Compos., 221. 1832 F

Calyptocarpus blepharolepis B. L. Robinson

Leaf blades 1–3(–6) cm. **Rays** often inconspicuous, laminae mostly 2–5 mm. **Cypselae** 2–4 mm; **pappi** of ± erect to spreading awns, the longer 1–2(–3+) mm. **2***n* = 24.

Flowering ± year round. Ruderal, lawns, along paths and roads, clays, sands, calcareous soils; 0–300 m; Ala., Fla., La., Miss., Tex.; Mexico; Central America.

294. **LASIANTHAEA** de Candolle in A. P. de Candolle and A. L. P. P. de Candolle, Prodr. 5: 607. 1836 • [Alteration of genus name *Lasianthus*; Greek *lasios*, hairy, and *anthos*, flower]

John L. Strother

Perennials [shrubs], 10–60+ [–700] cm (rootstocks bearing tubers). **Stems** erect, branched ± throughout. **Leaves** cauline; opposite; petiolate; blades (usually 3-nerved) rounded-deltate to

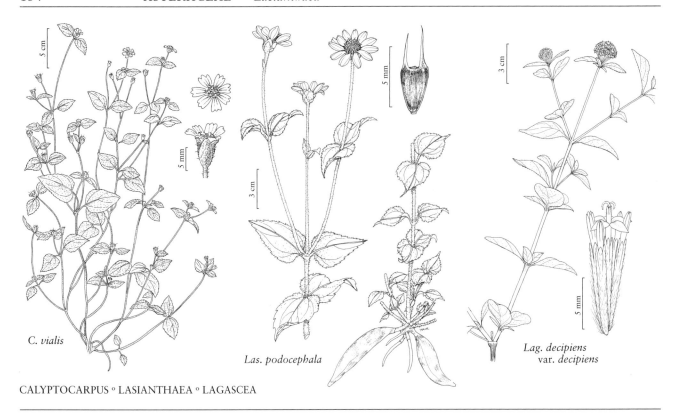

C. vialis

Las. podocephala

Lag. decipiens
var. decipiens

CALYPTOCARPUS ∘ LASIANTHAEA ∘ LAGASCEA

ovate or lance-ovate [lanceolate], bases cuneate to rounded or truncate, margins serrate, faces ± scabrous. **Heads** radiate, borne singly [in corymbiform arrays]. **Involucres** hemispheric [obconic], [4–]10–14[–25+] mm diam. **Phyllaries** persistent, 12–20 in 2–5 series (ovate to lanceolate, subequal or outer shorter [longer], herbaceous to papery). **Receptacles** convex, paleate (paleae membranous to indurate, conduplicate, apices acute [erose]). **Ray florets** [4–]11–14[30], pistillate, fertile; corollas yellow to orange [purplish]. **Disc florets** 20–100+[–200+], bisexual, fertile; corollas yellow to orange [purplish], tubes about equaling ampliate, cylindric throats, lobes 5, deltate. **Cypselae** 3-angled (ray) or strongly compressed or flattened (all with sharp edges, not truly winged); **pappi** persistent, of 2(–3) subulate scales or awns (continuous with margins of cypselae) plus 0–4 shorter scales. x = 10, 11?

Species 11 (1 in the flora): sw United States, Mexico.

SELECTED REFERENCE Becker, K. M. 1979. Monograph of the genus *Lasianthaea* (Asteraceae). Mem. New York Bot. Gard. 31(2): 1–64.

1. Lasianthaea podocephala (A. Gray) K. M. Becker, Mem. New York Bot. Gard. 31(2): 52. 1979 F

Verbesina podocephala A. Gray, Smithsonian Contr. Knowl. 5(6): 92. 1853

Leaf blades 2–8+ cm. **Cypselae** 2.5–4.5 mm; **pappi** 1–4 mm. $2n$ = ca. 80.

Flowering Jun–Oct. Open or shaded sites, grasslands, forests; 1200–2400 m; Ariz., N.Mex.; Mexico (Chihuahua, Sonora).

187m.9. ASTERACEAE Martinov (tribe HELIANTHEAE) subtribe HELIANTHINAE Cassini ex Dumortier, Fl. Belg., 71. 1827 (as Heliantheae)

Lagasceinae Bentham & Hooker f.

Annuals, perennials, subshrubs, or shrubs, (2–)20–500 cm. **Leaves** usually mostly cauline, rarely mostly basal; mostly opposite (distal often alternate) or mostly alternate, rarely whorled; petiolate or sessile; blades (often 3- or 5-nerved) orbiculate or deltate to lanceolate or linear (and intermediate shapes), sometimes lobed, ultimate margins entire or toothed, faces usually hairy (often hispid to scabrellous), often gland-dotted. **Heads** usually radiate, sometimes discoid, borne singly or in corymbiform (paniculiform or racemiform) arrays (glomerules or second-order heads of 1–2-flowered heads in *Lagascea*). **Calyculi** 0. **Involucres** cylindric to hemispheric or rotate. **Phyllaries** persistent, 4–45(–100+) in 1–6+ series (usually distinct, ± connate in *Lagascea*, broadly ovate to linear, subequal or unequal). **Receptacles** usually flat or convex, sometimes hemispheric or conic, usually paleate (epaleate in *Lagascea*; paleae usually falling, sometimes persistent, mostly oblong to linear, often conduplicate, herbaceous to scarious, apices sometimes ± 3-lobed or -toothed; becoming hardened perigynia completely investing cypselae in *Sclerocarpus*). **Ray florets** 0 or 5–30(–100+), neuter, or styliferous and sterile; corollas usually yellow to orange, sometimes partly or wholly brown to purple or red-brown to reddish. **Disc florets** 1 (*Lagascea*) or 10–200(–1000+), bisexual, fertile; corollas usually yellow to orange, sometimes brown to purple or reddish to red-brown, tubes shorter than or about equaling cylindric or funnelform throats, lobes 5, ± deltate (equal); anther thecae dark or pale; stigmatic papillae usually continuous. **Cypselae** often ± compressed, flattened, biconvex, clavate, columnar, obovoid, obpyramidal, or prismatic (lengths usually 1–2 times diams.), glabrous or hairy (faces and/or angles); **pappi** 0, or (often readily falling or fragile, sometimes persistent) usually of (1–)2(–8+) (sometimes aristate) scales and/or awns, sometimes coroniform.

Genera 17, species 363 (8 genera, 69 species, including 1 hybrid, in the flora): mostly subtropical, tropical, and warm-temperate New World.

Circumscription of Heliathinae adopted here (i.e., that of H. Robinson 1981) is narrower than traditional circumscriptions (e.g., T. F. Stuessy 1977[1978]), which included genera here treated in Ecliptinae, Galinsoginae, and Rudbeckiinae.

1. Phyllaries 4–5 in ± 1 series (connate); disc florets 1(–2; the 1–2-flowered heads aggregated in glomerules or second-order heads) . . . . 295. *Lagascea*, p. 136
1. Phyllaries 5–45(–100+) in (1–)2–6+ series (distinct); disc florets (10–)15–200(–1000+).
  2. Receptacles paleate (each palea completely investing and falling with a cypsela, each forming a hardened perigynium) . . . . 296. *Sclerocarpus*, p. 137
  2. Receptacles paleate (paleae sometimes conduplicate, ± enfolding cypselae, not forming perigynia).
    3. Heads borne singly (peduncles usually distally dilated, fistulose) . . . . 297. *Tithonia*, p. 138
    3. Heads borne singly or in corymbiform, paniculiform, racemiform, or thyrsiform arrays (peduncles rarely, if ever, notably dilated or fistulose).
      4. Cypselae flattened, thin-margined . . . . 298. *Simsia*, p. 140
      4. Cypselae ± compressed, biconvex, or 3- or 4-angled, often obpyramidal.
        5. Cypselae glabrous or glabrate; pappi 0.
          6. Shrubs (leaves often lobed, lobes usually 3–9, ± linear) . . . . 301. *Viguiera* (in part), p. 172

6. Annuals or perennials (leaves not lobed).
    7. Annuals; leaf blades lanceolate to linear; involucres 5–6 mm diam.; phyllaries 11–17 . . . . 299. *Helianthus* (in part), p. 141
    7. Annuals or perennials; leaf blades lance-linear, lanceolate, ovate, rhombic, or rhombic-ovate; involucres 6–14 mm diam.; phyllaries 14–25 . . . . 300. *Heliomeris*, p. 169

[5. Shifted to left margin.—Ed.]

5. Cypselae usually ± strigose, sometimes glabrous or glabrate; pappi of 2–6+, persistent, readily falling, or tardily falling scales.
    8. Shrubs; involucres 5–9 mm diam. . . . . 302. *Bahiopsis*, p. 174
    8. Annuals or perennials; involucres (5–)7–40+ mm diam.
        9. Pappi readily falling . . . . 299. *Helianthus* (in part), p. 141
        9. Pappi persistent or tardily falling . . . . 301. *Viguiera* (in part), p. 172

295. LAGASCEA Cavanilles, Anales Ci. Nat. 6: 331. 1803 (as Lagasca), name and orthography conserved • Doll's-head, acuate, silk-leaf, velvet-bush [For Mariano Lagasca y Segura, Spanish botanist at the Madrid Botanical Garden]

Elizabeth M. Harris

Shrubs [annuals, perennials], to 50–100[–300] cm. **Stems** erect [ascending to decumbent], branched from bases or ± throughout. **Leaves** cauline; opposite; petiolate [sessile]; blades 3-nerved, lance-ovate to ovate [lanceolate to oblanceolate], bases broadly cuneate [to sub-auriculate], margins ± serrate, faces usually sericeous to strigose or glabrate, often stipitate-glandular. **Heads** discoid, borne in headlike glomerules (of [8–]30–50+, 1(–2)[–8]-flowered heads, glomerules borne singly or in ± corymbiform [racemiform] arrays). **Involucres** cylindric, 1–2 mm diam. (glomerules of heads usually subtended by leaves or ± foliaceous bracts). **Phyllaries** persistent, 4–5[–8+] in ± 1 series (linear-attenuate, proximally connate, often 1 or more with 1[–3+] glands in abaxial face). **Receptacles** convex (often hirtellous), rarely paleate (paleae linear). **Ray florets** 0. **Disc florets** 1(–2)[–8], bisexual, fertile; corollas yellow [white, pink, or red], tubes shorter than cylindric to campanulate throats, lobes 5, lance-linear to lance-ovate [rounded-deltate] (often hairy). **Cypselae** (brown to black) narrowly cylindric to obovoid or clavate (faces glabrous or pilosulous, minutely grooved); **pappi** ± coroniform. $x = 17$.

Species 8 (1 in the flora): sw United States, Mexico, Central America (to Nicaragua); introduced in West Indies, South America, Asia (India, Java, Sri Lanka, Thailand), Africa, Pacific Islands (Hawaii).

Lagascea mollis Cavanilles, now a nearly pantropical weed, was collected at Apalachicola, Franklin County, Florida (*Chapman, n.d.,* presumably ca. 1860), evidently from an ephemeral population, probably from ballast.

SELECTED REFERENCES Harris, E. M. 1994. Developmental evidence for the derivation of syncephalia in *Lagascea* (Heliantheae; Asteraceae). Amer. J. Bot. 81: 1139–1148. Stuessy, T. F. 1976. A systematic review of the subtribe Lagasceinae (Compositae, Heliantheae). Amer. J. Bot. 63: 1289–1294. Stuessy, T. F. 1978. Revision of *Lagascea* (Compositae, Heliantheae). Fieldiana, Bot. 38: 75–133.

1. Lagascea decipiens Hemsley, Diagn. Pl. Nov. Mexic., 33. 1879 [F]

Varieties 2 (1 in the flora): sw United States, Mexico.

1a. Lagascea decipiens Hemsley var. **decipiens** [F]

Leaves 25–50(–75) × 15–30(–50+) mm. **Peduncles** (i.e., stalks supporting glomerules of heads) 5–55 mm. **Glomerules** hemispheric to subspheric, 10–17 × 20–35 mm. **Phyllaries** 8–10 mm, connate $^1/_2$–$^2/_3$ lengths, resinous glands to 3 mm. **Disc corollas** 10 mm. **Cypselae** 2–3 mm; **pappi** 0.1–1 mm. $2n = 34$.

Flowering spring–fall [year round in Mexico]. Dry canyonlands, granitic slopes, deciduous forests, thorn forests, oak grasslands; 900–1300 m; Ariz.; Mexico.

Pima and Santa Cruz counties represent the northernmost extension of the range of var. *decipiens*, which extends into Chihuahua and Sonora to northern Sinaloa, whence it is disjunct to Guerrero, Jalisco, Michoacán, and Oaxaca.

296. SCLEROCARPUS Jacquin, Icon. Pl. Rar. 1: 17, plate 176. 1781 • Mexican bonebract [Greek *skleros*, hard, and *karpos*, fruit, alluding to hardened paleae enfolding disc cypselae]

Neil A. Harriman

Annuals [perennials or subshrubs], 60–200 cm. **Stems** sprawling to erect, branched from bases or ± throughout. **Leaves** mostly cauline; opposite (proximal) or alternate; petiolate [sessile]; blades 3-nerved, ovate to rhombic or lanceolate, bases cuneate to rounded, margins coarsely toothed, faces glabrous or ± scabrous. **Heads** radiate, borne singly. **Involucres** rotate to hemispheric, 12–20 mm diam. **Phyllaries** persistent, 5–8[–21] in 1(–2) series (green, oblong to linear). **Receptacles** low-conic, paleate (paleae each closely investing subtended floret, each forming a hard perigynium around a fruit and shed with it). **Ray florets** 5–8, neuter; corollas (readily falling) yellow to orange. **Disc florets** 10–50, bisexual, fertile; corollas usually yellow to orange, sometimes with purple, tubes much shorter than cylindric throats, lobes 5, lance-linear. **Cypselae** (within perigynia) weakly compressed, obliquely obovoid or arcuate (glabrous); **pappi** 0, or coroniform (of distinct or connate scales or bristles). $x = 12$.

Species ca. 12 (1 in the flora): s United States, Mexico, Central America; one species in Old World tropics.

SELECTED REFERENCE Feddema, C. 1966. Systematic Studies in the Genus *Sclerocarpus* and the Genus *Aldama* (Compositae). Ph.D. dissertation. University of Michigan.

1. Sclerocarpus uniserialis (Hooker) Bentham & Hooker f. ex Hemsley, Biol. Cent.-Amer., Bot. 2: 164. 1881 [F]

Gymnopsis uniserialis Hooker, Icon. Pl. 2: plate 145. 1837; *Sclerocarpus uniserialis* var. *austrotexanus* B. L. Turner

Stems often densely hairy (hairs whitish, pustule-based, antrorse-appresssed, stiff). **Leaf blades** 2–5 cm, margins toothed mostly distally, apices acute. **Peduncles** 3–10 cm. **Ray laminae** rotundly elliptic, 5–15 mm. **Disc corollas** usually yellow with reddish veins (sometimes purple). **Cypselae:** perigynia brown, ± warty. $2n = 24$.

Flowering late summer–winter. Disturbed sites, caliche, limestone, and sandy soils; 0–300 m; Tex.; Mexico; Central America (Guatemala).

Of the four varieties sometimes recognized within *Sclerocarpus uniserialis*, only var. *uniserialis* and var. *austrotexanus* B. L. Turner occur within the flora area. Variety *austrotexanus* is said to differ by having the disc corollas longer and more uniformly purple and by growing mainly on dunes.

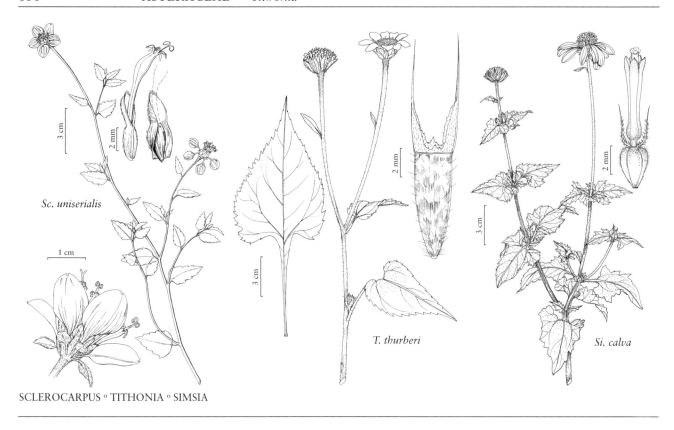

Sc. uniserialis

T. thurberi

Si. calva

SCLEROCARPUS ° TITHONIA ° SIMSIA

297. TITHONIA Desfontaines ex Jussieu, Gen. Pl., 189. 1789 • Sunflowerweed [From Greek mythology, *Tithonus*, son of Laomedon and consort of Aurora, symbolic of old age; perhaps alluding to gray to white induments of some plants]

John C. La Duke

Annuals, perennials, subshrubs, or shrubs [trees], 70–500[–700] cm. **Stems** erect, branched. **Leaves** all or mostly cauline; opposite (proximal) or mostly alternate; petiolate or sessile; blades often (1-), 3-, or 5-nerved, mostly deltate or pentagonal [lanceolate, linear], sometimes 3- or 5-lobed, bases ± truncate or auriculate [attenuate] (sometimes decurrent onto petioles), ultimate margins serrate to crenate, faces glabrate, ± hirsute, pilose, soft-pubescent, or villous, often gland-dotted. **Heads** borne singly (peduncles usually distally dilated, fistulose). **Involucres** campanulate to hemispheric, 10–20+ mm diam. **Phyllaries** persistent, 12–28+ in 2–5 series (linear to broadly rounded, unequal to subequal, apices acute to rounded). **Receptacles** hemispheric to convex, paleate (paleae persistent, embracing cypselae, striate, ± 3-toothed, middle teeth larger, stiff, acute or acuminate to aristate). **Ray florets** 8–30, neuter; corollas yellow or orange. **Disc florets** 40–120[–200+], bisexual, fertile; corollas yellow, tubes shorter than throats (bases of throats bulbous and hairy), lobes 5, ± triangular (anthers black, brown, or tan, bases cordate-sagittate, appendages ovate; style branches relatively slender, appendages penicillate or lanceolate to attenuate). **Cypselae** (black or brown) ± compressed or flattened, often 3- or 4-angled or biconvex, ± cuneiform in silhouette (sometimes with basal elaiosomes); **pappi** 0, or ± coroniform (of ± connate scales, 1–2 scales sometimes subulate to aristate). *x* = 17.

Species 11 (3 in the flora): sw United States, Mexico, Central America; introduced in se United States, West Indies, South America, and Old World.

SELECTED REFERENCE La Duke, J. C. 1982. Revision of *Tithonia*. Rhodora 84: 453–522.

1. Perennials, subshrubs, or shrubs; phyllaries 16–28; ray laminae 48–69 mm; disc florets 80–
    120+; cypselae 4–6 mm . . . . 1. *Tithonia diversifolia*
1. Annuals; phyllaries 12–16; ray laminae 9–15 or 20–33 mm; disc florets 40–60 or 60–90;
    cypselae 5–9 mm.
    2. Leaf blades ± deltate to pentagonal, often 3- or 5-lobed, abaxial faces soft-pubescent;
        rays usually orange, rarely yellow, laminae 20–33 × 6–17 mm . . . . 2. *Tithonia rotundifolia*
    2. Leaf blades mostly deltate, rarely, if ever, lobed, abaxial faces sparsely hirsute (hairs
        larger on veins); rays yellow, laminae 9–15 × 4–6 mm . . . . 3. *Tithonia thurberi*

1. Tithonia diversifolia (Hemsley) A. Gray, Proc. Amer. Acad. Arts 19: 5. 1883 • Tree marigold, Mexican sunflowerweed [I]

Mirasolia diversifolia Hemsley, Biol. Cent.-Amer., Bot. 2: 168, plate 47. 1881

Perennials, subshrubs, or shrubs, to 250(–500) cm. **Leaves:** petioles 2–6 cm; blades ± deltate to pentagonal, 7–33 × 7–22 cm, sometimes 3- or 5-lobed, abaxial faces glabrous to hispid-pilose. **Peduncles** 7–24 cm. **Phyllaries** 16–28 in (3–)4 series, oblong to ovate; outer 6–10 × 4–7 mm, apices rounded to acute, abaxial faces usually glabrous; inner 10–20 × 3–10 mm, apices rounded to acute, abaxial faces glabrous. **Paleae** 10–13 × 2–3 mm, mucros 1.5–2.5 mm. **Ray florets** 7–14; corollas yellow, laminae linear, 48–69 × 9–16 mm. **Disc florets** 80–120+. **Cypselae** 4–6 mm. $2n = 34$.

Flowering Sep–Jan. Disturbed sites; 0–20+ m; introduced; Fla., Tex.; Mexico; also introduced in West Indies, Central America, Asia, Africa, Atlantic Islands, Indian Ocean Islands, Pacific Islands, Australia.

Tithonia diversifolia is widely cultivated in warm climates and may persist after plantings; a report for Louisiana was not confirmed for this treatment.

2. Tithonia rotundifolia (Miller) S. F. Blake, Contr. Gray Herb. 52: 41. 1917 [I]

Tagetes rotundifolia Miller, Gard. Dict. ed. 8, Tagetes no. 4. 1768

Annuals, mostly 100–200(–400) cm. **Leaves:** petioles 3–8 cm; blades ± deltate to pentagonal, 7–38 × 7–30 cm, sometimes 3- or 5-lobed, abaxial faces glabrous or soft-pubescent. **Peduncles** 11–27 cm. **Phyllaries** 14–21 in (2–)3 series, lanceolate to linear; outer 17–30 × 4–7.5 mm, apices acute, abaxial faces usually densely puberulent; inner 16–28 × 5–8 mm, apices rounded to acute, abaxial faces usually densely puberulent. **Paleae** 11.5–15 × 2–3 mm, mucros 2.5–5.5 mm. **Ray florets** 8–13; corollas usually orange, sometimes yellow, laminae oval to oblong, 20–33 × 6–17 mm. **Disc florets** 60–90. **Cypselae** 5–7. $2n = 34$.

Flowering Nov.–Jan. Disturbed sites; 0–20+ m; introduced; Fla., La.; Mexico; West Indies; Central America.

Tithonia rotundifolia is widely cultivated in warm climates and may persist after plantings (e.g., in Asia).

3. Tithonia thurberi A. Gray, Proc. Amer. Acad. Arts 8: 655. 1873 • Arizona sunflowerweed [F]

Tithonia palmeri Rose

Annuals, 70–200 cm. **Leaves:** petioles 0–3.3 cm; blades deltate, 7–28 × 5.5–21.5 cm, abaxial faces sparsely hirsute (hairs larger on veins). **Peduncles** 17–36 cm. **Phyllaries** 13–16 in 3 series, linear to spatulate; outer 8–15 × 2.5–4.5 mm, apices acute, abaxial faces puberulent; inner 12–15.5 × 3.5–7 mm, apices acute, abaxial faces glabrous. **Paleae** 11–16.5 × 2–4 mm, mucros 3 mm. **Ray florets** 7–10; corollas yellow, laminae oblong, 9–15 × 4–6 mm. **Disc florets** 40–60. **Cypselae** 6–9 mm. $2n = 34$.

Flowering Aug–Sep. Disturbed areas, canyons, watercourses; 900–1000 m; Ariz.; Mexico (Sonora).

298. SIMSIA Persoon, Syn. Pl. 2: 478. 1807 • Bush sunflower [For John Sims, 1749–1831, British physician and botanist]

David M. Spooner

Annuals, perennials, or subshrubs [shrubs], 20–400 cm. **Stems** erect or ascending [decumbent], sparingly to freely branched. **Leaves** cauline; opposite (proximal) or alternate [whorled]; petiolate (petioles often ± winged, often with expanded bases, those bases sometimes fused to form nodal "discs") [sessile]; blades 3-nerved from bases, mostly deltate to ovate [linear], sometimes 3- [5-]lobed[pinnatifid], bases cordate to cuneate, ultimate margins entire or toothed, faces hirsute, hispid, pilose, puberulent, scabrous, or scabro-hispid [sericeous], often gland-dotted or ± stipitate-glandular to glandular-puberulent. **Heads** radiate [discoid], borne singly or in 2s or 3s, or in tight to loose, corymbiform [paniculiform] arrays. **Involucres** campanulate [ovoid-campanulate to urceolate], 5–16[–22] mm diam. **Phyllaries** persistent, [11–]13–43[–66] in 2–4 series (tightly appressed to broadly reflexed, unequal to subequal). **Receptacles** low-convex, paleate (paleae conduplicate, ± enclosing cypselae). **Ray florets** [0–]5–21[–45], styliferous and sterile; corollas orange-yellow [lemon-yellow, pink, purple, or white]. **Disc florets** [12–]13–154 [–172], bisexual, fertile; corollas concolorous with rays (usually turning purple apically), tubes (often glandular-hairy) shorter than throats, lobes 5, ± triangular (anthers black, yellow, or yellow proximally and bronze or purple distally; style branches relatively slender, apices sometimes attenuate). **Cypselae** flattened, thin-margined [thickened, biconvex] (shoulders minute to conspicuous, faces glabrous or hairy); **pappi** 0, or fragile or readily falling, of 2 ± subulate scales [plus 4–12 shorter scales]. *x* = 17.

Species 20 (2 in the flora): sw United States, Mexico, West Indies (Jamaica), Central America, South America.

SELECTED REFERENCE Spooner, D. M. 1990. Systematics of *Simsia* (Compositae–Heliantheae). Syst. Bot. Monogr. 30: 1–90.

1. Perennials or subshrubs (roots fusiform-thickened); ray florets 8–21, corollas light orange-yellow (abaxial faces often brown- or purple-lined, or wholly brown or purple); disc florets (26–)90–154; anthers usually yellow, rarely black . . . . 1. *Simsia calva*
1. Annuals (rarely persisting, taprooted or fibrous rooted); ray florets 5–10, corollas orange-yellow; disc florets 13–27; anthers yellow proximally, usually purple to bronze distally
. . . . 2. *Simsia lagasceiformis*

1. Simsia calva (A. Gray & Engelmann) A. Gray, Boston J. Nat. Hist. 6: 228. 1850 • Awnless bush sunflower

F

Barrattia calva A. Gray & Engelmann, Proc. Amer. Acad. Arts 1: 48. 1847

Perennials or subshrubs, 30–150 cm (roots ± fleshy, fusiform-thickened). **Leaves:** petiole bases dilated (pairs fused to form discs at nodes); blades ovate, 2–8 × 1.5–6 cm, sometimes 3-lobed. **Heads** usually borne singly, sometimes in 2s or 3s. **Peduncles** 3–30 cm. **Involucres** 10–12 × 7–16 mm. **Phyllaries** 21–43, subequal to unequal. **Ray florets** 8–21; corollas light orange-yellow (abaxial faces often brown- or purple-lined, or wholly brown or purple), laminae 5–16 mm. **Disc florets** (26–)90–154; anthers usually yellow, rarely black. **Cypselae** 3.5–5.7 mm; **pappi** 0 or to 4 mm. *2n* = 34.

Flowering year round. Sand to heavy clay soils, rock crevices, often limestone, prairies, thickets, oak savannas, along streams, roadsides, upland pine or pine-oak forests; 30–2400 m; N.Mex., Tex.; Mexico.

Simsia calva is widespread throughout central, southern, and southwestern Texas from the southern Texas Plains to the trans-Pecos mountains and into southeastern New Mexico.

Simsia calva is distinguished from *S. lagasceiformis* by its perennial habit, fusiform-thickened roots, petioles winged and fused at bases to form nodal discs, heads borne singly or in 2s or 3s, and anthers usually yellow, rarely black. The common name, awnless bush sunflower, is not truly appropriate. Most populations are epappose; some have minute scales, and some populations of *S. lagasceiformis* (normally pappose) are epappose.

2. **Simsia lagasceiformis** de Candolle in A. P. de Candolle and A. L. P. P. de Candolle, Prodr. 5: 577. 1836 (as lagascaeformis) • Annual bush sunflower

Simsia exaristata A. Gray

Annuals (rarely persisting), 20–400 cm (taprooted or fibrous rooted). **Leaves:** petiole bases rarely dilated (not fused to form nodal discs), blades ovate to deltate, 2–21 × 1–16 cm, rarely 3-lobed. **Heads** in tight to loose, corymbiform arrays. **Peduncles** 0.5–10 cm. **Involucres** 8–12 × 5–10 mm. **Phyllaries** 13–19, unequal. **Ray florets** 5–10; corollas orange-yellow, laminae 5.1–12 mm. **Disc florets** 13–27; anthers yellow proximally, usually purple to bronze distally. **Cypselae** 4.2–6 mm; **pappi** usually 2.5–4.6 mm, rarely 0. $2n = 34$.

Flowering year round (mostly Sep–Dec). Deserts, roadsides, open fields, borders of agricultural fields; 100–2200; Ariz., N.Mex., Tex.; Mexico; Central America (Guatemala).

299. **HELIANTHUS** Linnaeus, Sp. Pl. 2: 904. 1753; Gen. Pl. ed. 5, 386. 1754

• Sunflower, tournesol [Greek *helios*, sun, and *anthos*, flower, alluding to heads]

Edward E. Schilling

Annuals or perennials, (5–)20–300(–500) cm. **Stems** erect or ascending to decumbent or procumbent, usually branched distally. **Leaves** basal and/or cauline; opposite, or opposite (proximal) and alternate, or alternate; petiolate or sessile; blades usually 3-nerved (1-nerved in *H. eggertii*, *H. smithii*, and *H. maximiliani*), mostly deltate, lance-linear, lanceolate, lance-ovate, linear, or ovate, bases cordate to narrowly cuneate, margins usually entire or serrate, rarely lobed, faces glabrous or hairy, often gland-dotted. **Heads** usually radiate (sometimes discoid in *H. radula*), borne singly or in ± corymbiform, paniculiform, or spiciform arrays. **Involucres** usually ± hemispheric, sometimes campanulate or cylindric, 5–40+(–200+ in cultivars) mm diam. **Phyllaries** persistent, 11–40(–100+ in cultivars) in 2–3+ series (subequal to unequal). **Receptacles** flat to slightly convex (conic in *H. porteri*), paleate (paleae ± conduplicate, usually rectangular-oblong, usually ± 3-toothed, sometimes entire, apices sometimes reddish or purplish). **Ray florets** usually 5–30+(–100+ in cultivars), rarely 0, neuter; corollas usually yellow. **Disc florets** (15–)30–150+(–1000+ in cultivars), bisexual, fertile; corollas yellow or reddish (at least distally), tubes shorter than campanulate throats, lobes 5, triangular (style branches slender, appendages ± attenuate). **Cypselae** (usually purplish black, sometimes mottled) ± obpyramidal, ± compressed (glabrous, glabrate, or ± hairy); **pappi** 0 (*H. porteri*), or readily falling, of 2(–3) usually lanceolate, aristate, or erose scales (at the 2 principal angles, 1–5 mm) plus 0–8 usually shorter scales (0.2–2 mm). $x = 17$.

Species 52, including 1 hybrid (52 in the flora): North America, Mexico; introduced in the Old World.

The identification of sunflower species has long been problematic. C. B. Heiser et al. (1969) felt that the greatest contribution of their sustained efforts to understand sunflower taxonomy was not providing an easy way to identify sunflowers but rather an explanation for why they are so difficult. Taxonomic difficulties are based on a combination of factors, notably developmental and ecologic plasticity, the frequency of interspecific hybridization, and the presence of polyploidy. L. H. Rieseberg (1991) and Rieseberg et al. (1988, 1990) have subsequently documented the presence of species of homoploid hybrid origin to add to the complexity. The keys in this treatment have been modified somewhat with observations of features that require microscopic observation, such as the distribution of glands (sometimes referred to as "resin dots") and the color of anther appendages; they still are based heavily on those of Heiser et al. There will still be specimens, of hybrid origin or growing in unusual conditions or incompletely collected, that defy certain placement into a single species.   The taxonomic rank of certain taxa is also problematic, and for the most part the treatment of Heiser et al. has been followed, with relatively few exceptions. With the exception of *Helianthus* ×*laetiflorus*, which is widespread and commonly encountered away from either parental species, hybrids are not treated as separate entities. *Helianthus verticillatus* has been verified as a distinct entity, unlikely to be simply a hybrid. Whether *H. praetermissus* E. Watson, which continues to be known only from a single fragmentary specimen, represents a distinct species or an anomalous collection is still uncertain; it has not been included here.

SELECTED REFERENCES   Heiser, C. B. 1976. The Sunflower. Norman.   Heiser, C. B., D. M. Smith, S. Clevenger, and W. C. Martin. 1969. The North American sunflowers (*Helianthus*). Mem. Torrey Bot. Club 22(3).   Rogers, C. E., T. E. Thompson, and G. J. Seiler. 1982. Sunflower Species of the United States. Fargo.   Storbeck, T. A. 1984. A Taxonomic Study of *Helianthus* Series *Microcephali* and Allied Species. Ph.D. dissertation. Indiana University.

1. Annuals; paleae 3.5–4.6 mm, apices entire (1-toothed); disc corollas 2.8–3.5 mm; pappi 0 . . . . 1. *Helianthus porteri*
1. Annuals or perennials; paleae 5+ mm, apices usually 3-toothed, sometimes subentire or entire; disc corollas (3–)3.5–8.5 mm; pappi (readily falling) of 2(–3+) scales.
  2. Annuals (taprooted; disc corolla lobes reddish and style branches yellow); paleae entire or weakly 3-toothed (apices purplish, abaxial faces glabrous); cypselae glabrous (usually ± tuberculate) . . . . 2. *Helianthus agrestis*
  2. Annuals (taprooted; if disc corolla lobes reddish, style branches also reddish), or perennials (often rhizomatous or with crown buds, sometimes taprooted); paleae entire or ± 3-toothed (apices purplish, greenish, or yellow-brown, abaxial faces glabrous or hairy); cypselae glabrous or hairy (not tuberculate).
    3. Annuals or perennials (taprooted); leaves mostly alternate, petiolate (petiole lengths at least 1/5 blades); paleae (at least central ones) either bearded (with apical tufts of whitish hairs) or prominently 3-toothed (middle teeth relatively narrow, lengths 4 or more times width); disc corolla lobes and style branches usually reddish (rarely yellow in *H. annuus* and *H. debilis*).
      4. Plants 100–300 cm; leaf blades (at least larger) 10–40 cm wide, abaxial faces gland-dotted; phyllaries ovate to lance-ovate (larger usually 5–8 mm wide), apices narrowed abruptly (acute to acuminate).
        5. Stems (leaves, phyllaries) hispid; leaf margins usually serrate . . . . 3. *Helianthus annuus* (in part)
        5. Stems (leaves, phyllaries) densely silvery white tomentose or floccose; leaf margins usually entire, sometimes serrulate . . . . 4. *Helianthus argophyllus*
      4. Plants mostly 25–200 cm; leaf blades (larger) usually less than 12 cm wide (bases cuneate, truncate, or cordate), abaxial faces sometimes gland-dotted; phyllaries usually lanceolate to lance-ovate (usually less than 4 mm, sometimes to 5 mm, wide), apices narrowed gradually.
        6. Palea apices (at least central paleae) bearded (with tufts of whitish hairs).
          7. Stems (leaves, phyllaries) densely canescent . . . . 9. *Helianthus petiolaris* (in part)
          7. Stems (leaves, phyllaries) strigillose to hispid.
            8. Leaf blades usually deltate-lanceolate to lanceolate (lengths of larger usually 2+ times widths), bases truncate to cuneate; phyllary apices relatively short-attenuate . . . . 9. *Helianthus petiolaris* (in part)
            8. Leaf blades usually lanceolate to deltate-ovate or ovate (lengths of larger usually less than 2 times widths), bases truncate to cordate; phyllary apices relatively short- to long-attenuate.

9. Stems hispid; leaf margins subentire to serrulate; phyllary apices relatively long-attenuate (surpassing discs) . . . . . . 10. *Helianthus neglectus*
9. Stems hispid to hirsute; leaf margins serrate to serrulate; phyllary apices relatively short-attenuate (scarcely surpassing discs) . . . . 8. *Helianthus praecox* (in part)
6. Paleae apices glabrous or puberulent to hispid (not bearded).
   10. Leaves abaxially gland-dotted; phyllaries hirsute (paleae: middle teeth notably surpassing, often arching over discs).
      11. Leaf blades ovate to lance-ovate, usually serrate (discs, at least larger heads, 2+ cm diam.); cypselae 3.5–4.5 mm . . . . 5. *Helianthus bolanderi*
      11. Leaf blades lance-linear to lance-ovate, usually entire or subentire (discs usually less than 2 cm diam.); cypselae 2.7–3.5 mm . . . . 6. *Helianthus exilis*
   10. Leaves sometimes abaxially gland-dotted; phyllaries hairy or glabrous (paleae: middle teeth equaling or slightly surpassing discs).
      12. Phyllaries densely white-canescent . . . . 11. *Helianthus niveus* (in part)
      12. Phyllaries glabrous or sparsely hispid to hispidulous.
         13. Leaf blades (at least proximal) usually ovate or deltate, bases usually cordate to truncate (sometimes widely cuneate in *H. debilis* subsp. *debilis*); cypselae 2.5–3 mm, glabrous or sparsely hairy.
            14. Leaf blades deltate-ovate, lance-ovate, or ovate (not constricted near middles, gradually tapering to apices); phyllaries 8–17 × 1–3 mm, apices acute to relatively long-acuminate . . . . 7. *Helianthus debilis*
            14. Leaf blades deltate to ovate (sometimes constricted near middles); phyllaries 9–15 × 2–4 mm, apices relatively short-acuminate to short-attenuate . . . . 8. *Helianthus praecox* (in part)
         13. Leaf blades lanceolate to lance-ovate, bases cuneate; cypselae 3–9 mm, glabrous or villous to pilose.
            15. Phyllaries 17–33 mm (greatly surpassing discs, margins notably ciliate); cypselae 4.6–9 mm . . . . 12. *Helianthus anomalus*
            15. Phyllaries 6–19 mm (equaling or slightly surpassing discs, margins not notably ciliate); cypselae 3–5 mm.
               16. Stems densely hispid; cypselae 4–5 mm, pilose; pappi usually of 2 linear scales 1.6–2.5 mm plus 2–4 linear or ovate scales 05.1–1 mm . . . . 13. *Helianthus deserticola*
               16. Stems glabrate or ± hispid; cypselae 3–4 mm, glabrous; pappi usually of 2 lanceolate scales 2.5–2.9 mm . . . . 14. *Helianthus paradoxus*

[3. Shifted to left margin—Ed.]

3. Perennials (rhizomatous or with crown buds); leaves opposite or alternate, petiolate or sessile; paleae (at least central) glabrous or ± hispid to puberulent (not bearded) and entire or relatively weakly 3-toothed (if 3-toothed, lengths of middle teeth usually less than 4 times widths); disc corolla lobes yellow or reddish (if reddish, style branches yellow).
   17. Leaves (at flowering) mostly or all basal (cauline leaves abruptly smaller).
      18. Ray florets 0 or 2–8, laminae 1–2(–10) mm (inconspicuous, often tinged reddish) . . . . 16. *Helianthus radula*
      18. Ray florets 8–13(–25), laminae (7–)10–40 mm (yellow).
         19. Stems (and leaves) usually glabrous; disc corolla lobes yellow.
            20. Heads usually borne singly; phyllaries 8–16 × 3–5 mm . . . . 17. *Helianthus carnosus*
            20. Heads 3–12; phyllaries 5–11 × 1–2.5 mm . . . . 18. *Helianthus longifolius*
         19. Stems (and leaves) usually hispid or hirsute to scabrous; disc corolla lobes yellow or reddish.

21. Phyllaries lanceolate (the larger 1.5–2.5 mm wide); abaxial faces of leaves (and usually ray laminae) notably gland-dotted; disc corolla lobes yellow; cypselae 3–5 mm . . . . 29. *Helianthus occidentalis* (in part)

21. Phyllaries lanceolate to ovate (the larger 3+ mm wide); abaxial faces usually not gland-dotted (if gland-dotted, cypselae 5–6 mm); disc corolla lobes reddish or yellow.

    22. Leaves mostly basal (cauline relatively few, narrowly lanceolate to linear); phyllary apices acute to accuminate . . . . 15. *Helianthus heterophyllus*

    22. Leaves mostly basal or basal and cauline (at least proximal 1–2 cauline pairs well developed, mostly lanceolate, rhombic-ovate, or ovate, similar to basal leaves); phyllary apices usually obtuse to acute, sometimes mucronate.

        23. Perennials (rhizomatous); leaf blades lance-linear, lance-ovate, oblong-lanceolate, or rhombic-ovate, bases cuneate (onto winged petioles); cypselae 5–6 mm . . . . 30. *Helianthus pauciflorus* (in part)

        23. Perennials (with crown buds); leaf blades broadly lanceolate, ovate, or suborbiculate, bases cuneate, rounded, or truncate (usually abruptly contracted onto winged petioles); cypselae 2.8–4 mm.

            24. Leaves (at least basal) lanceolate to ovate (hairs on abaxial midribs 1+ mm); cauline leaves to 8 pairs proximal to heads; nonflowering stems usually absent . . . . 19. *Helianthus atrorubens*

            24. Leaves (at least basal) ovate to suborbiculate (hairs on abaxial midribs to 1 mm); cauline leaves 9+ pairs proximal to heads; nonflowering stems usually present . . . . . . 20. *Helianthus silphioides* (in part)

[17. Shifted to left margin—Ed.]

17. Leaves (at flowering) mostly cauline (not abruptly smaller distally).

  25. Disc corolla lobes reddish(at least at tips).

    26. Leaf blades lanceolate, lance-linear, or linear (lengths usually 10+ times widths).

      27. Stems glabrous (glaucous); leaf margins entire (flat) . . . . 40. *Helianthus salicifolius*

      27. Stems usually hairy; leaf margins entire (± revolute).

        28. Plants 50–150 cm (rhizomes absent or poorly developed); leaves 8–15 × 0.15–0.5(–1) cm . . . . 37. *Helianthus angustifolius* (in part)

        28. Plants 150–260 cm (rhizomes well developed); leaves 9–22 × 0.7–4 cm . . . . 38. *Helianthus simulans* (in part)

    26. Leaf blades deltate, deltate-ovate, lanceolate, lance-ovate, or linear (lengths seldom more than 5 times widths).

      29. Leaves sessile or nearly so, margins often irregularly toothed or lobed; ray laminae 8–11 mm.

        30. Plants 50–120 cm; stems glabrate to strigose or hispid; leaves (green or grayish, not bluish green) hispid . . . . 51. *Helianthus laciniatus*

        30. Plants 40–70 cm; stems glabrous or glabrate (glaucous); leaves (often bluish green) glabrous or glabrate (glaucous) . . . . 52. *Helianthus ciliaris*

      29. Leaves petiolate to subsessile, margins entire or serrate; ray laminae 15–37 mm.

        31. Phyllaries oblong, oblong-lanceolate, obovate, or ovate, 3–5 mm wide.

            32. Leaf blades usually broadly ovate to orbiculate, abaxial faces not gland-dotted . . . . 20. *Helianthus silphioides* (in part)

            32. Leaf blades lance-linear, lanceolate, lance-ovate, oblong-lanceolate, or rhombic-ovate, abaxial faces usually gland-dotted.

                33. Phyllaries ovate, apices acute, abaxial faces glabrate to hispid . . . . 30. *Helianthus pauciflorus* (in part)

                33. Phyllaries oblong-lanceolate, apices acuminate, abaxial faces usually hairy . . . . 31. *Helianthus ×laetiflorus* (in part)

        31. Phyllaries lanceolate, 1.5–3 mm wide.

34. Stems (usually reddish or purplish) hairy; leaves all or mostly opposite; disc corollas 5–6 mm (anthers reddish brown, appendages yellow or dull orange); cypselae 3–4 mm . . . . 47. *Helianthus gracilentus* (in part)
34. Stems (usually green) hairy; leaves usually opposite or alternate, rarely whorled; disc corollas 4–5 mm (anthers dark brown or black, appendages dark); cypselae 2.5–3 mm . . . . 39. *Helianthus floridanus* (in part)

[25. Shifted to left margin—Ed.]

25. Disc corolla lobes yellow.
    35. Phyllaries ovate to lanceolate, (3–)5–8 mm wide, apices abruptly attenuate (disc corolla throats notably bulbous at bases) . . . . 3. *Helianthus annuus* (in part)
    35. Phyllaries linear to lanceolate or lance-ovate, usually 2–4 mm wide, apices gradually narrowed (disc corolla throats not notably bulbous at bases).
        36. Stems glabrous or glabrate (at least proximal to arrays of heads, sometimes glaucous).
            37. Leaves usually whorled (3s–6s), sometimes opposite . . . . 43. *Helianthus verticillatus*
            37. Leaves opposite or alternate (not whorled).
                38. Leaves sessile (grayish green or bluish green), abaxial faces glabrous (glaucous).
                    39. Plants 20–30 cm; stems ascending to erect; ray laminae 7–9 mm; disc corollas 3–3.5 mm . . . . 50. *Helianthus arizonensis*
                    39. Plants (30–)100–220 cm; stems erect; ray laminae 15–20 mm; disc corollas 5–7 mm.
                        40. Leaves: adaxial faces glabrous or glabrate (smooth or slightly rough to touch); phyllaries 2–3 mm wide; rays 5–10 . . . . 25. *Helianthus laevigatus*
                        40. Leaves: adaxial faces strumose (rough to touch); phyllaries 3.5–4.5 mm wide; rays 10–18 . . . . 26. *Helianthus eggertii*
                38. Leaves sessile or petiolate (light to dark green, sometimes whitish abaxially, not grayish or bluish green); abaxial faces glabrous or hairy.
                    41. Leaf blades (1-nerved) lance-linear to lanceolate . . . . 28. *Helianthus smithii*
                    41. Leaf blades (3-nerved) lance-linear, lanceolate, lance-ovate, or ovate.
                        42. Involucres 5–7 or 8–9 mm diam; rays usually 5 or 8.
                            43. Leaves: abaxial faces (whitish) glabrous, not gland-dotted (glaucous) . . . . 23. *Helianthus glaucophyllus*
                            43. Leaves: abaxial faces (greenish) usually tomentulose, sometimes glabrate, densely gland-dotted . . . . 35. *Helianthus microcephalus*
                        42. Involucres (8–)9–28 mm diam.; rays (8–)10–21 (at least in larger heads).
                            44. Plants (not rhizomatous); phyllaries (paleae and ray laminae) gland-dotted; anthers dull orange or orange-brown, appendages dull orange to yellowish.
                                45. Perennials (with crown buds); phyllaries 4–8 mm . . . . 47. *Helianthus gracilentus* (in part)
                                45. Perennials (taproots stout, parsniplike); phyllaries 11–18 mm . . . . 48. *Helianthus cusickii* (in part)
                            44. Plants (rhizomatous); phyllaries sometimes gland-dotted; anthers usually dark brown to black, rarely reddish brown, appendages yellow, or dark or reddish brown.
                                46. Leaves sessile (3-nerved from bases), bases rounded to cordate . . . . 21. *Helianthus divaricatus*
                                46. Leaves sessile or petiolate (3-nerved distal to bases), bases ± cuneate (narrowing gradually).

47. Anther appendages yellow.
  48. Leaves petiolate (petioles 2.5–10 cm, lengths $^1/_2$+ blades), blades oblong-lanceolate or elliptic to ovate; phyllaries usually appressed, strongly unequal, not surpassing discs ... 29. *Helianthus occidentalis* (in part)
  48. Leaves petiolate or sessile (petioles 05–1.5 or 1–5 cm, lengths usually less than $^1/_4$ blades), blades lanceolate to lance-ovate; phyllaries usually loose, spreading, ± subequal.
    49. Leaves: petioles (1–)2–5 cm, blades 10–32 × (1.2–)4–9 cm, margins coarsely serrate .... 42. *Helianthus grosseserratus*
    49. Leaves: petioles 0.5–1.5 cm, blades 4–20 × 0.8–4 cm, margins entire or shallowly serrate .... 45. *Helianthus nuttallii* (in part)
47. Anther appendages dark or reddish brown.
  50. Peduncles (gland-dotted); phyllaries (strongly reflexed) 3–5 mm wide (notably surpassing discs), gland-dotted .... 46. *Helianthus californicus*
  50. Peduncles (not gland-dotted); phyllaries (sometimes reflexed) to 3 mm wide (sometimes surpassing discs), sometimes gland-dotted.
    51. Plants 100–200(–300) cm (producing tubers); leaves sessile or subsessile (petioles 0–1 cm) .... 36. *Helianthus schweinitzii* (in part)
    51. Plants 60–200 cm (not producing tubers); leaves petiolate (petioles 1–5 cm).
      52. Leaves: petioles 1–3 cm, blades moderately serrate or entire, abaxial faces usually densely gland-dotted; phyllaries (equaling or slightly surpassing discs): apices acute .... 27. *Helianthus strumosus*
      52. Leaves: petioles 2–5 cm, blades (at least larger leaves) moderately to notably serrate, abaxial faces usually sparsely gland-dotted; phyllaries (at least longer, usually surpassing discs, by $^1/_2$+ their lengths): apices acuminate .... 24. *Helianthus decapetalus*

[36. Shifted to left margin—Ed.]
36. Stems hairy (± throughout, not glaucous).
  53. Leaves all or mostly opposite, sessile, bases cordate .... 34. *Helianthus mollis*
  53. Leaves opposite or alternate, petiolate or sessile, bases mostly cuneate (not cordate).
    54. Leaf blades lanceolate to ovate (bases gradually narrowed onto petioles); phyllaries (squarrose to reflexed, notably surpassing discs): apices attenuate, abaxial faces densely gland-dotted (at least toward apices) .... 33. *Helianthus resinosus*
    54. Leaf blades lance-linear or lanceolate (and bases attenuate to truncate or rounded), or lance-ovate to ovate (and sessile or petioles narrowly winged); phyllaries (not reflexed): apices acute to attenuate, abaxial faces sometimes gland-dotted.
      55. Perennials (taproots parsniplike); anthers and appendages usually dull orange or orange-brown, sometimes yellowish................... 48. *Helianthus cusickii* (in part)
      55. Perennials (rhizomatous or with crown buds or slightly thickened taproots); anthers dark brown or black, appendages dark brown or black, or yellow.

[56. Shifted to left margin—Ed.]

56. Leaf blades (usually 1-nerved, conduplicate) entire; heads (1–)3–15 (borne singly or in racemiform or spiciform arrays) . . . . 44. *Helianthus maximiliani*
56. Leaf blades (3-nerved, not conduplicate) entire or serrate; heads (1–)3–16 (borne singly or in ± corymbiform, not racemiform or spiciform arrays).
  57. Perennials (taprooted); leaves all or mostly opposite, both faces gland-dotted (phyllaries, paleae, and ray laminae as well) . . . . 49. *Helianthus pumilus*
  57. Perennials (rhizomatous or with crown buds); leaves opposite or alternate, abaxial faces sometimes (adaxial never) gland-dotted (abaxial faces of phyllaries, paleae, and ray laminae sometimes gland-dotted).
    58. Phyllaries usually appressed, strongly unequal.
      59. Petioles 2.5–10 cm (lengths usually ¹/₂+ blades); leaf blades oblong-lanceolate, ovate, or elliptic; anther appendages yellow; cypselae 3–4 mm. . . . .
      . . . . 29. *Helianthus occidentalis* (in part)
      59. Petioles 1–5 cm (lengths usually less than ¹/₂ blades); leaf blades lanceolate to lance-ovate; anther appendages dark brown or black; cypselae (seldom formed) 4–5 mm . . . . 31. *Helianthus ×laetiflorus* (in part)
    58. Phyllaries usually loose or spreading, ± subequal.
      60. Leaves petiolate, petioles 2–8 cm; blades lanceolate to ovate, 7–15 cm wide; cypselae 5–7 mm (plants producing tubers, late in growing season) . . . . . 32. *Helianthus tuberosus*
      60. Leaves sessile or petiolate, petioles 0–2 cm; blades elliptic, lance-linear, lanceolate, lance-ovate, linear, or ovate, 0.15–4(–8) cm wide; cypselae 2–5 cm (plants sometimes producing tubers).
        61. Leaves petiolate, blade bases truncate to rounded . . . . 22. *Helianthus hirsutus*
        61. Leaves petiolate or sessile, blade bases cuneate (gradually narrowing).
          62. Leaf margins entire or subentire to serrulate (± flat); ray laminae not gland-dotted.
            63. Stems (usually reddish) erect; leaves subsessile or petiolate (petioles 0–1.2 cm, ciliate), abaxial faces scabrous or ± hirsute; anther appendages dark brown or black . . . . 41. *Helianthus giganteus*
            63. Stems (usually yellow-brown or greenish) erect; leaves petiolate (petioles 0.5–1.5 cm, not ciliate), abaxial faces hispid to villous or tomentose; anther appendages yellow . . . . 45. *Helianthus nuttallii* (in part)
          62. Leaf margins entire or subentire to serrulate (usually revolute); ray laminae gland-dotted abaxially.
            64. Involucres 9–10(–16) mm diam. (plants producing tubers) . . . .
            . . . . 36. *Helianthus schweinitzii* (in part)
            64. Involucres (7–)10–20 mm diam. (plants not producing tubers).
              65. Leaves lanceolate to elliptic or lance-ovate (lengths rarely more than 5 times widths), margins entire or serrulate (revolute and/or undulate); phyllary apices (at least outer) usually obtuse. . . . . 39. *Helianthus floridanus* (in part)
              65. Leaves lanceolate to linear (lengths more than 5 times widths), margins entire or subentire (not notably undulate); phyllary apices (at least outer) acute to acuminate.
                66. Plants 50–150+ cm (rhizomes absent or poorly developed); leaves 0.15–0.5(–1) cm wide . . . . 37. *Helianthus angustifolius* (in part)
                66. Plants 150–260 cm (rhizomes well developed); leaves 0.7–4 cm wide . . . . 38. *Helianthus simulans* (in part)

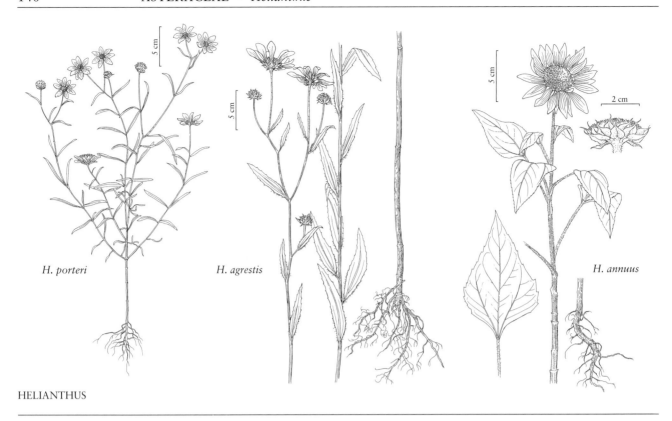

H. porteri   H. agrestis   H. annuus

HELIANTHUS

1. Helianthus porteri (A. Gray) Pruski, Castanea 63: 75. 1998 · Confederate daisy E F

Rudbeckia porteri A. Gray, Mem. Amer. Acad. Arts, n. s. 4: 83. 1849; *Heliomeris porteri* (A. Gray) Cockerell; *Viguiera porteri* (A. Gray) S. F. Blake

Annuals, 40–100 cm. **Stems** erect, usually sparsely strigose, sometimes hispid as well. **Leaves** mostly cauline; opposite (proximal) or alternate; petioles 0–0.2 cm; blades (± 3-nerved distal to bases) narrowly lanceolate to linear, 5–11.5 × 0.15–1 cm, bases cuneate, margins entire (often ± ciliate proximally), faces sparsely strigose or scabrous to glabrate, sparsely gland-dotted. **Heads** usually 5+. **Peduncles** 1–10 cm. **Involucres** hemispheric, 5–6 mm diam. **Phyllaries** 11–17, linear, 5.5–8 × 0.8–1.3 mm (margins sparsely hispido-ciliate, hairs 0.5–1.2 mm), apices acute to acuminate, abaxial faces glabrate, not gland-dotted. **Paleae** (ovate) 3.5–4.6 mm, entire (1-toothed). **Ray florets** 7–8; laminae (10–)15–20 mm. **Disc florets** 30+; corollas 2.8–3.5 mm, lobes yellow; anthers dark, appendages dark. **Cypselae** 2.2–2.3 mm, sparsely puberulent (bases and apices); **pappi** 0. $2n = 34$.

Flowering fall. Granite outcrops; 200–500+ m; Ala., Ga., N.C., S.C.

Helianthus porteri was established at Rocky Face Mountain, North Carolina, following its introduction as part of an ecologic experiment. Where it occurs, *H. porteri* produces conspicuous massed floral displays when in bloom. It is remarkably similar morphologically in its reduced habit, epappose cypselae, conic receptacles, and unlobed, mucronate pales to *Heliomeris* (*Viguiera* sect. *Heliomeris*), and it is treated under *Viguiera* in most southeastern United States treatments; its style appendages and chromosome number agree with molecular phylogenetic data in placing it within *Helianthus*.

2. Helianthus agrestis Pollard, Proc. Biol. Soc. Wash. 13: 184. 1900 · Southeastern sunflower E F

Annuals, 100(–200+) cm. **Stems** erect, glabrous or glabrate (glaucous). **Leaves** mostly cauline; mostly opposite; petioles 0.5–1.2 cm; blades (often 3-nerved distal to bases) lanceolate, 6–11 × 0.7–1.9 cm, bases cuneate, margins ± serrate, faces scabrous, not gland-dotted. **Heads** 1–15. **Peduncles** 2–8 cm. **Involucres** hemispheric, 10–15 mm diam. **Phyllaries** 15–25, lanceolate, 8–9 × 1.5–2 mm, (margins ciliate) apices acuminate, abaxial faces glabrate. **Paleae** 5–7, subentire to ± 3-toothed (apices purplish, glabrous). **Ray florets** ca. 12; laminae 12–25 mm. **Disc florets** 50+; corollas 3.5–4 mm, lobes reddish purple; anthers dark, appendages dark. **Cypselae** 2.5–3.2 mm, glabrous (± tuberculate); **pappi** of 2 aristate scales 1.5–2 mm. $2n = 34$.

Flowering late summer to fall. Mucky wet soils, marshes, pine flatwoods; 0–50+ m; Fla., Ga.

Distinctive among species of *Helianthus* with an annual habit in having a combination of reddish disc corolla lobes and yellow style branches, *H. agrestis* is also characterized by its glabrate stems and phyllaries and by relatively long hairs on the petioles and basal margins of the leaves.

3. **Helianthus annuus** Linnaeus, Sp. Pl. 2: 904. 1753 • Common sunflower [F]

Helianthus annuus subsp. *jaegeri* (Heiser) Heiser; *H. annuus* subsp. *lenticularis* (Douglas ex Lindley) Cockerell; *H. annuus* var. *lenticularis* (Douglas ex Lindley) Steyermark; *H. annuus* var. *macrocarpus* (de Candolle) Cockerell; *H. annuus* subsp. *texanus* Heiser; *H. aridus* Rydberg; *H. jaegeri* Heiser; *H. lenticularis* Douglas ex Lindley; *H. macrocarpus* de Candolle

Annuals, 100–300 cm. **Stems** erect, usually hispid. **Leaves** mostly cauline; mostly alternate; petioles 2–20 cm; blades lance-ovate to ovate, 10–40 × 5–40 cm, bases cuneate to subcordate or cordate, margins serrate, abaxial faces usually ± hispid, sometimes gland-dotted. **Heads** 1–9. **Peduncles** 2–20 cm. **Involucres** hemispheric or broader, 15–40(–200+) mm diam. **Phyllaries** 20–30 (–100+), ovate to lance-ovate, 13–25 × (3–)5–8 mm, (margins usually ciliate) apices abruptly narrowed, long-acuminate, abaxial faces usually hirsute to hispid, rarely glabrate or glabrous, usually gland-dotted. **Paleae** 9–11 mm, 3-toothed (middle teeth long-acuminate, glabrous or hispid). **Ray florets** (13–)17–30(–100+); laminae 25–50 mm. **Disc florets** 150+(–1000+); corollas 5–8 mm (throats ± bulbous at bases), lobes usually reddish, sometimes yellow; anthers brownish to black, appendages yellow or dark (style branches yellow). **Cypselae** (3–)4–5(–15) mm, glabrate; **pappi** of 2 lanceolate scales 2–3.5 mm plus 0–4 obtuse scales 0.5–1 mm. $2n = 34$.

Flowering summer–fall. Open areas; 0–3000 m; St. Pierre and Miquelon; Alta., B.C., Man., N.B., N.W.T., N.S., Ont., P.E.I., Que., Sask.; Ala., Ariz., Ark., Calif., Colo., Conn., Del., D.C., Fla., Ga., Idaho, Ill., Ind., Iowa, Kans., Ky., La., Maine, Md., Mass., Mich., Minn., Miss., Mo., Mont., Nebr., Nev., N.H., N.J., N.Mex., N.Y., N.C., N.Dak., Ohio, Okla., Oreg., Pa., R.I., S.C., S.Dak., Tenn., Tex., Utah, Vt., Va., Wash., W.Va., Wis., Wyo.; Mexico; intoduced nearly worldwide.

Helianthus annuus is widely distributed, including weedy, cultivated, and escaped plants. It is the only native North American species to become a major agronomic crop. Despite its considerable variability, attempts have failed to produce a widely adopted infraspecific system of classification. Forms with red-colored ray laminae, known from cultivation and occasionally seen escaped, trace their ancestry to a single original mutant plant. It hybridizes with many of the other annual species.

4. **Helianthus argophyllus** Torrey & A. Gray, Fl. N. Amer. 2: 318. 1842 • Silverleaf sunflower [E]

Annuals, 100–300 cm. **Stems** (silvery white) erect, usually tomentose or floccose. **Leaves** mostly cauline; mostly alternate; petioles 2–10 cm; blades ovate to lance-ovate, 15–25 × 10–20 cm, bases truncate to subcordate, margins entire or serrulate, abaxial faces usually floccose, sericeous, or tomentose, gland-dotted. **Heads** 1–5. **Peduncles** 2–8 cm. **Involucres** hemispheric, 20–30 mm diam. **Phyllaries** 20–35, ovate to lance-ovate, 15–18 × (2–)5–8 mm, apices long-attenuate, abaxial faces usually densely white-villous, gland-dotted. **Paleae** 10–11 mm, 3-toothed (apices glabrous or sparsely villous). **Ray florets** 15–20; laminae 20–30 mm. **Disc florets** 150+; corollas 6.5–7.5 mm, lobes reddish; anthers dark, appendages dark (style branches reddish). **Cypselae** 4–6 mm, glabrate; **pappi** of 2 aristate scales 2–2.7 mm. $2n = 34$.

Flowering late summer–fall. Open areas, sandy soils; 0–30+ m; Fla., N.C., Tex.; introduced in South America (Argentina), Africa, Australia.

Helianthus argophyllus hybridizes naturally with *H. annuus*.

5. **Helianthus bolanderi** A. Gray, Proc. Amer. Acad. Arts 6: 544. 1865 • Bolander's sunflower [E]

Annuals, 60–150 cm. **Stems** erect, hispid to hirsute. **Leaves** mostly cauline; mostly alternate; petioles 1–4 cm; blades lance-linear or lance-ovate to ovate, 3–15 × 2–6 cm, bases cuneate to truncate, margins usually serrate, abaxial faces sparsely hirsute, gland-dotted. **Heads** 1–3. **Peduncles** 3–13 cm. **Involucres** hemispheric, 17–25 mm diam. **Phyllaries** 10–18, usually lanceolate to lance-ovate, 9–27 × (3–)3.5–5 mm (often surpassing discs), apices gradually attenuate, abaxial faces hirsute. **Paleae** 9.5–10.5 mm, 3-toothed (middle teeth subulate, surpassing discs, apices glabrous). **Ray florets** 12–17; laminae 14–20 mm. **Disc florets** 75+ (discs usually 2+ cm diam.); corollas 5–7 mm, lobes usually reddish; anthers dark, appendages dark (style branches reddish). **Cypselae** 3.5–4.5 mm, glabrate; **pappi** of 2 lanceolate scales 1.7–3 mm. $2n = 34$.

Flowering summer–early fall. Grassy, often disturbed sites; 10–1200 m; Calif., Oreg.

Helianthus bolanderi and *H. exilis* form a closely related pair of sister species that share the distinctive feature of having the middle teeth of the paleae glabrous and relatively elongated, surpassing the disc florets. As treated here, *H. bolanderi* corresponds to the "valley weed race" (C. B. Heiser 1949; L. H. Rieseberg et al. 1988); it is separated from the "serpentine foothill race," here recognized as *H. exilis*. Heiser proposed that *H. bolanderi* originated through hybridization between *H. annuus* and *H. exilis*; molecular studies by Rieseberg et al. do not support this scheme. In an ironic twist, it appears that *H. bolanderi* may be undergoing "genetic assimilation" through hybridization with *H. annuus* (S. E. Carney et al. 2000).

6. **Helianthus exilis** A. Gray, Proc. Amer. Acad. Arts 6: 545. 1865 • Serpentine sunflower [C][E]

Annuals, 30–100 cm. Stems (often reddish) erect, hairy. Leaves mostly cauline; mostly alternate; petioles 0.7–2.5 cm; blades lance-linear to lance-ovate or ovate, 3–15 × 0.5–3 cm, bases cuneate, margins usually entire or shallowly serrate, abaxial faces gland-dotted. Heads 1–7. Peduncles (1–)3–13(–20) cm. Involucres hemispheric, 15–20 mm diam. Phyllaries 12–17, lanceolate, 8–17 × 3–4 mm, apices narrowed gradually, abaxial faces hirsute. Paleae 9.5–10 mm, 3-toothed (middle teeth surpassing discs, apices greenish or yellow-brown, glabrous). Ray florets 10–13; laminae 14–20 mm. Disc florets 50+; corollas 4–6 mm, lobes reddish; anthers reddish purple, appendages usually purplish (style branches reddish or yellow). Cypselae (2.5–)3–3.5(–4) mm, glabrate; pappi of 2 lanceolate scales 1.7–2.7 mm. 2*n* = 34.

Flowering summer–early fall. Gravelly streamsides on serpentine; of conservation concern; 100–1400 m; Calif.

Helianthus exilis is very similar morphologically to *H. bolanderi* and is often included in it; there are discrete differences between the two for fatty acid composition (C. E. Rogers et al. 1982) as well as some molecular markers (L. H. Rieseberg et al. 1988). It is listed as of concern by the California Native Plant Society.

7. **Helianthus debilis** Nuttall, Trans. Amer. Philos. Soc., n. s. 7: 367. 1841 [E]

Annuals or perennials, 30–200 cm (taprooted). Stems decumbent to erect, glabrous, hirsute, or puberulent. Leaves mostly cauline; mostly alternate; petioles 1–7 cm; blades deltate-ovate, lance-ovate, or ovate, 2.5–14 × 1.8–13 cm, bases cordate to truncate or broadly cuneate, margins subentire to serrate, abaxial faces glabrate to hispid, not gland-

dotted. Heads 1–3. Peduncles 9–50 cm. Involucres hemispheric, 10–22 mm diam. Phyllaries 20–30, lanceolate, 8–17 × 1–3 mm, apices acute to long-attenuate, abaxial faces glabrous or ± hispid, not gland-dotted. Paleae 7.5–8 mm, apices 3-toothed (middle teeth acuminate, usually glabrous or hispid, sometimes ± villous or bearded). Ray florets 11–20; laminae 12–23 mm. Disc florets 30+; corollas 4.5–5 mm, lobes usually reddish, sometimes yellow; anthers dark, appendages dark (style branches usually reddish, rarely yellow). Cypselae 2.5–3.2 mm, glabrous or sparsely hairy; pappi of 2 lanceolate or lance-linear scales 1.2–2.5 mm.

Subspecies 5 (5 in the flora): United States (Atlantic and Gulf coasts).

C. B. Heiser (1956) placed 8 subspecies in *Helianthus debilis*; he noted that alternative taxonomic treatments might recognize these in as many as three species, or expand the single species to include *H. petiolaris*. Later, Heiser et al. (1969) separated three of the subspecies as *H. praecox*. Isozyme data (R. P. Wain 1982, 1983; L. H. Rieseberg and M. F. Doyle 1989) show that all are closely related. Documented hybridization with *H. annuus* further complicates the situation. The treatment by Heiser et al. is followed here.

Helianthus debilis is adventive beyond the Atlantic and Gulf coasts of the United States.

1. Stems decumbent; peduncles 9–20(–22) cm.
   2. Stems glabrous or puberulent; leaf blades serrulate or shallowly, regularly serrate, abaxial faces sparsely, if at all, gland-dotted . . . . . .
   . . . . 7a. *Helianthus debilis* subsp. *debilis*
   2. Stems hirsute; leaf blades deeply, irregularly serrate, abaxial faces densely gland-dotted . . . . 7b. *Helianthus debilis* subsp. *vestitus*
1. Stems erect; peduncles (15–)20–50 cm.
   3. Leaves 8–14 cm; peduncles (relatively slender) 20–40 cm; discs 10–15(–17) mm diam. . . . . 7c. *Helianthus debilis* subsp. *silvestris*
   3. Leaves 2.5–9 cm; peduncles (not notably slender) 10–50 cm; discs 14–20 mm diam.
      4. Leaf blades usually deeply, irregularly serrate; peduncles 10–25(–30) cm; ray laminae 12–20(–22) mm . . . . . 7d. *Helianthus debilis* subsp. *tardiflorus*
      4. Leaf blades usually shallowly, regularly serrate; peduncles 25–50 cm; ray laminae (15–)20–23 mm . . . . 7d. *Helianthus debilis* subsp. *cucumerifolius*

7a. Helianthus debilis Nuttall subsp. **debilis** • Beach sunflower [E]

Stems decumbent, glabrous or puberulent. **Leaf blades** 3–10 cm, margins serrulate or shallowly, regularly serrate, abaxial faces sparsely, if at all, gland-dotted. **Peduncles** 10–20(–22) cm. **Ray laminae** 12–20 mm. **Discs** 11–14 mm diam. **2n** = 34.

Flowering year round. Coastal dunes; 0–10 m; Fla.

7b. Helianthus debilis Nuttall subsp. **vestitus** (E. Watson) Heiser, Madroño 13: 154. 1956 [C][E]

Helianthus vestitus E. Watson, Pap. Michigan Acad. Sci. 9: 347, plate 48. 1929; *H. debilis* Nuttall var. *vestitus* (E. Watson) Cronquist

Stems decumbent, hirsute. **Leaf blades** 4–8 cm, margins deeply, irregularly serrate, abaxial faces densely gland-dotted. **Peduncles** 9–15 cm. **Ray laminae** 12–16 mm. **Discs** 11–12 diam. **2n** = 34.

Flowering year round. Coastal dunes; of conservation concern; 0–10 m; Fla.

7c. Helianthus debilis Nuttall subsp. **silvestris** Heiser, Madroño 13: 158, fig. 6. 1956 [E]

Stems erect, glabrate. **Leaf blades** (6–)8–14 cm, margins serrate, abaxial faces densely gland-dotted. **Peduncles** (relatively slender) 20–40 cm. **Ray laminae** 15–23 mm. **Discs** 14–15(–16) mm diam. **2n** = 34.

Flowering late spring–fall. Open areas, sandy soils with pine and oak; 10–100+ m; Tex.

7d. Helianthus debilis Nuttall subsp. **tardiflorus** Heiser, Madroño 13: 154, figs. 5, 6. 1956 [E]

Helianthus debilis var. *tardiflorus* (Heiser) Cronquist

Stems erect, glabrous or sparsely hispid. **Leaf blades** 5–9(–10) cm, margins usually deeply, irregularly serrate, abaxial faces densely gland-dotted. **Peduncles** 10–25 cm. **Ray laminae** 14–16(–20) mm. **Discs** 14–15 mm diam. **2n** = 34.

Flowering spring–fall. Sandy beaches; 0–30+ m; Ala., Fla., Ga., Miss.

7e. Helianthus debilis Nuttall subsp. **cucumerifolius** (Torrey & A. Gray) Heiser, Madroño 13: 160. 1956 • Cucumberleaf sunflower [E]

Helianthus cucumerifolius Torrey & A. Gray, Fl. N. Amer. 2: 319. 1842; *H. debilis* var. *cucumerifolius* (Torrey & A. Gray) A. Gray

Stems erect, hispid. **Leaf blades** 4–9 cm, margins usually shallowly, regularly serrate, abaxial faces sometimes gland-dotted. **Peduncles** 25–50 cm. **Ray laminae** (15–)20–30 mm. **Discs** 16–18(–20) mm diam. **2n** = 34.

Flowering spring–fall. Open areas, sandy soils; 10–100+ m; Conn., Fla., Ga., La., Md., Mass., Mich., N.H., N.Y., N.C., Pa., R.I., S.C., Tex., Vt., Va., W.Va.

8. Helianthus praecox Engelmann & A. Gray, Boston J. Nat. Hist. 5: 221. 1847 • Texas sunflower [E]

Annuals, 40–150 cm. **Stems** erect to ± procumbent, hispid, hispid-hirsute, or hirsute. **Leaves** mostly cauline; mostly opposite; petioles 5–9 cm; blades deltate to ovate (sometimes constricted near middles, subsp. *praecox*), 3–9 × 2–7 cm, bases cordate, cuneate, or truncate, margins serrate to serrulate, abaxial faces hirsute to hispid, usually not gland-dotted. **Heads** 1–3. **Peduncles** 15–40 cm. **Involucres** hemispheric, 13–18 mm diam. **Phyllaries** 30–39, lanceolate, 9–15 × 2–4 mm, apices short-attenuate to short-acuminate (subsp. *praecox*), abaxial faces hispidulous to hispid-hirsute. **Paleae** 6.5–8 mm, 3-toothed (middle teeth acuminate, equaling or slightly surpassing discs, apices puberulent or bearded, hairs whitish). **Ray florets** 11–16; laminae 16–26 mm. **Disc florets** 35+; corollas 5–6 mm, lobes reddish; anthers dark, appendages purplish red (style branches reddish). **Cypselae** 2.5–3.3 mm, ± villous; **pappi** of 2 aristate scales 1.2–1.5 mm.

Subspecies 3 (3 in the flora): Texas.

The taxa of *Helianthus praecox* were placed as subspecies of *H. debilis* by C. B. Heiser (1956). They were separated later, primarily on the basis of crossing results (Heiser et al. 1969). The subspecies appear to form a grade between *H. debilis* and *H. petiolaris*.

1. Stems procumbent to erect, branches horizontal, moderately hispid or hirsute; phyllary apices short-acuminate; paleae: middle teeth puberulent . . . . 8a. *Helianthus praecox* subsp. *praecox*
1. Stems erect, branches ascending, moderately or densely hirsute or hispid-hirsute; phyllary apices short-attenuate; paleae: middle teeth white-villous ("bearded").

[2. Shifted to left margin—Ed.]
2. Stems moderately hirsute; leaves (at least larger) 3–7.5(–8) cm; peduncles 20–30(–35) cm; rays 11–13 .... 8b. *Helianthus praecox* subsp. *runyonii*
2. Stems densely hirsute or hispid-hirsute; leaves (larger) 8–9 cm; peduncles (25–)30–40 cm; rays usually 14–16 ... 8c. *Helianthus praecox* subsp. *hirtus*

8a. **Helianthus praecox** Engelmann & A. Gray subsp. **praecox** • Texas sunflower C E

Helianthus debilis Nuttall subsp. *praecox* (Engelmann & A. Gray) Heiser

Stems procumbent to erect, branches horizontal, moderately hispid or hirsute. **Leaves** (at least larger) 3–8 cm. **Peduncles** 15–30 cm. **Phyllary apices** short-acuminate. **Paleae:** middle teeth puberulent. **Rays** 14–15. **2n** = 34.

Flowering early summer–fall. Sandy soils; of conservation concern; 0–50 m; Tex.

8b. **Helianthus praecox** Engelmann & A. Gray subsp. **runyonii** (Heiser) Heiser, Mem. Torrey Bot. Club 22(3): 54. 1969 • Runyon's sunflower E

Helianthus debilis Nuttall subsp. *runyonii* Heiser, Madroño 13: 161, fig. 7. 1956; *H. praecox* var. *runyonii* (Heiser) B. L. Turner

Stems erect, branches ascending, moderately hirsute. **Leaves** (larger) 3–7.5(–8) cm. **Peduncles** 20–30(–35) cm. **Phyllary apices** short-attenuate. **Paleae:** middle teeth white-villous ("bearded"). **Rays** 11–13(–14). **2n** = 34.

Flowering spring–fall. Coastal prairies; 0–20+ m; Tex.

8c. **Helianthus praecox** Engelmann & A. Gray subsp. **hirtus** (Heiser) Heiser, Mem. Torrey Bot. Club 22(3): 54. 1969 • Dimmit sunflower C E

Helianthus debilis Nuttall subsp. *hirtus* Heiser, Madroño 13: 162, fig. 7. 1956

Stems erect, branches ascending, densely hirsute or hispid-hirsute. **Leaves** (larger) 8–9 cm. **Peduncles** (25–)30–40 cm. **Phyllary apices** short-attenuate. **Paleae:** middle teeth white-villous ("bearded"). **Rays** 14–16. **2n** = 34.

Flowering early summer–fall. Sandy soils; of conservation concern; 50–100+ m; Tex.

Subspecies *hirtus* is known from only a single population, near Carrizo Springs; it is genetically distinct from the two other subspecies (L. H. Rieseberg and M. F. Doyle 1989)

9. **Helianthus petiolaris** Nuttall, J. Acad. Nat. Sci. Philadelphia 2: 115. 1821 • Prairie sunflower E F

Annuals, 40–200 cm. **Stems** erect, usually densely canescent, hispid, or strigillose, rarely ± hirsute or glabrate. **Leaves** mostly cauline; mostly alternate; petioles 2–4 cm; blades (often bluish green) lanceolate to deltate-ovate or ovate, 4–15 × 1–8 cm, bases subcordate or truncate to cuneate, margins entire or ± serrate, abaxial faces strigose, sparsely to densely, or not at all, gland-dotted. **Heads** 1–5. **Peduncles** 4–15(–40) cm. **Involucres** ± hemispheric, 10–24 mm diam. **Phyllaries** 14–25, lance-linear to lanceolate to lance-ovate, 10–14 × 1–4(–5) mm, (margins sometimes ciliate) apices short-attenuate, abaxial faces usually hispidulous, rarely sparsely hirsute to glabrate. **Paleae** 4.5–7.5 mm, 3-toothed, middle teeth ± ciliate or bearded, hairs whitish, 0.5–0.7 mm. **Ray florets** 10–30; laminae 15–20 mm. **Disc florets** 50–100+; corollas 4.5–6 mm, lobes usually reddish, rarely yellow; anthers reddish to purplish, appendages purplish (style branches reddish). **Cypselae** 3–4.5 mm, ± villous; **pappi** of 2 aristate scales 1.5–3 mm plus 0–2 erose scales 0.3–0.5 mm. **2n** = 34.

Subspecies 2 (2 in the flora): w North America.

Helianthus petiolaris is adventive beyond western North America.

A third subspecies has yet to be named at that rank in *Helianthus petiolaris*; it has been called *H. petiolaris* var. *canescens* A. Gray. It differs in having stems, leaves, and phyllaries densely canescent and abaxial faces of leaves densely gland-dotted. It is additionally characterized by peduncles usually ebracteate, phyllaries 1–2 mm wide, disc corolla throats gradually narrowed distal to slight, not densely hairy basal bulges, and **2n** = 34. It flowers late spring through late summer and grows on sandy soils in open areas at (10–)1000–2300 m in Arizona, California, Nevada, New Mexico, and Texas and in Mexico. It was treated as *H. niveus* (Bentham) Brandegee subsp. *canescens* (A. Gray) Heiser by C. B. Heiser et al. (1969); molecular and morphologic data appear to favor a placement within *H. petiolaris*.

1. Stems usually hispidulous to strigillose; peduncles usually bractless; phyllaries 3–5 mm wide; disc corollas: throats abruptly narrowed distal to densely hairy basal bulbs ....
.... 9a. *Helianthus petiolaris* subsp. *petiolaris*
1. Stems usually ± hispid; peduncles usually each with leafy bract subtending head; phyllaries 2–3.5 mm wide; disc corollas: throats gradually narrowed distal to slight, not densely hairy, basal bulges ...
.... 9b. *Helianthus petiolaris* subsp. *fallax*

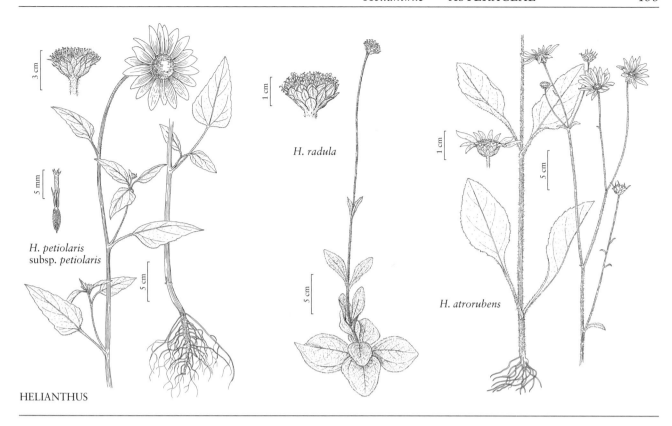

H. radula

H. petiolaris
subsp. *petiolaris*

H. atrorubens

HELIANTHUS

9a. Helianthus petiolaris Nuttall subsp. **petiolaris**
E F

Helianthus couplandii B. Boivin

Stems usually hispidulous to strigillose, rarely ± hirsute to glabrate. **Leaves:** abaxial faces sparsely, if at all, gland-dotted. **Peduncles** usually bractless. **Phyllaries** 3–5 mm wide. **Disc corollas:** throats abruptly narrowed distal to densely hairy basal bulbs. $2n = 34$.

Flowering early summer–fall. Dry open areas, usually sandy soils; 10–1500(–2000) m; Alta., Man., Ont., Sask.; Ariz., Ark., Calif., Colo., Conn., Del., D.C., Idaho, Ill., Ind., Iowa, Kans., La., Maine, Mass., Mich., Minn., Mo., Mont., Nebr., Nev., N.J., N.Mex., N.Y., N.C., N.Dak., Ohio, Okla., Oreg., Pa., S.C., S.Dak., Tenn., Tex., Utah, Va., Wash., W.Va., Wis., Wyo.

Subsp. *petiolaris* is native to western North America and adventive elsewhere (e.g., introduced in Ontario). Hybrids with with *H. annuus* and *H. debilis* have been reported.

9b. Helianthus petiolaris Nuttall subsp. **fallax** Heiser, Rhodora 60: 279, fig. 3. 1958 E

Stems usually ± hispid, rarely ± hirsute to glabrate. **Leaves:** abaxial faces sparsely, if at all, gland-dotted. **Peduncles** usually with leafy bracts subtending heads. **Phyllaries** 2–3.5 mm wide. **Disc corollas:** throats gradually narrowed distal to slight, not densely hairy basal bulges. $2n = 34$.

Flowering summer–fall. Dry, open areas; 1200–2500 m; Ariz., Colo., Nev., Utah.

10. Helianthus neglectus Heiser, Rhodora 60: 275, fig. 2. 1958 • Neglected sunflower C E

Annuals, 80–200 cm. **Stems** erect (leafy), ± hispid to glabrate. **Leaves** cauline; alternate; petioles 7–12 cm; blades deltate-ovate, 7–14 × 7.5–12.3 cm, bases usually ± cordate, sometimes truncate, margins subentire to serrulate, abaxial faces strigose, not gland-dotted. **Heads** 1–5. **Peduncles** 10–40 cm. **Involucres** ± hemispheric, (23–28 ×) 10–14 mm. **Phyllaries** 25–35, lanceolate, 14–25 × 2.3–4 mm (surpassing discs), apices long-attenuate, abaxial faces strigillose to hispid. **Paleae** 9–10 mm, 3-toothed (middle teeth ± ciliate or bearded, hairs whitish, 0.3–0.6 mm). **Ray florets** 21–31; laminae 29–39 mm. **Disc florets** 150+; corollas 6–6.5 mm, lobes reddish; anthers dark, appendages purplish (style branches reddish). **Cypselae** 4–5 mm, strigillose to glabrate; **pappi** of 2 aristate scales 2.8–3.2 mm. $2n = 34$.

Flowering summer–fall. Sandy soils, sand dunes; of conservation concern; 800–1100 m; N.Mex., Tex.

11. **Helianthus niveus** (Bentham) Brandegee, Proc. Calif. Acad. Sci., ser. 2, 2: 173. 1889 [C]

Encelia nivea Bentham, Bot. Voy. Sulphur, 27. 1844

Subspecies 2 (1 in the flora): sw United States, Mexico.

11a. **Helianthus niveus** (Bentham) Brandegee subsp. **tephrodes** (A. Gray) Heiser, Mem. Torrey Bot. Club 22(3): 43. 1969 • Algodones or Algodones dunes sunflower [C]

Helianthus tephrodes A. Gray in W. H. Emory, Rep. U.S. Mex. Bound. 2(1): 90. 1859

Annuals or perennials, 50–150 cm (taproots stout). **Stems** erect or decumbent, sericeo-villous (hairs white). **Leaves** mostly cauline; mostly alternate; petioles 1.5–3.5 cm; blades deltate or deltate-ovate, 3–7 × 2–4 cm, bases cuneate, margins entire or serrulate, abaxial faces densely sericeous, gland-dotted. **Heads** 1–3. **Peduncles** 4–17 cm. **Involucres** hemispheric, 8–28 mm diam. **Phyllaries** 18–21, lanceolate, 8–10 × 1–3.5 mm, apices shortly attenuate, abaxial faces densely white-canescent. **Paleae** 8–11 mm, subentire to 3-toothed. **Ray florets** 10–13; laminae 16–25 mm. **Disc florets** 50+; corollas 4.5–6 mm, lobes reddish; anthers brownish red, appendages reddish (style branches reddish). **Cypselae** 4–8 mm, ± villous; **pappi** of 2(–3) lanceolate scales 1.5–3 mm plus 2–4 obtuse scales 0.5–1 mm. $2n = 34$.

Flowering fall–spring. Sand hills; of conservation concern; 50–300 m; Ariz., Calif.; Mexico (Sonora).

Subspecies *tephrodes* differs from subsp. *niveus*, which is not known to occur north of Mexico, by having less densely hairy phyllaries and generally broader leaves (larger leaves more than 3 cm broad). Both differ from *H. petiolaris* subsp. *canescens* (*H. niveus* subsp. *canescens*) by lacking the conspicuous white-villous tufts ("beard") of the median cusp of the paleae in the center of the head that characterize *H. petiolaris*.

Subspecies *tephrodes* is in the Center for Plant Conservation's National Collection of Endangered Plants.

12. **Helianthus anomalus** S. F. Blake, J. Wash. Acad. Sci. 21: 333. 1931 • Anomalous sunflower [E]

Annuals, 25–60 cm. **Stems** erect, sparsely hispid. **Leaves** mostly cauline; mostly alternate; petioles 1.2–5 cm; blades ovate to lance-ovate, 4.5–13 × 1–4 cm, bases cuneate, margins entire, abaxial faces sparsely hispid, not gland-dotted. **Heads** usually borne singly. **Peduncles** 4–9 cm. **Involucres** hemispheric, 20–27 mm diam. **Phyllaries** 11–23, laceolate to lance-linear or lance-ovate, 17–33 × 1–2.9 mm (notably surpassing discs), (margins ± ciliate) apices attenuate, abaxial faces usually sparsely hispid to hispidulous. **Paleae** 7–8 mm, apices 3-toothed (middle teeth attenuate, apices purplish, ± hispid). **Ray florets** 7–12; laminae 18–37 mm. **Disc florets** 25+; corollas 6.8–7.5 mm, lobes reddish; anthers dark, appendages dark (style branches reddish). **Cypselae** 4.6–9 mm, appressed hairy; **pappi** of 2 linear scales 2–4.5 mm plus 4–6 scales 1–3.5 mm. $2n = 34$.

Flowering spring–fall. Sand dunes; 1300–1500 m; Ariz., Utah.

L. H. Rieseberg (1991) provided evidence that *Helianthus anomalus* is a hybrid species derived from *H. annuus* and *H. petiolaris*, and subsequent work has confirmed and extended that hypothesis.

13. **Helianthus deserticola** Heiser, Proc. Indiana Acad. Sci. 70: 209, fig. 1. 1960 (as deserticolus) • Desert sunflower [E]

Annuals, 10–40 cm. **Stems** (green or red) erect, densely hirsute and gland-dotted. **Leaves** mostly cauline; mostly alternate; petioles 1–6 cm; blades lanceolate to lance-ovate, 2.5–5 × 1–2 cm, bases cuneate, margins entire, abaxial faces hispid-hirsute, densely gland-dotted. **Heads** 1–5. **Peduncles** 1–3(–8) cm. **Involucres** hemispheric, 13–25 mm diam. **Phyllaries** 13–21, lanceolate to lance-ovate, 7–19 × 1.5–2 mm (equaling or slightly surpassing discs), (margins ciliate) apices acuminate, abaxial faces hispid (hairs erect, often 1+ mm) gland-dotted. **Paleae** 8–9 mm, ± 3-toothed (middle teeth equaling or slightly surpassing discs, apices yellowish brown, usually hispid). **Ray florets** 7–13; laminae 10–30 mm (abaxial faces gland-dotted). **Disc florets** 25+; corollas 5.5–6 mm, lobes reddish; anthers dark, appendages purplish (style branches reddish). **Cypselae** 4–5 mm, pilose; **pappi** usually of 2 linear scales 1.6–2.5 mm plus 4–6 linear or ovate, erose scales 0.5–1 mm. $2n = 34$.

Flowering spring–fall. Dry, open areas; 400–1500 m; Ariz., Nev., Utah.

Helianthus deserticola is relatively uncommon; it is not federally listed. Abundant subsessile glands ("resin dots") on stems, leaves, phyllaries, and abaxial faces of ray laminae help to distinguish *H. deserticola* from *H. anomalus*, with which it is sometimes lumped. L. H. Rieseberg (1991) demonstrated that it is of hybrid origin; parental species are *H. annuus* and *H. petiolaris*.

14. **Helianthus paradoxus** Heiser, Rhodora 60: 272, fig. 1. 1958 • Pecos or paradox sunflower C E

Annuals, 130–200 cm. **Stems** erect, glabrous or ± hispid. **Leaves** mostly cauline; opposite (proximal) or mostly alternate; petioles 1.5–6 cm; blades lanceolate to lance-ovate, 7–17.5 × 1.7–8.5 cm, bases cuneate, margins entire or (larger leaves) toothed, abaxial faces ± scabrous, not gland-dotted. **Heads** 1–5. **Peduncles** 12–18 cm. **Involucres** hemispheric, 15–20 mm diam. **Phyllaries** 15–25, lanceolate to lance-ovate, 6–19 × 0.7–4 mm (equaling or slightly surpassing discs), (margins ciliate) apices (spreading to recurved) acuminate, abaxial faces usually glabrate or sparsely hispid. **Paleae** 8–9 mm, apices 3-toothed (apices glabrous). **Ray florets** 12–20; laminae 20–30 mm. **Disc florets** 50+; corollas 5–5.5 mm, lobes reddish; anthers dark, appendages yellowish or dark (style branches reddish). **Cypselae** 3–4 mm, glabrous; **pappi** of 2 lanceolate scales 2.5–2.9 mm. $2n = 34$.

Flowering late summer–fall. Saturated saline soils, desert wetlands; of conservation concern; 1000–1200 m; N.Mex., Tex.

Helianthus paradoxus is listed by the U.S. Fish and Wildlife Service as an endangered species and is in the Center for Plant Conservation's National Collection of Endangered Plants. It is of hybrid origin; the parents are *H. annuus* and *H. petiolaris* (L. H. Rieseberg et al. 1990). It occupies a different habitat type than either parent (*H. annuus* usually on clay-based mesic soils and *H. petiolaris* usually on dry, sandy soils).

15. **Helianthus heterophyllus** Nuttall, J. Acad. Nat. Sci. Philadelphia 7: 74. 1834 • Wetland sunflower E

Helianthus elongatus Small

Perennials, 50–120 cm (with crown buds). **Stems** erect, usually hispid to ± hirsute. **Leaves** mostly basal; mostly opposite; petioles 0–3 cm (broadly winged); blades (3-nerved distal to bases) ovate or lanceolate to spatulate, 6–28 × 1.2–4.3 cm, bases cuneate, margins entire (often revolute), abaxial faces hispid to ± hirsute, not gland-dotted (cauline leaves relatively few, narrowly lanceolate to linear, much smaller). **Heads** 1–3(–5). **Peduncles** 10–15 cm. **Involucres** broadly hemispheric, 15–25 mm diam. **Phyllaries** 24–30, lanceolate to lance-ovate, 8–13 × 2–5 mm, (margins sometimes ciliate) apices acute to short-acuminate or acuminate, abaxial faces sparsely hispid to glabrate. **Paleae** 7–9 mm, 3-toothed (apices purplish). **Ray florets** 12–18; laminae 14–36 mm (abaxial faces not gland-dotted). **Disc florets** 100+; corollas 5.5–6.5 mm, lobes reddish; anthers purplish, appendages purplish (style branches usually reddish). **Cypselae** 4–5 mm, glabrate; **pappi** of 2 aristate scales 1.7–2.5 mm plus 1–3 deltate scales 0.5–1.5 mm. $2n = 34$.

Flowering late summer–fall. Wet sandy soils; 0–50+; Ala., Fla., Ga., La., Miss., N.C., S.C., Tex.

Helianthus heterophyllus is found on the Atlantic and Gulf coastal plains. The cauline leaves are usually abruptly reduced relative to the basal leaves; individuals sometimes have relatively large cauline leaves.

16. **Helianthus radula** (Pursh) Torrey & A. Gray, Fl. N. Amer. 2: 321. 1842 • Rayless or pineland sunflower E F

Rudbeckia radula Pursh, Fl. Amer. Sept. 2: 575. 1813

Perennials, 50–100 cm (with crown buds). **Stems** erect, distally densely hispid. **Leaves** mostly basal; opposite; petioles obscure (intergrading with blades); blades obovate to orbiculate, 4.6–14.5 × 2.1–12 cm, bases broadly cuneate to rounded, margins entire or serrulate, abaxial faces strigoso-hispid, not gland-dotted (cauline usually much smaller, alternate distally). **Heads** usually borne singly. **Peduncles** 10–20 cm. **Involucres** shallowly hemispheric, (15–25 ×) 5–8 mm. **Phyllaries** (often dark purple) 25–33, lanceolate to ovate, 10–14 × 3–5 mm, apices acute to acuminate, abaxial faces hispid or glabrous. **Paleae** 9–10 mm, subentire to 3-toothed (apices purplish, mucronate). **Ray florets** 0 or 2–8; laminae (sometimes purplish) 1–2(–10) mm. **Disc florets** 100–150+; corollas 7–8 mm, lobes reddish; anthers dark, appendages dark. **Cypselae** 3–4 mm, glabrate; **pappi** of 2 (often unequal) aristate scales 0.5–2.9 mm. $2n = 34$.

Flowering fall. Sandy, open pine barrens, flatwoods; 0–50+ m; Ala., Fla., Ga., La., Miss., S.C.

17. Helianthus carnosus Small, Torreya 2: 74. 1902

• Lakeside sunflower [C] [E]

Perennials, 10–60 cm, (with crown buds). Stems erect, (unbranched) glabrous, eglandular. Leaves mostly basal; opposite (proximal) or alternate; petioles obscure (intergrading with blades); blades lanceolate to lance-linear or linear, 10–25 × 0.5–1.5 cm, bases cuneate, margins entire, faces glabrous, not gland-dotted (cauline 1–5 pairs, smaller). Heads 1(–3). Peduncles 5–10 cm. Involucres broadly hemispheric, 17–25 mm diam. Phyllaries (green with purple veins) 25–30, ovate to lance-ovate, 8–16 × 3–5 mm, apices acuminate, abaxial faces usually glabrous, sometimes puberulent. Paleae 8–9 mm, 3-toothed (middle teeth long-acuminate). Ray florets 12–17; laminae 20–40 mm, (abaxial faces not gland-dotted). Disc florets 100+; corollas 6–6.5 mm, lobes yellow; anthers dark, appendages yellowish or dark. Cypselae 3 mm, glabrous; pappi of 2 aristate scales 1.5–2 mm plus 2–4 erose scales 0.3–0.6 mm. 2*n* = 34.

Flowering early–late summer. Wet prairies, flatwoods; of conservation concern; 0–10+ m; Fla.

Helianthus carnosus is distinctive in morphology and is known from relatively few counties in northeastern Florida.

18. Helianthus longifolius Pursh, Fl. Amer. Sept. 2: 571. 1813 • Longleaf sunflower [E]

Perennials, 10–30 cm (with crown buds). Stems (green or purplish) erect, glabrous. Leaves mostly basal; opposite; petioles 0–1 cm; blades linear to narrowly obovate, 13–30 × 0.7–2 cm, bases cuneate, margins entire or obscurely serrate, faces glabrous, not gland-dotted (cauline smaller). Heads 3–12. Peduncles 4–11 cm. Involucres hemispheric, 8–12 mm diam. Phyllaries 18–23, linear-linear, 5–11 × 1–2.5 mm, (margins ciliolate) apices ± attenuate, abaxial faces glabrate or glabrous. Paleae 6–8 mm, weakly 3-toothed (apices hairy). Ray florets 8–13; laminae 10–19 mm (abaxial faces not gland-dotted). Disc florets 35+; corollas 4.2–5.5 mm, lobes yellow; anthers dark, appendages dark. Cypselae 2–3 mm, glabrate; pappi of 2 aristate scales 1.4–2.5 mm. 2*n* = 34.

Flowering late summer–fall. Sandstone and granite outcrop edges; 100–600 m; Ala., Ga., N.C.

Helianthus longifolius is locally escaped from a planting in one county in North Carolina. It is locally abun-

dant where it occurs. It is not similar to or closely related to any other species; natural hybrids of *H. longifolius* with *H. atrorubens* and *H. occidentalis* are known.

19. Helianthus atrorubens Linnaeus, Sp. Pl. 2: 906. 1753 • Purpledisc sunflower [E] [F]

Helianthus atrorubens var. *alsodes* Fernald; *H. sparsifolius* Elliott

Perennials, 50–200 cm (with crown buds; nonflowering stems usually absent). Stems erect, proximally villous to strigoso-hispid, distally hispid or glabrate. Leaves mostly basal; opposite; petioles 4–25 cm (usually winged at least ¹/₂ their lengths); blades lanceolate to ovate, 7–26 × 3–10 cm, bases broadly cuneate to nearly truncate (often decurrent onto petioles), margins serrate to crenate, abaxial faces strigoso-hispid (hairs of midribs1+ mm), not gland-dotted (cauline to 8 pairs proximal to heads, ovate to elliptic, smaller). Heads (1–)3–15+. Peduncles 0.3–17 cm. Involucres broadly hemispheric, 9–16 mm diam. Phyllaries 15–22, broadly ovate to oblong, 7–9 × 4–5 mm, (margins ciliolate) apices obtuse to acute, sometimes mucronate, abaxial faces usually glabrous. Paleae 4–5.5 mm, ± 3-toothed to entire. Ray florets 10–15; laminae 15–22 mm (abaxial faces not gland-dotted). Disc florets 75+; corollas 4–6 mm, lobes reddish; anthers dark, appendages dark (style branches yellow). Cypselae 2.8–3 mm, glabrous or distally puberulent; pappi of 2 aristate scales 2.5–2.8 mm. 2*n* = 34.

Flowering late summer–fall. Open mixed woods, roadsides; 0–900 m; Ala., Fla., Ga., Ky., La., N.J., N.C., S.C., Tenn., Va.

Helianthus atrorubens and *H. silphioides* form a morphologically similar pair of species that share the distinctive feature of relatively broad, tightly appressed phyllaries with apices obtuse to acute. *Helianthus atrorubens* has a more easterly geographic distribution along the Piedmont and the Atlantic coastal plain and the southern Appalachian Mountains; distribution of *H. silphioides* is centered in the Ozark region. In general, *H. silphioides* differs in its usually well developed cauline leaves, and basal leaves with winged petioles that are less than half the total lengths. Involucres with phyllaries tightly appressed are also observed in *H. occidentalis* and *H. pauciflorus* subsp. *pauciflorus*, which differs in having the phyllary apices acute to acuminate as well as in usually having abaxial faces of leaves and ray laminae densely gland-dotted.

20. Helianthus silphioides Nuttall, Trans. Amer. Philos. Soc., n. s. 7: 366. 1841 • Rosinweed sunflower E

Helianthus kentuckiensis F. T. McFarland & W. A. Anderson

Perennials, 50–300 cm (with crown buds; nonflowering stems usually present). **Stems** erect, hispid to strigoso-hispid proximally, glabrate distally. **Leaves** basal and cauline; mostly opposite; petioles 0.1–5.5 cm; blades ovate to broadly ovate or suborbiculate, 7–15 × 4.5–15 cm, bases rounded to truncate or ± cuneate, margins entire or crenate to serrate, abaxial faces usually scabrous, sometimes ± strigose (hairs on midribs to 1 mm), not gland-dotted (cauline 9+ pairs proximal to heads, smaller). **Heads** 3–15+. **Peduncles** 0.5–10 cm. **Involucres** hemispheric, 10–20 mm diam. **Phyllaries** 16–23, oblong to obovate, 8–10 × 3–5 mm, (margins ciliolate) apices obtuse to acute, sometimes mucronate, abaxial faces glabrous or glabrate. **Paleae** 9–10 mm, entire or ± 3-toothed. **Ray florets** 8–13; laminae 15–20 mm (abaxial faces not gland-dotted). **Disc florets** 75+; corollas 6–7 mm, lobes reddish; anthers dark, appendages dark (style branches yellow). **Cypselae** 3–4 mm, glabrous or distally puberulent; **pappi** of 2 aristate scales 2.5–2.7 mm. **2n** = 34.

Flowering late summer–early fall. Open sites; 100–300 m; Ala., Ark., Ill., Ky., La., Miss., Mo., Okla., Tenn.

Similar to *Helianthus atrorubens*, *H. silphioides* is distinguished by shorter hairs (less than 2 mm) on stems proximally and on abaxial leaf midveins, and by petioles of basal leaves less than ¹/₂ lengths of blades and winged less than ¹/₂ their lengths. As befits the name, specimens of *H. silphioides* (and also *H. atrorubens*) are not infrequently misidentified as species of *Silphium*.

21. Helianthus divaricatus Linnaeus, Sp. Pl. 2: 906. 1753 • Woodland sunflower E

Helianthus divaricatus var. *angustifolius* Kuntze

Perennials, 20–150 cm (rhizomatous). **Stems** erect, usually glabrous, rarely ± hairy (often glaucous). **Leaves** cauline; opposite; sessile; blades (light to dark green, sometimes whitish abaxially, 3-nerved at bases) lanceolate to lance-ovate, 6–15 × 1–5 cm, bases rounded to cordate, margins subentire to serrate, abaxial faces sparsely hispid to hispidulous, gland-dotted. **Heads** 1–10. **Peduncles** 0.5–9 cm. **Involucres** hemispheric, 10–15 mm diam. **Phyllaries** 18–25, lanceolate, lance-linear, or lance-ovate, 6–12 × 2–2.5 mm, (margins ciliate) apices acuminate to attenuate, abaxial faces hispidulous to glabrate, not

gland-dotted. **Paleae** 5–8 mm, 3-toothed (apices ciliate). **Ray florets** 8–12; laminae 15–30 mm. **Disc florets** 40+; corollas 4.2–5.5 mm, lobes yellow; anthers usually dark brown to black, appendages yellow. **Cypselae** 3–3.6 mm, glabrate; **pappi** of 2 aristate scales 2.2–2.5 mm. **2n** = 34.

Flowering summer–early fall. Dry, open sites; 10–900+ m; Ont., Que.; Ala., Ark., Conn., Del., D.C., Fla., Ga., Ill., Ind., Iowa, Ky., La., Maine, Md., Mass., Mich., Miss., Mo., N.H., N.J., N.Y., N.C., Ohio, Okla., Pa., R.I., S.C., Tenn., Vt., Va., W.Va., Wis.

One of the earlier flowering perennial *Helianthus*, *H. divaricatus* resembles the tetraploid *H. hirsutus* but differs by its usually glabrous and often glaucous stems, sessile or subsessile leaves, and smaller reproductive organs (disc corollas, paleae, cypselae). Plants from the Ozark region of Arkansas have larger leaves and heads and may represent a polyploid form of *H. divaricatus*. Natural hybrids with *H. microcephalus* have been named *H. glaucus* Small (D. M. Smith and A. T. Guard 1958). Hybrids with other species differ from *H. divaricatus* in having short but distinct petioles, hairy stems, leaves with more rounded bases, and primary lateral leaf veins diverging in a subopposite manner distal to bases, rather than being strictly opposite and basal.

22. Helianthus hirsutus Rafinesque, Ann. Nat. 1: 14. 1820 • Hairy sunflower

Helianthus hirsutus var. *stenophyllus* Torrey & A. Gray; *H. hirsutus* var. *trachyphyllus* Torrey & A. Gray; *H. stenophyllus* (Torrey & A. Gray) E. Watson

Perennials, 100–200 cm (rhizomatous). **Stems** erect, hirsute. **Leaves** cauline; mostly opposite; petioles 0.4–2 cm; blades (3-nerved from bases) lanceolate to ovate, 6.5–18 × 1–8 cm, bases truncate to broadly rounded or cuneate, margins subentire to serrate (flat), abaxial faces ± hirsute, gland-dotted (adaxial not gland-dotted). **Heads** 1–7. **Peduncles** 1–5 cm. **Involucres** hemispheric, 10–25 mm diam. **Phyllaries** 18–25 (usually loose, spreading, not reflexed), lanceolate, 7–12 × 2.5–3.5 mm, (margins ciliate) apices acute to short-acuminate, abaxial faces not gland-dotted. **Paleae** 7–10 mm, 3-toothed (apices yellowish, hairy). **Ray florets** 10–15; laminae 15–20 mm. **Disc florets** 40+; corollas 5.5–6.5 mm, lobes yellow; anthers dark brown or black, appendages dark or yellowish. **Cypselae** 4–4.5 mm, glabrate or distally puberulent; **pappi** of 2 aristate scales 2.5–3.2 mm. **2n** = 68.

Flowering late summer–fall. Dry, open sites, woodland edges, roadsides; 10–900+ m; Ont.; Ala., Ark., Conn., Del., D.C., Fla., Ga., Ill., Ind., Iowa, Kans., Ky., La., Md., Mich., Minn., Miss., Mo., Nebr., N.Y., N.C., Ohio, Okla., Pa., S.C., Tenn., Tex., Va., W.Va., Wis.; Mexico (Coahuila, Nuevo León).

Helianthus hirsutus is distinguished from *H. strumosus* by hairy stems and usually yellow (as opposed to dark) anther appendages, and from *H. divaricatus* by petioles and leaf blades 3-nerved distal to bases. Mexican plants of *H. hirsutus* are sometimes labeled with the synonymous *H. leptocaulis* (S. Watson) S. F. Blake, and plants from Mexico and the southwestern United States often have leaf bases cuneate rather than truncate.

23. **Helianthus glaucophyllus** D. M. Smith, Brittonia 10: 192, fig. 1. 1958 • Whiteleaf sunflower [E]

Perennials, 100–200+ cm (rhizomatous). **Stems** erect, glabrous (glaucous). **Leaves** mostly cauline; opposite (proximal) or alternate; petioles 1–3 cm; blades (light to dark green, abaxially whitish, 3-nerved), lanceolate to lance-ovate, 9–18 × 2.5–7 cm, bases abruptly narrowed, margins serrate, faces glabrous (abaxial) or sparsely scabrous (adaxial), not gland-dotted (abaxial glaucous). **Heads** 3–15+. **Peduncles** 0.5–8 cm. **Involucres** hemispheric, 8–9 mm diam. **Phyllaries** 12–15, lanceolate to ovate, 8–9 × 1.8–3 mm, (margins ciliate) apices acuminate. **Paleae** 6.2–7.5 mm, 3-toothed (apices hairy). **Ray florets** 5–8; laminae 12–14 mm (abaxial faces not gland-dotted). **Disc florets** 20–35; corollas 5–6 mm, lobes yellow; anthers dark, appendages dark. **Cypselae** 3.5–3.8 mm, glabrous; **pappi** of 2 aristate scales 3.2–3.5 mm. $2n = 34$.

Flowering late summer–fall. Mesic woodlands; 700–1300 m; N.C., S.C., Tenn.

Helianthus glaucophyllus is found in the southern Blue Ridge Province. It is distinguished from *H. microcephalus*, with which it shares relatively small heads, by glabrous or sparsely scabrous, glaucous leaves. Its closest relative may be *H. decapetalus*, which has larger heads and, usually, longer phyllaries.

24. **Helianthus decapetalus** Linnaeus, Sp. Pl. 2: 905. 1753 • Thinleaf sunflower, hélianthe à dix rayons [E]

Helianthus trachelifolius Miller

Perennials, 60–200 cm (rhizomatous). **Stems** erect, glabrous. **Leaves** cauline; opposite (proximal or all) or alternate (distal); petioles (1–)2–5 cm; blades (green, 3-nerved distal to bases) lanceolate to ovate, 7–21 × 4–10 cm, bases rounded to cuneate (often shortly decurrent onto petioles, margins usually serrate (moderately to notably in larger leaves), abaxial faces ± scabro-hispidulous, relatively sparsely gland-dotted. **Heads** 3–6(–10). **Peduncles** 2–12 cm (not gland-dotted). **Involucres** hemispheric, 12–25 mm diam. **Phyllaries** 20–25 (often reflexed), lance-linear to lanceolate, 11–16 × 2–3 mm (sometimes leaflike, longest surpassing discs by 1/2+ their lengths), (margins ciliate) apices attenuate, abaxial faces strigillose to glabrate, not gland-dotted. **Paleae** 8–10 mm, 3-toothed. **Ray florets** 8–12; laminae 20–25 mm. **Disc florets** 40+; corollas 6.5–7.2 mm, lobes yellow; anthers usually dark brown to black (rarely reddish brown), appendages dark or reddish brown. **Cypselae** 3.5–5 mm; **pappi** of 2 aristate scales 3–4 mm. $2n = 34, 68$.

Flowering summer–fall. Mesic to wet woodland edges; 10–1200 m; N.B., Ont., Que.; Ala., Ark., Conn., Del., D.C., Ga., Ill., Ind., Iowa, Ky., La., Maine, Md., Mass., Mich., Miss., Mo., N.H., N.J., N.Y., N.C., Ohio, Okla., Pa., R.I., S.C., Tenn., Vt., Va., W.Va., Wis.

Helianthus decapetalus is sometimes confused with *Heliopsis helianthoides* because of shared habitats and superficial similarities. The tetraploid cytotype of *H. decapetalus* intergrades (and apparently hybridizes) with *H. strumosus*, particularly in the southern Appalachians; individual specimens can be difficult to place in one or the other species. In addition to morphologic differences, *H. decapetalus* usually occurs in more mesic habitats, particularly along watercourses; *H. strumosus* is found in drier sites such as roadside slopes. *Helianthus ×multiflorus* Linnaeus is a sterile hybrid, often with "doubled" heads (in which disc florets are replaced by ray florets); it is cultivated and is sometimes included within *H. decapetalus*, e.g., *H. decapetalus* var. *multiflorus* (Linnaeus) A. Gray; its parents are *H. decapetalus* and *H. annuus*.

25. **Helianthus laevigatus** Torrey & A. Gray, Fl. N. Amer. 2: 330. 1842 • Smooth sunflower [E]

Helianthus laevigatus subsp. *reindutus* E. S. Steele; *H. reindutus* (E. S. Steele) E. Watson

Perennials, 100–220 cm (rhizomatous). **Stems** erect, usually glabrous, sometimes proximally hirsute (glaucous). **Leaves** cauline; opposite (proximal) or alternate; sessile or subsessile; blades (grayish green or bluish green) lanceolate, 8–15 × 1.5–3.5 cm, bases ± cuneate, margins serrate to subentire, abaxial faces glabrous or glabrate (smooth or slightly rough to touch, glaucous). **Heads** 1–6. **Peduncles** 2–8 cm. cm. **Involucres** cylindric, 10–15 mm diam. **Phyllaries** 23–28, lanceolate, 6–13 × 2–3 mm, (margins sometimes ciliate) apices acuminate, abaxial faces glabrous, not gland-dotted. **Paleae** 8.5–10 mm, 3-toothed to subentire. **Ray florets** 5–10; laminae 15–20 mm. **Disc florets** 35+; corollas 5.5–6.5 mm, lobes yellow; anthers dark, appendages dark. **Cypselae** 4–5.5 mm, glabrous; **pappi** of 2 aristate scales 3.2–4.1 mm plus 0–1 deltate scales 0.5–1.1 mm. $2n = 68$.

Flowering late summer–fall. Shale barrens; 300–900+ m; N.C., S.C., Va., W.Va.

26. **Helianthus eggertii** Small, Fl. S.E. U.S., 1267, 1340. 1903 • Eggert's sunflower C E

Perennials, 30–200+ dm (rhizomatous). **Stems** erect, glabrous (distinctively bluish, glaucous). **Leaves** cauline; all or mostly opposite; sessile; blades (1-nerved) lanceolate to lance-ovate, 7–16.5 × 1.5–3.5 cm, bases cuneate, margins entire or serrulate, faces glabrous (abaxial) or strumose (adaxial), gland-dotted. **Heads** 1–5. **Peduncles** 1–4 cm. **Involucres** hemispheric, 10–25 mm diam. **Phyllaries** 30–38, lanceolate, 12–16 × 3.5–4.5 mm, (margins ciliate) apices acuminate, abaxial faces glabrate, not gland-dotted. **Paleae** 9–11 mm, entire or ± 3-toothed (apices deltate). **Ray florets** 10–18; laminae 15–20 mm. **Disc florets** 70+; corollas 5–7 mm, lobes yellow; anthers dark, appendages dark or ± yellowish. **Cypselae** 4–6 mm, glabrous; **pappi** of 2 aristate scales 3–4 mm plus 0–1 deltate scales 0.5–1 mm. $2n = 102$.

Flowering late summer–fall. Open barrens, open oak-hickory woodlands; of conservation concern; 100–300 m; Ala., Ky., Tenn.

Helianthus eggertii is similar to *H. strumosus* and *H. laevigatus;* it is distinguished by the distinctive blue coloration of stems and leaves and by the leaves usually 1-nerved, in contrast to the 3-nerved condition typical for the genus. It is listed federally as an endangered species; R. L. Jones (1994) found it to be more common than was previously thought, and it is now in the process of being removed from the federal list.

27. **Helianthus strumosus** Linnaeus, Sp. Pl. 2: 905. 1753 • Rough sunflower, hélianthe scrofuleux E

Helianthus montanus E. Watson; *H. saxicola* Small

Perennials, 100–200 cm (rhizomatous). **Stems** erect, glabrous or glabrate. **Leaves** cauline; mostly opposite, sometimes alternate (distal); petioles 1–3 cm; blades (light to dark green, 3-nerved distal to bases) lanceolate to lance-ovate or ovate, 7–18 × 2–10 cm, bases subcordate to ± cuneate, margins entire to ± serrate, abaxial faces glabrous or tomentulose, usually densely gland-dotted. **Heads** 3–15. **Peduncles** 1–9 cm. **Involucres** cylindric to hemispheric, 8–20 mm diam. **Phyllaries** 18–25 (erect, loose, or squarrose), lanceolate, 5.5–10 × 1.5–3 mm (equaling or slightly surpassing discs), (margins usually

ciliate) apices acute to acuminate, abaxial faces glabrous or hispidulous, usually not gland-dotted. **Paleae** 5.3–6.5 mm, 3-toothed. **Ray florets** 10–20; laminae ca. 12–20(–30) mm. **Disc florets** 35+; corollas 5.5–6.5 mm, lobes yellow; anthers dark, appendages dark or reddish brown. **Cypselae** 4–5.5 mm, glabrate; **pappi** of 2 aristate scales (1–)2–2.5 mm. $2n = 68, 102$.

Flowering late summer–fall. Woods, roadsides, prairies; 0–1500 m; Ont., Que.; Ala., Ark., Conn., Del., D.C., Fla., Ga., Ill., Ind., Iowa, Kans., Ky., La., Maine, Md., Mass., Mich., Minn., Miss., Mo., N.H., N.J., N.Y., N.C., N.Dak., Ohio, Okla., Pa., R.I., S.C., Tenn., Tex., Vt., Va., W.Va., Wis.

Helianthus strumosus is variable, particularly for leaf shape and indument; sufficient study has not been undertaken to evaluate whether regional forms deserve taxonomic recognition. Hybridization with other polyploids, including *H. decapetalus, H. hirsutus,* and *H. tuberosus,* further complicates the situation. It differs from *H. tuberosus* in having glabrous (or glabrate) stems and in lacking tubers.

28. **Helianthus smithii** Heiser, Rhodora 66: 346. 1964 • Smith's sunflower C E

Helianthus parviflorus Bernhardi ex Sprengel [not Kunth] var. *attenuatus* A. Gray in A. Gray et al., Syn. Fl. N. Amer. 1(2): 278. 1884, not *H. attenuatus* E. Watson 1929

Perennials, 70–150 cm (rhizomatous). **Stems** (usually purplish) erect, glabrous (glaucous). **Leaves** cauline; opposite (proximal) or alternate; petioles to 1.3 cm; blades (light to dark green, 1-nerved) lance-linear to lanceolate, 7–18 × 1–1.8 cm, bases ± cuneate, margins serrulate, abaxial faces ± hispid, gland-dotted. **Heads** 1–5. **Peduncles** 3–6 cm. **Involucres** hemispheric, 5–15 mm diam. **Phyllaries** 15–20, lanceolate, 6–9 × 2–2.7 mm, (margins ciliate) apices short-acuminate, abaxial faces glabrous or glabrate, not gland-dotted. **Paleae** 5.5–6.5 mm, 3-toothed (gland-dotted). **Ray florets** 6–9; laminae 10–15 mm. **Disc florets** 25+; corollas 5.5–6 mm, lobes yellow; anthers dark, appendages dark. **Cypselae** 2.8–3 mm, glabrate; **pappi** of 2 aristate scales 1.5–2 mm. $2n = 68$.

Flowering late summer–fall. Dry, open woods; of conservation concern; 100–300+ m; Ala., Ga., Tenn.

Helianthus smithii is known from relatively few sites. T. A. Storbeck (1984) concluded that it is a distinct species; he could not rule out the possibility that it might be a hybrid of *H. microcephalus* and *H. strumosus.* It may also be an extreme variant of *H. strumosus* that has relatively narrow leaves and small heads.

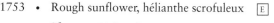

29. Helianthus occidentalis Riddell, W. J. Med. Phys. Sci. 9: 577. 1836 • Western sunflower [E]

Perennials, 60–150 cm (rhizomatous, sometimes stoloniferous as well). **Stems** (often reddish) erect, pilose to appressed-hairy proximally. **Leaves** mostly basal; opposite; petioles 2.5–10 cm; blades oblong-lanceolate or elliptic to ovate, 5–20 × 1.5–7 cm, bases cuneate, margins entire or serrulate, abaxial faces scabrous to hispidulous, gland-dotted (cauline usually smaller, rarely 2–4 pairs well developed). **Heads** 1–4(–12). **Peduncles** 1–14 cm. **Involucres** cylindric, 9–14 mm diam. **Phyllaries** 20–25, lanceolate, 4.5–7(–11) × 1.5–2.5 mm, (margins usually ciliate) apices acute to short-acuminate, abaxial faces glabrate to sparsely pilose (hairs moniliform), not gland-dotted. **Paleae** 5–7 mm, subentire to 3-toothed (shortly mucronate). **Ray florets** 8–14; laminae 18–22 mm (abaxial faces densely gland-dotted). **Disc florets** 50+; corollas 4.5–5.5 mm, lobes yellow; anthers dark brown or black, appendages yellow. **Cypselae** 3–4(–5) mm, sparsely villous or glabrate; **pappi** of 2 aristate scales 1.5–2.8 mm plus 0–4 deltate scales 0.4–1 mm. **2n** = 34 (subspecies unknown).

Subspecies 2 (2 in the flora): c, e United States.

1. Leaves entire or nearly so, scabrous or hirsute, rough to touch . . . . 29a. *Helianthus occidentalis* subsp. *occidentalis*
1. Leaves usually serrate, strigose or glabrous, smooth to touch . . . . 29b. *Helianthus occidentalis* subsp. *plantagineus*

29a. Helianthus occidentalis Riddell subsp. **occidentalis** [E]

Helianthus dowellianus M. A. Curtis; *H. occidentalis* var. *dowellianus* (M. A. Curtis) A. Gray

Leaves entire or nearly so, scabrous or hirsute, rough to touch.

Flowering late summer–fall. Fields, barrens; 10–600+ m; Ark., D.C., Fla., Ga., Ill., Ind., Iowa, Kans., Ky., Md., Mass., Mich., Mo., N.J., N.C., Ohio, Pa., S.C., Tenn., Va., W.Va., Wis.

In its typical form, *Helianthus occidentalis* is distinctive with its relatively few basal leaves and scapiform stems. Still unresolved is the status of *H. dowellianus*, which may deserve recognition at some level. These plants have the head morphology of *H. occidentalis* but have multiple pairs of well-developed cauline leaves. Plants with this morphology are found in the southern Appalachian region from Maryland to Georgia. According to C. B. Heiser et al. (1969), these plants may be hybrids of *H. occidentalis* and *H. atrorubens*. Natural hybrids of *H. occidentalis* with *H. mollis* have been called *H.* ×*cinereus* Torrey & A. Gray. Hybrids of *H. occidentalis* with other species are known.

29b. Helianthus occidentalis Riddell subsp. **plantagineus** (Torrey & A. Gray) Heiser, Mem. Torrey Bot. Club 22(3): 107. 1969 • Shinner's sunflower [E]

Helianthus occidentalis var. *plantagineus* Torrey & A. Gray, Fl. N. Amer. 2: 323. 1842

Leaves usually serrate, strigose or glabrous, smooth to touch.

Flowering summer–early fall. Dry, open areas; 10–100+ m; Ark., Tex.

30. Helianthus pauciflorus Nuttall, Gen. N. Amer. Pl. 2: 177. 1818 [E]

Perennials, 50–200 cm (rhizomatous). **Stems** erect, scabro-hispid. **Leaves** mostly basal; opposite, or opposite (proximal) and alternate; petioles 0–1 cm; blades lance-ovate to oblong-lanceolate, or rhombic-ovate to lance-linear, 5–27 × 2–6 cm, bases cuneate, margins serrate to subentire, abaxial faces sparsely hispid, sparsely, if at all, gland-dotted (cauline smaller). **Heads** 1–10. **Peduncles** 3–12 cm. **Involucres** hemispheric, 15–23 mm diam. **Phyllaries** 25–35, ovate, 6–10 × 3–5 mm, (margins ciliate) apices acute, abaxial faces glabrate to hispid. **Paleae** 9–10 mm, shallowly 3-toothed to subentire (apices ciliate, hairy). **Ray florets** 10–20; laminae 20–35 mm (abaxial faces usually gland-dotted). **Disc florets** 75+; corollas 6.5–7 mm, lobes usually reddish, sometimes yellow; anthers dark, appendages dark (style branches yellow). **Cypselae** 5–6 mm, glabrate; **pappi** of 2 aristate scales 4.5–5 mm plus 1–4 smaller aristate scales 0.3–2 mm. **2n** = 102 (subspecies unknown).

Subspecies 2 (2 in the flora): North America.

Variation within *Helianthus pauciflorus* is complex and is complicated by hybridization with *H. tuberosus*. Hybrids with *H. tuberosus* are known as *H.* ×*laetiflorus*, which is frequently cultivated.

1. Plants 80–200 cm; leaves usually alternate distally, blades oblong-lanceolate to lance-ovate, 8–27 cm, apices acuminate . . . . 29a. *Helianthus pauciflorus* subsp. *pauciflorus*
1. Plants 50–120 cm; leaves opposite, blades rhombic-ovate to lance-linear, 5–12 cm, apices acute or obtuse . . . . 29b. *Helianthus pauciflorus* subsp. *subrhomboideus*

30a. Helianthus pauciflorus Nuttall subsp. **pauciflorus**

• Stiff sunflower, hélianthe rigide [E]

Helianthus laetiflorus Persoon var. *rigidus* (Cassini) Fernald; *H. rigidus* (Cassini) Desfontaines

Plants 80–200 cm. **Leaves** usually alternate distally; blades oblong-lanceolate to lance-ovate, 8–27 cm, apices acuminate.

Flowering late summer–fall. Dry to moderately wet prairies, open places; 50–600+ m; Alta., B.C., Man., Ont., P.E.I., Que., Sask.; Ark., Conn., Ga., Ill., Ind., Iowa, Kans., Ky., Maine, Mass., Mich., Minn., Mo., Mont., Nebr., N.H., N.J., N.Dak., Okla., R.I., S.Dak., Tex., Vt., Va., Wis.

Subspecies *pauciflorus* is introduced in Ontario, and probably in all eastern Canada.

30b. Helianthus pauciflorus Nuttall subsp. **subrhomboideus** (Rydberg) O. Spring & E. E. Schilling, Biochem. Syst. & Ecol. 18: 22. 1990

• Hélianthe subrhomboïdal [E]

Helianthus subrhomboideus Rydberg, Mem. New York Bot. Gard. 1: 419. 1900; *H. laetiflorus* Persoon var. *subrhomboideus* (Rydberg) Fernald; *H. pauciflorus* var. *subrhomboideus* (Rydberg) Cronquist; *H. rigidus* (Cassini) Desfontaines subsp. *subrhomboideus* (Rydberg) Heiser; *H. rigidus* var. *subrhomboideus* (Rydberg) Cronquist

Plants 50–120 cm. **Leaves** opposite; blades rhombic-ovate to lance-linear, 5–12 cm, apices acute or obtuse.

Flowering summer–early fall. Dry, open places; 400–2400 m; Alta., Man., N.B., N.S., Ont., Que., Sask.; Colo., Conn., Ill., Ind., Iowa, Kans., Maine, Mass., Minn., Mo., Mont., Nebr., N.H., N.J., N.Mex., N.Dak., Pa., R.I., S.Dak., Tex., Vt., Wis., Wyo.

31. Helianthus ×laetiflorus Persoon, Syn. Pl. 2: 476. 1807 (as species) • Hélianthe à belles fleurs [E]

Helianthus severus E. Watson; *H. superbus* E. Watson

Perennials, 100–200 cm (rhizomatous). **Stems** erect, slightly hairy. **Leaves** basal and cauline; mostly opposite; petioles 1–5 cm; blades lanceolate to lance-ovate, 10–25 × 2–8 cm, bases cuneate, margins serrulate to serrate, abaxial faces sparsely hispid, gland-dotted. **Heads** 3–6. **Peduncles** 7–14 cm. **Involucres** hemispheric, 14–18 mm diam. **Phyllaries**

25–35, oblong-lanceolate, 7–12 × 3–3.5 mm (strongly unequal), apices acuminate, abaxial faces usually puberulent, gland-dotted. **Paleae** 9–10 mm, entire or 3-toothed (apices reddish). **Ray florets** 15–20; laminae 25–37 mm. **Disc florets** 75+; corollas 7–7.5 mm, lobes reddish (at least at tips); anthers dark, appendages dark to yellowish (style branches yellow). **Cypselae** (seldom formed) 4–5 mm, glabrate; **pappi** of 2 aristate scales 2.5–3.5 mm. $2n = 102$.

Flowering late summer–fall. Roadsides, fields; 0–300+ m; N.B., Nfld. and Labr. (Nfld.), Ont., Que.; Ala., Conn., Del., D.C., Ga., Ill., Ind., Iowa, Kans., Ky., Maine, Md., Mass., Mich., Minn., Mo., Mont., Nebr., N.H., N.J., N.Y., N.C., Ohio, Okla., Oreg., Pa., R.I., S.C., Tenn., Tex., Vt., Va., W.Va., Wis.

Helianthus ×laetiflorus is introduced in Ontario, Newfoundland, and probably in Quebec; it was excluded from British Columbia by G. W. Douglas et al. (1989–1994, vol. 1); and it was reported in New Brunswick by H. R. Hinds (2000).

Plants called *Helianthus ×laetiflorus* are usually interpreted to be hybrids and backcrosses of *H. tuberosus* and *H. pauciflorus*; they are widely cultivated and often escape. An alternative treatment has been to use the name *H. laetiflorus* for plants that are treated here as *H. pauciflorus*, sometimes with infraspecific taxa.

Helianthus atrorubens Lamarck 1789, not Linnaeus 1753, has been applied to plants here called *H. ×laetiflorus*.

SELECTED REFERENCE Clevenger, S. and C. B. Heiser. 1963. *Helianthus laetiflorus* and *Helianthus rigidus*—hybrids or species? Rhodora 65: 121–133.

32. Helianthus tuberosus Linnaeus, Sp. Pl. 2: 905. 1753

• Jerusalem artichoke, topinambour

Helianthus tomentosus Michaux; *H. tuberosus* var. *subcanescens* A. Gray

Perennials, 50–200+ cm (rhizomatous, producing tubers late in growing season). **Stems** erect, scabro-hispid to hirsute (sometimes glaucous). **Leaves** mostly cauline; opposite or alternate proximally, usually alternate distally; petioles 2–8 cm (often ± winged); blades (3-nerved from near bases) lanceolate to ovate, 10–23 × 7–15 cm, bases broadly to narrowly cuneate, margins entire or serrate (flat), abaxial faces puberulent or hirsutulous to tomentulose and gland-dotted (abaxial) or ± scabrous (adaxial). **Heads** 3–15. **Peduncles** 1–15 cm. **Involucres** hemispheric, (10–25 ×) 8–12 mm. **Phyllaries** (often dark green, drying nearly black) 22–35 (bases appressed, apices ± spreading, sometimes reflexed in fruit), lanceolate, 8.5–15 × 2–4 mm (subequal), (margins ciliate) apices acuminate, abaxial

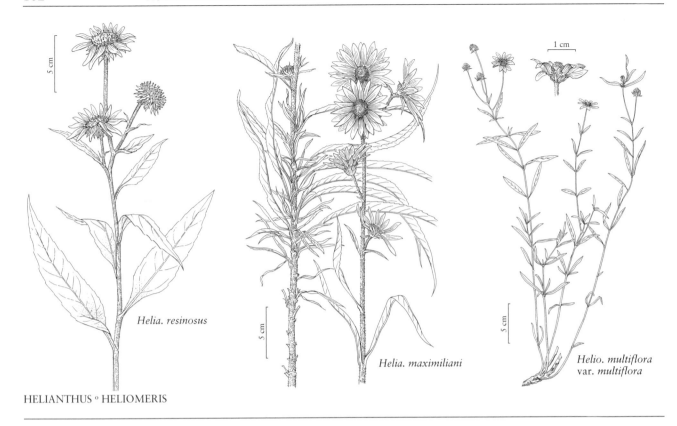

Helia. resinosus

Helia. maximiliani

Helio. multiflora
var. multiflora

HELIANTHUS ° HELIOMERIS

faces hispidulous or puberulent, gland-dotted. **Paleae** 8–9 mm, 3-toothed (apices hairy). **Ray florets** 10–20; laminae 25–40 mm. **Disc florets** 60+; corollas 6–7 mm, lobes yellow; anthers dark brown or black, appendages dark or yellowish. **Cypselae** 5–7 mm, glabrous or distally hairy; **pappi** of 2 aristate scales 1.9–3 mm plus 0–1 deltate scales 0.5–0.8 mm. $2n = 102$.

Flowering late summer–fall. Roadsides, fields, waste areas; 0–1000(–1500) m; Man., N.B., N.S., Ont., P.E.I., Que., Sask.; Ala., Ark., Colo., Conn., Del., D.C., Fla., Ga., Idaho, Ill., Ind., Iowa, Kans., Ky., La., Maine, Md., Mass., Mich., Minn., Miss., Mo., Nebr., N.H., N.J., N.Y., N.C., N.Dak., Ohio, Okla., Pa., R.I., S.C., S.Dak., Tenn., Tex., Utah, Vt., Va., Wash., W.Va., Wis., Wyo.; cultivated and adventive in Europe.

Helianthus tuberosus is variable, probably in part stemming from hybridization with other polyploids, including *H. pauciflorus*, *H. resinosus*, and *H. strumosus*. *Helianthus tuberosus* is so widely spread as a weedy species that its original distribution is difficult to discern. It has been used as a food plant for its tubers by native Americans (although not necessarily domesticated or even cultivated); it has been developed as a crop primarily in Europe, where it has become widely naturalized. The common name Jerusalem artichoke is a misnomer, as explained by C. B. Heiser (1976).

33. Helianthus resinosus Small, Fl. S.E. U.S., 1269, 1340. 1903 • Resindot sunflower [E] [F]

Perennials, 100–300 cm (rhizomatous). **Stems** (often reddish or purplish) erect, hirsute or villous. **Leaves** mostly cauline; mostly alternate; petioles 0.5–2 cm (broadly winged); blades lanceolate to ovate, 6.5–20 × 3.2–9 cm, bases gradually narrowed (onto petioles), margins entire or serrate, abaxial faces hirsute to tomentose, gland-dotted. **Heads** 1–5. **Peduncles** 1–3 cm. **Involucres** hemispheric, 15–27 mm diam. **Phyllaries** 20–30 (squarrose to reflexed, at least tips), lanceolate, 10–21 × 3–5 mm (notably surpassing discs), apices acute to attenuate, abaxial faces hispid to villous, densely gland-dotted (at least toward apices). **Paleae** 9–11 mm, 3-toothed (gland-dotted). **Ray florets** 10–20; laminae (often light yellow) 18–30 mm. **Disc florets** 90+; corollas 8–8.5 mm, lobes yellow; anthers dark, appendages dark. **Cypselae** 5–7 mm, glabrate; **pappi** of 2 aristate scales 2.4–2.6 mm. $2n = 102$.

Flowering late summer–fall. Roadsides, open areas; 0–900+ m; Ala., Fla., Ga., Miss., N.C., S.C.

Consistent with its epithet, *Helianthus resinosus* has abundant subsessile glandular hairs ("resin dots") on its leaves, phyllaries, paleae, and ray laminae. It intergrades and has been reported to hybridize with *H. tuberosus*. It is cultivated and usually sold under the long-misapplied name *H. tomentosus* Michaux.

34. Helianthus mollis Lamarck in J. Lamarck et al., Encycl. 3: 85. 1789 • Ashy sunflower E

Helianthus mollis var. *cordatus* S. Watson

Perennials, 50–150+ cm (rhizomatous). Stems erect, hirsute to villous. Leaves mostly cauline; mostly opposite (sometimes alternate among heads); sessile; blades (ashy or gray-green, 3-nerved distal to bases) lance-olate to broadly ovate, 5.5–14.5 × 1.8–6.5 cm, bases rounded to cordate, margins entire or serrulate, abaxial faces hispid to tomentose, gland-dotted. Heads 1–15. Peduncles 0.1–15 cm. Involucres broadly hemispheric, 12–25 mm diam. Phyllaries 30–40, lanceolate, (5–)10–16 × 2–3.5 mm, apices usually acute, sometimes acuminate, abaxial faces densely hispid to villous or tomentose, densely gland-dotted. Paleae (oblanceolate) 9–11 mm, entire (1-toothed, densely hairy, densely gland-dotted). Ray florets 17–22; laminae 25–30 mm (abaxial faces densely gland-dotted). Disc florets 75+; corollas 6–7.5 mm, lobes yellow; anthers dark, appendages dark. Cypselae 3.5–4 mm, distally villous; pappi of 2 aristate scales 2.8–3.2 mm. 2*n* = 34.

Flowering summer–early fall. Prairies, roadsides; 10–600+ m; Ont.; Ala., Ark., Conn., Del., D.C., Ga., Ill., Ind., Iowa, Kans., Ky., La., Maine, Md., Mass., Mich., Miss., Mo., Nebr., N.J., N.Y., N.C., Ohio, Okla., Pa., R.I., S.C., Tenn., Tex., Va., W.Va., Wis.

Helianthus mollis is introduced in Ontario and adventive in the eastern United States (e.g., Maine), where it is continuing to spread, particularly along roads. Natural hybrids between *H. mollis* and *H. occidentalis* have been named *H. cinereus* Torrey & A. Gray (R. C. Jackson and A. T. Guard 1957); they differ from *H. mollis* by having smaller heads with fewer ray florets and narrower leaves with cuneate bases. Hybrids of *H. mollis* with *H. giganteus* have been called *H. doronicoides* Lamarck (Jackson 1956).

35. Helianthus microcephalus Torrey & A. Gray, Fl. N. Amer. 2: 329. 1842 • Woodland sunflower E

Perennials, 20–200 cm (with crown buds). Stems erect, glabrous. Leaves cauline; opposite or alternate; petioles 0.3–3 cm; blades (greenish, at least abaxially, 3-nerved at bases) lanceolate, 7.2–15.5 × 1.3–4 cm, bases cuneate, margins entire or serrate, abaxial faces tomentulose, densely gland-dotted. Heads (1–)3–15+. Peduncles 1–3(–8) cm. Involucres cylindric, 5–7 mm diam. Phyllaries 12–17, lance-linear, 3–6.5 × 1.5–2.5 mm, (margins ciliate) apices acuminate, abaxial faces glabrate, not gland-dotted. Paleae 5–7 mm, 3-toothed (apices hairy). Ray florets 5–8; laminae 10–14 mm. Disc florets 15–22; corollas 4–5.5 mm, lobes yellow; anthers dark, appendages dark. Cypselae 3.5–4.2 mm, glabrous; pappi of 2 aristate scales 1.5–2.2 mm. 2*n* = 34.

Flowering late summer–fall. Open woodlands, shaded roadsides; 10–900 m; Ala., Ark., Conn., D.C., Fla., Ga., Ill., Ind., Iowa, Ky., La., Md., Mich., Minn., Miss., N.J., N.C., Ohio, Pa., S.C., Tenn., Va., W.Va.

Helianthus microcephalus is distinguished by its relatively small heads, which have relatively few phyllaries, ray florets, and disc florets, as well as the usually tomentulose abaxial faces of the leaves. Hybrids with *H. divaricatus* are known.

36. Helianthus schweinitzii Torrey & A. Gray, Fl. N. Amer. 2: 330. 1842 • Schweinitz's sunflower C E

Perennials, 100–200(–300) cm (rhizomatous, producing tubers). Stems (often reddish, leafy) erect, ± strigose proximally, strigose or glabrous distally. Leaves cauline; opposite (proximal) or alternate; petioles 0–1 cm; blades (light to dark green, 3-nerved distal to bases) lanceolate to lance-linear, 6–18 × 1–2 cm, bases cuneate, margins entire or subentire (revolute), faces hirsute to tomentose and gland-dotted (abaxial) or scabrous (adaxial). Heads 3–6. Peduncles 1.5–8 cm (not gland-dotted). Involucres hemispheric, 9–10(–16) mm diam. Phyllaries 16–20 (reflexed or loose), lanceolate or lance-ovate, 5–7 × 1.5–2.3 mm, apices acute to acuminate, abaxial faces hirsute (proximally) to glabrate, gland-dotted. Paleae 5.5–6.5 mm, 3-toothed (apices hairy). Ray florets 8–15; laminae 15–22 mm (abaxial faces gland-dotted). Disc florets 40+; corollas 5–6 mm, lobes yellow; anthers dark brown to black, appendages dark or reddish brown. Cypselae 3–4 mm, glabrate; pappi of 2 aristate scales 1.5–2 mm. 2*n* = 68.

Flowering fall. Clearings, woodland edges; of conservation concern; 20–100+ m; N.C., S.C.

Helianthus schweinitzii is distinguished by its relatively small heads and sessile to subsessile, narrowly lanceolate leaves with revolute margins. Chromosome number, typification issues, and identification of voucher specimens were updated by J. F. Matthews et al. (1997).

Helianthus schweinitzii is found on the Carolina piedmont. It was listed as endangered in the Federal Register on May 7, 1991, and is known from 16 populations. It is in the Center for Plant Conservation's National Collection of Endangered Plants.

37. Helianthus angustifolius Linnaeus, Sp. Pl. 2: 906. 1753 • Swamp sunflower E

Helianthus angustifolius var. *planifolius* Fernald

Perennials, 50–150+ cm (with crown buds, rhizomes absent or poorly developed). **Stems** erect, usually ± hairy. **Leaves** mostly cauline; opposite or alternate; sessile or subsessile; blades (3-nerved near bases) narrowly lanceolate to linear, 8–15 × 0.15–0.5(–1) cm, bases cuneate, margins entire (revolute, abaxial face sometimes obscured by rolled margins), abaxial faces strigose to hispid or hirsute, sometimes (adaxial not) gland-dotted. **Heads** 3–16. **Peduncles** 5–15 cm. **Involucres** shallowly hemispheric, 10–20 mm diam. **Phyllaries** 25–32 (loose), lanceolate, 4–9 × 1–2 mm, apices acute to slightly acuminate, abaxial faces usually scabrous, rarely glabrous, usually gland-dotted. **Paleae** (oblanceolate) 5.5–6.5 mm, entire or ± 3-toothed (apices purplish, mucronate, gland-dotted). **Ray florets** 10–20; laminae 10–20 mm (abaxial faces gland-dotted). **Disc florets** 75+; corollas 4–4.5 mm, lobes yellow; anthers dark brown or black, appendages dark (style branches usually yellow). **Cypselae** 2–3 mm, glabrate; **pappi** of 2 aristate scales 1.5–2.1 mm. **2n** = 34.

Flowering late summer–fall. Open to shaded, usually moist places; 10–700 m; Ala., Ark., Fla., Ga., Ind., Iowa, Ky., La., Md., Miss., Mo., N.J., N.C., Ohio, Okla., Pa., S.C., Tenn., Tex., Va., W.Va.

The distal parts of *Helianthus angustifolius* are identical in appearance to those of *H. simulans*; plants of the latter are taller (200+ cm) and more robust. The leaves are not only narrower in *H. angustifolius* but also tend to be more strongly revolute (abaxial faces sometimes scarcely visible).

38. Helianthus simulans E. Watson, Pap. Michigan Acad. Sci. 9: 363, plate 49. 1929 • Muck sunflower E

Perennials, 150–260 cm (rhizomatous). **Stems** erect, strigose or hirsute. **Leaves** basal and cauline; opposite or alternate; petioles 0.5–1 cm; blades (3-nerved distal to bases) linear-lanceolate to lanceolate, 9–22 cm × 0.7–4 cm, bases cuneate, margins entire or subentire (usually slightly revolute), abaxial faces hispidulous to tomentulose, gland-dotted. **Heads** 1–15. **Peduncles** 1–13 cm. **Involucres** ± hemispheric, 13–18 mm diam. **Phyllaries** 25–35 (loose), lance-linear to lanceolate, 6–17 × 1.5–3 mm, apices acuminate, abaxial faces glabrate to sparsely strigose,

gland-dotted. **Paleae** 5.5–6.5 mm, 3- (or 5-)toothed (distally strigose, gland-dotted). **Ray florets** 12–23; laminae 16–40 mm (abaxial faces gland-dotted). **Disc florets** 100+; corollas 5–5.5 mm, lobes yellow or reddish; anthers dark brown or black, appendages dark. **Cypselae** 2.5–3 mm, glabrous; **pappi** of 2 (usually unequal) aristate scales 1.5–2.4 mm. **2n** = 34.

Flowering fall. Wet soils, ditches, roadsides; 0–50+ m; Ala., Fla., Ga., La., S.C.

Helianthus simulans is similar to *H. angustifolius*, but much more robust with larger leaves, and to *H. floridanus*, but with longer and narrower leaves. Some workers have described *H. simulans* as having yellow disc corollas; this is variable within the species; individuals with reddish disc corolla lobes also occur. It is becoming popular as a garden plant, which may increase its range and may lead to an increase in the occurrence of hybrids with *H. angustifolius* and *H. floridanus*. It may be present in Arkansas and Texas.

39. Helianthus floridanus A. Gray ex Chapman, Fl. South. U.S. ed. 2, 629. 1883 • Florida sunflower E

Perennials, 100–200 cm (rhizomatous). **Stems** (green) erect, hispid to hispidulous. **Leaves** cauline; usually opposite or alternate, rarely whorled; petioles 0–1 cm; blades (1- or 3-nerved) lanceolate to elliptic or lance-ovate, 4–15 × 0.5–6 cm, bases rounded to cuneate, margins entire or serrulate (often undulate, often ± revolute), faces ± scabrous to tomentulose and gland-dotted (abaxial) or scabrous (adaxial). **Heads** 1–6. **Peduncles** 2–15 cm. **Involucres** ± hemispheric, 12–19 × 7–9 mm. **Phyllaries** 33–38, lanceolate, 5–9 × 1.5–3 mm, apices (at least outer) obtuse, abaxial faces hispidulous to hispid, gland-dotted. **Paleae** 4.5–6 mm, entire or weakly 3-toothed (apices purplish, glabrate to puberulent, often gland-dotted). **Ray florets** 10–20; laminae 20–25 mm (abaxial faces gland-dotted). **Disc florets** 90+; corollas 4–5 mm, lobes usually reddish, sometimes yellow; anthers dark brown or black, appendages dark (style branches yellow). **Cypselae** 2.5–3 mm, glabrous; **pappi** of 2 aristate scales 1.5–2.2 mm. **2n** = 34.

Flowering late summer to fall. Sandy, open areas; 0–50 m; Ala., Fla., Ga., La., N.C., S.C.

Helianthus floridanus is similar to and intergrades with *H. angustifolius*, which has narrower leaves, and *H. simulans*, which is more robust and has longer leaves without undulate margins.

40. Helianthus salicifolius A. Dietrich, Allg.
  Gartenzeitung 2: 337. 1834 [E]

Helianthus filiformis Small

Perennials, 150–250+ cm (rhizomatous). **Stems** (green or purplish) erect, glabrous (glaucous). **Leaves** cauline; alternate; sessile or subsessile; blades linear to lance-linear, 8–21 × 0.2–1.2 cm, bases ± attenuate, margins slightly serrulate to subentire (flat), abaxial faces glabrate, gland-dotted. **Heads** 6–15+. **Peduncles** 2–6 cm. **Involucres** campanulate, 10–18 mm diam. **Phyllaries** 40–50, linear to lance-linear, 12–20 × 1.8–2 mm, (margins ciliate) apices long-attenuate, abaxial faces glabrous or glabrate, not gland-dotted. **Paleae** 8–10 mm, entire or weakly 3-toothed (apices greenish, ± hirsute). **Ray florets** 10–20; laminae 28–35 mm (apices often bifid). **Disc florets** 50+; corollas 5.5–6 mm, lobes reddish; anthers dark, appendages dark (style branches yellow). **Cypselae** 4–6 mm, glabrous; **pappi** of 2 aristate scales 3–3.6 mm plus 2–8 lacerate scales 0.3–0.5 mm. **2*n* = 34.**

  Flowering late summer–fall. Limestone prairies; 100–300 m; Kans., Mo., Nebr., Okla., Tex.

Helianthus salicifolius is found chiefly in the region of the Ozark Plateau. It is cultivated and may occasionally escape. It was recorded from a single site in Chicago, Illinois, where it has now been extirpated.

41. Helianthus giganteus Linnaeus, Sp. Pl. 2: 905. 1753
  • Giant or tall sunflower, hélianthe géant [E]

Helianthus alienus E. Watson;
H. validus E. Watson

Perennials, 100–400 cm (rhizomatous). **Stems** (usually reddish) erect, hairy. **Leaves** cauline; mostly alternate; petioles 0–1.2 cm (ciliate); blades (3-nerved from bases) lanceolate to lance-ovate, 7–20 × 1.2–3.5 cm, bases cuneate, margins serrulate to subentire (± flat), abaxial faces scabrous or ± hirsute, sometimes gland-dotted. **Heads** 1–12. **Peduncles** 8–12 mm. **Involucres** hemispheric, 10–24 mm diam. **Phyllaries** 20–25 (loose or spreading), linear, 8–15 × 1.2–2 mm, (margins usually ciliate) apices acute to attenuate, abaxial faces sparsely strigose, not gland-dotted. **Paleae** 7–9 mm, entire or ± 3-toothed (apices acute, hairy). **Ray florets** 12–20; laminae (often pale yellow) 15–25 mm (abaxial faces not gland-dotted). **Disc florets** 60+; corollas 5–6 mm, lobes yellow; anthers dark brown or black, appendages dark. **Cypselae** 3–4 mm, glabrate; **pappi** of 2 aristate scales 2.4–3 mm. **2*n* = 34.**

  Flowering late summer–fall. Usually wet, open sites; 10–600+ m; N.B., Nfld. and Labr. (Nfld.), N.S., Ont., Que.; Conn., Del., D.C., Ga., Ill., Ind., Iowa, Kans., Ky., Maine, Md., Mass., Mich., Minn., Miss., N.J., N.Y., N.C., Ohio, Pa., S.C., Tenn., Vt., Va., W.Va., Wis.

Helianthus giganteus was probably introduced in Newfoundland. It intergrades with *H. nuttallii* and *H. grosseserratus*; it is distinctive in its hairy, reddish-colored stems, dark anther appendages, and prominent setae on the petioles. The natural hybrid of *H. giganteus* with *H. grosseserratus* has been called *H. luxurians* E. Watson (R. W. Long 1954). Hybrids of *H. giganteus* with *H. divaricatus* have been referred to as *H. ambiguus* (Torrey & A. Gray) Britton and *H. ×divariserratus* R. W. Long (Long 1954).

42. Helianthus grosseserratus M. Martens, Index
  Seminum (Louvain) 1839: unpaged. 1839 (as grosse-serratus) • Sawtooth sunflower, hélianthe à grosses dents [E]

Helianthus instabilis E. Watson

Perennials, 100–400(–500) cm (rhizomatous). **Stems** erect, glabrous proximally, distally glabrate, scabrellous, or strigillose (glaucous). **Leaves** cauline; opposite (proximal) or alternate; petioles (1–)2–5 cm; blades (light to dark green, 3-nerved distal to bases) lanceolate to lance-ovate, 10–32 × (1.2–)4–9 cm, bases cuneate, margins usually coarsely to shallowly serrate, rarely subentire (flat), abaxial faces puberulent to tomentulose, gland-dotted. **Heads** 3–15+. **Peduncles** 0.3–10 cm. **Involucres** broadly hemispheric, 15–25 mm diam. **Phyllaries** 25–30 (loose, spreading), lance-linear, 10–14 × 1.5–2.5 mm (subequal), (margins ± ciliate) apices attenuate, abaxial faces glabrous or puberulent, not gland-dotted. **Paleae** 7–8 mm, entire or 3-toothed (apices acuminate, ± hairy). **Ray florets** 14–20; laminae 23–40 mm. **Disc florets** 100+; corollas 5–6 mm, lobes yellow; anthers dark brown to black, appendages yellow. **Cypselae** 3–4 mm, glabrate; **pappi** of 2 aristate scales 1.9–2.5 mm. **2*n* = 34.**

  Flowering late summer–fall. Dry to wet prairies, other open sites; 10–300+ m; Ont., Que.; Ark., Conn., Del., D.C., Ga., Ill., Ind., Iowa, Kans., Ky., La., Maine, Md., Mass., Mich., Minn., Miss., Mo., Nebr., N.H., N.J., N.Y., N.C., N.Dak., Ohio, Okla., Pa., S.Dak., Tenn., Tex., Va., Wash., W.Va., Wis.

Helianthus grosseserratus is native to midwestern North America and has spread as a roadside weed into other areas, such as New England and the southeastern United States. It is introduced in Canada. Hybrids between *H. maximiliani* and *H. grosseserratus* are known as *H. ×intermedius* R. W. Long (R. W. Long 1954; Long 1966). Hybrids of *H. grosseserratus* with *H. salicifolius* have been described as *H. kellermannii* Britton

(Long 1955), and those with *H. mollis* as *H. brevifolius* E. Watson (R. C. Jackson and A. T. Guard 1957b).

43. Helianthus verticillatus Small, Bull. Torrey Bot. Club 25: 479. 1898 • Whorled sunflower [C] [E]

Perennials, 200–300 cm (rhizomatous). **Stems** erect, glabrous (glaucous). **Leaves** cauline; usually whorled (3s–6s), sometimes opposite; sessile or subsessile, petioles 0–1 cm; blades linear to lanceolate, 7.5–18.5 × 0.7–3 cm, bases cuneate, margins serrulate or entire (flat), faces puberulent (abaxial) or hispid-strumose (adaxial), gland-dotted. **Heads** (1–)3–7+. **Peduncles** 1–10 cm. **Involucres** broadly campanulate, 15–20 mm diam. **Phyllaries** 45–55, lance-linear, 12–18 × 2–2.5 mm, (margins ciliate) apices long-attenuate, abaxial faces glabrate to sparsely puberulent, gland-dotted. **Paleae** 5–8 mm, entire or 3-toothed (apices ciliate). **Ray florets** 10–19; laminae 20–26 mm. **Disc florets** 150+; corollas 5–7 mm, lobes yellow; anthers dark, appendages yellow. **Cypselae** 4–5 mm, glabrous; **pappi** of 2 aristate scales 2.2–2.5 mm. *2n* = 34.

Flowering late summer–fall. Wet prairies; of conservation concern; 100–300 m; Ala., Ga., Tenn.

Helianthus verticillatus is relatively uncommon and was recently rediscovered in scattered locations; it is a candidate for federal listing as an endangered species. It has been suggested to be a hybrid of *H. angustifolius* and *H. eggertii* or *H. grosseserratus*; no supporting evidence has been found to corroborate that suggestion.

44. Helianthus maximiliani Schrader, Index Seminum (Göttingen) 1834: unpaged. 1835 • Maximilian sunflower, hélianthe de Maximilien [F]

Helianthus dalyi Britton

Perennials, 50–300 cm (rhizomatous). **Stems** erect, 5–30 dm, scabrous to scabro-hispidulous. **Leaves** cauline; mostly alternate; petioles 0–2 cm; blades (light green to gray-green, 1-nerved, conduplicate) lanceolate, 10–30 × 2–5.5 cm, bases cuneate, margins usually entire, sometimes serrulate, abaxial faces scabrous to scabro-hispid, gland-dotted. **Heads** (1–)3–15 (often in racemiform to spiciform arrays). **Peduncles** 1–11 cm. **Involucres** hemispheric, 13–28 mm diam. **Phyllaries** 30–40, lanceolate, 14–20 × 2–3 mm, (margins ciliate) apices acute to attenuate, abaxial faces canescent, gland-dotted. **Paleae** 7–11 mm, entire or 3-toothed (apices greenish, mucronate, hairy). **Ray florets** 10–25; laminae (15–)25–40 mm. **Disc florets** 75+; corollas 5–7 mm, lobes yellow;

anthers dark brown or black; appendages usually yellow, sometimes partly dark. **Cypselae** 3–4 mm, glabrate; **pappi** of 2 aristate scales 3–4.1 mm. *2n* = 34.

Flowering late summer–fall. Prairies, fields, waste areas; 0–300(–2100+) m; Alta., B.C., Man., Ont., Que., Sask.; Ala., Ark., Calif., Colo., Conn., Del., D.C., Idaho, Ill., Ind., Iowa, Kans., Ky., Maine, Md., Mass., Mich., Minn., Miss., Mo., Nebr., N.J., N.Y., N.C., N.Dak., Ohio, Okla., Pa., S.C., S.Dak., Tenn., Tex., Utah, Va., Wash., W.Va., Wis., Wyo.; Mexico.

Helianthus maximiliani is introduced in eastern Ontario and in Quebec. It appears to be native to midcontinental prairie regions and has spread along railroads and highways into all areas of North America. Its wide dispersal may be aided by cultivation for its attractive, showy floral displays. In addition to the usually conduplicate, single-nerved leaves and spikelike arrangement of the heads, it is distinguished by the whitish-canescent indument of the leaves and stems and the long-attenuate phyllaries.

45. Helianthus nuttallii Torrey & A. Gray, Fl. N. Amer. 2: 324. 1842 • Nuttall's sunflower, hélianthe de Nuttall [E]

Perennials, 100–400 cm (rhizomatous). **Stems** (usually yellow-brown or greenish, sometimes glaucous) erect, glabrous, hispid, ± hirsute, or scabrous. **Leaves** cauline; all or mostly opposite to mostly alternate; petioles 0.5–1.5 cm; blades light to dark green, 3-nerved distal to bases) ± lanceolate to ± ovate, 4–20 × 0.8–4 cm, bases cuneate, margins entire or ± serrate (flat), abaxial faces hispid to hispidulous or tomentulose to villous-tomentose, gland-dotted. **Heads** 1–6. **Peduncles** 1–18 cm. **Involucres** hemispheric, 10–20 mm diam. **Phyllaries** 30–38 (loose, spreading), lanceolate to lance-ovate, 8–16 × 1.5–3 mm, (margins ciliate) apices acute to acuminate, abaxial faces usually ± strigose, rarely glabrate, not gland-dotted. **Paleae** 8–12 mm, 3-toothed. **Ray florets** 10–21; laminae 20–25 mm (abaxial faces not gland-dotted). **Disc florets** 60+; corollas 5–7 mm, lobes yellow; anthers dark brown to black, appendages yellow. **Cypselae** 3–5 mm, glabrate; **pappi** of 2 aristate scales 2.2–4.5 mm plus 0–2 lanceolate scales 0.5–1 mm. *2n* = 34 (subspecies unknown).

Subspecies 3 (3 in the flora): North America.

1. Leaves: abaxial faces tomentulose to villous-tomentose; phyllaries densely hairy (longer hairs 1+ mm) . . . . 45c. *Helianthus nuttallii* subsp. *parishii*
1. Leaves: abaxial faces hispid to hispidulous; phyllaries ± glabrate (longer hairs 0.5–1 mm).

[2. Shifted to left margin—Ed.]

2. Plants usually 200–300 cm; leaves mostly alternate, blades narrowly to broadly lanceolate, apices acute to acuminate . . . . 45a. *Helianthus nuttallii* subsp. *nuttallii*

2. Plants usually 100–250 cm; leaves all or mostly opposite, blades lanceolate to nearly ovate, apices acute to obtuse . . . . 45b. *Helianthus nuttallii* subsp. *rydbergii*

45a. Helianthus nuttallii Torrey & A. Gray subsp. nuttallii E

Helianthus nuttallii subsp. *canadensis* R. W. Long

Plants usually 200–300 cm. **Leaves** mostly alternate; blades narrowly to broadly lanceolate, apices acute to acuminate, abaxial faces hispid to hispidulous. **Phyllaries:** abaxial faces ± glabrate or sparsely hairy (longer hairs 0.5–1 mm).

Flowering late summer–fall. Moist open places, ditches, roadsides; (300–)1200–2700 m; Alta., B.C., Man., Sask.; Ariz., Calif., Colo., Idaho, Kans., Mo., Mont., Nebr., Nev., N.Mex., N.Dak., Okla., Oreg., S.Dak., Utah, Wash., Wyo.

Plants in the Dakotas and south-central and eastern Canada are intermediate between subsp. *nuttallii* and subsp. *rydbergii* and were placed by R. W. Long (1966) in subsp. *canadensis*; C. B. Heiser et al. (1969) returned this to subsp. *nuttallii* with the comment that it was hardly equivalent to other subspecies that they included in their treatment. The name *H. ×luxurians* may refer to hybrids between subsp. *nuttallii* and *H. giganteus* (Heiser et al.).

45b. Helianthus nuttallii Torrey & A. Gray subsp. rydbergii (Britton) R. W. Long, Brittonia 18: 77. 1966 • Rydberg's sunflower E

Helianthus rydbergii Britton, Man. Fl. N. States, 993. 1901

Plants usually 100–250 cm. **Leaves** all or mostly opposite; blades lanceolate to nearly ovate, apices acute to obtuse, abaxial faces hispid to hispidulous. **Phyllaries:** abaxial faces ± glabrate or sparsely hairy (longer hairs 0.5–1 mm).

Flowering late summer–fall. Open areas, sandy dry soils, wet places; 200–1000 m; Alta., Man., Ont., Que., Sask.; Colo., Iowa, Minn., Mont., Nebr., N.Dak., S.Dak.

45c. Helianthus nuttallii Torrey & A. Gray subsp. parishii (A. Gray) Heiser, Contr. Dudley Herb. 4: 316. 1955 • Parish's or Los Angeles sunflower C E

Helianthus parishii A. Gray, Proc. Amer. Acad. Arts 19: 7. 1883

Plants usually 200–400 cm. **Leaves** mostly alternate; blades lanceolate, apices acute to acuminate, abaxial faces tomentulose to villous-tomentose. **Phyllaries:** abaxial faces densely hairy (longer hairs 1+ mm).

Flowering late summer–fall. Marshes; of conservation concern; 10–500 m; Calif.

The wetland habitat for subsp. *parishii* has nearly disappeared through draining of marshes and channelization of watercourses. The subspecies was thought to be extinct until, relatively recently, claims have been made that a population of it has been found.

46. Helianthus californicus de Candolle in A. P. de Candolle and A. L. P. P. de Candolle, Prodr. 5: 589. 1836 • California sunflower

Perennials, 150–350 cm (rhizomatous). **Stems** erect, glabrous (glaucous). Leaves mostly cauline; mostly alternate; petioles 0–3 cm; blades (light to dark green, 3-nerved distal to bases), lanceolate, 10–20 × 2.5–6 cm, bases ± cuneate, margins entire or subentire, abaxial faces hispid to hirsute, gland-dotted. **Heads** 3–10. **Peduncles** (1–)3–15 cm (gland-dotted). **Involucres** broadly hemispheric, 10–25 mm diam. **Phyllaries** 25–35 (loose), lanceolate, 8–15 × 3–5 mm (notably surpassing discs), (margins sometimes ciliate) apices acute to attenuate, abaxial faces hispidulous to glabrate, gland-dotted. **Paleae** 10–11 mm, 3-toothed (apices acute, hispidulous). **Ray florets** 12–21; laminae 15–30 mm. **Disc florets** 50+; corollas 6–8 mm, lobes yellow; anthers reddish to dark, appendages reddish brown. **Cypselae** 4.5–5 mm, glabrate; **pappi** of 2 aristate scales 3–4 mm. $2n = 102$.

Flowering summer–fall. Stream banks, wet meadows, freshwater marshes; to 1600 m; Calif.; Mexico (Baja California).

47. Helianthus gracilentus A. Gray, Proc. Amer. Acad.
Arts 11: 77. 1876 • Slender sunflower

Perennials, 60–200 cm (taprooted).
Stems (reddish or purplish) erect,
hairy. **Leaves** cauline; all or mostly
opposite; petioles 0–3 cm; blades
(1- or 3-nerved) lanceolate to lance-
ovate, 5–11 × 2–3.5 cm, bases cuneate, margins entire or serrate,
abaxial faces hispid, gland-dotted.
Heads 1–5. **Peduncles** 2.5–30 cm.
Involucres hemispheric, 13–20 mm diam. **Phyllaries** 20–
25, lanceolate, 4–8 × 1.5–3 mm, (margins ciliate) apices
acute, abaxial faces hispidulous, gland-dotted. **Paleae**
8–9 mm, subentire to weakly 3-toothed (apices acute,
ciliate, gland-dotted). **Ray florets** 13–21; laminae 15–
25 mm (abaxial faces gland-dotted). **Disc florets** 50+;
corollas 5–6 mm, lobes reddish; anthers reddish brown,
appendages yellow or dull orange (style branches yellow).
Cypselae 3–4 mm, glabrate; **pappi** of 2 aristate scales
2.5–3 mm. **2n** = 34.

Flowering late spring–fall. Dry slopes; 0–1800 m;
Calif.; Mexico (Baja California).

48. Helianthus cusickii A. Gray, Proc. Amer. Acad. Arts
21: 413. 1886 • Cusick's sunflower E

Perennials, 60–120 cm (taproots
stout, parsniplike). **Stems** erect to
prostrate, glabrous or glabrate.
Leaves cauline; mostly opposite,
petioles 0–1 cm; blades (light to
dark green, 3-nerved at bases)
lance-linear to lanceolate, 5–15 ×
0.5–2.5 cm, bases cuneate, margins entire, abaxial faces glabrous
or scabro-hispid, gland-dotted. **Heads** 1–3. **Peduncles**
2–10 cm. **Involucres** hemispheric, 12–28 mm diam.
Phyllaries 12–18, lanceolate, 11–18 × 1.5–3 mm, apices
acute to acuminate, abaxial faces gland-dotted. **Paleae**
(lanceolate) 9–13 mm, subentire or 3-toothed (apices
sparsely hairy, sometimes gland-dotted). **Ray florets** 12–
16; laminae 20–35 mm (abaxial faces gland-dotted). **Disc
florets** 40+; corollas 6.5–7.5 mm, lobes yellow; anthers
usually dull orange or orange-brown, sometimes yellowish, appendages dull orange to yellowish. **Cypselae** 4–5
mm, glabrate; **pappi** of 2 aristate scales 3–4.2 mm. **2n** =
34.

Flowering late spring–summer. Dry slopes, open
woods; 600–2000 m; Calif., Idaho, Nev., Oreg., Wash.

49. Helianthus pumilus Nuttall, Trans. Amer. Philos.
Soc., n. s. 7: 366. 1841 • Little sunflower E

Perennials, 30–100 cm (taproots
slightly thickened). **Stems** erect,
hispid or strigose. **Leaves** cauline;
opposite; petioles 0.4–3 cm; blades
(ashy green, usually 3-nerved from
bases) lanceolate to ovate, 4–15 ×
1–5 cm, bases ± cuneate to truncate, margins entire or serrate,
faces strigoso-hispid, gland-dotted.
Heads 1–6. **Peduncles** 0.5–5.5 cm. **Involucres** cylindric,
7–14 mm diam. **Phyllaries** 15–25 (loosely appressed),
lanceolate to ovate, 3.5–7.5 × 2–3.5 mm, apices acute to
acuminate, abaxial faces ± hispid or strigose (hairs whitish), gland-dotted. **Paleae** 7–8 mm, entire or ± 3-toothed
(apices acute to acuminate, hispid). **Ray florets** 8–13;
laminae 15–20 mm (abaxial faces gland-dotted). **Disc
florets** 30+: corollas 5–6 mm, lobes yellow; anthers dark
brown to black, appendages dark. **Cypselae** 3–4 mm,
sparsely hairy; **pappi** of 2 aristate scales 4–4.5 mm plus
0–4 deltate scales 0.5–1 mm. **2n** = 34.

Flowering late summer. Dry, rocky soil in open areas;
1200–2700 m; Colo., Wyo.

50. Helianthus arizonensis R. C. Jackson, Brittonia 15:
266, fig. 2. 1963 • Arizona sunflower E

Perennials, 20–30 cm (roots creeping). **Stems** erect to ascending, glabrous (glaucous). **Leaves** cauline;
opposite; sessile; blades (bluish
green) lanceolate, 2–7 × 0.5–1.3
cm, bases cuneate, margins entire
(undulate), faces glabrous (glaucous).
Heads 1–7. **Peduncles** 2–7.5 cm.
Involucres hemispheric, 9–18 mm
diam. **Phyllaries** 16–19, lance-ovate, 3.5–7 × 2–2.8 mm,
(margins ciliate) apices acute, abaxial faces glabrate, sparingly gland-dotted. **Paleae** 5.5–6.5 mm, 3-toothed (apices
acute, gland-dotted). **Ray florets** 10–14; laminae 7–9
mm. **Disc florets** 30+; corollas 3–3.5 mm, lobes yellow;
anthers reddish brown, appendages brownish. **Cypselae**
2.8–3.2 mm, glabrous; **pappi** of 2 deltate, erose scales 1–
1.7 mm. **2n** = 34, 51.

Flowering summer–fall. Open pine woodlands; 1200–
2100 m; Ariz., N.Mex.

Helianthus arizonensis is morphologically similar to
the polyploid *H. ciliaris*; *H. arizonensis* is diploid (or triploid) and has entirely yellow disc corollas. It is known
from relatively few sites.

51. Helianthus laciniatus   A. Gray, Mem. Amer. Acad. Arts, n. s. 4: 84. 1849

Helianthus crenatus R. C. Jackson; *H. heiseri* R. C. Jackson

Perennials, 50–120(–200) cm. Stems erect, usually strigose or hispid to glabrate. Leaves cauline; opposite or alternate; sessile; blades (green or grayish, 1- or 3-nerved) lanceolate, 5–9 × 0.5–3.5 cm, bases ± cuneate, margins entire or irregularly toothed to lobed, faces strigose to strumose, gland-dotted (adaxial sometimes glaucous). Heads 1–9. Peduncles 4–13 cm. Involucres hemispheric, 10–24 mm diam. Phyllaries 16–21, lanceolate, 6–7.5 × 1.8–2.5 mm (often subequal), (margins ciliate) apices acute, abaxial faces hispidulous or strigose to glabrate, gland-dotted. Paleae 7–7.8 mm, entire or 3-toothed (apices obtuse to acute, hispid-ciliate to glabrate). Ray florets 14–20; laminae ca. 8–11 mm. Disc florets 40+; corollas 4.8–5.8 mm, lobes reddish; anthers purplish, appendages reddish (style branches yellow). Cypselae 2.7–3.5 mm, glabrate; pappi of 2(–3) aristate scales 1.4–2.5 mm. 2*n* = 34.

Flowering summer–fall. Open, dry, alkaline soils; 1000–1200 m; N.Mex., Tex.; Mexico.

Helianthus laciniatus reaches the northern extent of its range in New Mexico and Texas; it is relatively common in the Chihuahuan Desert areas of Mexico. It is similar to *H. ciliaris;* it usually has hairy stems as well as denser leaf indument that includes a greater number of subsessile, glandular hairs.

52. Helianthus ciliaris   de Candolle in A. P. de Candolle and A. L. P. P. de Candolle, Prodr. 5: 587. 1836
• Blueweed, blueweed sunflower, Texas blueweed

Perennials, 40–70 cm (rhizomatous or with creeping roots, often forming extensive colonies). Stems decumbent to ± erect, glabrous or glabrate (glaucous). Leaves cauline; mostly opposite; sessile; blades (often bluish green, 1- or 3-nerved) linear to lanceolate, 3–7.5 × 0.5–2.2 cm, bases ± cuneate, margins entire or serrate (usually ciliate and undulate), faces glabrous or glabrate to hispid. Heads 1–5. Peduncles (1–)3–13 cm. Involucres hemispheric, 12–25 mm diam. Phyllaries 16–19, ovate to lance-ovate, 3–8 × 2–3.5 mm, (margins ciliate) apices obtuse to acute, abaxial faces glabrate to ± strigose, not gland-dotted). Paleae 7–7.5 mm, subentire to 3-toothed (apices obtuse to acute, hairy, gland-dotted). Ray florets 10–18; laminae 8–9 mm. Disc florets 35+; corollas 4–6 mm, lobes reddish; anthers brownish red, appendages brownish red (style branches yellow). Cypselae 3–3.5 mm, glabrous; pappi of 2 aristate scales 1.2–1.5 mm. 2*n* = 68, 102.

Flowering summer–fall. Roadsides, ditches, cultivated fields, open drainage areas; 10–2600 m; Ariz., Calif., Colo., Ill., Kans., Nebr., Nev., N.Mex., Okla., Tex., Utah; Mexico (Chihuahua, Coahuila, Durango, San Luis Potosí, Sonora, Tamaulipas).

Helianthus ciliaris is considered a noxious weed in some states. It can propagate vegetatively from detached pieces of rhizome and spread aggressively, especially in cultivated fields. It has been noted to occur in Idaho and Washington, where control measures have been taken to eliminate it.

300. HELIOMERIS   Nuttall, Proc. Acad. Nat. Sci. Philadelphia 4: 19. 1848

• Golden-eye [Greek *helios*, sun, and *-merus*, part]

Edward E. Schilling

Viguiera Kunth sect. *Heliomeris* (Nuttall) S. F. Blake

Annuals or perennials, (2–)10–90(–120+) cm. Stems erect, branched distally or ± throughout. Leaves mostly cauline; opposite or alternate; sessile or subsessile; blades usually 1(–3)-nerved, elliptic, lance-linear, lanceolate, lance-ovate, linear, ovate, rhombic, or rhombic-ovate, margins entire (often revolute), faces hispid or strigose to stigillose, sometimes gland-dotted. Heads radiate, borne singly or in cymiform to paniculiform arrays. Involucres hemispheric or broader, 6–14 mm diam. Phyllaries persistent, 14–25 in ± 2–3 series (mostly lance-linear, herbaceous). Receptacles conic; paleate (paleae tan to brown, ovate to oblong-rectangular, conduplicate). Ray florets 5–15, neuter; corollas yellow (laminae elliptic, oblong, obovate, oval, or ovate). Disc florets 25–50+, bisexual, fertile; corollas yellow, tubes shorter than campanulate throats, lobes 5, triangular (style branches relatively slender, apices acute). **Cypselae** (black, mottled, or gray-striate) weakly 4-angled, ± obpyramidal (glabrous); **pappi** 0. x = 8.

Species 5 (4 in the flora): sw United States, Mexico.

Heliomeris has often been submerged within *Viguiera*; herbaceous phyllaries, epappose cypselae, and distinctive chromosome base number provide morphologic and genetic features that correlate with molecular phylogenetic studies to suggest that it be recognized as distinct. A species from granite outcrops of the southeastern United States, *Helianthus porteri*, shows remarkable morphologic similarity to *Heliomeris*, but cytologic, crossing, and molecular phylogenetic studies have provided abundant evidence that it is properly placed in *Helianthus*. W. F. Yates and C. B. Heiser (1979) provided a useful summary of *Heliomeris*, and their treatment is followed here.

SELECTED REFERENCE   Yates, W. F. and C. B. Heiser. 1979. Synopsis of *Heliomeris*. Proc. Indiana Acad. Sci. 88: 364–372.

1. Annuals; leaf margins conspicuously ciliate at least ³/₄+ their lengths (hairs often 0.5+ mm)
   . . . . 1. *Heliomeris hispida*
1. Annuals or perennials; leaf margins ciliate to ¹/₄ their lengths (hairs to 0.5 mm).
   2. Annuals; leaf blades (at least proximal) ovate to rhombic-ovate, abaxial faces not gland-dotted; peduncle lengths 2–5 times leafy portions of stems . . . . 2. *Heliomeris soliceps*
   2. Annuals or perennials; leaf blades (at least proximal) elliptic, lanceolate, lance-linear, lance-ovate, linear, or ovate, abaxial faces often gland-dotted; peduncle lengths ¹/₂₀–¹/₂ leafy portions of stems.
      3. Perennials (caudices woody), proximal and midstem leaves relatively broad (lengths 2–8 times widths) . . . . 4. *Heliomeris multiflora* (in part)
      3. Annuals (taprooted) or perennials (caudices woody), proximal and midstem leaves relatively narrow (lengths 6–30+ times widths).
         4. Annuals (taprooted), proximal and midstem leaves (40–)80–160 × 4–8(–12) mm . . . . 3. *Heliomeris longifolia* (in part)
         4. Annuals (taprooted), proximal and midstem leaves 10–70(–85) × 1.5–5 mm, or perennials (caudices woody), proximal and midstem leaves 10–90 × 2–20 mm.
            5. Annuals (taprooted, stem branching symmetric), proximal and midstem leaves 10–70(–85) × 1.5–5 mm, heads relatively small, involucres 6–9 diam.
               . . . . 3. *Heliomeris longifolia* (in part)
            5. Perennials (caudices woody, stem branching often asymmetric), proximal and midstem leaves 10–90 × 2–20 mm, heads relatively large, involucres 6–14 mm diam. . . . . 4. *Heliomeris multiflora* (in part)

1. **Heliomeris hispida** (A. Gray) Cockerell, Torreya 18: 183. 1918 • Hairy goldeneye

Heliomeris multiflora Nuttall var. *hispida* A. Gray, Smithsonian Contr. Knowl. 5(6): 87. 1853; *H. hispida* var. *ciliata* (B. L. Robinson & Greenman) Cockerell; *Viguiera ciliata* (B. L. Robinson & Greenman) S. F. Blake; *V. ciliata* var. *hispida* (A. Gray) S. F. Blake

Annuals, 10–90 cm (taprooted). **Stems** hispid. **Leaves** mostly alternate; blades lance-linear to linear, 20–100 × 1–3 mm, margins conspicuously ciliate at least ³/₄ their lengths, hairs often than 0.5+ mm, faces hispid, abaxial sometimes gland-dotted. **Heads**

(1–)3–15+. **Peduncles** 0.5–9 cm, lengths ¹/₁₀–¹/₃ leafy portions of stems. **Involucres** 9–14 mm diam. **Paleae** ovate to rectangular, 4.5–5.5 mm, prominently cuspidate. **Phyllaries** 5.5–10 mm. **Ray florets** 9–15; laminae oval to obovate, 6–13 mm. **Disc florets** 50+; corollas 2–3 mm. **Cypselae** black, 1.5–2.5 mm. $2n$ = 16.

Flowering Jul–Oct. Saline marshes and meadows; 1200–2200 m; Calif., N.Mex., Utah; Mexico (Chihuahua, Sonora).

Variety *ciliata* has been separated based on having less dense pubescence; W. F. Yates and C. B. Heiser (1979) felt that this is insufficient for taxonomic recognition, noting that the type collection of *H. hispida* includes both indument forms. *Heliomeris hispida* appears to be relatively rare.

2. **Heliomeris soliceps** (Barneby) W. F. Yates, Proc. Indiana Acad. Sci. 88: 370. 1979 • Tropical false goldeneye, paria sunflower [C] [E]

Viguiera soliceps Barneby, Leafl. W. Bot. 10: 316. 1966

Annuals, (2–)10–30(–40+) cm (taprooted). **Stems** strigillose. **Leaves** opposite (proximal) or alternate (distal); blades ovate to rhombic-ovate, 15–38 × 6–20+ mm, margins ciliate to ¼ their lengths, hairs mostly less than 0.5 mm, faces hispid or strigose (abaxial not gland-dotted). **Heads** 3–15+. **Peduncles** 7–28 cm, lengths 2–5 times leafy portions of stems. **Involucres** 7–10 mm diam. **Phyllaries** lanceolate, 5–6 mm. **Paleae** oblong, ca. 5 mm, cuspidate. **Ray florets** 10–12; laminae ovate, 10–15 mm (glabrous). **Disc florets** 25+; corollas 4 mm. **Cypselae** black or gray-striate, 2.8–3.3 mm.

Flowering May–Jun. Gumbo clay knolls and bluffs; of conservation concern; 1400–1500 m; Utah.

Heliomeris soliceps is characterized by the relatively long, scapiform peduncles rising beyond relatively short, leafy stems, and ovate to rhombic-ovate leaf blades that are relatively broader than those of most other members of the genus. Its blooming period is also significantly earlier than those of other members of *Heliomeris*. It is known from southern Utah, where it forms large populations.

3. **Heliomeris longifolia** (B. L. Robinson & Greenman) Cockerell, Torreya 18: 183. 1918 • Longleaf false goldeneye

Gymnolomia longifolia B. L. Robinson & Greenman, Proc. Boston Soc. Nat. Hist. 29: 92. 1899; *Viguiera longifolia* (B. L. Robinson & Greenman) S. F. Blake

Annuals, 7–15 cm (taprooted). **Stems** strigose. **Leaves** usually opposite proximally, sometimes alternate (distal); blades lance-linear to linear, 10–160 × 1.5–8(–12) mm, margins ciliate to ¼ their lengths, hairs mostly less than 0.5 mm, faces strigose to strigillose (hair bases narrowly tuberculate), abaxial often gland-dotted. **Heads** 6–25+. **Peduncles** 0.2–0.5(–2) cm, lengths ¹⁄₂₀–¹⁄₂ times leafy portions of stems. **Involucres** 6–14 mm diam. **Phyllaries** 3–6 mm. **Paleae** oblong, 4–4.5 mm, acute to cuspidate. **Ray florets** 12–14; laminae ± elliptic, 5–17 mm (glabrous or puberulent). **Disc florets** 50+; corollas 2.5–4 mm. **Cypselae** black or mottled, 1.8–2 mm.

Varieties 2 (2 in the flora): sw United States, Mexico.

1. Stems branching mostly in distal ¹⁄₂, bases usually 7+ mm diam.; leaves (40–)80–160 × 4–8(–12) mm; involucres 9–14 mm diam. . . . .
. . . . 3a. *Heliomeris longifolia* var. *longifolia*
1. Stems branching ± throughout, bases usually less than 5 mm diam.; leaves 10–70(–85) × 1.5–5 mm; involucres 6–9 mm diam. . . . .
. . . . 3b. *Heliomeris longifolia* var. *annua*

3a. **Heliomeris longifolia** (B. L. Robinson & Greenman) Cockerell var. **longifolia**

Stems branching mostly in distal ¹⁄₂, bases usually more than 7+ mm diam. **Leaves** (40–)80–160 × 4–8 (–12) mm. **Involucres** 9–14 mm diam. $2n = 16$.

Flowering Jun–Oct. Desert shrub, pinyon-juniper communities; 600–2300 m; Ariz., N.Mex, Tex.; Mexico.

3b. **Heliomeris longifolia** (B. L. Robinson & Greenman) Cockerell var. **annua** (M. E. Jones) W. F. Yates, Proc. Indiana Acad. Sci. 88: 369. 1979

Gymnolomia multiflora (Nuttall) Bentham & Hooker f. var. *annua* M. E. Jones, Proc. Calif. Acad. Sci., ser. 2, 5: 698. 1895; *Viguiera annua* (M. E. Jones) S. F. Blake

Stems branching ± throughout, bases usually less than 5 mm diam. **Leaves** 10–70(–85) × 1.5–5 mm. **Involucres** 6–9 mm diam. $2n = 16$.

Flowering Jun–Oct. Desert shrub, pinyon-juniper communities; 600–2700 m; Ariz., N.Mex, Tex.; Mexico (Chihuahua, Coahuila, Durango, Sonora).

4. **Heliomeris multiflora** Nuttall, J. Acad. Nat. Sci. Philadelphia, n. s. 1: 171. 1847 (as multiflorus) • Showy goldeneye [F]

Viguiera multiflora (Nuttall) S. F. Blake

Perennials, 20–120+ cm (caudices woody). **Stems** strigose or puberulent to glabrate. **Leaves** opposite or alternate; blades elliptic, lance-linear, lance-ovate, linear, or ovate, 10–90 × 2–28 mm, margins ciliate to ¼ their lengths, hairs mostly less than 0.5 mm, faces strigose, abaxial often gland-dotted. **Heads** 6–30+. **Peduncles** 1–15 cm, lengths ¹⁄₂₀–¹⁄₂ leafy portions of stems. **Involucres** 6–14 mm diam. **Phyllaries** 4–8 mm. **Paleae** oblong, 5–6 mm, cuspidate. **Ray florets** 5–14; laminae oval to oblong, 7–20 mm (gland-dotted along veins and abaxially). **Disc florets** 50+; corollas 3–4 mm. **Cypselae** black or gray-striate, 1.2–3 mm.

Heliomeris multiflora most frequently occurs in montane habitats. It produces a branched, woody caudex with adventitious buds, even in the first growing season. From the aerial parts of the plant there is little to distinguish it from *H. longifolia*, although it tends to have longer peduncles and broader leaves.

Varieties 3 (3 in the flora): sw United States, Mexico.

1. Leaves mostly alternate, blades elliptic to ovate, apices obtuse (often mucronate) . . . . 4c. *Heliomeris multiflora* var. *brevifolia*
1. Leaves mostly opposite, blades lance-ovate, lance-linear, or linear, apices acute.
    2. Leaves lance-ovate to lance-linear, 5–20 mm wide, margins flat . . . . 4a. *Heliomeris multiflora* var. *multiflora*
    2. Leaves lance-linear to linear, 2–5 mm wide, margins usually strongly revolute . . . . 4b. *Heliomeris multiflora* var. *nevadensis*

4a. Heliomeris multiflora Nuttall var. **multiflora**
• False goldeneye [F]

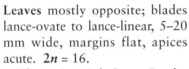

Leaves mostly opposite; blades lance-ovate to lance-linear, 5–20 mm wide, margins flat, apices acute. $2n = 16$.

Flowering Jul–Oct. Rocky slopes and valleys, open woods, roadsides; 1500–3200 m; Ariz., Colo., Idaho, Mont., Nev., N.Mex., Tex., Utah, Wyo.; Mexico.

4b. Heliomeris multiflora Nuttall var. **nevadensis** (A. Nelson) W. F. Yates, Proc. Indiana Acad. Sci. 88: 368. 1979 • Nevada goldeneye

Gymnolomia nevadensis A. Nelson, Bot. Gaz. 37: 271. 1904

Leaves mostly opposite; blades lance-linear to linear, 2–5 mm wide, margins usually strongly revolute, apices acute. $2n = 16, 32$

Flowering Jul–Oct. Dry, rocky slopes and valleys, roadsides; 1200–2400 m; Ariz., Calif., Nev., N.Mex., Utah; Mexico (Chihuahua, Durango, Jalisco, San Luis Potosí, Sonora, Zacatecas).

4c. Heliomeris multiflora Nuttall var. **brevifolia** (Greene ex Wooton & Standley) W. F. Yates, Proc. Indiana Acad. Sci. 88: 368. 1979 [E]

Gymnolomia brevifolia Greene ex Wooton & Standley, Contr. U.S. Natl. Herb. 16: 190. 1913; *Viguiera ovalis* S. F. Blake

Leaves mostly alternate; blades elliptic to ovate, 8–28 mm wide, margins usually flat or obscurely revolute, apices obtuse (often mucronate). $2n = 16$.

Flowering Jul–Oct. Canyons, woods, often in shade; 2400–3000 m; Ariz., N.Mex., Utah.

301. **VIGUIERA**   Kunth in A. von Humboldt et al., Nov. Gen. Sp. 4(fol.): 176. 1818; 4(qto.): 224, plate 379. 1820 • Goldeneye [For L. G. A. Viguier, 1790–1867, French physician]

Edward E. Schilling

Perennials or shrubs [annuals], 50–200 cm. **Stems** erect or ascending, unbranched or branched. **Leaves** mostly cauline; opposite (proximal) and/or alternate (distal); petiolate or ± sessile; blades often 3-nerved (from at or near bases), deltate, deltate-ovate, lance-linear, lanceolate, lance-ovate, linear, linear-filiform, rhombic-ovate, or ovate, sometimes pinnately lobed, ultimate margins entire or toothed, faces glabrous or hairy, often gland-dotted. **Heads** radiate, borne singly or (2–25+) in open to crowded, ± corymbiform arrays. **Involucres** hemispheric [campanulate], (5–18 ×) 7–15 mm. **Phyllaries** persistent, 10–30 in 2–5 series (subequal to unequal). **Receptacles** flat to convex, paleate (paleae tan, conduplicate, apices often 3-lobed or -toothed). **Ray florets** 6–18, neuter; corollas yellow (laminae 2–4-lobed). **Disc florets** 40–100+, bisexual, fertile; corollas yellow, tubes shorter than campanulate throats, lobes 5, triangular (style branches slender, apices acute or attenuate). **Cypselae** (purplish black, often mottled) ± compressed, often obpyramidal (usually strigose, sometimes glabrous); **pappi** 0, or persistent or tardily falling, of 2 lacerate, aristate scales (2.2–5.5 mm) plus [0–]2–4[–6] lacerate or aristate scales (0.5–1.3 mm). *x* = 17.

Species ca. 150 (3 in the flora): sw United States, Mexico, Central America, South America.

Viguiera has been long recognized as paraphyletic. Molecular studies are helping to clarify its phylogeny, and the corresponding systematic adjustments are beginning (E. E. Schilling and J. L. Panero 2002). Some groups have already been separated, including *Bahiopsis* and *Heliomeris*. It is likely that the genus will be narrowed to include only a single species, *V. dentata* (including *V. helianthoides* Kunth), and that new or resurrected genera will accommodate the remaining species. As currently circumscribed, the genus occurs primarily in Mexico and South America and reaches its northern limit of distribution in the southwestern United States. The Mexican (Chihuahua, Durango) *V. phenax* was collected a single time in Texas [where it was named *Helianthus ludens* Shinners and *Viguiera ludens* (Shinners) M. C. Johnston]; it apparently has not become established in the flora area.

SELECTED REFERENCE Blake, S. F. 1918. A revision of the genus *Viguiera*. Contr. Gray Herb. 54: 1–218.

1. Shrubs; leaves ovate (and shallowly to deeply lobed, lobes ± linear) or linear; pappi 0 . . .
. . . . 3. *Viguiera stenoloba*
1. Perennials; leaves deltate-ovate, lanceolate, lance-ovate, ovate, or rhombic-ovate (not lobed); pappi usually of 2 lacerate, aristate scales (2.2–5.5 mm) plus 2–4 lacerate or aristate scales (0.5–1.3 mm).
  2. Petioles (10–)20 mm; phyllary apices abruptly narrowed to acuminate or spatulate (phyllary bases broad, indurate, apices herbaceous); disc corollas 3–4 mm (staminal filaments hairy) . . . . 1. *Viguiera dentata*
  2. Petioles 1–8 mm; phyllary apices gradually narrowed; disc corollas 5.5–6.5 mm (staminal filaments glabrous) . . . . 2. *Viguiera cordifolia*

1. **Viguiera dentata** (Cavanilles) Sprengel, Syst. Veg. 3: 615. 1826 • Sunflower goldeneye, toothleaf F

Helianthus dentatus Cavanilles, Icon. 3: 10, plate 220. 1795

Perennials, 100–200 cm. **Leaves** opposite (proximal) or alternate (distal); petioles 10–55 mm; blades ovate or rhombic-ovate to lance-ovate or lanceolate, 3.5–12.5 × 1–8 cm, margins serrate or serrulate, faces strigose. **Heads** usually 3–9+ in ± corymbiform arrays. **Peduncles** 3–14 cm. **Involucres** 11–18 × 7–10 mm. **Phyllary apices** abruptly narrowed to acuminate or spatulate (phyllary bases broad, indurate, apices herbaceous). **Paleae** 6.5 mm, apices prominently cuspidate. **Ray florets** 10–14; tubes 1 mm, laminae 7–15 mm. **Disc florets** 50+; corollas 3–4 mm (staminal filaments hairy). **Cypselae** 3.5–3.8 mm, ± strigose; **pappi** of 2 lacerate, aristate scales 2.2–2.8 mm plus 2–4 lacerate scales 0.5–0.7 mm. **2*n*** = 34.

Flowering Jun–Oct. Dry slopes and canyons, fields, roadside ditches; 100–2300 m; Ariz., N.Mex., Tex.; Mexico; West Indies (Cuba); Central America.

Viguiera dentata is widespread and variable. It is unique in Helianthinae in the hairy staminal filaments. The name *V. dentata* var. *lancifolia* S. F. Blake has been used for plants from Mexico.

2. **Viguiera cordifolia** A. Gray, Smithsonian Contr. Knowl. 3(5): 107. 1852 • Heartleaf goldeneye

Perennials, 50–100 cm. **Leaves** opposite (proximal) or alternate (distal); petioles 1–8 mm; blades ovate to deltate-ovate or lance-olate (distal), 2–10 × 1.3–7 cm, margins serrate or serrulate, faces strigose. **Heads** borne singly or 2–9 in ± corymbiform arrays. **Peduncles** 1–15 cm. **Involucres** 11–15 × 7–15 mm. **Phyllary** apices gradually narrowed. **Paleae** 8–11 mm, apices acuminate, shortly cuspidate. **Ray florets** 6–8; tubes 1.5–2 mm, laminae 9–19 mm. **Disc florets** 40+; corollas 5.5–6.6 mm (staminal filaments glabrous). **Cypselae** 5–6.5 mm, ± strigose; **pappi** of 2 lacerate, aristate scales 5–5.5 mm plus 2–4 lacerate scales 0.5–1.3 mm. **2*n*** = 34.

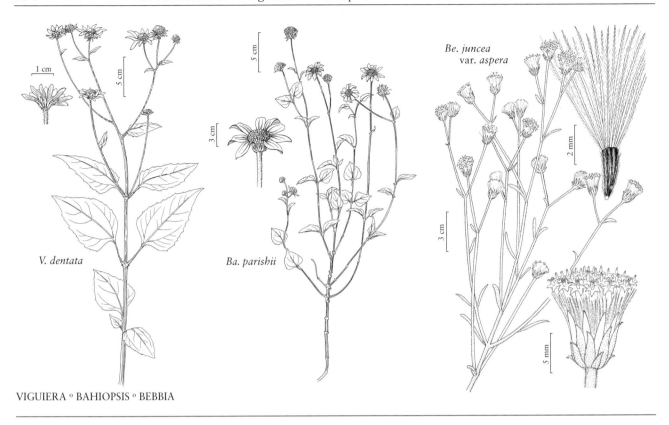

V. dentata

Ba. parishii

Be. juncea
var. *aspera*

VIGUIERA ◦ BAHIOPSIS ◦ BEBBIA

Flowering Jul–Sep. Dry slopes and canyons, mostly pine forests and limestone soils; 1300–2700 m; Ariz., N.Mex., Tex.; Mexico.

3. **Viguiera stenoloba** S. F. Blake, Contr. Gray Herb. 54: 97. 1918 • Skeletonleaf goldeneye

Heliomeris tenuifolia A. Gray, Mem. Amer. Acad. Arts, n. s. 4: 84. 1849, not *Viguiera tenuifolia* Gardner 1848

Shrubs, 50–150 cm. **Leaves** opposite or alternate; sessile or subsessile; blades ovate (and shallowly to deeply lobed, lobes 3–9, ± linear) or linear, 1.5–10.8 × 1–9 cm, ultimate margins entire (revolute), faces: abaxial loosely to densely strigillose (canescent) and gland-dotted, adaxial densely strigose to glabrate. **Heads** usually borne singly. **Peduncles** 2–5 cm. **Involucres** 5–12 × 7–9 mm. **Phyllaries** abruptly narrowed to acuminate or spatulate (phyllary bases broad, indurate, apices herbaceous). **Paleae** (oblong-rectangular) 4–5 mm, apices abruptly acuminate. **Ray florets** 13–18; tubes 0.9–1.1 mm, laminae 7–12 mm. **Disc florets** 100+; corollas 3.3–4.2 mm. **Cypselae** 2–3 mm, glabrous; **pappi** 0. $2n = 34, 68$.

Flowering May–Sep. Tamaulipan and Chihuahuan Desert scrub; 800–1900 m; N.Mex., Tex.; Mexico (Chihuahua, Coahuila, Durango, Nuevo León, San Luis Potosí, Tamaulipas, Zacatecas).

Viguiera stenoloba is a dominant shrub in some places in the trans-Pecos region of Texas. M. L. Butterwick (1975) suggested that the species is separable into two varieties; those have never received formal taxonomic recognition.

302. **BAHIOPSIS** Kellogg, Proc. Calif. Acad. Sci. 2: 35. 1863 • [Generic name *Bahia* and Greek *-opsis*, resembling]

Edward E. Schilling

Shrubs, 50–150 cm. **Stems** ascending to erect, much branched. **Leaves** mostly cauline; alternate or opposite; petiolate or sessile; blades usually 3-nerved (from at or near bases), deltate, deltate-ovate, lanceolate, lance-ovate, or ovate, margins entire or toothed to laciniate, faces hispid, sericeous, or strigillose, often resinous or gland-dotted. **Heads** radiate, borne singly or (3–25+)

in ± thyrsiform arrays. **Involucres** campulate to hemispheric, (9–14 ×) 4–7 mm. **Phyllaries** persistent, 16–28 in 2–3 series (unequal, bases ovate to lance-ovate, indurate, apices abruptly narrowed, herbaceous). **Receptacles** convex, paleate (paleae tan, conduplicate, apices acute). **Ray florets** 8–15, neuter; corollas yellow (2–3-lobed). **Disc florets** 40–50+, bisexual, fertile; corollas yellow, tubes shorter than campanulate throats, lobes 5, triangular (style branches relatively slender, apices acute). **Cypselae** (brown to black) ± compressed, ± 3- or 4- angled, often obpyramidal, ± strigose; **pappi** persistent, of 2(–6) lacerate, aristate scales (1–2.8 mm) plus (0–)2–6 lacerate scales (0.2–1 mm). *x* = 18.

Species 12 (3 in the flora): sw United States, nw Mexico.

Although traditionally included within *Viguiera*, the species of *Bahiopsis* differ by their distinctive base chromosome number and form a basally diverging clade in subtribe Helianthinae, based on molecular data. Morphologically, they are most similar to *Calanticaria*, another segregate of *Viguiera* that is Mexican in geographic distribution, from which they differ by their branched capitulescences and having phyllaries in which the apical herbaceous portions are longer.

SELECTED REFERENCE Schilling, E. E. 1990. Taxonomic revision of *Viguiera* subg. *Bahiopsis* (Asteraceae: Heliantheae). Madroño 37: 149–170.

1. Leaves lanceolate to lance-ovate, margins ± laciniate, abaxial faces strigillose (resinous, shiny) . . . . 1. *Bahiopsis laciniata*
1. Leaves deltate, deltate-ovate, or ovate, margins entire or serrate (not incised or laciniate-toothed), abaxial faces hispid, or sericeous (not shiny).
   2. Leaf blades 1–3.5 cm, margins usually toothed, faces: abaxial hispid and gland-dotted (and slightly reticulate), adaxial scabrous (bases of hairs notably enlarged) . . . . 2. *Bahiopsis parishii*
   2. Leaf blades 2.8–9 cm, margins entire, faces: abaxial sericeous and gland-dotted (and strongly reticulate, not shiny), adaxial sericeous (hair bases slightly enlarged) . . . . 3. *Bahiopsis reticulata*

1. Bahiopsis laciniata (A. Gray) E. E. Schilling & Panero, Bot. J. Linn. Soc. 140: 72. 2002 • San Diego sunflower, San Diego County viguiera

Viguiera laciniata A. Gray in W. H. Emory, Rep. U.S. Mex. Bound. 2(1): 89. 1859

Shrubs, 50–130 cm. **Leaves** mostly alternate; petioles (0–)3–7 mm; blades lanceolate to lance-ovate, 1.4–4.3 × 0.6–2.5 cm, margins ± laciniate, faces sparsely strigillose and gland-dotted or resinous (abaxial resinous, shiny). **Heads** (1–)3–9. **Peduncles** 0.5–5 cm. **Involucres** hemispheric, 8–12 × 6–9 mm. **Phyllaries** 20–25, 4–8 × 1.8–2.2 mm. **Paleae** lance-ovate, 4.5–7.3 mm (densely glandular). **Ray florets** 5–13; laminae 6–12 mm. **Disc florets** 50+; corollas 3.5–4.6 mm. **Cypselae** 1.7–3.5 mm (± strigose, margins ciliolate); **pappi** of 2 lacerate, aristate scales 1–2.5 mm plus 2–6 (easily detached) lacerate-erose scales 0.2–0.5 mm. *2n* = 36.

Flowering Jan–Jun(–Sep). Xeric scrub; 90–800 m; Calif.; Mexico (Baja California, Sonora).

Within the flora area, *Bahiopsis laciniata* is found only in San Diego County; its range extends south into the xeric Central Desert area of Baja California.

2. Bahiopsis parishii (Greene) E. E. Schilling & Panero, Bot. J. Linn. Soc. 140: 72. 2002 [F]

Viguiera parishii Greene, Bull. Torrey Bot. Club 9: 15. 1882; *V. deltoidea* A. Gray var. *parishii* (Greene) Vasey & Rose

Shrubs, 50–130 cm. **Leaves** mostly opposite, sometimes alternate (distal); petioles 2–8 mm; blades deltate to deltate-ovate, 1–3.5 × 1–3.5 cm, margins usually toothed, faces: abaxial hispid and gland-dotted (and slightly reticulate), adaxial scabrous, (bases of hairs notably enlarged). **Heads** (1–)3–5. **Peduncles** 1–15 cm. **Involucres** hemispheric, 8–13 × 5–9 mm. **Paleae** ovate to oblong, 5.5–7 mm. **Phyllaries** 16–28, 3–9 × 1.5–2 mm. **Ray florets** 8–15; laminae 10–15 mm. **Disc florets** 50+; corollas 3.5–5 mm. **Cypselae** 2.7–3.2 mm; **pappi** of 2 (–6) lacerate, aristate scales 2.1–2.5 mm plus (0–)2–6 lacerate scales 0.5–1 mm. *2n* = 36.

Flowering (Jan–Oct). Xeric scrub; 500–1500 m; Ariz., Calif., Nev.; Mexico (Baja California, Sonora).

Bahiopsis parishii is a diploid that is closely related to polyploids that have traditionally been recognized as varieties of *B.* (*Viguiera*) *deltoidea* and occur throughout the Baja California Peninsula in Mexico.

3. Bahiopsis reticulata (S. Watson) E. E. Schilling & Panero, Bot. J. Linn. Soc. 140: 72. 2002 • Death Valley goldeneye [E]

Viguiera reticulata S. Watson, Amer. Naturalist 7: 301. 1873

Shrubs, 50–150 cm. **Leaves** alternate or opposite; petioles 3.5–30 mm; blades broadly ovate, 2.8–9 × 2.3–6.5 cm, margins entire, faces: abaxial sericeous and gland-dotted (and strongly reticulate, not shiny), adaxial sericeous (hair bases slightly enlarged). **Heads** 3–9(–25). **Peduncles** 0.5–5 cm. **Involucres** campanulate to hemispheric, 10–15 × 6–9 mm. **Phyllaries** 18–25, 2.5–5 × 1.2–2 mm. **Paleae** oblanceolate, 4–8 mm. **Ray florets** 8–15; laminae 7–15 mm. **Disc florets** 40+; corollas 3–5.5 mm. **Cypselae** 2.5–4 mm; **pappi** of 2 lacerate, aristate scales 1.5–2.8 mm and 2–6 lacerate scales 0.6–1 mm. **2n** = 36.

Flowering Mar–Jun. Xeric scrub, desert grottoes and canyons; 600–1400 m; Calif., Nev.

187m.10. ASTERACEAE Martinov (tribe HELIANTHEAE) Cassini subtribe GALINSOGINAE Bentham & Hooker f., Gen. Pl. 2: 198. 1873 (as Galinsogeae)

Annuals, perennials, subshrubs, or shrubs, (2–)10–150[–300+] cm. **Leaves** mostly basal or mostly cauline; usually opposite (distal sometimes alternate); petiolate or sessile; blades (often 3- or 5-nerved) mostly deltate, elliptic, lanceolate, lance-ovate, linear, ovate, or rhombic, sometimes pinnately or palmately lobed, ultimate margins entire or ± toothed, faces glabrate, hispid, hispidulous, pilose, puberulent, scabrellous, scabrous, strigillose, or strigose, often gland-dotted or stipitate-glandular. **Heads** radiate or discoid (sometimes ± disciform in *Bebbia*), borne singly or in loose to tight, corymbiform or cymiform arrays. **Calyculi** 0. **Involucres** campanulate, cylindric, hemispheric, or ± obpyramidal. **Phyllaries** persistent or falling, 6–30+ in 2–5+ series, usually distinct, elliptic, lanceolate, lance-linear, lance-ovate, oblong, or ovate, subequal or unequal (outer longer or shorter). **Receptacles** flat, convex, or conic, paleate (paleae persistent or falling with cypselae, mostly oblong to lanceolate or linear, flat to conduplicate, herbaceous or scarious, often 2- or 3-lobed or -toothed). **Ray florets** 0, or (3–)6–21, pistillate, fertile; corollas yellow or whitish or pinkish [purplish]. **Disc florets** 5–150+, bisexual, fertile (peripheral sometimes pistillate in *Bebbia*); corollas yellow to orange, or whitish, pinkish, or purplish, tubes shorter than throats, lobes 5, deltate to lance-deltate or lanceolate (± equal [outer larger, e.g., some plants of *Tridax*]); anther thecae pale or slightly darkened; stigmatic papillae in 2 lines. **Cypselae** ± compressed, often clavate, obconic, or ovoid, or obpyramidal and 3-, 4-, or 5-angled, glabrous or sparsely to densely piloso-sericeous, sericeous, strigillose, or strigose; **pappi** 0, or persistent, of 1–10+ subulate to acerose scales, or 5–20 ± spatulate, entire to erose, fimbriate, or laciniate, sometimes aristate, scales in 1 series, or 15–25+[–40], plumose, setiform scales (or flattened bristles) in 1 series.

Genera 15, species 103 (4 genera, 8 species in the flora): subtropical, tropical, and warm-temperate New World.

All genera of Galinsoginae are centered in tropical and subtropical to warm-temperate North America and South America; relatively few extend into cool-temperate areas; fewer still are adventive in the Old World.

1. Phyllaries 10–20 in ± 2 series (outer 4 broadly lanceolate, foliaceous, notably larger than inner)... 304. *Tetragonotheca*, p. 178
1. Phyllaries 6–30+ in 2–5 series (subequal or unequal, outer smaller than inner).

[2. shifted to left margin—Ed.]

303. BEBBIA Greene, Bull. Calif. Acad. Sci. 1: 179. 1885 • Sweetbush [For Michael S. Bebb, 1833–1895, American botanist and willow specialist]

Molly A. Whalen

Subshrubs or shrubs, 50–140[–300] cm (often forming dense, rounded masses). **Stems** erect or spreading (glabrous or rough hairy [sparsely hirsute]). **Leaves** cauline; opposite (all or proximal) or alternate; sessile [petiolate]; blades linear to narrowly elliptic [± deltate, often hastate], sometimes lobed, ultimate margins entire [dentate], faces glabrous or ± scabrous [sparsely hirsute], sometimes gland-dotted or stipitate-glandular. **Heads** usually discoid (sometimes ± disciform), borne singly or in corymbiform arrays. **Involucres** ± campanulate or broader, 5–15 mm diam. **Phyllaries** persistent, 15–30[–40+] in 3–5 series (ovate or elliptic to lanceolate or lance-linear, strongly unequal; outer herbaceous, margins often chartaceous, ciliate, abaxial faces pubescent; inner more chartaceous, striate, margins ciliate, abaxial faces less pubescent to glabrate). **Receptacles** convex, paleate (paleae persistent, lanceolate to linear-lanceolate or linear-elliptic, conduplicate, each ± clasping a cypsela, scarious, striate, margins ciliate, abaxial faces glabrous or glabrate). **Ray florets** 0. **Disc florets** usually 20–50, bisexual, fertile (sometimes 2–8+ peripheral florets lack anthers); corollas yellow to orange, tubes shorter than narrowly funnelform throats, lobes 5, deltate. **Cypselae** ± compressed, ± obpyramidal, ± 3-angled, sericeous to strigillose; **pappi** of 15–25+ plumose, setiform scales in 1 series. *x* = 9.

Species 2 (1 in the flora): sw United States, n Mexico.

SELECTED REFERENCE Whalen, M. 1977. Taxonomy of *Bebbia* (Compositae: Heliantheae). Madroño 24: 112–123.

1. Bebbia juncea (Bentham) Greene, Bull. Calif. Acad. Sci. 1: 180. 1885 F

Carphephorus junceus Bentham, Bot. Voy. Sulphur, 21. 1844

Varieties 2 (1 in the flora): sw United States, n Mexico.
M. Whalen (1977) recognized two varieties of *Bebbia juncea*: var. *aspera*, which occurs in the flora area and in northern Mexico, and var. *juncea*, which occurs in Mexico (Baja California and Baja California Sur). Variety *juncea* has generally broader, obtuse, rounded or apiculate, ovate to elliptic phyllaries.

1a. Bebbia juncea (Bentham) Greene var. **aspera** Greene, Bull. Calif. Acad. Sci. 1: 180. 1885 F

Bebbia aspera (Greene) A. Nelson

Leaf blades 5–65 × 1–6(–12) mm, faces rough hairy to glabrate (hairs antrorse). **Phyllaries** usually lanceolate, lance-linear, or lance-elliptic, sometimes narrowly ovate, 0.9–7.5 × 0.5–2 mm, outer usually less than 1.6 mm wide, apices acute to blunt. **Corollas** 4.5–9 mm. **Pappi** 5–8 mm. *2n* = 18.

Flowering mostly year round. Rocky and sandy soils, dry hillsides, slopes, desert washes, canyons; 0–1300 m; Ariz., Calif., Nev., N.Mex., Tex., Utah; Mexico (Baja California, Sinaloa, Sonora).

304. TETRAGONOTHECA Linnaeus, Sp. Pl. 2: 903. 1753; Gen. Pl. ed. 5, 384. 1754

•   [Greek *tetra*, four, *gonio*, angle, and *theca*, container, alluding to quadrangular involucres]

John L. Strother

Perennials, mostly 30–120 cm. **Stems** erect. **Leaves** basal and/or cauline; mostly opposite; petiolate, subpetiolate, or sessile (bases sometimes connate-perfoliate); blades lanceolate, ovate, rhombic, or rounded-deltate, sometimes pinnatifid, ultimate margins usually toothed, faces glabrate, sparsely hispidulous, or puberulent, gland-dotted. **Heads** radiate, borne singly or in loose, corymbiform arrays. **Involucres** obpyramidal to hemispheric, 12–25+ mm diam. **Phyllaries** persistent, 10–25+ in ± 2 series (outer 4 broadly lanceolate, foliaceous, the inner ovate to lanceolate, smaller, more scarious, each subtending a ray floret). **Receptacles** conic, paleate (paleae persistent, lanceolate to lance-ovate, flat or weakly conduplicate, apices acute). **Ray florets** 6–21+, pistillate, fertile; corollas yellow (often with reddish nerves). **Disc florets** 25–150+, bisexual, fertile; corollas yellow, tubes (basally dilated) much shorter than ampliate, cylindric throats, lobes 5, deltate. **Cypselae** ± ovoid or plumply 4- or 5-angled, finely 32–40-ribbed, sparsely strigose or glabrous; **pappi** 0, or of 1–10+ subulate to acerose scales (to 0.5 mm), or of 16–30, ± spatulate scales (0.5–2 mm). $x = 17$.

Species 4 (4 in the flora): s United States, n Mexico.

SELECTED REFERENCE   Turner, B. L. and D. Dawson. 1980. Taxonomy of *Tetragonotheca* (Asteraceae–Heliantheae). Sida 8: 296–303.

1. Pappi of 16–30, ± obovate to spatulate scales 0.5–2 mm.
    2. Leaves mostly cauline (distal connate-perfoliate); ray florets 12–13, laminae 8–18 mm
       . . . . 1. *Tetragonotheca ludoviciana*
    2. Leaves mostly basal (cauline much smaller, none connate-perfoliate); ray florets 16–
       21+, laminae 20–30+ mm . . . . 2. *Tetragonotheca repanda*
1. Pappi 0, or of 1–10+ subulate to acerose scales to 0.5 mm.
    3. Stems sparsely tomentulose or glabrous (nodes usually auriculate-appendaged); ray
       laminae 10–18 mm; cypselae ± quadrate, 2.5–4 mm . . . . 3. *Tetragonotheca texana*
    3. Stems patently to retrorsely ± villous; ray laminae 20–40+ mm; cypselae ± ovoid, 4–6
       mm . . . . 4. *Tetragonotheca helianthoides*

1. Tetragonotheca ludoviciana (Torrey & A. Gray) A. Gray ex E. Hall, Pl. Tex., 13. 1873 E

Halea ludoviciana Torrey & A. Gray, Fl. N. Amer. 2: 304. 1842

Stems sparsely tomentulose or glabrous. **Leaves** mostly cauline (distal connate-perfoliate); blades rounded-deltate to ovate, 8–12 × 4–9 cm, coarsely toothed. **Ray florets** 12–13; laminae 8–18 mm. **Cypselae** ± quadrate, 4–6 mm; **pappi** of 16–30, obovate to spatulate scales 0.5–2 mm. $2n = 34$.

Flowering late spring–summer. Deep sandy soils with scrub oaks; 10–100 m; La., Tex.

2. Tetragonotheca repanda (Buckley) Small, Fl. S.E. U.S., 1340. 1903 E

Halea repanda Buckley, Proc. Acad. Nat. Sci. Philadelphia 13: 458. 1861

Stems sparsely tomentulose or glabrous. **Leaves** mostly basal (cauline much smaller); blades rounded-deltate to ovate, 5–12 × 3–7 cm, coarsely toothed. **Ray florets** 16–21; laminae 20–30+ mm. **Cypselae** ± quadrate, 4–5.5 mm; **pappi** of 20–30, obovate to spatulate scales 1–2 mm. $2n = 34$.

Flowering mostly spring–summer. Sandy soils; 0–100 m; Tex.

Tetragonotheca repanda is known from the Gulf Coastal Plain.

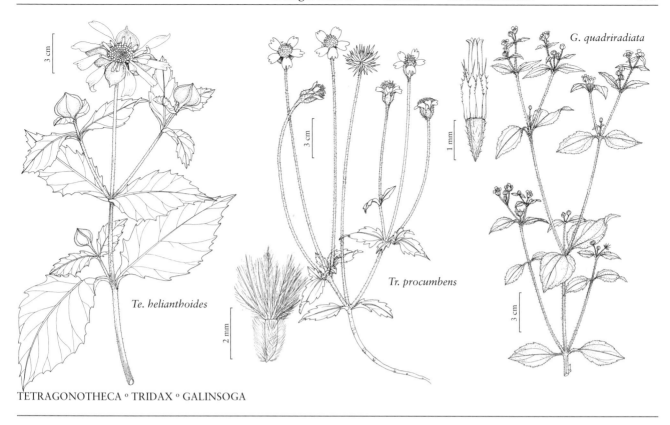

Te. helianthoides

Tr. procumbens

G. quadriradiata

TETRAGONOTHECA ° TRIDAX ° GALINSOGA

3. Tetragonotheca texana A. Gray & Engelmann, Proc. Amer. Acad. Arts 1: 48. 1847

Stems sparsely tomentulose or glabrous (nodes usually auriculate-appendaged). **Leaves** mostly cauline; blades ovate to lanceolate in outline, 3–7 × 1–3 cm, subentire to incised or irregularly pinnatifid. **Ray florets** 8–13; laminae 10–18 mm. **Cypselae** ± quadrate, 2.5–4 mm; **pappi** 0, or (often fragile) of 1–10+ subulate or acerose scales to 0.5(–1.5) mm. $2n = 34$.

Flowering spring–summer. Sandy soils, often with mesquite; 50–200 m; Tex.; Mexico (Coahuila, Nuevo León, Tamaulipas).

4. Tetragonotheca helianthoides Linnaeus, Sp. Pl. 2: 903. 1753 ⒠ Ⓕ

Stems patently to retrorsely ± villous. **Leaves** cauline, not connate-perfoliate; blades ovate to rhombic or lanceolate, 7–20 × 3–13 cm, coarsely toothed. **Ray florets** 6–14; laminae 15–40+ mm. **Cypselae** ± ovoid, 4–6 mm; **pappi** 0. $2n = 34$.

Flowering mid spring–early summer. Sandy soils, often in thickets; 0–400 m; Ala., Fla., Ga., Miss., N.C., S.C., Tenn., Va.

305. TRIDAX Linnaeus, Sp. Pl. 2: 900. 1753; Gen. Pl. ed. 5, 382. 1754 • [Etymology recondite; perhaps alluding to 3-lobed leaf blades characteristic of original species] Ⓘ

John L. Strother

Perennials [annuals], 10–40[–80+] cm. **Stems** procumbent to ascending. **Leaves** basal and/or cauline; opposite; petiolate [sessile]; blades deltate, lanceolate, lance-ovate, or ovate, often pinnately or palmately lobed, ultimate margins coarsely toothed to subentire, faces glabrate, hirsute, scabrellous, or strigillose. **Heads** radiate [discoid], borne singly [in ± cymiform arrays].

Involucres cylindric to hemispheric, 4–8 mm diam. **Phyllaries** persistent, 11–15+ in 2–3+ series (outer usually shorter, more herbaceous, the inner often scarious). **Receptacles** convex to conic, paleate (paleae ± persistent, lance-linear, scarious, weakly conduplicate, often apically toothed). **Ray florets** [0–]3–8[–13], pistillate, fertile; corollas pale yellow [white or purplish]. **Disc florets** [20–]40–80+, bisexual, fertile; corollas yellowish [whitish or purplish], tubes shorter than cylindric or funnelform throats, lobes 5, usually deltate [sometimes rays 0 and corollas of peripheral florets ± bilabiate]. **Cypselae** obconic to obpyramidal, 3-, 4-, or 5-angled, densely piloso-sericeous [glabrous, villous]; **pappi** [sometimes wanting in rays] persistent, of [10–]20 [–40], plumose [ciliate], setiform scales. $x = 10$.

Species ca. 26 (1 in the flora): introduced; mostly tropical, mostly New World.

SELECTED REFERENCE Powell, A. M. 1965. Taxonomy of *Tridax* (Compositae). Brittonia 17: 47–96.

1. Tridax procumbens Linnaeus, Sp. Pl. 2: 900. 1753

F I W

Leaves: petioles 1–10(–30) mm; blades 10–40(–120) × 5–20(–60) mm, often 3-lobed. **Cypselae** 2–2.5 mm; **pappi** 4–5(–7.5) mm. $2n = 36$.

Flowering Jan–Dec. Disturbed sites; 0–10 m; introduced; Fla., Tex.; Mexico.

Tridax procumbens is listed as a noxious weed for United States and is listed as a pest or noxious weed for nine or more states.

306. GALINSOGA Ruiz & Pavón, Fl. Peruv. Prodr., 110, plate 24. 1794 • Quickweed [For Mariano Martínez de Galinsoga, 1766–1797, court physician and director of the Botanic Garden, Madrid]

Judith M. Canne-Hilliker

Annuals, 2–62 cm. **Stems** erect. **Leaves** cauline; opposite; petiolate; blades (3-nerved) lanceolate to broadly ovate, margins entire or serrulate to serrate, faces glabrate to densely pilose. **Heads** radiate or discoid, in ± cymiform arrays. **Involucres** hemispheric to campanulate, 2.5–6 mm diam. **Phyllaries** persistent or falling, 6–9[–16] in 2[–3] series (elliptic, lance-ovate, oblong, or ovate, outer shorter, herbaceous or scarious, margins entire or minutely laciniate). **Receptacles** conic, paleate (paleae persistent or falling, scarious, proximal broadly elliptic to obovate, often connate at bases or nearly to apices, united in groups of 2–3 to adjacent proximal phyllary, each complex often enclosing and shed with a ray cypsela; distal persistent or falling, lanceolate to obovate, entire or 2- or 3-lobed, convex to conduplicate). **Ray florets** [0–](3–)5(–8)[–15], pistillate, fertile; corollas white or dull white to pinkish [purplish] (tubes pilose, laminae quadrate-obovate to oblong, lobes 0–3). **Disc florets** 5–50[–150], bisexual, fertile; corollas yellow, tubes (pilose) shorter than cylindric throats, lobes 5, deltate (anthers yellow; style-branch apices acute). **Cypselae** obconic to obpyramidal, glabrous or strigose (rays often shed with subtending phyllary plus 2–3 adjacent paleae); **pappi** 0, or persistent, of 5–20, white or gray, fimbriate, sometimes aristate scales. $x = 8$.

Species 15–33+ (2 in the flora): North America, Mexico, West Indies, Bermuda, Central America, South America; introduced in Europe, Asia, Africa, Atlantic Islands, Pacific Islands, Australia.

Galinsoga is closely related to *Sabazia* Cassini of Mexico and South America and to at least some members of *Alloispermum* Willdenow, primarily from South America. *Galinsoga* might best be treated as a single larger genus including *Alloispermum* and *Sabazia* (J. M. Canne 1978; J. L. Panero et al. 1999).

SELECTED REFERENCES Canne, J. M. 1977. A revision of the genus *Galinsoga* (Compositae: Heliantheae). Rhodora 79: 319–389.   Canne, J. M. 1978. Circumscription and generic relationships of *Galinsoga* (Compositae: Heliantheae). Madroño 25: 81–93.

1. Phyllaries persistent; inner paleae 3-lobed, lobes $^1/_3$+ total lengths, acute; disc pappi 0, or of 15–20 often gray, sometimes white, linear, fimbriate, obtuse scales 0.5–2 mm. ...... 1. *Galinsoga parviflora*
1. Phyllaries falling; inner paleae entire or 2- or 3-lobed, lobes to $^1/_3$ total lengths, blunt; disc pappi 0, or of 1–5 or 14–20 white, lanceolate to oblanceolate, fimbriate, sometimes aristate, scales. .... 2. *Galinsoga quadriradiata*

1. Galinsoga parviflora Cavanilles, Icon. 3: 41, plate 281. 1795   W

Plants 4–60 cm. **Leaf blades** 7–110 × 3–70(–80) mm. **Peduncles** 1–40 mm. **Involucres** campanulate, 2.5–5 mm diam. **Phyllaries** persistent. **Paleae:** outer persistent with distal phyllaries or falling, elliptic to obovate; inner usually persistent, lanceolate to ovate or obovate, 2–3.5 mm, 3-lobed, lobes to $^1/_3$+ total lengths, acute. **Ray florets** (3–)5(–8); corollas usually dull white (pink), laminae 0.5–1.8 × 0.7–1.5 mm. **Disc florets** 15–50. **Cypselae:** rays 1.5–2.5 mm; discs 1.3–2.5 mm, glabrous or strigose; **pappi:** rays 0 or of 5–10, laciniate scales 0.5–1 mm; discs 0, or of 15–20 white or gray, linear, fimbriate, obtuse (acute) scales 0.5–2 mm.

Varieties 2 (2 in the flora): North America; West Indies; Central America; South America; introduced in Europe; Asia; Africa; Australia.

1. Leaf blades lanceolate to broadly ovate, margins serrulate to serrate; arrays of heads loose; axillary peduncles longer than subtending bractlets.... .... 1a. *Galinsoga parviflora* var. *parviflora*
1. Leaf blades lanceolate, oblong, or ovate, margins usually entire or serrulate, sometimes serrate; arrays of heads congested; axillary peduncles shorter than subtending bractlets .... 1b. *Galinsoga parviflora* var. *semicalva*

1a. Galinsoga parviflora Cavanilles var. **parviflora**
I   W

Plants 10–60 cm. **Leaf blades** lanceolate to broadly ovate, 20–110 × 15–70 mm, margins serrulate to serrate. **Arrays** of heads loose. **Axillary peduncles** longer than subtending bractlets. $2n = 16$.

Flowering summer–fall. Disturbed soils, fields, orchards, gardens, lawns, roadsides; 0–2300 m; introduced; B.C., Man., N.B., Ont., Que.; Ariz., Calif., Colo., Conn., D.C., Ill., Ind., Iowa, Kans., Ky., Md., Mass., Mich., Minn., Mo., Nebr., N.J., N.Y., N.Dak., Ohio, Oreg., Pa., R.I., Tex., Vt., Va., W.Va., Wis.; Mexico; West Indies; Central America; South America; introduced in Europe; Asia; Africa; Australia.

1b. Galinsoga parviflora Cavanilles var. **semicalva**
A. Gray, Smithsonian Contr. Knowl. 5(6): 98. 1853

Galinsoga semicalva (A. Gray) H. St. John & D. White

Plants 4–35 cm. **Leaf blades** lanceolate to ovate, 7–55(–80) × 3–20 mm, margins usually entire or serrulate, sometimes serrate. **Arrays** of heads congested. **Axillary peduncles** shorter than subtending bractlets. $2n = 32$.

Flowering summer–fall. Rocky slopes; 1800–2700 m; Ariz., N. Mex., Tex.; Mexico.

Plants of var. *semicalva* are less branched and usually have narrower leaves and shorter axillary peduncles than those of var. *parviflora*. They are found above 1500 m as far south as Distrito Federal, Mexico. Morphologic intermediates between the diploid var. *parviflora* and the tetraploid var. *semicalva* occur throughout the range of var. *semicalva*.

2. **Galinsoga quadriradiata** Ruiz & Pavón, Syst. Veg.
Fl. Peruv. Chil. 1: 198. 1798 [F] [I] [W]

Galinsoga bicolorata H. St. John &
D. White; *G. caracasana* (de Candolle)
Schultz-Bipontinus; *G. ciliata*
(Rafinesque) S. F. Blake

Plants 8–62 cm. **Leaf blades** 20–
60 × 15–45 mm. **Peduncles** 5–20
mm. **Involucres** hemispheric to
campanulate, 3–6 mm diam. **Phyllaries** falling. **Paleae:** outer falling,
broadly elliptic to obovate, 2–3 mm; inner falling, linear
to lanceolate, 2–3 mm, entire or 2- or 3-lobed, lobes to
¹/₃ total lengths, blunt. **Ray florets** (4–)5(–8); corollas
usually white, sometimes pink, laminae 0.9–2.5 × 0.9–2
mm. **Disc florets** 15–35. **Cypselae:** rays 1.5–2 mm; discs
1.3–1.8 mm; **pappi:** rays of 6–15 fimbriate scales 0.5–1

mm; discs 0, or of usually 14–20, rarely 1–5, white, lanceolate to oblanceolate, fimbriate, sometimes aristate,
scales 0.2–1.7 mm. **2n** = 32 [48, 64].

Flowering and fruiting summer–fall. Disturbed soils,
fields, orchards, gardens, lawns, roadsides; 0–2500 m;
introduced; Alta., B.C., Man., N.B., N.S., Ont., P.E.I.,
Que., Sask.; Ala., Ark., Colo., Conn., Del., D.C., Fla.,
Ga., Ill., Ind., Iowa, Kans., Ky., Maine, Md., Mass., Mich.,
Minn., Mo., Nebr., N.H., N.J., N.Y., N.C., N.Dak., Ohio,
Pa., R.I., S.C., Tenn., Vt., Va., Wash., W.Va., Wis.;
Mexico; West Indies; Central America; South America;
introduced in Europe; Asia; Africa; Pacific Islands.

Tetraploids of *Galinsoga quadriradiata* are native to
Mexico. Higher polyploids are found in South America
and differ from the tetraploids by their coarsely crenate-
serrate leaves, cylindro-campanulate involucres, and usually reddish purple limbs of ray corollas that extend at
right angles to involucres.

187m.11. ASTERACEAE Martinov (tribe HELIANTHEAE) subtribe DIMERESIINAE H. Robinson,
Phytologia 41: 40. 1978 [E]

Annuals, 1–3(–5) cm. **Leaves** basal and/or cauline; opposite (crowded); subsessile or ± petiolate; blades oblanceolate to spatulate, margins entire, faces tomentose to glabrescent, often
gland-dotted. **Heads** discoid, usually (2–5) in congested clusters (clusters often crowded, total
heads top 100+). **Calyculi** 0. **Involucres** cylindric to obconic. **Phyllaries** persistent, 2–3 in 1
series (oblong, subequal, herbaceous, basally connate or distinct). **Receptacles** flat (minute),
epaleate. **Ray florets** 0. **Disc florets** 2–3, bisexual, fertile; corollas white, pink, or purple, tubes
shorter than funnelform throats, lobes 5, deltate-ovate; anther thecae pale; stigmatic papillae
in 2 lines. **Cypselae** cylindric to clavate, fine-ribbed, glabrous; **pappi** soon falling (all together),
of ca. 20, basally connate, setiform to subulate, ± plumose scales (or flattened bristles).

Genus 1, species 1: w United States.

The sole genus of Dimeresiinae was originally placed in a broadly circumscribed Inuleae.
H. Robinson (1981) suggested possible relationships with Galinsoginae and/or Helianthinae.
B. G. Baldwin et al. (2002) positioned *Dimeresia* in Chaenactideae with *Chaenactis* and
Orochaenactis as the other two genera in the tribe.

307. DIMERESIA A. Gray in A. Gray et al., Syn. Fl. N. Amer. ed. 2, 1(2): 448. 1886

• [Greek *dimeres*, in 2 parts or with 2 members, allusion unclear]   [E]

John L. Strother

Annuals, 1–3(–5) cm. **Stems** decumbent to erect. **Leaves** basal and/or cauline; opposite
(crowded, seemingly whorled); subsessile or ± petiolate; blades spatulate to oblanceolate,
entire. **Heads** discoid, in clusters (of 2–5, congested, total heads to 100+). **Involucres** cylindric to obconic, 2(–4) mm diam. **Phyllaries** 2(–3) in 1 series (basally distinct or connate,
oblong, herbaceous). **Ray florets** 0. **Disc florets** 2(–3), bisexual, fertile; corollas white, pink,
or purple, tubes slightly shorter than funnelform throats, lobes 5, deltate-ovate. **Receptacles**

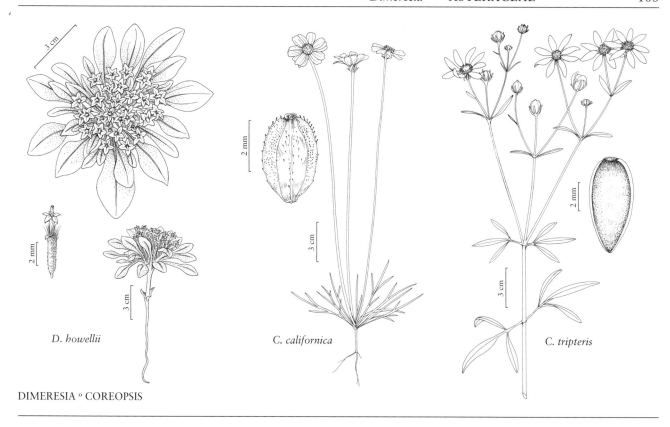

D. *howellii*   C. *californica*   C. *tripteris*

DIMERESIA ° COREOPSIS

minute, epaleate.   **Cypselae** cylindric to clavate, striate, glabrous; **pappi** falling, of ca. 20 basally connate, setiform to subulate, ± plumose scales (or flattened bristles). **x** = 7.

Species 1: w United States.

1. Dimeresia howellii A. Gray in A. Gray et al., Syn. Fl. N. Amer. ed. 2, 1(2): 449. 1886   E F

Leaf blades (2–)10–30+ mm, initially cobwebby, glabrescent, often gland-dotted. **Corollas** 5–6 mm. **Cypselae** 2–3 mm; **pappi** 3–4 mm (bristles falling together). **2*n*** = 14.

Flowering spring–summer. Open gravelly or sandy places, talus slopes screes, serpentine; 1100–2900 m; Calif., Idaho, Nev., Oreg.

187m.12. ASTERACEAE Martinov (tribe HELIANTHEAE) subtribe COREOPSIDINAE Lessing, Linnaea 5: 153. 1830 (as Coreopsideae)

Coreopsideae Lindley; *Petrobiinae* Bentham & Hooker f.

Annuals, perennials, subshrubs, or shrubs, 10–400 cm (sometimes rhizomatous or with cormiform bases, stoloniferous in *Coreopsis auriculata*). **Leaves** usually cauline (sometimes mostly basal); usually mostly opposite (distal sometimes alternate), rarely whorled; petiolate or sessile; blades (often pinnately or palmately lobed, sometimes compound), or lobes or leaflets, mostly deltate, elliptic, filiform, lanceolate, linear, oblanceolate, or ovate, ultimate margins entire or toothed, faces usually glabrous or glabrate, sometimes hairy (rarely, if ever, gland-dotted). **Heads** radiate or discoid, borne singly or in ± corymbiform, cymiform, or paniculiform arrays. **Calyculi** usually of 3–8(–21+) bractlets or bracts (usually ± herbaceous, usually shorter than phyllaries and/or reflexed, sometimes ± foliaceous and surpassing phyllaries). **Involucres** mostly campanulate, cylindric, turbinate, or hemispheric or broader. **Phyllaries** persistent, 3–34+ in ± 2 series (usually distinct, notably connate in *Thelesperma*), usually ± membranous, margins usually ± scarious. **Receptacles** mostly flat to convex, paleate (paleae often stramineous to white with red-brown to purplish striae, orbiculate to oblong or linear, ± flat to slightly cupped). **Ray florets** 0 or 1–21+ (more in cultivars), pistillate and fertile, or neuter, or styliferous and sterile; corollas usually yellow to orange (sometimes bicolored: yellow to orange with brown, red-brown, or purple), sometimes cyanic or white. **Disc florets** 3–150+, usually bisexual, fertile (functionally staminate in *Dicranocarpus*); corollas usually yellow to orange, sometimes brown, red-brown, or purple, tubes usually shorter than, sometimes equaling funnelform throats (longer than throats in some *Thelesperma* spp.), lobes (3–)5, ± deltate to lance-ovate; (staminal filaments hairy in *Cosmos*) anther thecae pale or dark; stigmatic papillae in 2 lines. **Cypselae** usually either obcompressed to obflattened, usually cuneate, linear, oblong, orbiculate, or ovate (bodies unequally 3–4-angled, margins often winged), or (all or inner) ± equally 4-angled and linear-fusiform, sometimes ± beaked, faces glabrous or hairy (often striate, tuberculate, or papillate); **pappi** 0, or persistent, of (1–)2–4(–8), usually ± barbellate awns or scales, rarely coroniform or of 1–2, smooth to ciliate or barbed awns or scales.

Genera 20, species 374 (7 genera, 69 species in the flora): mostly subtropical and warm-temperate New World and Old World.

The circumscription of Coreopsidinae followed here (H. Robinson 1981) is a bit narrower than the traditional one. The subtribe is remarkable among Heliantheae for having a relatively high number of species native in subtropical and tropical Old World, especially Africa. Distinctions among some genera (e.g., *Bidens*, *Coreopsis*, and *Cosmos*) are often subtle.

In Coreopsidinae, each involucre is subtended by a calyculus of more or less herbaceous (sometimes leaflike) bractlets or bracts (sometimes surpassing the phyllaries). In keys and descriptions here, shapes, heights, and diameters given for involucres are based on the phyllaries collectively (exclusive of calyculi) at flowering; the involucres are sometimes notably larger in fruit.

1. Phyllaries (excluding calyculi) 3–6 in 1(–2) series; disc florets 3–4+ (functionally staminate) . . . . 314. *Dicranocarpus*, p. 219
1. Phyllaries (3–)8–34+ in ± 2 series; disc florets 3–10, 10–20, or (5–)12–150+ (bisexual, fertile).
   2. Phyllaries connate ⅕–⅞+ their lengths . . . . 310. *Thelesperma*, p. 199
   2. Phyllaries usually distinct, rarely connate ± ¹⁄₁₀ their lengths.
      3. Cypselae (at least inner) ± 4-angled, ± linear-fusiform, often apically attenuate or beaked (none winged).
         4. Disc florets 10–20 (staminal filaments hairy near anthers); cypselae usually with 1 groove on each face . . . . 311. *Cosmos*, p. 203
         4. Disc florets (5–)12–150+ (staminal filaments not hairy); cypselae with 0 or 2 grooves on each face . . . . 312. *Bidens* (in part), p. 205
      3. Cypselae all ± obcompressed (sometimes winged).
         5. Annuals; ray florets 1–3 (laminae 1–2+ mm); cypselae mostly ellipsoid or obovoid (inner obovoid to obscurely urceolate, ± beaked) . . . . 313. *Heterosperma*, p. 218
         5. Annuals, perennials, subshrubs, or shrubs; ray florets usually 1, 2, 3, 5, 8, 13, or 21+, sometimes 0 (laminae mostly 4–30+ mm); cypselae mostly cuneate, linear, oblanceolate, oblong, ± orbiculate, or ovate (not beaked).

[6. Shifted to left margin—Ed.]

6. Cypselae rarely winged (margins sometimes thickened, winged in *B. aristosa* and *B. polylepis*); pappi usually of barbellate (rarely smooth) awns, sometimes 0 . . . . 312. *Bidens* (in part), p. 205

6. Cypselae (some or all) usually thin-margined or ± winged; pappi usually 0, sometimes coroniform, or of 2 bristly cusps or scales (in *Coreopsis*), or of 1–2 retrorsely barbellate awns (in *Coreocarpus*).

308. COREOPSIS Linnaeus, Sp. Pl. 2: 907. 1753; Gen. Pl. ed. 5, 388. 1754 • [Greek *korios*, bedbug, and *-opsis*, resembling, alluding to cypselae of original species]

John L. Strother

Annuals, perennials, subshrubs, or shrubs, 10–80(–200+) cm (often rhizomatous or with cormiform bases, stoloniferous in *C. auriculata*). **Stems** usually 1, erect, branched distally or ± throughout. **Leaves** basal, basal and cauline, or cauline; opposite or alternate or both; petiolate or sessile; blades simple and entire (dentate in *C. latifolia*), or ± pinnately or pedately lobed, faces glabrous or hairy. **Heads** radiate, borne singly or in open, ± corymbiform arrays. **Calyculi** of (3–)8+, distinct, ± herbaceous bractlets. **Involucres** ± globose to cylindric, 4–25+ mm diam. **Phyllaries** usually ± 8 in ± 2 series (usually distinct, rarely connate ± $^{1}/_{10}$ their lengths, mostly oblong to linear, ± membranous, margins ± scarious). **Receptacles** flat to convex, paleate; paleae falling, ovate to linear or subulate, ± flat, ± scarious (not adnate to and not falling with cypselae except in *C. bigelovii*). **Ray florets** mostly (5–)8(–12+, more in "double" cultivars), neuter, or styliferous and sterile, or pistillate and fertile; corollas usually yellow, sometimes red-brown to purple proximally, sometimes wholly purple or pink to white. **Disc florets** 8–150+, bisexual, fertile; corollas usually yellow, sometimes red-brown to purple at tips or throughout, tubes equaling or shorter than narrowly funnelform throats, lobes 4 or 5 (adaxial sinus seldom deeper than others). **Cypselae** obcompressed, ± orbiculate to ovate, oblong, or linear, usually thin-margined or winged, wings membranous to chartaceous or corky, entire or lobed to toothed, sometimes ciliolate; faces smooth or ± papillate to tuberculate; **pappi** 0, or persistent, of 2 bristly cusps or scales (sometimes pappi 0 and shoulders of cypsela wings ± bristly, pappus-like). *x* = 14.

Species ca. 35 (28 in the flora): mostly temperate North America, also tropical New World and Old World.

Early leaves of some coreopsises often differ from later leaves on individual plants. In such plants, early (proximal) leaves may be 1–3-pinnately or pedately lobed and 12–25+ cm long with 9–15+ orbiculate to lanceolate, linear, or filiform lobes and contrast markedly with later (distal) leaves 1–3 cm that are undivided or have 3–5+ lanceolate to linear or filiform lobes.

Cultivars (often "doubles" with multiple series of "ray" florets) derived from *Coreopsis auriculata*, *C. grandiflora*, *C. lanceolata*, and *C. tinctoria* are grown in public and residential gardens and are grown commercially for cut flowers.

SELECTED REFERENCES Jansen, R. K., E. B. Smith, and D. J. Crawford. 1987. A cladistic study of North American *Coreopsis* (Asteraceae: Heliantheae). Pl. Syst. Evol. 157: 73–84. Smith, E. B. 1976. A biosystematic survey of *Coreopsis* in eastern United

States and Canada. Sida 6: 123–215. Smith, E. B. 1984. Biosystematic study and typification of the Californian *Coreopsis* (Compositae) sections *Tuckermannia*, *Pugiopappus*, and *Euleptosyne*. Sida 10: 276–289.

1. Ray florets usually pistillate and fertile, sometimes styliferous and sterile.
    2. Perennials, subshrubs, or shrubs; ray laminae ± elliptic to oblong, broadest at or near middles . . . . 308a. *Coreopsis* sect. *Tuckermannia*, p. 186
    2. Annuals; ray laminae obovate to ± flabellate, broadest distal to middles.
        3. Cypselae of disc florets ciliolate; pappi of 2 lanceolate to subulate scales . . . . . .
        . . . . 308b. *Coreopsis* sect. *Pugiopappus*, p. 187
        3. Cypselae of disc florets not ciliolate; pappi 0 or coroniform (0.1–0.2 mm) . . . . .
        . . . . 308c. *Coreopsis* sect. *Leptosyne*, p. 188
1. Ray florets neuter (without styles).
    4. Leaf blades usually not lobed, rarely with 1–2 auricles at bases (margins dentate); ray florets 4–5; disc florets 10–15+ . . . . 308d. *Coreopsis* sect. *Silphidium*, p. 190
    4. Leaf blades sometimes lobed (ultimate margins not dentate); ray florets (5–)8(–13; more in "double" cultivars); disc florets 30–150+.
        5. Disc corolla lobes 5; style-branch apices usually ± penicillate, sometimes short-conic or truncate-mucronate.
            6. Ray laminae ± oblong-elliptic to suborbiculate, broadest at or near middles; paleae usually linear to subulate, often distally dilated . . . .
            . . . . 308e. *Coreopsis* sect. *Gyrophyllum*, p. 190
            6. Ray laminae ± cuneate, broadest distal to middles; paleae basally ± lanceolate to linear, distally attenuate (± filiform) . . . . 308f. *Coreopsis* sect. *Coreopsis*, p. 192
        5. Disc corolla lobes 4; style-branch apices ± truncate.
            7. Leaf blades sometimes with 1–2 lateral lobes; cypselae sometimes winged, wings ± pectinate or toothed . . . . 308g. *Coreopsis* sect. *Eublepharis*, p. 195
            7. Leaf blades usually 1–2(–3)-pinnately lobed; cypselae sometimes winged, wings entire . . . . 308h. *Coreopsis* sect. *Calliopsis*, p. 197

308a. Coreopsis Linnaeus sect. Tuckermannia (Nuttall) S. F. Blake, Proc. Amer. Acad. Arts 49: 340. 1913

Tuckermannia Nuttall, Trans. Amer. Philos. Soc., n. s. 7: 363. 1841

Perennials, subshrubs, or shrubs (often ± fleshy). **Leaves** cauline; alternate; blades (1–)2–3 -pinnately lobed. **Calyculi** of 4–12 bractlets. **Phyllaries** 8–14+. **Paleae** lanceolate to linear-oblong. **Ray florets** 11–21+, pistillate, fertile; laminae yellow, ± elliptic to oblong, each broadest at or near middle, apices entire or 2–3-toothed. **Disc florets** 80–150+; corollas yellow, lobes 5; style-branch apices ± truncate. **Cypselae** ± oblong, wings chartaceous to membranous, entire, not ciliolate, faces glabrous, smooth; **pappi** 0, or (1–)2 cusps 0.1–0.2+ mm (shoulders of wings sometimes bristly).

Species 2 (2 in the flora): coastal California, adjacent islands; nw Mexico (including Guadalupe Island).

1. Subshrubs or shrubs, (10–)30–250+ cm; heads (1–)8–12+ per stem; peduncles 4–12(–18+) cm; ray florets 8–13+ . . . . 1. *Coreopsis gigantea*
1. Perennials, 1–4(–8) dm; heads 1–2(–4+) per stem; peduncles (8–)15–30+ cm; ray florets 16–21+ . . . . 2. *Coreopsis maritima*

1. Coreopsis gigantea (Kellogg) H. M. Hall, Univ. Calif. Publ. Bot. 3: 142. 1907

Leptosyne gigantea Kellogg, Proc. Calif. Acad. Sci. 4: 198. 1872

Subshrubs or shrubs (± pachycaul, fleshy), (10–)30–250+ cm. **Leaf lobes** mostly filiform, 10–35(–75) × 0.5–1(–2) mm. **Heads** (1–)8–12+ per stem. **Peduncles** 4–12(–18+) cm. **Calyculi** of 4–12+ ovate to linear bractlets (5–)7–12+ mm. **Phyllaries** 8–14+, lance-ovate, 10–12+ mm. **Ray florets** 8–13+; laminae 20–35+ mm. **Disc corollas** 5–6 mm. **Cypselae** oblong-obovate, 5–7 mm. *2n* = 24.

Flowering Jan–Jul. Coastal bluffs, dunes; 0–200 m; Calif.; Mexico (Baja California).

2. Coreopsis maritima (Nuttall) Hooker f., Bot. Mag. 102: plate 6241. 1876 [E]

Tuckermannia maritima Nuttall, Trans. Amer. Philos. Soc., n. s. 7: 363. 1841

Perennials, 10–40(–80) cm. **Leaf lobes** mostly linear, 5–30(–45) × (0.5–)1–2(–4) mm. **Heads** 1–2 (–4+) per stem. **Peduncles** (8–)15– 30+ cm. **Calyculi** of 7–11+ oblong-deltate to linear bractlets (8–)12–20 (–30) mm. **Phyllaries** 12–13+, lanceolate, 12–20 mm. **Ray florets** 16–21+; laminae 20–35+ mm. **Disc corollas** 5.5–7 mm. **Cypselae** oblong-rectangular, 6–7 mm. *2n* = 24.

Flowering Feb–Jun. Coastal bluffs and dunes; 0–100 m; Calif.

308b. COREOPSIS Linnaeus sect. PUGIOPAPPUS (A. Gray) S. F. Blake, Proc. Amer. Acad. Arts 49: 340. 1913 [E]

Pugiopappus A. Gray in War Department [U.S.], Pacif. Railr. Rep. 4(5): 104. 1857

Annuals. Leaves mostly basal, seldom cauline; alternate; blades simple or (1–)2–3-pinnately lobed. **Calyculi** of 5–8 bractlets. **Phyllaries** (5–)8(–12). **Paleae** ± linear. **Ray florets** (3–)8, usually pistillate and fertile, sometimes styliferous and sterile; laminae yellow, ± obovate to flabellate, each broadest distal to middle, apices entire or 2–3-toothed. **Disc florets** (12–) 20–60(–100+); corollas yellow, lobes 5; style-branch apices ± truncate or deltoid. **Ray cypselae** ± rounded-quadrate to oblong, wings chartaceous to membranous, entire, spreading (or margins thin, winglike), not ciliolate, faces usually ± tuberculate, sometimes smooth; **pappi** 0. **Disc cypselae** spatulate to oblong, wings chartaceous to membranous, entire, spreading (or margins thin, winglike), ciliolate, faces usually ± tuberculate, sometimes smooth; **pappi** of 2 lanceolate to subulate scales 1–4(–5+) mm. *x* = 12.

Species 3 (3 in the flora): California.

1. Calyculi of ovate-deltate bractlets (3–)4–6+ mm; scales of pappi 3–4(–5) mm . . . . . . 3. *Coreopsis calliopsidea*
1. Calyculi of oblong to linear bractlets (1–)3–8+ mm; scales of pappi 1–3 mm.
   2. Ray laminae (5–)12–20+ mm; paleae basally adnate to and falling with cypselae; scales of pappi 2–3 mm . . . . 4. *Coreopsis bigelovii*
   2. Ray laminae 3.5–8 mm; paleae not adnate to and falling with cypselae; scales of pappi ca. 1 mm . . . . 5. *Coreopsis hamiltonii*

3. Coreopsis calliopsidea (de Candolle) A. Gray in W. H. Emory, Rep. U.S. Mex. Bound. 2(1): 90. 1859 [E]

Agarista calliopsidea de Candolle in A. P. de Candolle and A. L. P. P. de Candolle, Prodr. 5: 569. 1836

Annuals, (5–)12–30(–50) cm. **Leaf lobes** linear, 3–30 × 0.5–3 mm. **Peduncles** 5–25+ cm. **Calyculi** of 3–5 ovate-deltate bractlets (3–)4–6 mm. **Phyllaries** (5–)8, lance-ovate to lanceolate, 6–10+ mm. **Paleae** seldom basally adnate to and falling with disc cypselae. **Ray florets** (5–)8+; laminae 12–20(–30) mm. **Disc florets** (12–)40–100+; corollas 4–6.6 mm. **Ray cypselae** ± oblong, 4–6 mm. **Disc cypselae** oblanceolate to linear, 5–6 mm; **pappi** 3–4(–5) mm. $2n = 24$.

Flowering Feb–Jun. Alkaline playas, flood plains, deep sands; 100–1000 m; Calif.

4. Coreopsis bigelovii (A. Gray) Voss, Vilm. Blumengärtn. ed. 3, 1: 488. 1894 (as bigelowii) [E]

Pugiopappus bigelovii A. Gray in War Department [U.S.], Pacif. Railr. Rep. 4(5): 104. 1857

Annuals, 3–20(–40+) cm. **Leaf lobes** linear, 5–25 × 0.5–1(–2) mm. **Peduncles** 2–20(–35+) cm. **Calyculi** of (1–)5–8 linear bractlets (1–)3–8+ mm. **Phyllaries** (3–)8+, ± ovate to obovate, 6–10(–12+) mm. **Ray florets** (3–)8+; laminae (5–)12–20+ mm. **Disc florets** (5–)50–100+; corollas 2–4.5 mm. **Paleae** basally adnate to and falling with disc cypselae. **Ray cypselae** ± oblong, 5–6 mm. **Disc cypselae** oblanceolate to ± oblong, 5–6(–7) mm; **pappi** 2–3 mm. $2n = 24$.

Flowering Feb–Jun. Openings in chaparral, desert scrub, yellow-pine forests; 100–1800 m; Calif.

5. Coreopsis hamiltonii (Elmer) H. Sharsmith, Madroño 4: 214. 1938 [E]

Leptosyne hamiltonii Elmer, Bot. Gaz. 41: 323. 1906

Annuals, (3–)10–20+ cm. **Leaf lobes** oblong to linear, 1–8(–15) × 0.5–1 mm. **Peduncles** 3–9(–18) cm. **Calyculi** of 3–5 oblong to linear bractlets 1–3(–5) mm. **Phyllaries** 8, ± lance-ovate, 4–8 mm. **Ray florets** 8; laminae 3.5–8 mm. **Disc florets** ca. 40; corollas 2.5–3.5 mm. **Paleae** seldom basally adnate to and falling with disc cypselae. **Ray cypselae** narrowly obovate, 5–6 mm. **Disc cypselae** narrowly oblanceolate, 4–5 mm; **pappi** ca. 1 mm. $2n = 24$.

Flowering Mar–May. Talus slopes, rocky soils; 600–1300 m; Calif.

308c. Coreopsis Linnaeus sect. **Leptosyne** (de Candolle) O. Hoffmann in H. G. A. Engler and K. Prantl, Nat. Pflanzenfam. 54[IV,5]: 243. 1891

Leptosyne de Candolle in A. P. de Candolle and A. L. P. P. de Candolle, Prodr. 5: 531. 1836

Annuals. Leaves mostly basal, seldom cauline; mostly alternate, sometimes opposite; blades simple or 1–2-pinnately lobed. **Calyculi** of (3–)5–8 bractlets. **Phyllaries** 5–8. **Ray florets** (5–)8 (–12+), pistillate, fertile; laminae yellow, ± obovate or flabellate, each broadest distal to middle, apices entire or 2–3-toothed. **Disc florets** 8–60(–100+); corollas yellow (to orange), lobes 5; style-branch apices ± truncate or deltoid. **Paleae** oblong or lanceolate to linear. **Cypselae** ± oblong to spatulate, wings ± corky to chartaceous, entire, ± spreading, not ciliolate, faces smooth or ± tuberculate; **pappi** 0 or coroniform (0.1–0.2 mm). $x = 12$.

Species 3 (3 in the flora): California, nw Mexico.

The name *Coreopsis* sect. *Leptosyne* has been misattributed to S. F. Blake, Proc. Amer. Acad. Arts 49: 341. 1913.

1. Leaf blades simple or 1(–2)-pinnately lobed, terminal lobes spatulate to narrowly oblanceolate, 1–3 mm wide; calyculus bractlets ciliolate near bases .... 6. *Coreopsis stillmanii*
1. Leaf blades simple or pinnately lobed, terminal lobes filiform to linear, 0.5–1(–1.5) mm wide; calyculus bractlets not ciliolate.
    2. Cypselae marked adaxially with red dots or dashes near wing/body interface, wings ± corky-thickened, faces ± tuberculate and/or hirtellous .... 7. *Coreopsis californica*
    2. Cypselae not marked adaxially with red, wings usually thin, sometimes corky-thickened, faces usually smooth, rarely tuberculate, not hirtellous .... 8. *Coreopsis douglasii*

6. Coreopsis stillmanii (A. Gray) S. F. Blake, Proc. Amer. Acad. Arts 49: 342. 1913 ☐E☐

Leptosyne stillmanii A. Gray, J. Acad. Nat. Sci. Philadelphia 3: 91. 1855

Annuals, 5–20+ cm. **Leaf blades** simple or 1(–2)-pinnately lobed, terminal lobes spatulate to narrowly oblanceolate, 1–3 mm wide. **Peduncles** 5–12(–20+) cm. **Calyculi** of 5–7+ oblong to linear bractlets 4–8(–10) mm, ciliolate near bases. **Phyllaries** (5–)8, ovate to lance-ovate, 4–5(–7) mm. **Ray florets** 5–8; laminae 8–12(–15+) mm. **Disc florets** 12–30+; corollas 2.5–3.5 mm. **Cypselae** ± obovate, 4–5 mm, not marked adaxially with red, wings corky-thickened, faces smooth or ± tuberculate, not hirtellous. $2n = 24$.

Flowering Mar–May. Rocky slopes, serpentine ridges; 400–800 m; Calif.

7. Coreopsis californica (Nuttall) H. Sharsmith, Madroño 4: 217. 1938 ☐E☐☐F☐

Leptosyne californica Nuttall, Trans. Amer. Philos. Soc., n. s. 7: 363. 1841; *Coreopsis californica* subsp. *newberryi* (A. Gray) E. Murray; *C. californica* var. *newberryi* (A. Gray) E. B. Smith

Annuals, 5–20(–30+) cm. **Leaf blades** simple or 1(–2)-pinnately lobed, terminal lobes filiform, 0.5–1(–1.5) mm wide. **Peduncles** 5–15(–30+) cm. **Calyculi** of (3–)5–8 linear bractlets 3–7+ mm, not ciliolate near bases. **Phyllaries** 5–8, obovate to oblanceolate, 4–6(–7+) mm. **Ray florets** (5–)8(–12+); laminae 7–10(–15+) mm. **Disc florets** 20–60(–100+); corollas 2–3.6 mm. **Cypselae** ± oblong, (2.5–)3–4+ mm, marked adaxially with red dots or dashes, wings corky-thickened, faces ± tuberculate and/or hirtellous. $2n = 24$.

Flowering Feb–Jun. Openings in desert scrub, stabilized dunes; 300–1000 m; Ariz., Calif.

Coreopsis californica may occur sporadically in New Mexico.

8. Coreopsis douglasii (de Candolle) H. M. Hall, Univ. Calif. Publ. Bot. 3: 140. 1907 ☐E☐

Leptosyne douglasii de Candolle in A. P. de Candolle and A. L. P. P. de Candolle, Prodr. 5: 531. 1836

Annuals, 5–15(–25+) cm. **Leaf blades** simple or pinnately lobed, terminal lobes filiform, 0.5–1+ mm wide. **Peduncles** 5–15(–20+) cm. **Calyculi** of (3–)5 linear bractlets 3–5(–12) mm, not ciliolate near bases. **Phyllaries** (5–)8, ovate, 5–7+ mm. **Ray florets** (5–)8(–9+); laminae 5–12(–16+) mm. **Disc florets** 8–60(–100+); corollas 2.8–3.5 mm. **Cypselae** ± oblanceolate, 3.5–4+ mm, not marked adaxially with red dots or dashes, wings usually thin, sometimes corky-thickened, faces usually smooth, rarely tuberculate, not hirtellous. $2n = 24$.

Flowering Mar–May. Shale and serpentine slopes, sandy flats; (150–)300–1000 m; Calif.

308d. COREOPSIS Linnaeus sect. SILPHIDIUM   (Torrey & A. Gray) A. Gray in A. Gray et al., Syn. Fl. N. Amer. 1(2): 294. 1884   E

Coreopsis Linnaeus [unranked] *Silphidium* Torrey & A. Gray, Fl. N. Amer. 2: 341. 1842

Perennials. Leaves cauline, opposite; blades usually simple, sometimes with 1–2 auricles at base (margins dentate). **Calyculi** of 3–8+ bractlets. **Phyllaries** 5. **Ray florets** 4–5, neuter; laminae ± elliptic to oblong, broadest at or near middle, apices entire or 2–3-toothed. **Disc florets** 10–15+; corollas ochroleucous or pale yellow, lobes 5; style-branch apices ± conic-attenuate. **Paleae** ± linear. **Cypselae** ± linear, margins thin, not winged, not ciliolate, faces smooth, glabrous; **pappi** 0 or (1–)2 cusps (0.1–0.2+ mm). $x = 13$.

Species 1: se United States.

9. Coreopsis latifolia   Michaux, Fl. Bor.-Amer. 2: 137. 1803   E

Perennials, 90–150 cm. **Leaf blades** ovate, 6–15(–20+) × 2–10+ cm. **Peduncles** 1–5(–10) cm. **Calyculi** of 3–8+ linear bractlets 5–8 mm. **Phyllaries** 5, oblong, 8–12 mm. **Ray laminae** 10–18 mm. **Disc corollas** 7–8 mm. **Cypselae** 7–8 mm. $2n = 26$.

Flowering Aug–Sep. Shaded slopes in woods; 300–700 m; Ga., N.C., S.C., Tenn.

Coreopsis latifolia is in the Center for Plant Conservation's National Collection of Endangered Plants.

308e. COREOPSIS Linnaeus sect. GYROPHYLLUM   Nuttall, Trans. Amer. Philos. Soc., n. s. 7: 358. 1841   E

Perennials. Leaves cauline, opposite; blades usually 3-foliolate, sometimes simple or ± pinnately lobed (with 5+ leaflets), leaflets simple or each with 2–9+ lobes, or blades not 3-foliolate (then often with 3(–5+) lobes). **Calyculi** of 5–12+ bractlets. **Phyllaries** 8. **Ray florets** 8, neuter; laminae yellow, ± oblong-elliptic to suborbiculate, broadest at or near middle, apices entire or 2–3-toothed. **Disc florets** 25–80+; corollas yellow or ± red-brown to purple, lobes 5; style-branch apices ± penicillate. **Paleae** usually linear to subulate, often distally dilated. **Cypselae** obovate to oblong, wings corky to chartaceous, entire, ± spreading, not ciliolate, faces glabrous, smooth; **pappi** 0 or (1–)2 cusps (0.1–0.5+ mm; shoulders of wings sometimes bristly, pappus-like). $x = 13$.

Species 6 (6 in the flora): mostly c, e United States.

E. B. Smith (Bot. Gaz. 136: 83. 1975) used a superfluous name, sect. *Palmatae* (F. Boynton) E. B. Smith [based on *Coreopsis* (unranked) *Palmatae* F. Boynton in J. K. Small, Man. S.E. Fl., 1446. 1933], for sect. *Gyrophyllum*.

1. Blades of leaves not 3-foliolate, most 3(–5+)-lobed . . . . 10. *Coreopsis palmata*
1. Blades of leaves usually 3-foliolate, leaflets simple or ± pinnately or ± pedately lobed.
   2. Petioles 5–45 mm . . . . 11. *Coreopsis tripteris*
   2. Petioles 0–1 mm.

[3. Shifted to left margin.—Ed.]

3. Leaf blades or leaflets seldom lobed, lance-ovate to lanceolate or lance-linear, 2–30 mm wide.
    4. Leaf blades or leaflets (4–)6–15(–30+) mm wide . . . . 12. *Coreopsis major*
    4. Leaf blades or leaflets 2–7 mm wide . . . . 13. *Coreopsis delphiniifolia*
3. Leaf blades or leaflets often lobed, ultimate lobes oblong to linear or filiform, 0.5–2.5 mm wide.
    5. Internodes (± mid stem) 15–40(–60) mm; leaflets entire or each with 3–9+, ± linear to
       filiform lobes 5–25(–45+) × 0.5–1(–2) mm; disc corollas yellow . . . . 14. *Coreopsis verticillata*
    5. Internodes (± mid stem) 8–12(–16) mm; leaflets entire or each with 2–5+, ± oblong to
       linear lobes 15–30+ × –1.5(–2.5+) mm; disc corollas yellow or purple . . . . 15. *Coreopsis pulchra*

10. Coreopsis palmata Nuttall, Gen. N. Amer. Pl. 2: 180. 1818 [E]

Perennials, 30–80 cm. **Internodes** (± mid stem) 25–50 mm. **Leaves:** petioles 0–1 mm (or 5–25+ mm, winged, and scarcely distinct from blades); blades not 3-foliolate, most with 3(–5+), ± oblong to linear lobes (5–)15–40+ × 2–3(–7+) mm (sometimes some leaves not lobed). **Peduncles** 1–4+ cm. **Calyculi** of 9–12+ oblong to lanceolate bractlets 3–9+ mm. **Phyllaries** 8, ± oblong to nearly orbiculate, 6–10 mm. **Ray laminae** 15–25+ mm. **Disc florets** 60–80+; corollas yellow (sometimes drying blackish), 5–6.5 mm. **Cypselae** oblong, 5–6 mm. **2n = 26.**

Flowering Jun–Aug. Prairies, open woods; 100–300 m; Ark., Ill., Ind., Iowa, Kans., La., Mich., Minn., Mo., Nebr., Okla., S.Dak., Wis.

11. Coreopsis tripteris Linnaeus, Sp. Pl. 2: 908. 1753 [E] [F]

Coreopsis tripteris var. *deamii* Standley; *C. tripteris* var. *smithii* Sherff

Perennials, 90–180+ cm. **Internodes** (± mid stem) 3–7(–10+) cm. **Leaves:** petioles 5–45 mm; blades usually 3-foliolate, leaflets usually simple, sometimes ± pinnately lobed, ultimate blades ± lanceolate, 4–9(–12+) cm × (9–)12–35+ mm. **Peduncles** 2–5+ cm. **Calyculi** of 5–6 oblong bractlets 1.5–5 mm. **Phyllaries** 8, ± oblong to lance-oblong, 6–8 mm. **Ray laminae** 12–22+ mm. **Disc florets** 40–80+; corollas red-brown to purplish, 5–6 mm. **Cypselae** obovate to oblong, 4–5(–6) mm. **2n = 26.**

Flowering Jul–Sep. Heavy loams, moist sands, along streams, boggy meadows; 100–500 m; Ont., Que.; Ala., Ark., Conn., D.C., Fla., Ga., Ill., Ind., Iowa, Kans., Ky., La., Md., Mass., Mich., Miss., Mo., N.C., Ohio, Okla., Pa., R.I., S.C., Tenn., Tex., Vt., Va., W.Va., Wis.

12. Coreopsis major Walter, Fl. Carol., 214. 1788 [E]

Coreopsis major var. *rigida* (Nuttall) F. E. Boynton; *C. major* var. *stellata* (Nuttall) B. L. Robinson

Perennials, 30–90 cm. **Internodes** (± mid stem) 5–11 cm. **Leaves:** petioles 0–1 mm; blades simple or 3-foliolate, simple blades or leaflets ± lance-elliptic to narrowly lanceolate, 25–55(–100+) × (4–)6–15(–30+) mm (seldom lobed). **Peduncles** 2–6+ cm. **Calyculi** of 7–8 oblong to linear bractlets 3–6 mm. **Phyllaries** 8, lance-oblong to lance-ovate, 6–8 mm. **Ray laminae** 12–30+ mm. **Disc florets** 40–80+; corollas yellow (often drying blackish to purplish), 6–7 mm. **Cypselae** obovate to oblong, 4–5 mm. **2n = 26, 78, 104.**

Flowering May–Aug. Oak barrens, road clearings, wooded bottoms; 300–800 m; Ala., Fla., Ga., Ind., Ky., La., Mass., Miss., N.Y., N.C., Ohio, Pa., S.C., Tenn., Va., W.Va.

13. Coreopsis delphiniifolia Lamarck in J. Lamarck et al., Encycl. 2: 108. 1786 (as delphinifolia) [E]

Perennials, 30–90 cm. **Internodes** (± mid stem) 3–8 cm. **Leaves:** petioles 0–1 mm; blades simple or 3-foliolate, simple blades or leaflets usually narrowly lanceolate to lance-linear, 35–80 × 2–5(–7) mm (seldom lobed, sometimes parted into 2–3+ lance-linear to ± linear lobes). **Peduncles** 15–45+ mm. **Calyculi** of 8–10 linear bractlets 3–5(–7) mm. **Phyllaries** 8, oblong-ovate, 5–6+ mm. **Ray laminae** 15–25(–30) mm. **Disc florets** 25–60+; corollas yellow (often drying blackish), 5–6 mm. **Cypselae** oblong, 4.5–6 mm. **2n = 52, 78, 104.**

Flowering May–Jul(–Sep). Open woods, barrens, swamps; ca. 300 m; Ga., S.C.

Plants treated here as *Coreopsis delphiniifolia* are questionably distinct from *C. major*. In 1976, E. B. Smith suggested that members of the taxon he called *C. ×delphiniifolia* may be hybrids or progeny of hybrids involving *C. verticillata* and *C. tripteris* and, possibly, *C. major*. A problem with such an interpretation is that although all of the 35 or so records for *C. delphiniifolia* in the sense of Smith map at or near known localities for *C. major*, all but 2 are from well south of the known distribution of *C. verticillata* and only 3 are from near known localities for *C. tripteris*.

14. **Coreopsis verticillata** Linnaeus, Sp. Pl. 2: 907. 1753 Ⓔ

Perennials, 20–60+ cm. Internodes (± mid stem) 15–40(–60) mm. Leaves: petioles 0–1 mm; blades usually 3-foliolate, simple blades or leaflets entire or each ± pedately divided with 3–9+, ± linear to filiform lobes 5–25(–45+) × 0.5–1 (–2) mm. Peduncles 2–5(–8+) cm. Calyculi of 8 linear bractlets 3–5 (–9) mm. Phyllaries 8, lance-ovate, 6–7 mm. Ray laminae 12–20+ mm. Disc florets 30–40+; corollas yellow (sometimes drying blackish), 3.5–5.5 mm. Cypselae obovate to oblong, 3.5–4 mm. $2n = 26, 52, 78$.

Flowering Jun–Jul(–Aug). Oak wood, sandy soils; 10–500 m; Ont., Que.; Ala., Ark., Conn., D.C., Fl., Ga., Ky. Md., Mass., Miss., N.J., N.Y., N.C., Ohio, S.C., Tenn., Va., W.Va.

15. **Coreopsis pulchra** F. E. Boynton in J. K. Small, Fl. S.E. U.S., 1277, 1340. 1903 Ⓔ

Perennials, 20–40+ cm. Internodes (± mid stem) 8–12(–16) mm. Leaves: petioles 0–1 mm; blades usually 3-foliolate, simple blades or leaflets entire or each ± pedately lobed with 2–5+, ± oblong to linear lobes 15–30+ × 1–1.5(–2.5+) mm. Peduncles 2–5 (–8+) cm. Calyculi of 8 oblong to linear bractlets 1.5–3+ mm. Phyllaries 8, lance-ovate to oblong, 4–5 mm. Ray laminae 10–20+ mm. Disc florets 30–45+; corollas yellow (sometimes drying blackish) or purple, 3–4 mm. Cypselae obovate to oblong, 4–5 mm. $2n = 26$.

Flowering Jun–Sep. Openings in woods, sandstone outcrops; 300 m; Ala., Ga.

308f. COREOPSIS Linnaeus sect. COREOPSIS Ⓔ

Annuals or perennials. Leaves basal and cauline or mostly cauline, opposite; blades simple or ± pinnately or pedately lobed (with 3–9+ lobes). **Calyculi** of (6–)8(–9) bractlets. **Phyllaries** 8. **Ray florets** 8 (more in "double" cultivars), neuter; laminae yellow, sometimes each with red-brown or purple near base, ± cuneate, broadest distal to middles, apices usually ± 3-lobed (middle lobes entire or toothed). **Disc florets** (30–)60–150+; corollas yellow, apices yellow or red-brown to purple, lobes 5; style-branch apices ± penicillate to short-conic or truncate-mucronate. **Paleae** basally ± lanceolate to linear, distally attenuate (± filiform). **Cypselae** obovate to oblong (often cupped), margins ± winged (or wingless and margins adaxially ± corky), wings ± chartaceous, usually entire, rarely irregularly toothed to pectinate, ± spreading, not ciliolate, faces glabrous, smooth or ± papillate; **pappi** 0, or usually falling, of (1–)2 cusps or subulate to lanceolate scales (0.1–0.5+ mm). $x = 13$.

Species 7 (7 in the flora): mostly c, e North America.

1. Annuals (rarely persisting); ray laminae yellow, usually each with proximal red-brown to purple flecks, spot, or band.
    2. Leaves mostly in proximal $^3/_4$–$^7/_8$ of plant heights; calyculi of lance-deltate to linear bractlets 6–9(–12+) mm; disc corolla apices red-brown to purple; cypselae wingless
    . . . . 16. *Coreopsis basalis*
    2. Leaves mostly in proximal $^1/_5$–$^1/_3$(–$^1/_2$) of plant heights; calyculi of deltate to lance-oblong bractlets 4–6(–8) mm; disc corolla apices yellow; cypselae winged . . . . . 17. *Coreopsis nuecensis*
1. Perennials (sometimes flowering first year); ray laminae yellow (without red-brown or purple near bases).
    3. Plants usually stoloniferous; wings of cypselae 0 or involute, ± corky . . . . 18. *Coreopsis auriculata*
    3. Plants not stoloniferous (± rhizomatous or bases ± cormose); wings of cypselae ± spreading, ± chartaceous.

[4. Shifted to left margin.—Ed.]

4. Aerial nodes proximal to first peduncle usually 1–3(–5+); leaves mostly on proximal ¼–⅓(–½) of plant heights; cypselae (2.6–)3–4 mm .... 19. *Coreopsis lanceolata*
4. Aerial nodes proximal to first peduncle usually (5–)6–12+; leaves mostly on proximal ½–⅞ of plant heights; cypselae 2–3 mm.
    5. Blades of leaves rarely simple, usually irregularly 1(–2)-pinnately or ± -pedately lobed with (3–)5–9+ lobes, rarely simple, simple blades or terminal lobes narrowly lanceolate to linear or filiform, (0.5–)2–8 (–12+) mm wide .... 20. *Coreopsis grandiflora*
    5. Blades of leaves usually simple, rarely with 1–2(–3+) lateral lobes, simple blades or terminal lobes oblong-elliptic or ovate to elliptic or lanceolate, (2–)5–35 mm wide.
        6. Simple leaf blades or terminal lobes oblong-elliptic or ovate to elliptic, 45–95+ × 22–35 mm; phyllaries 9–12 mm .... 21. *Coreopsis intermedia*
        6. Simple leaf blades or terminal lobes lance-elliptic to oblanceolate or lanceolate, 15–60(–80) × 2–25(–35)mm; phyllaries 5–8+ mm .... 22. *Coreopsis pubescens*

16. Coreopsis basalis (A. Dietrich) S. F. Blake, Proc. Amer. Acad. Arts 51: 525. 1916 [E]

Calliopsis basalis A. Dietrich, Allg. Gartenzeitung 3: 329. 1835; *Coreopsis basalis* var. *wrightii* (A. Gray) S. F. Blake; *C. wrightii* (A. Gray) H. M. Parker ex E. B. Smith

Annuals, 10–50+ cm. **Aerial nodes** proximal to first peduncle usually 5–10+, distalmost 1–3 internodes 4–7(–10) cm. **Leaves:** basal and cauline on proximal ¾–⅞ of plant heights; petioles 8–35(–120) mm; blades simple or 1(–2)-pinnately lobed with 3–9+ lobes, simple blades or terminal lobes elliptic or lanceolate to oblanceolate or linear, 25–55+ × (1–)2–9(–20) mm. **Peduncles** 6–15+ cm. **Calyculi** of lance-deltate to linear bractlets 6–9(–12+) mm. **Phyllaries** lance-ovate, 7–9+ mm. **Ray laminae** yellow, usually each with a proximal, red-brown to purple spot or band, 15–20+ mm. **Disc corollas** 3–4 mm, apices red-brown to purple. **Cypselae** 1.2–1.8 mm, wingless (margins ± inrolled adaxially, ± corky). *2n* = 26.

Flowering Apr–Jun. Sandy soils in open, often disturbed, places; 10–300+ m; Ala., Ark., Fla., Ga., La., Miss., N.C., Okla., S.C., Tex.

Plants in the western part of the distribution of *Coreopsis basalis* usually have narrower lobes of leaf blades and narrower outer phyllaries; such plants have been treated as *C. wrightii* or as *C. basalis* var. *wrightii*.

17. Coreopsis nuecensis A. Heller, Contr. Herb. Franklin Marshall Coll. 1: 106. 1895 [E]

Coreopsis coronata Hooker, Bot. Mag. 63: plate 3460. 1836, not Linnaeus 1763; *C. nuecensoides* E. B. Smith

Annuals, 10–30(–50+) cm. **Aerial nodes** proximal to first peduncle usually 1–3(–5), distalmost 1–3 internodes 2–4(–9+) cm. **Leaves:** basal and cauline on proximal ⅕–⅓(–½) of plant heights; petioles 1–8+ cm; blades simple or 1(–2)-pinnately lobed, simple blades or terminal lobes ± elliptic to lanceolate, 10–30 × 5–15 mm. **Peduncles** (5–)10–25+ cm. **Calyculi** of ± deltate to lance-oblong bractlets 4–6(–8) mm. **Phyllaries** lance-oblong, 6–8(–10+) mm. **Ray laminae** yellow, usually each with (2–)4–8+ proximal, red-brown or purple flecks, 12–18+ mm. **Disc corollas** 2.5–3.2 mm, apices yellow. **Cypselae** 2.5–4 mm, wings spreading, ± chartaceous, entire. *2n* = 12, 14, 18, 20.

Flowering Mar–May. Sandy soils in post-oak and oak-mesquite woodlands; 10–200+ m; Fla., Tex.

Coreopsis nuecensis is probably in Mexico (Tamaulipas) and was collected once in Florida ("Escaped from cultivation"). Plants with glabrous phyllaries and chromosome numbers of *2n* = 18 and 20 and included here in *C. nuecensis* have been called *C. nuecensoides*.

18. Coreopsis auriculata Linnaeus, Sp. Pl. 2: 908. 1753 [E]

Perennials, 10–30(–60+) cm (often stoloniferous). **Aerial nodes** proximal to first peduncle usually 1–2 (–3), distalmost 1–3 internodes 1–4 (–8+) cm. **Leaves:** basal and cauline on proximal ¼–½ of plant heights; petioles 1–6(–10+) cm; blades simple or with 1–2+ lateral lobes, simple blades or terminal lobes suborbiculate or ovate-elliptic to lance-ovate, 15–55 (–75) × 9–25(–35) mm. **Peduncles** 8–20+ cm. **Calyculi** of

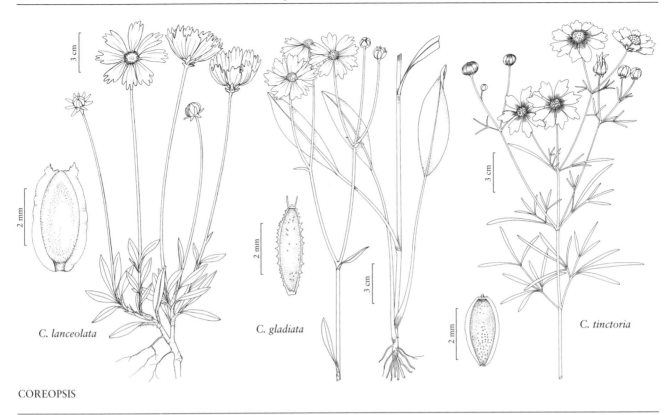

C. lanceolata

C. gladiata

C. tinctoria

COREOPSIS

ovate to oblong bractlets 5–10+ mm. **Phyllaries** lance-deltate to lance-ovate, 9–12 mm. **Ray laminae** yellow, 15–20+ mm. **Disc corollas** 3.5–4.5 mm, apices yellow. **Cypselae** 1.5–2.5 mm, wings 0 or involute and ± corky. $2n = 26$.

Flowering Apr–Jun. Roadsides and other openings in woods, pine barrens; 10–500+ m; Ala., Fla., Ga., Ky., La., Miss., N.C., S.C., Tenn., Va., W.Va.

19. Coreopsis lanceolata Linnaeus, Sp. Pl. 2: 908. 1753 E F

Perennials, 10–30(–60+) cm. **Aerial nodes** proximal to first peduncle usually 1–3(–5+), distalmost 1–3 internodes 1–2(–8+) cm. **Leaves:** basal and cauline on proximal $^1/_4$–$^1/_3$(–$^1/_2$) of plant heights; petioles 1–5(–8+) cm; blades simple or with 1–2+ lateral lobes, simple blades or terminal lobes lance-ovate or lanceolate to oblanceolate or lance-linear, 5–12 cm × 8–15(–18+) mm. **Peduncles** (8–)12–20(–35+) cm. **Calyculi** of lance-ovate to lance-linear or linear bractlets 4–8(–12) mm. **Phyllaries** deltate to lance-deltate, 8–12+ mm. **Ray laminae** yellow, 15–30+ mm. **Disc corollas** 6–7.5 mm, apices yellow. **Cypselae** (2.6–)3–4 mm, wings ± spreading, ± chartaceous, entire. $2n = 26$ (+ 0–4B).

Flowering (Mar–)May–Jul(–Aug). Sandy soils, ditches and roadsides, other disturbed sites; 30–500(–1000+) m; B.C., Ont.; Ala., Ark., Fla., Ga., Ill., Ind., Kans., Ky., La., Md., Mich., Miss., Mo., N.J., N.Mex., N.Y., N.C., Ohio, Okla., Pa., S.C., Tenn., Tex., Vt., Va., W.Va., Wis.

Plants that have been called *Coreopsis lanceolata* var. *villosa* Michaux often have 5+ aerial internodes 6+ cm long proximal to the first peduncle; they may merit recognition as a distinct taxon or may be hybrids (or derivatives) from crosses between *C. lanceolata* and *C. pubescens*.

20. Coreopsis grandiflora Hogg ex Sweet, Brit. Fl. Gard. 2: plate 175. 1826 E

Coreopsis grandiflora var. *harveyana* (A. Gray) Sherff; *C. grandiflora* var. *longipes* (Hooker) Torrey & A. Gray; *C. grandiflora* var. *saxicola* (Alexander) E. B. Smith; *C. saxicola* Alexander

Perennials, 40–60+ cm. **Aerial nodes** proximal to first peduncle usually 6–10+, distalmost 1–3 internodes 4–7+ cm. **Leaves:** mostly cauline on proximal $^2/_3$–$^7/_8$ of plant heights; petioles 0 or 10–35+ mm; blades usually 1(–2)-irregularly pinnately or ± pedately lobed with (3–)5–9+ lobes, rarely simple, simple blades or terminal lobes narrowly lanceolate to linear or filiform, 15–45(–90+) × (0.5–)2–8(–12+) mm. **Peduncles** 8–15(–25+) cm. **Calyculi** of lanceolate to linear bractlets 3.5–9+ mm. **Phyllaries** lance-ovate, 7–9(–12) mm. **Ray laminae** yellow, 12–25+ mm. **Disc corollas** 3.3–4.8 mm, apices yellow. **Cypselae** 2–3+ mm, wings spreading, ± chartaceous, entire or irregularly toothed to pectinate. **2n** = 26 (+ 0–2B).

Flowering May–Aug. Sandy soils, ditches and road-sides, other disturbed sites, granite and sandstone outcrops; 30–300+ m; Ont., Que.; Ala., Ark., Conn., Fla., Ga., Ill., Ind., Iowa, Kans., Ky., La., Maine, Md., Mich., Miss., Mo., N.J., N.Mex., N.Y., N.C., Ohio, Okla., Pa., R.I., S.C., Tenn., Tex., Vt., Va., W.Va., Wis.

Coreopsis grandiflora var. *inclinata* J. R. Allison from glades in Alabama may merit recognition.

21. Coreopsis intermedia Sherff, Bot. Gaz. 88: 299. 1929 [E]

Perennials, 60–90+ cm. **Aerial nodes** proximal to first peduncle usually (5–)6–12+, distalmost 1–3 internodes 4–12 cm. **Leaves:** mostly cauline on proximal ¹⁄₂–²⁄₃ of plant heights; petioles 0–35+ mm; blades usually simple, rarely with 1(–3+) lateral lobes, simple blades or terminal lobes oblong-elliptic or ovate to elliptic, 45–95+ × 22–35+ mm. **Peduncles** 12–25(–40) cm. **Calyculi** of lanceolate to lance-linear bractlets 6–9 mm. **Phyllaries** lance-ovate to lanceolate, 9–12 mm. **Ray laminae** yellow, 20–30+ mm. **Disc corollas** 3.4–4.4 mm, apices yellow. **Cypselae** 2.2–2.8 mm, wings ± spreading, ± chartaceous, entire. **2n** = 26

Flowering May–Jul. Sandy soils, openings in scrub-oak woodlands, roadsides and other disturbed sites; 60–300 m; Ark., La., Okla., Tex.

22. Coreopsis pubescens Elliott, Sketch Bot. S. Carolina 2: 441. 1823 [E]

Coreopsis pubescens var. *debilis* (Sherff) E. B. Smith; *C. pubescens* var. *robusta* A. Gray ex Eames

Perennials, 20–70(–90+) cm. **Aerial nodes** proximal to first peduncle usually (5–)6–12+, distalmost 1–3 internodes 4–6(–10+) cm. **Leaves:** mostly cauline on proximal ²⁄₃–⁷⁄₈ of plant heights; petioles 2–10(–25+) mm; blades usually simple, rarely with 1–2(–3+) lateral lobes, simple blades or terminal lobes lance-elliptic to oblanceolate or lanceolate, 15–60(–80) × 2–25(–35) mm. **Peduncles** (7–)12–15+ cm. **Calyculi** of lanceolate to lance-linear bractlets 3–7 mm. **Phyllaries** lance-ovate to lanceolate, 5–8+ mm. **Ray laminae** yellow, 12–15+ mm. **Disc corollas** 4.6–5.4 mm, apices yellow. **Cypselae** 2.5–3 mm, wings ± spreading, ± chartaceous, entire. **2n** = 26 (+ 0–2B).

Flowering (Apr–)Jun–Aug(–Sep). Sandy soils, granite outcrops, open pine-oak woods, ditches and roadsides, other disturbed sites; 30–1000+ m; Ala., Ark., Conn., Fla., Ga., Ill., Kans., Ky., La., Mass., Miss., Mo., N.C., Okla., S.C., Tenn., Tex., Va., W.Va.

Bushy to wiry, nearly glabrous plants with leaf blades mostly simple, mostly oblanceolate, and 2–15+ mm wide and with relatively small heads, florets, and fruits from coastal Mississippi and included here in *Coreopsis pubescens* may merit recognition as *C. debilis* Sherff or *C. pubescens* var. *debilis*.

308g. Coreopsis Linnaeus sect. **Eublepharis** Nuttall, Trans. Amer. Philos. Soc., n. s. 7: 359. 1841 [E]

Perennials (sometimes flowering first year; bases ± rhizomatous or cormose). **Leaves** basal and cauline, usually alternate, sometimes both alternate and opposite, or all opposite; blades mostly simple, sometimes with 1–2 lateral lobes (none in *C. nudata*). **Calyculi** of 5–8(–12+) bractlets. **Phyllaries** (7–)8(–9). **Ray florets** 8(–13), neuter; laminae yellow or purple to pink or white, ± cuneate, broadest distal to middles, usually 3-lobed (middle lobes sometimes toothed). **Disc florets** 40–120+; corollas ochroleucous to yellow or red-brown to purple or apices red-brown to purple, lobes 4; style-branch apices ± truncate. **Paleae** ± linear (not distally dilated or attenuate). **Cypselae** obovate or oblong to linear, margins winged or not, wings ± corky to chartaceous, ± pectinate, ± spreading, not ciliolate, faces glabrous, smooth or ± tuberculate; **pappi** 0, or (1–)2 cusps, awns, or scales (0.1–1+ mm). *x* = 13.

Species 4 (4 in the flora): e, se North America.

1. Leaves lacking blades (petioles ± terete, 5–30+ cm × 0.5–2+ mm, functionally equivalent to
   leaves) . . . . 23. *Coreopsis nudata*
1. Leaves with blades (petioles not 5–30+ cm × 0.5–2 mm).
   2. Ray corollas ± pinkish to white; cypselae 1.3–1.8 mm, not winged . . . . 24. *Coreopsis rosea*
   2. Ray corollas yellow; cypselae 3–4.5 mm, wings ± pectinate.
      3. Leaves all opposite; petioles ± ciliate; wings of cypselae 0.1 mm wide; pappi of 1–2
         cusps 0.1–0.2 mm . . . . 25. *Coreopsis integrifolia*
      3. Leaves usually all or mostly alternate, seldom some or all opposite; petioles not ciliate;
         wings of cypselae 0.2–0.4+ mm wide; pappi of 2, subulate scales 0.5–1+ mm. . . 26. *Coreopsis gladiata*

23. Coreopsis nudata Nuttall, Gen. N. Amer. Pl. 2: 180. 1818 E

Perennials, 40–60(–100+) cm. **Internodes** (± mid stem) 5–25 cm. **Leaves** basal and cauline; alternate; petioles ± terete, 5–30+ cm × 0.5–2+ mm, not ciliate; blades none. **Peduncles** 4–8+ cm. **Calyculi** of lance-deltate to lance-ovate bractlets 2–5 mm. **Phyllaries** lance-ovate, 7–8 mm. **Ray laminae** purplish to pinkish, 12–20(–30+) mm. **Disc florets** 60–120+; corollas ochroleucous to yellow, 3.5–4.5 mm. **Cypselae** narrowly oblong to linear, 2.5–3+ mm, winged, wings ± pectinate, 0.2–0.4+ mm wide; **pappi** of 2 subulate scales 1–1.5 mm. *2n* = 26.

Flowering Apr–May. Ditches, pond margins, swamps, in pine barrens; 0–50+ m; Ala., Fla., Ga., La., Miss.

24. Coreopsis rosea Nuttall, Gen. N. Amer. Pl. 2: 179. 1818 E

Perennials, 10–30(–60) cm. **Internodes** (± mid stem) 1–4(–5+) cm. **Leaves** mostly cauline; opposite; petioles 0–1 mm, ciliate or not; blades lance-linear to linear or filiform, 20–45(–60) × 1–2(–3+) mm, rarely with 1–2 lateral lobes. **Peduncles** 2–4(–6+) cm. **Calyculi** of oblong to linear bractlets 1.5–2+ mm. **Phyllaries** deltate-ovate, 4.5–5.5 mm. **Ray laminae** pinkish to white, 9–15+ mm. **Disc florets** 40–60+; corollas ochroleucous to yellow, 2.5–3 mm. **Cypselae** narrowly oblong, 1.3–1.8 mm, not winged; **pappi** 0. *2n* = 26.

Flowering Aug–Sep. Sandy shores, marsh edges, etc.; 0–50 m; N.S.; Del., Mass., N.J., Pa., R.I., S.C.

Occurrence of *Coreopsis rosea* in South Carolina may represent a human-mediated disjunction; the collection came from a "lime sink" near a trailer park close to a freeway.

25. Coreopsis integrifolia Poiret in J. Lamarck et al., Encycl., suppl. 2: 353. 1811 E

Perennials, 40–60 cm. **Internodes** (± mid stem) 3–8 cm. **Leaves** mostly cauline; opposite; petioles 5–15 mm, ciliate; blades elliptic to lanceolate, 20–45+ × 9–18+ mm, not lobed. **Peduncles** 3–8+ cm. **Calyculi** of ± oblong to linear bractlets 3–5 mm. **Phyllaries** lance-ovate, 7–8+ mm. **Ray laminae** yellow, 20–25+ mm. **Disc florets** 45–80+; corollas purplish or purple-tipped, 2.8–3.7 mm. **Cypselae** narrowly obovate, 4–4.5 mm, winged, wings 0.1 mm wide, ± pectinate; **pappi** of 2 cusps 0.1–0.2 mm. *2n* = 26.

Flowering Sep–Oct. Low woodlands, flood plains; 0–50 m; Fla., Ga., N.C., S.C.

26. Coreopsis gladiata Walter, Fl. Carol., 215. 1788 E F

Coreopsis falcata F. E. Boynton; *C. floridana* E. B. Smith; *C. gladiata* var. *linifolia* (Nuttall) Cronquist; *C. helianthoides* Beadle; *C. linifolia* Nuttall; *C. longifolia* Small

Perennials, 30–70 cm. **Internodes** (± mid stem) (1–)8–12+ cm. **Leaves** basal and cauline; usually alternate, sometimes opposite and alternate or all opposite; petioles 1–10(–15+) cm, not ciliate; blades elliptic or lanceolate to ± linear, (2–)4–8 (–15+) cm × (3–)7–12(–40+) mm, sometimes with 1–2 lateral lobes. **Peduncles** (1–)4–8(–10+) cm. **Calyculi** of broadly deltate-ovate to lance-ovate or lance-linear bractlets 2–6+ mm. **Phyllaries** lance-ovate to lanceolate, 6–9(–12) mm. **Ray laminae** yellow, (12–)15–20+ mm. **Disc florets** 40–80+; corollas purplish or purple-tipped, 2.8–4.8 mm. **Cypselae** ± obovate to oblong, 3–4 mm, winged, wings 0.2–0.4+ mm wide, ± pectinate; **pappi** of 2 subulate scales 0.5–1+ mm. *2n* = 26.

Flowering ± year round, mostly Sep–Oct. Peaty bogs, swamps, depressions, in pine barrens; 10–100+ m; Ala., Ark., Fla., Ga., La., Miss., N.C., S.C., Tex., Va.

As here circumscribed, *Coreopsis gladiata* includes plants that others (without agreement among themselves)

have treated as distinct species and/or varieties: *C. falcata* (leaves mostly basal and/or proximal, alternate, blades mostly lance-linear to linear, often 1–2 with 1–2 lateral lobes, faces not dotted; outer phyllaries lance-ovate to lance-linear; Ga., N.C., S.C.); *C. floridana* (leaves mostly basal and/or proximal, alternate, blades mostly lance-linear to linear, not lobed, not dotted; outer phyllaries broadly deltate-ovate; Fla.), *C. helianthoides* (leaves mostly cauline, alternate, blades mostly elliptic to lanceolate, not lobed, faces not dotted; outer phyllaries lance-ovate to lance-linear; Ala., Fla., Ga., La., Miss., N.C., S.C.), and *C. linifolia* (leaves basal and cauline or mostly cauline, alternate or opposite or both, blades mostly lance-linear to linear, seldom lobed, faces bearing dark or translucent dots; outer phyllaries lance-ovate to lance-linear; Ala., Fla., Ga., La., Miss., N.C., S.C., Tex., Va.).

308h. COREOPSIS Linnaeus sect. CALLIOPSIS (Reichenbach) Nuttall, Trans Amer. Philos. Soc., n. s. 7: 360. 1841

Calliopsis Reichenbach, Mag. Aesth. Bot. 1: plate 70. 1823

Annuals (sometimes persisting). **Leaves** basal and cauline or mostly cauline; usually opposite, rarely alternate; blades simple or 1–2+-pinnately lobed. **Calyculi** of 6–10 bractlets. **Phyllaries** 8. **Ray florets** (5–)8, neuter; laminae yellow or ± proximally red-brown to purple (sometimes almost wholly red-brown to purple), obovate to cuneate, broadest distal to middles, apices ± 3-lobed (middle lobes often toothed). **Disc florets** 40–80(–100+); corollas red-brown to purple (at least at apices), lobes 4; style-branch apices ± truncate. **Paleae** linear (not distally dilated or attenuate). **Cypselae** ± oblong to spatulate, margins winged or not, wings ± membranous, entire, ± spreading, not ciliolate, faces glabrous, smooth or tuberculate; **pappi** 0, or of (1–)2 cusps or subulate scales 0.1–1(–1.4+) mm. $x = 12$ (+ 0–2 Bs).

Species 3 (2 in the flora): North America, Mexico.

1. Blades of cauline leaves usually 1-pinnate or simple; ray corollas usually yellow throughout, rarely with proximal red-brown blotch; pappi of 2, subulate scales (0.2–)0.4–1(–1.4+) mm . . . . 27. *Coreopsis leavenworthii*
1. Blades of cauline leaves usually 1–2(–3)-pinnate, seldom simple; ray corollas usually yellow with red-brown blotch, sometimes red-brown in proximal 1/3–9/10 and distally yellow, rarely yellow throughout; pappi 0, or of 1–2, bristly cusps or subulate scales 0.1–1+ mm . . . 28. *Coreopsis tinctoria*

27. Coreopsis leavenworthii Torrey & A. Gray, Fl. N. Amer. 2: 346. 1842 [E]

Annuals, (10–)30–70(–150+) cm. **Leaves:** proximal blades usually 1(–2)-pinnate, terminal lobes ± elliptic or oblanceolate to lanceolate or linear, 15–30+ × 4–8(–12+) mm; cauline blades simple or 1(–3)-pinnate, simple blades or terminal lobes narrowly oblanceolate to linear, 8–35(–70+) × 1–3(–5) mm. **Peduncles** 2–5(–8) cm. **Calyculi** of deltate-ovate to oblong or linear bractlets 2–6+ mm. **Phyllaries** ± lance-oblong to lanceolate, (4–)5–9 mm. **Ray laminae** usually yellow throughout, rarely with proximal red-brown blotch, 8–15+ mm. **Disc corollas** 2.2–3.5 mm. **Cypselae** 2–3 mm, wings 0.2–0.8+ mm wide; **pappi** of 2 subulate scales (0.2–)0.4–1(–1.4+) mm. $2n = 24$ (+ 0–2 Bs).

Flowering year round, mostly May–Jul. Moist, sandy soils, flatwoods, ditches; 0–20+ m; Fla.

Plants here (and generally) treated as *Coreopsis leavenworthii* probably should be included within the circumscription adopted here for *C. tinctoria*.

28. Coreopsis tinctoria Nuttall, J. Acad. Nat. Sci. Philadelphia 2: 114. 1821 [F]

Coreopsis atkinsoniana Douglas ex Lindley; *C. cardaminefolia* Torrey & A. Gray; *C. tinctoria* var. *atkinsoniana* (Douglas ex Lindley) H. M. Parker ex E. B. Smith; *C. tinctoria* var. *similis* (F. E. Boynton) H. M. Parker ex E. B. Smith

Annuals, (10–)30–70(–150+) cm. **Leaves:** proximal blades usually 1(–3)-pinnate, terminal lobes lance-ovate to oblanceolate, 10–60 × 5–25 mm; cauline blades usually 1–2(–3)-pinnate, rarely simple, simple blades or terminal lobes lance-linear to linear or filiform, 10–45 × 0.5–2(–5+) mm. **Peduncles** 1–5(–15+) cm.

Calyculi of deltate-lanceolate bractlets 1–3+ mm. **Phyllaries** ± lance-oblong to lance-ovate, 4–7(–9) mm. **Ray laminae** usually yellow with red-brown blotch, sometimes red-brown in proximal ¹/₃–⁹/₁₀ and distally yellow, rarely yellow throughout, 12–18+ mm. **Disc corollas** 2.5–3+ mm. **Cypselae** 1.5–3(–4+) mm, wings 0 or 0.1–0.7+ mm wide; **pappi** 0, or of 1–2 cusps or subulate scales 0.1–1+ mm. $2n = 24$ (+ 0–2 Bs).

Flowering year round, mostly Jun–Aug. Moist, sandy or clay soils, sometimes alkaline flats, prairies, ditches, disturbed places; (0–)20–1500(–2000) m; Alta., B.C., Man., Ont., Que., Sask.; Ala., Ariz., Ark., Calif., Colo., Conn., Del., D.C., Fla., Ga., Idaho, Ill., Ind., Iowa, Kans., Ky., La., Maine, Md., Mass., Mich., Minn., Miss., Mo., Mont., Nebr., N.H., N.J., N.Mex., N.Y., N.C., N.Dak., Ohio, Okla., Oreg., Pa., R.I., S.C., S.Dak., Tenn., Tex., Vt., Va., Wash., W.Va., Wis., Wyo.; Mexico (Coahuila, Nuevo León, Tamaulipas).

Coreopsis tinctoria is widely grown in public and residential gardens, and commercially (for cut flowers), and has become widely established in the flora area.

As here circumscribed, *Coreopsis tinctoria* includes plants that others (without agreement among themselves) have treated as distinct species or infraspecific taxa: *C. atkinsoniana* (plants mostly 50–150+ cm, seldom branched from bases; cypselae 2.5–3 mm, "narrowly" winged; pappi 0.1–0.2 mm; mostly Idaho, Montana, Oregon, Washington), *C. cardaminefolia* (plants mostly 20–50 cm, seldom branched at bases; cypselae 2 mm, "narrowly to widely" winged; pappi 0 or 0.1–0.2 mm; mostly Arkansas, Kansas, Louisiana, Nebraska, Oklahoma, Texas), and *C. tinctoria* var. *similis* (plants mostly 10–30 cm, usually branched from bases; cypselae 2–3 mm, "widely" winged; pappi 0.2–1 mm; Texas and Mexico).

309. COREOCARPUS Bentham, Bot. Voy. Sulphur 28, plate 16. 1844 • [Greek *koreos*, bug, and *karpos*, fruit, alluding to pectinately winged cypselae of original species]

John L. Strother

Perennials [subshrubs], 10–70(–120+) cm. **Stems** 1, erect [sprawling], branched throughout. **Leaves** cauline; opposite; obscurely petiolate; blades 1–2-pinnately or -pedately lobed (ultimate lobes narrowly lanceolate to filiform [lanceolate to ovate]), ultimate margins entire, faces usually glabrous, rarely hairy [pilose]. **Heads** usually radiate, sometimes discoid, in open, corymbiform to paniculiform arrays [borne singly]. **Calyculi** of (0–)1–3+ linear to subulate, herbaceous bractlets. **Involucres** ± campanulate to turbinate, 3–5+ mm diam. **Phyllaries** persistent, 5–8[–13+] in ± 2 series, distinct, (green to stramineous with red-brown to purplish nerves) ovate to oblong, ± equal, membranous, margins scarious. **Receptacles** flat to convex, paleate; paleae falling, (stramineous with red-brown striae) lance-linear to lanceolate, membranous to scarious [adnate to cypselae]. **Ray florets** 0, or [1–]5–8, pistillate, fertile [styliferous and sterile or neuter]; corollas [orange] yellow to pale yellow or white [purplish]. **Disc florets** [5–]12–25[–30+], bisexual, fertile; corollas yellow with red-brown nerves, tubes shorter than narrowly funnelform throats, lobes 5, ± deltate. **Cypselae** (blackish or red-brown to stramineous) obcompressed, ± oblanceolate, faces smooth or ± granular-papillate, margins ± winged, wings ± corky, ± pectinate [entire]; **pappi** 0, or persistent, of 1–2 retrorsely [antrorsely] barbellate awns. $x = 12$.

Species ca. 9 (1 in the flora): sw United States, nw Mexico.

SELECTED REFERENCE Smith, E. B. 1989. A biosystematic study and revision of the genus *Coreocarpus* (Compositae). Syst. Bot. 14: 448–472.

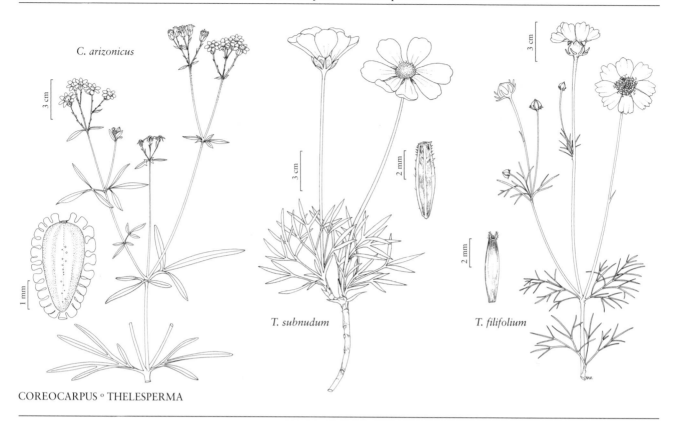

COREOCARPUS ° THELESPERMA

1. Coreocarpus arizonicus (A. Gray) S. F. Blake, Proc. Amer. Acad. Arts 49: 344. 1913 [F]

Leptosyne arizonica A. Gray, Proc. Amer. Acad. Arts 17: 218. 1882

Perennials (may flower first year). **Leaves** ± scattered over proximal ³⁄₄+ of plant height, internodes mostly 30–75+ mm; lobes mostly filiform to linear, sometimes lanceolate, (5–)15–35+ × 0.5–1(–3+) mm. **Ray laminae** pale yellow (fading to white), 2–4.5(–6+) mm. **Disc corollas** 3–4.5 mm. **Cypselae** 2–3(–5) mm; **pappi** 0, or 0.5–1 mm. $2n = 24$.

Flowering year round, following rains. Open sites, along streams, rocky canyons; 100–1500 m; Ariz.; Mexico (Chihuahua, Sinaloa, Sonora).

E. B. Smith (1989) treated plants of *Coreocarpus arizonicus* from the flora area and Mexico as var. *arizonicus* and other plants of the species from Mexico as var. *pubescens* (B. L. Robinson & Fernald) S. F. Blake.

310. **THELESPERMA** Lessing, Linnaea 6: 511. 1831 • [Greek *thele*, nipple, and *sperma*, seed, alluding to papillate cypselae of original species]

John L. Strother

Annuals, perennials, or subshrubs, 10–70+ cm. **Stems** usually 1, erect, branched distally or ± throughout. **Leaves** mostly basal, basal and cauline, or mostly cauline; mostly opposite (distal rarely alternate); blades usually 1(–3)-pinnately lobed (ultimate lobes oblanceolate to filiform), faces usually glabrous, rarely hairy. **Heads** radiate or discoid, borne singly or in loose, corymbiform arrays. **Calyculi** of 3–8+ distinct, usually spreading or reflexed, usually linear to subulate, herbaceous bractlets. **Involucres** hemispheric to urceolate, 4–15+ mm diam. **Phyllaries**

5–8 in ± 2 series, persistent, connate $^1/_5$–$^7/_8$+ their lengths, lance-ovate to ovate, ± equal, ± leathery to membranous, margins (of distinct apices) scarious. **Receptacles** flat to convex, paleate; paleae falling, (whitish with red-brown striae, each ± appressed to abaxial face of subtended cypsela), obovate to oblong, scarious. **Ray florets** 0 or ca. 8, neuter; corollas yellow or red-brown, or bicolored (yellow and red-brown). **Disc florets** 20–100+, bisexual, fertile; corollas yellow (with red-brown nerves) or red-brown, either with throats equal to or longer than 5, ± deltate, ± equal lobes, or with throats shorter than 5, ± lance-linear, ± unequal lobes (in either form, the abaxial sinus usually ± deeper than others). **Cypselae** (dark red-brown or stramineous) outer often arcuate, shorter, inner more columnar, usually some or all ± obcompressed (each usually shed together with its subtending palea), faces smooth or papillate to tuberculate or verrucate, margins sometimes ± winged; pappi 0, or persistent, of 2 retrorsely ciliate, subulate scales or awns. *x* = 12.

Species 10+ (9 in the flora): w North America, Mexico, South America.

SELECTED REFERENCES Hansen, C. J., L. Allphin, and M. D. Windham. 2002. Biosystematic analysis of the *Thelesperma subnudum* complex (Asteraceae). Sida 20: 71–96. Greer, L. F. 1997. *Thelesperma curvicarpum* (Asteraceae), an achene form in populations of *T. simplicifolium* var. *simplicifolium* and *T. filifolium* var. *filifolium*. Southw. Naturalist 42: 242–244.

1. Throats of disc corollas equal to or longer than lobes; pappi usually 0, rarely of 2 awns 0.1–0.3(–0.5) mm.
    2. Cauline leaves ± scattered over proximal $^3/_4$+ of plant heights, internodes mostly 45–95 mm . . . . 1. *Thelesperma simplicifolium*
    2. Cauline leaves ± crowded over proximal $^1/_4$–$^1/_2$ of plant heights, internodes mostly 5–25+ mm or 1–5(–35) mm.
        3. Leaf lobes mostly linear to filiform, 5–25(–45+) × 0.5(–1) mm; cypselae 2–3 mm. . . . . 2. *Thelesperma longipes*
        3. Leaf lobes mostly oblanceolate to linear, (5–)10–35(–45+) × (1–)2–3(–5) mm; cypselae 5–7 mm . . . . 3. *Thelesperma subnudum*
1. Throats of disc corollas shorter than lobes; pappi usually of 2 awns or scales (0.5–)1–3 mm, rarely 0.
    4. Perennials (sometimes flowering first year); calyculi of 3–5+, ovate to oblong bractlets 1–3 mm; ray florets 0, or 8, laminae 4–8(–12+) mm; cypselae (4–)5–8 mm.
        5. Plants 10–30(–50) cm; ray florets usually 8, rarely 0; disc corollas red-brown . . . . 4. *Thelesperma ambiguum*
        5. Plants (20–)30–80+ cm; ray florets 0; disc corollas yellow . . . . 5. *Thelesperma megapotamicum*
    4. Annuals (sometimes persisting); calyculi of 5–9, linear to narrowly triangular bractlets (2–)4–8+ mm; ray florets 8, laminae 10–20+ mm; cypselae 3.5–5.5+ mm.
        6. Ray laminae proximally to wholly red-brown to purplish, or each with a proximal red-brown spot; disc corollas red-brown to purplish.
            7. Calyculi of 5–6(–9), linear to narrowly triangular bractlets (margins hispido-ciliate); cypselae 3.5–4 mm. . . . . 6. *Thelesperma burridgeanum*
            7. Calyculi of 6–7+, narrowly triangular bractlets (margins not ciliate); cypselae 5–5.5+ mm. . . . . 7. *Thelesperma nuecense*
        6. Ray laminae yellow to golden (rarely proximally suffused with red-brown); disc corollas yellow or red-brown.
            8. Plants 30–70(–120+) cm; cauline leaves ± scattered over proximal $^9/_{10}$ of plant heights, internodes 35–100+ mm; disc corollas yellow, often with brown nerves . . . . 8. *Thelesperma flavodiscum*
            8. Plants 10–40(–70+) cm; cauline leaves crowded to ± scattered over proximal $^1/_2$–$^3/_4$+ of plant heights, internodes mostly 10–35(–50+ mm); disc corollas red-brown or yellow with red-brown nerves. . . . . 9. *Thelesperma filifolium*

1. Thelesperma simplicifolium (A. Gray) A. Gray, Hooker's J. Bot. Kew Gard. Misc. 1: 252. 1849

Cosmidium simplicifolium A. Gray, Mem. Amer. Acad. Arts, n. s. 4: 86. 1849; *Thelesperma curvicarpum* Melchert

Perennials or subshrubs (may flower first year), (20–)30–70+ cm. **Cauline leaves** ± scattered over proximal $^3/_4$+ of plant heights, internodes mostly 45–95 mm; lobes mostly linear to filiform, sometimes oblanceolate, (5–)15–45(–60) × 0.5–1(–2) mm. **Calyculi** of 6–8+ linear to subulate bractlets 1–3+ mm. **Ray florets** usually 8, rarely 0; laminae yellow, 9–15(–20+) mm. **Disc corollas** yellow with red-brown nerves, throats equal to or longer than lobes. **Cypselae** 3–4 mm; **pappi** usually 0, rarely 0.1–0.3+ mm. $2n = 20$.

Flowering Apr–Jul(–Oct). Openings in oak/juniper woodlands or desert scrub, usually on limestone; 100–1500 m; Tex.; Mexico (Chihuahua, Coahuila, Nuevo León, Tamaulipas).

Thelesperma simplicifolium is used in roadside plantings and may be encountered as a waif outside its natural range in the flora area (e.g., in California).

2. Thelesperma longipes A. Gray, Smithsonian Contr. Knowl. 3(5): 109. 1852

Perennials or subshrubs (may flower first year), 20–40+ cm. **Cauline leaves** mostly crowded over proximal $^1/_4$–$^1/_2$ of plant heights, internodes mostly 5–25+ mm; lobes mostly linear to filiform, 5–25 (–45+) × 0.5(–1) mm. **Calyculi** of 5 ovate bractlets 1–3 mm. **Ray florets** 0. **Disc corollas** yellow, sometimes with red-brown nerves, throats equal to or longer than lobes. **Cypselae** 2–3 mm; **pappi** usually 0, rarely 0.1–0.3+ mm. $2n = 40$.

Flowering (Mar–)Apr–Oct. Openings in desert scrub, limestone ridges; 500–2100 m; Ariz., N.Mex., Tex.; Mexico (Coahuila, Nuevo León, San Luis Potosí, Tamaulipas).

3. Thelesperma subnudum A. Gray, Proc. Amer. Acad. Arts 10: 72. 1874 E F

Thelesperma caespitosum Dorn; *T. marginatum* Rydberg; *T. pubescens* Dorn; *T. pubescens* var. *caespitosum* (Dorn) C. J. Hansen; *T. subnudum* var. *alpinum* S. L. Welsh; *T. subnudum* var. *marginatum* (Rydberg) Cronquist; *T. subnudum* var. *pubescens* (Dorn) S. L. Welsh; *T. windhamii* C. J. Hansen

Perennials, 10–30(–40+) cm (caudices often woody). **Cauline leaves** mostly crowded over proximal $^1/_4$(–$^1/_2$) of plant heights, internodes mostly 1–5 (–35 mm); lobes mostly oblanceolate to linear, (5–)10–35 (–45+) × (1–)2–3(–5) mm. **Calyculi** of 7–9 deltate to lance-linear bractlets 2–4+ mm. **Ray florets** 0, or 8; laminae yellow, (6–)12–20+ mm. **Disc corollas** yellow, sometimes with red-brown nerves, throats equal to or longer than lobes. **Cypselae** 5–7 mm; **pappi** usually 0, rarely 0.1–0.5 mm. $2n = 22, 24$.

Flowering May–Sep. Openings in pinyon/juniper or yellow-pine forests, often on talus or "balds"; 1000–2900 m; Alta.; Ariz., Colo., Mont., Nev., N.Mex., N.Dak., Utah, Wyo.

As here circumscribed, *Thelesperma subnudum* includes plants that have been treated as distinct species and/or varieties (some of which have been treated as of conservation concern). See C. J. Hansen et al. (2002) for an alternate interpretation of *T. subnudum*.

4. Thelesperma ambiguum A. Gray, Proc. Amer. Acad. Arts 19: 16. 1883

Thelesperma fraternum Shinners; *T. megapotamicum* (Sprengel) Kuntze var. *ambiguum* (A. Gray) Shinners

Perennials or subshrubs, 10–30 (–50) cm. **Cauline leaves** crowded to ± scattered over proximal $^1/_3$–$^1/_2$ of plant heights, internodes mostly 5–50(–80+) mm; lobes mostly linear to filiform, sometimes oblanceolate, (5–)15–75 × 0.5–1(–2) mm. **Calyculi** of 3–5+ ovate to oblong bractlets 1–3 mm. **Ray florets** usually 8, rarely 0; laminae yellow, 4–8(–12+) mm. **Disc corollas** red-brown, throats shorter than lobes. **Cypselae** 4–5+ mm; **pappi** usually 1.5–2+ mm, rarely 0. $2n = 44$.

Flowering Mar–Jun. Disturbed sites on sands or clays; 10–200 m; N.Mex., Tex.; Mexico (Nuevo León, Tamaulipas).

Thelesperma ambiguum is similar to *T. megapotamicum*; it may be better treated as a variety of the latter.

5. **Thelesperma megapotamicum**   (Sprengel) Kuntze,
   Revis. Gen. Pl. 3(2): 182. 1898 (as Thelespermum)

Bidens megapotamica Sprengel, Syst.
Veg. 3: 454. 1826; *Thelesperma
gracile* (Torrey) A. Gray

**Perennials or subshrubs, (20–)30–
80+ cm. Cauline leaves** ± scattered
over proximal $^1/_2$–$^3/_4$ of plant
heights, internodes mostly 40–100
mm; lobes mostly linear to filiform,
sometimes oblanceolate, 20–40
(–50+) × 0.5–1(–2.5) mm. **Calyculi** of 3–5+ ovate to
oblong bractlets 1–2(–3) mm. **Ray florets** 0. **Disc corollas**
yellow, often with red-brown nerves, throats shorter than
lobes. **Cypselae** 5–8 mm; **pappi** 1–2(–3) mm. $2n = 22, 44$.

Flowering (Apr–)May–Oct. Disturbed places on sands
or clays, oak/juniper woodlands, desert scrub, yellow-
pine forests; 300–2900 m; Ariz., Ark., Colo., Kans., Nebr.,
N.Mex., Okla., S.Dak.. Tex., Utah, Wyo.; Mexico (Chihuahua, Coahuila); South America.

Thelesperma megapotamicum sometimes persists
after plantings outside its natural range in the flora area
and may become established (e.g., in California).

6. **Thelesperma burridgeanum**   (Regel) S. F. Blake, Proc.
   Biol. Soc. Wash. 41: 146. 1928   E

Cosmidium burridgeanum Regel,
Index Seminum (St. Petersburg)
1857: 40. 1858

**Annuals, 30–40(–70+) cm. Cauline
leaves** crowded to ± scattered over
proximal $^1/_2$–$^3/_4$ of plant heights,
internodes mostly 5–40(–60+) mm;
lobes mostly linear to filiform,
sometimes oblanceolate, 5–25(–45)
× 0.5–1(–2+) mm. **Calyculi** of 5–6(–9) linear to narrowly
triangular bractlets 2–4 mm (margins hispido-ciliate). **Ray
florets** 8; laminae wholly red-brown to purplish or mostly
red-brown with distal margins yellow to orange, 7–9(–12+)
mm. **Disc corollas** red-brown to purplish, throats shorter
than lobes. **Cypselae** 3.5–4 mm; **pappi** 0.5–1+ mm.
$2n = 18$.

Flowering Apr–Jun. Disturbed sites on sands; 10–200
m; Tex.

Thelesperma burridgeanum may be a color-form of
T. filifolium.

7. **Thelesperma nuecense**   B. L. Turner, Rhodora 61: 243.
   1959   E

**Annuals, 30–70(–90+) cm. Cauline
leaves** ± scattered over proximal
$^1/_4$–$^1/_2$(–$^3/_4$) of plant heights, internodes mostly 4–10 cm; lobes mostly
linear to filiform, sometimes ± oblong, 25–50(–120) × 0.5–1+ mm.
Calyculi of 6–7+ narrowly triangular bractlets 2–4 mm. **Ray florets**
8; laminae yellow to golden, usu-
ally proximally, sometimes almost wholly, suffused with
red-brown or with a red-brown spot or band, 10–20+ mm.
Disc corollas red-brown, throats shorter than lobes.
Cypselae 5–5.5+ mm; **pappi** 0.5–1+ mm. $2n = 20$.

Flowering Mar–Jul. Disturbed sites on sands; 0–200
m; Tex.

Distinctions between *Thelesperma nuecense* and *T.
filifolium* are subtle; they may be better treated as one
species.

8. **Thelesperma flavodiscum**   (Shinners) B. L. Turner,
   Rhodora 61: 245. 1959   E

Thelesperma filifolium (Hooker)
A. Gray var. *flavodiscum* Shinners,
Field & Lab. 18: 98. 1950

**Annuals, 30–70(–120+) cm.
Cauline leaves** ± scattered over
proximal $^9/_{10}$ of plant heights, internodes mostly 35–100+ mm;
lobes mostly linear to filiform,
sometimes oblanceolate, (10–)25–
45+ × 0.5–1(–2) mm. **Calyculi** of 7–9 linear to narrowly
triangular bractlets 4–6(–8) mm. **Ray florets** 8; laminae
yellow, 12–20+ mm. **Disc corollas** yellow, often with
red-brown nerves, throats shorter than lobes. **Cypselae**
4–5 mm; **pappi** 0.5–1.5+ mm. $2n = 18, 20$.

Flowering May–Jun. On sands, post-oak savannas;
100–200 m; Tex.

Differences between *Thelesperma flavodiscum* and *T.
filifolium* are subtle; they may be better treated as one
species.

9. **Thelesperma filifolium**   (Hooker) A. Gray, Hooker's
   J. Bot. Kew Gard. Misc. 1: 252. 1849   F

Coreopsis filifolia Hooker, Bot.
Mag. 63: plate 3505. 1836;
Thelesperma filifolium var.
intermedium (Rydberg) Shinners;
T. intermedium Rydberg

Annuals (sometimes persisting),
10–40(–70+) cm. **Cauline leaves**
crowded to ± scattered over proximal $^1/_2$–$^3/_4$ of plant heights, internodes mostly 10–35(–50+) mm; lobes mostly linear to filiform, sometimes oblanceolate, 5–30(–55+) × 0.5–1(–3+)

mm. **Calyculi** of 7–8+ linear to narrowly triangular bractlets (2–)4–8+ mm. **Ray florets** 8; laminae yellow to golden yellow (sometimes proximally suffused with red-brown), 12–20+ mm. **Disc corollas** red-brown or yellow with red-brown nerves, throats shorter than lobes. **Cypselae** 3.5–4+ mm; **pappi** 0.5–1(–2+) mm. $2n = 16, 18$.

Flowering Mar–Aug(–Oct). Disturbed sites on clays or sandy soils, rocky slopes, often on limestone; 10–2200 m; Ark., Colo., Kans., La., Miss., Mo., Nebr., N.Mex., Okla., S.Dak., Tex., Wyo.; Mexico (Nuevo León).

As here circumscribed, *Thelesperma filifolium* includes plants that others have treated as a distinct species or variety: *T. intermedium* or *T. filifolium* var. *intermedium*, characterized as plants mostly 10–40 cm (versus taller); internodes "relatively short" (versus longer); calyculus bractlets mostly 1/4–1/2 lengths of phyllaries (versus more than 1/2 as long); ray corollas yellow (versus "golden yellow"); disc corollas yellow (versus sometimes red-brown); distribution mostly north and west of the typical form (Colorado, Kansas, Nebraska, New Mexico, w Oklahoma, South Dakota, c and w Texas, Wyoming).

According to A. Cronquist (1980), *Thelesperma trifidum* (Poiret) Britton has been misapplied (e.g., M. L. Fernald 1950) to *T. filifolium*.

311. COSMOS Cavanilles, Icon. 1: 9, plate 14. 1791 • [Greek *kosmos*, harmoniously ordered universe, or *kosmo*, ornament]

Robert W. Kiger

Annuals [perennials or subshrubs], 30–250 cm. **Stems** usually 1, erect or ascending, branched distally or ± throughout. **Leaves** mostly cauline; opposite; petiolate or sessile; blades usually 1–3-pinnately lobed [undivided], ultimate margins usually entire, faces usually glabrous, sometimes glabrate, hispid, puberulent, or scabridulous. **Heads** radiate, borne singly or in corymbiform arrays. **Calyculi** of [5–]8 basally connate, ± linear to subulate, herbaceous (striate) bractlets. **Involucres** hemispheric or subhemispheric [cylindric], 3–15 mm diam. **Phyllaries** persistent, [5–]8 in ± 2 series, distinct, lanceolate, lance-oblong, lance-ovate, or oblong, ± equal, membranous or herbaceous, margins ± scarious. **Receptacles** flat, paleate; paleae falling, linear, flat or slightly concave-convex, scarious (entire). **Ray florets** [0, 5] 8 (more in "double" cultivars), neuter; corollas white to pink or purple, or yellow to red-orange. **Disc florets** 10–20[–80+], bisexual, fertile; corollas yellow [orange] (at least distally), tubes shorter than funnelform throats, lobes 5, ± deltate (staminal filaments hairy near anthers; style branches linear, flattened, thicker distally, hirtellous, appendages relatively slender). **Cypselae** (dark brown or black) relatively slender, quadrangular-cylindric or -fusiform [outer somewhat obcompressed], sometimes slightly arcuate, attenuate-beaked, not winged [winged], faces glabrous or hispid to scabridulous or ± setose, sometimes papillate, usually with 1 groove; **pappi** persistent [falling], of 2–4[–8] retrorsely [antrorsely] barbed awns, sometimes 0. $x = 12$.

Species ca. 26 (4 in the flora): tropical and subtropical America, especially Mexico, widely introduced elsewhere.

SELECTED REFERENCES Pandey, A. K., S. Chopra, and R. P. Singh. 1986. Development and structure of seeds and fruits in Compositae: *Cosmos* species. J. Indian Bot. Soc. 65: 362–368. Sherff, E. E. 1955. *Cosmos*. In: N. L. Britton et al., eds. 1905+. North American Flora. 47+ vols. New York. Ser. 2, part. 2, pp. 130–146.

1. Ray laminae yellow to red-orange . . . . 4. *Cosmos sulphureus*
1. Ray laminae pink, purple, purplish, rose-pink, violet, or white.
   2. Leaves: ultimate lobes 2–10 mm wide . . . . 2. *Cosmos caudatus*
   2. Leaves: ultimate lobes to 1.5 mm wide.
      3. Ray laminae 15–50 mm . . . . 1. *Cosmos bipinnatus*
      3. Ray laminae 5–9 mm . . . . 3. *Cosmos parviflorus*

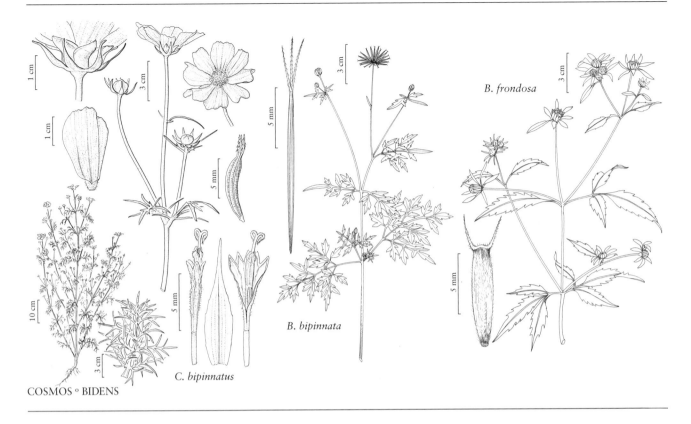

B. frondosa

B. bipinnata

C. bipinnatus

COSMOS ○ BIDENS

1. **Cosmos bipinnatus** Cavanilles, Icon. 1: 10, plate 14. 1791 [F]

Plants 30–200 cm, glabrous or sparsely puberulent, sometimes scabridulous. **Leaves:** petioles 0 or to 1 cm; blades 6–11 cm, ultimate lobes to 1.5 mm wide, margins entire, apices acute (indurate). **Peduncles** 10–20 cm. **Calyculi** of spreading, linear to lanceolate bractlets 6–13 mm, apices acuminate. **Involucres** 7–15 mm diam. **Phyllaries** erect, lanceolate to ovate-lanceolate, 7–13 mm, apices round or obtuse. **Ray corollas** white, pink, or purplish, laminae obovate to oblanceolate, 15–50 mm, apices ± truncate, dentate. **Disc corollas** 5–7 mm. **Cypselae** 7–16 mm, glabrous, papillose; **pappi** 0, or of 2–3 ascending to erect awns 1–3 mm. **2n** = 24.

Flowering summer–fall. Disturbed sites, roadsides; 0–1000 m; Ont., Que.; Ala., Ariz., Ark., Calif., Colo., Conn., Del., Fla., Ga., Ill., Kans., Ky., La., Maine, Md., Mass., Mich., Minn., Mo., N.H., N.J., N.Mex., N.Y., N.C., Ohio, Okla., Pa., R.I., S.C., Tenn., Tex., Utah, Va., W.Va., Wis., Wyo.; Mexico; introduced in West Indies, Central America, South America, Asia.

Cosmos bipinnatus is native to Mexico and the southwestern United States. A garden favorite, it has escaped and naturalized widely elsewhere in the flora area (and in warm climates almost worldwide), and it has been seeded along roadsides by some highway departments. Many cultivated races and hybrids differ considerably from the wild type described above, varying widely in stature and in coloration of both ray and disc corollas. Some plants in cultivation lack pappi; they are referable to var. *exaristatus* de Candolle, not treated formally here.

SELECTED REFERENCES Bate Smith, E. C. 1980. Astringent tannins of *Cosmos bipinnatus*. Phytochemistry 19: 982.   Pillai, A., S. K. Pillai, and O. Jacob. 1975. Embryogeny, histogenesis and apical meristems of *Cosmos bipinnatus* Cav. Acta Bot. Indica 3: 68–78.

2. **Cosmos caudatus** Kunth in A. von Humboldt et al., Nov. Gen. Sp. 4(fol.): 188. 1818; 4(qto.): 240. 1820 [I]

Plants 30–250 cm, glabrous or sparsely hispid. **Leaves:** petioles 1–7 cm; blades 10–20 cm, ultimate lobes 2–10 mm wide, margins spinulose-ciliate, apices acute, often mucronulate. **Peduncles** 10–30 cm. **Calyculi** of usually spreading, linear-subulate bractlets 6–10 mm, apices acuminate. **Involucres** 5–15 mm diam. **Phyllaries** erect, oblong-lanceolate, 7–11 mm, apices acute to obtuse. **Ray corollas** rose-pink to purple, laminae oblong-oblanceolate, 5–15 mm, apices obtusely 3-lobed. **Disc corollas** 5–6 mm. **Cypselae**

12–35 mm, glabrous or scabridulous proximally, setose distally; **pappi** of 2–3 widely divergent to reflexed awns 3–5 mm. $2n = 48$.

Flowering summer–fall. Disturbed sites; 0–10 m; introduced; Fla.; Mexico; West Indies; Central America; South America; also introduced in Asia, Pacific Islands.

Within the flora area, *Cosmos caudatus* is found only in the Florida Keys.

SELECTED REFERENCE Melchert, T. E. 1990. *Cosmos caudatus* (Asteraceae: Coreopsideae) in Mexico: A cytotaxonomic reappraisal. Phytologia 69: 200–215.

3. **Cosmos parviflorus** (Jacquin) Persoon, Syn. Pl. 2: 477. 1807 (as Cosmus)

Coreopsis parviflora Jacquin, Pl. Hort. Schoenbr. 3: 65, plate 374. 1798

Plants 30–90 cm, glabrous or sparsely pubescent. **Leaves:** petioles ca. 0.5 cm; blades 2.5–6.5 cm, ultimate lobes to 1 mm wide, margins usually spinulose-ciliate, apices acute to obtuse. **Peduncles** 10–30 cm. **Calyculi** of spreading to reflexed, linear-oblong to narrowly lanceolate bractlets 6–9 mm, apices acute. **Involucres** (3–)5–10(–15) mm diam. **Phyllaries** erect, oblong, 5–8 mm, apices acute. **Ray corollas** white to rose-pink or violet, laminae cuneate-obovate, 5–9 mm, apices dentate. **Disc corollas** 4–5 mm. **Cypselae** 9–16 mm, setulose; **pappi** of 2–4 erect awns 2–3 mm. $2n = 24$.

Flowering summer–fall. Open or forested slopes and canyons, sometimes in disturbed and cultivated areas; 100–3000 m; Ariz., Colo., Md., Mass., Mo., N.Mex., R.I., Tex.; Mexico.

Within the flora area, *Cosmos parviflorus* is native only in the southwestern United States.

4. **Cosmos sulphureus** Cavanilles, Icon. 1: 56, plate 79. 1791 ☐

Plants 30–200 cm, glabrous or sparsely pilose to hispid. **Leaves:** petioles 1–7 cm; blades 5–12(–25) cm, ultimate lobes 2–5 mm wide, margins sparsely spinulose-ciliate, apices apiculate. **Peduncles** 10–20 cm. **Calyculi** of spreading-ascending, linear-subulate bractlets 5–7 (–10) mm, apices acute; **Involucres** 6–10 mm diam. **Phyllaries** erect, oblong-lanceolate, 9–13(–18) mm, apices acute to rounded-obtuse. **Ray corollas** intensely yellow to red-orange, laminae obovate, 18–30 mm, apices ± truncate, denticulate. **Disc corollas** 6–7 mm. **Cypselae** 15–30 mm, usually hispidulous, rarely glabrous; **pappi** 0, or of 2–3 widely divergent awns 1–7 mm. $2n = 24, 48$.

Flowering summer–fall. Disturbed sites; 0–1000 m; introduced; Ala., Ark., Calif., Conn., Del., Fla., Ga., Ill., La., Md., Mich., Mo., N.Y., N.C., Okla., Pa., S.C., Tenn., Tex., Va.; Mexico; also introduced in West Indies, Central America, South America, Asia, Africa, Pacific Islands.

312. **BIDENS** Linnaeus, Sp. Pl. 2: 831. 1753; Gen. Pl. ed. 5, 362. 1754 • Beggar-ticks, bident, fourchette [Latin *bis*, two, and *dens*, tooth, alluding to 2-awned pappi of the original species]

John L. Strother

Ronald R. Weedon

Megalodonta Greene

Annuals or perennials [shrubs, vines], 5–400 cm. **Stems** usually 1, usually erect, (terete or 4-angled, often striate or sulcate) branched distally or ± throughout. **Leaves** usually cauline; usually opposite, rarely whorled, distal sometimes alternate; petiolate or sessile; blades simple, compound (leaflets petiolulate), or 1–3+-pinnatisect or -pinnately lobed (submerged leaves multifid in *B. beckii*, an aquatic), ultimate margins entire, dentate, laciniate, serrate, or toothed, faces usually glabrous, sometimes hirtellous, hispidulous, pilosulous, puberulent, scabrellous, or strigillose. **Heads** usually radiate or discoid, sometimes ± disciform, usually in corymbiform arrays, sometimes in 2s or 3s or borne singly. **Calyculi** of (3–)5–13(–21+) erect to spreading or reflexed, ± herbaceous (sometimes foliaceous) bractlets or bracts (sometimes surpassing phyllaries). **Involucres** mostly hemispheric or campanulate to cylindric, (1–)4–12(–25+) mm diam.

Phyllaries persistent, mostly (4–)8–21(–30+) in ± 2 series, usually distinct, sometimes connate 0.05–0.1 their lengths, mostly oblong or ovate to lance-oblong, chartaceous to membranous or scarious (usually striate with brownish nerves, margins usually hyaline). **Receptacles** flat or slightly convex, paleate; paleae usually falling, (usually stramineous, sometimes yellow to orange, with darker striae) ± flat to slightly navicular. **Ray florets** usually 1–21+ (often 3, 5, 8, or 13), sometimes 0, usually neuter, sometimes styliferous and sterile; corollas usually yellow, sometimes white or pinkish. **Disc florets** (5–)12–60(–150+), bisexual, fertile; corollas usually yellow to orange, sometimes whitish [purplish], tubes shorter than throats, lobes (3–)5, ± deltate (staminal filaments glabrous; style-branch appendages deltate or lanceolate to subulate). **Cypselae** usually obcompressed to flat, unequally 3–4-angled, and cuneate to oblanceolate or obovate, sometimes (all or inner) ± equally 4-angled and linear-fusiform, rarely subterete, margins (± corky-winged in *B. aristosa*, *B. cernua*, and *B. polylepis*) usually retrorsely, sometimes patently or antrorsely, barbed or ciliate, apices sometimes attenuate, not beaked [beaked], faces smooth, striate, or ± tuberculate, glabrous or hairy, each sometimes with 2 grooves; **pappi** 0, or persistent, of (1–)2–4(–8) usually retrorsely, sometimes antrorsely, barbellate or ciliate, rarely smooth, awns. *x* = 13.

Species 150–250+ (25 in the flora): widespread, especially in subtropical, tropical, and warm-temperate North America and South America.

Within species of *Bidens*, leaves may be simple, 1-pinnately compound, or 1–3-pinnatisect or -pinnately or -ternately lobed. For compound leaves, leaflets are described; for leaves mostly 1-pinnately lobed, primary lobes are described; for leaves 2–3-pinnatisect or -pinnately lobed, the ultimate lobes are described. Involucres in *Bidens* species are subtended by more or less herbaceous (sometimes foliaceous) bractlets or bracts; collectively, they constitute calyculi and they often well surpass the phyllaries. Together, phyllaries constitute involucres. In keys and descriptions here, shapes, heights, and diameters given for involucres are for involucres at flowering; involucres are sometimes notably larger in fruit. In some *Bidens* species, the outer and inner cypselae are obcompressed (unequally 3–4-angled) to flat, are mostly cuneate to obovate or nearly linear (broadest at or near apices), and are more or less similar except for sizes (outer smaller); in other species, the outer cypselae are obcompressed to flat and obovate to cuneate or linear and the inner are more or less equally 4-angled (in cross section) and linear-fusiform (thickest near their middles and more or less attenuate toward their tips).

SELECTED REFERENCES   Hall, G. W. 1967. A Biosystematic Study of the North American Complex of the Genus *Bidens* (Compositae). Ph.D. dissertation. Indiana University.   Sherff, E. E. 1937. The genus *Bidens*. Publ. Field Mus. Nat. Hist., Bot. Ser. 16.   Weedon, R. R. 1973. Taxonomy and Distribution of the Genus *Bidens* (Compositae) in the North-central Plains States. Ph.D. dissertation. University of Kansas.

1. Leaves (some or most) multifid (aquatics, multifid leaves usually submersed, ultimate lobes mostly 0.1–0.3 mm diam.); pappus awns 13–25(–40) mm . . . . 1. *Bidens beckii*
1. Leaves simple, 1-pinnately compound, or 1–3-pinnatisect or -pinnately lobed (plants sometimes rooted in mud, submersed leaves usually 0; leaflets or lobes mostly 0.5–40+ mm wide); pappus awns 0 or (0.1–)2–4(–8.5+) mm.
   2. All or inner cypselae ± equally 4-angled, mostly linear-fusiform (usually thickest in middle ¹/₂ and distally ± attenuate proximally and distally, each of 4 faces usually 2-grooved).
      3. Involucres ± campanulate to cylindric, 1–3(–4) mm diam.; disc florets (3–)5–9 (–13).
         4. Leaves: primary lobes lance-rhombic or ovate to lanceolate, 15–30(–50+) × 5–15(–30+) mm; pappi (1–)2–4 mm . . . . 6. *Bidens bigelovii* (in part)
         4. Leaves: primary or ultimate lobes ovate, oblanceolate, lanceolate, or spatulate to linear or filiform, 2–25+ × 0.5–5(–15) mm; pappi 0 or 1–3 mm.

5. Calyculus bractlets spatulate to linear (or 3–5+ lobed, some ± foliaceous),
   3–10(–25+) mm (usually surpassing phyllaries) . . . . 2. *Bidens lemmonii*
5. Calyculus bractlets linear to subulate or filiform (not foliaceous), 1–4 mm
   (seldom surpassing phyllaries).
     6. Leaves: ultimate lobes 0.5–1(–2) mm wide; cypselae usually glabrous
       . . . . 3. *Bidens heterosperma*
     6. Leaves: ultimate lobes (1–)2–3(–15) mm wide; cypselae ± hispidulous
       (at least distally) . . . . 4. *Bidens leptocephala*
3. Involucres campanulate to turbinate or hemispheric or broader, (3–)4–8 mm diam.;
   disc florets (5–)20–80+.
     7. Leaves: ultimate lobes ± linear, (2–)5–15+ × (0.5–)2–3+ mm . . . . 5. *Bidens tenuisecta*
     7. Leaves: blades simple, or primary or ultimate lobes lance-rhombic, lance-deltate,
       ovate, or lanceolate, (10–)15–30(–90+) × (5–)10–15(–40+) mm.
       8. Leaf blades either ovate to lanceolate and 30–70(–120) × 12–18(–45) mm,
         or 1-pinnately lobed, primary lobes 3–7, ovate to lanceolate, (10–)25–80+
         × (5–)10–40+ mm; disc florets 20–40(–80+) . . . . 8. *Bidens pilosa*
       8. Leaf blades rounded-deltate to ovate or lanceolate overall, (20–)25–90 ×
         15–35(–60+) mm, usually 2–3-pinnatisect, ultimate lobes lanceolate, lance-
         rhombic, obovate, or ovate, 15–30(–50+) × 5–15(–30+) mm; disc florets
         10–20(–30+).
         9. Outer cypselae 6–7 mm, inner cypselae 10–14 mm; pappi of 2(–3) awns
           . . . . 6. *Bidens bigelovii* (in part)
         9. Outer cypselae 7–15 mm, inner cypselae 12–18 mm; pappi of (2–)3–4
           awns . . . . 7. *Bidens bipinnata*
[2. Shifted to left margin.—Ed.]
2. All cypselae ± flattened (if 3–4-angled, 2 angles ± acute and 1–2 ± obtuse), usually cuneate,
   linear, oblanceolate, or obovate, seldom ± linear-fusiform (usually broadest in distal ¹⁄₃).
10. Blades of all or most leaves either 1-pinnately compound (leaflets 3–5+, petiolulate) or
   1–2-pinnately lobed (ultimate lobes lanceolate or oblanceolate to lance-linear or linear,
   usually 1–12 mm, sometimes to 25 mm wide); calyculus bractlets seldom foliaceous.
   11. Ray florets 0, or 1–3+, or 3–5, laminae 2–3.5 mm.
     12. Calyculus bractlets (3–)4(–5+), seldom ciliate; disc florets (10–)15–20+ . . . . 9. *Bidens discoidea*
     12. Calyculus bractlets 5–21, usually ciliate; disc florets 20–150.
       13. Leaves usually 3(–5)-foliolate (leaflets lanceolate to lance-ovate); calyculus
         bractlets (5–)8(–10) . . . . 10. *Bidens frondosa*
       13. Leaves usually laciniately 1-pinnatisect or 2–5-foliolate (leaflets or lobes
         ± lanceolate), rarely 2–3-pinnatisect; calyculus bractlets 10–16(–21) . . . 11. *Bidens vulgata*
   11. Ray florets (5–)8–13, laminae 10–30 mm.
     14. Phyllaries ovate to lanceolate (margins usually yellow); ray florets 5–6;
       margins of cypselae not barbed or ciliolate . . . . 12. *Bidens aurea*
     14. Phyllaries lance-linear, lanceolate, lance-ovate or oblong (margins not notably
       yellow); ray florets (5–)8–13; margins of cypselae usually barbed or ciliate
       (except *B. mitis*).
       15. Cypselae narrowly cuneate (lengths mostly 2.5–4 times widths) . . 13. *Bidens trichosperma*
       15. Cypselae broadly cuneate, oblanceolate, or obovate (lengths mostly 1.5–
         2.5 times widths).
         16. Cypselae 2.5–5 mm, margins (not winged) not barbed or ciliate, faces
           glabrous or hirtellous; pappi 0, or of 2 antrorsely barbed awns or
           ± deltate scales 0.5(–1) mm . . . . 14. *Bidens mitis* (in part)
         16. Cypselae (4–)5–8 mm, margins (often ± corky-winged) usually barbed
           or ciliate, faces glabrate, glabrous, or strigillose; pappi 0, or of 2
           (–4) antrorsely or retrorsely barbed awns or scales 0.1–6 mm.
           17. Calyculus bractlets 8–12(–16), (4–)5–7(–12) mm . . . . 15. *Bidens aristosa*
           17. Calyculus bractlets 12–21+, (6–)8–12(–20+) mm . . . . 16. *Bidens polylepis*
10. Blades of all or most leaves simple, margins ± dentate to serrate or margins coarsely
   incised (lobes 3–7+); calyculus bractlets or bracts often foliaceous.

[18. Shifted to left margin.—Ed.]

18. Leaves sessile.
    19. Calyculus bractlets (3–)8–12(–25+) mm, or (6–)10–12(–20+) mm, usually spreading to
        reflexed.
        20. Rays usually 6–8, sometimes 0, laminae 2–15(–18) mm; outer cypselae (3–)5–6+
            mm, inner 4–8 mm (margins ± thickened or winged); pappi of (2–)4 awns (1–)2–4
            mm (paleae stramineous to yellowish at tips) . . . .   17.   *Bidens cernua*
        20. Rays usually 7–8, rarely 0, laminae (10–)15–25(–30) mm; outer cypselae 6–8 mm,
            inner cypselae 8–10 mm (margins not notably thickened or winged); pappi of 2–4
            awns 3–5 mm (paleae usually orange at tips) . . . .   18.   *Bidens laevis*
    19. Calyculus bractlets (6–)10–30(–75+) mm, usually erect, sometimes spreading.
        21. Involucres (4–)7–10 mm diam.; calyculus bractlets erect; disc florets 15–30(–60);
            cypselae narrowly cuneate (faces ± 1-nerved and ± striate) . . . .   19.   *Bidens hyperborea* (in part)
        21. Involucres (3–)6–12(–15+) or 18–25 mm diam.; calyculus bractlets ± erect to spread-
            ing or reflexed; disc florets (5–)25–60(–150+); cypselae cuneate to oblanceolate or
            linear (faces usually smooth or weakly striate, sometimes tuberculate).
            22. Involucres 6–10 × 18–25 mm; ray florets usually 6–11 . . . .   20.   *Bidens amplissima* (in part)
            22. Involucres (4–)5–7(–12) × (3–)6–12(–15+) mm; ray florets usually 0,
                sometimes 1–5.
                23. Cypselae ± flattened, sometimes weakly 3(–4)-angled, faces usually smooth,
                    seldom notably tuberculate (disc corollas gradually ampliate, anthers
                    usually pale) . . . . 22.   *Bidens tripartita* (in part)
                23. Cypselae (at least inner) usually ± 4-angled, faces usually tuberculate (disc
                    corollas abruptly ampliate, anthers usually dark, blackish) . . .   23.   *Bidens connata* (in part)
18. Leaves petiolate, petioles (5–)10–25(–60) mm (sometimes ± winged).
    24. Ray florets 8–13 (laminae 12–25+ mm); cypselae 2.5–4.5 mm (margins not barbed
        or ciliate) . . . .   14.   *Bidens mitis* (in part)
    24. Ray florets 0 or 1–11 (laminae 2–12 mm); cypselae (3–)6–13 mm (margins sometimes
        barbed or ciliate).
        25. Involucres ± hemispheric or broader, 6–10 × 18–25+ mm; ray florets 6–11 (laminae
            4–8 mm) . . . .   20.   *Bidens amplissima* (in part)
        25. Involucres cylindric, campanulate, or hemispheric or broader, 4–10(–16) × (3–)5–
            8(–15) mm; ray florets 0 or 1–7 (laminae 2–12 mm).
            26. Involucres campanulate to hemispheric or broader; disc florets (5–)20–60
                (–150+); cypsela faces usually smooth or tuberculate (not notably striate).
                27. Cypselae ± flattened, sometimes weakly 3(–4)-angled, faces usually smooth,
                    seldom notably tuberculate (disc corollas gradually ampliate, anthers
                    usually pale) . . . . 22.   *Bidens tripartita* (in part)
                27. Cypselae (at least inner) usually ± 4-angled, faces usually tuberculate (disc
                    corollas abruptly ampliate, anthers usually dark, blackish) . . . .   23.   *Bidens connata* (in part)
            26. Involucres usually campanulate to cylindric, sometimes ± hemispheric; disc
                florets (6–)10–25(–60); cypsela faces usually ± striate, sometimes tuberculate.
                28. Petioles ± winged; involucres usually ± campanulate to cylindric,
                    sometimes ± hemispheric, 5–7(–9) × (4–)7–10 mm; disc florets 15–30(–60)
                    . . . .   19.   *Bidens hyperborea* (in part)
                28. Petioles not notably winged; involucres mostly campanulate to cylindric,
                    (4–)8–10(–16) × (4–)7–9(–12) mm; disc florets 6–20(–35+).
                    29. Cypselae narrowly cuneate to linear, margins evenly antrorsely
                        strigillose, faces ± evenly antrorsely strigillose . . . . 21.   *Bidens bidentoides*
                    29. Cypselae broadly to narrowly cuneate or oblanceolate to linear,
                        margins usually barbed or ciliate (proximally antrorsely, distally
                        retrorsely), faces glabrous or sparsely setulose.
                        30. Involucres (4–)8–10(–12) × (4–)7–9 mm; cypselae: outer 5.5–10
                            mm, inner 6.5–11 mm . . . .   24.   *Bidens eatonii*
                        30. Involucres (4–)6–7 × (4–)5–7+ mm; cypselae: outer 4–5(–6) mm,
                            inner (5–)6–7(–8+) mm . . . .   25.   *Bidens heterodoxa*

1. **Bidens beckii** Torrey ex Sprengel, Neue Entd. 2: 135. 1821 • Bident de Beck E

Megalodonta beckii (Torrey ex Sprengel) Greene

Perennials (perhaps flowering first year) to 200+ cm (aquatics). **Leaves** sessile; submersed blades multifid (2–3+-pinnatisect), ultimate lobes filiform, mostly 0.1–0.3 mm diam.; aerial blades ovate to lanceolate, 10–45+ × 5–20+ mm, bases cuneate, margins pectinately incised to serrate or entire, not ciliate, apices obtuse to acuminate or attenuate, faces glabrous. **Heads** usually borne singly. **Peduncles** (10–)20–100 mm. **Calyculi** of 5–6 usually spreading, oblong to obovate bractlets 5–8 mm, margins entire, not ciliate, abaxial faces usually glabrous. **Involucres** ± hemispheric, 7–12 × 12–15 mm. **Phyllaries** 7–8+, ovate, 7–10 mm. **Ray florets** 8; laminae yellow, 10–15 mm. **Disc florets** 10–30+; corollas pale yellow, 5–6 mm. **Cypselae** (outer and inner ± alike) yellowish to greenish brown, nearly terete or weakly 4-angled, ± linear, 10–15 mm, margins not barbed or ciliate, apices truncate, faces smooth or ± striate, glabrous; **pappi** of 2(–6) divergent to patent, retrorsely barbed (on distal ¹/₄ or so) awns 13–25(–40) mm. $2n = 26$.

Flowering Jul–Sep. Still or slow-moving waters; 0–300+ m; B.C., Man., N.B., N.S., Ont., Que., Sask.; Conn., Idaho, Ill., Ind., Iowa, Maine, Mass., Mich., Minn., Mo., Mont., N.H., N.J., N.Y., Ohio, Oreg., Pa., R.I., Vt., Wash., Wis.

M. L. Roberts (1985) presented a strong case for treating *Bidens beckii* as *Megalodonta beckii*.

SELECTED REFERENCE Roberts, M. L. 1985. The cytology, biology and systematics of *Megalodonta beckii* (Compositae). Aquat. Bot. 21: 99–110.

2. **Bidens lemmonii** A. Gray in A. Gray et al., Syn. Fl. N. Amer. 1(2): 297. 1884 (as lemmoni)

Annuals, (10–)15–25(–30+) cm. **Leaves:** petioles 10–20 mm; blades either oblanceolate to linear, 5–15+ × 1–2+ mm, or rounded-deltate overall, 10–25(–60+) × 15–25+ mm, (1–)2–3-pinnatisect, ultimate lobes oblanceolate to spatulate or linear, 5–15+ × 0.5–5 mm, bases ± cuneate, ultimate margins entire, sometimes ciliolate, apices obtuse to acute, faces glabrous. **Heads** usually borne singly, sometimes in open, ± corymbiform arrays. **Peduncles** 10–20(–90) mm. **Calyculi** of (1–)3–4 appressed to spreading, spatulate to linear bractlets or bracts 3–10(–25+) mm (sometimes foliaceous: pinnate, lobes 3–5+, linear), margins ciliolate, abaxial faces usually glabrous, sometimes sparsely hispidulous. **Involucres** ± campanulate to cylindric, 2–3(–8) × 2(–3)[–4] mm. **Phyllaries** (3–)5, oblong to lanceolate or linear, (2–)3–8 mm. **Ray florets** 0 or 1(–3+); laminae whitish, 1–1.5(–3+) mm. **Disc florets** (3–)5–9; corollas whitish to yellowish, 2–2.5 mm. **Cypselae:** outer red-brown (sometimes with lighter blotches), ± equally 4-angled, linear-fusiform, 5–6(–8) mm, margins not ciliate, apices ± attenuate, faces 2-grooved, usually glabrous; inner similar, 10–14 mm; **pappi** of 2–3 erect, retrorsely barbed awns 1–2(–3) mm.

Flowering Sep–Oct. Wettish spots on rocky slopes; 1400–2100 m; Ariz., N.Mex.; Mexico.

3. **Bidens heterosperma** A. Gray, Smithsonian Contr. Knowl. 5(6): 90. 1853

Annuals, (5–)10–30(–60+) cm. **Leaves:** petioles 3–20 mm; blades usually rounded-deltate overall, 10–30+ × 5–20+ mm, 1–2-pinnatisect, ultimate lobes linear to filiform, 2–25 × 0.5–1(–2) mm, bases obscurely cuneate, ultimate margins minutely, if at all, ciliate, apices obtuse to acute (sometimes apiculate), faces glabrous or sparsely hirtellous. **Heads** usually borne singly. **Peduncles** 10–50(–100) mm. **Calyculi** of (1–)3(–5), appressed to spreading, linear bractlets (1–)3(–4) mm, margins minutely, if at all, ciliate, abaxial faces usually glabrous, sometimes sparsely hispidulous. **Involucres** ± cylindric, 3.5–5 × (1–)2–3 mm. **Phyllaries** 5–6, lance-elliptic to lanceolate, 2.5–4(–5) mm. **Ray florets** 0 or 1–3+; laminae yellowish, 1–4 mm. **Disc florets** 5–9(–13); corollas yellowish, 1–1.5 mm. **Cypselae** red-brown to blackish (sometimes with lighter blotches), outer ± obcompressed and linear, 3–5 mm, margins not ciliate, apices ± truncate, faces 2-grooved, usually glabrous; inner similar, ± equally 4-angled, linear-fusiform, (6–)7–12 mm, apices ± attenuate; **pappi** 0, or of 2(–3) spreading to divergent, retrorsely barbed awns 1–2 mm.

Flowering Sep–Oct. Seeps on rocky slopes; 1500–2800 m; Ariz., Colo., N.Mex.; Mexico (Baja California, Chihuahua, Sinaloa, Sonora).

4. **Bidens leptocephala** Sherff, Bot. Gaz. 64: 22. 1917

Annuals, 10–25(–50) cm. **Leaves:** petioles 5–40 mm (usually hispido-ciliate); blades rounded-deltate overall, 15–50+ × 12–40+ mm, 1–2-pinnatisect, ultimate lobes ovate or lanceolate to linear, 3–25+ × (1–)2–3(–15) mm, bases ± cuneate, ultimate margins entire, ciliolate, apices acute to attenuate, faces glabrous or hirtellous. **Heads** usually borne singly, sometimes in open, ± corymbiform arrays. **Peduncles** (10–)30–80+ mm. **Calyculi** of 3–4(–6) ± appressed, subulate to filiform bractlets 1–3 mm, margins ciliolate, abaxial faces usually glabrous. **Involucres** ± cylindric, 2–3(–4+) × 1.5–2+ mm. **Phyllaries** 4–5(–7), oblong to lanceolate, (2–)3–4+ mm. **Ray florets** 0 or 2–3; laminae yellowish or whitish, 1–3 mm. **Disc florets** 5–13; corollas yellowish to whitish, 1.5–2 mm. **Cypselae:** outer red-brown, flat or unequally 4-angled, linear, 6–8 mm, margins not ciliate, apices somewhat attenuate, faces obscurely 2-grooved, antrorsely hispidulous toward tips; inner similar, blackish (sometimes with pale tips), ± equally 4-angled, linear-fusiform, 9–14 mm, margins not ciliate, apices attenuate, faces 2-grooved, antrorsely hispidulous (at least distally); **pappi** of 2(–3), erect, retrorsely barbed awns 1–3 mm.

Flowering Sep. Along streams; 900–1800 m; Ariz., N.Mex., Tex.; Mexico (Baja California, Chihuahua).

5. Bidens tenuisecta A. Gray, Mem. Amer. Acad. Arts, n. s. 4: 86. 1849

Annuals, (10–)20–40 cm. **Leaves:** petioles 5–40 mm; blades rounded-deltate to ovate overall, 20–50+ × 15–35+ mm, (1–)2–3-pinnatisect, ultimate lobes ± linear, (2–)5–15+ × (0.5–)2–3+ mm, bases ± cuneate, ultimate margins entire, sometimes ciliolate, apices acute to attenuate, faces usually glabrous, sometimes hispidulous. **Heads** usually borne singly, sometimes in 2s or 3s. **Peduncles** 20–30(–80) mm. **Calyculi** of 6–12 ± appressed, linear bractlets 5–7 mm, margins scabro-ciliate, abaxial faces usually pilosulous to hirsutulous, rarely glabrous. **Involucres** campanulate to hemispheric or broader, (5–)6–7 × 4–8 mm. **Phyllaries** 8, lance-linear, 5–7 mm. **Ray florets** 0 or 3–6; laminae yellow, 4–6 mm. **Disc florets** (5–)12–20(–45); corollas yellow, 2–3 mm. **Cypselae:** outer red-brown, flat or weakly 4-angled, linear, 6–8 mm, margins not ciliate, apices somewhat attenuate, faces obscurely 2-grooved, often tuberculate, glabrous or ± strigillose; inner similar, blackish (sometimes with pale tips), ± equally 4-angled, linear-fusiform, 8–15 mm, margins not ciliate, apices attenuate, faces 2-grooved, glabrous or, toward tips, strigillose; **pappi** of 2(–3), spreading to divergent, retrorsely barbed awns (1–)1.5–3 mm.

Flowering (Jul–)Aug–Sep(–Oct). Meadows, along streams; 1800–2600 m; Ariz., Colo., Idaho, N.Mex., Utah; Mexico (Chihuahua).

Reports of *Bidens tenuisecta* from Maryland, Massachusetts, and New York may be based on ephemeral introductions.

6. Bidens bigelovii A. Gray in W. H. Emory, Rep. U.S. Mex. Bound. 2(1): 91. 1859

Annuals, (10–)20–80+ cm. **Leaves:** petioles 5–25 mm; blades rounded-deltate overall, 25–90+ × 15–35+ mm, (1–)2(–3)-pinnatisect, ultimate lobes lance-rhombic or ovate to lanceolate, 15–30(–50+) × 5–15(–30+) mm, bases truncate to cuneate, ultimate margins entire or ± serrate to incised, usually ciliolate, apices obtuse to acuminate, faces glabrous. **Heads** usually borne singly, sometimes in open, ± corymbiform arrays. **Peduncles** (10–)30–50(–150+) mm. **Calyculi** of 8–13 usually spreading, narrowly lanceolate, oblanceolate, or subulate to linear bractlets 2–5 mm, margins entire, usually ciliate, abaxial faces usually glabrous. **Involucres** ± campanulate, 2.5–5+ × 1.5–3(–4) mm. **Phyllaries** 8–13+, lanceolate, (3–)4–6+ mm. **Ray florets** 0 or 1(–5+); laminae whitish, 1–3(–7) mm. **Disc florets** 13–25+; corollas yellowish, 1–2 mm. **Cypselae:** outer red-brown, obcompressed, cuneate, 6–7 mm, margins not ciliate, apices truncate, faces 2-grooved, antrorsely tuberculate-hirtellous; inner dark brown to blackish, ± equally 4-angled, linear-fusiform, 10–14 mm, margins not ciliate, apices ± attenuate, faces glabrous; **pappi** of 2(–3), erect, retrorsely barbed awns (1–)2–4 mm.

Flowering Sep. Along streams, other wettish sites; 900–2000 m; Ariz., Colo., N.Mex., Okla., Tex.; Mexico.

7. Bidens bipinnata Linnaeus, Sp. Pl. 2: 832. 1753 [F]

Bidens bipinnata var. *biternatoides* Sherff

Annuals, (15–)30–100(–150+) cm. **Leaves:** petioles 20–50 mm; blades rounded-deltate to ovate or lanceolate overall, (20–)30–70+ × (20–)30–60+ mm, (1–)2(–3)-pinnatisect, ultimate lobes obovate or lanceolate, 15–45+ × 10–25+ mm, bases truncate to cuneate, ultimate margins entire, sometimes ciliolate, apices rounded to acute or attenuate, faces usually glabrous, sometimes hirtellous. **Heads** usually borne singly, sometimes in ± corymbiform arrays. **Peduncles** (10–)20–50(–100) mm. **Calyculi** of (7–)8 (–10) linear bractlets 3–5 mm, ± appressed, margins ciliate, abaxial faces usually glabrous. **Involucres** ± campanulate, 5–7 × 3–4(–5) mm. **Phyllaries** 8–12, lanceolate to linear, 4–6 mm. **Ray florets** 0 or 3–5+; laminae yellowish or whitish, 1–2(–3) mm. **Disc florets** 10–20 (–30+); corollas yellowish to whitish, 2–3 mm. **Cypselae** red-brown, outer weakly obcompressed, 7–15 mm, inner ± 4-angled, linear to linear-fusiform, 12–18 mm, margins not ciliate, apices ± attenuate, faces 2-grooved, often tuberculate-hispidulous; **pappi** of (2–)3–4, erect to divergent, retrorsely barbed awns 2–4 mm. $2n = 24, 72$.

Flowering (Jun–)Aug–Sep(–Oct). Fields, forests, disturbed, wettish sites; 10–1800 m; N.B., Ont.; Ala., Ariz., Ark., Conn., D.C., Fla., Ga., Ill., Ind., Iowa, Kans., Ky., La., Md., Miss., Mo., Nebr., N.J., N.Mex., N.Y., N.C., Ohio, Okla., Pa., R.I., S.C., Tenn., Tex., Va., W.Va.; South America; Europe; Asia; Pacific Islands.

Bidens bipinnata is probably native in eastern Asia and introduced in South America, Europe, Asia, and Pacific Islands.

8. **Bidens pilosa** Linnaeus, Sp. Pl. 2: 832. 1753 • Bident poilu

Bidens alba (Linnaeus) de Candolle; *B. alba* var. *radiata* (Schultz-Bipontinus) R. E. Ballard; *B. odorata* Cavanilles; *B. pilosa* var. *radiata* (Schultz-Bipontinus) Schultz-Bipontinus

Annuals [perennials], (10–)30–60 (–180+)[–250] cm. **Leaves:** petioles 10–30(–70) mm; blades either ovate to lanceolate, 30–70(–120) × 12–18(–45) mm, or 1-pinnately lobed, primary lobes 3–7, ovate to lanceolate [linear], (10–)25–80+ × (5–)10–40+ mm [blades 2(–3)-pinnatisect], bases truncate to cuneate, ultimate margins serrate or entire, usually ciliate, apices acute to attenuate, faces pilosulous to sparsely hirtellous or glabrate. **Heads** usually borne singly, sometimes in open, ± corymbiform arrays. **Peduncles** 10–20(–90) mm. **Calyculi** (6–)7–9 (–13) ± appressed, spatulate to linear bractlets (3–)4–5 mm, margins ciliate, abaxial faces usually hispidulous to puberulent. **Involucres** turbinate to campanulate, 5–6 × (6–)7–8 mm. **Phyllaries** (7–)8–9(–13), lanceolate to oblanceolate, 4–6 mm. **Ray florets** 0 or (3–)5–8+; laminae whitish to pinkish [yellowish], 2–3 or 7–15+ mm. **Disc florets** 20–40(–80+); corollas yellowish, (2–)3–5 mm. **Cypselae:** outer red-brown, ± flat, linear to narrowly cuneate, (3–)4–5+ mm, margins antrorsely hispidulous, apices ± truncate or somewhat attenuate, faces obscurely 2-grooved, sometimes tuberculate-hispidulous; inner blackish, ± equally 4-angled, linear-fusiform, 7–16 mm, margins antrorsely hispidulous, apices ± attenuate, faces 2-grooved, tuberculate-hispidulous to sparsely strigillose; pappi 0, or of 2–3(–5), erect to divergent, retrorsely barbed awns (0.5–)2–4 mm. **2n = 24, 36, 48, 72.**

Flowering year round. Disturbed, wettish sites; 10–1900 m; Ont. Que.; Ala., Ariz., Calif., Conn., Fla., Ga., Ky., La., Md., Mass., Miss., Mo., N.Mex., N.C., Pa., S.C., Tex.; Mexico; Central America; introduced, Europe.

R. Ballard (1986) adopted a narrower circumscription of *Bidens pilosa* than that used here. He used: *B. pilosa* for plants with outer phyllaries 7–10, ray florets usually 0 (when present, laminae 2–3 mm), disc florets 35–75, pappi of 3(–5) awns 1–3 mm, and *2n = 72*; *B. alba* for plants with outer phyllaries (8–)12(–16), ray

florets 5–8 (laminae 5–16 mm), pappi of 2 awns 1–2 mm, and *2n = 48*; and *B. odorata* for plants with outer phyllaries (6–)8(–12), ray florets 5–8 (laminae 3–18 mm), disc florets 12–61, pappi 0, or of 1–2 awns 1–3 mm, and *2n = 24*.

SELECTED REFERENCE Ballard, R. 1986. *Bidens pilosa* complex (Asteraceae) in North and Central America. Amer. J. Bot. 73: 1452–1465.

9. **Bidens discoidea** (Torrey & A. Gray) Britton, Bull. Torrey Bot. Club 20: 281. 1893 • Bident discoïde [E]

Coreopsis discoidea Torrey & A. Gray, Fl. N. Amer. 2: 339. 1842

Annuals, (10–)20–60(–180) cm. **Leaves:** petioles 10–40(–60) mm; blades deltate to lance-ovate overall, 30–80(–100+) × 10–30(–80+) mm, usually 3-foliolate, leaflets petiolulate, lance-ovate to lanceolate, (10–)20–50(–100) × 5–20(–40) mm, bases cuneate, ultimate margins usually serrate, sometimes ciliate, apices acuminate to attenuate, faces glabrous or hirtellous. **Heads** usually borne singly, sometimes in 2s or 3s. **Peduncles** 10–20(–50+) mm. **Calyculi** of (3–)4(–5+) ± appressed, spatulate to linear, seldom foliaceous bractlets or bracts (3–)12–25 mm, margins seldom ciliate, abaxial faces usually glabrous. **Involucres** ± hemispheric, 4–5(–7) × 4–6(–9) mm. **Phyllaries** 5–7, oblong to lanceolate, 4–6+ mm. **Ray florets** 0. **Disc florets** (10–)15–20+; corollas ± orange, 1.5–2 mm. **Cypselae** blackish to red-brown or stramineous, ± flattened, linear to narrowly cuneate, outer 3–5 mm, inner 4–6+ mm, margins not distinctly ciliate, apices ± truncate to concave, faces ± tuberculate, antrorsely strigillose; pappi of 2 ± erect, antrorsely barbed or smooth awns (0.2–)1–2.4 mm. **2n = 24.**

Flowering Aug–Oct(–Nov). Ponds, swamps, other relatively wet sites; 10–300 m; N.B., N.S., Ont., Que.; Ala., Ark., Conn., Del., D.C., Fla., Ga., Ill., Ind., Iowa, Ky., La., Maine, Md., Mass., Mich., Minn., Miss., Mo., N.H., N.J., N.Y., N.C., Ohio, Okla., Pa., R.I., S.C., Tenn., Tex., Vt., Va., W.Va., Wis.

Bidens discoidea is not nearly as variable morphologically as the closely related *B. vulgata* or *B. frondosa*. It differs from *B. frondosa* in having smaller heads, cypselae with shorter, antrorsely barbed awns, fewer phyllaries and flowers per head, sparse indument, and leaves with fewer, more acuminate leaflets. Both *B. discoidea* and *B. frondosa* differ from *B. vulgata* in texture of flowers, in heads, and in shapes and surfaces of cypselae (M. L. Roberts 1982, 1983; M. G. Hickler 1999).

10. Bidens frondosa Linnaeus, Sp. Pl. 2: 832. 1753

• Bident feuillu F

Bidens frondosa var. *anomala* Porter ex Fernald; *B. frondosa* var. *caudata* Sherff; *B. frondosa* var. *pallida* (Wiegand) Wiegand; *B. frondosa* var. *stenodonta* Fernald & H. St. John; *B. melanocarpa* Wiegand

Annuals, (10–)20–60(–180) cm. **Leaves:** petioles 10–40(–60) mm; blades deltate to lance-ovate overall, 30–80(–150+) × 20–60(–100+) mm, 3(–5)-foliolate, leaflets petiolulate, lanceolate to lance-ovate, (15–)35–60(–120) × (5–)10–20(–30) mm, bases cuneate, margins dentate to serrate, sometimes ciliate, apices acuminate to attenuate, faces glabrous or hirtellous. **Heads** usually borne singly, sometimes in 2s or 3s or in open, corymbiform arrays. **Peduncles** 10–40(–80+) mm. **Calyculi** of (5–)8(–10) ascending to spreading, spatulate or oblanceolate to linear, sometimes ± foliaceous bractlets or bracts 5–20(–60) mm, margins usually ciliate, abaxial faces glabrous or hirtellous. **Involucres** campanulate to hemispheric or broader, 6–9 × 7–12 mm. **Phyllaries** 6–12, oblong or ovate to lance-ovate, 5–9 mm. **Ray florets** 0 or 1–3+; laminae golden yellow, 2–3.5 mm. **Disc florets** 20–60(–120+); corollas ± orange, 2.5–3+ mm. **Cypselae** blackish to brown or stramineous, ± obcompressed, obovate to cuneate, outer 5–7 mm, inner 7–10 mm, margins antrorsely or retrorsely barbed, apices ± truncate to concave, faces usually 1-nerved, sometimes tuberculate, glabrous or sparsely hirtellous; **pappi** of 2 ± erect to spreading, antrorsely or retrorsely barbed awns 2–5 mm. $2n$ = 24, 48, 72.

Flowering (Jun–)Aug–Sep(–Oct). Moist woods, meadows, thickets, fields, roadsides, railroads, borders of streams, ponds, sloughs, swamps, ditches; 10–2000 m; Alta., B.C., Man., N.B., Nfld. and Labr. (Nfld.), N.S., Ont., P.E.I., Que., Sask.; Ala., Alaska, Ariz., Ark., Calif., Colo., Conn., Del., D.C., Fla., Ga., Idaho, Ill., Ind., Iowa, Kans., Ky., La., Maine, Md., Mass., Mich., Minn., Miss., Mo., Nebr., Nev., N.H., N.J., N.Mex., N.Y., N.C., N.Dak., Ohio, Okla., Oreg., Pa., R.I., S.C., S.Dak., Tenn., Tex., Utah, Vt., Va., Wash., W.Va., Wis., Wyo.; Mexico; introduced, Europe.

Infusions and tinctures of *Bidens frondosa* are rated as outstanding herbal therapies for irritation, inflammation, pain, and bleeding of the urinary tract mucosa and are used for benign prostatic hypertrophy and increasing excretion of uric acid, decreasing the risk of gout attacks, as well as other medical uses (M. Moore 1993).

11. Bidens vulgata Greene, Pittonia 4: 72. 1899 (as vulgatus) • Bident vulgaire

Bidens frondosa Linnaeus var. *puberula* Wiegand; *B. puberula* (Wiegand) Rydberg; *B. vulgata* var. *dissectior* Sherff; *B. vulgata* var. *puberula* (Wiegand) Greene; *B. vulgata* var. *schizantha* Lunell

Annuals, (15–)30–50(–150) cm. **Leaves:** petioles 10–50 mm; blades ± deltate to ovate overall, 50–100(–150+) × (15–)30–80(–120+) mm, usually laciniately 1-pinnatisect or 3–5-foliolate, primary lobes or leaflets ± lanceolate, 20–80(–120) × 10–25(–40+) mm, blades rarely 2–3-pinnatisect, bases cuneate, ultimate margins dentate to serrate, little, if at all, ciliate, apices acute to attenuate, faces glabrous or ± hispidulous. **Heads** borne singly or in 2s or 3s or in open, corymbiform arrays, erect. **Peduncles** (10–)40–150+ mm. **Calyculi** of 10–16(–21) ascending to spreading, spatulate to linear, seldom foliaceous bractlets or bracts 10–20(–40) mm, margins usually hispid-ciliate, abaxial faces ± hispidulous. **Involucres** hemispheric or broader, 5–6 × 8–10 mm. **Phyllaries** 10–12, ovate to lanceolate, 6–9 mm. **Ray florets** 0 or 3–5+; laminae pale yellow, 2.5–3.5 mm. **Disc florets** 40–60(–150+); corollas yellow, 2.5–3.5 mm. **Cypselae** purplish, brown, olive, or stramineous, ± flattened, obovate to cuneate, outer 6–10 mm, inner 8–12 mm, margins (sometime ± winged) proximally antrorsely, distally retrorsely barbed, apices ± truncate, faces obscurely 1-nerved, sometimes tuberculate, glabrous or sparsely strigillose; **pappi** of 2 erect to divergent, retrorsely barbed awns 3–4(–7) mm. $2n$ = 24, 48.

Flowering Aug–Sep(–Oct). Ditches, shores of lakes and streams, swamps, marshes, moist woods, roadsides, railroads, fields, waste areas; 10–1000 m; Alta., B.C., Man., N.B., N.S., Ont., Que., Sask.; Ark., Calif., Colo., Conn., Del., D.C., Ga., Idaho, Ill., Ind., Iowa, Kans., Ky., Maine, Md., Mass., Mich., Minn., Miss., Mo., Mont., Nebr., Nev., N.H., N.J., N.Y., N.C., N.Dak., Ohio, Okla., Oreg., Pa., R.I., S.C., S.Dak., Tenn., Vt., Va., Wash., W.Va., Wis., Wyo.; introduced in Europe.

Bidens vulgata is similar to *B. frondosa*; it is more robust. Both are phenotypically plastic. Locally, *B. vulgata* often matures earlier than *B. frondosa*.

12. Bidens aurea (Aiton) Sherff, Bot. Gaz. 59: 313. 1915

Coreopsis aurea Aiton, Hort. Kew. 3: 252. 1789; *Bidens aurea* var. *wrightii* (A. Gray) Sherff; *B. ferulifolia* (Jacquin) de Candolle

Annuals [perennials], (10–)50–100[–250] cm. **Leaves:** petioles 20–40 mm; blades deltate or lanceolate to lance-linear overall, (30–)50–120(–220) × 10–50(–150) mm, sometimes either 1-pinnately lobed, primary lobes 3–5+, lanceolate to lance-linear, 30–120 × 3–25+ mm, or 2-pinnatisect, ultimate lobes linear, 5–30 × 1–2+ mm, bases truncate to cuneate, ultimate margins entire or serrate, seldom ciliate, apices obtuse to acuminate, faces glabrous or puberulent. **Heads** in ± corymbiform arrays. **Peduncles** (10–)40–60+ mm. **Calyculi** of 8–12(–17) usually erect, linear, seldom foliaceous bractlets 3–6 mm, margins ciliate, abaxial faces usually glabrous, sometimes hispidulous at bases. **Involucres** campanulate to hemispheric, 4–6 × 5–10 mm. **Phyllaries** 8–12, ovate to lanceolate, 4–5(–7) mm (margins usually yellow). **Ray florets** 5–6; laminae yellow, 10–30 mm. **Disc florets** 12–30(–60+); corollas yellowish, 3–4 mm. **Cypselae** dark brown to blackish, ± flattened, sometimes unequally 3–4-angled, ± linear, outer 4–6 mm, inner 5–7 mm, margins not barbed or ciliate, apices truncate, faces weakly 2-grooved, glabrous or sparsely strigillose, sometimes tuberculate; **pappi** 0, or of (1–)2(–4) erect to spreading, retrorsely barbed awns (1–)1.5–4 mm. *2n* = 24 (Mexico).

Flowering mostly Aug–Sep. Marshes, borders of streams, other wet sites; 900–2000 m; Ariz.; Mexico; Central America; introduced in Europe.

13. Bidens trichosperma (Michaux) Britton, Bull. Torrey Bot. Club 20: 281. 1893 [E]

Coreopsis trichosperma Michaux, Fl. Bor.-Amer. 2: 139. 1803; *Bidens coronata* (Linnaeus) Britton var. *brachyodonta* Fernald; *B. coronata* var. *tenuiloba* (A. Gray) Sherff; *B. coronata* var. *trichosperma* (Michaux) Fernald

Annuals (biennials), (15–)30–50 (–150+) cm. **Leaves:** petioles 5–20+ mm; blades ± deltate to ovate overall, 40–75(–150) × 10–45(–130) mm, usually laciniately 1–2-pinnatisect, ultimate lobes (3–)5–7+, oblanceolate or lanceolate to linear, (5–)10–30(–80+) × (1–)4–8(–12+) mm, bases cuneate, ultimate margins incised, dentate, serrate, or entire, little, if at all, ciliate, apices acute to attenuate, faces glabrous or ± hirtellous to strigillose. **Heads** usually in open, ± corymbiform arrays. **Peduncles** 20–150 mm. **Calyculi** of (6–)8(–11), ascending to spreading,

spatulate to linear, sometimes ± foliaceous bractlets or bracts 3–10(–18) mm, margins sometimes ciliate, abaxial faces glabrous. **Involucres** hemispheric or broader, 4–6(–8) × 6–12 mm. **Phyllaries** 6–8(–10), oblong, 3–8 mm. **Ray florets** (7–)8–9; laminae golden yellow, 10–30 mm. **Disc florets** 40–60(–80+); corollas yellow, 3–5 mm. **Cypselae** blackish or brown, flattened, narrowly cuneate, outer 3–6 mm, inner 5–9 mm (lengths mostly 2.5–4 times widths), margins antrorsely barbed or ciliate, apices ± truncate, faces obscurely 1-nerved, sometimes tuberculate, glabrous or sparsely hispidulous; **pappi** of 2 erect, ± patently barbed awns or scales (0.4–)1–2.5(–4) mm. *2n* = 24.

Flowering Aug–Oct. Marshes, estuaries; 0–300 m; Ont., Que.; Ala., Ark., Conn., Del., Fla., Ga., Ill., Ind., Iowa, Ky., Md., Mass., Mich., Minn., Miss., Mo., Nebr., N.J., N.Y., N.C., Ohio, Pa., R.I., S.C., S.Dak., Tenn., Va., W.Va., Wis.

Plants here called *Bidens trichosperma* have long been known as *B. coronata* (Linnaeus) Britton (or Britton ex Sherff). Alas, Britton's *B. coronata* (1913) is a later homonym of *B. coronata* Fischer ex Colla (1834) and cannot be used.

14. Bidens mitis (Michaux) Sherff, Bot. Gaz. 81: 43. 1926 [E]

Coreopsis mitis Michaux, Fl. Bor.-Amer. 2: 140. 1803

Annuals (sometimes persisting), (10–)30–100+ cm. **Leaves:** petioles 5–20(–30+) mm; blades ± deltate to ovate or lanceolate overall, 30–100 × 10–50+ mm, usually laciniately 1-pinnatisect, ultimate lobes 3(–7), ovate or lanceolate to linear, (5–)25–60(–70+) × (1–)5–10(–12+) mm, bases cuneate, ultimate margins serrate or entire, often ciliolate, apices acute to attenuate, faces glabrous or hirtellous. **Heads** usually in open, ± corymbiform arrays. **Peduncles** 20–120 mm. **Calyculi** of 7–10 spreading, spatulate to linear, seldom foliaceous (± fleshy) bractlets 5–10 mm, margins usually ciliate, abaxial faces glabrous or hirtellous. **Involucres** hemispheric or broader, 5–6 × 5–8 mm. **Phyllaries** (6–)8+, lance-ovate to lanceolate, 4–5 mm. **Ray florets** 8–13; laminae golden yellow, 12–25+ mm. **Disc florets** 25–50+; corollas yellowish, 2.5–3 mm. **Cypselae** blackish or brown, flattened or unequally (3–)4-angled, broadly cuneate, outer 2.5–4, inner 3.5–5 mm (lengths mostly 1.5–2.5 times widths), margins not barbed or ciliate, apices ± truncate, faces sometimes obscurely 1-nerved, sometimes tuberculate, glabrous or hirtellous; **pappi** 0, or of 2 erect, antrorsely barbed awns or ± deltate scales 0.5–(1) mm. *2n* = 24.

Flowering Sep–Nov. Marshes, borders of estuaries; 0–100(–300) m; Ala., Ark., Del., Fla., Ga., La., Md., Miss., Mo., N.J., N.C., S.C., Tenn., Tex., Va.

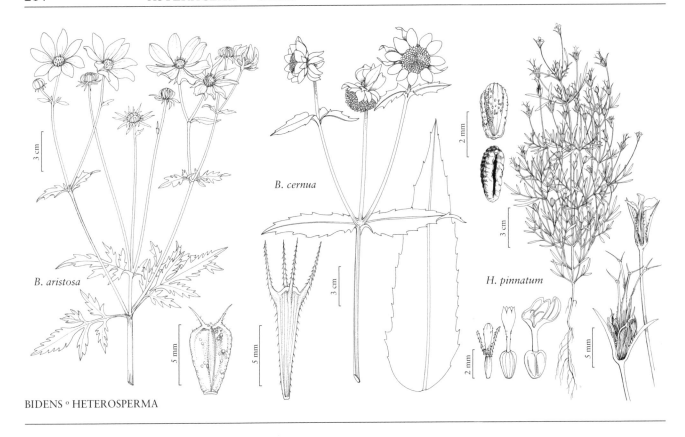

B. cernua

B. aristosa

H. pinnatum

BIDENS ○ HETEROSPERMA

15. Bidens aristosa (Michaux) Britton, Bull Torrey Bot. Club 20: 281. 1893 [E] [F]

Coreopsis aristosa Michaux, Fl. Bor.-Amer. 2: 140. 1803; *Bidens aristosa* var. *fritcheyi* Fernald; *B. aristosa* var. *mutica* (A. Gray) Gattinger ex Fernald

Annuals, (15–)30–60(–150) cm. **Leaves:** petioles 10–30 mm; blades lanceolate to lance-linear overall, 30–80(–150) × 10–30(–50) mm, usually laciniately pinnatisect, primary lobes 3–7+, 15–60+ × 2–20 mm, bases cuneate, ultimate margins laciniate to serrate, ciliate, apices acuminate, faces glabrous or sparsely scabrellous. **Heads** in ± corymbiform arrays. **Peduncles** (10–)30–80+ mm. **Calyculi** of 8–12(–16) usually spreading to reflexed, linear, seldom foliaceous bractlets (4–)5–7(–12) mm, margins entire, ciliate, abaxial faces glabrous or ± scabrellous. **Involucres** ± hemispheric or broader, 6–8(–12) × 10–15 mm. **Phyllaries** 7–8(–13), lance-ovate to lance-linear, 6–8(–12) mm. **Ray florets** (5–)8–10+; laminae yellow, 10–25 mm. **Disc florets** (12–)20–40+; corollas yellowish, 2–3+ mm. **Cypselae** red-brown to blackish, ± flattened (unequally 3–4-angled), broadly cuneate, outer (4–)5–6 mm, inner 5–7 mm (lengths mostly 1.5–2.5 times widths), margins (± corky-winged) patently to antrorsely barbed, apices truncate to concave, faces weakly striate, sometimes tuberculate, glabrous or sparsely strigillose; **pappi** 0, or

of 2(–4) spreading to divergent, antrorsely or retrorsely barbed awns (0.5–)2–4(–6) mm.

Flowering Aug–Oct(–Nov). Marshes, meadows, pine forests, disturbed sites; 0–300 m; Ont.; Ala., Ark., D.C., Ill., Ind., Ky., La., Md., Mich., Minn., Mo., N.C., S.C., Tenn., Tex., Va.

16. Bidens polylepis S. F. Blake, Proc. Biol. Soc. Wash. 35: 78. 1922 [E]

Coreopsis involucrata Nuttall, J. Acad. Nat. Sci. Philadelphia 7: 74. 1834, not *Bidens involucrata* Philippi 1891; *B. polylepis* var. *retrorsa* Sherff

Annuals (biennials), 30–100+ cm. **Leaves:** petioles 5–30 mm; blades ± deltate to ovate overall, 40–80 (–150+) × 25–50(–80+) mm, usually laciniately 1-pinnatisect, ultimate lobes 3(–7), lanceolate to linear, 25–60(–120+) × (3–)6–20(–25) mm, bases cuneate, ultimate margins incised, dentate, or serrate, ciliate, apices acute to attenuate, faces glabrous or hirtellous to scabrellous. **Heads** borne singly or in 2s or 3s or in open, ± corymbiform arrays. **Peduncles** 10–20(–50) mm. **Calyculi** of 12–21+ spreading to reflexed, linear, seldom foliaceous bractlets (6–)8–12(–20+) mm, margins coarsely ciliate, abaxial faces glabrous or scabrellous. **Involucres** turbinate to hemispheric, 4–5 × 6–10 mm. **Phyllaries** 6–8+, lance-ovate to lanceolate, 4–7 mm. **Ray florets** 8; laminae golden yellow, 10–25+ mm. **Disc florets** (40–)60–100+; corollas yellow, 2.5–3 mm. **Cypselae** blackish or brown to yellowish, ± flattened, obovate or oblanceolate to cuneate, outer 5–7 mm, inner 7–8 mm (lengths mostly 1.5–2.5 widths), margins (± corky-winged) antrorsely barbed or ciliate, apices ± truncate, faces glabrous or strigillose, sometimes tuberculate; **pappi** 0, or of 2 divergent, retrorsely barbed awns or ± deltate scales 0.1–0.5 (–1.5) mm. $2n = 24$.

Flowering Aug–Oct. Marshes, bogs, flood plains, disturbed sites; 10–1500 m; Ont.; Ala., Ark., Colo., Ill., Ind., Iowa, Kans., La., Md., Mich., Mo., Nebr., N.J., N.Mex., N.C., Okla., Pa., S.C., Tenn., Tex., Va.

17. **Bidens cernua** Linnaeus, Sp. Pl. 2: 832. 1753
   • Bident penché   F

Bidens cernua var. *elliptica* Wiegand; *B. cernua* var. *integra* Wiegand; *B. cernua* var. *minima* (Hudson) Pursh; *B. cernua* var. *oligodonta* Fernald & H. St. John; *B. cernua* var. *radiata* de Candolle; *B. filamentosa* Rydberg; *B. glaucescens* Greene; *B. gracilenta* Greene; *B. minima* Hudson; *B. prionophylla* Greene

Annuals, (2–)20–100(–400+) cm (stems sometimes rooting at proximal nodes). **Leaves** (rarely in 3s) sessile; blades lance-ovate or oblanceolate to lanceolate or linear, 40–100(–200+) × (2–)5–25(–45+) mm, bases cuneate to rounded, margins usually coarsely dentate to serrate, sometimes entire, sometimes ciliate, apices acute to acuminate, faces glabrous. **Heads** (erect or nodding at flowering, usually nodding in fruit) borne singly or in open, ± corymbiform arrays. **Peduncles** 10–40(–100+) mm. **Calyculi** of (3–)5–8(–10) spreading to reflexed, oblong to lance-linear, often ± foliaceous bractlets or bracts (3–)8–12(–25+) mm, margins usually ciliate, abaxial faces usually glabrous, bases sometimes hispidulous. **Involucres** ± hemispheric or broader, (3–)6–10 × (8–)12–20+ mm. **Phyllaries** 6–8+, ovate or lance-ovate to lanceolate, 2–10 mm. **Ray florets** usually 6–8, sometimes 0; laminae orange-yellow, 2–15(–18) mm. **Disc florets** (10–)40–100(–150+); corollas orange-yellow, 3–4 mm. **Cypselae** blackish or brown, usually ± flattened, sometimes ± 4-angled, ± cuneate, outer (3–)5–6+ mm, inner 4–8 mm, margins (± thickened or winged) retrorsely ciliate, apices ± truncate to convex, faces ± striate, glabrous or tuberculo-strigillose; **pappi** of (2–)4, ± erect, retrorsely barbed awns (1–)2–4 mm. $2n = 24, 48$.

Flowering late (Jul–)Aug–Sep(–Oct). Swamps, marshes, peat and sedge bogs, flood plains; 0–2300 m; Alta., B.C., Man., N.B., N.W.T., N.S., Ont., P.E.I., Que., Sask., Yukon; Ala., Alaska, Ariz., Ark., Calif., Colo., Conn., Del., D.C., Ga., Idaho, Ill., Ind., Iowa, Kans., Ky.,

La., Maine, Mass., Mich., Minn., Mo., Mont., Nebr., Nev., N.H., N.J., N.Mex., N.Y., N.C., Ohio, Okla., Oreg., Pa., R.I., S.Dak., Tenn., Tex., Utah, Vt., Va., Wash., W.Va., Wis., Wyo.; Europe, Asia.

Bidens cernua is used medicinally to treat urinary-tract infections.

18. **Bidens laevis** (Linnaeus) Britton, Sterns & Poggenberg, Prelim. Cat., 29. 1888

Helianthus laevis Linnaeus, Sp. Pl. 2: 906. 1753; *Bidens chrysanthemoides* Michaux; *B. helianthoides* Kunth; *B. nashii* Small; *B. quadriaristata* de Candolle

Annuals (sometimes persisting), (10–)20–60(–120+) cm. **Leaves** (sometimes in 3s or 4s) sessile; blades obovate or elliptic to lanceolate or linear, (20–)50–100(–160+) × (5–)10–25(–40+) mm, bases cuneate to rounded, margins usually coarsely dentate to serrate, sometimes ciliate, apices acute to acuminate, faces glabrous. **Heads** (erect at flowering, sometimes nodding in fruit) borne singly or in open, ± corymbiform arrays. **Peduncles** (10–)20–60 mm. **Calyculi** of 5–7(–9+) erect or spreading to reflexed, oblanceolate or lanceolate to linear, often foliaceous bractlets or bracts (6–)10–12(–20+) mm, margins usually ciliate, abaxial faces glabrous or bases hispidulous. **Involucres** turbinate to hemispheric or broader, (4–)6–8(–10+) × 8–12+ mm. **Phyllaries** 8–12, ovate or obovate to lance-oblong, (4–)6–8(–10+) mm (tips often orange to purplish, as are tips of paleae). **Ray florets** usually 7–8, rarely 0; laminae orange-yellow, (10–)15–25(–30) mm. **Disc florets** (25–)60–100(–150+); corollas yellow to orange-yellow, 3–6.5 mm. **Cypselae** blackish, red-brown, or stramineous, obcompressed, flattened or unequally 3–4-angled, ± cuneate, outer 6–8 mm, inner 8–10 mm, margins retrorsely ciliate or barbed, apices ± truncate to convex, faces ± 1-nerved, ± striate, glabrous; **pappi** of 2–4 ± erect, retrorsely barbed awns 3–5 mm. $2n = 22, 24$.

Flowering Aug–Oct(–Dec). Meadows, marshes, and margins of pools, streams, estuaries; 0–2800 m; Ala., Ariz., Ark., Calif., Colo., Conn., Del., D.C., Fla., Ga., Ind., Ky., La., Maine, Md., Mass., Miss., Mo., Nev., N.H., N.J., N.Mex., N.Y., N.C., Ohio, Pa., R.I., S.C., Tenn., Tex., Va., W.Va.; Mexico; Central America; South America; introduced Pacific Islands (Hawaii).

Bidens laevis and *B. cernua* are similar in gross appearance; specimens of one are sometimes misidentified as the other. Perhaps they represent extremes of a single species.

19. Bidens hyperborea Greene, Pittonia 4: 257. 1901

• Bident hyperboréal [E]

Bidens hyperborea var. *arcuans* Fernald; *B. hyperborea* var. *cathancensis* Fernald; *B. hyperborea* var. *colpophila* (Fernald & H. St. John) Fernald; *B. hyperborea* var. *gaspensis* Fernald; *B. hyperborea* var. *svensonii* Fassett

Annuals, (5–)10–30(–70+) cm. **Leaves** sessile or petioles 5–25 mm (± winged); blades oblanceolate to linear, 15–40(–100+) × (3–)6–10(–15+) mm, bases cuneate, margins entire or ± serrate, sometimes ciliate, apices rounded or obtuse to attenuate, faces glabrous. **Heads** borne singly or in 2s or 3s. **Peduncles** 10–60 mm. **Calyculi** of 2–3(–9) erect, oblanceolate to lance-linear (± foliaceous) bractlets or bracts (6–)10–20(–75+) mm, margins usually ciliolate, abaxial faces glabrous. **Involucres** usually ± campanulate to cylindric, sometimes ± hemispheric, 5–7(–9) × (4–)7–10 mm. **Phyllaries** 6–8(–12), oblong, 6–11 mm (tips usually yellow, as are tips of paleae). **Ray florets** 0 or 3–7; laminae pale yellow, 4–12 mm. **Disc florets** 15–30(–60); corollas yellow (often orange-tipped), 3–4 mm. **Cypselae** blackish to olive, flattened, narrowly cuneate, outer 4–9 mm, inner 5–10 mm, margins retrorsely barbed, apices ± truncate to convex, faces ± 1-nerved, ± striate, glabrous; **pappi** of (2–)3–4 ± erect, retrorsely barbed awns 2–5 mm.

Flowering Aug–Oct. Marshes, borders of estuaries; 0–10 m; N.B., N.S., Nunavut, Ont., Que.; Maine, Mass., N.H., N.Y.

20. Bidens amplissima Greene, Pittonia 4: 268. 1901

[C] [E]

Bidens cernua Linnaeus var. *elata* Torrey & A. Gray; *B. elata* (Torrey & A. Gray) Sherff

Annuals, (30–)50–120 cm. **Leaves** sessile or petioles 10–50 mm (winged); blades lance-elliptic to lance-oblong, 80–120(–250) × 30–100 mm (proximal usually with 1–2 lobes near bases), bases cuneate, margins laciniate to serrate, ciliate, apices acuminate, faces glabrous. **Heads** in ± corymbiform arrays. **Peduncles** (10–)30–70+ mm. **Calyculi** of 8–10 usually spreading, lance-linear to lanceolate, sometimes ± foliaceous bractlets or bracts 10–20(–75) mm, margins entire or incised, ciliate, abaxial faces ± hispid at bases. **Involucres** ± hemispheric or broader, 6–10 × 18–25 mm. **Phyllaries** 13–16+, oblong to lance-ovate, 6–10 mm. **Ray florets** 6–11; laminae yellow, 4–8 mm. **Disc florets** 50–100+; corollas yellow, 2.5–3.5 mm. **Cypselae** olivaceous green, ± flattened, unequally 3–4-angled, cuneate, outer

5–6 mm, inner 6–7 mm, margins antrorsely to retrorsely barbed, apices truncate to concave, faces weakly striate, glabrous or sparsely puberulent; **pappi** 0, or of (1–)3 (–4) suberect, retrorsely barbed awns 2–4 mm.

Flowering Aug–Oct. Seasonal wetlands, borders of ponds, lakes, streams, estuaries; of conservation concern; 0–100 m; B.C.; Wash.

For decades, *Bidens amplissima* was known only from Vancouver Island; it is now known from southwestern British Columbia and northwestern Washington (F. R. Ganders et al. 2003). Occurrences in southern Manitoba and northwestern Nebraska may represent ephemeral introductions. The restricted natural range of *B. amplissima* has stimulated interest in it, the addition of it to rare plant lists, and the consideration of need for protected sites where it grows.

SELECTED REFERENCE Ganders, F. R., B. Klinkenberg, and R. Klinkenberg. 2003. Taxonomy in conservation: The enigmatic Vancouver Island beggarticks. Davidsonia 14: 63–70.

21. Bidens bidentoides (Nuttall) Britton, Bull Torrey Bot. Club 20: 281. 1893 [E]

Diodonta bidentoides Nuttall, Trans. Amer. Philos. Soc., n. s. 7: 361. 1841; *Bidens bidentoides* var. *mariana* (S. F. Blake) Sherff; *B. mariana* S. F. Blake

Annuals, 10–90+ cm. **Leaves:** petioles 10–25 mm; blades lanceolate to lance-linear, 40–160 × 3–30+ mm, bases cuneate, margins entire or laciniate to serrate or denticulate, minutely, if at all, ciliate, apices acuminate to attenuate, faces glabrous. **Heads** usually borne singly, sometimes in open, ± corymbiform arrays. **Peduncles** 10–30(–60+) mm. **Calyculi** of 3–5 usually spreading, narrowly lanceolate to oblanceolate, sometimes foliaceous bractlets or bracts 10–30(–60) mm, margins entire, usually ciliate, abaxial faces usually glabrous. **Involucres** narrowly campanulate to cylindric, 9–16 × 5–8(–12) mm. **Phyllaries** 4–8+, ± oblong, 9–16 mm. **Ray florets** usually 0, sometimes 3–5+; laminae yellowish, 2–10 mm. **Disc florets** 6–20 (–30+); corollas yellowish, 3–6 mm. **Cypselae** red-brown, flattened, sometimes weakly 4-angled, narrowly cuneate to linear, outer 6–10 mm, inner 8–13 mm, margins evenly antrorsely strigillose, apices truncate, faces smooth or ± striate, ± evenly antrorsely strigillose; **pappi** of 2(–4) erect to spreading, antrorsely barbed awns (2–)3–9 mm.

Flowering Sep. Borders of streams, estuaries; 0–10+ m; Del., Md., N.J., N.Y., Pa.

22. **Bidens tripartita** Linnaeus, Sp. Pl. 2: 831. 1753
    • Bident à trois divisions

Bidens comosa (A. Gray) Wiegand

Annuals, (5–)20–70(–200) cm. **Leaves** sessile or petioles 5–15 (–35+) mm (± winged); blades ± elliptic to ovate or lanceolate, 40–80(–150+) × 15–40(–60+) mm, sometimes laciniately 1-pinnatisect with 1–4+ lobes near bases, bases cuneate, margins entire or dentate to serrate, usually ciliate, apices acute to acuminate, faces glabrous or hirtellous. **Heads** borne singly or in 2s or 3s. **Peduncles** 10–40(–80) mm. **Calyculi** of (2–)6–7(–10) ± spreading, oblanceolate or lanceolate to linear, ± foliaceous bractlets or bracts 7–35(–60) mm, margins (entire or serrate) sometimes sparsely ciliate, abaxial faces hispidulous near bases, distally glabrous. **Involucres** campanulate to hemispheric or broader, (4–)5–7(–12) × (3–)6–12(–15+) mm. **Phyllaries** (6–)7–8(–13), elliptic-ovate to lance-ovate, (4–)6–9(–12) mm. **Ray florets** usually 0, sometimes 1–5; laminae orange yellowish, 4–8 mm. **Disc florets** (5–)20–60(–150+); corollas pale yellow to orange, (2–)3–4 mm (gradually ampliate, anthers usually pale). **Cypselae** blackish to purplish or brown, ± flattened, sometimes weakly 3(–4)-angled, usually cuneate to linear, outer (3–)6–7(–10) mm, inner (4–)6–9(–11) mm, margins proximally antrorsely to patently, distally retrorsely, barbed, apices ± truncate to concave, faces ± 1-nerved, usually smooth, seldom notably tuberculate, glabrous or sparsely strigillose; **pappi** 0, or of (1–)3–3(–4+) ± erect to spreading, retrorsely barbed awns (0.2–)2–3(–6) mm. **2*n* = 48.**

Flowering Aug–Oct. Marshes and other wet sites; 10–1700 m; Alta., B.C., Man., Que.; Ala., Alaska, Calif., Colo., Conn., Del., D.C., Ga., Ill., Ind., Iowa, Kans., Ky., La., Maine, Md., Mass., Mich., Minn., Miss., Mo., Mont., Nebr., N.H., N.J., N.Mex., N.Y., N.C., N.Dak., Ohio, Okla., Oreg., Pa., R.I., S.C., S.Dak., Tenn., Tex., Utah, Vt., Va., Wash., W.Va., Wis., Wyo.; Europe; Asia; n Africa; introduced in Pacific Islands, Australia.

Plants with cypsela mid-nerves strongly developed (cypselae more or less strongly 4-angled and, often, tuberculate) that are treated below as *Bidens connata* have been included in *B. tripartita*, perhaps rightly so. And some botanists have included (or advocated inclusion of) *B. eatonii*, *B. heterodoxa*, and/or *B. infirma* in *B. tripartita*, as well, perhaps rightly so.

23. **Bidens connata** Muhlenberg ex Willdenow, Sp. Pl. 3: 1718. 1803 • Bident conné

Bidens connata var. *ambiversa* Fassett; *B. connata* var. *anomala* Farwell; *B. connata* var. *fallax* (Warnstorf) Sherff; *B. connata* var. *gracilipes* Fernald; *B. connata* var. *inundata* Fernald; *B. connata* var. *petiolata* (Nuttall) Farwell; *B. connata* var. *pinnata* S. Watson; *B. connata* var. *submutica* Fassett

Annuals, (5–)25–150(–200) cm. **Leaves:** sessile or petioles (± winged) 5–15(–35+) mm; blades ± elliptic to lanceolate, (20–)40–100(–200+) × 10–30(–70) mm, sometimes laciniately 1-pinnatisect with 1–4+ lobes near bases, bases cuneate, margins entire or dentate to serrate, usually ciliate, apices attenuate, faces glabrous or hirtellous. **Heads** borne singly or in 2s or 3s. **Peduncles** 10–35(–80) mm. **Calyculi** of 2–6+ spreading or ascending, lanceolate, linear, oblanceolate, oblong, or spatulate, ± foliaceous bractlets or bracts 10–25(–35+) mm, margins (entire or serrate) usually ciliate, abaxial faces hispidulous near bases, distally glabrous. **Involucres** campanulate to hemispheric or broader, (4–)6–12 × (3–)6–9(–15+) mm. **Phyllaries** (6–)8(–9), elliptic to oblong or ovate, (4–)5–6(–12) mm. **Ray florets** usually 0, sometimes 1–5+; laminae yellowish, 4–8 mm. **Disc florets** (5–)20–40(–60+); corollas pale yellow to orange, 2–3 mm (± abruptly ampliate, anthers usually dark, blackish). **Cypselae** blackish to purplish or brown, usually (at least inner) ± 4-angled, usually cuneate to linear, sometimes ± obpyramidal, outer (3–)4–7 mm, inner (4–)5–8 mm, margins proximally antrorsely to patently, distally retrorsely, barbed, apices ± truncate to concave, faces usually strongly 1-nerved, usually tuberculate, glabrous or setulose; **pappi** 0, or of (1–)2–4(–6), ± erect to spreading, antrorsely or retrorsely barbed awns (0.2–)2–5 mm. **2*n* = 48.**

Flowering Aug–Oct. Marshes and other wet sites; 10–1700 m; N.B., Nfld. and Labr. (Nfld.), N.S., Ont., P.E.I., Que.; Conn., Del., Ga., Ill., Ind., Iowa, Kans., Ky., Maine, Md., Mass., Mich., Minn., Mo., Mont., Nebr., N.H., N.J., N.Y., N.C., N.Dak., Ohio, Pa., R.I., S.C., S.Dak., Tenn., Va., W.Va., Wis.; Europe; Asia; n Africa; introduced in Pacific Islands, Australia.

Bidens connata may be better treated as part of *B. tripartita*.

24. Bidens eatonii Fernald, Rhodora 5: 92, plate 45, figs. 11–14. 1903 (as eatoni) · Bident d'Eaton E

Bidens eatonii var. *fallax* Fernald; *B. eatonii* var. *illicita* S. F. Blake; *B. eatonii* var. *interstes* (Fassett) Fassett; *B. eatonii* var. *kennebecensis* Fernald; *B. eatonii* var. *major* Fassett; *B. eatonii* var. *mutabilis* Fassett; *B. eatonii* var. *simulans* Fassett

Annuals, (10–)25–70(–150) cm. Leaves: petioles (narrowly, if at all, winged) 10–35+ mm; blades lance-ovate to lanceolate, (20–)35–100+ × 5–25+ mm, sometimes with 1–4 lobes near bases, bases cuneate, margins serrate, seldom ciliate, apices attenuate, faces glabrous. Heads borne singly or in 2s or 3s. Peduncles 10–35+ mm. Calyculi of (2–)3–5 ± erect, oblanceolate to linear, ± foliaceous bractlets or bracts 7–25(–70+) mm, margins sometimes obscurely ciliate, abaxial faces glabrous. Involucres usually cylindric to campanulate, rarely hemispheric, (4–)8–10(–12) × (4–)7–9 mm. Phyllaries 5–6, oblong to lance-ovate, 8–13 mm. Ray florets usually 0, sometimes 1–3+; laminae yellow-orange, 2–5 (–10) mm. Disc florets 7–30+; corollas yellowish, 2.5–3.5 mm. Cypselae blackish to purplish or red-brown, ± flattened, sometimes 3–4-angled, narrowly oblanceolate, outer 5.5–10 mm, inner 6.5–11 mm, margins barbed or ciliate (proximally antrorsely, distally retrorsely), apices ± truncate, faces ± 1-nerved, glabrous or sparsely setulose, often ± striate and/or tuberculate; pappi of 2–4(–5) erect to spreading, antrorsely or retrorsely barbed awns 3–4.5 mm. 2*n* = 48.

Flowering Sep–Oct. Borders of estuaries, marshes, tidal flats; 0–10 m; N.B., P.E.I., Que.; Conn., Maine, Mass., N.J., N.Y., R.I.

Bidens eatonii may be better treated as part of *B. tripartita*.

25. Bidens heterodoxa (Fernald) Fernald & H. St. John, Rhodora 17: 23. 1915 · Bident différent E

Bidens tripartita Linnaeus var. *heterodoxa* Fernald, Rhodora 15: 76. 1913; *B. heterodoxa* var. *agnostica* Fernald; *B. heterodoxa* var. *monardifolia* Fernald; *B. heterodoxa* var. *orthodoxa* Fernald & H. St. John; *B. infirma* Fernald

Annuals, (5–)25–35(–55+) cm. Leaves: petioles (narrowly, if at all, winged) 5–20(–35+) mm; blades lanceolate to lance-linear, (20–)30–50(–80+) × (3–)6–20(–30+) mm, sometimes laciniately 1-pinnatisect with 2–4+ lobes near bases, bases cuneate, margins incised to serrate, seldom ciliate, apices acute to attenuate, faces glabrous. Heads borne singly or in 2s or 3s. Peduncles 10–35(–55) mm. Calyculi of (2–)3–5(–7) ± erect, oblanceolate or spatulate to linear, ± foliaceous bractlets or bracts (6–)10–25 (–40+) mm, margins rarely ciliate, abaxial faces usually glabrous, sometimes sparsely hispidulous. Involucres usually obconic to broadly campanulate, rarely hemispheric, (4–)6–7 × (4–)5–7 mm. Phyllaries (4–)5–8, oblong to lance-ovate, (4–)5–8+ mm. Ray florets usually 0, sometimes 1–3+; laminae orange yellowish, 2–5 (–10) mm. Disc florets (7–)20–70+; corollas yellowish, 2.5–3.5 mm. Cypselae: outer red-brown, ± flattened, sometimes 3–4-angled, broadly cuneate, 4–5(–6) mm, inner blackish, narrowly cuneate, (5–)6–7(–8+) mm, margins glabrous or sparsely antrorsely strigillose, apices ± truncate to convex, faces ± 1-nerved, glabrous or sparsely setulose, usually smooth, sometimes ± tuberculate; pappi 0, or of 2(–4) spreading to divergent, antrorsely or retrorsely barbed awns (0.1–)1–3.5+ mm. 2*n* = 48.

Flowering Aug–Oct. Borders of estuaries, lakes, marshes, streams; 0–10 m; N.B., P.E.I., Que.; Conn.

Bidens heterodoxa may be better treated as part of *B. tripartita*.

313. HETEROSPERMA Cavanilles, Icon. 3: 34. 1795/1796 · [Greek, *heteros*, differing, and *sperma*, seed; probably alluding to the contrasting outer and inner cypselae]

Justin W. Allison

Annuals, 10–40(–70+) cm (taprooted). Stems 1 (bases relatively thick), ascending to erect, branched distally or ± throughout (striate). Leaves mostly cauline; opposite; obscurely petiolate or sessile; blades mostly pinnately lobed [undivided] (lobes 3–5, usually linear to filiform, sometimes lanceolate), ultimate margins usually entire, sometimes denticulate (often ciliate, at least proximally, apices acute), faces glabrous [hairy]. Heads radiate, borne singly or in loose, cymiform arrays. Calyculi of (1–)3–5+ erect, narrowly spatulate or linear to filiform, herbaceous bractlets (often surpassing phyllaries, margins usually hispid-ciliate). Involucres cylindric to obconic, 2–5 mm diam. (larger in fruit). Phyllaries 3–5+ in ± 2 series, distinct, erect, mostly oblong or ovate, equal, membranous. **Receptacles** concave to flat, paleate; paleae similar to phyllaries, oblong or oval to lanceolate or linear (± embracing cypselae, ± hyaline with brown striae, apices obtuse, apiculate). **Ray florets** 1–3[–8], pistillate, fertile; corollas pale yellow [orange]. **Disc florets** 3–10[–20+], bisexual, fertile; corollas ± yellow (at least distally), tubes shorter than funnelform throats, lobes 5, deltate. **Cypselae** (± dimorphic) outer obcompressed, ellipsoid or obovoid, winged (wings ± corky, often cucullate and/or pectinate), faces often corky-tuberculate, glabrous, inner ± obovoid, usually some (innermost) tapered to ± barbellate beaks; **pappi** 0, or persistent or tardily falling, of (1–)2–3 spreading to reflexed, retrorsely barbellate awns. *x* = 25.

Species 5–10 (1 in the flora): sw United States, Mexico, Central America, South America; introduced in West Indies.

1. Heterosperma pinnatum Cavanilles, Icon. 3: 34. 1795/1796 (as pinnata) [F]

Leaves 1–4+ cm, lobes 0.5–1(–3) mm wide. Ray laminae 1–2+ mm. Disc corollas ca. 2.5 mm. Cypselae 5–18 mm; pappi 0.5–1(–3) mm. $2n = 48, 50$.

Flowering Jun–Oct. Igneous and calcareous soils; 900–2500 m; Ariz., N. Mex., Tex.; Mexico; Central America.

Heterosperma pinnatum occurs throughout much of the southwestern United States. No specimens have been seen to substantiate claims of occurrences in Maryland and Massachusetts, where it may have been ephemeral. Calyculi of *H. pinnatum* have been interpreted as leaves, the heads being interpreted as terminal and sessile.

314. DICRANOCARPUS   A. Gray, Mem. Amer. Acad. Arts, n. s. 5: 322. 1854

• Pitchfork [Greek *di-*, two, *kranos*, skull or helmet, and *karpos*, fruit; probably alluding to the "two-horned" cypselae]

Justin W. Allison

Annuals, 10–70 cm (± succulent; taprooted). **Stems** 1, erect, branched dichotomously, divaricate (drying brittle). **Leaves** mostly cauline; opposite; obscurely petiolate; blades 1(–2)-pedately or -pinnately lobed (lobes filiform), ultimate margins entire, faces glabrous. **Heads** radiate, borne singly or (2–4) in cymiform arrays. **Calyculi** of (0–)1–3+ erect, linear to subulate, herbaceous bractlets. **Involucres** cylindric to obconic, 1.5–3 mm diam. (larger in fruit). **Phyllaries** persistent, 3–6 in 1(–2) series, lance-ovate to lanceolate, scarious, brown-nerved. **Receptacles** slightly convex, paleate; paleae linear to subulate, scarious. **Ray florets** 3–6, pistillate, fertile; corollas yellow. **Disc florets** 3–4+, functionally staminate; corollas yellow, tubes shorter than throats, lobes 5, triangular. **Cypselae** obcompressed, oblong to linear, ± ribbed, smooth or tuberculate, glabrous or densely papillate; **pappi** persistent, of 2 spreading to recurved, smooth or proximally papillate awns. *x* = 10.

Species 1: sw United States, n Mexico.

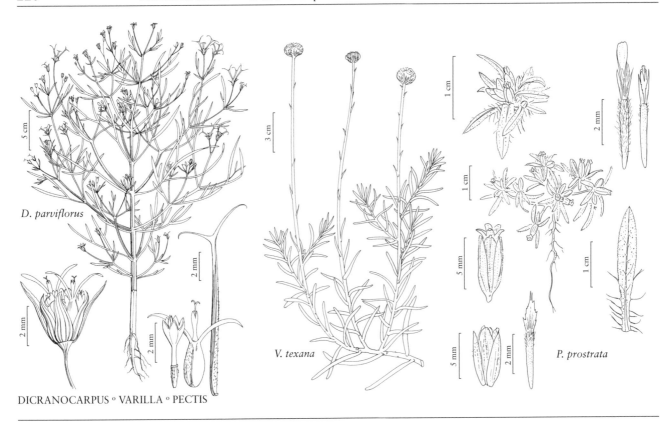

D. parviflorus

V. texana

P. prostrata

DICRANOCARPUS ○ VARILLA ○ PECTIS

1. Dicranocarpus parviflorus A. Gray, Mem. Amer. Acad. Arts, n. s. 5: 322. 1854 [F]

Heterosperma dicranocarpum A. Gray, Smithsonian Contr. Knowl. 3(5): 109.

Leaves 2–8 cm. **Ray corollas** 2–3 mm. **Disc corollas** 1–3 mm. **Cypselae** (3–)5–10 mm; **pappi** 2–5 mm. $2n = 20$.

Flowering Sep–Oct. Alkaline or gypseous soils; 900–1700 m;

N.Mex, Tex.; Mexico (Chihuahua, Durango, Nuevo León, San Luis Potosí).

Dicranocarpus parviflorus is a gypsophile that ranges throughout the Chihuahuan Desert.

187m.13. ASTERACEAE Martinov (tribe HELIANTHEAE) subtribe VARILLINAE B. L. Turner & A. M. Powell in V. H. Heywood et al., Biol. Chem. Compositae 2: 719. 1978

Subshrubs [shrubs], (10–)20–30+ cm (sometimes rhizomatous). **Leaves** cauline; alternate [opposite]; ± sessile; blades linear to filiform (usually ± terete), margins entire, faces glabrous (not gland-dotted). **Heads** discoid, borne singly [in corymbiform arrays]. **Calyculi** 0. **Involucres** obconic. **Phyllaries** persistent, [30–]45–60 in [2–]4–5 series (distinct, lance-linear, unequal, outer shorter, all resin-nerved). **Receptacles** conic, paleate (paleae falling, linear, flattish or weakly conduplicate, scarious, resin-nerved). **Ray florets** 0. **Disc florets** [25–]60–150+, bisexual, fertile; corollas yellow, tubes shorter than or about equaling campanulate to cylindric throats, lobes 5, deltate; anther thecae pale; stigmatic papillae in 2 lines. **Cypselae** columnar to clavate, 8–15-ribbed, glabrous; **pappi** 0 [bristles].

Genus 1, species 2 (1 in the flora): sw United States, Mexico.

Varillinae may be allied with Clappiinae (H. Robinson 1981) or with Jaumeinae (P. O. Karis and O. Ryding 1994).

315. VARILLA   A. Gray, Mem. Amer. Acad. Arts, n. s. 4: 106. 1849 • [Spanish *varilla*, rod or wand, a common name used in Mexico]

John L. Strother

Subshrubs [shrubs], (10–)20–30+ cm (± succulent, sometimes rhizomatous). **Stems** erect. **Leaves** cauline; mostly alternate [opposite]; ± sessile; blades linear to filiform (± terete) [lance-linear], margins entire, faces glabrous (not gland-dotted). **Heads** discoid, borne singly [in corymbiform arrays]. **Involucres** obconic, [3–]5–8 mm diam. **Phyllaries** persistent, [30–]45–60 in [2–]4–5 series ± appressed, lance-linear to subulate, unequal, resinous-nerved). **Receptacles** conic, paleate; paleae falling, linear, scarious, flat or weakly conduplicate, resinous-veined. **Ray florets** 0. **Disc florets** 60–150+, bisexual, fertile; corollas greenish yellow, tubes shorter than or about equaling campanulate to cylindric throats, lobes 5, narrowly deltate. **Cypselae** columnar to clavate, ribs 8–15, faces glabrous; **pappi** 0 [bristles]. $x = 18$.

Species 2 (1 in the flora): Texas, Mexico.

1. **Varilla texana**   A. Gray, Smithsonian Contr. Knowl. 3(5): 123. 1852   [F]

Leaf blades 10–30 × 1–2 mm. **Disc corollas** ca. 3 mm. **Cypselae** ca. 2 mm. $2n = 36$.

Flowering spring-early summer, fall. Usually on saline, clay or gypseous soils; 50–200 m; Tex.; Mexico (Coahuila, Nuevo Léon, Tamaulipas).

187m.14. ASTERACEAE   Martinov (tribe HELIANTHEAE) subtribe PECTIDINAE   Lessing, Linnaea 5: 134. 1830 (as Pectideae)

Tageteae Cassini

Annuals, perennials, subshrubs, or shrubs, (1–)5–50(–120+)[200+] cm. **Leaves** mostly cauline, sometimes basal or basal and cauline; mostly opposite (distal sometimes alternate) or mostly alternate; petiolate or sessile; blades usually elliptic, filiform, lanceolate, linear, or ovate, often 1(–2)-pinnately lobed, ultimate margins entire or toothed (often proximally bristly-ciliate), faces glabrous or puberulent to tomentose (usually bearing oil-glands, embedded pellucid glands filled with strong-scented, essential oils). **Heads** radiate or discoid (sometimes radiant, e.g., in *Thymophylla*), usually borne singly, sometimes in loose to crowded, corymbiform or cymiform arrays. **Calyculi** 0 or of 1–22 bractlets. **Involucres** campanulate, cylindric, hemispheric, obconic, or turbinate. **Phyllaries** falling or persistent, 3–21+ in 1–2 series (distinct or ± connate, mostly linear to lanceolate, usually subequal, usually some or all bearing pellucid oil-glands as in leaves). **Receptacles** flat or convex to hemispheric or conic, epaleate (receptacles sometimes pitted, pit margins sometimes fimbrillate to laciniate). **Ray florets** 0 or 1–21+, pistillate, fertile; corollas yellow to orange (sometimes with brown or mostly brown) or white to pink, magenta, or reddish purple. **Disc florets** 1–120+, usually bisexual and fertile [functionally staminate]; corollas yellow to orange or brown, or pinkish to purplish or reddish, tubes much shorter than to much longer than cylindric to funnelform throats, lobes 5, deltate to lance-linear (equal or unequal); anther thecae pale or ± darkened; stigmatic papillae in 2 lines. **Cypselae** mostly clavate to obpyramidal (lengths usually 2–3+ times diams.), glabrous or hairy; **pappi** usually persistent, of 20–50 bristles in 1–2+ series, or of 5–20 (sometimes aristate) scales in 1–2 series, or combinations of bristles and scales in 1–2+ series, rarely coroniform or 0.

Genera 22, species 225 (9 genera, 38 species in the flora): mostly subtropical, tropical, and warm-temperate, especially arid New World.

Circumscription of Pectidinae has been fairly constant for decades. The group is sometimes given tribal rank, as Tageteae. Here, we have followed H. Robinson (1981). The group is usually distinguished on the basis of schizogenous glands containing essential oils (mostly terpenes) in tissues of leaves and/or phyllaries (sometimes in tissues of other organs as well).

SELECTED REFERENCES Strother, J. L. 1969. Systematics of *Dyssodia* Cavanilles (Compositae: Tageteae). Univ. Calif. Publ. Bot. 48: 1–88. Strother, J. L. 1986. Renovation of *Dyssodia* (Compositae: Tageteae). Sida 11: 371–378.

1. Leaves opposite (blade margins proximally bristly-ciliate; ray florets borne on bases of subtending phyllaries; style branches of bisexual florets knoblike) . . . . 316. *Pectis*, p. 222
1. Leaves opposite or mostly alternate (blades often lobed, margins sometimes bristly-ciliate; ray florets borne on receptacles, not on bases of subtending phyllaries; style branches of bisexual florets linear).
  2. Phyllaries distinct to bases or nearly so.
    3. Calyculi of (0–)1–9, deltate, lanceolate, or linear bractlets; pappi wholly or partly of scales (individual scales sometimes comprising 5–10, basally connate bristles).
      4. Ray corollas yellow to orange . . . . 317. *Dyssodia*, p. 230
      4. Ray corollas whitish with pinkish or purplish stripes . . . . 318. *Nicolletia*, p. 231
    3. Calyculi 0; pappi wholly of distinct bristles.
      5. Heads radiate . . . . 319. *Chrysactinia*, p. 232
      5. Heads discoid . . . . 320. *Porophyllum*, p. 233
  2. Phyllaries ± connate $^{1}/_{3}$–$^{7}/_{8}$+ their lengths (margins of outer may be free to bases).
    6. Calyculi 0; pappi of 2–5(–10) elements in ± 1 series (usually 0–5+ oblong to lanceolate, erose-truncate to laciniate, plus 0–2+ longer, subulate or aristate, some or all sometimes connate) . . . . 321. *Tagetes*, p. 235
    6. Calyculi usually of (1–)5–8(–22) bractlets, rarely 0; pappi usually of 8–20 elements in 2 series, rarely coroniform.
      7. Bractlets of calyculi subulate or pectinate . . . . 322. *Dysodiopsis*, p. 237
      7. Bractlets of calyculi (deltate to subulate) not pectinate.
        8. Plants (20–)30–70+ cm; involucres (7–18 ×) 5–12 mm; phyllaries weakly connate $^{1}/_{3}$–$^{2}/_{3}$ their lengths . . . . 323. *Adenophyllum*, p. 237
        8. Plants mostly (1–)5–30 cm; involucres (4–6 ×) 2–7 mm; phyllaries strongly connate $^{2}/_{3}$–$^{7}/_{8}$ their lengths (margins of outer may be free to bases) . . . 324. *Thymophylla*, p. 239

316. PECTIS Linnaeus, Syst. Nat. ed. 10, 2: 1221. 1759 • [Greek *pecten*, comb, alluding to ciliate leaf margins]

David J. Keil

Annuals or perennials, 1–120 cm (herbage often lemon or spicy scented). **Stems** prostrate to erect, simple to much branched. **Leaves** cauline; opposite; usually sessile; blades mostly linear to elliptic, oblanceolate, oblong, or obovate, margins usually setose-ciliate (mostly near bases), faces glabrous or hairy (abaxial and/or margins dotted with oil-glands). **Heads** radiate, borne singly or in congested to open, cymiform arrays (peduncles usually bracteate). **Calyculi** 0. **Involucres** campanulate, cylindric, ellipsoid, or fusiform, 2–8+ mm diam. **Phyllaries** 3–15[–21] in 1 series (usually distinct, falling individually, each with a ray cypsela, sometimes cohering at bases, falling together as units enclosing all cypselae of a head; individually convex, indurate-keeled, narrowly to broadly hyaline-margined, apices often ciliolate, bearing oil-glands on margins and/or faces). **Receptacles** flat to hemispheric, smooth or pitted, epaleate. **Ray florets** 3–15[–21], pistillate, fertile (inserted on phyllary bases); corollas yellow, often reddened abaxially (often drying white to purplish; laminae elliptic, entire or 2- or 3-lobed, glabrous or proximally glandular-puberulent). **Disc florets** [1–]3–55[–100], usually bisexual; corollas yellow (sometimes drying white to purplish), tubes shorter than narrowly funnelform throats, lobes (4–)5, deltate to lance-ovate (lobes ± equal and corollas actinomorphic, not 2-lipped, or lobes unequal with 3 or 4 forming an adaxial lip opposite 1-lobed lip and corollas zygomorphic, 2-lipped, all lobes glabrous or proximally glandular-puberulent; anther bases rounded or subcordate, apical appendages rounded or emarginate; styles included to long-exserted, branches ± papillose knobs). **Cypselae** (blackish or dark brown) cylindric to narrowly clavate, ribbed or angled, puberulent to pilose; **pappi** persistent, usually of awns, bristles, or scales, sometimes coroniform. $x = 12$.

Species about 90 (13, including 1 hybrid, in the flora): North America, Mexico, West Indies, Central America, South America, Pacific Islands (Galapagos Islands, Hawaii).

Pectis is unusual among Compositae in having the C_4 photosynthetic pathway and the accompanying Kranz anatomy—leaves with vascular bundle sheath cells that contain numerous chloroplasts (B. N. Smith and B. L. Turner 1975). All of the closely related genera have the C_3 pathway. The selective advantage imparted by C_4 photosynthesis (greatly reduced photorespiration in bundle sheath cells) has enabled *Pectis* species to occupy a variety of hot, dry habitats including deserts, tropical and subtropical grasslands, arid scrublands, and tropical beaches. Some species of *Pectis* grow in hot desert areas of the western United States following summer precipitation.

Leaves and phyllaries in all *Pectis* species are dotted with embedded pellucid glands (here called oil-glands). In some species, the liquid within the schizogenous cavities includes a mixture of strongly scented monoterpenes; in other species the gland contents have little or no aroma. Herbage containing strongly scented essential oils are described as "lemon-scented" when citral is the predominant compound and "spicy-scented" when other oils are predominant. For most taxa the odor (or lack thereof) of the crushed herbage is a readily apparent field characteristic. Unfortunately, labels of most herbarium specimens lack information on odor. Intact glands on herbarium specimens may retain liquid contents for years; odors may change as the specimens age. On living specimens, the glands are translucent; on herbarium specimens, they tend to be golden brown or blackish. The glands probably function as a deterrent to herbivores, including insects.

Ray florets of *Pectis* are inserted directly onto the bases of the phyllaries, and the phyllaries and ray cypselae tend to fall together when the heads shatter at maturity. The numbers of phyllaries and ray florets per *Pectis* capitulum vary along the Fibonacci series and rarely deviate from the sequence (i.e., in *Pectis*, the modal numbers of phyllaries and rays are 3, 5, 8, 13, and 21).

Pectis coulteri Harvey & A. Gray has been attributed in floras to California and Arizona on the basis of the ambiguously labeled type collection ("California," without locality or date, *T. Coulter 331*, holotype TCD) and speculations as to its origin. This species is known to occur only from central Sonora to northern Sinaloa, Mexico, and has never been documented in the United States. The type was probably collected in Sonora (D. J. Keil 1975).

SELECTED REFERENCES Keil, D. J. 1975. Revision of *Pectis* sect. *Heteropectis* (Compositae: Tageteae). Madroño 23: 181–191. Keil, D. J. 1977. A revision of *Pectis* section *Pectothrix* (Compositae: Tageteae). Rhodora 79: 32–78. Keil, D. J. 1977b. Chromosome studies in North and Central American species of *Pectis* L. (Compositae: Tageteae). Rhodora 79: 79–94. Keil, D. J. 1978. Revision of *Pectis* section *Pectidium* (Compositae: Tageteae). Rhodora 80: 135–146. Keil, D. J. 1986. Synopsis of the Florida species of *Pectis* (Asteraceae). Sida 11: 385–395. Keil, D. J. 1996. *Pectis*. In: B. L. Turner, ed. 1996+. The Comps of Mexico: A Systematic Account of the Family Asteraceae. 2+ vols. Huntsville, Tex. Vol. 6, pp. 22–43. [Phytologia Mem. 10, 11.]

1. Ray florets (7–)8–13(–15).
   2. Heads in congested or open, cymiform arrays; peduncles 1–40 mm.
      3. Pappi of disc cypselae usually of 16–24 subplumose bristles, rarely coroniform;
         cypselae strigillose to short-pilose (hair tips curled, bulbous) . . . . 9. *Pectis papposa*
      3. Pappi of disc cypselae coroniform and/or of 1–7 scabrid awns or bristles; cypselae
         strigillose (hair tips straight, forked) . . . . 10. *Pectis angustifolia*
   2. Heads borne singly or in open, cymiform arrays; peduncles 20–160 mm.
      4. Perennials (rhizomatous); heads borne singly; peduncles 30–160 mm; ray florets
         (8–)13(–15) . . . . 7. *Pectis longipes*
      4. Annuals (taprooted); heads borne singly or in open, (leafy) cymiform arrays; ray
         florets 8(–13) . . . . 8. *Pectis rusbyi*
1. Ray florets 3–5.
   5. Disc corollas not 2-lipped (actinomorphic, lobes equal, 4 or all 5 each bearing an oil-
      gland); leaf margins with 0–1 pairs of setae; pappi coroniform or of awns.
      6. Perennials; ray corollas 6–11 mm . . . . 12. *Pectis imberbis*
      6. Annuals; ray corollas 2–3 mm . . . . 13. *Pectis linifolia*
   5. Disc corollas ± 2-lipped (zygomorphic, lobes not bearing oil-glands); leaf margins with
      1–12 pairs of setae; pappi usually of bristles, scales, or awns, sometimes coroniform
      plus 0–3 awns or bristles (*P. filipes*).
      7. Phyllaries coherent (at least at bases, falling together as units enclosing all cypselae
         of a head).
         8. Ray florets 3(–4) . . . . .2. *Pectis cylindrica*
         8. Ray florets 5.
            9. Peduncles 1–2 mm; pappi of lanceolate scales (ovaries usually forming
               cypselae) . . . . 1. *Pectis prostrata*
            9. Peduncles 5–25 mm; pappi of slender, aristate scales (pericarps darkening,
               not swelling, ovules abortive). . . . . 6. *Pectis ×floridana*
      7. Phyllaries distinct (spreading and falling individually, each with a ray cypsela).
       10. Leaves oblong-oblanceolate to obovate; herbage not scented . . . . 3. *Pectis humifusa*
       10. Leaves linear, linear-oblanceolate, narrowly elliptic, or narrowly linear; herb-
         age lemon-scented or spicy-scented.
         11. Peduncles 0–1 mm . . . . 5. *Pectis linearifolia*
         11. Peduncles 3–65 mm.
            12. Pappi of 0–5, antrorsely scabrid bristles or slender scales 1–2 mm plus
               0–5, entire or irregularly lacerate scales 0.2–0.7 mm; Florida . . . . 4. *Pectis glaucescens*
            12. Pappi of 0–3 awns 3–4 mm, usually with a crown of shorter scales; sw
               United States . . . . 11. *Pectis filipes*

1. Pectis prostrata Cavanilles, Icon. 4: 12, plate 324.
1797 • Spreading chinchweed [F]

Annuals, 1–30 cm (across); herbage not scented. **Stems** prostrate to ascending (often mat-forming, densely leafy, especially distally), puberulent (in lines or throughout). **Leaves** linear to narrowly oblanceolate, 10–40 × 1.5–7 mm, margins with 4–12 pairs of setae 1–3 mm, faces glabrous (abaxial densely dotted with round oil-glands 0.1–0.3 mm). **Heads** borne singly or in congested, (leafy) cymiform arrays. **Peduncles** 1–2 mm. **Involucres** campanulate, cylindric, or ellipsoid. **Phyllaries** coherent (falling together), oblong to obovate, 5–8 × 1–3 mm (often dotted in submarginal rows and sometimes along midribs with elliptic oil-glands 0.1–0.3 mm). **Ray florets** 5; corollas 2.5–3.5 mm (scarcely surpassing phyllaries). **Disc florets** 3–17; corollas 1.8–2.5 mm (2-lipped). **Cypselae** 2.5–4.5 mm, strigillose; **pappi** of 2 (ray) or 5 (disc) lanceolate scales 1.5–2.5 mm. 2*n* = 24.

Flowering Jul–Nov. Open sites in deserts, grasslands, oak-pine-juniper woodlands, roadsides; 0–2000 m; Ariz., Fla., La., N.Mex., Tex.; Mexico; West Indies; Central America.

The development of roads and highways has created ideal habitats for *Pectis prostrata*. Its range appears to be expanding along the coasts of Florida; it was discovered in Louisiana relatively recently. It can be expected to spread along the Gulf Coast and perhaps northward along the Atlantic Coast as well. Autogamy has apparently assisted *P. prostrata* to spread rapidly as suitable new habitats have become available.

2. **Pectis cylindrica** (Fernald) Rydberg in N. L. Britton et al., N. Amer. Fl. 34: 198. 1916 • Three-rayed or Sonoran chinchweed

Pectis prostrata Cavanilles var. *cylindrica* Fernald, Proc. Amer. Acad. Arts 33: 68. 1897

Annuals, 1–20 cm (across or high); herbage not scented. **Stems** prostrate to ascending (often mat-forming, densely leafy distally), puberulent (in decurrent lines or throughout) or glabrate. **Leaves** (bluish green) linear to linear-oblanceolate or narrowly oblong, 10–30 × 1.5–4 mm, margins with 2–5 pairs of setae 1–2 mm, faces glabrous (abaxial densely dotted with scattered, circular oil-glands 0.05–0.2 mm). **Heads** borne singly or in congested, (leafy) cymiform arrays. **Peduncles** 1–5 mm. **Involucres** cylindric. **Phyllaries** coherent (falling together), oblong to oblong-obovate, 6–10 × 2–3 mm (dotted with scattered, elliptic oil-glands 0.05–0.15 mm). **Ray florets** 3(–4); corollas 3–4 mm (scarcely surpassing phyllaries). **Disc florets** (3–)7–14; corollas 2.2–2.6 mm (2-lipped). **Cypselae** 4–5.5 mm, puberulent (distally glandular-puberulent); **pappi** of 2 (ray) and 5 (disc) lanceolate scales 1.5–3.5 mm. $2n = 48$.

Flowering Jul–Nov. Deserts, oak-juniper woodlands, grasslands, wash channels, mud flats, lawns, roadsides; 700–1500 m; Ariz., N.Mex., Tex.; Mexico (Baja California Sur, Chihuahua, Coahuila, Durango, Nuevo León, Sinaloa, Sonora).

Pectis cylindrica ($2n = 48$) is similar to *P. prostrata* ($2n = 24$); the two occasionally grow together (D. J. Keil 1975b). Some herbaria contain mixed collections of the two. No evidence is available of hybrids between the two.

3. **Pectis humifusa** Swartz, Prodr., 114. 1788 • Yerba de San Juan [1]

Annuals or perennials, 2–25 cm (across; bases often ± woody); herbage not scented. **Stems** prostrate (mat-forming, densely leafy), puberulent (in decurrent lines). **Leaves** oblong-oblanceolate to obovate, 3–17 × 1.5–4 mm wide, margins with 2–6 pairs of setae 1–2 mm, faces glabrous (dotted with scattered, round oil-glands 0.1–0.2 mm). **Heads** borne singly or in congested, (leafy) cymiform arrays. **Peduncles** 1–12 mm. **Involucres** campanulate. **Phyllaries** distinct, obovate, 4.5–6 × 2–4 mm (faces densely dotted with scattered, circular oil-glands 0.05–0.2 mm). **Ray florets** 5; corollas 3.5–5 mm. **Disc florets** 12–21; corollas 2.5–3.5 mm (2-lipped). **Cypselae** 2.5–4 mm, mostly puberulent (ray cypselae abaxially glabrous); **ray pappi** of 2–3

slender, aristate scales 1.5–2.5 mm plus 2–10 lacerate scales or bristles; **disc pappi** of 4–15, antrorsely scabrid bristles or aristate scales 2–3 mm plus 0–15 bristles or scales. $2n = 72$.

Flowering year round. Sandy soils; 0–10 m; Fla.; West Indies (Puerto Rico, Lesser Antilles), South America (Suriname).

Pectis humifusa has been reported once from Florida (D. J. Keil 1975c), where it is probably adventive. In the Lesser Antilles, it occurs most frequently in the salt spray zone near the seashore; on some islands, it occurs inland as well.

4. **Pectis glaucescens** (Cassini) D. J. Keil, Sida 11: 386. 1986 • Sand dune chinchweed

Chthonia glaucescens Cassini in F. Cuvier, Dict. Sci. Nat. ed. 2, 9: 173. 1817; *Pectis leptocephala* (Cassini) Urban; *P. lessingii* Fernald

Annuals or perennials, 2–50 cm (across or high); herbage spicy-scented. **Stems** prostrate to erect, usually sparsely to densely puberulent (sometimes in decurrent lines), sometimes glabrate. **Leaves** narrowly linear, 10–35 × 0.2–1.8 mm, margins with 1–5 pairs of setae 1–2 mm, faces glabrous (abaxial submarginally dotted with broadly elliptic to circular oil-glands 0.2–0.3 mm, sometimes with additional, scattered oil-glands). **Heads** borne singly or in diffuse, cymiform arrays. **Peduncles** filiform, (3–)7–35(–54) mm. **Involucres** cylindric. **Phyllaries** distinct, linear-oblanceolate, 4–5 × 0.8–1 mm (dotted with 1–2 elliptic, subapical oil-glands 0.2–0.3 mm, sometimes with additional, smaller submarginal or scattered oil-glands). **Ray florets** 5; corollas 3.5–5 mm. **Disc florets** 3–7; corollas 2–3 mm (2-lipped). **Cypselae** 2.5–3 mm, strigillose; **pappi** of 0–5, antrorsely scabrid bristles or slender scales 1–2 mm plus 0–5 entire or irregularly lacerate scales 0.2–0.7 mm. $2n = 48$.

Flowering year round. Sandy or gravelly soils, grassy areas, openings in pinelands, scrub, roadsides; 0–50 m; Fla.; West Indies (Bahamas, Hispaniola, Jamaica).

Pectis glaucescens is widespread in southern Florida and the Bahamas. Human disturbances, especially road constructions, have created habitats suitable for it. It grows most commonly on limestone soils in open, grassy sites. Occasionally, it is a lawn weed.

5. **Pectis linearifolia** Urban, Symb. Antill. 5: 276. 1907

• Florida chinchweed E

Annuals, 4–40 cm; herbage lemon-scented. **Stems** decumbent to erect (very leafy), glabrous or puberulent (in decurrent lines). **Leaves** linear, 10–50 × 1–3 mm, margins with 2–6 pairs of setae, faces glabrous (abaxial dotted submarginally with round oil-glands 0.2–0.5 mm). **Heads** borne singly or in congested, (leafy) cymiform arrays. **Peduncles** 0–1 mm. **Involucres** narrowly campanulate to cylindric. **Phyllaries** distinct, linear or linear-oblanceolate, 5–6 × 1–1.5 mm (dotted with scattered, elliptic oil-glands 0.4–0.5 mm). **Ray florets** 5; corollas 4.5–5.5 mm. **Disc florets** 4–10; corollas 2.5–3 mm (2-lipped). **Cypselae** 2.25–3.25 mm, puberulent; **pappi** (ray and disc similar) of 2–5 antrorsely barbed bristles or awns 1.5–2.5 mm plus shorter, barbellate scales. $2n = 48$.

Flowering mostly Aug–Dec. Open sites, sandy-gravelly soils with grasses and other herbs; 0–50 m; Fla.

Reports of *Pectis linearifolia* from Jamaica were based on misidentifications of diminutive, short-peduncled individuals of *P. glaucescens* (D. J. Keil 1986). In Florida, *P. linearifolia* and *P. glaucescens* are largely allopatric; they occasionally occur in mixed populations. No hybrids are known.

6. **Pectis ×floridana** D. J. Keil, Sida 11: 389. 1986 E

Annuals, 5–30 cm; herbage spicy-scented. **Stems** decumbent to erect, glabrous. **Leaves** linear, 15–35 × 1–2 mm, margins with 3–5 pairs of setae, faces glabrous (abaxial densely dotted with round oil-glands ca. 0.2 mm). **Heads** borne singly or in condensed, cymiform arrays. **Peduncles** 5–25 mm. **Involucres** cylindric to narrowly fusiform. **Phyllaries** coherent (falling together), linear or linear-oblanceolate, 5–7 × 1–2 mm (dotted with scattered, oval oil-glands 0.2–0.3 mm). **Ray florets** 5; corollas 3.5–4.2 mm. **Disc florets** 4–6; corollas 2.2–2.5 mm (2-lipped). **Cypselae** 3–3.5 mm (pericarps darkening, not swelling, ovules abortive), strigillose to short-pilose; **pappi** of 2 (ray) or 5 (disc) slender, antrorsely scabrid, aristate scales 2–2.5 mm, sometimes with additional shorter scales or bristles. $2n = 3x = 36$.

Flowering Sep–Dec. Roadsides; 0–50 m; Fla.

Pectis ×floridana is a triploid hybrid between *P. glaucescens* ($2n = 48$) and *P. prostrata* ($2n = 24$). The hybrids have been observed where the parental taxa grow together in roadside habitats in southern Florida. At the type locality, the hybrids were nearly as common as the parentals. Meiosis in the hybrids is very irregular; resulting pollen grains are malformed and variable in size and apparently all sterile. No fruits have been observed. Regeneration of the hybrids from season to season apparently requires new hybridization events.

The hybrids superficially resemble *Pectis linearifolia* from which they differ in having spicy-scented rather than lemon-scented herbage, longer peduncles, and strongly carinate phyllaries that cohere at the bases and fall together as a group.

7. **Pectis longipes** A. Gray, Smithsonian Contr. Knowl. 5(6): 69. 1853 • Longstalk chinchweed F

Perennials, 8–25 cm (rhizomes branched, 1–10 mm diam.); herbage lemon-scented or spicy-scented. **Stems** ascending to erect (very leafy), glabrous. **Leaves** linear to linear-oblanceolate, 10–55 × 1–3 mm, margins with 1–4 pairs of setae, faces glabrous (conspicuously dotted on margins with round oil-glands 0.3–0.5 mm). **Heads** borne singly. **Peduncles** (30–)50–160 mm. **Involucres** campanulate. **Phyllaries** distinct, linear, linear-oblanceolate, or linear-elliptic, 5–8 × 0.7–2 mm (dotted with 1–3, swollen, subterminal oil-glands 0.3–0.4 mm plus 1–3 pairs of narrow, submarginal oil-glands). **Ray florets** (8–)13 (–15); corollas 8–12 mm. **Disc florets** 25–50; corollas 4–6 mm (2-lipped). **Cypselae** 2.5–4.5 mm, strigillose (hairs tips acute or blunt); **ray pappi** of 1–2 awns 3–3.5 mm; **disc pappi** of 2–30 unequal bristles 3–5 mm. $2n = 24, 48$.

Flowering Apr–Nov. Grasslands, oak-juniper-mesquite woodlands; 900–1700 m; Ariz., N.Mex.; Mexico (Chihuahua, Durango, Sonora).

Pectis longipes has been listed from Texas in floras; I have seen no collections from that state.

Pectis longipes comprises two cytological races. Diploid, spicy-scented plants occur throughout the range. In southern Arizona, the diploid race is broadly sympatric but locally allopatric with a tetraploid, lemon-scented race. The tetraploid race is nested within the range of the diploids. The races are easily separable by odor, and although they are very similar morphologically, they can be separated also by statistically significant differences in floral dimensions and pollen size (M. A. Luckow 1983). Based upon those minute differences, the type collection is diploid. Because the races are so similar morphologically and because so many of the specimens of *P. longipes* in herbaria bear no indication of odor, I chose not to give the cytological races formal recognition.

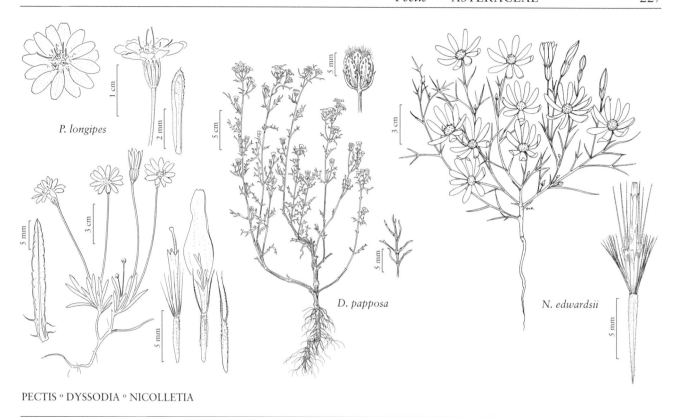

P. longipes

D. papposa

N. edwardsii

PECTIS ∘ DYSSODIA ∘ NICOLLETIA

8. Pectis rusbyi Greene ex A. Gray in A. Gray et al., Syn. Fl. N. Amer. 1(2): 361. 1884 • Rusby's chinchweed

Pectis palmeri S. Watson

Annuals, 5–50 cm (taprooted); herbage spicy-scented. **Stems** erect or ascending, glabrous or sparsely puberulent (in decurrent lines). **Leaves** linear to narrowly elliptic, 10–50 × 1–5 mm, margins with 1–3 pairs of setae, faces glabrous or sparsely puberulent (dotted on margins with round oil-glands 0.2–0.7 mm). **Heads** borne singly or in open, cymiform arrays. **Peduncles** 20–80 mm. **Involucres** campanulate. **Phyllaries** distinct, oblong or narrowly obovate, 4–7 × 1–2 mm (dotted with 0–2, subterminal oil-glands plus 2–4 pairs of inconspicuous, round to narrowly elliptic, submarginal oil-glands). **Ray florets** 8(–13); corollas 5–11 mm. **Disc florets** (7–)20–55; corollas 3.5–5 mm (2-lipped). **Cypselae** 3–4.5 mm, strigillose or short-pilose; **ray pappi** of 1–4, antrorsely barbed awns 1–4 mm or coroniform; **disc pappi** of 15–30, antrorsely barbed bristles 2.5–5 mm or coroniform. $2n$ = 24 (as *P. palmeri*).

Flowering Jul–Oct. Deserts, desert grasslands, arid scrub, dry woodlands; 600–1600 m; Ariz.; Mexico (Baja California Sur, Sinaloa, Sonora).

Pectis rusbyi is much less common in Arizona than *P. papposa* var. *papposa,* with which it sometimes grows.

9. Pectis papposa Harvey & A. Gray, Mem. Amer. Acad. Arts, n. s. 4: 62. 1849 • Common chinchweed

Annuals, 1–30 cm (often forming rounded bushes); herbage spicy-scented. **Stems** ascending, glabrous or puberulent. **Leaves** linear, 10–60 × 1–2 mm, margins with 1–3 pairs of setae, faces glabrous (dotted on margins with round to oval oil-glands 0.3–0.5 mm). **Heads** in congested or open, cymiform arrays. **Peduncles** 3–40 mm. **Involucres** campanulate to cylindric. **Phyllaries** distinct, linear, 3–8 × 0.5–1.7 mm (dotted with 1–5 subterminal oil-glands plus 2–5 pairs of submarginal oil-glands). **Ray florets** (7–)8(–10); corollas 3–8 mm. **Disc florets** 6–34; corollas 2–5.5 mm (weakly 2-lipped, glabrous or glandular-puberulent). **Cypselae** 2–5.5 mm, strigillose to short-pilose (hair tips curled, bulbous); **ray pappi** usually coroniform, rarely of 1+ awns or bristles 1–4 mm; **disc pappi** usually of 16–24, subplumose bristles 1.5–4 mm, rarely coroniform.

Varieties 2 (2 in the flora): sw United States, nw Mexico.

Pectis papposa generally flowers following summer monsoon rains in the desert of southwestern United States and northern Mexico. In favorable years, it becomes an aspect dominant, coloring wide areas of the desert with its bright yellow heads.

1. Phyllaries 3–5 mm; disc florets 6–14(–18); disc pappi 1–2.5 mm . . . . . . 9a. *Pectis papposa* var. *papposa*
1. Phyllaries 5–8 mm; disc florets 12–24(–34); disc pappi 2.5–4 mm . . . . 9b. *Pectis papposa* var. *grandis*

9a. Pectis papposa Harvey & A. Gray var. papposa

Plants 1–20 cm. Leaves 10–30 mm. Peduncles 3–10(–25) mm. Phyllaries 3–5 mm. Ray corollas 3–6 mm. Disc florets 6–14(–18); corollas 2–3.5(–4) mm. Cypselae 2–4.5 mm; disc pappi 1–2.5 mm. 2*n* = 24.

Flowering Jun–Dec. Deserts, desert grasslands, dry woodlands, roadsides; -60–1500 m; Ariz., Calif., Nev., N.Mex., Utah; Mexico (Baja California, Baja California Sur, Sinaloa, Sonora).

Variety *papposa* is widely distributed in the Sonoran and Mojave deserts and reaches the western fringes of the Chihuahuan Desert in southeastern Arizona and southwestern New Mexico, where it intergrades with var. *grandis* in some localities.

9b. Pectis papposa Harvey & A. Gray var. grandis
D. J. Keil, Brittonia 26: 32. 1974

Plants 5–30 cm. Leaves 20–60 mm. Peduncles 10–40 mm. Phyllaries 5–8 mm. Ray corollas 5–8 mm. Disc florets 12–24(–34); corollas 3–5 mm. Cypselae 3–5.5 mm; disc pappi 2.5–4 mm. 2*n* = 24.

Flowering Jun–Dec. Deserts, desert grasslands, dry woodlands, roadsides; 1000–2000 m; Ariz., N.Mex., Tex.; Mexico (Chihuahua, Coahuila).

Variety *grandis* is restricted for the most part to the Chihuahuan Desert with outlying populations in central Arizona. It often grows with *Pectis angustifolia* var. *angustifolia* in trans-Pecos Texas, especially on roadsides; occasional hybrids have been noted.

10. Pectis angustifolia Torrey, Ann. Lyceum Nat. Hist. New York 2: 214. 1827

Annuals or perennials, 1–20 cm (caudices slender, woody); herbage lemon-scented or spicy-scented. Stems erect or ascending, glabrous or puberulent. Leaves linear, 10–45 × 1–3 mm, margins with 2–5 pairs of bristles 1–2 mm, faces glabrous (dotted with oil-glands 0.2–0.7 mm). Heads in congested, cymiform arrays. Peduncles 1–20 mm. Involucres narrowly campanulate or cylindric. Phyllaries distinct, linear or narrowly oblanceolate, 2.5–5.5 × 0.5–1 mm (dotted subterminally with 1 or 2 swollen oil-glands and submarginally with 2–5 pairs of smaller oil-glands). Ray florets 8; corollas 3–5(–7) mm (glandular puberulent or nearly glabrous). Disc florets (7–)10–20; corollas 2.5–3.5 mm (sometimes weakly 2-lipped, glandular puberulent). Cypselae 2.5–4 mm, strigillose (hair tips straight, forked); pappi coroniform and/or of 0–7 scabrid awns or bristles 1–2 mm.

Varieties 3 (3 in the flora): w United States; Mexico.

1. Bases of distal leaves notably expanded (often partially concealing peduncles); pappi usually coroniform, 0.1–0.3 mm (plus 1–7 awns or bristles in some Texas populations); herbage lemon-scented . . . . 10a. *Pectis angustifolia* var. *angustifolia*
1. Bases of distal leaves seldom expanded; pappi either of 0–4 awns 0.5–1(–2) mm and sometimes additionally coroniform, or of 0–7 awns or bristles 1–3 mm and/or coroniform; herbage usually spicy-scented.
   2. Phyllaries widest near apices, each with subterminal oil gland 0.5–1 mm; pappi usually less than 1 mm; annuals or perennials; c Texas (Edwards Plateau) . . . . 10b. *Pectis angustifolia* var. *fastigiata*
   2. Phyllaries widest near middles, each with a subterminal oil-gland less than 0.5 mm; pappi often 1+ mm; annuals; s Texas, Mexico . . . . 10c. *Pectis angustifolia* var. *tenella*

10a. Pectis angustifolia Torrey var. angustifolia
• Limoncillo, lemonscent, crownseed pectis

Pectis angustifolia var. *subaristata* A. Gray

Annuals, 1–20 cm; herbage lemon-scented. Leaves (crowded at stems tips) 10–45 × 1–3 mm, bases of distal blades notably expanded. Peduncles mostly 1–10 mm (often wholly or partially concealed by bases of subtending leaves). Phyllaries linear, 2.5–5.5 mm, widths ± uniform throughout, each with 1–2 subterminal oil-glands 0.2–0.5 mm plus smaller, submarginal oil-glands. Cypselae 2.5–4 mm; pappi usually coroniform, 0.1–0.3 mm (plus 1–7 awns or bristles in some Texas populations). 2*n* = 24 (as *P.* cf. *texana*).

Flowering Jul–Oct. Deserts, grasslands, woodlands, sand bars, roadsides; 300–2300 m; Ariz., Colo., Kans., Nebr., N.Mex., Okla., Tex., Utah, Wyo.; Mexico (Chihuahua, Coahuila, Durango).

Variety *angustifolia* intergrades with var. *tenella* across a broad front from southern Texas to eastern Durango. It is locally sympatric and occasionally hybridizes with *Pectis papposa* var. *grandis* from southern New Mexico and western Texas to eastern Durango. Forms of var. *angustifolia* with a pappus of 1 or more awns in addition to the usual low crown have been called var. *subaristata* A. Gray and may represent introgression from either var. *tenella* or *P. papposa* var. *grandis*.

10b. Pectis angustifolia Torrey var. **fastigiata** (A. Gray) D. J. Keil, Rhodora 79: 60. 1977 • Texas chinchweed
E

Pectis fastigiata A. Gray, Mem. Amer. Acad. Arts, n. s. 4: 62. 1849; *P. texana* Cory

Annuals or perennials, 5–15 cm; herbage spicy-scented. **Leaves** 10–40 × 1–2 mm, bases of distal blades not notably expanded. **Peduncles** 3–30 mm. **Phyllaries** narrowly oblanceolate, 2.5–4.5 mm, widest near apices, each with a conspicuous, subterminal oil-gland 0.5–1 mm plus 1 or 2 pairs of smaller submarginal oil-glands. **Cypselae** 2.5–3.5 mm; **pappi** usually of 1–7 bristles or awns, sometimes coroniform, usually less than 1 mm. **2***n* = 24.

Flowering Jun–Nov. Open areas in grasslands and open woodlands, mostly on shallow limestone soils; 100–700 m; Tex.

Variety *fastigiata* is known only from the Edwards Plateau region of central Texas, mostly on limestone soils. It appears to intergrade in the western portion of its range with var. *angustifolia*.

10c. Pectis angustifolia Torrey var. **tenella** (de Candolle) D. J. Keil, Rhodora 79: 58. 1977 • Mexican chinchweed, low pectis

Pectis tenella de Candolle in A. P. de Candolle and A. L. P. P. de Candolle, Prodr. 5: 99. 1836

Annuals, 2–15 cm; herbage spicy-scented (lemon-scented in some Mexican populations). **Leaves** 10–40 mm, bases of distal blades not notably expanded. **Peduncles** 5–15 mm. **Phyllaries** linear, 3–5 mm, widest near middles, each with 1 subterminal oil-gland 0.2–0.5 mm plus smaller, submarginal oil-glands. **Cypselae** 2.5–4 mm; **pappi** of 0–7 awns or bristles 1–3 mm and/or coroniform. **2***n* = 24.

Flowering Jun–Dec. Deserts, grasslands, arid woodlands, shrublands, roadsides; 100–1100(–2500) m; N.Mex., Tex.; Mexico (Aguascalientes, Chihuahua, Coahuila, Durango, Nuevo León, San Luis Potosí, Tamaulipas, Veracruz, Zacatecas).

Variety *tenella* is widely distributed in the Chihuahuan Desert and adjacent regions. The isolated occurrence in northern New Mexico perhaps represents an introduction.

11. Pectis filipes Harvey & A. Gray, Mem. Amer. Acad. Arts, n. s. 4: 62. 1849 • Threadstalk chinchweed

Varieties 2 (1 in the flora): w United States, Mexico. Variety *filipes* is known only from Mexico.

11a. Pectis filipes Harvey & A. Gray var. **subnuda** Fernald, Proc. Amer. Acad. Arts 33: 76. 1897

Annuals, 5–40 cm; herbage spicy-scented. **Stems** erect or ascending, glabrous or puberulent. **Leaves** linear to narrowly elliptic, 10–60 × 0.5–5.5 mm, margins with 1–4 pairs of setae, faces glabrous or puberulent (dotted on margins with round oil-glands ca. 0.5 mm). **Heads** borne singly or in open, cymiform arrays. **Peduncles** 20–65 mm. **Involucres** cylindric. **Phyllaries** distinct, narrowly oblong to oblong-obovate, 3.5–6 mm (dotted with 0–2, subterminal oil-glands plus 0–3 pairs of submarginal oil-glands). **Ray florets** 5; corollas 4–9 mm. **Disc florets** 7–12; corollas 3–4 mm (weakly 2-lipped). **Cypselae** 2.5–4 mm, strigillose to short-pilose; **pappi** of 0–3 awns 3–4 mm usually with a shorter crown. **2***n* = 24.

Flowering Jun–Nov. Deserts, grasslands, pine-oak-juniper woodlands, shrublands; 1000–2000 m; Ariz., N.Mex., Tex.; Mexico (Chihuahua, Sonora).

Hybrids of var. *subnuda* and *Pectis papposa* var. *papposa* are known from southeastern Arizona. In northern Mexico, hybrids occur between *P. filipes* and other *Pectis* species.

12. Pectis imberbis A. Gray, Smithsonian Contr. Knowl. 5(6): 70. 1853 • Tall or beardless chinchweed

Perennials, 30–120 cm (caudices woody, 2–8+ mm diam.); herbage unscented. **Stems** erect, glabrous. **Leaves** narrowly linear, 10–50 × 1–2 mm (sometimes smaller, bractlike distally), margins with 0–1 pairs of setae, faces glabrous (abaxial dotted near each margin with a row of elliptic oil-glands ca. 0.3 mm). **Heads** borne singly or in open, cymiform arrays. **Peduncles** 10–80 mm. **Involucres** cylindric. **Phyllaries** distinct, linear-oblong, 5–9.5 × 1–1.5 mm (each dotted with 1–2 swollen, subapical oil-glands and a row of 2–3 linear, submarginal oil-glands on each side of mid-rib). **Ray florets** 5; corollas 6–11 mm (laminae often dotted near margins with inconspicuous oil-glands). **Disc florets** 4–7; corollas 3.7–6 mm (lobes 5, equal, each with 1 subterminal oil-gland). **Cypselae** 3.5–5 mm, puberulent (hair tips blunt); **pappi** of 1–3 stout awns 1–2 mm or coroniform. **2***n* = 24.

Flowering Aug–Oct. Pine-oak-juniper woodlands, grasslands, arid shrublands; 1000–1700 m; Ariz.; Mexico (Chihuahua, Sonora).

Pectis imberbis occurs in relatively small, widely separated populations. Overgrazing may be a factor in the scarcity of these plants. They are generally more than 25 cm before they begin to flower and may be unable to reproduce under grazing pressure.

13. Pectis linifolia Linnaeus, Syst. Nat. ed. 10, 2: 1221. 1759 • Romero-macho

Varieties 2 (1 in the flora): sw United States, Mexico, West Indies, Central America, South America, Pacific Islands.

13a. Pectis linifolia Linnaeus var. **linifolia**

Pectis linifolia var. *marginalis* Fernald; *P. punctata* Jacquin

Annuals, 10–100 cm; herbage unscented. **Stems** erect or ascending, glabrous. **Leaves** linear to linear-elliptic, 20–60 × 1–4.5 mm, margins with 0–1 pairs of setae, faces glabrous (abaxial dotted with scattered or marginal oil-glands, adaxial often with 1 elongate, medial, subterminal oil-gland). **Heads** borne singly or in diffuse, cymiform arrays.

Peduncles 10–30 mm. **Involucres** cylindric. **Phyllaries** distinct, linear, 4.5–7.5 × 0.5 mm (dotted submarginally with linear-elongate to elliptic oil-glands and subapically with 1+ elliptic oil-glands). **Ray florets** 5; corollas 2–3 mm. **Disc florets** 1–3; corollas 2–3.5 mm (lobes 5, equal, 4 or 5 each with 1 round oil-gland, throats sometimes with additional oil-glands). **Cypselae** 3.5–5 mm, strigillose or hirtellous (ray cypselae and subtending phyllaries widely spreading at maturity with pappi projecting downward); **pappi** of 1–4 stiff, often divaricate, awns 2–3 mm. $2n = 24$.

Flowering Aug–Dec. Desert scrub, arid grasslands; 100–1200 m; Ariz., Fla.; Mexico, West Indies, Central America, South America, Pacific Islands (Galapagos Islands, Hawaii).

Pectis linifolia is widespread with disjunctions among populations. Most of it is attributable to the widespread var. *linifolia*. Variety *hirtella* S. F. Blake is known only from southern Mexico. The wide distribution of var. *linifolia* is probably attributable to its adaptations for animal dispersal. The cypselae and spreading pappus awns resemble tiny grappling hooks. Field observations indicate that the cypselae readily cling to fabrics, and presumably they cling equally well to feathers and fur. The plants are autogamous and a single long-distance dispersal event is probably sufficient to establish a new population.

317. DYSSODIA Cavanilles, Descr. Pl., 202. 1802 • [Greek *dysodia*, a bad odor]

John L. Strother

Annuals [perennials], 10–30(–70+) cm. **Stems** erect to decumbent, branched from bases or throughout. **Leaves** cauline; mostly opposite (distal sometimes alternate); blades (1–)2–3-pinnatisect, primary lobes linear to linear-cuneate, ultimate margins entire or toothed, faces puberulent (little, if at all, setaceous at bases, on teeth, or at tips of lobes, oil-glands submarginal). **Heads** radiate, borne singly or in 2s or 3s [pseudocephalia]. **Calyculi** of [0] 1–9 ± linear bractlets (lengths 1/2–1 phyllaries, bearing oil-glands). **Involucres** turbinate to campanulate [hemispheric], 5–10 mm diam. **Phyllaries** persistent, 6–12 in ± 2 series (distinct to bases or nearly so, oval-oblanceolate, each bearing 1–7 round to elliptic oil-glands). **Receptacles** convex, ± pitted (socket margins fimbrillate to setose), epaleate. **Ray florets** usually 5–8, pistillate, fertile; corollas yellow-orange. **Disc florets** 12–50[–100+], bisexual, fertile; corollas yellow to orange, tubes shorter than cylindric throats, lobes 5, deltate to lance-deltate. **Cypselae** obpyramidal to obconic, subsericeous or glabrescent; **pappi** persistent, of 15–20 scales in ± 2 series (each scale comprising 5–10 basally connate, unequal bristles). $x = 13$.

Species 4 (1 in the flora): North America, Mexico, Central America; introduced in South America.

1. Dyssodia papposa (Ventenant) A. Hitchcock, Trans. Acad. Sci. St. Louis 5: 503. 1891 • Dogweed [F]

Tagetes papposa Ventenant, Descr. Pl. Nouv., plate 36. 1801

Leaves 15–50 × 10–40 mm overall, ultimate lobes (7–)11–15, 5–20 × 1–3 mm, glabrous or sparsely hairy, dotted with oil-glands. **Peduncles** 1–5(–10) mm. **Involucres** 6–10 mm. **Phyllaries** each bearing 1–7 oil-glands. **Ray laminae** 1.5–2.5 × 1–2 mm. **Disc corollas** ca. 3 mm. **Cypselae** 3–3.5 mm; **pappi** 1–3 mm. $2n = 26$.

Flowering summer–fall. Grasslands, open woodlands, often ruderal, fields, along roadways; 0–2000 m; Ont.; Ariz., Ark., Calif., Colo., Ill., Ind., Iowa, Kans., Ky., La., Maine, Mass., Mich., Minn., Mo., Mont., Nebr., Nev., N.H., N.Mex., N.Y., N.Dak., Ohio, Okla., Pa., S.Dak., Tenn., Tex., Utah, Vt., W.Va., Wis., Wyo.; Mexico; Central America; introduced in South America.

Records of *Dyssodia papposa* from Ontario and from California, Maine, Massachusetts, New Hampshire, New York, and Vermont evidently document local, probably ephemeral, introductions. In 1837, C. W. Short noted of *D. papposa* on a specimen label, "This plant is so abundant, and exhales an odor so unpleasant as to sicken the traveler over the western prairies of Illinois, in autumn."

318. NICOLLETIA A. Gray in J. C. Frémont, Rep. Exped. Rocky Mts., 315. 1845 • [For Jean Nicholas Nicollet, 1786–1843, "…who spent several years in exploring the country watered by the Mississippi and Missouri rivers, and who was employed by the United States Government in a survey of the region…." Quoted from protologue.]

John L. Strother

Annuals or perennials, (5–)10–50 cm, often glaucous. **Stems** erect or spreading, branched from bases or throughout. **Leaves** cauline; mostly alternate (1–4 proximal pairs opposite); blades mostly 1-pinnately lobed (lobes usually 3–11, bristle-tipped), ultimate margins entire, faces glabrous (often glaucous, oil-glands subterminal). **Heads** radiate, borne singly. **Calyculi** of 2–6 deltate to lanceolate bractlets (each usually bearing 1–5 oval to linear oil-glands). **Involucres** campanulate to turbinate or fusiform, 4–8(–10) mm diam. **Phyllaries** persistent, 6–12 in ± 2 series (distinct to bases, lanceolate or ovate to linear, most bearing 1–5 oil-glands). **Receptacles** convex to conic, ± pitted, epaleate. **Ray florets** 7–12, pistillate, fertile; corollas whitish with pinkish to purplish stripes. **Disc florets** 15–100+, bisexual, fertile; corollas yellow or purplish or whitish, tubes much shorter than cylindric throats, lobes 5, deltate to lanceolate. **Cypselae** clavate, sparsely puberulent (hairs usually reddish); **pappi** persistent, of 5(–6) fascicles of 7–15 barbellate bristles subtending and alternating with 5(–6) lanceolate, 1-aristate scales. $x = 10$.

Species 3 (2 in the flora): sw United States, n Mexico.

Nicolletia trifida Rydberg is known from Baja California and Baja California Sur, Mexico.

SELECTED REFERENCE Strother, J. L. 1978. Taxonomy and geography of *Nicolletia* (Compositae: Tageteae). Sida 7: 369–374.

1. Leaf lobes 5–11, rachis widths mostly 2–3 times lobe widths; ray laminae 4–8 mm; disc florets (30–)60–100+ .... 1. *Nicolletia occidentalis*
1. Leaf lobes mostly 3–5, rachis widths hardly greater than lobe widths; ray laminae 7–16 mm; disc florets 15–25(–50) .... 2. *Nicolletia edwardsii*

1. Nicolletia occidentalis A. Gray in J. C. Frémont, Rep. Exped. Rocky Mts., 316. 1845 • Hole-in-the-sand plant

Perennials, (5–)10–29 cm (deeply taprooted). **Stems** erect. **Leaf blades** 20–70 mm overall, lobes 5–11, ± quadrate to linear, rachis widths mostly 2–3 times lobe widths. **Peduncles** 2–10 mm. **Calyculi** of 2–4 bractlets 4–8 mm. **Involucres** turbinate to cylindric, 14–18 mm. **Phyllaries** 8–12, linear to ovate. **Ray florets** 8–12, laminae 4–9 × 2.5–4 mm. **Disc florets** (30–)60–100+; corollas yellow, purple-tipped, 8–9.5 mm. **Cypselae** 7–9 mm; **pappi:** bristles 3–7 mm, scales 6–8 mm. $2n = 20$.

Flowering spring. Deep sands of fans and floors of desert washes; 600–1400 m; Calif.; Mexico (Baja California).

2. Nicolletia edwardsii A. Gray, Smithsonian Contr. Knowl. 3(5): 119. 1852 [F]

Annuals or perennials, 3–25 cm. **Stems** erect to spreading. **Leaves** 10–60 mm overall, lobes 3–5, linear to filiform, rachis widths hardly greater than lobe widths. **Peduncles** 5–30 mm. **Calyculi** of 4–6 bractlets 1–2 mm. **Involucres** turbinate to fusiform, 11–15 mm. **Phyllaries** 6–8, lanceolate. **Ray florets** 7–9, laminae 7–16 × 3–6 mm. **Disc florets** 15–25(–50); corollas yellow, 7–8 mm. **Cypselae** 5–7 mm; **pappi:** bristles 3–5 mm, scales 5–7 mm. $2n = 20$.

Flowering mostly fall, following rains. Sandy, often calcareous or gypseous soils of fans, flats, or playas in desert scrub; 800–1700 m; Tex.; Mexico (Chihuahua, Coahuila, San Luis Potosí, Zacatecas).

319. CHRYSACTINIA A. Gray, Mem. Amer. Acad. Arts, n. s. 4: 93. 1849 • [Greek *chrysos*, gold, and *actinos*, ray]

John L. Strother

Subshrubs or shrubs, 10–40[–80] cm (evergreen). **Stems** erect, strictly branched. **Leaves** cauline; mostly alternate [opposite]; blades simple [pinnate], linear to clavate or acerose (± fleshy), margins entire, faces usually glabrous, sometimes puberulent (oil-glands marginal or submarginal). **Heads** radiate, borne singly. **Calyculi** 0. **Involucres** turbinate to hemispheric, 3.5–5 [–8] mm diam. **Phyllaries** persistent, 8–14 in ± 2 series (distinct to bases, linear to lance-linear [ovate], carinate, each usually bearing 1–5 oil-glands). **Receptacles** convex to hemispheric, ± pitted, rarely paleate (paleae readily falling). **Ray florets** 8(–13), pistillate, fertile; corollas bright yellow. **Disc florets** 15–70, bisexual, fertile; corollas yellow to orange, tubes much shorter than cylindro-funnelform throats, lobes 5, lance-deltate. **Cypselae** cylindric to fusiform, strigillose or glabrescent; **pappi** persistent, of [20–]30–40 bristles in ± 1 series. $x = 15$.

Species 5 (1 in the flora): sw United States, Mexico.

SELECTED REFERENCE Strother, J. L. 1977. Taxonomy of *Chrysactinia*, *Harnackia*, and *Lescaillea* (Compositae: Tageteae). Madroño 24: 129–139.

1. Chrysactinia mexicana A. Gray, Mem. Amer. Acad. Arts, n. s. 4: 93. 1849 [F]

Plants twiggy. **Leaf blades** 5–10 (–23) × 1–2 mm. **Peduncles** 15–75 mm (heads usually held well beyond foliage). **Ray laminae** 6–12 × 2–4 mm. **Disc corollas** 4.5–7 mm. **Cypselae** 3–4 mm; **pappi** 3–5.5 mm. $2n = 45$.

Flowering spring–fall. Mostly on limestone and derived soils; 200–2200+ m; N.Mex., Tex.; Mexico.

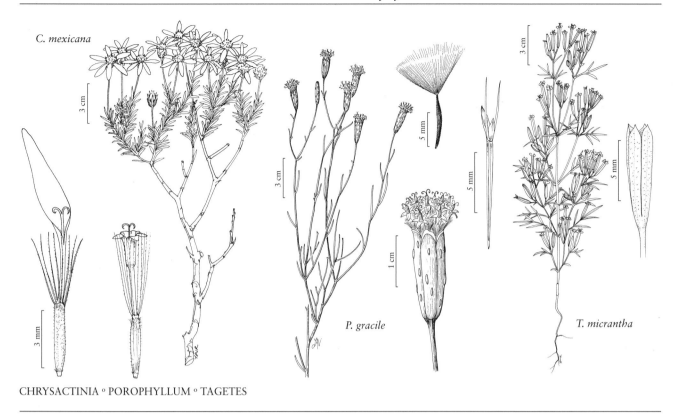

C. mexicana

P. gracile

T. micrantha

CHRYSACTINIA ∘ POROPHYLLUM ∘ TAGETES

320. **POROPHYLLUM** Guettard, Hist. Acad. Roy. Sci. Mém. Math. Phys. (Paris, 4to) 1750: 377. 1754 • [Greek *poros*, hole, and *phyllon*, leaf, alluding to gland-dotted leaf blades]

John L. Strother

Annuals, perennials, subshrubs, or shrubs, 10–120[–200+] cm. **Stems** erect, usually strictly branched. **Leaves** cauline; opposite or alternate; petiolate or sessile; blades orbiculate to linear or filiform, margins crenate or entire, faces usually glabrous (oil-glands scattered and/or at margins). **Heads** discoid, borne singly or in loose, ± corymbiform arrays. **Calyculi** 0. **Involucres** cylindric to campanulate [turbinate], 3–12 mm diam. **Phyllaries** 5–10 in ± 2 series (distinct to bases, oblong or lanceolate to linear, bearing oval to linear oil-glands). **Receptacles** convex to conic, pitted, epaleate. **Ray florets** 0. **Disc florets** (5–)10–80[–100+], bisexual, fertile; corollas usually yellow, sometimes whitish to greenish or purplish [brownish], tubes either very slender, much longer than funnelform throats, or stout, much shorter than narrowly cylindric throats, lobes 5, deltate to lanceolate (often unequal). **Cypselae** cylindric to fusiform, hirtellous to villous [glabrous]; **pappi** persistent, of 25–50(–100) coarse to fine bristles in 1–2+ series. $x = 12$.

Species ca. 25 (5 in the flora): sw United States, Mexico, West Indies (Antilles), Central America, South America.

SELECTED REFERENCE   Johnson, R. R. 1969. Monograph of the plant genus *Porophyllum* (Compositae: Helenieae). Univ. Kansas Sci. Bull. 48: 225–267.

1. Annuals; leaf blades oval or elliptic to obovate, 8–25+ mm wide . . . . 5. *Porophyllum ruderale*
1. Perennials, or subshrubs or shrubs; leaf blades linear to filiform, 1–5 mm wide.
   2. Corollas whitish or purplish . . . . 1. *Porophyllum gracile*
   2. Corollas yellow (sometimes tinged reddish).
     3. Subshrubs or shrubs, 20–60+ cm; phyllaries 7–10 . . . . 2. *Porophyllum scoparium*
     3. Perennials (rarely woody at bases), mostly 5–30 cm; phyllaries 5(–8).
       4. Internodes mostly 1–5(–12) mm; leaf blades 8–15 × 2 mm; pappi: longer bristles
         6–7 mm . . . . 3. *Porophyllum pygmaeum*

       4. Internodes mostly 10–20+ mm; leaf blades 20–60 × 1–2 mm; pappi: longer bristles 8–10 mm . . . . 4. *Porophyllum greggii*

1. Porophyllum gracile

Bentham, Bot. Voy. Sulphur, 29. 1844 [F]

Subshrubs or shrubs, 20–70 cm; internodes mostly 10–30 mm. **Leaf blades** linear to filiform, 20–50 × 1–5 mm. **Phyllaries** 5, oblong to linear, 9–15 mm. **Florets** (5–)12–30; corollas whitish to purplish, 6–9 mm. **Cypselae** 6–10 mm; **pappi:** longer bristles 7–9 mm. $2n = 48$.

Flowering late winter–early summer; fall. Sandy, gravelly, and rocky sites (usually granitic) with desert scrub; 200–2100 m; Ariz., Calif., N.Mex., Tex.; Mexico (Baja California, Sonora).

2. Porophyllum scoparium

A. Gray, Smithsonian Contr. Knowl. 3(5): 119. 1852

Subshrubs or shrubs, 20–60+ cm; internodes mostly 10–30 mm. **Leaf blades** mostly filiform, 10–40 × 1–2 mm. **Phyllaries** 7–10, oblanceolate to linear, 6–10 mm. **Florets** mostly 40–80+; corollas yellow, 5–8 mm. **Cypselae** 3.5–5.5 mm; **pappi:** longer bristles 6–7.5 mm. $2n = 24$.

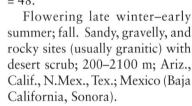
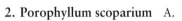

Flowering spring–fall. Sandy, gravelly, and rocky sites, calcareous or igneous soils, with desert scrub; 500–1400 m; N.Mex., Tex; Mexico (Chihuahua, Coahuila, Nuevo León).

3. Porophyllum pygmaeum

D. J. Keil & Morefield, Syst. Bot. 14: 583. 1989 [E]

Perennials, 5–15 cm; internodes mostly 1–5(–12) mm. **Leaf blades** mostly filiform, 8–15 × 1–2 mm. **Phyllaries** 5(–8), ovate to oblong or linear, 8–11 mm. **Florets** mostly 25–75+; corollas yellow, 7–9 mm. **Cypselae** 7–8 mm; **pappi:** longer bristles 6–7 mm. $2n = 24$.

Flowering spring–summer. Sandy soils, with desert scrub; 1300–2100 m; Nev.

4. Porophyllum greggii

A. Gray, Smithsonian Contr. Knowl. 3(5): 120. 1852 [E]

Perennials, 15–30 cm; internodes mostly 10–20+ mm. **Leaf blades** mostly filiform, 20–60 × 1–2 mm. **Phyllaries** 5, ovate to oblong or elliptic, 9–12 mm. **Florets** mostly 50–80+; corollas yellow, 7.5–12 mm. **Cypselae** 7–8 mm; **pappi:** longer bristles 8–10 mm. $2n = 72$.

Flowering spring–fall. Open, rocky sites, grasslands and desert scrub; 1200–2000 m; Tex.

5. Porophyllum ruderale

(Jacquin) Cassini in F. Cuvier, Dict. Sci. Nat. ed. 2, 43: 56. 1826

Kleinia ruderale Jacquin, Enum. Syst. Pl., 28. 1760

Varieties 2 (1 in the flora): sw United States, Mexico, West Indies, Central America, South America.

5a. Porophyllum ruderale

(Jacquin) Cassini var. **macrocephalum** (de Candolle) Cronquist, Madroño 20: 255. 1970

Porophyllum macrocephalum de Candolle in A. P. de Candolle and A. L. P. P. de Candolle, Prodr. 5: 648. 1836; *P. ruderale* subsp. *macrocephalum* (de Candolle) R. R. Johnson

Annuals, 15–50+ cm; internodes mostly 15–30+ mm. **Leaf blades** oval or elliptic to obovate, 15–35+ × 8–25+ mm. **Phyllaries** 5, linear, 15–25+ mm. **Florets** mostly

30–50+; corollas greenish to purplish, 9–12.5 mm. **Cypselae** 9–12 mm; **pappi:** longer bristles 7–10 mm. $2n = 22$.

Flowering late summer–fall. Ephemerally wet sites in desert mountains; 1000–1500 m; Ariz., N.Mex., Tex.; Mexico; West Indies (Antilles); Central America; South America.

321. TAGETES Linnaeus, Sp. Pl. 2: 887. 1753; Gen. Pl. ed. 5, 378. 1754 • Marigold

[Etymology unknown; possibly from New Latin *Tages*, an Etruscan god]

John L. Strother

Annuals, perennials, subshrubs, or shrubs [perennials], mostly 10–80(–200) cm. **Stems** erect, branched distally or ± throughout. **Leaves** cauline; mostly opposite (distal sometimes alternate); petiolate or sessile; blades mostly lanceolate to oblanceolate overall, usually 1–3-pinnately lobed or -pinnatisect, ultimate margins toothed or entire, faces glabrous or hairy (oil-glands scattered and/or submarginal). **Heads** radiate or discoid, borne singly or in ± corymbiform arrays. **Calyculi** 0. **Involucres** narrowly cylindric or fusiform to turbinate or broadly campanulate, 1–12+ mm diam. **Phyllaries** persistent, 3–21+ in 1–2 series (connate to $^7/_8$+ their lengths, usually streaked and/or dotted with oil-glands). **Receptacles** convex to conic, smooth or finely pitted, epaleate. **Ray florets** 0 or 1–8(–13+; to 100+ in "double" cultivars), pistillate, fertile (except "double" cultivars); corollas yellow or orange, red-brown (with or without yellow/orange), or white. **Disc florets** 6–120+, bisexual, fertile; corollas greenish yellow to orange, sometimes tipped with red or red-brown, tubes much longer than or about equaling funnelform throats, lobes 5, deltate to lance-linear (equal or 2 sinuses deeper than others). **Cypselae** narrowly obpyramidal or fusiform-terete, sometimes weakly flattened, glabrous or hairy; **pappi** persistent, of 2–5 (–10 dissimilar, distinct or connate scales in ± 1 series: 0–5+ oblong to lanceolate, erose-truncate or laciniate plus 0–2(–5) longer, subulate to aristate. $x = 12$.

Species 40+ (4 in the flora): tropical and warm-temperate America, especially Mexico; introduced in Old World.

Some *Tagetes* species (e.g., *T. erecta*) produce nematicidal thiophenes in their roots and have been shown to be effective controls for nematodes in diverse crops (cf., http://www.ncagr.com/agronomi/nnote1.htm).

Reports of "*Tagetes minima* L." for Pennsylvania (cf. http://plants.usda.gov) are evidently rooted in an error for *T. minuta*. Report of *T. pusilla* Kunth (= *T. filifolia* Lagasca) for Maryland (http://plants.usda.gov) was not verified for this treatment.

SELECTED REFERENCES Neher, R. T. 1966. Monograph of the Genus *Tagetes* (Compositae). Ph.D. thesis. Indiana University. Rydberg, P. A. 1915. *Tagetes*. In: N. L. Britton et al., eds. 1905+. North American Flora.... 47+ vols. New York. Vol. 34, pp. 148–159. Soule, J. A. 1993. Systematics of *Tagetes* (Asteraceae–Tageteae) (Mexico, Argentina). Ph.D. thesis. University of Texas.

1. Leaf lobes (or simple blades) linear to filiform, 10–25(–35+) × 0.5–1.5 mm . . . . 1. *Tagetes micrantha*
1. Leaf lobes lance-elliptic or lanceolate to lance-linear, 12–25(–50+) × (2–)4–8(–12+) mm.
   2. Perennials, subshrubs, or shrubs . . . . 2. *Tagetes lemmonii*
   2. Annuals.
      3. Heads in ± corymbiform clusters; peduncles 1–5+ mm; involucres 7–10+ × 1.5–3 mm . . . . 3. *Tagetes minuta*
      3. Heads borne ± singly; peduncles 30–100(–150+) mm; involucres 10–22 × (3–)5–12+ mm . . . . 4. *Tagetes erecta*

1. **Tagetes micrantha**   Cavanilles, Icon. 4: 31, plate 352. 1797 [F]

Annuals, 3–15(–35+) cm. **Leaf blades** 10–25(–35+) overall, lobes or leaflets 0 or 3–5, linear to filiform, 10–25(–35+) × 0.5–1.5+ mm. **Heads** borne ± singly. **Peduncles** 10–50+ mm. **Involucres** 9–12+ × 1–2 mm. **Ray florets** 0–1+; laminae ochroleucous to yellowish, ± oblong, 1.5–2.5 mm. **Disc florets** 5–6+; corollas 2.5–3.5 mm. **Cypselae** 7 mm; **pappi** of 2 ± subulate-aristate scales 4–5+ mm plus 2 distinct, oblong to ovate, ± erose scales 1.5–2.5 mm. **2n = 24.**

Flowering Aug–Sep. Disturbed sites, moist soils in meadows, juniper woodlands, pine forests; 1500–2600 m; Ariz., N.Mex., Tex.; Mexico

2. **Tagetes lemmonii**   A. Gray, Proc. Amer. Acad. Arts 19: 40. 1883 (as lemmoni)

Perennials, subshrubs, or shrubs 10–50(–100+) cm. **Leaf blades** 50–70(–120+) mm overall, lobes or leaflets (3–)5–7+, lance-elliptic to lanceolate, 15–30(–45) × 3–7(–10+) mm. **Heads** in ± corymbiform arrays. **Peduncles** 10–50 mm. **Involucres** 8–10 × 5–7 mm. **Ray florets** (3–)5–8+; laminae yellow, ± oblong, 9–15+ mm. **Disc florets** 12–30+; corollas 6–7 mm. **Cypselae** 5–6 mm; **pappi** of 0–5 lanceolate- to subulate-aristate scales 2.5–3+ mm plus 5+ lanceolate to oblong, ± erose scales 0.5–1 mm, distinct or connate, linear-oblong, ± erose scales 2–6+ mm. **2n = 24.**

Flowering (Aug–)Sep–Nov. Moist sites, grasslands, scrublands, woodlands, cliffs, streamsides; 1400–2500 m; Ariz.; Mexico (Sinaloa, Sonora).

3. **Tagetes minuta**   Linnaeus, Sp. Pl. 2: 887. 1753 [I]

Annuals, 30–100(–180+) cm. **Leaf blades** 80–150+ mm overall, lobes or leaflets 9–17+, narrowly lanceolate to lance-linear, 12–25(–50+) × (2–)4–7+ mm. **Heads** in ± corymbiform clusters. **Peduncles** 1–5+ mm. **Involucres** 7–10+ × 1.5–3 mm. **Ray florets** 1–3; laminae yellow, ± ovate to elliptic, 1–2 mm. **Disc florets** 3–5; corollas 3–4 mm. **Cypselae** (4.5–)6–7+ mm; **pappi** of 1–2 ± subulate scales 2–3+ mm plus 3–5 distinct, ovate to lanceolate scales 0.5–1 mm. **2n = 48.**

Flowering Sep–Oct. Disturbed sites; 0–100+ m; introduced; Ala., Calif., Conn., Fla., Ga., Md., Mass., N.C., Pa., S.C., Va.; South America.

Tagetes minuta is widely cultivated for use as a condiment and has become widely established or persists after plantings.

4. **Tagetes erecta**   Linnaeus, Sp. Pl. 2: 887. 1753
   • African or French marigold

Tagetes patula Linnaeus; *T. tenuifolia* Cavanilles

Annuals, 10–120+ cm. **Leaf blades** 30–120(–250+) mm overall, principal lobes/leaflets 9–25, lanceolate to lance-linear, 15–25 (–45+) × 3–8(–12+) mm. **Heads** borne ± singly. **Peduncles** 30–100(–150+) mm. **Involucres** 10–22+ × (3–)5–12+ mm. **Ray florets** (3–)5–8+ (to 100+ in "double" cultivars); laminae yellow to orange, red-brown (sometimes particolored: yellow/red-brown), or white (some cultivars), ± flabellate to oval-quadrate, (2–)12–18(–25+) mm. **Disc florets** (10–)50–120+; corollas 7–12+ mm. **Cypselae** 6–11 mm; **pappi** of 0–2, ± subulate-aristate scales 6–12+ mm plus 2–4 distinct or connate, linear-oblong, ± erose scales 2–6+ mm. **2n = 24, 48.**

Flowering ± year round. Disturbed sites; 0–1000+ m; Ala., Ark., Calif., Conn., Fla., Ill., Kans., Ky., La., Md., Mass., Mo., N.Y., N.C., Ohio, Okla., Pa., S.C., Utah, Vt., Va., Wis., Wyo.; Mexico.

Cultivars of *Tagetes erecta* are widely grown in gardens and, commercially, for cut flowers. They often persist after abandoned plantings. Tetraploid plants (2n = 48) with smaller involucres and wholly or partially red-brown corollas included here in *T. erecta* have been called *T. patula* by some botanists.

322. DYSODIOPSIS (A. Gray) Rydberg in N. L. Britton et al., N. Amer. Fl. 34: 170. 1915

  • [Generic name *Dyssodia* and Greek *-opsis*, resembling]   E

John L. Strother

Hymenatherum Cassini [unranked] *Dysodiopsis* A. Gray, Smithsonian Contr. Knowl. 3(5): 116. 1852

Annuals or perennials, 40–80+ cm. **Stems** erect, distally branched. **Leaves** cauline; mostly opposite (distal alternate); blades linear, margins coarsely toothed, faces glabrous (bases ± setaceous, oil-glands scattered along midveins). **Heads** radiate, in loose, corymbiform arrays or borne singly. **Calyculi** of 5–8 subulate or pinnatisect bractlets (bearing oil-glands). **Involucres** ± campanulate, 5–8 mm diam. **Phyllaries** persistent, 10–12 in 2 series (strongly connate, oblong to linear, margins of outer free to bases, faces of all or outer bearing submarginal and subapical oil-glands). **Receptacles** convex, pitted (socket margins fimbrillate), epaleate. **Ray florets** 7–12, pistillate, fertile; corollas lemon to greenish yellow. **Disc florets** 20–40+, bisexual, fertile; corollas dull yellow, tubes shorter than cylindro-funnelform throats, lobes 5, deltate. **Cypselae** obpyramidal, glabrous or sparsely strigillose; **pappi** persistent, of 10–12, 1(–3)-aristate scales in ± 2 series. $x = 13$.

Species 1: sc United States.

1. Dysodiopsis tagetoides (Torrey & A. Gray) Rydberg in N. L. Britton et al., N. Amer. Fl. 34: 171. 1915

E   F

Dyssodia tagetoides Torrey & A. Gray, Fl. N. Amer. 2: 361. 1842

Leaf blades 40–90 × 2–6 mm. **Peduncles** 30–50 mm. **Ray laminae** linear-elliptic, 10–15 × 3–6 mm. **Disc corollas** ca. 4 mm. **Cypselae** 3–3.5 mm; **pappi:** scales 1–2.5 mm. $2n = 26$.

Flowering late spring–summer. Limestone outcrops and derived soils; 100–700 m; Okla., Tex.

323. ADENOPHYLLUM Persoon, Syn. Pl. 2: 458. 1807 • [Greek *adeno*, gland, and *phyllon*, leaf]

John L. Strother

Annuals, perennials, or subshrubs [shrubs], (20–)30–70[–250+] cm. **Stems** erect, branched from bases or throughout. **Leaves** cauline; mostly opposite (distal sometimes alternate); petiolate or sessile; blades (or lobes) ovate, oblanceolate, or linear, 1-pinnately lobed, ultimate margins ± toothed (teeth often bristle-tipped), faces glabrous or ± hirtellous to scabrellous (oil-glands usually borne at bases of lobes and subterminally near apices). **Heads** usually radiate, rarely discoid, borne singly. **Calyculi** of 3–5 or 12–22 linear to subulate bractlets (sometimes with 1–2 lobes, often bristle-tipped, usually bearing oil-glands). **Involucres** campanulate to obconic

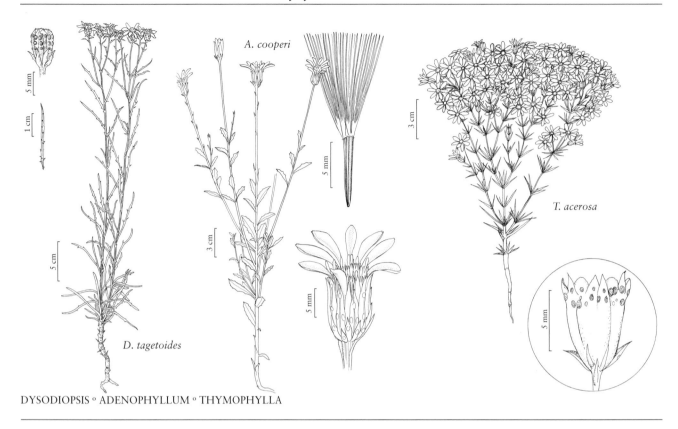

DYSODIOPSIS ∘ ADENOPHYLLUM ∘ THYMOPHYLLA

[hemispheric], [5–]7–20+ mm diam. **Phyllaries** persistent, 8–20+ in ± 2 series (weakly connate $^1/_3$–$^3/_4$ their lengths, separating in age, mostly lanceolate to linear, margins of outer distinct to bases or nearly so, oil-glands in faces and near margins). **Receptacles** convex, ± pitted (socket margins fimbrillate to setose), epaleate. **Ray florets** usually 7–16[–16], rarely 0, pistillate, fertile; corollas yellow to orange or red-orange. **Disc florets** 25–70[100+], bisexual, fertile; corollas yellow to yellow-orange, tubes shorter than cylindric throats, lobes 5, lance-ovate to lance-linear. **Cypselae** obpyramidal [obconic], glabrous or ± strigose or sericeous; **pappi** persistent, of 8–20 scales in 1–2 series (single scales muticous, or unequally 3-aristate, or comprising 5–11 basally connate bristles). x = 7, 13.

Species 10 (3 in the flora): sw United States, Mexico, Central America.

1. Annuals; calyculus bractlets eglandular; pappi: outer scales erose, inner each unequally 3-aristate .... 3. *Adenophyllum wrightii*
1. Perennials or subshrubs; calyculus bractlets bearing oil-glands; pappi: each scale comprising 5–11 basally connate bristles.
    2. Leaves 3–5-lobed (lobes linear to cuneate or oblanceolate, ultimate margins entire or toothed); disc florets 25–40; pappi of 8–12 scales .... 1. *Adenophyllum porophylloides*
    2. Leaves not lobed (blades ovate to oblanceolate, margins toothed); disc florets 50–80; pappi of 15–20 scales .... 2. *Adenophyllum cooperi*

1. **Adenophyllum porophylloides** (A. Gray) Strother, Sida 11: 377. 1986

Dyssodia porophylloides A. Gray, Mem. Amer. Acad. Arts, n. s. 5: 322. 1854

Perennials or subshrubs, 20–50+ cm. **Leaves** pinnately lobed, 15–40 mm, lobes 3–5, linear to cuneate or oblanceolate (oil-glands at base of each lobe and subterminal in tips). **Peduncles** 20–80 mm. **Calyculi** of 12–16 subulate bractlets 3–8 mm (bearing glands). **Involucres** obconic, 10–15 mm. **Phyllaries** 12–20, lanceolate, separating in fruit. **Ray florets** 10–14; corollas yellow, becoming red-orange; tubes ca. 2 mm, laminae 6 × 2 mm. **Disc florets** 25–40; corollas yellow-orange, 7–8 mm. **Cypselae** 5 mm; **pappi** of 8–12 scales 7–8 mm, each comprising 7–11 basally connate bristles. $2n = 26$.

Flowering spring and fall. Alluvial fans and rocky slopes in deserts; 0–1200 m; Ariz., Calif., Nev.; Mexico (Baja California, Sonora).

2. **Adenophyllum cooperi** (A. Gray) Strother, Sida 11: 376. 1986 [E] [F]

Dyssodia cooperi A. Gray, Proc. Amer. Acad. Arts 9: 201. 1874

Perennials or subshrubs, 30–50+ cm. **Leaves** mostly not lobed, blades oblanceolate to ovate, 8–25 × 4–8 mm (bases sometimes toothed or obscurely lobed, oil-glands 1–2 pairs near base, 1 near tip, each leaf). **Peduncles** 60–150 mm. **Calyculi** of 12–22 linear-attenuate, gland-bearing, bractlets 5–8 mm. **Involucres** campanulate to obconic, 15–18 mm. **Phyllaries** ca. 20, lanceolate to linear, separating in fruit. **Ray florets** (0–)7–13; corollas yellow-orange, becoming red-orange; tubes 5 mm, laminae 8–10 × 2.5–4 mm. **Disc florets** 50–80+; corollas yellow, 8–10 mm. **Cypselae** 5–7 mm; **pappi** of 15–20 scales 7–10 mm, each comprising 5–9, basally connate bristles. $2n = 26$.

Flowering spring–fall. Sandy and gravelly soils of washes and alluvial fans in deserts; 0–1300 m; Ariz., Calif., Nev.

3. **Adenophyllum wrightii** A. Gray, Smithsonian Contr. Knowl. 5(6): 92. 1853

Dyssodia neomexicana (A. Gray) B. L. Robinson

Varieties 2 (1 in the flora): sw United States, Mexico.

3a. **Adenophyllum wrightii** A. Gray var. **wrightii**

Annuals, to 70 cm. **Leaves** linear or pinnately lobed, 25–35 mm, lobes 5–9, linear (each with oil-gland at base). **Peduncles** 3–6 cm. **Calyculi** of 3–5 subulate bractlets 3–5 mm (eglandular). **Involucres** broadly campanulate to obconic, 7–8 mm. **Phyllaries** 8–13, narrowly to broadly oblanceolate, remaining connate in fruit. **Ray florets** usually 8; corollas yellow-orange; tubes 3 mm, laminae 2 × 1 mm. **Disc florets** 30–50; corollas pale yellow, often tipped with crimson, 4–5 mm. **Cypselae** 3.5–4.5 mm; **pappi** of 10 outer, erose scales 0.7–1 mm plus 10 inner, unequally 3-aristate scales 5–6.5 mm. $2n = 14$.

Flowering fall. Wettish sites in otherwise arid areas, coarse, sandy alluvium; 1500–2600 m; Ariz., N.Mex.; Mexico (Chihuahua).

324. **THYMOPHYLLA** Lagasca, Gen. Sp. Pl., 25. 1816 • [Greek *thymon*, thyme, and *phyllon*, leaf]

John L. Strother

Dyssodia Cavanilles sect. *Gnaphalopsis* (de Candolle) Strother; *Dyssodia* sect. *Hymenatherum* (Cassini) Strother; *Hymenatherum* Cassini

Annuals, perennials, subshrubs, or shrubs, (1–)5–30 cm. **Stems** erect to spreading or decumbent, branched from bases or ± throughout. **Leaves** cauline; opposite or alternate; petiolate or sessile; blades often pinnately lobed, blades or lobes spatulate to linear or filiform, ultimate margins entire or toothed, faces glabrous or arachnose, canescent, floccose, lanate, puberulent, or tomentose (bases little, if at all, bristly-ciliate, oil-glands scattered in laminae or submarginal). **Heads** usually radiate, sometimes discoid or radiant, borne singly. **Calyculi** 0 or of 1–8 deltate to linear bractlets (bearing oil-glands). **Involucres** campanulate to obconic, 2–7 mm diam. **Phyllaries** persistent, 8–13(–22) in ± 2 series (strongly connate ²/₃–⁷/₈+ their lengths, seldom with outer margins distinct more than ¹/₂ their lengths, exceptions: *T. aurea*, *T. pentachaeta* var. *belenidium*), usually bearing oil-glands. **Receptacles** convex, smooth or ± pitted, epaleate. **Ray florets** usually 5, 8, 13, or 21, pistillate, fertile; corollas usually yellow to orange, rarely white. **Disc florets** 16–100+, bisexual, fertile; corollas yellow to orange, tubes shorter than ± cylindric throats, lobes 5, deltate. **Cypselae** obpyramidal, obconic, or cylindro-clavate, glabrous or sparsely strigillose; **pappi** persistent, coroniform or of 10(–20) distinct scales (each scale erose, or 1–5-aristate, or a fascicle of 5–9 basally connate bristles). *x* = 8.

Species 13 (8 in the flora): sw United States, Mexico; introduced in West Indies, South America, Asia, Africa.

SELECTED REFERENCE   Strother, J. L. 1989. Chromosome numbers in *Thymophylla* (Compositae: Tageteae). Sida 13: 351–358.

1. Leaf blades not lobed (margins entire, toothed, or distally 3-fid); calyculi: lengths of bractlets usually ¹/₂+ phyllaries.
    2. Plants green, usually puberulent, sometimes glabrescent, rarely glabrous; leaves mostly opposite . . . . 4. *Thymophylla acerosa*
    2. Plants ashy white, arachnose, lanate, floccose, or tomentose; leaves mostly alternate.
        3. Stems spreading, often prostrate; leaf blades spatulate . . . . 2. *Thymophylla micropoides*
        3. Stems erect; leaf blades linear to filiform, sometimes distally 3-fid . . . . 3. *Thymophylla tephroleuca*
1. Leaf blades usually lobed (lobes 3–15+, linear to filiform, blades linear, usually entire in *T. tenuiloba* var. *wrightii*); calyculi 0, or lengths of bractlets less than ¹/₂ phyllaries.
    4. Perennials or subshrubs (sometimes flowering first year); leaves mostly opposite.
        5. Plants green, usually puberulent to canescent, sometimes glabrescent or glabrous; calyculi 0 or of 1–5 deltate bractlets . . . . 5. *Thymophylla pentachaeta*
        5. Plants ashy white, tomentose; calyculi of 1–3 subulate bractlets . . . . 1. *Thymophylla setifolia*
    4. Annuals (rarely persisting); leaves mostly alternate (sometimes opposite at 1–3 proximal nodes).
        6. Calyculi of 3–8 bractlets; disc florets 50–100+ . . . . 6. *Thymophylla tenuiloba*
        6. Calyculi 0, or of 1–2 bractlets; disc florets 25–45.
            7. Margins of outer phyllaries distinct nearly to bases . . . . 7. *Thymophylla aurea*
            7. Margins of outer phyllaries distinct less than ¹/₂+ their lengths . . . . 8. *Thymophylla concinna*

1. Thymophylla setifolia Lagasca, Gen. Sp. Pl., 25. 1816

Dyssodia setifolia (Lagasca) B. L. Robinson

Varieties 2 (1 in the flora): sw United States, Mexico.

1a. Thymophylla setifolia Lagasca var. **greggii** (A. Gray) Strother, Sida 21: 287. 2004

Thymophylla greggii A. Gray, Mem. Amer. Acad. Arts, n. s. 4: 92. 1849; *Dyssodia setifolia* (Lagasca) B. L. Robinson var. *greggii* (A. Gray) M. C. Johnston; *D. greggii* (A. Gray) B. L. Robinson

Perennials, ashy white, to 15 cm, tomentose. **Stems** spreading (mat-forming). **Leaves** opposite; blades (at least proximal) lobed (from near bases), 6–12 mm overall, lobes 3–7, stiff, filiform (distal blades often

acerose). **Peduncles** 30–40 mm, glabrous or glabrate. **Calyculi** of 1–3 subulate bractlets, lengths less than ¹/₂ phyllaries. **Involucres** obconic to campanulate, 3.5–4 mm. **Phyllaries** 9–15, margins of outer free less than ¹/₃ their lengths, abaxially glabrous. **Ray florets** 7–10; corollas bright yellow, laminae 2–3 × 0.8–1.2 mm. **Disc florets** 20–40; corollas dull yellow, 2–2.5 mm. **Cypselae** 1.5–2.2 mm; **pappi** coroniform (of connate scales) 0.3–0.5 mm.

Flowering summer. Rocky, calcareous outcrops and derived soils; 1200–1500+ m; N.Mex., Tex.; Mexico (Chihuahua, Coahuila, Nuevo León).

Thymophylla greggii var. *greggii* resulted from publication of *T. greggii* var. *radiata* A. Gray. *Dyssodia setifolia* var. *radiata* (A. Gray) Strother is illegitimate.

2. Thymophylla micropoides (de Candolle) Strother, Sida 11: 377. 1986

Gnaphalopsis micropoides de Candolle in A. P. de Candolle and A. L. P. P. de Candolle, Prodr. 7: 258. 1838; *Dyssodia micropoides* (de Candolle) Loesener

Perennials, ashy white, to 15+ cm, arachnose to floccose. **Stems** spreading, often prostrate. **Leaves** mostly alternate; blades not lobed, spatulate, 10–25 × 3–6 mm, margins entire or toothed. **Peduncles** 0–5+ mm, tomentose. **Calyculi** of 3–5 linear bractlets, lengths 1/2+ phyllaries. **Involucres** obconic, 6–7 mm. **Phyllaries** 12–14, margins of outer distinct less than 1/5 their lengths, abaxial faces tomentose. **Ray florets** 10–15; corollas bright yellow, becoming greenish, laminae 5 × 3 mm. **Disc florets** ca. 60; corollas yellow, 3 mm. **Cypselae** 2.3–3 mm; **pappi** of 5 erose scales to 1 mm alternating with 5 lanceolate, 1-aristate scales to 3 mm. $2n = 16$.

Flowering year round, following rains. Calcareous outcrops and derived soils; 50–200+ m; Tex.; Mexico (Coahuila, Nuevo León, Tamaulipas).

3. Thymophylla tephroleuca (S. F. Blake) Strother, Sida 11: 378. 1986 [C] [E]

Dyssodia tephroleuca S. F. Blake, J. Wash. Acad. Sci. 25: 320. 1935

Subshrubs or shrubs, ashy white, 10–30 cm, lanate to tomentose. **Stems** erect. **Leaves** all or mostly alternate; blades not lobed, linear to filiform, 10–15 × 0.3–1 mm, margins usually entire, sometimes 3-fid distally. **Peduncles** 10–30 mm, tomentose. **Calyculi** of 3–4 linear bractlets, lengths 1/2+ phyllaries. **Involucres** campanulate, 5–7 mm. **Phyllaries** 12–13, margins of outer distinct less than 1/4 their lengths, abaxial faces tomentose. **Ray florets** 12–13; corollas golden yellow, laminae 6–8 × 3–4 mm. **Disc florets** ca. 30; corollas yellow, 4.5–5 mm. **Cypselae** 3–4 mm; **pappi** of 10–11 subequal scales 4–5 mm (each 3–5-aristate). $2n = 16$.

Flowering early spring and late summer, following rains. Light, sandy soils; of conservation concern; 50–100 m; Tex.

Thymophylla tephroleuca probably grows also in adjacent Mexico. It is in the Center for Plant Conservation's National Collection of Endangered Plants.

4. Thymophylla acerosa (de Candolle) Strother, Sida 11: 376. 1986 [F]

Dyssodia acerosa de Candolle in A. P. de Candolle and A. L. P. P. de Candolle, Prodr. 5: 641. 1836

Subshrubs or shrubs, green, to 25 cm, usually puberulent, sometimes glabrescent, rarely glabrous. **Stems** erect (branched from bases). **Leaves** mostly opposite; blades not lobed, linear to acerose, 10–18 mm. **Peduncles** 0–10 mm, puberulent or glabrous. **Calyculi** of 5 lance-linear bractlets, lengths 1/2+ phyllaries. **Involucres** campanulate to cylindric, 5–7 mm. **Phyllaries** ca. 13, margins of outer distinct less than 1/6 their lengths, abaxial faces puberulent or glabrous. **Ray florets** 7–8; corollas lemon-yellow, laminae 5–6 × 2–3 mm. **Disc florets** 18–25+; corollas pale yellow, 3–4 mm. **Cypselae** 3–3.5 mm; **pappi** of ca. 20 scales (each a fascicle of 3–5 bristles), 3–4 mm. $2n = 16, 24$.

Flowering through the year, mostly summer–fall. Calcareous outcrops, gypseous soils; 1000–2000+ m; Ariz., Nev., N.Mex., Tex., Utah; Mexico.

5. Thymophylla pentachaeta (de Candolle) Small, Fl. S.E. U.S., 1295, 1341. 1903 [F]

Hymenatherum pentachaetum de Candolle in A. P. de Candolle and A. L. P. P. de Candolle, Prodr. 5: 642. 1836; *Dyssodia pentachaeta* (de Candolle) B. L. Robinson

Perennials or subshrubs, ± grayish to green, to 15(–25) cm, usually puberulent to canescent, sometimes glabrescent or glabrous. **Stems** erect or spreading. **Leaves** mostly opposite; blades mostly pinnately lobed, 6–28+ mm overall, lobes 3–11 linear to filiform (usually stiff, setiform). **Peduncles** 20–100 mm, puberulent or glabrous. **Calyculi** 0, or of 1–5 deltate bractlets, lengths less than 1/2 phyllaries. **Involucres** obconic to campanulate or hemispheric, 4–6 mm. **Phyllaries** 12–21, margins of outer distinct 1/5 to nearly all their lengths, abaxial faces puberulent or glabrous. **Ray florets** (8–)12–21; corollas yellow to orange-yellow, laminae 2–6(–8) × 1–3 mm. **Disc florets** 16–40 or 50–80; corollas yellow, 2–4 mm (tending to zygomorphy in peripheral florets in some plants). **Cypselae** 2–3 mm; **pappi** of 10 erose and/or aristate scales mostly 1–3 mm.

Varieties 4 (4 in the flora): sw United States, Mexico; introduced(?) in South America (Argentina).

1. Involucres cylindric, 2–3.5 mm diam.; disc florets 16–40 . . . . 5d. *Thymophylla pentachaeta* var. *hartwegii*
1. Involucres obconic, campanulate, or hemispheric, mostly 4–5 mm diam.; disc florets 50–80.
   2. Peduncles 20–50 mm; margins of outer phyllaries distinct almost to bases . . . . 5a. *Thymophylla pentachaeta* var. *belenidium*
   2. Peduncles (40–)50–100 mm; margins of outer phyllaries distinct less ca. ¹/₂ or ca. ¹/₃ their lengths.
      3. Leaf lobes mostly 9–11; phyllaries usually glabrous or glabrescent, rarely hairy . . . . 5b. *Thymophylla pentachaeta* var. *pentachaeta*
      3. Leaf lobes mostly 5–7(–10); phyllaries densely puberulent . . . . . . 5c. *Thymophylla pentachaeta* var. *puberula*

5a. Thymophylla pentachaeta   (de Candolle) Small var. **belenidium**   (de Candolle) Strother, Sida 11: 377. 1986

Dyssodia belenidium de Candolle in A. P. de Candolle and A. L. P. P. de Candolle, Prodr. 7: 292. 1838; *D. pentachaeta* (de Candolle) B. L. Robinson var. *belenidium* (de Candolle) Strother; *D. thurberi* (A. Gray) B. L. Robinson

Leaves 10–20 mm overall, lobes 3–7, subequal. **Peduncles** 20–50 mm. **Calyculi** of 3–5 bractlets. **Involucres** obconic to campanulate, 4.5–6 × 4–5 mm. **Phyllaries** ca. 13, margins of outer distinct almost to bases, abaxial faces glabrous (margins ciliate). **Disc florets** 50–70. **Pappi** usually of 10, 3-aristate scales, sometimes of 5 muticous scales alternating with 5, 3-aristate scales. $2n = 32$ (Argentina).

Flowering spring–summer. Sandy or rocky soils of desert washes and fans, commonly with limestone, often ruderal along roadways; 500–2500+ m; Ariz., Calif., Nev., N.Mex., Tex., Utah; Mexico; South America (Argentina).

5b. Thymophylla pentachaeta   (de Candolle) Small var. **pentachaeta**   F

Leaves 6–25 mm overall, lobes (3–)9–11, subequal. **Peduncles** 60–100 mm. **Calyculi** of 3–5 bractlets. **Involucres** obconic to campanulate, 4.3–6 × 4–5 mm. **Phyllaries** 12–21, margins of outer distinct ca. ¹/₂ their lengths, abaxial faces usually glabrescent or glabrous, rarely hairy. **Disc florets** 50–80. **Pappi** of 5 erose scales alternating with 5, 1–3-aristate scales. $2n = 32$.

Flowering early spring and late summer, following rains. Calcareous outcrops and derived soils, scrublands

or deserts, roadways, other disturbed sites; 50–700+ m; Tex.; Mexico.

5c. Thymophylla pentachaeta   (de Candolle) Small var. **puberula**   (Rydberg) Strother, Sida 11: 377. 1986

Thymophylla puberula Rydberg in N. L. Britton et al., N. Amer. Fl. 34: 177. 1915; *Dyssodia puberula* (Rydberg) Standley; *D. pentachaeta* (de Candolle) B. L. Robinson var. *puberula* (Rydberg) Strother

Leaves 10–28 mm overall, lobes 3–7(–10), subequal. **Peduncles** 40–100 mm. **Calyculi** 0, or of 1–3 bractlets. **Involucres** campanulate to hemispheric, 4.5–5.5 × 4–5 mm. **Phyllaries** 12–17, margins of outer distinct ca. ¹/₃ their lengths, abaxial faces densely puberulent. **Disc florets** 50–70. **Pappi** of 5 erose scales alternating with 5, 1–3-aristate scales.

Flowering early spring and late summer, following rains. Calcareous bluffs and slopes, scrublands, deserts; 300–2200 m; Tex.; Mexico.

5d. Thymophylla pentachaeta   (de Candolle) Small var. **hartwegii**   (A. Gray) Strother, Sida 11: 377. 1986

Hymenatherum hartwegii A. Gray, Smithsonian Contr. Knowl. 3(5): 117. 1852; *Dyssodia hartwegii* (A. Gray) B. L. Robinson; *D. pentachaeta* (de Candolle) B. L. Robinson var. *hartwegii* (A. Gray) Strother

Leaves 12–34 mm overall, lobes 3–7, unequal (terminal lobes usually notably longer than laterals). **Peduncles** 2–5 cm. **Calyculi** of 3–5 bractlets. **Involucres** cylindric, 3–5 × 2–3.5 mm. **Phyllaries** ca. 13, margins of outer distinct ¹/₃ or less their lengths, abaxial faces sparsely puberulous, often glabrescent. **Disc florets** 16–40. **Pappi** of 5 erose scales alternating with 5, 1–3-aristate scales.

Flowering early spring and late summer, following rains. Calcareous outcrops and bluffs, deserts; 1500–2500 m; Ariz., N.Mex., Tex.; Mexico.

6. Thymophylla tenuiloba   (de Candolle) Small, Fl. S.E. U.S., 1295, 1341. 1903

Hymenatherum tenuilobum de Candolle in A. P. de Candolle and A. L. P. P. de Candolle, Prodr. 5: 462. 1836; *Dyssodia tenuiloba* (de Candolle) B. L. Robinson

Annuals (rarely persisting), green, to 30 cm, glabrous or sparsely hirtellous. **Stems** decumbent to erect. **Leaves** mostly alternate (sometimes

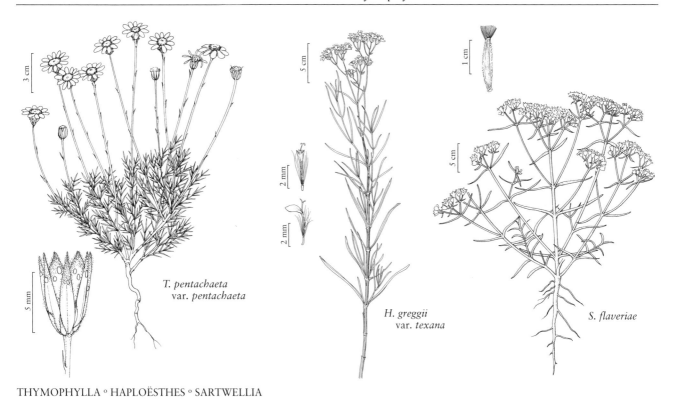

T. pentachaeta
var. pentachaeta

H. greggii
var. texana

S. flaveriae

THYMOPHYLLA ○ HAPLOËSTHES ○ SARTWELLIA

opposite at 1–3 proximal nodes); blades usually lobed, 12–28 mm overall, lobes 7–15, linear to filiform, pliable (blades mostly not lobed in var. *wrightii*. **Peduncles** 30–80 mm, glabrous or hirtellous. **Calyculi** of 3–8 deltate to subulate bractlets, lengths less than ¹⁄₂ phyllaries. **Involucres** obconic, 5–7 mm. **Phyllaries** 12–22, margins of outer distinct less than ¹⁄₅ their lengths, abaxial faces glabrous or sparsely hirtellous. **Ray florets** 10–21; corollas yellow-orange, laminae 4–10 × 1.5–3 mm. **Disc florets** 50–100+; corollas yellow, 2.5–4.5 mm (peripheral ± zygomorphic in some plants). **Cypselae** 2–3.5 mm; **pappi** of erose and/or aristate scales 0.7–3.5 mm.

Varieties 4 (4 in the flora): Texas, n Mexico.

1. Leaf blades seldom lobed (linear to spatulate, usually entire, rarely with 3–5 toothlike lobes) . . . . 6d. *Thymophylla tenuiloba* var. *wrightii*
1. Leaf blades lobed (lobes 7–15 linear).
    2. Pappi of 3–5-aristate scales . . . . . . 6a. *Thymophylla tenuiloba* var. *tenuiloba*
    2. Pappi of truncate to lanceolate and/or 1-aristate scales.
        3. Pappi of 5 erose scales alternating with 5 unequal, 1-aristate scales . . . . . 6b. *Thymophylla tenuiloba* var. *treculii*
        3. Pappi of 5–10 lanceolate to muticous scales, all ± erose or 1–3, 1-aristate . . . 6c. *Thymophylla tenuiloba* var. *texana*

6a. Thymophylla tenuiloba (de Candolle) Small var. **tenuiloba** • Dahlberg daisy, golden fleece

Leaves 7–15-lobed. **Pappi** of 10–12, 3–5-aristate scales 2–3.5 mm. $2n$ = 16, 24, 32, 40.

Flowering early spring–late fall. Heavy soils or loams or sands, often with limestone, frequently ruderal along roadways and in other disturbed places; 0–300 m; Tex.; Mexico (Tamaulipas); introduced in West Indies, Asia, Africa.

Variety *tenuiloba* is relatively recently adventive in Alabama, Florida, Louisiana, Mississippi, Bahamas, West Indies (Cuba), Asia, and Africa.

The names Dahlberg daisy and golden fleece are recent coinages applied to cultivars of var. *tenuiloba* in the horticultural trade.

6b. Thymophylla tenuiloba (de Candolle) Small var. **treculii** (A. Gray) Strother, Sida 11: 378. 1986

Hymenatherum treculii A. Gray, Proc. Amer. Acad. Arts 19: 42. 1883; *Dyssodia treculii* (A. Gray) B. L. Robinson; *D. tenuiloba* (de Candolle) B. L. Robinson var. *treculii* (A. Gray) Strother

Leaves 7–15-lobed. **Pappi** of 5 oblanceolate, erose scales 0.8–1.1 mm alternating with 5 unequal, 1-aristate scales 2.5–3 mm. $2n = 32, 40$.

Flowering early spring and late fall. Mostly sandy soils; 0–300 m; Tex.; Mexico (Coahuila, Nuevo León, Tamaulipas).

6c. Thymophylla tenuiloba (de Candolle) Small var. **texana** (Cory) Strother, Sida 11: 378. 1986

Dyssodia texana Cory, Rhodora 49: 162. 1947; *D. tenuiloba* (de Candolle) B. L. Robinson var. *texana* (Cory) Strother

Leaves 7–15-lobed. **Pappi** of 5–10 lanceolate to muticous scales 0.5–1.2 mm, all erose or 1–3, 1-aristate. $2n = 16$.

Flowering spring and fall. Limestone outcrops and derived soils in grasslands; 800–900 m; Tex.; Mexico (Coahuila, Nuevo León).

6d. Thymophylla tenuiloba (de Candolle) Small var. **wrightii** (A. Gray) Strother, Sida 11: 378. 1986 [E]

Hymenatherum wrightii A. Gray, Mem. Amer. Acad. Arts, n. s. 4: 89. 1849; *Dyssodia wrightii* (A. Gray) B. L. Robinson; *D. tenuiloba* (de Candolle) B. L. Robinson var. *wrightii* (A. Gray) Strother

Leaves linear to spatulate, usually entire, rarely with 1–5 toothlike lobes. **Pappi** of 10–12 unequal, aristate scales to 3 mm. $2n = 16$.

Flowering spring. Sandy soils, coastal grasslands, roadways; 0–30 m; Tex.

7. Thymophylla aurea (A. Gray) Greene in N. L. Britton and A. Brown, Ill. Fl. N. U.S. 3: 453. 1898

Lowellia aurea A. Gray, Mem. Amer. Acad. Arts, n. s. 4: 91. 1849; *Dyssodia aurea* (A. Gray) A. Nelson

Annuals, green, to 20(–30) cm, glabrous or puberulent. **Stems** erect (branched from bases). **Leaves** alternate; blades 15–40 mm overall, lobed, lobes 5–13, linear. **Peduncles** 10–70 mm, glabrous or puberulent. **Calyculi** 0, or of 1–2 subulate bractlets, lengths less than $^1/_2$ phyllaries. **Involucres** obconic to campanulate, 5–6 mm. **Phyllaries** 12–15, margins of outer distinct almost to bases, abaxial faces glabrous or puberulent. **Ray florets** 8–12; corollas bright yellow, laminae 4–6 × 2–3 mm. **Disc florets** 30–45; corollas yellow, 3 mm. **Cypselae** 3 mm; **pappi** either of erose scales 0.3–0.6 mm, or of 3–5-aristate scales 2–3 mm.

Varieties 2 (2 in the flora): sw United States, n Mexico.

The two varieties of *Thymophylla aurea* are sometimes found in mixed populations, which may include intermediate plants.

1. Pappi of 18–20, 3–5-aristate scales 2–3 mm . . . .
. . . . 7a. *Thymophylla aurea* var. *polychaeta*
1. Pappi of 8–10 erose scales 0.3–0.6 mm . . . . . .
. . . . 7b. *Thymophylla aurea* var. *aurea*

7a. Thymophylla aurea (A. Gray) Greene var. **polychaeta** (A. Gray) Strother, Sida 11: 376. 1986

Hymenatherum polychaetum A. Gray, Smithsonian Contr. Knowl. 3(5): 116. 1852; *Dyssodia aurea* (A. Gray) A. Nelson var. *polychaeta* (A. Gray) M. C. Johnston; *D. polychaeta* (A. Gray) B. L. Robinson; *Thymophylla polychaeta* (A. Gray) Small

Plants to 30 cm. **Peduncles** 20–70 mm. **Pappi** of 18–20, 3–5-aristate scales 2–3 mm. $2n = 16$.

Flowering late summer. Grasslands and desert scrublands; 1000–1500+ m; N.Mex., Tex.; Mexico (Chihuahua, Coahuila, Durango).

7b. Thymophylla aurea (A. Gray) Greene var. **aurea**

Plants to 20 cm. **Peduncles** 10–30 mm. **Pappi** of 8–10 erose scales 0.3–0.6 mm. $2n = 16$.

Flowering summer. Swales in grasslands; 1500–2000 m; Colo., Kans., N.Mex., Tex.; Mexico (Chihuahua).

8. **Thymophylla concinna**   (A. Gray) Strother, Sida 11: 376. 1986

Hymenatherum concinnum A. Gray in A. Gray et al., Syn. Fl. N. Amer. 1(2): 446. 1884; *Dyssodia concinna* (A. Gray) B. L. Robinson

Annuals, green, 3–12+ cm, glabrous. **Stems** spreading to ascending or erect. **Leaves** alternate; blades 10–15 mm overall, lobed, lobes (3–)5–9, linear to filiform. **Peduncles** 1–5+ mm, glabrous. **Calyculi** 0, or of 1–2 subulate bractlets, lengths less than ¹/₂ phyllaries. **Involucres** obconic to campanulate, 5–6.5 mm. **Phyllaries** 8–16, margins of outer distinct less than ¹/₂ their lengths, abaxial faces glabrous. **Ray florets** 9–12; corollas white to pale yellow, laminae 3–4 × 1–3.5 mm. **Disc florets** ca. 25; corollas pale yellow, 3–4.5 mm. **Cypselae** 2.5–3 mm; **pappi** of 10–15 scales (each comprising 5–9 basally connate bristles) 1–2.5 mm. **2*n*** = 16.

Flowering spring, following rains. Sandy washes and flats, in deserts; 50–100 m; Ariz.; Mexico (Sonora).

187m.15. ASTERACEAE Martinov (tribe HELIANTHEAE) subtribe FLAVERIINAE Lessing, Syn. Gen. Compos., 235. 1832 (as Flaverieae)

Annuals, perennials, or subshrubs, 10–80 cm (often ± succulent). **Leaves** cauline; opposite; petiolate or sessile; blades (1- or 3-nerved) oblong, oblong-ovate, lanceolate, linear, or filiform, margins entire or toothed, faces glabrous, glabrescent, or puberulent. **Heads** radiate or discoid, in glomerules or corymbiform to paniculiform arrays. **Calyculi** 0 or of 1–2 bractlets. **Involucres** campanulate, cylindric, obconic, turbinate, or urceolate. **Phyllaries** persistent, 2–8 in ± 1 series (distinct, linear to oblong, subequal, ± succulent to membranous or scarious). **Receptacles** flat to convex, epaleate. **Ray florets** 0, or 1–6, pistillate, fertile; corollas yellow. **Disc florets** 1–60[–100+], bisexual, fertile; corollas yellow, tubes shorter than or about equaling the funnelform to campanulate throats, lobes 5, deltate; anther thecae pale; stigmatic papillae in 2 lines. **Cypselae** cylindric to clavate or ± compresssed, linear-oblong to oblanceolate, 10–15-ribbed, glabrous or hairy; **pappi** 0, or persistent, coroniform, or of 2–5 ovate, scarious scales, or of 20–25+ unequal bristles, or of 5 scales plus 5 bristles (all in 1 series, sometimes connate).

Genera 3, species 27 (3 genera, 10 species in the flora): mostly subtropical, tropical, and warm-temperate New World.

As noted by H. Robinson (1981), traditionally, *Flaveria* and *Sartwellia* were treated in Helenieae, *Haploësthes* in Senecioneae.

1. Disc florets 18–30(–60); pappi of 20–25+ bristles . . . . 325. *Haploësthes*, p. 245
1. Disc florets 1–15; pappi 0, or coroniform, or wholly or partly of scales.
  2. Ray florets 3–5; pappi of 5 erose scales alternating with 5 setiform scales or bristles (sometimes all 10 elements basally connate) . . . . 326. *Sartwellia*, p. 246
  2. Ray florets 0 or 1; pappi 0, or coroniform, or of 2–4 hyaline scales . . . . 327. *Flaveria*, p. 247

325. HAPLOËSTHES   A. Gray, Mem. Amer. Acad. Arts, n. s. 4: 109. 1849 • [Greek *haploos*, simple, and *esthes*, raiment]

John L. Strother

Perennials or subshrubs, 10–80 cm (± succulent). **Stems** erect to sprawling, branched from bases or throughout. **Leaves** cauline; opposite; sessile; blades filiform to linear, margins entire, faces glabrous or glabrescent. **Heads** radiate, (5–25) in corymbiform arrays. **Involucres** obconic,

2.5–5+ mm diam. **Phyllaries** persistent, 4–5 in 1 series (lance-elliptic to ovate or orbiculate, membranous to subscarious). **Receptacles** convex, epaleate. **Ray florets** 3–6, pistillate, fertile; corollas yellow. **Disc florets** 18–30(–60)[–100+], bisexual, fertile; corollas yellow, tubes shorter than or equaling cylindric or funnelform throats, lobes 5, deltate to lance-deltate. **Cypselae** ± cylindric (10–15-ribbed); **pappi** persistent, of 20–25+ barbellulate bristles. $x = 18$.

Species 3 (1 in the flora): sw United States, Mexico.

SELECTED REFERENCE   Turner, B. L. 1975. Taxonomy of *Haploësthes* (Asteraceae–Senecioneae). Wrightia 5: 108–115.

1. **Haploësthes greggii**   A. Gray, Mem. Amer. Acad. Arts, n. s. 4: 109. 1849   F

Varieties 2 (1 in the flora): sw United States, Mexico.

1a. **Haploësthes greggii**   A. Gray var. **texana**   (J. M. Coulter) I. M. Johnston, J. Arnold Arbor. 22: 169. 1941   F

Haplopappus texanus J. M. Coulter, Contr. U.S. Natl. Herb. 1: 40. 1890 (as Aplopappus)

Leaf blades 20–80 × 0.5–4 mm. **Peduncles** 3–8(–15) mm. **Phyllaries** drying dark gray-green. **Ray laminae** elliptic to ovate, 2.5–3.5 × 1.3–2 mm. **Disc corollas** 3–4 mm. **Cypselae** 1.5–2.5 mm; **pappi** 2.5–3.5 mm. $2n = 36$.

Flowering sporadically year round following rains. Gypseous soils; 200–1000 m; Colo., Kans., N.Mex., Okla., Tex.; Mexico (Coahuila, Nuevo León).

Variety *texana* may be no longer present in Colorado and/or Kansas.

326. **SARTWELLIA**   A. Gray, Smithsonian Contr. Knowl. 3(5): 122, plate 6. 1852   •   [For Henry P. Sartwell, 1792–1867, "one of my earliest and most valued botanical correspondents...." Quoted from protologue.]

John L. Strother

Perennials, 10–30 cm (± succulent). **Stems** ascending to erect, branched from at or near bases. **Leaves** cauline; opposite; sessile; blades linear to filiform, margins entire, faces glabrous or glabrescent. **Heads** radiate, in tight, corymbiform arrays. **Involucres** campanulate, 2–3 mm diam. **Phyllaries** persistent, 5 in 1 series (usually yellowish, sometimes greenish, oblong to obovate, subequal, ± membranous). **Receptacles** flat, epaleate. **Ray florets** 3–5, pistillate, fertile; corollas yellow. **Disc florets** 5–15, bisexual, fertile; corollas yellow, tubes shorter than or equaling funnelform throats, lobes 5, ± deltate. **Cypselae** ± cylindric (10-ribbed); **pappi** persistent, usually of 5 muticous scales alternating with 5 subulate scales or coarse bristles, sometimes the 10 elements connate, forming a ± erose cup. $x = 18$.

Species 3 (1 in the flora): sw United States, Mexico.

SELECTED REFERENCE   Turner, B. L. 1971. Taxonomy of *Sartwellia* (Compositae Helenieae). Sida 4: 265–273.

1. Sartwellia flaveriae A. Gray, Smithsonian Contr. Knowl. 3(5): 122, plate 6. 1852 [E] [F]

Leaf blades 20–50 × 1–2.5 mm. **Ray laminae** ca. 2 mm. **Disc corollas** ca. 2 mm. **Cypselae** 1.5–2 mm; **pappi** ca. 0.5+ mm. $2n = 36$.

Flowering mostly late summer, rarely spring. Usually gypseous soils; 700–1900 m; N.Mex., Tex.

327. **FLAVERIA** Jussieu, Gen. Pl., 186. 1789 • [Latin *flavus*, yellow]

Sharon C. Yarborough

A. Michael Powell

Annuals, perennials, or subshrubs, to 200+ cm [trees to 400 cm] (usually ± succulent, herbage usually glaucous). **Stems** (often purplish) erect or decumbent, branched distally or ± throughout. **Leaves** cauline; opposite (decussate); petiolate or sessile (weakly connate to connate-perfoliate); blades (often 3-nerved) oblong-ovate to lanceolate or linear, margins entire, serrate, or spinulose-serrate, faces glabrous or short-pubescent. **Heads** radiate or discoid, usually in tight or loose aggregations in (often flat-topped) ± corymbiform arrays or glomerules. **Involucres** oblong, urceolate, cylindric, or turbinate, 0.5–2 mm diam. **Phyllaries** persistent, 2–6(–9) in ± 1 series (linear, concave, or boat-shaped, subequal). **Receptacles** convex, epaleate ("receptacles" of glomerules sometimes setose). **Ray florets** 0–1(–2), pistillate, fertile; corollas yellow or whitish (laminae inconspicuous). **Disc florets** 1–15, bisexual, fertile; corollas yellow, tubes shorter than to about equaling funnelform to campanulate throats, lobes 5, ± deltate. **Cypselae** (black) weakly compressed, narrowly oblanceolate or linear-oblong (usually 10-nerved, glabrous); **pappi** usually 0, sometimes persistent, of 2–4 hyaline scales, or coroniform (of connate scales). $x = 18$.

Species 21 (8 in the flora): United States, Mexico, West Indies (Greater Antilles), Central America, South America, Asia (India), Africa, Australia.

Members of *Flaveria* are frequently found in alkaline, saline, and gypseous soils, often in disturbed and moist areas. Heads of *Flaveria* may be either radiate or discoid; when both are present in the same capitulescence, the discoid heads tend to be central and the radiate heads peripheral. Many species of *Flaveria* have persistent sheathing leaf bases that ring the stems after the leaves have fallen.

Flaveria is notable because certain species exhibit C_3 photosynthesis, some C_3–C_4 (intermediate) photosynthesis, and others (*F. brownii, F. bidentis, F. campestris,* and *F. trinervia* in the United States) classic C_4 photosynthesis.

SELECTED REFERENCE Powell, A. M. 1978. Systematics of *Flaveria* (Flaveriinae–Asteraceae). Ann. Missouri Bot. Gard. 65: 590–636.

1. Pappi usually of 2–4 scales or coroniform, rarely 0.
   2. Leaf blades to 10–40 mm wide, bases connate-perfoliate; New Mexico, Texas . . . . 1. *Flaveria chlorifolia*
   2. Leaf blades 2–7 mm wide, bases weakly connate; Arizona . . . . 8. *Flaveria mcdougallii*
1. Pappi 0.
   3. Heads in tight, axillary glomerules ("receptacles" of glomerules setose) . . . . 7. *Flaveria trinervia*
   3. Heads usually in corymbiform, paniculiform, or spiciform arrays, seldom in tight, axillary glomerules ("receptacles" of glomerules not setose).

[4. Shifted to left margin.—Ed.]

4. Annuals.
   5. Heads in scorpioid cymiform arrays; ray laminae to 1 mm; Alabama, Florida, Georgia . . . . . 5. *Flaveria bidentis*
   5. Heads in corymbiform arrays; ray laminae 1.5–2.5 mm; sc, sw United States . . . . 6. *Flaveria campestris*
4. Perennials (annuals).
   6. Calyculus bractlets surpassing involucres (sw coast, Florida). . . . . 3. *Flaveria floridana*
   6. Calyculus bractlets shorter than involucres.
      7. Ray laminae oblong-elliptic, 2 mm; disc florets (5–)7–10; Texas . . . . 2. *Flaveria brownii*
      7. Ray laminae oval to obovate-spatulate, 2–3 mm; disc florets (2–)5–7(–8); Florida . . . . . . 4. *Flaveria linearis*

1. Flaveria chlorifolia A. Gray, Mem. Amer. Acad. Arts, n. s. 4: 88. 1849 (as chloraefolia) • Clasping yellowtops F

Perennials, to 200 cm (robust, glaucous). Stems erect to subdecumbent. Leaves sessile; blades oblong-ovate to lanceolate or hastate, (20–)30–100 × 10–40(–50) mm, bases connate-perfoliate, margins entire. Heads 25–150+ in paniculiform arrays. Calyculi of 1(–2) linear-lanceolate bractlets ca. 2 mm. Involucres oblong-urceolate, 5–7 mm. Phyllaries 5, oblong-lanceolate. Ray florets 0. Disc florets 9–14; corolla tubes 1.2–1.4 mm, throats narrowly funnelform, 2 mm. Cypselae narrowly oblanceolate to linear, 2.5–3 mm; pappi usually of 2–4 unequal, hyaline scales 0.2–1 mm, usually on 1 shoulder, rarely 0. 2*n* = 36.

Flowering Jul–Nov. Near saline or gypseous water sources: marshes, springs, creeks, rivers, irrigation canals, and roadside ditches; 700–1700 m; N.Mex., Tex.; Mexico (Chihuahua, Coahuila, Nuevo León).

Flaveria chlorifolia is widely distributed in the northern region of the Chihuahuan Desert and extends up the Pecos River drainage into central New Mexico and the Rio Grande River drainage to north of Las Cruces. The perfoliate leaves, succulent habit, relatively large, flat-topped clusters of discoid heads, and usually 2–4 pappus scales distinguish it. It may be considered primitive on the basis of habit, leaves, and pappus scales. It is not thought to be closely related to any other species of *Flaveria*.

2. Flaveria brownii A. M. Powell, Ann. Missouri Bot. Gard. 65: 611, fig. 4. 1979 • Brown's yellowtops E

Perennials (perhaps flowering first year), 15–70 cm (glabrate). Stems erect or decumbent. Leaves sessile; blades linear, 5–12 cm × 2–8 mm, bases barely connate, margins entire or weakly serrate. Heads 20–100+, in ± open, paniculiform-corymbiform arrays. Calyculi of 1–2 linear bractlets 1–2.5 mm. Involucres oblong-angular, 4–5 mm. Phyllaries 5–6, boat-shaped (phyllaries enclosing ray florets conspicuously keeled). Ray florets 0 or 1; laminae yellow, oblong-elliptic, 2 mm. Disc florets (5–)7–10; corolla tubes 0.8–1.2 mm, throats funnelform, 1.2–1.5 mm (distal ¹/₂ expanded). Cypselae linear, 1.5–2 mm (those of rays longer); pappi 0. 2*n* = 36.

Flowering Jun–Dec. Saline, sandy, and marshy areas of costal flats and islands; 0–30 m; Tex.

Flaveria brownii, which has radiate heads present among the discoid heads in each capitulescence, is superficially similar to the discoid *F. oppositifolia*. It occurs in the lower Gulf Coast region of Texas; the latter is Mexican.

3. Flaveria floridana J. R. Johnston, Proc. Amer. Acad. Arts 39: 291. 1903 • Florida yellowtops E

Perennials (or annuals), 50–120 cm (usually glabrous, sometimes sparsely pubescent distally on peduncles). Stems erect. Leaves sessile; blades usually linear or lanceolate, rarely subelliptic, 50–140 × 4–17 mm, bases ± connate, margins entire, minutely serrate, or spinulose-serrate. Heads 20–100+, in dense, scorpioid-cymiform arrays. Calyculi of 1–3 linear-lanceolate bractlets 4–6 mm (often surpassing involucres). Involucres suburceolate, oblong-angular, 3.6–4.5 mm. Phyllaries 5–6(–9), oblong or ovate-orbiculate. Ray florets 0 or 1; laminae yellow, ovate-spatulate, 2–2.8 × 1.5–2 mm. Disc florets 9–14; corolla tubes 0.8–1 mm, throats funnelform, 1.3–1.5 mm. Cypselae oblong-oblanceolate or linear, 1.2–1.8 mm; pappi 0. 2*n* = 36.

Flowering year round. Saline sand areas, beaches, brackish marshes, pinewoods; 0–10 m; Fla.

Flaveria floridana is known only from the lower Gulf coast of Florida; it is especially prevalent in the Tampa Bay area. The relatively long calyculus bractlets and modified peduncular leaves that extend beyond the heads distinguish it from *F. linearis*, the other perennial species in Florida. The closest relative of *F. floridana* is probably the similar *F. brownii* of the Texas Gulf coast.

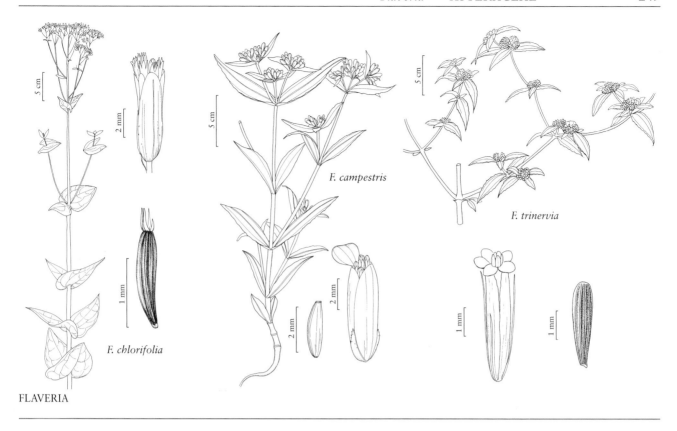

F. campestris

F. trinervia

F. chlorifolia

FLAVERIA

4. Flaveria linearis Lagasca, Gen. Sp. Pl., 33. 1816

• Narrowleaf yellowtops

Flaveria ×*latifolia* (J. R. Johnston) R. W. Long & E. L. Rhamstine

Perennials, 30–80 cm (glabrous or pubescent, mostly on distal peduncles). **Stems** erect. **Leaves** sessile; blades linear, 50–100(–130) × 1–4 (–15) mm, ± connate, margins entire or spinulose-serrulate. **Heads** 10–150+, in clusters in corymbiform-paniculiform arrays. **Calyculi** of 1–3 linear bractlets 1–2.5 mm. **Involucres** oblong-angular, 3.3–4.5 mm. **Phyllaries** 5 (–6), linear or oblong. **Ray florets** 0 or 1; laminae yellow, oval to obovate-spatulate, 2–3 mm. **Disc florets** (2–)5–7(–8); corolla tubes 0.8–1.2 mm, throats basally tubular, becoming funnelform-campanulate apically, 1–1.5 mm. **Cypselae** linear, 1.2–1.8 mm; **pappi** 0. $2n = 36$.

Flowering year round. Disturbed sites, beaches, hammocks, pinelands; 0–10(–20+) m; Fla.; Mexico (Quintana Roo, Yucatán); West Indies (Bahamas, Cuba).

Flaveria linearis is variable; it typically has linear leaves, calyculi of relatively short, linear bractlets, and oblong-angular involucres. The heads are relatively small with 5–8 florets, and throats of the disc corollas are tubular at the base, abruptly expanding distally to become funnelform-campanulate. Plants of this species, the most common *Flaveria* in Florida, occur throughout most of the Florida peninsula, often near the coast.

5. Flaveria bidentis (Linnaeus) Kuntze, Revis. Gen. Pl. 3([3]): 148. 1898 • Coastal plain yellowtops [I]

Ethulia bidentis Linnaeus, Mant. Pl. 1: 110. 1767

Annuals, to 100 cm (delicate or robust, sparsely villous). **Stems** erect. **Leaves** petiolate (proximal, petioles 3–15 mm) or sessile (distal); blades lanceolate-elliptic, 50–120(–180) × 10–25(–70) mm, bases (distal) connate, margins serrate or spinulose serrate. **Heads** 20–100+ in tight subglomerules in scorpioid, cymiform arrays. **Calyculi** of 1–2 linear bractlets 1–2 mm. **Involucres** oblong-angular, 5 mm. **Phyllaries** 3(–4), oblong. **Ray florets** 0 or 1; laminae pale yellow, ovate-oblique, to 1 mm (not or barely surpassing phyllaries). **Disc florets** (2–)3–8; corolla tubes ca. 0.8 mm, throats funnelform, 0.8 mm. **Cypselae** oblanceolate or subclavate, 2–2.5 mm (those of ray florets longer); **pappi** 0. $2n = 36$.

Flowering year round. Moist places, waste or disturbed ground, clay, gravel, or sand; 0–10 m; introduced; Ala., Fla., Ga., Mass.; West Indies; Central America; South America; also introduced in Europe (Great Britain) and Africa (Egypt and south).

In the United States, *Flaveria bidentis* often occurs on ballast and waste ground near wharfs and the shore, locations that suggest its introduction from sailing vessels. The species is widespread and well established in South America, where it is thought to be native. Its floral structure appears to be the least advanced among the annual species of *Flaveria*, the heads being in somewhat scorpioid arrays. Distinguishing characteristics are 3(–4) phyllaries, 2–8 florets per head, reduced ray corollas, pubescent stems, and lanceolate-elliptic, bluish green, sometimes pubescent leaves. *Flaveria bidentis* is probably most closely related to *F. campestris* and *F. trinervia*.

6. Flaveria campestris J. R. Johnston, Proc. Amer. Acad. Arts 39: 287. 1903 · Alkali yellowtops E F

Annuals, 18–70(–90) cm (delicate or robust, mostly glabrous). Stems usually erect (tufted-pubescent at nodes). Leaves sessile; blades linear-lanceolate to lanceolate, 30–90 × 6–22 mm, bases barely connate, margins strongly to weakly serrate or spinulose-serrate. Heads (10–)15–100+, in tight subglomerules in corymbiform arrays. Calyculi of 2 linear-lanceolate bractlets 1–3 mm. Involucres urceolate or oblong, 6–7 mm. Phyllaries 3, elliptic, obovate, or oblong-obovate. Ray florets 0 or 1; laminae yellow, ovate, 1.5–2.5 mm (apices notched). Disc florets 5–6(–8); corolla tubes 0.8–1.3 mm, throats funnelform, 1.2–1.7 mm. Cypselae oblanceolate, linear, or oblong, 2.8–3.6 mm (those of rays longer by ca. 0.2 mm); pappi 0. $2n = 36$ (18).

Flowering May–Nov. Saline soils, lake, pond, stream margins, flood plains, and disturbed pastures; 1000–1800 m; Ariz., Colo., Kans., Mo., N.Mex., Okla., Tex., Utah.

Flaveria campestris is distinguished by subglomerulate capitulescences with 5–6(–8) florets per head, subtended by 3 leafy bracts, keeled phyllaries, and linear-lanceolate leaves with serrate or spinulose-serrate margins, and its distribution. The prevalence of *F. campestris* fluctuates greatly from year to year, generally depending on the existence of wet, saline habitat.

7. Flaveria trinervia (Sprengel) C. Mohr, Contr. U.S. Natl. Herb. 6: 810. 1901 · Clustered yellowtops F

Odera trinervia Sprengel, Bot. Gart. Halle, 63. 1800

Annuals, to 200+ cm (delicate or robust, glabrate or glabrous). Stems erect. Leaves petiolate (proximal, petioles 10–20 mm) or sessile (distal); blades lanceolate or oblanceolate to elliptic or subovate, 30–150 × (7–)10–40 mm, bases (distal) connate, margins serrate, serrate-dentate, or spinulose-serrate. Heads 30–300+, in tight, axillary, sessile glomerules (receptacles of glomerules setose). Calyculi 0. Involucres oblong and cylindric or

angular, 3.8–4.5 mm. Phyllaries usually 2, oblong (closely investing and falling with mature cypselae). Ray florets 0–1; laminae pale yellow or whitish, oblique or suborbiculate, 0.5–1 mm. Disc florets 0–1(–2); corolla tubes 0.5–1.4 mm, throats campanulate, 0.5–0.8 mm. Cypselae oblanceoloid to subclavate, 2–2.6 mm (rays longer); pappi 0. $2n = 36$.

Flowering Mar–Dec. Near water, saline and gypseous areas; 0–1900 m; Ariz., Calif., Fla., Mass., Mo., N.Mex., Tex., Va.; West Indies; Central America (British Honduras); South America (Brazil, Ecuador, Peru, Venezuela); probably introduced in Asia (India, Middle East); Africa; Pacific Islands (Hawaii).

Flaveria trinervia is widespread and weedy; it often occurs in saline, gypseous, disturbed areas near permanent or ephemeral water sources in southern Florida and from Texas to southern California. It occurs also in scattered locations in some eastern states and has been reported from Alabama.

The heads of *Flaveria trinervia*, which usually contain just one floret, are either radiate or discoid; radiate heads tend to occur on the periphery of setose glomerules. Reduction of some of the floral features, including number of florets [0–1(–2)], phyllaries per head (2), and size of ray laminae, suggest that *F. trinervia* may be the most derived species in the genus.

8. Flaveria mcdougallii M. E. Theroux, Pinkava & D. J. Keil, Madroño 24: 13, fig. 1. 1977 · Mcdougall's yellowtops E

Perennials or subshrubs, 50 cm (glabrous). Stems erect. Leaves sessile; blades linear to narrowly linear-lanceolate, 50–110 × 2–7 mm, bases weakly connate, margins entire. Heads 50–200+, in clusters in compound, corymbiform or paniculiform arrays. Calyculi of 0–1 linear bractlets 1–3 mm. Involucres turbinate, 3 × 0.7 mm. Phyllaries 2–6, oblong. Ray florets 0. Disc florets 2–6 (exserted conspicuously from involucres); corolla tubes 1 mm, throats funnelform, 1.5 mm. Cypselae linear, 1.5 mm; pappi coroniform (fringed scales), ca. 0.3 mm. $2n = 36$.

Flowering Sep–Jan. Alkaline springs and seeps; 500–800 m; Ariz.

Flaveria mcdougallii is known only from four locations near shaded alkaline seeps and springs close to the Colorado River in the Grand Canyon of northern Arizona. Delimiting characteristics of *F. mcdougallii* include linear leaves with entire margins, pappi of crowns of scales, and the distribution. Morphologic character differences and experimental hybridization studies suggest that *F. mcdougallii* could well be assigned to a separate, monotypic genus.

187m.16. ASTERACEAE Martinov (tribe HELIANTHEAE) subtribe CLAPPIINAE H. Robinson, Phytologia 41: 39. 1978

Subshrubs or shrubs, 5–40 cm (often succulent or fleshy). **Leaves** mostly cauline; opposite (proximal) or distally or mostly alternate; sessile; blades mostly linear to filiform (terete or subterete), margins entire, faces glabrous. **Heads** radiate, borne singly. **Calyculi** 0. **Involucres** hemispheric or obconic. **Phyllaries** persistent, 12–16 in 2–4+ series (distinct, obovate or ovate to lanceolate or linear, unequal, striate with resinous nerves in *Clappia*). **Receptacles** convex, usually epaleate (paleae rarely scattered, subulate scales; receptacles sometimes with setiform or subulate enations). **Ray florets** 2–11 or 10–15, pistillate, fertile; corollas yellowish or yellow. **Disc florets** 20–100+, bisexual, fertile; corollas yellowish or yellow, tubes shorter than to longer than funnelform throats, lobes 5, deltate to narrowly lance-ovate; anther thecae pale; stigmatic papillae in 2 lines. **Cypselae** clavate to columnar (10–12-ribbed, sparsely hairy); **pappi** persistent, either of 12–25 subulate scales in 1 series, or of ca. 50 bristles or setiform scales in 3–4 series.

Genera 2, species 3 (2 genera, 3 species in the flora): sw United States, Mexico.

Traditionally, *Clappia* has been included in Helenieae and *Pseudoclappia* in Senecioneae. H. Robinson (1981) suggested that the two genera are related to *Varilla* (Varillinae).

1. Ray florets 10–15; pappi of 12–25 subulate scales in 1 series . . . . 328. *Clappia*, p. 251
1. Ray florets 2–4; pappi of ca. 50 bristles or setiform scales in 3–4 series . . . . 329. *Pseudoclappia*, p. 252

328. CLAPPIA A. Gray in W. H. Emory, Rep. U.S. Mex. Bound. 2(1): 93. 1859 • [For "Dr. Asahel Clapp, of New Albany, Indiana, one of the most zealous botanists of our Western States...." Quoted from protologue.]

John L. Strother

Subshrubs or shrubs, to 30+ cm (± succulent or fleshy). **Stems** mostly erect, branched from bases or throughout. **Leaves** cauline; opposite (proximally) or mostly alternate; sessile; blades linear (subterete), margins mostly entire, rarely distally 3-lobed, faces glabrous. **Heads** radiate, borne singly. **Involucres** hemispheric, 6–8 mm diam. **Phyllaries** 12–16 in 3–4+ series (broadly ovate to obovate, unequal, striate with resinous nerves). **Receptacles** convex, epaleate (± densely setose with subulate enations). **Ray florets** 10–15, pistillate, fertile; corollas yellow. **Disc florets** 30–100+, bisexual, fertile; corollas yellow; tubes about equaling or longer than narrowly funnelform throats; lobes 5, narrowly deltate. **Cypselae** columnar to clavate (10–12-ribbed, sparsely hairy on ribs); **pappi** persistent, of 12–25 subulate scales in 1 series. $x = 16$.

Species 1: Texas, Mexico.

1. **Clappia suaedifolia** A. Gray in W. H. Emory, Rep. U.S. Mex. Bound. 2(1): 93. 1859 (as suaedaefolia) F

Leaf blades 2–4+ cm. Heads held well beyond foliage. **Peduncles** 3–8+ cm, distally dilated, fistulose. **Ray corollas** 8–12 × 1–2 mm. **Cypselae** 3–4 mm; **pappi** 3–5 mm. $2n = 32$.

Flowering year round, mostly spring and fall. Low, often ± saline sites; 0–10+ m; Tex.; Mexico.

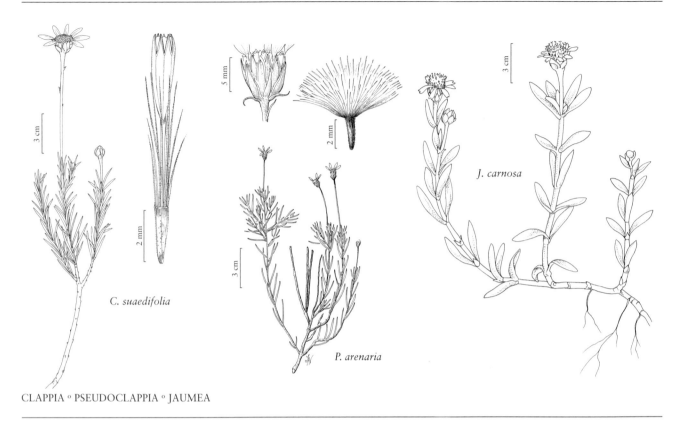

C. *suaedifolia*

J. *carnosa*

P. *arenaria*

CLAPPIA ∘ PSEUDOCLAPPIA ∘ JAUMEA

329. PSEUDOCLAPPIA   Rydberg, J. Wash. Acad. Sci. 13: 288. 1923   •   [Greek *pseudo*, resembling but not equaling, and generic name *Clappia*]

John L. Strother

Subshrubs or shrubs, 5–40+ cm (fleshy). **Stems** mostly erect, branched from bases. **Leaves** cauline, mostly alternate (proximal opposite); sessile; blades linear to filiform (terete), margins entire, faces glabrous. **Heads** radiate, borne singly. **Involucres** obconic, 4–6 mm diam. **Phyllaries** persistent, 12–16 in 2+ series (lanceolate to linear, subequal to unequal, succulent to membranous). **Receptacles** convex, usually epaleate, rarely with scattered, slender scales. **Ray florets** 2–11, pistillate, fertile; corollas yellowish. **Disc florets** 20–40+, bisexual, fertile; corollas yellowish, tubes shorter than or about equaling narrowly funnelform throats, lobes 5, deltate. **Cypselae** ± columnar (obscurely 10–12-ribbed, sparsely hairy on ribs); **pappi** ca. 50 fine to coarse, nearly smooth to barbellate, unequal bristles or setiform scales in 3–4 series (all ± connate at bases). $x = 18$.

Species 2 (2 in the flora): sw United States, n Mexico.

1. Rays 2–4+, laminae 5–11 mm; cypselae ca. 3 mm; pappi 4–6 mm . . . . 1. *Pseudoclappia arenaria*
1. Rays 6–11, laminae 10–13 mm; cypselae ca. 4 mm; pappi 8–10 mm . . . . 2. *Pseudoclappia watsonii*

1. Pseudoclappia arenaria   Rydberg, J. Wash. Acad. Sci. 13: 289. 1923   [F]

Leaf blades 10–35 mm. **Ray florets** 2–4+; laminae 5–11 mm. **Disc florets** 20–40+; corollas 6–7 mm. **Cypselae** ca. 3 mm; **pappi** 4–6 mm. $2n = 36$.

Flowering spring–summer. Clays or gypseous-sandy soils, probably seeps, springs; 1000–2000 m; N.Mex., Tex.; Mexico (Coahuila).

2. Pseudoclappia watsonii   A. M. Powell & B. L. Turner, Sida 6: 317, fig. 1. 1976   [E]

Leaf blades 12–17 mm. **Ray florets** 6–11; laminae 10–13 mm. **Disc florets** ca. 25; corollas 8–8.5 mm. **Cypselae** ca. 4 mm; **pappi** 8–10 mm. $2n$ = ca. 36.

Flowering spring–summer. Clays or gypseous soils, seeps, springs; 900 m; Tex.

187m.17. ASTERACEAE   Martinov (tribe HELIANTHEAE) subtribe JAUMEINAE   Bentham & Hooker f., Gen. Pl. 2: 199. 1873

Perennials, 5–30 cm (often ± succulent). **Leaves** mostly cauline; opposite; sessile; blades mostly linear, margins entire, faces glabrous. **Heads** radiate [discoid], borne singly. **Calyculi** 0. **Involucres** cylindric to obconic. **Phyllaries** persistent, 12–15 in 3+ series (distinct, ovate to lanceolate, unequal, succulent to membranous). **Receptacles** convex to conic, epaleate. **Ray florets** [0] 3–10, pistillate, fertile; corollas yellow. **Disc florets** 20–50+, bisexual, fertile; corollas yellow, tubes about equaling narrowly cylindric throats, lobes 5, deltate; anther thecae pale; stigmatic papillae in 2 lines (joining near tips of branches). **Cypselae** obconic to clavate or columnar, often 10-ribbed, glabrous; **pappi** usually 0, sometimes persistent, of 1–5 subulate scales.

Genus 1, species 2 (1 species in the flora): w North America, Mexico, South America.

Some species formerly included in *Jaumea* are now included in *Espejoa* or *Hypericophyllum* (in Chaenactidinae). H. Robinson (1981) allied Jaumeinae with Clappiinae and Flaveriinae.

330. JAUMEA   Persoon, Syn. Pl. 2: 397. 1807 •   [For J. H. Jaume St. Hilaire, 1772–1845, French botanist]

John L. Strother

Perennials, to 30+ cm (rhizomatous or stoloniferous). **Stems** sprawling, ascending, branched mostly from bases. **Leaves** opposite; sessile; blades spatulate to linear (succulent), margins entire, faces glabrous. **Heads** radiate [discoid], borne singly. **Involucres** cylindric to obconic, 4–8+ mm diam. **Phyllaries** 12–15 in 3+ series (ovate to lanceolate, unequal, succulent to membranous). **Receptacles** conic, epaleate. **Ray florets** [0] 3–10, pistillate, fertile; corollas yellow. **Disc florets** 20–50+, bisexual, fertile; corollas yellow, tubes about equaling narrow, cylindric throats, lobes 5, deltate. **Cypselae** obconic to clavate or columnar (10-ribbed, glabrous); **pappi** usually 0, rarely persistent, of 1–5 subulate scales. $x = 19$.

Species 2 (1 in the flora): w North America, nw Mexico, South America.

1. Jaumea carnosa (Lessing) A. Gray in C. Wilkes et al.,
U.S. Expl. Exped. 17: 360. 1874 [F]

Coinogyne carnosa Lessing, Linnaea
6: 520. 1831

Leaf blades 15–35+ mm. **Phyllaries** often purplish. **Ray laminae** 1–3+ mm. **Disc corollas** 6–7 mm. **Cypselae** 2–3 mm. $2n = 38$.

Flowering late summer–fall. Saline marshes, ocean bluffs; 0–20+ m; B.C.; Calif., Oreg., Wash.; Mexico (Baja California).

187m.18. ASTERACEAE   Martinov (tribe HELIANTHEAE) subtribe MADIINAE   Bentham & Hooker f.,
Gen. Pl. 2: 198.   1873 (as Madieae)

Annuals, perennials, subshrubs, or shrubs, (0.5–)1–250 cm. **Leaves** mostly basal, or basal and cauline, or mostly cauline; opposite and/or alternate; usually sessile, rarely petiolate; blades usually spatulate to oblanceolate or linear to linear-elliptic, sometimes pinnately lobed, ultimate margins entire or toothed, faces usually hairy and glandular (glands sometimes elaborate, apices sometimes gland- or spine-tipped). **Heads** usually radiate, rarely discoid, usually in corymbiform, glomerulate, paniculiform, spiciform, or ± umbelliform arrays, sometimes borne singly. **Calyculi** usually 0, sometimes 2–12+ bractlets (*Centromadia*, some members of *Hemizonia* and *Lagophylla*). **Involucres** campanulate, cylindric, ellipsoid, globose, fusiform, hemispheric, obconic, or urceolate. **Phyllaries** persistent or falling, 1–75+ in 1(–2) series or 0 (the "involucre" then interpreted as constituted of 1 series of receptacular paleae; phyllaries distinct or weakly connate, narrowly ovate to lanceolate or linear, equal or subequal, herbaceous, usually conduplicate, often each partly or wholly investing a subtended floret, apices often attenuate, faces usually glandular). **Receptacles** flat or convex to conic, ± paleate (paleae either in 1 series interior to ray florets, then usually connate and persistent, or 1 palea subtending each disc floret, then usually distinct and soon falling, or, in discoid heads, paleae functioning as phyllaries in 1 series and together constituting an "involucre"). **Ray florets** 0, or 1–75+, pistillate, fertile; corollas usually yellow or whitish, sometimes proximally yellow and distally whitish, often marked with red or purple or turning red on drying, rarely orange or red-orange (apices usually ± deeply divided, lobes usually 3, sometimes 2). **Disc florets** 1–200+, bisexual and fertile, or functionally staminate; corollas usually yellow, sometimes brown, orange, reddish, or whitish, tubes much shorter than to about equaling cylindric to funnelform throats, lobes 5, deltate; anther thecae usually dark (reddish to purple, often described as "black"), sometimes pale (yellow to brown); stigmatic papillae in 2 lines. **Cypselae** obcompressed, compressed, or terete, often obpyramidal, clavate, or fusiform, glabrous or hairy (ray and disc often different); **pappi** 0, or persistent or falling, of awns, bristles, or scales (sometimes in combination) in 1–2 series (the elements often ciliate or plumose).

Genera 24, species 121 (20 genera, 83 species in the flora): mostly w North America, South America, Pacific Islands (Hawaii).

Madiinae has long included the "tarweeds," which are centered in or restricted to the California Floristic Province, the silversword alliance of the Hawaiian Islands (*Argyroxiphium*, *Dubautia*, and *Wilkesia*), and *Raillardella* (in a broad sense), which was formerly included in

Senecioneae (Carlquist et al. 2003). Such a circumscription was used by H. Robinson (1981) and is followed here. In papers by B. G. Baldwin and B. Wessa (2000) and Baldwin et al. (2002), Madiinae were linked to genera traditionally included in Heliantheae (e.g., *Eriophyllum*, *Hulsea*, *Lasthenia*, *Venegasia*) and in Senecioneae (e.g., *Arnica*) and Madiinae plus Arnicinae, Baeriinae, Hulseinae, and Venegasiinae were included in tribe Madieae.

Key to genera of Madiinae here is based on key by B. G. Baldwin (in S. Carlquist et al. 2003).

1. Ray cypselae obcompressed (each mostly or completely enveloped by a phyllary; if rays 0, plants annuals, pappose).
  2. Annuals, 1–20 cm; disc florets 1(–2) . . . . 345. *Hemizonella*, p. 296
  2. Annuals or perennials, 2–150 cm; disc florets 3–120+.
    3. Perennials (rhizomatous); disc corollas white . . . . 343. *Holozonia*, p. 294
    3. Annuals; disc corollas yellow (sometimes reddish with age).
      4. Pappi of 10 apically obtuse scales. . . . . 332. *Achyrachaena*, p. 258
      4. Pappi 0 or of bristles or apically acute scales.
        5. Calyculi 0 or of 2–5 bractlets; ray florets 5; disc florets 6, functionally staminate; disc pappi 0 . . . . 334. *Lagophylla*, p. 260
        5. Calyculi 0; ray florets 0 or 3–27; disc florets 4–120+, bisexual; disc pappi usually of 2–32 bristles or scales, rarely 0 . . . . 335. *Layia*, p. 262
1. Ray cypselae usually compressed, ± terete, or ± 3-angled in cross section (if ± obcompressed, then each ± ¹⁄₂ enveloped by a phyllary; if rays 0, perennials, or annuals and epappose).
  6. Annuals (styles of discs hairy proximal to minute branches; receptacles paleate throughout, ray corollas white with abaxial purple lines, pappi of subulate, plumose scales) . . . . 333. *Blepharipappus*, p. 259
  6. Annuals, perennials, subshrubs, or shrubs (styles of discs glabrous proximal to branches).
    7. Perennials (± scapiform); disc pappi of subulate, ciliate-plumose scales. . . . 331. *Raillardella*, p. 256
    7. Annuals, perennials (leafy-stemmed), subshrubs, or shrubs; disc pappi 0 or of scales (scales seldom both subulate and ciliate-plumose).
      8. Annuals or perennials; peduncular bracts without terminal pit-glands, tack-glands, or spines; heads radiate or discoid; ray corollas yellow; ray cypselae usually compressed, rarely terete (cross sections usually ± 3-angled, then abaxial sides relatively broad, ± rounded, adaxial sides ± 2-faced, angles between those faces 15–70°; each ray cypsela usually completely or mostly enveloped by phyllary).
        9. Disc pappi 0 . . . . 350. *Madia*, p. 303
        9. Disc pappi of 5–21 scales (scales sometimes subulate to setiform, bristlelike).
          10. Heads discoid . . . . 349. *Carlquistia*, p. 302
          10. Heads (all or some) radiate.
            11. Perennials.
              12. Involucres campanulate to hemispheric; anthers ± dark purple; ray cypselae not beaked . . . . 344. *Kyhosia*, p. 295
              12. Involucres campanulate, ellipsoid, or globose; anthers yellow to brownish; ray cypselae beaked . . . . 347. *Anisocarpus*, p. 299
            11. Annuals.
              13. Anthers yellow to brownish . . . . 346. *Harmonia*, p. 297
              13. Anthers ± dark purple . . . . 348. *Jensia*, p. 301
      8. Annuals, subshrubs, or shrubs; peduncular bracts sometimes each with terminal pit-gland, tack-gland, or spine (or apiculus); heads radiate; ray corollas yellow, whitish, or rose; ray cypselae terete to subterete or ± obcompressed (cross sections nearly circular with adaxial sides ± flattened to slightly bulging, or ± 3-angled, then abaxial sides usually ± broadly 2-faced, angles between those faces usually 90+° and adaxial sides ± flattened to slightly bulging; in *Centromadia* spp., distal leaves spine-tipped, each cypsela ± enveloped by phyllary, cypselae sometimes compressed).

[14. Shifted to left margin—Ed.]

14. Annuals; leaves filiform to narrowly linear, margins often strongly revolute; peduncular bracts usually with tack-glands; ray corolla lobes (at least the lateral) often spreading (lengths often ¹/₂–⁵/₆ of total laminae).
    15. Ray cypselae beaked; tack-glands absent . . . . 336. *Osmadenia*, p. 269
    15. Ray cypselae not beaked; tack-glands present . . . . 337. *Calycadenia*, p. 270
14. Annuals, subshrubs, or shrubs; leaves linear or broader, margins seldom strongly revolute; peduncular bracts usually without tack-glands; ray corolla lobes ± parallel (lengths usually ¹/₁₀–¹/₂ of total laminae).
    16. Ray corollas usually white, sometimes yellow, often with abaxial purple lines; cypselae not beaked or each with an inconspicuous, straight beak (beak lengths less than diams.).
        17. Receptacles: paleae restricted to bases of outermost disc florets; disc florets bisexual; cypselae hairy . . . . 341. *Blepharizonia*, p. 289
        17. Receptacles: paleae throughout; disc florets functionally staminate; cypselae glabrous . . . . 342. *Hemizonia*, p. 291
    16. Ray corollas yellow (without abaxial purple lines); cypselae each with an adaxial, ascending beak (beak lengths greater than diams.).
        18. Peduncular bracts apiculate or each with an apical spine . . . . 338. *Centromadia*, p. 276
        18. Peduncular bracts not apiculate, without apical spines.
            19. Annuals; peduncular bracts each with an apical pit-gland; receptacles paleate throughout . . . . 340. *Holocarpha*, p. 287
            19. Annuals, subshrubs, or shrubs; peduncular bracts without pit-glands; receptacles: paleae usually restricted to bases of outermost disc florets (if in 2–3+ series, subshrubs or shrubs) . . . . 339. *Deinandra*, p. 280

331. RAILLARDELLA   (A. Gray) Bentham & Hooker f., Gen. Pl. 2: 442. 1873

• [*Raillardia*, orthographic variant of generic name *Railliardia*, and Latin *-ella*, diminutive]   E

Bruce G. Baldwin

John L. Strother

Railliardia Gaudichaud-Beaupré sect. *Raillardella* A. Gray, Proc. Amer. Acad. Arts 6: 550. 1865 (as Raillardia)

Perennials, 1–50+ cm (with branching caudices or rhizomes). **Stems** ± erect (± scapiform). **Leaves** mostly basal (in rosettes); opposite; sessile; blades lanceolate or oblanceolate to linear, margins usually entire, sometimes, denticulate, faces glabrous or sericeous or sparsely hirtellous and/or stipitate-glandular. **Heads** radiate or discoid, borne singly (on scapiform peduncles) or rarely in open, ± corymbiform arrays. **Peduncular bracts:** (usually 0) pit-glands, tack-glands, and/or spines 0. **Involucres** hemispheric or campanulate to cylindric, 3–25+ mm diam. **Phyllaries** 0 ("involucres" then consisting of 1 series of paleae), or 1–13 in 1 series (lanceolate or oblanceolate to linear, herbaceous, each usually ¹/₂+ enveloping a subtended ray ovary, abaxially hirsute and ± stipitate-glandular). **Receptacles** flat or convex, glabrous or setulose, paleate (paleae ± persistent, usually in 1 series between rays and discs, peripheral and constituting "involucres" in discoid heads, connate or distinct, sometimes overlapping, herbaceous to scarious). **Ray florets** 0 or 1–13, pistillate, fertile; corollas yellow to yellow-orange or orange to red-orange (not nerved with red to purple abaxially). **Disc florets** 7–80+, bisexual, fertile; corollas concolorous with rays, tubes shorter than funnelform throats, lobes 5, deltate (anthers yellowish; styles glabrous proximal to branches). **Ray cypselae** ± terete, ± clavate, basal attachments central, apices not beaked, faces strigose; **pappi** 0, or of 8–30 ± subulate, ciliate to plumose scales. **Disc cypselae** ± terete, ± clavate, (basal attachments central, apices not beaked, faces strigose); **pappi** of 8–30 ± subulate, ciliate to plumose scales. x = 17 or 18.

Species 3 (3 in the flora): w United States.

Raillardella represents an early diverging lineage of tarweeds (Carlquist et al. 2003); putatively ancestral morphologic, ecologic, and chromosomal characteristics are shared with *Arnica*, the sister group of Madiinae (B. G. Baldwin and B. L. Wessa 2000). Two species that have been included in *Raillardella* are more closely related to other genera (e.g., *Madia*, B. G. Baldwin 1996) and are here treated as *Anisocarpus scabridus* and *Carlquistia muirii*.

1. Leaves of basal rosettes glabrous; ray florets 6–13, corollas orange to red-orange; disc florets 45–80+ . . . . 1. *Raillardella pringlei*
1. Leaves of basal rosettes sericeous or stipitate-glandular; ray florets 0 or 1–7, corollas yellow to yellow-orange; disc florets 7–40+.
  2. Plants 1–15 cm; leaf blades sericeous (silvery), sometimes sparsely stipitate-glandular as well; ray florets 0 . . . . 2. *Raillardella argentea*
  2. Plants 6–50+ cm; leaf blades stipitate-glandular, sometimes sparsely hirtellous as well; ray florets 0 or 1–7 . . . . 3. *Raillardella scaposa*

1. Raillardella pringlei Greene, Bull. Torrey Bot. Club 9: 17. 1882 C E

Plants 25–50+ cm. **Leaf blades** lanceolate or oblanceolate to linear, margins entire or toothed, faces glabrous or ± stipitate-glandular (distal). **Ray florets** 6–13; corollas orange to red-orange, laminae 6–20+ mm. **Disc florets** 45–80+; corollas orange to red-orange, 8–11.5 mm. $2n$ = 34.

Flowering Jul–Oct. Meadows, stream banks, seeps; of conservation concern; 1200–2300 m; Calif.

Raillardella pringlei occurs on serpentine-derived soils in the Klamath Ranges.

2. Raillardella argentea (A. Gray) A. Gray in W. H. Brewer et al., Bot. California 1: 417. 1876 E F

Railliardia argentea A. Gray, Proc. Amer. Acad. Arts 6: 550. 1865 (as *Raillardia*)

Plants 1–15 cm. **Leaf blades** oblanceolate, margins entire or toothed, faces sericeous (silvery), sometimes sparsely stipitate-glandular as well. **Ray florets** 0. **Disc florets** 7–26; corollas yellow, 6–11 mm. $2n$ = 34, 36.

Flowering Jul–Sep. Dry, exposed, often gravelly sites; 1800–3900 m; Calif., Nev., Oreg.

Raillardella argentea occurs widely in the Sierra Nevada and southern Cascade Range and locally in the Klamath Ranges, San Bernardino Mountains (Mt. San Gorgonio), and western ranges of the Great Basin.

Artificial hybrids with *R. pringlei* are completely fertile and vigorous (D. W. Kyhos et al. 1990); the two species are not known to co-occur in nature.

3. Raillardella scaposa (A. Gray) A. Gray in W. H. Brewer et al., Bot. California 1: 417. 1876 E

Railliardia scaposa A. Gray, Proc. Amer. Acad. Arts 6: 551. 1865 (as *Raillardia*)

Plants 6–53 cm. **Leaf blades** lanceolate or oblanceolate to linear, margins entire, faces stipitate-glandular, sometimes sparsely scabrellous as well. **Ray florets** 0 or 1–7; corollas yellow to yellow-orange, laminae 5–25+ mm. **Disc florets** 7–44; corollas yellow to yellow-orange, 7.5–12 mm. $2n$ = 68, 70.

Flowering Jun–Sep. Wet to dry, often sandy sites; 2000–3500 m; Calif., Nev., Oreg.

Variation among populations of *Raillardella scaposa* spans the morphologic and ecologic divide between *R. argentea* and *R. pringlei*; *R. scaposa* is evidently an allopolyploid that descended from ancestors closely related to each of the other two species. *Raillardella scaposa* is widely distributed in the Sierra Nevada and southern Cascade Range, often near populations of *R. argentea*. Putative hybrids between *R. scaposa* and *R. argentea* have been noted (e.g., *R. Snow 293*, UC, from Tuolumne County, California).

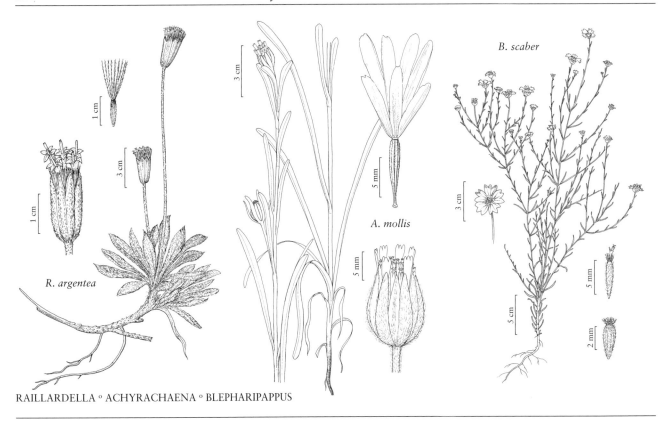

RAILLARDELLA ° ACHYRACHAENA ° BLEPHARIPAPPUS

332. ACHYRACHAENA   Schauer, Index Seminum (Bratislava) 1837: [3]. 1837   •   [Greek *achyron*, scale, and Latin *achaenium*, fruit, alluding to cypselae]

Bruce G. Baldwin

John L. Strother

Annuals, 4–62 cm. **Stems** erect, branched distally. **Leaves** mostly cauline; proximal opposite, distal alternate; ± sessile; blades linear, margins entire or toothed, faces hirsute or villous and (distal leaves) sparsely glandular-pubescent. **Heads** radiate, borne singly or in loose, ± corymbiform arrays. **Peduncular bracts:** pit-glands, tack-glands, and/or spines 0 at tips. **Involucres** ± campanulate to cylindric or ellipsoid, 3–12+ mm diam. **Phyllaries** 3–8 in 1 series (± lance-linear, herbaceous, each fully enveloping a ray ovary, abaxially hirsute or villous and sparsely glandular-pubescent or eglandular). **Receptacles** flat to convex (setulose), paleate (paleae falling, in 1 series between rays and disc, distinct). **Ray florets** 3–8, pistillate, fertile; corollas yellow, turning reddish (lobes ± parallel). **Disc florets** 4–35, bisexual and fertile; corollas yellow to reddish, tubes ± equaling narrowly funnelform throats, lobes 5, lance-deltate (anthers ± dark purple; styles glabrous proximal to branches). **Ray cypselae** (black) obcompressed, clavate (10-ribbed, straight, basal attachments centered, apices beakless, faces glabrous or ± scabrous); **pappi** 0. **Disc cypselae** (brown to black) similar to rays (± scabrous); **pappi** of 10 white, oblong scales (apices obtuse). $x = 8$.

Species 1: w North America, nw Mexico.

1. Achyrachaena mollis Schauer, Index Seminum (Bratislava) 1837: [3]. 1837 • Blow-wives F

Leaf blades 2–15 cm × 1–7+ mm. **Phyllaries** 10–20 mm. **Ray corollas** inconspicuous, laminae usually erect, sometimes spreading, 4–7 mm. **Disc corollas** 6–10 mm. **Cypselae** black, 4.5–9 mm; **pappi** (disc) of 5 outer scales 3–9 mm plus 5 inner scales 6–13 mm. **2*n* = 16.**

Flowering Mar–Jun. Grassy, often wet sites, usually with clay-rich soils; 0–900 m; Calif., Oreg.; Mexico (Baja California).

Achyrachaena mollis grows in low-elevation grasslands, savannas, and open woodlands throughout the California Floristic Province and occurs with vernal tarweeds of the genus *Layia*. During flowering, pappus scales of each disc floret form cylinders (imparting a honeycomb-like appearance to apices of heads); they are widely spread in fruiting heads. The species is self-compatible and is not known to hybridize with other tarweeds.

333. BLEPHARIPAPPUS Hooker, Fl. Bor.-Amer. 1: 316. 1833 • [Greek *blepharis*, eyelash, and *pappos*, pappus, alluding to ciliate pappus scales] E

Bruce G. Baldwin

John L. Strother

Annuals, 5–20(–40+) cm. **Stems** ± erect (scabrous and, sometimes, hirsute, usually stipitate-glandular distally). **Leaves** mostly cauline; proximal opposite, most alternate; sessile; blades narrowly spatulate to linear, margins entire, faces scabrous, hirsute, strigose, sericeous, or villous (distal leaves usually stipitate-glandular as well). **Heads** radiate, borne singly or in open, corymbiform arrays. **Peduncular bracts:** pit-glands, tack-glands, and/or spines 0 at tips. **Involucres** turbinate to campanulate or hemispheric, 3–6+ mm diam. **Phyllaries** (2–)3–5(–8) in 1 series, ± lanceolate or oblanceolate, herbaceous, each ± ½ investing subtended floret proximally, abaxially ± hirsute and/or stipitate-glandular. **Receptacles** convex, glabrous, paleate (paleae falling, subtending all or most disc florets, outer herbaceous, inner scarious). **Ray florets** (2–)3–5(–8), pistillate, fertile; corollas whitish (nerved with purple abaxially). **Disc florets** 6–25(–60+), bisexual, fertile; corollas whitish, tubes shorter than or about equaling funnelform throats, lobes 5, deltate (anthers ± dark purple; styles ± hairy proximal to branches, branches ca. 0.2 mm). **Cypselae** ± obconic or terete (basal attachments ± central, faces ± villous, apices not beaked); **pappi** 0 or of 12–18(–26) subulate fimbriate to ciliate or plumose scales. *x* = 8.

Species 1: nw North America.

Blepharipappus and members of *Layia* were once treated as congeneric; *Layia* appears to be more closely related to *Lagophylla* than to *Blepharipappus*, based on molecular phylogenetic data (S. Carlquist et al. 2003).

1. Blepharipappus scaber Hooker, Fl. Bor.-Amer. 1: 316. 1833 E F

Blepharipappus scaber subsp. *laevis* (A. Gray) D. D. Keck

Leaf blades 6–25+ × 0.5–1.5 mm. **Phyllaries** 3–8 mm. **Ray laminae** 2–11 mm. **Disc corollas** 2–3.5 mm. **Cypselae** 2–3.5 mm; **pappi** 0.1–2 mm. **2*n* = 16.**

Flowering Apr–Sep. Openings in sagebrush scrub, pinyon-juniper woodlands, yellow-pine forests; 300–2200 m; Calif., Idaho, Nev., Oreg., Wash.

Blepharipappus scaber is unusual among self-incompatible, continental tarweeds for occurring widely in western North America and having a relatively limited distribution in the California Floristic Province.

334. LAGOPHYLLA Nuttall, Trans. Amer. Philos. Soc., n. s. 7: 390. 1841 • [Greek *lago*, hare, and *phyllon*, leaf, alluding to sericeous leaves of original species] E

Bruce G. Baldwin

John L. Strother

Annuals, 8–100(–150) cm. **Stems** ± erect. **Leaves** mostly cauline; proximal opposite, most alternate; ± sessile; blades narrowly elliptic to linear or (proximal) oblanceolate to spatulate, margins usually entire (proximal sometimes toothed), faces hirsute to strigose, sericeous, or villous (all or distal sometimes stipitate-glandular as well). **Heads** radiate, in ± paniculiform arrays or in glomerules. **Peduncular bracts:** pit-glands, tack-glands, and/or spines 0. **Involucres** ± hemispheric or obovoid to obconic, 3–6+ mm diam. (sometimes subtended by calyculi of 2–5 bractlets). **Phyllaries** 5 in 1 series (linear to oblanceolate, herbaceous, each wholly enveloping a subtended ray ovary, abaxially piloso-hirsute to hirtellous or scabrellous). **Receptacles** flat to convex, densely hirtellous, paleate (paleae in rings between rays and discs, distinct or proximally connate, scarious). **Ray florets** 5, pistillate, fertile; corollas yellow (often nerved with red to purple abaxially). **Disc florets** 6, functionally staminate; corollas yellow, tubes shorter than funnelform throats, lobes 5, deltate (anthers ± dark purple; styles glabrous proximal to branches). **Ray cypselae** ± obcompressed (attachments basal, apices beakless, faces glabrous); **pappi** 0. **Disc cypselae** 0; **pappi** 0. *x* = 7.

Species 4 (4 in the flora): w North America.

Molecular phylogenetic data have indicated that *Lagophylla* is sister to *Layia* (S. Carlquist et al. 2003). Members of *Lagophylla* are easily overlooked in the field; leaves usually wither before flowering and heads often close at midday. Ray laminae often shrivel greatly in pressed specimens (making lengths difficult to assess). The species are moderately interfertile (W. C. Thompson 1983); most do not co-occur and no natural hybrids have been documented.

SELECTED REFERENCE Thompson, W. C. 1983. A Biosystematic Study of *Lagophylla* (Compositae: Heliantheae) and Related Taxa. Ph.D. dissertation. University of California, Davis.

1. Leaves grayish (the distal stipitate-glandular abaxially, glands whitish or yellowish; plants otherwise eglandular); heads in paniculiform arrays or in glomerules; calyculi of 2–5 bractlets;phyllaries piloso-hirsute on angles; ray laminae 3–6 mm (plants self-compatible) .... 4. *Lagophylla ramosissima*
1. Leaves green or gray-green (the distal eglandular or stipitate-glandular, glands yellow, golden, or purple; plants otherwise eglandular or stipitate-glandular on distal stems); heads in paniculiform arrays; calyculi 0 or of 2–5 bractlets; phyllaries hirtellous to scabrellous or piloso-hirsute on angles; ray laminae 7–13 mm (plants strongly or weakly self-incompatible).
    2. Branching excurrent; distal leaves stipitate-glandular, glands yellow or golden; calyculi of 3–5 bractlets; involucres obconic; phyllaries hirtellous to piloso-hirsute on angles, hairs 0.3–1+ mm, ± patent to antrorsely curved .... 2. *Lagophylla glandulosa*
    2. Branching ± pseudo-dichotomous (main stems ± zigzag); leaves usually eglandular (the distal eglandular or stipitate-glandular, glands mostly purple, some yellow); calyculi 0, or of 2–3 bractlets; involucres ± hemispheric to obovoid; phyllaries ± hirtellous to scabrellous or piloso-hirsute on angles, hairs 0.1–1+ mm, ± antrorsely curved.
        3. Phyllaries 4–5 mm, piloso-hirsute on angles, hairs 0.5–1+ mm; cypselae glossy ...... 1. *Lagophylla minor*
        3. Phyllaries 4–6.5 mm, ± hirtellous to scabrellous on angles, hairs 0.1–0.6 mm; cypselae dull .... 3. *Lagophylla dichotoma*

1. **Lagophylla minor** (D. D. Keck) D. D. Keck, Aliso 4: 105. 1958 [E]

Lagophylla dichotoma Bentham subsp. *minor* D. D. Keck, Madroño 3: 16. 1935

Plants 8–30+ cm (strongly self-incompatible); branching ± pseudo-dichotomous, distal stems eglandular or sparsely glandular. **Leaves:** blades green, usually eglandular (distal eglandular or stipitate-glandular, glands mostly purple, some yellow). **Heads** in paniculiform arrays. **Calyculi** 0. **Involucres** ± hemispheric to obovoid. **Phyllaries** 4–5 mm, piloso-hirsute on angles, hairs ± antrorsely curved, 0.5–1+ mm. **Ray laminae** 7–13 mm. **Cypselae** glossy. $2n = 14$.

Flowering Apr–Jun. Openings in chaparral and woodlands, serpentine slopes; 70–700 m; Calif.

Lagophylla minor is known only from serpentine of the North Inner Coast Ranges and in the northern Sierra Nevada foothills.

2. **Lagophylla glandulosa** A. Gray, Proc. Amer. Acad. Arts 17: 219. 1882 [E]

Lagophylla glandulosa subsp. *serrata* (Greene) D. D. Keck

Plants 10–100(–150) cm (weakly self-incompatible); branching excurrent, distal stems usually sparsely to densely stipitate-glandular, rarely eglandular. **Leaves:** blades green or gray-green, distal stipitate-glandular, glands yellow or golden. **Heads** in paniculiform arrays. **Calyculi** of 3–5 bractlets. **Involucres** obconic. **Phyllaries** 5–7 mm, hirtellous to piloso-hirsute on angles, hairs ± patent to antrorsely curved, 0.3–1+ mm. **Ray laminae** 7–13 mm. **Cypselae** glossy. $2n = 14$.

Flowering May–Nov. Grasslands, openings in chaparral and woodlands; 10–900 m; Calif.

Lagophylla glandulosa comprises spring and summer–fall flowering populations that occur widely in the northern Great Valley and surrounding foothills of the North Inner Coast Ranges, Cascade Range, and Sierra Nevada.

3. **Lagophylla dichotoma** Bentham, Pl. Hartw., 317. 1849 [E]

Plants 10–60+ cm (strongly self-incompatible); branching ± pseudo-dichotomous, distal stems eglandular or sparsely stipitate-glandular. **Leaves:** blades green, usually eglandular (distal eglandular or stipitate-glandular, glands mostly purple, some yellow). **Heads** in paniculiform arrays. **Calyculi** 0 or of 2–3 bractlets. **Involucres** ± hemispheric to obovoid. **Phyllaries** 4–6.5 mm, ± hirtellous to scabrellous on angles, hairs ± antrorsely curved, 0.1–0.6 mm. **Ray laminae** 7–13 mm. **Cypselae** dull (striate). $2n = 14$.

Flowering Apr–Jun. Grasslands, openings in woodlands; 50–900 m; Calif.

Lagophylla dichotoma occurs in the western Sierra Nevada foothills and adjacent eastern San Joaquin Valley and in the northern South Inner Coast Ranges (where plants are notably stipitate-glandular, unlike most Sierran and San Joaquin Valley collections).

4. **Lagophylla ramosissima** Nuttall, Trans. Amer. Philos. Soc., n. s. 7: 391. 1841 [E] [F]

Lagophylla congesta Greene; *L. ramosissima* subsp. *congesta* (Greene) D. D. Keck

Plants 10–100(–150) cm (plants self-compatible); branching excurrent or ± pseudo-dichotomous, distal stems eglandular. **Leaves:** blades grayish, mostly eglandular (distal stipitate-glandular abaxially, glands whitish or yellowish; plants otherwise eglandular). **Heads** in paniculiform arrays or in glomerules. **Calyculi** of 2–5 bractlets. **Involucres** obconic to obovoid. **Phyllaries** 4–7 mm, piloso-hirsute on angles, hairs ± patent to antrorsely curved, 0.5–1+ mm. **Ray laminae** 3–6 mm. **Cypselae** dull to ± glossy (weakly striate). $2n = 14$.

Flowering Apr–Oct. Grasslands, openings in chaparral, scrub, woodlands, and forests; 10–1800 m; Calif., Idaho, Mont., Nev., Oreg., Wash.

Lagophylla ramosissima occurs widely in dry, often disturbed or poor soils of the California Floristic Province, Great Basin, and Pacific Northwest. Plants with heads in glomerate arrays have been treated as *L. congesta* or *L. ramosissima* subsp. *congesta*; W. C. Thompson (1983, p. 21) concluded that *L. congesta* represents an "extreme morphological variant of *L. ramosissima*" unworthy of taxonomic recognition.

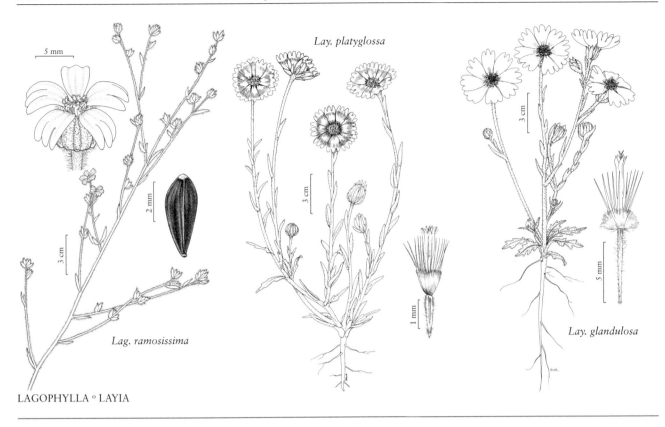

Lay. platyglossa

Lag. ramosissima

Lay. glandulosa

LAGOPHYLLA ∘ LAYIA

335. LAYIA   Hooker & Arnott ex de Candolle in A. P. de Candolle and A. L. P. P. de Candolle, Prodr. 7: 294. 1838, name conserved   •   [For George Tradescant Lay, a naturalist on Beechey's voyage (1825–1828)]

Bruce G. Baldwin

Susan J. Bainbridge

John L. Strother

Annuals, 2–60(–130) cm.   **Stems** usually ± erect (prostrate in *L. chrysanthemoides* and *L. platyglossa* of coastal bluffs).   **Leaves** mostly cauline; proximal opposite, most alternate; sessile; blades ovate, lanceolate, or oblanceolate to linear, sometimes 1–2-pinnatifid, ultimate margins toothed or entire, faces glabrous or hirsute to strigose (distal leaves sometimes stipitate-glandular as well).   **Heads** usually radiate (discoid in *L. discoidea*), borne singly or in ± corymbiform arrays.   **Peduncular bracts:** pit-glands, tack-glands, and/or spines 0.   **Involucres** ± hemispheric, campanulate, cylindric, ellipsoid, obconic, or urceolate, 2–15+ mm diam.   **Phyllaries** 0 (then outer paleae functioning as phyllaries, in *L. discoidea*), or 3–27 in 1(–2) series (lanceolate to lance-attenuate or oblanceolate, herbaceous, each usually wholly enveloping a subtended ray ovary, abaxially hirsute to strigose or scabrous, sometimes glandular).   **Receptacles** flat to convex, setulose, paleate (paleae falling, in 1 series, between rays and discs or subtending ± all disc florets, distinct, phyllary-like, more scarious).   **Ray florets** 0 or 3–27, pistillate, fertile; corollas yellow, cream, white, or bicolored.   **Disc florets** 5–120+, bisexual, fertile; corollas yellow, tubes shorter than funnelform throats, lobes 5, deltate (anthers ± dark purple or yellow to brownish; styles glabrous proximal to branches). **Ray cypselae** obcompressed, clavate (± arcuate to falcate, basal attachments central, apices beakless, faces glabrous or sparsely hairy); **pappi** 0. **Disc cypselae** ± clavate (usually ± strigose to sericeous, sometimes glabrous); **pappi** 0 or of 1–32 elliptic, lance-attenuate, ovate, setiform, or subulate, glabrous, scabrous, or plumose scales or bristles (often each basally villous and/or adaxially woolly). $x = 8$.

Species 14 (14 in the flora): w North America, nw Mexico.

Layia is evidently most closely related to *Lagophylla* based on molecular phylogenetic data. Species of *Layia* with $2n = 14$ constitute a clade that has been regarded as exemplary of geographic diversification (speciation) in plants (J. Clausen 1951; B. G. Baldwin, unpubl.). All members of *Layia* except *L. carnosa* and *L. hieracioides* are self-incompatible.

1. Disc pappi 0.
   2. Plants not glandular; paleae subtending ± all disc florets . . . . 2. *Layia chrysanthemoides* (in part)
   2. Plants glandular; paleae in 1 (involucre-like) series between ray and disc florets.
      3. Plants apple- or banana-scented; margins (basal leaves) usually denticulate to serrulate, seldom lobed; ray corollas white to cream . . . . 1. *Layia heterotricha* (in part)
      3. Plants unscented, or not apple- or banana-scented; margins (basal leaves) lobed; ray corollas white, yellow, or bicolored (proximally yellow, distally whitish).
         4. Stems usually purple-streaked; involucres campanulate to hemispheric, subglobose, or ± urceolate.
            5. Involucres campanulate to hemispheric or subglobose; ray florets 6–18, corollas yellow or bicolored (proximally yellow, distally whitish) . . . . .
. . . . 12. *Layia gaillardioides* (in part)
            5. Involucres ± urceolate; ray florets 13–27 (in 2 series), corollas bicolored
. . . . 5. *Layia jonesii* (in part)
         4. Stems usually not purple-streaked; involucres ± hemispheric.
            6. Plants strongly lemon- or acrid-scented; ray corollas white or yellow; anthers yellow or brownish . . . . 8. *Layia pentachaeta* (in part)
            6. Plants not strongly scented; ray corollas yellow or bicolored (proximally yellow, distally white); anthers ± dark purple (sometimes yellow to brownish in sw California) . . . . 4. *Layia platyglossa* (in part)
1. Disc pappi of 1–32 bristles or scales.
   7. Ray florets 0 . . . . 10. *Layia discoidea*
   7. Ray florets 3–27 (ray laminae sometimes inconspicuous).
      8. Plants not glandular; paleae subtending ± all disc florets.
         9. Disc pappi of unequal, subulate to setiform scales . . . . 2. *Layia chrysanthemoides* (in part)
         9. Disc pappi of ± equal, lance-attenuate scales . . . . 3. *Layia fremontii*
      8. Plants glandular; paleae in 1 (involucre-like) series between ray and disc florets.
         10. Disc pappi of elliptic, lance-linear, or ovate, non-plumose, non-woolly scales (bases sparsely setose), 0.5–3.5 mm.
            11. Ray corollas white; anthers yellow to brownish; ray cypselae sparsely hairy . . . . 6. *Layia leucopappa*
            11. Ray corollas yellow or bicolored (proximally yellow, distally white); anthers ± dark purple; ray cypselae glabrous or sparsely hairy.
               12. Stems usually purple-streaked; involucres ± urceolate; ray florets 13–27 (in 2 series); ray cypselae glabrous; pappus scales 0.5–2 mm . . .
. . . . 5. *Layia jonesii* (in part)
               12. Stems not purple-streaked; involucres hemispheric to ± urceolate; ray florets 6–15; ray cypselae glabrous or sparsely hairy; pappus scales 2–3.5 mm . . . . 7. *Layia munzii*

[10. Shifted to left margin—Ed.]

10. Disc pappi usually of bristles or setiform scales, if scales linear-attenuate to subulate, then proximally plumose and often proximally woolly adaxially, 1–7 mm.

    13. Plants usually not strongly scented (if not contacted); disc pappus bristles or scales mostly proximally plumose and adaxially woolly, sometimes ± scabrous throughout, if proximally plumose and not woolly, then linear-attenuate to subulate scales.

        14. Ray corollas usually white (sometimes yellow); disc pappi of 10–15 linear attenuate to subulate scales. . . . . 9. *Layia glandulosa*

        14. Ray corollas yellow or bicolored (proximally yellow, distally white); disc pappi of 14–32 bristles or setiform scales.

            15. Involucres usually ± hemispheric; phyllary apices often longer than folded bases; ray corollas yellow or bicolored (proximally yellow, distally white); anthers usually ± dark purple (sometimes yellow to brownish in sw California); disc pappus bristles or scales usually scabrous (sometimes proximally plumose and adaxially woolly in sw California) . . . . 4. *Layia platyglossa* (in part)

            15. Involucres ± ellipsoid to campanulate; phyllary apices usually shorter than folded bases; ray corollas yellow; anthers yellow to brownish; disc pappus bristles or scales proximally plumose and adaxially woolly . . . . 11. *Layia septentrionalis*

    13. Plants usually strongly scented, even without contact (except *L. carnosa*, fleshy-leaved plants of coastal dunes); disc pappus bristles or scales proximally plumose, seldom woolly adaxially (not linear-attenuate to subulate).

        16. Margins (basal leaves) usually denticulate or serrulate, seldom toothed or lobed; ray corollas white to cream (disc pappi readily falling as units) . . . . . . 1. *Layia heterotricha* (in part)

        16. Margins (basal leaves) lobed; ray corollas white, yellow, or bicolored (proximally yellow, distally whitish).

            17. Anthers yellow to brownish . . . . 8. *Layia pentachaeta* (in part)

            17. Anthers ± dark purple.

                18. Stems not purple-streaked; ray corollas white, laminae 1.5–3.5 mm; ray cypselae sparsely hairy . . . . 14. *Layia carnosa*

                18. Stems purple-streaked; ray corollas yellow or bicolored (proximally yellow, distally white), laminae 1–18 mm; ray cypselae glabrous.

                    19. Ray corollas yellow or bicolored (proximally yellow, distally whitish), 3.5–18 mm; disc pappi of 15–24 bristles or setiform scales (main stems ascending, not strictly erect) . . . . 12. *Layia gaillardioides* (in part)

                    19. Ray corollas yellow, 1–4 mm; disc pappi of 10–16 bristles or setiform scales (main stems strictly erect) . . . . 13. *Layia hieracioides*

1. Layia heterotricha (de Candolle) Hooker & Arnott, Bot. Beechey Voy., 358. 1839 [C][E]

Madaroglossa heterotricha de Candolle in A. P. de Candolle and A. L. P. P. de Candolle, Prodr. 5: 694. 1836

Plants 13–90 cm (self-incompatible); glandular, strongly apple- or banana-scented. **Stems** not purple-streaked. **Leaf blades** elliptic to ovate, 10–120 mm, margins (basal leaves) entire or shallowly toothed. **Involucres** hemispheric, 7–12 × 6–13+ mm. **Phyllaries** 7–13, apices usually shorter than folded bases. **Paleae** in 1 series between ray and disc florets. **Ray florets** 7–13; laminae white to cream, 5–24 mm. **Disc florets** 40–90+; corollas 4–7 mm; anthers yellow to brownish. **Ray cypselae** usually glabrous, sometimes sparsely hairy. **Disc pappi** 0, or (readily falling as units) of 14–20 white, ± equal bristles or setiform scales 3–6 mm, each proximally plumose, not adaxially woolly. *2n* = 16.

Flowering Apr–Jun. Grasslands, meadows, openings in woodlands, on clayey or sandy, sometimes ± alkaline soils; of conservation concern; 200–1800 m; Calif.

Layia heterotricha occurs in the South Coast Ranges, western Transverse Ranges, and Tehachapi Range. Molecular phylogenetic data have indicated that *L. heterotricha* is sister to all other members of *Layia* (B. G. Baldwin 1996). Weak, ± sterile artificial hybrids have been produced with other species of *Layia* (no natural hybrids have been reported; J. Clausen 1951).

2. **Layia chrysanthemoides** (de Candolle) A. Gray, Proc. Amer. Acad. Arts 7: 360. 1868 [E]

Oxyura chrysanthemoides de Candolle in A. P. de Candolle and A. L. P. P. de Candolle, Prodr. 5: 693. 1836; *Layia chrysanthemoides* subsp. *maritima* D. D. Keck

Plants 4–53 cm (self-incompatible); not glandular, not strongly scented. **Stems** not purple-streaked. **Leaf blades** lanceolate or oblanceolate to linear, 5–120 mm, margins (basal leaves) lobed to pinnatifid. **Involucres** hemispheric or depressed-hemispheric, 4–12 × 4–14+ mm. **Phyllaries** 6–16, apices often longer (sometimes shorter) than folded bases. **Paleae** subtending ± all disc florets. **Ray florets** 6–16; laminae usually proximally yellow, distally white or light yellow, rarely uniformly yellow throughout, 3–18 (–24) mm. **Disc florets** 28–100+; corollas 3–5 mm; anthers ± dark purple. **Ray cypselae** glabrous. **Disc pappi** 0 or of 2–18 tawny, subulate to setiform, unequal scales 1–4 mm, each ± scabrous, not adaxially woolly. $2n = 16$.

Flowering Mar–Jun. Grasslands, open woodlands, often valley bottoms, disturbed sites, edges of vernal pools, waterways, and salt marshes, usually on heavy soils, sometimes ± alkaline or saline; 0–800 m; Calif.

Layia chrysanthemoides occurs from the western Great Valley to the coast in northern and central California. Molecular and morphologic data have indicated that *L. chrysanthemoides* is most closely related to *L. fremontii* (B. G. Baldwin, unpubl.); the two species are reportedly highly interfertile (natural hybrids have not been reported; J. Clausen 1951).

3. **Layia fremontii** (Torrey & A. Gray) A. Gray, Mem. Amer. Acad. Arts, n. s. 4: 103. 1849 [E]

Calliachyris fremontii Torrey & A. Gray, Boston J. Nat. Hist. 5: 110. 1845

Plants 8–40 cm (self-incompatible); not glandular, not strongly scented. **Stems** not purple-streaked. **Leaf blades** lanceolate or oblanceolate to linear, 6–70(–90) mm, margins (basal leaves) lobed (pinnatifid). **Involucres** hemispheric to depressed-hemispheric, 4–11 × 3–11+ mm. **Phyllaries** 3–15, apices often longer (sometimes shorter) than folded bases. **Paleae** subtending ± all disc florets. **Ray florets** 3–15; laminae proximally yellow, distally white or light yellow, 5–18(–23) mm. **Disc florets** 4–100+; corollas 3.5–4.5 mm; anthers ± dark purple. **Ray cypselae** glabrous.

Disc pappi of 9–12 white to tawny, lance-attenuate, ± equal scales 2–5 mm, each neither plumose nor adaxially woolly. $2n = 14$.

Flowering Feb–May. Grasslands, meadows, open woodlands, disturbed sites, often valley bottoms, swales, edges of vernal pools, usually on heavy or shallow soils, sometimes serpentine; 10–800 m; Calif.

Layia fremontii occurs in the Great Valley and adjacent foothills of the Cascade Range and Sierra Nevada.

4. **Layia platyglossa** (Fischer & C. A. Meyer) A. Gray, Mem. Amer. Acad. Arts, n. s. 4: 103. 1849 • Tidy tips [F]

Callichroa platyglossa Fischer & C. A. Meyer, Index Seminum (St. Petersburg) 2: 31. 1836; *Layia platyglossa* A. Gray subsp. *campestris* D. D. Keck; *L. ziegleri* Munz

Plants 3–70 cm (self-incompatible); glandular, not strongly scented. **Stems** usually not purple-streaked. **Leaf blades** lanceolate or oblanceolate to linear, 4–100(–120) mm, margins (basal leaves) toothed to pinnatifid. **Involucres** hemispheric, 4–18 × 4–15+ mm. **Phyllaries** 5–18, apices often longer (sometimes shorter) than folded bases. **Paleae** in 1 series between ray and disc florets. **Ray florets** 5–18; laminae yellow or proximally yellow and distally white, 3–21 mm. **Disc florets** 6–120+; corollas 3.5–6 mm; anthers ± dark purple (in southwest of range, sometimes yellow or brownish). **Ray cypselae** glabrous or sparsely hairy. **Disc pappi** 0 or of 14–32 white to tawny, ± equal bristles or setiform scales 2–5 mm, each ± scabrous, usually not plumose and not adaxially woolly (in southwest of range, often proximally plumose and adaxially woolly). $2n = 14$.

Flowering Feb–Jul. Open, often grassy slopes or flats, disturbed sites, often on sandy or clayey soil, sometimes serpentine; 0–2000 m; Calif.; Mexico (Baja California).

Layia platyglossa occurs from the North Coast Ranges through central-western and southwestern California and west to the immediate coast and east into the central Great Valley. A specimen reported by S. L. Welsh et al. (1993) from dunes in Utah (*Harrison 2545*, BRY) may be mislabeled.

5. **Layia jonesii** A. Gray, Proc. Amer. Acad. Arts 19: 18. 1883 [C] [E]

Plants 7–55 cm (self-incompatible); glandular, not strongly scented. **Stems** usually purple-streaked. **Leaf blades** lanceolate or oblanceolate to linear, 10–70 (–90) mm, margins (basal leaves) lobed to pinnatifid. **Involucres** ± broadly urceolate, 4–8 × 4–8(–12) mm. **Phyllaries** 13–27, apices usually shorter, sometimes longer than folded bases. **Paleae** in 1 series between ray and disc florets. **Ray florets** 13–27 (in 2 series); laminae proximally yellow, distally white, 5–10(–14) mm. **Disc florets** 35–100+; corollas 3–5 mm; anthers ± dark purple. **Ray cypselae** glabrous. **Disc pappi** 0 or of 8–14 whitish, ± ovate or elliptic, ± equal scales 0.5–2 mm, not plumose, not adaxially woolly (bases sparsely setose). 2*n* = 14.

Flowering Mar–May. Grasslands, openings in chaparral, on clay or serpentine soils; of conservation concern; 0–400 m; Calif.

Layia jonesii occurs on the immediate coast from Cayucos to Morro Bay and in the Outer South Coast Ranges near San Luis Obispo. *Layia jonesii* is most closely related to *L. leucopappa* and *L. munzii* (B. G. Baldwin, unpubl.); the three species are interfertile (they do not co-occur; J. Clausen 1951).

6. **Layia leucopappa** D. D. Keck, Madroño 3: 17. 1935 [C] [E]

Plants 8–60 cm (self-incompatible); glandular, not strongly scented. **Stems** not purple-streaked. **Leaf blades** oblong to oblanceolate, 6–50(–80) mm, margins (basal leaves) toothed to lobed. **Involucres** hemispheric to depressed-hemispheric, 3.5–8 (–11) × 4–10(–13) mm. **Phyllaries** 6–15, apices longer or shorter than folded bases. **Paleae** in 1 series between ray and disc florets. **Ray florets** 6–15; laminae white, 3–12(–19) mm. **Disc florets** 20–100+; corollas 2.5–5 mm; anthers yellow to brownish. **Ray cypselae** sparsely hairy. **Disc pappi** 10–13 white, lanceolate, acuminate, ± equal scales 2–3.5 mm, not plumose, not adaxially woolly (bases sparsely setose). 2*n* = 14.

Flowering Mar–Apr. Grasslands, openings in chenopod scrub (historically), on sparsely-vegetated, clayey, subalkaline soils; of conservation concern; 100–400 m; Calif.

Layia leucopappa occurs in the Comanche and Tejon hills (western Tehachapi Range). Populations of the southern San Joaquin Valley (south and east of Bakersfield) have been extirpated.

7. **Layia munzii** D. D. Keck, Madroño 3: 16. 1935 [C] [E]

Plants 6–50 cm (self-incompatible); glandular, not strongly scented. **Stems** not purple-streaked. **Leaf blades** oblanceolate to linear, 5–60 mm, margins (basal leaves) lobed. **Involucres** hemispheric to ± urceolate, 5–9 × 5–10+ mm. **Phyllaries** 6–15, apices shorter or longer than folded bases. **Paleae** in 1 series between ray and disc florets. **Ray florets** 6–15; laminae proximally yellow, distally white, 3–14 mm. **Disc florets** 16–100+; corollas 3.5–5 mm; anthers ± dark purple. **Ray cypselae** glabrous or sparsely hairy. **Disc pappi** of 9–12 whitish, lance-linear, attenuate, ± equal scales 2–3.5 mm, each ± scabrous, not plumose, not adaxially woolly (bases sparsely setose). 2*n* = 14.

Flowering Mar–Apr. Open, ± alkaline flats, depressions, flood plains, in grasslands, saltbush scrub, on clayey soils; of conservation concern; 30–800 m; Calif.

Layia munzii occurs in the southern San Joaquin Valley and South Inner Coast Ranges. Populations are evident only in some wet years, when dense stands may be found, sometimes near *L. platyglossa* (natural hybrids have not been reported; artificial hybrids have reduced fertility; J. Clausen 1951).

8. **Layia pentachaeta** A. Gray in War Department [U.S.], Pacif. Railr. Rep. 4(5): 108. 1857 [E]

Plants 5–100 cm (self-incompatible); glandular, strongly lemon- or acrid-scented. **Stems** not purple-streaked. **Leaf blades** oblanceolate or lanceolate to linear, 6–110 mm, margins (basal leaves) 1–2-pinnatifid. **Involucres** ± hemispheric, 5–12 × 2–12+ mm. **Phyllaries** 4–14, apices shorter or longer than folded bases. **Paleae** in 1 series between ray and disc florets. **Ray florets** 3–14; laminae yellow or white, 3–26 mm. **Disc florets** 7–120+; corollas 3–6 mm; anthers yellow to brownish. **Ray cypselae** glabrous. **Disc pappi** 0 or of 1–22 whitish, ± equal, setiform scales 1.5–3.5 mm, each proximally plumose, seldom adaxially woolly.

Subspecies 2 (2 in the flora): California.

Molecular phylogenetic data have indicated that *Layia pentachaeta* is closely related to *L. glandulosa* and *L. discoidea* (B. G. Baldwin, unpubl.). Distributions of *L. pentachaeta* and *L. glandulosa* overlap broadly; the two species occur in different habitats and natural hybridization has been rarely detected (first-generation hybrids are of low to moderate fertility; J. Clausen 1951).

1. Ray corollas yellow . . . .
. . . . 8a.   *Layia pentachaeta* subsp. *pentachaeta*
1. Ray corollas white . . . .
. . . . 8b.   *Layia pentachaeta* subsp. *albida*

8   Layia pentachaeta A. Gray subsp. **pentachaeta** E

Ray corollas yellow. $2n = 16$.

Flowering Mar–Jun. Grasslands, openings in woodlands, chaparral, sandy or clayey soils, sometimes serpentine; 100–1200 m; Calif.

Subspecies *pentachaeta* occurs in the Sierra Nevada foothills north of the Tehachapi Range and in the San Joaquin Valley and South Inner Coast Ranges. Putative hybrids with *Layia pentachaeta* subsp. *albida* have been collected in the southern Sierra Nevada (e.g., *Keck 3258* in UC); artificial, highly fertile hybrids of the same combination have been produced; (J. Clausen 1951).

8b.   Layia pentachaeta A. Gray subsp. **albida**
D. D. Keck, Aliso 4: 107. 1958   E

Layia pentachaeta var. *albida* (D. D. Keck) Hoover

Ray corollas white. $2n = 16$.

Flowering Mar–May. Grasslands, openings in woodlands, saltbush scrub, clayey, subalkaline soils, sometimes sandy soils; 70–900 m; Calif.

Subspecies *albida* occurs in the southern Sierra Nevada and Tehachapi Range foothills, southern San Joaquin Valley, and South Inner Coast Ranges.

9.   Layia glandulosa (Hooker) Hooker & Arnott, Bot. Beechey Voy., 358. 1839   F

Blepharipappus glandulosa Hooker, Fl. Bor.-Amer. 1: 316. 1833; *Layia glandulosa* subsp. *lutea* D. D. Keck; *L. glandulosa* var. *lutea* (D. D. Keck) Hoover

Plants 3–60 cm (self-incompatible); glandular, often not strongly scented, sometimes spice-scented. **Stems** not purple-streaked (often uniformly dark purple). **Leaf blades** obovate to linear, 6–100 mm, margins (basal leaves) toothed to lobed. **Involucres** ± campanulate to hemispheric, 4–11 × 3–11+ mm. **Phyllaries** 3–14, apices often shorter (sometimes longer) than folded bases. **Paleae** in 1 series between ray

and disc florets. **Ray florets** 3–14; laminae often white, sometimes yellow or pale yellow, 3–22 mm. **Disc florets** 17–100+; corollas 3.5–6.5 mm; anthers yellow to brownish. **Ray cypselae** glabrous. **Disc pappi** of 10–15 usually white (rarely tawny), linear-attenuate to subulate, ± equal scales 2–5 mm, each proximally plumose and often adaxially woolly. $2n = 16$.

Flowering Feb–Jul. Openings in scrub, woodlands, forests, grasslands, and meadows, gravelly or sandy soils, sometimes dunes; 0–2700 m; Ariz., Calif., Idaho, N.Mex., Oreg., Utah, Wash.; Mexico (Baja California).

Layia glandulosa occurs in deserts of western North America, extending to the Pacific coast in central and southern California. As treated here (provisionally) and previously, *L. glandulosa* corresponds to a paraphyletic group; molecular phylogenetic data have indicated that *L. discoidea* is most closely related to a subset of lineages in *L. glandulosa*, including yellow-rayed populations previously recognized as subsp. *lutea* or var. *lutea* (B. G. Baldwin, unpubl.). Report of *L. glandulosa* from British Columbia has not been confirmed.

10.   Layia discoidea D. D. Keck, Aliso 4: 106. 1958
C E

Plants 3–20 cm (self-incompatible); glandular, not strongly scented. **Stems** not purple-streaked. **Leaf blades** oblanceolate or lanceolate to linear, 2–35 mm, margins (basal leaves) lobed. **Involucres** cylindric or narrowly obconic to campanulate, 4–7 × 2–6+ mm. **Phyllaries** 0 ("involucres" formed of "paleae"). **Paleae** in 1 series (interpreted as constituting the involucre). **Ray florets** 0. **Disc florets** 5–35+; corollas 2.5–4 mm; anthers yellow to brownish. **Ray cypselae** 0. **Disc pappi** of 8–15 whitish to tawny, lanceolate to subulate, ± equal (often apically or marginally notched) scales 0.5–1.5 mm, each ± plumose or villous, not adaxially woolly. $2n = 16$.

Flowering Apr–Jun. Open, ± barren slopes and terraces, in chaparral, woodlands, forest, and meadows, on serpentine soils, talus; of conservation concern; 800–1600 m; Calif.

Layia discoidea occurs in the South Inner Coast Ranges (Fresno and San Benito counties). Artificial hybrids with *L. glandulosa* are highly fertile (J. Clausen 1951).

11. Layia septentrionalis   D. D. Keck, Aliso 4: 106. 1958   C E

Plants 6–35 cm (self-incompatible); glandular, not strongly scented. **Stems** purple-streaked or not. **Leaf blades** oblanceolate or lanceolate to linear, 4–70 mm, margins (basal leaves) toothed to pinnatifid. **Involucres** ± ellipsoid to campanulate, 5–12 × 3–12+ mm. **Phyllaries** 5–9, apices usually shorter than folded bases. **Paleae** in 1 series between ray and disc florets. **Ray florets** 5–9; laminae yellow, 4–15 mm. **Disc florets** 10–65+, corollas 5–8 mm; anthers yellow to brownish. **Ray cypselae** glabrous or sparsely hairy. **Disc pappi** of 16–22 white, ± equal bristles or setiform scales 4–7 mm, each proximally plumose and adaxially woolly. $2n = 16$.

Flowering Apr–Jun. Grasslands, openings in chaparral, woodlands, on serpentine or sandy soils; of conservation concern; 100–900 m; Calif.

Layia septentrionalis occurs in the central and southern Inner North Coast Ranges and the Sutter Buttes (southern Sacramento Valley).

12. Layia gaillardioides   (Hooker & Arnott) de Candolle in A. P. de Candolle and A. L. P. P. de Candolle, Prodr. 7: 294. 1838   E

Tridax gaillardioides Hooker & Arnott, Bot. Beechey Voy., 148. 1833 (as galardioides)

Plants 6–60 cm (self-incompatible); not glandular, often strongly scented. **Stems** purple-streaked. **Leaf blades** oblanceolate or lanceolate to linear, 5–120 mm, margins (basal leaves) toothed to pinnatifid. **Involucres** subglobose to ± campanulate or hemispheric, 4–9+ × 4–12+ mm. **Phyllaries** 6–18, apices shorter than folded bases. **Paleae** in 1 series between ray and disc florets. **Ray florets** 6–18; laminae yellow or bicolored (proximally yellow, distally whitish or pale yellow), 3.5–18 mm. **Disc florets** 14–100+; corollas 3–5 mm; anthers ± dark purple. **Ray cypselae** glabrous. **Disc pappi** 0 or of 15–24 white to rufous or purplish, ± equal bristles or setiform scales 1–4 mm, each proximally plumose, not adaxially woolly. $2n = 16$.

Flowering Mar–Aug. Open or semishady sites in woodlands, grasslands, meadows, chaparral, or forests, often on serpentine or sandy soils; 0–1300 m; Calif.

Layia gaillardioides occurs on the North Coast and in the North Coast Ranges, San Francisco Bay area, and Inner South Coast Ranges, and in the Outer South Coast Ranges (near Cayucos). As treated here (provisionally)

and previously, *L. gaillardioides* is not monophyletic; molecular phylogenetic data have indicated that some lineages of *L. gaillardioides* are more closely related to *L. carnosa*, *L. hieracioides*, and/or *L. septentrionalis* than to one another (B. G. Baldwin, unpubl.).

13. Layia hieracioides   (de Candolle) Hooker & Arnott, Bot. Beechey Voy., 358. 1839   E

Madaroglossa hieracioides de Candolle in A. P. de Candolle and A. L. P. P. de Candolle, Prodr. 5: 694. 1836; *Layia paniculata* D. D. Keck

Plants 5–130 cm (self-compatible); glandular, usually strongly sweet- or pungent-scented. **Stems** purple-streaked. **Leaf blades** elliptic, lanceolate, linear, or oblanceolate, 7–150 mm, margins (basal leaves) lobed to toothed. **Involucres** ± ellipsoid to obconic, 4–9 × 4–9+ mm. **Phyllaries** 6–16, apices shorter than folded bases. **Paleae** in 1 series between ray and disc florets. **Ray florets** 6–16; laminae yellow, 1–4 mm. **Disc florets** 9–80; corollas 2.5–4.5 mm; anthers ± dark purple. **Ray cypselae** glabrous. **Disc pappi** of 10–16 white to rufous or purplish, ± equal bristles or setiform scales 2–4 mm, each proximally plumose, not adaxially woolly. $2n = 16, 32$.

Flowering Apr–Jul. Open, often grassy, or semishady sites in chaparral, woodlands, forests, and coastal scrub, often sandy soils, disturbed sites; 0–1200 m; Calif.

Layia hieracioides occurs in the southern San Francisco Bay area, South Coast Ranges, and Western Transverse Ranges. Diploids and tetraploids are morphologically similar and reportedly geographically distinct, with diploids documented from the San Francisco Bay area and tetraploids (= *L. paniculata*) documented to the south. Natural, ± sterile hybrids between tetraploid plants and *L. glandulosa* have been documented from the Central Coast (B. D. Tanowitz and J. W. Adams 1986; R. F. Hoover 8369, UC).

14. Layia carnosa   (Nuttall) Torrey & A. Gray, Fl. N. Amer. 2: 394. 1843   C E

Madaroglossa carnosa Nuttall, Trans. Amer. Philos. Soc., n. s. 7: 393. 1841

Plants 2–18 cm (self-compatible); glandular, not strongly scented. **Stems** not purple-streaked. **Leaf blades** oblong or ovate, 3–45 mm, margins (basal leaves) lobed. **Involucres** ± campanulate, 4–8 × 3–7+ mm. **Phyllaries** 4–10, apices shorter than folded bases. **Paleae** in 1 series between ray and disc florets.

Ray florets 4–10; laminae white, 1.5–3.5 mm. **Disc florets** 5–45; corollas 2–4 mm; anthers ± dark purple. **Ray cypselae** sparsely hairy. **Disc pappi** of 24–32 white to tawny or rufous, ± equal bristles or setiform scales 2.5–3.5 mm, each proximally plumose, not adaxially woolly. **2***n* = 16.

Flowering Apr–Jul. Coastal sand dunes; of conservation concern; 0–60 m; Calif.

Layia carnosa occurs on the Central Coast and North Coast and in the San Francisco Bay area.

336. OSMADENIA Nuttall, Trans. Amer. Philos. Soc., n. s. 7: 391. 1841 • [Greek *osma*, odor, and *aden*, gland, alluding to strong-scented, glandular herbage]

Robert L. Carr

Gerald D. Carr

Annuals, 5–40 cm (self-incompatible; herbage strongly scented). **Stems** erect (branched, branchlets relatively many, commonly spreading, threadlike, densely glandular, scabrous to shaggy-hairy). **Leaves** mostly cauline; alternate; sessile; blades linear, margins entire, faces hispidulous and glandular (proximally ± ciliate or shaggy-hairy). **Heads** radiate, in loose, cymiform arrays. **Peduncular bracts** with tack-glands 0 (strigillose, strongly glandular, margins often proximally pectinate, ciliate). **Involucres** ovoid, 2–4 mm diam. **Phyllaries** falling, 3–5 in 1 series (each partly enveloping a ray cypsela, apices acute to attenuate). **Receptacles** flat, glabrous, paleate (paleae persistent, in 1 series between rays and discs, connate, forming cups, distinct apices acute, often apiculate). **Ray florets** 3–5, pistillate, fertile; corollas wholly or mostly white (laminae 3-lobed, lobes sometimes each with prominent medial red blotch, overall often fading reddish, sinuses ± equaling laminae). **Disc florets** 3–10, bisexual, fertile; corollas usually wholly white, sometimes lobes each with prominent medial red blotch or reddish overall, tubes shorter than throats, lobes 5, deltate (anthers ± dark purple; styles glabrous proximal to branches). **Cypselae** (rays) ± obcompressed (± 3-angled, abaxial sides usually broadly 2-faced, angles between those faces usually 90+°, adaxial sides nearly flat), apices beaked, beaks off center, faces rugose, glabrous; or (discs) narrowly clavate, appressed-hairy; **pappi** (rays) 0, or (discs) of 4–5 lance-attenuate to aristate scales alternating with 4–5 shorter, ± fimbriate scales. *x* = 9.

Species 1: California, nw Mexico.

SELECTED REFERENCE Carr, G. D. 1977. A cytological conspectus of the genus *Calycadenia* (Asteraceae): An example of contrasting modes of evolution. Amer. J. Bot. 64: 694–703.

1. Osmadenia tenella Nuttall, Trans. Amer. Philos. Soc., n. s. 7: 392. 1841 F

Calycadenia tenella (Nuttall) Torrey & A. Gray

Leaves (proximal) 1–3(–5) cm, smaller distally. **Peduncular bracts** 4–7, narrowly lanceolate, 4–8 mm. **Phyllaries** 3.5–6.5 mm (paleae 4–6 mm) ± strigillose, strongly glandular, abaxial faces bristly, shaggy-hairy, especially margins and apices. **Ray corollas** tubes ca. 2.5 mm, laminae ca. 4 mm. **Disc corollas** 3–5 mm, lobes ± hirsute. **Cypselae** (rays) 1.8–2.7 mm, beaks 0.3 mm, or (discs) 2–2.5 mm; **pappi** of scales 4–6 mm plus scales ca. 1 mm. **2***n* = 18.

Flowering spring–summer. Barren, often rocky, exposed hillsides, canyons, chaparral, woodlands; 10–600(–1200) m; Calif.; Mexico (Baja California).

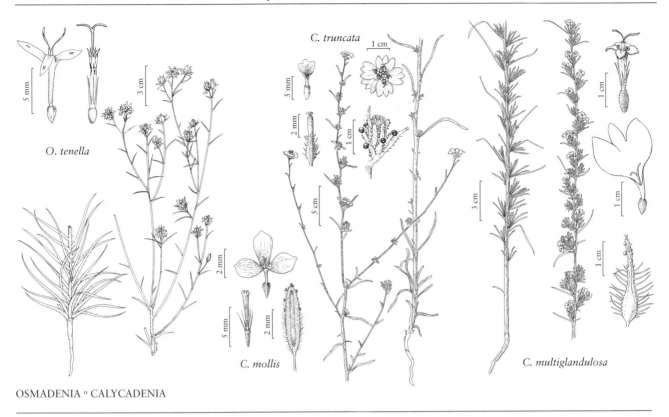

O. tenella

C. truncata

C. mollis

C. multiglandulosa

OSMADENIA ∘ CALYCADENIA

337. CALYCADENIA   de Candolle in A. P. de Candolle and A. L. P. P. de Candolle, Prodr. 5: 695. 1836 • [Greek *calyx*, cup, and *aden*, gland, alluding to tack-glands of peduncular bracts and/or phyllaries]   E

Robert L. Carr

Gerald D. Carr

Annuals, 5–120 cm (mostly self-incompatible). **Stems** erect to spreading (unbranched or branched). **Leaves** basal and cauline (at least basal mostly withering before flowering); usually mostly alternate (opposite in *C. oppositifolia*); sessile; blades linear (grasslike), margins entire or sparsely toothed (revolute), faces often scabrous. **Heads** radiate, usually borne singly or in open to dense, corymbiform or ± spiciform arrays, sometimes in axillary glomerules. **Peduncular bracts** (and reduced, distal leaves; peduncles often relatively short, bracts crowded, closely subtending phyllaries): tack-glands 0 or 1–6+ (at tips and, sometimes, on margins; margins sometimes thickened or rolled, often proximally pectinate and/or ciliate). **Phyllaries** falling or persistent, 1–6 in 1 series (each partly enveloping a ray cypsela, commonly with simple and/or tack-glands). **Receptacles** ± flat, hairy, paleate (paleae falling or persistent, in 1 series between rays and discs, ± connate, forming plicate cups, distinct apices usually acute to acuminate). **Ray florets** 1–6, pistillate, fertile; corollas white or cream to rose, or yellow, sometimes with proximal dark-red spot (laminae 3-lobed, lobes spreading or ± parallel, lengths $1/4$+ total laminae). **Disc florets** 1–25, usually bisexual and fertile; corollas white or cream to rose, or yellow, tubes shorter than throats, lobes 5, deltate to lance-ovate (anthers usually dark purple, rarely yellow or light brown). **Ray cypselae** usually obcompressed (± 3-angled, abaxial sides ± 2-faced, angles between those faces 90+°, adaxial sides ± flat, overall smooth or rough-wrinkled, glabrous or hairy); **pappi** 0. **Disc cypselae** narrowly clavate (usually angled in cross section, glabrous or hairy);

pappi usually of 6–13 subulate to lanceolate-aristate or shorter and broader scales (longer and shorter often alternating), rarely 0. $x = 4, 5, 6, 7$.

Species 10 (10 in the flora): w United States.

Calycadenia is nearly confined to the Central Valley and adjacent lower elevations of bordering mountain ranges of California. R. L. Carr and G. D. Carr (1993) recognized nine species; a tenth species is included here. All species except *C. hooveri* and *C. micrantha* are strongly self-incompatible. The genus appears to exhibit two contrasting modes of evolution (G. D. Carr 1977). One group of taxa (*C. pauciflora* and *C. fremontii* complex) comprises distinct chromosome races that lack clear and consistent morphologic differentiation. The chromosome races are differentiated primarily by reciprocal chromosome translocations and, in some cases, paracentric inversions (G. D. Carr 1975, 1975b, 1977, 1980; Carr and Carr 1983, 2000). In contrast, *C. multiglandulosa* and *C. hispida* (combined in this treatment) consist of distinct morphologic races (D. D. Keck 1960b) and lack chromosomal differentiation. Our incomplete understanding of the biology of this genus makes the production of a clear, concise, and practical taxonomic treatment difficult.

SELECTED REFERENCES Baldwin, B. G. and S. Markos. 1998. Phylogenetic utility of the external transcribed spacer (ETS) of 18S-26S rDNA: Congruence of ETS and ITS trees of *Calycadenia* (Compositae). Molec. Phylogen. Evol. 10: 449–463. Carr, G. D. 1977. A cytological conspectus of the genus *Calycadenia* (Asteraceae): An example of contrasting modes of evolution. Amer. J. Bot. 64: 694–703.

1. Peduncular bracts (and, sometimes, paleae): tack-glands (0–)2–6+ (some Butte County plants of *C. fremontii* and *C. pauciflora* with fewer tack-glands are usually less than 20 cm); ray laminae: central lobes equaling or narrower than laterals, widest near or beyond middles (except in *C. multiglandulosa*, with copious tack-glands on peduncular bracts and paleae).
    2. Leaves opposite (heads in congested, axillary and terminal glomerules, appearing whorled) . . . . 8. *Calycadenia oppositifolia*
    2. Leaves mostly alternate (sometimes opposite to beyond midstems in *C. multiglandulosa*; heads borne singly or in glomerules in corymbiform or spiciform arrays, not appearing whorled).
        3. Ray laminae: central lobes widest at bases (notably smaller and narrower than laterals) . . . . 7. *Calycadenia multiglandulosa*
        3. Ray laminae: central lobes widest near or beyond middles (± similar to laterals).
            4. Stems branched (plants usually without obvious axes, branches relatively many, slender, usually divergent, often zigzag); paleae 4–6 mm (receptacular cups ± clavate to fusiform, lengths ± 2 times diams.); ray florets 1(–2), corollas white to reddish; disc florets 2–5 . . . . 10. *Calycadenia pauciflora*
            4. Stems simple or branched (plants with main axes usually obvious, except in some populations centered in Butte, s Lake, and Sonoma counties, branches relatively few, ± rigid); paleae 3–7 mm (receptacular cups ± campanulate, lengths ± equaling diams.); ray florets (1–)2–6, corollas white or cream to pinkish, or yellow (laminae sometimes basally reddish); disc florets (4–)6–21 . . . . . 9. *Calycadenia fremontii*
1. Peduncular bracts: tack-glands (0–)1 (and terminal), or 3–6+ (and confined to thickened or rolled margins in *Calycadenia mollis*); paleae: tack-glands 0; ray laminae: central lobes usually notably smaller than laterals, widest at bases (widest beyond middles in *Calycadenia mollis*).
    5. Peduncular bracts: apices rounded, tack-glands mostly 3–6+ (confined to thickened or rolled margins); ray laminae: central lobes slightly narrower than laterals, widest beyond middles . . . . 1. *Calycadenia mollis*
    5. Peduncular bracts: apices concave, rounded, or truncate, tack-glands usually 1 (terminal); ray laminae: central lobes notably smaller than laterals, widest at bases.
        6. Corollas yellow; ray cypselae rough-wrinkled.
            7. Plants 20–120 cm (sometimes robust); ray laminae (4–)5–12 mm . . . . . 2. *Calycadenia truncata*
            7. Plants 10–50 cm (relatively slender); ray laminae 2–2.5(–3) mm . . . . 3. *Calycadenia micrantha*
        6. Corollas white (sometimes fading reddish); ray cypselae usually smooth (rough-wrinkled in some *C. hooveri*)

[8. Shifted to left margin.—Ed.]

8. Stems branched (branches flexible); ray florets (0–)1(–2); disc florets 1–2; ray cypselae glabrous . . . . 4. *Calycadenia hooveri*
8. Stems branched (branches rigid); ray florets 1–5; disc florets 4–15; ray cypselae densely appressed-hairy.
   9. Peduncular bracts lance-elliptic (flat distally), apices rounded to truncate . . . . 5. *Calycadenia villosa*
   9. Peduncular bracts narrowly lanceolate or lance-attenuate (± cylindric distally 1–3+ mm), apices truncate to strongly concave . . . . 6. *Calycadenia spicata*

1. **Calycadenia mollis** A. Gray, Proc. Amer. Acad. Arts 7: 360. 1868   E F

Hemizonia mollis (A. Gray) A. Gray

Plants 30–90 cm; self-incompatible. **Stems** usually simple, often zigzag-curved, ± hirsute, long-soft-hairy, ± glandular. **Leaves** mostly alternate, 2–8 cm (often longest at midstems), hirsute, ± long-hairy, ± glandular. **Heads** usually (3–10+) in dense, cymiform arrays. **Peduncular bracts** oblanceolate, 3–9 mm (flat, flexible, margins thickened or rolled, hirsute to hispid, usually bristly, especially margins, ± glandular), apices rounded, tack-glands usually 3–6+, confined to margins. **Phyllaries** 4–6 mm, abaxial faces spreading-hairy (hairs mixed, short, slender hairs and longer bristles), densely glandular, tack-glands usually 0. **Paleae** 4–6 mm (tack glands usually 0). **Ray florets** 1–4; corollas yellow, white, or rose, tubes ca. 2 mm (papillate, throats often with red "eyes"), laminae 5–6 mm (central lobes slightly narrower than laterals, nearly symmetric, widest beyond middles, sinuses usually ³/₄+ laminae). **Disc florets** 2–10; corollas yellow to white or rose, ca. 5 mm. **Ray cypselae** ca. 2.5 mm, usually rough-wrinkled, glabrous or glabrate. **Disc cypselae** ca. 2.5 mm, ± appressed-hairy; **pappi** 6–8 usually lanceolate-aristate (often 1–3 shorter, blunt) scales 2–5 mm. $2n = 14$.

Flowering spring–summer. Open, dry meadows, fields; 100–1500 m; Calif.

Calycadenia mollis grows in the Sierra Nevada foothills from Tuolumne County to Tulare County.

2. **Calycadenia truncata** de Candolle in A. P. de Candolle and A. L. P. P. de Candolle, Prodr. 5: 695. 1836   E F

Calycadenia truncata subsp. *scabrella* (Drew) D. D. Keck

Plants 20–120 cm (sometimes robust); self-incompatible (herbage glaucescent, acrid-scented). **Stems** (usually reddish) usually branched (branches 1–15+, remote, arcuate-ascending, slender, arising mostly distal to midpoints of main stems), glabrous or ± scabrous distally. **Leaves** mostly alternate, 2–10 cm (at least proximal), hispidulous to hispid and ± long-hairy (at least margins and adaxial faces), sometimes with tack-glands. **Heads** in spiciform arrays (usually l–3 per node). **Peduncular bracts** ± narrowly elliptic, 1–12 mm (± glabrous or hispidulous to hispid, sometimes ± bristly and/or pectinate-fimbriate), apices ± truncate, tack-glands (0–) 1 (terminal). **Phyllaries** 5–10 mm, abaxial faces glabrous or sparsely strigose and/or long-straight-hairy, especially distally, shaggy-long hairy on distal margins, tack-glands usually 0. **Paleae** 5–10 mm (tack-glands 0). **Ray florets** 3–6; corollas yellow, tubes ca. 2.5 mm, ± papillate, throats sometimes with red "eyes," laminae (4–)5–12 mm (central lobes smaller than laterals, symmetric, widest at bases, laterals asymmetric, sinuses less than ¹/₃ laminae). **Disc florets** 3–25 (ovaries often sterile or of low fertility); corollas yellow, 4–6 mm (sometimes papillate distally, lobes ± hirsute; anthers yellow, light brown, or dark purple). **Ray cypselae** ca. 3 mm, rough-wrinkled, glabrous. **Disc cypselae** 2–4 mm, glabrous or ± appressed-hairy; **pappi** 0 or of 7–12 oblong, fimbriate scales, to 1 mm. $2n = 14$.

Flowering spring–fall. Dry, open hillsides, rocky ridges, talus; 50–1500 m; Calif., Oreg.

Calycadenia truncata is the most widespread of all of the taxa. It ranges from Jackson and Josephine counties in southwestern Oregon [reported by D. D. Keck (1960b) from Douglas County] in the Coast Range to San Luis Obispo County, California, and in the Sierra Nevada foothills south to Fresno County. It is also localized in sites in the Sacramento Valley.

3. **Calycadenia micrantha** R. L. Carr & G. D. Carr, Sida 21: 261, figs. 1, 2B. 2004   E

Plants 10–50 cm (relatively slender); self-compatible. **Stems** (often purplish, especially distally, slender, usually less than 3 mm diam. at bases) branched (branches often relatively many, usually from near midstems, arcuate to ascending), glabrous. **Leaves** mostly alternate, 2–5 cm (proximal), hispidulous and ± long-hairy (especially margins and adaxial faces). **Heads** borne singly or in ± spiciform arrays (1–3 per node, ± sessile). **Peduncular bracts** lance-oblong to linear, 2–4 mm (nearly terete to strongly flattened, glabrous or hispid overall, sometimes pectinate-fimbriate), apices ± rounded, tack-glands usually 1 (terminal). **Phyllaries** 4–5 mm, abaxial faces glabrous or ± hispid distally or sparsely bristly, shaggy long-hairy on margins distally, tack glands 0(–1) (terminal). **Paleae** 5–6 mm. **Ray florets** 1–3(–6); corollas bright yellow, tubes 1–1.5 mm, laminae 2–2.5(–3) mm (central lobes smaller than laterals, oblong to narrowly triangular, symmetric, widest at bases, laterals asymmetric, sinuses ¹/₄ laminae). **Disc florets** 1–3; corollas yellowish, 3–4 mm. **Ray cypselae** ca. 3 mm, rough-wrinkled, glabrous. **Disc cypselae** ca. 3 mm (seldom formed, terete and tapered toward bases, smooth to ± ridged, glabrous); **pappi** 0. $2n = 14$.

Flowering usually Aug–fall. Dry, open, rocky ridges, hillsides and talus; 300–1500 m; Calif.

Calycadenia micrantha is most closely related to variants of *C. truncata* and is found in the North Coast Range from Trinity County to Lake and Colusa counties. *Calycadenia truncata* subsp. *microcephala* H. M. Hall ex D. D. Keck may be a synonym of *C. micrantha*. D. D. Keck (1960b) reported this to be on the east slope of the Santa Lucia Mountains of Monterey County.

4. Calycadenia hooveri G. D. Carr, Brittonia 27: 140, fig. 19. 1975 [E]

Plants 10–60 cm; self-compatible. **Stems** branched (branches relatively many, distal, filiform, flexible, minutely scabrous, glandular). **Leaves** mostly alternate, 1–6(–8) cm, ± thinly hispidulous and ± long-hairy (especially proximal margins and adaxial faces). **Heads** borne singly or in ± spiciform arrays (1–4 per node). **Peduncular bracts** subclaviform, 1–5 mm (hispidulous, sometimes ± pectinate-fimbriate), apices rounded, tack-glands 1 (terminal). **Phyllaries** 2.5–3.5 mm, abaxial faces ± hispidulous (hairs scattered, stout), ± shaggy long-hairy distally, especially margins, minutely glandular, tack-glands (0–)1 (terminal). **Paleae** 3–5 mm (vestiture similar to phyllaries, tack-glands 0). **Ray florets** (0–)1(–2); corollas white, tubes ca. 2 mm, laminae 2–3.5 mm (central lobes smaller than laterals, widest at bases, symmetric, sometimes 2-partite, laterals weakly asymmetric, sinuses ca. ²/₃ laminae). **Disc florets** 1–2; corollas white, 2.5–3.5 mm. **Ray cypselae** 1.5–2.5 mm, smooth to rugose, glabrous. **Disc cypselae** 2–3 mm, ± appressed-hairy; **pappi** of 6–13 lanceolate-aristate, scales 1.5–2.5 mm. $2n = 14$.

Flowering spring–summer. Rocky, exposed places; 100–400 m; Calif.

Calycadenia hooveri resembles variants of *Calycadenia pauciflora*; it is more closely related to *C. villosa* (G. D. Carr 1975b). *Calycadenia hooveri* is known only from the Sierra Nevada foothills of Calaveras, Mariposa, and Stanislaus counties.

5. Calycadenia villosa de Candolle in A. P. de Candolle and A. L. P. P. de Candolle, Prodr. 5: 695. 1836 [E]

Plants 10–40 cm; self-incompatible. **Stems** (± reddish distally) simple or branched (branches 1–4+ and ascending, or more numerous and spreading, rigid), hirsutulous-hispidulous, ± villous. **Leaves** (sometimes basal clusters persisting to flowering) mostly alternate, 2–5 cm, ± hispidulous (margins ± densely ciliate, especially proximally). **Heads** in ± spiciform arrays (1–3 per node). **Peduncular bracts** (crowded, closely investing, obscuring heads before flowering) lance-elliptic to linear (± flat distally), 3–10 (–15) mm (hirsutulous, bristly and/or pectinate-fimbriate, deeply grooved proximally), apices (often ± reddish) rounded to truncate, tack-glands (0–)1 (terminal). **Phyllaries** (± reddish distally) 5–6 mm, abaxial faces hirsutulous, usually bristly, ± shaggy long-hairy near distal margins, tack-glands 0. **Paleae** 5–6 mm. **Ray florets** 1–4; corollas white to pinkish, tubes ca. 2 mm (papillate), laminae ca. 5 mm (central lobes much narrower than laterals, widest at bases, symmetric, laterals asymmetric, sinuses ¹/₃–¹/₂ laminae). **Disc florets** 5–15; corollas white to pinkish, 5–6 mm. **Ray cypselae** ca. 2.5 mm, smooth, densely appressed-hairy. **Disc cypselae** ca. 2.5 mm, appressed-hairy, especially on angles; **pappi** of ca. 10 lanceolate-aristate scales ca. 4–6 mm. $2n = 14$.

Flowering spring–summer. Dry, rocky hills, ridges; 300–1100 m; Calif.

Calycadenia villosa occurs in Monterey and San Luis Obispo counties and is often localized. It is in the Center for Plant Conservation's National Collection of Endangered Plants.

6. Calycadenia spicata (Greene) Greene, Fl. Francisc. 4: 422. 1897

Hemizonia spicata Greene, Bull. Torrey Bot. Club 9: 16. 1882

Plants 20–60 cm; self-incompatible (herbage gray-green, pleasantly scented). **Stems** simple or branched (branches 1–6, ± ascending, mostly distal to midpoints, rigid), ± densely strigillose-hirsutulous, strigose (hairs longer, glandular distally). **Leaves** mostly alternate, 2–5 cm (often longest at midstems), hispidulous and ± long-hairy (especially proximal margins). **Heads** in glomerate-spiciform arrays (1–3+ per node). **Peduncular bracts** (15–20+, closely investing and concealing heads before flowering) narrowly lanceolate to lance-attenuate

(± cylindric distally 1–3+ mm), 3–7 mm (hispidulous, ± bristly and/or pectinate-fimbriate), apices truncate to strongly concave, tack-glands 1 (terminal). **Phyllaries** 6–9 mm (± prominently nerved), abaxial faces (often reddish) hispidulous, often long-hairy, especially distally, tack-glands 1–5+. **Paleae** 6–9 mm (tack-glands usually 0). **Ray florets** 1–5; corollas white, fading reddish, tubes 3–4 mm (± papillate, sometimes with red "eyes"), laminae 6–11 mm (central lobes smaller than laterals, widest at bases, symmetric, laterals asymmetric, sinuses $^1/_3$–$^2/_3$ laminae). **Disc florets** 4–11; corollas pinkish, 7–10 mm. **Ray cypselae** ca. 3 mm, smooth, densely appressed-hairy. **Disc cypselae** ca. 3 mm, appressed hairy; **pappi** of 9–16 lanceolate-aristate scales, mostly 3–6 mm. $2n = 8$.

Flowering summer. Dry, open meadows and hillsides; 50–1400 m; Calif.

Calycadenia spicata is found in the Sierra Nevada foothills from Butte County to Kern County. Populations are often localized.

7. **Calycadenia multiglandulosa** de Candolle in A. P. de Candolle and A. L. P. P. de Candolle, Prodr. 5: 695. 1836 E F

Calycadenia hispida (Greene) Greene; *C. hispida* subsp. *reducta* D. D. Keck; *C. multiglandulosa* subsp. *bicolor* (Greene) D. D. Keck; *C. multiglandulosa* subsp. *cephalotes* (de Candolle) D. D. Keck; *C. multiglandulosa* subsp. *robusta* D. D. Keck

Plants 10–70 cm; self-incompatible. **Stems** simple or branched (branches diffuse, relatively short, bracted), hirsutulous to strigose, often ± pilose, often glandular, especially distally. **Leaves** mostly alternate, sometimes opposite to beyond midstems, 3–8 cm, ± hispidulous and ± long-hairy (especially margins and adaxial faces). **Heads** borne singly or in spiciform (often congested) arrays. **Peduncular bracts** (often reddish) usually narrowly lanceolate, sometimes narrowed and ± cylindric toward apices, 4–20 mm hispidulous, ± bristly, especially along margins, ± glandular), apices acute, rounded, or truncate, tack-glands usually 2–6+. **Phyllaries** (± reddish) 4–10 mm, abaxial faces hispidulous, often with long bristles, shaggy long-hairy, especially distally and along margins, glandular, tack-glands (0–)l–15+. **Paleae** 4–10 mm. **Ray florets** 2–6; corollas white or cream to rose, or yellow, tubes ca. 3 mm (papillate), laminae 5–10 mm (central lobes smaller than laterals, symmetric, widest at bases, laterals strongly asymmetric, sinuses usually $^1/_3$–$^1/_2$ laminae). **Disc florets** 4–20: corollas white or cream to rose, or yellow, 6–10 mm. **Ray cypselae** ca. 2.5 mm, smooth, glabrous or ± appressed-hairy. **Disc cypselae** ca. 2.5 mm, nearly smooth to ± appressed-hairy; **pappi**

of ca. 11 lanceolate-aristate scales ca. 2.5–4.5 mm (2–3 shorter, blunt). $2n = l2$.

Flowering spring–fall. Dry, open valleys, hillsides, rocky ridges; 50–1000 m; Calif.

Calycadenia multiglandulosa is found in the Coast Range from Sonoma and mid-Napa counties to San Mateo, Santa Clara, and Monterey counties and in the Sierra Nevada foothills from southern Butte and Yuba counties to Tulare County.

Some populations of the variable *Calycadenia multiglandulosa* (especially the "robusta" forms) appear to approach *C. spicata* in certain characters. The most striking similarity involves presence of narrower peduncular bracts with ± cylindric and truncate tips; the same specimens retain distinct characteristics of *C. multiglandulosa*, such as multiple tack-glands on peduncular bracts (as opposed to single terminal glands in *C. spicata*), considerable reduction in overall glandularity, reduced presence of hairs on cypselae (especially rays), and fewer long hairs on the abaxial faces of the phyllaries and paleae (subspp. *bicolor*, *cephalotes*, and *robusta*, and *C. hispida* including subsp. *reducta*).

8. **Calycadenia oppositifolia** (Greene) Greene, Fl. Francisc. 4: 423. 1897 E

Hemizonia oppositifolia Greene, Bull. Torrey Bot. Club 9: 110. 1882

Plants 10–30 cm; self-incompatible. **Stems** simple or sparingly branched (very slender), strigose and sparsely spreading-hairy. **Leaves** mostly or all opposite, 1–5 cm (little reduced distally), strigillose and ± long-hairy (at least proximally). **Heads** in congested, axillary glomerules (appearing whorled). **Peduncular bracts** lanceolate (flat, stiff, sometimes ± cylindric near apices), 4–10 mm (hispidulous, ± bristly and/or pectinate-fimbriate, especially proximally), apices ± rounded, tack-glands (0–)1–5+. **Phyllaries** (often reddish) 4–7 mm, abaxial faces sometimes minutely scabrous, often sparsely bristly, tack-glands usually 0, sometimes 1+. **Paleae** 4–7 mm. **Ray florets** 2–4; corollas white to reddish, tubes ca. 2 mm, laminae 6–9 mm (central lobes ± equaling or narrower than laterals, elliptic, nearly symmetric, widest near middles, sinuses equaling laminae). **Disc florets** 4–20; corollas white to pink, ca. 6 mm. **Ray cypselae** ca. 3 mm, usually smooth, glabrous. **Disc cypselae** ca. 3 mm, ± appressed-hairy; **pappi** of 8–10 usually lanceolate, acuminate scales ca. 2 mm (often 2–4 shorter, blunt). $2n = 14$.

Flowering spring–early summer. Open, dry meadows, hillsides; 50–900 m; Calif.

Calycadenia oppositifolia is known only from the

Sierra Nevada foothills of Butte County. Its closest relative may be *C. multiglandulosa*.

9. Calycadenia fremontii A. Gray in W. H. Emory, Rep. U.S. Mex. Bound. 2(1): 100. 1859 [E]

Calycadenia ciliosa Greene; *C. elegans* Greene; *C. pauciflora* A. Gray var. *elegans* Jepson

Plants 10–100 cm; self-incompatible. **Stems** simple or branched (main axes usually obvious, branches relatively few, ± rigid), hispidulous to strigose (especially proximally, usually strigose with scattered, stouter hairs distally). **Leaves** mostly alternate, 2–8 cm (proximal), hispid to hispidulous and ± long-hairy (especially on margins and adaxial faces). **Heads** borne singly or in open, spiciform arrays (1–2+ per node). **Peduncular bracts** narrowly elliptic to oblanceolate, 2–10 mm (hispidulous, ± bristly and/or strongly pectinate-fimbriate) apices rounded or truncate, tack-glands 1–5+. **Phyllaries** 3–7 mm, abaxial faces ± hispidulous, ± bristly, shaggy long-hairy on distal margins, tack-glands (0–)1–5+. **Paleae** 3–7 mm (receptacular cups ± campanulate, lengths often ± equaling diams.). **Ray florets** (1–)2–6; corollas white or cream to pinkish, or yellow (sometimes basally reddish), tubes 1–2 mm (± papillate, throats sometimes with dark red "eyes"), laminae 4.5–7 mm (central lobes ± equaling or narrower than laterals, mostly obovate, widest beyond middles, sinuses nearly equaling laminae). **Disc florets** (4–)6–21; corollas white to pinkish, or yellow, 4–5 mm. **Ray cypselae** 2.5–4 mm, smooth, glabrous or glabrate. **Disc cypselae** 2–4 mm (often ± angular), appressed-hairy; **pappi** of 8–14 long-aristate and short-blunt, fimbriate scales 1–3 mm (alternating). *2n* = 12.

Flowering spring–fall. Open, dry meadows, hillsides, gravelly outwashes, rocky barrens in chaparral associations; 50–1400 m; Calif., Oreg.

Calycadenia fremontii is variable and not easily divisible. It ranges from Siskiyou County through Mendocino, Lake, and Butte counties in California with one outlying population near Grants Pass in southern Oregon. Genetic change and chromosome repatterning have produced an extremely complex taxon that (as here circumscribed) includes populations with individuals to 20 cm and extensively branched with white corollas as well as populations with individuals to 100 cm and unbranched or with relatively few branches with yellow corollas. Some relatively small forms with white corollas are more closely related to tall forms with yellow corollas than they are to each other. At present, there is no known combination of characters that allows reasonably consistent delimitation of these or other elements within this assemblage of populations. The present circumscription combines three previously described species, including one traditionally considered an element of *C. pauciflora*. Additionally, *C. fremontii* intergrades with *C. pauciflora* (see discussion under 10. *C. pauciflora*).

10. Calycadenia pauciflora A. Gray in W. H. Emory, Rep. U.S. Mex. Bound. 2(1): 100. 1859 [E]

Plants 10–50 cm; self-incompatible. **Stems** branched (plants without obvious main axes, branches usually relatively many, slender, usually divergent, often zigzag), finely strigose (some hairs longer). **Leaves** mostly alternate, 1–5 cm (proximal), ± hispidulous and ± long-hairy (proximal margins and adaxial faces). **Heads** borne singly or in open, spiciform arrays (1–4 per node). **Peduncular bracts** ± linear to narrowly elliptic to oblanceolate, 2–3 mm (hispidulous, ± bristly and/or pectinate-fimbriate), apices rounded to truncate, tack-glands 1–3+. **Phyllaries** 4–6 mm, abaxial faces scabrous, ± strongly bristly, ± shaggy long-hairy on distal margins, tack-glands (0–)1–7+. **Paleae** 4–6 mm (receptacular cups ± clavate to fusiform, lengths ± 2 times diams.). **Ray florets** 1(–2); corollas white to reddish, tubes ca. 2 mm (papillate), laminae 5–7 mm (central lobes equaling or slightly narrower than laterals, ± elliptic or distally broader and more rounded, symmetric or nearly so, widest near or beyond middles, sinuses nearly equaling laminae). **Disc florets** 2–5; corollas white to pinkish, ca. 5–6 mm. **Ray cypselae** 2.5–4 mm, smooth, usually glabrous, sometimes hairy distally, sometimes with slightly elevated areolae. **Disc cypselae** angular, 2.5–3.5 mm, appressed hairy, especially on angles; **pappi** of ca. 8–10 lanceolate-attenuate and shorter, blunt fimbriate scales ca. 1–2.5 mm (alternating). *2n* = 10, 12.

Flowering spring–late summer. Open, dry, usually rocky meadows, hillsides; 50–1000 m; Calif.

Calycadenia pauciflora is variable and closely related to and intergrades with *C. fremontii*. It is found mostly in the foothills of the Coast Ranges from Tehama County to Lake, Napa, and Yolo counties. It is also known from one location on the valley floor in Butte County.

The morphologic and chromosomal similarity of some populations of *Calycadenia ciliosa* to *C. fremontii* led R. L. Carr and G. D. Carr (1983) to merge the two species. Traditional treatments of these taxa and *C. pauciflora* have placed together populations that are far from homogeneous and may vary from one another in morphology and chromosome number and structure. For instance, *C. fremontii* and *C. pauciflora* are each known to include at least six different chromosome races. In this treatment, we have chosen to recognize the larger-headed members of *C. pauciflora* in the sense of G. D. Carr (1975) as *C. fremontii*. This reflects the current state of our knowledge, including molecular studies (B. G. Baldwin and S. Markos 1998). Molecular studies have provided a separate means of assessing relationships and insights into the evolution of the genus. Providing a practical taxonomic framework that reflects what is now known about the complicated morphologic and chromosomal patterns of variation of *C. pauciflora* and *C. fremontii* has proved to be an elusive goal.

338. CENTROMADIA   Greene, Fl. Francisc. 4: 424. 1897 • Spikeweed [Latin *centron*, prickle, and generic name *Madia*]

Bruce G. Baldwin

John L. Strother

Annuals, 10–120 cm. **Stems** ± erect to prostrate. **Leaves** mostly cauline (at flowering); proximal opposite (often in winter-spring rosettes), most alternate; ± sessile; blades oblanceolate to linear or lance-linear, proximal usually 1–2-pinnatifid, ultimate margins toothed or entire (sometimes bristly-ciliate), (apices of distal leaves usually spine-tipped) faces glabrous, scabroso-hirtellous, ± hirsute, or villous, often glandular as well. **Heads** radiate, borne in glomerules or ± spiciform-paniculiform or ± umbelliform arrays. **Peduncular bracts**: pit-glands and tack-glands 0 (apices usually spine-tipped, sometimes apiculate). **Involucres** ± obconic or urceolate, 3–8+ mm diam. (subtended by calyculi of 5–12+ usually spine-tipped bractlets). **Phyllaries** falling or persistent, 5–75+ in 1 series (lanceolate to lance-attenuate or oblanceolate, herbaceous, each usually ¹/₂ enveloping subtended ray floret proximally, abaxially scabroso-hirtellous, hirsute, or villous and/ or glandular, apices often spine-tipped). **Receptacles** flat to convex, setulose, paleate (paleae persistent, subtending all or most disc florets, distinct, phyllary-like, more scarious). **Ray florets** 5–75+, pistillate, fertile; corollas yellow. **Disc florets** 6–200+, usually functionally staminate, rarely bisexual and fertile; corollas yellow, tubes shorter than funnelform throats, lobes 5, deltate (anthers reddish to dark purple or yellow to brownish; styles glabrous proximal to branches). **Ray cypselae** ± compressed (abaxially gibbous, basal attachments basal or oblique, apices beaked or elevated adaxially, faces glabrous); **pappi** 0. **Disc cypselae** usually 0; **pappi** (of disc florets) 0 or of 3–12 linear, oblanceolate, or subulate scales. *x* = 13.

Species 4 (3 in the flora): w United States, nw Mexico.

Following B. G. Baldwin (1999b), *Centromadia* is treated here as distinct from *Hemizonia*; *Centromadia* is more closely related to *Calycadenia*, *Deinandra*, *Holocarpha*, and *Osmadenia* than to *Hemizonia* in the strict sense (S. Carlquist et al. 2003). Taxa of *Centromadia* are self-incompatible and of low to high interfertility (C. S. Venkatesh 1958). Most occur in somewhat poorly drained or alkaline sites.

SELECTED REFERENCE   Venkatesh, C. S.   1958. A cyto-genetic and evolutionary study of *Hemizonia* section *Centromadia* (Compositae). Evolution 39: 1236–1241.

1. Leaves not glandular; disc pappi 0 . . . . 1. *Centromadia pungens*
1. Leaves sometimes glandular; disc pappi of 3–5 linear to subulate scales, or 8–12 narrowly oblanceolate to linear scales.
  2. Leaves densely villous or hirsute, stipitate-glandular (glands yellow, brown, or black); anthers reddish to dark purple; disc pappi of 8–12 linear or narrowly oblanceolate scales . . . . 2. *Centromadia fitchii*
  2. Leaves glabrous, scabroso-hirtellous, ± hirsute to hirtellous, or villous, sometimes glandular (glands yellow); anthers yellow, brownish, or reddish to dark purple; disc pappi of 3–5 linear to subulate scales . . . . 3. *Centromadia parryi*

1. Centromadia pungens (Hooker & Arnott) Greene, Man. Bot. San Francisco, 196. 1894

Hartmannia pungens Hooker & Arnott, Bot. Beechey Voy., 357. 1839; *Hemizonia pungens* (Hooker & Arnott) Torrey & A. Gray

Plants 10–120 cm. **Leaves** glabrous (but for bristly-ciliate margins and midribs), scabrous, or hirsute (not glandular). **Involucres** 3–6 mm. **Anthers** yellow to brownish. **Pappi** 0.

Subspecies 2 (2 in the flora): w United States (plus New York), nw Mexico.

1. Leaves: margins bristly-ciliate, faces glabrous but for bristly midribs; paleae little, if at all, exserted, apices obtuse to acute, sometimes apiculate . . . . 1a. *Centromadia pungens* subsp. *laevis*
1. Leaves: margins sometimes ciliate, faces glabrous, scabrous, or hirsute, midribs sometimes bristly; paleae often conspicuously exserted, apices ± attenuate, spine-tipped . . . . 1b. *Centromadia pungens* subsp. *pungens*

1a. Centromadia pungens (Hooker & Arnott) Greene subsp. **laevis** (D. D. Keck) B. G. Baldwin, Novon 4: 467. 1999 [C]

Hemizonia pungens (Hooker & Arnott) Torrey & A. Gray subsp. *laevis* D. D. Keck, Madroño 3: 14. 1935; *H. laevis* (D. D. Keck) D. D. Keck

Leaves: margins bristly-ciliate, faces glabrous but for bristly midribs. **Paleae** little, if at all, exserted, apices obtuse to acute, sometimes apiculate. $2n = 18$.

Flowering Apr–Sep. Open, ± alkaline flats, depressions, stream banks and beds, grasslands, disturbed sites; of conservation concern; 90–500 m; Calif.; Mexico (Baja California).

Subspecies *laevis* is known only from the interior South Coast region and Peninsular Ranges.

1b. Centromadia pungens (Hooker & Arnott) Greene subsp. **pungens** • Common spikeweed [F]

Centromadia maritima Greene; *C. pungens* subsp. *maritima* (Greene) B. G. Baldwin; *C. pungens* subsp. *septentrionalis* (D. D. Keck) B. G. Baldwin; *Hemizonia pungens* (Hooker & Arnott) Torrey & A. Gray subsp. *maritima* (Greene) D. D. Keck; *H. pungens* subsp. *septentrionalis* D. D. Keck; *H. pungens* var. *septentrionalis* (D. D. Keck) Cronquist

Leaves: margins sometimes ciliate, faces glabrous, scabrous, or hirsute, midribs sometimes bristly. **Paleae** often exserted, apices ± attenuate, spine-tipped. $2n = 18, 20$.

Flowering Apr–Nov. Open, ± alkaline flats, depressions, waterway banks and beds, in grasslands, saltbush scrub, disturbed sites (e.g., fallow fields); 0–1200(–1800) m; Ariz., Calif., Idaho, Nev., N.Y., Oreg., Wash.; Mexico (Baja California).

Subspecies *pungens* is circumscribed broadly to include subsp. *maritima* and subsp. *septentrionalis* based on morphologic and molecular data (B. G. Baldwin, unpubl.). As treated here, *Centromadia pungens* subsp. *pungens* occurs widely in central and northern California, and it is putatively introduced in southwestern California and outside the state.

2. Centromadia fitchii (A. Gray) Greene, Man. Bot. San Francisco, 197. 1894 • Fitch's spikeweed [E]

Hemizonia fitchii A. Gray in War Department [U.S.], Pacif. Railr. Rep. 4(5): 109. 1857

Plants 5–50 cm. **Leaves** densely villous or hirsute, stipitate-glandular (glands yellow, brown, or black). **Involucres** 5–10 mm. **Anthers** reddish to dark purple. **Pappi** of 8–12 narrowly oblanceolate to linear scales. $2n = 26$.

Flowering May–Nov. Grasslands, ± alkaline plains, vernal pool edges and beds, disturbed sites (e.g., fallow fields), open woodlands, sometimes on serpentine; 0–1000 m; Calif., Oreg.

Centromadia fitchii occurs widely in northern and central California (except deserts) and southwestern Oregon and has putatively introduced populations in southwestern California. Unlike most other members of *Centromadia*, *C. fitchii* often occurs in upland, presumably well-drained sites. Putative natural hybrids with *C. parryi* subsp. *rudis* have been noted (e.g., *D. D. Keck 2573*, UC, from San Joaquin Co., California).

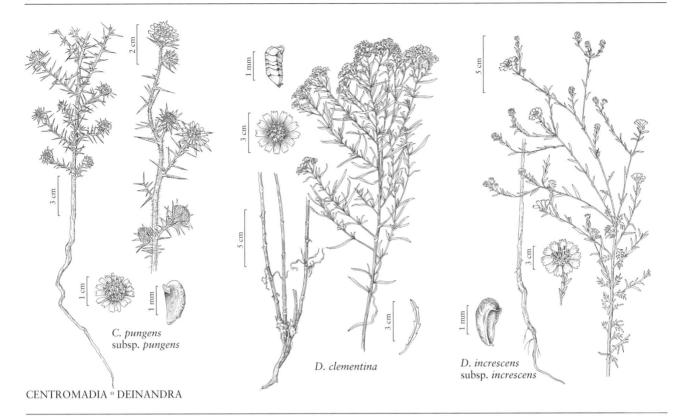

C. *pungens*
subsp. *pungens*

D. *clementina*

D. *increscens*
subsp. *increscens*

CENTROMADIA ○ DEINANDRA

3. Centromadia parryi (Greene) Greene, Man. Bot. San Francisco, 197. 1894

Hemizonia parryi Greene, Bull. Torrey Bot. Club 9: 16. 1882; *H. pungens* (Hooker & Arnott) Torrey & A. Gray var. *parryi* (Greene) H. M. Hall

Plants 10–70 cm. **Leaves** glabrous, scabroso-hirtellous, ± hirsute to hirtellous, or villous, glandular or not, glands yellow. **Involucres** 2.5–10 mm. **Anthers** yellow, brownish, or purple. **Pappi** of 3–5 linear or subulate scales. Subspecies 4 (4 in the flora): California, nw Mexico.

1. Leaves villous or ± hirsute, coarsely stipitate-glandular; paleae sometimes with 2 purple lines (along inner edges of scarious margins); anthers reddish to dark purple . . . .
. . . . 3a. *Centromadia parryi* subsp. *australis*
1. Leaves glabrous, scabroso-hirtellous, or ± hirsute to hirtellous, not villous, glandular or not; paleae lacking purple lines; anthers yellow or brownish.
    2. Leaves scabroso-hirtellous (and usually ± hirsute or setose-ciliate), not glandular or glands minute, sparsely and irregularly distributed; involucres 3.5–5 mm; ray laminae 2–3 mm . . . . 3d. *Centromadia parryi* subsp. *rudis*
    2. Leaves glabrous or hirsute to hirtellous, glandular or not; involucres 4–10 mm; ray laminae 2.5–5(–6) mm.

[3. Shifted to left margin—Ed.]
3. Peduncular bracts seldom glandular (sometimes with yellowish, minute, stipitate glands among eglandular hairs); ray laminae 2.5–3(–4.5) mm . . . . 3b. *Centromadia parryi* subsp. *congdonii*
3. Peduncular bracts coarsely glandular (glands yellow, sessile or stipitate); ray laminae 3–5(–6) mm . . . . 3c. *Centromadia parryi* subsp. *parryi*

3a. Centromadia parryi (Greene) Greene subsp. **australis** (D. D. Keck) B. G. Baldwin, Novon 9: 466. 1999 [C]

Hemizonia parryi Greene subsp. *australis* D. D. Keck, Madroño 3: 15. 1935; *H. australis* (D. D. Keck) D. D. Keck

Leaves (and peduncular bracts) villous or hirsute to hirtellous, coarsely stipitate-glandular (glands yellow). **Involucres** 2.5–6(–7) mm. **Paleae** sometimes with 2 purple lines (along inner edges of scarious margins). **Ray laminae** 2–4 mm. **Anthers** reddish to dark purple. $2n = 22$.

Flowering Jun–Oct. Inner edges of salt marshes, coastal grasslands, vernal pool edges and beds, openings in coastal scrub, often ± saline or alkaline sites; of conservation concern; 0–200 m; Calif.; Mexico (Baja California).

Subspecies *australis* occurs along the southern coast of California. Natural, ± sterile hybrids with *Deinandra fasciculata* have been documented from Santa Barbara County (B. D. Tanowitz 1977).

3b. Centromadia parryi (Greene) Greene subsp. **congdonii** (B. L. Robinson & Greenman) B. G. Baldwin, Novon 9: 466. 1999 C E

Hemizonia congdonii B. L. Robinson & Greenman, Bot. Gaz. 22: 169. 1896; *H. parryi* Greene subsp. *congdonii* (B. L. Robinson & Greenman) D. D. Keck; *H. parryi* var. *congdonii* (B. L. Robinson & Greenman) Hoover

Leaves (and peduncular bracts) glabrous or ± hirsute, seldom glandular (sometimes with yellowish, minute, stipitate glands among eglandular hairs). **Involucres** 4–9 mm. **Paleae** lacking purple lines. **Ray laminae** 2.5–3(–4.5) mm. **Anthers** yellow or brownish. $2n = 24$.

Flowering Jun–Oct. Terraces, swales, flood plains, often ± alkaline sites, grasslands, disturbed sites (e.g., fallow fields); of conservation concern; 0–300 m; Calif.

Subspecies *congdonii* is found in the South Coast Ranges and San Francisco Bay area. Molecular and morphologic studies have indicated that the southern and northern populations are evolutionarily distinct and may warrant treatment as distinct taxa (B. G. Baldwin, unpubl.).

3c. Centromadia parryi (Greene) Greene subsp. **parryi** C E

Leaves (and peduncular bracts) ± hirsute to hirtellous, usually coarsely, sometimes sparsely, glandular (glands yellow, sessile or stipitate). **Involucres** 5–10 mm. **Paleae** lacking purple lines. **Ray corolla laminae** 3–5(–6) mm. **Anthers** yellow or brownish. $2n = 24$.

Flowering Jun–Oct. Grasslands (often ± alkaline), coastal salt marshes, alkaline springs and seeps; of conservation concern; 0–400 m; Calif.

Subspecies *parryi* occurs in the northern San Francisco Bay area and southern North Coast Ranges. Putative natural hybrids between *Centromadia parryi* subsp. *parryi* and *C. parryi* subsp. *rudis* have been noted (e.g., *D. D. Keck 2598,* UC, from Sonoma County).

3d. Centromadia parryi (Greene) Greene subsp. **rudis** (Greene) B. G. Baldwin, Novon 4: 467. 1999 E

Centromadia rudis Greene, Man. Bot. San Francisco, 197. 1894; *Hemizonia parryi* Greene subsp. *rudis* (Greene) D. D. Keck

Leaves (and peduncular bracts) scabroso-hirtellous and usually ± hirsute or setose-ciliate, not glandular or with sparsely and irregularly distributed, minute, sessile or stipitate, yellowish glands. **Involucres** 3.5–5 mm. **Paleae** lacking purple lines. **Ray corolla laminae** 2–3 mm. **Anthers** yellow or brownish. $2n = 22$.

Flowering Jun–Oct. Grasslands, edges of marshes and vernal pools, disturbed sites (e.g., fallow fields); 0–500 m; Calif.

Subspecies *rudis* occurs in the southern Sacramento Valley, northern San Joaquin Valley, and, uncommonly, in the southern Inner North Coast Ranges.

339. DEINANDRA   Greene, Fl. Francisc. 4: 424. 1897   •   [No etymology stated in protologue; meaning uncertain]

Bruce G. Baldwin

John L. Strother

Annuals, subshrubs, or shrubs, 4–120(–150) cm. **Stems** usually ± erect, rarely decumbent. **Leaves** mostly cauline (annuals forming winter-spring rosettes, usually withering by flowering); proximal opposite, most alternate; sessile; blades oblanceolate to linear or lance-linear, usually pinnatifid or toothed, sometimes serrate or entire, faces often hirsute or hispid-hirsute, sometimes villous, pilose, pubescent, canescent, strigose, or scabrous (often sessile- or stipitate-glandular as well) or glabrous (with scabrous or hispid margins). **Heads** radiate, usually in corymbiform or ± paniculiform arrays, sometimes in racemiform arrays or in glomerules. **Peduncular bracts:** pit-glands, tack-glands, and spines 0. **Involucres** ± obconic, campanulate, hemispheric, or urceolate, 2–13+ mm diam. **Phyllaries** falling, 3–35 in 1 series (± lanceolate to lance-attenuate or oblanceolate, herbaceous, each usually ¹/₂ enveloping subtended ray floret proximally, abaxially ± hirsute and sessile- or stipitate-glandular). **Receptacles** flat to convex, glabrous or setulose, paleate (paleae falling, in 1 series between rays and disc in annuals, in 2–3+ series or subtending all or most disc florets in subshrubs and shrubs, connate or distinct, phyllary-like, more scarious). **Ray florets** 3–35, pistillate, fertile; corollas deep or pale yellow. **Disc florets** 3–70, usually functionally staminate, seldom bisexual and fertile; corollas yellow, tubes shorter than or about equaling funnelform throats, lobes 5, deltate (anthers usually reddish to dark purple or yellow, rarely maroon; styles glabrous proximal to branches). **Ray cypselae** slightly obcompressed (adaxial sides flatter than abaxials), clavate (abaxially gibbous, often ± arcuate, basal attachments oblique, apices ± beaked, beaks offset adaxially, ascending, faces glabrous); **pappi** 0. **Disc cypselae** usually 0; **pappi** (of disc florets) usually of 1–15 elliptic, lance-linear, lanceolate, linear, oblong, quadrate, setiform, or subulate, entire, erose, fimbriate, fringed, or laciniate scales, sometimes 0 or coroniform (crowns of ± linear, sometimes fimbriate scales). x = 12 or 13.

Species 21 (16 in the flora): w United States, nw Mexico.

Deinandra is treated here as distinct from *Hemizonia*; *Deinandra* is more closely related to *Calycadenia, Centromadia, Holocarpha,* and *Osmadenia* than to *Hemizonia* in the strict sense (S. Carlquist et al. 2003). As circumscribed here (following B. G. Baldwin 1999b), *Deinandra* comprises all taxa included in *Hemizonia* sect. *Madiomeris* by B. D. Tanowitz (1982) plus the shrubs and subshrubs constituting the informal "Fruticosae" or "Zonamra" (J. Clausen 1951; D. D. Keck 1959b). All are self-incompatible except *D. arida* and *D. mohavensis.* Five species (*D. frutescens, D. greeneana, D. martirensis, D. palmeri, D. streetsii*), none annuals, are known only from Baja California, Mexico. Most species north of Mexico are cross-incompatible or ± intersterile (J. Clausen 1951).

SELECTED REFERENCE Tanowitz, B. D. 1982. Taxonomy of *Hemizonia* sect. *Madiomeris* (Asteraceae: Madiinae). Syst. Bot. 7: 314–339.

1. Subshrubs or shrubs; paleae in 2 series or in 3+ series or throughout receptacles.
    2. Paleae in 2 series (1 between ray and disc florets, 1 between outermost and adjacent disc florets); ray florets (11–)13(–20); anthers reddish to dark purple . . . . 7. *Deinandra clementina*
    2. Paleae in 3+ series or throughout receptacles; ray florets (4–)8; anthers yellow or brownish . . . . 8. *Deinandra minthornii*
1. Annuals; paleae in 1 series (between ray and disc florets).
    3. Ray florets 3–5; disc florets 3–6; pappi of 5–12 scales.

4. Ray florets 3(–4); disc florets 3(–4) . . . . 12. *Deinandra lobbii*
4. Ray florets 5; disc florets 6.
    5. Phyllaries usually sessile-glandular, at least near margins, rarely stipitate-glandular (stalks shorter than glands); anthers reddish to dark purple.
        6. Bracts subtending heads usually overlapping at least proximal ¹/₂ of each involucre; phyllaries glandular near margins, sometimes with non-glandular, non-pustule-based hairs as well . . . . 1. *Deinandra fasciculata*
        6. Bracts subtending heads usually overlapping proximal 0–¹/₂ of each involucre; phyllaries ± evenly glandular and with pustule-based hairs, at least on midribs . . . . 11. *Deinandra pentactis*
    5. Phyllaries stipitate-glandular (stalks often equal to or longer than glands); anthers yellow or brownish (reddish to dark purple or maroon in some, mostly s Californian, *D. kelloggii*).
        7. Proximal leaves usually pinnatifid to toothed (rarely entire); heads in open, paniculiform arrays; pappi of linear to oblong, entire or fringed scales . . . . 13. *Deinandra kelloggii*
        7. Proximal leaves usually entire, sometimes serrate; heads usually in glomerules; pappi of irregular, erose scales. . . . . 5. *Deinandra mohavensis*

[3. Shifted to left margin—Ed.]

3. Ray florets (5–)8–35; disc florets 8–70; pappi 0, or of 1–14 scales, or coroniform (reputedly sometimes rudimentary).
    8. Anthers yellow or brownish.
        9. Faces of proximal leaves glabrous or hispid-hirsute and stipitate-glandular; disc florets 17–60; pappi usually 0, rarely of 1–5 linear to setiform scales 0.1–0.6 mm or, reputedly, rudimentary.
            10. Stems solid; leaves (proximal) hispid-hirsute and stipitate-glandular; disc florets 17–25 . . . . 4. *Deinandra arida*
            10. Stems fistulose; leaves (proximal) glabrous (margins and midribs sometimes scabrous or hispid); disc florets 28–60 . . . . 10. *Deinandra halliana*
        9. Faces of proximal leaves ± hirsute and stipitate-glandular; disc florets 10–21; pappi usually of 4–13 scales, sometimes coroniform (in *D. bacigalupii*), rarely 0 (in *Deinandra pallida*).
            11. Ray laminae pale yellow, 6–12 mm . . . . 14. *Deinandra pallida*
            11. Ray laminae deep yellow, 2–4 mm . . . . 15. *Deinandra bacigalupii*
    8. Anthers reddish to dark purple.
        12. Disc florets all or mostly bisexual . . . . 3. *Deinandra floribunda*
        12. Disc florets all or mostly functionally staminate.
            13. Ray florets 15–35; pappi 0, or coroniform (irregular crowns of entire, erose, or laciniate scales 0.1–0.9 mm) . . . . 16. *Deinandra corymbosa*
            13. Ray florets (7–)8–13; pappi of 4–14 elliptic, lanceolate, linear, oblong, or quadrate, barely fringed to erose scales 0.5–2 mm.
                14. Phyllaries sessile- and stipitate-glandular, mostly near proximal margins and on apices. . . . . 2. *Deinandra conjugens*
                14. Phyllaries ± evenly stipitate-glandular, including margins and apices.
                    15. Ray florets (7–)8(–10); disc florets 8–14(–15) (basal leaves sometimes present at flowering in South Coast Ranges) . . . . 6. *Deinandra paniculata*
                    15. Ray florets 8–13(–15); disc florets 11–32 . . . . 9. *Deinandra increscens*

1. **Deinandra fasciculata** (de Candolle) Greene, Fl. Francisc. 4: 424. 1897

Hartmannia fasciculata de Candolle in A. P. de Candolle and A. L. P. P. de Candolle, Prodr. 5: 693. 1836; *Hemizonia ramosissima* Bentham

Annuals, 4–100 cm. **Stems** ± solid. **Leaves:** proximal blades toothed, faces hirsute. **Heads** usually in glomerules or pairs, sometimes well separated, in racemiform or paniculiform arrays. **Bracts** subtending heads usually overlapping at least proximal ¹/₂ of each involucre. **Phyllaries** sessile-glandular near margins, sometimes with non-glandular, non-pustule-based hairs as well. **Paleae** in 1 series. **Ray florets** 5; laminae deep yellow, 6–14 mm. **Disc florets** 6, all or mostly functionally staminate; anthers reddish to dark purple. **Pappi** of 5–12 lanceolate to oblong or linear, entire or fringed scales 1–1.5 mm. *2n* = 24.

Flowering Apr–Sep. Grasslands, openings in chaparral, coastal scrub, and woodlands, vernal pool beds, disturbed sites (e.g., burns), often in sandy or clayey soils, sometimes serpentine; 0–1000 m; Calif.; Mexico (Baja California).

Deinandra fasciculata occurs in southwestern California, especially on immediate coast, and on southern Central Coast and in the western Outer South Coast Ranges.

2. **Deinandra conjugens** (D. D. Keck) B. G. Baldwin, Novon 9: 468. 1999 [C]

Hemizonia conjugens D. D. Keck, Aliso 4: 109. 1958

Annuals, 10–50 cm. **Stems** ± solid. **Leaves:** proximal blades pinnatifid to toothed, faces ± hirsute. **Heads** in glomerules or well separated, in paniculiform arrays. **Bracts** subtending heads usually overlapping proximal 0–¹/₂+ of each involucre. **Phyllaries** ± sessile- and stipitate-glandular, mostly near proximal margins and on apices, sometimes with non-glandular, non-pustule-based hairs as well. **Paleae** in 1 series. **Ray florets** 7–10; laminae deep yellow, 3–6 mm. **Disc florets** 13–21, all or mostly functionally staminate; anthers reddish to dark purple. **Pappi** of 6–9 scales 0.5–1 mm. *2n* = 24.

Flowering Apr–Jun. Grasslands, openings in coastal scrub, disturbed sites, clayey soils; of conservation concern; 20–300 m; Calif.; Mexico (Baja California).

Deinandra conjugens is known from the inner South Coast and in southwestern foothills of the Peninsular Ranges, near San Diego.

3. **Deinandra floribunda** (A. Gray) Davidson & Moxley, Fl. S. Calif., 401. 1923 [C]

Hemizonia floribunda A. Gray, Proc. Amer. Acad. Arts 11: 79. 1876

Annuals, 30–100 cm. **Stems** ± solid. **Leaves:** proximal blades entire or toothed, faces pilose and stipitate-glandular. **Heads** in racemiform-paniculiform arrays (often narrow; side branches strict, short). **Bracts** subtending heads overlapping proximal 0–¹/₂+ of each involucre or not. **Phyllaries** ± evenly stipitate-glandular, including margins and apices, usually with non-glandular, non-pustule-based hairs as well. **Paleae** in 1 series. **Ray florets** 13–20; laminae deep yellow, 4–7 mm. **Disc florets** 24–31, all or mostly bisexual; anthers reddish to dark purple. **Pappi** of 6–10 (often rufous and/or flecked), oblong to elliptic, fringed scales 0.7–1.5 mm. *2n* = 26.

Flowering Aug–Nov. Moist openings in chaparral, streambeds, disturbed sites; of conservation concern; 70–1200 m; Calif.; Mexico (Baja California).

Deinandra floribunda is known from valleys and hills of the southern Peninsular Ranges, within ca. 15 km of the Mexican border.

4. **Deinandra arida** (D. D. Keck) B. G. Baldwin, Novon 9: 467. 1999 [C] [E]

Hemizonia arida D. D. Keck, Aliso 4: 109. 1958

Annuals, 20–80 cm. **Stems** ± solid. **Leaves:** proximal blades toothed to entire, faces hispid-hirsute and stipitate-glandular. **Heads** in paniculiform arrays. **Bracts** subtending heads usually not overlapping each involucre. **Phyllaries** ± evenly stipitate-glandular, including margins and apices, with non-glandular, non-pustule-based hairs as well. **Paleae** in 1 series. **Ray florets** (4–)8(–10); laminae deep yellow, 5–7 mm. **Disc florets** 17–25, all or mostly functionally staminate; anthers yellow or brownish. **Pappi** usually 0, rarely of 1–5 linear to setiform scales 0.1–0.6 mm. *2n* = 24.

Flowering Apr–Nov. Washes, edges of springs and seeps, and adjacent slopes, cliffs, or ledges, often in ± alkaline, sandy, gravelly, or clayey soils; of conservation concern; 600–1000 m; Calif.

Deinandra arida is known only from Red Rock and Last Chance canyons and associated tributaries in the El Paso Mountains, western Mojave Desert. Depauperate specimens resemble *D. kelloggii*, which also has been documented from the Red Rock Canyon area.

5. **Deinandra mohavensis** (D. D. Keck) B. G. Baldwin, Novon 9: 469. 1999 [C] [E]

Hemizonia mohavensis D. D. Keck, Madroño 3: 9. 1935

Annuals, 10–100(–150) cm. **Stems** solid. **Leaves:** proximal blades usually entire, sometimes serrate, faces pubescent and stipitate-glandular. **Heads** usually in glomerules or ± crowded, sometimes in open, paniculiform arrays. **Bracts** subtending heads usually overlapping proximal 0–1/2+ of each involucre. **Phyllaries** ± evenly stipitate-glandular, including margins and apices, usually with non-glandular, non-pustule-based hairs as well. **Paleae** in 1 series. **Ray florets** 5; laminae deep yellow, 4–7 mm. **Disc florets** 6, functionally staminate or bisexual; anthers yellow or brownish. **Pappi** of 5–9 (often ± connate, irregular), subulate to quadrate, erose scales 0.1–0.6 mm. $2n = 22$.

Flowering (May–)Jul–Oct(–Jan). Edges of streams, springs, seeps, in swales, in openings on drier surrounding slopes of chaparral, desert scrub, or woodlands, clayey, silty, or gravelly soils; of conservation concern; 800–1600 m; Calif.

Deinandra mohavensis is known from the Peninsular Ranges (San Jacinto Mountains, Palomar Mountains, Anza Bench) and southernmost Sierra Nevada (Short Canyon, Jawbone Canyon region), and San Bernardino Mountains (where possibly extirpated). As noted by A. C. Sanders et al. (1997), *D. mohavensis* occurs in unusually mesic habitats for *Deinandra* (somewhat similar to situations where *D. arida* and *D. floribunda* occur) and has a highly disjunct distribution. It is in the Center for Plant Conservation's National Collection of Endangered Plants.

6. **Deinandra paniculata** (A. Gray) Davidson & Moxley, Fl. S. Calif., 401. 1923 [C]

Hemizonia paniculata A. Gray, Proc. Amer. Acad. Arts 19: 17. 1883; *Deinandra increscens* (H. M. Hall ex D. D. Keck) B. G. Baldwin subsp. *foliosa* (Hoover) B. G. Baldwin; *H. increscens* (H. M. Hall ex D. D. Keck) Tanowitz subsp. *foliosa* (Hoover) Tanowitz; *H. paniculata* subsp. *foliosa* Hoover

Annuals, 10–80(–150) cm. **Stems** ± solid. **Leaves:** (basal rosettes present at flowering in South Coast Ranges) proximal blades pinnatifid to toothed, faces hispid-hirsute and sometimes stipitate-glandular. **Heads** in paniculiform arrays. **Bracts** subtending heads usually overlapping proximal 0–1/2 of each involucre. **Phyllaries** ± evenly stipitate-glandular, including margins and apices, often with some non-glandular, non-pustule-based hairs as well. **Paleae** in 1 series. **Ray florets** (7–)8(–10); laminae deep yellow, 3–5(–6) mm. **Disc florets** 8–14(–15), all or mostly functionally staminate; anthers reddish to dark purple. **Pappi** of 6–12 oblong, fimbriate scales 1–2 mm. $2n = 24$.

Flowering May–Nov. Grasslands, openings in chaparral and woodlands, disturbed sites, often sandy soils; of conservation concern; 10–800 m; Calif.; Mexico (Baja California).

Deinandra paniculata occurs in dry foothills, mesas, and valleys of the southern Peninsular Ranges and South Coast and in the central South Coast Ranges. Plants from the South Coast Ranges, previously treated as *D. increscens* subsp. *foliosa*, may warrant recognition as a distinct taxon.

7. **Deinandra clementina** (Brandegee) B. G. Baldwin, Novon 9: 468. 1999 [C] [E] [F]

Hemizonia clementina Brandegee, Erythea 7: 70. 1899

Subshrubs or shrubs, 15–80 cm. **Stems** ± solid (floccose in some leaf axils). **Leaves:** proximal blades toothed or entire, faces pubescent to somewhat canescent, ± hirsute, strigose, or scabrous, often stipitate-glandular. **Heads** usually in crowded, corymbiform or paniculiform arrays. **Bracts** subtending heads usually overlapping proximal 0–1/2+ of each involucre. **Phyllaries** ± evenly stipitate-glandular (sometimes sparsely), including margins and apices, often with non-glandular, pustule-based or non-pustule-based hairs as well. **Paleae** in 2 series (1 between ray and disc florets, 1 between outermost and adjacent disc florets). **Ray florets** (11–)13(–20); laminae deep yellow, 4.5–7 mm. **Disc florets** 18–30, all or mostly functionally staminate; anthers reddish to dark purple. **Pappi** of 7–10(–15) lance-linear, fimbriate scales 1–3 mm. $2n = 24$.

Flowering Mar–Dec. Grassy areas, coastal scrub, semi-barren sites, edges of salt marshes, on terraces, slopes, and ridges, often in rocky or clayey soils; of conservation concern; 0–200 m; Calif.

Deinandra clementina is known only from six of the Channel Islands (Anacapa, San Clemente, San Nicolas, Santa Barbara, Santa Cruz, and Santa Catalina).

8. Deinandra minthornii (Jepson) B. G. Baldwin, Novon 9: 469. 1999 [C] [E]

Hemizonia minthornii Jepson, Man. Fl. Pl. Calif., 1092. 1925

Subshrubs or shrubs, 15–100 cm. **Stems** ± solid. **Leaves:** proximal blades pinnatifid to toothed, faces ± hirtellous, sometimes stipitate-glandular. **Heads** borne singly or in loose, racemiform or racemiform-paniculiform arrays. **Bracts** subtending heads usually overlapping proximal 0–¹/₂ of each involucre. **Phyllaries** ± evenly stipitate-glandular, including margins and apices, often with some non-glandular, non-pustule-based hairs as well. **Paleae** in 3+ series or throughout receptacles. **Ray florets** (4–)8; laminae deep yellow, 5.5–6.5 mm. **Disc florets** 18–23, all or mostly functionally staminate; anthers yellow or brownish. **Pappi** of 8–12 linear to lance-linear, entire or fringed scales 1–3 mm. $2n = 24$.

Flowering Jun–Nov. In chaparral, coastal scrub, often in sandstone crevices; of conservation concern; 200–800 m; Calif.

Deinandra minthornii occurs in the western Transverse Ranges (Santa Monica Mountains and Santa Susana Mountains). It is in the Center for Plant Conservation's National Collection of Endangered Plants.

9. Deinandra increscens (H. M. Hall ex D. D. Keck) B. G. Baldwin, Novon 9: 468. 1999 [E]

Hemizonia paniculata A. Gray subsp. *increscens* H. M. Hall ex D. D. Keck, Madroño 3: 11. 1935; *H. increscens* (H. M. Hall ex D. D. Keck) Tanowitz

Annuals, 6–100 cm. **Stems** ± solid. **Leaves:** proximal blades pinnatifid, faces ± hirsute or villous and sometimes stipitate-glandular. **Heads** usually in paniculiform arrays sometimes congested. **Bracts** subtending heads usually overlapping proximal 0–¹/₂ of each involucre. **Phyllaries** evenly stipitate-glandular, including margins and apices, usually with non-glandular, non-pustule-based hairs as well. **Paleae** in 1 series. **Ray florets** 8–13(–15); laminae pale to deep yellow, 4–8 mm. **Disc florets** 11–32, all or mostly functionally staminate; anthers reddish to dark purple. **Pappi** of 5–14 linear, lance-linear, or oblong, fimbriate to erose scales 1–2 mm.

Subspecies 2 (2 in the flora): California.

Extensive morphologic and ecologic variation within *Deinandra increscens* has been interpreted variously. Subspecies recognized here follow the treatment of B. D. Tanowitz (1982) except for return of Hoover's subsp. *foliosa* to *D. paniculata* A. Gray and recognition of *D. bacigalupii* B. G. Baldwin for members of subsp.

increscens in the sense of Tanowitz from the eastern San Francisco Bay Area.

1. Stems usually erect, sparsely to densely leafy; heads in open, paniculiform arrays . . . .
 . . . . 9a. *Deinandra increscens* subsp. *increscens*
1. Stems usually decumbent, densely leafy; heads ± congested . . . . . 9b. *Deinandra increscens* subsp. *villosa*

9a. Deinandra increscens (H. M. Hall ex D. D. Keck) B. G. Baldwin subsp. **increscens** [E] [F]

Stems usually erect, sparsely to densely leafy. **Heads** in open, paniculiform arrays. **Ray florets** 8–13(–15). **Disc florets** 11–32. $2n = 24$.

Flowering Apr–Nov. Grasslands, openings in woodlands, chaparral, and coastal scrub, disturbed sites (e.g., roadsides, fallow fields), often sandy or clayey soils, sometimes serpentine; 0–300 m; Calif.

Subspecies *increscens* occurs in foothills and valleys of the South Coast Ranges, on the immediate Central Coast and northern South Coast, and on Santa Rosa Island.

9b. Deinandra increscens (H. M. Hall ex D. D. Keck) B. G. Baldwin subsp. **villosa** (Tanowitz) B. G. Baldwin, Novon 9: 469. 1999 [C] [E]

Hemizonia increscens (H. M. Hall ex D. D. Keck) Tanowitz subsp. *villosa* Tanowitz, Syst. Bot. 7: 331. 1982

Stems usually decumbent, densely leafy. **Heads** ±congested. **Ray florets** (8–)12–13. **Disc florets** (12–)18–31. $2n = 24$.

Flowering Jun–Sep. Coastal bluffs, fields; of conservation concern; 30–50 m; Calif.

Subspecies *villosa* occurs on the northern South Coast, near Gaviota. It is in the Center for Plant Conservation's National Collection of Endangered Plants.

10. Deinandra halliana (D. D. Keck) B. G. Baldwin, Novon 9: 468. 1999 [C] [E]

Hemizonia halliana D. D. Keck, Madroño 3: 12. 1935

Annuals, 15–120 cm. **Stems** fistulose. **Leaves:** proximal blades entire or serrate, faces glabrous (margins and midribs sometimes scabrous or hispid). **Heads** in open, corymbiform, racemiform, or paniculiform arrays. **Bracts** subtending heads usually not overlapping involucres.

Phyllaries evenly stipitate-glandular, including margins and apices, usually with some non-glandular, non-pustule-based hairs as well. **Paleae** in 1 series. **Ray florets** (8–)10–14; laminae deep yellow, 5–10 mm. **Disc florets** 28–60, all functionally staminate; anthers yellow or brownish. **Pappi** 0 (reportedly sometimes rudimentary). $2n$ = 20.

Flowering Apr–May. Grasslands, edges of alkali sinks, open muddy slopes, heavy, ± alkaline, clayey soils; of conservation concern; 300–1000 m; Calif.

Deinandra halliana occurs in the Inner South Coast Ranges. Evident population sizes vary greatly from year to year; active plants may be absent in dry years and form dense stands in some wet years. *Madia radiata* is morphologically similar to *D. halliana*; the two species sometimes co-occur and can be difficult to distinguish without close examination of phyllaries and cypselae.

11. Deinandra pentactis (D. D. Keck) B. G. Baldwin, Novon 9: 469. 1999 [E]

Hemizonia lobbii Greene subsp. *pentactis* D. D. Keck, Madroño 3: 8. 1935; *H. pentactis* (D. D. Keck) D. D. Keck

Annuals, 4–75 cm. **Stems** ± solid. **Leaves:** proximal blades pinnatifid to toothed, faces hispid-hirsute and sometimes sessile- or short-stipitate-glandular, rarely glabrous. **Heads** in open, paniculiform arrays. **Bracts** subtending heads usually overlapping proximal 0–1/2 of each involucre. **Phyllaries** ± evenly sessile- or stipitate-glandular, including margins and apices, and usually with pustule-based hairs, at least on midribs. **Paleae** in 1 series. **Ray florets** 5; laminae deep yellow, 3–5 mm. **Disc florets** 6, all or mostly functionally staminate; anthers reddish to dark purple. **Pappi** of (5–)6–8(–11), linear or oblong, fimbriate to laciniate scales 0.8–1 mm. $2n$ = 22.

Flowering Apr–Oct. Grasslands, open woodlands, disturbed sites, sandy, loamy, or clayey soils; (0–)200–900 m; Calif.

Deinandra pentactis occurs in interior valleys and hills of the central South Coast Ranges and in the San Francisco Bay area (near Palo Alto), where presumably introduced. The closely related, morphologically similar, and slightly interfertile *D. lobbii* replaces *D. pentactis* in the northern South Coast Ranges.

12. Deinandra lobbii (Greene) Greene, Fl. Francisc. 4: 425. 1897 [E]

Hemizonia lobbii Greene, Bull. Torrey Bot. Club 9: 109. 1882

Annuals, 5–70 cm. **Stems** ± solid. **Leaves:** proximal blades pinnatifid to toothed, faces hispid-hirsute, sometimes sessile- or short-stipitate-glandular. **Heads** in paniculiform arrays. **Bracts** subtending heads overlapping proximal 0–1/2 of each involucre or not. **Phyllaries** ± evenly sessile- or short-stipitate-glandular, including margins and apices, and with pustule-based hairs, at least on midribs. **Paleae** in 1 series. **Ray florets** 3(–4); laminae deep yellow, 3–5 mm. **Disc florets** 3(–4), all or mostly functionally staminate; anthers reddish to dark purple. **Pappi** of (4–)6–8(–12) quadrate or oblong to lance-linear, fimbriate to laciniate scales 0.5–1 mm. $2n$ = 22.

Flowering May–Dec. Grasslands, open woodlands, sagebrush scrub, disturbed sites, often in sandy or clayey soils; 0–700(–1800) m; Calif.

Deinandra lobbii occurs on hills and flats of the northern South Coast Ranges and San Francisco Bay area and is disjunct in northeastern California, where putatively introduced. Populations in northeastern California occur at higher elevations than elsewhere (1000–1800 m).

13. Deinandra kelloggii (Greene) Greene, Fl. Francisc. 4: 424. 1897

Hemizonia kelloggii Greene, Bull. Torrey Bot. Club 10: 41. 1883

Annuals, 10–100(–150) cm. **Stems** ± solid. **Leaves:** proximal blades usually pinnatifid to toothed (rarely entire), faces hispid to hirsute and sometimes stipitate-glandular. **Heads** in open, paniculiform arrays. **Bracts** subtending heads usually not overlapping involucres. **Phyllaries** ± evenly stipitate-glandular, including margins and apices, with or without non-glandular, non-pustule-based hairs. **Paleae** in 1 series. **Ray florets** 5; laminae deep yellow, 4–8 mm. **Disc florets** 6, all or mostly functionally staminate; anthers usually yellow or brownish (reddish to dark purple or maroon in some, mostly southern Californian, *D. kelloggii*). **Pappi** of 6–12 linear to oblong, entire or fringed scales 1–2 mm. $2n$ = 18.

Flowering Mar–Nov. Grasslands, openings in woodlands or shrublands, barrens, disturbed sites, sandy or clayey soils; 0–900(–1200) m; Ariz., Calif.; Mexico (Baja California).

Deinandra kelloggii occurs in the San Joaquin Valley, eastern San Francisco Bay area, eastern Inner South Coast Ranges, and south-central Sierra Nevada foothills and in mostly interior, southwestern California. Plants in the Mojave and Sonoran deserts and Inner North Coast Ranges are presumably introductions. Plants morphologically intermediate between D. *kelloggii* and D. *pallida* (e.g., 5–7 ray florets) have been collected where ranges of the two, somewhat interfertile species meet near Porterville (e.g., *D. D. Keck 3335*, UC; *L. L. Donnelly 29*, UC).

14. Deinandra pallida   (D. D. Keck) B. G. Baldwin, Novon 9: 469. 1999   [E]

Hemizonia pallida D. D. Keck, Madroño 3: 8. 1935

Annuals, 9–100 cm. **Stems** ± solid or fistulose. **Leaves:** proximal blades pinnatifid to toothed, faces ± hirsute and sometimes sparsely stipitate-glandular as well. **Heads** in crowded to open, corymbiform or paniculiform arrays. **Bracts** subtending heads sometimes overlapping proximal 0–½ of each involucre. **Phyllaries** ± evenly and minutely stipitate-glandular, including margins and apices, with non-glandular, non-pustule-based hairs as well. **Paleae** in 1 series. **Ray florets** (7–)8–12; laminae pale yellow, 6–12 mm. **Disc florets** 10–21, all or mostly functionally staminate; anthers yellow or brownish. **Pappi** usually of 4–9 linear or oblong scales 0.8–1.1 mm, or of 1–5 subulate to setiform scales 0.1–0.9 mm, rarely 0. $2n = 18$.

Flowering Mar–Jul. Grasslands, open woodlands and shrublands, barrens, disturbed sites, sandy, silty, or clayey soils, often ± alkaline; 70–900 m; Calif.

Deinandra pallida occurs in the southern San Joaquin Valley and the bordering Inner South Coast Ranges and southern Sierra Nevada foothills.

15. Deinandra bacigalupii   B. G. Baldwin, Madroño 46: 55, fig. 1. 1999   [C] [E]

Annuals, 10–40 cm. **Stems** ± solid. **Leaves:** proximal blades entire or irregularly lobed, faces ± hirsute and stipitate-glandular. **Heads** in corymbiform or paniculiform arrays. **Bracts** subtending heads usually overlapping proximal 0–½ of each involucre. **Phyllaries** ± evenly stipitate-glandular, including margins and apices, with non-glandular, non-pustule-based hairs as well. **Paleae** in 1 series. **Ray florets** (6–)8(–9); laminae deep yellow, 2–4 mm. **Disc florets** (10–)15–18(–21), all or mostly functionally staminate;

anthers yellow or brownish. **Pappi** usually of 8–13, subulate to quadrate, fringed to deeply erose scales 0.1–0.8 mm, sometimes fimbriate crowns. $2n = 24$.

Flowering Jun–Oct. Alkaline meadows, edges of alkali barrens or sinks; of conservation concern; 100–200 m; Calif.

Deinandra bacigalupii occurs in the eastern San Francisco Bay area (Livermore Valley). Prior to recognition of *D. bacigalupii*, plants that constitute the species were regarded as northern outliers of *D. increscens* subsp. *increscens*.

16. Deinandra corymbosa   (de Candolle) B. G. Baldwin, Novon 9: 468. 1999   [E]

Hartmannia corymbosa de Candolle in A. P. de Candolle and A. L. L. P. de Candolle, Prodr. 5: 694. 1836; *Deinandra corymbosa* subsp. *macrocephala* (Nuttall) B. G. Baldwin; *Hemizonia corymbosa* (de Candolle) Torrey & A. Gray; *H. corymbosa* subsp. *macrocephala* (Nuttall) D. D. Keck

Annuals, 6–100 cm. **Stems** ± solid. **Leaves:** proximal blades pinnatifid, faces ± hirsute to villous and sometimes stipitate-glandular. **Heads** in corymbiform, racemiform, or paniculiform arrays or in glomerules. **Bracts** subtending heads often overlapping proximal 0–½+ of each involucre. **Phyllaries** evenly stipitate-glandular, including margins and apices, with non-glandular, non-pustule-based hairs as well. **Paleae** in 1 series. **Ray florets** 15–35; laminae deep yellow, 4–8 mm. **Disc florets** 24–70, all functionally staminate; anthers reddish to dark purple. **Pappi** 0, or coroniform (irregular crowns of entire, erose, or laciniate scales 0.1–0.9) mm. $2n = 20$.

Flowering Mar–Nov. Coastal grasslands, openings in coastal scrub or woods, dunes, disturbed sites (e.g., fallow fields), sandy or clayey soils; 0–600 m; Calif.

Deinandra corymbosa occurs on the Central Coast and the Northern Coast and in the Outer South Coast Ranges and San Francisco Bay area. Plants with relatively large heads in glomerules from the Central Coast south of Big Sur have been treated as subsp. *macrocephala*; no subspecies are recognized here because of wide variation in head size and arrangement along the Central Coast north and south of Big Sur.

340. HOLOCARPHA Greene, Fl. Francisc., 426. 1897 • [Greek *holo-*, whole, complete, and *karphos*, chaff, alluding to paleate receptacles] E

Bruce G. Baldwin

John L. Strother

Annuals, 10–120 cm. **Stems** ± erect. **Leaves** mostly cauline (at flowering); proximal opposite (forming winter–spring rosettes, usually withering before flowering), mostly alternate; sessile; blades linear to oblanceolate, margins serrate to serrulate or entire, faces hirsute to strigose, sericeous, or villous (distal leaves sometimes stipitate-glandular and/or gland-dotted as well, apices of distal leaves usually each with pit-gland). **Heads** radiate, borne singly or in ± corymbiform, paniculiform, racemiform, or spiciform arrays or in glomerules. **Peduncular bracts:** pit-glands 1 (terminal). **Involucres** ± obconic or campanulate to ± globose, 4–8+ mm diam. **Phyllaries** 3–16 in 1 series (elliptic, oblanceolate, or obovate, herbaceous, each usually $^1/_2$ enveloping a subtended ray floret proximally, abaxially with pit-gland-tipped processes and glabrous or hispid, hispidulous, puberulent, and/or sessile- or stipitate-glandular). **Receptacles** flat to convex, glabrous, paleate (paleae subtending all or most disc florets). **Ray florets** 3–16, pistillate, fertile; corollas yellow (without purplish nerves, lobes $^1/_8$–$^1/_2$ lengths of laminae, ± parallel). **Disc florets** 9–90, some bisexual and fertile, the rest functionally staminate; corollas yellow, tubes shorter than or about equaling funnelform throats, lobes 5, deltate (anthers yellow to brownish or reddish to dark purple; styles glabrous proximal to branches). **Ray cypselae** ± obcompressed (gibbous, basal attachments lateral, apices beaked, beaks adaxial, ascending, faces glabrous); **pappi** 0. **Disc cypselae** ± clavate (basal attachments central, apices beakless, faces glabrous); **pappi** 0. $x = 6$.

Species 4 (4 in the flora): California.

Molecular phylogenetic data have indicated that *Holocarpha* is most closely related to *Deinandra* (S. Carlquist et al. 2003). The low chromosome numbers in *Holocarpha* are probably the result of extensive, descending dysploidy. All are self-incompatible.

SELECTED REFERENCE Palmer, R. E. 1982. Ecological and Evolutionary Patterns in *Holocarpha* (Compositae, Madiinae). Ph.D. dissertation. University of California, Davis.

1. Anthers yellow to brownish.
  2. Stems resinous distally (not notably stipitate-glandular); involucres ± obconic to ± globose; phyllaries each bearing (0–)5–15(–20) gland-tipped processes and usually glabrous or minutely sessile- or stipitate-glandular, rarely hispid . . . . 1. *Holocarpha obconica*
  2. Stems notably stipitate-glandular; involucres campanulate to ± globose; phyllaries each bearing 25–50 gland-tipped processes and minutely sessile- or stipitate-glandular and puberulent or hispidulous . . . . 2. *Holocarpha heermannii*
1. Anthers reddish to dark purple.
  3. Heads borne singly or in glomerules or spiciform-glomerulate arrays; ray florets 8–16; disc florets 40–90 . . . . 3. *Holocarpha macradenia*
  3. Heads in ± racemiform to spiciform arrays; ray florets 3–7; disc florets 9–25 . . . . 4. *Holocarpha virgata*

1. **Holocarpha obconica** (J. C. Clausen & D. D. Keck) D. D. Keck, Aliso 4: 111. 1958 [E]

Hemizonia obconica J. C. Clausen & D. D. Keck, Madroño 3: 7. 1935; *H. vernalis* D. D. Keck; *Holocarpha obconica* subsp. *autumnalis* D. D. Keck

Plants 10–80(–120) cm; stems ± resinous distally (not notably stipitate-glandular). **Heads** in corymbiform or paniculiform arrays. **Involucres** ± obconic to ± globose. **Phyllaries** each bearing (0–)5–15(–20) gland-tipped processes and usually glabrous or minutely sessile- or stipitate-glandular (rarely ± hispid). **Ray florets** 4–9. **Disc florets** 11–21; anthers yellow to brownish. $2n = 12$.

Flowering Apr–Nov. Grasslands, savannas; 10–500 m; Calif.

Holocarpha obconica ranges from the eastern San Francisco Bay area and northwest San Joaquin Valley to the Inner South Coast Ranges and is disjunct in western foothills of the southern Sierra Nevada. Often, *H. obconica* occurs with or near *H. heermannii* or *H. virgata*. Crossing studies have shown that different populations of *H. obconica* retain high interfertility; crosses to other species of *Holocarpha* have failed or yielded sterile hybrids (J. Clausen 1951).

2. **Holocarpha heermannii** (Greene) D. D. Keck, Aliso 4: 112. 1958 [E]

Hemizonia heermannii Greene, Bull. Torrey Bot. Club 9: 15. 1882

Plants 10–120 cm; stems notably stipitate-glandular. **Heads** borne singly (at ends of branches) or in paniculiform or racemiform arrays. **Involucres** campanulate to ± globose. **Phyllaries** each bearing 25–50 gland-tipped processes and minutely sessile- or stipitate-glandular and puberulent or hispidulous. **Ray florets** 3–13. **Disc florets** 9–22; anthers yellow to brownish. $2n = 12$.

Flowering May–Nov. Grasslands, savannas; 10–1400 m; Calif.

Holocarpha heermannii occurs mostly in the eastern San Francisco Bay area and in the southern Sierra Nevada foothills and Tehachapi Range; there are fewer records from the San Joaquin Valley, central Sierra Nevada foothills, South Coast Ranges, and western Transverse Ranges. Sometimes, *H. heermannii* occurs with or near *H. obconica* or *H. virgata*. Crossing studies have indicated lack of crossability or intersterility of morphologically similar populations of *H. heermannii*; artificial hybrids of *H. heermannii* with some populations of *H. virgata* were marginally fertile (J. Clausen 1951).

3. **Holocarpha macradenia** (de Candolle) Greene, Fl. Francisc. 4: 426. 1897 [C] [E]

Hemizonia macradenia de Candolle in A. P. de Candolle and A. L. P. P. de Candolle, Prodr. 5: 693. 1836

Plants 10–50 cm; stems notably stipitate-glandular. **Heads** borne singly or in glomerules or spiciform-glomerulate arrays. **Involucres** ± globose. **Phyllaries** each bearing ± 25 gland-tipped processes and minutely sessile- or stipitate-glandular. **Ray florets** 8–16. **Disc florets** 40–90; anthers reddish to dark purple. $2n = 8$.

Flowering Jun–Nov. Grassy areas, clay soils; of conservation concern; 10–200 m; Calif.

Holocarpha macradenia occurs on the coast from near Santa Cruz to central Monterey Bay area (most populations in the San Francisco Bay area are extirpated). Populations are reportedly highly interfertile; crosses to *H. virgata* have sometimes yielded moderately to highly fertile hybrids (the two species do not co-occur). J. Clausen (1951) and R. E. Palmer (1982) suggested that *H. macradenia* is most closely related to *H. virgata*; that hypothesis has been confirmed by molecular data (B. G. Baldwin, unpubl.).

4. **Holocarpha virgata** (A. Gray) D. D. Keck, Aliso 4: 111. 1958 [E]

Hemizonia virgata A. Gray in W. H. Emory, Rep. U.S. Mex. Bound. 2(1): 100. 1859

Plants 20–120 cm.; stems usually notably stipitate-glandular, rarely merely ± resinous distally. **Heads** borne singly (at ends of branches) or in ± racemiform to spiciform arrays. **Involucres** ± obconic to campanulate. **Phyllaries** each bearing 5–20 gland-tipped processes and glabrous or minutely sessile- or stipitate-glandular. **Ray florets** 3–7. **Disc florets** 9–25; anthers reddish to dark purple. $2n = 8$.

Subspecies 2 (2 in the flora): California.

Holocarpha virgata occurs widely in the Great Valley and bordering foothills of the Coast Ranges and Sierra Nevada in northern and central California (subsp. *virgata*); it is disjunct in southwestern California (e.g., subsp. *elongata* near San Diego). Different populations are often intersterile and distinguished by chromosomal rearrangements.

1. Stem branches ± gracile, ± pliable; peduncles 0–5 (–15) cm . . . . . . 4a. *Holocarpha virgata* subsp. *elongata*
1. Stem branches ± virgate, ± rigid; peduncles 0–5+ cm . . . . 4b. *Holocarpha virgata* subsp. *virgata*

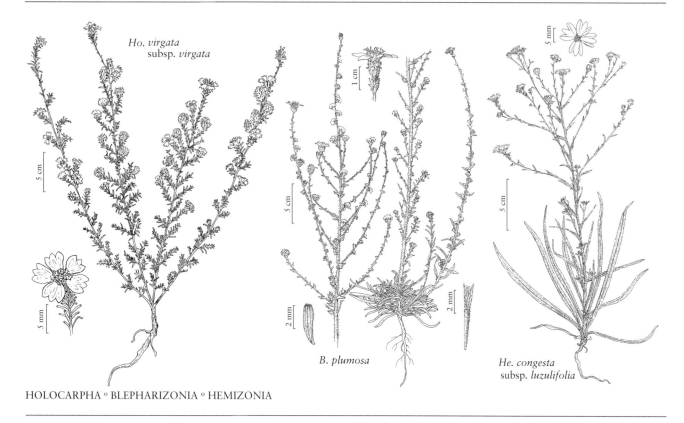

Ho. *virgata*
subsp. *virgata*

B. *plumosa*

He. *congesta*
subsp. *luzulifolia*

HOLOCARPHA ° BLEPHARIZONIA ° HEMIZONIA

4a. Holocarpha virgata (A. Gray) D. D. Keck subsp. **elongata** D. D. Keck, Aliso 4: 111. 1958 [C][E]

Stem branches ± gracile, pliable. Peduncles 0–5(–15) cm. $2n = 8$.

Flowering Jul–Nov. Grasslands, savannas; of conservation concern; 0–800(–1100) m; Calif.

4b. Holocarpha virgata (A. Gray) D. D. Keck subsp. **virgata** [E][F]

Stem branches ± virgate, rigid. Peduncles 0–5+ cm. $2n = 8$.

Flowering May–Nov. Grasslands, savannas; 0–900 m; Calif.

341. BLEPHARIZONIA (A. Gray) Greene, Bull. Calif. Acad. Sci. 1: 279. 1885 • [Greek *blepharis*, eyelash, and *zona*, girdle or ring; perhaps alluding to rings of ciliate pappus scales, or from generic names *Blepharipappus* and *Hemizonia*, alluding to resemblance] [E]

Bruce G. Baldwin

John L. Strother

Hemizonia de Candolle subg. *Blepharizonia* A. Gray, Proc. Amer. Acad. Arts 9: 192. 1874

Annuals, 10–180+ cm. **Stems** ± erect. **Leaves** basal and cauline; proximal opposite (forming winter–spring rosettes, often withering before flowering), most alternate; ± sessile; blades narrowly spatulate to linear or distal lanceolate, margins entire or serrate, faces hirsute to strigose, sericeous, pilose, or villous (distal leaves ± stipitate-glandular and/or bearing tack-glands as well).

Heads radiate, borne singly or in ± racemiform-paniculiform, paniculiform, or spiciform arrays. **Peduncular bracts:** tack-glands usually 1+ at tips and on margins. **Involucres** campanulate or obconic to ± globose, 4–9+ mm diam. **Phyllaries** 5–13 in 1 series, lanceolate or oblanceolate (± navicular), herbaceous, each ± 1/2 investing subtended ray floret proximally, abaxially canescent to hirsute or hispid to glabrate, usually bearing tack-glands as well. **Receptacles** flat, hirtellous, paleate (paleae falling, distinct, in 1 series, between rays and discs). **Ray florets** 5–13, pistillate, fertile; corollas whitish (often nerved with red to purple abaxially, lobes ± parallel, 1/4–1/2 lengths of laminae). **Disc florets** 5–35(–60+), bisexual, fertile; corollas whitish to purplish, tubes shorter than the funnelform throats, lobes 5, lanceolate (anthers ± dark purple; styles glabrous proximal to branches). **Cypselae** ± terete or ± obcompressed, basal attachments central, faces (10-ribbed) ± sericeous, apices not beaked; **pappi:** rays 0, or coroniform; discs 0, or coroniform, or of 12–20+, (reddish brown) subulate, ciliate to plumose scales. x = 14.

Species 2 (2 in the flora): California.

Blepharizonia is evidently the sister group of *Hemizonia*, based on molecular phylogenetic data (S. Carlquist et al. 2003). Members of both genera have $2n$ = 28. Summer–fall-flowering populations of *Hemizonia* with white corollas may be confused with *Blepharizonia* unless the foliar glands and heads are closely examined. Distributions of *Blepharizonia* and *Hemizonia* abut in the eastern San Francisco Bay area, northwestern San Joaquin Valley, and South Coast Ranges; hybridization between the two genera has not been reported.

SELECTED REFERENCES Baldwin, B. G., R. E. Preston, B. L. Wessa, and M. Wetherwax. 2001. A biosystematic and phylogenetic assessment of sympatric taxa in *Blepharizonia* (Compositae–Madiinae). Syst. Bot. 26: 184–194. Gregory, S. D., E. K. Espeland, T. M. Carlsen, and E. K. Bissell. 2001. Demography and population biology of a rare tarplant, *Blepharizonia plumosa* (Asteraceae), a California summer annual forb. Madroño 48: 271–285.

1. Herbage gray-green, sparsely, if at all, stipitate-glandular, apart from scattered tack-glands; heads usually borne in ± spiciform-paniculiform arrays (branches often arched-ascending); involucres usually ± canescent (usually also with scattered tack-glands); pappi 1.5–3 mm . . . . 1. *Blepharizonia plumosa*
1. Herbage yellowish green, usually densely stipitate-glandular distally and with scattered to abundant tack-glands; heads borne singly or in ± open, racemiform-paniculiform arrays (branches ± virgate); involucres ± glabrate or hirsute to hispid (usually also with scattered to abundant tack-glands); pappi 0 or 0.1–1 mm . . . . 2. *Blepharizonia laxa*

1. **Blepharizonia plumosa** (Kellogg) Greene, Bull. Calif. Acad. Sci. 1: 279. 1885 C E F

Calycadenia plumosa Kellogg, Proc. Calif. Acad. Sci. 5: 49. 1873; *Hemizonia plumosa* (Kellogg) A. Gray

Herbage gray-green, sparsely, if at all, stipitate-glandular, usually with scattered tack-glands as well. **Heads** usually in spiciform-paniculiform arrays (branches often arched-ascending). **Involucres** usually ± canescent (sometimes also with scattered, longer, glandless hairs), usually with scattered tack-glands as well. **Disc pappi** 1.5–3 mm. $2n$ = 28.

Flowering Jul–Nov. Dry slopes, in grasslands; of conservation concern; 0–500 m; Calif.

Blepharizonia plumosa is found in the eastern San Francisco Bay area and northwestern San Joaquin Valley. Northern and southern populations appear to be evolutionarily divergent and may warrant treatment as distinct taxa (B. G. Baldwin et al. 2001).

2. **Blepharizonia laxa** Greene, Bull. Calif. Acad. Sci. 1: 279. 1885 E

Hemizonia plumosa (Kellogg) A. Gray var. *subplumosa* A. Gray in A. Gray et al., Syn. Fl. N. Amer. 1(2): 312. 1884; *Blepharizonia plumosa* (Kellogg) Greene subsp. *viscida* D. D. Keck; *B. plumosa* (Kellogg) Greene var. *subplumosa* (A. Gray) Jepson

Herbage yellowish green, usually densely stipitate-glandular distally and with scattered to abundant tack-glands. **Heads** usually borne singly or in ± racemiform-paniculiform arrays (branches ± virgate). **Involucres** ± glabrate or hirsute to hispid, usually with scattered to abundant tack-glands as well. **Disc pappi** 0 or 0.1–1 mm. $2n$ = 28.

Flowering Jul–Nov. Openings in woodlands, chaparral, grasslands; 0–1500 m; Calif.

Blepharizonia laxa occurs widely in the Inner South Coast Ranges (rarely in the Outer South Coast Ranges) north to the eastern San Francisco Bay area and northwestern San Joaquin Valley. Most recent workers have treated the taxon as a subspecies or variety of *B.* *plumosa*; following B. G. Baldwin et al. (2001), we recognize *B. laxa* at species rank based on molecular and biosystematic evidence for extensive evolutionary divergence. Where *B. laxa* and *B. plumosa* co-occur, the two species remain distinct and appear to be ecologically divergent, in addition to being only minimally interfertile (Baldwin et al.).

342. HEMIZONIA de Candolle in A. P. de Candolle and A. L. P. P. de Candolle, Prodr. 5: 692. 1836 • [Greek *hemi-*, half, and *zona*, belt or girdle, alluding to cypselae half enfolded by phyllaries] E

Bruce G. Baldwin

John L. Strother

Annuals, 5–80 cm. **Stems** ± erect. **Leaves** mostly cauline; proximal opposite (forming winter-spring rosettes, sometimes present at flowering), mostly alternate; sessile; blades narrowly elliptic to linear or lance-linear, margins serrulate or entire, faces puberulent, hispidulous, hirsute, strigose, sericeous, or villous (distal leaves sometimes stipitate-glandular as well). **Heads** radiate, borne singly or in ± paniculiform, racemiform, or spiciform arrays or in glomerules. **Peduncular bracts:** pit-glands, tack-glands, and/or spines 0. **Involucres** hemispheric to ± urceolate or globose, 3–8+ mm diam. (usually subtended by calyculi of 5–7 bractlets in *H. congesta* subsp. *calyculata*). **Phyllaries** 5–14 in 1 series (linear to lanceolate or oblanceolate, herbaceous, each usually ¹/₂ enveloping a subtended ray floret proximally, abaxially pubescent to hirsute or villous, and stipitate-glandular). **Receptacles** flat to conic, glabrous, paleate (paleae connate, forming cells around all or most individual disc florets, scarious, ± deliquescent). **Ray florets** 5–14, pistillate, fertile; corollas white or yellow (often purple-veined abaxially). **Disc florets** 5–60+, functionally staminate; corollas white or yellow, tubes shorter than funnelform throats, lobes 5, deltate (anthers ± dark purple; styles glabrous proximal to branches). **Ray cypselae** ± obcompressed, abaxially gibbous, basal attachments oblique, apices sometimes beaked, beaks inconspicuous, straight, diameters greater than lengths, faces glabrous); **pappi** 0. **Disc cypselae** 0; **pappi** 0. $x = 14$.

Species 1: w North America.

Hemizonia as often circumscribed (to include *Centromadia* and *Deinandra*) is polyphyletic; molecular phylogenetic studies have indicated that *Centromadia* and *Deinandra* are more closely related to *Calycadenia*, *Holocarpha*, and *Osmadenia* than to *Hemizonia* in the strict sense, which is most closely related to *Blepharizonia*, also with $2n = 28$ (S. Carlquist et al. 2003). Crosses between taxa of *Hemizonia* have yielded hybrids of minimal to full fertility; natural hybrids also have been documented (J. Clausen 1951). All taxa in *Hemizonia* are self-incompatible. The name *H. congesta* subsp. *congesta* has been sometimes misapplied to plants treated here as *H. congesta* subsp. *lutescens*; the type of *H. congesta* corresponds to plants that have been called *H. leucocephala*, a synonym of *H. congesta* subsp. *congesta* (B. G. Baldwin et al. 2001).

SELECTED REFERENCE Babcock, E. B. and H. M. Hall. 1924. *Hemizonia congesta*: A genetic, ecologic, and taxonomic study of the hay-field tarweeds. Univ. Calif. Publ. Bot. 13: 15–100.

1. Hemizonia congesta de Candolle in A. P. de Candolle and A. L. P. P. de Candolle, Prodr. 5: 692. 1836 [E]

Leaf blades 50–175+ × 2–8(–12+) mm. **Involucres** hemispheric to ± urceolate or globose, 3.5–12 × 3–8+ mm. **Ray laminae** 5–12 mm. **Disc corollas** 2.5–3.5 mm. **Cypselae** black, 2–3.5 mm.

Subspecies 6 (6 in the flora): w United States.

1. Heads in spiciform-racemiform to spiciform-paniculiform arrays (heads ± sessile); ray laminae white, abaxially purple-veined.
  2. Peduncular bracts not or barely surpassing phyllaries; calyculi 0; phyllaries 3–7 mm, apices usually shorter than bodies; cypsela widths 0.6–0.75 times lengths . . . .
    . . . . 1a. *Hemizonia congesta* subsp. *clevelandii*
  2. Peduncular bracts (and calyculi) often notably surpassing phyllaries; calyculi usually of 5–7, sometimes 0, bractlets; phyllaries 6–12 mm, apices usually longer than bodies; cypsela widths 0.5–0.6 times lengths . . . .
    . . . . 1b. *Hemizonia congesta* subsp. *calyculata*
1. Heads in racemiform or paniculiform arrays or in glomerules; ray laminae white or yellow, abaxially purple-veined or not.
  3. Leaves usually puberulent and eglandular throughout, rarely the distal villous and glandular; heads in paniculiform arrays; ray laminae white, not purple-veined . . . .
    . . . . 1c. *Hemizonia congesta* subsp. *tracyi*
  3. Leaves pubescent, sericeous, or villous, all or the distal glandular; heads in paniculiform arrays or in glomerules; ray laminae white or yellow, abaxially purple-veined.
    4. Leaves ± villous (the distal with hairs at margins often longer); some or all heads usually in glomerules, sometimes in corymbiform or paniculiform arrays; phyllaries 6–10 mm, apices usually longer than bodies; ray laminae white . . . .
      . . . . . . 1d. *Hemizonia congesta* subsp. *congesta*
    4. Leaves pubescent, villous, or sericeous (the distal with hairs at margins not notably longer); heads in paniculiform arrays; phyllaries 3–7(–8) mm, apices usually shorter than bodies; ray laminae white or yellow.
      5. Ray laminae white . . . .
        . . . . 1e. *Hemizonia congesta* subsp. *luzulifolia*
      5. Ray laminae yellow . . . .
        . . . . 1f. *Hemizonia congesta* subsp. *lutescens*

1a. Hemizonia congesta de Candolle subsp. **clevelandii** (Greene) Babcock & H. M. Hall, Univ. Calif. Publ. Bot. 13: 48. 1924 [E]

Hemizonia clevelandii Greene, Bull. Torrey Bot. Club 9: 109. 1882

Leaves hirsute to villous and often hispidulous or strigose, all or distal usually stipitate-glandular. **Heads** in spiciform-racemiform to spiciform-paniculiform arrays. **Peduncles** 0 or 1–4 mm, bracts not or barely surpassing phyllaries. **Calyculi** 0. **Phyllaries** 3–7 mm, apices usually shorter than bodies. **Ray florets** 5–8; laminae white, abaxially purple-veined. **Cypsela widths** 0.6–0.75 times lengths. $2n = 28$.

Flowering Jun–Nov. Dry sites, grasslands, openings in chaparral and woodlands; 40–1600 m; Calif., Oreg.

Subspecies *clevelandii* occurs in interior habitats from southern North Coast Ranges in California to northern Klamath Ranges in southwestern Oregon.

1b. Hemizonia congesta de Candolle subsp. **calyculata** Babcock & H. M. Hall, Univ. Calif. Publ. Bot. 13: 42. 1924 [C][E]

Hemizonia calyculata (Babcock & H. M. Hall) D. D. Keck; *Hemizonia congesta* var. *calyculata* (Babcock & H. M. Hall) Jepson

Leaves hispidulous or strigose and often hirsute, all or distal densely stipitate-glandular. **Heads** in spiciform-racemiform to spiciform-paniculiform arrays. **Peduncles** 0 or 1–3 mm, bracts (and calyculi) surpassing phyllaries. **Calyculi** usually of 5–7, sometimes 0, bractlets. **Phyllaries** 6–12 mm, apices usually longer than bodies. **Ray florets** 5–8; laminae white, abaxially purple-veined. **Cypsela widths** 0.5–0.6 times lengths. $2n = 28$.

Flowering Jul–Nov. Clay soils, grasslands, openings in woodlands; of conservation concern; 200–1400 m; Calif.

Subspecies *calyculata* occurs in the North Inner Coast Ranges and northern Sacramento Valley.

1c. Hemizonia congesta de Candolle subsp. **tracyi**
Babcock & H. M. Hall, Univ. Calif. Publ. Bot. 13: 46.
1924 C E

Hemizonia congesta var. *tracyi* (Babcock & H. M. Hall) Jepson; *H. tracyi* (Babcock & H. M. Hall) D. D. Keck

Leaves usually puberulent, hispidulous, or strigillose and often eglandular throughout, distal rarely villous and glandular. **Heads** in paniculiform arrays. **Peduncles** 0 or 1–18 mm, bracts not surpassing phyllaries. **Calyculi** 0. **Phyllaries** 3–10.5 mm, apices usually slightly shorter than bodies. **Ray florets** 5–8(–13); laminae white, not purple-veined. **Cypsela widths** 0.5–0.63 times lengths. $2n = 28$.

Flowering May–Nov. Grassy slopes, valleys, stream banks, openings in woodlands, forests, or chaparral; of conservation concern; 0–1200 m; Calif.

Subspecies *tracyi* occurs in the north and central North Coast Ranges and on the outer North Coast.

1d. Hemizonia congesta de Candolle subsp. **congesta**
C E

Hemizonia congesta subsp. *leucocephala* (Tanowitz) D. J. Keil; *H. leucocephala* Tanowitz

Leaves villous (the distal often with hairs longer at margins), all or distal glandular, often sparsely (rarely eglandular). **Heads** (some or all) usually in glomerules, sometimes in corymbiform or paniculiform arrays. **Peduncles** 0 or 1–4 mm, bracts not surpassing phyllaries. **Calyculi** 0. **Phyllaries** 6–10 mm, apices usually longer than bodies. **Ray florets** 5–13; laminae white, abaxially purple-veined. **Cypsela widths** 0.5–0.6 times lengths. $2n = 28$.

Flowering May–Nov. Grasslands, meadows, marsh edges; of conservation concern; 0–100 m; Calif.

Subspecies *congesta* occurs in the San Francisco Bay area and North Coast Ranges. Natural hybrids between subsp. *congesta* and subsp. *lutescens* have been reported (E. B. Babcock and H. M. Hall 1924).

1e. Hemizonia congesta de Candolle subsp. **luzulifolia**
(de Candolle) Babcock & H. M. Hall, Univ. Calif. Publ. Bot. 13: 43. 1924 • Hayfield tarweed E F

Hemizonia luzulifolia de Candolle in A. P. de Candolle and A. L. P. P. de Candolle, Prodr. 5: 692. 1836 (as luzulaefolia); *H. congesta* var. *luzulifolia* (de Candolle) Jepson; *H. luzulifolia* subsp. *rudis* (Bentham) D. D. Keck

Leaves pubescent, villous, or sericeous (distal with hairs not notably longer at margins), all or distal glandular. **Heads** in paniculiform arrays. **Peduncles** 0 or 1–30 mm, bracts not surpassing phyllaries. **Calyculi** 0. **Phyllaries** 3.5–6.5 mm, apices usually shorter than bodies. **Ray florets** 5–11; laminae white, abaxially purple-veined. **Cypsela widths** 0.5–0.6 times lengths. $2n = 28$.

Flowering Mar–Dec. Disturbed sites, grassy slopes, valley bottoms, openings in chaparral and woodlands, often clayey soils, serpentine; 0–1000 m; Calif.

Subspecies *luzulifolia* occurs from the North Inner Coast Ranges and northern Great Valley through the San Francisco Bay area to the southern South Coast Ranges. Natural hybridization between subsp. *luzulifolia* and subsp. *lutescens* has been reported (E. B. Babcock and H. M. Hall 1924).

1f. Hemizonia congesta de Candolle subsp. **lutescens**
(Greene) Babcock & H. M. Hall, Univ. Calif. Publ. Bot. 13: 38. 1924 E

Hemizonia luzulifolia de Candolle var. *lutescens* Greene, Bull. Torrey Bot. Club 9: 16. 1882; *H. citrina* Greene; *H. congesta* var. *lutescens* (Greene) Jepson; *H. congesta* subsp. *vernalis* (D. D. Keck) Tanowitz; *H. lutescens* (Greene) D. D. Keck; *H. luzulifolia* de Candolle var. *citrina* (Greene) Jepson; *H. multicaulis* Hooker & Arnott; *H. multicaulis* subsp. *vernalis* D. D. Keck

Leaves pubescent, villous, or sericeous (distal with hairs not notably longer at margins), all or distal glandular. **Heads** in paniculiform arrays. **Peduncles** 0 or 1–20 mm, bracts not surpassing phyllaries. **Calyculi** 0. **Phyllaries** 4.5–6.5(–8) mm, apices usually shorter than bodies. **Ray florets** 5–14; laminae yellow, abaxially purple-veined. **Cypsela widths** 0.4–0.63 times lengths. $2n = 28$.

Flowering Apr–Dec. Grassy slopes, flats, barrens, openings in woodlands or chaparral, often on serpentine; 0–500 m; Calif.

Subspecies *lutescens* occurs in the southern, outer North Coast Ranges and San Francisco Bay area and on the immediate coast, mostly north of the Golden Gate.

343. **HOLOZONIA**   Greene, Bull Torrey Bot. Club 9: 122. 1882 • [Greek *holos*, whole or entire, and *zona*, belt or girdle; alluding to each phyllary fully (or mostly) investing a ray ovary (cypsela), in contrast to the half-invested cypselae of *Hemizonia*]   E

Bruce G. Baldwin

John L. Strother

Perennials, 30–150 cm (rhizomatous). **Stems** (aerial) ± erect. **Leaves** mostly cauline; proximal opposite (basally connate), distal alternate; sessile; blades lanceolate to linear, margins entire, faces hirsute and (distal leaves) glandular-hirtellous (glands cup-shaped). **Heads** radiate, borne singly or in loose, corymbiform arrays (peduncles filiform). **Peduncular bracts:** pit-glands, tack-glands, and/or spines 0. **Involucres** ± obconic or turbinate, 2–4+ mm diam. **Receptacles** flat to convex, glabrous or setulose, paleate (paleae falling, in 1 series between rays and discs, connate). **Phyllaries** 4–10 in 1 series (each mostly or wholly enveloping a ray ovary, ± lance-linear, herbaceous, abaxially hirsute, sometimes glandular-hirtellous, glands cup-shaped). **Ray florets** 4–10, pistillate, fertile; corollas whitish (abaxially purplish-veined). **Disc florets** 9–28, functionally staminate; corollas white (pubescent), tubes shorter than funnelform throats, lobes 5, deltate (anthers ± dark purple; styles glabrous proximal to branches). **Cypselae** (black) obcompressed, ± clavate (basal attachments centered, apices beakless, areolae broadly cupulate, faces glabrous); **pappi** (rays) 0 or coroniform (0.1–0.3 mm), or (discs) 0 or (readily falling) of 1–5 subulate scales. *x* = 14.

Species 1: California.

Holozonia has been treated as congeneric with *Hemizonia* and with *Lagophylla*. Like *Lagophylla*, *Holozonia* has functionally staminate disc florets, cup-shaped glands, and obcompressed cypselae, each completely or mostly invested by a phyllary. Additional morphologic considerations and biosystematic studies led W. C. Thompson (1983) to reject the hypothesis of a close relationship between *Holozonia* and *Lagophylla*. Molecular phylogenetic data are in keeping with a closer relationship of *Holozonia* to other continental tarweeds with white corollas and *x* = 14 (i.e., *Blepharizonia* and *Hemizonia*), than to *Lagophylla* (S. Carlquist et al. 2003).

SELECTED REFERENCE   Thompson, W. C. 1983. A Biosystematic Study of *Lagophylla* (Compositae: Heliantheae) and Related Taxa. Ph.D. dissertation. University of California, Davis.

1. **Holozonia filipes** Greene, Bull Torrey Bot. Club 9: 122. 1882   E F

Leaf blades (proximal) 3–10 cm × 2–8 mm. **Phyllaries** 3–5 mm. **Disc corollas** 3–4.5 mm. **Cypselae** 2.5–3.5 mm. *2n* = 28.

Flowering Jun–Oct. Banks and dry beds of streams and pools, often in rocky sites or alkaline clays; 30–600 m; Calif.

Holozonia filipes is the only perennial, continental tarweed with white corollas or filiform peduncles and is unique among the perennials for occurring in low-elevation, summer-hot, interior habitats. Like most of the perennial, continental tarweeds, *H. filipes* is self-incompatible and has a geographic distribution marked by major disjunctions between some populations.

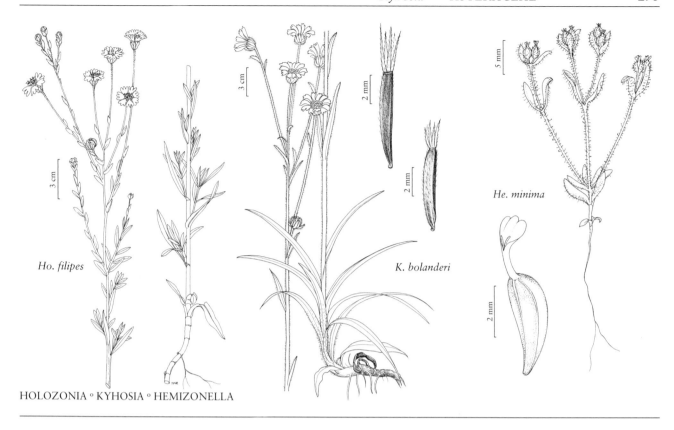

He. minima

Ho. filipes

K. bolanderi

HOLOZONIA ° KYHOSIA ° HEMIZONELLA

344. KYHOSIA   B. G. Baldwin, Novon 9: 465. 1999 • [For Donald William Kyhos, b. 1929, Californian botanist]   E

Bruce G. Baldwin

John L. Strother

Perennials, 50–120 cm (rhizomatous; self-incompatible). **Stems** (aerial) erect. **Leaves** basal and cauline (at flowering); proximal opposite (basalmost in rosettes), distal alternate; sessile; blades lance-linear to linear, margins entire, faces hirsute and (distal leaves) glandular-hirtellous. **Heads** radiate, borne singly or in loose, ± corymbiform arrays. **Peduncular bracts:** pit-glands, tack-glands, and/or spines 0. **Involucres** ± campanulate to hemispheric, 6–12+ mm diam. **Phyllaries** 8–12 in 1 series (each mostly or wholly enveloping a ray ovary, lanceolate to lance-linear, herbaceous, abaxially hirsute and glandular-hirtellous). **Receptacles** flat to convex, glabrous, paleate (paleae falling, in 1 series between rays and discs, weakly connate or distinct). **Ray florets** 8–12, pistillate, fertile; corollas bright yellow. **Disc florets** 28–65, bisexual and fertile; corollas bright yellow, pubescent, tubes shorter than funnelform throats, lobes 5, deltate (anthers ± dark purple; styles glabrous proximal to branches). **Ray cypselae** (black) compressed, clavate (arcuate, basal attachments centered, apices beakless, faces glabrous or hispidulous); **pappi** 0, or coroniform. **Disc cypselae** (brown to black) ± terete (straight or arcuate, faces hispidulous, otherwise similar to rays); **pappi** of 5–10 (stramineous to purplish) lanceolate to subulate, ciliate to plumose scales. $x = 6$.

Species 1: w United States.

Kyhosia has been treated as congeneric with *Madia,* which is evidently more closely related to the Hawaiian silversword alliance (*Argyroxiphium, Dubautia,* and *Wilkesia*) than to *Kyhosia*

(B. G. Baldwin 1996). Head architecture is similar in *Kyhosia* and *Argyroxiphium*, as noted by S. Carlquist (1959). *Kyhosia* is the only perennial tarweed with $2n = 12$.

1. Kyhosia bolanderi (A. Gray) B. G. Baldwin, Novon 9: 466. 1999 [E] [F]

Anisocarpus bolanderi A. Gray, Proc. Amer. Acad. Arts 7: 360. 1868; *Madia bolanderi* (A. Gray) A. Gray

Leaf blades (proximal) 5–35 cm × 4–15 mm. **Phyllaries** 7–14 mm. **Disc corollas** 5–8 mm. **Ray cypselae** 5–7 mm; **pappi** 0 or to 0.7 mm. **Disc cypselae** 5–9 mm; **pappi** 1–5 mm. $2n = 12$.

Flowering Jul–Sep. Meadows, stream banks; 1000–2600 m; Calif., Nev., Oreg.

Kyhosia bolanderi is unusual among continental tarweeds for occurring in wet, montane habitats, in the North Coast Ranges, Klamath Ranges, and Sierra Nevada of the California Floristic Province. Ecologically, *K. bolanderi* is similar to another self-incompatible, robust, rhizomatous, perennial tarweed, *Raillardella pringlei*, which occurs with *K. bolanderi* in the Klamath Ranges.

345. HEMIZONELLA   (A. Gray) A. Gray, Proc. Amer. Acad. Arts 9: 189. 1874

· [Generic name *Hemizonia* and Latin *-ella*, diminutive]   [E]

Bruce G. Baldwin

John L. Strother

Hemizonia de Candolle [unranked] *Hemizonella* A. Gray, Proc. Amer. Acad. Arts 6: 548. 1865

Annuals, 1–20 cm. **Stems** ± erect (branches often divaricate, hispidulous, stipitate-glandular). **Leaves** basal and/or cauline; proximal opposite, distal alternate (often in clusters of 2–3+ immediately proximal to branches); sessile; blades linear, margins entire or toothed, faces hirsute and (distal leaves) glandular-puberulent. **Heads** radiate, borne singly or in corymbiform arrays or glomerules (peduncles filiform). **Peduncular bracts:** pit-glands, tack-glands, or spines 0. **Involucres** ± obovoid, (1–)2–4 mm diam. **Phyllaries** 3–5 in 1 series (each mostly or wholly enveloping a ray ovary, ± oblanceolate, herbaceous, abaxially hirsute and glandular-hirtellous). **Receptacles** flat to convex, glabrous or sparsely setulose, paleate (paleae falling, in 1 series between rays and discs, connate). **Ray florets** 3–5, pistillate, fertile; corollas pale yellow. **Disc florets** 1–2, bisexual, fertile; corollas pale yellow, (pubescent) tubes about equaling funnelform throats, lobes 5, deltate (anthers yellow; styles glabrous proximal to branches). **Ray cypselae** (black) obcompressed (arcuate, basal attachments centered, apices minutely beaked, beaks straight, oriented adaxially, 0.1–0.15 mm, faces sparsely hispidulous or glabrate); **pappi** 0. **Disc cypselae** (black) ± terete, clavate (± hispidulous, apices minutely beaked, beaks straight, oriented vertically, 0.1–0.15 mm); **pappi** 0. $x = 21–22$.

Species 1: w North America.

Hemizonella is treated as distinct from *Madia* based on morphologic, cytologic, and molecular evidence (S. Carlquist et al. 2003). Unlike members of *Madia*, *Hemizonella* has obcompressed ray cypselae and hairy (ray and disc) fruits. D. D. Keck (1949) suggested that a somatic count of $2n = 32$ for *H. minima* was evidence for a close relationship between *Hemizonella* and the morphologically similar *Madia exigua* ($2n = 32$). Keck's count contrasts with a subsequent meiotic count of $2n = 21–22$ II for *H. minima*. *Hemizonella* was resolved as the sister-group of

Kyhosia in a phylogenetic analysis of nuclear ribosomal DNA (B. G. Baldwin 1996). The molecular data and morphologic resemblance of *Hemizonella* to *M. exigua* are consistent with the possibility that *Hemizonella* may be an ancient allopolyploid descended from a hybrid between ancestors with *n* = 6 (like *Kyhosia*) and *n* = 16 (like *M. exigua*).

SELECTED REFERENCE Keck, D. D. 1949. *Hemizonella* becomes a *Madia*. Madroño 10: 22.

1. Hemizonella minima (A. Gray) A. Gray, Proc. Amer. Acad. Arts 9: 189. 1874 E F

Hemizonia minima A. Gray, Proc. Amer. Acad. Arts 6: 548. 1865; *Madia minima* (A. Gray) D. D. Keck

Leaf blades 5–25 × (0.5–)1–3 mm. **Involucres** (1–)2–4 × (1–)2–4 mm. **Disc corollas** 1–2.5 mm. **Ray cypselae** 1.8–2.8 mm; disc cypselae 1.8–2.8 mm. **2n** = 42–44.

Flowering Apr–Aug. Gravelly or rocky, usually open sites, scrublands, meadows, forests; 300–2900 m; B.C.; Calif., Idaho, Mont., Nev., Oreg., Wash.

Hemizonella minima is self-compatible, like most other tarweeds that are distributed widely in western North America. *Taylor 8147* (JEPS), from the Sierra Nevada (Tioga Peak, Mono County, California), is an unusually late-flowering (15 Sep 1982) and high-elevation (2900 m) collection.

346. HARMONIA B. G. Baldwin, Novon 9: 463. 1999 • [For Harvey Monroe Hall, 1874–1932, Californian botanist] E

Bruce G. Baldwin

John L. Strother

Annuals, 5–40 cm. **Stems** erect. **Leaves** mostly cauline; proximal opposite, distal alternate; sessile; blades linear, margins entire or toothed, faces usually hirsute, sometimes minutely stipitate-glandular as well (glands usually black, sometimes yellowish). **Heads** radiate, borne singly or in loose, ± umbelliform to corymbiform arrays. **Peduncular bracts:** pit-glands, tack-glands, and spines 0. **Involucres** obovoid to obconic, 2–5+ mm diam. **Phyllaries** 3–8 in 1 series (lanceolate to oblanceolate, herbaceous, strongly conduplicate, each wholly enveloping a ray ovary). **Receptacles** flat to convex, glabrous or setulose, paleate (paleae falling, in 1 series between rays and discs, distinct or weakly connate, phyllary-like, more scarious). **Ray florets** usually 3–8, pistillate, fertile; corollas bright yellow (laminae flabelliform to obovate). **Disc florets** 7–30, bisexual and fertile or functionally staminate (sometimes in same head); corollas bright yellow, tubes shorter than narrowly funnelform throats, lobes 5, deltate (anthers yellowish to brownish; styles glabrous proximal to branches). **Ray cypselae** (black) terete to ± compressed, weakly arcuate, gibbous or not, beaked or beakless, glabrous; **pappi** 0 or of 3–12 lanceolate to subulate, fimbrillate to plumose scales. **Disc cypselae** (black) ± terete, ± clavate, glabrous or hairy; **pappi** of 7–11 lance-attenuate, linear, oblong, quadrate, or subulate, fimbriate, fimbrillate, or plumose scales. ***x*** = 9.

Species 5 (5 in the flora): California.

Members of *Harmonia* occur in mountains of northwestern California, as far south as the northern San Francisco Bay area. All but *H. nutans* are known only from serpentine exposures and are probably descended from a common, serpentine-endemic ancestor (B. G. Baldwin 2001). *Harmonia* has been treated in *Madia*, which is more closely related to *Carlquistia* than to

Harmonia (B. G. Baldwin 1996).   See Baldwin (2001) for discussion of phylogeny within *Harmonia*.

SELECTED REFERENCE   Baldwin, B. G. 2001. *Harmonia guggolziorum* (Compositae–Madiinae), a new tarweed from ultramafics of southern Mendocino County, California.   Madroño 48: 293–297.

1. Heads usually reflexed in bud and fruit; ray pappi 0; disc pappi of lance-attenuate, fimbrillate scales 2–3.7 mm . . . . 1. *Harmonia nutans*
1. Heads usually erect in bud and fruit; ray pappi 0.2–1.5 mm; disc pappi of lanceolate, linear, oblong, quadrate, or subulate, fimbriate or plumose scales 0.2–3.5 mm.
   2. Leaves ± evenly distributed on stems; ray cypselae strongly gibbous (bowed abaxially), beaked (beaks 0.4–0.5 mm); disc florets functionally staminate . . . . 3. *Harmonia doris-nilesiae*
   2. Leaves unevenly distributed, mostly along primary (central) stems and immediately proximal to branches supporting heads; ray cypselae slightly or not gibbous, beakless (pappi elevated adaxially 0.1–0.2 mm in *H. stebbinsii*); disc florets (some or all) bisexual, fertile.
      3. Phyllaries densely white-villous near folded edges; ray cypselae slightly gibbous; disc pappi of linear-attenuate to subulate, plumose scales 1.2–3.5 mm . . . . . . 4. *Harmonia stebbinsii*
      3. Phyllaries hirsute and/or hirtellous near folded edges; ray cypselae not gibbous; disc pappi of lanceolate, linear, oblong, or quadrate, fimbriate scales 0.2–0.8 mm.
         4. Proximal, unbranched portions of primary (central) stems usually shorter than branches supporting heads (distal leaves of primary stems densely congested); disc pappi of oblong or quadrate scales 0.2–0.5 mm . . . . 5. *Harmonia hallii*
         4. Proximal, unbranched portions of primary (central) stems usually longer than branches supporting heads (distal leaves of primary stems not densely congested); disc pappi of lanceolate to linear scales 0.6–0.8 mm . . . . 2. *Harmonia guggolziorum*

1. Harmonia nutans (Greene) B. G. Baldwin, Novon 9: 464. 1999 C E

Callichroa nutans Greene, Pittonia 2: 227. 1892; *Layia nutans* (Greene) Jepson; *Madia nutans* (Greene) D. D. Keck

Plants 5–25 cm; proximal unbranched portions of primary stems shorter or longer than branches supporting heads. **Leaves** ± evenly distributed on stems. **Heads** usually reflexed in bud and fruit. **Phyllaries** 4–8, hirsute and/or hirtellous near folded edges. **Ray florets** 4–8; laminae 3–7 mm. **Disc florets** 7–30, bisexual, fertile. **Ray cypselae** not gibbous, beakless; **pappi** 0. **Disc cypselae** 2.3–4.5 mm; **pappi** of 9–11 lance-attenuate, fimbrillate scales 2–3.7 mm. $2n = 18$.

Flowering Apr–Jun. Rocky soils, openings in chaparral or woodlands; of conservation concern; 100–1000 m; Calif.

Harmonia nutans occurs in the southern North Coast Ranges and northern San Francisco Bay area, mostly on volcanic substrates. Based on molecular phylogenetic data, *H. nutans* evidently represents the sister-group to the serpentine-endemic species of *Harmonia* (B. G. Baldwin 2001).

2. Harmonia guggolziorum B. G. Baldwin, Madroño 48: 293, figs. 1, 2. 2002 C E

Plants 10–30 cm; proximal unbranched portions of primary stems usually longer than branches supporting heads. **Leaves** mostly on primary stems (distal leaves of primary stems not congested) and immediately proximal to branches supporting heads. **Heads** usually erect in bud and fruit. **Phyllaries** 3–6, hirsute and/or hirtellous near folded edges. **Ray florets** 3–6; corolla laminae 4–5 mm. **Disc florets** 8–13, some or all bisexual, fertile. **Ray cypselae** not gibbous, beakless; **pappi** to 0.5 mm. **Disc cypselae** 3–3.5 mm; **pappi** of 9–11 lanceolate to linear, fimbriate scales 0.6–0.8 mm. $2n = 18$.

Flowering Apr–May. Serpentine slopes; of conservation concern; 100–200 m; Calif.

Harmonia guggolziorum occurs in the southern Inner North Coast Ranges. Molecular phylogenetic analyses have indicated a sister-group relationship between *H. guggolziorum* and the other serpentine-endemic species of *Harmonia* (*H. doris-nilesiae*, *H. hallii*, and *H. stebbinsii*; B. G. Baldwin 2001).

3. **Harmonia doris-nilesiae** (T. W. Nelson & J. P. Nelson) B. G. Baldwin, Novon 9: 464. 1999 [C][E]

Madia doris-nilesiae T. W. Nelson & J. P. Nelson, Brittonia 37: 394, fig. 1. 1985

Plants 9–40 cm; proximal unbranched portions of primary stems shorter or longer than branches supporting heads. Leaves ± evenly distributed on stems. Heads usually erect in bud and fruit. Phyllaries 4–8, densely white-villous near folded edges. Ray florets 4–8; laminae 2.5–7 mm. Disc florets 8–20, functionally staminate. Ray cypselae strongly gibbous, beaked (beaks 0.4–0.5 mm); pappi to 0.9 mm. Disc cypselae 0; pappi (of disc florets) of 7–10 lance-attenuate to subulate, plumose scales 0.2–0.9 mm. $2n = 18$.

Flowering May–Jun. Serpentine slopes; of conservation concern; 800–1600 m; Calif.

Harmonia doris-nilesiae occurs in the southern Klamath Ranges. Sometimes, *H. doris-nilesiae* and *H. stebbinsii* occur in proximity (e.g., *V. Parker 757, 759*; JEPS); hybrids have not been reported.

4. **Harmonia stebbinsii** (T. W. Nelson & J. P. Nelson) B. G. Baldwin, Novon 9: 464. 1999 [C][E]

Madia stebbinsii T. W. Nelson & J. P. Nelson, Brittonia 32: 323, fig. 1. 1980

Plants 5–27 cm; proximal, unbranched portions of primary stems usually shorter than branches supporting heads. Leaves mostly on primary stems (distal leaves of primary stems congested) and immediately proximal to branches supporting heads. Heads usually erect in bud and fruit. Phyllaries 4–6, densely white-villous near folded edges. Ray florets 4–6; laminae 4–6.5 mm. Disc florets 8–20, outer bisexual and fertile, inner functionally staminate. Ray cypselae slightly gibbous, beakless (elevated 0.1–0.2 mm adaxially); pappi 0.5–1.5 mm. Disc cypselae 2–4 mm; pappi of 8–10 linear-attenuate to subulate, plumose scales 1.2–3.5 mm. $2n = 18$.

Flowering May–Jul. Serpentine slopes; of conservation concern; 1100–1600 m; Calif.

Harmonia stebbinsii occurs in the southern Klamath Ranges and North Inner Coast Ranges. Based on molecular phylogenetic data, *H. stebbinsii* appears to be most closely related to *H. hallii* (B. G. Baldwin 2001).

5. **Harmonia hallii** (D. D. Keck) B. G. Baldwin, Novon 9: 464. 1999 [C][E][F]

Madia hallii D. D. Keck, Madroño 3: 5. 1935

Plants 5–18 cm; proximal, unbranched portions of primary stems usually shorter than branches supporting heads. Leaves mostly on primary stems (distal leaves of primary stems densely congested) and immediately proximal to branches supporting heads. Heads usually erect in bud and fruit. Phyllaries 3–6, hirsute and/or hirtellous near folded edges. Ray florets 3–6; laminae 2–5 mm. Disc florets 8–20, mostly bisexual and fertile, sometimes functionally staminate. Ray cypselae not gibbous, beakless; pappi 0.2–0.5 mm. Disc cypselae 2.8–3.2 mm; pappi of 8–10 oblong or quadrate, fimbriate scales 0.2–0.5 mm. $2n = 18$.

Flowering Apr–Jun. Serpentine openings in chaparral; of conservation concern; 500–1000 m; Calif.

Harmonia hallii occurs in the southern Inner North Coast Ranges.

347. **ANISOCARPUS** Nuttall, Trans. Amer. Philos. Soc., n. s. 7: 388. 1841 • [Greek *anisos*, unequal or dissimilar, and *karpos*, fruit, alluding to contrasting ray (fertile) and disc (sterile) ovaries in type species] [E]

Bruce G. Baldwin

John L. Strother

Perennials, 10–80 cm. Stems erect, branched from bases or throughout. Leaves basal and cauline; proximal opposite (sometimes rosettes), distal alternate; ± sessile; blades oblong to linear, lance-linear, or oblanceolate, margins entire or toothed, faces hirsute to strigose or pubescent and (distal leaves) stipitate-glandular. Heads radiate or discoid, borne singly or in corymbiform or racemiform arrays. Peduncular bracts: pit-glands, tack-glands, and/or spines 0 at tips. Involucres

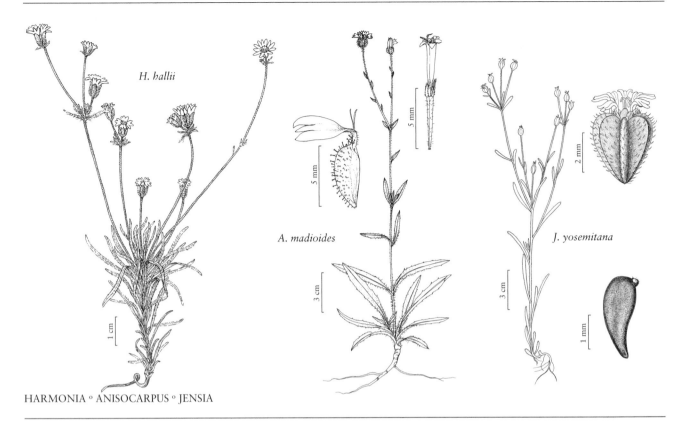

HARMONIA ○ ANISOCARPUS ○ JENSIA

± globose or broadly ellipsoid to campanulate, 4–6+ mm diam. **Phyllaries** 0 (see paleae at receptacles) or falling, 1–3 or 7–15 in 1 series (lanceolate to lance-attenuate or oblanceolate, herbaceous, each ¹/₂ or fully enveloping subtended ray floret proximally, ciliolate, abaxially stipitate-glandular, sometimes hirtellous). **Receptacles** flat to convex, glabrous or setulose, paleate (paleae falling, in 1 series, between rays and discs, usually connate, sometimes distinct, phyllary-like, more scarious; in discoid heads, functionally an "involucre"). **Ray florets** 0, 1–3, or 8–15, pistillate, fertile; corollas yellow. **Disc florets** 5–30, bisexual and fertile, or functionally staminate; corollas yellow, tubes shorter than funnelform throats, lobes 5, deltate (styles glabrous proximal to branches; anthers yellow). **Ray cypselae** (black or grayish) compressed or ± obcompressed, clavate, ± arcuate (basal attachments centered, apices beaked, beaks offset adaxially, 0.2–0.3 mm, faces glabrous or hairy); **pappi** 0 or coroniform. **Disc cypselae** (black or grayish) ± terete, clavate (± straight, faces hairy); **pappi** of 5–8 or 11–21 lanceolate, linear, quadrate, or subulate, ciliolate-plumose, erose, or fimbrillate scales. *x* = 7.

Species 2 (2 in the flora): w North America.

Following B. G. Baldwin (1999b), *Anisocarpus* comprises two species that have resided in different genera and different tribes. *Anisocarpus scabridus* was placed in *Raillardella* or *Raillardiopsis*—both regarded as members of Senecioneae until S. Carlquist's (1959) anatomic studies. *Anisocarpus madioides* was treated in *Madia* and included in D. D. Keck's (1959) informal "section Anisocarpus," along with the other pappose species of *Madia* in the sense of Keck (now treated in *Harmonia*, *Jensia*, and *Kyhosia*). Molecular phylogenetic data support a sister-group relationship between *A. madioides* and *A. scabridus* (Baldwin 1996).

Anisocarpus madioides and *A. scabridus* are the only perennial herbaceous tarweeds that combine non-scapiform capitulescences, radiate heads (at least in part), ellipsoid or spheric involucres, and yellow anthers, and are the only perennials in Madiinae with 2*n* = 14. The two species are highly distinct ecologically: *A. madioides* occurs in low- to mid-elevation, forest and woodland understories; *A. scabridus* occurs on high-elevation, exposed scree slopes and ridges.

Artificial hybrids between the two species are vigorous, easily produced (with *Anisocarpus madioides* as seed parent), and largely pollen-sterile, except for large, diploid grains. B. G. Baldwin used diploid pollen from an F₁ hybrid between *A. madioides* and *A. scabridus* in an artificial hybridization with a member of the closely-related Hawaiian silversword alliance, *Dubautia knudsenii* Hillebrand, a wet-forest tree from Kaua'i (see M. Barrier et al. 1999).

1. Leaves dark green, 40–130 mm, margins entire or toothed, apices acute; involucres ± globose, 4–6 mm; ray florets 8–15; disc florets functionally staminate; cypselae compressed; disc pappi of 5–8 linear, lanceolate, or quadrate, fimbrillate or erose scales 0.2–1.5 mm . . . . 1. *Anisocarpus madioides*
1. Leaves blue-green to grayish blue-green, 10–30 mm, margins entire, apices acute or obtuse; involucres broadly ellipsoid to campanulate, 6–12 mm; ray florets 0 or 1–3; disc florets bisexual, fertile; ray cypselae ± compressed; disc pappi of 11–21 subulate, ciliate-plumose scales 4–7 mm . . . . 2. *Anisocarpus scabridus*

1. Anisocarpus madioides Nuttall, Trans. Amer. Philos. Soc., n. s. 7: 388. 1841 E F

Madia madioides (Nuttall) Greene

Plants 15–80 cm. **Leaf blades** dark green, 40–130 × 5–15 mm, margins entire or toothed, apices acute. **Involucres** ± globose, 4–6 mm. **Ray florets** 7–15. **Disc florets** 5–30, functionally staminate. **Ray cypselae** compressed, 3–5 mm. **Disc pappi** 5–8 linear, lanceolate, or quadrate, fimbrillate or erose scales 0.2–1.5 mm. 2*n* = 14.

Flowering Apr–Sep. Forests, woodlands; 10–1300 m; B.C.; Calif., Oreg., Wash.

Like some other self-compatible tarweeds, *Anisocarpus madioides* ranges beyond the California Floristic Province and is the most widespread perennial in Madiinae. It occurs in the Pacific coast ranges from central California (Santa Lucia Range) north to Vancouver Island, British Columbia; outliers include populations in the Feather River region of the northern Sierra Nevada and in the Agua Tibia Mountains of southern California.

2. Anisocarpus scabridus (Eastwood) B. G. Baldwin, Novon 9: 463. 1999 C E

Raillardella scabrida Eastwood, Bull. Torrey Bot. Club 32: 216. 1905; *Raillardiopsis scabrida* (Eastwood) Rydberg

Plants 10–50 cm (stems hirsute and stipitate-glandular). **Leaf blades** blue-green to grayish blue-green, 10–30 × 1–4 mm, margins entire, apices acute or obtuse. **Involucres** broadly ellipsoid to campanulate, 6–12 mm. **Ray florets** 0 or 1–3. **Disc florets** 5–23, bisexual, fertile. **Ray cypselae** ± obcompressed, 6–8 mm. **Disc pappi** of 11–21 subulate, ciliate-plumose scales 4–7 mm. 2*n* = 14.

Flowering Jun–Sep. Open ridges and slopes, on metamorphics; of conservation concern; 1600–2400 m; Calif.

Anisocarpus scabridus occurs mostly in the Inner North Coast Ranges, with interior outliers in the southern Cascade Range.

348. JENSIA B. G. Baldwin, Novon 9: 464. 1999 • [For Jens Christian Clausen, 1891–1969, Californian botanist] E

Bruce G. Baldwin

John L. Strother

Annuals, 5–60 cm (self-incompatible). **Stems** erect. **Leaves** mostly cauline; proximal opposite (often crowded), distal alternate; sessile; blades spatulate to linear, margins entire or toothed, faces hirsute to strigose (distal leaves sometimes stipitate-glandular as well). **Heads** radiate, in

± umbelliform arrays. **Peduncular bracts:** pit-glands, tack-glands, and/or spines 0. **Involucres** ± obconic or urceolate to globose, 3–5 mm diam. **Phyllaries** 2–12 in 1 series (lanceolate to lance-attenuate, herbaceous, each usually wholly enveloping a ray ovary, abaxially hirsute, hair tips ± uncinate). **Receptacles** flat to convex, glabrous or setulose, paleate (paleae falling, in 1 series between rays and discs, connate, herbaceous to ± scarious). **Ray florets** 2–12, pistillate, fertile; corollas yellow, sometimes purple-veined abaxially. **Disc florets** 1–65, functionally staminate; corollas yellow, tubes shorter than funnelform throats, lobes 5, deltate (anthers ± dark purple; styles glabrous proximal to branches). **Ray cypselae** compressed, clavate, arcuate, basal attachments oblique, faces glabrous, apices beaked. **Ray pappi** crowns of scales (0.1–1 mm); **disc pappi** of 5–7 (white or purple-tipped) subulate, crisped, ciliolate scales (2.5–3 mm). $x = 8$.

Species 2 (2 in the flora): California.

Recognition of *Jensia* is based on evidence that *Madia* in the sense of D. D. Keck (1959) is not monophyletic; the epappose annuals constituting *Madia* in the restricted sense are more closely related to *Carlquistia* than to *Jensia* (B. G. Baldwin 1996).

1. Plants 6–60 cm; involucres urceolate or globose; ray florets 5–12, laminae 4–10 mm; disc florets 16–65 . . . . 1. *Jensia rammii*
1. Plants 5–25 cm; involucres broadly obconic; ray florets 2–8, laminae 0.5–3 mm; disc florets 1–7 . . . . 2. *Jensia yosemitana*

1. Jensia rammii (Greene) B. G. Baldwin, Novon 9: 465. 1999 E

Madia rammii Greene, Bull. Calif. Acad. Sci. 1: 90. 1885

Plants 6–60 cm. **Leaves:** blades 15–100 × 1–3 mm. **Involucres** urceolate or globose, (3–)4–5 mm. **Ray florets** 5–12; laminae 4–10 mm. **Disc florets** 16–65. $2n = 16$.

Flowering Apr–Jul. Grassy slopes, openings in woodlands and forests; 400–1100(–1600) m; Calif.

Jensia rammii occurs in the western foothills and ponderosa-pine belt of the northern Sierra Nevada, often in clayey soils.

2. Jensia yosemitana (Parry ex A. Gray) B. G. Baldwin, Novon 9: 465. 1999 C E F

Madia yosemitana Parry ex A. Gray, Proc. Amer. Acad. Arts 17: 219. 1882

Plants 5–15(–25) cm. **Leaves:** blades 10–50 × 1–2 mm. **Involucres** broadly obconic, 2.5–4 mm. **Ray florets** 2–8; laminae 0.5–3 mm. **Disc florets** 1–7. $2n = 16$.

Flowering May–Aug. Sandy places, meadows; of conservation concern; 1200–2300 m; Calif.

Jensia yosemitana is known from widely scattered sites on the western slopes of the Sierra Nevada, at higher elevations than most populations of *J. rammii*.

349. CARLQUISTIA B. G. Baldwin, Novon 9: 463. 1999 · [For Sherwin Carlquist, b. 1930, Californian botanist] E

Bruce G. Baldwin

John L. Strother

Perennials, 7–54 cm (rhizomatous, often matted). **Stems** (aerial) erect. **Leaves** mostly cauline; proximal opposite, distal alternate; sessile; blades lanceolate to linear, margins entire, faces hirsute to villous and glandular-pubescent. **Heads** discoid, borne singly or in loose, corymbiform arrays. **Peduncular bracts:** pit-glands, tack-glands, and/or spines 0 at tips. **Involucres** ± campanulate, 5–10+ mm diam. **Phyllaries** (modified paleae) (5–)7–16 in 1 series, lanceolate to lance-linear, herbaceous, abaxially hirsute and glandular-pubescent. **Receptacles** flat, setulose, epaleate (except for bracts constituting "involucres"). **Ray florets** 0. **Disc florets** 7–29, bisexual, fertile; corollas yellow, tubes shorter than or about equaling funnelform throats, lobes 5, deltate (anthers yellow to brownish; styles glabrous proximal to branches). **Cypselae** ± terete (apices not beaked, faces scabrellous); **pappi** of 9–17 white to mauve or tawny, subulate, ± plumose scales (flattened bristles). $x = 8$.

Species 1: California.

Carlquistia resembles another montane, perennial tarweed, *Anisocarpus scabridus* ($2n = 14$), and the two taxa were long treated as congeneric, in *Raillardella* or *Raillardiopsis*. Molecular phylogenetic evidence for a sister-group relationship between *Carlquistia* and *Madia* in the strict sense is consistent with $x = 8$ in both groups (B. G. Baldwin 1996). *Carlquistia* is treated as distinct from *Madia* on the basis of morphologic and ecologic disparity between the two groups.

Carlquistia is closely related to the Hawaiian silversword alliance (*Argyroxiphium* de Candolle, *Dubautia* Gaudichaud-Beaupré, and *Wilkesia* A. Gray); vigorous, largely sterile hybrids have been produced between *C. muirii* and *Dubautia laevigata* A. Gray, a large shrub of mesic forests in Kauai (B. G. Baldwin et al. 1991). Vigorous hybrids of low fertility also have been produced between *C. muirii* and the perennial California tarweeds *Anisocarpus scabridus* and *Kyhosia bolanderi* (D. W. Kyhos et al. 1990). Diploid pollen from a hybrid between *Carlquistia* and *Kyhosia* was successfully used in a cross with *Dubautia scabra* (de Candolle) D. D. Keck (G. D. Carr et al. 1996).

SELECTED REFERENCE Baldwin, B. G. and D. W. Kyhos. 1990. A systematic and biogeographic review of *Raillardiopsis* [*Raillardella*] *muirii*, with special reference to a disjunct California Coast Range population. Madroño 37: 43–54.

1. Carlquistia muirii (A. Gray) B. G. Baldwin, Novon 9: 463. 1999 Ⓒ Ⓔ Ⓕ

Raillardella muirii A. Gray in W. H. Brewer et al., Bot. California 1: 618. 1876; *Raillardiopsis muirii* (A. Gray) Rydberg

Leaf blades 9–42 × 2–4 mm. **Involucres** 8–13 × 5–10+ mm. **Disc corollas** 6.5–10 mm. **Cypselae** black, 4–7.5 mm; **pappi** 5–11 mm. $2n = 16$.

Flowering Jun–Oct. On granitic soils, openings in forests; of conservation concern; 1100–2500 m; Calif.

Carlquistia muirii is known only from granitic substrates in the southern Sierra Nevada and Santa Lucia Range (Ventana Double Cone). Major disjunctions among the populations of *C. muirii*, a self-incompatible tarweed, may be explained by vicariance, ancient dispersal, or both (B. G. Baldwin and D. W. Kyhos 1990).

350. MADIA Molina, Sag. Stor. Nat. Chili, 136, 354. 1782 • [From native name in Chile]

Bruce G. Baldwin

John L. Strother

Annuals, 5–250 cm. **Stems** erect. **Leaves** mostly cauline (at flowering) proximal opposite (often in rosettes), distal alternate; sessile; blades lanceolate or oblong-linear to linear, margins usually entire, sometimes toothed, faces hirsute to strigose, usually glandular-pubescent as well. **Heads** usually radiate (sometimes discoid in *M. glomerata*), in corymbiform, paniculiform, racemiform, or spiciform arrays or in glomerules. **Peduncular bracts:** pit-glands, tack-glands, and/or spines 0. **Involucres** ellipsoid, depressed-globose, globose, obconic, ovoid, or urceolate, 1–10+ mm diam.

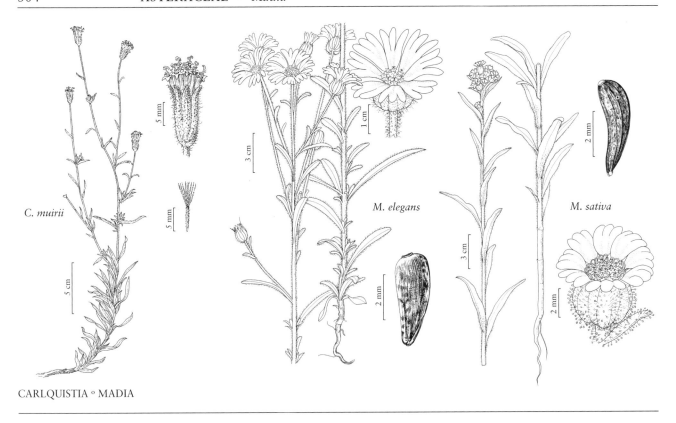

CARLQUISTIA ∘ MADIA

Phyllaries 0 (then outer paleae functioning as phyllaries, sometimes in *M. glomerata*), or 1–22 in 1 series (lance-linear to lance-attenuate or oblanceolate, herbaceous, each mostly or wholly enveloping a subtended ray ovary, abaxially hirsute and, usually, glandular). **Receptacles** flat to convex, glabrous or setulose, paleate (paleae persistent or falling readily, in 1 series between rays and discs, ± connate or distinct, phyllary-like, more scarious). **Ray florets** 0 (sometimes in *M. glomerata*), or 1–22, pistillate, fertile; corollas yellowish (with maroon bases sometimes in *M. elegans*; purplish red sometimes in *M. sativa*). **Disc florets** 1–80+, bisexual and fertile or functionally staminate; corollas usually yellow, sometimes purplish, tubes shorter than or about equaling funnelform throats, lobes 5, deltate (anthers ± dark purple or yellow to brownish; styles glabrous proximal to branches). **Ray cypselae** compressed, ± 3-angled, or rarely terete, clavate (often arcuate, basal attachments central or offset, apices sometimes beaked, faces glabrous); **pappi** 0. **Disc cypselae** similar, sometimes obovoid (often ± straight, basal attachments central, apices not beaked), sometimes 0; **pappi** 0. $x = 8$.

Species 10 (10 in the flora): North America, South America, Pacific Islands (Hawaii, probably introduced).

Madia is more narrowly circumscribed here than in previous treatments by D. D. Keck (1959) and others. Molecular phylogenetic data have indicated that *Madia* in those earlier senses is not monophyletic (B. G. Baldwin 1996). As treated here, *Madia* comprises all members of Keck's informal "section Madia" except *M. minima* (= *Hemizonella*) (Baldwin 1999b). Most species are reportedly either cross-incompatible or intersterile (J. Clausen 1951).

1. Ray laminae 6–19 mm; paleae readily falling; anthers yellow to brownish; ray cypselae beaked (beaks adaxially offset, curved) . . . . 1. *Madia radiata*
1. Ray laminae 0.7–20 mm (rays sometimes 0 in *M. glomerata*); paleae mostly persistent; anthers yellow to brownish or ± dark purple; ray cypselae sometimes beaked.

[2. Shifted to left margin—Ed.]

2. Plants self-incompatible (heads showy) or self-compatible (heads not showy); disc florets functionally staminate (cypselae 0, ovary walls remaining pallid, membranous).
   3. Ray laminae greenish yellow, 4–11 mm; anthers ± dark purple; ray cypselae ± 3-angled (abaxial sides broadly rounded, adaxial sides 2-faced, angles between those faces ca. 70°), glossy . . . . 5. *Madia citriodora*
   3. Ray laminae bright yellow (sometimes with maroon bases), 4–20 mm; anthers yellow to brownish or ± dark purple; ray cypselae compressed (abaxial sides slightly rounded, adaxial sides 2-faced, angles between those faces 15–45°), dull or glossy . . . . 6. *Madia elegans*
2. Plants self-compatible (heads not showy); disc florets bisexual (forming cypselae, ovary walls becoming dark, rigid).
   4. Heads usually in glomerules, sometimes in corymbiform or paniculiform arrays; involucres narrowly ovoid or ellipsoid; ray florets 0 or 1–3. . . . . 2. *Madia glomerata*
   4. Heads in crowded or open, corymbiform, paniculiform, racemiform, or spiciform arrays; involucres depressed-globose, globose, obovoid, ovoid, or urceolate; ray florets (1–)3–13(–14).
      5. Stems 1–30(–60) cm; phyllaries glandular-pubescent (glands golden yellow), apices ± erect, often sulcate; anthers yellow to brownish.
         6. Heads in ± spiciform arrays; involucres globose or ovoid, 6–8 mm; disc florets 5–15, corollas pubescent; ray cypselae sometimes purple-mottled, beakless . . . 3. *Madia subspicata*
         6. Heads in racemiform or paniculiform arrays (peduncles filiform); involucres depressed-globose, 3–5 mm; disc florets 1(–2), corollas glabrous; ray cypselae black, beaked (beaks adaxially offset, curved) . . . . 4. *Madia exigua*
      5. Stems 10–200 cm; phyllaries glandular-pubescent (glands black, purple, or yellow), apices ± reflexed, flat; anthers ± dark purple.
         7. Ray cypselae black or purple, terete, glossy . . . . 8. *Madia anomala*
         7. Ray cypselae black, purple, or mottled, compressed, dull or glossy.
            8. Stems glandular-pubescent throughout; heads in racemiform, paniculiform, or spiciform arrays; involucres 6–16 mm; mostly coastal . . . . 10. *Madia sativa*
            8. Stems distally glandular-pubescent; heads in corymbiform, racemiform, or paniculiform arrays; involucres 5–10 mm; mostly away from immediate coast.
               9. Lateral branches often surpassing main stems (in large plants); ray laminae 6–8 mm . . . . 7. *Madia citrigracilis*
               9. Lateral branches seldom surpassing main stems; ray laminae 1.5–8 mm . . . . 9. *Madia gracilis*

1. **Madia radiata** Kellogg, Proc. Calif. Acad. Sci. Sci. 4: 190. 1870 [C] [E]

Plants 10–90 cm; self-incompatible (heads showy). **Stems** glandular-pubescent, glands yel-lowish or purple, lateral branches often surpassing main stems. **Leaf blades** lanceolate to linear, 2–10 cm × 4–15 mm. **Heads** in open, ± corymbiform arrays. **Involucres** depressed-globose, 4–7 mm. **Phyllaries** pilose to hispid (hairs uncinate) and glandular-pubescent, glands yellowish or purple, apices ± erect or reflexed, flat. **Paleae** readily falling, distinct. **Ray florets** 8–16; corollas golden yellow, laminae 6–19 mm. **Disc florets** 18–65, bisexual, fertile; corollas 3.5–5.5 mm, pubescent; anthers yellow to brownish. **Ray cypselae** black, purple, or mottled, dull or glossy, compressed (strongly arcuate), beaked (beaks adaxially offset, curved). **Disc cypselae** similar, not beaked. $2n = 16$.

Flowering Mar–May. Grasslands, openings in woodlands or chaparral, disturbed sites, usually heavy, clayey soils, often from decomposed shale; of conservation concern; 20–1200 m; Calif.

Madia radiata occurs in the Inner South Coast Ranges and, locally, in the eastern San Francisco Bay area. It sometimes co-occurs with *Deinandra halliana*; the two species are morphologically similar.

2. Madia glomerata Hooker, Fl. Bor.-Amer. 2: 24. 1834

E

Plants 5–120 cm, self-compatible (heads not showy). **Stems** proximally villous to hispid, glandular-pubescent distally, glands yellowish or black, lateral branches sometimes surpassing main stems. **Leaf blades** linear to lance-linear, 2–10 cm × 2–7 mm. **Heads** usually in crowded glomerules, sometimes in corymbiform or paniculiform arrays. **Involucres** narrowly ovoid or ellipsoid, 5.5–9 mm. **Phyllaries** ± pilose and glandular-pubescent, glands yellowish or black, apices erect or reflexed, ± flat. **Paleae** mostly persistent, distinct. **Ray florets** 0 or 1–3; corollas greenish yellow to purplish, laminae 1–3 mm. **Disc florets** 1–5(–12), bisexual, fertile; corollas 3–4.5 mm, pubescent; anthers ± dark purple. **Ray cypselae** black, dull, compressed, beakless. **Disc cypselae** similar. $2n = 28$.

Flowering Jun–Sep. Openings in grasslands, meadows, swales, shrublands, woodlands, forests, edges of marshes, lakes, or watercourses, disturbed sites, often in coarse, sandy or gravelly soils; 0–3100 m; Alta., B.C., Man., Ont., Que., Sask., Yukon; Alaska, Ariz., Calif., Colo., Conn., Idaho, Iowa, Maine, Mich., Minn., Mont., Nev., N.Dak., N.Mex., Oreg., S.Dak., Utah, Vt., Wash., Wyo.

Madia glomerata has the most extensive North American distribution of any species in Madiinae. At southern latitudes, *M. glomerata* occurs mostly in montane settings. Occurrences in eastern North America are mostly local and widely scattered.

3. Madia subspicata D. D. Keck, Publ. Carnegie Inst. Wash. 564: 45. 1945   E

Plants 5–60 cm, self-compatible (heads not showy). **Stems** proximally ± villous, distally glandular-pubescent, glands yellowish, lateral branches not surpassing main stems. **Leaf blades** linear to lance-linear, 2–7 cm × 1–5 mm. **Heads** in spiciform or spiciform-racemiform arrays (peduncles 0 or lengths usually less than 2 times heads). **Involucres** globose or ovoid, 6–8 mm. **Phyllaries** ± hirsute and thick-stalked-glandular as well, glands golden yellow, apices ± erect, sulcate or flat. **Paleae** mostly persistent, distinct or connate less than 1/2 their lengths. **Ray florets** 5–8; corollas pale yellow, laminae 1–2.5 mm. **Disc florets** 5–15, bisexual, fertile; corollas 3–3.5 mm, pubescent; anthers yellow to brownish. **Ray cypselae** black or brown, sometimes purple-mottled, dull, compressed, ± clavate, beakless. **Disc cypselae** similar. $2n = 16$.

Flowering Apr–Jun. Grasslands and open woodlands, often in shade; 50–800 m; Calif.

Madia subspicata occurs locally in the central and northern Sierra Nevada foothills, sometimes with the morphologically similar *M. gracilis.*

4. Madia exigua (Smith) A. Gray, Proc. Amer. Acad. Arts 8: 391. 1872

Sclerocarpus exigua Smith in A. Rees, Cycl. 31: Sclerocarpus no. 3. 1815

Plants 1–30(–50) cm, self-compatible (heads not showy). **Stems** hirsute and glandular-pubescent, glands yellowish or purple, lateral branches seldom surpassing main stems. **Leaf blades** linear, 0.2–4 cm × 0.5–2 mm. **Heads** in open, corymbiform arrays (peduncles ± filiform). **Involucres** depressed-globose, 2.5–5 mm. **Phyllaries** ± hirsute and glandular-pubescent as well, glands golden yellow, apices ± erect, sulcate. **Paleae** mostly persistent, connate 1/2+ their lengths. **Ray florets** 1–8; corollas pale yellow, laminae 0.7–1 mm. **Disc florets** 1(–2), bisexual, fertile; corollas 1–1.8 mm, glabrous; anthers yellow to brownish. **Ray cypselae** black or brown, dull, compressed (strongly arcuate), beaked (beaks adaxially offset, curved). **Disc cypselae** obovoid, weakly compressed. $2n = 32$.

Flowering Apr–Jul. Openings in grasslands, meadows, shrublands, woodlands, and forests, disturbed sites, often sandy, gravelly, or clayey soils, sometimes serpentine; 30–2500 m; B.C.; Calif., Idaho, Mont., Nev., Oreg., Wash.; Mexico (Baja California).

Madia exigua occurs in seasonally dry situations in much of western North America outside the warm deserts. Morphologically, *M. exigua* is somewhat similar to *Hemizonella minima*, which (unlike *M. exigua*) has subumbellate arrays of heads and obcompressed, sparsely hairy ray cypselae.

5. Madia citriodora Greene, Bull. Torrey Bot. Club 9: 63. 1882   E

Plants 10–70 cm, self-compatible (heads not showy). **Stems** proximally villous to hirsute, distally glandular-pubescent, glands purple, lateral branches often surpassing main stems. **Leaf blades** linear, 2–9 cm × 1–10 mm. **Heads** in open, corymbiform arrays. **Involucres** ± ovoid to hemispheric, 6–8 mm. **Phyllaries** ± villous or hirsute, glandular-pubescent as well (often sparsely), glands purple, apices usually ± erect, flat. **Paleae** mostly persistent, mostly connate 1/2+ their lengths. **Ray florets** 5–12; corollas greenish yellow, laminae 4–11 mm. **Disc florets** 8–50+, functionally staminate; corollas 2–3 mm, pubescent; anthers ± dark purple. **Ray cypselae** black or brown, sometimes mottled, glossy, ± 3-angled (abaxial sides rounded, adaxial sides 2-faced, angles between those faces ca. 70°), beakless (or nearly so). **Disc cypselae** 0. **2n** = 16.

Flowering Apr–Jul. Openings in woodlands, forests, and shrublands, disturbed sites, stream banks, often in dry, stony or clayey soils; 30–1600 m; Calif., Nev., Oreg., Wash.

Madia citriodora occurs in northern California, northwestern Nevada, Oregon, and Washington, sometimes with (and often confused with) *M. gracilis*.

6. Madia elegans D. Don ex Lindley, Edwards's Bot. Reg. 17: plate 1458. 1831 • Common madia [F]

Madia elegans subsp. *densifolia* (Greene) D. D. Keck; *M. elegans* subsp. *vernalis* D. D. Keck; *M. elegans* subsp. *wheeleri* (A. Gray) D. D. Keck

Plants 6–250 cm, self-incompatible (heads showy). **Stems** proximally villous to hirsute, distally glandular-pubescent, glands yellowish, purple, or black, lateral branches sometimes surpassing main stems. **Leaf blades** lanceolate to linear, 3–20 cm × 2–20 mm. **Heads** in open, corymbiform arrays. **Involucres** ± globose to campanulate, 4.5–12 mm. **Phyllaries** ± hirsute or villous, usually glandular-pubescent as well, glands yellowish, purple, or black, apices erect or reflexed, flat. **Paleae** mostly persistent, mostly connate ¹/₂+ their lengths. **Ray florets** (2–)5–22; corollas bright yellow (sometimes with maroon bases), laminae 4–20 mm. **Disc florets** 25–80+, functionally staminate; corollas 2.5–5 mm, pubescent; anthers yellow to brownish or ± dark purple. **Ray cypselae** black or brown, sometimes mottled, dull, compressed or ± 3-angled (slightly rounded abaxially, angled 15–45° adaxially), beakless (or nearly so). **Disc cypselae** 0. **2n** = 16.

Flowering Apr–Nov. Grasslands, meadows, open sites in shrublands, woodlands, and forests, disturbed sites, often in coarse or clayey soils, sometimes serpentine; 0–3400 m; Calif., Nev., Oreg., Wash.; Mexico (Baja California).

Madia elegans occurs widely in California outside the deserts and in southwestern Oregon and locally in western Nevada and Washington. It is unusually variable in morphology, ecology, and phenology. Molecular data have indicated that D. D. Keck's (1959) infraspecific taxonomy for *M. elegans* needs revision. Putative natural (sterile) hybrids with *M. sativa* have been collected (e.g., *D. D. Keck 2647*, UC, from northern California).

7. Madia citrigracilis D. D. Keck, Publ. Carnegie Inst. Wash. 564: 44. 1945 [E]

Plants (10–)25–60 cm, self-compatible (heads not showy). **Stems** hirsute to villous, distally glandular-pubescent, glands yellowish, purple, or black, lateral branches often surpassing main stems (in large plants). **Leaf blades** lanceolate to linear-oblong or linear, 3–15 cm × 2–14 mm. **Heads** in open, corymbiform arrays. **Involucres** ± globose to ovoid or obovoid, 6–8 mm. **Phyllaries** hirsute and glandular-pubescent, glands yellowish, purple, or black, apices erect or ± reflexed, flat. **Ray florets** 5–8(–14); corollas pale yellow or greenish yellow, laminae 6–8 mm. **Disc florets** 3–10(–30), bisexual, fertile; corollas 2.5–3.5 mm, pubescent; anthers ± dark purple. **Paleae** mostly persistent, mostly connate ¹/₂+ their lengths. **Ray cypselae** black or brown, sometimes mottled, dull, compressed, beakless (or nearly so). **Disc cypselae** similar. **2n** = 48.

Flowering Jun–Aug. Openings in shrublands, woodlands, and forests; 1400–2700 m; Calif., Nev.

Madia citrigracilis occurs in northeastern California and northern Nevada. Small or young plants can be exceedingly difficult to distinguish morphologically from *M. gracilis*, which is one of two putative parental species suggested by J. Clausen et al. (1945) to have been involved in the hybrid (allopolyploid) origin of *M. citrigracilis*. Molecular data reinforce Clausen et al.'s hypothesis that *M. citrigracilis* is evolutionarily distinct from *M. gracilis*.

8. Madia anomala Greene, Bull. Calif. Acad. Sci. 1: 91. 1885 [E]

Plants (10–)20–55 cm, self-compatible (heads not showy). **Stems** hirsute, distally glandular-pubescent, glands yellowish, purple, or black, lateral branches rarely surpassing main stems. **Leaf blades** linear, 2–10 cm × 2–7 mm. **Heads** in open, racemiform or paniculiform arrays. **Involucres** globose or depressed-globose, 6–10 mm. **Phyllaries** hirsute and glandular-pubescent, glands yellowish, purple, or black, apices erect or ± reflexed, flat. **Paleae** mostly persistent, connate ¹/₄–¹/₂+ their lengths. **Ray florets** 3–8; corollas greenish yellow, laminae 3–4.5 mm. **Disc florets** 3–8, bisexual, fertile; corollas 3.5–4 mm, pubescent; anthers ± dark purple. **Ray cypselae** black or purple, glossy, ± terete, beakless. **Disc cypselae** similar. **2n** = 32.

Flowering Apr–Jun. Open, often grassy slopes in woodlands and chaparral; 0–500 m; Calif.

Madia anomala occurs locally in the North Coast Ranges, San Francisco Bay area, and Sutter Buttes (southern Sacramento Valley), sometimes with the morphologically similar *M. gracilis*.

9. **Madia gracilis** (Smith) D. D. Keck, Madroño 5: 169. 1940 • Gumweed

Sclerocarpus gracilis Smith in A. Rees, Cycl. 31: Sclerocarpus no. 2. 1815; *Madia gracilis* subsp. *collina* D. D. Keck; *M. gracilis* subsp. *pilosa* D. D. Keck

Plants 6–100 cm, self-compatible (heads not showy). **Stems** proximally pilose to hirsute, distally glandular-pubescent, glands yellowish, purple, or black, lateral branches seldom surpassing main stems. **Leaf blades** oblong to linear, 1–10 (–15) cm × 1–8(–10) mm. **Heads** in ± open, paniculiform or racemiform arrays. **Involucres** depressed-globose to urceolate, 5–10 mm. **Phyllaries** sometimes hirsute, always finely or coarsely glandular-pubescent, glands yellowish, purple, or black, apices erect or ± reflexed, flat. **Paleae** mostly persistent, connate ¹/₂+ their lengths. **Ray florets** 3–10; corollas lemon yellow or greenish yellow, laminae 1.5–8 mm. **Disc florets** 2–16+, bisexual, fertile; corollas 2.5–5 mm, pubescent; anthers ± dark purple. **Ray cypselae** black, purple, or mottled, dull, compressed, beakless (or nearly so). **Disc cypselae** similar. **2n** = 32, 48.

Flowering Apr–Aug. Open or partially shaded slopes or flats in grasslands, meadows, shrublands, woodlands, and forests, disturbed sites, stream banks, roadsides, coarse to fine textured soils, sometimes serpentine; 0–2500 m; B.C.; Calif., Idaho, Mont., Nev., Oreg., Utah, Wash.; Mexico (Baja California).

Madia gracilis occurs widely in California (except the warm deserts), is scattered across much of Nevada, Oregon, and Washington (outside the driest regions), and extends into southernmost British Columbia, northwestern Montana, and northern Utah. Near the coast, *M. gracilis* sometimes co-occurs with *M. sativa*; the two species are partially interfertile (*M. gracilis* tends to flower earlier than *M. sativa*; J. Clausen 1951). Reported occurrences of *M. gracilis* in Maine and South America have not been confirmed.

10. **Madia sativa** Molina, Sag. Stor. Nat. Chili, 136. 1782 • Chile tarweed F

Madia capitata Nuttall

Plants (0.3–)35–100(–240) cm, self-compatible (heads not showy). **Stems** hirsute and glandular-pubescent, glands yellowish, purple, or black, lateral branches rarely surpassing main stems. **Leaf blades** broadly lanceolate to linear-oblong or linear, 2–18 cm × 3–18 (–29) mm. **Heads** in usually crowded, paniculiform, racemiform, or spiciform arrays. **Involucres** ovoid to urceolate, 6–16 mm. **Phyllaries** hirsute and glandular-pubescent, glands yellowish, purple, or black, apices erect or ± reflexed, flat. **Paleae** mostly persistent, connate ¹/₂+ their lengths. **Ray florets** (5–)8–13; corollas greenish yellow or sometimes purplish red abaxially or throughout, laminae 1.5–4 mm. **Disc florets** 11–14, bisexual, fertile; corollas 2–5 mm, pubescent; anthers ± dark purple. **Ray cypselae** black or brown, sometimes mottled, dull, compressed, beakless. **Disc cypselae** similar. **2n** = 32.

Flowering May–Oct. Grasslands, openings in shrublands and woods, disturbed sites, stream banks, roadsides; 0–1000 m; B.C., Calif., Oreg., Wash.; South America (Argentina, Chile); Pacific Islands (Hawaii, probably introduced).

In North America, *Madia sativa* occurs on the Pacific Coast from California to British Columbia, sporadically in coastal ranges, and rarely eastward. Reports of *M. sativa* from Ontario and Quebec and from Alaska, Connecticut, Georgia, Idaho, Indiana, Iowa, Maine, Maryland, Massachusetts, New York, North Carolina, Pennsylvania, Vermont, and Wisconsin are putative waifs or misidentified *M. glomerata*. Molecular data and greenhouse studies have indicated that plants referable to *M. capitata* and *M. sativa* in California are not distinct (B. G. Baldwin, unpubl.). Sampled populations of *M. sativa* (including *M. capitata*) from California are somewhat divergent in DNA sequences from sampled Chilean populations, in apparent conflict with earlier suggestions that *M. sativa* was recently introduced to North America from South America by Europeans (Baldwin, unpubl.). *Madia sativa* has been cultivated for seed-oil in South America, Europe, Africa, and Asia Minor (E. Zardini 1992).

187m.19. ASTERACEAE Martinov (tribe HELIANTHEAE) subtribe HYMENOPAPPINAE Rydberg in N. L. Britton et al., N. Amer. Fl. 34: 43. 1914 (as Hymenopappanae)

Annuals, biennials, or perennials, (5–)20–150 cm. **Leaves** mostly basal or basal and cauline; alternate; petiolate or sessile; blades deltate to lanceolate overall, usually 1–2-pinnately or -palmati-pinnately lobed, lobes usually filiform, ultimate margins entire or toothed, faces glabrous or hairy, often tomentose, usually gland-dotted). **Heads** usually discoid, rarely radiate, usually in ± corymbiform arrays. **Calyculi** 0. **Involucres** obconic to hemispheric. **Phyllaries** persistent or tardily falling, 5–13+ in 2+ series (distinct, often yellowish, whitish, or purplish, orbiculate to lance-linear, subequal, often ± petaloid, margins usually notably membranous or scarious). **Receptacles** flat or convex, usually epaleate (paleae scarious, conduplicate, each ± investing subtended floret in *H. newberryi*). **Ray florets** usually 0, sometimes 8, pistillate, fertile; corollas white or whitish. **Disc florets** 12–70+, bisexual, fertile; corollas yellow or whitish to purplish, tubes longer than or about equaling funnelform or abruptly dilated, campanulate throats, lobes 5, ± deltate (reflexed at anthesis); anther thecae pale; stigmatic papillae in 2 lines. **Cypselae** obconic to obpyramidal, usually 4-angled and 12–16-ribbed, glabrous or hairy; **pappi** 0 or of 12–22 orbiculate to spatulate scales.

Genera 5, species 28 (1 genus, 10 species in the flora): North America, Mexico.

H. Robinson (1981) suggested that Hymenopappinae is related to Gaillardiinae and other "core" groups of traditional Helenieae.

351. HYMENOPAPPUS L'Héritier, Hymenopappus, plate. 1788 • [Greek *hymen*, membrane, and *pappos*, pappus, alluding to membranous pappus scales] E

John L. Strother

Biennials or perennials, to 120 cm. **Stems** erect, often scapiform (1 stem per crown in biennials, usually 3+ in perennials). **Leaves** mostly basal or basal and cauline; alternate; petiolate; blades usually 1–2+ times pinnately lobed (lobes usually ± filiform), ultimate margins usually entire, rarely toothed, faces glabrous or hairy, often tomentose, usually gland-dotted as well. **Heads** radiate or discoid, in ± corymbiform arrays. **Involucres** obconic to hemispheric, 4–15+ mm diam. **Phyllaries** 5–13+ in 2–3+ series, subequal, usually membranous- or scarious-margined (often yellowish, whitish, or purplish). **Receptacles** flat or convex, epaleate (except *H. newberryi*). **Ray florets** 0 or 8, pistillate, fertile; corollas white to ochroleucous. **Disc florets** 12–70+, bisexual, fertile; corollas usually yellow to ochroleucous or whitish to purplish, tubes longer than or about equaling funnelform or abruptly dilated, campanulate throats, lobes 5 (reflexed at anthesis). **Cypselae** obpyramidal (4-, sometimes 5-angled, each face usually 1–4-ribbed, glabrous or hirtellous to villous); **pappi** 0 or of 12–22 orbiculate to spatulate scales. x = 17.

Species 11 (10 in the flora): North America, Mexico.

My treatment of *Hymenopappus* is essentially an adaptation of that by B. L. Turner (1956).

SELECTED REFERENCE Turner, B. L. 1956. A cytotaxonomic study of the genus *Hymenopappus* (Compositae). Rhodora 58: 163–186, 208–242, 259–269, 295–308.

1. Perennials (usually 3+ aerial stems from crown); cauline leaves mostly 0–12; heads 1–20 (–50) per stem; disc corolla throat lengths 2–8 times lobes.
    2. Ray florets 8.
        3. Basal leaves 12–25 cm; peduncles 6–15 cm; pappi 0, or 0.01–0.1 mm (receptacles paleate) . . . . 1. *Hymenopappus newberryi*
        3. Basal leaves 8–14 cm; peduncles 3–8 cm; pappi 0.4–0.6 mm . . . . 2. *Hymenopappus radiatus*
    2. Ray florets 0.
        4. Leaf blades simple or 1-pinnate; cypselae glabrous or sparsely hirtellous . . . . . .
            . . . . 3. *Hymenopappus mexicanus*
        4. Leaf blades 2-pinnate; cypselae hirtellous to villous . . . . 4. *Hymenopappus filifolius*
1. Biennials (usually 1 stem from crown); cauline leaves (2–)10–100; heads (15–)20–200 per stem; disc corolla throat lengths 1–2(–3) times lobes.
    5. Ray florets 8 . . . . 5. *Hymenopappus biennis*
    5. Ray florets 0.
        6. Basal leaves simple or 1–2-pinnate, lobes (1–)5–30 mm wide; disc corollas whitish or purplish (not yellowish), tubes 2–3 mm.
            7. Basal leaves simple or 1-pinnate; disc corollas usually purplish, rarely whitish; pappi 0.3–2 mm . . . . 6. *Hymenopappus artemisiifolius*
            7. Basal leaves 1–2-pinnate; disc corollas whitish; pappi 0.1–1 mm . . . .
                . . . . 7. *Hymenopappus scabiosaeus*
        6. Basal leaves 2-pinnate, lobes 0.5–6 mm wide; disc corollas whitish or yellowish, tubes 1.5–2.2 mm.
            8. Lobes of basal leaves 1–6 mm wide; disc corollas yellowish; pappi 0.5–1.5 mm
                . . . . 8. *Hymenopappus flavescens*
            8. Lobes of basal leaves 0.5–1.5 mm wide; disc corollas whitish; pappi 1–1.5(–2) mm.
                9. Disc corollas 3–4 mm, throats ± funnelform . . . . 9. *Hymenopappus carrizoanus*
                9. Disc corollas 2.5–3 mm, throats ± campanulate . . . . 10. *Hymenopappus tenuifolius*

1. Hymenopappus newberryi (A. Gray ex Porter & J. M. Coulter) I. M. Johnston, Contr. Gray Herb. 68: 96. 1923 [E] [F]

Leucampyx newberryi A. Gray ex Porter and J. M. Coulter, Syn. Fl. Colorado, 77. 1874

Perennials, 20–60 cm. **Leaves:** basal 2-pinnate, 12–25 cm, lobes 5–15(–25+) × 0.5–1.5(–3+) mm; cauline 1–3(–5). **Heads** 3–8 per stem. **Peduncles** 6–15 cm. **Phyllaries** whitish or yellowish, 8–10 × 4–7 mm. **Ray florets** 8; corollas white or pinkish, laminae 14–20 mm. **Disc florets** 60–150+; corollas yellowish, 3.5–4 mm, tubes 1.5–2 mm, throats campanulate, 1.5–2 mm, lengths 2–3 times lobes. **Cypselae** 3.5–4 mm, glabrous; **pappi** 0 or of 1–12+ scales 0.01–0.1 mm. **2n** = 34.

Flowering Jun–Sep. Clays or igneous soils, openings in aspen, pine, and spruce woodlands; 2100–3000 m; Colo., N.Mex.

2. Hymenopappus radiatus Rose, Contr. U.S. Natl. Herb. 1: 122. 1891 (as radiata) [E]

Perennials, 30–45 cm. **Leaves:** basal 2-pinnate, 8–14 cm, lobes 5–15+ × 1–2 mm; cauline 2–3. **Heads** 6–8 per stem. **Peduncles** 3–8 cm. **Phyllaries** yellowish, 7–9 × 3–5 mm. **Ray florets** 8; corollas white, laminae 14–16 mm. **Disc florets** 30–50+; corollas yellowish, 3–4 mm, tubes 1.5–2 mm, throats campanulate, 1.5 mm, lengths 3 times lobes. **Cypselae** 4–4.5 mm, ± hirtellous to villous; **pappi** of 14–18 scales 0.4–0.6 mm. **2n** = 34.

Flowering May–Jul. Sandy soils, pine woodlands; 2100–2700+ m; Ariz.

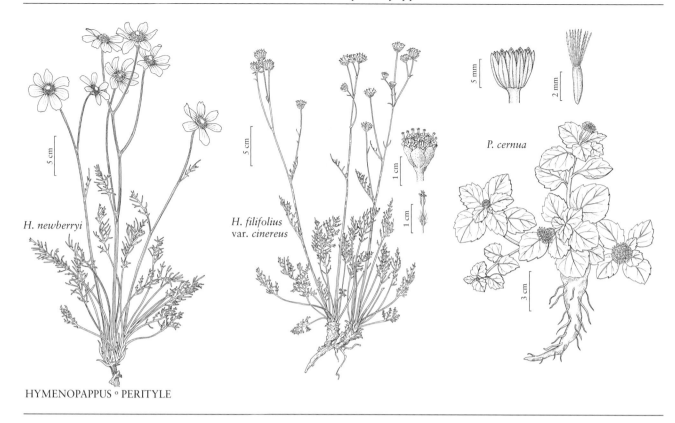

H. *newberryi*

H. *filifolius* var. *cinereus*

P. cernua

HYMENOPAPPUS ° PERITYLE

3. Hymenopappus mexicanus A. Gray, Proc. Amer. Acad. Arts 19: 29. 1883

Perennials, 20–90 cm. **Leaves:** basal simple or 1-pinnate, 6–20 cm, lobes 5–15+ × 1–9 mm; cauline 0 or 3–6. **Heads** 1–8(–20+) per stem. **Peduncles** 1–10 cm. **Phyllaries** yellowish to whitish or red-tinged, 6–9 × 2–3 mm. **Ray florets** 0. **Disc florets** 20–40; corollas yellow, 3–4.5 mm, tubes 2–2.5 mm, throats campanulate, 1–1.5 mm, lengths 3–4 times lobes. **Cypselae** 4–6 mm, glabrous or sparsely hirtellous; **pappi** of 12–20 scales 0.1–0.4 mm. $2n = 34$.

Flowering Jun–Oct. Openings in pine, spruce, and aspen woodlands; 2100–3000 m; Ariz., N.Mex.; Mexico.

4. Hymenopappus filifolius Hooker, Fl. Bor.-Amer. 1: 317. 1833

Perennials, 5–10 cm. **Leaves:** basal 2-pinnate, 3–20 cm, lobes 2–50 × 0.5–2(–3) mm; cauline 0–12. **Heads** 1–60 per stem. **Peduncles** 1–16 cm. **Phyllaries** whitish to yellowish, 3–14 × 2–5 mm. **Ray florets** 0. **Disc florets** 10–80; corollas usually yellowish, sometimes whitish, 2.2–7 mm, tubes 1–2.5 mm, throats campanulate, 1–3 mm, lengths 2–8 times

lobes. **Cypselae** 3–7 mm, ± hirtellous to villous; **pappi** of 12–22 scales 0.1–3 mm.

Varieties 13 (13 in the flora): North America, n Mexico.

Some plants are morphologically intermediate between some pairs of varieties of *Hymenopappus filifolius* as treated by B. L. Turner (1956).

1. Axils of basal leaves glabrous or sparsely tomentose; cauline leaves 0 or 1(–2).
  2. Peduncles 1–2 cm; corollas 2–3 mm; anthers 2 mm; cypselae 3.5–4 mm . . . .
    . . . . 4a. *Hymenopappus filifolius* var. *parvulus*
  2. Peduncles 2–10 cm; corollas 3–5 mm; anthers 2.5–3 mm; cypselae 4–6 mm . . . .
    . . . . 4b. *Hymenopappus filifolius* var. *nudipes*
1. Axils of basal leaves ± densely tomentose; cauline leaves (0–)2–12.
  3. Terminal lobes of basal leaves 2–6 mm; phyllaries 3–7 mm; florets 10–30; corollas 2–3 mm.
    4. Plants 5–20 cm; cauline leaves 0–3; pappi 0.1–0.8 mm . . . .
      . . . . 4c. *Hymenopappus filifolius* var. *luteus*
    4. Plants 20–35 cm; cauline leaves 4–9; pappi 0.5–1.5 mm . . . .
      . . . 4d. *Hymenopappus filifolius* var. *pauciflorus*
  3. Terminal lobes of basal leaves (2–)3–50 mm; phyllaries (5–)6–14 mm; florets (15–)20–70+; corollas (2.5–)3–7 mm.

[5. Shifted to left margin—Ed.]

5. Corollas 4–7 mm, throat lengths (2–)3–8 times lobes; anthers 3–4 mm; cypselae 5–7 mm; pappi (1.2–)1.5–3 mm.
  6. Peduncles 8–16 cm; corollas whitish; cypsela hairs 0.5–1 mm . . . . . . 4e. *Hymenopappus filifolius* var. *eriopodus*
  6. Peduncles 2–12 cm; corollas yellowish; cypsela hairs 1–2 mm.
    7. Cauline leaves (1–)2–6; terminal lobes of basal leaves 5–30 mm; phyllaries 8–12(–14) mm . . . . 4f. *Hymenopappus filifolius* var. *megacecphalus*
    7. Cauline leaves 0–3; terminal lobes of basal leaves 3–15 mm; phyllaries 6–10 mm . . . . 4g. *Hymenopappus filifolius* var. *lugens*
5. Corollas 2.5–4.5 mm, throat lengths 2–4(–5) times lobes; anthers 2–3 mm; cypselae 4–6 mm; pappi 0.2–2(–3) mm.
  8. Cauline leaves (2–)3–8; heads 5–60; corolla throats 1.3–1.8 mm . . . 4h. *Hymenopappus filifolius* var. *polycephalus*
  8. Cauline leaves 0–5(–12); heads 1–10(–40); corolla throats 1.5–2.5 mm.
    9. Terminal lobes of basal leaves 3–6 mm; cauline leaves 5–12 . . . . . 4i. *Hymenopappus filifolius* var. *tomentosus*
    9. Terminal lobes of basal leaves 2–50 mm; cauline leaves 0–5.
      10. Plants 30–100 cm; terminal lobes of basal leaves 15–50 mm; pappi 0.2–1 mm . . . . 4j. *Hymenopappus filifolius* var. *filifolius*
      10. Plants 5–50 cm; terminal lobes of basal leaves 2–30 mm; pappi (0.6–)1–3 mm.
        11. Peduncles 3–15 cm; cypsela hairs 0.1–1 mm; pappi 1.5–3 mm . . . . 4k. *Hymenopappus filifolius* var. *nanus*
        11. Peduncles 1–6 cm; cypsela hairs 1–3 mm; pappi 0.6–2 mm.
          12. Cauline leaves 0–2; cypsela hairs 1 mm; pappi 0.6–1.3 mm . . . . 4l. *Hymenopappus filifolius* var. *idahoensis*
          12. Cauline leaves (0–)2–4; cypsela hairs 1–3 mm; pappi 1.5–2 mm . . . . . . 4m. *Hymenopappus filifolius* var. *cinereus*

4a. Hymenopappus filifolius Hooker var. **parvulus** (Greene) B. L. Turner, Rhodora 58: 216. 1956 [E]

Hymenopappus parvulus Greene, Pl. Baker. 3: 30. 1901

Stems 10–25 cm. **Leaves:** basal axils glabrous or sparsely tomentose, terminal lobes 3–20 mm; cauline 0–1(–2). **Heads** 3–10(–15). **Peduncles** 1–2 cm. **Phyllaries** 4–8 mm. **Florets** 10–20; corollas yellowish, 2–3 mm, throats 1.5–1.8 mm, lengths 2–3 times lobes; anthers 2 mm. **Cypselae** 3.5–4 mm, hairs 1 mm; **pappi** 0.6–1.8 mm. **2*n*** = 34.

Flowering Jun–Aug. Sands, gravels; 1600–2600 m; Colo.

4b. Hymenopappus filifolius Hooker var. **nudipes** (Maguire) B. L. Turner, Rhodora 58: 219. 1956 [E]

Hymenopappus nudipes Maguire, Amer. Midl. Naturalist 37: 143. 1947

Stems 5–45 cm. **Leaves:** basal axils glabrous or sparsely tomentose, terminal lobes 6–40 mm; cauline 0–1(–2). **Heads** 1–12. **Peduncles** 2–10 cm. **Phyllaries** 6–10 mm. **Florets** 10–35; corollas yellowish, 3–5 mm, throats 1.5–2.5 mm, lengths 2–4 (–5) times lobes; anthers 2.5–3 mm. **Cypselae** 4–6 mm, hairs 0.5–1.2 mm; **pappi** 1.5–2.5 mm. **2*n*** = 34.

Flowering Jun–Jul. Sands, gravels, clays; 1800–3500 m; Utah, Wyo.

4c. Hymenopappus filifolius Hooker var. **luteus** (Nuttall) B. L. Turner, Rhodora 58: 221. 1956 [E]

Hymenopappus luteus Nuttall, Trans. Amer. Philos. Soc., n. s. 7: 374. 1841

Stems 5–20+ cm. **Leaves:** basal axils ± densely tomentose, terminal lobes 2–6 mm; cauline 0–3(–4). **Heads** (5–)10–50. **Peduncles** 1–7 cm. **Phyllaries** 5–7 mm. **Florets** 15–25; corollas yellowish, 2–3 (–3.5) mm, throats 1–2 mm, lengths (2–)3–4 times lobes; anthers 2 mm. **Cypselae** 4–5 mm, hairs 1–2 mm; **pappi** 0.1–0.8 mm. **2*n*** = 34.

Flowering Jun–Jul. Sands, gravels, calcareous; 1600–2200 m; Colo., Utah, Wyo.

4d. Hymenopappus filifolius Hooker var. **pauciflorus**
(I. M. Johnston) B. L. Turner, Rhodora 58: 224. 1956
[E]

Hymenopappus pauciflorus
I. M. Johnston, Contr. Gray Herb.
68: 97. 1923

Stems 20–35 cm. **Leaves:** basal
axils ± densely tomentose, terminal
lobes 2–6 mm; cauline (3–)4–9.
Heads 2–15(–30). **Peduncles** 2–6
cm. **Phyllaries** 3–6 mm. **Florets**
10–30; corollas yellowish, 2–3
mm, throats 1–1.6 mm, lengths 2–3 times lobes; anthers
1.6–2 mm. **Cypselae** 3–4.5 mm, hairs 1–2 mm; **pappi**
0.5–1.5 mm. *2n* = 34.

Flowering May–Sep. Sands; 1100–1700 m; Ariz.,
Colo., Utah.

4e. Hymenopappus filifolius Hooker var. **eriopodus**
(A. Nelson) B. L. Turner, Rhodora 58: 225. 1956 [E]

Hymenopappus eriopodus
A. Nelson, Bot. Gaz. 37: 274. 1904
(as eripoda)

Stems 40–80 cm. **Leaves:** basal
axils ± densely tomentose, terminal lobes 6–40 mm; cauline 2–7.
Heads 3–8. **Peduncles** 8–16 cm.
Phyllaries 7–10 mm. **Florets** 30–
60; corollas whitish, 4–5 mm,
throats 2–3 mm, lengths 3–5 times lobes; anthers 3–3.5
mm. **Cypselae** 5–6 mm, hairs 0.5–1.5 mm; **pappi** 1.5–2
mm.

Flowering May–Jul. Limestone outcrops; 1500–3000
m; Calif., Nev., Utah.

4f. Hymenopappus filifolius Hooker var.
megacephalus B. L. Turner, Rhodora 58: 227, fig. 36.
1956 [E]

Stems 30–70 cm. **Leaves:** basal
axils ± densely tomentose, terminal
lobes 5–30 mm; cauline (1–)2–6.
Heads 3–10(–14). **Peduncles** 2–
10 cm. **Phyllaries** 8–12(–14) mm.
Florets 25–60; corollas yellowish,
4–7 mm, throats 2–5 mm, lengths
2–5+ times lobes; anthers 3–4 mm.
Cypselae 5–7 mm, hairs 1–2 mm;
pappi (1–)1.5–3 mm. *2n* = 34.

Flowering May–Nov. Sands, gravels; 300–1700 m;
Ariz., Calif., Colo., Nev., Utah.

4g. Hymenopappus filifolius Hooker var. **lugens**
(Greene) Jepson, Man. Fl. Pl. Calif., 1128. 1925

Hymenopappus lugens Greene,
Pittonia 4: 43. 1899

Stems (20–)30–60 cm. **Leaves:**
basal axils ± densely tomentose,
terminal lobes 3–15 mm; cauline
0–3. **Heads** 3–8. **Peduncles** 2–12
cm. **Phyllaries** 6–10 mm. **Florets**
20–70+; corollas yellowish or
whitish, 4–6 mm, throats 2–3 mm,
lengths (2.2–)3–8 times lobes; anthers 3–3.5 mm.
Cypselae 5–6 mm, hairs 1–2 mm; **pappi** 1.2–2.5 mm. *2n*
= 34, 68.

Flowering May–Sep. Sands, clays, rocks (limestones,
schists, volcanics); 1500–2700 m; Ariz., Calif., Nev.,
N.Mex., Utah; Mexico (Baja California).

4h. Hymenopappus filifolius Hooker var.
polycephalus (Osterhout) B. L. Turner, Rhodora 58:
235. 1956 [E]

Hymenopappus polycephalus
Osterhout, Torreya 18: 90. 1918

Stems (20–)30–60 cm. **Leaves:**
basal axils ± densely tomentose,
terminal lobes (4–)10–30 mm;
cauline (2–)3–8. **Heads** 5–60.
Peduncles 2–8 cm. **Phyllaries** 5–8
mm. **Florets** 20–50; corollas yellowish, 3–3.5 mm, throats 1.3–1.8
mm, lengths 2–3 times lobes; anthers 2.5–3 mm. **Cypselae**
4–5 mm, hairs 1–2 mm; **pappi** 0.6–2 mm. *2n* = 68.

Flowering Jun–Aug. Hills, prairies, sands, clays; 600–
2100+ m; Alta., Sask.; Colo., Kans., Mont., Nebr.,
N.Dak., S.Dak., Wyo.

4i. Hymenopappus filifolius Hooker var. **tomentosus**
(Rydberg) B. L. Turner, Rhodora 58: 237. 1956 [E]

Hymenopappus tomentosus
Rydberg, Bull. Torrey Bot. Club 27:
633. 1900

Stems 30–60(–70) cm. **Leaves:**
basal axils ± densely tomentose, terminal lobes 3–6 mm; cauline 5–12.
Heads 4–14. **Peduncles** 1–6 cm.
Phyllaries 7–10 mm. **Florets** 30–
40; corollas yellowish, 3–4.5 mm,
throats 1.5–2.5 mm, lengths 3–4(–6) times lobes; anthers
2.5–3 mm. **Cypselae** 4.5–5 mm, hairs 0.5–1 mm; **pappi**
0.8–1.8 mm. *2n* = 34.

Flowering May–Jun. Sands, rocky soils, limestones;
1000–2100 m; Utah.

4j. Hymenopappus filifolius Hooker var. **filifolius**   E

Stems 30–100 cm. **Leaves:** basal axils ± densely tomentose, terminal lobes 15–50 mm; cauline (0–) 2–5. **Heads** (2–)6–40. **Peduncles** 1–15 cm. **Phyllaries** 6–10 mm. **Florets** 15–45; corollas yellowish, 2.5–4.5 mm, throats 1.5–2.5 mm, lengths 3–5 times lobes; anthers 2.5–3 mm. **Cypselae** 4.5–6 mm, hairs 0.4–1 mm; **pappi** 0.2–1 mm. *2n* = 34.

Flowering May–Sep. Sands, clays, gravels; 100–300+ m; Idaho, Oreg., Wash.

4k. Hymenopappus filifolius Hooker var. **nanus** (Rydberg) B. L. Turner, Rhodora 58: 240. 1956   E

Hymenopappus nanus Rydberg in N. L. Britton et al., N. Amer. Fl. 34: 53. 1914

Stems 5–50 cm. **Leaves:** basal axils ± densely tomentose, terminal lobes 3–15 mm; cauline 0–2 (–3). **Heads** 1–6. **Peduncles** 3–15 cm. **Phyllaries** 6–9 mm. **Florets** 30–60; corollas pale yellowish, 3–4 mm, throats 1.5–2 mm, lengths 2–4 times lobes; anthers 2.5–3 mm. **Cypselae** 4.5–5.5 mm, hairs 0.1–1(–2) mm; **pappi** (1–)1.5–3 mm. *2n* = 34.

Flowering May–Jul. Limestones, shales; 1500–3000 m; Ariz., Calif., Nev., Utah.

4l. Hymenopappus filifolius Hooker var. **idahoensis** B. L. Turner, Rhodora 58: 250, fig. 28. 1956   E

Stems 10–45 cm. **Leaves:** basal axils ± densely tomentose, terminal lobes 6–30 mm; cauline 0–2. **Heads** 4–14. **Peduncles** 1–6 cm. **Phyllaries** 7–9 mm. **Florets** 25–70; corollas yellowish, 3.2–4.5 mm, throats 1.5–2 mm, lengths 3–4 times lobes; anthers 2.5–3 mm. **Cypselae** 4.5–6 mm, hairs 1 mm; **pappi** 0.6–1.3 mm. *2n* = 34.

Flowering Jun–Jul. Gravelly or rocky sites; 1200–1600 m; Idaho.

4m. Hymenopappus filifolius Hooker var. **cinereus** (Rydberg) I. M. Johnston, Contr. Gray Herb. 68: 98. 1923   E F

Hymenopappus cinereus Rydberg, Bull. Torrey Bot. Club 27: 634. 1900

Stems 15–40 cm. **Leaves:** basal axils ± densely tomentose, terminal lobes 2–8(–15) mm; cauline (0–)2–4. **Heads** 1–6. **Peduncles** 1–6 cm. **Phyllaries** 6–9 mm. **Florets** 25–40; corollas yellowish or whitish, 3–4.5 mm, throats 1.5–2.5 mm, lengths 2.5–4 times lobes; anthers 2–3 mm. **Cypselae** 4–6 mm, hairs 1–3 mm; **pappi** 1.5–2 mm. *2n* = 34.

Flowering May–Sep. Open sites; 1700–3000 m; Ariz., Colo., N.Mex., Tex., Utah.

5. Hymenopappus biennis B. L. Turner, Rhodora 58: 256, figs. 24, 38. 1956   E

Biennials, 60–100 cm. **Leaves:** basal 2-pinnate, 6–16 cm, lobes 6–20 × 1(–3) mm; cauline 10–40. **Heads** 20–40 per stem. **Peduncles** 1–6 cm. **Phyllaries** yellowish, 5–8 × 3–5 mm. **Ray florets** 8; corollas white, laminae (6–)14–16 mm. **Disc florets** 32–50+; corollas yellowish, 3–3.5 mm, tubes 1.5 mm, throats campanulate, 1.5–2 mm, lengths 2–3 times lobes. **Cypselae** 4 mm, glabrous; **pappi** 0 or of scales 0.1–0.2 mm. *2n* = 34.

Flowering Jul–Oct. Limestone soils; 2000–3000 m; N.Mex., Tex.

6. Hymenopappus artemisiifolius de Candolle in A. P. de Candolle and A. L. P. P. de Candolle, Prodr. 5: 658. 1836 (as artemisiaefolius)   E

Biennials, 40–100 cm. **Leaves:** basal simple (5–20+ mm wide) or 1-pinnate, 8–20 cm, lobes 5–20+ × 1–5(–20+) mm; cauline 2–16. **Heads** 20–60 per stem. **Peduncles** 1–6 cm. **Phyllaries** whitish (sometimes red-tinged) or yellowish, 5–12 × 2–7 mm. **Ray florets** 0. **Disc florets** 40–60; corollas white to purplish or yellowish to reddish, 3–5 mm, tubes 2–3 mm, throats funnelform, 1–1.5 mm, lengths 1–1.5 times lobes. **Cypselae** 3.5–5 mm, hirtellous or villous; **pappi** of 16–18 scales 0.3–1(–1.5) or 1.5–2 mm.

Varieties 2 (2 in the flora): sc United States.

1. Cauline leaves (6–)8–16; phyllaries 6–12 mm;
corollas whitish to purplish; pappi 0.3–1(–1.5) mm
. . . . 6a. *Hymenopappus artemisiifolius*
var. *artemisiifolius*

1. Cauline leaves 2–8; phyllaries 5–7 mm; corollas
yellowish to reddish; pappi 1.5–2 mm . . . . . .
. . . . 6b. *Hymenopappus artemisiifolius*
var. *riograndensis*

6a. Hymenopappus artemisiifolius de Candolle var. artemisiifolius E

Cauline leaves (6–)8–16. Peduncles 1–4(–6) cm. Phyllaries usually whitish, sometimes red-tinged, 6–12 × 3–7 mm. Disc florets 20–30+; corollas whitish to purplish, 3.5–5 mm, tubes 2.5–3 mm, throats funnelform, 1–1.5 mm, lengths 1–1.5 times lobes. Cypselae 3.5–4 mm, hirtellous; pappi 0.3–1(–1.5) mm. *2n* = 34.

Flowering Mar–May. Sandy soils, pine and oak woodlands; 10–300+ m; Ark., La., Tex.

6b. Hymenopappus artemisiifolius de Candolle var. riograndensis B. L. Turner, Rhodora 58: 305, fig. 47. 1956 E

Cauline leaves 2–8. Peduncles 2–5 cm. Phyllaries yellowish, 5–7 × 2–4 mm. Disc florets 40–60; corollas yellowish to reddish, 3–3.5 mm, tubes 2 mm, throats funnelform, 1.5 mm, lengths 1–1.2 times lobes. Cypselae 4–5 mm, ± villous; pappi 1.5–2 mm. *2n* = 34.

Flowering Mar–May. Sandy-clay or gravelly soils; 0–50+ m; Tex.

7. Hymenopappus scabiosaeus L'Héritier, Hymenopappus, plate. 1788

Biennials, 40–150 cm. Leaves: basal simple (5–20 mm wide) or 1(–2)-pinnate, 5–25 cm, lobes (3–)5–20+ × (1–)2–8+ mm; cauline 15–50. Heads (20–)40–100 per stem. Peduncles 1–5 cm. Phyllaries white to yellowish, 5–15 × 2–8 mm. Ray florets 0. Disc florets 20–80; corollas whitish, 3–5.5 mm, tubes 2–3 mm, throats funnelform, 1–3 mm, lengths ± 1 times lobes. Cypselae 3–5 mm, ± hirtellous; pappi of 14–18 scales 0.1–1 mm.

Varieties 2 (2 in the flora): c, se United States, n Mexico.

Morphologic distinctions between varieties of *Hymenopappus scabiosaeus* as treated by B. L. Turner (1956) fail for some specimens.

1. Peduncles usually subtended by ± membranous
bracts 5–14 × 3–10 mm; phyllaries whitish, 7–15
× 4–8 mm . . . . 7a. *Hymenopappus scabiosaeus*
var. *scabiosaeus*

1. Peduncles not subtended by bracts, or bracts not
membranous, 3–5+ × 1–3+ mm; phyllaries whitish to yellowish, 5–9 × 2–4 mm . . . .
. . . . . . 7a. *Hymenopappus scabiosaeus* var. *corymbosus*

7a. Hymenopappus scabiosaeus L'Héritier var. scabiosaeus E

Peduncles usually subtended by ± membranous bracts 5–14 × 3–10 mm. Phyllaries whitish, 7–15 × 4–8 mm.

Flowering Apr–Jun. Sandy soils, openings in pine and/or oak woodlands; 10–300+ m; Ala., Ark., Fla., Ga., Ill., Ind., Miss., Mo., Okla., S.C.

7b. Hymenopappus scabiosaeus L'Héritier var. corymbosus (Torrey & A. Gray) B. L. Turner, Rhodora 58: 297. 1956

Hymenopappus corymbosus Torrey & A. Gray, Fl. N. Amer. 2: 372. 1842

Peduncles not subtended by bracts or bracts not membranous, 3–5+ × 1–3+ mm. Phyllaries whitish to yellowish, 5–9 × 2–4 mm. *2n* = 34.

Flowering Mar–Jul. Clays and/or calcareous soils, limestone, grasslands; 10–300 m; Kans., Nebr., Okla., Tex.; Mexico (Coahuila).

8. Hymenopappus flavescens A. Gray, Mem. Amer. Acad. Arts, n. s. 4: 97. 1849

Biennials, 30–90 cm. Leaves: basal 2-pinnate, 6–15 cm, lobes 3–15(–20+) × 1–6 mm; cauline 10–50. Heads 15–100 per stem. Peduncles 1–6 cm. Phyllaries yellowish, 4–8 × 2–4 mm. Ray florets 0. Disc florets 20–40; corollas yellow, 2.5–3.5 mm, tubes 1.5–2 mm, throats campanulate, 0.8–1.5 mm, lengths 1–1.5 times lobes. Cypselae 3–4.5 mm, ± villous; pappi of 16–22 scales 0.5–1.5 mm.

Varieties 2 (2 in the flora): c United States, n Mexico.

1. Lobes of basal leaves (1–)2–6 mm wide, abaxial
   faces glabrous or less hairy than adaxial; pappi
   0.5–1(–1.5) mm . . . . 8a. *Hymenopappus flavescens*
   var. *flavescens*
1. Lobes of basal leaves 1–2 mm wide, abaxial and
   adaxial faces ± equally hairy; pappi 1–1.5 mm . .
   . . . . 8b. *Hymenopappus flavescens*
   var. *canotomentosus*

8a. Hymenopappus flavescens A. Gray var. **flavescens** Ⓔ

Lobes of basal leaves (1–)2–6 mm wide, abaxial faces glabrous or less hairy than the adaxial. **Pappi** 0.5–1 (–1.2) mm.

Flowering May–Sep. Sandy soils; 1200–2100+ m; Colo., Kans., N.Mex., Okla., Tex.

8b. Hymenopappus flavescens A. Gray var. **canotomentosus** A. Gray, Smithsonian Contr. Knowl. 5(6): 94. 1853 (as cano-tomentosus)

Lobes of basal leaves 1–2 mm wide, abaxial and adaxial faces ± equally hairy. **Pappi** 1–1.5 mm.

Flowering Apr–May, Jul–Aug. Sandy or gravelly, often limestone soils; 1100–2100+ m; N.Mex., Tex.; Mexico (Chihuahua).

9. Hymenopappus carrizoanus B. L. Turner, Phytologia 67: 294, fig. 1. 1989 Ⓒ Ⓔ

Biennials (perhaps flowering first year), 45–150 cm. **Leaves:** basal 2-pinnate, (3–)8–12 cm, 2-pinnate, lobes 4–10 × 0.5–1 mm; cauline 12–25+. **Heads** 45–60 per stem. **Peduncles** (1–)2–5 cm. **Phyllaries** gray-green, (3.5–)5–6 × (1.5–)2.5–3.5 mm. **Ray florets** 0. **Disc florets** 20–40; corollas whitish (drying pinkish), 3–4 mm, tubes 1.5–2 mm, throats funnelform, 1–1.5 mm, lengths 1–2 times lobes. **Cypselae** 4 mm, ± hirtellous to villous; **pappi** of 16–20 scales 1–1.5 mm.

Flowering Apr–Jun. Sandy soils, oak woodlands; of conservation concern; 50–100+ m; Tex.

10. Hymenopappus tenuifolius Pursh, Fl. Amer. Sept. 2: 742. 1813 Ⓔ

Biennials, 40–150 cm. **Leaves** 2-pinnate, 8–15 cm, lobes (3–)5–15+ × 0.5–1.5+ mm; cauline 8–30. **Heads** 20–200 per stem. **Peduncles** 1–5 cm. **Phyllaries** yellowish, 5–8 × 2–4 mm. **Ray florets** 0. **Disc florets** 25–50; corollas whitish, 2.5–3 mm, tubes 1.5–2.2 mm, throats campanulate, 0.8–1.5 mm, lengths 1.5–2 times lobes. **Cypselae** 3.5–4.5 mm, ± villous; **pappi** of 16–18 scales 1–1.5(–2) mm. $2n = 34$.

Flowering (Apr–)May–Jul(–Aug). Sandy, gravelly, or silty soils, limestone; 50–2100 m; Colo., Kans., Nebr., N.Mex., Okla., S.Dak., Tex., Wyo.

187m.20. ASTERACEAE Martinov (tribe HELIANTHEAE) subtribe PERITYLINAE Rydberg in N. L. Britton et al., N. Amer. Fl. 34: 11. 1914 (as Peritylanae)

Amauriinae Rydberg

Annuals, perennials, subshrubs, or shrubs, 2–45(–75) cm. **Leaves** cauline; opposite or alternate; petiolate or sessile; blades mostly deltate to ovate or lanceolate overall, sometimes cordate, linear, reniform, rhombic, orbiculate, or ovate, often 3-lobed to pinnatisect, ultimate margins subentire to doubly serrate, faces glabrous or hairy, usually gland-dotted. **Heads** radiate or discoid, borne singly or in corymbiform arrays. **Calyculi** 0. **Involucres** cylindric to campanulate or hemispheric. **Phyllaries** persistent, 8–21 in 1(–3) series (distinct or connate, ovate to obovate or linear, subequal, often conduplicate, herbaceous, margins scarious). **Receptacles** flat or convex, epaleate. **Ray florets** 0 or 8–12 pistillate, fertile; corollas yellow or white. **Disc florets** 5–100+, bisexual, fertile; corollas yellow or white, tubes shorter than or about equaling cylindric to funnelform or campanulate throats, lobes 4, deltate (equal); anther thecae pale; stigmatic papillae in 2 lines. **Cypselae** linear or cuneiform to oblong or oblanceolate, strongly flattened or weakly 3–4-angled (callous-margined, usually ciliate); **pappi** 0, or persistent or falling, vestigial, or of 8–12 linear to subulate, laciniate or fimbriate scales, sometimes 1–2(–4) scales aristate, or of 1–2(–6) or 8–35 bristles.

Genera 5, species 67 (2 genera, 36 species in the flora): sw United States, Mexico, South America (1 species).

Peritylinae appears to be relatively isolated among epaleate Heliantheae (H. Robinson 1981). B. G. Baldwin et al. (2002) included *Lycapsus* Philippi in Peritylinae.

1. Leaf blades usually lobed (mostly 5–30 mm); phyllaries 8–16 in 2–3 series, distinct . . . 352. *Perityle*, p. 317
1. Leaf blades usually triangular-hastate to narrowly deltate, seldom notably lobed (30–120 mm); phyllaries 15–21 in 1(–2) series, wholly or partially connate . . . . 353. *Pericome*, p. 334

352. PERITYLE Bentham, Bot. Voy. Sulphur, 23, plate 15. 1844 • Rock daisy [Greek *peri*, around, and *tyle*, a callus, alluding to calloused cypselae margins]

Sharon C. Yarborough

A. Michael Powell

Annuals, perennials, subshrubs, or shrubs, 2–45(–75) cm (glabrous or hairy, often gland-dotted). **Stems** erect to pendent. **Leaves** mostly cauline; often proximally opposite, distally alternate; petiolate or sessile; blades usually 3-lobed, ultimate margins entire, toothed, or lobed. **Heads** radiate or discoid, borne singly or in corymbiform arrays. **Involucres** campanulate, cylindric, funnelform, or hemispheric, 3–15 mm diam. **Phyllaries** persistent, 5–28 in (1–)2(–3) series (distinct, linear to ovate, equal or subequal, flat or keeled, glabrous or hairy, apices obtuse, acute, or attenuate). **Receptacles** flat or convex, pitted, epaleate. **Ray florets** 0, or (1–)3–18, pistillate, fertile; corollas cream, yellow, or white (showy or rudimentary). **Disc florets** 5–200, bisexual, fertile; corollas cream, yellow, or white, tubes shorter than or nearly equal to cylindric, funnelform, or campanulate throats, lobes 4, ± deltate (acute; stamens 4; style branches flattened, linear, usually tapering to fine, minutely hairy tips). **Cypselae** (black) flattened to subcylindric, linear to oblanceolate or obovate (margins ± calloused and glabrous, hairy, or ciliate, faces usually hairy, sometimes glabrous); **pappi** 0, or persistent or falling, of 1–35 bristles plus callous crowns or hyaline scales. $x = 17, 19$.

Species 66 (35 in the flora): United States, Mexico, South America.

Most species of *Perityle* are rock-dwelling subshrubs exhibiting geographic speciation; most are found in the eroded mountain and basin region of the southwestern United States.

1. Pappi (disc florets) usually each of 8–35 unequal, antrorsely barbellulate bristles 0.5–3.5 mm (in *P. vitreomontana* all but 2–3 may be reduced to vestigial nubs); New Mexico, Texas . . . . 352a. *Perityle* sect. *Pappothrix*, p. 318
1. Pappi (disc florets) 0, or of 1–2(–6) bristles, sometimes plus callous crowns or scales; Arizona, California, Nevada, New Mexico, Texas, and Utah.
   2. Pappi (disc florets) 0, or of 1–2(–3) unequal to subequal, ± barbellate bristles 0.5–4 (–6) mm plus hyaline, laciniate scales; margins of cypselae usually densely ciliate . . . . 352b. *Perityle* sect. *Perityle*, p. 320
   2. Pappi (disc florets) 0, or of 1–2(–6) equal or subequal, antrorsely barbellulate or subplumose bristles 0.5–4.6 mm plus callous crowns or hyaline scales; margins of cypselae usually sparsely short-hairy . . . . 352c. *Perityle* sect. *Laphamia*, p. 323

352a. PERITYLE Bentham sect. PAPPOTHRIX (A. Gray) A. M. Powell, Sida 3: 277. 1968 [E]

Laphamia A. Gray sect. *Pappothrix* A. Gray, Smithsonian Contr. Knowl. 3(5): 99. 1852; *Pappothrix* (A. Gray) Rydberg

Perennials or subshrubs (usually in rock crevices, stems often brittle). **Involucres** cylindric, narrowly funnelform, or campanulate, 6–12 × 3–15 mm. **Ray florets** 0. **Disc florets** 5–150; corollas cream, yellow, or white. **Cypselae** oblong to narrowly oblanceolate, partially flattened to subcylindric, usually somewhat rounded or angled on 1 or both faces, margins inconspicuously calloused, almost glabrous or short-hairy; **pappi** of 8–35 unequal, antrorsely barbellulate bristles 0.5–3.5 mm (in *P. vitreomontana* all but 2–3 may be reduced to vestigial nubs). *x* = 17.

Species 6 (6 in the flora): sw United States.

SELECTED REFERENCE   Powell, A. M. 1969. Taxonomy of *Perityle*, section *Pappothrix* (Compositae–Peritylanae). Rhodora 71: 58–93.

1. Disc florets 20–150; corollas yellow; New Mexico . . . . 1. *Perityle cernua*
1. Disc florets 5–16; corollas cream, yellow, or white; New Mexico, Texas.
   2. Disc florets 5–6(–8); phyllaries 5–6; plants usually glabrate or puberulent, sometimes hirtellous; New Mexico, Texas . . . . 2. *Perityle quinqueflora*
   2. Disc florets 8–16; phyllaries 8–16; plants hirsute, pilose, puberulent, or tomentose-canescent; Texas.
      3. Leaf blades deltate to ovate overall, 3-lobed . . . . 4. *Perityle fosteri*
      3. Leaf blades ovate, ovate-deltate, subcordate, subdeltate, subovate, or subreniform (not 3-lobed, margins usually toothed).
         4. Corollas yellow or white; pappi of (16–)20–30 bristles; plants hirsute, pilose, or puberulent . . . . 3. *Perityle rupestris*
         4. Corollas white; pappi of 2–20 bristles; plants pilose or tomentose-canescent.
            5. Plants pilose; pappi of 2–6 bristles plus vestigial nubs between bristles, or of 10–20 bristles . . . . 5. *Perityle vitreomontana*
            5. Plants densely tomentose-canescent; pappi of (10–)14–20 bristles . . . . . . 6. *Perityle cinerea*

1. Perityle cernua (Greene) Shinners, SouthW. Naturalist 4: 204. 1959 • Organ Mountain rock daisy [C][E][F]

Laphamia cernua Greene, Bull. Torrey Bot. Club 25: 122, plate 333, figs. 3, 4. 1898

Plants 3–12(–20) cm (closely appressed to rocks); glabrate or with transparent, multicellular hairs on petioles, peduncles, and phyllaries. **Leaves:** petioles 7–25 mm; blades ovate-reniform to ovate-cordate, 10–40 × 8–40 mm, margins unevenly serrate-dentate. **Heads** borne singly (nodding or erect), 10–12 × 12–14 mm. **Peduncles** 10–19 mm. **Involucres** broadly campanulate. **Phyllaries** 18–28, linear-lanceolate, 6–9 × 1.2–2 mm. **Disc florets** 20–150; corollas yellow, tubes 1.2–2 mm, throats narrowly funnelform, 2–3.3 mm, lobes 0.7–0.8 mm. **Cypselae** 2.1–3 mm; **pappi** of 25–35 bristles 0.5–1.5(–2) mm. **2n** = 34 + 1 frag.

Flowering spring–fall. Limestone or igneous cliffs and slopes; of conservation concern; 1600–3000 m; N.Mex.

Perityle cernua is known only from the Organ Mountains of Doña Ana County.

2. Perityle quinqueflora (Steyermark) Shinners, SouthW. Naturalist 4: 205. 1959 • Five-flower rock daisy [E]

Laphamia quinqueflora Steyermark, Ann. Missouri Bot. Gard. 19: 392. 1932; *Pappothrix quinqueflora* (Steyermark) Everly

Plants 7–30 cm (often densely leafy); usually glabrate or puberulent, sometimes hirtellous. **Leaves:** petioles 5–8(–12) mm; blades reniform to cordate, 8–20(–33) × 8–20(–33) mm, margins subentire, dentate-toothed, or shallow-lobed. **Heads** in corymbiform arrays, 7–8(–9) × 2–3 mm. **Peduncles** 2–10(–15) mm. **Involucres** cylindric to narrowly campanulate. **Phyllaries** 5–6, linear-lanceolate, 5–6(–9) × (1–)1.3–1.7 mm. **Disc florets** 5–6(–8); corollas yellow, tubes 1–1.3 mm, throats tubular, 2–2.3(–2.8) mm, lobes 0.5–1 mm. **Cypselae** 1.9–2(–2.9) mm; **pappi** of 25–30 bristles 0.6–2.5(–3.5) mm. **2n** = 34.

Flowering spring–fall. Crevices of limestone (rarely igneous) rock in high canyons, bluffs, and caprock; 1200–2500 m; N.Mex., Tex.

Perityle quinqueflora grows in mountains of west-central trans-Pecos Texas and adjacent New Mexico. Possibly *P. quinqueflora* and the similar *P. rupestris* var. *rupestris* have both evolved from a widespread common ancestor resembling *P. cernua*.

3. **Perityle rupestris** (A. Gray) Shinners, SouthW. Naturalist 4: 204. 1959 • Leafy rock daisy E F

Laphamia rupestris A. Gray, Smithsonian Contr. Knowl. 3(5): 100, plate 9, fig. A. 1852; *Pappothrix rupestris* (A. Gray) Rydberg

Plants 4–25(–35)+ cm (profusely branched, leafy); hirsute, pilose, or puberulent. **Leaves:** petioles 3–13 mm; blades ovate, subcordate, subdeltate, or subreniform, 5–17(–22) × 5–17(–30) mm, margins almost entire to irregularly serrate or serrate-lobed (faces pilose to hirsute). **Heads** borne singly or in corymbiform arrays, 6–8 × 3–5 mm. **Peduncles** 2–13(–20) mm. **Involucres** subcylindric, subfunnelform, or narrowly campanulate. **Phyllaries** 8–10(–14), narrowly lanceolate to oblanceolate, 3.5–7 × 1–2 mm. **Disc florets** 8–16; corollas yellow or white, tubes 0.8–1.3 mm, throats tubular to subfunnelform, 1.8–3 mm, lobes 0.5–1.3 mm. **Cypselae** (1.8–)2–2.5 mm; **pappi** of (16–)20–30 bristles 1–3(–3.5) mm.

Varieties 2 (2 in the flora): Texas.

1. Corollas white . . . . . 3a. *Perityle rupestris* var. *albiflora*
1. Corollas yellow . . . . 3b. *Perityle rupestris* var. *rupestris*

3a. **Perityle rupestris** (A. Gray) Shinners var. **albiflora** A. M. Powell, Sida 3: 180. 1967 • White leafy rock daisy E

Disc florets 8–18; corollas white. **Pappi** of (16–)18–25 bristles 1–2 mm. *2n* = 34, 68, 170.

Flowering spring–fall. Crevices of igneous bluffs and boulders; 1200–1900 m; Tex.

Variety *albiflora* is found throughout much of Brewster County, eastern Presidio County, and extreme southern Jeff Davis County.

3b. **Perityle rupestris** (A. Gray) Shinners var. **rupestris**
• Yellow leafy rock daisy F

Disc florets (7–)8–12; corollas yellow. **Pappi** of 20–30 bristles 2–3(–3.5) mm. *2n* = 34, 68, 136.

Flowering spring–fall. Crevices of igneous boulders and bluffs; 1500–2600 m; Tex.

Variety *rupestris* is found throughout Jeff Davis County and possibly in northern Presidio County.

4. **Perityle fosteri** A. M. Powell, Madroño 30: 217, fig. 1. 1983 • Foster's rock daisy C E

Plants 6–15 cm (forming mats on rock faces, profusely branched, leafy); short-hairy. **Leaves:** petioles 4–8 mm; blades ± ovate to deltate overall, 5–12 × 5–12 mm, 3-lobed, ultimate margins entire. **Heads** borne singly or in corymbiform arrays, 6–7 × ca. 4 mm. **Peduncles** 2–5 mm. **Involucres** funnelform-cylindric. **Phyllaries** ca. 8, linear-lanceolate, 5–6 × 0.8–1.3 mm. **Disc florets** ca. 10; corollas cream to pale yellow, tubes 1 mm, throats tubular-funnelform, 3 mm, lobes 0.5 mm. **Cypselae** 1.5–1.6 mm; **pappi** of ca. 20 bristles 1–2 mm. *2n* = 34.

Flowering spring–fall. Rock faces in protected canyons; of conservation concern; 1500–1660 m; Tex.

Perityle fosteri, possibly a close relative of *P. rupestris*, is known only from Culberson County.

5. **Perityle vitreomontana** Warnock, SouthW. Naturalist 12: 475, fig. 1. 1967 • Glass Mountain rock daisy C E

Plants 3–10 cm (densely leafy); pilose. **Leaves:** petioles 2–4 mm; blades subovate to ovate-deltate, 5–7(–10) × 5–8(–12) mm, margins obtuse-serrate to incised-lobed. **Heads** usually in corymbiform arrays, rarely borne singly, 6–7 × 3–4 mm. **Peduncles** 1–3(–8) mm. **Involucres** narrowly campanulate. **Phyllaries** 9–12, linear-lanceolate, 4–5 × 0.8–1.2 mm. **Disc florets** 11–16; corollas white, tubes 1 mm, throats tubular, 2.5 mm, lobes 1 mm. **Cypselae** 1.9–2 mm; **pappi** of 2–6 bristles 0.5–2 mm plus vestigial nubs between bristles, or of 10–20 bristles 0.1–2 mm. *2n* = 34.

Flowering spring–fall. Limestone bluffs; of conservation concern; 1500–2200 m; Tex.

Perityle vitreomontana is known only from in the Glass Mountains of Brewster County. Although *P. vitreomontana* resembles *P. bisetosa*, close study of the

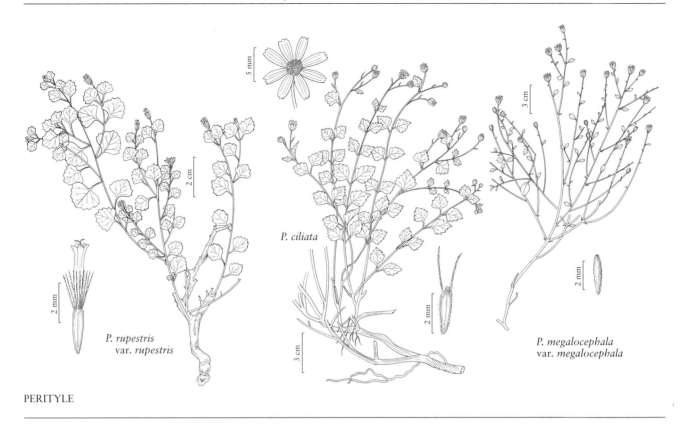

P. ciliata

P. rupestris
var. *rupestris*

P. megalocephala
var. *megalocephala*

PERITYLE

pappus reveals that it belongs in sect. *Pappothrix*. Leaf and flower morphology suggest the relationship of *P. vitreomontana* to *P. rupestris* var. *albiflora*.

6. **Perityle cinerea** (A. Gray) A. M. Powell, Sida 3: 278. 1968 · Gray rock daisy [C][E]

Laphamia cinerea A. Gray in W. H. Emory, Rep. U.S. Mex. Bound. 2(1): 82. 1859; *Pappothrix cinerea* (A. Gray) Rydberg

Plants 8–25(–45) cm (profusely branched, densely leafy); densely tomentose-canescent. **Leaves:** petioles 6–10 mm; blades usually broadly ovate to subdeltate, rarely subreniform, 6–15 × 6–18 mm, margins deeply serrate (veins conspicuous). **Heads** borne singly or in corymbiform arrays, 8–10 × 4–7 mm. **Peduncles** 3–10 mm. **Involucres** narrowly campanulate. **Phyllaries** 8–16, linear-lanceolate to oblanceolate, 4–7 × 1–2 mm. **Disc florets** 10–15; corollas white, tubes 1–1.6 mm, throats tubular, 2.4–3.2 mm, lobes 0.6–1.2 mm. **Cypselae** 2.6–3(–3.5) mm; **pappi** of (10–)14–20 bristles 1–3 mm. $2n$ = ca. 136.

Flowering spring–fall. Crevices, limestone caprock of mesas; of conservation concern; 800–1300 m; Tex.

Perityle cinerea is found in Pecos, Terrell, and Upton counties. The relatively short, appressed habit seen in some populations may be the result of browsing by sheep. *Perityle vitreomontana* and *P. rupestris* var. *albiflora* are thought to be closely related to the uniquely tomentose-canescent *P. cinerea*.

352b. PERITYLE Bentham sect. PERITYLE

Annuals, perennials, or subshrubs (delicate or robust, sometimes weedy, in rock crevices or soil). **Involucres** campanulate to hemispheric, 3–11 × 4–14 mm. **Ray florets** 0 or (1–)8–18; corollas yellow or white. **Disc florets** 20–100+; corollas yellow (sometimes pink or purple tinged). **Cypselae** linear, linear-elliptic, linear-oblong, oblanceolate, oblong, ovate, or subcuneate, usually flattened (outer often 3-angled), margins usually prominently calloused, sometimes thin (not calloused), usually ± densely ciliate; **pappi** 0 or of 1–2(–3+), unequal to subequal, ± barbellate bristles 0.5–4(–6) mm, sometimes plus crowns of hyaline, laciniate scales. x = 17, 19.

Species 30 (7 in the flora): United States, Mexico, South America.

SELECTED REFERENCE   Powell, A. M. 1974. Taxonomy of *Perityle*, section *Perityle* (Compositae–Peritylanae). Rhodora 76: 229–306.

1. Ray florets 0 (Brewster and Terrell counties, Texas) . . . . 13. *Perityle aglossa*
1. Ray florets usually 6–18 (sometimes 0 or 1–6 in *P. parryi* from w Presidio County, Texas).
  2. Ray and disc corollas yellow; Texas.
    3. Leaves usually cordate to subreniform, margins usually irregularly dentate, laciniate, or ± 3-lobed, seldom more divided (perennials or subshrubs in rock crevices or in soil) . . . . 11. *Perityle parryi*
    3. Leaves usually ternately lobed or ± cruciform, lobes usually again ternately lobed (perennials in soil) . . . . 12. *Perityle vaseyi*
  2. Ray corollas white, disc corollas yellow (sometimes purple tinged); Arizona, California, Nevada, New Mexico, Utah.
    4. Annuals (sometimes persisting).
      5. Cypselae (1.5–)2–3 mm, margins thin (not calloused); pappi 0, or of single, antrorsely to retrorsely barbellate bristles 1–3 mm plus crowns of hyaline, laciniate scales . . . . 7. *Perityle emoryi*
      5. Cypselae 1.5–2 mm, margins prominently calloused; pappi of 2 unequal, antrorsely barbellate bristles 0.8–1.2 mm (the longest) plus crowns of hyaline, laciniate scales . . . . 8. *Perityle microglossa*
    4. Subshrubs.
      6. Leaf blades deltate-ovate to ovate-rhombic, margins usually entire or serrate to serrate-crenate, sometimes shallow-lobed; c Arizona . . . . 9. *Perityle ciliata*
      6. Leaf blades pedately lobed (lobes spatulate or linear) or 2–3-pinnatifid (lobes linear-filiform); s Arizona, New Mexico . . . . 10. *Perityle coronopifolia*

7. Perityle emoryi Torrey in W. H. Emory, Not. Milit. Reconn., 142. 1848 • Emory's rock daisy

Annuals (sometimes persisting), 2–60 cm (delicate or robust, stems relatively few to many, erect or spreading); puberulent to hirsute, glandular-pubescent. **Leaves:** petioles 3–45 mm; blades ovate, cordate, suborbiculate, or triangular, 17–60 × 10–50 mm, margins deeply toothed, lobed, cleft, or divided, lobes indented to irregularly dissected. **Heads** borne singly or in corymbiform arrays, 4–10 × 4–10 mm. **Peduncles** 1–70 mm. **Involucres** campanulate to hemispheric. **Phyllaries** 10–20, lanceolate or oblanceolate to ovate-lanceolate, 4–6 × 1–2 mm. **Ray florets** usually 8–14, rarely rudimentary or 0; corollas white, laminae oblong, 1–4(–6) × 1–3 mm. **Disc florets** 40–100+; corollas yellow, tubes 0.7–1.3 mm, throats tubular to tubular-funnelform, 0.8–1.3 mm, lobes 0.1–0.2 mm. **Cypselae** suboblong, oblanceolate, or subcuneate, (1.5–)2–3 mm, margins thin (not calloused), long- or short-ciliate; **pappi** 0 or of 1 antrorsely to retrorsely barbellate bristles 1–3 mm plus crowns of hyaline, laciniate scales. *2n* = 65–72 or 100–116.

Flowering year round (depending on latitude). Coastal bluffs, desert plains, slopes, washes; 10–1500 m; Ariz., Calif., Nev., Utah; Mexico; South America (Chile, Peru).

Perityle emoryi is a widespread polyploid of diverse habitats and is often weedy. It is variable; none of the variation appears to have population significance and does not require taxonomic recognition. The range of *P. emoryi* appears to be gradually expanding.

8. Perityle microglossa Bentham, Bot. Voy. Sulphur, 119. 1844 • Short-ray rock daisy

Varieties 2 (1 in the flora): sc United States, Mexico.

8a. Perityle microglossa Bentham var. microglossa

Perityle microglossa var. *effusa* A. Gray

Annuals, 20–60 cm (stems erect or decumbent); usually copiously glandular-pubescent, rarely puberulent with scattered glandular hairs. **Leaves:** petioles 5–30(–40) mm; blades often turning purplish (variable in size and shape), cordate, ovate, or broadly ovate-cordate to subreniform, or subdeltate, usually 20–60 × 20–50 mm, margins singly or double crenate, even or irregular to strongly 3-lobed or cleft, pedately divided to subhastate (faces puberulent, glandular-puberulent, or glabrous. **Heads** borne singly or in corymbiform arrays, 3.5–4.5(–6) × 4–6.5(–7) mm. **Peduncles** 6–70 mm. **Involucres**

campanulate. **Phyllaries** 14–30, narrowly to broadly lanceolate, 2.5–3 × 0.6–1 mm. **Ray florets** 5–13; corollas white, laminae oblong, 1.5–3.5 mm. **Disc florets** 40–100; corollas yellow, tubes 0.5–0.7 mm, throats tubular-funnelform, 0.5–0.8 mm, lobes 0.1–0.2 mm. **Cypselae** usually linear-oblong to linear-elliptic, rarely narrowly obovate, 1.5–2 mm, margins usually prominently calloused, ciliate; **pappi** of 2 unequal, antrorsely barbellate bristles 0.5–1.2 mm plus crowns of hyaline, laciniate scales. $2n = 68, 102$.

Flowering year around. Disturbed sites; 0–1000 m; Ariz., Tex.; Mexico.

The only known occurrence of *Perityle microglossa* from Arizona is the type of *P. microglossa* var. *effusa* (Santa Catalina Mountains, Pima County, 1882). Variety *microglossa* occurs as an introduced weed in extreme southern Texas and northern Mexico. Its shorter ligules and more glandular nature distinguish it from var. *saxosa*, which occurs only in Mexico. Cypsela and pappus characters, and distribution, distinguish *P. microglossa* from *P. emoryi*.

9. Perityle ciliata (L. H. Dewey) Rydberg in N. L. Britton et al., N. Amer. Fl. 34: 17. 1914 · Fringed rock daisy

E F

Laphamia ciliata L. H. Dewey, Bot. Gaz. 20: 425. 1895

Subshrubs, 15–30 cm (in rock crevices, stems relatively many, erect to pendulous); sparsely to densely short-hairy, glandular. **Leaves:** petioles 2–15 mm; blades deltate-ovate to ovate-rhombic, 6–23 × 5–24 mm, margins usually entire or serrate to serrate-crenate, sometimes shallow-lobed. **Heads** in corymbiform arrays, 5–7 × 5.5–7 mm. **Peduncles** 5–25 mm. **Involucres** campanulate. **Phyllaries** 13–20, linear-lanceolate to narrow-ovate, 4–5.5 × 1–2 mm. **Ray florets** 6–10; corollas white, sometimes pink tinged, laminae broadly oblong to oblong-elliptic, 3–7 × 1.5–3 mm. **Disc florets** 30–40; corollas yellow, often purple tinged, tubes 0.8–1 mm, throats tubular to tubular-funnelform, 1.2–1.4 mm, lobes 0.3–0.4 mm. **Cypselae** linear-oblong to oblanceolate, 2–2.8 mm, margins prominently calloused, long-ciliate; **pappi** of 2(–3+) barbellulate bristles 1.5–2.5 mm plus crowns of hyaline, laciniate scales. $2n = 34$.

Flowering spring–fall. In rock crevices; 1100–2500 m; Ariz.

Perityle ciliata is found only in the mountains of central Arizona in Apache, Coconino, Gila, Mohave, and Yavapai counties. It appears to be most closely related to *P. coronopifolia*.

10. Perityle coronopifolia A. Gray, Smithsonian Contr. Knowl. 5(6): 82. 1853 · Crow-foot rock daisy

Laphamia coronopifolia (A. Gray) Hemsley

Subshrubs, 6–36 cm (in rock crevices, stems relatively many, erect or pendulous, very leafy); sparsely to densely grayish hairy. **Leaves:** petioles 2–8(–12) mm; blades pedately 3-lobed (lobes spatulate or linear), or 2–3-pinnatifid (lobes linear-filiform), 4–30 × 4–20 mm, ultimate margins entire. **Heads** (2–5) in corymbiform arrays, 5–6.5 × 5–6 mm. **Peduncles** 7–15 mm. **Involucres** campanulate. **Phyllaries** 2–16, linear-lanceolate to narrow-ovate, 3.5–5 × 0.5–1.5 mm. **Ray florets** 8–12; corollas white, laminae broadly oblong or oblong-elliptic to subspatulate, 3–7 × 2–3 mm. **Disc florets** 30–40; corollas yellow, often purple tinged, tubes 0.8–1 mm, throats tubular, tubular-funnelform, or tubular-campanulate, 1–1.3 mm, lobes 0.3–0.4 mm. **Cypselae** linear-oblong to narrowly oblanceolate, 1.8–2.5 mm, margins usually prominently calloused, sometimes thin, usually ciliate; **pappi** of 2 (–3+) barbellulate bristles 1.5–2.5 mm plus crowns of hyaline, laciniate scales. $2n = 34$.

Flowering spring–fall. Rock and cliff faces; 1000–2600 m; Ariz., N.Mex.; Mexico (Chihuahua).

Perityle coronopifolia is widespread in south-central and southeastern Arizona, and southwestern and south-central New Mexico. The combination of white rays, often pinnatifid leaves, and perennial habit distinguish it.

11. Perityle parryi A. Gray, Smithsonian Contr. Knowl. 5(6): 106. 1853 · Parry's rock daisy

Laphamia parryi (A. Gray) Bentham & Hooker f.

Perennials or subshrubs, 10–75 cm (in rock crevices, or in soil); puberulent, villous, or rarely tomentose. **Leaves:** petioles 12–40 mm; blades usually cordate to subreniform, 8–40(–50) × 8–50(–60) mm, margins irregularly dentate, laciniate, or ± 3-lobed, seldom more divided. **Heads** borne singly or (2–3) in corymbiform arrays, 8–10 × 7–14 mm. **Peduncles** 10–70 mm. **Involucres** campanulate to hemispheric. **Phyllaries** 28–36, lanceolate to linear, seldom oblanceolate, 5–9 × 0.5–1.6 mm, apices acute to short- or long-attenuate. **Ray florets** usually 12–18 (sometimes 0 or 1–6 in plants from w Presidio County); corollas yellow, laminae oblong, 4–10 × 2–4 mm. **Disc florets** 60–100; corollas yellow, tubes 1–1.3 mm, throats tubular, 2.5–3.2 mm, lobes 0.4–0.6 mm. **Cypselae** linear-elliptic to narrowly oblanceolate, (2–)3–4 mm, margins usually prominently calloused (sometimes thin), usually densely ciliate, sometimes short-hairy; **pappi** 0 or of 1(–2) usually antrorsely, sometimes erectly or retrorsely barbellate, bristles (1–)3–4(–6) mm plus crowns of hyaline, laciniate scales. $2n = 34$.

Flowering year around. Igneous or limestone rock or soil; 700–2000 m; Tex.; Mexico (Chihuahua).

Perityle parryi occurs in southern Presidio and Brewster counties. Atypical populations with pappus bristles resembling those of *P. vaseyi* and/or with ray florets present or absent may be found in western Presidio County.

12. **Perityle vaseyi** J. M. Coulter, Contr. U.S. Natl. Herb. 1: 42. 1890 • Vasey's rock daisy

Perennials or subshrubs, (10–)15–75 cm (in soil, tap roots fleshy, stems erect or spreading); glandular-pubescent. **Leaves:** petioles 10–35 mm; blades usually palmately 3-lobed or ± cruciform, 6–35 × 10–50 mm, lobes usually again ternately dissected, cleft, or parted. **Heads** borne singly or (2–3) in corymbiform arrays, 8–10 × 10–13 mm. **Peduncles** 10–80 mm. **Involucres** campanulate to hemispheric. **Phyllaries** 20–30, oblanceolate or lanceolate to linear-lanceolate, 6–7 × 1–2 mm, apices usually acute, sometimes short-attenuate. **Ray florets** 12–18; corollas yellow, laminae oblong, 4–10 × 2–4 mm. **Disc florets** 50–100; corollas yellow, tubes 1.2–1.5 mm, throats tubular to broadly tubular, 2.2–3 mm, lobes 0.4–0.6 mm. **Cypselae** linear-elliptic to narrowly oblanceolate, (2–)3–4 mm, margins usually prominently calloused, sometimes thin, usually densely ciliate, sometimes short-hairy; **pappi** 0 or of 1(–2) retrorsely or erectly barbellate (at least near tips) bristles 1–4 mm plus crowns of hyaline, laciniate scales. $2n = 34$.

Flowering year around. Desert soils, especially gypsiferous clays; 700–1800 m; Tex.; Mexico (Chihuahua, Coahuila).

Perityle vaseyi is always found growing in soil, often on roadsides and in road cuts in the southwestern Big Bend area. The more deeply dissected leaves, shorter pappus bristles with barbellate tips, and broader phyllaries help distinguish it from the soil-growing form of *P. parryi*.

13. **Perityle aglossa** A. Gray, Smithsonian Contr. Knowl. 5(6): 107. 1853 • Rayless rock daisy, bluff rock daisy

Laphamia aglossa (A. Gray) Bentham & Hooker f.

Perennials or subshrubs, 15–45 cm (in rock crevices); glandular-pubescent. **Leaves:** petioles 8–23 mm; blades ovate, subcordate, or subdeltate, 6–20 × 5–25 mm, margins irregularly lobed, laciniate, serrate-dentate, or shallowly 3-lobed. **Heads** borne singly or (2–3) in corymbiform arrays, 8–11 × 4–8 mm. **Peduncles** 10–25 mm. **Involucres** campanulate. **Phyllaries** 14–24, linear to linear-lanceolate, (5–)6–10 × 0.5–1 mm, apices long-attenuate. **Ray florets** 0. **Disc florets** 20–60; corollas yellow, often pink or purple tinged, tubes 1–1.7 mm, throats tubular, 2.1–3.2 mm, lobes 0.4–0.6 mm. **Cypselae** oblong to narrowly oblanceolate, (1–)2–3.2 mm, margins prominently calloused, densely ciliate; **pappi** of 1(–2) antrorsely barbellate bristles 3.5–6 mm plus crowns of hyaline, laciniate scales. $2n = 34$.

Flowering spring–fall. Limestone rock crevices; 600–1300 m; Tex.; Mexico (Coahuila).

Perityle aglossa is consistently discoid and occurs in limestone crevices along and near the Rio Grande in Brewster and Terrell counties. A combination of subtle characteristics including shallower leaf lobing, narrower shape and more attenuate apices of the phyllaries, shorter cypselae, and proportionally longer pappus bristles distinguish these plants from the rayless ones of *P. parryi*.

352c. PERITYLE Bentham sect. LAPHAMIA (A. Gray) A. M. Powell, Sida 3: 277. 1968

Laphamia A. Gray, Smithsonian Contr. Knowl. 3(5): 99, plate 9. 1852

Perennials or subshrubs (in rock crevices). **Involucres** funnelform, campanulate, or hemispheric, 5–11 × 3.5–14 mm. **Ray florets** 0, or 3–5, or 6–18; corollas yellow. **Disc florets** 9–200; corollas yellow or white. **Cypselae** linear, linear-lanceolate, oblanceolate or oblong to oblanceolate, flattened, often obscurely rounded or angled on 1 or both surfaces, margins usually inconspicuously calloused and sparsely short-hairy; **pappi** 0 or of 1–2(–6) equal or subequal, smooth, antrorsely barbellulate, or subplumose bristles 0.5–4.6 mm plus callous crowns or hyaline scales. $x = 17$.

Species 29 (22 in the flora): United States, Mexico.

SELECTED REFERENCE Powell, A. M. 1973b. Taxonomy of *Perityle*, section *Laphamia* (Compositae–Peritylanae). Sida 5: 61–128.

1. Corollas white; rays 0; plants 2–12 cm; Texas (Big Bend area).
   2. Pappi 0 or callous crowns (bristles 0); plants densely scabrous-pubescent . . . . 14. *Perityle warnockii*
   2. Pappi usually of 2–3(–5) bristles; plants glabrate, minutely hairy, or scabrous-hispidulous . . . . 15. *Perityle bisetosa*
1. Corollas yellow; rays 0, or 3–5, or 6–18; plants (4–)12–75 cm; Arizona, California, Nevada, New Mexico, Texas, or Utah.
   3. Rays 3–5 or 6–18.
      4. Leaf blades broadly ovate, ovate-deltate, ovate-lanceolate, or ovate-rhombic; disc florets 11–22; Texas.
         5. Disc florets 11–13; Hueco Mountains . . . . 34. *Perityle huecoensis*
         5. Disc florets 16–22; Edwards Plateau . . . . 31. *Perityle lindheimeri*
      4. Leaf blades ± cruciform, subdeltate, or subovate, usually 3-partite or lobed, divided, or pinnate; disc florets 30–200; Arizona, Nevada, New Mexico, Utah.
         6. Leaf blades usually ± cruciform, sometimes lobed or pinnate; New Mexico . . . . 33. *Perityle staurophylla* (in part)
         6. Leaf blades subdeltate, suborbiculate, or subovate, or deeply lobed, 3-foliolate, or 2–3-pinnate; Arizona, Nevada, Utah.
            7. Disc florets 60–80; leaf blades subdeltate, suborbiculate, or subovate, margins serrate or ± lobed; Nevada, Utah. . . . . 16. *Perityle stansburii*
            7. Disc florets 75–200; leaf blades 3-foliolate, or 2–3-pinnate; Arizona.
               8. Leaf blades 2–3-pinnate (lobes linear to flat-filiform, scarcely broadened distally) . . . . 25. *Perityle saxicola*
               8. Leaf blades usually 3-foliolate (lobes much broadened distally) . . . . 26. *Perityle gilensis*
   3. Rays usually 0 (sometimes 1–2, reduced, in *P. ambrosiifolia*).
      9. Plants long-hairy, pilose, or villous.
         10. Leaf blades deeply dissected (lobes relatively numerous, irregular).
            11. Phyllaries 4–6.4(–8) mm (apices acute); disc corolla throats broadly tubular to subfunnelform; pappus bristles delicate . . . . 28. *Perityle lemmonii* (in part)
            11. Phyllaries 6–9 or (5–)7–13 mm (apices usually attenuate, sometimes acute); disc corolla throats narrowly tubular to narrowly funnelform; pappus bristles moderately stout.
               12. Leaf blades 15–30 mm wide; phyllaries 14–20; Arizona . . . . . 29. *Perityle ambrosiifolia*
               12. Leaf blades 4–15 mm wide; phyllaries 12–15; Texas . . . . 30. *Perityle dissecta*
         10. Leaf blades entire, serrate, or shallow-lobed.
            13. Leaf margins entire or with 1–3 lobes per side (leaves all alternate) . . . . . 22. *Perityle villosa*
            13. Leaf margins serrate (leaves opposite or alternate).
               14. Pappi usually callous crowns (bristles 0); California . . . . 21. *Perityle inyoensis*
               14. Pappus usually of single, delicate bristles 1–3 mm; Arizona, New Mexico . . . . 28. *Perityle lemmonii* (in part)
      9. Plants glabrous, or hirtellous, hispidulous, or puberulent, or short-hairy, or short-tomentose.
         15. Plants densely short-hairy or short-tomentose.
            16. Leaf margins serrate; plants densely short-hairy; nw Arizona, Nevada, Utah . . . . 23. *Perityle tenella*
            16. Leaf margins usually shallow-crenate; plants densely short-tomentose; sw Arizona . . . . 27. *Perityle ajoensis*
         15. Plants glabrous, or hirtellous, hispidulous, or puberulent.
            17. Leaf blades 2–8(–15) mm (stems sparsely leafy).
               18. Pappus bristles usually 1–3(–4), rarely 0; se Utah . . . . 19. *Perityle specuicola*
               18. Pappus bristles usually 0, sometimes 1–2; California, s Nevada.

19. Heads borne singly or (2–3) in loose, corymbiform arrays; peduncles 10–45(–80) mm; leaf blades usually elliptic, lanceolate, lance-ovate, ovate, or suborbiculate (sometimes linear in var. *oligophylla*)............................. 17. *Perityle megalocephala*
19. Heads (2–6) in tight, corymbiform arrays; peduncles 1–10 mm; leaf blades linear................................. 18. *Perityle intricata*

[17. Shifted to left margin.—Ed.]

17. Leaf blades 3–30 mm (stems moderately to densely leafy).
  20. Pappi usually 0, sometimes 1–2 bristles; plants glabrate or puberulent; Texas... 32. *Perityle angustifolia*
  20. Pappi usually of 1–3(–6) bristles, sometimes 0; plants glabrous, hirtellous, or short-hairy; Arizona, Nevada.
    21. Leaf blades usually deeply divided (lobes 3, relatively broad to linear, usually secondarily lobed, cleft, or parted, often ± cruciform, sometimes almost entire); sc Arizona... 33. *Perityle staurophylla* (in part)
    21. Leaf blades entire, toothed, or lobed; Arizona, Nevada.
      22. Plants 4–12(–15+) cm; petioles 5–30 mm (equal to or longer than blades; leaves usually glabrous); se Arizona .... 35. *Perityle cochisensis*
      22. Plants 10–30(–45) cm (petioles shorter than blades; leaves hairy); nw Arizona, se Nevada.
        23. Leaf blades usually 3-lobed to ± cruciform; nw Arizona, se Nevada.... 20. *Perityle gracilis*
        23. Leaf blades elliptic, lanceolate, or ovate, usually entire or 2–6-toothed; nw Arizona .... 24. *Perityle congesta*

14. **Perityle warnockii** A. M. Powell, Sida 3: 177, fig. 1. 1966 • Warnock's rock daisy C E

Perennials or subshrubs, 2–10 cm; densely scabrous-pubescent. **Leaves:** petioles 2–5(–8) mm; blades lance-ovate, ovate, or suborbiculate, 6–10 × 4–8(–10) mm, margins serrate. **Heads** borne singly (often partially obscured by leaves), 7–10 × 4–7 mm. **Peduncles** 1–7 mm. **Involucres** campanulate. **Phyllaries** 9–10, linear to lanceolate, 5–7 × 1.2–2.2 mm. **Ray florets** 0. **Disc florets** 9–12; corollas bright white, tubes 0.6–1 mm, throats subfunnelform, 2–2.5 mm, lobes 1–1.6 mm. **Cypselae** oblong to oblanceolate, 2.2–2.8 mm, margins notably calloused, glabrous; **pappi** callous crowns, bristles 0. $2n = 34$.

Flowering spring–fall. Limestone; of conservation concern; 500–700 m; Tex.

Perityle warnockii is known only from the type locality in Val Verde County. The relationship of *P. warnockii* to *P. bisetosa* is evidenced by its similar though often reduced vegetative and floral morphology. The two species may have been derived separately from a common ancestor, perhaps a white-flowered member of sect. *Pappothrix* such as *P. vitreomontana*.

15. **Perityle bisetosa** (Torrey ex A. Gray) Shinners, SouthW. Naturalist 4: 204. 1959 • Two-bristle rock daisy C

Laphamia bisetosa Torrey ex A. Gray, Smithsonian Contr. Knowl. 5(6): 106. 1853

Perennials or subshrubs, 2–12 cm; glabrate, glabrous, minutely hairy, or scabrous-hispidulous. **Leaves:** petioles (0–)1–7 mm; blades lanceolate, lance-ovate, ovate, or suborbiculate, 4–8 × 2–8 mm, margins entire, serrate, or shallow-lobed. **Heads** usually borne singly (often obscured by the dense leaves), 7–8(–10) × 3.5–7 mm. **Peduncles** 0–3 mm. **Involucres** narrowly campanulate. **Phyllaries** 8–10(–12), linear to broadly or narrowly lanceolate, 4.5–6 × 1–2 mm. **Ray florets** 0. **Disc florets** 11–17; corollas white, tubes 0.8–1 mm, throats tubular to subfunnelform, 1.5–3 mm, lobes 0.8–1 mm. **Cypselae** oblong to oblanceolate, 1.8–2.2(–2.5) mm, margins notably calloused, glabrous; **pappi** of 2–3(–5) ± flattened bristles 1.5–2.3 mm.

Varieties 4 (3 in the flora): Texas; Mexico.

1. Petioles 0–2 mm (glabrate); leaf margins entire or subserrate......... 15b. *Perityle bisetosa* var. *bisetosa*
1. Petioles 2–7 mm (glabrous, minutely hairy, or scabrous-hispidulous); leaf margins usually serrate or lobed.
  2. Leaves glabrous or minutely hairy........ .... 5a. *Perityle bisetosa* var. *appressa*
  2. Leaves scabrous-hispidulous .... 5c. *Perityle bisetosa* var. *scalaris*

15a. Perityle bisetosa (Torrey ex A. Gray) Shinners var. **appressa** A. M. Powell, Sida 3: 178, fig. 2. 1967
C

Plants 4–8 cm. **Leaves** mostly alternate; petioles 3–7 mm; blades ovate, margins with 2–3 serrations or lobes, faces glabrous or minutely hairy; **pappi** of 2–3(–5) bristles (1–3 reduced). **2*n*** = 34.

Flowering spring–fall. Caprock bluffs; of conservation concern; 1200 m; Tex.; Mexico.

Variety *appressa* is restricted to caprock bluffs in southeastern Brewster County and adjacent Mexico.

15b. Perityle bisetosa (Torrey ex A. Gray) Shinners var. **bisetosa** C E

Plants 2–7(–10) cm. **Leaves** usually opposite; petioles 0–2 mm; blades ovate or ovate-lanceolate to lanceolate, margins entire or subserrate, faces glabrate; **pappi** of 2 bristles. **2*n*** = 34.

Flowering spring–fall. Limestone caprocks; of conservation concern; 1200–1300 m; Tex.

Variety *bisetosa* appears to be restricted to a relatively small area in extreme southern Pecos County and adjacent Brewster County.

15c. Perityle bisetosa (Torrey ex A. Gray) Shinners var. **scalaris** A. M. Powell, Sida 3: 177, fig. 3. 1967
C E

Plants 6–12 cm. **Leaves** proximally opposite, distally alternate; petioles 2–4 mm; blades broadly ovate to suborbiculate, margins with 1–2 shallow serrations or lobes, faces scabrous-hispidulous; **pappi** of 2–3(–4) bristles (1–2 reduced). **2*n*** = ca. 204.

Flowering spring–fall. Crevices of limestone bluffs; of conservation concern;1200–1600 m; Tex.

Variety *scalaris* is known only from at or near the type locality east of the Dead Horse Mountains in southeastern Brewster County.

16. Perityle stansburii (A. Gray) J. F. Macbride, Contr. Gray Herb. 56: 39. 1918 • Stansbury's rock daisy
E

Laphamia stansburii A. Gray, Smithsonian Contr. Knowl. 3(5): 101. 1852

Perennials or subshrubs, 7–45 cm (often dense clumps to 60 cm across); hirtellous. **Leaves:** petioles 3–10 mm; blades subdeltate, suborbiculate, or subovate, 3–14 × 3–15 mm, margins usually 2–5-lobed or serrate, sometimes subentire or 3-lobed. **Heads** borne singly or in corymbiform arrays, 7–8 × 5–9 mm. **Peduncles** 5–60 mm. **Involucres** campanulate. **Phyllaries** 14–22, lanceolate to broadly oblanceolate, 5–6 × 1–2 mm. **Ray florets** 6–14; corollas yellow, laminae 3–6 × 1.2–3 mm. **Disc florets** 60–80; corollas yellow, tubes 1.2–1.5 mm, throats tubular to subfunnelform, 2.4–3 mm, lobes 0.4–0.6 mm. **Cypselae** narrowly oblanceolate, 2–3.5 mm, margins thin-calloused, short-hairy; **pappi** of single, ± stout bristles 2.5–4 mm plus crowns of vestigial, hyaline scales. **2*n*** = 34.

Flowering spring–fall. Rock crevices; 1200–2600 m; Nev., Utah.

Perityle stansburii with its large habit, relatively broad leaves, radiate heads, bristle pappus, chromosome number, and wide distribution, fills most expectations as the ancestral taxon of the group of related species called the "southwestern alliance." This natural assemblage of taxa, which is thought to have evolved through geographic displacement and subsequent genetic differentiation, includes *P. congesta, P. gracilis, P. intricata, P. inyoensis, P. megalocephala, P. specuicola, P. tenella,* and *P. villosa.* These taxa, which are found mostly to the south and west of *P. stansburii,* all have rayless heads and may or may not have pappus bristles. In western and northwestern Utah and adjacent Nevada, *P. stansburii* occurs in crevices of rock exposures.

17. Perityle megalocephala (S. Watson) J. F. Macbride, Contr. Gray Herb. 56: 39. 1918 • Nevada rock daisy
E F

Laphamia megalocephala S. Watson, Amer. Naturalist 7: 301. 1873

Perennials or subshrubs, 15–55 cm (often dense, profusely branched clumps, sparsely leafy); densely hirtellous. **Leaves:** petioles 1–6 mm; blades usually elliptic, lanceolate, lance-ovate, ovate, or suborbiculate, sometimes linear, 4–8(–15) × 1–10(–12) mm, margins entire, irregularly and sparsely serrate, or serrate-lobed.

Heads borne singly or (2–3) in loose, corymbiform arrays, 6–9(–10) × 5–6(–8) mm. **Peduncles** 10–45(–80) mm. **Involucres** campanulate. **Phyllaries** 14–20, lanceolate to suboblanceolate, 5–6 × 1.3–1.9 mm. **Ray florets** 0. **Disc florets** 45–60; corollas yellow, tubes 1–1.6 mm, throats tubular to subfunnelform, 1.6–2.2 mm, lobes 0.4–0.7 mm. **Cypselae** narrowly oblanceolate to suboblanceolate, 2.5–3 mm, margins thin-calloused, short-hairy; **pappi** 0, or of single bristles.

Varieties 2 (2 in the flora): w United States.

1. Leaf blades ovate, lanceolate-ovate or elliptic to suborbiculate . . . .
. . . . 17a. *Perityle megalocephala* var. *megalocephala*
1. Leaf blades linear to lanceolate . . . .
. . . . 17b. *Perityle megalocephala* var. *oligophylla*

17a. Perityle megalocephala (S. Watson) J. F. Macbride var. **megalocephala** [E] [F]

Plants 30–55 cm. **Leaves:** petioles 1–6 mm; blades ovate, lanceolate-ovate or elliptic to suborbiculate, 4–8(–12) × 3–9(–12) mm. **Pappi** callous crowns or of vestigial, hyaline scales. $2n = 34$.

Flowering spring–fall. Crevices and rocky slopes in arid mountains; 1400–3000 m; Calif., Nev.

Variety *megalocephala* is relatively common, growing usually in volcanic, also granitic and limestone rock in arid mountains of western and south-central Nevada and adjacent California. The lack of rays and pappus bristles separates var. *megalocephala* from the similar *Perityle stansburii.*

17b. Perityle megalocephala (S. Watson) J. F. Macbride var. **oligophylla** A. M. Powell, Sida 5: 91, fig. 6. 1973 • Small-leaved rock daisy [E]

Plants 15–35 cm. **Leaves:** petioles 0–4 mm; blades linear to lanceolate, 4–10(–15) × 1–4(–6) mm. **Pappi** 0, or of single bristles (sometimes plus callous crowns or vestigial, hyaline scales). $2n = 68$.

Flowering spring–fall (depending on rains). Rocky slopes; 1300–2500 m; Calif.

Variety *oligophylla* is ephemeral in northern exposures or protected areas in mountains in the Death Valley region. At least some leaves are linear and the heads and habit are generally smaller than in var. *megalocephala.*

18. Perityle intricata (Brandegee) Shinners, SouthW. Naturalist 4: 204. 1959 • Desert rock daisy [E]

Laphamia intricata Brandegee, Bot. Gaz. 27: 450. 1899; *L. megalocephala* S. Watson subsp. *intricata* (Brandegee) D. D. Keck; *Perityle megalocephala* (S. Watson) J. F. Macbride var. *intricata* (Brandegee) A. M. Powell

Perennials or subshrubs, 13–35 (–40) cm; hirtellous. **Leaves:** petioles usually 0 mm; blades linear, 3–8(–15) × 0.2–1 mm; margins usually entire, rarely 1–3-lobed. **Heads** (2–6) in tight, corymbiform arrays, 4–6.5 × 3.5–5.5 mm. **Peduncles** 1–10 mm. **Involucres** campanulate. **Phyllaries** 3–4, linear or lanceolate to oblanceolate, 5(–5) × 0.5–1.2 mm. **Ray florets** 0. **Disc florets** 24–36; corollas yellow, tubes 0.5–1 mm, throats tubular, 1–1.2, lobes 0.4–0.6 mm. **Cypselae** oblanceolate, (1.5–)2.2–2.8 mm; margins moderately calloused, moderately to densely coarse-ciliate; **pappi** usually 0, sometimes of 1–2 bristles 1–2 mm, usually plus crowns of cilialike scales, rarely callous crowns. $2n = 38 \pm 2$.

Flowering spring–fall (depending on rains). Limestone crevices, dry mountain slopes and canyons; 800–1700 m; Calif., Nev.

Perityle intricata has been treated as a subspecies or variety of *P. megalocephala.* Now that more collections are available for study, it is evident that this taxon should be treated as a species. The linear leaves, tightly aggregated heads on relatively short peduncles, and densely coarse-ciliate cypsela margins separate it from *P. megalocephala* with its broader leaves, loosely aggregated heads on longer peduncles, and short-hairy cypsela margins. Where the two distributions overlap (southern Nye County), they are reported to appear distinct in the field, with *P. intricata* occurring at lower elevations than *P. megalocephala.*

19. Perityle speciicola S. L. Welsh & Neese, Great Basin Naturalist 43: 373. 1983 • Hanging-garden or alcove rock daisy [C] [E]

Perennials or subshrubs, 30–75 cm (sprawling or pendulous, much branched, sparsely leafy); hispidulous. **Leaves** (mostly alternate): petioles 1–3(–8) mm; blades ovate to elliptic, 2–8 × 1–5 mm, margins entire. **Heads** borne singly or in corymbiform arrays, 3.5–5 × 5–6 mm. **Peduncles** 4–70 mm. **Involucres** campanulate. **Phyllaries** 11–16, lanceolate to oblanceolate, 3.5–5 × 0.5–1 mm. **Ray florets** 0. **Disc florets** 30–60; corollas yellow, tubes 0.5–0.8 mm, throats broadly tubular, 1.3–1.8 mm, lobes 0.4–0.6 mm. **Cypselae** narrowly oblanceolate, 3–3.8 mm, faces glabrous, margins thin-calloused, relatively long-ciliate;

pappi usually of 1–3(–4) unequal bristles 1–2.5 mm, rarely 0.

Flowering spring–fall. Rock crevices and faces along rivers and seeps; of conservation concern; 1100–2300 m; Utah.

Perityle specuicola, which resembles *P. megalocephala* and *P. stansburii* in having relatively small, sparse leaves, is found only in "hanging gardens" along streams and near seeps in Grand and San Juan counties. The stems of this rayless species are usually sprawling and long-pendent, the cypsela margins are thin and long-ciliate, and the pappi are of 1–4 bristles.

20. **Perityle gracilis** (M. E. Jones) Rydberg in N. L. Britton et al., N. Amer. Fl. 34: 19. 1914 • Slender or three-lobed rock daisy C E

Laphamia gracilis M. E. Jones, Proc. Calif. Acad. Sci., ser. 2, 5: 703. 1895

Perennials or subshrubs, 10–20 (–25) cm (stems erect, spreading, or pendent); glabrate or densely puberulent. **Leaves:** petioles 2–5 (–10) mm; blades palmately 3-lobed, or ± cruciform, sometimes entire, or with 1–3 lobes or teeth per side, 3–10(–15) × 2–15 mm, ultimate margins sometimes irregularly lobed or toothed. **Heads** borne singly or (2–6+) in corymbiform arrays, 5–7 × 4–5.5 mm. **Peduncles** 5–15 mm. **Involucres** campanulate. **Phyllaries** (8–)12–17, linear-lanceolate to lanceolate, 3.5–5.5 × 0.7–1.2 mm. **Ray florets** 0. **Disc florets** 30–42; corollas yellow, tubes 0.8–1.1 mm, throats ± tubular, (1.3–)1.6–2 mm, lobes 0.4–0.6 mm. **Cypselae** oblong-elliptic, 1.8–2.5(–2.8) mm, margins thin-calloused, short-hairy; **pappi** 0 or of 1(–3) stout bristles to 2 mm, usually plus callous crowns or vestigial, hyaline scales. 2*n* = 68.

Flowering spring–fall. Crevices of limestone rock faces; of conservation concern; 1200–2200 m; Ariz., Nev.

Perityle gracilis is known from southeastern Nevada in Meadow Valley Wash and in the Sheep Range. The only Arizona collection is the type (*M. E. Jones* in 1894, Kaibab Plateau). The 3-fid leaves with cuneate bases set *P. gracilis* apart from the similar *P. congesta* and *P. tenella*.

21. **Perityle inyoensis** (Ferris) A. M. Powell, Sida 3: 278. 1968 • Inyo rock daisy C E

Laphamia inyoensis Ferris, Contr. Dudley Herb. 5: 104, fig. 2. 1958

Perennials or subshrubs, 12–25 cm; pilose-villous indument intermixed with short glandular hairs. **Leaves** (opposite or alternate): petioles 5–20(–50) mm; blades orbiculate, ovate, or ovate-deltate, 8–18 (–21) × 6–12(–15) mm, margins serrate to serrate-lobed. **Heads** borne singly or (2–3) in

corymbiform arrays, 7–8.5(–9) × 5–7(–8) mm. **Peduncles** 8–40 mm. **Involucres** campanulate. **Phyllaries** 14–21, linear-lanceolate to sub-lanceolate, 5.5–6.5 × 1.1–1.5 mm. **Ray** florets 0. **Disc florets** 35–60; corollas yellow, tubes 1.4–1.6 mm, throats subtubular to subfunnelform, 2–2.4 mm, lobes 0.6–0.7 mm. **Cypselae** narrowly oblanceolate to suboblanceolate, (2.5–)3–3.5 mm, margins thin-calloused, short-hairy; **pappi** usually inconspicuous, callous crowns, rarely of minute scales. 2*n* = 36.

Flowering spring–fall. Dry, rocky slopes; of conservation concern; 1800–2800 m; Calif.

Perityle inyoensis occurs in the southern Inyo Mountains. It is allied to *P. megalocephala*, which occurs at intermediate and lower elevations to the north and east of *P. inyoensis*.

22. **Perityle villosa** (S. F. Blake) Shinners, SouthW. Naturalist 4: 205. 1959 • Hanaupah rock daisy C E

Laphamia villosa S. F. Blake, Proc. Biol. Soc. Wash. 45: 142. 1932

Perennials or subshrubs, 13–20 cm; villous. **Leaves** (all alternate): petioles 3–6 mm; blades ovate and entire, or ovate-cuneate and 1–3-lobed, 3–16 × 4–10 mm, ultimate margins entire. **Heads** borne singly or (2–3) in corymbiform arrays, 7.5–9.5 × 5–7 mm. **Peduncles** 10–20(–25) mm. **Involucres** campanulate. **Phyllaries** 13–23, linear-lanceolate to oblong-lanceolate, 4–7 × 1–1.5 mm. **Ray florets** 0. **Disc florets** 30–75; corollas yellow, tubes 1–1.7 mm, throats 1.8–3 mm, tubular to subfunnelform, lobes 0.7–0.8 mm. **Cypselae** linear to oblanceolate, 3–3.5 mm, margins thin-calloused, short-hairy; **pappi** 0, or of 1–2, equal or unequal bristles 1–2 mm plus callous crowns. 2*n* = ca. 3*x* = 51.

Flowering spring–fall. Rock crevices; of conservation concern; 1700–2600 m; Calif.

Perityle villosa occurs in the Inyo, Panamint, and Grapevine mountains in and near the Death Valley region. Two distinct morphotypes characterized by either entire or lobed leaves appear to exist in Hanaupah Canyon in the Panamint Mountains. Some chromosomal evidence suggests that the variability of *P. villosa* within that population may result from the co-occurrence there of diploids, triploids, and tetraploids.

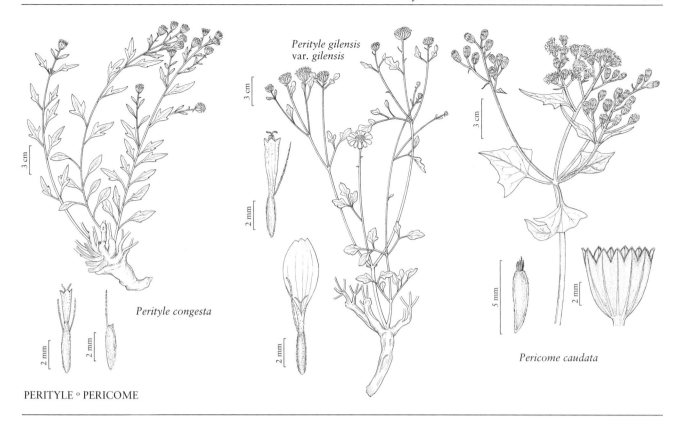

Perityle gilensis var. *gilensis*

Perityle congesta

Pericome caudata

PERITYLE ∘ PERICOME

23. Perityle tenella (M. E. Jones) J. F. Macbride, Contr.
Gray Herb. 56: 39. 1918 • Dixie or Springdale rock
daisy C E

Laphamia palmeri A. Gray var.
tenella M. E. Jones, Proc. Calif.
Acad. Sci., ser. 2, 5: 703. 1895;
L. palmeri A. Gray 1878, not
Perityle palmeri S. Watson 1889

Perennials or subshrubs, (5–)15–
30(–35) cm (stems erect to pendulous often densely leafy); usually
densely short-hairy. **Leaves:** peti-
oles 2–8 mm; blades usually broadly deltate-ovate, sometimes subovate or subcordate, 5–18(–25) × 4–18(–20) mm,
(bases truncate) margins usually ± serrate, rarely laciniate.
Heads borne singly or (2–10) in corymbiform arrays, 6–7
× 4–6 mm. **Peduncles** 4–25 mm. **Involucres** campanulate.
Phyllaries 13–20, linear to narrowly or broadly lanceolate,
4–5(–6) × 0.8–1.3 mm. **Ray florets** 0. **Disc florets** 23–
35(–48); corollas yellow, tubes (1–)1.2–1.6 mm, throats
tubular to narrowly funnelform, 1.2–1.6(–2.2) mm, lobes
0.3–0.7 mm. **Cypselae** narrowly oblanceolate, 2.5–3.3 mm,
margins obviously calloused, minutely hairy; **pappi** 0 or of
1(–3) moderately stout bristles 2–2.8(–4.1) mm plus callous crowns. **2*n*** = 32.

Flowering spring–fall. Rocky slopes or crevices in
limestone or sandstone rock faces; of conservation concern; 600–2700 m; Ariz., Nev., Utah.

Similarities of *Perityle tenella* to *P. congesta* are discussed
under the latter. *Perityle tenella* occurs in northwestern
Arizona, southeastern Nevada, and southwestern Utah.

24. Perityle congesta (M. E. Jones) Shinners, SouthW.
Naturalist 4: 204. 1959 • Kaibab or Grand Canyon
rock daisy C E F

Laphamia congesta M. E. Jones,
Proc. Calif. Acad. Sci., ser. 2, 5:
703. 1895

Perennials or subshrubs, 10–30
(–45) cm (densely clumped, stems
spreading or pendent); hirtellous.
Leaves: petioles 2.5–10(–15) mm
(shorter than blades); blades elliptic,
lanceolate, or ovate, 5–17(–20) × 4–
8(–10) mm, (bases cuneate) margins usually entire or with 2
(–6) teeth. **Heads** borne singly or (2–8) in corymbiform
arrays, 5–6(–6.5) × 3.5–4.5(–5) mm. **Peduncles** 3–17 mm.
Involucres campanulate. **Phyllaries** 12–16, narrowly lanceolate,
oblong-lanceolate, or lanceolate, 3.2–4.5 × 0.8–1.4 mm. **Ray
florets** 0. **Disc florets** 30–40; corollas yellow, tubes 0.8–1 mm,
throats tubular to subfunnelform, 1.2–1.5 mm, lobes 0.3–0.5
mm. **Cypselae** narrowly oblanceolate, (2–)2.5–3 mm,
margins thin-calloused, short-hairy; **pappi** of 1–2 subequal or
unequal bristles 2–2.7 mm. **2*n*** = 32.

Flowering spring–fall. Crevices of limestone cliffs and
bluffs; of conservation concern; 900–2700 m; Ariz.

The leaf blades of *Perityle congesta* usually have a
tooth on each margin, cuneate bases, and unobtrusive

veins abaxially; those of *P. tenella* usually have serrate margins, truncate bases, and raised veins abaxially. *Perityle congesta* is found on both rims and extending into the Grand Canyon and northward almost to the Utah border. Available specimens suggest that *P. congesta* and *P. tenella* intergrade where their ranges overlap in northwestern Arizona on the Colorado Plateau. Specimens from the Colorado Plateau have intermediate leaf shapes and induments.

25. Perityle saxicola (Eastwood) Shinners, SouthW. Naturalist 4: 204. 1959 • Roosevelt Dam rock daisy [C] [E]

Laphamia saxicola Eastwood, Proc. Calif. Acad. Sci., ser. 4, 20: 159. 1931

Perennials or subshrubs, 20–40 cm; glabrate or puberulent. **Leaves:** petioles 10–30 mm; blades (2–)3-pinnate, 15–55 × 10–50 mm, lobes linear to flat-filiform, scarcely broadened distally. **Heads** borne singly or (2–4) in loose, corymbiform arrays, 7.5–10 × 10–14 mm. **Peduncles** 1–3 mm. **Involucres** broadly campanulate to hemispheric. **Phyllaries** 14–25, linear-lanceolate to lanceolate, 5–7 × 0.5–1 mm. **Ray florets** 10–17; corollas yellow, laminae oblong to oblong-ovate, 4–6 × 2–3.2 mm. **Disc florets** 75–200; corollas yellow, tubes 1–1.2 mm, throats narrowly tubular to narrowly funnelform, 2–2.5 mm, lobes ca. 0.7 mm. **Cypselae** narrowly oblanceolate to narrowly oblong-elliptic, 2.8–3.2 mm, margins notably calloused, sparsely short-hairy; **pappi** 0 or of 1(–2) bristles (2–)4–5 mm. **2*n* = 34.**

Flowering spring–fall. Crevices of igneous rocks; of conservation concern; 700–1100 m; Ariz.

Perityle saxicola occurs in the area of Tonto National Monument and Roosevelt Dam in central Arizona. The 2–3-pinnately dissected leaves with long-linear to flat-filiform lobes distinguish it from the similar *P. gilensis* with its broader leaf lobes.

26. Perityle gilensis (M. E. Jones) J. F. Macbride, Contr. Gray Herb. 56: 39. 1918 [C] [E] [F]

Laphamia gilensis M. E. Jones, Zoë 2: 15. 1891

Perennials or subshrubs, 22–70 cm (often in dense clumps, stems upright to pendent or spreading); glabrous or puberulent, often glandular. **Leaves:** petioles 9–45 (–80) mm; blades usually pinnately 3-foliolate (proximal) or 3-lobed, 14–30 × 4–20 mm, lobes much broadened distally, often cruciform, sometimes 3-lobed to divided,

ultimate margins entire or lobed. **Heads** borne singly or (2–4) in corymbiform arrays, 8–11 × 9–12 mm. **Peduncles** 1–4 mm. **Involucres** broadly campanulate to hemispheric. **Phyllaries** 20–30, linear-lanceolate to oblanceolate, 4–7 × 0.8–1.5 mm. **Ray florets** 8–18; corollas yellow, laminae oblong to subovate, 4–10 × 2–4 mm. **Disc florets** 70–200; corollas yellow, tubes 1.5–2 mm, throats subtubular to narrowly funnelform, 2.5–3.5 mm, lobes 0.7–0.8 mm. **Cypselae** narrowly oblanceolate to narrow oblong-elliptic, 2.6–4 mm, margins notably calloused, sparsely short-hairy, sometimes glandular; **pappi** of 1(–3) bristles 3–4.6 mm, sometimes plus hyaline scales.

Varieties 2 (2 in the flora): Arizona.

1. Leaves: petioles 9–25(–30) mm, lobes of mid and distal leaves elliptic to ovate . . . . 26a. *Perityle gilensis* var. *gilensis*
1. Leaves: petioles 25–45(–80) mm, lobes of mid and distal leaves linear to long-spatulate . . . . 26b. *Perityle gilensis* var. *salensis*

26a. Perityle gilensis (M. E. Jones) J. F. Macbride var. **gilensis** • Gila rock daisy [C] [E] [F]

Plants upright to spreading, 22–40 cm. **Leaves:** petioles 9–25(–30) mm; lobes of mid and distal leaves elliptic to ovate, to 1 × 1 cm. **2*n* = 68–72.**

Flowering spring–fall. Crevices igneous bluffs and boulders especially canyon walls; of conservation concern; 500–1300 m; Ariz.

Proximal leaves of vars. *gilensis* and *salensis* may appear similar; lobes of the mid and distal leaves are elliptic to ovate in var. *gilensis* and linear to long-spatulate in var. *salensis*.

26b. Perityle gilensis (M. E. Jones) J. F. Macbride var. **salensis** A. M. Powell, Sida 5: 104, fig 7. 1973 • Salt River rock daisy [C] [E]

Laphamia gilensis M. E. Jones subsp. *longilobus* W. E. Niles; *Perityle gilensis* var. *longilobus* (W. E. Niles) A. M. Powell & Yarborough

Plants trailing, 30–70 cm. **Leaves:** petioles 25–45(–80) cm, lobes of mid and distal leaves linear to long-spatulate, to 6 × 0.5(–1) cm. **2*n* = 34.**

Flowering spring–fall. Crevices and ledges, igneous canyon walls; of conservation concern; 900–2000 m; Ariz.

Variety *salensis* is known only from Salt River Canyon.

27. **Perityle ajoensis** Todsen, J. Arizona Acad. Sci. 9: 35. 1974 • Ajo rock daisy [C] [E]

Perennials or subshrubs, to 30 cm; usually densely short-tomentose. **Leaves** (proximally opposite, distally alternate): petioles 3–10 mm; blades deltate, ovate, or suborbiculate, 5–15 × 5–15 mm, margins shallow-crenate to dentate. **Heads** borne singly or (2–10) in loose, corymbiform arrays, 7–9 × 6–9 mm. **Peduncles** 5–15 mm. **Involucres** campanulate to hemispheric. **Phyllaries** 15–20, oblanceolate to broadly lanceolate, 4–5 × 0.8–1.8 mm. **Ray florets** 0. **Disc florets** 20–45; corollas yellow, tubes 1–1.2 mm, throats tubular to narrowly funnelform, 1.3–1.6 mm, lobes 0.4–0.6 mm. **Cypselae** narrowly suboblong to oblanceolate, 2.5–3.5 mm, margins prominently calloused, short-hairy; **pappi** of 1 (–2) stout bristles 1.8–2.8 mm. $2n = 34$.

Flowering spring–fall. Crevices of rocky canyon walls and cliff faces; of conservation concern; 800–1200 m; Ariz.

Perityle ajoensis occurs in the Ajo Mountains of Organ Pipe Cactus National Monument in Pima County.

28. **Perityle lemmonii** (A. Gray) J. F. Macbride, Contr. Gray Herb. 56: 39. 1918 (as lemmoni) • Lemmon's rock daisy

Laphamia lemmonii A. Gray, Proc. Amer. Acad. Arts 16: 101. 1880 (as lemmoni); *L. dissecta* Torrey subsp. *lemmonii* (A. Gray) W. E. Niles

Perennials or subshrubs, 6–17(–23) cm (densely clumped, stems leafy); usually densely pilose to villous, sometimes glabrate. **Leaves** (opposite or alternate): petioles 3–8(–10) mm (usually shorter than blades); blades ovate to ovate-deltate (margins crenate, lacerate, laciniate, or serrate), or 3–5-lobed or pinnately divided (lobes crenate, lacerate, laciniate, lobed, or serrate), 6–18 × 7–20 mm. **Heads** borne singly or (2–3) in corymbiform arrays (often obscured by leaves), 7–10 × 5–9 mm. **Peduncles** 1–5 (–10) mm. **Involucres** campanulate. **Phyllaries** 10–16, linear-lanceolate, lanceolate, to oblanceolate, 4–6.4(–8) × 1–1.7 mm (apices acute). **Ray florets** 0. **Disc florets** 20–40; corollas yellow, often tinged with purple, tubes 1–1.9 mm, throats broadly tubular to subfunnelform, 2–2.5 (–2.9) mm, lobes 0.4–0.6 mm. **Cypselae** narrowly oblong to narrowly oblanceolate, 2.5–3.2(–3.6) mm, margins thin-calloused, short-hairy; **pappi** usually of 1(–2) delicate bristles 1–3(–4) mm, rarely plus vestigial, hyaline scales, sometimes 0. $2n = 34$.

Flowering spring–fall. Crevices of granitic boulders and cliffs; 600–2300 m; Ariz., N.Mex.; Mexico (Chihuahua).

Perityle lemmonii, which occurs in the mountain areas of southeastern Arizona and southwestern New

Mexico, has variable leaf morphology that is often consistent within geographic populations. Its former inclusion within *P. dissecta* is discussed under that species.

29. **Perityle ambrosiifolia** Greene ex A. M. Powell & Yarborough, Phytologia 76: 325, fig. 1. 1994 • Lace-leaved rock daisy [C] [E]

Perennials or subshrubs, 10–30 cm (stems brittle, densely leafy); usually villous, often with glandular hairs, sometimes pilose. **Leaves:** petioles 5–10 mm; blades 3-partite or compound-pinnatifid, 15–35 × 15–30 mm, lobes lobed, cleft, parted, or divided, ultimate margins crenate. **Heads** in corymbiform arrays, 7–10 × 6–11 mm. **Peduncles** 3–10 mm. **Involucres** campanulate. **Phyllaries** 14–20, linear to linear-lanceolate, 6–9 × 0.5–1.2(–2) mm (apices usually short-attenuate, sometimes acute). **Ray florets** usually 0 (sometimes 1–2 reduced rays in isolated heads, laminae color unknown, 3–5 × 1.5–2 mm). **Disc florets** 25–45; corollas yellow, tubes 1–1.2 mm, throats tubular to narrowly funnelform, 2–2.5 mm, lobes 0.6–0.8 mm. **Cypselae** narrowly oblanceolate, 3–4 mm, margins thin-calloused, short-hairy; **pappi** usually 0, sometimes of 1–3 moderately stout bristles 2.8–4.5 mm, often plus hyaline, laciniate scales. $2n =$ ca. 17.

Flowering spring–fall. Rock crevices, cliff faces, and canyons; of conservation concern; 1000–1500 m; Ariz.

Perityle ambrosiifolia occurs in the vicinity of Clifton and Morenci in Greenlee County. Most heads are discoid; 1 or 2 ray florets sometimes appear on isolated heads; color of the laminae is not known; only dried specimens without the ray color noted have been examined. The species was first recognized by E. L. Greene in 1900; no record exists that the species as proposed by him was formally published. *Perityle ambrosiifolia* is morphologically and geographically distinct from *P. lemmonii* and may have resulted from either intrasectional or intersectional hybridization between two of several possible taxa.

30. **Perityle dissecta** (Torrey) A. Gray in A. Gray et al., Syn. Fl. N. Amer. 1(2): 320. 1884 • Slim-lobe or dissected rock daisy [C]

Laphamia dissecta Torrey, Smithsonian Contr. Knowl. 5(6): 81. 1853

Perennials or subshrubs, 7–20 cm (densely leafy); pilose. **Leaves:** petioles (3–)4–10 mm; blades lanceolate to deltate, 4–15 × 4–15 mm, usually 1–3-pinnately divided, lobes irregular, rounded, ultimate margins laciniate. **Heads** borne singly or in corymbiform arrays, 8–10 × 4–5 mm. **Peduncles** 2–8 (–17) mm. **Involucres** campanulate. **Phyllaries** 12–15, linear to linear-lanceolate, (5–)7–13 × 0.5–0.9 mm (apices usually attenuate, sometimes acute). **Ray florets** 0. **Disc florets** 20–30; corollas yellow, tubes 1–1.5 mm, throats usually narrowly tubular, (2–)2.5–2.8 mm, lobes 0.7–1 mm. **Cypselae** oblong to narrowly oblanceolate, (2.8–)3–3.8 mm, margins moderately calloused, short-hairy; **pappi** 0, or of 1(–4) moderately stout bristles (1–)2.2–3 mm. $2n = 34, 68$.

Flowering spring–fall. Limestone walls of canyons; of conservation concern; 800–1200 m; Tex.; Mexico (Chihuahua).

Perityle dissecta occurs in southern Brewster and Presidio counties. It and *P. lemmonii* have been considered conspecific by some taxonomists because of similarities in gross leaf morphology. The differences in phyllaries, disc corollas, and pappus bristles, as well as distinct distributions, support recognizing them as separate species.

31. **Perityle lindheimeri** (A. Gray) Shinners, SouthW. Naturalist 4: 204. 1959 • Lindheimer's rock daisy

Laphamia lindheimeri A. Gray, Smithsonian Contr. Knowl. 3(5): 101. 1852; *L. rotundata* Rydberg; *Perityle rotundata* (Rydberg) Shinners

Perennials or subshrubs, (10–)18–45(–60) cm; glabrate or sparsely hairy. **Leaves:** petioles 4–15 mm; blades broadly ovate to ovate-lanceolate, 10–40 × 10–35 mm, margins entire or serrate-lobed. **Heads** in corymbiform arrays, 6–7 × 4–6 mm. **Peduncles** 3–20 mm. **Involucres** narrowly campanulate. **Phyllaries** 12–14, narrowly lanceolate to oblanceolate, 3–4 × 2 mm. **Ray florets** 3–5; corollas yellow, laminae oblong to suborbiculate, 2.5–3 × 1.5–2 mm. **Disc florets** 16–22; corollas yellow, tubes 0.8–1.5 mm, throats narrowly campanulate, 1–1.5 mm, lobes 0.5 mm. **Cypselae** linear-oblong to narrowly oblanceolate, 2–2.8 mm, margins thin-calloused, minutely hairy; **pappi** 0, or of 1(–2) bristles 0.5–1.8 mm.

Varieties 2 (2 in the flora): Texas.

A specimen resembling *Perityle lindheimeri* was collected in 1936 near Ruidosa in Lincoln County, New Mexico. The identity and exact location of that seemingly out-of-place collection remain to be determined.

1. Pappi usually 0, sometimes of single bristles 0.5–1(–1.5) mm, usually plus obscure, callous crowns; w Edwards Plateau . . . . 31a. *Perityle lindheimeri* var. *halimifolia*
1. Pappi usually of 1(–2) bristles 0.5–1.8 mm plus vestigial, laciniate scales or obscure callous crowns, rarely 0; c, e Edwards Plateau . . . . 31b. *Perityle lindheimeri* var. *lindheimeri*

31a. **Perityle lindheimeri** (A. Gray) Shinners var. **halimifolia** (A. Gray) A. M. Powell, Sida 3: 278. 1968 E

Laphamia halimifolia A. Gray, Smithsonian Contr. Knowl. 3(5): 100, plate 9, fig. B. 1852; *Perityle halimifolia* (A. Gray) Shinners

Pappi usually 0, sometimes of single bristles 0.5–1(–1.5) mm, usually plus obscure, callous crowns. $2n = 34$.

Flowering Apr–Jun. Cretaceous limestone outcrops; 300–800 m; Tex.

Variety *halimifolia* is known only from Cretaceous limestone exposures in Val Verde and Crockett counties on the western edge of the Edwards Plateau. The pappus generally lacks bristles and the plants usually are smaller and somewhat depauperate in habit when compared with var. *lindheimeri*.

31b. **Perityle lindheimeri** (A. Gray) Shinners var. **lindheimeri** E

Pappi usually of 1(–2) bristles 0.5–1.8 mm plus vestigial, laciniate scales or obscure callous crowns, rarely 0.

Flowering spring–fall. Crevices of cretaceous limestone rock, often beside streams and springs; 300–700 m; Tex. $2n = 34$.

Variety *lindheimeri* is widely distributed on the Edwards Plateau. The more numerous, smaller, radiate heads and ovate leaves distinguish it from the related *Perityle angustifolia*, which has larger, discoid heads and linear to lanceolate leaves.

32. **Perityle angustifolia** (A. Gray) Shinners, SouthW. Naturalist 4: 204. 1959 • Narrow-leaved rock daisy E

Laphamia angustifolia A. Gray, Smithsonian Contr. Knowl. 3(5): 100. 1852; *L. angustifolia* subsp. *laciniata* (A. Gray) W. E. Niles

Perennials or subshrubs, 8–35(–70) cm (stems densely leafy); glabrate or puberulent. **Leaves:** petioles 0–12 mm; blades usually lanceolate, linear, or oblanceolate, rarely ovate, usually 3- or 5-lobed (lobes acute or attenuate; bases cuneate), (5–)15–30 × 2–15 mm, ultimate margins entire, subentire, or laciniate. **Heads** borne singly or in loose, corymbiform arrays, 5.5–7(–9) × 3.5–5.5 (–6) mm. **Peduncles** 6–50 mm. **Involucres** narrowly campanulate to campanulate. **Phyllaries** 12–15, narrowly to broadly lanceolate, 4–5 × 1–2 mm. **Ray florets** 0. **Disc florets** 20–33(–50); corollas yellow, tubes 1–1.5 mm, throats narrowly campanulate, 1–1.5 mm, lobes 0.5–1 mm. **Cypselae** narrowly oblanceolate to linear-oblong, 2–3 mm, margins thin-calloused, puberulent; **pappi** usually 0, sometimes 1–2 bristles 1.5–2 mm. $2n = 34$, ca. 136.

Flowering spring–fall. Limestone crevices; 300–1300 m; Tex.

Perityle angustifolia occurs in crevices of Cretaceous limestone in west Texas near the Rio Grande and along the eroded, western edge of the Edwards Plateau.

33. Perityle staurophylla (Barneby) Shinners, SouthW. Naturalist 4: 205. 1959 • New Mexico rock daisy   [E]

Laphamia staurophylla Barneby, Leafl. W. Bot. 8: 168, fig. s.n. [p. 169]. 1957

Perennials or subshrubs, 15–40 cm (often forming dense clumps, densely leafy); glabrous or puberulent. **Leaves:** petioles 10–25(–40) mm; blades usually deeply divided, 5–30 × 7–25 mm, lobes 3, relatively broad to linear, usually secondarily lobed, cleft, or parted, often ± cruciform, sometimes almost entire. **Heads** usually in corymbiform arrays, sometimes borne singly, 6–7.5 × 4.5–6 mm. **Peduncles** 4–25 mm. **Involucres** campanulate. **Phyllaries** 13–16(–22), sublanceolate, 3.4–4 × 0.7–1.1(–1.8) mm. **Ray florets** 0, or 4–8; corollas yellow, laminae oblong, 3.6–4.8 × 1.2–2.2 mm. **Disc florets** 40–50; corollas yellow, tubes 1.5–1.6 mm, throats subtubular to narrowly funnelform, 1.5–2 mm, lobes 0.6–0.8 mm. **Cypselae** oblanceolate, 1.8–2.3(–2.7) mm, margins thin-calloused, short-hairy; **pappi** of 2–3(–6) bristles 1.8–2.5 mm (at least the longer), often plus minute, hyaline scales.

Varieties 2 (2 in the flora): New Mexico.

1. Leaves: lobes usually linear to filiform; ray florets 0 (n end of San Andres Mountains) . . . . 33a. *Perityle staurophylla* var. *homoflora*
1. Leaves: lobes relatively broad to filiform; ray florets 4–8 (not n end of San Andres Mountains) . . . . 33b. *Perityle staurophylla* var. *staurophylla*

33a. Perityle staurophylla (Barneby) Shinners var. homoflora Todsen, Madroño 30: 116. 1983   [C]

Leaves: lobes usually linear to filiform. **Ray florets** 0. $2n = 34$.

Flowering spring–fall. Crevices in limestone and granite cliffs; of conservation concern; 1600–2000 m; N.Mex.

Variety *homoflora* occurs at the northern end of the San Andres Mountains (Rhodes Canyon and northward) in south-central New Mexico. A waxy, white

coating on the cypselae, which was included by the describing author as a distinguishing characteristic, also occurs in some populations of var. *staurophylla* and other species of *Perityle* as well.

33b. Perityle staurophylla (Barneby) Shinners var. staurophylla

Leaves: lobes relatively broad to filiform. **Ray florets** 4–8. $2n = 17$.

Flowering spring–fall. Crevices in limestone and granite cliffs; 1500–3000 m; N.Mex.

Variety *staurophylla* occurs in the Caballo, Fra Cristobal, Sacramento, and San Andres mountains of south-central New Mexico.

34. Perityle huecoensis A. M. Powell, Madroño 30: 219, fig. 2. 1983 • Hueco rock daisy   [C]

Perennials or subshrubs, 10–20 cm; densely short-hairy. **Leaves:** petioles 10–80 mm; blades broadly ovate, ovate-deltate, or ovate-rhombic, 5–12(–15) × 4–10(–15) mm, margins entire or serrate-lobed. **Heads** borne singly or in corym-biform arrays, 5.5–6 × 4–6 mm. **Peduncles** 5–15 mm. **Involucres** funnelform. **Phyllaries** 8–10, lanceolate to oblanceolate, 3–4.5 × 0.6–1.2 mm. **Ray florets** 3–5; corollas yellow, laminae oblong, 2.5–3 × 1.5–2 mm. **Disc florets** 11–13; corollas yellow, tubes 1.3–1.5 mm, throats broadly campanulate-funnelform, 0.8–1.2 mm, lobes 0.5–0.7 mm. **Cypselae** linear-lanceolate, 2.2–3.2 mm, margins usually thin-calloused, short-hairy; **pappi** usually of 1–3, antrorsely barbellate bristles 1–2.5 mm plus 0–4 shorter bristles, sometimes plus hyaline, laciniate scales. $2n = 34$.

Flowering spring–fall. Limestone cliffs; of conservation concern; 1400–1500 m; Tex.; Mexico (Chihuahua).

Perityle huecoensis is found in the United States only in the Hueco Mountains of El Paso County, generally growing on north or northeast facing slopes. Plants in the nearby Sierra Juarez in Mexico tend to lack bristles in pappi on the disc florets and to be taller with larger leaves compared to plants in the United States populations.

35. Perityle cochisensis (W. E. Niles) A. M. Powell, Sida 5: 123. 1973 • Cochise rock daisy [C] [E]

Laphamia cochisensis W. E. Niles, Mem. New York Bot. Gard. 21(1): 47, fig. 28. 1970

Perennials or subshrubs, 4–12 (–15+) cm (usually erect, compact, sometimes spreading, moderately to densely leafy); usually glabrate, sometimes sparsely or densely short-hairy, rarely sparsely villous. **Leaves:** petioles 5–30 mm (usually equal to or longer than blades); blades deltate-orbiculate, ovate, or subdeltate, 5–27 × 6–27 mm, irregularly lobed, lobes 3–5, ultimate margins serrate or lobed. **Heads** usually borne singly, 8–10 × ca. 10 mm. **Peduncles** 5–10 mm. **Involucres** broadly campanulate to hemispheric. **Phyllaries** 12–22, lanceolate to ovate, 4.5–6.5 × 1–2.5 mm. **Ray florets** 0. **Disc florets** 30–45(–60); corollas yellow, tubes 0.8–1 mm, throats funnelform to narrowly campanulate, 1.8–2 mm, lobes 0.6–0.8 mm. **Cypselae** oblong to narrowly oblanceolate, 2.5–3 mm, margins thin-calloused, sparsely short-ciliate; **pappi** usually of 1(–2) unequal, distally subplumose bristles 2–2.5 mm, sometimes 0. $2n = 34$.

Flowering spring–fall. Granite rock crevices in canyons; of conservation concern; 1500–2100 m; Ariz.

Perityle cochisensis is known only from the Chiricahua Mountains of southeastern Arizona. Some collections of *P. lemmonii* from the Big Hatchet Mountains of Hidalgo County in southwestern New Mexico may have petioles approaching the length of the blades as in *P. cochisensis*. The densely pilose or villous induments of those New Mexican plants is similar to that of *P. lemmonii*.

353. PERICOME A. Gray, Smithsonian Contr. Knowl. 5(6): 81. 1853 • [Greek *peri*, around, and *come*, tuft of hairs, alluding to ciliate margins of cypselae]

Sharon C. Yarborough

A. Michael Powell

Perennials or subshrubs, (20–)50–150+ cm. **Stems** multiple, spreading, branched. **Leaves** mostly cauline; mostly opposite, distal sometimes alternate; petiolate; blades (3-nerved) usually deltate-hastate basally, sometimes subhastate, ovate, or cordate, margins entire, cleft, toothed, or 3–5-lobed (apices long-attenuate, caudate), faces sparsely to densely puberulent, usually copiously gland-dotted (distal leaves rarely tomentulose). **Heads** discoid, in corymbiform or compound-corymbiform arrays. **Involucres** turbinate-campanulate, 4–10 mm diam. **Phyllaries** 16–24 in 1 series (connate ¹/₂+ their lengths, linear, subequal). **Receptacles** convex, pitted, epaleate. **Disc florets** 30–70, bisexual, fertile; corollas yellow (glabrate or densely gland-dotted), tubes mostly shorter than abruptly dilated, cylindric throats, lobes 4, triangular. **Cypselae** (black with pale margins) flattened, oblong-oblanceolate to narrowly linear-oblanceolate (margins calloused or thin, densely ciliate); **pappi** 0 or, persistent or falling, vestigial, or of 8–12 distinct or partially connate, fimbriate to laciniate scales (sometimes 1–4 scales aristate) plus 1–2(–6) ciliolate bristles. $x = 18$.

Species 2 (1 in the flora): sw United States, n Mexico.

The close relationship of *Pericome* to *Perityle* is apparent in the similarity of general vegetative and floral morphology, particularly in the similarities of cypselae and pappus structure. The shrubby habit and capitulescence of densely aggregated heads with united involucral bracts of *Pericome* distinguish it from *Perityle*. *Pericome* is widespread in Upper Sonoran and Montane zones of the southwestern United States.

SELECTED REFERENCE Powell, A. M. 1973. Taxonomy of *Pericome* (Compositae–Peritylinae). SouthW. Naturalist 18: 335–339.

1. Pericome caudata A. Gray, Smithsonian Contr. Knowl. 5(6): 82. 1853 • Tail-leaf pericome, taper leaf, yerba del chivato F

Pericome caudata var. *glandulosa* (Goodman) H. D. Harrington; *P. glandulosa* Goodman

Stems striate, terete, glabrous or hairy, often densely puberulent to tomentulose distally, sometimes gland-dotted. **Leaves:** petioles (5–)10–45 mm; blades (2–)3.5–12 (–15) × 1–12 cm. **Heads** 3–30+, usually tightly clustered. **Peduncles** 0.5–4 cm. **Involucres** 4.5–10 × 4–10 mm. **Phyllaries** 0.5–1 mm wide, apices attenuate. **Disc corollas:** tubes 1–3.5 mm, throats 2–5.5 mm, lobes 0.5–1 mm. **Cypselae** 3–5 mm; **pappi** crowns to ca. 1 mm plus 0–2 bristles 1–4.5 mm. $2n = 36$.

Flowering spring–fall. Among rocks, boulders, on talus slopes, bluffs, crags, canyons, disturbed roadsides, in volcanic, limestone, and sandstone substrates; 1400–3300 m; Ariz., Calif., Colo., Nev., N.Mex., Okla., Tex.; Mexico (Chihuahua).

Populations of *Pericome caudata* may vary in leaf shape, head size, and indument, particularly in some far western and eastern populations. Central populations (e.g., in Colorado, New Mexico, and Arizona) tend to have relatively large, deltate-hastate leaves with long-attenuate tips and capitulescences of medium-sized heads. Specimens from Nevada and California often have smaller, ovate or cordate distal leaves with short-attenuate tips, as well as larger and fewer heads per capitulescence. Oklahoma specimens are often coarsely pubescent with copious glands and have leaves similar to the Nevada and California populations. This variability does not appear to warrant taxonomic distinction, nor does the presence or absence of pappus bristles appear to be taxonomically significant.

187m.21. ASTERACEAE Martinov (tribe Heliantheae subtribe BAERIINAE Bentham & Hooker f., Gen. Pl. 2: 200. 1873 (as Baerieae)

Eriophyllinae Rydberg

Annuals, perennials, subshrubs, or shrubs, 1–200 cm. **Leaves** basal, basal and cauline, or mostly cauline; mostly opposite (*Lasthenia*) or mostly alternate; usually sessile, sometimes obscurely petiolate; blades (often 1–2 times pinnately lobed) or lobes often linear, ultimate margins entire or toothed, faces often ± woolly to tomentose, sometimes glabrate or glabrous, often gland-dotted. **Heads** radiate, discoid, or disciform, borne singly or in corymbiform, glomerate, or paniculiform arrays. **Calyculi** 0. **Involucres** ovoid or obconic to campanulate or hemispheric. **Phyllaries** persistent, mostly 3–18 in 1–2 series, (erect or reflexed in fruit) distinct or connate, mostly elliptic, lanceolate, ovate, or obovate, usually ± equal, mostly herbaceous, sometimes indurate (at least proximally), flat or weakly cupped at bases, sometimes scarious-margined, often woolly to tomentose, sometimes glabrate or glabrous. **Receptacles** flat, convex, hemispheric, or conic (smooth, knobby, or pitted, glabrous or hairy), usually epaleate (paleae usually 0, rare in *Eriophyllum*). **Ray florets** 0 or 4–21, pistillate, fertile (3–8 peripheral florets pistillate, fertile, corollas tubular in *Amblyopappus* and *Monolopia congdonii*); corollas yellow to orange, often darker proximally, sometimes purplish (usually ± bilabiate in *Monolopia*). **Disc florets** 2–300, bisexual, fertile; corollas yellow to orange, tubes shorter than or about equaling funnelform or campanulate throats, lobes 4–5, deltate, glabrous or papillate; anther thecae usually pale; stigmatic papillae in 2 lines. **Cypselae** clavate or obovoid to terete, or obpyramidal, sometimes compressed or obcompressed, glabrous, hairy, or papillate (compressed, callous-margined, and ciliolate in *Eatonella*, *Lasthenia chrysantha*, and *Monolopia congdonii*; sometimes winged in *Monolopia*); **pappi** 0 or of 1–12+ aristate, erose, laciniate, or truncate scales or awns in 1–2 series (often 2 sorts of scales in combination on 1 cypsela).

Genera 9, species 44 (7 genera, 41 species in the flora): w North America, Mexico, w South America.

Members of Baeriinae are found mostly in western North America; there are disjuncts in western South America. H. Robinson (1981) treated Baeriinae as a relatively isolated element among epaleate subtribes of Heliantheae. B. G. Baldwin (in Baldwin et al. 2002) included Baeriinae within Madieae.

1. Leaves all or mostly opposite . . . . 354. *Lasthenia,* p. 336
1. Leaves mostly alternate (proximal sometimes opposite).
   2. Leaves glabrous (often granular-glandular, not woolly) . . . . 355. *Amblyopappus,* p. 348
   2. Leaves ± woolly or tomentose (usually stems and/or phyllaries as well).
      3. Heads ± disciform (4–7 peripheral florets pistillate, corollas with laminae ca. 0.5 mm) . . . . 357. *Monolopia* (in part), p. 349
      3. Heads usually radiate, sometimes discoid.
         4. Phyllaries usually reflexed in fruit (ray laminae 2–2.5 mm; pappi of 2, ± laciniate to nearly entire, often uniaristate scales 1.5–3 mm) . . . . 356. *Eatonella,* p. 348
         4. Phyllaries usually ± erect in fruit.
            5. Ray florets 7–11 (corollas obscurely bilabiate, laminae each opposed by adaxial lobe or tooth) . . . . 357. *Monolopia* (in part), p. 349
            5. Ray florets 0 or 3–13 (corollas not bilabiate).
               6. Annuals; ray cypselae obcompressed (disc corollas with rings of hairs at bases of limbs) . . . . 358. *Pseudobahia,* p. 351
               6. Annuals, perennials, subshrubs, or shrubs; ray cypselae usually prismatic, 4–5-angled (disc corollas without rings of hairs).
                  7. Annuals, perennials, subshrubs, or shrubs; phyllaries 4–13(–15) in 1+ series; pappi 0, or coroniform, or of 6–12+ (distinct) scales (in 1–2 similar or contrasting series) . . . . 359. *Eriophyllum,* p. 353
                  7. Subshrubs; phyllaries 8–16+ in ± 2 series; pappi of 2–6+ (unequal or 2 opposite, longer, and ± equal, basally connate, oblong to subulate) scales (tips acute to erose) . . . . 360. *Constancea,* p. 362

354. LASTHENIA Cassini, Opusc. Phytol. 3: 88. 1834 • Goldfields [Greek, for a student of Plato, said to have been a woman who dressed as a man]

Raymund Chan

Robert Ornduff†

Annuals (perennials), to 40(–60) cm (taprooted or roots fibrous, fleshy and clustered in *Lasthenia californica* subsp. *bakeri*). **Stems** usually erect, sometimes decumbent, prostrate, or sprawling, simple or branched (usually distally, often proximally in decumbent plants). **Leaves** mostly cauline; opposite; petiolate or sessile; blades usually linear, often 1(–2)-pinnately lobed, ultimate margins entire or toothed, faces glabrous or hairy. **Heads** usually radiate, sometimes ± disciform (in *L. glaberrima* and *L. microglossa*), borne singly or in ± corymbiform arrays. **Involucres** obconic to hemispheric, 3–5+ mm diam. **Phyllaries** usually persistent, sometimes falling with cypselae, 4–18 in 1(–2) series (usually ± erect in fruit, distinct or ± connate, narrowly oblong to broadly ovate, mostly herbaceous, bases flat or weakly cupped, faces not woolly, except sometimes in *L. minor* and *L. platycarpha*). **Receptacles** hemispheric to narrowly conic or subulate, smooth, papillate, or pitted, glabrous or hairy, epaleate. **Ray florets** 4–16, pistillate, fertile; corollas yellow to orangish (sometimes white in *L. debilis*, often somewhat darker proximally, laminae rarely lacking in *L. glaberrima* and *L. microglossa*). **Disc florets** 5–100+, bisexual, fertile; corollas yellow to orangish (sometimes white in *L. debilis*), tubes shorter than or about equaling funnelform or campanulate throats, lobes (4–)5, deltate. **Cypselae** (black to gray) usually cylindric to obovoid, glabrous or hairy, sometimes papillate (flattened, margins fringed with blunt, curved hairs in *L. chrysantha*); **pappi** 0, or of 1–12 erose, fimbriate, or laciniate, truncate or aristate scales (sometimes 2 kinds in combination on single cypselae). *x* = 8.

Species 18 (17 in the flora): w North America, nw Mexico, South America (Chile).

Lasthenias occur in a wide variety of habitats; some are particularly conspicuous members of vernal-pool floras. The characteristic rich, golden yellow color of *Lasthenia gracilis* can be seen to cover thousands of hectares of grasslands and open woodlands in early spring, giving the genus its common name. Relatively few taxa are widely distributed; most have relatively restricted distributions. Some are considered to be of conservation concern.

Lasthenia glaberrima, L. kunthii, L. maritima, and *L. microglossa* are self-pollinating; the rest are self-incompatible, obligate outcrossers. *Lasthenia ornduffii* and two subspecies of *L. californica* are perennial; the rest are spring annuals. It is not uncommon to find two *Lasthenia* species growing more or less sympatrically (in discrete populations).

Most lasthenias show a wide latitude of morphologic response to environmental conditions. Growth of individual plants of *Lasthenia* is robust in good conditions; in unfavorable conditions, single stems terminating in relatively small heads with relatively few florets are produced. Other morphologic characters such as the degree of leaf dissection, leaf margin, and pappus elements sometimes are plastic.

Circumscriptions of taxa here are based on R. Ornduff (1966b) with realignments proposed by R. Chan (2000). Sections are characterized by morphology, chromosome numbers, biochemistry, ecology, and molecular data.

SELECTED REFERENCES Chan, R. 2000. Molecular Systematics of the Goldfield Genus *Lasthenia* (Compositae: Heliantheae Sensu Lato). Ph.D. dissertation. University of California, Berkeley. Chan, R., B. G. Baldwin, and R. Ornduff. 2001. Goldfields revisited: A molecular phylogenetic perspective on the evolution of *Lasthenia* (Compositae: Heliantheae sensu lato). Int. J. Pl. Sci. 162: 1347–1360. Ornduff, R. 1966b. A biosystematic survey of the goldfield genus *Lasthenia* (Compositae: Helenieae). Univ. Calif. Publ. Bot. 40: 1–92.

1. Phyllaries connate ²/₃+ their lengths.
   2. Cypselae epappose . . . . 354c. *Lasthenia* sect. *Hologymne*, p. 342
   2. Cypselae pappose . . . . 354d. *Lasthenia* sect. *Lasthenia*, p. 344
1. Phyllaries distinct or connate ¹/₄–¹/₂ their lengths.
   3. Leaves entire or ± toothed (not pinnatifid); corolla floral pigments turning deep red in dilute aqueous alkali.
      4. Ray laminae 2.5–16 mm; receptacles usually narrowly conic to conic (subulate in *L. leptalea*, ray laminae 2.5–5 mm) . . . . 354a. *Lasthenia* sect. *Amphiachaenia*, p. 338
      4. Ray laminae 0–1 mm (and anther appendages eglandular), or 3–5 mm; receptacles subulate . . . . 354b. *Lasthenia* sect. *Burrielia*, p. 341
   3. Leaves (especially midstem) usually 1(–2)-pinnately lobed or -pinnatifid, sometimes entire; corolla floral pigments remaining yellow in dilute aqueous alkali.
      5. Cypselae to 1.5 mm . . . . 354e. *Lasthenia* sect. *Ornduffia*, p. 344
      5. Cypselae 1.5–3.5 mm.
         6. Cypselae pappose (pappi of 4–6 lanceolate to ovate, aristate scales) . . . . 354f. *Lasthenia* sect. *Platycarpha*, p. 346
         6. Cypselae epappose or pappose (pappi of 2–12 lanceolate, ovate, subulate, aristate scales, sometimes plus 4–5+ shorter, truncate, fimbriate, or laciniate scales) . . . . 354g. *Lasthenia* sect. *Ptilomeris*, p. 346

354a. Lasthenia Cassini sect. Amphiachaenia (Nuttall) R. Chan, Madroño 48: 205. 2002

Burrielia de Candolle sect. *Amphiachaenia* Nuttall, Trans. Amer. Philos. Soc., n. s. 7: 381. 1841; *Baeria* Fischer & C. A. Meyer

Annuals or perennials. Leaves entire or ± toothed. **Involucres** campanulate to depressed-hemispheric, hemispheric, or obconic. **Phyllaries** ± persistent or falling with cypselae, distinct. **Receptacles** narrowly conic to conic (subulate in *L. leptalea*), muricate or papillate, glabrous. **Ray corollas** bright yellow, laminae 2.5–16 mm, floral pigments turning deep red in dilute aqueous alkali. **Disc corolla** lobes 5; anther appendages deltate to sublanceolate or subulate (broadened distal to bases); style apices ± deltate with apical tufts of hairs and subapical fringes of shorter hairs. **Cypselae** to 4 mm; epappose or pappose. $2n$ = 16, 32, 48.

Species 4 (4 in the flora): w United States, nw Mexico.

Perennial lasthenias are found only in sect. *Amphiachaenia*. Epappose plants of sect. *Amphiachaenia* are more likely to be found in coastal populations and may constitute entire local populations.

SELECTED REFERENCES   Chan, R. 2001. Taxonomic changes and a new species in *Lasthenia* sect. *Amphiachaenia* (Compositae: Heliantheae sensu lato). Madroño 48: 205–210.   Chan, R., B. G. Baldwin, and R. Ornduff. 2002. Cryptic goldfields: A molecular phylogenetic reinvestigation of *Lasthenia californica* sensu lato and close relatives (Compositae: Heliantheae sensu lato). Amer. J. Bot. 89: 1103–1112.

1. Stems glabrous proximally, villous distally; phyllaries usually 4–6, glabrous but for apices; receptacles subulate, papillate; anther appendages subulate; interior w California . . . . . 3. *Lasthenia leptalea*
1. Stems ± hairy throughout, sometimes more so distally; phyllaries 4–16, hairy; receptacles conic, muricate; anther appendages deltate to sublanceolate; Arizona, California, Oregon.
   2. Perennials; coastal Oregon . . . . 4. *Lasthenia ornduffii*
   2. Annuals or perennials; coastal or inland Arizona, California, Oregon.
     3. Annuals; leaves entire; pappi usually of (2–)4(–6) opaque, white (aging to brown), ovate-lanceolate, aristate scales, sometimes 0; Arizona, California (including Channel Islands) . . . . 2. *Lasthenia gracilis*
     3. Annuals or perennials (if perennial, only in coastal California); leaves entire or with 3–5+ teeth; pappi 0, or of 1–7 translucent (rarely opaque), brown (rarely white), linear to subulate, aristate scales; n California, s Oregon . . . . 1. *Lasthenia californica*

1. Lasthenia californica de Candolle ex Lindley, Edwards's Bot. Reg. 21: sub plate 1780. 1835 [E] [F]

Baeria chrysostoma Fischer & C. A. Meyer; *Lasthenia chrysostoma* (Fischer & C. A. Meyer) Greene; *L. hirsutula* Greene

Annuals or perennials, to 40 cm (cespitose). **Stems** erect or decumbent, branched proximally or distally, ± hairy. **Leaves** linear to oblanceolate or oblong, 8–210 × 1–5.5(–15) mm, (± fleshy in coastal forms) margins entire or with 3–5+ teeth, faces glabrous or ± hairy. **Involucres** campanulate to depressed-hemispheric or hemispheric, 5–14 mm. **Phyllaries** (persistent or falling with cypselae) 4–16 (in 1–2 series), elliptic to ovate or lanceolate to oblong, hairy. **Receptacles** conic, muricate, glabrous. **Ray florets** 6–16; laminae linear to oblong, 5–18 mm. **Anther appendages** deltate to sublanceolate.

Cypselae black to gray or silver-gray, linear to narrowly clavate, to 4 mm, glabrous or hairy; **pappi** 0, or of 1–7 translucent (rarely opaque), brown (rarely white), linear to subulate, aristate scales.

Subspecies 3 (3 in the flora): w United States.

Plants of *Lasthenia californica*, especially those in coastal populations, have the largest, showiest heads in the genus. Report of *L. californica* from Massachusetts was not confirmed for this study

1. Annuals; coastal or inland; n California, s Oregon
   . . . . 1b. *Lasthenia californica* subsp. *californica*
1. Perennials (sometimes flowering first year); coastal; California.
   2. Roots fleshy, clustered; stems erect, branched distally; leaf blades 1–2+ mm wide . . . .
     . . . . 1a. *Lasthenia californica* subsp. *bakeri*
   2. Roots usually not fleshy or clustered; stems decumbent, branched proximally; leaf blades 1.5–5.5(–15) mm wide . . . .
     . . . . 1c. *Lasthenia californica* subsp. *macrantha*

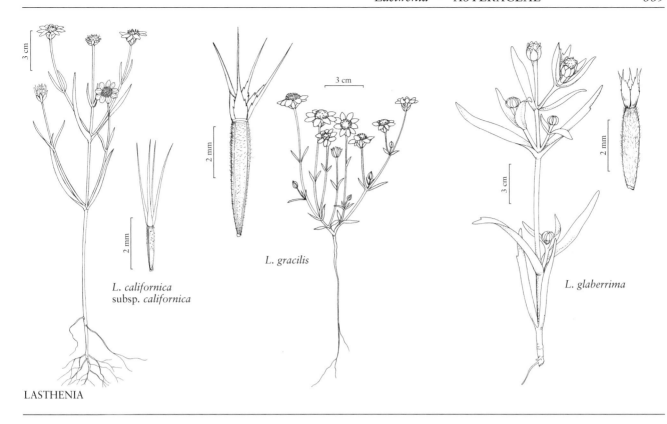

L. californica
subsp. californica

L. gracilis

L. glaberrima

LASTHENIA

1a. Lasthenia californica de Candolle ex Lindley subsp. **bakeri** (J. T. Howell) R. Chan, Madroño 48: 208. 2002 • Baker's goldfields [C][E]

Baeria bakeri J. T. Howell, Leafl. W. Bot. 1: 7. 1932; *B. macrantha* (A. Gray) A. Gray var. *bakeri* (J. T. Howell) D. D. Keck; *Lasthenia macrantha* (A. Gray) Greene subsp. *bakeri* (J. T. Howell) Ornduff

Perennials (sometimes flowering first year). **Roots** fleshy, clustered. **Stems** erect, branched distally, ± hairy. **Leaves** mostly basal; blades linear to oblong, 20–210 × 1–2+ mm, not fleshy, margins entire, faces glabrous or ± hairy. **Involucres** campanulate to depressed-hemispheric, 9–14 mm. **Phyllaries** ± persistent, 13–16 in 2 series, elliptic to ovate. **Ray florets** 8–16; laminae elliptic to oblong, 5–16 mm. **Anther appendages** deltate to sublanceolate. **Cypselae** silver-gray, linear to narrowly clavate, to 4 mm, glabrous; **pappi** 0, or of 1–4 translucent, brown, subulate, aristate scales (sometimes variable or 0 within heads). **2n = 48.**

Flowering year round (mostly May–Jun). Grasslands, woodlands, coastal; of conservation concern; 0–500 m; Calif.

Subspecies *bakeri* is found along the coast in central California. Its most distinctive feature is its fleshy, clustered roots.

1b. Lasthenia californica de Candolle ex Lindley subsp. **californica** • California goldfields [E][F]

Annuals. Roots not fleshy, not clustered. **Stems** usually erect, sometimes decumbent (especially in coastal forms), usually branched distally, sometimes proximally, ± hairy, usually more so distally. **Leaves** mostly cauline; blades linear to oblanceolate, 8–40(–70) × 1–6 mm, ± fleshy in coastal forms, margins entire or with 3–5+ teeth (coastal forms), faces ± hairy. **Involucres** campanulate or hemispheric, 5–10 mm. **Phyllaries** persistent or falling with cypselae, 4–13 in 1 series, ovate-lanceolate to oblong. **Ray florets** 6–13; laminae oblong, 5–10 mm. **Anther appendages** deltate. **Cypselae** black to gray, ± clavate, to 3 mm, glabrous or hairy; **pappi** usually of 1–7 translucent (rarely opaque), brown (rarely white), linear to subulate, aristate scales (rarely variable or 0 within heads). **2n = 16, 32, 48.**

Flowering Feb–Jun. Mostly open sites (virtually every habitat but desert); 0–1500 m; Calif., Oreg.

Subspecies *californica* sometimes occurs with other *Lasthenia* taxa. R. Ornduff (1993) included *L. gracilis* within his circumscription of subsp. *californica*. Coastal forms of subsp. *californica* tend to have shorter, wider, toothed, fleshy leaves and larger heads; they can be distinguished from *L. gracilis* by their translucent, brown, linear to subulate, aristate pappus scales and more northern distribution. Sympatric epappose plants of subsp. *californica* and *L. gracilis* are not easily distinguished morphologically; molecular markers (R. Chan et al. 2001) show them to be distinct taxa.

1c. Lasthenia californica de Candolle ex Lindley subsp. **macrantha** (A. Gray) R. Chan, Madroño 48: 208. 2002 • Perennial goldfields [C] [E]

Burrielia chrysostoma (Fischer & C. A. Meyer) Torrey & A. Gray var. *macrantha* A. Gray in War Department [U.S.], Pacif. Railr. Rep. 4(5): 106. 1857; *Baeria macrantha* (A. Gray) A. Gray; *B. macrantha* var. *pauciaristata* A. Gray; *B. macrantha* var. *thalassophila* J. T. Howell; *Lasthenia macrantha* (A. Gray) Greene

Perennials (sometimes flowering first year; cespitose). **Roots** usually not fleshy or clustered. **Stems** decumbent, branched proximally, ± hairy. **Leaves** mostly basal; blades linear to oblong, 28–88 × 1.5–5.5(–15) mm, fleshy, margins entire or with 3–5+ teeth (ciliate), faces glabrous or ± hairy. **Involucres** campanulate to depressed-hemispheric, 9–14 mm. **Phyllaries** ± persistent, 9–16 in 2 series, elliptic to ovate. **Ray florets** 8–16; laminae elliptic to oblong, 6–18 mm. **Anther appendages** deltate to sublanceolate. **Cypselae** silvergray, linear to narrowly clavate, to 4 mm, glabrous; **pappi** 0, or of 1–4 translucent, brown, subulate, aristate scales (sometimes variable within heads). $2n = 48$.

Flowering year round (mostly May–Aug). Grasslands, dunes, woodlands, immediate coast; of conservation concern; 0–500 m; Calif.

Subspecies *macrantha* is morphologically similar to *Lasthenia ornduffii*; their ranges are allopatric. Plants of subsp. *macrantha* found in San Luis Obispo County may have fleshy roots approaching those of subsp. *bakeri*.

2. Lasthenia gracilis (de Candolle) Greene, Man. Bot. San Francisco, 206. 1894 • Common goldfields [F]

Burrielia gracilis de Candolle in A. P. de Candolle and A. L. P. P. de Candolle, Prodr. 5: 664. 1836; *Baeria chrysostoma* Fischer & C. A. Meyer subsp. *gracilis* (de Candolle) Ferris; *B. gracilis* (de Candolle) A. Gray

Annuals, to 40 cm. **Stems** erect or decumbent, usually branched distally (sometimes proximally in desert forms), ± hairy (more so distally). **Leaves** linear to oblanceolate, 8–40

(–70) × 1–3(–6) mm, (± fleshy in coastal forms) margins entire or with 3–5+ teeth, faces ± hairy. **Involucres** campanulate or hemispheric, 5–10 mm. **Phyllaries** (persistent or falling with cypselae) 4–13 (in 1 series), ovate-lanceolate to oblong, ± hairy. **Receptacles** conic, muricate, glabrous. **Ray florets** 6–13; laminae oblong, 5–10 mm. **Anther appendages** deltate. **Cypselae** black to gray, ± linear, to 3 mm, glabrous or hairy; **pappi** usually of (2–)4(–6) opaque, white (aging to brown), ovate-lanceolate, aristate scales, sometimes 0. $2n = 16, 32$.

Flowering Feb–Jun. Mostly open sites (virtually all habitats); 0–1500 m; Ariz., Calif.; Mexico (Baja California, Sonora).

D. D. Keck (1959c) said of *Lasthenia gracilis* (as *Baeria chrysostoma* subsp. *gracilis*), "The most abundant composite in the state [of California]." It is widespread throughout California, central Arizona, the Channel Islands, Guadalupe Island, and Baja California. It is variable and is sometimes similar to *L. californica* subsp. *californica*, from which it differs in its opaque, white, ovate-lanceolate, aristate pappus scales. Coastal forms of *L. gracilis* tend to have shorter, wider, fleshy leaves. *Lasthenia gracilis* often occurs with other *Lasthenia* species.

3. Lasthenia leptalea (A. Gray) Ornduff, Univ. Calif. Publ. Bot. 40: 63. 1966 • Salinas Valley goldfields [E]

Burrielia leptalea A. Gray, Proc. Amer. Acad. Arts 6: 546. 1865; *Baeria leptalea* (A. Gray) A. Gray

Annuals, to 15 cm. **Stems** erect (peduncles sometimes sinuous), branched distally, glabrous proximally, villous distally. **Leaves** linear, 3–20 × 0.5–1 mm, (± fleshy) margins entire, faces sparsely hairy. **Involucres** obconic to campanulate, 4–6 mm. **Phyllaries** (± persistent) usually 4–6 (in 1 series), elliptic to ovate, glabrous but for hairy apices. **Receptacles** subulate, papillate, glabrous. **Ray florets** 6–9; corolla laminae broadly elliptic, 2.5–5 mm. **Anther appendages** subulate. **Cypselae** gray, narrowly clavate, to 2 mm, sparsely hairy; **pappi** usually of 1–4 translucent, white to yellowish, subulate, aristate scales (sometimes 0 in some florets within heads). $2n = 16$.

Flowering Feb–Apr. Open areas of oak woodlands; 0–700 m; Calif.

Lasthenia leptalea grows in southern Monterey and northern San Luis Obispo counties. Originally assigned by R. Ornduff (1966b) to sect. *Burrielia*, *L. leptalea* is morphologically similar to *L. gracilis*, from which it can be distinguished by its subulate anther appendages and phyllaries that are hairy only at their tips.

4. **Lasthenia ornduffii** R. Chan, Madroño 48: 209. 2002 • Ornduff's goldfields C E

Lasthenia macrantha (A. Gray) Greene subsp. *prisca* Ornduff, Madroño 21: 96. 1971

Perennials, 4–28 cm. **Stems** decumbent, branched proximally, ± hairy throughout, more so distally. **Leaves** linear to oblong, 6–40 × 1.8–5 mm, (± fleshy) margins entire or with 3–5+ teeth, faces glabrous or ± hairy. **Involucres** campanulate to depressed-hemispheric, 5–14 mm. **Phyllaries** (± persistent) 8–14 (in 2 series), elliptic to ovate, ± hairy. **Receptacles** conic, muricate, glabrous. **Ray florets** 8–15; laminae elliptic to oblong, 5–9 mm. **Anther appendages** deltate to sublanceolate. **Cypselae** silver-gray, linear to narrowly clavate, to 4 mm, glabrous; **pappi** 0, or of 1–4 translucent, brown, subulate, aristate scales (often variable within heads). $2n = 32$.

Flowering year round (mostly May–Aug). Coastal bluffs; of conservation concern; 0–500 m; Oreg.

Lasthenia ornduffii is known only from six or so populations in grasslands along the immediate coast in Curry County. The plants are usually scapiform.

354b. LASTHENIA Cassini sect. BURRIELIA (de Candolle) Ornduff, Univ. Calif. Publ. Bot. 40: 62. 1966 E

Burrielia de Candolle in A. P. de Candolle and A. L. P. P. de Candolle, Prodr. 5: 663. 1836; *Baeria* Fischer & C. A. Meyer sect. *Burrielia* (de Candolle) H. M. Hall

Annuals. Leaves ± entire or toothed (not pinnatifid). **Involucres** campanulate to cylindric. **Phyllaries** falling with cypselae, distinct. **Receptacles** subulate, papillate, glabrous. **Ray corollas** yellow, laminae 0–1 or 3–5 mm, floral pigments turning deep red in dilute aqueous alkali. **Disc corolla** lobes 4–5; anther appendages subulate or deltate (with 1–4 wartlike glands in *L. debilis*); style apices lanceolate and glabrous, or ± deltate with apical tufts of hairs and subapical fringes of shorter hairs. **Cypselae** to 5 mm; epappose or pappose. $2n = 8, 24$.

Species 2 (2 in the flora): California.

Within *Lasthenia*, only members of sects. *Amphiachaenia* and *Burrielia* have floral anthochlor pigments that turn red in aqueous alkali. Unlike most lasthenias, the two members of *Lasthenia* sect. *Burrielia* are inconspicuous and usually occur in relatively moist, shaded woodland areas. Both have relatively thick stems, relatively broad, obscurely toothed leaves, relatively few rays, phyllaries that fall with cypselae, and similar induments.

1. Ray laminae 3–5 mm; disc corolla lobes 5; anther appendages deltate . . . . 5. *Lasthenia debilis*
1. Ray laminae 0–1 mm; disc corolla lobes 4(–5); anther appendages subulate . . . . . . 6. *Lasthenia microglossa*

5. **Lasthenia debilis** (Greene ex A. Gray) Ornduff, Univ. Calif. Publ. Bot. 40: 63. 1966 • Greene's goldfields E

Baeria debilis Greene ex A. Gray in A. Gray et al., Syn. Fl. N. Amer. 1(2): 325. 1884

Annuals, to 30 cm. **Stems** erect, usually branched proximally, sometimes distally, villous. **Leaves** linear to linear-oblong, 10–80 × 1–5.5 mm, margins usually entire, sometimes 1–2-toothed (teeth to 2 mm), faces hairy. **Heads** showy. **Involucres** campanulate to obconic, 5–7 mm. **Phyllaries** 5, obovate to ovate, slightly hairy. **Receptacles** subulate, papillate, glabrous. **Ray florets** 5–10; (corollas yellow or white) laminae broadly elliptic to oblong, 3–5 mm. **Disc corolla** lobes 5. **Anther appendages** deltate (with 1–4 wartlike glands; style apices ± deltate with apical tufts of hairs and subapical fringes of shorter hairs). **Cypselae** black, ± linear, to 3 mm, hairy; **pappi** 0, or of 2–4 brown or white, ovate or lanceolate, aristate scales. $2n = 8$.

Flowering Mar–May. Shaded or open, moist woodland slopes; 0–500 m; Calif.

Lasthenia debilis is inconspicuous and has pale green foliage and lightly pigmented, relatively short rays. Populations with white or nearly white rays, uncharacteristic for lasthenias, are known.

6. **Lasthenia microglossa**   (de Candolle) Greene, Man. Bot. San Francisco, 205. 1894 • Small-ray goldfields [E]

Burrielia microglossa de Candolle in A. P. de Candolle and A. L. P. P. de Candolle, Prodr. 5: 664. 1836; *Baeria microglossa* (de Candolle) Greene

Annuals, to 25 cm. **Stems** erect or sprawling, branched distally, villous, especially distally. **Leaves** linear or subulate, 15–80 × 1.5–2(–4) mm, margins ± entire, faces moderately hairy. **Involucres** cylindric to narrowly obconic, 6–8.5 mm. **Phyllaries** 4, elliptic to oblong, hairy. **Receptacles** subulate, papillate, glabrous. **Ray florets** 4; (corollas yellow) laminae lance-elliptic, (0–)0.5–1 mm. **Disc corolla** lobes 4(–5). **Anther appendages** subulate (without wartlike glands; style apices lanceolate, glabrous). **Cypselae** black, ± linear, to 5 mm, hairy; **pappi** 0, or of 1–4 brown or white, linear to ovate or lanceolate, aristate scales (sometimes variable within heads). **2*n*** = 24.

Flowering Mar–May. Shaded areas, woodlands, chaparral, deserts; 0–1000 m; Calif.

Because the rays are inconspicuous, plants of the self-pollinating *Lasthenia microglossa* are easy to overlook. Besides growing in habitats similar to those occupied by *L. debilis*, *L. microglossa* is found also beneath chaparral shrubs and extends into desert areas, where it grows near rocks that may provide favorable shade and moisture conditions.

354c. **Lasthenia** Cassini sect. **Hologymne** (Bartling) Nuttall, Trans. Amer. Philos. Soc., n. s. 7: 383. 1841   [E]

Hologymne Bartling, Index Seminum (Göttingen) 1837: 4. 1837; *Crockeria* Greene ex A. Gray

Annuals. **Leaves** entire (blades ± fleshy, bases sometimes connate and sheathing). **Involucres** hemispheric. **Phyllaries** persistent, connate ²/₃+ their lengths (forming cups, glabrous but for distinct apices). **Receptacles** ± conic, papillate or warty, glabrous or sparsely hairy. **Ray corollas** yellow, laminae 4–14 mm, floral pigments remaining yellow in dilute aqueous alkali. **Disc corolla** lobes 5; anther appendages deltate or broadly ovate; style apices deltate with apical tufts of hairs and subapical fringes of shorter hairs. **Cypselae** 2–3.5 mm; epappose. **2*n*** = 14.

Species 3 (3 in the flora): California.

Members of sect. *Hologymne* have a high salt tolerance and usually occur in moderately to strongly alkaline habitats, such as coastal salt marshes, around vernal pools, and in wet areas around alkali flats (i.e., hard-packed, poorly aerated, saline clays). The herbage of these plants also becomes succulent in response to increasing soil salinity as the soil dries up. All three species show a narrow range of morphologic characteristics except for their variable cypselae.

1. Cypselae black, strongly flattened, obovoid, margins ciliate (hairs stiff, blunt), faces glabrous or scabrous . . . . 7. *Lasthenia chrysantha*
1. Cypselae black, ± flattened and obovate to oblong, or gray, ± clavate or obovoid, margins sometimes ciliate, faces glabrous or hairy (sometimes papillate).
    2. Cypselae ± flattened and obovate to oblong, to 2.5 mm, faces ± hairy (hairs curved) and papillate . . . . 8. *Lasthenia ferrisiae*
    2. Cypselae ± clavate or obovoid, to 3.5 mm, faces glabrous or hairy and papillate (papillae rusty or yellowish, wartlike) . . . . 9. *Lasthenia glabrata*

7. Lasthenia chrysantha (Greene ex A. Gray) Greene, Man. Bot. San Francisco, 204. 1894 • Alkali-sink goldfields E

Crockeria chrysantha Greene ex A. Gray in A. Gray et al., Syn. Fl. N. Amer. 1(2): 445. 1884

Annuals, to 28 cm. **Stems** erect, branched distally, glabrous or ± hairy. **Leaves** linear, 10–80 × 1–2 (–3+) mm, margins entire, faces glabrous or sparsely hairy. **Involucres** hemispheric, 5–7 mm. **Phyllaries** 8–14, ± deltate (distinct tips ± deltate), glabrous but for apices. **Receptacles** ± conic, warty, glabrous or sparsely hairy. **Ray florets** 6–10; (corollas lemon- or golden yellow) laminae narrowly oblong, 6–7 mm. **Anther appendages** deltate or broadly ovate. **Cypselae** black, strongly flattened, obovoid, 2–3 mm, margins ciliate (hairs white or stramineous, stiff, blunt, curved), faces glabrous or scabrous; **pappi** 0. $2n = 14$.

Flowering Feb–Apr. Vernal pools and wet alkali flats; 0–100 m; Calif.

Lasthenia chrysantha has the narrowest ecologic range within *Lasthenia* sect. *Hologymne* and is mostly limited to alkali flats in the San Joaquin Valley.

8. Lasthenia ferrisiae Ornduff, Univ. Calif. Publ. Bot. 40: 74, figs. 4n,o, 23. 1966 • Ferris's goldfields E

Annuals, to 40 cm. **Stems** erect, branched distally, glabrous or slightly hairy. **Leaves** linear, 10–80 × 1–3+ mm, margins entire, faces glabrous. **Involucres** hemispheric, 5–10 mm. **Phyllaries** 6–14, ± lanceolate (distinct tips ± deltate), glabrous but for apices. **Receptacles** ± conic, papillate, glabrous. **Ray florets** 6–13; (corollas golden yellow to yellow) laminae ± oblong, 6–10 mm. **Anther appendages** deltate or broadly ovate. **Cypselae** black, ± clavate, ± flattened, 2–2.5 mm, margins ciliate (hairs whitish or stramineous, curved), faces ± hairy and papillate; **pappi** 0. $2n = 14$.

Flowering Feb–May. Vernal pools and wet alkali flats; 0–700 m; Calif.

R. Ornduff (1966b) suggested that *Lasthenia ferrisiae* may have originated from hybridization between *L. chrysantha* and *L. glabrata* subsp. *coulteri*; it has morphologic features that are intermediate between the two.

9. Lasthenia glabrata Lindley, Edwards's Bot. Reg. 21: plate 1780. 1835 E

Annuals, to 60 cm. **Stems** erect, branched distally, glabrous or slightly hairy. **Leaves** linear or subulate, 40–150 × 2–3+ mm, margins entire, faces glabrous. **Involucres** hemispheric, 5–10 mm. **Phyllaries** 10–14, ± lanceolate (distinct tips ± deltate), glabrous but for apices. **Receptacles** ± conic, papillate, glabrous or sparsely hairy. **Ray florets** 7–15; (corollas yellow) laminae oblong, 4–14 mm. **Anther appendages** deltate or broadly ovate. **Cypselae** gray, clavate or obovoid, 2–3.5 mm, margins not ciliate, faces glabrous, or hairy and papillate (papillae rusty or yellowish, wartlike); **pappi** 0.

Subspecies 2 (2 in the flora): California.

Historically, aboriginal Californians used fruits and leaves of *Lasthenia glabrata* for food. The subspecies are allopatric and almost identical except for their cypselae.

1. Cypselae glabrous, not papillate . . . .
. . . . 9a. *Lasthenia glabrata* subsp. *glabrata*
1. Cypselae ± hairy and papillate . . . .
. . . . 9b. *Lasthenia glabrata* subsp. *coulteri*

9a. Lasthenia glabrata Lindley subsp. **glabrata**
  • Yellow-ray goldfields E

Lasthenia glabrata var. *californica* Jepson

Cypselae glabrous, not papillate. $2n = 14$.

Flowering Mar–May. Salt marshes, vernal pools, wet, alkali flats; 0–600 m; Calif.

9b. Lasthenia glabrata Lindley subsp. **coulteri** (A. Gray) Ornduff, Univ. Calif. Publ. Bot. 40: 74. 1966 • Coulter's goldfields E

Lasthenia glabrata var. *coulteri* A. Gray in A. Gray et al., Syn. Fl. N. Amer. 1(2): 324. 1884; *L. coulteri* (A. Gray) Greene

Cypselae ± hairy and papillate (papillae rusty or yellowish, wartlike). $2n = 14$.

Flowering Apr–May. Salt marshes, vernal pools, and wet, alkali flats; 0–1000 m; Calif.

354d. LASTHENIA Cassini sect. LASTHENIA

Annuals. Leaves entire. **Involucres** hemispheric to obconic. **Phyllaries** persistent, connate $^2/_3$+ their lengths. **Receptacles** conic, papillate, glabrous. **Ray corollas** yellow, laminae (0–)0.5–2 mm, floral pigments remaining yellow in dilute aqueous alkali. **Disc corolla** lobes 4(–5); anther appendages oblong or obovate; style apices ± deltate, glabrous or glabrate. **Cypselae** to 4 mm; pappose. **2n = 10.**

Species 2 (1 in the flora): w North America, South America (Chile).

Section *Lasthenia* has an amphitropical distribution: *L. glaberrima* ranges from southwestern Washington to central California; *L. kunthii* Hooker & Arnott is known only from central Chile (P. H. Raven 1963c). The species are very similar morphologically, and conclusive identification of specimens sometimes is difficult without knowing where the plant was collected. Both *L. glaberrima* and *L. kunthii* are self-pollinating and are found in vernal pools or other seasonally wet habitats.

10. **Lasthenia glaberrima** de Candolle in A. P. de Candolle and A. L. P. P. de Candolle, Prodr. 5: 664. 1836 • Smooth goldfields E F

Annuals, to 35 cm. **Stems** erect or sprawling, branched distally, glabrous. **Leaves** linear, 30–100 × 2–4(-6+) mm, (bases often connate and sheathing, apices blunt) margins entire, faces glabrous. **Involucres** hemispheric or campanulate, 5–7 mm. **Phyllaries** 5–10, ± lanceolate, hairy. **Receptacles** conic, papillate, glabrous. **Ray florets** 6–13; (corollas pale yellow) laminae ovate-elliptic, (0–)0.5–2 mm. **Anther append-**ages oblong or obovate. **Cypselae** grayish, somewhat flattened, ± linear, to 4 mm, hairy; **pappi** of 5–10 subulate to lanceolate, often fimbriate, laciniate, or aristate scales. **2n = 10.**

Flowering Mar–Jul. Vernal pools, wet meadows; 0–900 m; B.C.; Calif., Oreg., Wash.

354e. LASTHENIA Cassini sect. ORNDUFFIA R. Chan, Madroño 48: 38. 2001 E

Annuals. Leaves usually pinnatifid, sometimes entire. **Involucres** hemispheric or obconic. **Phyllaries** persistent, usually distinct (connate $^1/_4$–$^1/_2$ their lengths in *L. conjugens*). **Receptacles** conic or dome-shaped, glabrous or hairy. **Ray corollas** yellow, laminae 5–10 mm, floral pigments remaining yellow in dilute aqueous alkali. **Disc corolla** lobes 5; anther appendices linear, obovate, or ovate (style apices deltate with apical tufts of hairs and subapical fringes of shorter hairs). **Cypselae** to 1.5 mm; epappose or pappose. **2n = 12.**

Species 3 (3 in the flora): California.

Members of sect. *Ornduffia* are morphologically similar; they can be distinguished by pappus and phyllary differences. They are typically found in vernal pools; the plants may start growing as submerged aquatics and flower as succulent xerophytes. Leaves of this section are usually pinnatifid; undivided leaves are sometimes produced in crowded conditions with poor light.

R. Ornduff (1966b) originally included the three members of *Lasthenia* sect. *Ornduffia* in sect. *Ptilomeris*. Molecular studies (R. Chan 2000) showed that the three species form a distinct, monophyletic group.

SELECTED REFERENCE Crawford, D. J. and R. Ornduff. 1989. Enzyme electrophoresis and evolutionary relationships among three species of *Lasthenia* (Asteraceae: Heliantheae). Amer. J. Bot. 76: 289–296.

1. Phyllaries 12–18 (connate ¹/₄–¹/₂ their lengths) . . . . 12. *Lasthenia conjugens*
1. Phyllaries 7–16 (distinct).
    2. Pappi usually of 1(–2) aristate scales plus 3–6+ shorter, ± subulate scales . . . . 11. *Lasthenia burke*
    2. Pappi usually of (3–)4(–5) subulate, aristate scales plus 3–5+ shorter, ± subulate scales
        or teeth, rarely of aristate scales only or 0 . . . . 13. *Lasthenia fremontii*

11. Lasthenia burkei (Greene) Greene, Man. Bot. San Francisco, 204. 1894 • Burke's goldfields C E

Baeria burkei Greene, Bull. Calif. Acad. Sci. 2: 151. 1887

Annuals, to 30 cm. **Stems** erect, branched distally, hairy. **Leaves** linear, 10–50 × 1–2+ mm (simple blades or single lobes), margins entire or pinnately lobed, faces glabrous or ± hairy. **Involucres** hemispheric or obconic, 4–6 mm. **Phyllaries** 7–16 (distinct), ovate, hairy. **Receptacles** conic or dome-shaped, muriculate, glabrous or hairy. **Ray florets** 8–13; laminae oblong to oval, to 6 mm. **Anther appendages** linear to ± ovate. **Cypselae** black or gray, clavate, to 1.5 mm, hairy; **pappi** usually of 1(–2) aristate scales plus 3–6+ shorter, ± subulate scales. **2*n* = 12.**

Flowering Jun. Vernal pools and wet meadows; of conservation concern; 0–500 m; Calif.

Lasthenia burkei is known only from relatively few populations in the coast ranges north of San Francisco Bay. It is allopatric from other members of *Lasthenia* sect. *Ornduffia*. It is in the Center for Plant Conservation's National Collection of Endangered Plants.

SELECTED REFERENCE Ornduff, R. 1969. The origin and relationship of *Lasthenia burkei* (Compositae). Amer. J. Bot. 56: 1042–1047.

12. Lasthenia conjugens Greene, Pittonia 1: 221. 1888 • Contra Costa goldfields C E

Baeria fremontii (Torrey ex A. Gray) A. Gray var. *conjugens* (Greene) Ferris

Annuals, to 40 cm. **Stems** erect, branched distally, glabrous or sparsely hairy. **Leaves** linear, 10–80 × 1–2+ mm (simple blades or single lobes), margins entire or pinnately lobed, faces glabrous. **Involucres** hemispheric or obconic, 6–10 mm. **Phyllaries** 12–18 (connate ¹/₄–¹/₂ their lengths), ± lanceolate,

puberulent to glabrate. **Receptacles** dome-shaped or obconic, densely hairy. **Ray florets** 6–13; laminae oblong to oval, 5–10 mm. **Anther appendages** linear to ± ovate. **Cypselae** black or gray (reflective), clavate, to 1.5 mm, glabrous; **pappi** 0. **2*n* = 12.**

Flowering Mar–Jun. Vernal pools and wet meadows; of conservation concern; 0–100 m; Calif.

Lasthenia conjugens often germinates under water in vernal pools; first leaves are submerged and entire; later, aerial leaves are pinnatifid.

13. Lasthenia fremontii (Torrey ex A. Gray) Greene, Man. Bot. San Francisco, 204. 1894 (as fremonti) • Fremont's goldfields E

Dichaeta fremontii Torrey ex A. Gray, Mem. Amer. Acad. Arts, n. s. 4: 102. 1849; *Baeria fremontii* (Torrey ex A. Gray) A. Gray

Annuals, to 35 cm. **Stems** erect, branched proximally, glabrous proximally, ± hairy distally. **Leaves** linear, 10–60 × 1–2+ mm (simple blades or single lobes), margins entire or with 1–3 pairs of linear lobes, faces glabrous or sparsely hairy. **Involucres** hemispheric or obconic, 4–6 mm. **Phyllaries** 8–16 (distinct), ovate, hairy. **Receptacles** dome-shaped, muricate, hairy. **Ray florets** 6–13; laminae oblong to oval, 5–7 mm. **Anther appendages** linear to narrowly ovate. **Cypselae** black or gray, clavate, to 1.5 mm, usually scabrous, rarely glabrous; **pappi** usually of (3–)4(–5) subulate, aristate scales plus 3–5+ shorter, ± subulate scales or teeth, rarely of aristate scales only, or 0. **2*n* = 12.**

Flowering Mar–May. Vernal pools and wet meadows; 0–700 m; Calif.

Epappose plants of *Lasthenia fremontii* are rarely found and then only in the southernmost part of its range in Tulare and San Luis Obispo counties, where they occur together with pappose plants. The ranges of *L. fremontii* and *L. conjugens* overlap slightly; the two species have not been found growing together.

354f. Lasthenia Cassini sect. Platycarpha (H. M. Hall) Ornduff, Univ. Calif. Publ. Bot. 40: 66. 1966 [E]

Baeria Fischer & C. A. Meyer sect. *Platycarpha* H. M. Hall in N. L. Britton et al., N. Amer. Fl. 34: 76. 1914

Annuals. Leaves usually with 1–2 pairs of linear lobes, sometimes entire. **Involucres** obconic. **Phyllaries** persistent, distinct. **Receptacles** conic, muricate, glabrous or sparsely hairy. **Ray corollas** yellow, laminae 7–8 mm, floral pigments remaining yellow in dilute aqueous alkali. **Disc corolla** lobes 5; anther appendages deltate; style apices deltate with apical tufts of hairs and subapical fringes of shorter hairs. **Cypselae** 1.5–3.5 mm; pappose. *2n* = 8.

Species 1: California.

14. Lasthenia platycarpha (A. Gray) Greene, Man. Bot. San Francisco, 205. 1894 • Alkali goldfields [E]

Burrielia platycarpha A. Gray in W. H. Emory, Rep. U.S. Mex. Bound. 2(1): 97. 1859; *Baeria platycarpha* (A. Gray) A. Gray

Annuals, to 30 cm. **Stems** erect, branched proximally, glabrous or woolly to villous, especially distally. **Leaves** linear, 10–60 × 1–2+ mm (simple blades or single lobes), margins usually lobed, sometimes entire, faces glabrous or hairy. **Involucres** obconic, 6–8 mm. **Phyllaries** 6–9, elliptic to ovate, glabrous or villous. **Receptacles** conic, muricate, glabrous or sparsely hairy. **Ray florets** 6–13; laminae elliptic, 7–8 mm. **Anther appendages** deltate. **Cypselae** black to gray, narrowly clavate, 1.5–3.5 mm, hairy; **pappi** of 4–6 white or yellowish, lanceolate to ovate, aristate scales. *2n* = 8.

Flowering Mar–Apr. Alkali flats; 0–100+ m; Calif.

Lasthenia platycarpha is known only from highly saline soils and is frequently found with species of sects. *Hologymne* and *Ornduffia*, particularly *L. fremontii*. In the northern part of its range, *L. platycarpha* is more robust and has longer pappus scales and more densely pubescent peduncles. Plants with entire leaves resemble species of sect. *Amphiachaenia*; *L. platycarpha* does not have anthochlor pigments that turn red in aqueous alkali.

354g. Lasthenia Cassini sect. Ptilomeris (Nuttall) Ornduff, Univ. Calif. Publ. Bot. 40: 75. 1966

Ptilomeris Nuttall, Trans. Amer. Philos. Soc., n. s. 7: 381. 1841; *Baeria* Fischer & C. A. Meyer sect. *Ptilomeris* (Nuttall) H. M. Hall

Annuals. Leaves usually pinnately lobed or pinnatifid, sometimes entire. **Involucres** hemispheric to obconic. **Phyllaries** falling with cypselae, distinct. **Receptacles** conic, smooth, muricate, or pitted, glabrous or hairy. **Ray corollas** yellow, laminae 1–10 mm, floral pigments remaining yellow in dilute aqueous alkali. **Disc corolla** lobes 5; anther appendages ovate to obovate; style apices ± deltate, glabrous or with apical tufts of hairs and subapical fringes of shorter hairs. **Cypselae** 1.5–3 mm; epappose or pappose. *2n* = 8, 10.

Species 3 (3 in the flora): w North America, nw Mexico.

1. Stems usually glandular-puberulent (herbage sweetly scented) . . . . 15. *Lasthenia coronaria*
1. Stems glabrous or hairy (at nodes and distally), or ± woolly (herbage not sweetly scented).
 2. Ray laminae 1–3 mm; pappi 0, or of 4–6(–12) aristate scales plus 4–5+ shorter, laciniate scales; coastal and offshore; British Columbia, California, Oregon, Washington . . . . 16. *Lasthenia maritima*
 2. Ray laminae 4–8 mm; pappi 0, or of 2–3(–4) aristate scales plus 4–5+ shorter, fimbriate scales; coastal and inland; c California . . . . 17. *Lasthenia minor*

15. Lasthenia coronaria (Nuttall) Ornduff, Univ. Calif. Publ. Bot. 40: 76. 1966 • Crowned or royal goldfields

Ptilomeris coronaria Nuttall, Trans. Amer. Philos. Soc., n. s. 7: 382. 1841; *Baeria californica* (Hooker) K. L. Chambers; *B. coronaria* (Nuttall) A. Gray

Annuals, to 40 cm (herbage sweetly scented). **Stems** erect, branched distally, usually glandular-puberulent (often with longer non-glandular hairs as well). **Leaves** linear, 15–60 × 0.5–5 mm, (not fleshy) margins entire or 1–2-pinnately lobed, faces hairy. **Involucres** hemispheric to obconic, 4–7 mm. **Phyllaries** 6–14, lanceolate to ovate, hairy. **Receptacles** conic, smooth, muricate, or pitted, hairy. **Ray florets** 6–15; (corollas yellow) laminae linear-oblong or oblong, 3–10 mm. **Anther appendages** elliptic, acute (style apices ± deltate with apical tufts of hairs and subapical fringes of shorter hairs). **Cypselae** black, linear to narrowly clavate, to 2.5 mm, hairy; **pappi** usually of 5–6+ lanceolate to ovate scales (1–5 uniaristate), sometimes of 4–5 subulate, aristate scales, or 0. $2n = 8, 10$.

Flowering Mar–May. Sunny, open grassy areas; 0–700 m; Calif.; Mexico (Baja California).

Pappus, head size, and branching pattern vary in *Lasthenia coronaria*. Two types of pappi are often found within a head and sometimes in different individuals of a population. The most distinctive feature of this species is its glandular herbage, which produces a characteristic sweet scent not present in any other lasthenia.

16. Lasthenia maritima (A. Gray) M. C. Vasey, Madroño 32: 139. 1985 • Maritime or seaside goldfields [E]

Burrielia maritima A. Gray, Proc. Amer. Acad. Arts. 7: 358. 1868; *Baeria maritima* (A. Gray) A. Gray; *B. minor* (de Candolle) Ferris subsp. *maritima* (A. Gray) Ferris; *Lasthenia minor* (de Candolle) Ornduff subsp. *maritima* (A. Gray) Ornduff

Annuals, to 25 cm (herbage not sweetly scented). **Stems** usually prostrate or decumbent, rarely erect, branched proximally, glabrous or hairy at nodes and distally. **Leaves** linear to oblanceolate, 10–90 × 2–12 mm, (fleshy) margins entire or lobed, faces glabrous. **Involucres** hemispheric, 4–7 mm. **Phyllaries** 6–14, lanceolate to ovate, hairy (especially at margins and midribs). **Receptacles** conic, muricate, glabrous. **Ray florets** 7–12; (corollas light to golden yellow) laminae oblong, 1–3 mm. **Anther appendages** ± oblong, obtuse (style apices ± deltate, glabrous or with apical tufts of hairs and subapical fringes of shorter hairs). **Cypselae** gray, linear to narrowly clavate, (2–)2.5–3 mm, ± hairy; **pappi** 0, or of 4–6(–12) brown, lanceolate or subulate, aristate scales plus 4–5+ shorter, laciniate scales. $2n = 8$.

Flowering May–Jul. Seabird roosting sites, coastal headlands, offshore rocks, islands; 0–100 m; B.C.; Calif., Oreg., Wash.

Lasthenia maritima is a self-pollinating, "guano endemic" of seabird nesting grounds. It is typically found on offshore islands and rocks from the Farallon Islands, California, to the northern tip of Vancouver Island, British Columbia and rarely occurs on the mainland.

17. Lasthenia minor (de Candolle) Ornduff, Univ. Calif. Publ. Bot. 40: 80. 1966 • Coastal goldfields [E]

Monolopia minor de Candolle in A. P. de Candolle and A. L. P. P. de Candolle, Prodr. 6: 74. 1838; *Baeria minor* (de Candolle) Ferris

Annuals, to 35 cm (herbage not sweetly scented). **Stems** erect, branched distally, ± woolly, especially distally. **Leaves** broadly to narrowly linear, 20–120 × 1–10 mm, margins entire or irregularly toothed or lobed (lobes to 1.5 mm), faces glabrous or villous. **Involucres** hemispheric, 4–6 mm. **Phyllaries** 7–14, oblong to ovate, hairy (especially at margins). **Receptacles** conic, muricate, glabrous. **Ray florets** (8–)13; (corollas light to golden yellow) laminae oblong, 4–8 mm. **Anther appendages** ovate or elliptic, acute (style apices ± deltate with apical tufts of hairs and subapical fringes of shorter hairs). **Cypselae** black, narrowly clavate, 2–2.5 mm, glabrous or hairy; **pappi** 0, or of 2–3(–4) brown or white, lanceolate or subulate, aristate scales plus 4–5+ shorter, ± truncate, fimbriate scales. $2n = 8$.

Flowering Mar–Jun. Grasslands, coastal and inland; 0–700 m; Calif.

Lasthenia minor is variable; coastal plants tend to have lower stature and broader leaves and tend to flower later in the year. Inland populations were once common in the San Joaquin Valley and are now harder to find as grasslands become cultivated, grazed, or built upon. Epappose plants are common and sometimes comprise entire populations.

355. AMBLYOPAPPUS   Hooker & Arnott, J. Bot. (Hooker) 3: 321. 1841   •   [Greek *ambly-*, blunt, and *pappos*, pappus]

John L. Strother

Annuals, to 40 cm (sweet scented). **Stems** erect, branched. **Leaves** cauline; opposite (proximal) or alternate (mostly); sessile; blades mostly linear, sometimes pinnately lobed, ultimate margins entire, faces glabrous (often granular-glandular). **Heads** discoid or disciform, in corymbiform or paniculiform arrays. **Involucres** ovoid to campanulate, 3–4+ mm diam. **Phyllaries** persistent, 3–6+ in 1–2 series (± erect in fruit, distinct, obovate, herbaceous to membranous or scarious, flat or weakly cupped at bases, glabrous or granular-glandular). **Receptacles** conic, pitted or smooth, glabrous, epaleate. **Ray florets** 0 (or inconspicuous; peripheral 3–8 florets pistillate, fertile; corollas yellow, tubular or nearly so). **Disc florets** 2–25+, bisexual, fertile; corollas yellow, tubes about equaling campanulate throats, lobes 5, ± deltate. **Cypselae** mostly obpyramidal, (3–)4-angled, hairy; **pappi** persistent, of 7–10+ oblong to spatulate scales. $x = 8$.

Species 1: California, nw Mexico, South America.

1. Amblyopappus pusillus   Hooker & Arnott, J. Bot. (Hooker) 3: 321. 1841   [F]

Leaves 1–3+ cm, often ± fleshy. **Ray or pistillate corollas** 0.6–1 mm (obscurely zygomorphic). **Disc corollas** 0.6–1 mm. **Cypselae** black, 1.5–2 mm; **pappi** of white or purplish, apically ± muticous scales 0.5–0.8 mm. $2n = 16$.

Flowering spring. Coastal bluffs, dunes, beaches; 0–40+ m; Calif.; Mexico (Baja California); South America.

356. EATONELLA   A. Gray, Proc. Amer. Acad. Arts 19: 19. 1883   •   [For Daniel Cady Eaton, 1834–1885, American botanist]   [E]

John L. Strother

Annuals, 1–3(–5+) cm. **Stems** decumbent to erect, often branched from bases (woolly). **Leaves** basal (mostly) and cauline; opposite (proximal) or alternate (mostly); petiolate or sessile; blades spatulate to oblanceolate, margins entire, faces woolly-tomentose. **Heads** radiate, borne singly. **Involucres** campanulate, 3–5 mm diam. **Phyllaries** persistent, 8–13 in (1–)2 series (reflexed in fruit, distinct, linear to oblong, herbaceous, ± conduplicate, woolly). **Receptacles** flat or convex, pitted or smooth, glabrous, epaleate. **Ray florets** 8–13, pistillate, fertile; corollas yellowish or purplish. **Disc florets** 7–12+, bisexual, fertile; corollas yellowish, tubes shorter than or about equaling campanulate throats, lobes 4–5, ± deltate. **Cypselae** compressed, linear-oblanceolate, (callous-margined) ciliate; **pappi** persistent, of 2 ± laciniate to nearly entire, often uniaristate scales. $x = 19$.

Species 1: w North America.

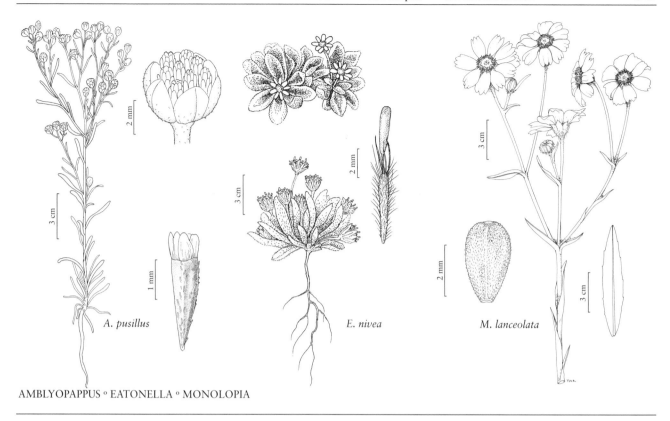

AMBLYOPAPPUS ° EATONELLA ° MONOLOPIA

A. pusillus   *E. nivea*   *M. lanceolata*

1. Eatonella nivea (D. C. Eaton) A. Gray, Proc. Amer. Acad. Arts 19: 19. 1883 ⊡Ⓔ ⊡Ⓕ

Burrielia nivea D. C. Eaton in S. Watson, Botany (Fortieth Parallel), 174, plate 18, figs. 6–14. 1871

Stems usually crowded in tufts. **Leaves** 10–2(0–30) × 3–8(–12) mm. **Peduncles** 1–35+ mm. **Involucres** 4–6+ mm. **Ray corollas** inconspicuous, laminae ± erect, 2–2.5 mm. **Disc corollas** 1.5–3 mm. **Cypselae** black, 2.5–3.5 mm, hairs on margins 0.6–1.2 mm; **pappus scales** 1.5–2.5 mm. $2n$ = 38.

Flowering spring. Sandy or gravelly soils, often with sagebrush scrub; 800–3100 m; Calif., Idaho, Nev., Oreg.

357. MONOLOPIA   de Candolle in A. P. de Candolle and A. L. L. P. de Candolle, Prodr. 6: 74. 1838 • [Greek *monos*, single, and *lopos*, husk, alluding to phyllaries] Ⓔ

Dale E. Johnson

Annuals, to 60 cm. **Stems** erect, branched. **Leaves** cauline; mostly alternate, proximal sometimes opposite; petiolate or sessile; blades narrowly oblong, oblanceolate, or lance-linear, margins sinuate-dentate or entire, faces loosely to densely woolly. **Heads** usually radiate (sometimes obscurely disciform in *M. congdonii,* usually borne singly (in corymbiform arrays in *M. congdonii*). **Involucres** hemispheric, 5–13 mm diam. **Phyllaries** persistent, 4–11 in 1 series (usually spreading, reflexed in fruit in *M. congdonii,* distinct or connate with distinct tips, elliptic to oblanceolate, herbaceous to membranous or indurate, bases flat or cupped, apices acute to acuminate, abaxial faces woolly, usually with black hairs toward tips). **Receptacles** convex to conic, smooth or pitted, glabrous, epaleate. **Ray florets** 4(–7, inconspicuous, corollas ± lacking laminae) or (7–)8(–11), pistillate, fertile; corollas usually yellow, sometimes cream in *M. major* (each lamina opposed by an adaxial lobule or tooth). **Disc florets** 20–100, bisexual, fertile; corollas yellow (glandular or with nonglandular hairs); tubes cylindric to cylindro-funnelform, about equaling gradually or abruptly dilated, narrowly to broadly funnelform throats, lobes (4–)5, ± deltate (anther appendages usually widest at bases, glandular; style-branch appendages deltate). **Cypselae** obcompressed or ± prismatic, 3-angled (ray or peripheral) or 2- or 4-angled (disc); **pappi** usually 0 (persistent, of 2–4+ spatulate, erose to laciniate scales in *M. congdonii*). $x = 13$.

Species 5 (5 in the flora): California.

1. Ray lamina apices entire or obscurely to slightly lobed (mid lobes shorter than outers); disc cypselae ± prismatic (not obcompressed).
    2. Cypselae ca. 2 mm, glabrous or sparsely hairy (branches spreading, usually distal) . . . . 4. *Monolopia gracilens*
    2. Cypselae 2.5–3 mm, uniformly gray-strigose (branches erect) . . . . 5. *Monolopia stricta*
1. Ray lamina apices usually ± equally 3-lobed (laminae sometimes inconspicuous); disc cypselae obcompressed.
    3. Ray laminae ca. 0.5 mm (inconspicuous); disc cypselae 2-angled (pappi of 2–7 scales) . . . . 1. *Monolopia congdonii*
    3. Ray laminae 8–20 mm; disc cypselae 4-angled (pappi 0).
        4. Phyllaries distinct or connate (± ¹/₂ their lengths); cypselae uniformly gray-strigose . . . . 2. *Monolopia lanceolata*
        4. Phyllaries connate (± ³/₄ their lengths, forming cups with distinct triangular lobes); cypselae glabrous or apically hairy . . . . 3. *Monolopia major*

1. Monolopia congdonii (A. Gray) B. G. Baldwin, Novon 9: 460. 1999 [E]

Eatonella congdonii A. Gray, Proc. Amer. Acad. Arts 19: 20. 1883 (as congdoni); *Lembertia congdonii* (A. Gray) Greene

Leaves narrowly oblong (10–45 mm, margins shallowly sinuate-dentate or entire, faces loosely woolly). **Peduncles** 2–20 mm. **Involucres** ± 4.5 mm. **Phyllaries** 4–7, distinct, elliptic to oblanceolate (reflexed in fruit), apices acute. **Ray florets** 4–7; laminae 0.5 mm, ± equally 3-lobed (inconspicuous). **Disc florets** 20–40 (corolla lobes 4, with glandular hairs; anther appendages widest at middles). **Cypselae** obcompressed, ± 3 mm, 3-angled (peripheral), 2-angled (disc), hairy on margins (pappi of 2–7 spatulate, erose to laciniate scales ± 1 mm). $2n = 22$.

Flowering Feb–May. Sandy areas in grasslands or on alkali sinks; 90–700 m; Calif.

2. Monolopia lanceolata Nuttall, Proc. Acad. Nat. Sci. Philadelphia 4: 21. 1848 [E][F]

Leaves oblanceolate to lance-linear. **Peduncles** 10–130 mm. **Involucres** 6–10 mm. **Phyllaries** ± 8, usually distinct, sometimes connate to ¹/₂ their lengths, elliptic to oblanceolate, apices acuminate. **Ray florets** ± 8; corollas yellow, laminae 10–20 mm, ± equally 3-lobed. **Disc florets** 60–100. **Cypselae** obcompressed, 2–4 mm, 3-angled (ray), 4-angled (disc), uniformly gray-strigose. $2n = 20$.

Flowering Feb–Jun. Clayey areas in grasslands or openings in chaparral or oak or foothill woodlands; 50–1600 m; Calif.

3. **Monolopia major** de Candolle in A. P. de Candolle and A. L. L. P. de Candolle, Prodr. 6: 74. 1838 E

Leaves oblanceolate to lance-linear. **Peduncles** 10–130 mm. **Involucres** 8–13 mm. **Phyllaries** ± 8, connate ± ³/₄ their lengths (forming cups with distinct, triangular lobes), oblanceolate, apices acute. **Ray florets** ± 8; corollas usually yellow, sometimes cream, laminae 8–20 mm, ± equally 3-lobed. **Disc florets** 60–100. **Cypselae** obcompressed, 2.5–4 mm, 3-angled (ray), 4-angled (disc), glabrous or apically hairy. $2n = 24$.

Flowering Feb–Jul. Clayey areas in grasslands; 10–1100 m; Calif.

4. **Monolopia gracilens** A. Gray, Proc. Amer. Acad. Arts 19: 20. 1883 E

Leaves oblanceolate to lance-linear. **Peduncles** 20–120 mm. **Involucres** 5–7 mm. **Phyllaries** 7–11, distinct, elliptic to oblanceolate, apices acuminate. **Ray florets** 7–11; corollas yellow, laminae 5–10 mm, entire or obscurely 3-lobed (mid lobes shorter than outers). **Disc florets** 20–40. **Cypselae** ± prismatic (not obcompressed), ± 2 mm, 3-angled (ray) or 4-angled (disc), glabrate. $2n = 26$.

Flowering Mar–Jul. Serpentinitic areas in grasslands or openings in chaparral or oak woodlands; 100–1200 m; Calif.

5. **Monolopia stricta** Crum, Madroño 5: 258. 1940 E

Leaves oblanceolate to lance-linear. **Peduncles** 30–50 mm. **Involucres** 5–7 mm. **Phyllaries** ± 8, distinct, oblanceolate, apices acute. **Ray florets** 8; corollas yellow, laminae 2–7 mm (w San Joaquin Valley) or 9–17 mm (se San Joaquin Valley), entire or slightly lobed (mid lobes shorter than outers). **Disc florets** 20–40. **Cypselae** ± prismatic, not obcompressed, 2.5–3 mm, 3-angled (ray) or 4-angled (disc), uniformly gray-strigose. $2n = 26$.

Flowering Feb–May. Clayey areas in grasslands or openings in chaparral or oak woodlands; 50–800 m; Calif.

358. **PSEUDOBAHIA** (A. Gray) Rydberg in N. L. Britton et al., N. Amer. Fl. 34: 83. 1915

• Sunburst [Greek *pseudes*, false, and generic name *Bahia*] E

Dale E. Johnson

Monolopia de Candolle sect. *Pseudobahia* A. Gray in W. H. Brewer et al., Bot. California 1: 383. 1876 (as Pseudo-Bahia)

Annuals, to 30(–70) cm. **Stems** erect, usually branched. **Leaves** cauline; mostly alternate; usually petiolate; blades usually 1–2-pinnately lobed, sometimes 3-lobed or entire, faces sparsely to moderately woolly. **Heads** radiate, borne singly. **Involucres** hemispheric, 5–9 mm diam. **Phyllaries** persistent, (3–)8 in 1 series (± erect in fruit, connate at bases or to ¹/₂ their lengths, elliptic, lanceolate, or oblanceolate, equal, margins somewhat hyaline, apices acute to acuminate, abaxial faces usually woolly). **Receptacles** conic to hemispheric, pitted or smooth, glabrous, epaleate. **Ray florets** (3–)8, pistillate, fertile; corollas yellow (with rings of hairs at bases of laminae). **Disc florets** 8–25+, bisexual, fertile; corollas yellow (with rings of hairs at bases of limbs), tubes shorter than funnelform or cylindric throats, lobes 5, ± deltate (anther appendages deltate, widest at bases, glandular; style-branch appendages deltate). **Cypselae** ± obcompressed, 3–4-angled and obpyramidal to clavate, hairy or glabrate; **pappi** 0, or coroniform (minute scales). $x = 4$.

Species 3 (3 in the flora): California.

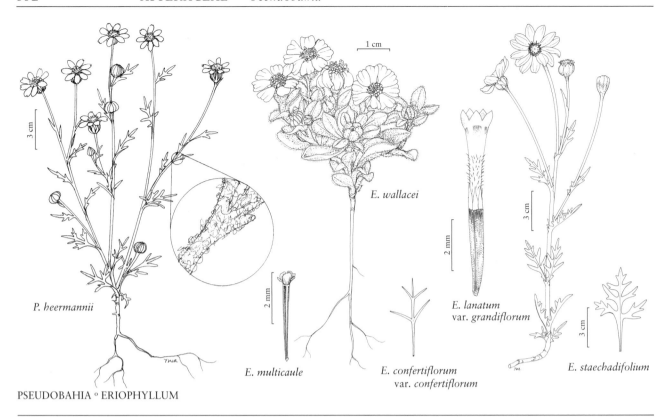

E. wallacei

E. lanatum
var. *grandiflorum*

P. heermannii

E. multicaule

E. confertiflorum
var. *confertiflorum*

E. staechadifolium

PSEUDOBAHIA ° ERIOPHYLLUM

B. G. Baldwin and B. L. Wessa (2000) found that *Pseudobahia bahiifolia* and *P. peirsonii* nest within a clade of *Eriophyllum* and *Syntrichopappus* species; circumscriptions of these genera are likely to change.

1. Leaves entire or 3-lobed . . . . 1. *Pseudobahia bahiifolia*
1. Leaves mostly 1–2-pinnately lobed.
   2. Leaves mostly 1-pinnately lobed; phyllaries connate to ½ their lengths . . . . . . 2. *Pseudobahia heermannii*
   2. Leaves mostly 2-pinnately lobed (except in smaller plants); phyllaries connate at bases
   . . . . 3. *Pseudobahia peirsonii*

1. **Pseudobahia bahiifolia** (Bentham) Rydberg in N. L. Britton et al., N. Amer. Fl. 34: 83. 1915 (as bahiaefolia)
  • Hartweg's golden sunburst [C] [E]

Monolopia bahiifolia Bentham, Pl. Hartweg., 317. 1849 (as bahiaefolia)

Plants to 20 cm. **Leaves** 8–25 mm (linear to oblanceolate), usually 3-lobed. **Peduncles** 20–50 mm. **Phyllaries** 3–8, connate at bases, elliptic to lanceolate. **Ray florets** 3–8. **Disc florets** 8–25+; tubes cylindric, abruptly dilated (throats broadly funnelform), lobes glabrous. **Cypselae** 1.5–2.5 mm. $2n = 8$. —

Flowering Mar–May. Grasslands or openings in foothill woodlands; of conservation concern; 100–200 m; Calif.

2. **Pseudobahia heermannii** (Durand) Rydberg in N. L. Britton et al., N. Amer. Fl. 34: 83. 1915 • Foothill sunburst, brittlestem [E] [F]

Monolopia heermannii Durand, J. Acad. Nat. Sci. Philadelphia, n. s. 3: 93. 1855

Plants to 30 cm. **Leaves** 10–30 mm, mostly 1-pinnately lobed (lobes 0.5–1.5 mm wide), distal sometimes entire. **Peduncles** 20–50 mm. **Phyllaries** ± 8, connate to ½ their lengths, elliptic to lanceolate (each pair with semicircular crest proximal to sinus, hard at maturity, ± enclosing cypselae). **Ray florets** ± 8. **Disc florets** 8–25+; tubes cylindric, abruptly dilated (throats broadly funnelform), lobes glabrous. **Cypselae** 2–2.5 mm. $2n = 6 + 0$–$2I$, $8 + 0$–$1I$ or 0–$1B$, $10 + 0$–$1I$.

Flowering Mar–Jun. Grasslands or sandy or rocky openings in chaparral, foothill woodlands, yellow pine forests; 100–1600 m; Calif.

3. **Pseudobahia peirsonii**   Munz, Aliso 2: 84, plate 12, fig. F, plate 13, fig. D. 1949 • San Joaquin adobe sunburst C E

Plants to 70 cm. **Leaves** 20–60 mm, mostly 2-pinnately (1-pinnately in smaller plants) lobed (lobes 1–5 mm wide). **Peduncles** 20–80 mm. **Phyllaries** ± 8, connate at bases, oblanceolate. **Ray florets** ± 8. **Disc florets** 25+; tubes cylindro-funnelform, gradually dilated, lobes with glandular hairs. **Cypselae** ± 3 mm. $2n = 16$.

Flowering Mar–May. Bare, clayey areas in grasslands; of conservation concern; 100–900 m; Calif.

Within populations, some individuals of *Pseudobahia peirsonii* have ray corollas with ultraviolet-reflective tips, others not.

359. **ERIOPHYLLUM**   Lagasca, Gen. Sp. Pl., 28. 1816 • Woolly sunflower [Greek *erion*, wool, and *phyllon*, leaf]

Dale E. Johnson

John S. Mooring

Annuals, perennials, subshrubs, or shrubs, 1–200 cm. **Stems** erect or decumbent, usually branched (proximally, distally, or ± throughout). **Leaves** mostly cauline; mostly alternate (proximal sometimes opposite); petiolate or sessile; blades usually 1–2(–3)-pinnately lobed, ultimate margins toothed, serrate, or entire, faces usually densely to sparsely woolly (abaxial or both, adaxial sometimes glabrescent). **Heads** radiate or discoid, borne singly or in loose to tight, corymbiform or compound-corymbiform arrays. **Involucres** campanulate to hemispheric, 3–12+ mm diam. **Phyllaries** persistent, 4–13(–15) in 1+ series (± erect in fruit, distinct or basally connate, lanceolate to oblanceolate, herbaceous or indurate, slightly to deeply concave, usually carinate, margins sometimes scarious, abaxial faces densely to sparsely woolly). **Receptacles** flat or convex to conic, smooth or pitted, glabrous, usually epaleate (with 1–6 hyaline paleae in *E. ambiguum*, obscurely setose in *E. mohavense*). **Ray florets** 0, or 4–13(–15), pistillate, fertile; corollas yellow or white (sometimes with reddish veins in *E. lanosum).* **Disc florets** (3–)10–300, bisexual, fertile; corollas yellow, tubes shorter than or about equaling funnelform throats, lobes 5, deltate. **Cypselae** linear-clavate to prismatic, flattened or 3-angled in rays, 4(–5)-angled in discs, hairy or glabrous; **pappi** 0, or persistent, of 6–12+ (distinct) erose to laciniate or aristate scales (in 1–2 similar or contrasting series), or ± coroniform. $x = 8$.

Species 13 (13 in the flora): w North America, nw Mexico.

Eriophyllum encompasses taxa that occur in seashore, chaparral, grassland, desert, forest, and alpine communities. Their disparateness encouraged taxonomic multiplication. Between 1890 and 1937, about 157 designations under the genus existed (L. Constance 1937). Constance reduced that taxonomic thicket to six annual and five perennial species; the perennial species *E. lanatum* consisted of ten varieties. Base diploid chromosome numbers for *Eriophyllum* (in the sense of Constance) species are $x = 4, 5,$ and 7 for the annuals, and $x = 8,$ 15, and 19 for the perennials (S. Carlquist 1956; J. S. Mooring 1997, 2001, 2002). Possibly, $x = 15$ and $x = 19$ represent paleopolyploidy. Only *E. mohavense* remains uncounted. B. G. Baldwin (1999) linked Mooring's (1997) report of $n = 19$ in *E. nevinii* to chromosomal, morphologic, and rDNA evidence, and erected the genus *Constancea* on that species. *Eriophyllum* (in the sense of Constance) seems most closely related to the annuals *Pseudobahia* $(x = 3, 4, 8)$ and *Syntrichopappus* $(x = 6, 7)$. *Eriophyllum* (in the sense of Baldwin), *Pseudobahia*, and *Syntrichopappus* constitute a clade, and nomenclatural changes are necessary for a monophyletic classification (Baldwin and B. L. Wessa 2000; Baldwin et al. 2002). Mooring (1997) hypothesized a descending dysploidy phylogeny in *Eriophyllum* (in the sense of Constance) from *E. nevinii*. Baldwin et al. (2002, p. 174) stated that *E. nevinii* "is an evolutionary outlier (although probably not ancestral) to" *Eriophyllum* in the sense of Constance.

Natural intertaxon hybrids have been reported for the perennial species (L. Constance 1937; J. S. Mooring 1994) but not for the annuals. Experimental hybridizations have produced sterile hybrids between the annual *E. congdonii* and the perennial *E. lanatum*. Experimental crosses among seven of the annual species produced fertile hybrids between two morphologically similar species; the other combinations either failed or produced sterile hybrids (Mooring 2002).

D. P. Tibor (2001) cited nine taxa of *Eriophyllum* as rare or endangered: the perennials *E. confertiflorum* var. *tanacetiflorum*, *E. jepsonii*, *E. latilobum*, *E. (Constancea) nevinii*, and *E. lanatum* vars. *hallii* and *obovatum*, and the annuals *E. congdonii*, *E. mohavense*, and *E. nubigenum*.

Here, accounts of the annuals are by D. E. Johnson, the perennials by J. S. Mooring.

SELECTED REFERENCES Constance, L. 1937. A systematic study of the genus *Eriophyllum* Lag. Univ. Calif. Publ. Bot. 18: 69–136. Mooring, J. S. 1997. A new base chromosome number and phylogeny for *Eriophyllum* (Asteraceae, Helenieae). Madroño 44: 364–373. Mooring, J. S. 2002. Experimental hybridizations of *Eriophyllum* annuals (Asteraceae, Helenieae). Amer. J. Bot. 89: 1973–1983.

1. Perennials (sometimes flowering in first year), subshrubs, or shrubs, (10–)20–150 cm.
  2. Perennials or subshrubs; heads borne singly or 2–5(–10) per array; peduncles (2–)3–30 cm; ray laminae 6–20 mm.
    3. Peduncles 3–30 cm; involucres 6–15 mm diam.; ray florets 0, or 5–13(–15), laminae 6–20 mm; disc florets 20–300; British Columbia, w United States . . . . . 9. *Eriophyllum lanatum*
    3. Peduncles 2–7 cm; involucres 5–6(–7) mm diam.; ray florets (6–)8–10(–13), laminae 6–10 mm; disc florets 40–75; California. . . . . 11. *Eriophyllum latilobum*
  2. Subshrubs or shrubs; heads 2–30+, or 40–80+, per array; peduncles 0–10(–14) cm; ray laminae 2–10 mm.
    4. Subshrubs, 30–150 cm; phyllaries 8–11; ray florets 0, or 6–9, laminae 3–5 mm; coastal dunes and bluffs . . . . 13. *Eriophyllum staechadifolium*
    4. Shrubs, 20–100 cm; phyllaries 4–8; ray florets 0, or 4–8, laminae 2–10 mm; mostly not coastal.
      5. Heads 2–4 per array; peduncles mostly 5–10(–14) cm; ray florets 6–8, laminae 6–10 mm; California (Inner Coast Ranges) . . . . 12. *Eriophyllum jepsonii*
      5. Heads (3–)4–7(–10) or 10–30+ per array; peduncles mostly 0–2.5(–9) cm; ray florets 0 or 4–6(–8), laminae 2–5 mm; California, Mexico . . . . . 10. *Eriophyllum confertiflorum*
1. Annuals, 1–15(–30) cm.
  6. Peduncles ± 0.1 cm.
    7. Ray florets 5–7 . . . . 1. *Eriophyllum multicaule*
    7. Ray florets 0.
      8. Leaf margins weakly, if at all, revolute, apices acute; phyllaries 3–4 . . . 2. *Eriophyllum mohavense*
      8. Leaf margins strongly revolute, apices ± rounded; phyllaries 6–8 . . . . . . 3. *Eriophyllum pringlei*
  6. Peduncles 1–10 cm.
    9. Ray laminae cream or yellow, or white with red veins (lobes of disc corollas glandular; anther appendages subulate).
      10. Ray laminae white with red veins; cypselae 2.5–4.5 mm; pappi of 5 subulate scales 1.5–2.5 mm plus 4 oblong scales ± 0.5 mm . . . . 4. *Eriophyllum lanosum*
      10. Ray laminae usually cream or yellow, sometimes white with red veins; cypselae ± 2 mm; pappi usually of 6–10 ± oblong scales 0.4–0.8 mm, rarely 0 . . . 5. *Eriophyllum wallacei*
    9. Ray laminae yellow (lobes of disc corollas not glandular; anther appendages deltate).
      11. Anther appendages not glandular; pappi 0, or of 6–10 scales 0.1–0.5 mm . . . . 6. *Eriophyllum ambiguum*
      11. Anther appendages glandular; pappi of 8–10 lanceolate to spatulate (± unequal) scales 0.5–1.5 mm, or of 3–5 ± spatulate scales 0.5–1 mm plus 3–5 lanceolate scales 1.5–2 mm.

[12. Shifted to left margin.—Ed.]

12. Stems ascending; ray laminae ± 1 mm (inconspicuous); 1800–2500 m . . . . 7. *Eriophyllum nubigenum*
12. Stems ± spreading; ray laminae 3–5 mm; 500–1900 m . . . . 8. *Eriophyllum congdonii*

1. **Eriophyllum multicaule** (de Candolle) A. Gray, Proc. Amer. Acad. Arts 19: 24. 1883 • Manystem woolly sunflower E F

Actinolepis multicaulis de Candolle in A. P. de Candolle and A. L. P. P. de Candolle, Prodr. 5: 656. 1836

Annuals, 2–15 cm. **Stems** spreading. **Leaves:** blades ± cuneate, ± 10 mm, usually 3-lobed, ultimate margins entire, plane (apices ± rounded), faces ± woolly. **Heads** borne singly or in clusters. **Peduncles** ± 0.1 cm. **Involucres** campanulate to hemispheric, 2–3+ mm diam. **Phyllaries** 5–7, distinct. **Ray florets** 5–7; laminae yellow, ± 2 mm. **Disc florets** 10–20; corollas ± 2 mm (tubes cylindric, throats broadly cylindric, abruptly dilated, lobes glandular; anther appendages narrowly deltate, not glandular). **Cypselae** ± 2 mm; **pappi** usually of 10–15 ± lanceolate scales ± 1 mm, rarely 0. $2n = 14$.

Flowering Mar–Jul. Sandy openings in chaparral or coastal scrublands; 0–1600 m; Calif.

2. **Eriophyllum mohavense** (I. M. Johnston) Jepson, Man. Fl. Pl. Calif., 1117. 1925 • Mohave or Barstow woolly sunflower C E

Eremonanus mohavensis I. M. Johnston, Contr. Gray Herb. 68: 101. 1923

Annuals, 1–2.5 cm. **Stems** erect to spreading. **Leaves:** blades spatulate to cuneate, 3.5–10 mm, sometimes 1–3-lobed (each lobe terminating in short, sharp cusp), ultimate margins entire, weakly, if at all, revolute (apices acute), faces loosely woolly. **Heads** borne singly or in clusters. **Peduncles** ± 0.1 cm. **Involucres** cylindric to obconic, 1–2+ mm diam. **Phyllaries** 3–4, distinct. **Ray florets** 0. **Disc florets** ± 3; corollas ± 2 mm (tubes cylindric, throats broadly cylindric, abruptly dilated, lobes glandular; anther appendages deltate, tapering distally, not glandular). **Cypselae** 2–2.5 mm; **pappi** of 12–14 linear to spatulate scales ± 1.5 mm.

Flowering Apr–May. Sandy or gravelly areas, creosote-bush scrublands; of conservation concern; 500–800 m; Calif.

3. **Eriophyllum pringlei** A. Gray, Proc. Amer. Acad. Arts 19: 25. 1883 • Pringle's woolly sunflower

Annuals, 1–5 cm. **Stems** ± spreading. **Leaves:** blades ± cuneate, 3–10 mm, usually 3-lobed, ultimate margins entire, strongly revolute (apices ± rounded), faces woolly. **Heads** borne singly or in clusters. **Peduncles** ± 0.1 cm. **Involucres** broadly campanulate to hemispheric, 3–5+ mm diam. **Phyllaries** 6–8, distinct. **Ray florets** 0. **Disc florets** 10–20; corollas ± 2 mm (tubes cylindric, throats broadly cylindric, abruptly dilated, lobes glandular; anther appendages deltate, tapering distally, not glandular). **Cypselae** 1.5–2 mm; **pappi** of 5–12 ± oblanceolate scales ± 1 mm. $2n = 14 + 0–1$ B or $0–1$ I, 16.

Flowering Feb–Jul. Sandy or gravelly openings in desert, creosote-bush, or sagebrush scrublands, or chaparral; 300–2200 m; Ariz., Calif., Nev.; Mexico (Baja California).

4. **Eriophyllum lanosum** (A. Gray) A. Gray, Proc. Amer. Acad. Arts 19: 25. 1883 • White easterbonnets or woolly daisy

Burrielia lanosa A. Gray in War Department [U.S.], Pacif. Railr. Rep. 4(5): 107. 1857; *Antheropeas lanosum* (A. Gray) Rydberg

Annuals, 3–15 cm. **Stems** decumbent to ascending. **Leaves:** blades oblanceolate to linear, 5–20 mm, rarely lobed, ultimate margins, usually entire, plane (apices acute), faces sparsely woolly. **Heads** borne singly. **Peduncles** 1–5 cm. **Involucres** campanulate to obconic, 3–5 mm diam. **Phyllaries** 8–10, distinct. **Ray florets** 8–10; laminae white with red veins, 3–5 mm. **Disc florets** 10–20; corollas 2–3 mm (tubes cylindric, throats funnelform, gradually dilated, lobes glandular; anther appendages subulate, not glandular). **Cypselae** 2.5–4.5 mm; **pappi** of 5 subulate scales 1.5–2.5 mm plus 4–5 oblong scales ± 0.5 mm. $2n = 8$.

Flowering Feb–May. Sandy or gravelly openings, desert scrublands; 70–1400 m; Ariz., Calif., Nev., N.Mex., Utah; Mexico (Baja California).

5. Eriophyllum wallacei (A. Gray) A. Gray, Proc. Amer. Acad. Arts 19: 25. 1883 • Woolly easterbonnets, Wallace's woolly daisy F

Bahia wallacei A. Gray in War Department [U.S.], Pacif. Railr. Rep. 4(5): 105. 1857; *Antheropeas wallacei* (A. Gray) Rydberg; *Eriophyllum wallacei* var. *rubellum* (A. Gray) A. Gray

Annuals, 1–15 cm. **Stems** erect to spreading or ascending. **Leaf blades** obovate to spatulate, 7–20 mm, sometimes 3-lobed, ultimate margins entire, plane (apices ± rounded), faces ± woolly. **Heads** usually borne singly. **Peduncles** 1–3 cm. **Involucres** broadly campanulate, 4–6 mm diam. **Phyllaries** 5–10, distinct. **Ray florets** 5–10; laminae usually cream or yellow, sometimes white with red veins, 3–4 mm. **Disc florets** 20–30; corollas 2–3 mm (tubes cylindric, throats funnelform, gradually dilated, lobes glandular; anther appendages subulate, not glandular). **Cypselae** ± 2 mm; **pappi** usually of 6–10 ± oblong scales 0.4–0.8 mm, rarely 0. $2n = 10 + 0–1$ I or 0–3 B.

Flowering Dec–Jul. Sandy or gravelly openings, creosote-bush or sagebrush scrublands, Joshua Tree or pinyon-juniper woodlands, or chaparral; 30–2400 m; Ariz., Calif., Nev., Utah; Mexico (Baja California).

6. Eriophyllum ambiguum (A. Gray) A. Gray, Proc. Amer. Acad. Arts 19: 26. 1883 E

Lasthenia ambigua A. Gray, Proc. Amer. Acad. Arts 6: 547. 1865

Annuals, 5–30 cm. **Stems** decumbent to ascending. **Leaves:** blades oblong to oblanceolate, 5–40 mm, margins entire or toothed, plane (apices rounded to acute), faces lanate or tomentose to tomentulose. **Heads** borne singly. **Peduncles** 1–8 cm. **Involucres** hemispheric to obconic, 5–6 mm diam. **Phyllaries** 6–10, ± connate. **Ray florets** 6–10; laminae yellow, 2–10 mm. **Disc florets** 10–30; corollas 1.3–3 mm (tubes cylindric, throats broadly funnelform, abruptly dilated, lobes glabrous or hairy, hairs 1-celled, not glandular; anther appendages deltate, widest at bases, not glandular). **Cypselae** 2.2–3 mm; **pappi** 0 or of 6–10 ± ovate to cuneate, entire, erose, or laciniate scales 0.1–0.5 mm.

Varieties 2 (2 in the flora): w United States.

1. Disc corolla lobes glabrous; pappi of 6–10 erose to laciniate scales 0.2–0.5 mm . . . . 6a. *Eriophyllum ambiguum* var. *ambiguum*
1. Disc corolla lobes hairy (hairs 1-celled); pappi 0, or of 6–10 entire scales 0.1–0.2 mm . . . . 6b. *Eriophyllum ambiguum* var. *paleaceum*

6a. Eriophyllum ambiguum (A. Gray) A. Gray var. **ambiguum** • Beautiful woolly sunflower E

Disc florets 10–20; corolla lobes glabrous. **Pappi** of 6–10 erose to laciniate scales 0.2–0.5 mm. $2n = 14$.

Flowering Mar–Jun. Sandy or grassy openings in chaparral; 200–1900 m; Calif.

6b. Eriophyllum ambiguum (A. Gray) A. Gray var. **paleaceum** (Brandegee) Ferris, Contr. Dudley Herb. 5: 100. 1958 E

Eriophyllum paleaceum Brandegee, Bot. Gaz. 27: 450. 1899

Disc florets 15–30; corolla lobes hairy (hairs 1-celled). **Pappi** 0 or of 6–10 entire scales 0.1–0.2 mm. $2n = 14$.

Flowering Jan–Jul. Sandy or rocky openings, creosote-bush scrublands, or Joshua tree or pinyon-juniper woodlands; 100–2800 m; Calif., Nev.

7. Eriophyllum nubigenum Greene ex A. Gray, Proc. Amer. Acad. Arts 19: 25. 1883 • Yosemite woolly sunflower C E

Annuals, 5–15 cm. **Stems** ascending. **Leaves:** blades oblanceolate, 1–2 cm, margins entire, plane (apices acute), faces woolly. **Heads** borne singly or in loose arrays. **Peduncles** ± 1 cm. **Involucres** cylindric, 3–4+ mm diam. **Phyllaries** 4–6, distinct. **Ray florets** 4–6; laminae yellow, ± 1 mm (inconspicuous). **Disc florets** 10–20; corollas ± 2 mm (tubes cylindric, throats broadly funnelform, abruptly dilated, lobes not glandular; anther appendages deltate, glandular). **Cypselae** 2.5–3 mm; **pappi** of 8–10 lanceolate to spatulate (± unequal) scales 0.5–1.5 mm. $2n = 14$.

Flowering Jun–Jul. Gravelly or rocky openings, yellow pine or red fir forests; of conservation concern; 1800–2500 m; Calif.

8. **Eriophyllum congdonii** Brandegee, Bot. Gaz. 27: 449. 1899 (as congdoni) [C] [E]

Annuals, 10–30 cm. Stems ± spreading. Leaves: oblanceolate, 10–40 mm, rarely 2–3-lobed, ultimate margins usually entire, plane (apices acute), faces ± woolly. Heads borne singly. Peduncles 3–10 cm. Involucres campanulate, 3–5 mm diam. Phyllaries 8–10, distinct. Ray florets 8–10; laminae yellow, 3–5 mm. Disc florets 10–30; corollas 2–3 mm (tubes cylindric, throats funnelform, gradually dilated, lobes not glandular; anther appendages deltate, glandular). Cypselae 2.5–3 mm; pappi of 3–5 ± spatulate scales 0.5–1 mm plus 3–5 lanceolate scales 1.5–2 mm. $2n = 14$.

Flowering Mar–Jun. Rocky openings, foothill woodlands or yellow pine forests; of conservation concern; 500–1900 m; Calif.

9. **Eriophyllum lanatum** (Pursh) J. Forbes, Hort. Woburn., 183. 1833 • Common woolly sunflower [F]

Actinella lanata Pursh, Fl. Amer. Sept. 2: 560. 1813

Perennials or subshrubs, 10–100 cm (sometimes flowering first year). Stems erect to decumbent (usually woolly). Leaves (proximal usually alternate): blades mostly lanceolate to oblanceolate, 1–8 cm, often 1–2(–3)-pinnately lobed, ultimate margins toothed, serrate, or entire, revolute or plane, faces hairy, often woolly (more densely abaxially, sometimes glabrate adaxially; distal leaves reduced in size and lobing). Heads borne singly or (2–5+) in corymbiform arrays. Peduncles mostly 3–30 cm. Involucres campanulate to hemispheric, 6–15 mm diam. Phyllaries 5–13(–15), distinct or connate at bases (lanceolate to ovate, carinate or plane). Ray florets 0 or 5–13(–15); laminae golden yellow to yellow, 6–20 (× 2–7) mm. Disc florets 20–300; corollas 2.5–5 mm (tubes usually glandular or glandular-hairy, glabrous in var. *hallii*). Cypselae 2–5 mm; pappi usually of 6–12 ovate or cuneate to lanceolate or lance-linear (often unequal), erose or lacerate scales 0.3–2 mm, sometimes coroniform, rarely 0.

Varieties 10 (10 in the flora): w North America, Mexico (probably extinct).

Eriophyllum lanatum is a polyploid complex of intergrading regional facies treated here as varieties. Artificial hybridization studies show that strong barriers to interbreeding exist among the varieties at the diploid level (J. S. Mooring 2001). In nature, morphologically intermediate polyploid populations often occur in regions where the ranges of the varieties approach one another. Edaphic factors and light intensity also make identification more difficult by strongly influencing leaf morphology and sizes of structures. For example, cultivated individuals of var. *achillioides* may have laciniately toothed rather than pinnatifid leaves. Rarely, plants of different varieties maintain their identity while growing side by side. In some instances, one is diploid and the other tetraploid; in others both are diploid. Varieties *arachnoideum*, *croceum*, *grandiflorum*, and *obovatum* apparently form natural hybrids with *E. confertiflorum* var. *confertiflorum*; past hybridizations may have resulted in the origin of *E. latilobum* and *E. jepsonii* (L. Constance 1937; P. A. Munz 1959; Mooring 1994) and *E. confertiflorum* var. *tanacetiflorum* (Mooring 1994).

Our treatment of *Eriophyllum lanatum* closely follows that of L. Constance (1937), which was done without benefit of cytogeographic studies. The key is to modal populations of the varieties, usually based on living plants.

Some varieties have been introduced into cultivation as ornamentals.

SELECTED REFERENCES Mooring, J. S. 1975. A cytogeographic study of *Eriophyllum lanatum* (Compositae, Helenieae). Amer. J. Bot. 62: 1027–1037. Mooring, J. S. 2001. Barriers to interbreeding in the *Eriophyllum lanatum* (Asteraceae, Helenieae) species complex. Amer. J. Bot. 88: 285–312.

1. Ray florets 8–9 (disc corolla tubes glabrous; s California) . . . . . . 9j. *Eriophyllum lanatum* var. *hallii*
1. Ray florets usually (5–)8–15, rarely 0 (disc corolla tubes glandular or glandular-hairy).
  2. Pappi to 0.2 mm, sometimes coroniform, rarely 0.
    3. Proximal leaves pinnately lobed, lobes 3–5, abaxial faces loosely cobwebby-woolly; California, Coast Ranges . . . .
    . . . . 9c. *Eriophyllum lanatum* var. *arachnoideum*
    3. Proximal leaves coarsely serrate or shallowly lobed distally, abaxial faces densely silky-woolly; California, Sierra Nevada . . . . 9d. *Eriophyllum lanatum* var. *croceum*
  2. Pappi (0.3–)0.5–2 mm.
    4. Proximal leaves 1–2(–3)-pinnately lobed.
      5. Proximal leaves 1-pinnately lobed; peduncles mostly 5–30 cm.
        6. Proximal leaves usually lobed (beginning in proximal 1/2 of leaf); peduncles mostly 10–30 cm; California, sw Oregon . . . . 9b. *Eriophyllum lanatum* var. *grandiflorum* (in part)
        6. Proximal leaves usually lobed (beginning in distal 1/2 of leaf); peduncles mostly 5–15(–20) cm; Idaho, Montana, ne Oregon, e Washington . . . . 9h. *Eriophyllum lanatum* var. *lanatum* (in part)

5. Proximal leaves 1–2(–3)-pinnately lobed (sinuses almost to midribs); peduncles mostly 3–15 cm.

  7. Perennials (often clumped); proximal leaves usually mostly opposite, 1(–2)-pinnately lobed, lobes ± linear (with 0–2 teeth, ultimate margins not crispate), abaxial faces usually densely gray-woolly; heads 2–5 per array or borne singly; ray laminae 8–20 mm; British Columbia, sw Oregon, Washington . . . .
. . . . 9i. *Eriophyllum lanatum* var. *leucophyllum*

  7. Perennials (sometimes flowering first year); proximal leaves usually mostly alternate, 1–2(–3)-pinnately lobed, lobes usually ± triangular (with 0–4 teeth, ultimate margins often crispate), abaxial faces woolly; heads usually 3–8 per array; ray laminae usually 6–9(–16) mm; California, Nevada, s Oregon . . . . . .
. . . . 9a. *Eriophyllum lanatum* var. *achillioides*

[4. Shifted to left margin.—Ed.]

4. Proximal leaves laciniately toothed, serrate, or entire.

  8. Leaves: margins revolute, abaxial faces woolly, adaxial tufted-woolly or glabrate.

    9. Peduncles mostly 10–30 cm; cypselae (2.4–)2.5–3.5(–4) mm; California, sw Oregon . . . . . .
. . . . 9b. *Eriophyllum lanatum* var. *grandiflorum* (in part)

    9. Peduncles mostly 5–15(–20) cm; cypselae (3.4–)3.5–5 mm; Idaho, Montana, ne Oregon, e Washington . . . .
. . . . 9h. *Eriophyllum lanatum* var. *lanatum* (in part)

  8. Leaves: margins usually plane, sometimes slightly revolute in var. *integrifolium*, abaxial and adaxial faces woolly.

    10. Involucres 6–10 mm diam.; ray florets (5–)8 (–10) . . . . 9g. *Eriophyllum lanatum* var. *integrifolium*

    10. Involucres 10–15 mm diam.; ray florets 10–13(–15).

      11. Involucres 10–12 mm diam.; ray laminae 6–7 mm; s California . . . .
. . . . 9f. *Eriophyllum lanatum* var. *obovatum*

      11. Involucres 12–15 mm diam.; ray laminae 7–10(–15) mm; n California, s Oregon . . . . 9e. *Eriophyllum lanatum* var. *lanceolatum*

9a. Eriophyllum lanatum (Pursh) J. Forbes var. **achillioides** (de Candolle) Jepson, Man. Fl. Pl. Calif., 1118. 1925 (as achillaeoides) E

Bahia achillioides de Candolle in A. P. de Candolle and A. L. P. P. de Candolle, Prodr. 5: 657. 1836 (as achillaeoides); *Eriophyllum lanatum* var. *aphanactis* J. T. Howell

Perennials (sometimes flowering first year; ± taprooted). **Proximal leaves** usually mostly alternate (opposite in ne California); blades oblanceolate to ovate, 1–2(–3)-pinnately lobed (lobes usually ± triangular), ultimate margins often toothed (teeth 1–4, 1–2 mm), often crispate, usually revolute, abaxial faces woolly, adaxial less so or glabrate. **Heads** usually 3–8 per array, sometimes borne singly (n California, s Oregon). **Peduncles** mostly 3–10 cm. **Involucres** usually 8–10 mm diam. **Ray florets** usually 9–13, sometimes 0; laminae usually 6–9(–16) mm. **Cypselae** 2–3(–3.3) mm; **pappi** (0.3–)0.5–0.9(–1) mm. $2n = 16, 32$.

Flowering Apr–Jun. Dry sites, chaparral, forests; 30–1300 m; Calif., Nev., Oreg.

In northern California, a taxonomically perplexing series of intermediate populations connects vars. *arachnoideum*, *achillioides*, and *grandiflorum*. Polyploids characterize the regions where the varieties meet. Putative *arachnoideum-achillioides* hybrid derivatives, compared to var. *achillioides*, have smaller, broader, thinner, and more shallowly lobed leaves, and shorter pappi. Putative *achillioides-grandiflorum* hybrid derivatives, compared to var. *achillioides*, have longer peduncles, larger heads (often borne singly), and longer ray laminae. Variety *aphanactis* ($2n = 32$), an Inner Coast Range local, rayless variant, is probably of *achillioides-grandiflorum* parentage. Varieties *achillioides*, *arachnoideum*, *grandiflorum*, and *lanceolatum* overlap in northwestern California and southwestern Oregon. L. Constance (1937, p. 89), referring to the Klamath area, observed "that it seems quite impossible to assign many specimens definitely to any one subspecific category." Variety *achillioides* passes gradually through northern California and southern Oregon populations into var. *leucophyllum* west of the Cascade Mountains. Oregon specimens with ternately lobed leaves and larger heads often borne singly (*Eriophyllum ternatum* Greene) are part of the continuum.

L. Constance (1937, p. 87) had var. *achillioides*, although "uncommon in the Sierra Nevada, occurring as far south as Mariposa County."

9b. Eriophyllum lanatum (Pursh) J. Forbes var. **grandiflorum** (A. Gray) Jepson, Fl. W. Calif., 524. 1901 [F]

Bahia lanata de Candolle var. *grandiflora* A. Gray in W. H. Brewer et al., Bot. California 1: 381. 1876

Perennials (sometimes flowering first year; ± taprooted). **Proximal leaves** mostly alternate; blades linear to lanceolate or oblanceolate, pinnately lobed (usually beginning in proximal part of leaf, lobes 5–7), ultimate margins entire or laciniately toothed (teeth 1–5), serrate, or entire, revolute, abaxial faces woolly, adaxial less so or glabrate. **Heads** 2–6 per array or borne singly. **Peduncles** mostly 10–30 cm. **Involucres** 10–15 mm diam. **Ray florets** usually 10–13(–15), rarely 0; laminae 10–20 mm. **Cypselae** (2.4–)2.5–3.5(–4) mm; **pappi** (0.5–)0.8–1.2(–2) mm. **2n** = 16, 32, 48, 64.

Flowering Apr–Jun. Dry sites, grasslands, lower montane forests; 30–1300 m; Calif., Oreg.; Mexico (probably extinct).

Edward Palmer collected an eriophyllum believed to represent variety *grandiflorum* on Guadalupe Island, Mexico, in 1875. It was last reported in 1893 and is presumed extirpated by goats (R. V. Moran 1996).

L. Constance (1937) observed that var. *grandiflorum* merges with var. *croceum* at higher elevations. We have noted that the two taxa can occur side by side and maintain their identities where ploidy level differs.

9c. Eriophyllum lanatum (Pursh) J. Forbes var. **arachnoideum** (Fischer & Avé-Lallemant) Jepson, Man. Fl. Pl. Calif., 1119. 1925 [E]

Bahia arachnoidea Fischer & Avé-Lallemant, Index Seminum (St. Petersburg) 9: 63. 1842

Perennials (clumped; ± taprooted or roots branched). **Proximal leaves** mostly alternate; blades rhombic to oblanceolate, pinnately lobed (usually in distal part of leaf, lobes 3–5, oblong or triangular), ultimate margins entire or dentate, revolute, abaxial faces loosely cobwebby-woolly, adaxial glabrate. **Heads** 2–4 per array or borne singly. **Peduncles** mostly 3–10 cm. **Involucres** 9–15 mm diam. **Ray florets** 8–15; laminae 8–10 mm. **Cypselae** (1.8–)2–3.2(–4) mm; **pappi** 0–0.2 mm. **2n** = 16, 32.

Flowering Apr–Jul. Ocean bluffs, inland sites, with conifers; 0–700 m; Calif.

Variety *arachnoideum* merges with var. *grandiflorum* in coastal northern California, where intermediates with pappi to 1.3 mm may be found. Variety *arachnoideum* is morphologically similar to var. *croceum*; their distributions are disjunct.

9d. Eriophyllum lanatum (Pursh) J. Forbes var. **croceum** (Greene) Jepson, Man. Fl. Pl. Calif., 1118. 1925 [E]

Eriophyllum croceum Greene, Erythea 3: 124. 1895

Perennials (sometimes stoloniferous; roots branched). **Proximal leaves** mostly alternate; blades narrow-oblanceolate to obovate, coarsely serrate or shallowly lobed distally, ultimate margins toothed or entire, revolute, abaxial faces densely silky-woolly, adaxial glabrate. **Heads** 2–4 per array or borne singly. **Peduncles** 3–8 cm. **Involucres** 10–12 mm diam. **Ray florets** 10–14; laminae 8–10 mm. **Cypselae** 2–2.5(–4) mm; **pappi** 0–0.1 mm. **2n** = 16, 32.

Flowering May–Jun. Usually with conifers; 1000–1700 m; Calif.

Variety *croceum*, one of the less frequently collected taxa of the complex, closely resembles var. *arachnoideum*.

9e. Eriophyllum lanatum (Pursh) J. Forbes var. **lanceolatum** (Howell) Jepson, Man. Fl. Pl. Calif., 1118. 1925 [E]

Eriophyllum lanceolatum Howell, Fl. N.W. Amer., 355. 1900

Perennials (± taprooted). **Proximal leaves** alternate; blades lanceolate to ovate, margins coarsely serrate or entire, plane, faces densely woolly. **Heads** usually borne singly, sometimes 2–3 per array. **Peduncles** mostly 3–10(–15) cm. **Involucres** 12–15 mm diam. **Ray florets** 10–15; laminae 7–10(–15) mm. **Cypselae** (1.9–)2.5–3.2 mm; **pappi** 0.5–1.4 mm. **2n** = 16, 32.

Flowering May–Aug. Dry, rocky sites, oak/conifer forests; 200–2200 m; Calif., Oreg.

Although individuals of var. *lanceolatum* usually differ strikingly from most members of the other northern Californian varieties of the complex, sometimes they are similar. For example, L. Constance (1937) cited specimens of var. *lanceolatum* with features of varieties *integrifolium* or *achillioides* and we have found that most members of var. *lanceolatum* closely resemble plants of var. *obovatum* of southern California.

9f. Eriophyllum lanatum (Pursh) J. Forbes var.
obovatum (Greene) H. M. Hall, Univ. Calif. Publ.
Bot. 3: 186. 1907 [C] [E]

Eriophyllum obovatum Greene,
Erythea 3: 123. 1895

Perennials (± taprooted). **Proximal
leaves** alternate; blades oblanceolate, margins entire or coarsely
serrate, plane, faces densely woolly.
Heads borne singly or 2–3 per
array. **Peduncles** mostly 4–10(–15)
cm. **Involucres** 10–12 mm diam.
Ray florets 10–13; laminae 6–7 mm. **Cypselae** 2.5–4 mm;
pappi 0.4–1.2 mm. $2n$ =16.

Flowering Jun–Jul. Conifer forests; of conservation
concern; 1300–2500 m; Calif.

Variety *obovatum* is known only from the Greenhorn
and the San Bernardino mountains (M. Skinner and B.
Pavlik 1994). It is apparently entirely diploid (J. S.
Mooring 1975) and is suspected of forming natural
hybrids with *Eriophyllum confertiflorum* (L. Constance
1937). The scarcely distinguishable var. *lanceolatum* is
disjunct by 800 km northward.

9g. Eriophyllum lanatum (Pursh) J. Forbes var.
integrifolium (Hooker) Smiley, Univ. Calif. Publ.
Bot. 9: 378. 1921 [E]

Trichophyllum integrifolium
Hooker, Fl. Bor.-Amer. 1: 316.
1833; *Eriophyllum lanatum* var.
cuneatum (Kellogg) Jepson;
E. lanatum var. *monoense*
(Rydberg) Jepson

Shrubs (multistemmed; taprooted).
Proximal leaves alternate; blades
cuneate, oblanceolate, or obovate,
margins entire or ± toothed to lobed (lobes 3–5), plane or
slightly revolute, faces (persistently) woolly. **Heads** usually borne singly. **Peduncles** mostly 3–10 cm. **Involucres**
6–10 mm diam. **Ray florets** (5–)8(–10); laminae 6–12 mm.
Cypselae (2.3–)3–4(–5) mm; **pappi** (0.3–)0.5–1(–2) mm
(scales equal or unequal). $2n$ = 16, 32, 48, 64.

Flowering Jun–Aug. Dry or rocky sites, sagebrush,
conifer forests alpine fell fields; 1200–3500 m; Calif.,
Idaho, Mont., Nev., Oreg., Utah, Wash., Wyo.

L. Constance (1937) sorted the regional races of this
complex, some taxonomically recognized, into five geographic groups. P. A. Munz (1959) referred high elevation populations in California and adjacent states to var.
monoense (Rydberg) Jepson. Variety *integrifolium* intergrades with var. *grandiflorum*. Constance and also
L. Abrams and R. S. Ferris (1923–1960, vol. 4) applied
var. *cuneatum* (Kellogg) Jepson to individuals of what
appear to be vars. *grandiflorum* × *integrifolium* parentage as well as to variants of vars. *achillioides*,

grandiflorum, and *integrifolium*. Variety *integrifolium*
intergrades with var. *lanatum* in Oregon and
Washington near the Columbia River. The intermediate
populations that have been analyzed are polyploid (J. S.
Mooring 2001).

9h. Eriophyllum lanatum (Pursh) J. Forbes var.
lanatum [E]

Perennials (± taprooted). **Proximal
leaves** alternate; blades mostly
oblanceolate, 1-pinnately lobed
(usually beginning in distal ¹/₂ of
leaf, lobes 5–7), ultimate margins
usually toothed or serrate, rarely
entire, revolute, abaxial faces
densely woolly, adaxial less so.
Heads 2–5 per array or borne sin-
gly. **Peduncles** mostly 5–15(–20) cm. **Involucres** 10–15 mm
diam. **Ray florets** (8–)11–13; laminae 10–20 mm. **Cypselae**
(3.4–)3.5–5 mm; **pappi** 0.4–1.8 mm. $2n$ = 16, 32.

Flowering May–Jun. Dry, rocky sites, sagebrush, forests; 200–1100 m; Idaho, Mont., Oreg., Wash.

Variety *lanatum* intergrades with var. *leucophyllum*
near the Cascade Mountains in Washington and Oregon,
and to the south and west with var. *integrifolium* (see
above). It closely resembles var. *grandiflorum*; their
distributions are disjunct.

9i. Eriophyllum lanatum (Pursh) J. Forbes var.
leucophyllum (de Candolle) W. R. Carter, Prelim.
Cat. Fl. Vancouver, 82. 1921 (as leucophylla) [E]

Bahia leucophylla de Candolle in
A. P. de Candolle and A. L. P. P. de
Candolle, Prodr. 5: 657. 1836

Perennials (often clumped; roots
multi-branched). **Proximal leaves**
(at least on flowering stems)
mostly opposite; blades lanceolate
to oblanceolate, 1–(2-)pinnately
lobed (lobes ± linear), ultimate
margins entire or toothed (teeth 1–2), usually strongly
revolute (not crispate), abaxial faces usually densely gray-
woolly, adaxial glabrate. **Heads** 2–5 per array or borne
singly. **Peduncles** mostly 3–14 cm. **Involucres** usually
7–11 mm diam. **Ray florets** 8–13, sometimes 0; laminae 8–20 mm. **Cypselae** 2–3(–4) mm; **pappi** (0.3–)0.5–
1(–1.5) mm. $2n$ = 16, 32.

Flowering May–Aug. Moist meadows, dry and rocky
coastal or inland sites; 0–800 m; B.C.; Oreg., Wash.

Variety *leucophyllum* intergrades so smoothly with
var. *achillioides* in southern Oregon that placement of
some specimens becomes arbitrary. Eastward, variety
leucophyllum passes into var. *lanatum* along the Columbia River near The Dalles. Polyploid populations of var.
leucophyllum occur where it abuts vars. *achillioides* and
lanatum (J. S. Mooring 1975, 2001). L. Constance (1937)

and A. Cronquist (1955) submerged var. *leucophyllum* in var. *lanatum*. Modal populations of the two varieties differ sharply in respect to roots, leaves, and cypselae.

9j. Eriophyllum lanatum (Pursh) J. Forbes var. **hallii** Constance, Proc. Natl. Acad. Sci. U.S.A. 20: 411. 1934 • Hall's woolly sunflower C E

Perennials (± taprooted). **Proximal leaves** opposite; blades ovate, 1-pinnately lobed, ultimate margins slightly revolute, abaxial faces sparsely woolly, adaxial glabrate. **Heads** 2–4 per array or borne singly. **Peduncles** mostly 5–12 cm. **Involucres** 7–8 mm diam. **Ray florets** 8–9; laminae 10–13 mm (disc corolla tubes glabrous). **Cypselae** 3.5–5 mm; **pappi** 0.7–2 mm. **2n** = 16.

Flowering Jun–Jul. Dry sites, woodlands; of conservation concern; 1200–1500 m; Calif.

Variety *hallii* is known only from two populations in Kern and in Santa Barbara counties and is threatened by cattle (D. P. Tibor 2001).

10. Eriophyllum confertiflorum (de Candolle) A. Gray, Proc. Amer. Acad. Arts 19: 25. 1883 • Golden or yellow yarrow F

Bahia confertiflora de Candolle in A. P. de Candolle and A. L. P. P. de Candolle, Prodr. 5: 657. 1836

Shrubs, 20–50(–100) cm. **Stems** erect (woolly). **Leaves** (proximal usually mostly alternate): blades cuneate to obovate, 0.5–5 cm (the longer 1–2-pinnately lobed, lobes linear, the shorter 3–5-lobed or -toothed), ultimate margins entire, revolute, abaxial faces woolly, adaxial usually glabrate (distal leaves gradually reduced in size and lobing). **Heads** (3–)4–7(–10) or 10–30+ per array. **Peduncles** mostly 0–2.5(–9) cm. **Involucres** campanulate, 3–7 mm diam. **Phyllaries** 4–7, distinct or connate at bases (broadly elliptic or ovate, carinate, apices obtuse or acute). **Ray florets** 0 or 4–6(–8); laminae golden yellow, 2–5 (× 1.5–3) mm. **Disc florets** 10–75; corollas 2–4 mm. **Cypselae** 1.5–3.5 mm; **pappi** of 2–14 cuneate to oblanceolate (subequal), erose scales 0.5–1.3 mm.

Varieties 2 (2 in the flora): California, Mexico.

SELECTED REFERENCE Mooring, J. S. 1994. A cytogenetic study of *Eriophyllum confertiflorum* (Compositae, Helenieae). Amer. J. Bot. 81: 919–926.

1. Heads 10–30+ per array; peduncles 0–1(–2) cm; involucres 3–5 mm diam.; disc corollas 2–3 mm . . . . 10a. *Eriophyllum confertiflorum* var. *confertiflorum*

1. Heads (3–)4–7(–10) per array; peduncles 0–2.5 (–9) cm; involucres 5–7 mm diam.; disc corollas 3.5–4 mm . . . . 10b. *Eriophyllum confertiflorum* var. *tanacetiflorum*

10a. Eriophyllum confertiflorum (de Candolle) A. Gray var. **confertiflorum** F

Eriophyllum confertiflorum var. *latum* H. M. Hall; *E. confertiflorum* var. *laxiflorum* A. Gray

Leaf blades cuneate to obovate, 0.5–5 cm, the larger 1–2-pinnately lobed, lobes linear, the smaller 3–5-lobed or -toothed. **Heads** 10–30+ per array. **Peduncles** 0–1(–2) cm. **Involucres** 3–5 mm diam. **Phyllaries** 4–6, broadly elliptic, apices obtuse. **Ray florets** usually 4–6, rarely 0; laminae 2–4 mm. **Disc florets** 10–35; corollas 2–3 mm. **Cypselae** 1.5–3 mm; **pappi** 0.5–1 mm. **2n** = 16, 32, 48, 64.

Flowering Mar–Jul. Dry, sandy to rocky sites, coastal scrub to forest communities; 0–3500 m; Calif.; Mexico.

Members of var. *confertiflorum* complex hybridize with at least four varieties of *Eriophyllum lanatum*. *Eriophyllum latilobum*, *E. jepsonii*, *E. confertiflorum* var. *tanacetiflorum*, and some populations of *E. confertiflorum* var. *confertiflorum* with larger, broader leaves and longer peduncles may have resulted from such hybridizations (L. Constance 1937; P. A. Munz 1959; J. S. Mooring 1994).

Some southern California Indian tribes used var. *confertiflorum* as a remedy for rheumatism (E. C. Jaeger 1941). It has been introduced into cultivation as an ornamental.

10b. Eriophyllum confertiflorum (de Candolle) A. Gray var. **tanacetiflorum** (Greene) Jepson, Man. Fl. Pl. Calif., 1116. 1925 C E

Eriophyllum tanacetiflorum Greene, Pittonia 2: 21. 1889

Leaf blades spatulate to obovate, 2–5 cm, 3–5-lobed, lobes ± oblong. **Heads** (3–)4–7(–10) per array. **Peduncles** 0–2.5(–9) cm. **Involucres** 5–7 mm diam. **Phyllaries** 5–7(–8), ovate, apices obtuse or acute. **Ray florets** 0 or 4–6; laminae 4–5 mm. **Disc florets** 35–75; corollas 3.5–4 mm. **Cypselae** 2–3.5 mm; **pappi** 0.5–1.3 mm. **2n** = 64.

Flowering May–Jul. Dry sites, oak woodlands; of conservation concern; 300–1000 m; Calif.

Variety *tanacetiflorum* is known only from the central Sierra Nevada foothills. At least one population is sympatric with *Eriophyllum lanatum* and *E. confertiflorum* var. *confertiflorum*. It may deserve specific recognition.

11. **Eriophyllum latilobum** Rydberg in N. L. Britton et al., N. Amer. Fl. 34: 94. 1914 • San Mateo woolly sunflower [C] [E]

Perennials or subshrubs, 30–60 (–90) cm. Stems erect (woolly). Leaves (proximal alternate): blades rhombic to obovate, 2–6 cm, 1-pinnately lobed (lobes 3–5, ± triangular), ultimate margins entire, revolute, abaxial faces loosely woolly, adaxial glabrate (distal leaves gradually reduced in size and lobing). Heads 2–5(–10) in corymbiform arrays. Peduncles mostly 2–7 cm. Involucres broadly campanulate, 5–6(–7) mm diam. Phyllaries 6–10, distinct (ovate, carinate, apices acute). Ray florets (6–)8–10(–13); laminae yellow, 6–10 (× 4.5–5) mm. Disc florets 40–75; corollas 3–4.5 mm. Cypselae 2–3 mm; pappi of 4 cuneate to oblanceolate scales 0.3–1 mm alternating with 4 shorter scales (in disc cypselae, disc pappi longer than ray pappi). 2*n* = 32.

Flowering Apr–Jun. Usually shady sites, oak woodlands; of conservation concern; 100–200 m; Calif.

Eriophyllum latilobum combines characteristics of *E. confertiflorum* var. *confertiflorum*, with which it sometimes mingles, and *E. lanatum* var. *arachnoideum*. It may have originated by hybridization between them (L. Constance 1937; P. A. Munz 1959).

12. **Eriophyllum jepsonii** Greene, Pittonia 2: 165. 1891 • Jepson's woolly sunflower [C] [E]

Shrubs, 50–80 cm. Stems erect (densely white-woolly). Leaves (proximal alternate): blades ovate, 3–6 cm, 1-pinnatately lobed (lobes 5–7, linear), ultimate margins entire, revolute, abaxial faces woolly, adaxial glabrate (distal leaves gradually reduced in size and lobing). Heads 2–4 per array. Peduncles mostly 5–10(–14) cm. Involucres broadly campanulate, 5–7 mm diam. Phyllaries 6–8, distinct (ovate, carinate, apices acute). Ray florets 6–8; laminae yellow, 6–10 (× 4–5) mm. Disc florets 35–50; corollas 3–5 mm. Cypselae 2–3 mm; pappi of 8 lanceolate to lance-linear, erose scales 1–1.5 mm. 2*n* = 64.

Flowering Mar–May. Dry sites, oak woodlands; of conservation concern; 200–500 m; Calif.

Eriophyllum jepsonii occurs in Inner Coast Ranges and D. P. Tibor (2001) treated it as uncommon. It combines characteristics of *E. lanatum* and *E. confertiflorum* var. *confertiflorum*. Populations have been found as close as 10 m to either species. L. Constance (1937, p. 106) commented that it might be an "evolutionary link" between them.

13. **Eriophyllum staechadifolium** Lagasca, Gen. Sp. Pl., 28. 1816 • Seaside woolly sunflower, lizard tail [E] [F]

Eriophyllum staechadifolium var. *artemisiifolium* (Lessing) J. F. Macbride; *E. staechadifolium* var. *depressum* Greene

Subshrubs, 30–150 cm. Stems ± erect. Leaves (proximal usually alternate, sometimes opposite): blades linear, lanceolate, or ovate, 3–7 cm, usually 1–2-pinnately lobed, ultimate margins entire or toothed, revolute, abaxial faces white-woolly, adaxial glabrate (distal leaves usually similar). Heads 40–80+ per array. Peduncles mostly 0–1(–1.5) cm. Involucres campanulate, 2–3+ mm diam. Phyllaries 8–11, distinct (oblong, carinate, apices obtuse or acute). Ray florets 0 or 6–9; laminae yellow, 3–5 (× 2–3) mm. Disc florets 30–40; corollas 4 mm. Cypselae 3–4 mm; pappi of 8–12 oblong or oblanceolate scales ca. 1 mm. 2*n* = 30.

Flowering Apr–Sep. Coastal dunes and bluffs; 0–100 m; Calif., Oreg.

Morphologically, *Eriophyllum staechadifolium* appears to connect *E. confertiflorum* var. *confertiflorum* and *E.* (*Constancea*) *nevinii*.

360. CONSTANCEA B. G. Baldwin, Madroño 46: 159. 2000 • [For Lincoln Constance, 1909–2001, Californian botanist] [E]

Bruce G. Baldwin

John L. Strother

Subshrubs, 50–150(–200) cm. Stems decumbent to ± erect, branched from bases or throughout (densely white-tomentose to glabrate). Leaves cauline; alternate; petiolate; blades broadly ovate, 1–2-pinnately lobed (lobes oblong or oblanceolate to linear), ultimate margins entire (somewhat revolute, apices ± rounded), faces white-tomentose (adaxial often glabrescent). Heads

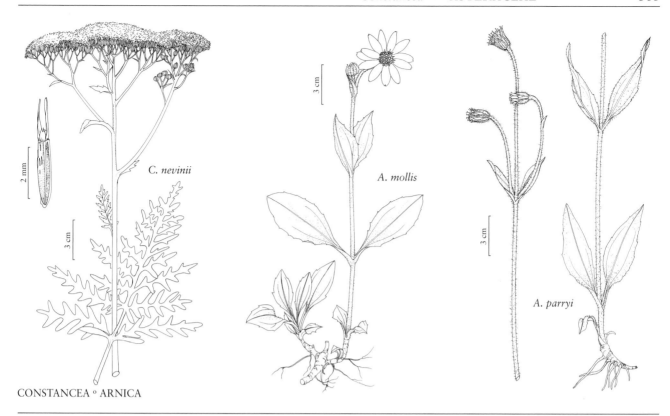

CONSTANCEA ° ARNICA

radiate, (50–100+) in corymbiform or paniculiform arrays. **Involucres** cylindric to campanulate, 3–5 mm diam. **Phyllaries** persistent, 8–16 in ± 2 series (± erect in fruit, distinct, oblong to linear, ± herbaceous, ± keeled). **Receptacles** flat to convex, shallowly pitted, glabrous, epaleate. **Ray florets** 4–9, pistillate, fertile; corollas yellow. **Disc florets** 10–25+, bisexual, fertile; corollas yellow, tubes shorter than campanulate or narrowly funnelform throats, lobes 5, deltate. **Cypselae** (blackish, dull) obpyramidal to clavate, ± scabrellous to glabrate; **pappi** persistent, of 2–6+ unequal (or 2 opposite, longer, and ± equal), basally connate, oblong to subulate scales (tips acute to erose). $x = 19$.

Species 1: California.

Recognition of *Constancea* is based on morphologic, cytologic, and molecular evidence for polyphyly of *Eriophyllum* if *E. nevinii* is included. *Constancea* has characteristics found in other, closely related $x = 19$ "helenioid" genera that are unusual or absent in the clade represented by *Eriophyllum*, *Pseudobahia*, and *Syntrichopappus*, such as leaves with well-developed petioles, phyllaries in more than one series and more than the number of ray florets, and pappus scales unequal or a longer pair opposite and ± equal. The previously suggested close relationship between *Constancea* and another coastal subshrub, *E. staechadifolium*, is untenable.

1. **Constancea nevinii** (A. Gray) B. G. Baldwin, Madroño 46: 160. 2000 [C] [E] [F]

Eriophyllum nevinii A. Gray in A. Gray et al., Syn. Fl. N. Amer. ed. 2, 1(2): 452. 1886

Leaves 8–25 cm. **Peduncles** 1–5 mm. **Phyllaries** 4–6 mm. **Ray laminae** 2–3 × 1–2 mm. **Disc corollas** 2–4 mm. **Cypselae** 2–3 mm; **pappus scales** 0.5–2.5 mm. $2n = 38$.

Flowering Apr–Sep. Coastal bluffs, cliff faces; of conservation concern; 0–200 m; Calif.

Constancea nevinii is known from San Clemente, Santa Barbara, and Santa Catalina islands.

187m.22. ASTERACEAE   Martinov (tribe HELIANTHEAE) subtribe CHAENACTIDINAE   Rydberg in N. L. Britton et al., N. Amer. Fl. 34: 63.   1914 (as Chaenactidanae)

Bahiinae Rydberg; Palafoxiinae Rydberg

Annuals, biennials, perennials, subshrubs, shrubs, or treelets, 1–80(–300) cm (some rhizomatous or with woody caudices).   **Leaves** basal, basal and cauline, or cauline; opposite or alternate; usually petiolate, sometimes sessile; blades mostly cordate, deltate, elliptic, lanceolate, linear, oblanceolate, oblong, obovate, ovate, rhombic, or spatulate, often 1–2-pinnately or -ternately lobed (lobes mostly filiform to linear, lanceolate, or oblanceolate), ultimate margins entire or toothed, faces ± lanate to woolly, or hispidulous to scabrellous or strigose, sometimes glabrescent, sometimes stipitate-glandular, rarely glabrous, often gland-dotted.   **Heads** usually radiate, sometimes discoid, rarely radiant (*Chaenactis* spp.), borne singly, or in clusters, or in corymbiform, cymiform, or paniculiform arrays.   **Calyculi** usually 0 (sometimes 1–3 bractlets in *Schkuhria*).   **Involucres** campanulate to narrowly cylindric or obconic, or hemispheric to rotate.   **Phyllaries** usually persistent, 4–60+ in 1–4+ series (usually erect at flowering, usually reflexed in fruit, usually distinct, basally coherent in some *Arnica* spp., mostly lance-ovate, linear, or ovate, unequal to subequal, mostly herbaceous to membranous, outer foliaceous in *Venegasia*).   **Receptacles** conic, flat, or hemispheric, often knobby or pitted (sometimes hairy or gland-dotted), usually epaleate (paleae sometimes subulate scales in *Chaenactis*; in *Bartlettia*, linear, membranous, hairy scales, each appressed to and shed with an associated cypsela may be interpreted as paleae).   **Ray florets** 0 (corollas of peripheral florets sometimes notably larger than those of the inner, then zygomorphic, showy, and somewhat raylike, e.g., *Chaenactis* spp.) or (1–)4–21(–60+), pistillate, fertile; corollas yellow, orange, red, purplish, or whitish (persistent, marcescent in *Arnica dealbata*).   **Disc florets** 4–200+, usually bisexual, fertile (functionally staminate in *Arnica dealbata*); corollas yellow, orange, purplish, or whitish, usually hairy and/or gland-dotted or stipitate-glandular, tubes shorter than to longer than cylindric or campanulate to funnelform throats, lobes 5, deltate to linear; anther thecae pale; stigmatic papillae in 2 lines.   **Cypselae** mostly obpyramidal and 4–5-angled, sometimes clavate, columnar, cylindric, or fusiform, sometimes compressed and ± linear (e.g., *Chaenactis*, *Hulsea*) or obcompressed (e.g., *Arnica dealbata*, *Bartlettia*; lengths usually 3+ times diams.), faces usually hairy and/or gland-dotted; **pappi** falling, fragile, or persistent, usually of 6–80 distinct or basally connate bristles, or of 4–20 distinct (or basally connate) scales, all, some, or none aristate, sometimes pappi none.

Genera 25, species 129 (19 genera, 83 species in the flora): North America, Mexico, Central America, South America (*Bahia*).

The circumscription of Chaenactidinae used here (that of H. Robinson 1981) contrasts sharply with that of B. G. Baldwin (Baldwin and B. L. Wessa 2000; Baldwin et al. 2002).   Baldwin included *Chaenactis*, *Dimeresia* (here segregated in Dimeresiinae), and *Orochaenactis* in his Chaenactidinae (as sole subtribe in his Chaenactideae) and placed other genera of Chaenactidinae in the sense of Robinson as follows: *Amauriopsis*, *Bahia*, *Bartlettia*, *Chamaechaenactis*, *Florestina*, *Hymenothrix*, *Palafoxia*, *Peucephyllum*, *Picradeniopsis*, *Platyschkuhria*, *Psathyrotopsis*, and *Schkuhria* in Bahiinae (in Bahieae); *Arnica* in Arnicinae (in Madieae); *Hulsea* in Hulseinae (in Madieae); *Syntrichopappus* in Baeriinae (in Madieae); *Venegasia* in Venegasiinae (in Madieae); and *Jamesianthus* in Pectidinae (in Tageteae).

1. Pappi wholly or partially of bristles (pappi 0 and ray corollas whitish with red veins in *Syntrichopappus lemmonii*).
  2. Leaves all or mostly opposite.
    3. Phyllaries 8–23 in (1–)2 series (subequal); pappi persistent, of 10–50 bristles . . . . 361. *Arnica* (in part), p. 366
    3. Phyllaries 14–18+ in ± 3 series (unequal); pappi fragile, of 6–8+ bristles. . . . 362. *Jamesianthus*, p. 377
  2. Leaves mostly alternate (sometimes mostly basal).
    4. Shrubs or treelets; leaf blades linear-filiform . . . . 363. *Peucephyllum*, p. 378
    4. Annuals or perennials; leaf blades mostly deltate, elliptic, or rhombic.
      5. Cypselae obcompressed (each shed together with a subtending, linear, membranous scale, margins ciliate) . . . . 364. *Bartlettia*, p. 378
      5. Cypselae clavate, fusiform, obconic, or obpyramidal (margins not ciliate).
        6. Ray florets 5–8; pappi 0, or ± fragile (falling in whole of fragmented rings), of 25–40 basally connate or coherent bristles in 1–2 series . . . . 365. *Syntrichopappus*, p. 379
        6. Ray florets 0; pappi persistent, of 70–90 distinct, unequal bristles in 2–3 series . . . . 366. *Psathyrotopsis*, p. 380
1. Pappi none or wholly of scales (all, some, or no scales aristate).
  7. Leaves all or mostly cauline, usually all or mostly opposite (distal sometimes, usually in *Bahia*, alternate).
    8. Leaves woolly; rays 5–12 (corollas persistent, marcescent); disc florets usually functionally staminate; cypselae obovoid or plumply fusiform, smooth or ca. 20-ribbed . . . . 361. *Arnica* (in part), p. 366
    8. Leaves usually ± hairy (not woolly, hairs white, straight, often bulbous, conic, or fusiform, 0.1–0.8 mm), sometimes glabrous, sometimes stipitate-glandular, usually gland-dotted; rays 0, or 1–15 (corollas withering, not marcescent); disc florets bisexual, fertile; cypselae obpyramidal, mostly 4-angled.
      9. Phyllaries 4–9(–12, margins often purplish or yellowish) . . . . 367. *Schkuhria*, p. 381
      9. Phyllaries 6–18+ (margins rarely purplish, not yellowish).
        10. Annuals, biennials, or perennials, 10–80+ cm; leaves all or mostly opposite (if perennials, blades lanceolate to oblong, 2–20+ mm wide), or all or mostly alternate . . . . 368. *Bahia*, p. 383
        10. Perennials, 3–20+ cm; leaves all or mostly opposite (blades or lobes lanceolate to lance-linear, mostly 1–8 mm wide) . . . . 369. *Picradeniopsis*, p. 384
  7. Leaves all or mostly basal, or mostly cauline, mostly alternate (proximal sometimes opposite).
    11. Phyllaries 20–40+ in 3–4+ series (unequal, outer spreading or reflexed at flowering, rotund to broadly ovate, foliaceous) . . . . 370. *Venegasia*, p. 385
    11. Phyllaries mostly 4–21(–60+ in *Hulsea*; subequal to unequal, all ± erect at flowering, mostly lanceolate, linear, oblanceolate, obovate, or spatulate, not foliaceous).
      12. Disc corolla lobes lance-linear, lance-oblong, or linear (lengths mostly 2+ times widths).
        13. Pappus scales 12–18 . . . . 371. *Hymenothrix*, p. 387
        13. Pappus scales 4–12.
          14. Leaf blades broadly lanceolate to linear (not lobed); corollas usually pinkish or purplish, sometimes whitish; cypselae densely to sparsely hairy (hairs straight) . . . . 372. *Palafoxia*, p. 388
          14. Leaf blades (at least mid-cauline) 3- or 5-lobed or -foliolate (blades or leaflets broadly to narrowly oblong to ovate); corollas whitish; cypselae sparsely hairy (hairs curled) . . . . 373. *Florestina*, p. 392
      12. Disc corolla lobes mostly deltate, lance-deltate, lanceolate, or ovate (lengths mostly 1–2 times widths, sometimes longer in zygomorphic corollas of some *Chaenactis* spp.).
        15. Leaves sparsely to densely hirtellous, scabrellous, or strigose (hairs white, straight, often conic or fusiform, 0.1–0.8 mm), sometimes glabrescent, sometimes stipitate-glandular, usually gland-dotted; cypselae obpyramidal, 4-angled.

16. Leaves mostly cauline, blades deltate, oblong, or ovate overall, usually
    1–2-ternately lobed . . . . 374. *Amauriopsis*, p. 392
16. Leaves mostly basal or basal and cauline, blades cordate, elliptic,
    lanceolate, ovate or rounded.
    17. Plants mostly 10–50+ cm; leaf blades mostly lanceolate, sometimes
        ± ovate; heads borne singly or in corymbiform to paniculiform
        arrays; rays 6–12 . . . . 375. *Platyschkuhria*, p. 394
    17. Plants 2–7(–9) cm (10–20+ cm across); leaf blades cordate, elliptic,
        ovate, or rounded; heads borne singly; rays 0 . . . . . 376. *Chamaechaenactis*, p. 395
        [15. Shifted to left margin.—Ed.]
15. Leaves mostly sparsely lanate, tomentose, or woolly (hairs crisped, tangled or matted, mostly
    0.8+ mm), sometimes glabrescent, or finely granular-pubescent (hairs bulbous, less than 0.2
    mm), or glabrous, sometimes stipitate-glandular or gland-dotted; cypselae clavate, ± cylin-
    dric, or obconic, often compressed and ± linear, obscurely, if at all, 4-angled (sometimes
    obscurely 8–20-angled).
    18. Ray florets 9–60+ . . . . 377. *Hulsea*, p. 396
    18. Ray florets 0 (corollas of peripheral florets sometimes zygomorphic and larger than
        inner, heads radiant).
        19. Heads borne singly or in ± cymiform arrays; disc florets 8–70+ . . . . 378. *Chaenactis*, p. 400
        19. Heads borne singly or (2–5) in clusters (at stem tips); disc florets 4–9 . . . . 379. *Orochaenactis*, p. 414

361. ARNICA Linnaeus, Sp. Pl. 2: 884. 1753; Gen. Pl. ed. 5, 376. 1754 • [Ancient Latin or Greek plant name]

Steven J. Wolf

Perennials, 5–100 cm (rhizomes relatively long and thin; caudices woody, relatively short and thick). **Stems** erect, simple or branched. **Leaves** basal (sterile basal rosettes often present) and/or cauline; mostly opposite (usually 1–10 pairs, distalmost sometimes alternate and usually smaller); petiolate or sessile; blades mostly cordate, deltate, elliptic, lanceolate, linear, oblanceolate, oblong, obovate, ovate, or spatulate, margins entire or toothed (usually dentate, denticulate, or serrate, sometimes crenate or slightly lobed), faces glabrous, hirsute, hispidulous, pilose, puberulent, scabrous, tomentose, villous, or woolly, often stipitate-glandular as well. **Heads** radiate or discoid, borne singly or in cymiform or corymbiform arrays. **Involucres** campanulate, hemispheric, or turbinate, mostly 6–20+ mm diam. **Phyllaries** persistent, 5–23 in (1–)2 series. **Receptacles** convex, smooth or pitted, epaleate. **Ray florets** 0, or 5–22, pistillate, fertile; corollas yellow to orange. **Disc florets** 10–120, usually bisexual and fertile (functionally staminate in *A. dealbata*); corollas usually yellow, rarely cream, tubes shorter than funnelform throats, lobes 5, ± deltate (anthers usually yellow, purple in *A. lessingii* and *A. unalaschcensis*). **Cypselae** (gray or brown to black) ± conic, fusiform, or obovoid-cylindric, nerves 5–10(–20), faces hairy, glandular, or glabrous; **pappi** usually persistent, of 10–50 white or stramineous to tawny, fine, barbellate or subplumose to plumose bristles (0 in *A. dealbata*). $x = 19$.

Species 29 (26 in the flora): North America, Mexico, Europe, Asia (Japan, Russia).

Arnica is circumboreal, predominantly montane, and exhibits maximum species diversity in western North America. It includes common and very widespread species as well as relatively uncommon, narrow endemics. Polyploidy and apomixis are common in the genus, resulting in considerable morphologic variability. *Arnica montana* from Europe has been used medicinally for centuries, and unsubstantiated claims have been made regarding the medicinal properties of some North American species.

Here, in key leads and descriptions, numbers of pairs of leaves refer to flowering shoots.

SELECTED REFERENCES Downie, S. R. and K. E. Denford. 1988. Taxonomy of *Arnica* (Asteraceae) subgenus *Arctica*. Rhodora 90: 245–275. Gruezo, W. S. and K. E. Denford. 1994. Taxonomy of *Arnica* L. subgenus *Chamissonis* Maguire (Asteraceae). Asia Life Sci. 3: 89–212. Maguire, B. 1943. A monograph of the genus *Arnica*. Brittonia 4: 386–510. Wolf, S. J and K. E. Denford. 1984. Taxonomy of *Arnica* (Compositae) subgenus *Austromontana*. Rhodora 86: 239–309.

1. Pappi 0 (ray corollas marcescent; disc florets functionally staminate) . . . . 20. *Arnica dealbata*
1. Pappi of 10–50, fine bristles.
    2. Anthers purple.
        3. Leaves mostly cauline; heads erect; pappus bristles subplumose . . . . 3. *Arnica unalaschcensis*
        3. Leaves mostly basal; heads nodding; pappus bristles usually barbellate . . . . 2. *Arnica lessingii*
    2. Anthers yellow.
        4. Ray florets 0 (disc corollas sometimes dilated in *A. parryi*, resembling rays).
            5. Cauline leaves usually sessile, rarely broadly petiolate.
                6. Leaves: (reticulate venation inconspicuous) margins ± entire, abaxial and adaxial faces sparsely to densely pilose and stipitate-glandular; disc corollas cream . . . . 26. *Arnica viscosa*
                6. Leaves: (reticulate venation conspicuous) margins irregularly and coarsely serrate, abaxial faces pilose and stipitate-glandular, adaxial glabrate to stipitate-glandular; disc corollas yellow . . . . 23. *Arnica venosa*
            5. Cauline leaves petiolate (petioles sometimes relatively broad).
                7. Heads 1–14 (nodding in bud); phyllaries linear to narrowly lanceolate; pappi usually stramineous, rarely tawny, bristles barbellate to ± subplumose . . . . 14. *Arnica parryi*
                7. Heads 1–30 (erect); phyllaries lanceolate, ovate, or ovate-lanceolate; pappi white, bristles usually barbellate, sometimes subplumose.
                    8. Leaves lanceolate, ovate, or subcordate (petioles usually narrow or scarcely winged); cypselae stipitate-glandular and hirsute (hairs duplex) . . . . 21. *Arnica discoidea*
                    8. Leaves elliptic-ovate to spatulate (petioles relatively broad); cypselae sparsely stipitate-glandular (lacking duplex hairs) . . . . 22. *Arnica spathulata*
        4. Ray florets 5–22.
            9. Leaves mostly basal (blades usually with 3, 5, or 7 prominent, subparallel veins; cauline leaves 0 or 1–2 and much reduced).
                10. Leaves broadly elliptic, ovate, or rhombic; involucres campanulate; rays yellow; e United States . . . . 1. *Arnica acaulis*
                10. Leaves oblanceolate, oblong, oval, or spatulate; involucres hemispheric; rays yellow-orange; Canada, w United States.
                    11. Leaf blades 4.5–20 × 0.5–2.5 cm (axils with tufts of brown wool) . . . . . 6. *Arnica fulgens*
                    11. Leaf blades 3.5–14.5 × 0.6–2.4 cm (axils lacking tufts of brown wool) . . . . . 7. *Arnica sororia*
            9. Leaves mostly cauline, sometimes also basal (blades lacking prominent, subparallel veins).
                12. Cauline leaves (at least at mid stems and exclusive of distal reduced leaves) sessile.
                    13. Pappi white, bristles barbellate.
                        14. Leaf blades ovate to elliptic-lanceolate, 2–10 × 1–6 cm, margins serrate to dentate . . . . 24. *Arnica latifolia*
                        14. Leaf blades oblanceolate to spatulate, 2–7 × 0.5–2.5 cm, margins usually entire, sometimes denticulate . . . . 10. *Arnica rydbergii*
                    13. Pappi stramineous to tawny, bristles usually subplumose to plumose sometimes barbellate.
                        15. Cauline leaves (2–)3(–4) pairs (pappus bristles with deep, amberlike deposits) . . . . 12. *Arnica mollis*
                        15. Cauline leaves usually 4–10 pairs (pappus bristles lacking deep, amberlike deposits).

16. Phyllaries 8–23, lanceolate to linear (apices with conspicuous
tufts of white hairs) . . . . 11. *Arnica chamissonis*
16. Phyllaries 8–20, narrowly to broadly lanceolate (apices lacking
conspicuous tufts of white hairs).
    17. Leaves lance-elliptic, obovate, or ovate, margins subentire to
dentate-serrate (basal leaves present at flowering) . . . 15. *Arnica lanceolata*
    17. Leaves lanceolate to lance-elliptic, margins usually entire
(basal leaves withered by flowering) . . . . 16. *Arnica longifolia*
[12. Shifted to left margin.—Ed.]
12. Cauline leaves (at least mid-stem pair) petiolate (petioles sometimes broad).
  18. Cauline leaf blades elliptic, lanceolate, linear, oblanceolate, ovate, or spatulate (lengths
2.5–10 times widths; basal leaves withered by flowering).
    19. Heads 1–8 (erect).
      20. Leaves 3–7 pairs (petioles relatively narrow, lengths ± equaling blades). . . . . 5. *Arnica lonchophylla*
      20. Leaves 1–5 pairs (petioles relatively broadly to narrowly winged, lengths shorter
than blades) . . . . 4. *Arnica angustifolia*
    19. Heads 1–3 (usually nodding).
      21. Leaves: margins usually entire, rarely denticulate or slightly undulate, faces
± densely stipitate-glandular; cypselae ± glandular; Alberta . . . . 9. *Arnica louiseana*
      21. Leaves: margins usually denticulate to dentate, rarely entire, faces rarely stipi-
tate-glandular; cypselae rarely glandular; e Canada or Alaska, nw Canada (not
Alberta) . . . . 8. *Arnica griscomii*
  18. Cauline leaf blades cordate, deltate, elliptic, ovate, ovate-lanceolate, or subcordate
(lengths 1–3 times widths; basal leaves present at flowering).
    22. Pappi stramineous to tawny, bristles subplumose (sterile basal leaf rosettes
lacking) . . . . 13. *Arnica ovata*
    22. Pappi white, bristles usually barbellate, rarely subplumose (sterile basal leaf
rosettes present).
      23. Leaves (relatively thick, often succulent and reddish) glabrous or scabrous; heads
1(–3; often nodding in bud; serpentine in nw California, sw Oregon) . . . . 18. *Arnica cernua*
      23. Leaves (relatively thin, not succulent nor reddish) hairy; heads 1 or 3–15 (erect
in bud; plants not of serpentine).
        24. Stems usually much branched distally (often in dense clumps); heads
(1–)5–15; cypselae black . . . . 25. *Arnica gracilis*
        24. Stems usually simple (not in dense clumps); heads 1 or 3–5(–10); cypselae
gray.
          25. Leaves cordate, ovate, or subcordate, margins dentate; involucres broadly
campanulate . . . . 17. *Arnica cordifolia*
          25. Leaves elliptic to ovate, margins entire or denticulate; involucres
campanulate-turbinate . . . . 19. *Arnica nevadensis*

1. Arnica acaulis (Walter) Britton, Sterns & Poggenburg,
Prelim. Cat., 30. 1888 • Common leopardbane [E]

Doronicum acaule Walter, Fl.
Carol., 205. 1788

Plants 20–80 cm. **Stems** simple.
Leaves 2–4(–6) pairs, nearly all
basal; sessile; blades (with 3, 5, or
7 prominent, subparallel veins)
broadly elliptic, ovate, or rhombic,
4–15 × 1.5–8 cm, margins mostly
entire, apices mostly obtuse, faces
hirsute and stipitate-glandular (cauline leaves 0 or 1–2,
distalmost often alternate, much reduced). **Heads** 3–

20. Involucres campanulate. **Phyllaries** 12–18, narrowly
ovate. **Ray florets** 10–16; corollas yellow. **Disc florets:**
corollas yellow; anthers yellow. **Cypselae** dark gray to
dark brown, 5–7 mm, glabrous or sparsely stipitate-
glandular; **pappi** white, bristles barbellate. *2n* = 38.

Flowering Apr–Jun. Sandy pine woods and clearings,
often in damp soils, chiefly on Coastal Plain; 0–300 m;
Del., D.C., Fla., Ga., Md., N.J., N.C., Pa., S.C., Va.

2. Arnica lessingii (Torrey & A. Gray) Greene, Pittonia 4: 167. 1900 • Nodding arnica

Arnica angustifolia Vahl var. *lessingii* Torrey & A. Gray, Fl. N. Amer. 2: 449. 1843; *A. lessingii* subsp. *norbergii* Hultén & Maguire

Plants 8–35 cm. **Stems** (1–3 together) usually simple, rarely branched (moderately to densely hairy, hairs translucent with purple septa). **Leaves** (2–)3–5 (–6) pairs, usually crowded toward stem bases; petiolate or sessile (petioles short-winged); blades elliptic, lanceolate, or broadly oblanceolate, 3–10 × 1–2.5 cm, margins usually entire proximal to mid blade, subentire to denticulate distally, apices acute to obtuse, faces: abaxial paler and often glabrous, adaxial slightly hairy. **Heads** 1 (nodding). **Involucres** turbinate-campanulate (bases moderately villous, hairs brownish). **Phyllaries** 11–15, lanceolate to elliptic (apices obtuse, tips blunt, callous). **Ray florets** 8–13; corollas yellow (laminae 14–20 mm). **Disc florets:** corollas yellow; anthers dark purple. **Cypselae** brown, 5–6 mm, sparsely strigose (hairs simple, apically pointed) or glabrous; **pappi** tawny, bristles barbellate. $2n$ = 38, 57, 76.

Flowering Jun–Sep. Coastal tundra to alpine slopes; 0–1500 m; B.C., N.W.T., Yukon; Alaska; Asia (Russian Far East).

3. Arnica unalaschcensis Lessing, Linnaea 6: 238. 1831 • Alaska arnica

Plants 8–30 cm. **Stems** simple (sparsely to moderately hairy, hairs translucent with purple septa). **Leaves** 2–5 pairs, mostly cauline; petiolate (proximal) or sessile; blades elliptic-oblanceolate, obovate or spatulate, (3–)5–12 × 1.5–3 cm, margins usually entire proximal to mid blade, serrate to dentate distally, apices acute to obtuse or abruptly pointed, faces usually glabrate, sometimes scabrous to sparsely stipitate-glandular. **Heads** 1 (erect). **Involucres** hemispheric or turbinate-hemispheric (bases moderately villous). **Phyllaries** 14–24, lanceolate (apices obtuse, tips blunt, callous). **Ray florets** 6–17; corollas yellow (laminae 12–17 mm). **Disc florets:** corollas yellow; anthers dark purple. **Cypselae** brown, 4–5 mm, shortly hispidulous (hairs duplex, apically forked) and stipitate-glandular; **pappi** tawny, bristles subplumose. $2n$ = 38.

Flowering Jun–Aug. Coastal tundra to alpine slopes; 0–1500 m; Alaska; Asia (Japan, Russian Far East).

4. Arnica angustifolia Vahl in G. C. Oeder et al., Fl. Dan. 9(26): 5, plate 1524. 1816 • Narrowleaf arnica

Arnica alpina subsp. *angustifolia* (Vahl) Maguire; *A. alpina* (Linnaeus) Olin & Ladau var. *angustifolia* (Vahl) Fernald

Plants 5–40 cm. **Stems** usually simple, rarely branched. **Leaves** 1–5 pairs, mostly cauline; petiolate (petioles broadly to narrowly winged, shorter than blades); blades broadly lanceolate to linear, 2–20 × 0.3–4 cm, margins entire or irregularly denticulate or dentate, apices acute or acuminate, faces glabrous or densely woolly-villous and stipitate-glandular. **Heads** 1–3(–5; erect). **Involucres** hemispheric. **Phyllaries** 9–22, usually narrowly to broadly lanceolate, sometimes oblanceolate. **Ray florets** 6–16; corollas yellow. **Disc florets:** corollas yellow; anthers yellow. **Cypselae** brown, 3–8 mm, densely hirsute, rarely glandular; **pappi** white, bristles barbellate.

Subspecies 2 (2 in the flora): North America; Europe; Asia.

1. Leaves linear to broadly lanceolate, faces glabrous or moderately villous . . . . 4a. *Arnica angustifolia* subsp. *angustifolia*
1. Leaves narrowly lanceolate, faces densely white-woolly-villous . . . . 4b. *Arnica angustifolia* subsp. *tomentosa*

4a. Arnica angustifolia Vahl subsp. **angustifolia** • Arnica à feuilles étroites

Arnica alpina (Linnaeus) Olin & Ladau subsp. *attenuata* (Greene) Maguire; *A. alpina* var. *attenuata* (Greene) Ediger & T. M. Barkley; *A. alpina* subsp. *iljinii* Maguire; *A. alpina* var. *linearis* Hultén; *A. alpina* subsp. *sornborgeri* (Fernald) Maguire; *A. alpina* var. *vestita* Hultén; *A. attenuata* Greene; *A. plantaginea* Pursh; *A. sornborgeri* Fernald; *A. terrae-novae* Fernald

Leaves linear to broadly lanceolate, faces glabrous or moderately villous. **Involucres:** bases densely pilose, sometimes stipitate-glandular as well. $2n$ = 38, 57, 76, 95.

Flowering May–Sep. Alpine slopes and ridges to Arctic tundra; 0–1500 m; Greenland; Alta., B.C., Man., Nfld. and Labr., N.W.T., Nunavut, Ont., Que., Sask., Yukon; Alaska, Mont.; Europe; Asia.

4b. Arnica angustifolia Vahl subsp. **tomentosa** (Macoun) G. W. Douglas & Ruyle-Douglas, Canad. J. Bot. 56: 1710. 1978 [E]

Arnica tomentosa J. M. Macoun, Ottawa Naturalist 13: 166. 1899; *A. alpina* (Linnaeus) Olin & Ladau subsp. *tomentosa* (J. M. Macoun) Maguire; *A. alpina* var. *tomentosa* (J. M. Macoun) Cronquist

Leaves narrowly lanceolate, faces densely white-woolly-villous. **Involucres:** densely woolly-villous and densely stipitate-glandular. $2n = 57, 76$.

Flowering Jul–Aug. Alpine slopes and subalpine meadows; 500–2500 m; Alta., B.C., Nfld. and Labr. (Nfld.), N.W.T., Yukon; Mont.

5. Arnica lonchophylla Greene, Pittonia 4: 164. 1900 • Longleaf arnica, northern arnica, arnica lonchophylle [E]

Arnica alpina (Linnaeus) Olin & Ladau subsp. *lonchophylla* (Greene) G. W. Douglas & Ruyle-Douglas; *A. angustifolia* Vahl subsp. *lonchophylla* (Greene) G. W. Douglas & Ruyle-Douglas; *A. arnoglossa* Greene; *A. chionopappa* Fernald; *A. gaspensis* Fernald; *A. lonchophylla* subsp. *arnoglossa* (Greene) Maguire; *A. lonchophylla* subsp. *chionopappa* (Fernald) Maguire

Plants 12–50 cm. **Stems** usually simple, rarely branched. **Leaves** 3–7 pairs, mostly cauline; petiolate (petioles, at least proximal, relatively narrow, ± equaling blades); blades narrowly to broadly lanceolate or ovate, 3.5–14 × 0.5–3.7 cm, margins regularly dentate or denticulate, apices acute, faces glabrous or moderately pilose, sparsely to densely stipitate-glandular. **Heads** 3–8 (erect). **Involucres** campanulate-turbinate. **Phyllaries** 6–14, lanceolate (apices acute). **Ray florets** 6–17; corollas yellow. **Disc florets:** corollas yellow; anthers yellow. **Cypselae** gray to brown, 3–6 mm, densely hirsute, sometimes stipitate-glandular; **pappi** white, bristles barbellate. $2n = 38, 57, 76$.

Flowering June–Aug. Dry to mesic, open montane slopes, open woodlands, stream gravels, shorelines, calcareous rocky outcrops, to lowland Arctic tundra; 0–1500 m; Alta., B.C., Man., N.B., Nfld. and Labr. (Nfld.), N.W.T., N.S., Nunavut, Ont., Que., Sask., Yukon; Alaska, S.Dak., Wyo.

6. Arnica fulgens Pursh, Fl. Amer. Sept. 2: 527. 1813 • Foothill arnica [E]

Arnica monocephala Rydberg; *A. pedunculata* Rydberg

Plants 10–75 cm. **Stems** solitary, simple. **Leaves** 3–5 pairs, crowded toward stem bases (some or all axils with dense tufts of brown wool); petiolate (petioles at least basal leaves, narrow or broadly winged); blades (with 3 or 5 prominent, subparallel veins) usually narrowly oblanceolate to oblong, rarely oval or broadly spatulate, 4.5–20 × 0.5–2.5 cm, margins denticulate, apices obtuse, faces moderately uniformly hairy, stipitate-glandular. **Heads** 1(–3). **Involucres** broadly hemispheric. **Phyllaries** 13–21, elliptic-oblong or narrowly to broadly lanceolate. **Ray florets** 8–16; corollas yellow-orange. **Disc florets:** corollas yellow; anthers yellow. **Cypselae** brown, 3.5–7 mm, densely hirsute, sometimes sparingly stipitate-glandular; **pappi** usually white, sometimes tawny, bristles barbellate. $2n = 38, 57$.

Flowering May–Jul. Prairies and grasslands to montane conifer forests; 500–3000 m; Alta., B.C., Man., Sask.; Calif., Colo., Idaho, Mont., Nev., N.Dak., Oreg., S.Dak., Utah, Wash., Wyo.

7. Arnica sororia Greene, Ottawa Naturalist 23: 213. 1910 • Twin arnica [E]

Arnica fulgens Pursh var. *sororia* (Greene) G. W. Douglas & Ruyle-Douglas

Plants 15–50 cm. **Stems** (1 or relatively few) simple or branched. **Leaves** 3–6 pairs, crowded toward stem bases (axils lacking tufts of brown wool); petiolate (petioles, at least basal, narrowly winged); blades (with 3 or 5, prominent, subparallel veins) oblanceolate to narrowly oblong, 3.5–14.5 × 0.6–2.4 cm, margins usually entire, rarely denticulate, apices obtuse, faces uniformly hairy, stipitate-glandular. **Heads** 1–5. **Involucres** hemispheric. **Phyllaries** 13–20, usually narrowly, sometimes broadly, lanceolate. **Ray florets** 9–17; corollas yellow-orange. **Disc florets:** corollas yellow; anthers yellow. **Cypselae** brown, 3.5–5 mm, densely hirsute, sometimes sparingly glandular as well; **pappi** white, bristles barbellate. $2n = 38$.

Flowering May–Jul. Prairies and grasslands to montane conifer forests; 500–1500 m; Alta., B.C., Sask.; Calif., Idaho, Mont., Nev., Oreg., Utah, Wash., Wyo.

8. **Arnica griscomii** Fernald, Rhodora 26: 105, plate 143, fig. 7. 1924 • Snow arnica

Arnica louiseana Farr subsp. *griscomii* (Fernald) Maguire

Plants 5–40 cm. **Stems** usually simple, rarely branched. **Leaves** 1–4 pairs, mostly cauline; petiolate; blades elliptic, lanceolate, oblanceolate, ovate, or spatulate, 1.2–10 × 0.5–3.5 cm, margins usually denticulate to dentate, rarely entire, apices acute to obtuse or abruptly pointed, faces usually glabrous or sparingly hispidulous-puberulent, rarely stipitate-glandular. **Heads** usually 1, sometimes 2–3 (erect or nodding). **Involucres** hemispheric to campanulate-turbinate (bases sparsely to densely pilose, hairs white or yellow). **Phyllaries** 9–20, lanceolate to oblanceolate, not stipitate-glandular. **Ray florets** 6–17; corollas yellow. **Disc florets:** corollas yellow; anthers yellow. **Cypselae** brown, 2.5–6 mm, abaxially glabrous, adaxially sparsely hirsute or glabrous throughout, rarely glandular; **pappi** white, bristles barbellate.

Subspecies 2 (2 in the flora): North America; Asia (e Russia).

1. Involucre bases sparsely to moderately white-pilose; Newfoundland, Québec.... 8a. *Arnica griscomii* subsp. *griscomii*
1. Involucre bases densely yellow-pilose; w Northwest Territories to n British Columbia, Alaska .... 8b. *Arnica griscomii* subsp. *frigida*

8a. **Arnica griscomii** Fernald subsp. **griscomii**
• Griscom's arnica, arnica de Griscom [E]

Involucre bases sparsely to moderately white-pilose. **Phyllaries** broadly lanceolate to oblanceolate. **2n** = 76.

Flowering Jun–Jul. Sea-cliffs, limestone barrens, hornblende schists and dry schistose talus in alpine areas; 10–1100 m; Nfld. and Labr. (Nfld.), Que.

8b. **Arnica griscomii** Fernald subsp. **frigida** (C. A. Meyer ex Iljin) S. J. Wolf, Taxon 38: 142 • Snow arnica

Arnica frigida C. A. Meyer ex Iljin, Trudy Bot. Muz. 19: 112, plate 11, fig. 3. 1926; *Arnica louiseana* Farr subsp. *frigida* (C. A. Meyer ex Iljin) Maguire; *A. louiseana* var. *frigida* (C. A. Meyer ex Iljin) S. L. Welsh; *A. louiseana* var. *mendenhallii* (Rydberg) Maguire; *A. louiseana* var. *pilosa* Maguire

Involucre bases densely yellow-pilose. **Phyllaries** lanceolate. **2n** = 38, 57, 76, 95.

Flowering Jul–Aug. Alpine meadows, tundras, and calcareous rocky outcrops; 0–1500 m; B.C., N.W.T., Yukon; Alaska; Asia (e Russia).

9. **Arnica louiseana** Farr, Ottawa Naturalist 20: 109. 1906 • Lake Louise arnica, snow arnica [E]

Plants 5–20 cm. **Stems** simple. **Leaves** 1–3 pairs, mostly cauline (shorter plants often with leaves crowed mostly toward bases); petiolate; blades elliptic, oblong, or ovate-lanceolate, 1.5–7.5 × 0.5–2 cm, margins usually entire, rarely denticulate or slightly undulate, apices usually obtuse, sometimes acute or acuminate, faces glabrous or hispidulous-puberulent, ± densely stipitate-glandular. **Heads** usually 1, sometimes 2–3 (nodding at flowering). **Involucres** campanulate-turbinate. **Phyllaries** 10–20, narrowly lanceolate (stipitate-glandular). **Ray florets** 7–10; corollas yellow. **Disc florets:** corollas yellow; anthers yellow. **Cypselae** brown, 3–5 mm, abaxially glabrous, adaxially sparsely hirsute or glabrous throughout, usually stipitate-glandular toward apices, sometimes densely stipitate-glandular throughout; **pappi** white, bristles barbellate. **2n** = 76, 95.

Flowering Jul–Aug. Exposed tundra slopes and calcareous rock slides; 1800–2100 m; Alta., B.C.

10. **Arnica rydbergii** Greene, Pittonia 4: 36. 1899
• Rydberg's arnica [E]

Plants 8–35 cm. **Stems** (branches from rhizomes) simple. **Leaves** 2–4 pairs (basal often persistent on sterile rosettes), mostly cauline; petiolate (at least basal) or sessile; blades oblanceolate to spatulate, 2–7 × 0.5–2.5 cm, margins usually entire, sometimes denticulate, rarely dentate, apices usually acute, sometimes obtuse, faces sparsely puberulent, moderately pubescent distally, stipitate-glandular. **Heads** 1–3(–5). **Involucres** turbinate-campanulate. **Phyllaries** 9–15, linear-oblong to narrowly lanceolate. **Ray florets** 6–10; corollas yellow. **Disc florets:** corollas yellow; anthers yellow. **Cypselae** brown, 4–7 mm, densely hirsute (not stipitate-glandular); **pappi** white, bristles barbellate. $2n = 38, 76$.

Flowering Jul–Aug. Dry to mesic, exposed, rocky alpine slopes and ridges or alpine meadows; 1200–3400 m; Alta., B.C.; Colo., Idaho, Mont., Oreg., Utah, Wash., Wyo.

11. Arnica chamissonis Lessing, Linnaea 6: 238. 1831

• Chamisso arnica, leafy arnica, arnica de Chamisso E

Arnica bernardina Greene; *A. chamissonis* var. *bernardina* (Greene) Jepson ex Maguire; *A. chamissonis* subsp. *foliosa* (Nuttall) Maguire; *A. chamissonis* var. *foliosa* (Nuttall) Maguire; *A. chamissonis* subsp. *incana* (A. Gray) Maguire; *A. chamissonis* var. *incana* (A. Gray) Hultén; *A. chamissonis* var. *interior* Maguire; *A. chamissonis* var. *jepsoniana* Maguire; *A. foliosa* Nuttall

Plants 20–80(–150) cm. **Stems** usually branched from mid heights or distally. **Leaves** 4–10 pairs, mostly cauline (evenly distributed; basal leaves often withered by flowering, 1–2 pairs, subsessile to short-petiolate); sessile (proximalmost with membranous connate-sheathing bases); blades lance-elliptic, broadly oblanceolate, or oblong, 5–20 × 2–6(–8) cm, margins entire or remotely denticulate to prominently dentate, apices acute, faces nearly glabrous or puberulent to sparsely or densely white-tomentose-pilose. **Heads** (1–)3–10(–16). **Involucres** campanulate (rarely hemispheric). **Phyllaries** 8–23, nearly linear to narrowly lanceolate (apices each with conspicuous tuft of white hairs). **Ray florets** 8–20; corollas yellow. **Disc florets:** corollas yellow; anthers yellow. **Cypselae** gray to brown, 3–8 mm, subglabrous to sparsely hirsutulous, stipitate-glandular; **pappi** stramineous, bristles barbellate to subplumose. $2n = 38, 57, 76$.

Flowering Apr–Sep. Moist meadows and conifer forests, stream banks, late snow-melt areas, often montane to subalpine; 0–3500 m; Alta., B.C., Man., N.W.T., Ont., Que., Sask., Yukon; Alaska, Ariz., Calif., Colo., Idaho, Mont., Nev., N.Mex., Oreg., Utah, Wash., Wyo.

12. Arnica mollis Hooker, Fl. Bor.-Amer. 1: 331. 1834

• Hairy arnica E F

Plants 15–70 cm. **Stems** (forming clumps) simple or branched among heads. **Leaves** (2–)3(–4) pairs, mostly cauline (basal sometimes present); petiolate (petioles relatively short, broad-winged) or subsessile; blades broadly elliptic, lance-elliptic, or narrowly to broadly lanceolate, 4–20 × 1–4 cm, margins entire or irregularly denticulate, apices acute, faces sparsely to moderately hairy (hairs relatively short to long, stipitate glands or soft, silky). **Heads** 1 or 3–7. **Involucres** hemispheric to campanulate. **Phyllaries** 10–22, usually broadly lanceolate, rarely narrowly lanceolate or oblanceolate. **Ray florets** 10–22; corollas yellow. **Disc florets:** corollas yellow; anthers yellow. **Cypselae** grayish brown to black, 4–8 mm, mostly stipitate-glandular, sparsely hirsutulous (hairs white to brownish, simple or bifid); **pappi** tawny, bristles plumose (with deep, amberlike deposits). $2n = 38, 57, 76, 95, 114, 133, 152$.

Flowering Jun–Sep. Moist meadows and conifer forests, stream banks, late snow-melt areas, montane to subalpine; 1000–4000 m; Alta., B.C., N.W.T., Yukon; Alaska, Calif., Colo., Idaho, Mont., Nev., Oreg., Utah, Wash., Wyo.

13. Arnica ovata Greene, Pittonia 4: 161. 1900

• Sticky leaf arnica E

Arnica diversifolia Greene; *A. latifolia* Bongard var. *viscidula* A. Gray

Plants 10–50 cm. **Stems** (forming clumps) simple or branched among heads. **Leaves** 2–3(–4) pairs (basal 1–2 pairs usually withered by flowering, petiolate, petioles broadly winged, blades round-ovate, relatively small; sterile rosettes lacking), mostly cauline; petiolate (at least middle pair, petioles broadly to narrowly winged); blades broadly deltate to ovate, 4–8 × 2–6 cm (middle pair largest), margins irregularly denticulate to coarsely dentate-serrate, apices acute, faces puberulent (hairs minute) and stipitate-glandular. **Heads** 1–3(–5). **Involucres** usually narrowly turbinate, rarely narrowly campanulate. **Phyllaries** 9–20, linear to narrowly lanceolate. **Ray florets** 8–16, yellow. **Disc florets:** corollas yellow; anthers yellow. **Cypselae** brown to black, 5–7 mm, sparsely to moderately pilose and stipitate-glandular; **pappi** stramineous to tawny, bristles subplumose. $2n = 57, 76$.

Flowering Jul–Sep. Moist meadows and conifer forests, stream banks, late snow-melt areas, montane to subalpine; 200–3600 m; Alta., B.C., Yukon; Alaska, Calif., Colo., Idaho, Mont., Oreg., Utah, Wash., Wyo.

14. Arnica parryi A. Gray, Amer. Naturalist 8: 213. 1874 • Parry's arnica, nodding arnica [E] [F]

Arnica angustifolia Vahl subsp. *eradiata* A. Gray, Proc. Acad. Nat. Sci. Philadelphia 15: 68. 1864; *A. parryi* A. Gray subsp. *sonnei* (Greene) Maguire; *A. parryi* var. *sonnei* (Greene) Cronquist

Plants 15–50(–60) cm. **Stems** simple or branched among heads. **Leaves** (2–)3–4 pairs, basal and cauline (basal usually withered by flowering, petiolate or subsessile, blades ovate or nearly round to oblong-ovate, much smaller than cauline; sterile rosettes often present, proximal cauline often crowded toward stems bases); petiolate; blades broadly to narrowly lanceolate, oblong-ovate, ovate-lanceolate, or round, 4–22 × 1–6 cm, (bases obtuse or cuneate to truncate) margins usually entire, sometimes sparsely denticulate, faces scantily to moderately pilose (hairs white), adaxial sparsely stipitate-glandular (distal cauline leaves sessile, extremely reduced, nearly linear to narrowly elliptic-lanceolate). **Heads** (1–)3–9(–14; often nodding in bud). **Involucres** turbinate to narrowly campanulate. **Phyllaries** 8–20, linear to narrowly lanceolate. **Ray florets** usually 0 (sometimes peripheral florets pistillate; corollas yellow, laminae rudimentary). **Disc florets** 20–50; corollas yellow (1 or more lobes sometimes expanded, giving appearance of laminae); anthers yellow. **Cypselae** brown to black, 4–7 mm, glabrous or sparsely stipitate-glandular to densely hirsute; **pappi** usually stramineous, rarely tawny, bristles barbellate to ± subplumose. $2n = 38, 57, 76$.

Flowering May–Sep. Open conifer forests to alpine meadows; 500–3800 m; Alta., B.C., Yukon; Calif., Colo., Idaho, Mont., Nev., Oreg., Utah, Wash., Wyo.

15. Arnica lanceolata Nuttall, Trans. Amer. Philos. Soc., n. s. 7: 407. 1841 [E]

Plants (5–)20–80 cm. **Stems** solitary to densely clumped, simple or branched among heads. **Leaves** 4–8(–10) pairs, mostly cauline; usually sessile (sometimes with partly connate-sheathing bases), sometimes petiolate (proximalmost); blades lance-elliptic, lanceolate, oblanceolate, obovate, or ovate, 4–12(–20) × (1–)2–6(–8) cm, margins subentire to dentate-serrate, apices usually acute, rarely obtuse, faces glabrate to pilose. **Heads** (1–)3–10(–20). **Involucres** narrowly campanulate to turbinate. **Phyllaries**

8–15(–19), narrowly to broadly lanceolate. **Ray florets** (5–)7–17(–20); corollas yellow. **Disc florets:** corollas yellow; anthers yellow. **Cypselae** brown, 4–8 mm, sparsely hirsutulous, stipitate-glandular; **pappi** tawny, bristles subplumose.

Subspecies 2 (2 in the flora): North America.

1. Cauline leaves mostly sessile (proximalmost pair petiolate); e North America . . . . 15a. *Arnica lanceolata* subsp. *lanceolata*
1. Cauline leaves sessile (including proximalmost pair, bases sometimes partly connate-sheathing); w North America . . . . 15b. *Arnica lanceolata* subsp. *prima*

15a. Arnica lanceolata Nuttall subsp. **lanceolata**
• Lanceleaf arnica, arnica à aigrettes brunes [E]

Arnica mollis Hooker var. *petiolaris* Fernald

Leaves: basal petiolate, blades elliptic-lanceolate or oblanceolate, margins subentire, denticulate, or dentate-serrate; cauline mostly sessile (proximalmost pair petiolate), blades lanceolate to elliptic-lanceolate or oblanceolate, margins subentire, irregularly denticulate, or serrate, faces glabrate to scantily pilose. **Phyllaries** narrowly to broadly lanceolate. $2n = 76$.

Flowering Jun–Sep. Moist areas, stream banks, gravel bars, to subalpine meadows; 0–1000 m; N.B., Que.; Maine, N.H.

15b. Arnica lanceolata Nuttall subsp. **prima** (Maguire) Strother & S. J. Wolf, Novon 16: 2006
• Clasping arnica [E]

Arnica amplexifolia Rydberg subsp. *prima* Maguire, Madroño 6: 154. 1942; *A. amplexicaulis* Nuttall; *A. amplexicaulis* var. *piperi* H. St. John & F. A. Warren; *A. amplexicaulis* subsp. *prima* (Maguire) Maguire; *A. amplexicaulis* var. *prima* (Maguire) B. Boivin

Leaves: basal commonly withered by flowering, sessile or subsessile to petiolate, blades usually oblong to narrowly lanceolate or oblanceolate, rarely ovate, margins subentire to prominently dentate-serrate; cauline sessile (bases sometimes partly connate-sheathing), blades narrowly to broadly lance-elliptic or ovate, margins subentire to serrate-dentate, faces sparsely to densely pilose. **Phyllaries** narrowly lanceolate. $2n = 38, 57, 76$.

Flowering May–Sep. Moist areas, along stream banks, snow-melt areas, montane to alpine meadows;

0–3000 m; Alta., B.C., N.W.T., Yukon; Alaska, Calif., Idaho, Mont., Nev., Oreg., Utah, Wash., Wyo.

The name *Arnica amplexifolia* Rydberg (1900c) was a superfluous (illegitimate) substitution for *A. amplexicaulis* Nuttall (1841).

16. Arnica longifolia D. C. Eaton in S. Watson, Botany (Fortieth Parallel), 186. 1871 · Spearleaf arnica E

Arnica longifolia subsp. *myriadenia* (Piper) Maguire; *A. myriadenia* Piper

Plants 30–60(–110) cm. **Stems** (often relatively numerous, clustered in clonal patches) simple. **Leaves** 5–7 pairs, mostly cauline (basal leaves usually withered by flowering); sessile or subsessile (proximalmost with connate-sheathing bases); blades lanceolate to lance-elliptic, 5–12(–15) × 1.5–3.5 cm, margins usually entire, rarely denticulate, apices acute to acuminate, faces scabrid-puberulent, sometimes glandular (distal leaves not much reduced). **Heads** 3–20(–35). **Involucres** turbinate-campanulate. **Phyllaries** 11–15(–20), narrow to broadly lanceolate. **Ray florets** 6–15; corollas yellow. **Disc florets** 6–11; corollas yellow; anthers yellow. **Cypselae** brown to black, 3–7 mm, glabrous or sparsely hirsutulous, stipitate-glandular; **pappi** stramineous to tawny, bristles barbellate to subplumose. $2n = 57, 76$.

Flowering Jun–Oct. Montane coniferous forests to alpine, usually moist areas, stream banks or late snow-melt areas; 1300–3700 m; Alta., B.C.; Calif., Colo., Idaho, Mont., Nev., Oreg., Utah, Wash., Wyo.

17. Arnica cordifolia Hooker, Fl. Bor.-Amer. 1: 331. 1834 · Heartleaf arnica E F

Arnica cordifolia var. *pumila* (Rydberg) Maguire; *A. paniculata* A. Nelson; *A. whitneyi* Fernald

Plants 10–40(–70) cm. **Stems** usually simple, sometimes branched. **Leaves** 2–4(–6) pairs, mostly cauline (basal often persistent on sterile rosettes); petiolate; blades cordate or subcordate to ovate, 3–10 × 2–10 cm, margins dentate to coarsely dentate, apices acute to rounded, faces puberulent to sparsely villous, sometimes stipitate-glandular (especially adaxial). **Heads** 1, or 3–5(–10). **Involucres** broadly campanulate. **Phyllaries** 10–20, narrowly ovate to lanceolate. **Ray florets** 6–13; corollas yellow. **Disc florets:** corollas yellow; anthers yellow. **Cypselae** dark gray, 5–10 mm, sparsely to densely hirsute (hairs duplex),

sometimes stipitate-glandular as well; **pappi** white, bristles barbellate. $2n = 38, 57, 76, 95, 114$.

Flowering May–Jul. Conifer forests to subalpine meadows; 500–3000 m; Alta., B.C., Man., N.W.T., Ont., Sask., Yukon; Ariz., Calif., Colo., Idaho, Mich., Mont., Nev., N.Mex., Oreg., S.Dak., Utah, Wash., Wyo.

18. Arnica cernua Howell, Fl. N.W. Amer., 373. 1900 (as cernuua) · Serpentine arnica

Arnica chandleri Rydberg

Plants 10–30 cm. **Stems** (often reddish purple) mostly simple. **Leaves** 3–4 pairs, mostly cauline (basal often persistent on sterile rosettes); petiolate (petioles usually narrow, proximalmost sometimes broadly winged); blades (often reddish purple) usually elliptic to ovate, sometimes subcordate, 1.5–8 × 1.5–4 cm (often relatively thick, nearly succulent), margins usually entire or serrate, sometimes crenate or slightly lobed, apices acute to rounded, faces glabrous or scabrous. **Heads** usually 1, sometimes 2–3 (often nodding in bud). **Involucres** campanulate-turbinate. **Phyllaries** 8–14, ovate to broadly lanceolate. **Ray florets** 5–10; corollas yellow to slightly orange. **Disc florets:** corollas yellow; anthers yellow. **Cypselae** dark gray, 6–8 mm, sparsely to densely hirsute (hairs duplex); **pappi** white, bristles usually barbellate, sometimes subplumose. $2n = 38$.

Flowering Apr–Jun. Serpentine soils, conifer forests; 500–1500 m; Calif., Oreg.

19. Arnica nevadensis A. Gray, Proc. Amer. Acad. Arts 19: 55. 1883 · Nevada arnica, Sierra arnica E

Arnica tomentella Greene

Plants 10–50 cm. **Stems** simple. **Leaves** 2–3 pairs, mostly cauline (basal leaves often persistent on sterile rosettes); petiolate (petioles narrowly to broadly winged); blades elliptic to ovate, 3–8 × 2–4 cm, margins entire or denticulate, apices acute to rounded, faces stipitate-glandular. **Heads** 1(–3). **Involucres** campanulate-turbinate. **Phyllaries** 10–16, oblanceolate. **Ray florets** 6–14; corollas yellow. **Disc florets:** corollas yellow; anthers yellow. **Cypselae** dark gray, 6–9 mm, stipitate-glandular; **pappi** white or tawny, bristles barbellate to subplumose. $2n = 76$.

Flowering Jul–Aug. Coniferous forests, meadows, rocky slopes; 1500–3000 m; Calif., Nev., Oreg., Wash.

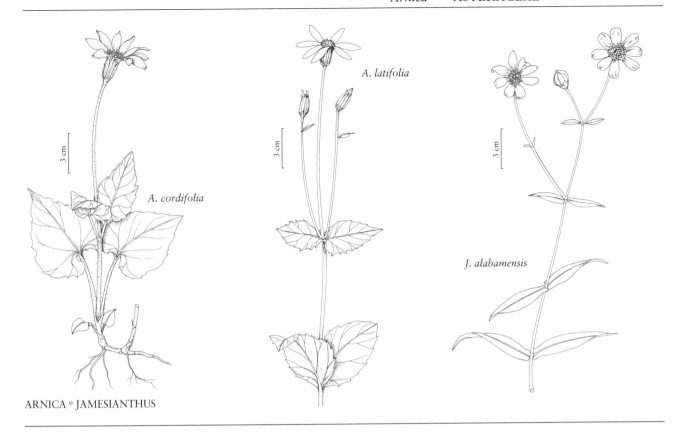

A. latifolia

A. cordifolia

J. alabamensis

ARNICA ∘ JAMESIANTHUS

20. Arnica dealbata (A. Gray) B. G. Baldwin, Novon 9: 460. 1999 • Mock leopardbane E

Whitneya dealbata A. Gray, Proc. Amer. Acad. Arts 6: 550. 1865

Plants 15–35 cm. **Stems** mostly simple. **Leaves** 3–8 pairs, mostly crowded toward bases; petiolate (at least proximal); blades elliptic to oblanceolate or obovate, 5–10 × 0.5–3.5 cm, margins entire, apices acute, faces densely hairy (hairs relatively short, curly) and sessile-glandular (distal leaves sessile, much reduced). **Heads** 1–6. **Involucres** campanulate to hemispheric. **Phyllaries** 5–12 (in 1 series), oblong or elliptic. **Ray florets** 5–12; corollas yellow. **Disc florets** (functionally staminate): corollas yellow; anthers yellow (styles rarely exserted). **Cypselae** brown to black, 4–10 mm, sparsely to densely hirsute (hairs usually simple, rarely duplex) and sparsely stipitate-glandular; **pappi** 0. **2n** = 38, 76.

Flowering Jun–Aug. Open forests, meadows, slopes; 1200–2400 m; Calif.

Although *Arnica dealbata* lacks pappi, morphologic, cytologic, and molecular data (B. G. Baldwin and B. L. Wessa 2000; Baldwin et al. 2002) clearly support its inclusion in *Arnica*.

21. Arnica discoidea Bentham, Pl. Hartw., 319. 1849 • Rayless arnica E

Arnica alata Rydberg; *A. cordifolia* Hooker var. *eradiata* A. Gray; *A. discoidea* var. *alata* (Rydberg) Cronquist; *A. discoidea* var. *eradiata* (A. Gray) Cronquist; *A. parviflora* A. Gray

Plants 15–60 cm. **Stems** mostly simple or distally branched. **Leaves** 3–7 pairs, sometimes crowded at stem bases (basal leaves often persistent on sterile rosettes); petiolate (petioles narrow, 1.5–8 cm, often broadly winged on distal reduced leaves); blades usually ovate to broadly lanceolate, seldom subcordate, 2–12 × 1–7 cm, margins usually serrate to coarsely dentate or crenate, rarely subentire, faces glabrate to pilose, stipitate-glandular. **Heads** 3–10(–30; erect). **Involucres** turbinate-campanulate. **Phyllaries** 8–15, ovate-lanceolate to narrowly lanceolate. **Ray florets** usually 0 (corollas of peripheral florets rarely dilated, resembling rays). **Disc florets** 20–50; corollas yellow; anthers yellow. **Cypselae** dark gray, 6–8 mm, hirsute (hairs duplex) and stipitate-glandular; **pappi** white, bristles usually barbellate, sometimes subplumose. **2n** = 38, 57, 76.

Flowering May–Jul. Chaparral, oak and pine forests; 100–1500 m; Calif., Nev., Oreg., Wash.

22. Arnica spathulata   Greene, Pittonia 3: 103. 1896
• Klamath arnica   E

Arnica eastwoodiae Rydberg;
A. spathulata subsp. *eastwoodiae*
(Rydberg) Maguire; *A. spathulata*
var. *eastwoodiae* (Rydberg) Ediger
& T. M. Barkley

Plants 15–45+ cm. **Stems** simple
or branched. **Leaves** 3–5 pairs,
sometimes crowded toward bases
(basal leaves often persistent on
sterile rosettes); petiolate (petioles mostly broadly
winged, 1–9 × 0.2–1.5 cm); blades elliptic-ovate to
spatulate, 2–8 × 1–4 cm, margins subentire to mostly
irregularly dentate, apices acute, faces sparsely to densely
villous, stipitate-glandular. **Heads** 1, or 3–9(–25).
Involucres turbinate-campanulate. **Phyllaries** 8–15,
broadly to narrowly lanceolate. **Ray florets** 0. **Disc
florets** 15–50; corollas yellow; anthers yellow. **Cypselae**
black, 5–10 mm, sparsely stipitate-glandular; **pappi**
white, bristles barbellate. **2*n*** = 38, 76.

Flowering Apr–Jul. Usually serpentine soils, open,
dry oak–conifer forests, disturbed areas; 200–1500 m;
Calif., Oreg.

23. Arnica venosa   H. M. Hall, Univ. Calif. Publ. Bot. 6:
174. 1915 • Shasta County arnica   E

Plants 20–60 cm. **Stems** simple
or branched (prominently ribbed;
caudices woody). **Leaves** 6–10 pairs,
cauline (basal leaves withered by
flowering; proximal cauline
scalelike, middle leaves largest,
distal reduced, bractlike) usually
broadly sessile, rarely broadly
petiolate; blades (3- or 5-nerved,
strongly reticulate-veined) ovate-elliptic to broadly
lanceolate, 3–7 × 1.5–4 cm (firm), margins irregularly
and coarsely serrate, apices acute to obtuse, faces: abaxial
pilose, stipitate-glandular (especially on veins), adaxial
glabrate to stipitate-glandular. **Heads** 1. **Involucres**
turbinate-campanulate. **Phyllaries** 8–19, ovate to
broadly lanceolate. **Ray florets** 0. **Disc florets** 30–60;
corollas yellow; anthers yellow. **Cypselae** dark gray,
6–8 mm, densely hirsute (hairs duplex); **pappi** white,
bristles barbellate. **2*n*** = 38.

Flowering May–Jun. Open, often disturbed, oak-pine
forests; 400–1400 m; Calif.

Arnica venosa is known only from Shasta and Trinity
counties.

24. Arnica latifolia   Bongard, Mém. Acad. Imp. Sci.
St.-Pétersbourg, Sér. 6, Sci. Math. 2: 147. 1832
• Broadleaf arnica   E F

Plants 10–50 cm. **Stems** usually
simple, sometimes branched
distally. **Leaves** 2–4(–6) pairs,
cauline (basal leaves often
persistent on sterile rosettes),
petiolate (proximal, petioles
relatively short, broadly winged)
or sessile (mid and distal); blades
lance-elliptic to ovate, 2–10 × 1–6
cm, margins serrate to dentate, apices obtuse to acute,
faces glabrous or sparsely villous. **Heads** 1 or 3–5(–9).
Involucres narrowly turbinate. **Phyllaries** 8–20,
lanceolate to oblanceolate. **Ray florets** 8–15; corollas
yellow. **Disc florets** 20–90; corollas yellow; anthers
yellow. **Cypselae** dark brown, 5–9 mm, sparsely villous;
pappi white, bristles barbellate. **2*n*** = 38, 76.

Flowering Jul–Aug. Relatively moist, montane conifer
forests to subalpine meadows; 500–3300 m; Alta., B.C.,
N.W.T., Yukon; Alaska, Calif., Colo., Idaho, Mont.,
Oreg., Utah, Wash., Wyo.

25. Arnica gracilis   Rydberg, Bull. Torrey Bot. Club 24:
297. 1897 • Smallhead arnica   E

Arnica latifolia Bongard var. *gracilis*
(Rydberg) Cronquist

Plants 10–30 cm. **Stems** (often 5–10
in dense clumps) branched distally. **Leaves** 2–3 pairs, mostly
cauline (basal leaves often
persistent on sterile rosettes,
similar to cauline leaves); petiolate
(petioles narrowly to broadly
winged, distal pair of leaves often reduced, connate-
perfoliate); blades ovate to ovate-lanceolate, 2–6 × 1–3
cm, margins irregularly serrate to subentire, apices acute,
faces: abaxial glandular, adaxial stipitate-glandular.
Heads (1–)5–15. **Involucres** turbinate-campanulate.
Phyllaries 10–16, ovate-lanceolate. **Ray florets** 5–12;
corollas yellow. **Disc florets:** corollas yellow; anthers
yellow. **Cypselae** black, 4.5–7 mm, hairy (hairs duplex)
and stipitate-glandular; **pappi** white, bristles barbellate.
2*n* = 57, 76.

Flowering Jul–Aug. Dry, exposed, rocky, alpine
slopes, sometimes subalpine meadows; 1200–2500 m;
Alta., B.C.; Idaho, Mont., Oreg., Utah, Wash., Wyo.

26. Arnica viscosa A. Gray, Proc. Amer. Acad. Arts 13: 374. 1878 • Mt. Shasta arnica [E]

Chrysopsis shastensis Jepson; *Raillardella paniculata* Greene

Plants 20–50 cm. **Stems** branched (prominently ribbed, strongly stipitate-glandular; caudices woody). **Leaves** 5–10 pairs (main stems, 2–6 pairs on branches; basal withered by flowering), cauline; sessile; blades obovate-oblong to ovate-oblong, (1–)2–4(–5) × 1–3 cm, margins ± entire, apices acute, faces sparsely to densely pilose, densely stipitate-glandular. **Heads** 10–20. **Involucres** narrowly turbinate. **Phyllaries** 10–20, broadly lanceolate (stipitate-glandular). **Ray florets** 0. **Disc florets** 10–30; corollas cream; anthers yellow. **Cypselae** dark gray, 4.5–6.5 mm, stipitate-glandular; **pappi** usually white, rarely tawny, bristles barbellate to subplumose. *2n* = 38.

Flowering Aug–Sep. Open, rocky subalpine to alpine; 1700–2500 m; Calif., Oreg.

362. JAMESIANTHUS S. F. Blake & Sherff, Publ. Field Mus. Nat. Hist., Bot. Ser. 22: 399. 1940 • [For Robert Leslie James, 1897–1977, American botanist and historian, and Greek *anthos*, flower] [E]

John L. Strother

Perennials, 60–150+ cm (fibrous-rooted). **Leaves** cauline; opposite; petiolate or sessile; blades ± lanceolate to narrowly trullate, (bases sometimes ± auriculate) margins entire or denticulate, faces: abaxial glabrous, adaxial minutely scabrellous to hirtellous. **Heads** radiate, borne singly or in open, corymbiform arrays. **Involucres** ± campanulate or broader, 9–12+ mm diam. **Phyllaries** persistent, 14–18+ in ± 3 series (± oblong, ovate, ovate-oblong, or ovate-attenuate, unequal, outer shorter). **Receptacles** flat to convex, epaleate. **Ray florets** 6–8, pistillate, fertile; corollas yellow. **Disc florets** 20–30, all bisexual or inner functionally staminate; corollas yellow, tubes shorter than to equaling ± campanulate to funnelform throats, lobes 5, ovate-deltate to deltate (style branches stout, appendages deltate). **Ray cypselae** ± obovoid, weakly obcompressed, 16–24-nerved, ± hispidulous; **pappi** 0, or fragile, of 6–8+ barbellate bristles (borne on crowns). **Disc cypselae** ± ellipsoid to clavate, ± hispidulous; **pappi** 0, or fragile, of 6–8+ barbellulate bristles (borne on crowns).

Species 1: Alabama.

1. Jamesianthus alabamensis S. F. Blake & Sherff, Publ. Field Mus. Nat. Hist., Bot. Ser. 22: 402. 1940 [E] [F]

Stems stipitate-glandular distally, internodes 4–6(–10) cm. **Leaf blades** 50–90 × 10–25(–40) mm. **Peduncles** 4–8+ cm. **Phyllaries:** outer ± oblong to ovate, 2–4+ mm, inner ovate-oblong to ovate-attenuate, 7–9 mm; all ± stipitate-glandular abaxially. **Ray laminae** 10–15+ mm. **Disc corollas** 5–6 mm, tubes stipitate-glandular. **Cypselae** 3–4 mm; **pappus bristles** 4–5 mm.

Flowering Aug–Oct. Shaded places, limestone outcrops; 70–200 m; Ala.

363. PEUCEPHYLLUM   A. Gray in W. H. Emory, Rep. U.S. Mex. Bound. 2(1): 74. 1859

• [Greek *peuke*, pine or fir, and *phyllon*, leaf]

John L. Strother

Shrubs or treelets, mostly 100–300 cm. **Stems** (1–5+), branched. **Leaves** cauline; alternate; sessile; blades linear-filiform, rarely with 1–2 lateral lobes, ultimate margins entire, faces glabrous, gland-dotted. **Heads** discoid, borne singly. **Involucres** turbinate to campanulate, 6–12 mm diam. **Phyllaries** persistent or tardily falling, 8–18 in ± 2 series (linear to lanceolate, outer intergrading with subtending leaves). **Receptacles** flat, pitted or knobby, epaleate. **Ray florets** 0. **Disc florets** 12–21, bisexual, fertile; corollas creamy yellow, distally purplish (stipitate-glandular), tubes shorter than cylindric throats, lobes 5, erect, deltate (style-branch appendages rounded-truncate, papillate). **Cypselae** obconic to obpyramidal, hirsute (hairs tawny to reddish); **pappi** persistent, of 30–60 bristles subtending 15–20 subulate-aristate scales, or of ca. 120 bristles. *x* = 10.

Species 1: sw United States, nw Mexico.

1. Peucephyllum schottii   A. Gray in W. H. Emory, Rep.
   U.S. Mex. Bound. 2(1): 74. 1859

Leaves 5–20(–50) × 1–2 mm. **Peduncles** 8–25 mm, glabrous or stipitate-glandular. **Disc corollas** 6.5–8.5 mm. **Cypselae** 3.2–4 mm; **pappus elements** 2–6 mm. *2n* = 40.

Flowering spring (following rains). Desert scrubs, soils from granitics, limestones, sandstones, volcanics; -50–1400 m; Ariz., Calif., Nev., Utah; Mexico (Baja California, Baja California Sur, Sonora).

364. BARTLETTIA   A. Gray, Pl. Nov. Thurb., 323. 1854 • [For John Russell Bartlett, 1805– 1886, United States Commissioner of the U.S.-Mexico Boundary Survey]

John L. Strother

Annuals, mostly 5–25 cm. **Stems** erect to spreading, branched from bases. **Leaves** mostly basal; subopposite or alternate; petiolate; blades ± deltate to elliptic, sometimes 3-lobed, ultimate margins ± serrate, faces sparsely hirsute or glabrate. **Heads** radiate, borne singly. **Involucres** turbinate to hemispheric, mostly 7–9 mm diam. **Receptacles** convex, knobby, epaleate. **Phyllaries** 8–22 in 2–3 series (lanceolate to linear, unequal, glandular-puberulent and sparsely strigose). **Ray florets** 5–13, pistillate, fertile; corollas golden yellow. **Disc florets** 30–90, bisexual, fertile; corollas yellow, sometimes suffused with purple, tubes (slender) longer than funnelform throats, lobes 5, spreading, deltate (style-branch apices truncate, penicillate). **Cypselae** obcompressed (each shed with subtending, linear, membranous scale, scales sometimes interpreted as paleae), margins ciliate; **pappi** fragile, of 17–22 bristles. *x* = 11.

Species 1: sw United States, Mexico.

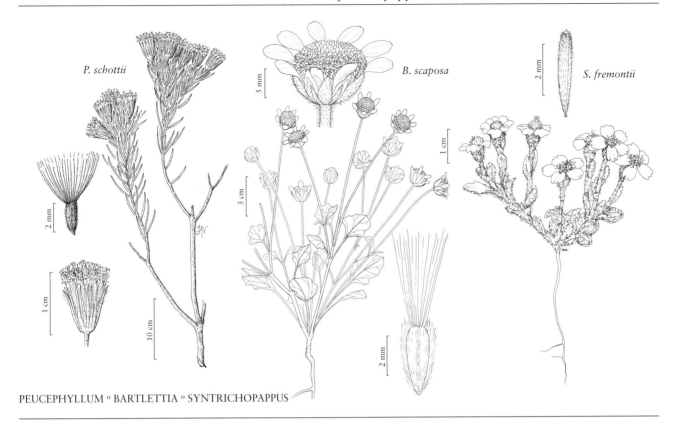

P. schottii   *B. scaposa*   *S. fremontii*

PEUCEPHYLLUM ° BARTLETTIA ° SYNTRICHOPAPPUS

1. **Bartlettia scaposa** A. Gray, Pl. Nov. Thurb., 323. 1854

Leaf blades 8–25 × 9–23 mm. **Peduncles** scapiform, 3–20 cm. **Ray corollas:** tubes 3–3.5 mm, laminae 7–8 × 3–4 mm. **Disc corollas** 4–7.5 mm. **Cypselae** 2.5–4 mm; **pappus bristles** 4–7 mm. $2n = 22$.

Flowering summer–fall, following rains. Gravelly or adobe soils, grasslands, playas; 1000–1900 m; N.Mex., Tex.; Mexico.

365. **SYNTRICHOPAPPUS** A. Gray in War Department [U.S.], Pacif. Railr. Rep. 4(5): 106, plate 15. 1857 • Fremont's-gold [Greek *syn,* united, *trichos,* hair, and *pappos,* pappus]

Dale E. Johnson

Annuals, 1–10 cm. **Stems** decumbent or erect, branched from bases or throughout. **Leaves** mostly cauline; mostly alternate (proximal opposite); petiolate or sessile; blades either linear and margins entire, or narrowly cuneate to spatulate and margins usually distally 3-lobed, faces sparsely tomentose to moderately woolly or glabrescent. **Heads** radiate, borne singly or in loose, corymbiform arrays. **Involucres** subcylindric, 2–5+ mm diam. **Phyllaries** persistent, 5–8 in 1 series (spreading, not reflexed in fruit, distinct, oblanceolate, subequal, alternate ones scarious-margined, apices acute). **Receptacles** convex, ± knobby or smooth, epaleate. **Ray**

florets 5–8, pistillate, fertile; corollas either yellow, or white or pinkish with reddish veins. **Disc florets** 10–20+, bisexual, fertile; corollas yellow, tubes shorter than narrowly funnelform throats, lobes 5, ovate-deltate (glabrous; anther appendages narrowly deltate, not glandular; style appendages narrowly triangular). **Cypselae** narrowly obconic, clavate, or fusiform, glabrous or sparsely to densely hispidulous; **pappi** 0, or ± fragile (falling in whole or fragmented rings), of 20–40 basally connate or coherent, ± barbellate bristles or subulate scales. $x = 7$.

Species 2 (2 in the flora): w United States, nw Mexico.

B. G. Baldwin and B. L. Wessa (2000) found that *Syntrichopappus lemmonii* nests within a clade of *Eriophyllum* and *Pseudobahia* species; circumscriptions of these genera are likely to change.

1. Ray laminae yellow . . . . 1. *Syntrichopappus fremontii*
1. Ray laminae white or pinkish with reddish veins . . . . 2. *Syntrichopappus lemmonii*

1. **Syntrichopappus fremontii** A. Gray in War Department [U.S.], Pacif. Railr. Rep. 4(5): 106, plate 15. 1857 • Yellowray Fremont's-gold, yellow syntrichopappus F

Stems decumbent or erect. **Leaves:** blades narrowly cuneate to spatulate, 5–20 mm, margins usually distally 3-lobed, sometimes entire (revolute). **Peduncles** 3–25 mm. **Involucres** 5–7 mm. **Phyllaries** ± 5. **Rays** yellow, laminae 3–5 mm. **Cypselae** 2–3.5 mm, sparsely to densely hispidulous; **pappi** of 30–40 bristles ± 2 mm. $2n = 12$.

Flowering Mar–Jun. Sandy or gravelly openings in deserts, creosote-bush or sagebrush scrublands, or Joshua Tree or pinyon-juniper woodlands; 600–2500 m; Ariz., Calif., Nev., Utah; Mexico (Baja California).

2. **Syntrichopappus lemmonii** (A. Gray) A. Gray, Proc. Amer. Acad. Arts 19: 20. 1883 (as lemmoni) • Pinkray Fremont's-gold, Lemmon's syntrichopappus E

Actinolepis lemmonii A. Gray, Notes Compositae, 101. 1880 (as lemmoni)

Stems erect. **Leaves:** blades linear, 3–8 mm, margins entire. **Peduncles** 3–5 mm. **Involucres** 4–5 mm. **Phyllaries** 5–8. **Rays** white or pinkish with reddish veins, laminae 2–3 mm. **Cypselae** 2–2.5 mm, glabrous or sparsely hispidulous; **pappi** 0 or of 20–30 bristles ± 1 mm. $2n = 14$.

Flowering Apr–May. Sandy or gravelly openings in chaparral or Joshua Tree or pinyon-juniper woodlands; 900–1500 m; Calif.

366. **PSATHYROTOPSIS** Rydberg in N. L. Britton et al., N. Amer. Fl. 34: 360. 1927 • [Genus *Psathyrotes* and Greek *-opsis*, appearance]

John L. Strother

Pseudobartlettia Rydberg

Perennials, 8–40 cm (sometimes functionally annual; caudices often ± woody). **Stems** 1–3+ (scapiform), erect, branched from bases or throughout. **Leaves** mostly basal; alternate; petiolate; blades rounded-deltate, lanceolate, or rhombic, margins crenate, serrate, or subentire, faces ± tomentose and stipitate-glandular. **Heads** discoid, borne singly or in corymbiform arrays. **Involucres** campanulate to hemispheric, 4–6(–10) mm diam. **Phyllaries** persistent, 8–21 in 2–3 series (lanceolate to linear, subequal, sparsely tomentulose and stipitate-glandular). **Receptacles** flat, obscurely pitted or knobby, epaleate. **Ray florets** 0. **Disc florets** 16–50, bisexual, fertile; corollas ochroleucous to bright yellow, (cylindro-funnelform) tubes much shorter than throats, lobes 5, (erect or spreading) deltate (style-branch appendages subulate to conic, shaggy-papillate). **Cypselae** narrowly obconic; **pappi** persistent, of 70–90 unequal bristles in 2–3 series. $x = 19$.

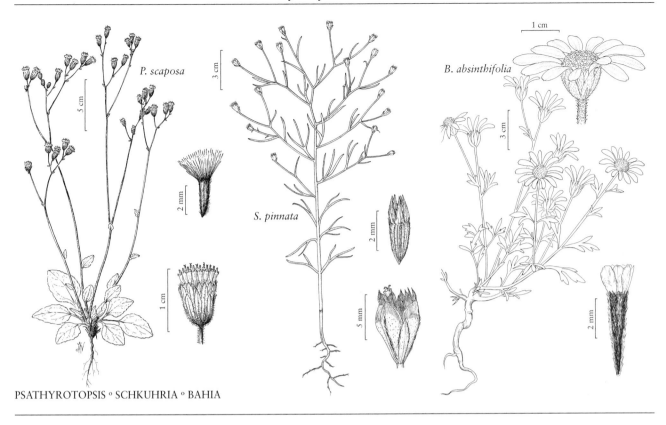

PSATHYROTOPSIS ° SCHKUHRIA ° BAHIA

Species 3 (1 in the flora): sw United States, Mexico.

SELECTED REFERENCE Strother, J. L. and G. E. Pilz. 1975. Taxonomy of *Psathyrotes* (Compositae: Senecioneae). Madroño 23: 24–40.

1. Psathyrotopsis scaposa (A. Gray) H. Robinson, Phytologia 45: 459. 1980

Psathyrotes scaposa A. Gray, Smithsonian Contr. Knowl. 5(6): 100, plate 13. 1853; *Pseudobartlettia scaposa* (A. Gray) Rydberg

Leaf blades 1–6 × 1–4 cm. **Peduncles** (3–)5–20 mm, stipitate-glandular (at least distally). **Disc corollas** ca. 5 mm. **Cypselae** 2.5–3.5 mm; **pappus bristles** 1–3 mm. $2n = 38$.

Flowering mostly spring and fall (following rains).

Gypseous and calcareous soils in creosote bush scrublands; 700–1300 m; N.Mex., Tex.; Mexico (Chihuahua).

367. SCHKUHRIA Roth, Catal. Bot. 1: 116. 1797, name conserved • [For Christian Schkuhr, 1741–1811, German botanist]

John L. Strother

Annuals, 5–50+ cm. **Stems** erect, simple or branched (usually distally). **Leaves** mostly cauline; mostly opposite (distal sometimes alternate); petiolate or sessile; blades linear to filiform or 1–2-pinnately lobed (lobes 3–7+, linear to filiform), ultimate margins entire, faces sparsely hairy (hairs bulbous, conic, or fusiform, 0.1–0.3 mm), glabrescent, usually gland-dotted. **Heads** radiate or discoid, usually in loose, corymbiform to paniculiform arrays. **Involucres** obconic-obpyramidal to ± turbinate or hemispheric, 3–6 mm diam. (subtended by 0–3 bractlets). **Phyllaries** persistent, 4–6+ in 1–2 series (spreading or reflexed in fruit, distinct, obovate to oblanceolate, subequal, thin-herbaceous to membranous, margins membranous, often purplish or yellowish). **Receptacles** convex, pitted, epaleate. **Ray florets** 0 or 1–2, pistillate, fertile; corollas yellow or white. **Disc florets** 2–8(–30+), bisexual, fertile; corollas yellow to yellowish, sometimes red-tipped (± stipitate-glandular), tubes shorter than or about equaling campanulate to funnelform throats, lobes 5, deltate. **Cypselae** narrowly obpyramidal, 4-angled, finely nerved, ± hirsutulous to villous (at least at bases and/or on angles); **pappi** persistent, of 8+ spatulate to lanceolate (basally and/or medially thickened, distally and/or laterally scarious) scales (all, some, none, or the alternate ones ± aristate). $x = 11$.

Species 2 or perhaps more (2 in the flora): sw United States, Mexico, Central America, South America.

According to B. G. Baldwin et al. (2002), among others, *Schkuhria* is closely related to *Achyropappus* Kunth and to *Bahia*.

SELECTED REFERENCE   Heiser, C. B. 1945. A revision of the genus *Schkuhria*. Ann. Missouri Bot. Gard. 32: 265–278.

1.   Phyllaries ± hirsutulous and gland-dotted; ray florets 0; disc florets 15–30 . . . . 1. *Schkuhria multiflora*
1.   Phyllaries gland-dotted, otherwise usually glabrous; ray florets usually 1–2, sometimes 0; disc florets 2–6(–8+) . . . . 2. *Schkuhria pinnata*

1. Schkuhria multiflora Hooker & Arnott, J. Bot. (Hooker) 3: 322. 1841

Bahia neomexicana (A. Gray) A. Gray; *Schkuhria neomexicana* A. Gray

Plants mostly 3–12(–25+) cm. **Stems** ± decumbent-ascending to erect. **Leaves** mostly 1–3 cm; blades linear or lobed (lobes 3–7+, linear to filiform), faces puberulent and gland-dotted. **Peduncles** mostly 5–25 mm. **Involucres** ± obconic, 5–6 mm. **Phyllaries** 7–10+, green to purple, weakly carinate, oblanceolate to obovate, ± hirsutulous and gland-dotted. **Ray florets** 0. **Disc florets** 15–30+; corollas yellowish, 1–2 mm. **Cypselae** blackish to buff, 3 mm, hispidulous to villous, especially on angles and/or at bases; **pappi** of 8 white to tawny or purplish, obovate-rounded or oblanceolate to ± quadrate, apically truncate or acute scales 1–2 mm. $2n = 22$.

Flowering Aug–Oct. Roadsides, sandy slopes, and washes; 1600–2500 m; Ariz., Calif., Colo., N.Mex., Tex.; Mexico (Chihuahua); South America.

2. Schkuhria pinnata (Lamarck) Kuntze ex Thellung, Repert. Spec. Nov. Regni Veg. 11: 308. 1912   F

Pectis pinnata Lamarck, J. Hist. Nat. 2: 150, plate 31. 1792; *Schkuhria anthemoidea* (de Candolle) J. M. Coulter; *S. anthemoidea* var. *wislizeni* (A. Gray) Heiser; *S. anthemoidea* var. *wrightii* (A. Gray) Heiser; *S. virgata* (La Llave & Lexarza) de Candolle; *S. wislizeni* A. Gray; *S. wislizeni* var. *frustrata* S. F. Blake; *S. wislizeni* var. *wrightii* (A. Gray) S. F. Blake

Plants (10–)25–40(–70+) cm. **Stems** usually strictly erect. **Leaves** mostly 10–25(–40) mm; blades linear or lobed (lobes 3–7+, linear to filiform), faces scaberulous (hairs ± conic, 0.1 mm) and gland-dotted. **Peduncles** mostly (3–)8–25(–30+) mm. **Involucres** obconic or obpyramidal, 4–6+ mm (each often subtended by 1–3, lanceolate bractlets). **Phyllaries** 4–6, green to purple, weakly carinate, oblanceolate to obovate, gland-dotted, otherwise usually glabrous. **Ray florets** usually 1(–2), sometimes 0; corollas yellow to white, laminae 0.8–1.2+ mm. **Disc florets** 2–6(–8+); corollas yellow (sometimes with purple), 1.5–2 mm. **Cypselae** blackish to buff, 3–4 mm, hirsutulous to villous, especially on angles; **pappi** of 8 white to tawny or purplish, obovate-rounded or elliptic to lanceolate scales 1–2.5 mm (sometimes some or all ± aristate). $2n = 20, 22, 40$.

Flowering Sep–Oct. Roadsides, pastures, wooded slopes; 1500–2100 m; Ariz., N.Mex., Tex.; Mexico; Central America; South America.

Schkuhria pinnata has been noted as persisting after plantings in Maine. *Schkuhria pinnata* (Lamarck) Kuntze var. *wislizeni* (A. Gray) B. L. Turner is a superfluous, illegitimate, name intended to refer to North American plants of *S. pinnata*.

SELECTED REFERENCE Turner, B. L. 1996. Taxonomy and nomenclature of *Schkuhria pinnata* (Asteraceae, Helenieae). Phytologia 79: 364–368.

368. BAHIA Lagasca, Gen. Sp. Pl., 30. 1816 • [For J. F. Bahí, 1775–1841, professor of botany at Barcelona]

John L. Strother

Annuals, biennials, or perennials, 10–80+ cm. **Stems** erect or spreading, branched from bases or throughout. **Leaves** mostly cauline; all or mostly opposite or all or mostly alternate; petiolate or sessile; blades usually 1–2-ternately lobed (blades or lobes mostly filiform, lanceolate, linar, oblanceolate, oblong, or ovate), ultimate margins toothed or entire, faces sparsely to densely hairy (hairs white, straight, conic or fusiform, 0.1–0.3 or 0.3–0.8 mm), often gland-dotted as well. **Heads** radiate, borne singly or in loose, corymbiform arrays. **Involucres** ± hemispheric or broader, 6–14+ mm diam. **Phyllaries** persistent, 8–18+ in ± 2 series (reflexed in fruit, distinct, subequal or outer smaller, mostly lanceolate or oblanceolate, thin-herbaceous, margins membranous, rarely purplish). **Receptacles** convex, smooth or knobby, epaleate. **Ray florets** 5–15, pistillate, fertile; corollas yellow (sometimes pale). **Disc florets** 25–120+, bisexual, fertile; corollas yellow (hairy), tubes shorter than or about equaling cylindric or campanulate to funnelform throats, lobes 5, deltate to lance-ovate. **Cypselae** obpyramidal, 4-angled, ± hirtellous to ± sericeous; **pappi** persistent, of 6–12 distinct, spatulate or oblanceolate to ovate or quadrate, basally and/or medially thickened, distally and/or laterally scarious scales in ± 1 series (apices usually muticous, sometimes some or all ± aristate). *x* = 12.

Species ca. 10 (4 in the flora): sw United States, Mexico, South America.

Bahia dissecta is treated here in *Amauriopsis*; *B. oppositifolia* and *B. woodhousei* are treated in *Picradeniopsis*; *B. nudicaulis* is treated in *Platyschkuhria*. These departures from the treatment of *Bahia* in a broad sense by W. L. Ellison (1964) are consistent with findings by B. G. Baldwin et al. (2002).

SELECTED REFERENCE Ellison, W. L. 1964. A systematic study of the genus *Bahia* (Compositae). Rhodora 66: 67–86; 177–215; 281–311.

1. Leaves all or mostly opposite (distal sometimes alternate).
    2. Annuals; leaf lobes filiform to linear, 0.5–1(–2+) mm wide, faces sparsely scabrellous; pappi 0.5–1 mm . . . . 1. *Bahia bigelovii*
    2. Perennials; leaf lobes lanceolate to oblong, 2–5(–20+) mm wide, faces densely scabrellocanescent; pappi 1–1.5 mm . . . . 2. *Bahia absinthifolia*
1. Leaves all or mostly alternate (proximal sometimes opposite).
    3. Leaf lobes filiform to linear, 0.5–1.5(–3+) mm wide; pappus scales (outer cypselae) ± ovate to quadrate, 0.5–1 mm, apices ± muticous or (innermost cypselae) lanceolate to lance-subulate, 1–2.5 mm, apices ± aristate . . . . 3. *Bahia biternata*
    3. Leaf lobes oblong or ovate to oblanceolate, (1–)2–6(–8+) mm wide; pappus scales (all cypselae) spatulate to oblanceolate, 1–1.5 mm, apices ± muticous. . . . . 4. *Bahia pedata*

1. Bahia bigelovii A. Gray in W. H. Emory, Rep. U.S. Mex. Bound. 2(1): 96. 1859 [E]

Annuals, 10–30+ cm. **Stems** mostly erect. **Leaves** all or mostly opposite (distal sometimes alternate); blades simple or 1–2-ternately lobed, lobes filiform to linear, 10–25(–50+) × 0.5–1(–2+) mm, faces sparsely scabrellous, usually gland-dotted as well. **Involucres** 3–4+ × 7–9+ mm. **Ray florets** 5–9(–13+); corolla laminae 4–5+ mm. **Disc florets** 25–50+; corollas 1.5–3 mm. **Cypselae** 2–3.5+ mm, faces ± hirtellous; **pappi** of ± ovate to quadrate, apically ± muticous scales 0.5–1 mm. $2n = 22$.

Flowering Jun–Oct. Sandy soils; 1500–1900 m; Tex.

2. Bahia absinthifolia Bentham, Pl. Hartw., 18. 1839

Bahia dealbata A. Gray; *B. absinthifolia* var. *dealbata* (A. Gray) A. Gray

Perennials, 10–40 cm. **Stems** spreading to erect. **Leaves** all or mostly opposite (distal sometimes alternate); blades simple or 1–2-ternately lobed, lobes lanceolate to oblong, 10–25(–55+) × 2–5(–20+) mm, faces ± densely scabrello-canescent, usually gland-dotted as well. **Involucres** 5–7+ × 9–14+ mm. **Ray florets** 8–13+; corolla laminae 6–15+ mm. **Disc florets** 60–80 (–120+); corollas 3–4 mm. **Cypselae** 3–4.5+ mm, faces hirtellous to ± strigose; **pappi** of ± spatulate to obovate, apically ± muticous scales 1–1.5 mm. $2n = 24, 48$.

Flowering Mar–Nov. Sandy soils with mesquite or desert scrubs, calcareous places; 600–1700 m; Ariz., N.Mex., Tex.; Mexico.

3. Bahia biternata A. Gray, Smithsonian Contr. Knowl. 5(6): 95. 1853 [E]

Annuals, 8–40+ cm. **Stems** mostly erect. **Leaves** all or mostly alternate (proximal sometimes opposite); blades simple or 1–2-ternately lobed, lobes filiform to linear, 5–12(–25+) × 0.5–1.5(–3+) mm, faces sparsely scabrellous, usually gland-dotted as well. **Involucres** 3–4+ × 6–9+ mm. **Ray florets** 8–13+; corolla laminae 3–4(–10+) mm. **Disc florets** 30–60+; corollas 2.5–3.5 mm. **Cypselae** 2–3+ mm, faces ± hirtellous; **pappi** of (outer cypselae) ± ovate to quadrate, apically ± muticous scales 0.5–1 mm or (innermost cypselae) lanceolate to lance-subulate, apically ± aristate scales 1–2.5 mm.

Flowering (May–)Sep–Oct. Granite outcrops; 1200–1600 m; Ariz., N.Mex.

4. Bahia pedata A. Gray, Smithsonian Contr. Knowl. 3(5): 123. 1852

Annuals or biennials, 15–70 (–120+) cm. **Stems** mostly erect. **Leaves** all or mostly alternate (proximalmost sometimes opposite); blades simple or 1–2-ternately lobed, lobes ovate or obovate to lanceolate or oblanceolate, 2–25 × 1–6(–8) mm, faces sparsely scabrellous, usually gland-dotted as well. **Involucres** 4–6+ × 8–12+ mm. **Ray florets** 10–15; corolla laminae 5–6(–10+) mm. **Disc florets** 40–80+; corollas 3–3.5 mm. **Cypselae** 2.5–3.5+ mm, faces ± hirtellous; **pappi** of ± spatulate to oblanceolate, apically ± muticous scales 1–1.5 mm. $2n = 24$.

Flowering (May–)Jul–Aug(–Nov). Sandy soils, limestone, openings in grasslands and pinyon-juniper woodlands; 900–2000 m; N.Mex., Tex.; Mexico (Chihuahua, Coahuila).

369. PICRADENIOPSIS Rydberg ex Britton, Man. Fl. N. States, 1008. 1901 • [Generic name *Picradenia* and Greek -*opsis*, resembling] [E]

John L. Strother

Perennials, 3–20+ cm (± rhizomatous, sometimes forming colonies). **Stems** ± erect or spreading, branched ± throughout. **Leaves** cauline; all or mostly opposite; petiolate or sessile; blades mostly ternately lobed (blades or lobes lanceolate to lance-linear), ultimate margins entire, faces sparsely to densely scabrellous (hairs white, straight, conic or fusiform, 0.1–0.4 mm) and gland-dotted. **Heads** radiate, borne singly or (3–6) in loose, corymbiform arrays. **Involucres** turbinate or obconic to ± hemispheric, 5–6+ mm diam. **Phyllaries** persistent, 6–10 in 1–2 series (reflexed in fruit, distinct, oblanceolate, subequal, herbaceous, margins membranous, not purplish). **Receptacles** ± convex, ± pitted, epaleate. **Ray florets** 3–8, pistillate, fertile; corollas usually yellow, sometimes ochroleucous. **Disc florets** 20–40+, bisexual, fertile; corollas yellow to orange (gland-dotted), tubes about equaling funnelform to campanulate throats, lobes 5, deltate to lance-deltate. **Cypselae** narrowly obpyramidal, 4-angled, finely nerved, shaggily hairy (at least at bases) and/or gland-dotted; **pappi** persistent, of 8–10 (distinct) ovate or elliptic to lanceolate or linear-subulate (basally and/or medially thickened, distally and/or laterally scarious) scales in 1 series (weakly, if at all, aristate). *x* = 12.

Species 2 (2 in the flora): c, w United States.

According to T. F. Stuessy et al. (1973), among others, *Picradeniopsis* is closely allied to *Bahia*. Indeed, the species are often treated as members of *Bahia*. Where they occur together, the two species of *Picradeniopsis* may produce more or less sterile hybrids.

SELECTED REFERENCE Stuessy, T. F., R. S. Irving, and W. L. Ellison. 1973. Hybridization and evolution in *Picradeniopsis* (Compositae). Brittonia 25: 40–56.

1. Cypselae usually gland-dotted, seldom hirsutulous; scales of pappi usually ovate or elliptic to obovate, sometimes lanceolate . . . . 1. *Picradeniopsis oppositifolia*
1. Cypselae seldom gland-dotted, usually hirsutulous; scales of pappi usually lanceolate to linear-subulate . . . . 2. *Picradeniopsis woodhousei*

1. Picradeniopsis oppositifolia (Nuttall) Rydberg in N. L. Britton, Man. Fl. N. States, 1008. 1901 E F

Trichophyllum oppositifolium Nuttall, Gen. N. Amer. Pl. 2: 167. 1818; *Bahia oppositifolia* (Nuttall) de Candolle

Plants ± erect or spreading, 3–15 (–20+) cm. **Leaf lobes** lanceolate to linear, 10–25+ × 1–3(–8) mm, faces ± canescent-scabrellous and gland-dotted. **Involucres** broadly turbinate to hemispheric, 5–6+ × 6–9+ mm. **Ray florets** 3–5(–6); corollas pale yellow, laminae 3–5 mm. **Disc florets** 30–60+; corollas 3.5–5 mm. **Cypselae** 3–5 mm, usually gland-dotted, seldom hirsutulous; **pappus scales** usually ovate or elliptic to obovate, sometimes lanceolate, 0.5–1.5 mm. *2n* = 48.

Flowering Jun–Oct. Roadsides, saline flats, shale sites; 900–2500 m; Ariz., Colo., Kans., Mont., Nebr., N.Mex., N.Dak., Okla., S.Dak., Tex., Wyo.

2. Picradeniopsis woodhousei (A. Gray) Rydberg, Bull. Torrey Bot. Club 37: 333. 1910 E

Achyropappus woodhousei A. Gray, Proc. Amer. Acad. Arts 6: 546. 1866; *Bahia woodhousei* (A. Gray) A. Gray

Plants ± erect, 3–15(–18+) cm. **Leaf lobes** narrowly lanceolate to linear, 8–20+ × 1–2(–5) mm, faces sparsely scabrellous and gland-dotted. **Involucres** ± turbinate or obconic to hemispheric, 5–6+ × 4–6+ mm. **Ray florets** 3–8; corollas pale yellow (sometimes ochroleucous), laminae 2–5 mm. **Disc florets** 20–35+; corollas 3–3.5 mm. **Cypselae** 3–4 mm, seldom gland-dotted, usually hirsutulous (at least at bases); **pappus scales** usually lanceolate to linear-subulate, 0.8–1.2(–2) mm. *2n* = 24.

Flowering May–Sep. Roadsides, etc., on sandy, silty, or loamy soils of grasslands, plains, or prairies; 1000–1700 m; Ariz., Colo., Kans., N.Mex., Okla., Tex.

370. VENEGASIA de Candolle in A. P. de Candolle and A. L. P. P. de Candolle, Prodr. 6: 43. 1838 • [For "rev. Patr. Mich. Venegas Hispani," 1680–1764, "qui primus notitiam naturalem civilemque circà Californiam scripsit"]

John L. Strother

Subshrubs or shrubs, 50–150(–250+) cm. **Stems** erect. **Leaves** cauline; opposite (proximal) or alternate (most); petiolate; blades (3-nerved) rounded-deltate or cordate to ovate, margins toothed or subentire, faces: abaxial puberulent and/or gland-dotted. **Heads** radiate, borne

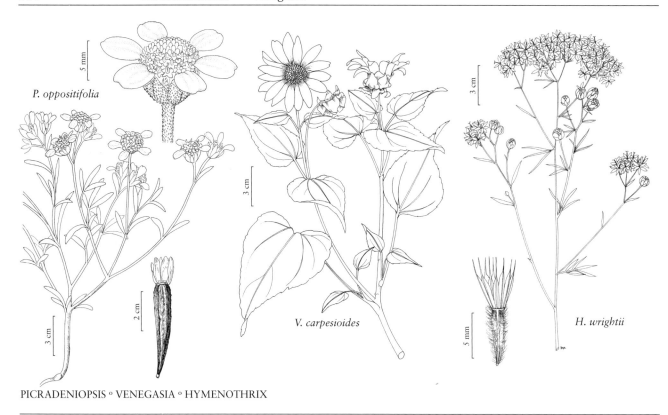

P. oppositifolia

V. carpesioides

H. wrightii

PICRADENIOPSIS ∘ VENEGASIA ∘ HYMENOTHRIX

singly or in ± corymbiform arrays. **Involucres** hemispheric to globose, 12–25+ mm diam. **Phyllaries** persistent (all or outer), 20–40+ in 3–4+ series (at least outer reflexed in fruit, outer rotund to oblong, herbaceous-foliaceous, inner oblanceolate, membranous or scarious, basally indurate). **Receptacles** flat or convex, pitted (pubescent), epaleate. **Ray florets** 12–34, pistillate, fertile; corollas yellow. **Disc florets** 35–150+; bisexual, fertile; corollas yellow, tubes (densely glandular-hairy at bases) shorter than campanulate to narrowly urceolate throats, lobes 5, ovate-deltate. **Cypselae** clavate or fusiform, sometimes arcuate, obscurely 4–5-angled, ca. 20-nerved, glabrous, minutely papillose; **pappi** 0. $x = 19$.

Species 1: California, nw Mexico.

SELECTED REFERENCE Turner, B. L. and D. Zippin. 1992. Taxonomic study of *Venegasia* (Asteraceae: Helenieae). Sida 15: 223–229.

1. **Venegasia carpesioides** de Candolle in A. P. de Candolle and A. L. P. P. de Candolle, Prodr. 6: 43. 1838

   F

Leaf blades mostly 3–12+ × 2–10+ cm. **Ray laminae** 15–30 mm. **Disc corollas** 5–6 mm. **Cypselae** 2–3 mm. **2n** = 38.

Flowering spring. Usually in shade in canyons and along streams; 0–1000 m; Calif.; Mexico (Baja California).

371. HYMENOTHRIX A. Gray, Mem. Amer. Acad. Arts, n. s. 4: 102. 1849 • [Greek *hymen*, membrane, and *thrix*, hair, possibly alluding to scarious-aristate pappus scales]

John L. Strother

Annuals, biennials, or perennials, 10–70[–150] cm. **Stems** erect, branched. **Leaves** basal and cauline; mostly alternate (proximal opposite); petiolate; blades 2-ternately [pinnately] lobed, ultimate margins entire, faces usually ± scabrellous (hairs white, antrorse, 0.1–0.5 mm), sometimes glabrate, sometimes gland-dotted. **Heads** radiate or discoid, in corymbiform to paniculiform arrays. **Involucres** obconic to hemispheric, 4–8+ mm diam. **Phyllaries** persistent, 8–16 in 2–3 series (reflexed in fruit, obovate or oblanceolate to linear, subequal, sometimes subtended by 1–3 bractlets, thin-herbaceous to membranous, margins scarious, often tinged with purple). **Receptacles** flat or convex, knobby or smooth, epaleate. **Ray florets** 0, or 3–8+, pistillate, fertile; corollas yellowish. **Disc florets** 10–30+, bisexual, fertile; corollas yellowish or whitish or pinkish to purplish (sparsely glandular-hairy), tubes about equaling cylindric throats or much longer than broadly funnelform throats, lobes 5, lance-oblong or lance-linear. **Cypselae** obpyramidal, 4–5-angled, faces ribbed and sparsely hairy (hairs straight); **pappi** persistent, of 12–18 narrowly lanceolate to subulate, medially thickened, scarious-margined scales in 1 series, some or all ± aristate. $x = 12$.

Species 5 (3 in the flora): sw United States, Mexico.

SELECTED REFERENCE Turner, B. L. 1962. Taxonomy of *Hymenothrix* (Helenieae, Compositae). Brittonia 14: 101–120.

1. Phyllaries (8–)12–16, mostly oblong to ovate or obovate; corollas white or pinkish to purplish, lobe lengths 2–3+ times throats; anthers pinkish to purplish . . . . 3. *Hymenothrix wrightii*
1. Phyllaries 8(–12), mostly linear to narrowly oblanceolate; corollas creamy to bright yellow, lobe lengths 0.3–1 times throats; anthers yellowish.
  2. Rays 3–8; corollas of disc florets mostly 5–7 mm . . . . 1. *Hymenothrix wislizeni*
  2. Rays 0; corollas of disc florets mostly 4–4.5 mm . . . . 2. *Hymenothrix loomisii*

1. Hymenothrix wislizeni A. Gray, Mem. Amer. Acad. Arts, n. s. 4: 102. 1849

Annuals, 30–70 cm. **Phyllaries** 8 (–12), mostly linear to narrowly oblanceolate. **Rays** 3–8. **Disc florets** 15–30; corollas creamy to bright yellow, mostly 5–7 mm, lobe lengths 0.3–1 times throats; anthers yellowish. **Cypselae** 3–5 mm; pappus scales 1.5–6 mm. $2n = 24, 48$.

Flowering summer–fall. Open slopes and washes, sandy or gravelly soils; 600–1500 m; Ariz., N.Mex.; Mexico (Chihuahua, Sonora).

2. Hymenothrix loomisii S. F. Blake, Proc. Biol. Soc. Wash. 40: 49. 1927   E

Annuals or biennials, 30–70 cm. **Phyllaries** 8(–12), mostly linear to narrowly oblanceolate. **Rays** 0. **Disc florets** 10–20; corollas creamy to bright yellow, mostly 4–4.5 mm, lobe lengths 0.3–1 times throats; anthers yellowish. **Cypselae** 2.5–4+ mm; **pappus scales** 3–5 mm. $2n = 24$.

Flowering summer–fall. Open, often disturbed, places; 900–2100 m; Ariz., Calif., Nev.

Occurrences of *Hymenothrix loomisii* in California have been geographically scattered, sporadic, and non-persistent introductions.

3. Hymenothrix wrightii   A. Gray, Smithsonian Contr. Knowl. 5(6): 97. 1853   F

Annuals or perennials, 3–6 dm. **Phyllaries** (8–)12–16, mostly oblong to ovate or obovate. **Rays** 0. **Disc florets** 15–30; corollas white or pinkish to purplish, 5–6 mm, lobe lengths 2–3+ times throats; anthers pinkish to purplish. **Cypselae** 4–5 mm; **pappus** scales 4–6 mm. $2n$ = 24.

Flowering mostly late summer–fall. Rocky places, often with pinyons and junipers; 900–2300 m; Ariz., Calif., N.Mex., Tex.; Mexico (Baja California, Chihuahua, Sonora).

372. PALAFOXIA   Lagasca, Gen. Sp. Pl., 26. 1816   •   [For General José Palafox, 1776–1847, Spanish patriot]

John L. Strother

Annuals or perennials, 10–80(–150+) cm. **Stems** erect, branched (mostly distally). **Leaves** cauline; opposite (proximal) or alternate (mostly); petiolate or sessile; blades (1- or 3-nerved) broadly lanceolate to linear, margins entire, faces sparsely to moderately scabrellous to strigillose or appressed-puberulent (hairs 0.1–0.8+ mm), not gland-dotted. **Heads** radiate, discoid, disciform, or radiant, in ± corymbiform arrays. **Involucres** narrowly cylindric or turbinate to hemispheric, 3–15+ mm diam. **Phyllaries** persistent or falling (in fruit), 8–14 in 2–3 series (erect in flowering, reflexed in fruit, purplish, or whitish, linear to oblanceolate, subequal or unequal, herbaceous to membranous). **Ray florets** 0, or 3–13, pistillate, fertile; corollas white or pinkish to purplish. **Disc florets** 12–40(–90), bisexual, fertile; corollas pinkish to purplish or whitish (glabrous or sparsely hairy, sometimes stipitate-glandular), tubes shorter than cylindric throats or much longer than short, broadly funnelform throats, lobes 5, lance-linear or linear (equal or unequal). **Receptacles** flat, pitted or knobby, epaleate. **Cypselae** obpyramidal, 4-angled, densely to sparsely hairy (hairs straight); **pappi** usually persistent, of 4–10 rotund and callous-thickened, or spatulate to subulate, medially thickened, laterally scarious scales in 1–2 series (all, some, or none aristate, all alike or pappi of outer cypselae shorter, rarely 0). x = 12.

Species 12 (10 in the flora): s United States, Mexico.

Circumscriptions of species adopted here closely follow those of B. L. Turner and M. I. Morris (1976). Some of their circumscriptions, especially those of *Palafoxia callosa*, *P. rosea*, and *P. texana*, merit reconsideration and possibly different interpretations. Their infraspecific taxa are not formally recognized here.

SELECTED REFERENCE   Turner, B. L. and M. I. Morris. 1976. Systematics of *Palafoxia* (Asteraceae: Helenieae). Rhodora 78: 567–628.

1. Ray florets 3–13.
  2. Leaf blades narrowly lanceolate to linear, 2–6 mm wide; phyllaries 6–8 mm; cypselae 5–6 mm. . . . . 3. *Palafoxia reverchonii*
  2. Leaf blades broadly to narrowly lanceolate, 3–25 mm wide; phyllaries 7–16+ mm; cypselae 6–9 mm.
    3. Stems usually stipitate-glandular ± throughout; phyllaries mostly 3–5 mm wide; ray corollas 10–20 mm, laminae 9–14 mm. . . . . 1. *Palafoxia hookeriana*
    3. Stems stipitate-glandular only distally (mainly on peduncles), if at all; phyllaries mostly 1.5–3 mm wide; ray corollas 15–25 mm, laminae 5–8(–12) mm . . . . 2. *Palafoxia sphacelata*

1. Ray florets 0 (peripheral florets sometimes with enlarged, ± zygomorphic corollas).
    4. Corolla throats ± cylindric, longer than lobes.
        5. Perennials, subshrubs, or shrubs; leaf blades narrowly elliptic to ovate; phyllaries 5–9 mm; cypselae 6–8 mm .... 4. *Palafoxia feayi*
        5. Annuals (rarely persisting); leaf blades narrowly lanceolate to linear; phyllaries 10–25 mm; cypselae 10–16 mm .... 5. *Palafoxia arida*
    4. Corolla throats ± funnelform, shorter than lobes.
        6. Perennials (rhizomatous); phyllaries unequal (outer shorter)............ 6. *Palafoxia integrifolia*
        6. Annuals (rarely persisting, becoming ± shrubby, not rhizomatous); phyllaries ± equal.
            7. Involucres ± cylindric; phyllaries 10–15 mm; corollas of outer florets ± zygomorphic .... 7. *Palafoxia riograndensis*
            7. Involucres ± turbinate to campanulate; phyllaries 3–10 mm; corollas of all florets ± actinomorphic.
                8. Phyllaries 3–5 mm; corollas 5–6 mm; pappus scales mostly 0.3–1 mm ... 8. *Palafoxia callosa*
                8. Phyllaries 5–10 mm; corollas 7–10 mm; pappus scales (1.5–)3–8 mm.
                    9. Leaf blades narrowly lanceolate to linear, 2–6(–12) mm wide; phyllaries usually stipitate-glandular .... 9. *Palafoxia rosea*
                    9. Leaf blades mostly lanceolate, (5–)10–20 mm wide; phyllaries rarely stipitate-glandular .... 10. *Palafoxia texana*

1. Palafoxia hookeriana Torrey & A. Gray, Fl. N. Amer. 2: 368. 1842 [E]

Annuals, 25–180 cm. **Stems** usually stipitate-glandular throughout. **Leaf blades** broadly to narrowly lanceolate, 50–100 × 3–25 mm. **Involucres** broadly turbinate. **Phyllaries** 7–16+ × 3–5 mm, ± equal, scabrous and/or stipitate-glandular. **Ray florets** 5–13; corollas 10–20 mm, laminae 9–14 mm. **Disc florets** 25–90; corollas ± actinomorphic, 10–12 mm, throats ± funnelform, shorter than lobes. **Cypselae** 6–9 mm; **pappus scales** of inner cypselae 5–8 mm. $2n = 24$.

Flowering late summer–fall. Sandy soils; 10–200+ m; Tex.

Plants of *Palafoxia hookeriana* with stems stipitate-glandular distally (versus throughout), phyllaries 7–12 (versus 10–16) mm, and florets 25–50 (versus 50–90) per head have been treated as var. *minor* Shinners.

2. Palafoxia sphacelata (Nuttall ex Torrey) Cory, Rhodora 48: 86. 1946 [F]

Stevia sphacelata Nuttall ex Torrey, Ann. Lyceum Nat. Hist. New York 2: 214. 1827

Annuals, 10–90 cm. **Stems** usually proximally hispid, distally stipitate-glandular. **Leaf blades** broadly to narrowly lanceolate, 30–90 × 3–20 mm. **Involucres** broadly to narrowly turbinate. **Phyllaries** 9–12+ × 1.5–3 mm, ± equal, ± scabrous to hispid and/or stipitate-

glandular. **Ray florets** 3–5; corollas 15–25 mm, laminae 5–12 mm. **Disc florets** 15–35; corollas ± actinomorphic, 10–14 mm, throats ± funnelform, shorter than lobes. **Cypselae** 6–9 mm; **pappus scales** of inner cypselae 7–9 mm. $2n = 24$.

Flowering summer–fall. Sandy soils; 1000–1800 m; Colo., Kans., Nebr., N.Mex., Okla., Tex.; Mexico (Chihuahua).

3. Palafoxia reverchonii (Bush) Cory, Rhodora, 48: 86. 1946 [E]

Othake reverchonii Bush, Trans. Acad. Sci. St. Louis 14: 180. 1904 (as reverchoni)

Annuals, 10–90 cm. **Stems** usually proximally appressed-pubescent, distally stipitate-glandular. **Leaf blades** narrowly lanceolate, 30–90 × 2–6 mm. **Involucres** turbinate. **Phyllaries** 6–8 × 1.5–3 mm, ± equal, ± scabrous to puberulous and stipitate-glandular. **Ray florets** 3–8; corollas 15–25 mm, laminae 6–12 mm. **Disc florets** 12–30; corollas ± actinomorphic, 5–8 mm, throats ± funnelform, shorter than lobes. **Cypselae** 5–6 mm; **pappus scales** of inner cypselae 3–6 mm. $2n = 24$.

Flowering late summer–fall. Sandy soils; 10–100+ m; Tex.

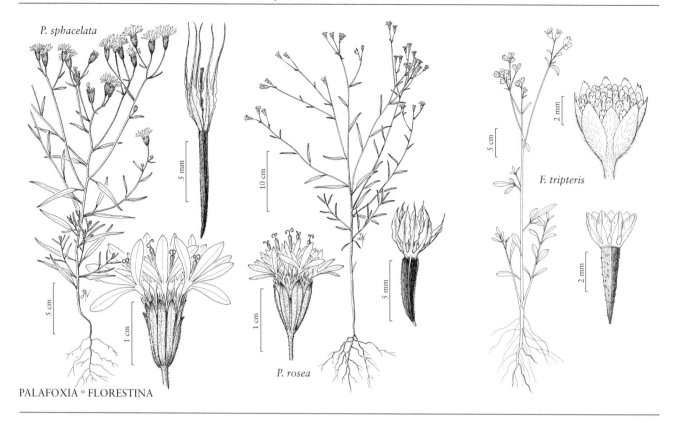

P. sphacelata

P. rosea

F. tripteris

PALAFOXIA ○ FLORESTINA

4. Palafoxia feayi A. Gray, Proc. Amer. Acad. Arts 12: 59. 1877 [E]

Perennials or subshrubs, 100–300 cm. **Stems** strigillose to nearly glabrous, not stipitate-glandular. **Leaf blades** narrowly elliptic, broadly oblong, or ovate, 20–60 × 5–25 mm. **Involucres** campanulate or broadly to narrowly turbinate. **Phyllaries** 5–9 × 1–1.5 mm, ± equal, ± strigillose. **Ray florets** 0. **Disc florets** 15–30; corollas ± actinomorphic, 10–14 mm, throats ± cylindric, longer than lobes. **Cypselae** 6–8 mm; **pappus scales** of inner cypselae 1.5–2 mm. $2n = 24$.

Flowering (spring–)late summer–fall. Sandy soils; 0–10 m; Fla.

5. Palafoxia arida B. L. Turner & M. I. Morris, Madroño 23: 79. 1975

Palafoxia arida var. *gigantea* (M. E. Jones) B. L. Turner & M. I. Morris

Annuals, 10–200 cm. **Stems** scabrous to nearly glabrous, sometimes distally stipitate-glandular. **Leaf blades** lance-linear to linear, 20–120 × 2–55 mm. **Involucres** cylindric to narrowly turbinate. **Phyllaries** ± equal, 10–25 × 1–2.5 mm, ± scabrous, sometimes stipitate-glandular. **Ray florets** 0. **Disc florets** 9–40; corollas ± actinomorphic, 9–13 mm, throats ± cylindric, longer than lobes. **Cypselae** 10–16 mm; **pappus scales** of inner cypselae 8–12 mm. $2n = 24$.

Flowering late winter–spring(–summer). Sandy soils; 30–1000 m; Ariz., Calif., Nev., Utah; Mexico (Baja California, Sonora).

Plants of *Palafoxia arida* 90–150 cm with phyllaries 16–25 mm (from dunes west of Yuma, Arizona) have been treated as var. *gigantea*. Plants referable to *P. arida* have been named *P. linearis* (Cavanilles) Lagasca (including var. *gigantea* M. E. Jones) in other floras; *P. linearis* is a Mexican species (see B. L. Turner and M. I. Morris 1976).

6. Palafoxia integrifolia (Nuttall) Torrey & A. Gray, Fl. N. Amer. 2: 369. 1842 [E]

Polypteris integrifolia Nuttall, Gen. N. Amer. Pl. 2: 139. 1818

Subshrubs or shrubs, 30–150 cm. **Stems** strigillose to glabrate, not stipitate-glandular. **Leaf blades** lanceolate to linear, 25–90 × 2–13 mm. **Involucres** broadly turbinate. **Phyllaries** 8–11 × 1–3.5 mm, unequal, glabrous or glabrate. **Ray florets** 0. **Disc florets** 10–26; corollas ± actinomorphic, 7–13 mm, throats ± funnelform, shorter than lobes.

Cypselae 5–6 mm; **pappus scales** of inner cypselae 4–7 mm. **2*n* = 24.**

Flowering (spring–)late summer–fall. Sandy soils; 0–50+ m; Fla., Ga.

7. Palafoxia riograndensis Cory, Rhodora 48: 84. 1946

Annuals, 30–60 cm. **Stems** scabrous to glabrate, not stipitate-glandular. **Leaf blades** lance-linear, 30–70 × 2–8 mm. **Involucres** ± cylindric. **Phyllaries** 10–15 × 1–2 mm, ± equal, ± scabrous, not stipitate-glandular. **Ray florets** 0. **Disc florets** 8–25; corollas zygomorphic (outer) to actinomorphic (inner), 6–9 mm, throats ± funnelform, shorter than lobes. **Cypselae** 7–12 mm; **pappus scales** of inner cypselae 5–8 mm. **2*n* = 24.**

Flowering (spring–)late summer–fall. Sandy or silty soils; 50–200+ m; Tex.; Mexico (Chihuahua, Coahuila).

8. Palafoxia callosa (Nuttall) Torrey & A. Gray, Fl. N. Amer. 2: 369. 1842

Stevia callosa Nuttall, J. Acad. Nat. Sci. Philadelphia 2: 121. 1821

Annuals, 20–60 cm. **Stems** scabrous to glabrate, not stipitate-glandular. **Leaf blades** linear, 20–70 × 1–4 mm. **Involucres** ± turbinate. **Phyllaries** 3–5 × 1+ mm, ± equal, ± strigillose, not stipitate-glandular. **Ray florets** 0. **Disc florets** 5–30; corollas actinomorphic, 5–6 mm, throats ± funnelform, shorter than lobes. **Cypselae** 3–5 mm; **pappus scales** of inner cypselae 0.3–1 mm. **2*n* = 20.**

Flowering summer–fall. Rocky, calcareous soils; 50–300 m; Ark., Mo., Tex.; Mexico (Coahuila).

Some depauperate specimens of *Palafoxia rosea* (e.g., Oklahoma, Beckham Co., 17 Oct 1936, *Eskero 1502,* US) closely resemble *P. callosa.*

9. Palafoxia rosea (Bush) Cory, Rhodora 48: 86. 1946
E F

Othake roseum Bush, Trans. Acad. Sci. St. Louis 14: 175. 1904; *Palafoxia rosea* var. *macrolepis* (Rydberg) B. L. Turner & M. I. Morris

Annuals, 10–50 cm. **Stems** scabrous to glabrate, not stipitate-glandular. **Leaf blades** linear-lanceolate, 30–60 × 2–6(–10) mm. **Involucres** ± turbinate. **Phyllaries** 5–10 × 1–2.5 mm, ± equal, usually scabrellous and stipitate-glandular. **Ray** florets 0. **Disc florets** 5–30; corollas actinomorphic, 7–10 mm, throats ± funnelform, shorter than lobes. **Cypselae** 5–8 mm; **pappus scales** of inner cypselae (1.5–)3–8 mm. **2*n* = 20.**

Flowering spring–fall. Sandy soils; 100–1500 m; Colo., Kans., N.Mex., Okla., Tex., Wyo.

Plants of *Palafoxia rosea* with phyllaries 7–10 (versus 5–7) mm and pappus scales 3–8 (versus 1–3) mm have been treated as var. *macrolepis.* Some specimens are not readily assignable to either variety, and some are not readily assignable to either *P. rosea* or *P. texana.* Other specimens referred here to *P. rosea* closely resemble *P. callosa.*

10. Palafoxia texana de Candolle in A. P. de Candolle and A. L. P. P. de Candolle, Prodr. 5: 125. 1836

Palafoxia texana var. *ambigua* (Shinners) B. L. Turner & M. I. Morris

Annuals or perennials (subshrubs), 20–80 cm. **Stems** ± scabrous, not stipitate-glandular. **Leaf blades** ovate-lanceolate to linear-lanceolate, 30–80 × (5–)10–20 mm. **Involucres** ± turbinate. **Phyllaries** 5–8 × 1–2.5 mm, ± equal, scabrellous, rarely stipitate-glandular. **Ray florets** 0. **Disc florets** 10–25; corollas actinomorphic, 7–10 mm, throats ± funnelform, shorter than lobes. **Cypselae** 4–6 mm; **pappus scales** of inner cypselae 2–6 mm. **2*n* = 22.**

Flowering spring–fall. Calcareous, usually rocky or gravelly soils; 10–100+ m; Fla., La., Tex.; Mexico (Coahuila, Nuevo León, Tamaulipas).

Plants of *Palafoxia texana* with non-stipitate-glandular phyllaries (versus stipitate-glandular) and pappus scales (on inner cypselae) 1–4 (versus 3–6) mm have been treated as var. *ambigua.* Some specimens are not readily assignable to either variety, and some are not readily assignable to either *P. rosea* or *P. texana.*

373. FLORESTINA   Cassini, Bull. Sci. Soc. Philom. Paris 1817: 11. 1817 • [No etymology given; possibly from Latin *floreus*, of flowers, and *tina*, a wine vessel; perhaps alluding to sometimes purplish involucres]

John L. Strother

Annuals [perennials], 10–60 cm. **Stems** erect, branched (mostly distally). **Leaves** cauline; opposite (proximal) or alternate (mostly); mostly petiolate; blades (3- or 5-nerved) simple (proximal and distal) or (mid-cauline) 3- or 5-lobed or -foliolate; blades or leaflets broadly to narrowly oblong to ovate, ultimate margins entire, faces sparsely to moderately scabrous (hairs white, antrorse, 0.3–0.7 mm), usually gland-dotted as well. **Heads** discoid, in ± corymbiform to paniculiform arrays. **Involucres** obconic to subhemispheric, 3–6+ mm diam. **Phyllaries** falling, 4–14 in 1–2 series (obovate to oblanceolate, subequal, thin-herbaceous, margins scarious, sometimes purplish). **Receptacles** concave to convex, knobby or smooth (sometimes minutely setose or hairy), epaleate. **Ray florets** 0. **Disc florets** 10–30+, bisexual, fertile; corollas whitish, sparsely glandular-hairy, tubes longer than funnelform throats, lobes 5, lance-linear (often unequal). **Cypselae** obpyramidal, 4-angled, faces sometimes ribbed, sparsely hairy (hairs curled); **pappi** persistent, of 8–12 (distinct) spatulate to lanceolate [suborbiculate], medially thickened, laterally scarious scales in 1 series (some, all, or the alternate aristate). *x* = 12.

Species 8 (1 in the flora): Texas, Mexico, Central America.

Florestinas are closely related to, perhaps congeneric with, palafoxias.

SELECTED REFERENCE   Turner, B. L. 1963. Taxonomy of *Florestina* (Helenieae, Compositae). Brittonia 15: 27–46.

1. **Florestina tripteris**   de Candolle in A. P. de Candolle and A. L. P. P. de Candolle, Prodr. 5: 655. 1836 [F]

Leaf blades or leaflets mostly 2–5 × 1–2 cm. **Disc corollas** 4–5 mm. **Cypselae** 3–6 mm; **pappus scales** 0.3–2 mm. *2n* = 20.

Flowering year round, mostly summer–fall. Open, often disturbed sites; 0–900 m; Tex.; Mexico (Chihuahua, Coahuila, Nuevo León, Tamaulipas).

374. AMAURIOPSIS   Rydberg in N. L. Britton et al., N. Amer. Fl. 34: 37. 1914 • [Generic name *Amauria*, and Greek *-opsis*, resembling]

John L. Strother

Biennials (sometimes flowering in first year, sometimes persisting), 10–80+ cm. **Stems** erect, usually branched distally, sometimes from bases. **Leaves** usually basal and cauline; alternate; petiolate; blades deltate to ovate or oblong (in broad outline), usually 1–2-ternately lobed (lobes mostly oblong to obovate), ultimate margins entire or toothed, faces ± scabrellous to hirtellous, usually gland-dotted. **Heads** radiate, in loose, corymbiform arrays. **Involucres** ± hemispheric or broader, 10–18+ mm diam. **Phyllaries** persistent, 12–21+ in ± 2 series (reflexed in fruit, distinct, ± oblanceolate or lanceolate to lance-attenuate, subequal, thin-herbaceous, margins membranous, sometimes purplish, apices usually attenuate to ± caudate). **Receptacles**

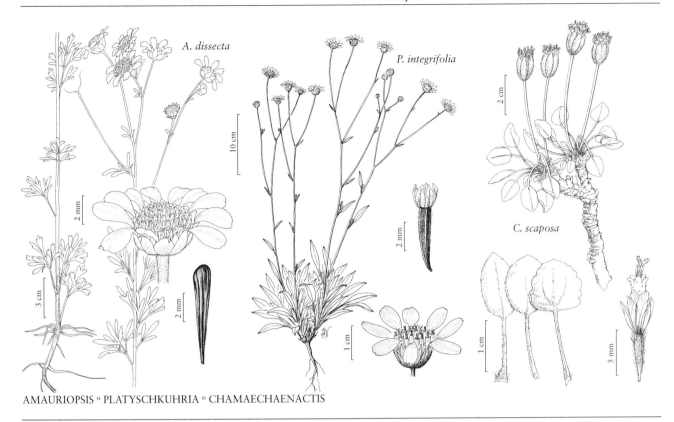

AMAURIOPSIS ° PLATYSCHKUHRIA ° CHAMAECHAENACTIS

convex, pitted, epaleate. **Ray florets** 10–21+, pistillate, fertile; corollas yellow. **Disc florets** 30–80+, bisexual, fertile; corollas yellow, tubes longer than funnelform throats, lobes 5, ± lanceolate (usually longer than throats). **Cypselae** obpyramidal, 4-angled, usually glabrous, sometimes hirtellous; **pappi** 0. x = 12.

Species 1: w United States, n Mexico.

Resurrection of *Amauriopsis*, which was included in *Bahia* Lagasca by W. L. Ellison (1964), is predicated largely on the work of B. G. Baldwin et al. (2002).

SELECTED REFERENCE Ellison, W. L. 1964. A systematic study of the genus *Bahia* (Compositae). Rhodora 66: 67–86; 177–215; 281–311.

1. **Amauriopsis dissecta** (A. Gray) Rydberg in N. L. Britton et al., N. Amer. Fl. 34: 37. 1914 F

Amauria dissecta A. Gray, Mem. Amer. Acad. Arts, n. s. 4: 104. 1849; *Bahia dissecta* (A. Gray) Britton

Leaf lobes (3–)7–25(–75+), ovate to oblong or lanceolate to lance-linear, 4–12(–25+) × (1–)2–6(–12+) mm. **Involucres** 5–7+ × 10–18+ mm. **Ray laminae** 5–10+ mm. **Disc corollas** 2.5–3(–4+) mm, ± stipitate-glandular. **Cypselae** 2.5–4+ mm. $2n$ (= 3x) = 36.

Flowering (Jun–)Aug–Sep(–Oct). Usually on sandy or gravelly soils, openings in pinyon-juniper, yellow pine, or spruce-fir forests; 1600–2900 m; Ariz., Calif., Colo., Nev., N.Mex., Tex., Utah, Wyo.; Mexico (Baja California, Chihuahua, Coahuila, Sonora).

W. L. Ellison (1964) reported "n = 18, metaphase II," based on observation of meiosis in a microsporocyte. D. J. Keil et al. (1988) noted that anthers are often abortive and meiosis is irregular and that seed set is usually high in *A. dissecta*. Keil et al. suggested that *A. dissecta* is an apomictic triploid derived from a cross between a diploid plant with n = x = 12 and a tetraploid plant (or unreduced gamete) with n = $2x$ = 24 so that for *A. dissecta* $2n$ = $3x$ = 36.

Some plants (from Arizona, Baja California, and New Mexico) with poorly developed ray corollas and ± hairy cypselae bearing pappi of 1–13 lance-linear to lance-subulate scales have been identified as belonging to *Amauriopsis dissecta*. Such plants may be hybrids resulting from crosses between *A. dissecta* and some other pappose species.

375. PLATYSCHKUHRIA   (A. Gray) Rydberg, Bull. Torrey Bot. Club 33: 154. 1906

* [Greek *platys*, broad, and genus *Schkuhria*]   E

John L. Strother

Schkuhria Roth sect. *Platyschkuhria* A. Gray, Amer. Naturalist 8: 213. 1874

Perennials, 10–50+ cm (bases ± woody, branched caudices). **Stems** ± erect. **Leaves** basal and cauline; mostly alternate; petiolate; blades mostly lanceolate, sometimes ± ovate, margins en- tire, faces sparsely to densely scabrellous (hairs white, straight, conic or fusiform, 0.1–0.8 mm) and gland-dotted. **Heads** radiate, borne singly or (3–11) in loose, corymbiform to paniculiform arrays. **Involucres** campanulate to ± hemispheric, 12–25+ mm diam. **Phyllaries** 9–21 in ± 2 series, distinct, subequal, oblong or elliptic to lanceolate, herbaceous, membranous-margined (not purplish-margined), persistent, reflexed in fruit. **Receptacles** ± convex, ± pitted, epaleate. **Ray florets** 6–12, pistillate, fertile; corollas yellow. **Disc florets** 25–80+, bisexual, fertile; corollas yellow to orange (gland-dotted), tubes about equaling funnelform to campanulate throats, lobes 5, deltate to lance-deltate. **Cypselae** narrowly obpyramidal, 4-angled, finely nerved, ± hirsutulous (at least on angles), not gland-dotted; **pappi** of 8–16, (distinct) lance-elliptic to lance-subulate (basally and/or medially thickened, distally and/or laterally scarious) scales in 1 series (weakly, if at all, aristate). *x* = 12.

Species 1: w United States.

SELECTED REFERENCES   Brown, G. K. 1983. Chromosome numbers in *Platyschkuhria* Rydberg (Compositae) and their systematic significance. Amer. J. Bot. 70: 591–601. Ellison, W. L. 1971. Taxonomy of *Platyschkuhria* (Compositae). Brittonia 23: 269–279.

1. Platyschkuhria integrifolia   (A. Gray) Rydberg, Bull. Torrey Bot Club 33: 155. 1906   E F

Schkuhria integrifolia A. Gray, Amer. Naturalist 8: 213. 1874; *Bahia nudicaulis* A. Gray 1883, not *B. integrifolia* (Hooker) de Candolle 1836; *B. nudicaulis* var. *desertorum* (M. E. Jones) Cronquist; *B. nudicaulis* var. *oblongifolia* (A. Gray) Cronquist; *B. nudicaulis* var. *ourolepis* (S. F. Blake) Cronquist; *Platyschkuhria integrifolia* var. *desertorum* (M. E. Jones) W. L. Ellison; *P. integrifolia* var. *oblongifolia* (A. Gray) W. L. Ellison; *P. integrifolia* var. *ourolepis* (S. F. Blake) W. L. Ellison

Leaf blades 2–10 cm × 5–35+ mm. **Involucres** 9–12+ × 12–25+ mm. **Ray corolla** laminae 6–16 mm. **Disc corollas** 3–6(–7) mm. **Cypselae** (1–)3–5(–8) mm; pappi 0.6–3+ mm. **2*n*** = 24, 48, 60, 72.

Flowering May–Jul. Seleniferous clays, shaley slopes; 1100–2200 m; Ariz., Colo., Mont., N.Mex., Utah, Wyo.

Specimens of *Platyschkuhria integrifolia* with stems leafy over ³⁄₄+ of their lengths have been called var. *oblongifolia* (mostly from "four-corners" area of Arizona, Colorado, New Mexico, and Utah). Those with leaves mostly on proximal ¹⁄₁₀–³⁄₄ their stem lengths, peduncles little or not at all stipitate-glandular, and phyllary tips ± caudate have been called var. *ourolepis* (mostly from Uinta Basin, Utah). Those with leaves mostly on proximal ¹⁄₁₀–³⁄₄ their stem lengths, leaf blades widest beyond their middles and rounded at tips, peduncles ± stipitate-glandular, and phyllary tips acute to acuminate have been called var. *nudicaulis* (mostly from Montana and Wyoming). Those with leaves mostly on proximal ¹⁄₁₀–³⁄₄ their stem lengths, leaf blades widest proximal to their middles and acute at tips, peduncles ± stipitate-glandular, and phyllary tips obtuse to acute or acuminate have been called var. *desertorum* (mostly from western Colorado and eastern Utah).

376. CHAMAECHAENACTIS Rydberg, Bull. Torrey Bot. Club 33: 155. 1906 • Fullstem

[Greek, *chamae-*, creeping, low, on the ground, and generic name *Chaenactis*] E

James D. Morefield

Perennials, 2–7(–9) cm (to 10–20+ cm across). **Stems** mostly subterranean (caudices relatively thick, branched; aerial stems essentially peduncles). **Leaves** mostly basal; alternate; petiolate; blades (1- or 3-nerved) cordate, elliptic, ovate, or rounded, margins entire or distally ± crenate, revolute to ± plane, faces ± strigose and gland-dotted, adaxial sometimes glabrescent. **Heads** discoid, borne singly. **Involucres** ± obconic, 6–15 (10–23 pressed) mm diam. **Phyllaries** 11–15 in 2 series (± erect, sometimes spreading in senescence, oblong to oblanceolate, ± unequal, herbaceous, abaxially densely villous, obscurely glandular). **Receptacles** convex, knobby, epaleate. **Ray florets** 0. **Disc florets** 10–30+, bisexual, fertile; corollas white to pinkish, tubes shorter than ± cylindric throats, lobes 5, ± deltate (style branches stigmatic in 2 lines, appendages linear-oblong, blunt). **Cypselae** clavate, ± quadrangular with 8–12 obscure nerves, densely piloso-strigose, eglandular; **pappi** persistent, of 7–11, (distinct) oblanceolate to narrowly spatulate, erose scales in 2 series (midnerves prominent). *x* = 16.

Species 1: w North America.

Chamaechaenactis has been considered a close relative of *Chaenactis*, which has superficially similar heads. Nuclear rDNA evidence (B. G. Baldwin et al. 2002) suggests closer relationship to *Bartlettia*, *Hymenopappus*, and other members of Bahiinae sensu Baldwin. Such a placement was also suggested by Rydberg when he segregated *Chamaechaenactis* from *Chaenactis*.

SELECTED REFERENCE Preece, S. J. and B. L. Turner. 1953. A taxonomic study of the genus *Chamaechaenactis* Rydberg (Compositae). Madroño 12: 97–103.

1. **Chamaechaenactis scaposa** (Eastwood) Rydberg, Bull. Torrey Bot. Club 33: 156. 1906 • Fullstem E F

Chaenactis scaposa Eastwood, Zoë 2: 231. 1891; *Actinella carnosa* A. Nelson; *Chamaechaenactis scaposa* var. *parva* Preece & B. L. Turner

Plants densely cespitose, ± pulvinate; taproots deep; caudices thickly branched. **Leaves** strictly basal (old bases marcescent); longest petioles 3–40 mm, lengths mostly 1–2 times blades, proximally dilated and ± chartaceous; blades: largest 4–15(–18) × 3–13(–15) mm, coriaceous. **Peduncles** (1–)5–30+, ascending to erect, (0–)1–5(–7) cm, ± strigose, usually villous (at least distally). **Phyllaries**: longest 9–17 mm; outer shorter than inner. **Corollas** 5–9 mm. **Cypselae** 5–8 mm; **pappi**: longest scales 4–7 mm. 2*n* = 32.

Flowering late Apr–early Jul. Dry, open, relatively barren silty to clay soils from shale (sometimes petroleum-bearing), sandstone, marl, or limestone, often armored by rocky, sandy, or gravelly overburden, usually in pinyon-juniper woodlands; 1400–2600 m; Ariz., Colo., N.Mex., Utah, Wyo.

Chamaechaenactis scaposa is grown and sold for use in rock gardens; it is considered a difficult horticultural subject. It is native to the Green, San Juan, and upper Colorado river basins in eastern Utah and western Colorado, extending to northeastern Arizona, northwestern New Mexico, and southwestern Wyoming. A previous report of *C. scaposa* from Idaho was in error.

Plants of *Chamaechaenactis scaposa* to the east and south tend to have heads and flowers larger, and leaves larger, more frequently ovate-cordate and distally crenate (var. *scaposa*), than plants to the west and north (var. *parva*), which also tend to have narrower and uniformly entire leaves (S. J. Preece and B. L. Turner 1953). Some recent collections from higher-elevation limestone sites at the western edge of the range of the species are reduced to a degree not originally included in var. *parva*, with heads ± sessile, embedded among dense leaves with blades scarcely reaching 5 mm. The two varieties appear to represent points within a broad, intermingled, clinal pattern of variation; they do not merit formal taxonomic recognition.

377. HULSEA   Torrey & A. Gray in War Department [U.S.], Pacif. Railr. Rep. 6(3): 77, plate 13. 1858 • Alpinegold [For Gilbert White Hulse, 1807–1883, physician and plant collector]

Dieter H. Wilken

Annuals or perennials, 5–150 cm. **Stems** erect, simple or branched (aerial shoots often from subterranean caudices). **Leaves** basal or basal and cauline, proximally whorled, distally alternate; petiolate (at least basal) or sessile; blades mostly lance-linear, oblanceolate, ovate, or spatulate, margins entire, lobed, or toothed, faces thinly lanate to densely woolly (hairs crisped, tangled, or matted, usually 0.8+ mm) and/or gland-dotted, glandular-puberulent, glandular-villous, or stipitate-glandular. **Heads** radiate, borne singly or in corymbiform arrays. **Involucres** obconic to hemispheric, 8–26 mm diam. **Phyllaries** persistent, 14–35(–60+) in 2–4 series (reflexed in fruit, mostly lance-ovate to lanceolate or linear, herbaceous). **Receptacles** flat, knobby or pitted, epaleate. **Ray florets** 9–60+, pistillate, fertile; corollas yellow, orange, or red. **Disc florets** 20–200+, bisexual, fertile; corollas yellow to orange (glabrous or sparsely hairy), tubes shorter than or about equaling cylindric throats, lobes 5, deltate to lance-ovate. **Cypselae** clavate to linear, compressed (lenticular in cross section), silky-hairy; **pappi** of 4 (distinct) quadrate to spatulate, equal, subequal, or unequal, erose to laciniate scales (the alternate alike). $x = 19$.

Species 7 (7 in the flora): w United States, nw Mexico.

SELECTED REFERENCE   Wilken, D. H. 1975. A systematic study of *Hulsea* (Asteraceae). Brittonia 27: 228–244.

1. Basal and proximal cauline leaves gray to grayish green, lanate to woolly (hairs mostly eglandular).
  2. Leaves basal and cauline, blades broadly oblanceolate to spatulate, 6–10 cm; heads 2–5; cypselae 4–6 mm . . . . 3. *Hulsea californica*
  2. Leaves mostly basal, blades spatulate, 1–9 cm; heads 1–2; cypselae 5–10 mm . . . . 7. *Hulsea vestita*
1. Basal and proximal cauline leaves ± green, mostly glandular-puberulent or glandular villous (sometimes sparsely lanate as well, sometimes woolly in *H. nana*).
  3. Leaves mostly basal (cauline none or relatively few); heads 1–2 (per basal leaf rosette).
    4. Plants (10–)20–40 cm; leaf margins sinuate, lobed, or toothed (lobes or teeth mostly triangular); phyllary apices acute to attenuate; ray florets 28–59 . . . . 1. *Hulsea algida*
    4. Plants 5–15(–20) cm; leaf margins lobed (lobes mostly oblong); phyllary apices acuminate to acute; ray florets 12–30 . . . . 6. *Hulsea nana*
  3. Leaves mostly cauline (basal usually present as well); heads 3–5.
    5. Ray laminae red to reddish purple (narrowly oblong to linear, ciliate) . . . . . 4. *Hulsea heterochroma*
    5. Ray laminae yellow (narrowly elliptic to narrowly lance-oblong).
      6. Ray florets 10–23, corolla tubes hairy . . . . 2. *Hulsea brevifolia*
      6. Ray florets 20–35, corolla tubes glabrous . . . . 5. *Hulsea mexicana*

1. Hulsea algida A. Gray, Proc. Amer. Acad. Arts 6: 547. 1865 • Pacific or alpine alpinegold E

Hulsea caespitosa A. Nelson & P. B. Kennedy; *H. carnosa* Rydberg; *H. nevadensis* Gandoger

Perennials, (10–)20–40 cm. **Stems** 1–15+, sparsely lanate. **Leaves** mostly basal (cauline relatively few); blades green, narrowly oblanceolate to narrowly spatulate, 5–11 cm, margins sinuate, lobed, or toothed (lobes or teeth triangular), faces sparsely to moderately lanate and glandular-puberulent; distal cauline leaves oblong to narrowly lanceolate, much reduced. **Heads** 1. **Involucres** obconic to hemispheric, 12–25 mm diam. **Phyllaries** 8–15 mm, outer lanceolate, apices acute to attenuate. **Ray florets** 28–59; corolla tubes glabrous, laminae yellow, 8–12 mm. **Disc corollas** yellow. **Cypselae** 6–10 mm; **pappus scales** subequal, 1–1.5 mm. $2n = 38$.

Flowering summer. Subalpine to alpine rocky slopes, talus, on various substrates; 2700–4000 m; Calif., Idaho, Mont., Nev., Oreg., Wyo.

Plants of *Hulsea algida* from the Wallowa Mountains of northeastern Oregon tend to have smaller, more lanate leaves than those elsewhere and have heads slightly larger than those of *H. nana*. They deserve further study.

2. **Hulsea brevifolia** A. Gray, Proc. Amer. Acad. Arts 7: 359. 1867 • Shortleaf alpinegold E

Perennials, 30–60 cm. **Stems** 1–7, sparsely lanate. **Leaves** mostly cauline; blades green, lanceolate to narrowly oblanceolate, 5–11 cm, margins sinuate to toothed (teeth triangular, obtuse), faces sparsely to moderately glandular-villous; distal cauline leaves ± oblanceolate, gradually reduced. **Heads** 3–4. **Involucres** obconic to hemispheric, 10–16 mm diam. **Phyllaries** 8–10 mm, outer lanceolate, apices acuminate. **Ray florets** 10–23; corolla tubes hairy, laminae yellow, 10–14 mm. **Disc corollas** yellow. **Cypselae** 6–8 mm; **pappus scales** unequal, shorter 1–2 mm, longer 2–3 mm. $2n = 38$.

Flowering late spring–summer. Common after fires, open sites in coniferous forests, mostly coarse-grained volcanic or granitic soils; 1500–2700 m; Calif.

Hulsea brevifolia grows in the Sierra Nevada.

3. **Hulsea californica** Torrey & A. Gray in War Department [U.S.], Pacif. Railr. Rep. 6(3): 77. 1858 • San Diego alpinegold C E

Biennials or perennials, 40–120 cm. **Stems** (1–)3–10, leafy, lanate to woolly. **Leaves** basal and cauline; blades gray to grayish green, broadly oblanceolate to spatulate, 6–10 cm, margins undulate to weakly lobed, faces densely lanate; distal cauline leaves lanceolate to narrowly obovate, gradually reduced. **Heads** 2–5. **Involucres** broadly conic to hemispheric, 15–26 mm diam. **Phyllaries** 9–14 mm, outer lanceolate, apices attenuate. **Ray florets** 22–40; corolla tubes glabrous, laminae yellow, 9–12 mm. **Disc corollas** yellow. **Cypselae** 4–6 mm; **pappus scales** subequal, 1–2 mm. $2n = 38$.

Flowering late spring–summer. Often common after fires, open sites in chaparral and woodlands, mostly rocky, metamorphic soils; of conservation concern; 1000–2000 m; Calif.

Hulsea californica grows in the eastern Peninsular Ranges, San Diego County.

4. **Hulsea heterochroma** A. Gray, Proc. Amer. Acad. Arts 7: 359. 1867 • Redray alpinegold E F

Biennials or perennials, 50–150 cm. **Stems** 1–3, glandular. **Leaves** mostly cauline; blades green, broadly oblanceolate to spatulate, 9–21 cm, margins dentate, faces glandular-villous; distal cauline leaves lanceolate to ovate, gradually reduced. **Heads** 3–5. **Involucres** obconic to hemispheric, 11–18 mm diam. **Phyllaries** 10–25 mm, outer narrowly lanceolate, apices attenuate. **Ray florets** 28–75+; corolla tubes hairy, laminae red to reddish purple (narrowly oblong to linear, ciliate), 7–15 mm. **Disc corollas** yellow (lobes and veins red-tinged). **Cypselae** 6–8 mm; **pappus scales** unequal, shorter 1–2 mm, longer 2–3 mm. $2n = 38$.

Flowering late spring–summer. Often common after fires, chaparral and woodlands, rocky or gravelly soils from various substrates; 300–2600 m; Ariz., Calif., Nev., Utah.

5. **Hulsea mexicana** Rydberg in N. L. Britton et al., N. Amer. Fl. 34: 41. 1914 • Mexican alpinegold

Annuals or biennials, (20–)40–100 cm. **Stems** 1–3, glandular. **Leaves** mostly cauline; blades green, oblanceolate to spatulate, 5–14 cm, margins dentate, faces glandular-villous; distal cauline leaves broadly oblanceolate to ovate, gradually reduced. **Heads** 3–5. **Involucres** hemispheric, 15–27 mm diam. **Phyllaries** 8–12 mm, outer narrowly lanceolate, apices acuminate. **Ray florets** 20–35; corolla tubes glabrous; laminae yellow (narrowly elliptic to narrowly oblong-lanceolate), 10–12 mm. **Disc corollas** yellow. **Cypselae** 4–6 mm; **pappus scales** subequal, 1–2 mm. $2n = 38$.

Flowering late spring–summer. Common after fires, chaparral and woodlands; 1000–1100 m; Calif.; Mexico (Baja California).

In the flora area, *Hulsea mexicana* grows in southern San Diego County.

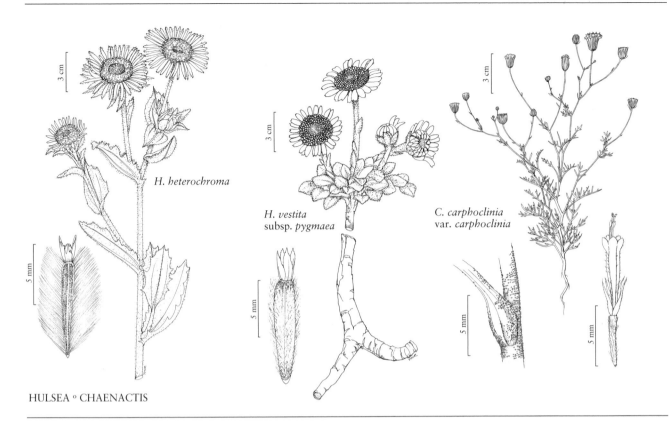

H. heterochroma

H. vestita
subsp. pygmaea

C. carphoclinia
var. carphoclinia

HULSEA ° CHAENACTIS

6. **Hulsea nana** A. Gray in War Department [U.S.], Pacif.
Railr. Rep. 6(3): 76, plate 13. 1858 • Dwarf
alpinegold E

Hulsea nana var. *larsenii* A. Gray;
H. vulcanica Gandoger

Perennials, 5–15(–20) cm. Stems
1–7, glandular-puberulent and
sparsely lanate. Leaves mostly
basal; blades narrowly spatulate,
2–6 cm, margins lobed or toothed
(lobes or teeth mostly oblong),
faces sparsely lanate to woolly;
distal cauline leaves narrowly lanceolate, much reduced.
Heads 1. Involucres obconic, 8–12 mm diam. Phyllaries 8–12 mm, outer narrowly obovate to oblong-
lanceolate, apices acuminate to acute. Ray florets 12–
30; corolla tubes glabrous, laminae yellow, 6–10 mm.
Disc corollas yellow. Cypselae 6–8 mm; pappus scales
subequal, 1–2 mm. 2*n* = 38.

Flowering summer. Subalpine to alpine rocky slopes,
taluses, mostly volcanic substrates; 2400–3000 m;
Calif., Oreg., Wash.

Densely lanate or woolly plants of *Hulsea nana* are
referable to var. *larsenii*. Such plants may occur in distinct populations but can be found together with sparsely
lanate and strictly glandular plants. The distribution of
lanate to woolly plants appears associated with higher
levels of insolation.

7. **Hulsea vestita** A. Gray, Proc. Amer. Acad. Arts 6: 547.
1865 • Pumice alpinegold E F

Perennials, 5–100 cm. Stems 3–10,
lanate to woolly. Leaves mostly
basal; blades gray to grayish green,
spatulate, 1–9 cm, margins entire
or undulate, dentate, or lobed, faces
densely lanate to woolly; distal
cauline leaves mostly lanceolate to
linear, gradually reduced or none.
Heads 1–3. Involucres obconic to
hemispheric, 8–18 mm diam. Phyllaries 7–12 mm, outer
oblong-lanceolate to narrowly oblanceolate, apices acuminate to acute. Ray florets 9–32; corolla tubes glabrous,
laminae yellow to reddish orange, 6–18 mm. Disc corollas
yellow to orange. Cypselae 5–10 mm; pappus scales equal
to subequal, 1–2 mm.

Subspecies 6 (6 in the flora): w United States.

1. Proximal leaves: margins dentate to lobed; ray and
   disc corollas orange to reddish orange.
   2. Leaf faces lanate; outer phyllaries narrowly
      lanceolate; cypselae 5–7 mm . . . .
      . . . . 7e. *Hulsea vestita* subsp. *parryi*
   2. Leaf faces: abaxial lanate, adaxial glandular-
      puberulent; outer phyllaries narrowly oblan-
      ceolate to obovate; cypselae 7–9 mm . . . . . .
      . . . . 7f. *Hulsea vestita* subsp. *pygmaea*

1. Proximal leaves: margins entire, undulate, or lobed (dentate to lobed in subsp. *inyoensis*); ray and disc corollas yellow, sometimes red-tinged.
   3. Plants 10–30 cm; cauline leaves (0 or 1–2 proximal to mid stems) narrowly lanceolate . . . .
       . . . . 7a. *Hulsea vestita* subsp. *vestita*
   3. Plants 20–100 cm; cauline leaves (usually 3+ proximal to mid stems) lanceolate to ovate.
      4. Proximal leaves: margins dentate to lobed; ray florets 18–32, laminae usually yellow, rarely red-tinged, 12–18 mm . . . .
        . . . . 7d. *Hulsea vestita* subsp. *inyoensis*
      4. Proximal leaves: margins entire, undulate, or lobed; ray florets 16–25, laminae usually yellow, sometimes red-tinged distally, 6–10 mm.
        5. Cauline leaves (on proximal ¼ of stems) lance-oblong to ovate (1.5–3 cm), margins entire or undulate . . . .
         . . . . 7b. *Hulsea vestita* subsp. *callicarpha*
        5. Cauline leaves (on proximal ½ of stems) lanceolate to ovate (2.5–3.5 cm), margins undulate to lobed . . . .
         . . . . 7c. *Hulsea vestita* subsp. *gabrielensis*

7a. Hulsea vestita A. Gray subsp. **vestita**   E

Plants 10–30(–40) cm. **Leaves:** proximal blades 2–6 cm, margins entire or ± undulate, faces densely lanate to woolly; cauline leaves (0 or 1–2 proximal to mid stems) narrowly lanceolate. **Heads** 1(–2). **Involucres** obconic, 10–12 mm diam. **Phyllaries** 8–11 mm, outer narrowly obovate to oblong-lanceolate, apices acuminate. **Ray florets** 12–21; laminae usually yellow, sometimes red-tinged, 8–10 mm. **Disc corollas** yellow. **Cypselae** 5–7 mm; **pappus scales** subequal, 1–2 mm. **2*n*** = 38.

Flowering late spring–summer. Rocky slopes, talus, and flats, alpine or coniferous forest, on various substrates; 2000–3000 m; Calif.

Subspecies *vestita* grows in and near the Sierra Nevada.

7b. Hulsea vestita A. Gray subsp. **callicarpha** (H. M. Hall) Wilken, Aliso 7: 413. 1972 • Beautiful hulsea   E

Hulsea vestita var. *callicarpha* H. M. Hall, Univ. Calif. Publ. Bot. 1: 129. 1902; *H. callicarpha* (H. M. Hall) S. Watson ex Rydberg

Plants 40–100 cm. **Leaves:** proximal blades 3–8 cm, margins entire or undulate (sometimes weakly lobed near apices), faces densely lanate to woolly; cauline leaves (on proximal ¼ of stems) lance-oblong to ovate (1.5–3 cm). **Heads** 1–3. **Involucres** obconic, 10–15 mm diam. **Phyllaries** 8–12 mm, outer narrowly obovate to lanceolate, apices acuminate. **Ray florets** 16–25; laminae yellow, sometimes red-tinged distally, 8–10 mm. **Disc corollas** yellow. **Cypselae** 5–7 mm; **pappus scales** subequal, 1.5–2 mm. **2*n*** = 38.

Flowering late spring–summer. Rocky slopes, talus, and flats, conifer forests, sometimes chaparral, on granitic and metamorphic substrates; 1300–2500 m; Calif.

Subspecies *callicarpha* grows in the Palomar, San Jacinto, and Santa Rosa mountains.

7c. Hulsea vestita A. Gray subsp. **gabrielensis** Wilken, Madroño 24: 53. 1977   E

Plants 30–70 cm. **Leaves:** proximal blades 3–8 cm, margins undulate or weakly lobed, faces densely lanate to woolly; cauline leaves (on proximal ½ of stems) lanceolate to ovate (2.5–3.5 cm). **Heads** 1–3. **Involucres** obconic to hemispheric, 13–18 mm diam. **Phyllaries** 8–10 mm, outer lanceolate to narrowly obovate, apices acuminate. **Ray florets** 16–23; laminae usually yellow, sometimes red-tinged distally, 6–10 mm. **Disc corollas** yellow (lobes often red-tinged). **Cypselae** 6–8 mm; **pappus scales** subequal, 1.5–2 mm.

Flowering late spring–summer. Rocky slopes, talus, and flats, mostly coniferous forest, sometimes chaparral, on granitic and metamorphic substrates; 1200–2100 m; Calif.

Subspecies *gabrielensis* grows in the San Emigdio and San Gabriel mountains.

7d. Hulsea vestita A. Gray subsp. **inyoensis** (D. D. Keck) Wilken, Aliso 7: 413. 1972   C E

Hulsea californica Torrey & A. Gray var. *inyoensis* D. D. Keck, Aliso 4: 101. 1958; *H. inyoensis* (D. D. Keck) Munz

Plants 30–70 cm. **Leaves:** proximal blades 5–9 cm, margins dentate to lobed, faces densely lanate to woolly; cauline leaves (on proximal ½ of stems) lanceolate. **Heads** 1(–2). **Involucres** obconic to hemispheric, 10–20 mm diam. **Phyllaries** obovate to oblong-lanceolate, 10–12 mm, apices acuminate. **Ray florets** 18–32; laminae usually yellow, rarely red-tinged, 12–18 mm. **Disc corollas** yellow. **Cypselae** 6–8 mm; **pappus scales** subequal, 1–2 mm. **2*n*** = 38.

Flowering late spring–summer. Rocky slopes, talus, and flats, desert shrublands and conifer woodlands, on volcanic or metamorphic substrates; of conservation concern; 1700–3000 m; Calif., Nev.

Subspecies *inyoensis* grows in the desert mountains of Inyo County and on the eastern slope of the Sierra Nevada of Mono County, California, and in Nye County, Nevada.

7e. Hulsea vestita A. Gray subsp. **parryi** (A. Gray) Wilken, Aliso 7: 413. 1972 · Parry's alpinegold [E]

Hulsea parryi A. Gray, Proc. Amer. Acad. Arts 12: 59. 1876

Plants 10–40(–50) cm. **Leaves:** proximal blades 1–5 cm, margins dentate to lobed, faces densely lanate to woolly; cauline leaves lanceolate. **Heads** 1. **Involucres** obconic, 8–10 mm diam. **Phyllaries** 7–10 mm, outer narrowly lanceolate, apices acuminate. **Ray florets** 10–16; laminae orange to reddish orange, 5–7 mm. **Disc corollas** orange. **Cypselae** 5–7 mm; **pappus scales** subequal, 1–2 mm. $2n = 38$.

Flowering late spring–summer. Rocky slopes, talus, and flats, montane to subalpine coniferous forests, sometimes chaparral, on granitic and metamorphic substrates; 1200–2800 m; Calif.

Plants on limestone substrates from the northeastern San Bernardino Mountains have broadly spatulate basal leaves with undulate margins. The heads and florets are typical of subsp. *parryi*. These plants are retained within subsp. *parryi*; they deserve further study.

7f. Hulsea vestita A. Gray subsp. **pygmaea** (A. Gray) Wilken, Aliso 7: 413. 1972 · Pygmy alpinegold [C] [E] [F]

Hulsea vestita var. *pygmaea* A. Gray in A. Gray et al., Syn. Fl. N. Amer. 1(2): 343. 1884

Plants 5–15 cm. **Leaves:** proximal blades spatulate, 1–4 cm, margins dentate to lobed, faces: abaxial densely lanate to woolly, adaxial glandular-puberulent. **Heads** 1. **Involucres** obconic to hemispheric, 15–18 mm diam. **Phyllaries** narrowly oblanceolate to obovate, 8–10 mm, (margins scarious) apices acuminate. **Ray florets** 9–20; laminae orange to reddish orange, 7–10 mm. **Disc corollas** orange. **Cypselae** 7–9 mm; **pappus scales** equal, 1.5–2 mm. $2n = 38$.

Flowering late spring–summer. Alpine to subalpine rocky slopes and talus, granitic or volcanic substrates; of conservation concern; 3000–3900 m; Calif.

Subspecies *pygmaea* grows in the San Bernardino Mountains and the southern Sierra Nevada.

378. **CHAENACTIS** de Candolle in A. P. de Candolle and A. L. P. P. de Candolle, Prodr. 5: 659. 1836 · Pincushion, dustymaidens [Greek *chaino*, to gape, and *aktis*, ray, alluding to enlarged peripheral corollas of type species]

James D. Morefield

Annuals, biennials, perennials, or subshrubs, (2–)5–70(–200) cm (taprooted). **Stems** erect to prostrate, usually branched. **Leaves** basal and/or cauline (smaller and sparser distally except in *C. cusickii*); alternate; usually petiolate; blades deltate, elliptic, linear, oblanceolate, or ovate (plane or ± 3-dimensional), (0–)1–4-pinnately (rarely -subpalmately) lobed, ultimate margins entire, faces glabrous or hairy, often stipitate-glandular or gland-dotted. **Heads** discoid or ± radiant, borne singly or in (terminal) ± cymiform arrays (erect in bud except *C. macrantha*). **Involucres** hemispheric to obconic or broadly cylindric, (3–)5–15[–25] mm diam. **Phyllaries** 5–21+ in 1–2(–3) series (subequal to unequal). **Receptacles** convex to ± flat, pitted and/or knobby, usually epaleate (paleae 3–10+ in *C. carphoclinia*). **Ray florets** 0 (sometimes simulated by enlarged peripheral disc corollas). **Disc florets** 8–70+, bisexual, fertile (diurnal with anthers exserted except in *C. macrantha*); corollas white, pinkish, cream, or yellow, tubes shorter than cylindric or funnelform throats, lobes 5, deltate to ± lanceolate (sometimes enlarged, unequal; style-branch appendages blunt, obscure). **Cypselae** clavate to ± cylindric or compressed, obscurely 8–20-angled, faces scabrous and strigose to densely sericeous (usually eglandular); **pappi** usually persistent, of (1–)4–20, distinct, ± erose scales in 1–4 series (equal or unequal, outer then shorter, scales usually fewer and/or shorter on peripheral cypselae, midnerves obscure), sometimes 0 or coroniform. $x = ?$ ($n = 6, 8$, plus polyploids and dysploid numbers).

Species 18 (17 in the flora): w North America, nw Mexico.

Chaenactis species grow in arid to alpine or Mediterranean climates, usually in open, unstable or early seral habitats (loose sand, scree, talus, shrink-swell clay, fire-adapted vegetation, recent disturbances). Some annual species have been grown in gardens in the eastern United States.

Differences in induments are key to distinguishing some *Chaenactis* species. Unless otherwise noted in descriptions and key leads, assume for any given plant that indument of proximal leaves and adjacent proximal portions of stems ("proximal indument") is similar; indument of distal leaves is, likewise, similar to indument of stem portions from which they arise; peduncle indument is denser distally than proximally; and distal peduncle indument is similar to proximal phyllary indument. Unless otherwise noted, phyllary traits apply to the outer series only. Pappus scales may be equal, subequal, or unequal; unequal scales may intergrade (here said to be in gradually unequal series) or may form two, more or less uniform, shorter and longer series (here said to be in abruptly unequal series); subequal scales are said to be in subequal series. Excellent illustrations of most *Chaenactis* species appeared in L. Abrams and R. S. Ferris (1923–1960, vol. 4) and A. Cronquist (1955).

Section *Acarphaea* is distinctive by its farinose indument and base chromosome number of 8, among other traits; it could prove to be a separate, convergent genus. Natural and artificial hybrids have been documented among some members of sect. *Chaenactis* (see further discussion there). Reports of hybrids among species of the other two sections are few and doubtful. *Chaenactis* appears to be most closely related to the monotypic *Dimeresia* and *Orochaenactis*, which B. G. Baldwin et al. (2002) treated together as a narrowly circumscribed tribe, Chaenactideae.

SELECTED REFERENCES Mooring, J. S. 1965. Chromosome studies in *Chaenactis* and *Chamaechaenactis* (Compositae, Helenieae). Brittonia 17: 17–25. Stockwell, P. 1940. A revision of the genus *Chaenactis*. Contr. Dudley Herb. 3: 89–167.

1. Annuals; proximal indument predominantly ± farinose, not arachnoid; largest leaf blades (2–)3–4-pinnately lobed. . . . . 378a. *Chaenactis* sect. *Acarphaea*, p. 401
1. Annuals, biennials, perennials, or subshrubs; proximal indument predominantly arachnoid, lanuginose, pannose, stipitate-glandular, or glabrescent, not farinose; largest leaf blades (0–)1–2-pinnately or -subpalmately lobed.
   2. Biennials, perennials, or subshrubs (rarely flowering first year); pappi of (8–)10–20 scales in 2–4 equal or gradually unequal series; leaf blades gland-dotted beneath indument . . . . 378b. *Chaenactis* sect. *Macrocarphus*, p. 403
   2. Annuals; pappi usually of (1–)4–8(–14) scales in 1, 2 abruptly unequal, or 2–3 gradually unequal series, sometimes 0 or coroniform; leaf blades not gland-dotted (except *C. macrantha*) . . . . 378c. *Chaenactis* sect. *Chaenactis*, p. 408

378a. CHAENACTIS de Candolle sect. ACARPHAEA (Harvey & A. Gray) A. Gray, Proc. Amer. Acad. Arts 10: 74. 1874

Acarphaea Harvey & A. Gray, Mem. Amer. Acad. Arts, n. s. 4: 98. 1849

Annuals; proximal indument predominantly ± farinose, not arachnoid. **Stems** usually 1, erect; branches mainly distal. **Leaves:** largest blades deltate to ± ovate [broadly elliptic], (2–)3–4-pinnately lobed, not gland-dotted. **Heads** discoid (mostly 3–20+ per stem). **Peduncles** ascending to erect. **Phyllaries:** outer acute to acuminate-aristate. **Florets:** corollas white to pinkish, actinomorphic, ± equal. **Cypselae** ± terete or compressed; **pappi** 0, coroniform (of ± 10 scales), or of 4(–5) [10–16] scales in 1 [2–3] series. $x = 8$.

Species 3 (2 in the flora): sw United States, nw Mexico.

1. Phyllaries: apices (all or inner) usually reddish, acuminate, aristate, terete; receptacles: paleae (0–)3–10+ (± phyllary-like, apices visible among mature floret buds); leaves: ultimate lobes ± terete; cypselae ± terete; involucres mostly 5–10 mm diam. . . . . 2. *Chaenactis carphoclinia*
1. Phyllaries: apices (all) ± green, acute or scarcely acuminate, not aristate, ± plane; receptacles: paleae 0; leaves: ultimate lobes plane or involute; cypselae compressed; involucres mostly 10–25 mm diam.
    2. Pappi 0 or coroniform (of ± 10 scales, longest 0.1–0.5 mm); involucres mostly 10–15 mm diam.; phyllaries ± densely villous, not or sparsely glandular . . . . 1. *Chaenactis artemisiifolia*
    2. Pappi of 10–16 scales in 2–3 subequal series, longest scales 4–6 mm; involucres mostly 20–25 mm diam.; phyllaries ± glandular, not or sparsely villous . . . .
        . . . . [*Chaenactis lacera*, see 1. *C. artemisiifolia*]

1. **Chaenactis artemisiifolia**   (Harvey & A. Gray) A. Gray, Proc. Amer. Acad. Arts 10: 74. 1874 (as artemisiaefolia) • White pincushion

Acarphaea artemisiifolia Harvey & A. Gray, Mem. Amer. Acad. Arts, n. s. 4: 98. 1849 (as artemisiaefolia)

Plants (15–)25–90(–200) cm. **Leaves** basal (withering) and cauline, 3–15(–20) cm; largest blades ± plane, not succulent; primary lobes mostly 5–10 pairs, ultimate lobes ± crowded, antrorse, lanceolate to elliptic, plane. **Peduncles** 1.5–6 cm. **Involucres** ± hemispheric, mostly 10–15 mm diam. **Phyllaries:** longest 7–10(–12) mm, ± densely villous, not or sparsely glandular; apices (all) erect, ± green, acute or scarcely acuminate, not aristate, ± plane. **Receptacles:** paleae 0. **Corollas** 5–7 mm. **Cypselae** compressed, 4–7 mm; **pappi** 0 or coroniform (of ± 10 scales, longest 0.1–0.5 mm). **2*n* = 16.**

Flowering Apr–early Jul. Dry canyons, open slopes, often over granitoid rocks, locally abundant in chaparral burns or other recovering disturbances; 80–1600 m; Calif.; Mexico (Baja California).

In the flora area, *Chaenactis artemisiifolia* is known from the Transverse and Peninsular ranges and seaward valleys of southwestern California. It is fire-adapted; its germination is significantly enhanced by exposure to biomass smoke (J. E. Keeley and C. J. Fotheringham 1998).

Chaenactis lacera Greene, the eighteenth species of the genus, is known from coastal portions (including islands) of the western Vizcaíno Desert in Baja California and Baja California Sur, Mexico. Forms of *C. artemisiifolia* sometimes resemble *C. lacera* in coastal southern California (P. Stockwell 1940), where *C. lacera* could eventually be introduced. Besides the key characteristics above, *C. lacera* differs from *C. artemisiifolia* by its largest leaf blades broadly ± elliptic, 2–3-pinnately lobed, ultimate lobes remote, recurved to retrorse, ± linear, involute (leaf blades appearing ± skeletal).

2. **Chaenactis carphoclinia**   A. Gray in W. H. Emory, Rep. U.S. Mex. Bound. 2(1): 94. 1859 • Pebble or straw-bed pincushion F

Plants (5–)10–30(–60) cm. **Leaves** basal (often withering) and, usually, cauline, 1–6(–10) cm; largest blades ± 3-dimensional, not to somewhat succulent; primary lobes mostly 2–7(–10) pairs, ultimate lobes ± crowded to remote, antrorse, linear, terete. **Peduncles** 2–6 cm. **Involucres** obconic to ± cylindric or hemispheric, mostly 5–10 mm diam. **Phyllaries:** longest 7–10 mm, ± granular-glandular and villous; apices (all or inner) erect to incurved, usually reddish, acuminate, aristate, terete. **Receptacles:** paleae (0–)3–10+ (persistent, ± phyllary-like, apices visible among mature floret buds). **Corollas** 4–6 mm. **Cypselae** ± terete, 3–4.5 mm; **pappi** usually of 4(–5) scales, longest 3–5 mm.

Varieties 2 (2 in the flora): sw United States, nw Mexico.

Though occasionally suspected (P. Stockwell 1940), there are no confirmed natural or artificial hybrids between *Chaenactis carphoclinia* and any other member of the genus. The presence of paleae on the receptacle of *C. carphoclinia* is unique in *Chaenactis*.

1. Leaves basal and cauline, longest 1–6(–7) cm; petioles scarcely dilated proximally, ± herbaceous . . . . 2a. *Chaenactis carphoclinia* var. *carphoclinia*
1. Leaves ± basal, longest 7–10 cm; petioles dilated proximally, ± indurate . . . . 2b. *Chaenactis carphoclinia* var. *peirsonii*

2a. Chaenactis carphoclinia A. Gray var. **carphoclinia**

Chaenactis carphoclinia var. *attenuata* (A. Gray) M. E. Jones

Plants (5–)10–30(–40) cm. **Leaves** basal (± withering) and cauline, longest 1–6(–7) cm; petioles scarcely dilated proximally, ± herbaceous. $2n = 16$.

Flowering Jan–Jun. Open, rocky or gravelly (sometimes sandy) desert slopes and flats, shrublands; -90–1900 m; Ariz., Calif., Nev., N.Mex., Utah; Mexico (Baja California, Sonora).

Variety *carphoclinia* is one of the most abundant spring wildflowers in the Sonoran Desert; it extends to the southern Great Basin and northwestern Chihuahuan Desert. It is reported to be eaten by desert tortoises (*Gopherus agassizii* Cooper). Some stunted forms have been named var. *attenuata*; such forms recur in the most arid (or otherwise severe) habitats throughout the range of the species.

2b. Chaenactis carphoclinia A. Gray var. **peirsonii** (Jepson) Munz, Man. S. Calif. Bot. 601. 1935

• Peirson pincushion [C] [E]

Chaenactis peirsonii Jepson, Madroño 1: 259. 1929

Plants mostly 40–60 cm. **Leaves** ± basal, longest 7–10 cm; petioles dilated proximally, ± indurate.

Flowering Mar–Apr. Open, rocky or gravelly desert slopes and flats, shrublands; of conservation concern; 0–200 m; Calif.

Variety *peirsonii* is known from the eastern Santa Rosa Mountains and adjacent Salton Sea area near the common border of Imperial, Riverside, and San Diego counties. Unlike var. *attenuata*, in which it has sometimes been included, var. *peirsonii* is geographically and morphologically well defined.

378b. CHAENACTIS de Candolle sect. MACROCARPHUS (Nuttall) Torrey & A. Gray, Fl. N. Amer. 2: 371. 1842

Macrocarphus Nuttall, Trans. Amer. Philos. Soc., n. s. 7: 376. 1841

Biennials, perennials, or subshrubs (rarely flowering first year); proximal indument predominantly arachnoid, lanuginose, or pannose (rarely glabrescent), not farinose. **Stems** 1–25+, prostrate to erect; branches mainly proximal. **Leaves:** largest blades deltate, elliptic, linear, or ovate, (0–)1–2-pinnately or -subpalmately lobed, gland-dotted beneath indument. **Heads** discoid. **Peduncles** prostrate to erect. **Phyllaries:** outer ± blunt. **Corollas** white to pinkish or cream, actinomorphic, ± equal. **Cypselae** ± terete; **pappi** of (8–)10–20 scales in 2–4 equal or gradually unequal series. $x = 6$.

Species 8 (8 in the flora): w North America, nw Mexico.

Species of sect. *Macrocarphus* occur mainly in montane to alpine habitats; all except *Chaenactis douglasii* are narrowly distributed. With *C. douglasii* here broadly defined, all the species of sect. *Macrocarphus* are sharply distinct.

1. Leaves ± cauline and, often, basal; plants not or scarcely cespitose, not matted; heads (1–)2–25+ per stem.
   2. Subshrubs (usually); proximal indument (especially stems) persistent, whitish, densely lanuginose or pannose; largest leaf blades deltate to ovate, ± plane (California).
      3. Phyllaries: longest 10–13 mm, outer predominantly arachnoid to closely lanuginose (sparsely, if at all, stipitate-glandular) . . . . 3. *Chaenactis parishii*
      3. Phyllaries: longest 14–18 mm, outer predominantly stipitate-glandular (other indument none or sparse) . . . . 4. *Chaenactis suffrutescens* (in part)
   2. Biennials or perennials (rarely slightly woody or flowering first year); proximal indument ± thinning with age, grayish, arachnoid to ± lanuginose; largest leaf blades ± elliptic or lanceolate to ovate, plane or ± 3-dimensional.

4. Outer phyllaries usually densely, sometimes sparsely or obscurely, stipitate-glandular and, often, arachnoid, lanuginose, and/or villous; largest leaf blades ± 3-dimensional, usually 2-pinnately lobed, primary lobes ± congested, ultimate lobes ± involute and/or twisted .... 6. *Chaenactis douglasii* (in part)

4. Outer phyllaries closely lanuginose, not stipitate-glandular; largest leaf blades ± plane, 1-pinnately lobed, lobes remote, ± plane (Washington) .... 8. *Chaenactis thompsonii*

1. Leaves ± basal; plants cespitose or ± matted; heads 1(–3) per stem.

5. Outer phyllaries predominantly arachnoid, sericeous, or ± lanuginose (sparsely, if at all, stipitate-glandular).

6. Cypselae sparsely glandular amidst other indument; largest leaf blades 2-pinnately lobed (± 3-dimensional, primary lobes 4–12 pairs, peduncles mostly ascending to erect) .... 6. *Chaenactis douglasii* (in part)

6. Cypselae eglandular; largest leaf blades (0–)1(–2)-pinnately or -subpalmately lobed (± plane, and/or primary lobes 0–4 pairs, and/or peduncles mostly prostrate).

7. Longest pappus scales 2.5–4.5 mm (lengths 0.4–0.8 times corollas); leaf blades ± plane; peduncles mostly ascending to erect; Idaho .... 9. *Chaenactis evermannii*

7. Longest pappus scales 5–8 mm (lengths 0.9–1 times corollas); leaf blades ± plane or 3-dimensional; peduncles mostly prostrate; California, Nevada ...... 10. *Chaenactis alpigena*

5. Outer phyllaries predominantly or evidently stipitate-glandular (other indument none, sparse, or ± arachnoid).

8. Largest leaf blades deltate to ovate, ± plane, ultimate lobes ± plane.

9. Plants 2–10(–12) cm; leaves 2.5–5 cm; longest phyllaries 9–12(–14) mm; corollas 5.5–8 mm; longest pappus scales 3–5 mm .... 5. *Chaenactis nevadensis*

9. Plants (10–)25–45(–60) cm; leaves 5–10 cm; longest phyllaries 14–18 mm; corollas 8.5–10 mm; longest pappus scales 7–9 mm ...... 4. *Chaenactis suffrutescens* (in part)

8. Largest leaf blades linear-cylindric to ± elliptic or slightly ovate, ± 3-dimensional, ultimate lobes ± involute and/or twisted.

10. Largest leaf blades ± elliptic to slightly ovate, primary lobes (4–)5–9(–12) pairs, ± congested, scarcely imbricate; not s California .... 6. *Chaenactis douglasii* (in part)

10. Largest leaf blades linear-cylindric to ± fusiform, primary lobes (7–)10–18+ pairs, ± imbricate; s California .... 7. *Chaenactis santolinoides*

3. **Chaenactis parishii** A. Gray, Proc. Amer. Acad. Arts 20: 299. 1885 • Parish chaenactis [C]

Subshrubs, (10–)20–40(–60) cm (not cespitose or matted); proximal indument (especially of stems) persistent, whitish, densely lanuginose or pannose. **Stems** mostly 5–15+, erect. **Leaves** mostly cauline, (1–)2–5 cm; largest blades lance-ovate or deltate, ± plane, 1-pinnately lobed; lobes mostly 2–5 pairs, remote, ± plane. **Heads** mostly 1–3 per stem. **Peduncles** ascending to erect, 2–8(–20) cm. **Involucres** ± obconic. **Phyllaries:** longest 10–13 mm; outer predominantly arachnoid to closely lanuginose (sparsely, if at all, stipitate-glandular), apices ± squarrose, pliant. **Corollas** 7–8.5 mm. **Cypselae** 4–7 mm; **pappi:** longest scales 6–8 mm. $2n = 12$.

Flowering May–Jul. Open rocky to sandy soils in low montane chaparral; of conservation concern; 1300–2500 m; Calif.; Mexico (Baja California).

Chaenactis parishii is sometimes cultivated in rock gardens. It is known from small, isolated populations in the higher Peninsular Ranges of Riverside and San Diego counties and adjacent Baja California. *Chaenactis parishii* and *C. suffrutescens* form a species pair well marked by the (usually) subshrubby habit, proximal indument persistent, white, felty, heads relatively large, and largest leaf blades lance-ovate to deltate.

4. **Chaenactis suffrutescens** A. Gray, Notes Compositae, 100. 1880 • Shasta pincushion [C][E]

Chaenactis suffrutescens var. *incana* Stockwell

Perennials or subshrubs (10–)25–45(–60) cm (sometimes cespitose, not matted); proximal indument (especially stems) persistent, whitish, densely lanuginose or pannose. **Stems** mostly 5–15+, usually erect. **Leaves** usually ± cauline, sometimes ± basal, 5–10 cm; largest blades lance-ovate or deltate, ± plane, 1–2-pinnately lobed; primary lobes mostly 2–5 pairs, ± remote, ultimate lobes ± plane. **Heads** mostly 1–3 per stem. **Peduncles** ascending to erect, (5–)10–20 cm. **Involucres** ± cylindric. **Phyllaries:** longest 14–18 mm; outer predominantly stipitate-glandular (other indument none or sparse), apices ± squarrose, pliant. **Corollas** 8.5–10 mm. **Cypselae** 7–9 mm; **pappi:** longest scales 7–9 mm. **2n** = 12.

Flowering May–early Sep. Loose scree, sand, rocky soils, slopes, drainages, usually on serpentine or other ultramafic deposits; of conservation concern; 700–2300 m; Calif.

Chaenactis suffrutescens is known from the southern and eastern Klamath Ranges and northern Coast Ranges of Trinity and Siskiyou counties (nearly to Oregon). It is sometimes cultivated in rock gardens and may be found outside its native range. Small forms of *C. suffrutescens* from southern Trinity County approach *C. nevadensis* in habit (see discussion there).

5. **Chaenactis nevadensis**   (Kellogg) A. Gray in W. H. Brewer et al., Bot. Calif. 1: 391. 1876 • Sierra pincushion, Nevada dustymaidens   E

Hymenopappus nevadensis Kellogg, Proc. Calif. Acad. Sci. 5: 46. 1873

Perennials, 2–10(–12) cm (cespitose or ± matted); proximal indument ± thinning with age, whitish, lanuginose. **Stems** mostly 10–20+, decumbent to ± erect. **Leaves** ± basal, 2.5–5 cm; largest blades ovate to deltate, ± plane, (1–)2-pinnately lobed; primary lobes mostly 2–4 pairs, ± congested, ultimate lobes ± plane. **Heads** 1(–2) per stem. **Peduncles** mostly ascending to erect, (0.5–)3–11 cm. **Involucres** obconic to ± cylindric. **Phyllaries:** longest 9–12(–14) mm; outer stipitate-glandular, apices erect, ± rigid. **Corollas** 5.5–8 mm. **Cypselae** 5.5–7.5 mm; **pappi:** longest scales 3–5 mm. **2n** = 12.

Flowering Jul–mid Sep. Loose sandy or gravelly, mainly volcanic soils or scree (rarely on serpentine), openings in or above subalpine conifer forests; 1900–3200 m; Calif., Nev.

Chaenactis nevadensis is known mainly from the northern Sierra Nevada and southern Cascade Range (Shasta to Placer counties, California; Washoe County, Nevada). It was recently discovered disjunct on ultramafic rocks of Bully Choop Mountain west of Redding, California, where it approaches small forms of *C. suffrutescens* in habit (see discussion there). It is sometimes cultivated in rock-gardens and may be found beyond its native range. *Chaenactis nevadensis* and *C. suffrutescens* appear to be sister or ancestor-derivative species. I have seen no evidence to support reports that *C. nevadensis* intergrades with *C. alpigena* (P. Stockwell 1940, as *C. nevadensis* var. *mainsiana*), with *C. douglasii* var. *alpina* (M. Graf 1999), or with any other taxon.

6. **Chaenactis douglasii**   (Hooker) Hooker & Arnott, Bot. Beechey Voy., 354. 1839 • Hoary pincushion, Douglas's dustymaiden   E F

Hymenopappus douglasii Hooker, Fl. Bor.-Amer. 1: 316. 1833; *Macrocarphus douglasii* (Hooker) Nuttall

Biennials or perennials, (2–)5–50(–60) cm (rarely slightly woody or flowering first year, sometimes cespitose or ± matted); proximal indument thinning with age, grayish, mostly arachnoid-sericeous to thinly lanuginose. **Stems** 1–25+, erect to spreading. **Leaves** basal, or basal (sometimes withering) and ± cauline, (1–)2–12(–15) cm; largest blades ± elliptic or slightly lanceolate to ovate, ± 3-dimensional, usually 2-pinnately lobed; primary lobes (4–)5–9(–12) pairs, ± congested, scarcely imbricate, ultimate lobes ± involute and/or twisted. **Heads** 1–25+ per stem. **Peduncles** mostly ascending to erect, 1–10 cm. **Involucres** obconic to ± hemispheric. **Phyllaries:** longest 9–15(–17) mm; outer usually stipitate-glandular (sometimes sparsely or obscurely, rarely eglandular) and, often, arachnoid to lanuginose and, sometimes, sparsely villous, apices usually ± squarrose, pliant. **Corollas** 5–8 mm. **Cypselae** 5–8 mm (usually sparsely glandular amidst other indument); **pappi:** longest scales 3–6 mm.

Varieties 2 (2 in the flora): North America.

Chaenactis douglasii is widespread and variable (see discussion under var. *douglasii*).

1. Leaves basal (sometimes withering) and ± cauline; plants not or scarcely cespitose, not matted; stems usually 1–5(–12); heads (1–)2–25+ per stem . . . . 6a. *Chaenactis douglasii* var. *douglasii*
1. Leaves strictly basal; plants cespitose or ± matted; stems (1–)10–25+; heads 1(–2) per stem . . . . 6b. *Chaenactis douglasii* var. *alpina*

6a. **Chaenactis douglasii** (Hooker) Hooker & Arnott var. **douglasii** • False yarrow   E F

Chaenactis angustifolia Greene; *C. douglasii* var. *achilleifolia* (Hooker & Arnott) A. Gray; *C. douglasii* var. *glandulosa* Cronquist; *C. douglasii* var. *montana* M. E. Jones; *C. douglasii* var. *rubricaulis* (Rydberg) Ferris; *C. pedicularia* Greene; *C. pumila* Greene; *C. ramosa* Stockwell

Biennials or perennials, (3–)8–50(–60) cm (sometimes slightly woody or flowering first year, not or scarcely cespitose, not matted). **Stems** usually 1–5(–12). **Leaves** basal (sometimes withering) and ± cauline, 1.5–12(–15) cm, indument ± persistent. **Heads** (1–)2–25+ per stem.

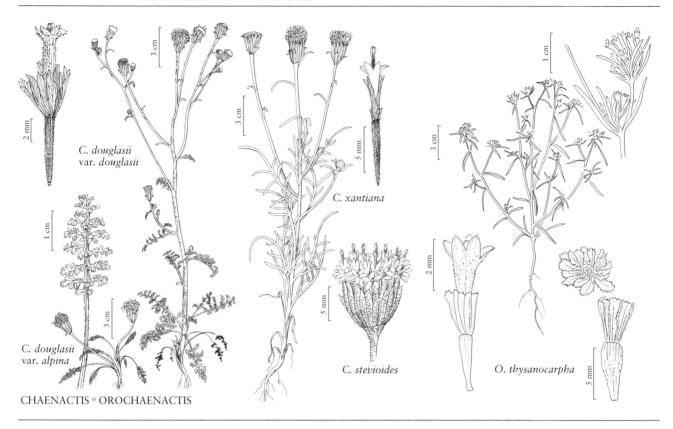

C. douglasii
var. douglasii

C. douglasii
var. alpina

CHAENACTIS ° OROCHAENACTIS

C. xantiana

C. stevioides

O. thysanocarpha

Phyllaries: longest 9–15(–17) mm; outer ± stipitate-glandular and, often, arachnoid to lanuginose and, sometimes, sparsely villous. $2n$ = 12, 24, 36 (plus dysploid numbers and triploids).

Flowering May–Sep. Open sandy, gravelly, or rocky substrates in shrublands, woodlands, forests, often unstable or recently disturbed sites; 400–3500 m; Alta., B.C.; Ariz., Calif., Colo., Idaho, Mont., Nev., N.Mex., N.Dak., Oreg., S.Dak., Utah, Wash., Wyo.

Variety *douglasii* is known in and east of the Sierra Nevada and Cascade Ranges, to the western edge of the northern Great Plains, and in the Klamath Ranges and North Coast Ranges of northwestern California (where its lowest elevations occur). It is known from most mountains within that range; northward it also descends to the higher plains and valley floors. A report of var. *douglasii* from Saskatchewan was not verified by me. Seeds of var. *douglasii* are commercially available and are planted for revegetation or horticultural purposes and it may be found outside its native range.

Four ploidy levels were documented across the range of var. *douglasii* by J. S. Mooring (1965, 1980) and P. H. Raven and D. W. Kyhos (1961). Mooring's (1980) data suggested that diploids are more frequent at higher elevations in major mountain chains and are more often mixed with polyploids at lower elevations or in areas of Pleistocene or later disturbance. The chromosome numbers listed above are predominant; Mooring (1965, 1980, 1992) also documented $2n$ = 14, 15, 26, or 30 as regular bivalents. Additional numbers, involving various combinations of bivalents with univalents, tetravalents, and/or extra chromosomes, have been observed.

Most of the diploid elements of var. *douglasii* are distinctive and are connected by a morphologically continuous series of polyploids (usually assigned to var. *achilleifolia*). Some diploid forms (including var. *rubricaulis* and *Chaenactis ramosa*) appear repeatedly and discontinuously in suitable habitats. In particular, forms named var. *montana* seem to arise wherever the species reaches sufficient elevation. Such populations have no historic or genetic cohesion to justify their recognition as a collective taxon, even though their reduced stature may become genetically fixed in each instance. (Variety *alpina*, recognized below with hesitation, may be just an extreme such case.) *Chaenactis angustifolia* of the northern Rocky Mountains and possibly *C. pedicularia* of Colorado are more distinctive morphologically and geographically, and merit further study and possible varietal status.

SELECTED REFERENCE   Mooring, J. S. 1980. A cytogeographic study of *Chaenactis douglasii* (Compositae, Helenieae). Amer. J. Bot. 67: 1304–1319.

6b. Chaenactis douglasii (Hooker) Hooker & Arnott var. **alpina** A. Gray in A. Gray et al., Syn. Fl. N. Amer. 1(2): 341. 1884 • Alpine dustymaidens or pincushion E F

Chaenactis alpina (A. Gray) M. E. Jones; *C. alpina* var. *leucopsis* (Greene) Stockwell; *C. alpina* var. *rubella* (Greene) Stockwell; *C. panamintensis* Stockwell

Perennials, mostly (2–)5–10(–20) cm (cespitose to ± matted). **Stems** (1–)10–25+. **Leaves** strictly basal, to (1–)2–6 cm, indument persistent or glabrate. **Heads** 1(–2) per stem. **Phyllaries:** longest 9–12 mm; outer ± stipitate-glandular (rarely eglandular) and, often, arachnoid to lanuginose. $2n = 12$.

Flowering Jul–Sep. Rocky or gravelly alpine ridges, talus, fell-fields, rock crevices; 2700–4000 m; Calif., Colo., Idaho, Mont., Oreg., Utah, Wyo.

In California, relatively few populations of *Chaenactis douglasii* approach typical var. *alpina*. A report of var. *alpina* from the Spring Mountains of Clark County, Nevada (P. Stockwell 1940, as *C. panamintensis*), could not be confirmed and was likely based on stunted var. *douglasii*; it should be sought in the Carson, Ruby, and Snake ranges of northern Nevada. A report of var. *alpina* from southern British Columbia has not been verified by me. Variety *alpina* is sometimes cultivated, especially for rock gardens, and it may be encountered outside its native range.

Local populations of these largely alpine plants can appear highly distinctive and uniform, leading to their past recognition as a separate species. Across the range, most such populations intergrade continuously with local populations of var. *douglasii*, as is often evident among specimens of the same gathering. The relative development of roots, stems, and branches appears highly mutable throughout *C. douglasii*.

Chaenactis pumila Greene was assigned to the variety or species *alpina* by R. D. Dorn (1988c), P. Stockwell (1940), and others; it is a small form of var. *douglasii*, as assigned by D. C. Eaton in S. Watson (1871).

7. Chaenactis santolinoides Greene, Bull. Torrey Bot. Club 9: 17. 1882 • Santolina pincushion C E

Perennials, 10–25(–35) cm (cespitose or ± matted); proximal indument thinning with age, whitish to grayish, lanuginose. **Stems** mostly 5–15+, erect to ± spreading. **Leaves** basal, (1–)3–11 cm; largest blades linear-cylindric to ± fusiform, 3-dimensional, 1–2-pinnately lobed; primary lobes (7–)10–18+ pairs, ± imbricate, ultimate lobes ± involute, twisted. **Heads**

1(–3) per stem. **Peduncles** mostly ascending to erect, mostly 8–25 cm. **Involucres** obconic to ± cylindric. **Phyllaries:** longest 8–13 mm; outer evidently stipitate-glandular and, sometimes, ± arachnoid, apices erect, ± rigid. **Corollas** 5–7 mm. **Cypselae** 4–6 mm; **pappi:** longest scales 3–4.5 mm. $2n = 12$.

Flowering (Mar–)May–Jul. Exposed sandy to rocky summits, ridges, scree, talus, openings in or above conifer forests, sometimes road cuts or other recent disturbances; of conservation concern; (1100–)1500–2800 m; Calif.

Chaenactis santolinoides is known from the southern Sierra Nevada and Transverse Ranges. It is sometimes cultivated in rock gardens and may be found beyond its native range. It may be relatively recently derived from an isolated segment of *C. douglasii* var. *alpina*. The morphology of *C. panamintensis* (here assigned to *C. douglasii* var. *alpina*) suggests past convergence toward, or genetic influence from, *C. santolinoides*.

8. Chaenactis thompsonii Cronquist in C. L. Hitchcock et al., Vasc. Pl. Pacif. N.W. 5: 123, fig. [p. 125]. 1955 • Thompson's pincushion C E

Perennials, 10–30 cm (not or scarcely cespitose, not matted); proximal indument thinning with age, grayish, arachnoid-sericeous to thinly lanuginose. **Stems** mostly 5–15+, ascending to erect. **Leaves** mostly cauline, 2–5 cm; largest blades ± elliptic, ± plane, 1-pinnately lobed; lobes mostly 2–5 pairs, remote, ± plane. **Heads** mostly 1–3 per stem. **Peduncles** ascending to erect, 2–5 cm. **Involucres** ± obconic. **Phyllaries:** longest (10–)12–15 mm; outer closely lanuginose, not stipitate-glandular, apices erect, ± rigid. **Corollas** 7–9 mm. **Cypselae** 7–9 mm (eglandular); **pappi:** longest scales 3.5–5 mm.

Flowering Jun–Aug. Rocky or gravelly serpentine slopes, scree, talus, openings in or above conifer forests; of conservation concern; (900–)1200–2200 m; Wash.

Chaenactis thompsonii appears to be sister to *C. evermannii*; it is known from the mountains of central and northwestern Washington. The similar habits of *C. thompsonii* and *C. ramosa* (= *C. douglasii* var. *douglasii*) appear to result from convergent evolution in the distinctive habitat of their type localities (Wenatchee Mountains), not from a close genetic relationship as suggested by Cronquist.

9. **Chaenactis evermannii** Greene, Leafl. Bot. Observ. Crit. 2: 224. 1912 • Evermann's pincushion E

Chaenactis nevadensis (Kellogg) A. Gray var. *mainsiana* (A. Nelson & J. F. Macbride) Stockwell

Perennials, mostly 6–12 cm (cespitose or ± matted); proximal indument thinning with age, grayish, mostly arachnoid-sericeous to thinly lanuginose. **Stems** mostly 5–20+, ascending to erect. **Leaves** basal, 1–5 cm; largest blades broadly ± elliptic, ± plane, 1-pinnately lobed; lobes 2–5 pairs, remote, ± plane. **Heads** 1(–3) per stem. **Peduncles** mostly ascending to erect, 5–10 cm. **Involucres** ± obconic. **Phyllaries:** longest 8–12 mm; outer predominantly closely lanuginose, sparsely, if at all, stipitate-glandular, apices erect, ± rigid. **Corollas** 5–6.5 mm. **Cypselae** 5–6.5 mm (eglandular); **pappi:** longest scales 2.5–4.5 mm (lengths 0.4–0.8 times corollas). $2n = 12$.

Flowering Jun–Aug. Subalpine, usually decomposing granitic sand or gravel slopes, ridges, scree, talus, openings in or above conifer forests; 1200–3000 m; Idaho.

Chaenactis evermannii is known from mountains of central Idaho. Reports of it from Washington and California/Nevada (P. Stockwell 1940, some as *C. nevadensis* var. *mainsiana*) were based on specimens of *C. thompsonii* and *C. alpigena*, respectively; all three species are closely related.

10. **Chaenactis alpigena** Sharsmith, Contr. Dudley Herb. 4: 319, plate 12, fig. a. 1955 • Sharsmith or southern Sierra or mountain pincushion E

Perennials, 2–7 cm (cespitose or ± matted); proximal indument thinning with age, grayish to yellowish, lanuginose (sometimes glabrescent). **Stems** mostly 5–15+, erect to prostrate. **Leaves** basal, 1–2.5(–3.5) cm; largest blades linear or broadly elliptic to slightly obovate, ± plane or 3-dimensional, (0–)1–2-pinnately or -subpalmately lobed; primary lobes 2–7 pairs, ± congested, ultimate lobes ± plane to involute and twisted. **Heads** 1 per stem. **Peduncles** mostly prostrate, (0.5–)2–7(–10) cm. **Involucres** obconic to ± cylindric. **Phyllaries:** longest 9–14 mm; outer ± lanuginose, not stipitate-glandular, apices erect, ± rigid. **Corollas** 5.5–8 mm. **Cypselae** 5–8 mm (eglandular); **pappi:** longest scales 5–8 mm (lengths 0.9–1 times corollas).

Flowering Jul–Sep. Open, loose, subalpine to alpine granitic sand, gravel, scree; 2200–3900 m; Calif., Nev.

Chaenactis alpigena is known from the central and southern Sierra Nevada and adjacent northern White Mountains. It is sometimes cultivated in rock gardens and may be found beyond its native range. It appears to be sister to *C. thompsonii* and/or *C. evermannii*.

Chaenactis alpigena shows clinal variation; northward the leaves tend to be ± plane, (0–)1-subpalmately lobed, lobes 2–5 pairs, ± plane; southward the leaves tend to be ± 3-dimensional, 1–2-pinnately lobed, primary lobes 4–7 pairs, ultimate lobes ± involute, twisted. Leaves of intermediate forms tend to be ± plane, 1-pinnately lobed, lobes mostly 4–5 pairs, ± plane. The variation is gradual and does not support recognition of infraspecific taxa. Throughout its range, occasional plants of *C. alpigena* are green and glabrate.

378c. **CHAENACTIS** de Candolle sect. **CHAENACTIS**

Annuals; proximal indument predominantly arachnoid, lanuginose, stipitate-glandular, or glabrescent, not farinose. **Stems** mostly 1–5(–12), usually erect to ascending; branches proximal and/or distal. **Leaves:** largest blades usually ± oblanceolate to elliptic or linear (± ovate in some *C. macrantha*), (0–)1–2-pinnately lobed, not gland-dotted (except *C. macrantha*). **Heads** discoid or ± radiant. **Peduncles** erect to ascending (by flowering). **Phyllaries:** outer usually ± blunt, sometimes acute. **Florets:** corollas white to pinkish, cream, or bright to dark yellow, actinomorphic and ± equal, or peripheral enlarged (and often zygomorphic). **Cypselae** ± terete (except *C. nevii*); **pappi** 0, coroniform (of ± 10 scales), or of (1–)4–8(–14) scales in 1 series, in 2, abruptly unequal series, or in 2–3, gradually unequal series. $x = 6$.

Species 7 (7 in the flora): sw United States, nw Mexico.

Taxa of sect. *Chaenactis* grow mainly in valley and foothill habitats; all except *Chaenactis macrantha*, *C. cusickii*, and *C. nevii* appear to form occasional natural hybrids where sympatric. *Chaenactis macrantha* and *C. cusickii* may each warrant a separate monotypic section and here probably render sect. *Chaenactis* polyphyletic. Both species share similarities with *C. xantiana*.

The species with zygomorphic and enlarged peripheral corollas (*Chaenactis glabriuscula, C. stevioides,* and *C. fremontii*) are closely related, and identification of some specimens requires careful evaluation, especially when fresh corolla color is not evident. D. W. Kyhos (1965) presented strong evidence that the species with corollas white and $2n = 10$ (*C. stevioides* and *C. fremontii*) evolved separately from within *C. glabriuscula* (corollas yellow, $2n = 12$) via chromosomal rearrangements. That study did not sample very well the large amount of variation in *C. glabriuscula*, nor did it rule out involvement of other taxa such as *C. xantiana*. *Chaenactis xantiana* (corollas white, $2n = 14$) is known to form hybrids with *C. glabriuscula* that also exhibit $2n = 10$ in some individuals (P. Stockwell 1940).

A series of natural hybrids and backcrosses (keyed below), involving various combinations of the four species named in the preceding paragraph, appears to occur in the inner South Coast Ranges and adjacent southern and western San Joaquin Valley of west-central California. Where *Chaenactis xantiana* or *C. glabriuscula* var. *heterocarpha* is involved, a partial second series of pappus scales is usually evident. Identification of specimens from this area can be difficult.

1. Corollas bright to dark yellow; mainly west of deserts.
    2. Peripheral corollas actinomorphic, scarcely enlarged (relative to inner); cypselae compressed; pappi 0 or coroniform (of ± 10 scales, longest 0.1–0.5 mm); Oregon . . . . . . 14. *Chaenactis nevii*
    2. Peripheral corollas ± zygomorphic, enlarged (relative to inner); cypselae ± terete; pappi of (1–)4, or (5–)8, scales, longest scales (1–)2–8 mm; California . . . . 15. *Chaenactis glabriuscula*
1. Corollas white to pinkish, cream, or pale yellow; mainly in and near deserts.
    3. Peripheral corollas zygomorphic, enlarged; pappi of (1–)4(–5) scales, usually in 1 series (rarely with partial outer series in *Chaenactis stevioides*); peduncles usually stipitate-glandular distally (at least early, sometimes glabrescent by fruit).
        4. Pappi: longest scales 1.5–6 mm, lengths mostly 0.3–0.9 times corollas (apices hidden among corollas at flowering); proximal indument ± arachnoid-sericeous (tardily glabrescent except around nodes); leaf blades mostly 1–2-pinnately lobed, primary lobes mostly 4–8 pairs; outer phyllaries usually stipitate-glandular and/or ± arachnoid in fruit, apices blunt . . . . 16. *Chaenactis stevioides*
        4. Pappi: longest scales 6–8.5 mm, lengths 1–1.3 times corollas (apices visible among corollas at flowering); proximal indument glabrescent (early ± arachnoid, glabrous by flowering); leaf blades 0–1-pinnately lobed, lobes 1–2(–5) pairs; outer phyllaries usually glabrescent in fruit, apices acute . . . . 17. *Chaenactis fremontii*
    3. Peripheral corollas actinomorphic, not or scarcely enlarged; pappi of (6–)8–14 scales in 2 abruptly unequal series, or in 2–3 gradually unequal series; peduncles ± arachnoid to lanuginose distally, not stipitate-glandular.
        5. Corollas (nocturnal) 9–12(–15) mm (lengths 1.8–2.2 times cypselae; anthers ± included); leaf blades not succulent . . . . 11. *Chaenactis macrantha*
        5. Corollas (diurnal) 3–10 mm (± equaling cypselae; anthers exserted); leaf blades ± succulent.
            6. Longest phyllaries 10–18 mm (surpassed by florets), outer distally tomentulose-puberulent in fruit (proximally glabrous), not stipitate-glandular . . . . . . 13. *Chaenactis xantiana*
            6. Longest phyllaries 6–9(–10) mm (or surpassing florets), outer (uniformly) ± arachnoid and/or stipitate-glandular or glabrescent.
                7. Pappi of (8–)10–14 scales in 2–3 gradually unequal series; leaf blades ± oblanceolate, not lobed (margins entire or distally ± crenate); peduncles: bracts 1–2, leaflike, surpassing heads; outer phyllaries sparsely arachnoid to glabrescent, not stipitate-glandular (inner apically brownish villosulous) . . . . 12. *Chaenactis cusickii*
                7. Pappi of 6–8 scales in 2, abruptly unequal series; leaves mostly ± elliptic, mostly 1–2-pinnately lobed; peduncles: bracts 0 or 1–2 ± subulate, surpassed by heads; outer phyllaries usually stipitate-glandular, other indument various (apices of inner not villosulous) [hybrids among *Chaenactis fremontii, C. glabriuscula, C. stevioides,* and *C. xantiana*; see discussion of *C.* sect. *Chaenactis*].

11. Chaenactis macrantha D. C. Eaton in S. Watson, Botany (Fortieth Parallel), 171, plate 18, figs. 1–5. 1871 • Showy or bighead dustymaidens, Mojave pincushion [E]

Plants 5–25(–35) cm; proximal indument grayish, arachnoid-sericeous to closely lanuginose (sometimes tardily glabrescent). **Stems** mostly 1–5; branches mainly proximal. **Leaves** basal (withering) and cauline, 1.5–7 cm; largest blades ± elliptic to ovate, ± plane, not succulent, 1(–2)-pinnately lobed (± gland-dotted beneath indument); primary lobes mostly 2–5 pairs, ± remote, ultimate lobes ± plane. **Heads** (± radiant, nocturnally), mostly 1–5(–7) per stem (nodding in bud). **Peduncles** 1.5–8 cm, arachnoid-sericeous to thinly lanuginose distally, not stipitate-glandular. **Involucres** ± obconic to broadly cylindric. **Phyllaries:** longest 12–18 mm; outer arachnoid-sericeous to thinly lanuginose in fruit, not stipitate-glandular, apices ± squarrose, blunt, pliant. **Florets:** corollas (nocturnal) white to pinkish or cream, 9–12(–15) mm (lengths 1.8–2.2 times cypselae; anthers ± included); peripheral corollas nocturnally spreading, actinomorphic, scarcely enlarged. **Cypselae** 5–6(–7) mm; **pappi** of 8 scales in 2, abruptly unequal series, longest scales 5–7 mm. $2n = 12$.

Flowering Mar–early Jul. Open, loose, light-colored, silty, usually calcareous or alkaline, desert soils, often covered by or mixed with gravel; 600–2200 m; Ariz., Calif., Idaho, Nev., Oreg., Utah.

Though the derived floral features of *Chaenactis macrantha* obscure its relationships, it may represent a link between sect. *Chaenactis* (annuals; pappus scales in regular, often strongly reduced series) and sect. *Macrocarphus* (leaf blades gland-dotted). Resemblance of its heads, leaves, and indument to those of *C. thompsonii* and relatives is striking. It appears to form no natural hybrids, perhaps because of its nocturnal corollas.

12. Chaenactis cusickii A. Gray in A. Gray et al., Syn. Fl. N. Amer. ed. 2. 1(2): 452. 1886 • Morning brides, Cusick's pincushion [C][E]

Plants 3.5–10(–15) cm; proximal indument grayish, sparsely arachnoid, glabrescent. **Stems** mostly 1–5; branches mainly proximal. **Leaves** basal (withering) and cauline (not notably smaller or sparser distally), 1–4 cm; largest blades ± oblanceolate, plane, ± succulent, not lobed (margins entire or distally ± crenate). **Heads** mostly 1–5(–12) per stem. **Peduncles** 0.5–2.5 cm, glabrescent distally (sparsely

arachnoid early; bracts 1–2, leaflike, surpassing heads). **Involucres** ± hemispheric to campanulate. **Phyllaries:** longest 6–9(–10) mm (surpassed by florets); outer (uniformly) sparsely arachnoid to glabrescent in fruit, not stipitate-glandular (inner apically brownish-villosulous), apices usually erect, blunt, ± rigid. **Florets:** corollas (diurnal) white to pinkish, 3–7 mm (± equaling cypselae; anthers exserted); peripheral corollas ± erect, actinomorphic, scarcely enlarged. **Cypselae** 4–6 mm; **pappi** of (8–)10–14 scales in 2–3 gradually unequal series, longest scales 1.5–3.5 mm. $2n = 12$.

Flowering late Apr–Jun. Light-colored shrink-swell clay soils from volcanic ash and tuff, semiarid shrublands; of conservation concern; 700–1300(–1800) m; Idaho, Oreg.

Chaenactis cusickii is known from Malheur County, Oregon, and adjacent Owyhee and Canyon counties, Idaho. Its relationship to other species is obscure.

SELECTED REFERENCE Moseley, R. K. 1994. The Status and Distribution of Cusick's False Yarrow (*Chaenactis cusickii*) in Idaho. Boise.

13. Chaenactis xantiana A. Gray, Proc. Amer. Acad. Arts 6: 545. 1865 • Fleshy pincushion, fleshcolor pincushion, Xantus pincushion [E][F]

Plants 10–40 cm; proximal indument grayish, sparsely arachnoid, early glabrescent (usually glabrous by flowering). **Stems** mostly 1–5(–12); branches proximal and/or distal. **Leaves** basal (withering) and cauline, (1–)2–6 cm; largest blades linear or ± elliptic, ± plane or terete, ± succulent, 0–1-pinnately lobed; lobes 1–2(–5) pairs, remote, ± terete. **Heads** mostly 1–5(–7) per stem. **Peduncles** 1–5 (–8) cm, glabrous (and ± expanded) distally. **Involucres** broadly obconic to campanulate. **Phyllaries:** longest 10–18 mm (surpassed by florets); outer distally tomentulose-puberulent in fruit (proximally glabrous, not stipitate-glandular), apices ± squarrose, blunt, pliant. **Florets:** corollas (diurnal) dirty-whitish to pinkish, 6–10 mm (± equal to cypsela lengths, anthers exserted); peripheral corollas erect to ascending, actinomorphic, scarcely enlarged. **Cypselae** 5–9 mm; **pappi** of 8 scales in 2, abruptly unequal series, longest scales 5–9 mm. $2n = 14$.

Flowering late Mar–Jul. Open, deep, loose sandy (rarely gravelly) soils, arid and semiarid shrublands, chaparral; (100–)300–2500 m; Ariz., Calif., Nev., Oreg.

14. Chaenactis nevii A. Gray, Proc. Amer Acad. Arts 19: 30. 1883 • John Day pincushion E

Plants 10–30 cm; proximal indument greenish to grayish, stipitate-glandular and, sometimes, sparsely arachnoid or villous. **Stems** mostly 1–3; branches mainly distal. **Leaves** basal (withering) and cauline, 2–5 cm; largest blades ± elliptic, ± plane to 3-dimensional, not succulent, 1–2-pinnately lobed; primary lobes mostly 3–8 pairs, ± remote, ultimate lobes ± plane to involute. **Heads** mostly 3–9 per stem. **Peduncles** 1–6 cm, distally stipitate-glandular and, sometimes, ± arachnoid to villous. **Involucres** ± hemispheric to campanulate. **Phyllaries:** longest 6–9 mm; outer predominantly stipitate-glandular and, sometimes, ± arachnoid to villous in fruit, apices ± erect, acute, rigid. **Florets:** corollas bright yellow, 4–6.5 mm; peripheral corollas ± erect, actinomorphic, scarcely enlarged. **Cypselae** 3.5–6 mm (compressed); **pappi** 0 or coroniform (of ± 10 scales, longest 0.1–0.5 mm). $2n = 12$.

Flowering late Apr–mid July. Nearly barren, heavy clay soils from volcanic ash and tuff, sometimes disturbed or moist sites; 400–1000 m; Oreg.

Chaenactis nevii is known from the John Day Basin area in Gilliam, Grant, Jefferson, Wasco, and Wheeler counties. This odd and isolated species combines traits of *C. artemisiifolia* (sect. *Acarphaea*) and *C. glabriuscula*, and might be descended from their common ancestor.

15. Chaenactis glabriuscula de Candolle in A. P. de Candolle and A. L. P. P. de Candolle, Prodr. 5: 659. 1836 • Yellow pincushion

Plants 6–60 cm; proximal indument grayish to whitish, arachnoid to densely lanuginose, or glabrescent. **Stems** mostly 1–5(–12; sometimes ± horizontal); branches proximal and, often, distal. **Leaves** basal (often withering) and cauline, 1–10 cm; largest blades linear or ± elliptic, plane to 3-dimensional, succulent or not, (0–)1–2-pinnately lobed; primary lobes 1–7 pairs, remote to ± congested, ultimate lobes ± plane, involute, twisted, and/or terete. **Heads** (± radiant) mostly 1–20+ per stem. **Peduncles** 1–20(–30) cm, distally stipitate-glandular, ± villous, arachnoid-sericeous, lanuginose, and/or glabrescent. **Involucres** ± hemispheric to obconic or broadly cylindric. **Phyllaries:** longest 4.5–10 mm; outer stipitate-glandular, ± villous, arachnoid-sericeous, lanuginose, and/or glabrescent in fruit, apices erect, blunt, ± rigid. **Florets:** corollas bright to dark yellow, 4–8 mm (inner); peripheral corollas spreading, ± zygomorphic, enlarged. **Cypselae** 3–9 mm (± terete);

pappi of (1–)4 scales in 1 series, or of (5–)8 scales in 2, abruptly unequal series, longest scales (1–)2–8 mm. $2n = 12$.

Varieties 5 (5 in the flora): sw United States, nw Mexico.

The diverse and intergrading forms here included in *Chaenactis glabriuscula* have been divided by P. Stockwell (1940) and subsequent workers into as many as four species and ten varieties. *Chaenactis glabriuscula* is known from the southern two-thirds of the Californian Floristic Province and adjacent desert edges. It has been reported in Massachusetts as a garden escape (variety unspecified); it is not expected to persist there outside cultivation.

Complete interfertility among the taxa recognized here as *Chaenactis glabriuscula* vars. *glabriuscula*, *megacephala*, and *lanosa* was demonstrated by P. Stockwell (1940). Intraspecific crosses involving *C. glabriuscula* var. *orcuttiana* were much less successful; *C. glabriuscula* var. *heterocarpha* was not tested. As noted by W. J. Hooker and G. A. W. Arnott ([1830–]1841) and D. W. Kyhos (1965), some forms of *C. glabriuscula* differ from *C. stevioides* or *C. fremontii* only in corolla color, which can be lost in older or poorly preserved specimens.

1. Pappi of (5–)8 scales in 2, abruptly unequal series . . . . 15a. *Chaenactis glabriuscula* var. *heterocarpha*
1. Pappi of (1–)4(–5) scales in 1 series.
  2. Proximal indument whitish, ± densely lanuginose; leaves ± basal (persistent) . . . . 15c. *Chaenactis glabriuscula* var. *lanosa*
  2. Proximal indument grayish, ± arachnoid, or glabrescent; leaves basal (withering) and cauline.
    3. Largest leaf blades succulent, mostly 2-pinnately lobed; stems ± horizontal . . . . 15e. *Chaenactis glabriuscula* var. *orcuttiana*
    3. Largest leaf blades scarcely succulent, 1(–2)-pinnately lobed; stems ascending to erect.
      4. Longest phyllaries 7–9 × 2–3 mm, outer glabrescent in fruit; pappi: lengths of longest scales ± 0.9(–1) times corollas . . . . . . 15b. *Chaenactis glabriuscula* var. *megacephala*
      4. Longest phyllaries 5–7 × 1–2 mm, outer (at least medially) ± arachnoid-sericeous and, often, ± stipitate-glandular in fruit; pappi: lengths of longest scales 0.4–0.7 times corollas . . . . 15d. *Chaenactis glabriuscula* var. *glabriuscula*

15a. Chaenactis glabriuscula de Candolle var. **heterocarpha** (Torrey & A. Gray) H. M. Hall, Univ. Calif. Publ. Bot. 3: 190. 1907 • Inner Coast Range chaenactis [E]

Chaenactis heterocarpha Torrey & A. Gray, Mem. Amer. Acad. Arts, n. s. 4: 98. 1849; *C. glabriuscula* var. *gracilenta* (Greene) D. D. Keck; *C. heterocarpha* var. *tanacetifolia* (A. Gray) A. Gray; *C. tanacetifolia* A. Gray; *C. tanacetifolia* var. *gracilenta* (Greene) Stockwell

Plants 6–40 cm; proximal indument grayish, arachnoid to sparsely lanuginose. **Stems** mostly 1–5, erect to spreading; branches proximal and, often, distal. **Leaves** basal (often withering) and ± cauline, 1–6(–10) cm; largest blades ± 3-dimensional, scarcely succulent, usually 2-pinnately lobed; primary lobes 2–7 pairs, remote to ± congested, ultimate lobes ± involute and/or twisted. **Heads** mostly (1–)2–5 per stem. **Peduncles** 2–7(–9) cm. **Involucres** ± hemispheric to broadly cylindric. **Phyllaries:** longest 6–10 × 1.5–2.5 mm; outer glabrescent to sericeous and, often, ± glandular in fruit. **Florets:** inner corollas 4.5–7 mm. **Cypselae** 4–9 mm; **pappi** of (5–)8 scales in 2, abruptly unequal series, longest scales mostly 2–7 mm, lengths 0.4–0.9 (–1) times corollas. $2n = 12$.

Flowering Mar–Jun. Slopes, ridges, mostly over serpentine, openings in chaparral, woodlands; 100–1500 m; Calif.

Variety *heterocarpha* is known mainly from west-central and northwestern California. It intergrades to minor degrees with vars. *megacephala*, *lanosa*, and *glabriuscula* at its southern limits. Most of the specimens from the Tehachapi Mountains region assigned to this taxon appear to represent hybrids between *C. xantiana* and vars. *megacephala* or *glabriuscula* (see sectional discussion). *Chaenactis tanacetifolia* is an occasional form with leaves mostly basal, persistent, and densely arachnoid, and lobes more congested; its taxonomic significance remains to be demonstrated.

15b. Chaenactis glabriuscula de Candolle var. **megacephala** A. Gray in War Department [U.S.], Pacif. Railr. Rep. 4(5): 104. 1857 • Yellow pincushion [E]

Plants 15–40 cm; proximal indument grayish, arachnoid, early glabrescent. **Stems** mostly 1–5, ascending to erect; branches mainly proximal. **Leaves** basal (withering) and cauline, 1–8 cm; largest blades ± plane, scarcely succulent, usually 1-pinnately lobed; lobes 2–7 pairs, ± remote, ± plane or terete. **Heads** mostly 1–3 per stem. **Peduncles** 5–20 cm. **Involucres** ± hemispheric to broadly cylindric. **Phyllaries:** longest 7–9 × 2–3 mm; outer glabrescent in fruit. **Florets:** inner corollas 5–8 mm. **Cypselae** 5–8.5 mm; **pappi** of 4(–5) scales in 1 series, longest scales mostly 5–8 mm, lengths ± 0.9(–1) times corollas.

Flowering Mar–Jun. Dry, often sandy slopes, openings in chaparral, woodlands; 300–1500 m; Calif.

Variety *megacephala* is known mainly from the southern Sierra Nevada foothills and adjacent San Joaquin Valley and Transverse Ranges; it intergrades extensively with vars. *lanosa* and *glabriuscula*. A specimen cited by P. Stockwell (1940) from Bingen, Klickitat County, Washington (*Suksdorf*, May 1907) was not checked; it is either misidentified or from an introduction that did not persist.

15c. Chaenactis glabriuscula de Candolle var. **lanosa** (de Candolle) H. M. Hall, Univ. Calif. Publ. Bot. 3: 192. 1907 • Sand buttons, South Coast Ranges pincushion [E]

Chaenactis lanosa de Candolle in A. P. de Candolle and A. L. P. P. de Candolle, Prodr. 5: 659. 1836; *C. glabriuscula* var. *denudata* (Nuttall) Munz

Plants 8–15(–35) cm; proximal indument whitish, ± densely lanuginose. **Stems** mostly 1–12, decumbent to erect; branches none or strictly proximal. **Leaves** ± basal (persistent), 2–10 cm; largest blades ± plane or terete, scarcely succulent, 0–1-pinnately lobed; lobes 1–2(–5) pairs, remote, ± plane to terete. **Heads** 1(–3) per stem. **Peduncles** 8–20(–30) cm. **Involucres** ± hemispheric to obconic. **Phyllaries:** longest 6–8 × 1–2 mm; outer ± densely lanuginose in fruit. **Florets:** inner corollas 5–6.5 mm. **Cypselae** 4–6 mm; **pappi** of 4 scales in 1 series, longest scales mostly 4–6 mm, lengths 0.8–0.9(–1) times corollas. $2n = 12$.

Flowering Mar–June. Deep loose sand (typical forms; intermediate forms also in habitats of other vars.), openings in chaparral; 10–700(–1400) m; Calif.

Variety *lanosa* is known mainly from the Coast Ranges of central California. Most (but not all) specimens from farther south (Pacific coast and Santa Rosa Island) and southeast (Transverse Ranges to Sonoran Desert edges) are intermediate with vars. *megacephala* or *glabriuscula*; these are sometimes recognized as var. *denudata*. Northward, intermediates with var. *heterocarpha* are known.

15d. Chaenactis glabriuscula de Candolle var. **glabriuscula**

Chaenactis glabriuscula var. *curta* (A. Gray) Jepson; *C. glabriuscula* var. *tenuifolia* (Nuttall) H. M. Hall; *C. tenuifolia* Nuttall

Plants 10–60 cm; proximal indument grayish, ± arachnoid. **Stems** mostly 1–5, ascending to erect; branches proximal and, often, distal. **Leaves** basal (withering) and cauline, 3–8 cm; largest blades ± plane to 3-dimensional, scarcely succulent, 1(–2)-pinnately lobed; primary lobes 2–7 pairs, remote to ± congested, ultimate lobes plane, twisted, involute, or terete. **Heads** 2–20+ per stem. **Peduncles** 1–4(–10) cm. **Involucres** ± hemispheric to obconic. **Phyllaries:** longest 5–7 × 1–2 mm; outer (at least medially) ± arachnoid-sericeous and, often, ± stipitate-glandular in fruit. **Florets:** inner corollas 4–6 mm. **Cypselae** 3–5.5 mm; **pappi** of (1–)4 scales in 1 series, longest scales mostly 2–4 mm, lengths 0.4–0.7 times corollas. *2n* = 12.

Flowering Feb–Jun. Dry slopes, sandy places, openings in chaparral, woodlands; 100–2300 m; Calif.; Mexico (Baja California).

Variety *glabriuscula* is known mainly in and west of the Peninsular Ranges and adjacent desert edges, inland from the coast; it also extends to Santa Rosa Island. Northward it intergrades with vars. *megacephala* and *lanosa*; near the coast it intergrades with var. *orcuttiana*. Forms sometimes recognized as var. *tenuifolia* are distinctive in the filiform leaf segments but merge seamlessly with the remainder of var. *glabriuscula*.

15e. Chaenactis glabriuscula de Candolle var. **orcuttiana** (Greene) H. M. Hall, Univ. Calif. Publ. Bot. 3: 192. 1907 • Orcutt's pincushion C

Chaenactis tenuifolia Nuttall var. *orcuttiana* Greene, W. Amer. Sci. 3: 157. 1887

Plants 10–30 cm; proximal indument grayish, sparsely arachnoid. **Stems** mostly 3–7(–12), ± horizontal; branches proximal and distal. **Leaves** basal (withering) and cauline, mostly 2–5 cm; largest blades ± plane to 3-dimensional, succulent, mostly 2-pinnately lobed; primary lobes 2–7 pairs, ± congested, ultimate lobes ± plane to involute. **Heads** mostly 2–5 per stem. **Peduncles** 1–4 cm. **Involucres** ± hemispheric. **Phyllaries:** longest 4.5–6.5(–9) × 1–2 mm; outer predominantly stipitate-glandular and ± villous in fruit. **Florets:** inner corollas 4.5–5.5 mm. **Cypselae** 4–5 mm; **pappi** of 4 scales in 1 series, longest scales mostly 1–2.5 mm, lengths 0.2–0.5 times corollas. *2n* = 12.

Flowering late Apr–Jun. Coastal dunes and bluffs; of conservation concern; 0–100 m; Calif.; Mexico (Baja California).

Variety *orcuttiana* is known along the coast from near Santa Barbara to northwestern Baja California. It appears to intergrade with var. *glabriuscula* shortly inland.

16. Chaenactis stevioides Hooker & Arnott, Bot. Beechey Voy., 353. 1839 • Desert or Esteve or broad-flower pincushion F

Chaenactis furcata Stockwell; *C. gillespiei* Stockwell; *C. latifolia* Stockwell; *C. mexicana* Stockwell; *C. stevioides* var. *brachypappa* (A. Gray) H. M. Hall; *C. stevioides* var. *thornberi* Stockwell

Plants 5–30(–45) cm; proximal indument grayish, ± arachnoid-sericeous (tardily glabrescent except around nodes). **Stems** 1–12 (sometimes decumbent); branches proximal and/or distal. **Leaves** basal (usually withering) and ± cauline, 1–8(–10) cm; largest blades ± elliptic, ± 3-dimensional, usually not succulent, mostly 1–2-pinnately lobed; primary lobes 4–8 pairs, remote or ± congested, ultimate lobes ± involute and/or twisted. **Heads** (± radiant) mostly 3–20+ per stem. **Peduncles** 1–5(–10) cm, usually stipitate-glandular distally and, often, ± arachnoid. **Involucres** ± hemispheric to obconic (bases green, rounded in fruit). **Phyllaries:** longest 5.5–8(–10) mm; outer stipitate-glandular and/or ± arachnoid in fruit, apices erect, blunt, ± rigid. **Florets:** corollas white to pinkish, cream, or pale yellow, 4.5–6.5 mm (inner); peripheral corollas spreading, zygomorphic, enlarged. **Cypselae** (3–)4–6.5 mm; **pappi** of (1–)4(–5) scales, usually in 1 series, rarely with partial outer, abruptly unequal series, longest scales 1.5–6 mm, lengths mostly 0.3–0.9 times corollas (apices hidden among corollas at flowering). *2n* = 10.

Flowering Feb–Jun. Open, arid or semiarid, sandy or gravelly slopes and flats, shrublands; -30–2100(–2300) m; Ariz., Calif., Colo., Idaho, Nev., N.Mex., Oreg., Utah, Wyo.; Mexico (Baja California, Sonora).

Chaenactis stevioides is found throughout the southwestern deserts; it is among the most abundant spring wildflowers in the higher Mojave Desert and southern Great Basin. It also extends seaward into west-central California. It has been reported in New York as a garden escape; it is not expected to persist there outside cultivation.

Chaenactis stevioides varies in more or less concentric zones. Plants from the core zone (centered on the Great Basin and Mojave Desert) typically have pappi and phyllaries relatively short and phyllaries predominantly stipitate-glandular (var. *brachypappa*). Surrounding this zone to the southwest, southeast, and northeast are plants with pappi and phyllaries relatively long and phyllaries evidently or predominantly lanuginose (var. *stevioides*). Scattered on the periphery in central Arizona, Baja California, and west-central and southwestern California (where hybrids may be involved; see sectional discussion) are mesophytic forms with relatively long and/or broad leaf divisions, corollas varying from white to pale yellow, and pappi and phyllaries like those of var. *brachypappa* (var. *thornberi*, *C. gillespiei*). An unnamed form with leaves arachnoid but otherwise like *C. fremontii* occurs around sand dunes in the Mojave Desert. *Chaenactis furcata* and *C. latifolia* are forms possibly influenced by *C. fremontii* genes, unusual substrates, or pathogens. Traits of all the above taxa are inconsistent within populations, and/or recurrent or recombinant elsewhere in the range of *C. stevioides*.

17. **Chaenactis fremontii** A. Gray, Proc. Amer. Acad. Arts 19: 30. 1883 (as fremonti) • Fremont or desert pincushion, pincushion flower

Plants 10–30(–40) cm; proximal indument glabrescent (early ± arachnoid, glabrous by flowering). Stems mostly 1–12; branches mainly proximal. Leaves basal (withering) and ± cauline, 1–7 (–10) cm; largest blades linear and terete or ± elliptic and plane, ± succulent, 0–1-pinnately lobed; lobes 1–2(–5) pairs, remote, ± terete. Heads (± radiant) mostly 1–5 per stem. Peduncles 2–8(–10) cm, distally usually ± stipitate-glandular and, sometimes, ± arachnoid (at least early, often glabrescent by fruit). Involucres ± hemispheric to obconic (bases pale and ± truncate in fruit). Phyllaries: longest 8–10(–12) mm; outer usually glabrescent in fruit, apices erect, acute, ± rigid. Florets: corollas white to pinkish, 5–8 mm (inner); peripheral corollas spreading, zygomorphic, enlarged. Cypselae (3–)6–8 mm; pappi of (1–)4(–5) scales in 1 series, longest scales 6–8.5 mm, lengths 1–1.3 times corollas (apices visible among corollas at flowering). $2n$ = 10.

Flowering Mar–May. Sandy or gravelly soils, warm deserts, often growing through shrubs; -10–1600 m; Ariz., Calif., Nev., Utah; Mexico (Baja California).

Chaenactis fremontii is often the most abundant spring wildflower in the lower Mojave and northern Sonoran deserts, where it is reported to be a significant food source for desert tortoises (*Gopherus agassizii* Cooper). It also extends seaward into the southern San Joaquin Valley area of west-central California, often as hybrids with other taxa (see sectional discussion).

The involucre bases described above are characteristic of *Chaenactis fremontii* and can help separate it from some forms of *C. stevioides*.

379. **OROCHAENACTIS** Coville, Contr. U.S. Natl. Herb. 4: 134, plate 10. 1893
  • California mountain-pincushion [Greek *horos*, mountain, and generic name *Chaenactis*] E

David J. Keil

Annuals, 1–25 cm. **Stems** erect to spreading, simple or ± branched (very slender). **Leaves** cauline; opposite (proximal) or alternate; obscurely petiolate or sessile; blades linear to narrowly oblong-oblanceolate, margins entire, faces tomentose to glabrate, gland-dotted. **Heads** discoid, borne singly or (2–5) in clusters (at stem tips). **Involucres** cylindric or narrowly turbinate, 3–4(–6) mm diam. **Phyllaries** persistent, 4–7 in ± 1 series (distinct, oblong, unequal, glandular-puberulent). **Receptacles** ± flat, ± knobby, epaleate. **Ray florets** 0. **Disc florets** 4–9; corollas yellow, lobes 5, short-triangular (anther bases short-sagittate, appendages short-triangular; style branches linear, tapering). **Cypselae** terete, linear-clavate, narrowly ribbed, minutely glandular-puberulent, glabrescent; **pappi** falling (in rings), of 11–17 (basally connate) oblanceolate, fringed, obtuse scales. x = 9.

Species 1: California.

1. **Orochaenactis thysanocarpha** (A. Gray) Coville, Contr. U.S. Natl. Herb. 4: 134. 1893 E F

Chaenactis thysanocarpha A. Gray, Proc. Amer. Acad. Arts 19: 30. 1883

Leaves 1–4 cm, bases tapered, apices obtuse. Heads 6–8 mm. Phyllaries ± purple, 3–6 mm. Disc corollas 3–5 mm, minutely glandular. Cypselae 2.5–4 mm; pappi 1.5–2.5 mm. $2n$ = 18.

Flowering Jun–Aug. Open forests, dry meadows, gravelly slopes; 2100–3800 m; Calif.

187m.23. Asteraceae Martinov (tribe Heliantheae) subtribe Gaillardiinae Lessing, Linnaea 6: 516. 1831 (as Galardieae)

Annuals, biennials, perennials, subshrubs, or shrubs, 3–160 cm. **Leaves** basal, basal and cauline, or cauline; alternate; petiolate or sessile; blades mostly elliptic, filiform, lanceolate, linear, oblong, ovate, or spatulate, often 1–2-pinnately or -ternately lobed, ultimate margins entire or toothed, faces glabrous or hairy, usually gland-dotted. **Heads** mostly radiate or discoid (rarely radiant), usually borne singly, sometimes in glomerules or corymbiform arrays. **Calyculi** 0. **Involucres** mostly campanulate, cylindric, hemispheric, obconic, or ± rotate. **Phyllaries** persistent, 5–50+ in 2–3+ series, distinct or outer connate, mostly elliptic, lanceolate, linear, oblanceolate, oblong, or ovate, subequal or unequal (erect or reflexed in fruit, usually herbaceous to scarious, margins seldom scarious). **Receptacles** mostly conic, convex, flat, globose, hemispheric, or ovoid, epaleate (smooth or slightly to deeply pitted, sometimes with setiform enations, sometimes gland-dotted; outer disc florets rarely subtended by paleae in *Amblyolepis*). **Ray florets** 0, or 1–55 (in 1–3 series), pistillate and fertile, or styliferous and sterile, or neuter; corollas mostly yellow to orange, sometimes wholly or partly purple, red-brown, or reddish. **Disc florets** 5–250+, usually bisexual and fertile, rarely functionally staminate; corollas yellow, orange, or wholly or partly purple, red-brown, or reddish, tubes much shorter than to about equaling slightly to abruptly and greatly dilated, funnelform to campanulate or urceolate throats, lobes 5, deltate to lance-deltate (equal or unequal, often attenuate, glabrous, papillate, or hairy, hairs sometimes moniliform); anther thecae pale; stigmatic papillae in 2 lines. **Cypselae** usually obpyramidal, sometimes clavate, columnar, or obconic (lengths usually 1–2.5, rarely to 3.5 diams.), mostly densely and shaggily hairy, sometimes sparsely hairy or glabrate; **pappi** usually persistent, of 2–12 scales, sometimes 0 or of 35–150 bristles (*Psathyrotes*).

Genera 12, species 123 (11 genera, 73 species in the flora): mostly sw United States, Mexico, South America.

The circumscription of Gaillardiinae adopted here is that of H. Robinson (1981). In studies of epaleate Heliantheae, B. G. Baldwin adopted a narrower circumscription of the subtribe (Baldwin and B. L. Wessa 2000; Baldwin et al. 2002). Gaillardiinae in sense of Baldwin includes *Balduina*, *Gaillardia*, and *Helenium*; dispositions by Baldwin of other genera of Gaillardiinae in the broad sense include: *Plateilema* in Plateileminae, *Psathyrotes* and *Trichoptilium* in Psathyrotinae, and *Amblyolepis*, *Baileya*, *Hymenoxys*, *Psilostrophe*, and *Tetraneuris* in Tetraneuriinae (all in Helenieae sensu Baldwin).

Psilostrophinae B. L. Turner & A. M. Powell is a superfluous name (circumscription included type of Riddelliinae O. Hoffmann).

1. Pappi of 35–150 distinct or basally connate bristles in 1–4 series . . . . 380. *Psathyrotes*, p. 416
1. Pappi 0 or of 2–12 scales.
    2. Pappi of 5 ovate to flabellate, deeply and finely lacerate scales (each seemingly constituted of 8–15+ connate bristles) . . . . 381. *Trichoptilium*, p. 418
    2. Pappi 0 or of 2–12 ovate or spatulate to lanceolate, entire, erose, or coarsely lacerate, often attenuate or uniaristate scales.
        3. Receptacles deeply pitted (each cypsela nested within a 5–6-sided cell) . . . . . . 382. *Balduina*, p. 419
        3. Receptacles smooth or ± pitted (sometimes with scattered subulate to setiform enations, cypselae not nested within cells; outer disc florets rarely subtended by paleae in *Amblyolepis*).

[4. Shifted to left margin.—Ed.]

4. Phyllaries 17–21 in 2 series (inner hyaline, scalelike; herbage notably sweet scented)....
.... 383. *Amblyolepis*, p. 420
4. Phyllaries 5–50 in 2–3 series (inner herbaceous to scarious or scarious-margined; herbage not notably sweet-scented).
    5. Phyllaries usually strongly reflexed in fruit; receptacles mostly globose (sometimes with setiform enations); disc corollas often brown-purple to red-brown or tipped with brown-purple to red-brown (tubes much shorter than abruptly much-dilated, urceolate to campanulate throats, lobes often shaggily hairy, hairs ± moniliform).
        6. Stems not winged (receptacles usually with setiform enations; style-branch apices ± attenuate) .... 384. *Gaillardia*, p. 421
        6. Stems often winged (by decurrent leaf bases; receptacles rarely with setiform enations; style-branch apices penicillate or truncate).................... 385. *Helenium*, p. 426
    5. Phyllaries mostly spreading to erect in fruit; receptacles flat, conic, domed, hemispheric, or ovoid (smooth or pitted, without setiform enations); disc corollas usually uniformly yellow to cream or sometimes purplish to reddish (tubes much shorter than to about equaling slightly dilated, funnelform to cylindric throats, lobes not shaggily hairy with moniliform hairs).
        7. Ray florets usually 3–16 (corollas withering, falling early or tardily).
            8. Leaf blades sometimes pinnately lobed (lobes mostly filiform, linear, or oblong); phyllaries: outer connate or distinct, inner distinct .... 386. *Hymenoxys*, p. 435
            8. Leaf blades pinnately lobed (lobes mostly deltate to obovate); phyllaries: all basally connate .... 387. *Plateilema*, p. 444
        7. Ray florets usually 1–55 (corollas marcescent).
            9. Pappi 0.. 388. *Baileya*, p. 444
            9. Pappi of 4–8 scales.
                10. Plants mostly scapiform (non-scapiform annuals in *T. linearifolia*); heads mostly borne singly; involucres hemispheric to rotate; rays 0 or 7–27; disc florets 20–250+ .... 389. *Tetraneuris*, p. 447
                10. Plants not scapiform; heads usually in close corymbiform or glomerulate clusters; involucres mostly campanulate, cylindric, or obconic; rays 1–8; disc florets 5–25+.................................... 390. *Psilostrophe*, p. 453

380. PSATHYROTES A. Gray, Smithsonian Contr. Knowl. 5(6): 100. 1853 • [Greek *psathyros*, brittle, fragile, looseness; allusion unclear]

John L. Strother

Bulbostylis de Candolle sect. *Psathyrotus* Nuttall, Proc. Acad. Nat. Sci. Philadelphia 4: 22. 1848

Annuals or perennials, 5–40+ cm. **Stems** erect or spreading, usually much branched. **Leaves** basal and cauline; alternate, petiolate; blades rounded-deltate, reniform, rounded-rhombic, or suborbiculate, margins toothed or entire, faces usually lanate, pilose, or tomentose, sometimes glandular-pubescent and/or furfuraceous as well. **Heads** discoid, borne singly. **Involucres** campanulate, cylindric, or turbinate, 3–6+ mm diam. **Phyllaries** falling or persistent, 8–24 in 2 (contrasting) series (distinct, lance-linear, lanceolate, obovate, or spatulate, tips erect or reflexed to squarrose). **Receptacles** flat to convex, smooth or ± pitted, epaleate. **Ray florets** 0. **Disc florets** bisexual, fertile; corollas yellowish, tubes much shorter than cylindric throats, lobes 5, erect, deltate (style-branch apices rounded-truncate). **Cypselae** cylindro-fusiform to obpyramidal; **pappi** of 35–150 distinct or basally connate bristles in 1–4 series. $x = 17$.

Species 3 (3 in the flora): sw United States, nw Mexico.

SELECTED REFERENCE Strother, J. L. and G. E. Pilz. 1975. Taxonomy of *Psathyrotes* (Compositae: Senecioneae). Madroño 23: 24–40.

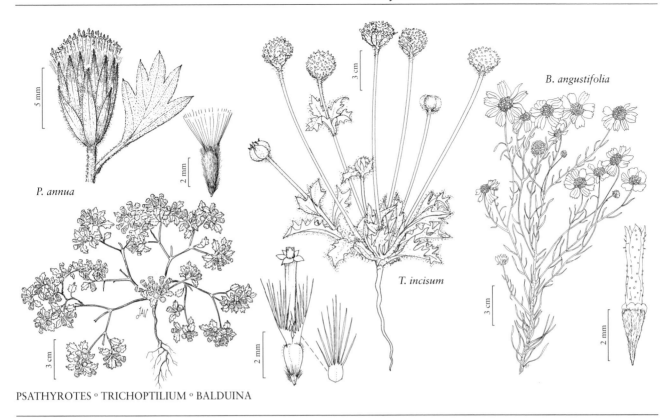

P. annua

T. incisum

B. angustifolia

PSATHYROTES ∘ TRICHOPTILIUM ∘ BALDUINA

1. Margins of all leaves entire (margins and veins bearing straight, multicellular hairs 2–3+ mm); cypselae 4–5 mm . . . . 3. *Psathyrotes pilifera*
1. Margins of at least some leaves toothed (faces hairy, hairs not 2–3+ mm); cypselae 1.5–3.5 mm.
  2. Plants lanate and furfuraceous; outer phyllaries spatulate to obovate, tips squarrose to reflexed; florets 16–32; pappi of 120–140 bristles . . . . 1. *Psathyrotes ramosissima*
  2. Plants sparsely tomentose and furfuraceous; outer phyllaries lance-linear, tips erect; florets 10–20; pappi of 35–50 bristles . . . . 2. *Psathyrotes annua*

1. **Psathyrotes ramosissima** (Torrey) A. Gray, Proc. Amer. Acad. Arts 7: 363. 1868 • Turtleback

Tetradymia ramosissima Torrey in W. H. Emory, Not. Milit. Reconn., 145. 1848

Annuals or perennials, 3–30 cm; lanate, furfuraceous. **Stems** erect and spreading, much branched. **Leaf blades** rounded-deltate to suborbiculate, 8–25 × 8–30 mm, margins toothed. **Peduncles** 3–50 mm. **Involucres** broadly turbinate to campanulate, 6–10 mm. **Phyllaries** 15–24, outer persistent, 5–6, apically spatulate, tips reflexed to squarrose, inner falling, 10–18, apically lanceolate, tips erect. **Florets** 16–32; corollas pale yellow, 4.5–5 mm. **Cypselae** 1.5–3.5 mm; **pappi** of 120–140 subequal bristles in 2–4 series, 3–4 mm. *2n* = 34.

Flowering mostly in spring, otherwise sporadically, following rains. Sandy soils and desert pavements; -30–900 m; Ariz., Calif., Nev.; Mexico (Baja California, Sonora).

2. **Psathyrotes annua** (Nuttall) A. Gray, Smithsonian Contr. Knowl. 5(6): 100. 1853 [E] [F]

Bulbostylis annua Nuttall, Proc. Acad. Nat. Sci. Philadelphia 4: 22. 1848

Annuals (rarely persisting), 4–15 cm; glandular-pubescent, weakly tomentose, furfuraceous. **Stems** erect and spreading, much branched. **Leaf blades** rounded-deltate to reniform, 4–18 × 6–26 mm, margins of some or all toothed. **Peduncles** 1–28 mm. **Involucres** turbinate, 6–9 mm. **Phyllaries** 13–19, outer falling or persistent, 5–6, lanceolate to lance-linear, tips erect, inner falling, 8–13, lance-linear, tips erect. **Florets** 10–21; corollas yellow, often purplish at tips, 4–5 mm. **Cypselae** 2–3 mm; **pappi** of 35–50, subequal bristles in 1 series, 1–4 mm. *2n* = 34.

Flowering spring–fall. Alkaline soils of washes and playas; 800–2000 m; Ariz., Calif., Idaho, Nev., Utah.

3. Psathyrotes pilifera A. Gray, Proc. Amer. Acad. Arts 19: 50. 1883 [E]

Annuals, 6–15 cm, glandular-pubescent, pilose, and furfuraceous. **Stems** erect and spreading. **Leaf blades** rounded-rhombic, 7–14 × 6–14 mm, margins entire (faces, at least abaxial, with straight, multicellular hairs 2–3+ mm). **Peduncles** 4–20 mm. **Involucres** narrowly cylindric, 7–9 mm. **Phyllaries** ca. 13, outer persistent, 5, lanceolate, tips erect, inner falling, 8, lance-linear, tips erect. **Florets**

9–14; corollas yellow, often purplish at tips, ca. 8 mm. **Cypselae** 4–5 mm; **pappi** of ca. 150 subequal bristles in 2–4 series, 6–7 mm. *2n* = 34.

Flowering sporadically following rains. Reddish, alkaline soils; 500–1400 m; Ariz., Nev., Utah.

381. TRICHOPTILIUM A. Gray in W. H. Emory, Rep. U.S. Mex. Bound. 2(1): 97. 1859

• [Greek *tricho-*, hairlike, and *ptilon*, feather, alluding to pappus scales]

John L. Strother

Annuals or perennials, 3–20(–30+) cm. **Stems** erect or ascending, branched from bases or throughout. **Leaves** basal and cauline; mostly alternate; petiolate; blades oblanceolate to spatulate or ovate, margins usually sharply and coarsely toothed or lobed, faces sparsely floccose to densely woolly, gland-dotted. **Heads** discoid, borne singly. **Involucres** hemispheric to campanulate, 6–12 mm diam. **Phyllaries** persistent, 18–25 in 2+ series (erect or spreading in fruit, distinct, lance-ovate to lanceolate, subequal, thin-herbaceous, loosely woolly, the inner more scarious, less woolly). **Receptacles** convex to conic, smooth or finely pitted, epaleate. **Ray florets** 0. **Disc florets** 30–100+, bisexual, fertile; corollas yellow, sometimes drying pinkish distally (outer sometimes larger and zygomorphic), tubes much shorter than narrowly funnelform throats, lobes 5, ovate to deltate (equal or unequal). **Cypselae** obpyramidal, glabrous or shaggily hairy; **pappi** of 5 scarious, deeply lacerate scales (each seemingly constituted of 8–15+ connate bristles). *x* = 13.

Species 1: sw United States, nw Mexico.

1. Trichoptilium incisum (A. Gray) A. Gray in W. H. Emory, Rep. U.S. Mex. Bound. 2(1): 97. 1859 [F]

Psathyrotes incisa A. Gray, Pl. Nov. Thurb., 322. 1854

Leaf blades 1–5 cm × 3–12 mm. **Disc corollas** ca. 4 mm. **Cypselae** ca. 3 mm; **pappi** 3–3.5 mm. *2n* = 26.

Flowering Jan–May. Sandy or gravelly soils, deserts; 0–700 m; Ariz., Calif., Nev.; Mexico (Baja California, Baja California Sur).

382. BALDUINA Nuttall, Gen. N. Amer. Pl. 2: 175. 1818, name conserved

• [For William Baldwin, 1779–1819, American botanist] E

Brian R. Keener

Annuals or perennials, 30–100 cm (gland-dotted). **Stems** erect, simple or branched (terete or angled), glabrous or hairy. **Leaves** cauline; alternate; sessile; blades linear, linear-spatulate, or spatulate, margins entire, faces glabrous or ± hairy. **Heads** borne singly or in loose, corymbiform to paniculiform arrays. **Involucres** hemispheric to broadly campanulate, 6–25 mm diam. **Phyllaries** persistent, 30–50 in 3–5 series (distinct, lanceolate to ovate, unequal). **Receptacles** slightly convex, deeply pitted (pits collectively forming honeycombed structure, each pit ± enveloping a disc floret, pit borders irregularly toothed to entire), epaleate (pits sometimes interpreted as formed from connate paleae). **Ray florets** 8–22, neuter; corollas yellow (tubes glabrous or hairy, laminae spatulate, abaxially hairy, adaxially glabrous). **Disc florets** 30–190, bisexual, fertile; corollas yellow, orange-red, or purple (at least tips), tubes shorter/longer than funnelform throats, lobes 5, deltate (abaxially hairy; stamens pale yellow; style-branch appendages subulate, papillate). **Cypselae** obconic, villous; **pappi** persistent, of 7–12 scales. *x* = 18.

Species 3 (3 in the flora): se United States.

SELECTED REFERENCE Parker, E. S. and S. B. Jones. 1975. A systematic study of the genus *Balduina* (Compositae, Heliantheae). Brittonia 27: 355–361.

1. Annuals (taprooted); cauline leaves linear, 0.5–1.9(–2.5) mm wide; outer phyllaries lanceolate, 0.6–1.7 mm wide, apices acuminate; pappus scales obovate to orbiculate, 0.3–0.6 mm . . . . 1. *Balduina angustifolia*
1. Perennials (fibrous rooted); cauline leaves linear-spatulate, 2–7 mm wide; outer phyllaries ovate, 1.7–3.1 mm wide, apices acute; pappus scales lanceolate, 1.3–2.2 mm.
   2. Basal leaves spatulate, 2.8–10.4 cm; disc corollas yellow to orange-red . . . . 2. *Balduina uniflora*
   2. Basal leaves linear-spatulate, 7–32 cm; disc corollas dark purple (at least tips) . . . . .
   . . . . 3. *Balduina atropurpurea*

1. Balduina angustifolia (Pursh) B. L. Robinson, Proc. Amer. Acad. Arts 47: 215. 1911 E F

Buphthalmum angustifolium Pursh, Fl. Amer. Sept. 2: 564. 1813; *Actinospermum angustifolium* (Pursh) Torrey & A. Gray

Annuals (taprooted). **Stems** 1–20+, branched. **Leaves** abaxially glabrous or sparsely hairy, adaxially glabrous; basal leaves (absent at flowering) linear-spatulate, 1.5–6 × 0.05–0.19(–0.25) cm, proximal and mid cauline leaves linear, 1.5–4.6 × 0.02–0.07 cm, distal cauline leaves similar, smaller and becoming reduced, distalmost bractlike. **Peduncles** 1–11.5 cm, sparsely hairy. **Heads** 1–20+. **Involucres** 6–15 mm diam. **Phyllaries** abaxially glabrous; outer proximally yellow, distally green, lanceolate, 2.4–6.7 mm, apices acuminate-aristate; inner lanceolate, 4.2–7 mm, apices acuminate-aristate. **Receptacles:** pit borders spinulose-cuspidate at angles. **Ray florets** 5–13; tubes 2–3 mm, laminae 8–18 × 2.2–6.5 mm. **Disc florets** 30–90; corollas yellow, 3.5–4.5

mm; style branches yellow. **Cypselae** 1–2 mm; **pappi** of obovate to orbiculate, entire scales 0.3–0.6 mm. **2n** = 36.

Flowering summer–fall (year round south). Dry sandy soils, especially in pinelands; 0–100 m; Ala., Fla., Ga., Miss.

2. Balduina uniflora Nuttall, Gen. N. Amer. Pl. 2: 175. 1818 E

Perennials (fibrous rooted). **Stems** 1–4+, branched. **Leaves** abaxially hairy, adaxially glabrous or sparsely hairy; basal (present or absent at flowering) spatulate, 2.8–10.4 × 0.5–1.3 cm; proximal cauline linear-spatulate, 1.5–5.5 × 0.2–0.7 cm; mid and distal cauline similar, smaller, distalmost bractlike. **Peduncles** 13.5–24 cm, hairy. **Heads** 1–4. **Involucres** 15–25 mm diam. **Phyllaries** abaxially hairy; outer light green proximally, dark green distally, ovate, 4–7.2 × 2.1–2.9 mm, apices broadly acuminate; inner yellow, lanceolate, 5.1–11 × 1.5–2 mm, apices acuminate. **Receptacles:** pit borders toothed to almost entire at angles. **Ray florets** 8–22; corolla tubes 3–5 mm, laminae 12–30 × 3.2–5.3 mm. **Disc florets** 50–180; corollas yellow to orange-red, 5.5–8.2 mm; style branches yellow. **Cypselae** 1.3–2.2 mm; **pappi** of lanceolate, entire or apically lacerate scales 1.3–2.2 mm. 2*n* = 72.

Flowering summer–fall. Moist to dry pinelands and savannas; 0–100 m; Ala., Fla., Ga., La., Miss., N.C., S.C.

3. **Balduina atropurpurea** R. M. Harper, Bull. Torrey Bot. Club 28: 483. 1901 (as Baldwinia) E

Perennials (fibrous rooted). **Stems** 1–4+, branched. **Leaves** glabrous or sparsely hairy adaxially, hairy abaxially; basal present or absent at flowering, linear-spatulate, 7–32 × 0.4–1 cm; proximal cauline linear-spatulate, 3.8–6.2 × 0.2–0.6 cm; mid and distal cauline similar, smaller, distalmost bractlike. **Peduncles** 5.5–20 cm, hairy. **Heads** 1–4. **Involucres** 15–25 mm diam. **Phyllaries** abaxially hairy; outer green to purple, ovate, 2.9–5.4 mm, apices acute; inner purple, lanceolate, 4.5–7.6 mm, apices acuminate. **Receptacles:** pit borders toothed to almost entire at angles. **Ray florets** 8–22, tubes 3–5 mm, laminae 16–27 × 3.5–5.5 mm. **Disc florets** 40–190; corollas purple (at least tips), 4–7 mm; style branches purple. **Cypselae** 1.3–2.2 mm; **pappi** of lanceolate, entire or apically lacerate scales 1.5–2.1 mm. 2*n* = 36.

Flowering summer–fall. Pitcher plant bogs, wet pinelands and savannas; 0–100 m; Fla., Ga., N.C., S.C.

383. AMBLYOLEPIS de Candolle in A. P. de Candolle and A. L. P. P. de Candolle, Prodr. 5: 667. 1836 • Huisache-daisy [Greek *ambly*, blunt, and *lepis*, scale]

Mark W. Bierner

Helenium Linnaeus sect. *Amblyolepis* (de Candolle) Bentham

Annuals, 10–60 cm (herbage sweet scented). **Stems** (1–10) erect or ± decumbent, unbranched or sparingly branched distally, sparsely to densely pilose. **Leaves** cauline; alternate; petiolate; blades lanceolate, oblanceolate, ovate, or spatulate, margins entire, faces usually sparsely to moderately pilose (especially on margins), sometimes glabrate or glabrous. **Heads** radiate, borne singly or (2–45) in paniculiform arrays. **Involucres** hemispheric to globose, 12–20 mm diam. **Phyllaries** persistent, 17–21 in 2 series (outer usually distinct, sometimes connate proximally, green, narrowly elliptic to lanceolate, herbaceous, moderately to densely pilose, especially bases and margins; inner phyllaries distinct, obovate, hyaline, scalelike, glabrous). **Receptacles** globose to ovoid, ± pitted, usually epaleate (outer disc florets rarely subtended by paleae, central disc florets usually subtended by persistent enations). **Ray florets** 8–13(–20), pistillate, fertile; corollas yellow. **Disc florets** 20–50, bisexual, fertile; corollas yellow, tubes shorter than narrowly funnelform to cylindric-campanulate throats, lobes 5, ± deltate (glabrous). **Cypselae** narrowly obconic, prominantly 10-ribbed, densely pubescent; **pappi** persistent, of 5–6 ovate to obovate scales. *x* = 19.

Species 1: Texas, n Mexico.

SELECTED REFERENCE Bierner, M. W. 1990. Present status of *Amblyolepis* (Asteraceae: Heliantheae). Madroño 37: 133–140.

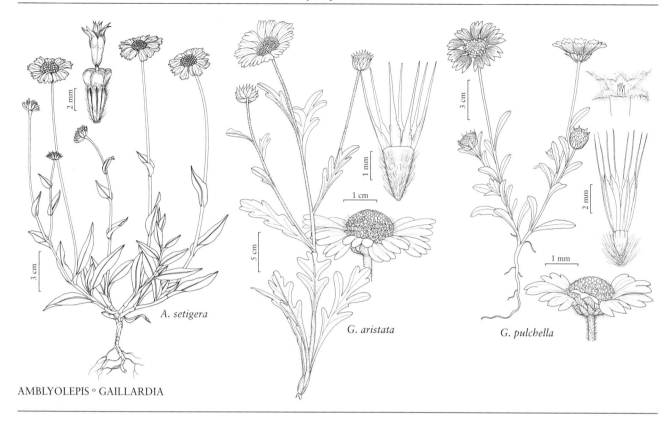

A. setigera

G. aristata

G. pulchella

AMBLYOLEPIS ° GAILLARDIA

1. Amblyolepis setigera de Candolle in A. P. de Candolle and A. L. P. P. de Candolle, Prodr. 5: 668. 1836 [F]

Helenium setigerum (de Candolle) Britton & Rusby

Leaves: basal and proximal spatulate to oblanceolate, bases narrowed; mid ovate to lanceolate, bases usually semiclasping; distal leaves similar, smaller. **Peduncles** (4–)8–20 cm, expanded distally, sparsely pilose proximally, moderately to densely pilose distally. **Involucres** hemispheric to globose, 9–17 × 12–20 mm. **Phyllaries:** outer usually 8, inner 9–13. **Ray corollas** 10–22 × 4.5–9.6(–11.6) mm. **Disc corollas** 5.1–7 mm. **Cypselae** 3–4.5 mm; **pappus scales** 2–3(–4.1) mm, apices usually rounded to acute or acuminate, rarely cuspidate, not aristate. $2n$ = 34(?), 36, 37 (18II + 1I), 38, 39 (19II + 1I), 39 (18II + 3I ?), 40.

Flowering (Jan–)Mar–May(–Sep). Roadsides, open fields; 10–900 m; Tex.; Mexico (Coahuila, Nuevo León, Tamaulipas).

384. GAILLARDIA Fougeroux, Observ. Phys. 29: 55. 1786 (as Gaillarda); Hist. Acad. Roy. Sci. Mém. Math. Phys. (Paris, 4to) 1786: 5. 1788 (corrected) • Blanket flower, firewheel [For M. Gaillard de Merentonneau (or Charentonneau?), eighteenth-century French patron of botanists]

John L. Strother

Annuals, perennials, or subshrubs, 10–80+ cm (sometimes rhizomatous). **Stems** usually erect, usually branched (from bases or throughout). **Leaves** basal, basal and cauline, or mostly cauline; alternate; petiolate or sessile; blades mostly elliptic, lanceolate, linear, oblong, obovate, ovate, or spatulate, often pinnately lobed, ultimate margins toothed or entire, faces usually scabrellous to villous (hairs simple or jointed), rarely glabrous, usually notably gland-dotted (except in 10. *G. suavis*). **Heads** usually radiate or discoid (sometimes radiant), borne singly. **Involucres** hemispheric to nearly rotate, mostly 9–20+ mm diam. **Phyllaries** persistent, 14–40+ in 2–3+ series (reflexed in fruit, usually distinct, oblanceolate, lanceolate, narrowly triangular, or linear, subequal to unequal, usually chartaceous proximally, distally herbaceous, apices often attenuate). **Receptacles** convex to hemispheric, smooth or pitted (often with scattered, hard, setiform enations), epaleate. **Ray florets** 0, or 5–15+, usually neuter, rarely pistillate and fertile or styliferous and sterile; corollas yellow to white or orange to red, purplish, or brown, often bicolored (sometimes "ray" corollas funnelform, 5-lobed, heads radiant). **Disc florets** 20–100+, bisexual, fertile; corollas yellow or orange to red, purplish, or brown (distally ± shaggily hairy, hairs jointed), tubes much shorter than ampliate, campanulate to cylindric or urceolate throats, lobes 5, broadly deltate or ovate (± equal, sometimes attenuate or attenuate-terete apically; style-branch apices ± attenuate, usually hispidulous). **Cypselae** obpyramidal to clavate, ± 4-angled (lengths 1–1.5+ diams.), sparsely to densely hairy (at least proximally, hairs straight); **pappi** usually persistent, of 6–10+ medially thickened, laterally scarious scales in 1–2 series (all, some, or none aristate), rarely 0. $x = 19$.

Species 15–17 or more (11 in the flora): North America, Mexico, South America (mostly Argentina).

Some horticultural gaillardias may be derived from hybrids (e.g., *Gaillardia* ×*grandiflora* van Houtte) involving *G. aristata* and some other species (probably *G. pulchella*). Such horticultural plants sometimes persist after cultivation or occur sporadically as waifs in places well beyond the "natural" ranges of the "parent" species; e.g., scattered localities in Arizona and California.

SELECTED REFERENCE Biddulph, S. F. 1944. A revision of the genus *Gaillardia*. Res. Stud. State Coll. Wash. 12: 195–256.

1. Leaves all basal (sometimes extending onto proximal ¹/₃+ of stems in desert annual *G. arizonica*), blades glabrous or sparsely villous (hairs jointed).
   2. Ray florets 0 or 7–10, corollas red to purple; disc corollas pinkish or purplish; pappus scales aristate, 6–9 mm . . . . 10. *Gaillardia suavis*
   2. Ray florets 10–13, corollas yellow to orange; disc corollas yellow; pappus scales muticous or aristate, 2–5 mm . . . . 11. *Gaillardia arizonica*
1. Leaves basal and cauline or all cauline (restricted to proximal ¹/₃ or less of plants in some perennials; mostly cauline in annuals), blades hispidulous to scabrellous or strigillose to villous (hairs simple or jointed).
   3. Receptacles rarely setose (setae 0.1–1 mm); lobes of disc corollas attenuate-terete (bearing jointed hairs to 0.3 mm) . . . . 9. *Gaillardia aestivalis*
   3. Receptacles usually setose (setae 0.3–6 mm); lobes of disc corollas ovate to deltate, sometimes attenuate (usually bearing jointed hairs 0.3+ mm, hairs shorter in *G. coahuilensis*).
      4. Annuals or perennials (seldom from woody caudices); cypselae hairy (hairs inserted at bases and sometimes on angles, not on faces).
         5. Perennials (sometimes flowering first year); leaves basal and cauline; receptacular setae (2–)3–6+ mm . . . . 5. *Gaillardia aristata*
         5. Annuals (seldom persisting); leaves all or mostly cauline; receptacular setae mostly 0.3–3 mm.
            6. Hairs on disc corollas to 0.3 mm . . . . 6. *Gaillardia coahuilensis*
            6. Hairs on disc corollas 0.3+ mm.
               7. Ray corollas usually bicolored (brown-purple to red proximally, tipped with yellow or orange), rarely unicolored; lobes of disc corollas deltate to ovate, often attenuate, 1–3+ mm; cypselae all alike . . . . 7. *Gaillardia pulchella*

7. Ray corollas usually uniformly dark red to purple, rarely yellow; lobes of disc corollas mostly ovate-deltate to lanceolate, 0.5–1 mm; cypselae dimorphic (outer longer than inner and with shorter pappus scales) .... 8. *Gaillardia amblyodon*

[4. Shifted to left margin.—Ed.]

4. Perennials (often with woody caudices, seldom flowering first year); cypselae hairy (hairs inserted at bases and on angles and faces).

    8. Leaves (most or all) usually pinnatifid, rarely entire; disc corollas usually purplish or purple-tipped, rarely wholly yellow .... 1. *Gaillardia pinnatifida*

    8. Leaves (most or all) entire or toothed; disc corollas yellow or purplish.

        9. Leaf blades linear, 3–5(–8+) mm wide, faces sparsely hispidulous or glabrate; disc corollas usually purplish .... 2. *Gaillardia multiceps*

        9. Leaf blades mostly oblanceolate to spatulate, usually (at least the larger) 10–25 mm wide, faces closely puberulent or sericeous to villous; disc corollas yellow.

            10. Leaves mostly basal, sometimes some to mid stems; peduncles 10–30 cm; cypselae mostly 1–2 mm .... 3. *Gaillardia parryi*

            10. Leaves mostly cauline; peduncles 2–6(–10) cm; cypselae mostly 3–4 mm .... 4. *Gaillardia spathulata*

1. Gaillardia pinnatifida Torrey, Ann. Lyceum Nat. Hist. New York 2: 214. 1827

Gaillardia flava Rydberg; *G. gracilis* A. Nelson; *G. mearnsii* Rydberg; *G. pinnatifida* var. *linearis* (Rydberg) Biddulph

Perennials (sometimes flowering first year, sometimes rhizomatous), (5–)15–35+ cm. **Leaves** basal and cauline, mostly restricted to proximal ¹/₃–¹/₂; petiolar bases 0–5+ cm; blades mostly oblanceolate to spatulate, 3–6(–12) cm × 3–18(–30) mm, margins mostly pinnatifid, distal sometimes toothed or entire (rarely all linear, entire), faces closely strigillose to shaggily villous. **Peduncles** (4–)8–25+ cm. **Phyllaries** 20–30, ovate- to lanceolate-attenuate, 7–12+ mm, hispidulous to villous (hairs often jointed). **Receptacular setae** usually 1–3 mm, rarely wanting. **Ray florets** 0 or 5–14; corollas usually uniformly yellow, sometimes proximally and/or abaxially reddish, 10–25+ mm. **Disc florets** (30–)60–100+; corollas usually proximally ochroleucous or yellow and distally purplish, rarely wholly yellow, tubes 0.8–1 mm, throats campanulate to plumply urceolate, 3–4.5 mm, lobes broadly deltate to deltate-ovate, 0.5–1 mm, jointed hairs 0.3+ mm. **Cypselae** obpyramidal, 1–3 mm, hairs 1–2 mm, inserted at bases and on angles and faces; **pappi** of 8–11 lanceolate, aristate scales 3–7 mm (scarious bases 1.5–4 × 0.5–1.5 mm). *2n* = 34.

Flowering Mar–Oct, mostly May–Jul. Clays or sandy soils, often disturbed places, in grasslands, desert scrublands, or pinyon woodlands; 900–2000 m; Ariz., Colo., Nev., N.Mex., Okla., Tex., Utah; Mexico (Chihuahua, Coahuila, Durango, Sonora).

Some plants, especially from Arizona, included here in *Gaillardia pinnatifida*, have mostly narrow, undivided leaf blades (mostly 3–8+ mm wide, villous to sparsely strigillose; var. *linearis*) and intergrade with similar plants called *G. multiceps*, which have sparsely and minutely hispidulous or glabrate leaf blades.

Plants from Utah with yellow disc corollas and densely gland-dotted leaves, included here in *Gaillardia pinnatifida*, have been recognized as *G. flava*.

2. Gaillardia multiceps Greene, Bull. Torrey Bot. Club 24: 512. 1897 [E]

Gaillardia multiceps var. *microcephala* B. L. Turner

Perennials or subshrubs, 30–45+ cm. **Leaves** cauline; petiolar bases essentially none; blades narrowly spatulate or linear, 2–6 cm × 3–5(–8+) mm, margins entire, faces sparsely and minutely hispidulous or glabrate. **Peduncles** 2–5+ cm. **Phyllaries** 20–26, lanceolate- to narrowly triangular-attenuate, 6–10 mm, ± sericeous to villous, not ciliate with jointed hairs. **Receptacular setae** 0.5–2.5 mm. **Ray florets** 8; corollas yellow, 13–20 mm. **Disc florets** 50–80+; corollas proximally yellow, distally purplish, tubes 1–1.5 mm, throats campanulate, 4–4.5 mm, lobes ovate-deltate, 0.8 mm, jointed hairs mostly 0.3+ mm. **Cypselae** obpyramidal, 2.5–3 mm, hairs 2.5–3 mm, inserted at bases and on angles and faces; **pappi** of 10 lanceolate, aristate scales 6–9 mm (scarious bases 4–5 × 0.8–1.5 mm). *2n* = 68, ca. 102.

Flowering Jul–Aug. Gypseous soils, including dunes; 1100–1800 m; Ariz., N.Mex., Tex.

See comments under 1. *Gaillardia pinnatifida*.

3. Gaillardia parryi Greene, Bull. Torrey Bot. Club 27: 512. 1897 E

Gaillardia acaulis A. Gray, Proc. Amer. Acad. Arts 10: 73. 1874, not Pursh 1813

Perennials, 10–35 cm. **Leaves** mostly basal, sometimes to midstems; petiolar bases 3–5+ cm; blades elliptic-obovate to ± spatulate, 3–5+ cm × 5–25 mm, margins weakly lobed or entire, faces closely puberulent. **Peduncles** 10–30 cm. **Phyllaries** 22–26+, lance-attenuate, 9–12 mm, villous with jointed hairs. **Receptacular setae** 2–4 mm. **Ray florets** 8–14; corollas yellow, 14–17+ mm. **Disc florets** 50–100+; corollas yellow, tubes 1–1.2 mm, throats cylindric, 4–5.5 mm, lobes ovate, 0.7–1.3 mm, jointed hairs 0.3+ mm. **Cypselae** ± obpyramidal, 1–2 mm, hairs 1+ mm, inserted at bases and on angles and faces; **pappi** of 10 ovate to lanceolate, aristate scales 5–6.5+ mm (scarious bases 3–4.5 × 1–1.2 mm). $2n = 68$.

Flowering May–Jun(–Sep). Clays or sandy soils, often disturbed places, with sagebrush, pinyons, or yellow pines; 1200–2000 m; Ariz., Utah.

4. Gaillardia spathulata A. Gray, Proc. Amer. Acad. Arts 12: 59. 1876 E

Perennials, 5–35 cm. **Leaves** mostly cauline; petiolar bases 2–8 cm; blades obovate to oblanceolate or spatulate, mostly 2–5(–8) cm × 5–25 mm, margins usually entire, sometimes obscurely lobed or toothed, faces sparsely villous. **Peduncles** 2–6(–10) cm. **Phyllaries** 18–24+, lanceolate-ovate to lanceolate-attenuate, 8–12 mm, villous with jointed hairs. **Receptacular setae** 0.3–2 mm. **Ray florets** 7–10; corollas yellow, 12–16 mm. **Disc florets** 60–100+; corollas yellow, tubes 0.8–1.2 mm, throats cylindric to urceolate, 4–4.5 mm, lobes ovate-deltate, 0.8–1.2 mm, jointed hairs 0.3+ mm. **Cypselae** obpyramidal, 3–4 mm, hairs 1.5–2 mm, inserted at bases and on angles and faces; **pappi** of 10 lanceolate, aristate scales 5–6.5+ mm (scarious bases 3.5–4.5 × 0.8–1.5 mm). $2n = 34$.

Flowering May–Jun(–Oct). Clays or sandy soils, often rocky places, deserts; 1200–2400 m; Colo., Utah.

5. Gaillardia aristata Pursh, Fl. Amer. Sept. 2: 573. 1813 (as Galardia) E F

Perennials (sometimes flowering first year), 20–80 cm. **Leaves** basal and cauline or cauline; petiolar bases 5–15 cm; blades oblanceolate to lanceolate, 5–15 cm × 5–30(–40) mm, margins raggedly pinnately lobed to toothed or entire, faces scabrellous and/or sparsely to densely villous (hairs jointed). **Peduncles** (5–)20–35+ cm. **Phyllaries** 24–40+ ovate to lance-attenuate, 10–15+ mm, ciliate with jointed hairs (also strigose and gland-dotted). **Receptacular setae** 2–6 mm. **Ray florets** (6–)12–18+; corollas yellow or yellow/purple, rarely tubular and 5-lobed, usually distally laminate and 3-lobed, 15–35+ mm. **Disc florets** 60–120+; corollas usually purple or purple-tipped, sometimes yellow, tubes 0.5–1.5 mm, throats cylindric to urceolate, 4.5–5.5 mm, lobes lance-ovate to triangular-attenuate, 1–2 mm, jointed hairs 0.3+ mm. **Cypselae** clavate (outer) to obpyramidal (inner), 2.5–6 mm, hairs 1.5–2.5 mm, inserted at bases; **pappi** of 8 ovate to lanceolate, aristate scales 5–6 mm (scarious bases 1.5–3 × 0.4–1.5 mm). $2n = 34, 68$.

Flowering May–Sep. Open places, usually among aspens or pines, or with sagebrush, often dry, sandy benches or bars; 200–2900 m; Alta., B.C., Man., N.W.T., Sask., Yukon; Colo., Conn., Idaho, Mass., Minn., Mont., N.H., N.Dak., Oreg., S.Dak., Utah, Wash., Wyo.

6. Gaillardia coahuilensis B. L. Turner, SouthW. Naturalist 21: 539. 1977

Annuals, 20–80 cm. **Leaves** basal and cauline; petiolar bases 0–3+ cm; blades ovate to lanceolate, 3–8 cm × 5–25 mm, (bases of distal usually clasping) margins of proximal often lobed or toothed, of distal usually entire, rarely lobed, faces sparsely hispidulous. **Peduncles** 5–20 cm. **Phyllaries** 14–22, lanceolate-acuminate to lanceolate-attenuate, 5–12 mm, closely hispidulous strigillose, not ciliate with jointed hairs. **Receptacular setae** 2–4 mm. **Ray florets** 5–13; corollas proximally reddish or orange to yellow, distally orange to yellow, 10–20 mm. **Disc florets** 40–100+; corollas proximally yellow or reddish, distally reddish or yellow, tubes 0.6–1 mm, throats campanulate, 3–4 mm, lobes deltate, 1–1.2 mm, jointed hairs to 0.3 mm. **Cypselae** obpyramidal, 1–3 mm, hairs 1–3 mm, inserted at bases; **pappi** of 8–9 ovate to lanceolate, aristate scales 2–5+ mm (scarious bases 1–2.5 × 0.4–1 mm). $2n = 34$.

Flowering Apr–Jun. Calcareous soils; 300–1200+ m; Tex.; Mexico (Coahuila).

Specimens of *Gaillardia coahuilensis* were earlier treated as members of *G. mexicana* A. Gray, a Mexican species. Plants of *G. coahuilensis* seem much closer to *G. pulchella* than to *G. mexicana* in most attributes and might be included within the circumscription of G. *pulchella* but for the short hairs on the corollas.

7. Gaillardia pulchella Fougeroux, Hist. Acad. Roy. Sci. Mém. Math. Phys. (Paris, 4to) 1786: 5, fig. 1. 1788

• Firewheel, Indian blanket [F]

Gaillardia drummondii (Hooker) de Candolle; *G. neomexicana* A. Nelson; *G. picta* D. Don; *G. pulchella* var. *australis* B. L. Turner & M. Whalen; *G. pulchella* var. *picta* (D. Don) A. Gray

Annuals (sometimes persisting), 5–35(–60+) cm. **Leaves** cauline; petiolar bases 0–3+ cm; blades linear, oblong, or spatulate, 1–5(–12) cm × 4–12(–35) mm, (bases of distal ± clasping) margins usually entire, sometimes toothed or lobed, faces closely strigillose or hirtellous to ± villous (hairs jointed). **Peduncles** 3–10(–20) cm. **Phyllaries** 18–28+, narrowly triangular- to linear-attenuate, 6–14+ mm, usually ciliate with jointed hairs. **Receptacular setae** 1.5–3 mm. **Ray florets** usually 8–14, rarely 0; corollas usually reddish to purplish proximally, yellow to orange distally, rarely yellow, reddish, or purplish throughout, 13–30+ mm. **Disc florets** 40–100+; corollas yellowish to purple or brown, often bicolored, tubes 0.8–1.2 mm, throats campanulate to urceolate, 3–4 mm, lobes deltate to ovate, often attenuate, 1–3+ mm, jointed hairs 0.3+ mm. **Cypselae** obpyramidal, 2–2.5 mm, hairs 1.5–2 mm, inserted at bases and on angles; **pappi** of 7–8 deltate to lanceolate, aristate scales 4–7 mm (scarious bases 1–2.5 × 0.7–1.3 mm). $2n = 34$.

Flowering Jan–Dec, mostly May–Aug. Sandy or calcareous soils, often disturbed places, mostly in grasslands or open places; 0–1800 m; Ala., Ariz., Colo., Conn., Fla., Ga., Kans., Maine, Miss., Mo., Nebr., N.H., N.Mex., N.C., Okla., S.C., S.Dak., Tex., Vt.; Mexico (Chihuahua, Coahuila, Nuevo León, Sonora, and Tamaulipas).

Spring-flowering plants with most of their proximal leaves pinnately lobed or coarsely toothed, mainly found in south-central Texas, have been treated as var. *australis* (B. L. Turner and M. Whalen 1975). Plants from near or on beaches of the Atlantic and the Gulf of Mexico, usually with somewhat fleshy leaves and often persisting for more than one year, have been distinguished as var. *picta*.

Cultivars of *Gaillardia pulchella* (or of hybrids between *G. pulchella* and *G. aristata*) are used horticulturally.

SELECTED REFERENCE Turner, B. L. and M. Whalen. 1975. Taxonomic study of *Gaillardia pulchella* (Asteraceae–Heliantheae). Wrightia 5: 189–192.

8. Gaillardia amblyodon Gay, Ann. Sci. Nat., Bot., sér. 2, 12: 62. 1839 [E]

Annuals, 20–45+ cm. **Leaves** cauline; petiolar bases 0–3+ cm; blades oblanceolate, oblong, or spatulate, 2–6 cm × 6–24 mm, (bases of distal ± clasping) margins usually entire or obscurely toothed, rarely lobed, faces scabrellous and/or ± villous (hairs scattered, jointed). **Peduncles** 5–10(–15) cm. **Phyllaries** 22–32, lanceolate to lance-attenuate, 7–15 mm, ciliate with jointed hairs. **Receptacular setae** 2.5–3 mm. **Ray florets** 8–12; corollas mostly bright red to dark purple, rarely yellow, 15–25 mm. **Disc florets** 30–60+; corollas yellow with purple tips, tubes 1.5 mm, throats narrowly cylindric, 4.5 mm, lobes ovate-deltate, 0.5–1 mm, jointed hairs more than 0.3 mm. **Cypselae** dimorphic: the peripheral usually ± obcompressed-clavate, 2–4+ mm, glabrous or nearly so, with **pappi** of (6–)8+, muticous scales 0.1–0.3+ mm; the inner obpyramidal, 2–3 mm, hairs 1–3 mm, inserted at bases and on angles, with **pappi** of 8–12 lanceolate, aristate scales 5–6 mm (scarious bases 2.5–3 × 0.5–1 mm). $2n = 34$.

Flowering Apr–Jul. Open places on sandy soils; 10–100+ m; Tex.

Gaillardia amblyodon may prove to be not distinct from *G. pulchella* at species rank.

9. Gaillardia aestivalis (Walter) H. Rock, Rhodora 58: 315. 1956 (as aestivale) [E]

Helenium aestivale Walter, Fl. Carol., 210. 1788; *Gaillardia aestivalis* var. *flavovirens* (C. Mohr) Cronquist; *G. chrysantha* Small; *G. fastigiata* Greene; *G. lanceolata* Michaux; *G. lanceolata* var. *fastigiata* (Greene) Waterfall; *G. lanceolata* var. *flavovirens* C. Mohr; *G. lutea* Greene

Perennials (sometimes flowering first year) 10–60+ cm (sometimes rhizomatous). **Leaves** cauline; petiolar bases 0–3+ cm; blades narrowly elliptic, linear, obovate, or spatulate, 15–60 × 3–12(–22) mm, (bases of distal often clasping) margins remotely toothed or entire, faces usually closely scabrellous (hairs sometimes crisped). **Peduncles** 1–10(–20) cm. **Phyllaries** 15–26, lance-ovate to lanceolate, 6–14+ mm, scabrellous. **Receptacular setae** 0 or 0.1–0.5+ mm. **Ray florets** usually 6–12(–15), sometimes 0; corollas pinkish to purple or yellow to cream or white, 13–25 mm. **Disc florets** 20–60(–100+); corollas yellow or purple to purple-brown or bicolored, tubes 0.5–1 mm, throats stoutly cylindric to campanulate or urceolate, 3.5–5 mm, lobes attenuate-terete, 1.5–3 mm, jointed hairs to 0.3 mm. **Cypselae** 1.5–2 mm, hairs 1 mm, inserted at bases and on angles and faces; **pappi** of 8–10 lanceolate, aristate scales 5–7 mm (scarious bases 2–3.5 × 0.6–1 mm). $2n = 34, 68$.

Flowering Jun–Oct. Open places in grasslands, pinelands; 10–1200 m; Ala., Ark., Fla., Ga., Kans., La., Miss., Mo., N.C., Okla., S.C., Tex.

Plants with both ray and disc corollas wholly or mostly yellow to cream or white have been treated as a distinct species (*Gaillardia lutea*) or as varieties [*G. aestivalis* var. *chrysantha* (Small) Cronquist and *G. aestivalis* var. *winkleri* (Cory) B. L. Turner]. B. L. Turner (1979) formally recognized three varieties of *G. aestivalis* in the broad sense: var. *aestivalis* with purple discs, var. *chrysantha* with yellow discs, and var. *winkleri* with white discs. Although plants with yellow or white corollas may occur as local populations and may be distinguished by different ploidy levels, failing one or more additional distinguishing morphologic traits, they do not, in my opinion, merit formal taxonomic recognition.

10. Gaillardia suavis (A. Gray & Engelmann) Britton & Rusby, Trans. New York Acad. Sci. 7: 11. 1887

Agassizia suavis A. Gray & Engelmann, Proc. Amer. Acad. Arts 1: 49. 1847

Perennials (sometimes flowering first year), 20–80 cm. **Leaves** wholly basal or nearly so; petiolar bases 0–3+ cm; blades spatulate to oblanceolate, 3–10(–15) cm × 5–30(–60) mm, pinnatifid, toothed, or entire, glabrous or very sparsely villous with jointed hairs (obscurely, if at all, gland-dotted). **Peduncles** 15–75+ cm. **Phyllaries** 22–32, ovate to lance-attenuate, 6–8 mm, nearly glabrous or villous, or at least ciliate, with jointed hairs. **Receptacular setae** 0 or 0.1–0.5 mm. **Ray florets** 0 or 7–10; corollas purplish to red, 10–15 mm.

Disc florets 40–100+; corollas pinkish to purplish, tubes 1.2–2 mm, throats urceolate, 3–4 mm, lobes narrowly triangular, 1–1.2 mm, jointed hairs 0.3+ mm (style-branch appendages glabrous or glabrate). **Cypselae** obpyramidal, 2 mm, hairs 2+ mm, inserted at bases and on angles; **pappi** of 8–10 lanceolate, oblanceolate, or ovate, aristate scales 6–9 mm (scarious bases 4–5 × 1.4–2 mm). $2n = 38$, ca. 72.

Flowering Mar–Jun. Calcareous or sandy soils of prairies, desert scrubs, or juniper woodlands; 30–800 m; Kans., Okla., Tex.; Mexico (Coahuila, Nuevo León, Tamaulipas).

11. Gaillardia arizonica A. Gray in A. Gray et al., Syn. Fl. N. Amer. 1(2): 353. 1884

Gaillardia arizonica var. *pringlei* (Rydberg) S. F. Blake

Annuals, 5–40 cm. **Leaves** mostly basal, mostly restricted to proximal $^1/_5(-^1/_3+)$; petiolar bases 0–2+ cm; blades oblanceolate, obovate, or spatulate, 3–6(–9) cm × 5–25 mm, mostly raggedly pinnately lobed, sometimes entire, sparsely to densely villous with jointed hairs. **Peduncles** 5–35 cm. **Phyllaries** 14–20, lanceolate to lance-attenuate, 6–12+ mm, ciliate with jointed hairs. **Receptacular setae** 0 or 0.3–0.8 mm. **Ray florets** 10–13; corollas yellow to orange, 17–22 mm. **Disc florets** 40–100+; corollas yellow, tubes 0.6–0.8 mm, throats campanulate, 2.8–3 mm, lobes deltate, 0.5–0.7 mm, jointed hairs to 0.3 mm. **Cypselae** obpyramidal, 2 mm, hairs 2 mm, inserted at bases and on angles and faces; **pappi** of 8–10 obovate or ovate to lanceolate, aristate or muticous scales 2–5 mm (if aristate, scarious bases 2–3.5 × 1.3–1.6 mm). $2n = 34$.

Flowering Mar–May. Sandy desert washes and fans; 300–1200 m; Ariz., Nev., Utah; Mexico (Sonora).

385. HELENIUM Linnaeus, Sp. Pl. 2: 886. 1753; Gen. Pl. ed. 5, 377. 1754 (as Helenia)

• Sneezeweed [For Helen of Troy]

Mark W. Bierner

Annuals or perennials, 10–160 cm. **Stems** 1(–10), erect, simple or branched distally (usually ± winged by decurrent leaf bases), glabrous or sparsely to densely hairy. **Leaves** usually mostly cauline; mostly alternate (proximal sometimes opposite); petiolate or sessile; blades mostly elliptic, lanceolate, linear, oblanceolate, oblong, ovate, or spatulate, often pinnately lobed or pinnatifid, ultimate margins entire or toothed, faces glabrous or sparsely to densely hairy, gland-dotted. **Heads** radiate or discoid, borne singly or (2–300+) in paniculiform to corymbiform arrays. **Involucres** ± globose, hemispheric, obconic, or ovoid, 4–34 mm diam. **Phyllaries**

persistent, 9–34(–40+) in (1–)2[–3] series (distinct or proximally connate, usually reflexed in fruit, glabrous or sparsely to densely hairy, gland-dotted). **Receptacles** conic, ± globose, hemispheric, or ovoid, pitted (glabrous), epaleate (rarely bearing setiform enations). **Ray florets** 0 or 7–34, pistillate and fertile or neuter; corollas yellow, yellow with purple streaks, reddish brown to red proximally and yellow distally, or reddish brown to red or purple throughout. **Disc florets** 75–1000+, bisexual, fertile; corollas yellow, purple, or yellow to yellow-green proximally and yellow-brown, brown or reddish brown to purple distally, tubes shorter than throats, lobes 4 or 5, ± deltate (style-branch apices penicillate or truncate). **Cypselae** mostly obpyramidal, 4–5-angled, glabrous or sparsely to densely hairy; **pappi** persistent, of 5–12 entire or ± lacerate, sometimes aristate scales. *x* = 17.

Species ca. 32 (18 in the flora): North America, Mexico, West Indies (Cuba), Central America, South America.

Helenium polyphyllum Small is apparently a hybrid between *H. autumnale* and *H. flexuosum*.

SELECTED REFERENCES Bierner, M. W. 1972. Taxonomy of *Helenium* sect. *Tetrodus* and a conspectus of North American *Helenium* (Compositae). Brittonia 24: 331–355. Bierner, M. W. 1989. Taxonomy of *Helenium* sect. *Amarum* (Asteraceae). Sida 13: 453–459. Rock, H. F. L. 1957. A revision of the vernal species of *Helenium* (Compositae). Rhodora 59: 101–116, 128–158, 168–178, 203–216.

1. Stems not winged .... 1. *Helenium amarum*
1. Stems ± winged (by decurrent leaf bases).
  2. Ray florets usually 8–34, neuter (if rays 0, perennials from e or se United States).
    3. Heads 1(–3); disc corollas mostly yellow, sometimes yellow-brown distally.
      4. Pappus scales deeply lacerate; e Texas, sw Louisiana .... 2. *Helenium drummondii*
      4. Pappus scales entire or slightly lacerate (never deeply lacerate); se Louisiana,
        s Mississippi, s Alabama, s Georgia, se South Carolina, s North Carolina, Florida.
        5. Leaves of basal rosettes usually pinnatifid; peduncles hairy; cypselae hairy
          .... 3. *Helenium pinnatifidum*
        5. Leaves of basal rosettes usually entire; peduncles usually glabrous, sometimes sparsely hairy; cypselae glabrous .... 4. *Helenium vernale*
    3. Heads 1–80+; disc corollas usually purple, sometimes red-brown or yellow-brown distally or throughout.
      6. Heads (1–)5–50(–80); pappi of aristate scales .... 5. *Helenium flexuosum*
      6. Heads 1–20; pappi of non-aristate scales.
        7. Stems densely hairy proximally; leaves moderately to densely hairy; pappi 0.3–0.5 mm; nc Arkansas .... 6. *Helenium campestre*
        7. Stems glabrous proximally; leaves glabrous or sparsely hairy; pappi 1–1.7 mm; se United States .... 7. *Helenium brevifolium*
  2. Ray florets usually 7–30, pistillate, fertile (if rays 0, plants either perennials from w United States or annuals).
    8. Perennials.
      9. Stems unbranched or sparingly branched distally.
        10. Peduncles usually densely hairy; pappi (2.5–)3–4.5 mm .... 11. *Helenium bolanderi*
        10. Peduncles sparsely to moderately hairy; pappi 1.3–2.2(–2.7) mm....
          .... 12. *Helenium bigelovii*
      9. Stems strongly branched distally.
        11. Rays usually 13–15, sometimes 0, corollas 3.8–10 mm; disc corollas yellow to reddish brown or purple distally, lobes 4(–5) .... 10. *Helenium puberulum* (in part)
        11. Rays 8–21, corollas 10–23 mm; disc corollas yellow to yellow-brown distally, lobes 5.
          12. Leaves: basal blades oblanceolate, obovate, or spatulate (withered by flowering), mid blades usually dentate, distal blades entire or dentate; United States and s Canada .... 8. *Helenium autumnale*

12. Leaves: basal blades oblanceolate (usually present at flowering), mid
    blades usually entire, sometimes dentate; swampy meadows, Missouri,
    Virginia . . . . 9. *Helenium virginicum*

[8. Shifted to left margin.—Ed.]

8. Annuals.
   13. Pappi of aristate scales.
       14. Disc corolla lobes 4(–5); California . . . . 10. *Helenium puberulum* (in part)
       14. Disc corolla lobes 5; Arizona, Texas.
           15. Stems weakly winged (leaf bases scarcely decurrent from node to node); cypselae
               2–2.9 mm; n Arizona . . . . 13. *Helenium arizonicum*
           15. Stems narrowly winged (leaf bases decurrent from node to node); cypselae 1–
               1.2 mm; s Texas . . . . 14. *Helenium linifolium*
   13. Pappi of non-aristate scales.
       16. Involucres globoid and distal leaves entire . . . . 15. *Helenium elegans*
       16. Involucres usually ovoid to obconic (if globoid, distal leaves serrate to undulate-
           serrate).
           17. Distal leaves serrate to undulate-serrate; disc corolla lobes (4–)5 . . . . . .
               . . . . 16. *Helenium microcephalum*
           17. Distal leaves entire; disc corolla lobes 4.
               18. Rays 10–15 . . . . 17. *Helenium quadridentatum*
               18. Rays 0 . . . . 18. *Helenium thurberi*

1. **Helenium amarum** (Rafinesque) H. Rock, Rhodora
59: 131. 1957 • Yellowdicks, fiveleaf sneezeweed

Gaillardia amara Rafinesque, Fl.
Ludov., 69. 1817 (as Galardia);
Helenium tenuifolium Nuttall

Annuals, 10–100 cm. **Stems** 1–3
(–15), usually branched distally,
not winged, glabrous or sparsely
hairy. **Leaves** glabrous or sparsely
hairy; basal blades linear to ovate,
entire or pinnately toothed or
lobed to pinnatifid; mid and distal blades linear, entire.
Heads (1–)10–150(–250+) per plant, in paniculiform
arrays. **Peduncles** 3–11 cm, sparsely hairy. **Involucres**
hemispheric to globose or globoid, 5–9 × 6–10 mm. **Phyllaries** moderately to densely hairy. **Ray florets** 8–10, pistillate, fertile; corollas yellow, 4.5–14 × 2–10 mm. **Disc
florets** 75–250+; corollas yellow proximally, yellow to
yellow-brown or purple distally, 1.6–2.7 × 0.8–1.2 mm,
lobes 5. **Cypselae** 0.7–1.3 mm, moderately to densely
hairy; **pappi** of 6–8 entire, aristate scales 1–1.8 mm.
Varieties 2 (2 in the flora): se United States, n Mexico.

1. Basal and proximal cauline leaves usually withered by flowering; basal blades usually entire or
   pinnately toothed, sometimes pinnatifid; proximal
   blades usually entire, sometimes pinnately toothed;
   disc corollas yellow to yellow-brown distally
   . . . . 1a. *Helenium amarum* var. *amarum*
1. Basal and proximal cauline leaves often present
   at, sometimes withered by, flowering; basal blades
   pinnatifid; proximal blades entire or pinnately
   toothed or lobed to pinnatifid; disc corollas purple
   distally . . . . 1b. *Helenium amarum* var. *badium*

1a. **Helenium amarum** (Rafinesque) H. Rock var.
amarum E

Plants (10–)20–60(–100) cm.
Leaves: basal and proximal
cauline usually withered by flowering; basal blades linear to ovate,
usually entire or pinnately toothed
sometimes pinnatifid; proximal
blades usually linear, usually entire, sometimes pinnately toothed.
Ray corollas 6.5–14 × 4–10 mm.
Disc florets 75–150+; corollas yellow to yellow-brown
distally, 2–2.7 mm. **Cypselae** 0.9–1.3 mm; **pappi** of 6–8
scales 1.2–1.8 mm. 2*n* = 30.

Flowering (Feb–)Jul–Oct(–Dec). Roadsides and fields;
20–400 m; Ala., Ark., Calif., D.C., Fla., Ga., Ind., Kans.,
La., Mich., Miss., Mo., N.C., Okla., Pa., S.C., Tenn.,
Tex., Va.

Variety *amarum* is introduced in California.

1b. **Helenium amarum** (Rafinesque) H. Rock var.
badium (A. Gray ex S. Watson) Waterfall, Rhodora
62: 321. 1960 • Basin sneezeweed

Helenium tenuifolium Nuttall var.
badium A. Gray ex S. Watson, Proc.
Amer. Acad. Arts 18: 108. 1883; *H.
badium* (A. Gray ex S. Watson)
Greene

Plants 10–50(–80) cm. **Leaves:**
basal and proximal often present
at, sometimes withered by, flowering; basal blades pinnatifid;
proximal blades entire or pinnately toothed or lobed to

pinnatifid. **Ray corollas** 4.5–14 × 2–9 mm. **Disc florets** 100–250+; corollas purple distally, 1.6–2.5 mm. **Cypselae** 0.7–1 mm; **pappi** of usually 6 scales 1–1.5 mm. **2***n* = 30.

Flowering (Mar–)Apr–Jul(–Nov). Roadsides and fields; 200–1500 m; Okla., Tex.; Mexico (Chihuahua, Coahuila).

2. **Helenium drummondii** H. Rock, Rhodora 59: 173. 1957 • Fringed sneezeweed E

Leptopoda fimbriatum Torrey & A. Gray, Fl. N. Amer. 2: 387. 1842, not (Michaux) Eaton 1829

Perennials, 20–60 cm. **Stems** usually 1, usually unbranched distally, moderately to strongly winged, glabrous proximally, glabrous or sparsely hairy distally. **Leaves** glabrous; basal blades narrowly obovate to narrowly oblanceolate, usually entire, sometimes undulate-serrate; proximal and mid blades narrowly lanceolate, usually entire, sometimes undulate-serrate; distal blades lance-linear, usually entire, rarely undulate-serrate. **Heads** 1(–3) per plant, usually borne singly. **Peduncles** 10–30 cm, sparsely hairy. **Involucres** hemispheric to globoid, 10–18 × 15–25 mm. **Phyllaries** sparsely to moderately hairy. **Ray florets** 13–30, neuter; corollas yellow, 14–25 × 4–10 mm. **Disc florets** 250–650(–1000+); corollas yellow, 4.3–5.4 mm, lobes 5. **Cypselae** 1.2–1.5 mm, moderately hairy; **pappi** of 5–12 deeply lacerate, non-aristate scales 2–3.7 mm. **2***n* = 32.

Flowering (Mar–)Apr(–May). Ditches, other moist areas such as wet woods, bogs, and swamp edges; 10–50 m; Ark., La., Tex.

3. **Helenium pinnatifidum** (Schweinitz ex Nuttall) Rydberg in N. L. Britton et al., N. Amer. Fl. 34: 130. 1915 • Southeastern sneezeweed E

Leptopoda pinnatifida Schweinitz ex Nuttall, Trans. Amer. Philos. Soc., n. s. 7: 372. 1841

Perennials, 30–80 cm. **Stems** usually 1, usually unbranched distally, weakly winged, glabrous or sparsely to moderately hairy proximally, sparsely to moderately hairy distally. **Leaves** usually glabrous, rarely sparsely hairy; basal blades obovate to oblanceolate, usually pinnatifid, sometimes undulate to undulate-serrate, rarely entire; proximal and mid blades linear-oblanceolate to lance-linear, entire or undulate-serrate; distal blades lance-linear, entire. **Heads** 1(–3) per plant, usually borne singly. **Peduncles** 3–20 cm, moderately to densely hairy. **Involucres** hemispheric, 12–20 × 15–30 mm. **Phyllaries** moderately hairy. **Ray**

florets 13–34, neuter; corollas yellow, 15–22 × 5–8 mm. **Disc florets** 250–650(–800+); corollas yellow proximally, yellow to yellow-brown distally, 4–5.5 mm, lobes 5. **Cypselae** 1.2–1.4 mm, moderately hairy; **pappi** of 8–11 entire or slightly lacerate, non-aristate scales 1.2–1.5 mm. **2***n* = 32, 34.

Flowering (Feb–)Mar–Apr(–May). Ditches, other moist areas such as wet woods, bogs, and swamp edges; 10–50 m; Fla., Ga., N.C., S.C.

4. **Helenium vernale** Walter, Fl. Carol., 210. 1788 • Savanna or spring sneezeweed E F

Perennials, 30–80 cm. **Stems** usually 1, usually unbranched distally, weakly winged, glabrous proximally, glabrous or sparsely hairy distally. **Leaves** usually glabrous, rarely sparsely hairy; basal blades obovate to narrowly oblanceolate, usually entire or undulate to un-dulate-serrate; proximal and mid blades usually narrowly lanceolate to narrowly oblanceolate, usually entire, sometimes toothed; distal blades lance-linear, entire. **Heads** 1(–3) per plant, usually borne singly. **Peduncles** 2–23 cm, usually glabrous, sometimes sparsely hairy. **Involucres** hemispheric, 10–17 × 15–27 mm. **Phyllaries** glabrous or sparsely hairy. **Ray florets** 13–30, neuter; corollas yellow, 15–21 × 5–10 mm. **Disc florets** 150–650(–800+); corollas yellow proximally, yellow to yellow-brown distally, 4.6–6 mm, lobes 5. **Cypselae** 1–1.5 mm, glabrous; **pappi** of usually 8 entire or lacerate, non-aristate scales 1.5–2 mm. **2***n* = 34.

Flowering Mar–May. Ditches, other moist areas such as wet woods, bogs, and swamp edges; 10–50 m; Ala., Fla., Ga., La., Miss., N.C., S.C.

The name *Helenium nuttallii* A. Gray is illegitimate; it has been applied to plants treated here as *H. vernale*.

5. **Helenium flexuosum** Rafinesque, New Fl. 4: 81. 1838 • Purplehead sneezeweed, hélénie nudiflore E

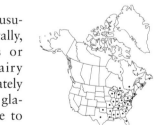

Helenium nudiflorum Nuttall

Perennials, 30–100 cm. **Stems** usually 1, usually branched distally, strongly winged, glabrous or sparsely hairy proximally, sparsely to moderately hairy distally. **Leaves** glabrous or sparsely to moderately hairy; basal blades oblanceolate to obovate or spatulate, entire or serrate; proximal and mid blades oblanceolate to lanceolate, entire or toothed; distal blades lanceolate to lance-linear, entire. **Heads** (1–)5–50(–80+) per plant, in paniculiform arrays. **Peduncles** 3–10 cm, sparsely to moderately hairy. **Involucres** globoid to

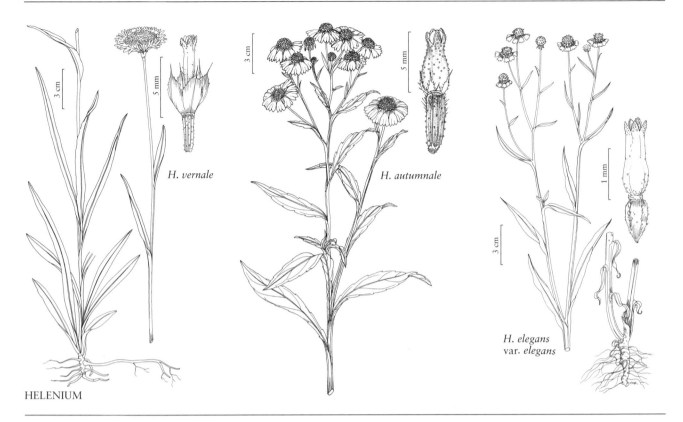

HELENIUM

usually ovoid, 8–17 × 9–17 mm. **Phyllaries** (connate proximally) moderately to densely hairy. **Ray florets** usually 8–13, neuter, sometimes 0; corollas yellow to reddish brown to red to purple, 10–20 × 5–10 mm. **Disc florets** 250–500(–700+); corollas yellow proximally and purple distally or purple throughout, 2.3–3.7 mm, lobes 4(–5). **Cypselae** 1–1.2 mm, moderately hairy; **pappi** of 5–6 entire, aristate scales 0.6–1(–1.7) mm. **2*n* = 28.**

Flowering (Apr–)May–Jul(–Oct). Ditches, fields, washes, along streams; 10–900 m; N.S., Ont., Que.; Ala., Ark., Fla., Ga., Ill., Ind., Ky., La., Md., Mass., Mich., Miss., Mo., N.J., N.C., Ohio, Okla., Pa., S.C., Tenn., Tex., Va., W.Va., Wis.

It has been suggested that *Helenium flexuosum* may be of hybrid origin from *H. campestre* (perennial, 2*n* = 28) and some member of *Helenium* sect. *Tetrodus* (H. F. L. Rock 1957). The only member of sect. *Tetrodus* that has a chromosome number of 2*n* = 28 is *H. linifolium*, an annual from southern Texas.

6. Helenium campestre Small, Fl. S.E. U.S., 1291, 1341. 1903 • Oldfield sneezeweed [E]

Perennials, 40–100 cm. **Stems** usually 1, branched distally, moderately to strongly winged, densely (the proximal) to moderately hairy. **Leaves** moderately to densely hairy; basal blades oblanceolate to obovate to spatulate, entire or undulate-serrate; proximal and mid blades oblanceolate to lanceolate, usually entire, sometimes undulate-serrate; distal blades lanceolate to lance-linear, entire. **Heads** 3–20 per plant, in paniculiform to corymbiform arrays. **Peduncles** 3–11 cm, moderately to densely hairy. **Involucres** globoid to ovoid, 9–17 × 10–20 mm. **Phyllaries** (barely connate proximally) moderately to densely hairy. **Ray florets** 9–15, neuter; corollas yellow, 20–30 × 8–15 mm. **Disc florets** 150–500(–700+); corollas yellow proximally, reddish brown to purple distally, 3–5 mm, lobes 5. **Cypselae** 1–1.4 mm, sparsely to moderately hairy; **pappi** of 6–7, entire, non-aristate scales 0.3–0.5 mm. **2*n* = 28.**

Flowering May–Jun(–Jul). Ditches, fields, in washes, along streams; 100–200 m; Ark.

7. **Helenium brevifolium** (Nuttall) Alph. Wood, Amer. Bot. Fl., 182. 1870 • Shortleaf sneezeweed E

Leptopoda brevifolia Nuttall, Trans. Amer. Philos. Soc., n. s. 7: 373. 1841; *Helenium curtisii* A. Gray

Perennials, 20–100 cm. **Stems** usually 1, unbranched or sparingly branched distally, weakly to moderately winged, glabrous proximally, sparsely to moderately hairy distally. **Leaves** glabrous or sparsely hairy; basal blades oblanceolate to spatulate, entire or undulate to undulate-serrate; proximal and mid blades oblanceolate to lanceolate, usually entire, sometimes undulate to undulate-serrate; distal blades lance-linear, entire. **Heads** 1–10 per plant, borne singly or in paniculiform to corymbiform arrays. **Peduncles** (3–)6–11(–15) cm, sparsely to moderately hairy. **Involucres** hemispheric to globoid, 10–15 × 10–20 mm. **Phyllaries** (connate proximally) sparsely to moderately hairy. **Ray florets** 9–24, neuter; corollas yellow, 12.5–21 × 4–8 mm. **Disc florets** 200–500(–800+); corollas yellow proximally, usually purple to yellow or yellow-brown distally, 4–5.6 mm, lobes 5. **Cypselae** 1–1.4 mm, sparsely to moderately hairy; **pappi** of 6–8 entire, non-aristate scales 1–1.7 mm. **2n** = 26, 28.

Flowering Mar–May. Ditches, fields, other moist areas such as wet woods, bogs, and swamp edges; 0–80 m; Ala., Fla., Ga., La., Miss., N.C., Va.

8. **Helenium autumnale** Linnaeus, Sp. Pl. 2: 886. 1753 • Common sneezeweed, hélénie automnale E F

Helenium autumnale var. *canaliculatum* (Lamarck) Torrey & A. Gray; *H. autumnale* var. *fylesii* B. Boivin; *H. autumnale* var. *grandiflorum* Torrey & A. Gray; *H. autumnale* var. *montanum* (Nuttall) Fernald; *H. autumnale* var. *parviflorum* (Nuttall) Fernald; *H. latifolium* Miller; *H. parviflorum* Nuttall

Perennials, 50–130 cm. **Stems** 1(–7), branched distally, strongly winged, sparsely to densely hairy proximally, moderately to densely hairy distally. **Leaves** usually moderately to densely hairy, sometimes glabrous; basal blades (withered by flowering) lanceolate, oblanceolate, or obovate, entire or weakly lobed; proximal and mid blades obovate to oblanceolate, usually dentate or entire; distal blades oblanceolate to lanceolate, entire or dentate. **Heads** 5–70(–100+) per plant, in paniculiform arrays. **Peduncles** 3–10 cm, moderately to densely hairy. **Involucres** globoid, 8–20 × 8–23 mm. **Phyllaries** (connate proximally) moderately to densely hairy. **Ray**

florets 8–21, pistillate, fertile; corollas yellow, 10–23 × 4–10 mm. **Disc florets** 200–400(–800+); corollas yellow proximally, yellow to yellow-brown distally, 2.4–4 mm, lobes 5. **Cypselae** 1–2 mm, sparsely to moderately hairy; **pappi** of 5–7 entire, aristate scales (0.5–)0.9–1.5(–1.8) mm. **2n** = 32, 34, 36.

Flowering (Jul–)Aug–Oct(–Nov). Roadsides, fields, along streams, ditches, seepage areas, around ponds and lakes; 20–2600 m; Alta., B.C., Man., N.W.T., Ont., Que., Sask.; Ala., Ariz., Ark., Calif., Colo., Del., D.C., Fla., Ga., Idaho, Ill., Ind., Iowa, Kans., Ky., La., Md., Mass., Mich., Miss., Mo., Mont., Nebr., Nev., N.J., N.Mex., N.Y., N.C., N.Dak., Ohio, Okla., Oreg., Pa., S.C., Tenn., Tex., Utah, Va., Wash., W.Va., Wyo.

9. **Helenium virginicum** S. F. Blake, Claytonia 3: 13. 1936 • Virginia sneezeweed C E

Perennials, 30–130 cm. **Stems** 1 (–2), branched distally, strongly winged, glabrous or sparsely to moderately hairy proximally, glabrous or sparsely hairy distally. **Leaves** glabrous or sparsely to densely hairy; basal blades oblanceolate, weakly to strongly lobed; proximal and mid blades oblanceolate to lanceolate, usually entire, sometimes dentate; distal blades usually lanceolate, entire. **Heads** 2–25+ per plant, in paniculiform arrays. **Peduncles** 2–8 cm, sparsely to moderately hairy. **Involucres** globoid, 8–15 × 10–16 mm. **Phyllaries** (connate proximally) sparsely to moderately hairy. **Ray florets** 8–13, pistillate, fertile; corollas yellow, 10–15 × 4–10 mm. **Disc florets** 200–400+; corollas yellow proximally, yellow to yellow-brown distally, 2.5–3.5 mm, lobes 5. **Cypselae** 1.8–2.3 mm, moderately hairy; **pappi** of 5–6 entire, aristate scales 1.3–2 mm.

Flowering Jul–Sep. Around ponds, lakes, and bogs, swampy meadows; of conservation concern; 300–500 m; Mo., Va.

M. C. Simurda and J. S. Knox (2000) presented ITS sequence data that placed a population of *Helenium virginicum* from the Ozark highlands of southern Missouri in a monophyletic group with six populations of *H. virginicum* from the Shenandoah Valley of Virginia. R. L. Rimer and J. W. Summers (Missouri Department of Conservation, pers. comm.) located 42 *H. virginicum* populations in six counties in the Ozark highlands of Missouri. G. A. Yatskievych (Missouri Botanical Garden, pers. comm. to editors) has stated, ". . . the existence of this taxon in Missouri is no longer a matter of a single odd population, but probably rather another case of a taxon with two disjunct centers following dissection of a range during the Pleistocene glaciation."

Helenium virginicum is in the Center for Plant Conservation's National Collection of Endangered Plants.

SELECTED REFERENCE Simurda, M. C., and J. S. Knox. 2000. ITS sequence evidence for the disjunct distribution between Virginia and Missouri of the narrow endemic *Helenium virginicum*. J. Torrey Bot. Soc. 127: 316–323.

10. **Helenium puberulum** de Candolle in A. P. de Candolle and A. L. P. P. de Candolle, Prodr. 5: 667. 1836 • Rosilla

Annuals or perennials, 50–160 cm. **Stems** usually 1, branched distally, strongly winged, glabrous proximally, sparsely hairy distally. **Leaves** glabrous or sparsely hairy; basal blades oblanceolate to oblong-elliptic, entire; proximal and mid blades lanceolate to oblong-elliptic, entire; distal blades lance-linear, entire. **Heads** 4–20(–30) per plant, in paniculiform arrays. **Peduncles** (6–)9–17(–23) cm, sparsely to moderately hairy. **Involucres** globose to depressed globose, 9–15 × 9–17(–19) mm. **Phyllaries** (distinct or connate proximally) moderately hairy. **Ray florets** 0, or 13–15, pistillate, fertile; corollas yellow, 3.8–10 × 2–4 mm. **Disc florets** 300–500(–1000+); corollas yellow proximally, yellow to reddish brown to purple distally, (1.6–)1.9–2.7 mm, lobes 4(–5). **Cypselae** 1.2–1.9 mm, moderately hairy; **pappi** of 5–6 entire, aristate scales 0.4–1 mm. $2n = 58$.

Flowering (Mar–)Jun–Aug(–Nov). Along streams, ditches, seepage areas, around ponds and lakes, forests, woodlands; 0–1200 m; Calif.; Mexico (Baja California).

Helenium puberulum may be of amphidiploid origin from hybridization between *H. bigelovii* (perennial, radiate, $2n = 32$) and *H. thurberi* (annual, rayless, $2n = 26$) (M. W. Bierner 1972).

11. **Helenium bolanderi** A. Gray, Proc. Amer. Acad. Arts 7: 358. 1868 • Coastal sneezeweed E

Perennials, 25–140 cm. **Stems** usually 1, usually unbranched distally, weakly winged, glabrous proximally, sparsely to moderately hairy distally. **Leaves** glabrous or sparsely hairy; basal and proximal blades spatulate, entire; mid and distal blades ovate to oblanceolate to oblong-elliptic, entire. **Heads** 1(–3) per plant, usually borne singly. **Peduncles** 10–30 (–50) cm, usually densely hairy (tomentose near involucres). **Involucres** depressed hemispheric to globoid, 14–24 × 18–34 mm. **Phyllaries** (connate proximally) usually tomentose. **Ray florets** 15–30, pistillate, fertile; corollas yellow, (12–)16–28(–37) × 6–13(–25) mm. **Disc florets** 300–750(–1000+); corollas yellow proximally, yellow to brown to purple distally, 4–5(–5.2) mm, lobes

5. **Cypselae** 1.5–2.5 mm, moderately to densely hairy; **pappi** of 6–8 entire, aristate scales (2.5–)3–4.5 mm. $2n = 32$.

Flowering Jun–Sep. Bogs, seepage areas, wet meadows; 30–200 m; Calif., Oreg.

12. **Helenium bigelovii** Torrey & A. Gray in War Department [U.S.], Pacif. Railr. Rep. 4(5): 107. 1857 • Bigelow's sneezeweed E

Perennials, 30–130 cm. **Stems** 1–3(–10), unbranched or sparingly branched distally, weakly to moderately winged, glabrous proximally, glabrous or sparsely hairy distally. **Leaves** glabrous or sparsely hairy; basal blades oblanceolate to oblong-elliptic, entire; proximal and mid blades lanceolate to oblong-elliptic, entire; distal blades linear to lanceolate, entire. **Heads** 1–20 per plant, borne singly or in paniculiform arrays. **Peduncles** (6–)10–30 cm, sparsely to moderately hairy. **Involucres** hemispheric to globoid, 12–20 × (14–)17–22(–25) mm. **Phyllaries** (connate proximally) moderately to densely hairy. **Ray florets** 14–20, pistillate, fertile; corollas yellow, 13–25 × 5–10(–12) mm. **Disc florets** 250–500(–800+); corollas yellow proximally, yellow to brown to purple distally, 3–4.4(–4.8) mm, lobes 5. **Cypselae** 1.8–2.4 mm, moderately hairy; **pappi** of 6–8 entire, aristate scales 1.3–2.2(–2.7) mm. $2n = 32$.

Flowering (May–)Jul–Aug(–Oct). Bogs and swamps, around ponds and lakes, along streams, in moist meadows; 60–3400 m; Calif., Oreg.

13. **Helenium arizonicum** S. F. Blake, J. Wash. Acad. Sci. 27: 388. 1937 • Arizona sneezeweed E

Annuals, 30–70 cm. **Stems** usually 1, sparingly branched distally, weakly winged, glabrous or sparsely hairy. **Leaves** glabrous or sparsely hairy; blades all lance-linear, entire or coarsely toothed. **Heads** 1–15(–25) per plant, in paniculiform arrays. **Peduncles** 4–13 cm, glabrous or sparsely to moderately hairy. **Involucres** globoid to ovoid, 11–16 × 14–20 mm. **Phyllaries** sparsely to densely hairy. **Ray florets** 12–18, pistillate, fertile; corollas usually yellow, sometimes yellow with purple streaks, 10–20 × 5–13 mm. **Disc florets** 300–500(–700+); corollas yellow proximally, yellow to purple to dark blue-purple distally, 3.2–3.6 mm, lobes 5. **Cypselae** 2–2.9 mm, moderately hairy; **pappi** of 6–7 entire, aristate scales 1.5–2 mm. $2n = 30$.

Flowering Jul–Sep. Pine forests, edges of wet places such as bogs, ponds, lakes, ditches; 1800–2400 m; Ariz.

14. Helenium linifolium Rydberg in N. L. Britton et al., N. Amer. Fl. 34: 125. 1915 • Slimleaf sneezeweed E

Annuals, 20–80 cm. **Stems** usually 1, branched distally, narrowly winged, sparsely hairy proximally, moderately hairy distally. **Leaves** glabrous or sparsely hairy (basal withered by flowering); proximal blades lance-linear, usually coarsely or irregularly laciniate; mid and distal blades lance-linear to linear, entire. **Heads** 5–50(–100+) per plant, in paniculiform arrays. **Peduncles** 2–8 cm, sparsely to moderately hairy. **Involucres** globose to slightly ovoid, 6–11(–13) × 6–11(–13) mm. **Phyllaries** moderately hairy. **Ray florets** 8–9, pistillate, fertile; corollas yellow throughout or red proximally and yellow distally or red throughout, 6–12 × 3–8 mm. **Disc florets** 150–300(–500+); corollas yellow-green proximally, brown to purple distally, 2.5–3.2 mm, lobes 5. **Cypselae** 1–1.2 mm, moderately hairy; **pappi** of 6(–7) entire, aristate scales 2–2.5 mm. 2*n* = 28.

Flowering (Feb–)Mar–May(–Jul). Sandy soils, ditches, open pastures, edges of ponds and streams; 10–200 m; Tex.

15. Helenium elegans de Candolle in A. P. de Candolle and A. L. P. P. de Candolle, Prodr. 5: 667. 1836 • Pretty sneezeweed F

Annuals, 20–120 cm. **Stems** usually 1, branched distally, moderately to strongly winged, glabrous or sparsely to moderately hairy. **Leaves** glabrous or sparsely hairy (basal withered by flowering); proximal and mid blades lanceolate to narrowly elliptic, entire, serrate, or irregularly toothed or lobed; distal blades lanceolate to linear, entire. **Heads** 20–200+ per plant, in paniculiform arrays. **Peduncles** 2–14 cm, moderately to densely hairy. **Involucres** globoid, 5–11 × 5–12(–13) mm. **Phyllaries** (connate proximally) moderately to densely hairy. **Ray florets** 10–17, pistillate, fertile; corollas yellow throughout or red-brown to red proximally and yellow distally, 5.5–15(–19) × 2.2–5.5(–8.7) mm. **Disc florets** 150–700+; corollas yellow proximally, yellow to brown to usually reddish brown distally, 1.5–2.3(–2.5) mm, lobes 5. **Cypselae** 0.6–1 mm, moderately hairy; **pappi** of 6–7 entire, non-aristate scales 0.05–0.3(–0.4) mm.

Varieties 2 (2 in the flora): sc United States, n Mexico.

1. Proximal leaves usually entire, rarely serrate; involucres 5–8 × 5–8 mm ...... 15a. *Helenium elegans* var. *elegans*
1. Proximal leaves usually irregularly toothed or lobed; involucres 7–11 × 8–12(–13) mm .... 15b. *Helenium elegans* var. *amphibolum*

15a. Helenium elegans de Candolle var. **elegans** E F

Plants (20–)30–100(–120) cm. **Leaves:** proximal blades usually entire, rarely serrate; mid and distal leaves entire. **Heads** 20–100 (–200+). **Peduncles** 2–5(–7) cm. **Involucres** 5–8 × 5–8 mm. **Ray florets** 10–14; corollas 5.5–11.5 × 2.7–5.2 mm. **Disc florets** 150–300(–500+). **Cypselae** 0.6–0.9 mm; **pappi** 0.05–0.2 mm. 2*n* = 26.

Flowering (May–)Jun–Aug (sometimes again in fall). Calcareous soils, ditches, washes, along streams; 100–600 m; Ark., Okla., Tex.

15b. Helenium elegans de Candolle var. **amphibolum** (A. Gray) Bierner, Sida 5: 47. 1972

Helenium amphibolum A. Gray, Proc. Amer. Acad. Arts 9: 202. 1874

Plants 25–70 cm. **Leaves:** proximal blades usually irregularly toothed or lobed; mid and distal leaves usually entire. **Heads** 20–50. **Peduncles** 3–14 cm. **Involucres** 7–11 × 8–12(–13) mm. **Ray florets** 13–17; corollas 6.3–15(–19) × 2.2–5.5(–8.7) mm. **Disc florets** 250–500(–700+). **Cypselae** 0.7–1 mm; **pappi** 0.1–0.3(–0.4) mm. 2*n* = 26.

Flowering (Feb–)Mar–Jun(–Sep). Calcareous soils, ditches, washes, along streams; 300–1900 m; Tex.; Mexico (Coahuila, Nuevo León, San Luis Potosí, Tamaulipas).

16. Helenium microcephalum de Candolle in A. P. de Candolle and A. L. P. P. de Candolle, Prodr. 5: 667. 1836 • Smallhead sneezeweed

Annuals, 25–120 cm. **Stems** 1 (2–10), branched distally, moderately to strongly winged, glabrous or sparsely hairy. **Leaves** glabrous or sparsely to moderately hairy (basal withered by flowering); proximal and mid blades narrowly elliptic to narrowly oblong-elliptic, usually serrate to undulate-serrate, sometimes deeply toothed or laciniate;

distal blades narrowly elliptic to narrowly oblong-elliptic, serrate to undulate-serrate. **Heads** 20–300+ per plant, in paniculiform arrays. **Peduncles** 1.8–8 cm, moderately hairy. **Involucres** globoid to usually ovoid, 4–12 × 4–12 mm. **Phyllaries** (connate proximally) moderately hairy. **Ray florets** 7–13, pistillate, fertile; corollas usually yellow throughout, sometimes reddish proximally and yellow distally, 2.6–9.1 × 1.5–6.7 mm. **Disc florets** 120–400+; corollas yellow to yellow-green proximally, brown to reddish brown distally, 1.2–2.4 mm, lobes (4–)5. **Cypselae** 0.7–1.4 mm, moderately hairy; **pappi** of usually 6 entire, non-aristate scales (0.2–)0.3–0.7 mm.

Varieties 2 (2 in the flora): sw, sc United States, n Mexico.

1. Peduncles 1–3.5 cm, projecting terminal heads barely beyond foliage; involucres 4–8 × 4–8 mm . . . . 16a. *Helenium microcephalum* var. *microcephalum*
1. Peduncles 1.5–8 cm, usually projecting terminal heads well beyond foliage; involucres 8–12 × 8–12 mm . . . . 16b. *Helenium microcephalum* var. *ooclinium*

16a. Helenium microcephalum de Candolle var. microcephalum

Plants 30–120 cm. **Leaves:** proximal and mid blades serrate to undulate-serrate, glabrous or sparsely hairy; distal blades glabrous or sparsely hairy. **Heads** 20–150(–300+) per plant. **Peduncles** 1–3.5 cm, projecting terminal heads barely beyond foliage. **Involucres** 4–8 × 4–8 mm. **Ray florets** 7–13; corollas yellow, 2.6–4.8 × 1.6–3.9 mm. **Disc florets** 120–300+; corollas 1.2–1.6 mm, lobes (4–)5. **Pappi** (0.2–)0.3–0.4(–0.5) mm. 2*n* = 26.

Flowering (Feb–)Jun–Jul(–Nov). Sandy or clay soils, ditches, washes, around ponds and lakes, along streams; 0–1500 m; Ariz., N.Mex., Okla., Tex.; Mexico (Coahuila, Nuevo León, and Tamaulipas).

16b. Helenium microcephalum de Candolle var. ooclinium (A. Gray) Bierner, Sida 5: 47. 1972

Helenium ooclinium A. Gray, Proc. Amer. Acad. Arts 9: 202. 1874

Plants 25–80 cm. **Leaves:** proximal and mid blades serrate to undulate-serrate to occasionally deeply toothed or laciniate, glabrate; distal blades glabrous or sparsely to moderately hairy. **Heads** 20–100+ per plant. **Peduncles** 1.5–8 cm, usually projecting terminal heads well beyond foliage. **Involucres** 8–12 × 8–12 mm. **Ray**

florets 9–13; corollas usually yellow throughout, sometimes reddish proximally and yellow distally, 2.7–9.1 × 1.5–6.7 mm. **Disc florets** 200–400+; corollas 1.7–2.4 mm, lobes 5. **Pappi** 0.4–0.7 mm. 2*n* = 26.

Flowering (Mar–)Apr–Jun(–Dec). Sandy or clay soils, ditches, washes, around ponds and lakes, along streams; 500–1900 m; Tex.; Mexico (Coahuila, Nuevo León, Durango, San Luis Potosí)

17. Helenium quadridentatum Labillardière, Actes Soc. Hist. Nat. Paris 1: 22, plate 4. 1792 • Longdisk sneezeweed

Annuals, 30–100 cm. **Stems** usually 1, branched distally, moderately to strongly winged, glabrous or sparsely hairy. **Leaves** glabrous or sparsely hairy; blades narrowly elliptic to lance-linear, usually entire; proximal blades sometimes pinnately lobed. **Heads** 5–15 (–50+) per plant, in paniculiform arrays. **Peduncles** 3–13 cm, sparsely to moderately hairy. **Involucres** ovoid to conic, 7–11(–14) × 6–11 mm. **Phyllaries** (connate proximally) sparsely to densely hairy. **Ray florets** 10–15, pistillate, fertile; corollas yellow, 7–12.4 × 4–8.8 mm. **Disc florets** 225–500+; corollas yellow proximally, yellow to brown to reddish brown distally, 1.1–1.8 mm, lobes 4. **Cypselae** 0.6–1.2 mm, moderately hairy; **pappi** of usually 6 entire, non-aristate scales 0.1–0.3 mm. 2*n* = 26.

Flowering Mar–Jun (year round). Sandy or clay soils, ditches, around ponds and lakes, along streams; 0–1800 m; Ala., La., Miss., Pa., Tex.; Mexico; West Indies (Cuba); Central America (Belize, Guatemala).

Helenium quadridentatum is introduced in Pennsylvania.

18. Helenium thurberi A. Gray, Proc. Amer. Acad. Arts 19: 32. 1883 • Thurber's sneezeweed

Annuals, 40–100 cm. **Stems** usually 1, branched distally, moderately winged, sparsely hairy proximally, moderately hairy distally. **Leaves** glabrous or sparsely hairy (basal leaves withered by flowering); proximal blades lanceolate to lance-linear, usually irregularly laciniate; mid and distal blades lance-linear, entire. **Heads** 10–75(–120+) per plant, in paniculiform arrays. **Peduncles** 3–10 cm, moderately hairy. **Involucres** ovoid to conic, 6–10 × 5–10 mm. **Phyllaries** (strongly connate proximally), moderately hairy. **Ray florets** 0. **Disc florets** 225–500+; corollas yellow-green proximally, brown to reddish brown distally,

1–1.3 mm, lobes 4. **Cypselae** 0.9–1.2(–1.5) mm, moderately hairy; **pappi** of 6(–7) entire, non-aristate scales 0.2–0.5 mm. $2n = 26$.

Flowering (Mar–)May–Jul(–Sep). Ditches, around ponds, along streams, moist fields; 30–1600 m; Ariz.; Mexico.

386. HYMENOXYS Cassini in F. Cuvier, Dict. Sci. Nat. ed. 2, 55: 278. 1828

• Bitterweed, rubberweed [Greek *hymen*, membrane, and *oxys*, sharp, alluding to aristate pappus scales]

Mark W. Bierner

Dugaldia Cassini; *Plummera* A. Gray

Annuals, biennials, or perennials, 5–150 cm (sometimes with ± branched, woody caudices or stout rhizomes). **Stems** 1–30+, erect, unbranched or branched, green throughout to purple-red-tinted proximally or distally to purple-red-tinted throughout, glabrous or ± hairy. **Leaves** alternate; blades simple or 1–2-pinnately lobed, ultimate margins entire or toothed, faces glabrous or hairy, usually ± gland-dotted (often in pits). **Heads** radiate [discoid], borne singly or in paniculiform to corymbiform arrays. **Involucres** subhemispheric, hemispheric, globoid, campanulate, or urceolate, (2.5–)6–32 mm diam. **Phyllaries** persistent (or inner falling), usually (6–)16–30(–40) in 2 series and unequal, sometimes 28–50 in 2–3 series and subequal (usually spreading to erect in fruit). **Receptacles** usually hemispheric, globoid, ovoid, or conic (flat in *H. ambigens*), smooth or pitted, epaleate. **Ray florets** usually (3–)8–13(–16), sometimes 14–34 [0], pistillate, fertile; corollas (usually withering after flowering, falling early or tardily) yellow or yellow-orange to orange (laminae fan-shaped, lobes 3–5). **Disc florets** usually 25–150(–400+), usually bisexual and fertile (6–15, functionally staminate in *H. ambigens*); corollas yellow to yellow-brown proximally, yellow distally, tubes shorter than cylindric to cylindric-campanulate throats, lobes 5, ± deltate. **Cypselae** obconic or obpyramidal, glabrous or hairy; **pappi** 0, or persistent, of 2–11(–15) usually aristate scales. $x = 15$.

Species 25 (17 in the flora): North America, Mexico, Central America, South America.

Subgenus *Hymenoxys*, which occurs only in South America, contains four species, three of which are discoid. Subgenus *Phileozera* (Buckley) Cockerell includes *Hymenoxys odorata* and *H. chrysanthemoides* (Kunth) de Candolle (Mexico) (M. W. Bierner 2001). Subgenus *Plummera* (A. Gray) Bierner comprises only *Hymenoxys ambigens* (Bierner 1994, 2001). Subgenus *Dugaldia* (Cassini) Bierner includes *Hymenoxys hoopesii*, *H. integrifolia* (Kunth) Bierner (Mexico and Guatemala), and *H. pinetorum* (Standley) Bierner (Mexico; Bierner 1994, 2001). Subgenus *Rydbergia* (Greene) Bierner comprises *Hymenoxys brandegeei*, *H. grandiflora*, and *H. insignis* (Mexico) (Bierner 2001, 2005). *Hymenoxys bigelovii* is the sole member of subg. *Macdougalia* (A. Heller) Bierner (Bierner 2001, 2004). *Hymenoxys richardsonii*, *H. subintegra*, *H. cooperi*, *H. lemmonii*, *H. rusbyi*, *H. jamesii*, and *H. brachyactis* belong to subg. *Picradenia* (Hooker) Cockerell (Bierner 2001).

J. L. Anderson et al. (1996) and M. W. Bierner and R. K. Jansen (1998) provided evidence that *Hymenoxys helenioides* is a hybrid between *H. hoopesii* and *H. richardsonii* var. *floribunda*. Bierner (2001) recognized *H. helenioides* as a species because he was unable to determine whether all of the plants were F₁ hybrids or at least some of them had given rise to breeding populations.

Plants identified as *Hymenoxys anthemoides*, a discoid, South American annual that resembles *H. odorata*, were reported as "adventive on ballast," Mobile County, Alabama, by C. T. Mohr (1901).

SELECTED REFERENCES   Bierner, M. W. 1974. A systematic study of *Dugaldia* (Compositae). Brittonia 26: 385–392.   Bierner, M. W. 1994. Submersion of *Dugaldia* and *Plummera* in *Hymenoxys* (Asteraceae: Heliantheae: Gaillardiinae). Sida 16: 1–8.   Bierner, M. W. 2001. Taxonomy of *Hymenoxys* subgenus *Picradenia* and a conspectus of the subgenera of *Hymenoxys* (Asteraceae, Helenieae, Tetraneurinae). Lundellia 4: 37–63.   Bierner, M. W. 2004. Taxonomy of *Hymenoxys* subgenus *Macdougalia* (Asteraceae, Helenieae, Tetraneurinae). Sida 21: 657–663.   Bierner, M. W. 2005. Taxonomy of *Hymenoxys* subgenus *Rydbergia* (Asteraceae, Helenieae, Tetraneurinae). Lundellia 8: 28–37.   Bierner, M. W. and R. K. Jansen. 1998. Systematic implications of DNA restriction site variation in *Hymenoxys* and *Tetraneuris* (Asteraceae, Helenieae, Gaillardiinae). Lundellia 1: 17–26.   Spring, O., B. Zitterell-Haid, M. W. Bierner, and T. J. Mabry. 1994. Chemistry of glandular trichomes in *Hymenoxys* and related genera. Biochem. Syst. & Ecol. 22: 171–195.   Wagner, W. L., R. Fletcher, and R. K. Shannon. 1999. Another look at *Hymenoxys* subgenus *Plummera* (Asteraceae: Heliantheae: Gaillardiinae) from Arizona and New Mexico. Brittonia 51: 79–86.

1. Annuals.
   2. Plants 5–10(–15) cm (delicate); ray corollas 2–3 × 0.7–1 mm (not surpassing phyllaries; Texas) . . . . 1. *Hymenoxys texana*
   2. Plants 10–80 cm (robust); ray corollas 8.5–11 × 3–5.5 mm (well surpassing phyllaries; widespread) . . . . 2. *Hymenoxys odorata*
1. Biennials or perennials.
   3. Disc florets 6–15 (functionally staminate; receptacles flat) . . . . 3. *Hymenoxys ambigens*
   3. Disc florets 25–400+ (bisexual; receptacles hemispheric to globoid, ovoid or conic).
      4. Phyllaries 22–50 in 2–3 series, subequal.
         5. Plants 30–100 cm; leaf blades simple; phyllaries 36–50 in 2 series; ray corollas usually yellow-orange to orange . . . . 4. *Hymenoxys hoopesii*
         5. Plants 8–30 cm; leaf blades simple and/or lobed; phyllaries 22–48+ in 2–3 series; ray corollas yellow.
            6. Basal leaves simple or (some, not all) lobed (lobes 3–7); involucres 13–16 × 19–23 mm; Arizona, Colorado, New Mexico . . . . 5. *Hymenoxys brandegeei*
            6. Basal leaves lobed (lobes 3–15); involucres 15–25 × 18–30 mm; Colorado, Idaho, Montana, Utah, Wyoming . . . . 6. *Hymenoxys grandiflora*
      4. Phyllaries (11–)16–30(–40) in 2 series, unequal.
         7. Outer phyllaries basally connate to $^1/_5$ their lengths; inner phyllaries narrowly lanceolate to oblanceolate, 8.5–12.6 mm, apices aristate . . . . 7. *Hymenoxys bigelovii*
         7. Outer phyllaries basally connate $^1/_4$–$^2/_3$ their lengths; inner phyllaries obovate to oblanceolate, 2.8–8 mm, apices usually mucronate, sometimes acuminate.
            8. Stems 1–20(–30+; plants usually with highly branched, woody caudices; basal leaf bases densely long-villous-woolly) . . . . 8. *Hymenoxys richardsonii*
            8. Stems 1–10(–20; plants often with sparingly or moderately branched, woody caudices; basal leaf bases sparsely, if at all, long-villous-woolly).
               9. Involucres 4–8 mm diam.
                  10. Perennials (polycarpic); mid blades simple or lobed (lobes 3, terminal lobes 2–4 mm wide) . . . . 15. *Hymenoxys rusbyi*
                  10. Biennials or perennials (monocarpic); mid blades lobed (lobes 3–7, terminal lobes 0.8–2.2 mm wide).
                     11. Stems usually purple-red-tinted proximally; disc corollas 2.4–3 mm; pappi 0.8–2.1 mm; Mogollon Plateau area, c Arizona . . . . 16. *Hymenoxys jamesii*
                     11. Stems green throughout; disc corollas 3.1–4.2 mm; pappi 2.5–3 mm; c New Mexico . . . . 17. *Hymenoxys brachyactis*
               9. Involucres (8–)10–18 mm diam.
                  12. Stems (and leaves) usually densely sericeous (mainly Kaibab Plateau in Arizona, Utah) . . . . 9. *Hymenoxys subintegra*
                  12. Stems (and leaves) sometimes densely hairy (only *H. cooperi* and *H. richardsonii* var. *floribunda* on Kaibab Plateau in Arizona, Utah).

[13. Shifted to left margin.—Ed.]

13. Outer phyllaries 5(–8); s Arizona (mostly Huachuca Mountains) . . . . 10. *Hymenoxys quinquesquamata*
13. Outer phyllaries 7–15; not s Arizona.
    14. Leaf blades lobed (lobes 3–23); outer phyllaries basally connate ¹/₂–²/₃ their lengths
    . . . . 11. *Hymenoxys vaseyi*
    14. Leaf blades simple or lobed (lobes 3–13); outer phyllaries basally connate ¹/₄–¹/₂ their lengths.
        15. Mid blades usually lobed (lobes 3, terminal lobes 2–5.5 mm wide), sometimes not lobed; outer phyllaries (weakly keeled) basally connate ¹/₄–¹/₃ their lengths; ray corollas yellow to yellow-orange . . . . 12. *Hymenoxys helenioides*
        15. Mid blades lobed (lobes 3–7, terminal lobes 0.7–2.5 mm wide); outer phyllaries (weakly to strongly keeled) basally connate ¹/₄–¹/₂ their lengths; ray corollas yellow.
            16. Biennials or perennials, (10–)20–80(–100) cm (monocarpic); leaves ± hairy
            . . . . 13. *Hymenoxys cooperi*
            16. Perennials, 30–50 cm (polycarpic); leaves glabrous or sparsely hairy . . . . . .
            . . . . 14. *Hymenoxys lemmonii*

1. **Hymenoxys texana** (J. M. Coulter & Rose) Cockerell, Bull. Torrey Bot. Club 31: 499. 1904 • Prairiedawn
C E

Actinella texana J. M. Coulter & Rose, Bot. Gaz. 16: 27. 1891

Annuals, 5–10(–15) cm (delicate). **Stems** 1–7, purple-red-tinted proximally or throughout, unbranched or branched distally, glabrous or sparsely hairy. **Leaves:** blades simple or lobed (lobes 3–5, ultimate margins entire or pinnately toothed), glabrous or sparsely hairy, weakly to moderately gland-dotted; mid leaves sometimes lobed (terminal lobes 1–2.5 mm wide). **Heads** 1–30 per plant, borne singly or in paniculiform arrays. **Peduncles** 0.5–1.5 cm, glabrous or sparsely hairy. **Involucres** campanulate to ± urceolate, 5–7 × 6–8 mm. **Phyllaries** in 2 series, unequal; outer 6–8, distinct or weakly basally connate, obovate or oblanceolate to lanceolate, 4–4.7, apices obtuse to acute; inner 6–8, obovate to ovate, 4–4.5 mm, apices usually acute. **Ray florets** 6–8; corollas yellow, 2–3 × 0.7–1 mm (not surpassing phyllaries). **Disc florets** 40–75+; corollas 1.5–2.1 mm. **Cypselae** obpyramidal to narrowly obpyramidal, 1.5–1.8 mm; **pappi** of ca. 5 obovate, aristate scales 1.4–1.8 mm. $2n = 6, 16$.

Flowering Mar–Apr. Open grassy areas; of conservation concern; 30–50 m; Tex.

Hymenoxys texana is known only from open grassy areas at relatively few localities within 150 km or so of Houston.

Morphology and DNA restriction site data clearly associate this species with *Hymenoxys*; its relationships to other taxa in the genus are unclear (M. W. Bierner and R. K. Jansen 1998). Furthermore, unlike *Hymenoxys* and like *Tetraneuris*, *H. texana* possesses monoterpene glycosides and lacks seco-pseudoguaianolides (O. Spring et al. 1994). Also, its very unusual chromosome num-

bers of $2n = 6$ and 16 (J. L. Strother and L. E. Brown 1988) are unlike any reported from other taxa in either *Hymenoxys* or *Tetraneuris* (mainly $2n = 30$ with some dysploidy and polyploidy; Bierner 1994). Despite the conflicting data, Bierner and Jansen, and Bierner (2001) maintained it in *Hymenoxys*, but separated it from the other members of the genus in its own subgenus.

Hymenoxys texana is in the Center for Plant Conservation's National Collection of Endangered Plants.

2. **Hymenoxys odorata** de Candolle in A. P. de Candolle and A. L. P. P. de Candolle, Prodr. 5: 661. 1836 • Western bitterweed, bitter rubberweed F

Annuals, 10–80 cm (robust). **Stems** 1–25, usually purple-red-tinted proximally, sometimes throughout, often branched throughout, ± hairy. **Leaves:** blades simple or lobed (lobes 3–19+), ± hairy, gland-dotted; mid leaves lobed (lobes 5–11+, terminal lobes 0.3–1 mm wide). **Heads** 15–350+ per plant in paniculiform arrays. **Peduncles** 2–12 cm, sparsely hairy. **Involucres** subhemispheric to campanulate to ± urceolate, 6–10 × 7–12.5 mm. **Phyllaries** in 2 series, unequal; outer 8–13, basally connate ¹/₄–¹/₃ their lengths, obovate to oblanceolate, 3.5–5.2 mm, apices acuminate; inner 8–13, obovate, 3.8–5.9 mm, apices acuminate to acute. **Ray florets** 8–13; corollas yellow, 8.5–11 × 3–5.5 mm. **Disc florets** 50–150+; corollas 2.6–4.1 mm. **Cypselae** narrowly obpyramidal, 1.7–2.5 mm; **pappi** of 5–6 obovate, aristate scales 1.6–2.3 mm. $2n = 22, 24, 28, 30$.

Flowering (Feb–)Apr–Jun(–Aug). Roadsides, open flats, mesquite and creosote-bush flats, ditches and drainage areas, stream banks and bottoms, 60–1500 m; Ariz., Calif., Colo., Kans., N.Mex., Okla., Tex.; Mexico (Chihuahua, Coahuila, Durango, Nuevo León, San Luis Potosí, Sonora, Tamaulipas).

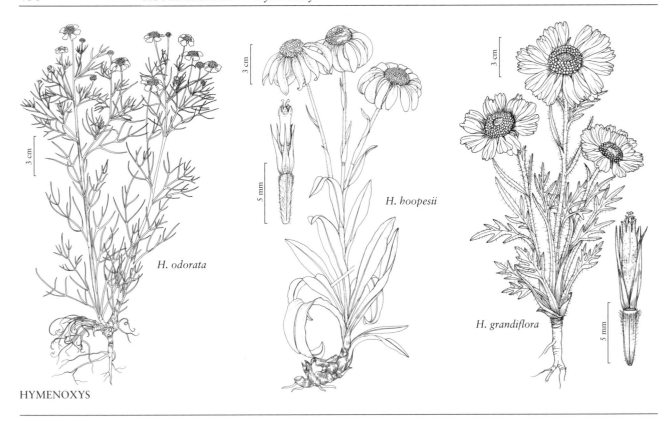

H. odorata

H. hoopesii

H. grandiflora

HYMENOXYS

3. Hymenoxys ambigens (S. F. Blake) Bierner, Sida 16: 6. 1994 • Pinaleno Mountains rubberweed C E

Plummera ambigens S. F. Blake, J. Wash. Acad. Sci. 19: 276, fig. 1a–i. 1929

Perennials, 30–150 cm (polycarpic, with sparingly branched, woody caudices). **Stems** 1–3, usually purple-red-tinted proximally, branched distally, glabrous or ± hairy. **Leaves:** blades simple or lobed (lobes 3–19+), glabrous or sparsely hairy, gland-dotted; mid leaves lobed (lobes 3–19+, terminal lobes 0.7–1.3 mm wide). **Heads** 25–400+ per plant, in paniculiform to corymbiform arrays. **Peduncles** 0.4–2 cm, glabrous or ± hairy. **Involucres** campanulate to narrowly campanulate, 5–7.5 × 2.2–4.5 mm. **Phyllaries** in 2 series, unequal; outer 3–5, basally connate ½–¾ their lengths, obovate to oblanceolate, 4–6 mm, apices acuminate, acute, obtuse, or rounded; inner 3–5, obovate to oblanceolate, 2.5–4.5 mm, apices acuminate, acute, obtuse, or rounded, sometimes mucronate. **Ray florets** 3–5; corollas yellow, extending beyond phyllaries, 3.4–6 × 1.7–4 mm. **Disc florets** 6–15 (functionally staminate); corollas 3–4 mm. **Cypselae** obconic, 1.7–3.5 mm (hairy, hairs straight or wavy); **pappi** 0, or of 2–6 ± lanceolate scales 1–2.5 mm.

Varieties 3 (3 in the flora): sw United States.

1. Apices of inner phyllaries acuminate to mucronate; pappus scales 3–6 (sometimes 0 on disc florets); hairs on cypselae usually straight, sometimes slightly wavy; Arizona, Mescal, Pinaleno, and Santa Teresa mountains .... 3a. *Hymenoxys ambigens* var. *ambigens*

1. Apices of inner phyllaries rounded or obtuse to weakly acute or mucronate; pappus scales 0 or 2–4; hairs on cypselae moderately to strongly wavy; Arizona or New Mexico.

  2. Plants 30–150 cm; outer phyllaries 3–4; hairs on cypselae strongly wavy; Arizona, Chiricahua, Dos Cabezas, Dragoon, Little Dragoon, and Mule Mountains.... 3b. *Hymenoxys ambigens* var. *floribunda*

  2. Plants 30–40+ cm; outer phyllaries 5; hairs on cypselae moderately wavy; New Mexico, Animas and Peloncillo mountains ....

  ...... 3c. *Hymenoxys ambigens* var. *neomexicana*

3a. Hymenoxys ambigens (S. F. Blake) Bierner var. **ambigens** [C] [E]

Plants 30–150 cm. **Heads** (50–)100–400+. **Phyllaries:** outer 4, basally connate ¹/₂–²/₃ their lengths, 4–5 mm, apices obtuse to acute; inner 4, 3.2–4.5 mm, apices acuminate to mucronate. **Ray florets** 4; corollas 4–5.5 × 2.3–4 mm. **Disc florets** 7–11. **Cypselae** 2.5–3 mm, hairs straight to slightly wavy; **pappi** usually of 3–6 scales 1.2–2.5 mm, sometimes 0 on disc florets. **2n** = 30.

Flowering Jun–Nov. Oak-juniper woodlands; of conservation concern; 1300–2200 m; Ariz.

Variety *ambigens* is known from the Mescal, Pinaleno, and Santa Teresa mountains.

3b. Hymenoxys ambigens (S. F. Blake) Bierner var. **floribunda** (A. Gray) W. L. Wagner, Brittonia 51: 84. 1999 [C] [E]

Plummera floribunda A. Gray, Proc. Amer. Acad. Arts 17: 215. 1882; *Hymenoxys microcephala* Bierner

Plants 30–150 cm. **Heads** (25–)100–350+. **Phyllaries:** outer 3–4, basally connate ³/₄ their lengths, 4–6 mm, apices rounded to usually obtuse to sometimes acute; inner 3–4, 3.2–4 mm, apices rounded or obtuse to weakly acute or mucronate. **Ray florets** 3–4; corollas 4.5–6 × 2–2.5 mm. **Disc florets** 6–13. **Cypselae** 3–3.5 mm, hairs strongly wavy; **pappi** 0 or of 2–4 scales 1–2 mm. **2n** = 30.

Flowering Jun–Oct. Savannas, oak-juniper woodlands, pine forests; of conservation concern; 1500–2300 m; Ariz.

Variety *floribunda* is known from the Chiricahua, Dos Cabezas, Dragoon, Little Dragoon, and Mule mountains.

3c. Hymenoxys ambigens (S. F. Blake) Bierner var. **neomexicana** W. L. Wagner, Brittonia 51: 85, figs. 1, 2G–L. 1999 [C] [F]

Plants 30–40+ cm. **Heads** 35–100+. **Phyllaries:** outer 5, basally connate ²/₃–³/₄ their lengths, 4–4.5 mm, apices obtuse to acuminate to acute; inner 5, 2.5–3 mm, apices rounded to obtuse to weakly acute or mucronate. **Ray florets** 4–5; corollas 3.4–4.2 × 1.7–2.2 mm. **Disc florets** 8–15. **Cypselae** 1.7–2.3 mm, hairs moderately wavy; **pappi** usually of 2–4 scales 1.2–2 mm, sometimes 0 on disc florets.

Flowering Jul–Sep. Open woodlands; of conservation concern; 1600–1900 m; N.Mex.

Variety *neomexicana* is known from the Animas and Peloncillo mountains.

4. Hymenoxys hoopesii (A. Gray) Bierner, Sida 16: 6. 1994 • Owl's-claws, orange-sneezeweed [E] [F]

Helenium hoopesii A. Gray, Proc. Acad. Nat. Sci. Philadelphia 15: 65. 1864; *Dugaldia hoopesii* (A. Gray) Rydberg

Perennials, 30–100 cm (polycarpic, often with blackish, stout rhizomes). **Stems** 1–4, green throughout to often purple-red-tinted proximally to purple-red-tinted throughout, branched distally, glabrous or ± hairy. **Leaves:** blades simple, glabrous or ± hairy, gland-dotted; mid leaves simple. **Heads** 1–12 per plant, in paniculiform to corymbiform arrays. **Peduncles** 3–16 cm, ± hairy, tomentose distally near involucres. **Involucres** hemispheric to broadly campanulate, 12–17 × 19–26 mm. **Phyllaries** in 2 series, subequal; outer 16–24, basally connate ¹/₄–¹/₃ their lengths, ovate to lanceolate, 11–16 mm, apices acute; inner 20–26, obovate to elliptic, 7.5–8.5 mm, apices acuminate to acute. **Ray florets** 14–26; corollas usually yellow-orange to orange, 21–45 × 4.2–8.5 mm. **Disc florets** 100–325+; corollas 4.2–5.4 mm. **Cypselae** obpyramidal to narrowly obpyramidal, 3.5–4.5 mm; **pappi** of 5–7 lanceolate to lanceolate-acuminate scales 2.9–4.1 mm. **2n** = 30.

Flowering (May–)Jul–Aug(–Nov). Mountain meadows, open forests, along streams; 1500–3700 m; Ariz., Calif., Colo., Idaho, Nev., N.Mex., Oreg., Utah, Wyo.

5. Hymenoxys brandegeei (Porter ex A. Gray) K. F. Parker, Madroño 10: 159. 1950 (as brandegei) • Brandegee's rubberweed, western bitterweed [E]

Actinella brandegeei Porter ex A. Gray, Proc. Amer. Acad. Arts 13: 373. 1878 (as brandegei), based on *A. grandiflora* Torrey & A. Gray var. *glabrata* Porter in T. C. Porter and J. M. Coulter, Syn. Fl. Colorado, 76. 1874; *Tetraneuris brandegeei* (Porter ex A. Gray) K. F. Parker

Perennials, 8–24 cm (polycarpic, often with sparingly branched, woody caudices). **Stems** 1–3(–10), green throughout to purple-red-tinted proximally or distally to throughout, unbranched distally, ± hairy. **Leaves:** blades simple or lobed (lobes 3–7), ± hairy, gland-dotted (basal leaf bases sparsely, if at all, long-villous-woolly); mid leaves usually simple, sometimes lobed (lobes 3, terminal lobes 1.5–3.5 mm wide). **Heads** 1–3(–10) per plant, borne singly. **Peduncles** 2–5 cm, ± hairy, tomentose distally near involucres. **Involucres** hemispheric to subglobose, 13–16 × 19–23 mm. **Phyllaries** in 2–3 series, subequal; outer 10–20, basally connate ¹/₅–¹/₃ their lengths, lanceolate, 10–13.5 mm, apices rounded to acute; inner 12–20+, lanceolate to elliptic to obovate to oblanceolate, 6–10 mm, apices acuminate. **Ray florets** 14–23; corollas yellow, 14–23 × 3.5–7.5 mm. **Disc florets** 150–250+; corollas 4.1–5.2 mm. **Cypselae** obpyramidal to narrowly obpyramidal, 2.8–3.1 mm; **pappi** of 5–6 lanceolate-aristate scales (2.7–)4.1–4.3 mm. **2n = 30.**

Flowering Jul–Sep. Meadows, often above timberline; 2800–4100 m; Ariz., Colo., N.Mex.

6. **Hymenoxys grandiflora** (Torrey & A. Gray) K. F. Parker, Madroño 10: 159. 1950 · Graylocks rubberweed, four-nerved daisy [E] [F]

Actinella grandiflora Torrey & A. Gray, Boston J. Nat. Hist. 5: 109. 1845; *Tetraneuris grandiflora* (Torrey & A. Gray) K. F. Parker

Perennials, 8–30 cm (polycarpic, often with sparingly branched, woody caudices). **Stems** 1–10, green throughout to purple-red-tinted proximally or distally to purple-red-tinted throughout, usually unbranched distally, ± hairy. **Leaves:** blades simple or lobed (lobes 3–15), ± hairy, gland-dotted (basal leaf bases, sparsely, if at all, long-villous-woolly); mid leaves simple or lobed (lobes 3–7, terminal lobes 1–2.5 mm wide). **Heads** 1–10 per plant, usually borne singly. **Peduncles** 1–10 cm, ± hairy, tomentose distally near involucres. **Involucres** hemispheric to subglobose, 15–25 × 18–30 mm. **Phyllaries** in 2–3 series, subequal; outer 16–24, basally connate ¹/₅–¹/₄ their lengths, lanceolate, 9–15 mm, apices acute; inner 16–24+, lanceolate to oblanceolate, 8–12 mm, apices acuminate to acute. **Ray florets** 15–34(–44); corollas yellow, 16–30 × 4–8 mm. **Disc florets** 150–400+; corollas 5–6 mm. **Cypselae** obpyramidal to narrowly obpyramidal, 3.3–3.7 mm; **pappi** of 5–7 lanceolate, aristate scales 4.5–5.3 mm. **2n = 30.**

Flowering Jun–Aug. Meadows, often above timberline; 2600–4300 m; Colo., Idaho, Mont., Utah, Wyo.

7. **Hymenoxys bigelovii** (A. Gray) K. F. Parker, Madroño 10: 159. 1950 · Bigelow's rubberweed [E]

Actinella bigelovii A. Gray, Smithsonian Contr. Knowl. 5(6): 96. 1853

Perennials, 20–70 cm (polycarpic, often with sparingly branched, woody caudices). **Stems** 1–5, green throughout or purple-red-tinted distally to throughout, usually unbranched distally, ± hairy (often tomentose proximally). **Leaves:** blades usually simple, rarely lobed (lobes 3), glabrous or ± hairy, eglandular or sparsely gland-dotted (basal leaf bases ± long-villous-woolly); mid leaves usually simple, rarely lobed (lobes 3, terminal lobes 1.5–3 mm wide). **Heads** 1–5 per plant, usually borne singly, sometimes in paniculiform arrays. **Peduncles** (1.5–)6–20(–29) cm, ± hairy, densely tomentose distally near involucres. **Involucres** hemispheric to broadly campanulate, 13–20 × 23–32 mm. **Phyllaries** in 2 series, unequal; outer 13–19, basally connate only slightly to ¹/₅ their lengths, lanceolate to narrowly lanceolate to obovate to oblanceolate, 7–11 mm, apices acuminate to acute; inner 13–18, narrowly lanceolate to oblanceolate, 8.5–12.6 mm, apices aristate. **Ray florets** 13–15; corollas yellow, 13–26 × 5.4–9.5 mm. **Disc florets** 100–250+; corollas 5.7–7.4 mm. **Cypselae** narrowly obpyramidal, 4.2–4.7 mm; **pappi** of 9–11(–15) obovate to oblanceolate, often aristate scales 4.7–7.3 mm. **2n = 30.**

Flowering May–Jun. Roadsides, edges of juniper-pine and pine forests; 1300–2500 m; Ariz., N.Mex.

8. **Hymenoxys richardsonii** (Hooker) Cockerell, Bull. Torrey Bot. Club 31: 468. 1904 (as richardsoni) · Richardson's bitterweed, pingue rubberweed [E]

Picradenia richardsonii Hooker, Fl. Bor.-Amer. 1: 317, plate 108. 1833 (as richardsoni)

Perennials, 7–34 cm (polycarpic; usually with highly branched woody caudices). **Stems** 1–20 (–30+), usually green throughout, rarely purple-red-tinted proximally, branched distally, glabrous or ± hairy. **Leaves:** blades simple or lobed (lobes 3–7), glabrous or ± hairy, gland-dotted (basal leaf bases densely long-villous-woolly); mid leaves lobed (lobes 3–5, terminal lobes 0.7–2 mm wide). **Heads** 5–300+ per plant, in corymbiform arrays. **Peduncles** 1–5 cm, ± hairy. **Involucres** campanulate, 7–11 × 7–14 mm. **Phyllaries** in 2 series, unequal; outer 8–13, obovate to ovate to lanceolate, basally connate ¹/₂ their lengths, 3.5–8 mm, apices acuminate to acute; inner 8–18, obovate, 3–8 mm, apices mucronate. **Ray florets** 7–14; corollas yellow, 7–17 × 3–6 mm. **Disc florets** 25–75+; corollas 3–5 mm. **Cypselae** narrowly obpyramidal, 2–3 mm; **pappi** of 4–8 obovate to lanceolate, aristate scales 1.8–3.5 mm. **2*n*** = 30.

Varieties 2 (2 in the flora): w North America.

1. Stems 1–10; leaves ± hairy; heads 5–35+ per plant; involucres 8–11 × 9–14 mm; outer phyllaries 9–13, weakly to moderately keeled; ray corollas 11–17 mm . . . . 8a. *Hymenoxys richardsonii* var. *richardsonii*
1. Stems (2–)5–20(–30+); leaves glabrous or sparsely hairy; heads (30–)80–250(–300+) per plant; involucres 7–8 × 7–9 mm; outer phyllaries 8–9, strongly keeled; ray corollas 7–11 mm . . . . 8b. *Hymenoxys richardsonii* var. *floribunda*

8a. Hymenoxys richardsonii (Hooker) Cockerell var. **richardsonii** E

Plants 7–24 cm. **Stems** 1–10, ± hairy. **Leaves** ± hairy. **Heads** 5–35+ per plant. **Involucres** 8–11 × 9–14 mm. **Phyllaries:** outer 9–13, weakly to moderately keeled, lanceolate, 5–8 mm, apices acuminate to acute; inner 9–18, 4–8 mm. **Ray florets** 8–14; corollas 11–17 mm. **Disc florets** 30–75+; corollas 3–5 mm. **Cypselae** 2–3 mm; **pappi** of 5–6(–8) usually lanceolate, aristate scales 2–3.5 mm. **2*n*** = 30.

Flowering (May–)Jun–Jul(–Sep). Roadsides, open areas, edges of forests; (1000–)1500–2500(–3400) m; Alta., Sask.; Colo., Mont., N.Dak., Utah, Wyo.

8b. Hymenoxys richardsonii (Hooker) Cockerell var. **floribunda** (A. Gray) K. F. Parker, Madroño 10: 159. 1950 • Colorado rubberweed, olivaceous bitterweed E

Actinella richardsonii (Hooker) Nuttall var. *floribunda* A. Gray, Mem. Amer. Acad. Arts, n. s. 4: 101. 1849

Plants 19–34 cm. **Stems** (2–)5–20(–30+), glabrous or ± hairy. **Leaves** glabrous or sparsely hairy. **Heads** (30–)80–250(–300+) per plant. **Involucres** 7–8 × 7–9 mm. **Phyllaries:** outer 8–9, strongly keeled, obovate, ovate, or lanceolate, 3.5–6 mm, apices acute; inner 8–12, 3–5 mm. **Ray florets** 7–9; corollas 7–11 mm. **Disc florets** 25–50+; corollas 3–4 mm. **Cypselae** 2 mm; **pappi** of (4–)5 obovate, aristate scales 1.8–2.8 mm. **2*n*** = 30.

Flowering (May–)Jul–Aug(–Oct). Roadsides, open areas, edges of forests; (1600–)1800–2700(–3400) m; Ariz., Colo., N.Mex., Tex., Utah.

9. Hymenoxys subintegra Cockerell, Bull. Torrey Bot. Club 31: 480. 1904 • Arizona rubberweed E

Actinea subintegra (Cockerell) S. F. Blake

Biennials or perennials, 30–60 cm (monocarpic; sometimes with sparingly branched, woody caudices). **Stems** 1(–5), green throughout or sometimes purple-red-tinted proximally, branched distally, usually densely sericeous. **Leaves:** blades simple or lobed (lobes 3), usually densely sericeous, gland-dotted; mid leaves simple or lobed (lobes 3, terminal lobes 1.8–3.5 mm wide). **Heads** 10–85+ per plant, in paniculiform to corymbiform arrays. **Peduncles** 3–9 cm, usually densely sericeous. **Involucres** hemispheric to campanulate, 8–10 × 10–13 mm. **Phyllaries** in 2 series, unequal; outer 10–14(–18), basally connate 1/4–1/2 their lengths, lanceolate, obovate, or ovate 5–7 mm, apices acuminate to acute; inner phyllaries 13–16(–22), obovate, 4–5 mm, apices mucronate. **Ray florets** 10–16; corollas yellow, 10–15 × 4.5–6.5 mm. **Disc florets** 50–100+; corollas 3.5–4 mm. **Cypselae** narrowly obpyramidal, 2.1–3 mm; **pappi** of 5(–6) obovate, aristate scales 2.5–3 mm. **2*n*** = 30.

Flowering (Jun–)Jul–Aug(–Sep). Roadsides, open areas, edges of forests; 2100–2800 m; Ariz., Utah.

Hymenoxys subintegra grows mainly on the Kaibab Plateau north of the Grand Canyon in Arizona and in western Kane County, Utah.

10. Hymenoxys quinquesquamata Rydberg in N. L. Britton et al., N. Amer. Fl. 34: 114. 1915 • Rincon rubberweed or bitterweed E

Perennials, 30–100 cm (polycarpic; sometimes with sparingly branched, woody caudices). **Stems** 1(–3), green throughout or purple-red-tinted proximally, branched distally, ± hairy. **Leaves:** blades simple or lobed (lobes 3–17), glabrous or sparsely hairy, gland-dotted; mid leaves lobed (lobes 5–11, terminal lobes 0.8–2 mm wide). **Heads** 5–50+ per plant, in paniculiform to corymbiform arrays. **Peduncles** 2.5–7 cm, ± hairy. **Involucres** campanulate, 8–10 × 9–12 mm. **Phyllaries** in 2 series, unequal; outer 5(–8), basally connate 1/3–1/2 their lengths, obovate to ovate, 6–7 mm, apices acuminate; inner 6–8, obovate, 6.5–8 mm, apices mucronate. **Ray florets** 5–8; corollas yellow, 14–18 × 7–12 mm. **Disc florets** 30–100+; corollas 3.5–4.3 mm. **Cypselae** obpyramidal, 2.5–3.9 mm; **pappi** of 5–6 obovate, sometimes aristate, scales 1.2–2.5 mm. **2*n*** = 30.

Flowering (Jun–)Jul–Sep(–Oct). Open areas, edges of pine-oak forests; 1500–2500 m; Ariz.

Hymenoxys quinquesquamata is known from southern Arizona, mostly Huachuca Mountains.

11. **Hymenoxys vaseyi** (A. Gray) Cockerell, Bull. Torrey Bot. Club 31: 493. 1904 • Vasey's rubberweed or bitterweed C E

Actinella vaseyi A. Gray, Proc. Amer. Acad. Arts 17: 219. 1882

Perennials, 20–60 cm (polycarpic; often with moderately branched, woody caudices). **Stems** 1–10, green throughout or sometimes purple-red-tinted proximally, sparingly or moderately branched distally, ± hairy. **Leaves:** blades usually lobed (lobes 3–23), glabrous or sparsely hairy, gland-dotted (basal leaf bases sparsely, if at all, long-villous-woolly); mid leaves lobed (lobes 5–21, terminal lobes 0.8–1.2 mm wide). **Heads** 10–50+ per plant, in paniculiform to corymbiform arrays. **Peduncles** 1.5–5.5 cm, ± hairy. **Involucres** campanulate, 8–9 × 8–11 mm. **Phyllaries** in 2 series, unequal; outer 8–11, basally connate 1/2–2/3 their lengths, obovate, 4.5–6 mm, apices acute; inner 8–12, obovate, 4.5–6 mm, apices mucronate. **Ray florets** 8–11; corollas yellow, 9–15 × 4–7 mm. **Disc florets** 25–80+; corollas 3.5–4.2 mm. **Cypselae** narrowly obpyramidal, 2–2.7 mm; **pappi** of 5–6 obovate, aristate scales 1.7–2.7 mm. *2n* = 30.

Flowering (Jun–)Jul–Sep. Open areas, edges of forests; of conservation concern; 1600–2100(–2500) m; N.Mex., Tex.

12. **Hymenoxys helenioides** (Rydberg) Cockerell, Bull. Torrey Bot. Club 31: 481. 1904 • Intermountain rubberweed or bitterweed E

Picradenia helenioides Rydberg, Bull. Torrey Bot. Club 28: 21. 1901

Perennials, 20–50 cm (polycarpic; usually with moderately branched, woody caudices). **Stems** 1–5(–10+), green throughout or lightly purple-red-tinted proximally, branched distally, ± hairy. **Leaves:** blades simple or lobed (lobes 3), glabrous or sparsely hairy, gland-dotted (basal leaf bases sparsely, if at all, long-villous-woolly); mid leaves usually lobed (lobes 3, terminal lobes 2–5.5 mm wide), sometimes simple. **Heads** 5–50+ per plant, in paniculiform to corymbiform arrays. **Peduncles** (2.5–)4–7.5 cm, ± hairy. **Involucres** subhemispheric to hemispheric, 9–12 × 12–18 mm. **Phyllaries** in 2 series, unequal; outer 10–15, basally connate 1/4–1/3 their lengths (weakly keeled), lanceolate, 7.5–11 mm, apices acuminate; inner 13(–17), obovate, 5–8 mm, apices acuminate to mucronate. **Ray florets** 10–16; corollas yellow to yellow-orange, 17–31 × 5–11 mm. **Disc florets** 50–150+; corollas 3.5–5.5 mm. **Cypselae** narrowly obpyramidal, 2.5–3.5 mm; **pappi** of 5–7 obovate to lanceolate, aristate scales 2.5–4 mm. *2n* = 30.

Flowering Jun–Aug. Roadsides, open areas, edges of forests; 2200–3000 m; Ariz., Colo., Utah.

13. **Hymenoxys cooperi** (A. Gray) Cockerell, Bull. Torrey Bot. Club 31: 494. 1904 • Cooper's rubberweed or bitterweed E

Actinella cooperi A. Gray, Proc. Amer. Acad. Arts 7: 359. 1868; *Hymenoxys cooperi* var. *canescens* (D. C. Eaton) K. F. Parker

Biennials or perennials, (10–)20–80(–100) cm (monocarpic; sometimes with sparingly branched, woody caudices). **Stems** 1–3(–15), usually purple-red-tinted proximally, sometimes purple-red-tinted throughout, branched distally, ± hairy. **Leaves:** blades simple or lobed (lobes 3–9), ± hairy, gland-dotted (basal leaf bases ± long-villous-woolly); mid leaves lobed (lobes 3–5, terminal lobes 1–2.5 mm wide). **Heads** (1–)7–45(–80) per plant, in paniculiform to corymbiform arrays. **Peduncles** (2–)3.5–8(–13) cm, ± hairy. **Involucres** subhemispheric to hemispheric, 8–10 × 10–17 mm. **Phyllaries** in 2 series, unequal; outer 8–15, basally connate 1/3–1/2 their lengths (moderately to strongly keeled), lanceolate, 4.5–8.9 mm, apices acuminate to acute; inner 14–22, obovate to oblanceolate, 4.1–6.8 mm, apices acuminate to mucronate. **Ray florets** 9–14; corollas yellow, 10.2–17(–21.5) × 4–7.5(–9.2) mm. **Disc florets** 30–150+; corollas 2.7–4.8 mm. **Cypselae** narrowly obpyramidal, 1.7–3.7 mm; **pappi** of 5–6(–8) obovate, aristate scales 1.3–3.3 mm. *2n* = 30.

Flowering May–Jun(–Sep). Roadsides, open areas, edges of juniper-pine forests; (1000–)1500–2500(–3500) m; Ariz., Calif., Idaho, Nev., Oreg., Utah.

Some plants of *Hymenoxys cooperi*, mostly 10–40 cm with mostly 1–6 heads, have been called var. *canescens*.

14. **Hymenoxys lemmonii** (Greene) Cockerell, Bull. Torrey Bot. Club 31: 477. 1904 (as lemmoni)
• Lemmon's rubberweed or bitterweed [E]

Picradenia lemmonii Greene, Pittonia 3: 272. 1898 (as lemmoni)

Perennials, 30–50 cm (polycarpic; often with moderately branched, woody caudices). **Stems** 3–10 (–15), green throughout or purple-red-tinted proximally, branched distally, glabrous or ± sparsely hairy. **Leaves:** blades simple or lobed (lobes 3–13), glabrous or sparsely hairy, gland-dotted; mid leaves lobed (lobes 3–7, terminal lobes 1.5–2.3 mm wide). **Heads** 10–85+ per plant, in paniculiform to corymbiform arrays. **Peduncles** (1–)2–4.5 cm, glabrous or ± hairy. **Involucres** subhemispheric to campanulate, 8–11 × 12–15 mm. **Phyllaries** in 2 series, unequal; outer 10–12, basally connate ¼–½ their lengths (moderately to strongly keeled), ovate to lanceolate, 4.5–7 mm, apices acuminate to acute; inner 13–18(–28), obovate, 4–6 mm, apices usually mucronate. **Ray florets** 9–12; corollas yellow, 10–16 × 4–6.5 mm. **Disc florets** 50–125+; corollas 3.5–4.2 mm. **Cypselae** obpyramidal to narrowly obpyramidal, 2.5–3.5 mm; **pappi** of 5(–6) obovate, aristate scales (1.5–)2.1–2.8 mm. *2n* = 30.

Flowering (Jun–)Jul–Aug(–Sep). Roadsides, open areas, meadows, on slopes, along drainages and streams; (800–)1400–2200(–3200) m; Calif., Nev., Utah.

15. **Hymenoxys rusbyi** (A. Gray) Cockerell, Bull. Torrey Bot. Club 31: 496. 1904 • Rusby's rubberweed or bitterweed [E]

Actinella rusbyi A. Gray, Proc. Amer. Acad. Arts 19: 33. 1883

Perennials, 30–150 cm (polycarpic, often with moderately branched, woody caudices). **Stems** 1–5(–15), usually purple-red-tinted proximally, branched distally, glabrous or ± hairy. **Leaves:** blades simple or lobed (lobes 3), glabrous, gland-dotted; mid leaves simple or lobed (lobes 3, terminal lobes 2–4 mm wide). **Heads** 50–250+ per plant, in corymbiform arrays. **Peduncles** 1–3.2 cm, glabrous or ± hairy. **Involucres** campanulate to urceolate, 6–8 × 6–7 mm. **Phyllaries** in 2 series, unequal; outer 7–9, basally connate ¼–⅓ their lengths, lanceolate, 4.5–6 mm, apices usually rounded; inner 7–11, obovate, 3.5–4.5 mm, apices mucronate. **Ray florets** 6–8; corollas yellow, 4.5–8 × 2.5–5 mm. **Disc florets** 25–50+; corollas 2.7–3.2 mm. **Cypselae** narrowly obpyramidal, 2–3 mm; **pappi** usually of 5 obovate, sometimes aristate, scales 0.9–1.8 mm. *2n* = 30.

Flowering Jul–Aug(–Oct). Roadsides, open areas, edges of forests; 1600–2100 m; Ariz., N.Mex.

16. **Hymenoxys jamesii** Bierner, Madroño 40: 43. 1993 • James's rubberweed [C][E]

Biennials or perennials, 30–120 cm (monocarpic; sometimes with sparingly branched, woody caudices). **Stems** 1–4(–20), usually purple-red-tinted proximally, branched distally, ± hairy. **Leaves:** blades simple or lobed (lobes 3–9), ± hairy, gland-dotted; mid leaves lobed (lobes 3–5, terminal lobes 0.8–2.2 mm wide). **Heads** (30–)50–180(–330) per plant, in paniculiform to corymbiform arrays. **Peduncles** 1.3–5 cm, ± hairy. **Involucres** subhemispheric to campanulate, 4–7 × 4–8 mm. **Phyllaries** in 2 series, unequal; outer 8–10, basally connate ⅓ their lengths, ovate to lanceolate, 2.9–5.5 mm, apices acuminate to acute; inner 8–14, obovate, 2.8–4.2 mm, apices mucronate. **Ray florets** 7–9; corollas yellow, 4.2–8 × 2.3–5 mm. **Disc florets** 30–75+; corollas 2.4–3 mm. **Cypselae** narrowly obpyramidal, 1.6–2.1 mm; **pappi** of 4–5(–8) usually obovate, sometimes aristate, scales 0.8–2.1 mm. *2n* = 30.

Flowering (Jun–)Jul–Aug(–Sep). Roadsides, open areas, edges of juniper-pine forests; of conservation concern; 2000–2400 m; Ariz.

Hymenoxys jamesii is known from the Mogollon Plateau area.

17. **Hymenoxys brachyactis** Wooton & Standley, Contr. U.S. Natl. Herb. 16: 192. 1913 • East view rubberweed, tall bitterweed [E]

Biennials or perennials, 30–60 cm (monocarpic; sometimes with sparingly branched, woody caudices). **Stems** 1–4(–8), green throughout, branched distally, glabrous or sparsely hairy. **Leaves:** blades simple or lobed (lobes 3–9), glabrous, gland-dotted; mid leaves lobed (lobes 3–7, terminal lobes 0.8–1 mm wide). **Heads** (7–)40–150(–250) per plant, in paniculiform to corymbiform arrays. **Peduncles** 1.2–2 cm, glabrous or sparsely hairy. **Involucres** usually urceolate, sometimes campanulate, 6–8 × 5–6 mm. **Phyllaries** in 2 series, unequal; outer 8(–9), basally connate ½ their lengths, lanceolate, 4–6 mm, apices acuminate to acute; inner 8(–10), obovate, 4.5–5 mm, apices mucronate. **Ray florets** 8(–9); corollas yellow, 7–8.5 × 3–4(–4.8) mm. **Disc florets** 25–60+; corollas 3.1–4.2 mm. **Cypselae** narrowly obpyramidal, 2.1–2.5 mm; **pappi** of 5(–7) obovate to lanceolate, aristate scales 2.5–3 mm.

Flowering Jul–Sep. Roadsides, open areas, edges of pine forests; 2000–2500 m; N.Mex.

387. **PLATEILEMA**   (A. Gray) Cockerell, Bull. Torrey Bot. Club 31: 462. 1904   •   [Greek *platys*, broad, and *eilema*, envelope, alluding to broad phyllaries]

John L. Strother

Actinella Nuttall sect. *Plateilema* A. Gray, Proc. Amer. Acad. Arts 19: 31. 1883

Perennials, mostly 5–8+ cm. **Stems** erect, branched mostly from bases. **Leaves** mostly basal; alternate; obscurely petiolate; blades obovate to oblanceolate, pinnately toothed or lobed (lobes deltate to obovate), ultimate margins toothed or entire, faces sparsely hispid (hairs white, coarse) or glabrate, not gland-dotted. **Heads** radiate, borne singly. **Involucres** broadly turbinate, mostly 7–9 mm diam. **Phyllaries** persistent, 8–12 in 2 series (spreading or erect in fruit, basally connate, gray-green, ± ovate, sparsely hispid). **Receptacles** convex, ± pitted, epaleate. **Ray florets** 8–12, pistillate, fertile; corollas yellowish with 4–5+ purplish nerves (apices 3–4-lobed). **Disc florets** 40–60+, bisexual, fertile; corollas yellowish to purplish, tubes much shorter than ampliate, cylindric throats, lobes 5, ± deltate (± equal, usually hispidulous and/or glandular-puberulent). **Cypselae** obpyramidal, ± 4–5-angled (lengths 1.5–3 times widths), ± hispid on angles; **pappi** persistent, of 4–5 brownish, moderately lacerate to erose or truncate scales.

Species 1: Texas, n Mexico.

SELECTED REFERENCE   Turner, B. L. 2000. *Plateilema* (Asteraceae: Helenieae) a new generic report for the United States. Sida 19: 185–187.

1. **Plateilema palmeri**   (A. Gray) Cockerell, Bull. Torrey Bot. Club 31: 462. 1904   [F]

Actinella palmeri A. Gray, Proc. Amer. Acad. Arts 19: 31. 1883

Leaf blades 3–5 cm × 12–18 mm overall. **Phyllaries** 8–10 mm. **Ray laminae** 7–8 mm. **Disc corollas:** tubes 0.8–1 mm, throats 2.5–3 mm, lobes 0.5–0.8 mm. **Cypselae** 3–4 mm; **pappus scales** 1.5–2 mm.

Flowering Apr. Limestone soils; 1100–1200 m; Tex; Mexico (Coahuila, Nuevo León).

388. **BAILEYA**   Harvey & A. Gray ex Torrey in W. H. Emory, Not. Milit. Recon., 143. 1848   •   [For Jacob Whitman Bailey, 1811–1857, researcher of diatomaceous algae at the U.S. Military Academy]

M. W. Turner

Annuals, biennials, or perennials, 15–100 cm. **Stems** erect, branched from bases or ± throughout, floccose-woolly. **Leaves** basal and cauline; alternate; petiolate (basal) or sessile (cauline); blades lance-linear to broadly ovate, sometimes pinnately lobed or pinnatifid, ultimate margins entire, faces usually floccose-woolly. **Heads** radiate, borne singly or in cymiform arrays. **Involucres** campanulate to hemispheric, 5–25 mm diam. **Phyllaries** persistent, 8–13 or 21–34 in 2 series (spreading to erect in fruit, distinct, mostly lance-linear, subequal, floccose-tomentose). **Receptacles** flat to convex, shallowly pitted, epaleate. **Ray florets** 5–7 or 20–55, pistillate,

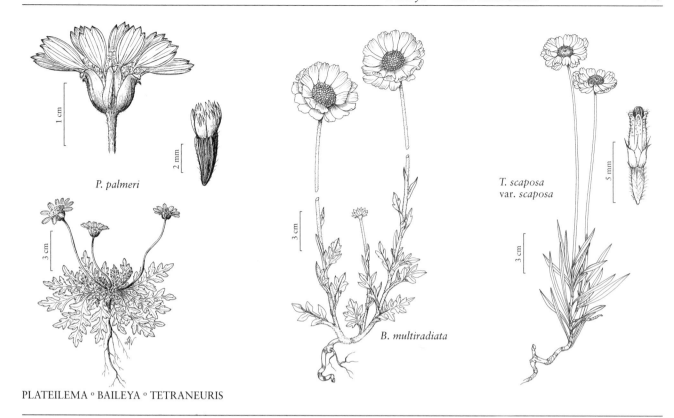

P. palmeri

T. scaposa
var. scaposa

B. multiradiata

PLATEILEMA ° BAILEYA ° TETRANEURIS

fertile; corollas (usually marcescent) yellow. **Disc florets** 10–20 or 40–100+, bisexual, fertile; corollas yellow (hairy), tubes shorter than throats, lobes 5, ± deltate (anther appendages ovate; style-branch apices truncate to acute). **Cypselae** narrowly obpyramidal, weakly ribbed or striate, glandular-pubescent; **pappi** usually 0 (rarely of scales). $x = 16$.

Species 3 (3 in the flora): sw United States, n Mexico.

Baileya is perhaps best known by *B. multiradiata*, which is the most widely distributed, most abundant, and (usually) earliest blooming of the three species in the genus. The large-headed vernal form is particularly attractive; this, together with its long flowering season and its drought tolerance, have given the plant recognition in horticultural circles.

SELECTED REFERENCES Brown, R. C. 1974. Biosystematics of *Baileya* and *Psilostrophe* (Compositae, Helenieae). Ph.D. dissertation. Arizona State University. Turner, M. W. 1993. Systematic study of the genus *Baileya* (Asteraceae: Helenieae). Sida 15: 491–508.

1. Heads in loose, cymiform arrays; rays mostly 5–7 . . . . 3. *Baileya pauciradiata*
1. Heads borne singly; rays 20–55.
    2. Peduncles (at least in vernal forms) 10–30 cm; style-branch apices truncate to slightly rounded . . . . 1. *Baileya multiradiata*
    2. Peduncles 3–12 cm; style-branch apices ± acute . . . . 2. *Baileya pleniradiata*

1. **Baileya multiradiata** Harvey & A. Gray, Mem. Amer. Acad. Arts, n. s. 4: 106. 1849 • Desert marigold F

Baileya australis Rydberg; *B. multiradiata* var. *nudicaulis* A. Gray; *B. pleniradiata* Harvey & A. Gray var. *multiradiata* (Harvey & A. Gray) Kearney; *B. thurberi* Rydberg

Plants mostly 20–100 cm. **Leaves:** basal (rosette) leaves mostly 3–10 × 1–5 cm; petioles 1–4 cm; blades ovate, usually pedately to pinnately lobed (cauline leaves in vernal forms much reduced; autumnal forms often leafy throughout and cauline leaves not much reduced). **Heads** borne singly. **Peduncles** 10–30 cm (vernal forms, often shorter on autumnal forms). **Involucres** hemispheric (vernal forms), mostly 5–10 × 10–25 mm. **Phyllaries** mostly 21–34, floccose-tomentose. **Rays** (vernal forms) mostly 34–55; laminae linear-oblanceolate, mostly 10–20 × 5 mm, apices moderately to deeply 3-toothed. **Disc florets** (vernal forms) 100+; corollas 4 mm, tubes 1 mm, lobes 0.25 mm; style-branch apices truncate to slightly rounded. **Cypselae** 4 mm. $2n = 32$.

Flowering Mar–Nov (depending on rains). Stony slopes, mesas, and sandy plains; 100–2000 m; Ariz., Calif., Nev., N.Mex., Tex., Utah; Mexico (Aguascalientes, Chihuahua, Coahuila, Durango, Sonora).

Baileya multiradiata is an attractive and bountiful wild flower over a large part of the desert Southwest. It has been touted as a promising plant for landscaping, and research on its nursery production has begun (D. J. Cotter et al. 1980, 1982).

Both *Baileya multiradiata* and *B. pleniradiata* produce an antineoplastic pseudoguaianolide, radiatin, which might prove useful in cancer therapy (J. J. Einck et al. 1978). In addition, the antibiotic sesquiterpene lactone, baileyolin, from *B. multiradiata* inhibits tumor formation (X. A. Dominguez et al. 1977).

Baileya multiradiata is reportedly toxic to livestock, especially to sheep and goats, where losses as high as 25% have been reported on overgrazed rangeland in Texas (D. W. Hill et al. 1979, 1980). Cattle and horses seem to be unaffected, or at least poisoning of these animals has gone unreported. The chemical agent responsible is believed to be hymenoxon, a sesquiterpene lactone originally found in the genus *Hymenoxys*, where it is also toxic.

The poorly known desert marigold moth, *Schinia minima* (Grote), appears to be endemic on *Baileya multiradiata*, using the heads of this species for its larval development (T. G. Myles and B. F. Binder 1990).

The autumnal blossoms of *Baileya multiradiata*, with smaller heads, fewer rays, and shorter peduncles, greatly resemble those of *B. pleniradiata*. This has caused much confusion in the distinction between these two species.

The shape of the style apex is a useful character to distinguish between fall-blooming specimens.

2. **Baileya pleniradiata** Harvey & A. Gray, Mem. Amer. Acad. Arts, n. s. 4: 105. 1849

Baileya multiradiata Harvey & A. Gray var. *perennis* (A. Nelson) Kittell; *B. multiradiata* var. *pleniradiata* (Harvey & A. Gray) Coville; *B. nervosa* M. E. Jones; *B. perennis* (A. Nelson) Rydberg; *B. pleniradiata* var. *perennis* A. Nelson

Plants mostly 15–45 cm. **Leaves:** basal rosettes not persistent; basal leaves mostly 2–7 × 0.5–2.5 cm; petioles 1–3 cm; blades oblanceolate, often pinnately lobed; cauline leaves gradually reduced distally. **Heads** borne singly. **Peduncles** 3–12 cm. **Involucres** hemispheric, mostly 5–8 × 7–12 mm **Phyllaries** mostly 21–34, floccose-tomentose. **Rays** mostly 20–40; laminae elliptic to obovate, mostly 7–10 × 4-7 mm, apices shallowly 3-toothed. **Disc florets** mostly 40–50; corollas 3 mm, tubes 0.3 mm, lobes 0.25 mm; style-branch apices acute. **Cypselae** 3 mm. $2n = 32$.

Flowering Mar–Nov (depending on rains). Sandy plains and mesas; 100–2000 m; Ariz., Calif., Nev., Utah; Mexico (Baja California, Sonora).

Baileya pleniradiata is superficially similar to *B. multiradiata* and autumnal forms of the latter have often been misidentified as the former (see discussion under *B. multiradiata*). The two species occasionally occur together or in proximity; hybrids have not been noted. Style appendages readily allow distinction between the two taxa.

3. **Baileya pauciradiata** Harvey & A. Gray, Mem. Amer. Acad. Arts, n. s. 4: 105. 1849

Plants mostly 20–75 cm. **Leaves:** basal leaves (not persistent) mostly 5–12 × 0.5–1 cm; petioles often indistinct; blades linear-oblanceolate to oblanceolate, sometimes pinnately lobed; cauline leaves linear, seldom lobed, often not reduced distally. **Heads** in loose, cymiform arrays. **Peduncles** 2–5 cm. **Involucres** campanulate, mostly 5–8 × 5–8 mm. **Phyllaries** mostly 8–13, floccose-tomentose. **Rays** mostly 5–7; laminae elliptic to ovate, mostly 5–8 × 4–6 mm, apices shallowly to indistinctly 3-toothed. **Disc florets** mostly 10–20; corollas 2.5–3 mm, tubes 0.4 mm, lobes 0.25 mm; style-branch apices truncate. **Cypselae** 3 mm. $2n = 32$.

Flowering Mar–May. Sandy soils, deserts; 60–700 m; Ariz., Calif.; Mexico (Baja California, Sonora).

The most distinct species of the genus, *Baileya pauciradiata* has the most limited distribution and is, perhaps, the least abundant.

Baileya pauciradiata also possesses cytotoxic sesquiterpene lactones, namely odoratin and paucin (J. J. Hoffmann et al. 1978).

Baileya pauciradiata is connected to the moth *Schinia pallicincta* Smith, which is closely related to the species of noctuid moth that inhabits *B. multiradiata*. *Schinia pallicincta* occupies mostly sand dunes of southern

California and uses *B. pauciradiata* as a primary feeding source (T. G. Myles and B. F. Binder 1990).

Baileya pauciradiata has occasionally been found growing with or near *B. pleniradiata*; natural hybrids between them have not been recorded, nor have synthetic crosses been successful (R. C. Brown 1974). Brown supported the genetic isolation of the species from the other taxa by demonstrating complete barriers to gene exchange.

389. **TETRANEURIS** Greene, Pittonia 3: 265. 1898 • Bitterweed [Greek *tetra*, four, and *neuron*, nerve, alluding to venation of ray floret corollas]

Mark W. Bierner

Billie L. Turner

Annuals or perennials, 2–50+ cm (perennials with ± branched, woody caudices). **Stems** 1–60, erect or ± decumbent, unbranched or sparingly branched distally. **Leaves** all basal, or basal-proximal, or basal and cauline; alternate; petiolate or sessile; blades (usually with distinct midribs) mostly oblanceolate to linear or filiform, sometimes lobed, ultimate margins usually entire, sometimes toothed, faces glabrous or ± hairy, eglandular or ± gland-dotted. **Heads** radiate or discoid, borne singly or in paniculiform to corymbiform or fastigiate arrays. **Involucres** hemispheric to campanulate 6–20 mm diam. **Phyllaries** 11–60+ in 3 series (mostly spreading to erect in fruit, distinct, herbaceous; outer with or without scarious margins, abaxial faces ± hairy; mid usually same number as, alternating with, and similar to outer, almost always with ± scarious margins; inner narrower than others, margins scarious). **Receptacles** hemispheric to conic, shallowly pitted or smooth, epaleate. **Ray florets** 0 or 7–27, pistillate, fertile; corollas (usually marcescent) yellow (laminae fan-shaped to oblanceolate, usually 3-lobed). **Disc florets** 20–250+, bisexual, fertile; corollas yellow proximally, yellow or purplish distally, tubes shorter than cylindric to cylindro-campanulate throats, lobes 5, ± deltate. **Cypselae** ± obpyramidal, moderately to densely hairy; **pappi** persistent, of 4–8 usually aristate, scales. $x = 15$.

Species 9 (9 in the flora): North America, Mexico.

Here, strigoso-canescent refers to often silvery induments (hairs tightly appressed, relatively short, 1–1.5 mm) of *Tetraneuris argentea* and *T. acaulis* var. *acaulis*. Such induments contrast with lanuginose or sericeous induments (hairs spreading, ca 1.5–3 mm) in other taxa, including *T. ivesiana* and the other three varieties of *T. acaulis*.

SELECTED REFERENCES Bierner, M. W. and R. K. Jansen. 1998. Systematic implications of DNA restriction site variation in *Hymenoxys* and *Tetraneuris* (Asteraceae, Helenieae, Gaillardiinae). Lundellia 1: 17–26. Bierner, M. W. and B. L. Turner. 2003. Taxonomy of *Tetraneuris* (Asteraceae: Helenieae: Tetraneurinae). Lundellia 6: 44–96. Greene, E. L. 1898. Studies in the Compositae—VII. 1. Some helenioid genera. *Tetraneuris*. Pittonia 3: 264–270. Rydberg, P. A. 1915b. *Tetraneuris*. In: N. L. Britton et al., eds. 1905+. North American Flora.... 47+ vols. New York. Vol. 34, 100–108.

1. Annuals . . . . 1. *Tetraneuris linearifolia*
1. Perennials (with caudices).
  2. Caudices: branches not notably thickened distally; leaves all basal-proximal (not tightly clustered, internodes often evident).
    3. Stems erect; leaves sparsely to densely hairy, not woolly; Colorado, Kansas, Nebraska, New Mexico, Oklahoma, Texas . . . . 2. *Tetraneuris scaposa*
    3. Stems erect or ± decumbent; leaves moderately to densely woolly; s Texas . . . . 3. *Tetraneuris turneri*

[2. Shifted to left margin.—Ed.]

2. Caudices: branches notably thickened distally; leaves all basal or basal and cauline (basal leaves tightly clustered, internodes usually not evident).

    4. Leaves basal and cauline.

        5. Leaf blades ± densely strigoso-canescent . . . . 4. *Tetraneuris argentea*

        5. Leaf blades glabrous or sparsely to moderately lanuginose to sericeous . . . . . 5. *Tetraneuris ivesiana*

    4. Leaves all basal.

        6. Ray florets 0 (leaves densely hairy, not strigoso-canescent; Yavapai County, Arizona) . . . . 6. *Tetraneuris verdiensis*

        6. Ray florets usually 7–27 (if 0, leaves strigoso-canescent; widespread).

            7. Rays 14–27; pappus scales usually non-aristate; Ontario, Illinois, Ohio . . . .
. . . . 7. *Tetraneuris herbacea*

            7. Rays 7–15(–21); pappus scales aristate; North Dakota to Texas and west.

                8. Leaves (midribs usually distinct) glabrous or usually sparsely, sometimes moderately or densely hairy (not strigoso-canescent or sericeous), gland-dotted; outer phyllaries 4–8, margins (0.3–)0.5–1.2 mm wide, usually conspicuously scarious . . . . 8. *Tetraneuris torreyana*

                8. Leaves (midribs not very distinct) glabrous or ± hairy, often lanuginose, sericeous, or strigoso-canescent, eglandular or ± gland-dotted; outer phyllaries 6–12, margins 0–0.4 mm wide, sometimes slightly scarious . . . . 9. *Tetraneuris acaulis*

1. Tetraneuris linearifolia (Hooker) Greene, Pittonia 3: 269. 1898

Hymenoxys linearifolia Hooker, Icon. Pl. 2: plate 146. 1837

Annuals, 16–50+ cm. **Caudices** none. **Stems** 1–10, erect or ± decumbent, leafy, unbranched or branched distally. **Leaves** basal and cauline, ± hairy, ± gland-dotted; basal blades spatulate to oblanceolate, entire or with 2–6 teeth or lobes; proximal cauline blades narrowly spatulate or oblanceolate to linear-oblanceolate, entire or with 2–6 teeth or lobes; mid blades oblanceolate to linear-oblanceolate to linear, 0.9–9.5(–16) mm wide, entire or with 1–2 teeth or lobes; distal blades narrowly oblanceolate to linear-oblanceolate to linear, entire or with 1–2 teeth or lobes. **Heads** 8–50(–80) per plant, borne singly or in corymbiform arrays. **Peduncles** 8–29 cm, ± hairy. **Involucres** 5–10 × 7–15 mm. **Outer phyllaries** 8–21, 2.4–5.5 mm, margins 0–0.2 mm wide, not scarious or scarious, abaxial faces ± hairy. **Ray florets** 9–25; corollas 8.2–16.8 mm. **Disc florets** 50–200+; corollas yellow, 1.6–3 mm. **Cypselae** 1.5–2.6 mm; **pappi** of 4–8 obovate, often aristate scales 1–2.5 mm.

Varieties 2 (2 in the flora): sw United States, n Mexico.

Plants of *Tetraneuris linearifolia* morphologically intermediate between the two varieties are found in Bee, Duval, Live Oak, and McMullen counties, Texas.

1. Stems erect, branched distally; leaves ± hairy, basal blades entire or with 2(–6) teeth or lobes, proximal cauline blades entire or with 1–2 teeth or lobes, mid blades (0.9–)1.4–3(–4.8) mm wide; involucres (7–)8–11(–12) mm diam.; pappus scales aristate; Kansas, New Mexico, Oklahoma, Texas, on limestone-derived soils . . . . 1a. *Tetraneuris linearifolia* var. *linearifolia*

1. Stems erect or ± decumbent, ± branched distally; leaves usually densely hairy, basal blades with 2–6 teeth or lobes, proximal cauline blades with 2(–6) teeth or lobes, mid blades (2.5–)3.9–9.5(–16) mm wide; involucres 12–15 mm diam.; pappus scales not aristate; s Texas on sand . . . . 1b. *Tetraneuris linearifolia* var. *arenicola*

1a. Tetraneuris linearifolia (Hooker) Greene var. **linearifolia**

Tetraneuris linearifolia subsp. *dodgei* Cockerell; *T. linearifolia* var. *latior* Cockerell; *T. oblongifolia* Greene

Plants 16–40+ cm. **Stems** 1(–6), erect, branched distally. **Leaves** ± hairy, sparsely to densely gland-dotted; basal blades spatulate to oblanceolate, entire or with 2(–6) teeth or lobes; proximal cauline blades narrowly spatulate to linear-oblanceolate, entire or with 1–2 teeth or lobes; mid blades linear-oblanceolate to linear, (0.9–)1.4–3(–4.8) mm wide. **Peduncles** (8–)9–14(–19) cm, ± hairy. **Involucres** (5–)6–7.8(–9) × (7–)8–11(–12) mm. **Outer phyllaries** 8–16, margins often scarious, abaxial faces ± hairy. **Ray florets** 9–20. **Cypselae** 1.5–2.1(–2.5) mm; **pappi** of 4–7 aristate scales 1.2–1.9 (–2.5) mm. $2n = 28, 30$.

Flowering (Jan–)Apr–Jul(–Dec). Roadsides, hillsides, pastures, open areas, edges of woods; 30–2700 m; Kans., N.Mex., Okla., Tex.; Mexico (Coahuila, Nuevo León, Tamaulipas).

1b. Tetraneuris linearifolia (Hooker) Greene var. **arenicola** Bierner, Sida 15: 237. 1992

Plants 21–50+ cm. **Stems** (1–)4–10, erect or ± decumbent, unbranched or branched distally. **Leaves** usually densely hairy, sparsely to moderately glanddotted; basal blades usually spatulate, usually with 2–6 teeth or lobes; proximal cauline blades usually oblanceolate, usually with 2(–6) teeth or lobes; mid blades usually oblanceolate, (2.5–)3.9–9.5(–16) mm wide. **Peduncles** (10–)13–22(–29) cm, densely hairy. **Involucres** 8–10 × 12–15 mm. **Outer phyllaries** 13–21, margins usually not scarious, abaxial faces densely hairy. **Ray florets** 14–25. **Cypselae** 1.9–2.6 mm; **pappi** of 5–8 non-aristate scales 1–1.7 mm. *2n* = 30.

Flowering (Feb–)Mar–Apr(–Jun; also Dec). Roadsides and pastures; 50–200 m; Tex.

Variety *arenicola* is known only from the vicinity of Brooks, Hidalgo, Jim Hogg, and Zapata counties. It grows in sand; var. *linearifolia* grows in limestone-derived soils.

2. Tetraneuris scaposa (de Candolle) Greene, Pittonia 3: 266. 1898 [F]

Cephalophora scaposa de Candolle in A. P. de Candolle and A. L. P. P. de Candolle, Prodr. 5: 663. 1836; *Hymenoxys scaposa* (de Candolle) K. F. Parker

Perennials, 14–40+ cm. **Caudices** often highly branched, branches not notably thickened distally. **Stems** 1–50, erect, sometimes branched near bases, sometimes densely woolly among proximal leaves. **Leaves** all basal-proximal (new leaves not tightly clustered); blades spatulate to oblanceolate to linear-oblanceolate or linear, entire or with 2 teeth or lobes, sparsely to densely hairy, ± gland-dotted. **Heads** 1–50 per plant, borne singly or in fastigiate arrays. **Peduncles** 12–40 cm, ± hairy. **Involucres** 5–10 × 7–12 mm. **Outer phyllaries** 8–16, 3.8–6.6 mm, margins 0–0.2(–0.4) mm wide, sometimes to often slightly scarious, abaxial faces ± hairy. **Ray florets** 12–26; corollas 7.4–22 mm. **Disc florets** 25–180+; corollas yellow, 2.5–3.5 mm. **Cypselae** 2–3 mm; **pappi** of 5–7 obovate to oblanceolate, aristate scales 1.6–2.3 mm.

Varieties 2 (2 in the flora): c, w United States, Mexico.

The name "*Hymenoxys scaposa* (de Candolle) K. F. Parker var. *glabra* (Nuttall) K. F. Parker" appears never to have been published. Parker did make the combination *Hymenoxys acaulis* (Pursh) K. F. Parker var. *glabra* (A. Gray) K. F. Parker based on *Actinella scaposa* (de Candolle) Nuttall var. *glabra* A. Gray, which she mistakenly called *Actinea scaposa* Nuttall var. *glabra* A. Gray.

1. Stems not densely woolly among proximal leaves; leaves crowded (internodes not evident); Colorado, Kansas, Nebraska, New Mexico, Oklahoma, Texas . . . . 2a. *Tetraneuris scaposa* var. *scaposa*
1. Stems densely woolly among proximal leaves; leaves not crowded (internodes evident); Texas . . . . 2b. *Tetraneuris scaposa* var. *argyrocaulon*

2a. Tetraneuris scaposa (de Candolle) Greene var. **scaposa** [F]

Hymenoxys glabra (Nuttall) Shinners; *H. scaposa* (de Candolle) K. F. Parker var. *linearis* (Nuttall) K. F. Parker; *H. scaposa* var. *villosa* Shinners; *Tetraneuris angustata* Greene; *T. angustifolia* Rydberg; *T. fastigiata* Greene; *T. glabra* (Nuttall) Greene; *T. linearis* (Nuttall) Greene; *T. scaposa* var. *linearis* (Nuttall) K. F. Parker; *T. scaposa* var. *villosa* (Shinners) Shinners; *T. stenophylla* Rydberg

Plants 14–40+ cm. **Stems** 1–30(–50), not densely woolly among proximal leaves. **Leaves** crowded (internodes not evident), blades linear, linear-oblanceolate, oblanceolate, or spatulate, usually densely gland-dotted. **Involucres** 5–10 × 7–12 mm. **Ray corollas** 9.5–22 mm. **Disc corollas** 2.8–3.5 mm. *2n* = 30, 60.

Flowering (Jan–)Mar–Jun(–Dec). Roadsides, hillsides, pastures, open areas, edges of woods; 300–2400 m; Colo., Kans., Nebr., N.Mex., Okla., Tex.; Mexico (Chihuahua, Coahuila, Nuevo León, San Luis Potosí, Tamaulipas, Zacatecas).

2b. Tetraneuris scaposa (de Candolle) Greene var. **argyrocaulon** (K. F. Parker) K. F. Parker, Phytologia 45: 467. 1980

Hymenoxys scaposa (de Candolle) K. F. Parker var. *argyrocaulon* K. F. Parker, Phytologia 20: 192. 1970

Plants 17–40+ cm. **Stems** 1–20, densely woolly among proximal leaves. **Leaves** not crowded (internodes evident); blades linear, linear-oblanceolate, or oblanceolate, ± gland-dotted. **Involucres** 6–10 × 9–12 mm. **Ray corollas** (7.4–)9.5–14 mm. **Disc corollas** 2.5–3 mm. *2n* = 30.

Flowering Feb–Apr(–Jun; also Nov). Roadsides, pastures; 100–200 m; Tex.; Mexico (Tamaulipas).

3. **Tetraneuris turneri** (K. F. Parker) K. F. Parker, Phytologia 45: 467. 1980

Hymenoxys turneri K. F. Parker, Phytologia 20: 192. 1970

Perennials, 22–50+ cm. Caudices ± branched, branches not notably thickened distally. **Stems** 1–20, erect or ± decumbent, sometimes branched near bases, usually densely woolly or long-hairy among proximal leaves. **Leaves** all basal-proximal, new leaves not tightly clustered; blades linear-oblanceolate, oblanceolate, or spatulate, entire or basal and proximal with 2(–4) teeth or lobes, moderately to densely woolly, usually densely gland-dotted. **Heads** 1–20 per plant, borne singly. **Peduncles** 20–50 cm, moderately to densely hairy. **Involucres** 8–15 × 12–20 mm. **Outer phyllaries** 12–16, 5.2–7.8(–10) mm, margins 0–0.2 mm wide, sometimes slightly scarious, abaxial faces ± hairy. **Ray florets** 12–24; corollas 14–20 mm. **Disc florets** 100–250+; corollas yellow proximally, yellow to purplish distally, 3.5–3.8 mm. **Cypselae** 2.6–3.4 mm; **pappi** of 6–7 lanceolate to obovate or oblanceolate, often aristate scales 3–4 mm. $2n = 90$.

Flowering (Feb–)Mar–Apr(–Jun). Roadsides, pastures; 50–200 m; Tex.; Mexico (Coahuila).

4. **Tetraneuris argentea** (A. Gray) Greene, Pittonia 3: 269. 1898 E

Actinella argentea A. Gray, Mem. Amer. Acad. Arts, n. s. 4: 100. 1849; *Hymenoxys argentea* (A. Gray) K. F. Parker; *Tetraneuris formosa* Greene ex Wooton & Standley; *T. leptoclada* (A. Gray) Greene; *T. trinervata* Greene

Perennials, 6–25(–42+) cm. Caudices ± branched, branches notably thickened distally. **Stems** 1–8(–12), erect, leafy, ± branched distally. **Leaves** basal and cauline, entire, usually densely strigoso-canescent, ± gland-dotted; basal leaves tightly clustered, blades spatulate to oblanceolate; proximal cauline blades oblanceolate; mid blades oblanceolate to linear-oblanceolate, 2.5–5 mm wide; distal blades linear-lanceolate to linear. **Heads** 1–10(–30) per plant (1–7 per stem), borne singly or in paniculiform to corymbiform arrays. **Peduncles** 2–13.5 cm, usually densely hairy. **Involucres** (5–)7–10 × 8–15 mm. **Outer phyllaries** 8–11, 4–6.5 mm, margins 0–0.3 mm wide, usually scarious, abaxial faces densely hairy. **Ray florets** 8–14; corollas 11.8–17 mm. **Disc florets** 25–75(–100+); corollas yellow proximally, yellow or rarely purplish

distally, 2.8–3.8 mm. **Cypselae** 2.4–3.1 mm; **pappi** of 5–6 obovate, aristate scales 2.1–3.5 mm. $2n = 30, 60$.

Flowering (Apr–)May–Aug(–Sep). Roadsides, hillsides, open treeless areas, edges of woods; (1500–)2000–2400(–3000) m; Ariz., N.Mex.

5. **Tetraneuris ivesiana** Greene, Pittonia 3: 269. 1898 E

Hymenoxys acaulis (Pursh) K. F. Parker var. *ivesiana* (Greene) K. F. Parker; *H. argentea* (A. Gray) K. F. Parker var. *ivesiana* (Greene) Cronquist; *H. ivesiana* (Greene) K. F. Parker; *Tetraneuris intermedia* Greene; *T. mancosensis* A. Nelson; *T. pilosa* Greene

Perennials, 10–26+ cm. Caudices ± branched, branches notably thickened distally. **Stems** (1–)5–20(–30), erect, leafy, ± branched distally. **Leaves** basal and cauline, entire, glabrous or sparsely to moderately lanuginose to sericeous, usually densely gland-dotted; basal leaves tightly clustered, blades linear-oblanceolate; proximal cauline blades linear-oblanceolate; mid blades linear-oblanceolate to linear, 1.7–2 mm wide; distal blades linear-oblanceolate to linear. **Heads** (1–)5–30(–40) per plant (1–5 per stem), borne singly or in paniculiform to corymbiform arrays. **Peduncles** 5–18.5 cm, ± hairy, usually densely hairy distally. **Involucres** 8–12 × 13–17 mm. **Outer phyllaries** 7–12, 5–6.4 mm, margins 0–0.3(–0.4) mm wide, usually scarious, abaxial faces ± hairy. **Ray florets** 7–10; corollas 10–20 mm. **Disc florets** (40–)80–150+; corollas yellow, 3.3–4.5 mm. **Cypselae** 3–4.1 mm; **pappi** of 5–7 lanceolate or obovate to oblanceolate, aristate scales 2.7–4.5 mm. $2n = 28, 30, 56, 60$.

Flowering Apr–Aug. Roadsides, hillsides, open treeless areas, edges of woods; 1300–2900 m; Ariz., Colo., N.Mex., Utah.

6. **Tetraneuris verdiensis** R. A. Denham & B. L. Turner, Phytologia 81: 5, fig. 1. 1996 C E

Perennials, 5–15+ cm. Caudices ± branched, branches notably thickened distally. **Stems** 1–15, erect, unbranched. **Leaves** all basal (new leaves tightly clustered); blades obovate or oblanceolate to linear-oblanceolate, entire, densely hairy (not strigoso-canescent), ± gland-dotted. **Heads** 1–15 per plant, borne singly. **Peduncles** 4–15 cm, ± hairy. **Involucres** 5–9 × 6–10 mm. **Outer phyllaries** 8–10, 4–5(–6) mm, margins 0–0.2 mm wide, sometimes slightly scarious, abaxial faces densely hairy. **Ray florets** 0. **Disc florets** 20–40; corollas yellow, 2.5–4 mm. **Cypselae** 2.8–

4 mm; **pappi** of 5–6 ovate to lanceolate or obovate to oblanceolate, aristate scales 2.5–3.5 mm.

Flowering May. Low gypsum hills; of conservation concern; ca. 1000 m; Ariz.

Tetraneuris verdiensis is found only at or near the type locality, east of Camp Verde, Yavapai County.

7. **Tetraneuris herbacea** Greene, Pittonia 3: 268. 1898 E

Actinea herbacea (Greene) B. L. Robinson; *Hymenoxys acaulis* (Pursh) K. F. Parker var. *glabra* (A. Gray) K. F. Parker; *H. herbacea* (Greene) Cronquist

Perennials, 6–35+ cm. Caudices ± branched, branches notably thickened distally. **Stems** 1–10, erect, unbranched. **Leaves** all basal, new leaves tightly clustered; blades spatulate or oblanceolate to linear-oblanceolate, entire, glabrous or sparsely to rarely moderately hairy, moderately to densely gland-dotted. **Heads** 1–10 per plant, borne singly. **Peduncles** 5–35 cm, ± hairy proximally, densely hairy distally. **Involucres** 10–15 × 12–20 mm. **Outer phyllaries** 8–12, 5–7.5 mm, margins 0–0.2 mm wide, sometimes slightly scarious, abaxial faces ± hairy. **Ray florets** 14–27; corollas 11.8–20 mm. **Disc florets** 50–100+; corollas yellow, 3.4–4 mm. **Cypselae** 2.4–3.4 mm; **pappi** of 4–7 obovate, usually non-aristate scales 1.9–2.2 mm. *2n* = 28.

Flowering (Apr–)May–Jun(–Oct). Alvars (limestone flats), openings in woods; 200 m; Ont.; Ill., Ohio.

Tetraneuris herbacea is in the Center for Plant Conservation's National Collection of Endangered Plants.

8. **Tetraneuris torreyana** (Nuttall) Greene, Pittonia 3: 265. 1898 E

Actinella torreyana Nuttall, Trans. Amer. Philos. Soc., n. s. 7: 379. 1841; *Hymenoxys depressa* (Torrey & A. Gray) S. L. Welsh & Reveal; *H. lapidicola* S. L. Welsh & Neese; *H. torreyana* (Nuttall) K. F. Parker; *Tetraneuris depressa* (Torrey & A. Gray) Greene

Perennials, 2–15+ cm. Caudices ± branched, branches notably thickened distally. **Stems** 1–30(–40), erect, unbranched. **Leaves** all basal, new leaves tightly clustered; blades (midribs usually pronounced) spatulate or oblanceolate to linear-oblanceolate, entire, glabrous or usually sparsely, sometimes moderately to densely, hairy, densely gland-dotted. **Heads** 1–10(–40) per plant, borne singly. **Peduncles** (0.5–)2–15 cm, ± hairy proximally, almost always densely lanate distally. **Involucres** 10–12 × 12–15 mm. **Outer phyllar-**

ies 4–8, 6–8.5 mm, margins (0.3–)0.5–1.2 mm wide, usually conspicuously scarious, abaxial faces ± hairy. **Ray florets** 7–14; corollas 11–17 mm. **Disc florets** 25–150+; corollas yellow, 4–4.7 mm. **Cypselae** 3–4 mm; **pappi** of 5–7 obovate to oblanceolate, aristate scales 3–4(–4.5) mm.

Flowering May–Jun(–Sep). Roadsides, hillsides, slopes, open treeless areas, edges of woods; (1300–)1700–3100 m; Colo., Mont., Utah, Wyo.

9. **Tetraneuris acaulis** (Pursh) Greene, Pittonia 3: 265. 1898

Gaillardia acaulis Pursh, Fl. Amer. Sept. 2: 743. 1813 (as Galardia); *Hymenoxys acaulis* (Pursh) K. F. Parker E F

Perennials, 2–30+ cm. Caudices ± branched, branches notably thickened distally. **Stems** 1–35 (–60), erect, unbranched. **Leaves** all basal, new leaves tightly clustered; blades spatulate or oblanceolate to linear-oblanceolate, entire, glabrous or ± hairy, often lanuginose, sericeous, or strigoso-canescent, eglandular or ± gland-dotted. **Heads** 1–35(–60) per plant, borne singly. **Peduncles** 0.5–30 cm, ± hairy. **Involucres** 7–12 × 8–16 mm. **Outer phyllaries** 6–12, 3.9–9(–11.5) mm, margins 0–0.4 mm wide, sometimes slightly scarious, abaxial faces ± hairy. **Ray florets** 8–15(–21); corollas 9–19 mm. **Disc florets** 25–200+; corollas yellow, 2.7–4.3 mm. **Cypselae** 2–4 mm; **pappi** of 5–8 obovate to oblanceolate, aristate scales 2.2–3.7 mm.

Varieties 4 (4 in the flora): c, w North America.

1. Leaves sparsely to densely hairy, often lanuginose, sometimes sericeous, not strigoso-canescent; peduncles 0.5–8(–12) cm; outer phyllaries 6.8–9 (–11.5) mm; growing at (2100–)3000–3900 m . . . . 9a. *Tetraneuris acaulis* var. *caespitosa*
1. Leaves sparsely to densely hairy, often strigoso-canescent; peduncles (1–)5–20(–30) cm; outer phyllaries 4–7.5 mm; growing at 700–2900(–3500) m.
    2. Leaf blades usually densely, sometimes sparsely, strigoso-canescent; e of continental divide (except for 2 reports from nw Idaho) . . . . 9b. *Tetraneuris acaulis* var. *acaulis*
    2. Leaf blades glabrous or sparsely to moderately to sometimes densely hairy, not strigoso-canescent; w of continental divide.
        3. Leaf blades glabrous or usually sparsely to moderately or sometimes densely hairy, densely gland-dotted; n Arizona, se California, w Colorado, s Idaho, e Nevada, Utah . . . . 9c. *Tetraneuris acaulis* var. *arizonica*
        3. Leaf blades glabrous or sparsely to moderately hairy, at least some (usually most or all) eglandular or sparsely gland-dotted; w Colorado, c, ne Utah, and sw Wyoming . . . . 9d. *Tetraneuris acaulis* var. *epunctata*

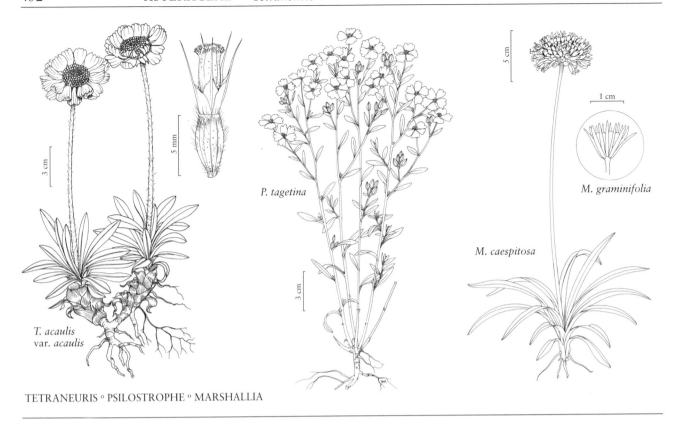

T. acaulis
var. acaulis

P. tagetina

M. graminifolia

M. caespitosa

TETRANEURIS ° PSILOSTROPHE ° MARSHALLIA

9a. Tetraneuris acaulis (Pursh) Greene var. **caespitosa**
A. Nelson, Bot. Gaz. 28: 127. 1899 [E]

Hymenoxys acaulis (Pursh) K. F.
Parker var. *caespitosa* (A. Nelson)
K. F. Parker; *Tetraneuris brevifolia*
Greene; *T. lanata* Greene; *T. lanigera*
Daniels

Plants 2–8(–12+) cm. **Leaves:**
blades spatulate or oblanceolate to
linear-oblanceolate, sparsely to
densely hairy, often lanuginose,
sometimes sericeous, sparsely to sometimes moderately
or densely gland-dotted. **Heads** 1–35(–60) per plant.
Peduncles 0.5–8(–12) cm. **Outer phyllaries** 6–12, 6.8–
9(–11.5) mm, margins 0–0.2 mm wide, sometimes slightly
scarious, abaxial faces usually densely hairy. **Ray florets**
9–15(–21); corollas 9–13 mm. **Cypselae** 2–3(–4) mm;
pappi 2.5–3.5 mm. $2n$ = 28, 30.

Flowering (May–)Jun–Aug(–Sep). Edges of montane
forests, alpine meadows above timberline, open savannas; (2100–)3000–3900 m; Colo., N.Mex., Wyo.

9b. Tetraneuris acaulis (Pursh) Greene var. **acaulis**
[E] [F]

Tetraneuris eradiata A. Nelson;
T. incana A. Nelson; *T. pygmaea*
(A. Gray) Wooton & Standley; *T.
septentrionalis* Rydberg; *T. simplex*
A. Nelson

Plants (3–)10–20(–30+) cm.
Leaves: blades spatulate or oblanceolate to linear-oblanceolate, usually densely, sometimes sparsely,
strigoso-canescent, sparsely to moderately or densely
gland-dotted. **Heads** 1–10(–15) per plant. **Peduncles**
(2–)8–20(–30) cm. **Outer phyllaries** 6–10, 4–7 mm,
margins 0–0.2(–0.4) mm wide, sometimes slightly scarious, abaxial faces usually densely hairy. **Ray florets** 8–
14; corollas 9–15 mm. **Cypselae** 2.5–3 mm; **pappi** 2.2–
2.9 mm. $2n$ = 28, 30, 56, 60.

Flowering (Apr–)May–Jul(–Oct). Roadsides, hillsides,
grasslands, edges of woods; 700–2000(–3500) m; Alta.,
Sask.; Colo., Idaho, Kans., Mont., Nebr., N.Mex.,
N.Dak., Okla., S.Dak., Tex., Wyo.

Some plants of var. *acaulis* from Colorado and
southeastern Wyoming have considerably less hairy and/
or more densely gland-dotted leaves, making them easy
to confuse with var. *arizonica*. As to distribution, label
on a collection at LL reads, "From W. A. Curtis,
Lewistown [Idaho], without definite locality but

probably collected in that vicinity." Also, the holotype and isotype of *Tetraneuris septentrionalis* were collected in "Palouse County, Idaho." These are the only collections we have seen that place var. *acaulis* west of the continental divide.

9c. Tetraneuris acaulis (Pursh) Greene var. **arizonica** (Greene) K. F. Parker, Phytologia 45: 467. 1980 [E]

Tetraneuris arizonica Greene, Pittonia 3: 266. 1898; *Hymenoxys acaulis* (Pursh) K. F. Parker var. *arizonica* (Greene) K. F. Parker; *H. acaulis* var. *nana* S. L. Welsh; *T. acaulis* var. *nana* (S. L. Welsh) Kartesz & Gandhi

Plants (2–)6–15(–30+) cm. **Leaves:** blades spatulate or oblanceolate to linear-oblanceolate, glabrous or moderately to densely hairy (not strigoso-canescent or sericeous), densely gland-dotted. **Heads** 1–20(–35) per plant. **Peduncles** (1–)5–15(–30) cm. **Outer phyllaries** 7–10, 3.9–6.5 mm, margins 0–0.3(–0.4) mm wide, often slightly scarious, abaxial faces sparsely to usually moderately to sometimes densely hairy. **Ray florets** 9–15; corollas 11–17 mm. **Cypselae** 3–3.7 mm; **pappi** 2–3.5 mm. *2n* = 28, 30, 56, 60.

Flowering (Apr–)May–Jul(–Sep). Roadsides, hillsides, grasslands, edges of woods, aspen meadows; 1300–2900 m; Ariz., Calif., Colo., Idaho, Nev., Utah.

9d. Tetraneuris acaulis (Pursh) Greene var. **epunctata** (A. Nelson) Kartesz & Gandhi, Phytologia 78: 2. 1995 [E]

Tetraneuris epunctata A. Nelson, Bot. Gaz. 37: 275. 1904; *Hymenoxys acaulis* (Pursh) K. F. Parker var. *epunctata* (A. Nelson) Cronquist; *T. crandallii* Rydberg

Plants (2–)6–15(–25+) cm. **Leaves:** blades oblanceolate to linear-oblanceolate, glabrous or sparsely to moderately hairy (not strigoso-canescent or sericeous), some (usually most or all) eglandular or sparsely gland-dotted. **Heads** 1–10 (–15) per plant. **Peduncles** (1–)5–15(–25) cm. **Outer phyllaries** 6–12, 5–7.5 mm, margins 0–0.2 mm wide, sometimes slightly scarious, abaxial faces ± hairy. **Ray florets** 10–14; corollas 11–19 mm. **Cypselae** 3–4 mm; **pappi** 2.8–3.7 mm. *2n* = 28, 56.

Flowering May–Jul(–Sep). Roadsides, hillsides, edges of woods; 1600–2700(–3500) m; Colo., Utah, Wyo.

Some plants of *Hymenoxys acaulis* from south-central to northeastern Utah and northwestern Colorado have moderately gland-dotted leaf blades. We interpret them to be intermediates between var. *epunctata* and var. *arizonica*. Plants referable to var. *epunctata* have eglandular or very sparsely gland-dotted blades; those referable to var. *arizonica* have densely gland-dotted blades.

390. PSILOSTROPHE de Candolle in A. P. de Candolle and A. L. P. de Candolle, Prodr. 7: 261. 1838 • [Greek *psilo-*, bare, and *strophe*, turn; perhaps alluding to epaleate receptacles, which differ from paleate receptacles in *Parthenium*, to which de Candolle compared *Psilostrophe*]

John L. Strother

Biennials, perennials, subshrubs, or shrubs (rarely flowering first year), 8–60+ cm. **Stems** erect to spreading, branched from bases or throughout (not scapiform). **Leaves** basal and cauline or all cauline; alternate; petiolate or sessile; blades spatulate to oblanceolate or linear, margins usually entire (sometimes toothed or lobed on larger rosette leaves), faces densely to sparsely arachno-villous or ± strigillose, often gland-dotted as well. **Heads** radiate, usually in compact, corymbiform arrays or glomerulate clusters (borne singly in *P. cooperi*). **Involucres** cylindric to campanulate or obconic, 2–7 mm diam. **Phyllaries** persistent, 5–12 in 1–2 series (erect in fruit, distinct, oblong to lanceolate, bases ± indurate). **Receptacles** flat or convex, smooth or ± pitted (without setiform enations, sometimes gland-dotted), epaleate. **Ray florets** 1–8, pistillate, fertile; corollas yellow to orange (marcescent, spreading or reflexed in fruit). **Disc florets** 5–25+, bisexual, fertile; corollas yellow to orange, tubes shorter than narrowly cylindric throats, lobes 5, deltate (equal, papillate abaxially). **Cypselae** cylindric to clavate or obpyramidal, sometimes weakly obcompressed (ray), all striate-ribbed, usually glabrous, sometimes gland-dotted (villous in *P. gnaphalodes*, sometimes hirtellous in *P. tagetina*); **pappi** of 4–8 ± oblong or elliptic to lanceolate or lance-subulate, entire (lacerate in *P. gnaphalodes*) scales (without prominent midribs). $x = 16$.

Species 7 (6 in the flora): United States, Mexico.

My treatment of psilostrophes closely follows that by R. C. Brown (1978). *Psilostrophe mexicana* R. C. Brown is known from Chihuahua and Durango.

SELECTED REFERENCES Brown, R. C. 1978. Biosystematics of *Psilostrophe* (Compositae: Helenieae). II. Artificial hybridization and systematic treatment. Madroño 25: 187–201. Turner, B. L., M. W. Turner, and J. C. Crutchfield. 1988. Populational analyses and new combinations in *Psilostrophe tagetina* and *P. gnaphalodes* (Asteraceae, Heliantheae). Phytologia 65: 231–240.

1. Subshrubs or shrubs; stems pannose (white); heads borne singly; peduncles 35–60(–80+) mm. . . . . 1. *Psilostrophe cooperi*
1. Biennials or perennials; stems arachno-villous (gray to gray-green) or strigillose (greenish); heads in corymbiform arrays; peduncles 0.5–25(–50) mm.
   2. Stems ± strigillose, (greenish, sometimes ± villous in proximal axils); ray laminae reflexed in fruit . . . . 2. *Psilostrophe sparsiflora*
   2. Stems mostly arachno-villous (gray to gray-green); ray laminae spreading in fruit.
      3. Involucres 7–9(–12) mm; disc florets (10–)12–15(–20); pappus scales oblong to ovate, 1.5–2 mm. . . . . 3. *Psilostrophe bakeri*
      3. Involucres 4–6(–8) mm; disc florets 5–8(–12); pappus scales elliptic or lanceolate to linear-subulate, 2–3+ mm.
         4. Cypselae villous . . . . 4. *Psilostrophe gnaphalodes*
         4. Cypselae usually glabrous, sometimes hirtellous and/or gland-dotted.
            5. Peduncles (3–)12–20(–40) mm; ray laminae (5–)7–14+ mm . . . . . . 5. *Psilostrophe tagetina*
            5. Peduncles (0.5–)1–3(–5+) mm; ray laminae 3–4(–6) mm . . . . 6. *Psilostrophe villosa*

1. Psilostrophe cooperi (A. Gray) Greene, Pittonia 2: 176. 1891

Riddellia cooperi A. Gray, Proc. Amer. Acad. Arts 7: 358. 1868

Shrubs or subshrubs, (15–)25–30(–50+) cm. **Stems** pannose (white). **Heads** borne singly. **Peduncles** 35–60(–80+) mm. **Involucres** 6–8 mm. **Rays** 3–6; laminae (8–)12–20 mm, spreading to reflexed in fruit. **Disc florets** (6–)10–17(–25). **Cypselae** usually glabrous, sometimes gland-dotted; **pappi** of 4–6 oblong to ± lanceolate scales 2–2.5 mm. $2n = 32$.

Flowering (Feb–)Apr–Jun(–Dec). Creosote-bush scrub, sandy washes; 100–2100 m; Ariz., Calif., Nev., Utah; Mexico (Baja California, Sonora).

2. Psilostrophe sparsiflora (A. Gray) A. Nelson, Proc. Biol. Soc. Wash. 16: 23. 1903 [E]

Riddellia tagetina Nuttall var. *sparsiflora* A. Gray in A. Gray et al., Syn. Fl. N. Amer. 1(2): 318. 1884

Biennials or perennials (rarely flowering first year), 10–40+ cm. **Stems** ± strigillose (greenish, sometimes ± villous in proximal axils). **Heads** in ± corymbiform arrays. **Peduncles** 8–12(–25+) mm. **Involucres** 4–6 mm. **Ray florets** (1–)2–3(–4); corolla laminae (6–)8–10+ mm, reflexed in fruit. **Disc florets** 7–9. **Cypselae** usually glabrous, sometimes gland-dotted; **pappi** of 4–5 lanceolate to lance-subulate scales 2–3 mm. $2n = 32$.

Flowering May–Sep. Grasslands, pinyon-juniper scrub, sagebrush scrub, yellow pine forests; 900–2200 m; Ariz., N.Mex., Utah.

3. **Psilostrophe bakeri** Greene, Pl. Baker. 3: 29. 1901
E

Biennials or perennials (rarely flowering first year), 5–12(–20+) cm. **Stems** arachno-villous (gray to gray-green). **Heads** in ± loose, corymbiform arrays. **Peduncles** 15–35(–50) mm. **Involucres** 7–9 (–12) mm. **Rays** 4–5(–8); laminae 8–15+ mm, spreading in fruit. **Disc florets** (10–)12–15(–20). **Cypselae** usually glabrous, sometimes gland-dotted; **pappi** of 4–5 ± oblong scales 1.5–2 mm. $2n = 32$.

Flowering May–Jun. Rocky, often alkaline, slopes, knolls; 1300–2000 m; Colo., Idaho.

4. **Psilostrophe gnaphalodes** de Candolle in A. P. de Candolle and A. L. P. P. de Candolle, Prodr. 7: 261. 1838

Biennials or perennials (rarely flowering first year), (15–)25–40 (–50+) cm. **Stems** arachno-villous (gray to gray-green). **Heads** in ± loose to crowded, corymbiform arrays. **Peduncles** (5–)10–25 mm. **Involucres** 5–6 mm. **Rays** (2–)3 (–4); laminae 4–6+ mm, spreading in fruit. **Disc florets** 5–9(–12). **Cypselae** villous; **pappi** of 4–5 lance-subulate scales 2–2.5 mm (scales ± villous abaxially and margins ± lacerate). $2n = 32$.

Flowering Mar–Jun(–Dec). Creosote-bush scrub, desert flats, dry banks, limestone soils; 300–1400+ m; Tex.; Mexico (Coahuila, Durango, Nuevo León, San Luis Potosí, Tamaulipas, Zacatecas).

5. **Psilostrophe tagetina** (Nuttall) Greene, Pittonia 2: 176. 1891 F

Riddellia tagetina Nuttall, Trans. Amer. Philos. Soc., n. s. 7: 371. 1841 (as tagetinae); *Psilostrophe tagetina* var. *grandiflora* (Rydberg) Heiser; *P. tagetina* var. *lanata* A. Nelson

Biennials or perennials (rarely flowering first year), 10–30(–60+) cm. **Stems** arachno-villous (gray to gray-green). **Heads** in ± loose to crowded, corymbiform arrays. **Peduncles** (3–)12–20(–40) mm. **Involucres** 4–8 mm. **Rays** 3–4(–6); laminae (3–)7–14+ mm, spreading in fruit. **Disc florets** 6–9(–12). **Cypselae** usually glabrous, sometimes hirtellous and/or gland-dotted; **pappi** of (4–)6–8 elliptic or lanceolate to lance-subulate scales 2–3+ mm. $2n = 32$.

Flowering (Apr–)May–Sep(–Dec). Desert scrub, grasslands, limestone soils, saline flats, sandy soils; (300–)600–2400 m; Ariz., N.Mex., Tex., Utah; Mexico (Chihuahua, Coahuila).

See comments under 6. *Psilostrophe villosa*.

6. **Psilostrophe villosa** Rydberg ex Britton, Man. Fl. N. States, 1006. 1901 E

Psilostrophe tagetina (Nuttall) Greene var. *cerifera* (A. Nelson) B. L. Turner

Biennials or perennials (rarely flowering first year), 10–45(–60+) cm. **Stems** arachno-villous (gray to gray-green). **Heads** in ± crowded, corymbiform arrays. **Peduncles** (0.5–)1–3(–5+) mm. **Involucres** 4–5(–7) mm. **Rays** (2–)3(–5); laminae 3–4 (–6) mm, spreading in fruit. **Disc florets** 5–8(–12). **Cypselae** usually glabrous, sometimes gland-dotted; **pappi** of 4–6 lanceolate to lance-subulate scales 2.5–3+ mm. $2n = 32$.

Flowering (Mar–)May–Aug(–Oct). Grasslands, limestone soils, gypseous soils, caliche; 300–900 m; Colo., Kans., N.Mex., Okla., Tex.

In trans-Pecos Texas, *Psilostrophe villosa* has been collected within 10 km or so of the Rio Grande; it probably occurs in Mexico. Within 100 km or so of the New Mexico-Texas border, *P. villosa* and *P. tagetina* intergrade; B. L. Turner et al. (1988) treated *P. villosa* as *P. tagetina* var. *cerifera*.

187m.24. Asteraceae Martinov (tribe Heliantheae) subtribe Marshalliinae H. Robinson, Phytologia 41: 42. 1978 [E]

Perennials, (10–)20–80(–120) cm. **Leaves** mostly basal and/or cauline; alternate; petiolate or sessile; blades 1- or 3-nerved, mostly elliptic, linear, oblanceolate, ovate, or spatulate, margins entire, faces usually glabrous, gland-dotted. **Heads** discoid, borne singly or in open, corymbiform arrays. **Calyculi** 0. **Involucres** turbinate to obconic or hemispheric. **Phyllaries** persistent, 10–20 in 2+ series (distinct, narrowly oblong to linear-elliptic or lance-subulate, subequal, herbaceous, basally thickened-keeled, sometimes callous- or scarious-margined). **Receptacles** convex to conic, paleate (paleae mostly linear to filiform, herbaceous). **Ray florets** 0. **Disc florets** (20–)45–90(–160+), bisexual, fertile; corollas lavender, pink, purple, or white, tubes much longer than to equaling funnelform throats, lobes 5, lance-linear to linear-oblong; anther thecae cream to purple; stigmatic papillae in 2 lines. **Cypselae** stoutly obpyramidal, 5(–6)-angled, hairy; **pappi** persistent, of 5(–6) lance-deltate to lance-attenuate, scarious (often hyaline) scales.

Genus 1, species 7 (7 species in the flora): c, e United States.

Marshalliinae was related to epaleate subtribes of Heliantheae by H. Robinson (1981), rather than to Eupatorieae, as suggested by T. F. Stuessy (1977[1978]) and by B. L. Turner and A. M. Powell (1977[1978]). B. G. Baldwin and B. L. Wessa (2000) suggested that *Marshallia* may be sister to Gaillardiinae and results presented by Baldwin et al. (2002) showed *Marshallia* nested in Gaillardiinae, near *Pelucha* S. Watson, *Psathyrotes*, and *Trichoptilium*.

391. MARSHALLIA Schreber, Gen. Pl. 2: 810. 1791, name conserved • Barbara's buttons [For Moses Marshall, 1758–1813, American botanist, nephew of and assistant to Humphrey Marshall] [E]

Linda E. Watson

Perennials, (10–)20–80(–120) cm (fibrous-rooted, usually with caudices, sometimes rhizomatous). **Stems** erect, sometimes branched (striate). **Leaves** basal and/or cauline; alternate; petiolate or sessile; blades (1- or 3-nerved) elliptic, linear, linear-elliptic, linear-oblanceolate, oblanceolate, ovate, or spatulate, margins entire, faces glabrous. **Heads** discoid, borne singly or in open, corymbiform arrays (peduncles relatively long). **Involucres** hemispheric to obconic, 10–25+ mm diam. **Phyllaries** persistent, 12–24 in ± 2 series (lanceolate, linear, oblong, ovate, or rhombic, subequal, herbaceous, midnerves often keeled, proximal margins often scarious). **Receptacles** conic or convex (usually hollow), paleate (paleae linear to spatulate, distally herbaceous). **Ray florets** 0. **Disc florets** (20–)45–90(–160+), bisexual, fertile; corollas pale lavender, pink, or purple, or white, (usually hairy) tubes longer than to nearly equaling throats, lobes 5, lance-linear to linear-oblong (often contorted). **Cypselae** turbinate or clavate (5-angled, ribs 10, often hairy); **pappi** persistent, of 5(–6) scarious to membranous (hyaline, sometimes scaberulous) scales. $x = 9$.

Species 7 (7 in the flora): s, c, se United States.

SELECTED REFERENCES Channell, R. B. 1957. A revision of the genus *Marshallia* (Compositae). Contr. Gray Herb. 81: 41–132. Watson, L. E. and J. R. Estes. 1990. Biosystematic and phenetic analysis of *Marshallia* (Asteraceae). Syst. Bot. 15: 403–414.

1. Leaves lance-elliptic, lance-linear, linear, or oblanceolate (lengths mostly 13–30 times widths), widths mostly 2–13 mm.
    2. Corollas usually pale lavender to purple, rarely white; flowering Aug–Sep . . . 7. *Marshallia graminifolia*
    2. Corollas usually white, sometimes pale lavender; flowering May–Jun.
        3. Heads 4–10, 10–25 mm diam . . . . 5. *Marshallia ramosa*
        3. Heads borne singly, 20–32 mm diam . . . . 6. *Marshallia caespitosa*
1. Leaves elliptic, lanceolate, oblanceolate, ovate, or spatulate (lengths mostly 3–13 widths), widths 8–33 mm.
    4. Leaves mostly cauline, blades ovate . . . . 1. *Marshallia trinervia*
    4. Leaves mostly basal, blades elliptic, lanceolate, oblanceolate, or spatulate.
        5. Phyllary apices obtuse; corollas white . . . . 4. *Marshallia obovata*
        5. Phyllary apices acute; corollas pink.
            6. Heads 1(–2), 26–45 mm diam . . . . 2. *Marshallia grandiflora*
            6. Heads 2–5(–10), 22–37 mm diam . . . . 3. *Marshallia mohrii*

1. Marshallia trinervia (Walter) Trelease, Rep. (Annual) Arkansas Geol. Surv. 1888(4): 196. 1891 [E]

Athanasia trinervia Walter, Fl. Carol., 201. 1788

Plants 40–60(–80) cm (rhizomatous). **Leaves** mostly cauline (not notably reduced distally); proximal petiolate; blades strongly 3-nerved, ovate, 4–10 cm × 11–30 mm. **Heads** 1–5, 20–33 mm diam. **Peduncles** 5–15+ cm. **Phyllaries** 8–12 × 1.5–3 mm, apices acute. **Paleae** ± linear, apices acute. **Corollas** pink, lobes 3.5–7.5 × 0.5–1 mm. **Pappi:** scale margins serrulate. **2***n* = 18.

Flowering May–Jun. Understories, mixed hardwood forests, along streams, slightly disturbed sites; 100–500 m; Ala., Ga., La., Miss., N.C., S.C., Tenn.

2. Marshallia grandiflora Beadle & F. E. Boynton, Biltmore Bot. Stud. 1: 7, plate 1. 1901 [C][E]

Plants 20–90 cm. **Leaves** mostly basal; basal petiolate (distal sessile, reduced); blades 3-nerved, ± elliptic, broadly oblanceolate, or spatulate, 12–25 cm × 14–19 mm. **Heads** 1–2, 26–45 mm diam. **Peduncles** 10–20 cm. **Phyllaries** 8–11 × 2–4 mm, apices acute. **Paleae** ± linear, apices acute. **Corollas** pink, lobes 3–7 × 0.5–0.8 mm. **Pappi:** scale margins serrulate. **2***n* = 18.

Flowering May–Jun. Stream banks, gravel bars; of conservation concern; 500–1500 m; Ky., N.C., Pa., Tenn., W.Va.

Marshallia grandiflora grows in the Appalachian Mountains. It is in the Center for Plant Conservation's National Collection of Endangered Plants.

3. Marshallia mohrii Beadle & F. E. Boynton, Biltmore Bot. Stud. 1: 8, plate 3. 1901 [E]

Plants 30–70 cm (aerial stems branched distally). **Leaves** mostly basal (distally reduced); basal petiolate; blades 3-nerved, narrowly elliptic to spatulate, 6–25 cm × 8–3 mm. **Heads** 2–5(–10), 22–37 mm diam. **Peduncles** 10–20 cm. **Phyllaries** 8–11 × 2–3.5 mm, apices acute. **Paleae** ± linear, apices acute. **Corollas** pink, lobes 5–7 × 1 mm. **Pappi:** scale margins irregular, broken, serrulate. **2***n* = 36.

Flowering May–Jun. Meadows, roadsides; 500–1000 m; Ala., Ga.

Marshallia mohrii grows in the southern Appalachians. It may be an allotetraploid derivative of diploids, probably *M. grandiflora* and either *M. trinervia* or *M. graminifolia* subsp. *tenuifolia*

SELECTED REFERENCE Watson, L. E., W. J. Elisens, and J. R. Estes. 1991. Cytogenetic and electrophoretic evidence for the allotetraploid origin of *Marshallia mohrii* (Asteraceae). Amer. J. Bot. 78: 408–416.

4. Marshallia obovata (Walter) Beadle & F. E. Boynton, Biltmore Bot. Stud. 1: 5. 1901 [E]

Athanasia obovata Walter, Fl. Carol., 201. 1788; *Marshallia obovata* var. *platyphylla* (M. A. Curtis) Beadle & F. E. Boynton; *M. obovata* var. *scaposa* Channell

Plants 10–60 cm. **Leaves** mostly basal (distal slightly reduced); basal petiolate; blades 3-nerved, elliptic, oblanceolate, or spatulate, 5–10 cm × 5–15 mm. **Heads** 1–2, 20–30 mm diam. **Peduncles** 20–40 cm. **Phyllaries** 5–10 × 2–3 mm, apices obtuse. **Paleae** linear-spatulate, apices obtuse. **Corollas** white, lobes 2.5–6 × 0.5–1 mm. **Pappi:** scale margins entire or denticulate. **2***n* = 18.

Flowering May–Jun. Understories, pine forests; 100–1000 m; Ala., Fla., Ga., N.C., S.C., Va.

Marshallia obovata grows on the piedmont and on the Atlantic coastal plain. Scapiform plants with the leafy parts of stems less than ¼ lengths of peduncles are sometimes known as var. *scaposa*. They occur along the inner Atlantic coastal plain.

5. Marshallia ramosa Beadle & F. E. Boynton, Biltmore Bot. Stud. 1: 8, plate 2. 1901 C E

Plants 20–60 cm. **Leaves** basal and cauline; basal petiolate; blades 3-nerved, linear, 6–18 cm × 2–7 mm. **Heads** (2–)4–10(–20), 10–25 mm diam. **Peduncles** 6–12 cm. **Phyllaries** 5–8 × 1.5–2 mm, (margins often winged proximally) apices obtuse to acute, often mucronulate. **Paleae** ± linear, apices obtuse, mucronulate to ± subulate. **Corollas** usually white, sometimes pale lavender, lobes 4–6 × 1 mm. **Pappi:** scales margins entire or denticulate. *2n* = 18.

Flowering May–Jun. Sandstone outcrops, pine savannas, mixed hardwoods and pines; of conservation concern; 100–200 m; Fla., Ga.

Marshallia ramosa is known only from sandstone outcrops in southeastern Georgia and in the Florida panhandle. It is associated with pine savannas in Florida and with mixed hardwoods and pines in Georgia.

6. Marshallia caespitosa Nuttall ex de Candolle in A. P. de Candolle and A. L. P. P. de Candolle, Prodr. 5: 680. 1836 E F

Marshallia caespitosa var. *signata* Beadle & F. E. Boynton

Plants (10–)20–50(–80) cm (aerial stems unbranched, usually clustered). **Leaves** mostly basal; basal petiolate; blades 3-nerved, linear, linear-elliptic, or linear-oblanceolate, 5–15 cm × 2.5–9.5(–15) mm. **Heads** borne singly, 20–32 mm diam. **Peduncles** 10–28 cm. **Phyllaries** 9–12 × 1–3 mm, (margins sometimes narrowly-winged proximally) apices obtuse to acute, often mucronate. **Paleae** linear, apices acute or mucronate. **Corollas** usually white, sometimes pale lavender, lobes 4–6 × 0.5–1 mm. **Pappi:** scale margins irregularly fimbriate. *2n* = 18, 36.

Flowering May–Jun. Limestone outcrops, sandy soils; 100–1000 m; Ark., Kans., La., Mo., Okla., Tex.

Scapiform to subscapiform plants of *Marshallia caespitosa* with relatively few cauline leaves and heads borne singly occur mostly in Oklahoma and southeast Texas and Louisiana. Plants with more or less leafy stems and one to relatively many heads are sometimes recognized as var. *signata*; they grow mostly in central Texas.

7. Marshallia graminifolia (Walter) Small, Bull. Torrey Bot. Club 25: 482. 1898 E F

Athanasia graminifolia Walter, Fl. Carol., 200. 1788; *Marshallia graminifolia* subsp. *tenuifolia* (Rafinesque) L. E. Watson; *M. lacinarioides* Small; *M. williamsonii* Small

Plants 30–80(–120) cm (aerial stems simple or branched near or proximal to middles). **Leaves** basal and cauline; basal petiolate; blades ± 3-nerved, oblanceolate, (2–)4–25(–34) cm × 2–13+ mm (cauline sessile, ascending, linear, sharply differentiated from basal). **Heads** 1–34, 20–30 mm diam. **Peduncles** 5–50 cm. **Phyllaries** 4–9.5 × 0.5–2.5 mm, (margins ± winged proximally) apices subulate. **Paleae** linear-attenuate, apices subulate (often gland-dotted). **Corollas** usually pale lavender to purple, rarely white, lobes 2.5–6 × 0.5–1 mm. **Pappi:** scale margins entire or denticulate. *2n* = 18.

Flowering Aug–Sep. Bogs and pocosins; 100–500 m; Ala., Fla., Ga., La., Miss., N.C., S.C., Tex.

Marshallia graminifolia grows on the Atlantic and Gulf coastal plains from the Carolinas to Texas, often in bogs with pitcher plants, sundews, and orchids in open areas of pine savannas.

Typical *Marshallia graminifolia* has undifferentiated, ascending, and firm leaves. Plants with thinner basal leaves that tend to spread and are more strongly differentiated from the cauline leaves have been called subsp. *tenuifolia*; they grow mostly in Florida and Georgia and along the Gulf coastal plain to Texas.

SELECTED REFERENCE Watson, L. E., W. J. Elisens, and J. R. Estes. 1994. Genetic differentiation in populations of *Marshallia graminifolia* s.lat. (Asteraceae). Biochem. Syst. & Ecol. 22: 577–582.

187n.   ASTERACEAE Martinov tribe EUPATORIEAE Cassini, J. Phys. Chim. Hist. Nat. Arts 88: 202. 1819

Annuals, biennials, perennials, subshrubs, shrubs, or vines [trees]. **Leaves** usually cauline, sometimes basal or basal and cauline; usually opposite, sometimes whorled or alternate; usually petiolate, sometimes sessile; blade margins entire, toothed, lobed, or dissected. **Heads** homogamous (usually discoid [radiant]), usually in corymbiform, paniculiform, racemiform, or spiciform arrays, sometimes borne singly or in glomerules. **Calyculi** 0. **Phyllaries** usually persistent (readily falling), usually in 2–8+ series, distinct, and unequal, sometimes in 1–2 series, distinct, and subequal to equal, usually herbaceous to chartaceous, margins and/or apices sometimes scarious (abaxial faces often striate-nerved). **Receptacles** usually flat to convex, sometimes spheric or conic, usually epaleate, rarely paleate (paleae readily falling). **Ray florets** 0. **Disc florets** bisexual, fertile; corollas white, ochroleucous, or pink to purplish, not yellow, not 2-lipped (sometimes ± zygomorphic), lobes (4–)5, usually ± deltate to lance-ovate, sometimes lanceolate to lance-linear; anther bases obtuse, rounded, or truncate, not tailed, apical appendages usually ovate to lanceolate, sometimes 0; styles abaxially papillate to hirsutulous (usually distally, sometimes at bases), branches ± linear, adaxially stigmatic in 2 lines from bases to appendages, appendages usually terete to clavate (lengths often 2–5+ times lengths of stigmatic lines), usually papillate. **Cypselae** usually ± monomorphic within heads, usually columnar to fusiform, sometimes prismatic or compressed to flattened, rarely, if ever, beaked, bodies often 10-ribbed or (4–)5-angled, smooth or papillate to rugose between ribs or angles (glabrous or hairy); **pappi** (rarely 0) usually persistent, usually of fine to coarse, barbellulate to plumose bristles, sometimes of scales (scales often aristate) or awns, sometimes of bristles and scales.

Genera 170, species 2400 (27 genera, 159 species in the flora): mostly subtropics, tropics, and warm-temperate New World, also in Old World.

In a survey of Compositae, G. Bentham (1873) noted 35 genera and 750 or so species for Eupatorieae; he treated more than 50% of those species as belonging within one genus, *Eupatorium*. The current view of circumscriptions of most genera within Eupatorieae has stemmed largely from the work of H. Robinson, which was summarized by R. M. King and Robinson (1987), who reported 45 species for *Eupatorium* (i.e., ca. 2 % of the total species in the tribe).

Authors of molecular studies have repeatedly found Eupatorieae to be a coherent clade "nested" within Heliantheae (broad sense) (e.g., R. K. Jansen et al. 1990). Some of those authors have suggested inclusion of Eupatorieae within Heliantheae in the broad sense as a subtribe; others have suggested break-up of Heliantheae into a dozen or so tribes (e.g., J. L. Panero and V. A. Funk 2002).

SELECTED REFERENCES   Gaiser, L. O. Studies in the Kuhniinae (Eupatorieae). II. J. Arnold Arbor. 35: 87–133.   King, R. M. and H. Robinson. 1987. The genera of the Eupatorieae (Asteraceae). Monogr. Syst. Bot. Missouri Bot. Gard. 22.   Schilling, E. E., J. L. Panero, and P. B. Cox. 1999. Chloroplast DNA restriction site data support a narrowed interpretation of *Eupatorium* (Asteraceae). Pl. Syst. Evol. 219: 209–223.   Schmidt, G. J. and E. E. Schilling. 2000. Phylogeny and biogeography of *Eupatorium* (Asteraceae: Eupatorieae) based on nuclear ITS sequence data. Amer. J. Bot. 87: 716–726.

1. Involucres narrowly cylindric, (1–)2–3 mm diam.; phyllaries 4 or 5(–6) in ± 1–2 series; florets 4 or 5(–6).
    2. Subshrubs or shrubs; phyllaries 5(–6); florets 5(–6). . . . . 397. *Stevia*, p. 483
    2. Vines; phyllaries 4; florets 4 . . . . 417. *Mikania*, p. 545
1. Involucres campanulate, cylindric, ellipsoid, hemispheric, or obconic, (2–)3–7(–25) mm diam.; phyllaries (5–)8–45(–65+) in (1–)2–8+ series; florets (3–)10–125(–200+).

[3. Shifted to left margin.—Ed.]

3. Cypselae 8–10-ribbed.

    4. Pappi of 0–5+, muticous, erose, lacerate, or lanceolate to subulate scales (1–4 mm) plus [5–]9–12+, aristate scales (10–15 mm) . . . . 398. *Carphochaete*, p. 486

    4. Pappi of 10–100+ bristles.

        5. Leaves basal or basal and cauline (cauline mostly sessile).

            6. Heads usually in spiciform or racemiform, rarely corymbiform or thyrsiform, arrays; receptacles epaleate; pappi of 12–40 coarsely barbellate to plumose bristles . . . . 410. *Liatris*, p. 512

            6. Heads in corymbiform to paniculiform arrays; receptacles sometimes (at least partially) paleate; pappi of 35–40 barbellulate to barbellate (subequal) bristles . . . . 411. *Carphephorus*, p. 535

        5. Leaves mostly cauline (at flowering; mostly petiolate, sometimes sessile).

            7. Leaves all or mostly alternate (at flowering) . . . . 412. *Garberia*, p. 538

            7. Leaves all or mostly opposite.

                8. Leaf blades deltate, lance-elliptic, lance-linear, lanceolate, lance-ovate, lance-rhombic, linear, oblong, obovate, ovate, rhombic-ovate, spatulate, or suborbiculate, margins crenate, dentate, entire, laciniate-dentate, lobed, or serrate; style bases enlarged, hairy . . . . 403. *Brickellia*, p. 491

                8. Leaf blades linear (distal sometimes scalelike), margins entire; style bases not enlarged, glabrous . . . . 406. *Asanthus*, p. 509

3. Cypselae (3–)4–5(–8)-ribbed.

    9. Pappi usually 0 or of 2–6(–12), muticous or aristate to subulate scales plus 0–6(–12), setiform scales or bristles, rarely coroniform (*Ageratum*) or of 1–5 ± glandular setae (*Hartwrightia*).

        10. Pappi usually 0, rarely 1–5 ± glandular setae (*Hartwrightia*).

            11. Leaves basal and cauline, mostly alternate; cypselae obpyramidal (gland-dotted) . . . . 413. *Hartwrightia*, p. 540

            11. Leaves cauline, all or mostly opposite; cypselae prismatic (not gland-dotted).

                12. Heads in dense to open, cymiform or corymbiform arrays; phyllaries 30–40; style branches ± linear to clavate (distally dilated) . . . . . 396. *Ageratum* (in part), p. 481

                12. Heads in tight, corymbiform to subcapitate arrays or borne singly; phyllaries 10–30; style branches ± filiform or linear-filiform (little, if at all, distally dilated).

                    13. Leaves sessile; phyllaries not notably nerved; receptacles epaleate; style bases not enlarged . . . . 400. *Shinnersia*, p. 488

                    13. Leaves petiolate or sessile; phyllaries 2- or 3-nerved; receptacles paleate (paleae similar to inner phyllaries); style bases enlarged . . . . 402. *Isocarpha*, p. 490

        10. Pappi usually of 2–6(–12) muticous or aristate to subulate scales plus 0–6 (–12), setiform scales or bristles, rarely coroniform (*Ageratum*).

            14. Phyllaries unequal; receptacles flat to convex (not warty).

                15. Leaves mostly sessile (or nearly so), blades linear; cypselae ± fusiform . . . . 407. *Malperia*, p. 509

                15. Leaves petiolate, blades ovate, deltate, or rhombic to lanceolate; cypselae prismatic . . . . 408. *Pleurocoronis*, p. 510

            14. Phyllaries ± equal; receptacles convex to conic or hemispheric (sometimes warty).

                16. Leaves whorled (4 or 6 per node), blades linear; heads borne singly . . . . 401. *Sclerolepis*, p. 488

                16. Leaves mostly opposite (distal sometimes alternate), blades elliptic, lanceolate, or oblong; heads usually in cymiform to corymbiform arrays, sometimes borne singly.

                    17. Leaves petiolate; involucres 3–6 mm diam.; phyllaries usually 2-nerved; pappi usually of 5–6 aristate scales, rarely coroniform . . . . 396. *Ageratum* (in part), p. 481

                    17. Leaves sessile; involucres 3–4(–5) mm diam.; phyllaries obscurely 3–4-nerved; pappi of 2–6 setiform scales . . . . 399. *Trichocoronis*, p. 487

[9. Shifted to left margin.—Ed.]
9. Pappi of (5–)10–80+ barbellulate, barbellate, or plumose bristles or setiform scales.
  18. Involucres cylindric (3–4+ mm diam); pappus bristles plumose (basally coherent or connate, falling together or in groups) . . . . 409. *Carminatia*, p. 511
  18. Involucres usually obconic to hemispheric, sometimes campanulate, cylindric, or ellipsoid (2–7 mm diam.); pappus bristles smooth to barbellulate or barbellate (not plumose).
    19. Phyllaries ± equal.
      20. Receptacles conic . . . . 394. *Conoclinium*, p. 478
      20. Receptacles flat or convex.
        21. Phyllaries 2- or 3-nerved, or not notably nerved, or pinnately nerved; style bases usually puberulent (glabrous in *Eupatorium capillifolium*); cypselae usually gland-dotted . . . . 392. *Eupatorium* (in part), p. 462
        21. Phyllaries 3-nerved, or 0- or 2-nerved; style bases glabrous; cypselae sometimes gland-dotted.
          22. Involucres 2–3 mm diam.; phyllaries 7–16 in 1–2 series; florets 3–13 . . . . 415. *Koanophyllon* (in part), p. 542
          22. Involucres 3–6 mm diam.; phyllaries ca. 30 in 2–3 series; florets 10–60 . . . . 418. *Ageratina*, p. 547
    19. Phyllaries unequal (outer shorter).
      23. Style bases usually puberulent (glabrous in *Eupatorium capillifolium*); cypselae usually glabrous and gland-dotted, sometimes scabrellous on ribs.
        24. Leaves mostly opposite (sometimes whorled, distal sometimes alternate) . . . . 392. *Eupatorium* (in part), p. 462
        24. Leaves mostly whorled (3–7 per node), rarely opposite . . . . 393. *Eutrochium*, p. 474
      23. Style bases usually glabrous (hirsute in *Flyriella*); cypselae glabrous or hirsute, hirtellous, hispidulous, hispidulo-strigose, puberulent, or scabrellous (sometimes gland-dotted).
        25. Annuals or perennials; involucres 2–5+ mm diam.; florets 10–30.
          26. Perennials, 20–60 cm (viscid); corollas white to ochroleucous, throats ± cylindric (± contracted distally, lengths 4–6 times diams.) . . . . . . 405. *Flyriella*, p. 507
          26. Annuals or perennials, 30–120+ cm (not viscid, stems usually puberulent, hairs curled); corollas bluish, pinkish, purplish, or white, throats funnelform (not contracted distally, lengths 2.5–4 times diams.) . . . . 414. *Fleischmannia*, p. 540
        25. Perennials, subshrubs, or shrubs; involucres (2–)4–7 mm diam.; florets (3–)25–50.
          27. Phyllaries usually readily falling, 18–65+ in 4–6+ series, 3–5-nerved; cypselae (3–)5-ribbed, scabrellous, usually gland-dotted . . . . 416. *Chromolaena*, p. 544
          27. Phyllaries usually persistent, 7–35 in (1–)2–4 series, 2- or 4-nerved, 3-nerved, or obscurely nerved; cypselae 5(–7)-ribbed, hispidulous, hispidulo-strigose, puberulent, or sparsely scabrellous (sometimes gland-dotted).
            28. Phyllaries 2- or 4-nerved; corollas white to yellowish white; pappi readily falling or fragile . . . . 404. *Brickelliastrum*, p. 507
            28. Phyllaries 3-nerved or obscurely nerved; corollas usually blue, lavender, or pinkish, sometimes white; pappi persistent.
              29. Involucres 5–7 mm diam.; phyllaries 30–35; florets 30–50 . . . . 395. *Tamaulipa*, p. 480
              29. Involucres 2–3 mm diam.; phyllaries 7–16; florets 3–13 . . . . 415. *Koanophyllon* (in part), p. 542

392. **EUPATORIUM** Linnaeus, Sp. Pl. 2: 836. 1753; Gen. Pl. ed. 5, 363. 1754 • [For Mithridates Eupator, King of Pontus, 132–63 B.C.]

Kunsiri Chaw Siripun

Edward E. Schilling

Perennials, 30–200 cm. **Stems** erect, usually not branched proximal to arrays of heads (from caudices or rhizomes). **Leaves** mostly cauline; usually opposite (rarely whorled, distal sometimes alternate); petiolate or sessile; blades usually 3-nerved from or distal to bases, or pinnately nerved, mostly deltate or ovate to lanceolate or linear (and intermediate shapes, sometimes elliptic, oblong, rhombic, or suborbiculate, sometimes pinnatifid, 1–2-pinnately, ternately, or palmately lobed), ultimate margins entire or toothed, faces glabrous or puberulent, pubescent, scabrous, or setulose, usually gland-dotted. **Heads** discoid, in corymbiform or diffuse to dense, paniculiform arrays. **Involucres** obconic to ellipsoid, 1–3(–5+) mm diam. **Phyllaries** persistent, 7–15+ in 2–3(–4+) series, (usually green) 2–3-nerved, or not notably nerved, or pinnately nerved, elliptic, lanceolate, oblong, or obovate, usually unequal, sometimes ± equal (margins scarious, hyaline, apices rounded to acute or acuminate sometimes mucronate, faces usually puberulent or villous, usually gland-dotted, rarely glabrous). **Receptacles** flat or convex, epaleate. **Florets** (3–)5(–15+); corollas usually white, rarely pinkish, throats funnelform to campanulate, lobes 5, triangular; styles: bases sometimes enlarged, usually puberulent (glabrous in *E. capillifolium*), branches mostly filiform. **Cypselae** (brownish to black) prismatic, 5-ribbed, usually glabrous, usually gland-dotted; **pappi** persistent, of 20–50 (whitish) barbellulate bristles in 1 series. *x* = 10.

Species 41+ (24 species, including 2 hybrids, in the flora): e North America, Europe, e Asia.

Eupatorium is treated here in a restricted circumscription, following R. M. King and H. Robinson (1987) in excluding genera that traditionally have been included in a broad *Eupatorium* (e.g., *Ageratina, Chromolaena, Critonia, Conoclinium, Fleischmannia, Koanophyllon, Tamaulipa*); *Eutrochium* (*Eupatorium* sect. *Verticillatum*) is also excluded here.

Species identification within *Eupatorium* is sometimes complicated; polyploidy and apomixis have contributed to the complications. Some species include both sexual diploid and apomictic polyploid plants or populations. V. I. Sullivan (1972) made important contributions to understanding *Eupatorium* in North America by showing that some fairly distinct, sexual diploid species may include apomictic polyploid plants or populations that do not differ greatly from the diploids. Other apomictic polyploids appear to be intermediate morphologically between pairs of diploid or diploid/polyploid species and were proposed by Sullivan to have originated from interspecific hybridization. Distinction and level of recognition of hybrid apomictic taxa have a large arbitrary component, in part because some apomicts appear to be ephemeral and others may be relatively stable and in part because differences in the relative genomic contributions of the progenitors through dosage effects or backcrossing may affect whether an apomict is morphologically distinctive or part of a continuous series of variation.

SELECTED REFERENCE Sullivan, V. I. 1972. Investigations of the Breeding Systems, Formation of Auto- and Alloploids and the Reticulate Pattern of Hybridization in North American *Eupatorium* (Compositae). Ph.D. dissertation. Florida State University.

1. Leaves (at least the principal) pinnatifid, or pinnately or ternately lobed, or palmately 3(–5)-lobed.
    2. Leaves (at least larger proximal) palmately 3(–5)-lobed (lobes relatively broad, margins serrate; corollas usually pinkish) .... 4. *Eupatorium cannabinum*
    2. Leaves pinnatifid or pinnately or ternately lobed (lobes relatively narrow; corollas usually white).

3. Leaves (mostly ternately or 1–2-pinnately lobed): blades or lobes linear; heads in subcorymbiform to subpaniculiform arrays; florets 7–9 . . . . 19. *Eupatorium ×pinnatifidum*
3. Leaves (usually pinnately or ternately lobed or pinnatifid): blades or lobes linear; heads in paniculiform arrays; florets usually 5.
    4. Stems glabrous (gland-dotted, branches supporting heads recurved and secund) . . . . 11. *Eupatorium leptophyllum*
    4. Stems puberulent (branches supporting heads not recurved or secund).
        5. Leaf blades or lobes 0.2–0.5(–1) mm wide (margins strongly revolute); phyllaries usually glabrate or glabrous, usually not gland-dotted . . . . 5. *Eupatorium capillifolium*
        5. Leaf blades or lobes 0.5–2.5(–4) mm wide; phyllaries usually puberulent (mostly on midveins), usually gland-dotted . . . . 6. *Eupatorium compositifolium*
1. Leaves simple (not lobed, margins crenate, entire, laciniate, serrate, or serrulate).
    6. Leaves petiolate (petioles 10–30 mm).
        7. Leaf blades deltate to rhombic (held vertically; stems green); florets 5 . . . 14. *Eupatorium mikanioides*
        7. Leaf blades lanceolate (held horizontally; stems sometimes reddish to purplish); florets 9–15 . . . . 23. *Eupatorium serotinum*
    6. Leaves sessile or subsessile (petioles 0 or 1–10 mm).
        8. Florets 7–11 (leaf bases connate-perfoliate) . . . . 16. *Eupatorium perfoliatum*
        8. Florets (4–)5–8 or 9–14 (leaf bases not connate-perfoliate).
            9. Florets usually 9–14 . . . . 20. *Eupatorium resinosum*
            9. Florets (4–)5–8.
                10. Phyllary apices (usually white) acuminate to attenuate.
                    11. Leaf bases (especially larger leaves) clasping to narrowly connate-perfoliate (leaf faces rugose); florets 5–8 . . . . 7. *Eupatorium ×cordigerum* (in part)
                    11. Leaf bases attenuate, cuneate, or rounded (leaf faces relatively smooth); florets (4–)5.
                        12. Leaves 4–10(–15) mm wide . . . . 12. *Eupatorium leucolepis*
                        12. Leaves mostly 10–45 mm wide.
                            13. Leaves little, if at all, gland-dotted; phyllaries glabrous, not gland-dotted (the larger 8–10 mm) . . . . 17. *Eupatorium petaloideum*
                            13. Leaves usually gland-dotted; phyllaries puberulent to villous (at least toward bases and on midveins), gland-dotted (the larger 5–10 mm).
                                14. Leaf blades elliptic to oblanceolate (lengths mostly 3–4 times widths), bases narrowly cuneate (sometimes oblique); phyllaries linear (larger 6–9 mm) . . . . 1. *Eupatorium album*
                                14. Leaf blades elliptic, lanceolate, or lance-ovate (lengths mostly 1–2 times widths), bases rounded to rounded-cuneate; phyllaries oblong to lance-oblong (larger 5–7 mm) . . . . 18. *Eupatorium pilosum* (in part)
                10. Phyllary apices (usually with narrow white margins, sometimes pigmented) acute, obtuse, or rounded.
                    15. Stems glabrous or glabrate proximally (and leaves pinnately nerved, 70–150 mm); leaves: bases truncate to somewhat rounded, faces glabrate (scattered, fine hairs) . . . . 24. *Eupatorium sessilifolium*
                    15. Stems usually pilose, puberulent, or pubescent proximally or throughout (if glabrous, leaves 3-nerved from bases, 30–50 × 5–13 mm, bases narrowly cuneate); leaves (3-nerved from or distal to bases or pinnately nerved): bases usually narrowly to broadly cuneate, rounded, subcordate, or truncate, sometimes clasping to perfoliate, faces glabrous, puberulent, scabrous, or villous.

[16. Shifted to left margin. —Ed.]

16. Leaf blades deltate, elliptic, lanceolate, lance-ovate, ovate, or suborbiculate (usually broadest proximal to middles), bases usually broadly cuneate, rounded, subcordate, or truncate, sometimes clasping to perfoliate.

 17. Leaf bases (especially larger leaves) clasping to narrowly connate-perfoliate (faces conspicuously rugose); florets 5–8 . . . . 7. *Eupatorium* ×*cordigerum* (in part)

 17. Leaf bases broadly cuneate, rounded, subcordate, or truncate (faces smooth to somewhat rugose); florets (4–)5.

  18. Leaves pinnately nerved, blades usually 50–100 mm . . . . 8. *Eupatorium godfreyanum*

  18. Leaves 3-nerved, blades 15–50(–90) mm.

   19. Leaf blades elliptic, lanceolate, or lance-ovate (lengths mostly 2–2.5 times widths, usually with purple borders, sometimes visible only in live plants; distal leaves sometimes alternate). . . . . 18. *Eupatorium pilosum* (in part)

   19. Leaf blades usually deltate to suborbiculate, sometimes ovate (lengths mostly 1–2 times widths; distal leaves sometimes alternate). . . . . 21. *Eupatorium rotundifolium*

16. Leaf blades elliptic, lance-elliptic, lanceolate, lance-oblong, linear, oblanceolate, or oblong (usually broadest near or distal to middles), bases cuneate.

 20. Leaves strongly to obscurely 3-nerved from bases.

  21. Stems usually glabrous or glabrate (at least proximally), sometimes pilose; leaves 30–50 mm, faces glabrous abaxially . . . . 10. *Eupatorium lancifolium*

  21. Stems pubescent or pilose proximally or throughout; leaves 20–60(–120, the larger usually 50+) mm, faces puberulent, scabrous, or villous abaxially.

   22. Leaves: margins entire proximally, serrate distally (teeth sharp, antrorse; midribs and 2 major lateral veins prominent). . . . . 2. *Eupatorium altissimum*

   22. Leaves: margins usually laciniate-serrate (teeth ± divergent; 2 major lateral veins usually obscure) . . . . 9. *Eupatorium hyssopifolium* (in part)

 20. Leaves 3-nerved distal to bases, or pinnately nerved, or relatively narrow and obscurely 3-nerved from bases.

  23. Leaves usually in whorls (3s or 4s), sometimes opposite, blades lance-linear to linear (lengths 6–40 times widths). 9. *Eupatorium hyssopifolium* (in part)

  23. Leaves usually opposite (sometimes whorled in *E. semiserratum*), blades elliptic, lance-elliptic, lance-oblong, oblanceolate, or oblong (lengths 2.5–7 times widths).

   24. Leaves (spreading): blades adaxially finely puberulent or villous (stems from short caudices or rhizomes).

    25. Stems densely branched distally; leaf blades (30–)50–70 mm, margins usually serrate; phyllaries elliptic (larger 2.5–3 mm); corollas 2.5–3 mm . . . . 22. *Eupatorium semiserratum*

    25. Stems usually branched at or near bases; leaf blades mostly 20–35(–45) mm, margins weakly serrate to subentire; phyllaries lanceolate (larger usually 4–5 mm); corollas 3–3.5 mm . . . . 13. *Eupatorium linearifolium*

   24. Leaves (ascending, recurved, or spreading): blades adaxially glabrous or glabrate (stems from tuberous-thickened rhizomes, not usually apparent until flowering is well under way).

    26. Leaves (usually spreading or ascending): blades 15–50 × (5–)10–20 mm . . . . 3. *Eupatorium anomalum*

    26. Leaves (usually recurved): blades 20–80 × 5–10(–20) mm . . . . 15. *Eupatorium mohrii*

1. Eupatorium album Linnaeus, Mant. Pl., 111. 1767; Syst. Nat. ed. 12, 2: 536. 1767 • White thoroughwort [E]

Perennials, 40–100+ cm. **Stems** (from short caudices or stout rhizomes) single, sparsely branched distally, pubescent throughout. **Leaves** usually opposite (distal sometimes alternate); sessile; blades ± 3-nerved or pinnately nerved, elliptic to oblanceolate, 30–120 × 10–40 mm (lengths mostly 3–4 times widths), bases narrowly cuneate (sometimes oblique), margins subentire, serrate, or serrulate, apices rounded to acute, faces setulose (denser on midribs and veinlets) to glabrate, usually gland-dotted. **Heads** in corymbiform arrays. **Phyllaries** 9–15 in 2–4 series, linear, 1.5–9 × 0.6–1 mm, apices acuminate to attenuate, sometimes mucronate, abaxial faces pubescent throughout, gland-dotted. **Florets** (4–)5; corollas 4–4.5 mm. **Cypselae** 2.5–3.5 mm; **pappi** of 40–50 bristles 3.5–4.5 mm.

Varieties 3 (3 in the flora): e North America.

Eupatorium album is represented by widespread diploid populations and by polyploid, apomictic populations, segregated as varieties, that may have arisen by hybridization with other species. Molecular data suggest that var. *vaseyi* arose via hybridization between var. *album* and *E. sessilifolium.* They also suggest that *E. album* var. *subvenosum* arose from hybridization between *E. album* var. *album* and *E. serotinum.*

1. Leaf blades: margins coarsely serrate, apices obtuse to rounded, faces pubescent. . . . .
. . . . 1a. *Eupatorium album* var. *album*
1. Leaf blades: margins subentire, serrulate, or serrate, apices acute, faces pubescent to glabrate.
  2. Leaf blades 3-nerved from bases, 40–70 × 10–20 mm, margins subentire to serrulate. . . .
. . . . 1b. *Eupatorium album* var. *subvenosum*
  2. Leaf blades 3-nerved distal to bases, 50–110 × 20–40 mm, margins evenly serrate. . . . . .
. . . . 1c. *Eupatorium album* var. *vaseyi*

1a. Eupatorium album Linnaeus var. **album** [E]

Eupatorium album var. *glandulosum* (Michaux) de Candolle

Leaf blades 3-nerved distal to bases, 5–9(–11) × 1–2.5(–4) mm, margins coarsely serrate, apices obtuse to rounded, faces pubescent. $2n = 20$.

Flowering Jul–Sep. Dry, open or wooded, disturbed sites, sandy pinelands; 10–300 m;

Ala., Conn., Del., Fla., Ga., La., Md., Miss., N.J., N.Y., N.C., Ohio, Pa., S.C., Tenn., Va., W.Va.

1b. Eupatorium album Linnaeus var. **subvenosum** A. Gray in A. Gray et al., Syn. Fl. N. Amer. 1(2): 98. 1884 [E]

Leaf blades 3-nerved from bases, 40–70 × 10–20 mm, margins subentire to serrulate, apices acute, faces pubescent to glabrate.

Flowering Jul–Sep. Dry, open pine barrens and sandy shores; 0–10+ m; Del., D.C., N.J., N.Y.

1c. Eupatorium album Linnaeus var. **vaseyi** (Porter) Cronquist, Brittonia 29: 220. 1977 [E]

Eupatorium vaseyi Porter, Bull. Torrey Bot. Club 19: 128. 1892; *E. album* var. *monardifolium* Fernald & Griscom; *E. fernaldii* R. K. Godfrey

Leaf blades 3-nerved distal to bases, 50–110 × 20–40 mm, margins evenly serrate, apices acute, faces pubescent to glabrate. $2n = 30$.

Flowering Jul–Sep. Dry to moist upland woods; 20–300 m; Ala., Del., D.C., Ga., Ky., Md., N.C., Pa., S.C., Tenn., Va., W.Va.

2. Eupatorium altissimum Linnaeus, Sp. Pl. 2: 837. 1753 • Tall thoroughwort [E]

Perennials, 50–150+ cm. **Stems** (from short caudices or stout rhizomes) single, sparsely branched distally, pubescent throughout (nodes sometimes with galls). **Leaves** usually opposite (nodes often appearing leafy, lateral buds producing 2+ pairs of leaves); sessile or subsessile; blades strongly 3-nerved from bases, lance-elliptic to oblanceolate, 50–120 × 5–20 mm, bases ± cuneate, margins entire proximally, serrate distally, apices acuminate, faces puberulent or villous, gland-dotted. **Heads** in corymbiform arrays. **Phyllaries** 8–10 in 2–3 series, oblong, 1–4 × 0.5–1.5 mm, (bases tapered) apices rounded to acute (not mucronate), abaxial faces pubescent throughout. **Florets** 5; corollas 3–3.5 mm. **Cypselae** 2–3 mm; **pappi** of 30–40 bristles 3.5–4 mm. $2n = 20, 30, 40$.

Flowering Jul–Sep. Clearings, open woods, thickets; 20–400 m; Ont.; Ala., Ark., Conn., Fla., Ga., Ill., Ind.,

Iowa, Kans., Ky., La., Md., Mass., Mich., Minn., Miss., Mo., Nebr., N.J., N.Y., N.C., Ohio, Okla., Pa., S.C., Tenn., Tex., Va., W.Va., Wis.

Eupatorium altissimum occurs in sexual diploid populations in the Ozark region of Missouri and Arkansas, and as apomictic polyploids elsewhere throughout its range. It occurs almost exclusively on limestone soils, where it is often accompanied by (and sometimes misidentified as) *Brickellia eupatorioides*, which has 10-ribbed cypselae and plumose pappus bristles. *Eupatorium altissimum* hybridizes with *E. serotinum*.

3. Eupatorium anomalum Nash, Bull. Torrey Bot. Club 23: 106. 1896 • Florida thoroughwort C E

Perennials, 80–150+ cm. Stems (from tuberous rhizomes) single, densely branched distally (shoots often develop from lateral buds), puberulent throughout (denser distally). Leaves usually opposite (distal sometimes alternate); sessile or subsessile; blades 3-nerved distal to bases, elliptic to oblong, 15–50 × (5–)10–20 mm, bases cuneate, margins entire or serrate (unevenly toothed), apices rounded to acute, faces puberulent (abaxial) or glabrate (adaxial), gland-dotted (both). Heads in corymbiform arrays. Phyllaries 8–12 in 2–3 series, oblong to lanceolate, 2.5–5 × 0.5–0.7 mm, apices rounded to acute, abaxial faces puberulent throughout. Florets 5; corollas 3–3.5 mm. Cypselae 1.8–3 mm; pappi of 20–35 bristles 3.5–4.5 mm. 2*n* = 20, 30, 40.

Flowering Jul–Aug. Wet, low ground, flatwoods; of conservation concern; 10–100 m; Ala., Fla., Ga., N.C., S.C.

Eupatorium anomalum has been proposed to be intermediate in morphology between *E. rotundifolium* and *E. mohrii* and, possibly, a hybrid derivative of that pairing. Molecular data suggest that its derivation is from hybridization between *E. serotinum* and *E. mohrii*.

4. Eupatorium cannabinum Linnaeus, Sp. Pl. 2: 838. 1753 • Hemp agrimony I

Perennials, 30–150 cm. Stems (from short rhizomes) single, branched distally, puberulent. Leaves opposite; subsessile or petiolate; blades palmately 3(–5)-lobed (at least larger proximal, lobes relatively broad), blades (or lobes) lanceolate to lance-ovate, 50–100 × 20–40 mm, margins serrate, apices rounded to acute, faces puberulent, gland-dotted. Heads in dense, corymbiform arrays. Phyllaries 8–10 in 2–3 series, oblong, 4.5–6 × 1.5–2 mm, apices rounded, abaxial faces puberulent, gland-dotted. Florets (4–)5(–6); corollas (usually pinkish) 2–2.5 mm. Cypselae 2–3 mm; pappi of 20–30 bristles 3–5 mm. 2*n* = 20.

Flowering Jul–Sep. Disturbed sites; 10–100 m; introduced; B.C.; N.Y., Pa., Va.; Europe.

Eupatorium cannabinum is a garden escape; it is native to Europe and may be established in British Columbia; it is only casually adventive elsewhere in the flora area.

5. Eupatorium capillifolium (Lamarck) Small, Mem. Torrey Bot. Club 5: 311. 1894 • Dogfennel E

Artemisia capillifolia Lamarck in J. Lamarck et al., Encycl. 1: 267. 1783

Perennials, 50–200 cm. Stems (from short caudices) multiple, branched distally, puberulent throughout. Leaves opposite (proximal) or alternate (nodes often appearing leafy because of development of leaves on lateral buds without axis elongation); sessile; blades (often ternately lobed) or lobes linear, 5–100 × 0.2–0.5(–1) mm, bases ± cuneate, margins entire (strongly revolute), apices rounded to acute, faces glabrate, gland-dotted. Heads in dense, paniculiform arrays. Phyllaries 8–10 in 2–3 series, oblong, 0.5–2.5 × 0.2–0.5 mm, apices acuminate and mucronate, abaxial faces glabrous or glabrate, not or little, if at all, gland-dotted. Florets 5; corollas 2–2.5 mm. Cypselae 1–1.7 mm; pappi of 20–30 bristles 2–2.5 mm. 2*n* = 20.

Flowering Aug–Oct. Old fields, open sites, roadsides, flatwoods; 10–300 m; Ala., Ark., Conn., Del., Fla., Ga., Ky., La., Md., Mass., Miss., Mo., N.J., N.C., Okla., Pa., S.C., Tenn., Tex., Va., W.Va.

6. Eupatorium compositifolium Walter, Fl. Carol., 199. 1788 • Yankeeweed E

Perennials, 50–200 cm. Stems (from short caudices) single, branched distally, puberulent throughout (lateral buds often develop fertile and infertile branches). Leaves opposite (proximal) or alternate (lateral buds dormant or producing 1 pair of leaves); sessile; blades (often pinnately or ternately lobed) or lobes pinnately nerved, linear, 20–80 × 0.5–2.5(–4) mm, bases cuneate, margins entire, apices acuminate, faces puberulent, gland-dotted. Heads in paniculiform arrays. Phyllaries 7–10 in 2–3 series, (sometimes purple) elliptic to oblong, 1–3 × 0.5–

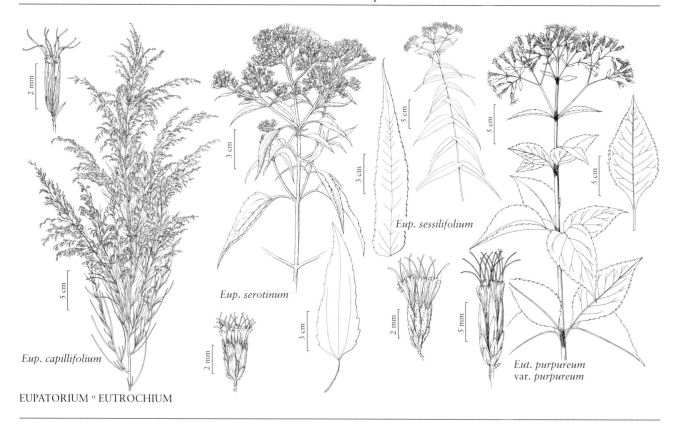

Eup. capillifolium

Eup. serotinum

Eup. sessilifolium

Eut. purpureum
var. purpureum

EUPATORIUM ◦ EUTROCHIUM

0.8 mm, (margins hyaline) apices acuminate and mucronate, abaxial faces puberulent (mostly on midveins), usually gland-dotted. **Florets** 5; corollas (sometimes with purple throats) 2.5–3 mm. **Cypselae** 1–1.7 mm; **pappi** of 20–30 bristles 3–3.5 mm. $2n = 20$.

Flowering Aug–Oct. Open or slightly shaded, wet sites, sand dunes, disturbed areas, roadsides, flatwoods; 10–200+ m; Ala., Ark., Fla., Ga., Ky., La., Miss., N.C., Okla., S.C., Tenn., Tex., Va.

7. Eupatorium ×cordigerum (Fernald) Fernald, Rhodora 47: 192. 1945 (as species) E

Eupatorium rotundifolium Linnaeus var. *cordigerum* Fernald, Rhodora 45: 477. 1943

Perennials, 40–100+ cm. **Stems** (from short rhizomes) single, branched distally, densely pilose. **Leaves** usually opposite (distal sometimes alternate); sessile; blades ± 3-nerved or pinnately nerved, ovate, 3–9.8 × 1.6–5.7 cm, bases clasping or narrowly perfoliate, margins serrate, apices acute, faces (rugose) sparsely pilose (adaxial) or densely pilose (abaxial) and densely gland-dotted. **Heads** in corymbiform arrays. **Phyllaries** 7–10 in 2–3 series, 4–12 × 1–4 mm, apices (usually white) acuminate to acute, abaxial faces pilose, gland-dotted. **Florets** 5–8; corollas

2.5–3 mm. **Cypselae** 2–3 mm; **pappi** of 30–40 bristles 3.5–4.5 mm. $2n = 30, 40$.

Flowering Jul–Aug. Margins of woods, moist areas; 10–100+ m; Ark., Miss., N.J., N.C., S.C., Va.

Eupatorium ×cordigerum is an apomictic, polyploid hybrid of *E. rotundifolium* and *E. perfoliatum* that occurs in disturbed areas where those two species occur together. It is not uncommonly encountered; it is unclear how long it persists.

8. Eupatorium godfreyanum Cronquist, Brittonia 37: 238, fig. 1. 1985 • Godfrey's thoroughwort E

Perennials, 60–100+ cm. **Stems** (from short caudices) single or multiple, sparsely branched distally, puberulent throughout. **Leaves** usually opposite (distal sometimes alternate, spreading or horizontal); sessile; blades pinnately nerved, elliptic or lance-ovate to narrowly ovate, 50–100 × 15–40 mm, bases rounded to cuneate (not connate-perfoliate), margins serrate, apices acute, faces puberulent or villous, gland-dotted. **Heads** in corymbiform arrays. **Phyllaries** 7–10 in 2–3 series, lanceolate (tapering at tips), 2–6 × 1–1.5 mm, apices acute, not mucronate, abaxial faces puberulent, gland-dotted. **Florets** (4–)5; corollas 3–3.5 mm. **Cypselae** 2–3 mm; **pappi** of 20–50 bristles 3.5–4.5 mm. $2n = 30, 40$.

Flowering Jul–Sep. Dry, open, disturbed sites, edges of deciduous woods; 20–300+ m; Ky., Md., N.J., N.C., Ohio, Va., W.Va.

Eupatorium godfreyanum is an apomictic polyploid derivative that includes genomes from *E. rotundifolium* and *E. sessilifolium*. Although it is relatively narrow in distribution, it is known to occur in localities where both progenitor species are absent and it seems to be persistent where it occurs. *Eupatorium vaseyi* Porter has been misapplied to *E. godfreyanum*.

9. Eupatorium hyssopifolium Linnaeus, Sp. Pl. 2: 836. 1753 (as hyssopifolia) • Hyssopleaf thoroughwort E

Perennials, 50–100+ cm. **Stems** (from short caudices or rhizomes) single, sparsely branched distally, pubescent throughout. **Leaves** usually opposite or whorled (distal sometimes alternate, spreading or horizontal); simple, sessile; blades 3-nerved from bases (laterals sometimes weak), lance-linear, lance-oblong, or linear, 20–60 × 2–15 mm (lengths mostly 6–40 times widths), bases cuneate, margins entire, laciniate-serrate, or serrate, apices acute, faces scabrous (at least abaxial). **Heads** in corymbiform arrays. **Phyllaries** 8–10 in 2–3 series, elliptic to oblong, 1.5–5 × 1–1.5 mm, apices obtuse to acute (not mucronate), abaxial faces pubescent throughout (more densely distally). **Florets** 5; corollas 3–3.5 mm. **Cypselae** 2–3 mm; **pappi** of 20–30 bristles 3.5–4 mm.

Varieties 2 (2 in the flora): e North America.

Eupatorium hyssopifolium includes both diploid and polyploid cytotypes and presents a complex situation taxonomically. The diploids are placed here in var. *hyssopifolium*, which also includes polyploids and is characterized by relatively narrow leaves. Variety *laciniatum* has broader leaves and molecular data suggest that it arose through hybridization between var. *hyssopifolium* and *E. serotinum*; the name *E. torreyanum* has been applied to plants of similar morphology, but these are hybrid derivatives of *E. serotinum* and *E. mohrii*. It has also been suggested that var. *hyssopifolium* has hybridized with *E. album* var. *album* to form *E. saltuense*, and with diploids of *E. linearifolium* to form polyploid populations included here under *E. linearifolium*. *Eupatorium hyssopifolium* is characterized by a tendency for the leaves to be whorled, in 3s and 4s, and for axillary buds to produce leaves without expansion of shoots; the combination of whorled leaves and suppressed axillary shoots gives the nodes a somewhat tufted or verticillate appearance.

1. Leaves mostly linear, 2–5 mm wide, margins usually entire or obscurely serrulate. . . . .
   . . . . . 9a. *Eupatorium hyssopifolium* var. *hyssopifolium*
1. Leaves mostly lance-linear to lanceolate, 5–15 mm wide, margins usually laciniately toothed. . . . .
   . . . . 9b. *Eupatorium hyssopifolium* var. *laciniatum*

9a. Eupatorium hyssopifolium Linnaeus var. **hyssopifolium** E

Eupatorium hyssopifolium var. *calcaratum* Fernald & B. G. Schubert; *E. lecheifolium* Greene

Leaves mostly linear, 2–5 mm wide, margins usually entire or obscurely serrulate. $2n = 20, 30$.

Flowering Aug–Oct. Dry, open, disturbed sites, along roadsides; 10–200+ m; Ala., Ark., Conn., Del., Fla., Ga., Ill., Ky., La., Md., Mass., Miss., Mo., N.J., N.Y., N.C., Ohio, Pa., R.I., S.C., Tenn., Tex., Va., W.Va., Wis.

9b. Eupatorium hyssopifolium Linnaeus var. **laciniatum** A. Gray in A. Gray et al., Syn. Fl. N. Amer. 1(2): 98. 1884 E

Eupatorium torreyanum Short & R. Peter

Leaves mostly lance-linear to lanceolate, 5–15 mm wide, margins usually laciniately toothed. $2n = 30, 40$.

Flowering Aug–Oct. Dry, open fields; 10–200+ m; Ala., Ark., Conn., Del., Fla., Ga., Ky., La., Md., Miss., N.J., N.Y., N.C., Ohio, Pa., R.I., S.C., Tenn., Va., W.Va.

10. Eupatorium lancifolium (Torrey & A. Gray) Small, Fl. S.E. U.S., 1167. 1903 • Lanceleaf thoroughwort E

Eupatorium parviflorum Elliott [not Aublet] var. *lancifolium* Torrey & A. Gray, Fl. N. Amer. 2: 85. 1841; *E. semiserratum* de Candolle var. *lancifolium* (Torrey & A. Gray) A. Gray

Perennials, 40–100+ cm. **Stems** (from short rhizomes) single, branched distally, usually glabrous or glabrate, sometimes pilose. **Leaves** usually opposite (distal sometimes alternate); sessile or subsessile (petioles to 5 mm); blades 3-nerved from bases, lanceolate to lance-linear, 30–50 × 5–13 mm, bases narrowly cuneate, margins serrate, apices acute, faces glabrous adaxially, sparsely pilose abaxially, densely gland-dotted. **Heads** in corymbiform arrays. **Phyllaries** 7–10 in 2–3 series, elliptic, 1–4.5 × 0.2–0.8 mm, apices rounded, abaxial faces pilose, gland-dotted. **Florets** 5; corollas 2.5–3 mm. **Cypselae** 2.5–2.8 mm; **pappi** of 30–40 bristles 3.5–4 mm. *2n* = 20.

Flowering Aug–Sep. Dry, rolling terrain, clay soils, shade to sun, shortleaf pine and oak woods; 50–200+ m; Ala., Ark., La., Tex.

Eupatorium lancifolium has been combined with *E. semiserratum* or listed within it as a subspecies; it differs in its habitat as well as in having leaves that are somewhat smaller, typically a dull blue-green (in contrast to yellow-green in *E. semiserratum*), and 3-nerved from bases of blades.

11. **Eupatorium leptophyllum** de Candolle in A. P. de Candolle and A. L. P. P. de Candolle, Prodr. 5: 176. 1836 • False fennel E

Eupatorium capillifolium (Lamarck) Small var. *leptophyllum* (de Candolle) H. E. Ahles

Perennials, 50–200 cm. **Stems** (from short caudices) single, branched distally, glabrous throughout, gland-dotted. **Leaves** opposite (proximal) or alternate (nodes often appearing leafy by development of leaves on lateral buds without axis elongation); petiolate or sessile; blades (usually pinnately or ternately lobed) or lobes 1-nerved or pinnately nerved, linear, 20–100 × 0.2–0.5(–1) mm, bases slightly expanded, margins entire, apices rounded to acute, faces sparsely puberulent to glabrate, gland-dotted. **Heads** in paniculiform arrays (branches supporting heads recurved, secund). **Phyllaries** 6–10 in 1–2 series, lanceolate, 1.5–3 × 0.5–0.7 mm, apices acuminate, mucronate, abaxial faces glabrous or glabrate (not gland-dotted). **Florets** 5; corollas 2–2.5 mm. **Cypselae** 1–1.5 mm; **pappi** of 15–30 bristles 2–2.5 mm. *2n* = 20.

Flowering Aug–Oct. Pond margins, wet, low places, shallow water; 10–50+ m; Ala., Fla., Ga., Miss., N.C., S.C.

12. **Eupatorium leucolepis** (de Candolle) Torrey & A. Gray, Fl. N. Amer. 2: 84. 1841 • Justiceweed, white-bracted thoroughwort E

Eupatorium glaucescens Elliott var. *leucolepis* de Candolle in A. P. de Candolle and A. L. P. P. de Candolle, Prodr. 5: 177. 1836

Perennials, 40–100+ cm. **Stems** (from short caudices or stout rhizomes) single, sparsely branched distally, puberulent throughout (more densely distally and among heads). **Leaves** usually opposite (distal sometimes alternate, ascending to vertical); sessile; blades pinnately nerved, lance-oblong to linear-oblong, 20–60 × 4–10(–15) mm, bases rounded to cuneate (not connate-perfoliate), margins entire or serrate, apices acute, faces villous (abaxial), scabrous (adaxial), gland-dotted. **Heads** in corymbiform arrays. **Phyllaries** 8–10 in 2–3 series, narrowly elliptic, 2.5–8 × 0.8–1.2 mm, acuminate to attenuate, mucronate, abaxial faces puberulent, gland-dotted. **Florets** 5; corollas 3–3.5 mm. **Cypselae** 2–3 mm; **pappi** of 30–40 bristles 4.5–5 mm.

Varieties 2 (2 in the flora): e North America.

Eupatorium leucolepis is distinct morphologically by its acuminate to attenuate phyllaries, linear and usually plicate leaves, and phyllotaxy that is almost always strictly opposite to the arrays, with well-separated nodes. *Eupatorium leucolepis* has recently been shown to include two distinct species, an unnamed diploid that is endemic to Carolina Bay habitats and a series of relatively widespread and mostly polyploid, apomictic populations that include the type. It was recognized after preparation of this treatment that *E. novae-angliae* (*E. leucolepis* var. *novae-angliae*) was derived from hybridization between the unnamed diploid and *E. perfoliatum*, and is here recognized as a distinct species; it is not directly related genetically to *E. leucolepis* and thus is not appropriately classified as a variety of it.

1. Leaf blades strongly folded along midribs, curved, 4–10 mm wide. . . . . 12a. *Eupatorium leucolepis* var. *leucolepis*
1. Leaf blades weakly folded along midribs, not curved, 10–15 mm wide. . . . . 12b. *Eupatorium leucolepis* var. *novae-angliae*

12a. Eupatorium leucolepis (de Candolle) Torrey & A. Gray var. **leucolepis** [E]

Leaf blades strongly folded along midribs, curved, 4–10 mm wide. **2n** = 20, 30, 40.

Flowering Jul–Sep. Pine barrens, wet meadows, margins of ponds, sandy soils; 20–100+ m; Ala., Del., Fla., Ga., La., Md., Miss., N.J., N.Y., N.C., Pa., S.C., Tenn., Tex., Va.

12b. Eupatorium leucolepis (de Candolle) Torrey & A. Gray var. **novae-angliae** Fernald, Rhodora 39: 453, plate 485, figs. 3–5. 1937 [C] [E]

Leaf blades weakly folded along midribs, not curved, 10–15 mm wide.

Flowering Jul–Sep. Ponds; of conservation concern; 20–100+ m; Mass., R.I.

As discussed above, the appropriate name for this entity is *Eupatorium novae-angliae* (Fernald) V. Sullivan ex A. Haines & Sorie, Botanical Notes 11: 2. 2005.

Variety *novae-angliae* is in the Center for Plant Conservation's National Collection of Endangered Plants.

13. Eupatorium linearifolium Walter, Fl. Carol., 199. 1788 [E]

Eupatorium cuneifolium Willdenow; *E. glaucescens* Elliott; *E. tortifolium* Chapman

Perennials, 30–100+ cm. **Stems** (from short caudices) single or multiple, branched at or near bases, pubescent throughout. **Leaves** usually opposite (distal sometimes alternate); sessile or subsessile; blades 3-nerved distal to bases, oblong to lance-oblong, 20–45 × 5–10 mm, bases cuneate, margins entire or serrate (teeth mostly proximal), apices acute, faces finely puberulent, gland-dotted. **Heads** in corymbiform arrays. **Phyllaries** 8–10 in 1–2 series, lanceolate (tapering toward apices), 2–5 × 0.5–1 mm, apices rounded to acute, abaxial faces puberulent, gland-dotted. **Florets** 5; corollas 3–3.5 mm. **Cypselae** 2.5–3 mm; **pappi** of 30–40 bristles 3–5 mm. **2n** = 20, 30, 40.

Flowering Jul–Sep. Dry, sandy soils, pine and oak woods, old fields; 20–100+ m; Ala., Del., Fla., Ga., Miss., N.C., S.C., Tex.

Plants treated here as *Eupatorium linearifolium* were long treated under the name *E. cuneifolium*; the latter name was superfluous when published (K. N. Gandhi and R. D. Thomas 1991). Because there appears to be a continuous range of variation between diploids that were

referred to by V. I. Sullivan (1972) as *E. cuneifolium* and the series of putative hybrids (with *E. hyssopifolium* suggested as the other parent) that she called *E. linearifolium*, these are combined here. The tendency for the plants to branch at or near the bases is distinctive within *Eupatorium*.

14. Eupatorium mikanioides Chapman, Fl. South. U.S., 195. 1860 • Semaphore thoroughwort [E]

Perennials, 50–100+ cm. **Stems** (from short rhizomes, green) single, branched distally, puberulent throughout. **Leaves** opposite (± vertical); petiolate (petioles 10–30 mm); blades ± 3-nerved distal to bases, deltate to rhombic, 25–80 × 20–60 mm, bases broadly cuneate, margins usually serrate (teeth uneven), sometimes crenate, apices acute, faces glabrate to sparsely puberulent, gland-dotted. **Heads** in corymbiform arrays. **Phyllaries** 8–12 in 2–3 series, elliptic to oblong, 1.5–5 × 0.8–1 mm, apices acute to acuminate, abaxial faces puberulent, gland-dotted. **Florets** 5; corollas 3.5–4 mm. **Cypselae** 1.2–1.5 mm; **pappi** of 20–35 bristles 3.5–4.5 mm. **2n** = 20.

Flowering Jul–Sep. Moist or wet, low, often saline places; 0–10+ m; Fla.

15. Eupatorium mohrii Greene, Contr. U.S. Natl. Herb. 6: 762, plate 11. 1901 • Mohr's thoroughwort [E]

Eupatorium recurvans Small

Perennials, 30–100+ cm. **Stems** (from tuberous rhizomes) multiple, densely branched distally, puberulent throughout. **Leaves** usually opposite (distal sometimes alternate); sessile or subsessile; blades 3-nerved distal to bases, oblanceolate, 20–80 × 5–10(–20) mm, bases cuneate, margins serrate proximally, entire distally, apices acute, faces puberulent (abaxial), glabrous or glabrate (adaxial), gland-dotted. **Heads** in corymbiform arrays. **Phyllaries** 7–10 in 1–2 series, oblanceolate, 1–3 × 0.2–0.5 mm, apices rounded, abaxial faces puberulent, gland-dotted. **Florets** 5; corollas 2–4 mm. **Cypselae** 1–2 mm; **pappi** of 20–30 bristles 2.5–3 mm. **2n** = 20, 30, 40.

Flowering Jul–Sep. Moist, low ground, margins of ponds, sandy soils; 10–100+ m; Ala., Fla., Ga., La., Miss., N.C., S.C., Tex., Va.

As treated here, *Eupatorium mohrii* includes both sexual diploid (sometimes segregated as *E. recurvans*) and apomictic polyploid populations that are suggested by molecular data to be autoploids. *Eupatorium saltuense* refers to hybrids shown by molecular data to involve *E. mohrii* and *E. serotinum*. Molecular data also document frequent hybridization with *E. rotundifolium*.

16. **Eupatorium perfoliatum** Linnaeus, Sp. Pl. 2: 838. 1753 • Common boneset, eupatoire perfoliée [E]

Eupatorium chapmanii Small; *E. perfoliatum* var. *colophilum* Fernald & Griscom; *E. perfoliatum* var. *cuneatum* Engelmann

Perennials, 40–100+ cm. **Stems** (from short caudices) single, sparsely branched distally, puberulent throughout. **Leaves** usually opposite (sometimes whorled); sessile; blades pinnately nerved, oblong (tapering toward apices), 50–150+ × 15–40 mm, bases connate-perfoliate, margins serrate, apices acute, faces pilose, gland-dotted (abaxial), glabrate, not gland-dotted (adaxial). **Heads** in corymbiform arrays. **Phyllaries** 7–10 in 1–2 series, oblong, 2–4.5 × 0.6–1 mm, apices (whitish) acute to acuminate, abaxial faces villous or puberulent, gland-dotted. **Florets** 7–11; corollas 2.5–3 mm. **Cypselae** 1.5–2 mm; **pappi** of 20–30 bristles 3–3.5 mm. *2n* = 20.

Flowering Aug–Oct. Moist, low ground, marshes, roadsides, swamps, wet pastures; 10–500+ m; Man., N.B., N.S, Ont., P.E.I., Que.; Ala., Ark., Conn., Del., Fla., Ga., Ill., Ind., Iowa, Kans., Ky., La., Maine, Md., Mass., Mich., Minn., Miss., Mo., Nebr., N.H., N.J., N.Y., N.C., N.Dak., Ohio, Okla., Pa., R.I., S.C., S.Dak., Tenn., Tex., Vt., Va., W.Va., Wis.

Eupatorium perfoliatum is both distinctive, with its perfoliate leaf bases, and extraordinarily widespread. Hybrids between *E. perfoliatum* and other species of *Eupatorium* have been reported. The name *E. resinosum* var. *kentuckiense* Fernald is based on hybrids between *E. perfoliatum* and *E. serotinum* (= *E.* ×*truncatum* Muhlenberg ex Willdenow). *Eupatorium perfoliatum* var. *colophilum* refers to populations characterized by narrower, more leathery leaves found in the extreme northeastern portion of the range.

17. **Eupatorium petaloideum** Britton, Bull. Torrey Bot. Club 24: 492. 1897 (as petalodium) • Showy white thoroughwort [E]

Eupatorium album Linnaeus var. *petaloideum* (Britton) R. K. Godfrey ex D. B. Ward

Perennials, 30–70 cm. **Stems** (from short, stout rhizomes) single, branched distally, puberulent. **Leaves** usually opposite (distal sometimes alternate); sessile; blades pinnately nerved, elliptic to ovate, 20–80 × 10–30 mm, bases cuneate, margins serrate to crenate-serrate, apices rounded to acute, faces sparsely villous (abaxial), glabrate (adaxial), little, if at all, gland-dotted. **Heads** in corymbiform arrays. **Phyllaries** 9–15 in 2–4 series, linear, 4–10 × 0.6–1 mm, apices (white) acuminate to attenuate, strongly mucronate, abaxial faces glabrous, not gland-dotted. **Florets** (4–)5; corollas 3–3.5 mm. **Cypselae** 2.5–3 mm; **pappi** of 40–50 bristles 3.5–4 mm. *2n* = 20.

Flowering Jul–Sep. Upland scrub oak and longleaf pinewoods, fine textured, loamy soils; 20–100+ m; Ala., Fla., Ga., Miss.

Eupatorium petaloideum is commonly combined with *E. album*; it has recently been separated as a distinct variety. V. I. Sullivan (1972) found the two taxa to be distinct chemically and to occupy different habitats. In both, the involucral bracts are white and long-acuminate or mucronate; those of *E. petaloideum* are almost entirely devoid of any pubescence including glands and those of *E. album* have both simple and glandular hairs.

18. **Eupatorium pilosum** Walter, Fl. Carol., 199. 1788 • Rough boneset [E]

Eupatorium rotundifolium Linnaeus var. *saundersii* (Porter ex Britton) Cronquist; *E. verbenifolium* Michaux

Perennials, 30–100+ cm. **Stems** (from short rhizomes) single, sparsely branched distally, densely puberulent to pilose throughout. **Leaves** usually opposite (distal sometimes alternate); simple, sessile or subsessile; blades ± 3-nerved (distal to bases), elliptic, lanceolate, or lance-ovate, 30–90 × 20–45 mm (lengths mostly 2–2.5 times widths), bases rounded to rounded-cuneate, margins unevenly serrate, apices acute to attenuate, faces puberulent, gland-dotted. **Heads** in corymbiform arrays. **Phyllaries** 7–10 in 2–3 series, oblong to lance-oblong, 2–7 × 1–1.5 mm, apices (sometimes whitish) acuminate to acute, mucronate, abaxial faces puberulent, gland-dotted. **Florets** 5; corollas 3.5–4 mm. **Cypselae** 3–4 mm; **pappi** of 30–50 bristles 4–5 mm. *2n* = 20, 30, 40.

Flowering Jul–Sep. Moist, low ground, margins of ponds, sandy soils, savannas, ditches; 20–100+ m; Ala., Conn., Del., Fla., Ga., Ky., La., Md., Mass., Miss., N.J., N.Y., N.C., Pa., R.I., S.C., Tenn., Va., W.Va.

Eupatorium pilosum has been most commonly treated as a variety of *E. rotundifolium*; it is distinct morphologically by its leaves, which are narrower and 3-nerved distal to bases, rather than 3-nerved from bases. Distal leaves of *E. pilosum* tend to differ from the proximal by being alternate with entire margins.

19. **Eupatorium ×pinnatifidum** Elliott, Sketch Bot. S. Carolina 2: 295. 1823 (as species)

Eupatorium eugenei Small; *E. pectinatum* Small

Perennials, 40–100+ cm. **Stems** (from short caudices) single, densely branched distally, puberulent throughout. **Leaves** opposite (proximal) or alternate (nodes often appearing leafy, lateral buds develop leaves without axis elongation); sessile; blades (usually 1–2-pinnately or ternately lobed) or lobes pinnately nerved or 1-nerved, linear, 20–100 × 0.5–1.5 mm, bases narrowly cuneate to attenuate, margins entire or serrate, apices acute, faces puberulent, gland-dotted. **Heads** in subcorymbiform to subpaniculiform arrays. **Phyllaries** 10–12 in 2–3 series, elliptic to oblong, 1–3.5 × 0.5–1 mm, apices acute to acuminate, shortly mucronate, abaxial faces puberulent, gland-dotted. **Florets** 7–9; corollas 2–2.5 mm. **Cypselae** 1–1.5 mm; **pappi** of 20–30 bristles 2.5–3 mm. *2n* = 20.

Flowering Aug–Oct. Dry to wet areas, roadsides, around ponds, open pine woods; 20–100 m; Ala., Fla., La., Miss., S.C., Tex., Va.

Eupatorium ×pinnatifidum has been suggested to represent a series of recurrent hybrids between *E. capillifolium* or *E. compositifolium* and *E. perfoliatum*; molecular data, in addition to confirming that this combination occurs, also show that *E. serotinum* rather than *E. perfoliatum* is the second parent in at least some plants included in *E. ×pinnatifidum.*

20. **Eupatorium resinosum** Torrey ex de Candolle in A. P. de Candolle and A. L. P. P. de Candolle, Prodr. 5: 176. 1836 · Pine barren thoroughwort E

Perennials, 40–100+ cm. **Stems** (from short rhizomes) single, branched distally (among heads), puberulent throughout. **Leaves** usually opposite (distal sometimes alternate); sessile; blades pinnately nerved, narrowly elliptic, 40–100+ × 5–20 mm, bases cuneate (not perfoliate), margins serrate, apices acute to acuminate, faces puberulent, gland-dotted. **Heads** in corymbiform arrays. **Phyllaries** 10–12 in 2–3 series, elliptic to oblong, 1.5–4 × 0.8–1 mm, apices broadly acute to acute, abaxial faces puberulent, gland-dotted. **Florets** 9–14; corollas 3–3.5 mm. **Cypselae** 2–3 mm; **pappi** of 20–30 bristles 3–3.5 mm. *2n* = 20.

Flowering Aug–Sep. Moist, low ground, pocosins, bogs, cutover woodlands, pine barrens, acid soils; 20–100+ m; N.J., N.C., S.C.

Eupatorium resinosum is in the Center for Plant Conservation's National Collection of Endangered Plants.

21. **Eupatorium rotundifolium** Linnaeus, Sp. Pl. 2: 837. 1753 · Roundleaf thoroughwort E

Perennials, 40–100+ cm. **Stems** (from short rhizomes) single, sparsely branched distally, puberulent. **Leaves** usually opposite (distal sometimes alternate, lateral buds dormant or producing 1 pair of leaves); sessile or subsessile; blades ± 3-nerved, usually ± deltate to orbiculate, sometimes ovate, 15–50(–70) × 15–40(–60) mm (lengths mostly 1–2 times widths), bases broadly rounded to truncate, margins crenate to serrate, apices acute, faces puberulent to villous, gland-dotted. **Heads** in corymbiform arrays. **Phyllaries** 8–10 in 2–3 series, elliptic to oblanceolate, 2–6 × 0.5–1.5 mm, apices acute, abaxial faces puberulent, gland-dotted. **Florets** 5; corollas 3–4 mm. **Cypselae** 2–3 mm; **pappi** of 30–40 bristles 3.5–4 mm.

Varieties 3 (3 in the flora): e North America.

Eupatorium rotundifolium is one of the more variable species of the genus; its apparent propensity to hybridize with other species has led it to be described as a compilospecies and has complicated its delimitation. Hybrids with *E. perfoliatum* and *E. sessilifolium* are sufficiently distinctive to be treated here as distinctive entities (*E. ×cordigerum* and *E. godfreyanum*, respectively), and the varieties *ovatum* and *scabridum* have also been proposed to be of hybrid origin. The boundaries between varieties are indistinct and individual specimens may be difficult to place with confidence.

1. Leaf blades 3-nerved from bases, bases subtruncate to broadly cuneate, margins crenate ..... 21b. *Eupatorium rotundifolium* var. *rotundifolium*
1. Leaf blades 3-nerved distal to bases, bases broadly cuneate to cuneate, margins serrate.
　2. Leaf blades 30–70 × 30–60 mm, broadest near middles .... 21a. *Eupatorium rotundifolium* var. *ovatum*
　2. Leaf blades 20–50 × 15–30 mm, broadest near bases .... 21c. *Eupatorium rotundifolium* var. *scabridum*

21a. **Eupatorium rotundifolium** Linnaeus var. **ovatum** (Bigelow) Torrey ex de Candolle in A. P. de Candolle and A. L. P. P. de Candolle, Prodr. 5: 178. 1836 E

Eupatorium ovatum Bigelow, Fl. Boston. ed. 2, 296. 1824; *E. pubescens* Muehlenberg ex Willdenow

Leaf blades 3-nerved distal to bases, 30–70 × 30–60 mm, broadest near middles, bases broadly cuneate to cuneate, margins serrate. *2n* = 20, 30.

Flowering Jul–Aug. Moist, low ground, roadsides,

sandy soils; 10–200+ m; Ala., Ark., Conn., Del., D.C., Fla., Ga., Ky., La., Md., Mass., Miss., N.H., N.J., N.Y., N.C., Ohio, Pa., R.I., S.C., Tenn., Va., W.Va.

Variety *ovatum* is an apomictic polyploid that may be a derivative of *Eupatorium rotundifolium* and *E. sessilifolium*. Morphologically, it is closer to *E. rotundifolium* than to *E. sessilifolium*; the name *E. pubescens* has priority at species rank.

21b. Eupatorium rotundifolium Linnaeus var. rotundifolium E

Leaf blades 3-nerved from bases, 20–43 × 13–35 mm, broadest near bases, bases broadly subtruncate to broadly cuneate, margins crenate. $2n = 20, 30$.

Flowering Jul–Aug. Moist, low ground, roadsides, margins of ponds, sandy soils; 10–200+ m; Ala., Ark., Del., Fla., Ga., Ind., Ky., La., Md., Mass., Miss., N.J., N.C., Ohio, Okla., Pa., R.I., S.C., Tenn., Tex., Va., W.Va.

Variety *rotundifolium* includes both sexual diploids and apomictic polyploids, which are morphologically not distinguishable. The sexual diploids are somewhat more restricted geographically, occurring in a band across northern Florida and southern Georgia.

21c. Eupatorium rotundifolium Linnaeus var. scabridum (Elliott) A. Gray in A. Gray et al., Syn. Fl. N. Amer. 1(2): 99. 1884 E

Eupatorium scabridum Elliott, Sketch Bot. S. Carolina 2: 299. 1823

Leaf blades 3-nerved distal to bases, 20–50 × 15–30 mm, broadest near bases, bases broadly cuneate to cuneate, margins serrate. $2n = 20, 30$.

Flowering Jul–Aug. Disturbed sites; 10–100+ m; Ala., Ark., Fla., Ga., La., Miss., Mo., Okla., S.C., Tex.

Variety *scabridum* includes apomictic polyploids and has been suggested to have been derived from hybridization between var. *rotundifolium* and *Eupatorium semiserratum*.

22. Eupatorium semiserratum de Candolle in A. P. de Candolle and A. L. P. P. de Candolle, Prodr. 5: 177. 1836 • Smallflower thoroughwort E

Eupatorium cuneifolium Willdenow var. *semiserratum* (de Candolle) Fernald & Griscom

Perennials, 40–100+ cm. **Stems** (from short rhizomes) single, densely branched distally, puberulent throughout. **Leaves** usually opposite (distal sometimes alternate, lateral buds dormant or producing 1 pair of leaves); simple, sessile or subsessile; blades ± 3-nerved distal to bases, elliptic to lance-elliptic, (30–)50–70 × 5–25 mm (lengths mostly 2–5 times widths), bases narrowly cuneate, margins usually serrate, apices acute, faces puberulent or villous, densely gland-dotted. **Heads** in corymbiform arrays. **Phyllaries** 7–10 in 2–3 series, elliptic, 1–3 × 0.5–1 mm, apices rounded to acute (not mucronate), abaxial faces puberulent, gland-dotted. **Florets** 5; corollas 2.5–3 mm. **Cypselae** 1.5–2 mm; **pappi** of 30–40 bristles 2.5–3 mm. $2n = 20$.

Flowering Aug–Sep. Moist to boggy, sandy, peaty soils, margins of pine flatwoods, gum swamps, bayheads, disturbed sites, roadsides; 10–100+ m; Ala., Ark., Fla., Ga., La., Miss., N.C., S.C., Tenn.

Eupatorium semiserratum has been included within *E. glaucescens* (*E. cuneifolium*); it is distinguished by its consistently smaller heads and stems that branch only within the capitulescences. It differs from the similar and sometimes sympatric *E. lancifolium* by its smaller heads, leaves 3-nerved distal to bases (rather than at bases), as well as preference for wetter habitats. It has been proposed that *E. rotundifolium* var. *scabridum* (*E. pubescens*) represents hybrids between *E. semiserratum* and *E. rotundifolium*; it also apparently hybridizes with *E. hyssopifolium*.

23. Eupatorium serotinum Michaux, Fl. Bor.-Amer. 2: 100. 1803 • Lateflowering thoroughwort E F

Perennials, 50–150+ cm. **Stems** (from short caudices) single, sparsely to densely branched distally, pubescent throughout (sometimes reddish to purplish). **Leaves** opposite; petiolate (petioles 10–25 mm); blades ± 3-nerved, lanceolate, 20–90+ × 5–40 mm, bases rounded to slightly oblique, margins entire or serrate, apices acute, faces puberulent, gland-dotted. **Heads** in corymbiform arrays. **Phyllaries** 8–12 in 1–2 series, elliptic to oblong, 1–3 × 0.5–1 mm, apices slightly rounded to acute, abaxial faces puberulent, gland-dotted. **Florets** 9–15; corollas 2.5–3 mm. **Cypselae** 1–1.5 mm; **pappi** of 20–30 bristles 2–2.5 mm. $2n = 20$.

Flowering Aug–Oct. Moist or dry, open sites, roadsides; 10–400+ m; Ala., Ark., Conn., Del., Fla., Ga., Ill., Ind., Iowa, Kans., Ky., La., Md., Mass., Mich., Minn., Miss., Mo., Nebr., N.J., N.Y., N.C., Ohio, Okla., Pa., R.I., S.C., Tenn., Tex., Va., W.Va., Wis.

Eupatorium serotinum has a wide distribution and is often abundant where it occurs. It includes only sexual, diploid populations. It is known to hybridize with *E. perfoliatum*. An introduction of *E. serotinum* in southeastern Ontario apparently is local and may not have persisted.

24. Eupatorium sessilifolium Linnaeus, Sp. Pl. 2: 837. 1753 • Upland boneset E F

Perennials, 50–100+ cm. **Stems** (from short caudices or rhizomes) single, sparsely branched distally, 5–10+ dm, glabrous or glabrate proximally, puberulent distally (among heads). **Leaves** usually opposite (distal sometimes alternate); simple, sessile; blades pinnately nerved, lanceolate to oblong, 70–150+ × 10–30 mm (lengths mostly 2–5 times

widths), bases truncate to somewhat rounded, margins serrate, apices narrowly acute to acuminate, faces glabrate (scattered, fine hairs), gland-dotted. **Heads** in corymbiform arrays. **Phyllaries** 10–15 in 2–3 series, elliptic to oblong, 2–5 × 0.5–1.5 mm, apices rounded to acute (not mucronate), abaxial faces villous to puberulent, gland-dotted. **Florets** 5; corollas 3–3.5 mm. **Cypselae** 2–3 mm; **pappi** of 30–40 bristles 3–4 mm. *2n* = 20, 30.

Flowering Jul–Sep. Dry, open, edges of mesic woods; 20–300+ m; Ala., Ark., Conn., Del., Ga., Ill., Ind., Iowa, Kans., Ky., Md., Mass., Mich., Minn., Miss., Mo., N.H., N.J., N.Y., N.C., Ohio, Pa., R.I., S.C., Tenn., Vt., Va., W.Va., Wis.

393. EUTROCHIUM Rafinesque, New Fl. 4: 78. 1838 • Joepyeweed, eupatoire [Greek *eu-*, well, truly, and *trocho-*, wheel-like, alluding to whorled leaves] E

Eric E. Lamont

Eupatoriadelphus R. M. King & H. Robinson; *Eupatorium* Linnaeus sect. *Verticillatum* de Candolle

Perennials, 30–350+ cm. **Stems** (sometimes ± purple, sometimes glaucous, internodes usually shorter than leaves) erect, unbranched. **Leaves** mostly cauline; mostly whorled (3–7 per node), rarely opposite; petiolate; blades pinnately veined or ± 3-nerved from at or near bases, deltate-ovate, lance-elliptic, lanceolate, lance-ovate, or ovate (bases gradually or abruptly tapered), margins serrate (usually with gland at apex of each tooth, apices acuminate), abaxial faces usually gland-dotted and hirsute, puberulent, pubescent, scabrous, or stipitate-glandular, sometimes glabrate, adaxial faces mostly puberulent to scabrous-hirsute and glabrescent, sometimes glabrate or glabrous. **Heads** discoid, in (flat-topped or convex to rounded) compound, corymbiform arrays. **Involucres** cylindric, 2.5–7 mm diam. **Phyllaries** persistent, 10–22 in 5–6 series, (pale pink to purple, rarely white, tightly appressed) striate or 1-nerved, mostly lance-ovate to lanceolate, unequal (outer 2–3 obtuse, densely pubescent or glabrescent, often gland-dotted, innermost glabrous or glabrescent). **Receptacles** flat or convex, epaleate. **Florets** 4–22; corollas usually purplish or pinkish, rarely white, throats funnelform, lobes 5, ovate to deltate; styles: bases enlarged, puberulent, branches ± filiform to clavate (slightly dilated or flattened distally, papillose). **Cypselae** (dark brown to black or yellowish brown) prismatic, 5-ribbed, usually gland-dotted, sometimes scabrellous on ribs; **pappi** persistent, of 25–40 (cream to pinkish purple) barbellate bristles in 1 series. *x* = 10.

Species 5 (5 in the flora): North America.

Detailed distribution maps for taxa of *Eutrochium* are available but treated as taxa under Eupatorium (E. E. Lamont 1995).

SELECTED REFERENCES Lamont, E. E. 1995. Taxonomy of *Eupatorium* section *Verticillata* (Asteraceae). Mem. New York Bot. Gard. 72: 1–66. Mackenzie, K. K. 1920. Scientific names applicable to our purple-flowered eupatoriums. Rhodora 22: 157–165. Wiegand, K. M. and C. A. Weatherby. 1937. The nomenclature of the verticillate eupatoria. Rhodora 39: 297–306.

1. Leaves usually ± 3-nerved and bases of blades abruptly contracted to petioles; florets (4–)5–9(–10) (near coast from South Carolina to New Hampshire and Nova Scotia). . . . . . . 1. *Eutrochium dubium*
1. Leaves either pinnately veined or bases of blades gradually narrowed to petioles or both; florets 4–10 or 8–22.
    2. Heads in ± flat-topped arrays; florets 8–22 . . . . 2. *Eutrochium maculatum*
    2. Heads in convex to rounded arrays; florets 4–10.
        3. Stems usually glandular-pubescent throughout, sometimes densely puberulent and sparingly glandular; abaxial leaf faces ± glandular-pubescent and sparsely hirsute; s Appalachian Mountains of sw Virginia, e Kentucky, e Tennessee, and w North Carolina . . . . 3. *Eutrochium steelei*
        3. Stems glabrous proximal to heads; abaxial leaf faces sparingly and minutely gland-dotted; relatively widespread in e North America.
            4. Stems purple at nodes, usually solid, rarely ± hollow near bases. . . . . 4. *Eutrochium purpureum*
            4. Stems purplish throughout (and glaucous), hollow. . . . . 5. *Eutrochium fistulosum*

1. Eutrochium dubium (Willdenow ex Poiret) E. E. Lamont, Sida 21: 901. 2004 • Coastal plain joepyeweed, joepye thoroughwort [E]

Eupatorium dubium Willdenow ex Poiret in J. Lamarck et al., Encycl., suppl. 2: 606. 1812; *Eupatoriadelphus dubius* (Willdenow ex Poiret) R. M. King & H. Robinson; *Eupatorium americanum* Hill; *E. ternifolium* Elliott

Plants 40–120(–170) cm. **Stems** usually purple-spotted, sometimes uniformly purple, solid, glabrous proximally, glandular-puberulent distally. **Leaves** mostly in 3s–4s; petioles 7–25 mm, glabrous; blades usually ± 3-nerved, rarely pinnately veined, deltate-ovate or ovate to lance-ovate, 5–16 × 2–8 cm, relatively thick and firm (often rugose), bases usually abruptly contracted to petioles, margins coarsely serrate, abaxial faces densely gland-dotted and sparingly hirsute (at least on midribs and main veins), adaxial faces scabrous, glabrescent. **Heads** usually in ± convex, sometimes flat-topped, corymbiform arrays. **Involucres** often purplish, 6.5–9 × 2.5–5 mm. **Phyllaries** glabrous or sparsely hairy. **Florets** (4–)5–9(–10); corollas usually dark purple, rarely pinkish or white, 4.5–7 mm. **Cypselae** 3–4.5 mm. 2*n* = 20.

Flowering late summer–early fall. Moist habitats, in sandy or gravelly, acid soils, open sun or partial shade; 0–50+ m; N.S.; Conn., Del., Maine, Md., Mass., N.H., N.J., N.Y., N.C., Pa., R.I., S.C., Vt., Va.

Eutrochium dubium is restricted to the coastal plain, and inland along major river systems, of eastern North America. Identification of *E. dubium* has been difficult for field botanists, as evidenced by misidentified herbarium specimens. The most prominent character distinguishing *E. dubium* from its relatives (especially *E. maculatum*) is the 3-nerved leaf venation: the proximalmost pair of lateral veins are more prominent and more prolonged than the others. This character is best observed on abaxial leaf faces.

2. Eutrochium maculatum (Linnaeus) E. E. Lamont, Sida 21: 902. 2004 • Spotted joepyeweed, eupatoire maculée [E]

Eupatorium maculatum Linnaeus, Cent. Pl. I, 27. 1755; *Eupatoriadelphus maculatus* (Linnaeus) R. M. King & H. Robinson; *Eupatorium purpureum* Linnaeus subsp. *maculatum* (Linnaeus) Á. Löve & D. Löve; *E. purpureum* var. *maculatum* (Linnaeus) Voss; *E. trifoliatum* Linnaeus var. *maculatum* (Linnaeus) Farwell

Plants 60–200 cm. **Stems** usually purple-spotted, sometimes uniformly purple, usually solid, sometimes hollow near bases, glabrous proximally to densely puberulent throughout, glandular-puberulent distally. **Leaves** in (3s–)4s–5s(–6s); petioles 5–20 mm, glabrous or pubescent; blades pinnately veined, lance-elliptic to lanceolate or lance-ovate, mostly (6–)8–23(–30) ×

(1.5–)2–7(–9) cm, bases gradually or abruptly tapered, margins sharply serrate or doubly serrate, abaxial faces gland-dotted and densely pubescent to glabrate, adaxial faces sparingly hairy or glabrous. **Heads** in flat-topped, corymbiform arrays. **Involucres** often purplish, 6.5–9 × 3.5–7 mm. **Phyllaries** glabrous or densely pubescent. **Florets** (8–)9–20(–22); corollas purplish, 4.5–7.5 mm. **Cypselae** 3–5 mm. $2n = 20$.

Varieties 3 (3 in the flora): North America.

Eutrochium maculatum has the widest geographic distribution and greatest morphologic variability among species in the genus. The three varieties recognized here show intergradation where the ranges overlap.

1. Stems densely puberulent throughout; leaves densely pubescent on abaxial faces, relatively thick; mostly w of Mississippi River . . . . 2c. *Eutrochium maculatum* var. *bruneri*
1. Stems glabrous proximally, puberulent among heads; leaves usually glabrate (seldom strongly puberulent) on abaxial faces, relatively thin; mostly e of Mississippi River.
    2. Distalmost whorls of leaves subtending heads equaling or surpassing arrays of heads; maritime Canada, adjacent New England to ne Minnesota, n Michigan, and s Ontario . . . . 2b. *Eutrochium maculatum* var. *foliosum*
    2. Distalmost whorls of leaves subtending heads not equaling arrays of heads; Newfoundland to sw Ontario, s to n and w New England, Pennsylvania, North Carolina, Ohio, Indiana, Illinois, and Iowa . . . . 2a. *Eutrochium maculatum* var. *maculatum*

2a. Eutrochium maculatum (Linnaeus) E. E. Lamont var. **maculatum** E

Stems puberulent only near summits. **Leaves**: petioles glabrous; blades lance-elliptic to lanceolate or lance-ovate, 8–25 × 2.5–9 cm, relatively thin, bases gradually narrowed, abaxial faces usually glabrate, seldom strongly puberulent; distalmost whorls of leaves directly subtending heads not surpassing heads (blades 1.5–2.5 × 0.2–0.5 cm). **Phyllaries** mostly glabrous, outer often hairy.

Flowering late summer–early fall. Moist sites, calcareous soils, open sun or partial shade; 10–1800 m; St. Pierre and Miquelon; N.B., Nfld. and Labr. (Nfld.), N.S., Ont., P.E.I., Que.; Conn., Ill., Ind., Iowa, Maine, Mass., Mich., Minn., N.H., N.J., N.Y., N.C., Ohio, Pa., Vt., Va., W.Va., Wis.

2b. Eutrochium maculatum (Linnaeus) E. E. Lamont var. **foliosum** (Fernald) E. E. Lamont, Sida 21: 902. 2004 E

Eupatorium purpureum Linnaeus var. *foliosum* Fernald, Rhodora 10: 86. 1908; *E. bruneri* A. Gray var. *foliosum* (Fernald) House; *E. maculatum* Linnaeus var. *foliosum* (Fernald) Wiegand; *E. trifoliatum* Linnaeus var. *foliosum* (Fernald) Farwell

Stems puberulent only near summits. **Leaves**: petioles glabrous; blades lance-elliptic to lanceolate or lance-ovate, 8–25 × 2.5–9 cm, relatively thin, bases gradually narrowed, abaxial faces usually glabrate, seldom strongly puberulent; distalmost whorls of leaves directly subtending heads equaling or surpassing heads (blades 8–20 × 2–4 cm). **Phyllaries** mostly glabrous, outer often hairy.

Flowering late summer–early fall. Moist sites, calcareous soils, open sun or partial shade, upland edges of spruce-fir swamps; 0–200+ m; N.B., Nfld. and Labr. (Nfld.), N.S., Ont., P.E.I., Que.; Maine, Mich., Minn., N.H., N.Y., Vt., Wis.

2c. Eutrochium maculatum (Linnaeus) E. E. Lamont var. **bruneri** (A. Gray) E. E. Lamont, Sida 21: 902. 2004 E

Eupatorium bruneri A. Gray in A. Gray et al., Syn. Fl. N. Amer. 1(2): 96. 1884; *Eupatoriadelphus maculatus* (Linnaeus) R. M. King & H. Robinson var. *bruneri* (A. Gray) R. M. King & H. Robinson; *Eupatorium atromontanum* A. Nelson; *E. maculatum* Linnaeus subsp. *bruneri* (A. Gray) G. W. Douglas; *E. maculatum* var. *bruneri* (A. Gray) Breitung; *E. purpureum* Linnaeus var. *bruneri* (A. Gray) B. L. Robinson; *E. rydbergii* Britton; *E. trifoliatum* Linnaeus var. *bruneri* (A. Gray) Farwell

Plants 60–150 cm. **Stems** densely puberulent throughout (glabrescent at bases). **Leaves**: petioles pubescent to glabrescent; blades lance-elliptic to lanceolate or lance-ovate, usually 6–17 × 1.5–5(–7) cm, relatively firm, bases gradually or abruptly narrowed, abaxial faces densely hairy (hairs relatively short, spreading); distalmost whorls of leaves directly subtending heads, not surpassing heads. **Phyllaries** mostly glabrescent, outer sometimes densely pubescent.

Flowering summer–fall. Stream and canal banks, wet meadows, bogs, and seeps, calcareous soils; 200–2500 m; Alta., B.C., Man., Ont., Sask.; Ariz., Colo., Idaho, Iowa, Kans., Mich., Minn., Mo., Nebr., N.Mex., N.Dak., S.Dak., Utah, Wash., Wis., Wyo.

3. Eutrochium steelei (E. E. Lamont) E. E. Lamont, Sida 21: 902. 2004 E

Eupatorium steelei E. E. Lamont, Brittonia 42: 279, fig. 1. 1990; *Eupatoriadelphus steelei* (E. E. Lamont) G. J. Schmidt & E. E. Schilling

Plants 60–200 cm. **Stems** usually greenish purple, sometimes evenly purplish, solid, usually glandular-pubescent throughout, sometimes densely puberulent and sparingly glandular. **Leaves** mostly in 3s–4s; petioles (0.7–)1.3–2.8(–3.6) mm, glabrate to densely ciliate; blades pinnately veined, lance-ovate or ovate to deltate-ovate, mostly 7–30 × 2.5–18 cm, relatively firm, bases abruptly or gradually tapered, margins sharply serrate, abaxial faces ± glandular-pubescent and sparsely hirsute (at least midribs and main veins), adaxial faces scabrous-hirsute, glabrescent. **Heads** in loose, convex, compound corymbiform arrays. **Involucres** often purplish, 6.5–9 × 3.5–5 mm. **Phyllaries** mostly glabrous, outer sometimes hairy on midveins. **Florets** (5–)6–9(–10); corollas usually pale pinkish or purplish, 4.5–7 mm. **Cypselae** 3–4.5 mm. $2n = 20$.

Flowering late summer–early fall. Gravelly embankments, open woods, thickets; 700–1500 m; Ky., N.C., Tenn., Va.

Eutrochium steelei is known from the Blue Ridge Province of western North Carolina and eastern Tennessee, the Ridge and Valley Province of southwestern Virginia, and the Appalachian Plateaus Province of eastern Kentucky.

4. Eutrochium purpureum (Linnaeus) E. E. Lamont, Sida 21: 902. 2004 • Sweetscented or sweet joepyeweed E

Eupatorium purpureum Linnaeus, Sp. Pl. 2: 838. 1753; *Eupatoriadelphus purpureus* (Linnaeus) R. M. King & H. Robinson

Plants 30–200 cm. **Stems** usually dark purple at nodes, usually otherwise greenish, rarely purplish green, usually solid, rarely ± hollow near bases, glabrous proximally, ± glandular-puberulent distally and among heads. **Leaves** mostly in 3s–4s(–5s); petioles 5–15(–20) mm, glabrous or sparingly puberulent, rarely ciliate; blades pinnately veined, lance-ovate or ovate to deltate-ovate, mostly (7–)9–26(–30) × (2.5–)3–15(–18) cm, bases abruptly or gradually tapered, margins coarsely serrate, abaxial faces sparingly and minutely gland-dotted and densely pubescent to glabrate, adaxial faces sparingly

puberulent and glabrescent or glabrous. **Heads** in loose, convex, compound corymbiform arrays. **Involucres** often purplish, 6.5–9 × 2.5–5 mm. **Phyllaries** usually glabrous, sometimes sparsely hairy. **Florets** (4–)5–7(–8); corollas usually pale pinkish or purplish, 4.5–7 mm. **Cypselae** 3–4.5 mm.

Varieties 2 (2 in the flora): North America.

Eutrochium purpureum is morphologically variable and is known to hybridize with all other species in the genus (E. E. Lamont 1995). Historically, more than a dozen infraspecific taxa have been recognized; the extent of intergradation and the lack of correlation among varying traits tend to make recognition of more than two varieties impractical.

1. Leaves: abaxial faces usually glabrous, sometimes sparsely to densely hairy along major veins (hairs fine, simple, 1-cellular); s New Hampshire to n Florida, w to se Minnesota, Iowa, and e Oklahoma. . . . . 4a. *Eutrochium purpureum* var. *purpureum*
1. Leaves: abaxial faces densely and persistently puberulent to villous; s Minnesota, Iowa, and ne Kansas to s Wisconsin, n Illinois, and n Missouri. . . . . 4b. *Eutrochium purpureum* var. *holzingeri*

4a. Eutrochium purpureum (Linnaeus) E. E. Lamont var. **purpureum** E F

Eupatorium amoenum Pursh; *E. falcatum* Michaux; *E. fuscorubrum* Walter; *E. harnedii* Steele ex Harned; *E. purpureum* Linnaeus var. *album* Barratt; *E. purpureum* var. *amoenum* (Pursh) A. Gray; *E. purpureum* var. *falcatum* (Michaux) Britton; *E. purpureum* var. *ovatum* Alph. Wood; *E. purpureum* var. *verticillatum* (Lamarck) Alph. Wood; *E. trifoliatum* Linnaeus; *E. trifoliatum* var. *amoenum* (Pursh) Farwell

Leaves: petioles glabrous; abaxial faces usually glabrous, sometimes sparsely to densely hairy along major veins (hairs fine, simple, 1-cellular). $2n = 20$.

Flowering late summer–early fall. Rich, open deciduous woodlands, woodland borders, and thickets, wooded ravines, near bases of slopes bordering wet grounds, partial or deep shade; 10–1200 m; Ont.; Ala., Ark., Conn., Del., Fla., Ga., Ill., Ind., Iowa, Kans., Ky., La., Md., Mass., Mich., Minn., Miss., Mo., Nebr., N.H., N.J., N.Y., N.C., Ohio, Okla., Pa., R.I., S.C., Tenn., Vt., Va., W.Va., Wis.

East of the Mississippi River, abaxial leaf faces of var. *purpureum* tend to be glabrous and rarely pubescent along the major veins. West of the Mississippi (and in the xeric sandhills of South Carolina), abaxial leaf faces are commonly densely pubescent along the major veins.

4b. Eutrochium purpureum (Linnaeus) E. E. Lamont var. **holzingeri** (Rydberg) E. E. Lamont, Sida 21: 902. 2004 [E]

Eupatorium holzingeri Rydberg, Brittonia 1: 97. 1931; *E. purpureum* Linnaeus var. *holzingeri* (Rydberg) E. E. Lamont

Leaves: petioles glabrous or sparingly puberulent or ciliate; abaxial faces densely and persistently puberulent to villous.

Flowering late summer–early fall. Steep slopes of rocky, deciduous woodlands, woodland borders, shade; 100–300 m; Ark., Ill., Iowa, Kans., Minn., Mo., Nebr., Wis.

5. Eutrochium fistulosum (Barratt) E. E. Lamont, Sida 21: 901. 2004 • Trumpetweed, hollow joepyeweed [E]

Eupatorium fistulosum Barratt, Eupatoria Verticillata, no. 1. 1841; *Eupatoriadelphus fistulosus* (Barratt) R. M. King & H. Robinson; *Eupatorium purpureum* Linnaeus var. *angustifolium* Torrey & A. Gray

Plants 60–350+ cm. **Stems** usually purple throughout, sometimes greenish or purple-spotted, hollow proximally, usually glabrous proximally (rarely pubescent toward bases when young), ± glandular-puberulent distally and among heads (glaucous throughout, at least when fresh). **Leaves** mostly in 4s–6s(–7s); petioles (5–)10–30(–50) mm, glabrous; blades pinnately veined, narrowly to broadly lanceolate, mostly (8–)12–25(–28) × (1.5–)2–6(–9) cm, bases gradually tapered, margins finely serrate (teeth rounded, blunt), abaxial faces sparingly and minutely ± scabrellous to glabrate, adaxial faces glabrous or sparingly puberulent. **Heads** in convex to rounded (dome-shaped), compound corymbiform arrays (ultimately broadly cylindric). **Involucres** often purplish, 6.5–9 × 2.5–5 mm. **Phyllaries** glabrous or outer with hairs on midveins. **Florets** (4–)5–7; corollas usually pale pinkish or purplish, 4.5–6 mm. **Cypselae** 3–4.5 mm. $2n = 20$.

Flowering late summer–early fall. Wet lowlands, alluvial woods, along streams, moist meadows, bogs, marshes with permanently saturated or seasonally flooded organic soils, open sun or partial shade; 10–1400+ m; Ala., Ark., Conn., Del., Fla., Ga., Ill., Ind., Ky., La., Maine, Md., Mass., Mich., Miss., Mo., N.H., N.J., N.Y., N.C., Ohio, Okla., Pa., R.I., S.C., Tenn., Tex., Va., W.Va.

In the field, *Eutrochium fistulosum* is the most distinct species of the genus; herbarium specimens do not always document the distinguishing characteristics: heights commonly surpassing 2 m, proximal stem diameters usually 2 cm or greater, stems strongly glaucous (mostly hollow, sometimes distally hollow), leaves commonly 6 or 7 per node, arrays of heads commonly 30 × 22 cm.

394. **CONOCLINIUM** de Candolle in A. P. de Candolle and A. L. P. P. de Candolle, Prodr. 5: 135. 1836 • [Greek *konos*, cone, and *kline*, bed, alluding to conic receptacles]

Thomas F. Patterson

Guy L. Nesom

Perennials, 50–200 cm (usually rhizomatous, colonial, rhizomes relatively slender). **Stems** erect to decumbent (sometimes rooting at proximal nodes), not much branched distal to bases. **Leaves** cauline; opposite; petiolate; blades usually 3-nerved from bases, deltate, oblanceolate, ovate, ovate-deltate, oblong, or triangular, margins dentate or lobed (to dissected in *C. dissectum*), faces glabrate to puberulous, villosulous, or hispidulous, gland-dotted. **Heads** discoid, in tight, corymbiform arrays. **Involucres** hemispheric, 3–6 mm diam. **Phyllaries** persistent, ca. 25 in 2–3 series, obscurely 2–3-nerved, lanceolate to linear, ± equal (herbaceous). **Receptacles** conic, epaleate. **Florets** 35–70+; corollas usually blue to purple or violet, rarely white, throats narrowly funnelform (lengths ca. 4 times diams.); styles: bases not enlarged, glabrous, branches filiform to linear-clavate. **Cypselae** prismatic (bases narrowed), 5-ribbed, glabrous or sparsely gland-dotted and/or hispidulous; **pappi** persistent, of ca. 30 barbellate bristles in 1 series. $x = 10$.

Species 4 (3 in the flora): sw, e North America, Mexico.

SELECTED REFERENCE Patterson, T. F. 1994. A Taxonomic Revision of *Conoclinium* DC. (Asteraceae, Eupatorieae). M.S. thesis. University of Texas.

1. Leaf blades deeply dissected or lobed (lobes dentate, apices sharply pointed or rounded)
. . . . 1. *Conoclinium dissectum*
1. Leaf blades serrate to dentate or crenate (not dissected or lobed).
    2. Leaf blades usually oblong to oblong-lanceolate, sometimes triangular, bases truncate to cordate (often subauriculate), apices rounded to obtuse; tips of pappus bristles dilated
. . . . 2. *Conoclinium betonicifolium*
    2. Leaf blades triangular to deltate or ovate, bases usually cuneate to truncate, rarely subcordate (not subauriculate), apices acute; tips of pappus bristles not dilated. . . .
. . . . 3. *Conoclinium coelestinum*

1. **Conoclinium dissectum** A. Gray, Smithsonian Contr. Knowl. 3(5): 88. 1852 • Palm-leaf mistflower

Conoclinium greggii (A. Gray) Small; *Eupatorium greggii* A. Gray

Stems erect (often from knotty crowns, sometimes basally lignescent). **Leaf blades** ovate-deltate to ovate, 1.5–4 cm, bases attenuate, margins dissected or lobed, apices pointed or rounded. **Phyllaries** 3.5–5 mm. **Corollas** blue to lavender or purple, 2.5–3.5 mm. **Cypselae** 1.8–2.5 mm, hispidulous; **pappi:** bristle tips not dilated. *2n* = 20.

Flowering Apr–Aug, Oct–Nov. Waterways, depressions and ditches, dry sandy or rocky soil, mesquite, creosote bush-mesquite; 400–1400 m; Ariz., N.Mex., Tex.; Mexico (Chihuahua, Coahuila, Durango, Nuevo León, San Luis Potosí, Sonora, Zacatecas).

2. **Conoclinium betonicifolium** (Miller) R. M. King & H. Robinson, Phytologia 19: 300. 1970 (as betonicaefolium) • Betony-leaf mistflower

Eupatorium betonicifolium Miller, Gard. Dict. ed. 8, Eupatorium no. 9. 1768

Stems decumbent or procumbent (rooting at nodes). **Leaf blades** usually oblong to oblong-lanceolate or oblanceolate to ovate, sometimes triangular, 2.5–5(–11) cm, bases cordate, obtuse, or truncate (often subauriculate), margins usually crenate to dentate, rarely entire, apices rounded to obtuse. **Phyllaries** ca. 4 mm. **Corollas** blue to purple, 2.5–3 mm. **Cypselae** 1.2–1.8 mm, glabrous; **pappi:** bristle tips dilated.

Varieties 2 (2 in the flora): Texas; Mexico.

1. Leaf blades mostly oblong to oblong-lanceolate, bases cordate to truncate (often subauriculate), margins crenate; coastal habitats . . . .
. . . 2a. *Conoclinium betonicifolium* var. *betonicifolium*
1. Leaf blades mostly oblanceolate to ovate, bases cordate, obtuse, or truncate, margins crenate or entire; inland habitats along streams and around lakes. . . . . 2b. *Conoclinium betonicifolium* var. *integrifolium*

2a. **Conoclinium betonicifolium** (Miller) R. M. King & H. Robinson var. **betonicifolium**

Conoclinium betonicum de Candolle; *Eupatorium betonicum* (de Candolle) Hemsley

Leaf blades mostly oblong to oblong-lanceolate, bases cordate to truncate (often subauriculate), margins crenate.

Flowering Apr–Jun, Sep–Oct. Coastal dunes, beaches, sandy loam, roadside ditches, edges of woods, salt marshes; 0–10 m; Tex.; Mexico.

2b. **Conoclinium betonicifolium** (Miller) R. M. King & H. Robinson var. **integrifolium** (A. Gray) T. F. Patterson, Sida 21: 1641. 2005

Conoclinium betonicum de Candolle var. *integrifolium* A. Gray, Smithsonian Contr. Knowl. 3(5): 88. 1852; *C. integrifolium* (A. Gray) Small

Leaf blades mostly oblanceolate to ovate, bases cordate, obtuse, or truncate, margins crenate or entire.

Flowering Apr–Jun, Sep–Oct. Along streams, around lakes, marsh edges, saline soils near irrigation, old fields; 50–1000 m; Tex.; Mexico.

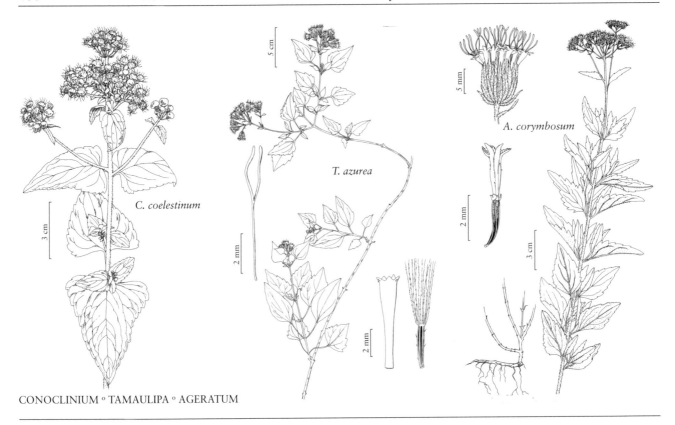

CONOCLINIUM ○ TAMAULIPA ○ AGERATUM

3. **Conoclinium coelestinum** (Linnaeus) de Candolle in A. P. de Candolle and A. L. P. P. de Candolle, Prodr. 5: 135. 1836 • Blue mistflower E F

Eupatorium coelestinum Linnaeus, Sp. Pl. 2: 838. 1753; *Conoclinium dichotomum* Chapman

Stems usually erect, sometimes decumbent or procumbent (rooting at nodes). **Leaf blades** triangular to deltate or ovate, 2–7(–13) cm, bases usually cuneate to truncate, rarely subcordate, margins serrate to serrate-dentate or crenate, apices acute. **Phyllaries** 3.5–4 mm. **Corollas** blue to blue-violet or rosy-violet, (1.6–)2–2.5 mm. **Cypselae** 1–1.5 mm, glabrous; **pappi:** bristle tips not dilated. **2***n* = 20.

Flowering (Jun–)Aug–Nov. Stream banks, lakeshores, coastal and inland sands, pine-oak woodlands, low woods, flood plains, ditches, ravines, bogs, wet slopes, pine savannas, disturbed sites, roadsides, along railroads; 50–400 m; Ont.; Ala., Ark., Del., Fla., Ga., Ill., Ind., Kans., Ky., La., Md., Mich., Miss., Mo., Nebr., N.J., N.Y., N.C., Ohio, Okla., Pa., S.C., Tenn., Tex., Va., W.Va.

395. **TAMAULIPA** R. M. King & H. Robinson, Phytologia 22: 154. 1971 • [Alluding to the Tamaulipan Desert region, to which the species is restricted]

Guy L. Nesom

Shrubs, (100–)200–300 cm (sometimes subscandent, climbing and sprawling over other plants). **Stems** erect or clambering, intricately branched (often brittle, glabrous or nearly so, not viscid). **Leaves** cauline; opposite; petiolate; blades usually 3-nerved from bases, deltate, margins toothed, faces puberulent to velutinous (not gland-dotted). **Heads** discoid, in corymbiform arrays. **Involucres** obconic to hemispheric, 5–7 mm diam. **Phyllaries** persistent,

30–35 in 2–3+ series, obscurely nerved, lanceolate to subulate, unequal (herbaceous to chartaceous or membranous). **Receptacles** convex to conic, epaleate. **Florets** 30–50+; corollas usually blue to lavender, sometimes white, throats narrowly funnelform (lengths 1.5–2 times diams.); styles: bases not enlarged, glabrous, branches linear-filiform. **Cypselae** prismatic, 5–6-ribbed, sparsely scabrellous; **pappi** persistent, of ca. 35 barbellate bristles in 1 series. $x = 10$.

Species 1: Texas, ne Mexico.

1. Tamaulipa azurea (A. P. de Candolle) R. M. King & H. Robinson, Phytologia 22: 154. 1971 • Blueweed, blue boneset F

Eupatorium azureum A. P. de Candolle in A. P. de Candolle and A. L. P. P. de Candolle, Prodr. 5: 168. 1836

Leaf blades mostly 20–50+ × 15–40+ mm. **Involucres** 6–8 mm. **Corollas** ca. 6 mm. **Cypselae** ca. 2.5 mm; **pappi** ca. 4 mm. $2n = 20$.

Flowering Feb–Jun(–Jul). Shrublands, often with acacias, palm groves, clay or clay-loam soils; 0–10 m; Tex.; Mexico.

King and Robinson hypothesized that *Tamaulipa azurea* is closely related to *Conoclinium*; B. L. Turner (1972) observed that it may be closer to *Chromolaena*.

396. AGERATUM Linnaeus, Sp. Pl. 2: 839. 1753; Gen. Pl. ed. 5, 363. 1754 • [Greek *a*, not, and *geras*, old age, apparently alluding to long-lasting nature of flowers]

Guy L. Nesom

Annuals and perennials, mostly 20–120 cm. **Stems** often decumbent (rooting at proximal nodes), sparsely to densely branched. **Leaves** cauline; all or mostly opposite; petiolate; blades mostly 1-nerved, deltate to ovate, or elliptic to lanceolate, margins entire or toothed, faces glabrous or ± pilose, puberulent, or strigoso-hispid, sometimes gland-dotted. **Heads** discoid, in dense to open, cymiform to corymbiform arrays. **Involucres** campanulate, 3–6 mm. **Phyllaries** persistent, 30–40 in 2–3 series, usually 2-nerved, lanceolate, ± equal (often indurate, margins scarious). **Receptacles** conic, epaleate [paleate]. **Florets** 20–125; corollas white or bluish to lavender, throats ± campanulate (lengths 2 times diams.); styles: bases not enlarged, glabrous, branches ± linear to clavate (usually papillose and dilated distally). **Cypselae** prismatic, 4–5-ribbed, glabrous or sparsely strigoso-hispidulous; **pappi** persistent, of 5–6 aristate scales, or coroniform, or 0. $x = 10$.

Species ca. 40 (4 in the flora): United States, Mexico, Central America; 2 species widespread as adventives.

SELECTED REFERENCE Johnson, M. F. 1971. A monograph of the genus *Ageratum* L. (Compositae–Eupatorieae). Ann. Missouri Bot. Gard. 58: 6–88.

1. Plants colonial; stems and leaves glabrous or glabrate . . . . 1. *Ageratum maritimum*
1. Plants not colonial; stems and leaves hairy.
 2. Stems puberulent to minutely strigoso-hispid; cypselae glabrous . . . . 2. *Ageratum corymbosum*
 2. Stems sparsely to densely pilose (usually in combination with other forms of vestiture); cypselae sparsely strigoso-hispidulous.

[3. Shifted to left margin.—Ed.]

3. Peduncles minutely puberulent and sparsely to densely pilose, eglandular; phyllaries oblong-lanceolate, abruptly tapering to subulate tips 0.5–1 mm, glabrous or sparsely pilose, margins often ciliate, abaxial faces eglandular . . . . 3. *Ageratum conyzoides*

3. Peduncles mixed pilose, stipitate-glandular, and viscid-puberulent; phyllaries narrowly lanceolate, gradually tapering to indurate-subulate tips 0.8–2 mm, margins not ciliate or inconspicuously ciliate, abaxial faces stipitate-glandular and sparsely to densely pilose . . . .
. . . . 4. *Ageratum houstonianum*

1. **Ageratum maritimum** Kunth in A. von Humboldt et al., Nov. Gen. Sp. 4(fol.): 117. 1818; 4(qto.) 150. 1820 • Cape Sable whiteweed

Ageratum littorale A. Gray; *A. littorale* var. *hondurense* B. L. Robinson

Annuals or perennials, 10–50 cm (semisucculent, rhizomatous, forming colonies). **Stems** decumbent to straggling or creeping (rooting at nodes), glabrous but for puberulous-pilose nodes. **Leaf blades** deltate-ovate to oblong, mostly 0.8–4 × 0.5–3 cm, (fleshy) margins toothed, faces glabrous or glabrate. **Peduncles** glabrous or glabrate. **Involucres** ca. 3 × 3–4 mm. **Phyllaries** elliptic-lanceolate, glabrous or glabrate, tips abruptly tapered to nearly obtuse. **Corollas** lavender or blue to white. **Cypselae** glabrous; **pappi** usually blunt coronas ca. 0.1 mm, rarely of separate scales.

Flowering year round. Beach sand and nearby thickets, coral soils, salt marshes, hammocks, roadsides; 0–10 m; Fla.; Mexico (Quintana Roo); West Indies (Cuba, Hispaniola); Central America (Belize).

Plants from Florida (*Ageratum littorale*, the type from Florida) are described here. Plants of the West Indies and Mexico (broadening the species concept to *A. maritimum*, the type from Cuba) have various elaborations of vestiture and a more conspicuous pappus–coronas with even to laciniate margins or rings of nearly separate scales mostly 0.2–1.5 mm. In addition to the distinctive relatively small, glabrous or glabrate leaves, plants of *A. maritimum* are characterized by heads in clusters, usually held well beyond the leaves.

2. **Ageratum corymbosum** Zuccagni, Cent. Observ. Bot., no. 85. 1806 • Flat-top whiteweed [F]

Ageratum corymbosum var. *jaliscense* B. L. Robinson; *A. salicifolium* Hemsley; *A. strictum* Hemsley

Perennials or subshrubs, 30–100 cm (fibrous-rooted). **Stems** erect to basally decumbent, puberulent to minutely strigoso-hispid. **Leaf blades** ovate to rhombic-lanceolate, 3–8 × 1–3.5 cm, margins toothed, abaxial faces usually puberulent, sometimes minutely strigoso-hispid, densely gland-dotted. **Peduncles** finely puberulent (not pilose), eglandular. **Involucres** 5–6 mm. **Phyllaries** narrowly lanceolate (0.4–0.7 mm wide, innermost often 1–1.5 mm longer than outer), finely puberulent, eglandular, tips green or purplish, filiform. **Corollas** usually blue to lavender, sometimes white. **Cypselae** glabrous; **pappi** usually crowns of connate scales with erose margins or tubular portions longer than divisions, rarely with 1 or more awnlike lobes. $2n = 20, 30, 40$.

Flowering Jul–Oct. Crevices, ledges, cliffs, other rocky sites in canyons, along streams, in desert grasslands, oak-agave, oak, oak-juniper, and pine-oak woodlands; (900–)1200–1900 m; Ariz., N.Mex.; Mexico.

Habitat information came mostly from collections from Sonora and Chihuahua, Mexico. *Ageratum corymbosum* grows in all Mexican states except for the extreme southeast. It has been included in various summaries as occurring in Texas; as noted by D. S. Correll and M. C. Johnston (1970), those records apparently were based on a collection by Charles Wright from southwestern New Mexico.

Varieties and forms of *Ageratum corymbosum* have been recognized (e.g., M. F. Johnson 1971; R. McVaugh 1984). McVaugh wryly noted that extremes of these intergrading infraspecific entities "can be recognized with a little imagination." The form that reaches the United States (with ovate-lanceolate leaves) is var. *jaliscense.*

3. **Ageratum conyzoides** Linnaeus, Sp. Pl. 2: 839. 1753 • Tropical whiteweed [I]

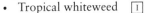

Ageratum latifolium Cavanilles

Annuals, perennials, or subshrubs, 20–150 cm (fibrous-rooted). **Stems** erect, sparsely to densely villous. **Leaf blades** ovate to elliptic-oblong, 2–8 × 1–5 cm, margins toothed, abaxial faces sparsely pilose and gland-dotted. **Peduncles** minutely puberulent and sparsely to densely pilose, eglandular. **Involucres** 3–3.5 × 4–5 mm. **Phyllaries** oblong-lanceolate (0.8–1.2 mm wide), glabrous or sparsely pilose (margins often ciliate), eglandular, tips abruptly tapering, subulate, 0.5–1 mm. **Corollas** usually blue to lavender, sometimes white. **Cypselae** sparsely strigoso-hispidulous; **pappi** usually of scales 0.5–1.5(–3) mm, sometimes with tapering setae, rarely 0. $2n = 20, 40$.

Flowering Jul–Aug. Disturbed sites, mostly coastal; 0–20 m; introduced; Ala., Calif., Conn., Fla., Ga., Ky., Md., Miss., Mo., N.C.; South America; introduced, Mexico; West Indies; Central America; Pacific Islands (Hawaii).

Ageratum conyzoides is apparently native to South America. North American plants were escapes and naturalized from cultivation.

4. Ageratum houstonianum Miller, Gard. Dict. ed. 8, Ageratum no. 2. 1768 • Bluemink [1]

Ageratum conyzoides Linnaeus var. *mexicanum* (Sims) de Candolle

Annuals, 30–80 cm (fibrous-rooted). **Stems** erect to decumbent, sparsely to densely pilose. **Leaf blades** deltate to ovate, mostly 3–8 × 2.5–4 cm, margins toothed, abaxial faces sparsely to densely pilose, not evidently gland-dotted. **Peduncles** viscid-puberulent, pilose, and stipitate-glandular. **Involucres** ca. 4 × 5–6 mm. **Phyllaries** narrowly lanceolate (0.6–1 mm wide), stipitate-glandular, sparsely to densely pilose, eciliate or inconspicuously ciliate, tips gradually tapering, indurate-subulate, 0.8–2 mm. **Corollas** usually lavender, rarely white. **Cypselae** sparsely strigoso-hispidulous; **pappi** of 5 distinct, oblong scales 2–3 mm. $2n = 20$.

Flowering Jun–Aug. Disturbed sites, mostly coastal; 0–20 m; introduced; Ala., Conn., Fla., Ga., Mass., N.C., S.C., Tex.; Mexico; Central America; introduced, Pacific Islands (Hawaii).

Ageratum houstonianum is apparently native to southeastern Mexico and Central America; the North American plants are escapes and naturalized from cultivars. M. F. Johnson (1971) observed that forma *isochroum* (B. L. Robinson) M. F. Johnson (type from the state of Veracruz, Mexico) sometimes may be nearly eglandular.

397. STEVIA Cavanilles, Icon. 4: 32, plates 354, 355. 1797 • Candyleaf [For Pedro Jaime Esteve (Stevius), d. 1556, noted medical practitioner and botany professor of Valencia, Spain]

Guy L. Nesom

Annuals, perennials, subshrubs, or shrubs, 50–120+ cm. **Stems** erect, usually branched. **Leaves** cauline; all or mostly opposite or mostly alternate; petiolate or sessile; blades 1- or 3-nerved, deltate, lanceolate, lance-elliptic, lance-linear, elliptic-oblong, linear, ovate, or trullate, margins entire or serrate, serrulate, or toothed, faces hirtellous, puberulent, glabrescent, or glabrous (sometimes shiny), sometimes gland-dotted. **Heads** discoid, in loose to dense, corymbiform arrays. **Involucres** ± cylindric, (1–)2–3 mm diam. **Phyllaries** persistent, 5(–6) in ± 1 series, 2–3-nerved, ovate or oblong to lanceolate or linear, ± equal (herbaceous). **Receptacles** flat or convex, epaleate. **Florets** 5(–6); corollas purple to pink or white, throats narrowly funnelform (lengths 3–4 times diams., ± hirtellous inside); styles: bases sometimes enlarged, glabrous (sometimes papillose), branches filiform. **Cypselae** columnar to prismatic or fusiform, 5-ribbed, gland-dotted and/or scabrellous; **pappi** (sometimes ± vestigial on 1–2 cypselae of each head) persistent, of 5, (distinct or connate) muticous scales plus 0–5[–30+] subulate-aristate to setiform scales or bristles. $x = 11, 12, 17$.

Species ca. 240 (7 in the flora): w United States, Mexico, Central America, South America.

SELECTED REFERENCE Grashoff, J. L. 1972. A Systematic Study of the North and Central American Species of *Stevia*. Ph.D. dissertation. University of Texas.

1. Annuals . . . . 1. *Stevia micrantha*
1. Perennials, subshrubs, or shrubs.

[2. Shifted to left margin.—Ed.]

2. Subshrubs or shrubs.
    3. Phyllaries densely stipitate-glandular and sparsely villous to hispid; corolla lobes hispidulous . . . . 2. *Stevia lemmonii*
    3. Phyllaries glabrous, usually viscid-shiny (from resinous exudate of ± evident sunken glands), not villous or hispid; corolla lobes glabrous . . . . 3. *Stevia salicifolia*
2. Perennials.
    4. Leaves mostly opposite, blades lanceolate, lance-ovate, oblanceolate, ovate, or trullate; phyllary apices rounded to blunt or obtuse; pappi usually shorter than corollas (sometimes coroniform).
        5. Leaves petiolate (raised venation usually including only midvein and primary laterals); involucres 4–5.5(–6) mm . . . . 4. *Stevia ovata*
        5. Leaves sessile or nearly so (raised venation reticulate, including secondary and tertiary veins); involucres (5.5–)6–8.5 mm . . . . 5. *Stevia plummerae*
    4. Leaves mostly alternate (often with abundant axillary clusters of smaller leaves), blades lance-linear, linear, linear-oblong, narrowly oblanceolate, or oblong; phyllary apices acute to acuminate; pappi usually equaling corollas (sometimes coroniform or 0).
        6. Heads in ± congested, compact clusters; peduncles 0 or 1–4 mm; involucres 5–6(–7) mm; phyllaries sessile-glandular, sparsely villosulous . . . . 6. *Stevia serrata*
        6. Heads borne in open, subcorymbiform arrays; peduncles mostly 5–30 mm; involucres 6–8 mm; phyllaries sessile- and stipitate-glandular (not villous) . . . . 7. *Stevia viscida*

1. Stevia micrantha Lagasca, Gen. Sp. Pl., 27. 1816 • Annual candyleaf

Annuals, 10–35(–40) cm. **Leaves** mostly opposite (distal often alternate); petioles (0–)4–11 mm; blades ovate to ovate-deltate or trullate, 1–3.5(–5) cm, margins serrate. **Heads** in open, lax, subcorymbiform or paniculiform (sometimes nearly cylindric) arrays. **Peduncles** (1–)3–10 mm, stipitate-glandular. **Involucres** 6–8 mm. **Phyllaries** stipitate- and sessile-glandular, apices acute. **Corollas** white, lobes sparsely hirsute. **Pappi** surpassing corollas. **2***n* = 22.

Flowering (Aug–)Sep–Oct. Grassy openings, damp and shady, oak, oak-walnut, oak-pine, and pinyon pine-ponderosa pine woodlands; 1800–2700 m; Ariz., N.Mex.; Mexico.

2. Stevia lemmonii A. Gray, Proc. Amer. Acad. Arts 17: 204. 1882 (as lemmoni) • Lemmon's candyleaf

Shrubs or subshrubs, 40–100 cm. **Leaves** opposite; petioles (0–)3–6 mm; blades elliptic-oblong to lance-elliptic, 3–8 cm, margins serrate or entire. **Heads** in ± congested, compact clusters. **Peduncles** 0 or 1–2 mm, hirsutulous. **Involucres** 5–6 mm. **Phyllaries** densely stipitate-glandular and sparsely villous to hispid, apices acute. **Corollas** white, lobes hispidulous. **Pappi** coroniform (scales shorter than corollas).

Flowering (Feb–)Apr–May(–Sep). Rocky slopes, road cuts, streamsides, oak-juniper, pine-oak, and cottonwood-sycamore woodlands; 900–1700 m; Ariz.; Mexico (Chihuahua, Sinaloa, Sonora).

Stevia hispidula (Grashoff) B. L. Turner, based on *S. lemmonii* var. *hispidula* Grashoff, is illegitimate; it is a later homonym of *S. hispidula* de Candolle.

3. Stevia salicifolia Cavanilles, Icon. 4: 32, plate 354. 1797 • Willow-leaf candyleaf

Varieties 3 (1 in the flora): Texas, Mexico.

3a. Stevia salicifolia Cavanilles var. **salicifolia**

Shrubs, 10–50(–80) cm (glabrous). **Leaves** opposite; petioles 0 or 2–10 mm; blades lance-linear to lanceolate or narrowly elliptic, 3–9 cm, (bases gradually narrowed) margins entire or serrate to serrulate. **Heads** borne in tightly clustered arrays, usually in discrete groups of 10–25. **Peduncles** 0–4 mm, glabrous or sparsely puberulent. **Phyllaries** 5.5–6.5(–7) mm, glabrous (usually viscid-shiny from resinous exudate of ± evident sunken glands), apices acute to obtuse. **Corollas** white, lobes glabrous. **Pappi** usually coroniform and shorter than corollas, sometimes equaling corollas. **2***n* = 24.

Flowering (Jul–)Aug–Oct (n Mexico). Oak and oak-pine woodlands, rocky sites, crevices, boulder pockets; 1500–2700 m; N.Mex.; Mexico.

Stevia salicifolia is variable. Some 14 specific and infraspecific names were regarded as synonyms of var.

salicifolia by J. L. Grashoff (1972). The variety differs from other members of *Stevia* in the flora area by its stem and leaf surfaces (glabrous-shiny from viscid exudate of sunken glands); other taxa in the flora area usually have sparsely hairy, not shiny, stems and leaves. The variety is known in the United States from a single locality (Hidalgo County); from Mexico, it closely approaches Texas and Arizona (B. L. Turner 1996+, vol. 2) and it seems likely that it may eventually be discovered in those states.

4. Stevia ovata Willdenow, Enum. Pl., 855. 1809 • Round-leaf candyleaf

Varieties 4 (1 in the flora): Texas, Mexico.

4a. Stevia ovata Willdenow var. **texana** Grashoff, Brittonia 26: 367 fig. 11. 1974 • Texas candyleaf

Perennials, 40–80 cm. **Leaves** mostly opposite, distal sometimes alternate; petioles (0–)2–10 mm; blades (raised venation mostly including only midvein and primary laterals) ovate to trullate, mostly 3–6 cm, margins serrate. **Heads** in ± congested, compact clusters. **Peduncles** 0 or 1–2 mm, sessile-glandular and finely villous. **Involucres** 4–5.5 (–6) mm. **Phyllaries** sessile-glandular and finely villous, apices rounded to blunt or obtuse. **Corollas** white or light pink, lobes sessile-glandular. **Pappi** shorter than corollas.

Flowering Jun–Sep. Rocky sites; 1700–2400 m; Tex.; Mexico (Coahuila).

The type of *Stevia rhombifolia* Kunth, a name that has been used for plants of var. *texana*, is referable to var. *ovata*.

5. Stevia plummerae A. Gray, Proc. Amer. Acad. Arts 17: 204. 1882 • Plummer's candyleaf

Varieties 2 (1 in the flora): sw United States, Mexico.

5a. Stevia plummerae A. Gray var. **plummerae**

Stevia plummerae var. *alba* A. Gray

Perennials, 30–80 cm. **Leaves** mostly opposite (distal sometimes alternate); petioles 0–3(–5) mm; blades (raised venation reticulate, including secondary and tertiary veins) lanceolate, lance-ovate, or oblanceolate, (2–)3–10 cm, margins coarsely serrate. **Heads** in ± congested, compact clusters. **Peduncles** 0 or 1–4 mm, villous-puberulent. **Involucres** (5.5–)6–8.5 mm. **Phyllaries** eglandular or very sparsely glandular, sparsely

villous-puberulent, apices usually rounded to blunt or obtuse, sometimes acute. **Corollas** whitish, pale rose, pink, or red, lobes sparsely glandular and hispid-villous. **Pappi** shorter than corollas, sometimes coroniform. $2n$ = 34, 42.

Flowering (Aug–)Sep–Oct. Canyon walls, open slopes, oak to ponderosa pine or ponderosa pine-Douglas fir woodlands; 2100–2800 m; Ariz., N.Mex.; Mexico (Chihuahua, Durango, Sonora).

6. Stevia serrata Cavanilles, Icon. 4: 33, plate 355. 1797 • Saw-tooth candyleaf [F]

Stevia serrata var. *haplopappa* B. L. Robinson; *S. serrata* var. *ivifolia* (Willdenow) B. L. Robinson

Perennials, 40–100 cm. **Leaves** mostly alternate (at least not regularly opposite, crowded, with axillary clusters of smaller leaves); petioles 0; blades (3-nerved) narrowly lanceolate to lance-linear, 1.5–4 cm, margins serrulate. **Heads** borne in ± congested, compact clusters. **Peduncles** 0 or 1–4 mm, sessile-glandular, villous-puberulent. **Involucres** 5–6 (–7) mm. **Phyllaries** sessile-glandular, sparsely villosulous, apices acute to acuminate. **Corollas** white or pink, lobes sparsely sessile-glandular, finely villous-hirsute. **Pappi** usually ± equaling corollas, sometimes coroniform or 0. $2n$ = (22–)34(–54) univalents, less often 17 pairs.

Flowering (Jul–)Aug–Oct. Roadsides, disturbed sites, oak-grasslands, oak-pine grasslands, and oak, mixed conifer-oak, mixed pine, ponderosa pine-Douglas fir, pine-fir-aspen, spruce-Douglas fir, and fir-hemlock woodlands; 1700–2700 m; Ariz., N.Mex., Tex.; Mexico.

Some collections of *Stevia serrata* from Cochise and Graham counties, Arizona, were annotated by J. L. Grashoff as "*S. serrata* > *plummerae*"; in leaf arrangement and morphology (venation, margin, and shape), they appear to be similar to typical *S. serrata* from the same area.

7. Stevia viscida Kunth in A. von Humboldt et al., Nov. Gen. Sp. 4(fol.): 110, plate 351. 1818; 4(qto.): 140. 1820 • Viscid candyleaf

Perennials, 40–100 cm. **Leaves** mostly alternate (usually crowded with axillary clusters of smaller leaves); petioles 0; blades (3-nerved) linear-oblong, linear-oblanceolate, or linear, mostly 2–5 cm, margins shallowly toothed distally or entire. **Heads** in open, subcorymbiform arrays. **Peduncles** mostly 5–30 mm, viscid-puberulent and

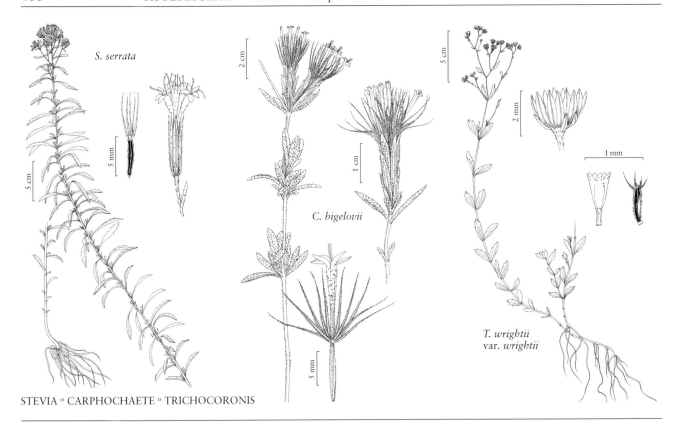

S. serrata

C. bigelovii

T. wrightii
var. wrightii

STEVIA ∘ CARPHOCHAETE ∘ TRICHOCORONIS

stipitate-glandular. **Involucres** 6–8 mm, stipitate- and sessile-glandular. **Phyllaries** stipitate- and sessile-glandular, apices acute to acuminate. **Corollas** white or pink, purplish, or rose, lobes glandular and sparsely hispid-villous. **Pappi** ± equaling corollas. $2n = 22, 33, 44$.

Flowering Aug–Oct(–Nov). Roadsides, pastures, other disturbed sites; 1500–2000 m; Ariz., Tex.; Mexico; Central America.

398. CARPHOCHAETE A. Gray, Mem. Amer. Acad. Arts, n. s. 4: 65. 1849 • [Greek *karphos*, chaff, and *chaite*, long bristle]

David J. Keil

Subshrubs or shrubs, (8–)20–45(–120+)[–300] cm. **Stems** erect [decumbent], branched from bases and/or ± throughout. **Leaves** cauline; opposite [alternate]; sessile; blades 1(–3)-nerved, narrowly elliptic or oblanceolate to linear, faces glabrate or puberulent, gland-dotted (in pits). **Heads** discoid, usually borne singly or in pairs [in ± cymiform arrays]. **Involucres** cylindric [turbinate], [2–]3–5 mm diam. **Phyllaries** persistent (at least outer), 8–12+ in 3–4+ series, not notably nerved, deltate-ovate to oblong, lanceolate, or linear, unequal. **Receptacles** flat to slightly convex (glabrous), epaleate [partially paleate]. **Florets** 3–4[–6+]; corollas white or pink to purple, throats narrowly funnelform (sometimes puberulent inside), lobes 5, narrowly triangular to linear (adaxially densely papillose); styles: bases enlarged, glabrous, branches ± filiform (appendages cylindric or slightly flattened, densely papillose). **Cypselae** subcylindric, [4–]8–10-ribbed, ± hirtellous; **pappi** persistent, of 0–5+ muticous, erose, lacerate, or lanceolate to subulate scales (1–4 mm) plus [5–]9–12+ aristate scales (10–15 mm) [coroniform]. $x = 11, 12$.

Species 7 (1 in the flora): sw United States, Mexico.

SELECTED REFERENCE Turner, B. L. 1987. Taxonomy of *Carphochaete* (Asteraceae–Eupatorieae). Phytologia 64: 145–162.

1. Carphochaete bigelovii A. Gray, Smithsonian Contr. Knowl. 3(5): 89. 1852 • Bristlehead F

Stems usually stiff, erect. **Leaves** often fascicled on older growth; blades 5–35 mm. **Heads** terminal (± pedunculate) or in leaf axils (sessile). **Involucres** 12–20 mm. **Phyllaries:** gland-dotted, sometimes puberulent as well, margins narrowly hyaline, apices acute. **Corollas** 15–20 mm, throats usually purplish, lobes creamy white. **Cypselae** yellow-green to golden brown, 11–14 mm; **pappus scales** mostly brownish to purplish with colorless, hyaline margins, aristae of the longer 1–3+ mm, scabrous. $2n = 22$.

Flowering Jan–Jun. Sandy soils, rock outcrops in grasslands, chaparral, pine-oak woodlands; 900–2200 m; Ariz., N.Mex., Tex.; Mexico (Chihuahua, Coahuila, Sonora).

399. TRICHOCORONIS A. Gray, Mem. Amer. Acad. Arts., n. s. 4: 65. 1849 • [Greek *trichos*, hair, and *koronos*, crown, apparently alluding to setiform pappus elements]

Guy L. Nesom

Annuals or perennials, 10–30 cm (aquatic or subaquatic). **Stems** decumbent to erect, branched from bases or ± throughout (sometimes rooting at proximal nodes). **Leaves** cauline; mostly opposite (distal sometimes alternate); sessile; blades usually 1–3-nerved from near bases, oblong to elliptic or lanceolate, margins usually ± dentate (at least distally), faces glabrous or sparsely pilose to pilosulous and glabrescent, sometimes minutely gland-dotted. **Heads** discoid, borne singly or in loose or open, corymbiform arrays. **Involucres** mostly hemispheric or broader, 3–4(–5) mm diam. **Phyllaries** persistent, 18–30 in 2–3 series, obscurely 3–4-nerved, lanceolate, ± equal (herbaceous). **Receptacles** convex to conic, epaleate (warty). **Florets** 75–125; corollas whitish or pinkish to purplish, throats tubular-funnelform (lengths ca. 2 times diams.); styles: bases not enlarged, glabrous, branches linear-filiform (distally densely papillose). **Cypselae** weakly fusiform, 4–5-ribbed, scabrellous on ribs; **pappi** persistent, of 2–6+ laciniate or fimbriate to setiform scales or coarsely barbellate bristles. $x = 15$.

Species 2 (1 in the flora): Texas, Mexico; introduced in California.

SELECTED REFERENCE King, R. M. and H. Robinson. 1970b. A monograph of the genus *Trichocoronis*. Phytologia 19: 497–500.

1. Trichocoronis wrightii (Torrey & A. Gray) A. Gray, Mem. Amer. Acad. Arts, n. s. 4: 65. 1849 • Limestone bughead

Ageratum wrightii Torrey & A. Gray, Proc. Amer. Acad. Arts 1: 46. 1847

Varieties 2 (1 in the flora): Texas, Mexico; introduced in California.

1a. Trichocoronis wrightii (Torrey & A. Gray) A. Gray var. **wrightii** F

Trichocoronis riparia (Greene) Greene

Plants sometimes colonial. **Leaves** mostly 10–25 mm. **Involucres** 1.8–2.5 mm. **Corollas** ca. 1 mm. **Cypselae** ca. 1 mm. $2n = 30$.

Flowering Feb–Jul(–Oct). Edges of ponds, streams, ditches, wet depressions in clay and sand, prairies, brushlands, roadsides; 0–20 m; Calif., Tex.; Mexico.

The California populations of *Trichocoronis wrightii* have been described as "adventive" and "native to Texas." *Trichocoronis wrightii* var. *wigginsii* R. M. King & H. Robinson is regarded as endemic to Sierra de la Giganta in Baja California Sur. It differs from var. *wrightii* in its taller, mostly unbranched stems, greater number of flowers per head, and less thickened carpopodial cells (King and Robinson 1970b).

400. SHINNERSIA R. M. King & H. Robinson, Phytologia 19: 297. 1970 • [For Lloyd Herbert Shinners, 1918–1971, botanist, long at Southern Methodist University, founder of the journal *Sida*]

Guy L. Nesom

Perennials, 10–30 cm (aquatic). **Stems** ascending, unbranched or branched (rooting at proximal nodes). **Leaves** cauline; opposite; sessile; blades 1- or 3-nerved, broadly oblanceolate or 3-lobed, ultimate margins coarsely toothed, faces sometimes pilose and glabrescent or glabrous, gland-dotted. **Heads** discoid, borne singly. **Involucres** hemispheric or broader, 6–9 mm diam. **Phyllaries** persistent, 25–30 in 2–3+ series, not notably nerved, ovate to lanceolate or linear, ± equal (herbaceous to membranous). **Receptacles** convex to conic, epaleate (warty). **Florets** 75–100+; corollas white, throats broadly campanulate (notably wider than tubes, lengths ca. 1 times diams.); styles: bases not enlarged, glabrous, branches linear-filiform. **Cypselae** prismatic, 4–5-ribbed, sparsely gland-dotted and sparsely hairy (hairs forked); **pappi** 0. $x = 15$?

Species 1: Texas, n Mexico.

1. Shinnersia rivularis (A. Gray) R. M. King & H. Robinson, Phytologia 19: 297. 1970 • Rio Grande bugheal [F]

Trichocoronis rivularis A. Gray, Mem. Amer. Acad. Arts, n. s. 4: 66. 1849

Plants rooted in muck; rhizomes relatively long. **Stems** commonly submerged except for terminal 10 cm or less. **Leaves** mostly 2–4 cm. **Heads** borne out of water. **Corollas** ca. 2 mm. **Cypselae** ca. 2 mm. $2n = 60$.

Flowering (Dec–)Mar–May(–Aug). Water along streams, in slow-moving water (to 1 m deep) arising from calcareous outcrops; 50–3000 m; Tex.; Mexico (Coahuila, Nuevo León).

Shinnersia rivularis has been treated within *Trichocoronis*. The genera differ in features noted in the couplet below. *Sclerolepis*, of the southeastern United States, apparently also is closely related, and its recognition as separate reinforces recognition of *Shinnersia*.

a. Heads borne singly, from axils of primary leaves; involucres 5–6 mm; corollas infundibular, tubes relatively long; cypselae sparsely gland-dotted and with forked hairs; carpopodia oblique; pappi 0. . . . . *Shinnersia*

b. Heads usually in loose or open, corymbiform, bracteate arrays; involucres 2–3 mm; corollas tubular-funnelform, tubes relatively short; cypselae eglandular, scabrellous on angles, without forked hairs; carpopodia at right angles; pappi persistent, of 2–6 setiform scales. . . . . *Trichocoronis*

401. SCLEROLEPIS Cassini, Bull. Sci. Soc. Philom. Paris 1816: 198. 1816 • Bogbutton [Greek *scleros*, hard, and *lepis*, scale, alluding to pappus] [E]

Eric E. Lamont

Perennials, 10–30(–60) cm (rhizomatous; aquatic to subaquatic). **Stems** decumbent or erect (sometimes floating), simple or branched distal to bases (terete, glabrous; internodes usually shorter than leaves). **Leaves** cauline; whorled (4–6 per node); sessile; blades 1-nerved, linear,

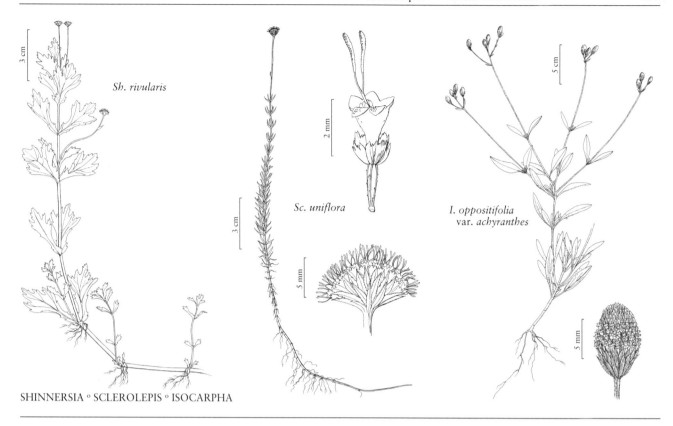

SHINNERSIA ∘ SCLEROLEPIS ∘ ISOCARPHA

margins entire (apices callus-tipped), faces glabrous or sparsely gland-dotted. **Heads** discoid, borne singly (peduncles ebracteate). **Involucres** campanulate to subhemispheric, ca. 10 mm diam. **Phyllaries** persistent, 22–30+ in 2–3+ series, not notably nerved, linear to broadly lanceolate, unequal to ± equal (glabrous). **Receptacles** conic or hemispheric, epaleate. **Florets** ca. 50; corollas pink-lavender to flesh-colored, throats funnelform (sparsely stipitate-glandular), lobes 5, deltate; styles: bases not enlarged, glabrous, branches filiform to linear-clavate. **Cypselae** prismatic, 5-ribbed, gland-dotted; **pappi** persistent, of 5 muticous scales. *x* = 15.

Species 1: e United States.

SELECTED REFERENCES Crow, G. E. and C. B. Hellquist. 2000. Aquatic and Wetland Plants of Northeastern North America.... 2 vols. Madison. Vol. 1, pp. 414–415. Mehrhoff, L. J. 1983. *Sclerolepis uniflora* (Compositae) in New England. Rhodora 85: 433–438.

1. **Sclerolepis uniflora** (Walter) Britton, Sterns & Poggenburg, Prelim. Cat., 25. 1888 • Pink bogbutton, one-flowered sclerolepis E F

Ethulia uniflora Walter, Fl. Carol., 195. 1788

Plants usually creeping and mat-forming; rhizomes relatively slender. **Stems:** submersed flexuous, to 60 cm; emersed ± ascending, 10–30(–45) cm. **Leaf blades** 7–23 × 0.3–2 mm. **Peduncles** 2–6 cm. **Phyllaries** 2–4 mm, margins minutely serrulate, faces glabrous. **Corollas** 2–3 mm (style branches conspicuously exserted). **Cypselae** 2–3.5 mm; **pappi:** scales indurate, broadly oblong, margins crenulate, apices obtuse, erose. *2n* = 30.

Flowering summer–fall. In still water, sandy or peaty shorelines of ponds, bogs, swamps, wet flatwoods, low savannas, wet depressions, ditches; 0–100+ m; Ala., Del., Fla., Ga., Md., Mass., N.H., N.J., N.C., R.I., S.C., Va.

Sclerolepis uniflora is uncommon north of North Carolina, being restricted to the Delmarva Peninsula in Delaware, Maryland, and Virginia; in New Jersey it occurs in the southern part of the pine barrens and in Cape May County. Two populations of *S. uniflora* are known in New England: one in southern New Hampshire, the other on the Massachusetts-Rhode Island state border.

The submersed, aquatic growth form of *Sclerolepis uniflora* has been rarely reported or collected. The submersed form is entirely vegetative; the stems are longer and the leaves more flaccid, elongated, and loosely arranged than in emergent plants. As water levels drop, the horizontal stems become anchored in exposed substrates and develop upright shoots of the terrestrial form, producing solitary, terminal heads of flowers.

402. ISOCARPHA R. Brown, Trans. Linn. Soc. London 12: 110. 1817 • Pearlhead

[Greek *iso*-, same, and *carphos*, small dry body, evidently alluding to uniform receptacular paleae]

Guy L. Nesom

Perennials or subshrubs [annuals], 10–150 cm. **Stems** usually erect, sometimes basally decumbent [creeping]. **Leaves** cauline; mostly opposite (distal sometimes alternate) [alternate]; petiolate or sessile; blades ± 3-nerved, elliptic to narrowly lance-elliptic [ovate or lanceolate to linear], margins entire or serrulate [dentate], faces glabrate to weakly hirsutulous or pilosulous, gland-dotted. **Heads** discoid, borne singly or in tight, corymbiform to subcapitate [paniculiform] arrays. **Involucres** ± obconic, 4–6 mm diam. **Phyllaries** persistent, 10–15+ in 2+ series, [1–]2–3[–6]-nerved, ovate or elliptic to lanceolate or linear, ± equal. **Receptacles** conic to columnar, paleate (paleae similar to inner phyllaries). **Florets** 60–200+; corollas white or pinkish, throats funnelform (lengths ca. 2 times diams.) [cylindric or campanulate], lobes 5, ± deltate; styles: bases enlarged, glabrous [papillose], branches filiform. **Cypselae** prismatic, 5-ribbed, glabrous [pubescent]; **pappi** 0. $x = 10$.

Species 5 (1 in the flora): s, c United States, Mexico, West Indies, Central America, South America.

SELECTED REFERENCE Keil, D. J. and T. F. Stuessy. 1981. Systematics of *Isocarpha* (Compositae: Eupatorieae). Syst. Bot. 6: 258–287.

1. Isocarpha oppositifolia (Linnaeus) Cassini in
  F. Cuvier, Dict. Sci. Nat. ed. 2, 24: 19. 1822
  • Rio Grande pearlhead

Santolina oppositifolia Linnaeus, Syst. Nat. ed. 10, 2: 1207. 1759

Varieties 2 (1 in the flora): Texas, Mexico, Central America, South America.

1a. Isocarpha oppositifolia (Linnaeus) Cassini var.
  achyranthes (de Candolle) D. J. Keil & Stuessy, Syst.
  Bot. 6: 280. 1981 F

Dunantia achyranthes de Candolle in A. P. de Candolle and A. L. P. P. de Candolle, Prodr. 5: 627. 1836

Stems: nodes without axillary fascicles. **Leaves** mostly sessile, sometimes ± petiolate; blades mostly 2–3(–5) cm × 5–8(–15) mm, without basal auricles. **Heads** pedunculate (peduncles 2–5 cm) or subsessile (in groups of 2–5). **Corollas** ca. 1.5 mm. **Cypselae** ca. 1.5 mm. $2n = 20$.

Flowering (Jun–)Oct–Dec. Thickets, clay dunes, shrublands; 0–10 m; Tex.; Mexico; Central America; South America.

The Texas plants are known only from around Brownsville in Cameron County. D. J. Keil and T. F. Stuessy (1981) identified them as var. *oppositifolia*, noting that they are "probably introduced" there. Distribution of the species continues from Cameron County southward along the Gulf coast of Mexico in Tamaulipas and Veracruz, and the Texas plants are similar to the Mexican ones, which have been identified as var. *achyranthes*. Abundant Texas material, as described above, fits the description of var. *achyranthes*, as delimited by Keil and Stuessy.

403. BRICKELLIA Elliott, Sketch Bot. S. Carolina 2: 290. 1823, name conserved · [For John Brickell, 1748–1809, Irish-born physician and naturalist who settled in Georgia (not John Brickell, 1710?–1745, Irish naturalist who visited North Carolina ca. 1729–1731 and published on the natural history of North Carolina in 1737)]

Randall W. Scott

Annuals, perennials, subshrubs, or shrubs, (12–)30–120(–200) cm. **Stems** mostly erect, often much branched (sometimes virgate, often striate). **Leaves** cauline; opposite or alternate; petiolate or sessile; blades usually 3-nerved from bases, deltate, lance-elliptic, lance-linear, lanceolate, lance-ovate, lance-rhombic, linear, oblong, obovate, ovate, rhombic-ovate, spatulate, or suborbiculate, margins mostly crenate, dentate, entire, laciniate-dentate, lobed, serrate, or toothed, faces glabrous (sometimes shiny) or glandular-puberulent, strigose, or tomentose, sometimes gland-dotted. **Heads** usually in corymbiform, sometimes cymiform, paniculiform, or racemiform, arrays, rarely borne singly. **Involucres** cylindric to obconic or campanulate, 5–12 mm diam. **Phyllaries** persistent, (10–)14–45(–60) in 3–7(–9) series, usually (4–)5–6 (–16)-striate or -nerved, linear or lanceolate to oblanceolate or oblong, usually unequal (usually chartaceous, sometimes herbaceous). **Receptacles** flat to convex, epaleate. **Florets** (3–)8–45 (–90); corollas usually white or whitish to cream, sometimes greenish, purplish or yellowish, throats mostly cylindric to narrowly funnelform (lengths 3–5 times diams.); styles: bases enlarged, hairy, branches narrowly clavate (± dilated distally). **Cypselae** narrowly prismatic, 10-ribbed, glabrous or hairy to glabrate, often gland-dotted; **pappi** persistent, of 10–80 usually smooth or barbellulate to barbellate, sometimes plumose or subplumose bristles in 1 series. $x = 9$.

Species ca. 100 (32 in the flora): North America, Mexico, Central America.

SELECTED REFERENCE Robinson, B. L. 1917. A monograph of the genus *Brickellia*. Mem. Gray Herb. 1: 3–151.

1. Pappi usually of plumose or subplumose, sometimes barbellate, bristles.
  2. Phyllaries puberulent, often densely gland-dotted as well . . . . 13. *Brickellia eupatorioides*
  2. Phyllaries glabrous or puberulent (not gland-dotted).
    3. Stems (at least distally) pubescent (lacking relatively long, gland-tipped hairs); florets 9–12. . . . . 5. *Brickellia brachyphylla*
    3. Stems (at least distally) stipitate-glandular (hairs relatively long); florets 11–22 . . . . 18. *Brickellia hinckleyi*
1. Pappi usually of smooth or barbellulate to barbellate, sometimes subplumose, bristles.
  4. Florets (40–)45–90.
    5. Stems and leaves densely white-tomentose . . . . 19. *Brickellia incana*
    5. Stems and leaves glabrous or glandular-puberulent, puberulent, pubescent, stipitate-glandular, strigose, villous, or viscid-glandular, often gland-dotted as well (not tomentose).
      6. Shrubs; leaf margins usually sharply dentate or dentate-serrate, rarely entire; outer phyllaries ovate, lance-ovate, or lance-linear . . . . 2. *Brickellia atractyloides*
      6. Perennials (caudices woody); leaf margins crenate, crenate-dentate, dentate, or serrate; outer phyllaries lance-ovate to narrowly lanceolate.
        7. Peduncles 0–2 mm; involucres 14–18 mm; cypselae 5.5–7 mm . . . . . . 17. *Brickellia greenei*
        7. Peduncles 4–50 mm; involucres 7–14 mm; cypselae 4–5 mm.
          8. Petioles 10–70 mm; peduncles 4–30 mm; involucres 7–12 mm; phyllaries 30–40; corollas 6.5–7.5 mm . . . . 16. *Brickellia grandiflora* (in part)
          8. Petioles 3–20 mm; peduncles 20–50 mm; involucres 12–14 mm; phyllaries 40–60; corollas 9–10 mm . . . . 30. *Brickellia simplex*

[4. Shifted to left margin.—Ed.]

4. Florets 3–30(–50).
    9. Florets 3–7.
        10. Leaf blades lance-elliptic, lance-ovate, lanceolate, lance-linear, or linear, margins
            entire or nearly so, faces gland-dotted (and shiny), often puberulent as well . . . 23. *Brickellia longifolia*
        10. Leaf blades lance-ovate to lanceolate, margins dentate to serrate, faces gland-
            dotted and sparsely pubescent (not shiny) . . . . 20. *Brickellia knappiana*
    9. Florets 8–30(–50).
        11. Petioles 4–70 mm.
            12. Leaf apices acuminate or long-acuminate to attenuate.
                13. Phyllaries, outer: apices obtuse to acute (cypselae 2–2.5 mm) . . . . 29. *Brickellia rusbyi*
                13. Phyllaries, outer: apices acute, acuminate, or mucronate.
                    14. Shrubs; leaf margins subentire or ± toothed (usually with 1–3 sets of
                        sharp teeth near bases); phyllaries 17–22 . . . . 9. *Brickellia coulteri*
                    14. Perennials; leaf margins crenate, dentate, or serrate; phyllaries 28–40.
                        15. Heads (erect in fruit) in corymbiform or paniculiform arrays;
                            corollas pale yellow-green, often purple-tinged, 6–7.2 mm; pappi
                            of 38–46 purple-tinged bristles . . . . 8. *Brickellia cordifolia*
                        15. Heads (nodding in flower and fruit) in loose, corymbiform or
                            paniculiform arrays; corollas pale yellow-green; pappi of 20–30
                            white bristles . . . . 16. *Brickellia grandiflora* (in part)
            12. Leaf apices acute to obtuse or rounded.
                16. Perennials; stems and peduncles stipitate-glandular.
                    17. Leaf bases rounded to cuneate; petioles 4–10 mm; phyllaries: outer
                        ovate to lance-ovate or lanceolate (equaling or surpassing inner,
                        glabrous or sparsely glandular-hirtellous) . . . . 7. *Brickellia chenopodina*
                    17. Leaf bases truncate to cordate; petioles 20–35 mm; phyllaries: outer
                        ovate to lanceolate (shorter than inner, glandular-hirtellous) . . . 14. *Brickellia floribunda*
                16. Shrubs; stems and peduncles pubescent or glandular-pubescent.
                    18. Leaves: bases cuneate, margins laciniate-dentate; peduncles 5–40 mm
                        . . . . 3. *Brickellia baccharidea*
                    18. Leaves: bases cordate to truncate, margins crenate to serrate; peduncles
                        1–5 mm . . . . 6. *Brickellia californica*
        11. Petioles 0–5 mm.
            19. Leaves opposite.
                20. Stems and leaves mostly puberulent to pubescent, sometimes gland-dotted
                    (not glandular-pubescent or stipitate-glandular).
                    21. Perennials mostly 30–90 cm; leaf blades oblong, obovate, or ovate,
                        (15–)30–80 × 10–35 mm . . . . 4. *Brickellia betonicifolia*
                    21. Perennials 12–30 cm; leaf blades deltate-ovate to lanceolate, 7–30
                        × 4–28 mm . . . . 27. *Brickellia parvula*
                20. Stems and leaves pubescent, puberulent, or tomentulose and gland-dotted,
                    glandular-pubescent, or stipitate-glandular.
                    22. Shrubs; leaf blades mostly 3–15 mm; peduncles 0–3 mm.
                        23. Phyllaries 20–24; florets 8–12 . . . . 12. *Brickellia desertorum* (in part)
                        23. Phyllaries 35–40; florets 18–33 . . . . 32. *Brickellia veronicifolia*
                    22. Perennials; leaf blades (2–)15–160 mm; peduncles (1–)3–150 mm (if
                        less than 3 mm, bracteate).
                        24. Leaf blades oblong to linear, 2–10 mm wide, margins entire or
                            dentate; peduncles (20–)30–150 mm . . . . 31. *Brickellia venosa*
                        24. Leaf blades elliptic, lanceolate, oblong, obovate, ovate, or
                            subdeltate, mostly 10–75 mm wide, margins crenate to serrate;
                            peduncles 1–15(–25) mm.

25. Peduncles (bracteate) 1–8 mm; involucres 13–15 mm; phyllaries 34–40 . . . . 28. *Brickellia pringlei*
25. Peduncles 2–15(–25) mm; involucres 8–13 mm; phyllaries 18–24.
    26. Leaf bases rounded to cordate (clasping); peduncles densely stipitate-glandular . . . . 1. *Brickellia amplexicaulis*
    26. Leaf bases ± cuneate; peduncles densely tomentose, often gland-dotted as well.
        27. Phyllaries 18–22 in 5–7 series (lengths in outer 2–3 series no more than ¹/₂ inner) . . . . . . 10. *Brickellia cylindracea* (in part)
        27. Phyllaries 18–20 in 3–4(–5) series (lengths in outer series grading to inner) . . . . 22. *Brickellia lemmonii* (in part)
        [19. Shifted to left margin.—Ed.]
19. Leaves alternate.
  28. Leaf blades oblong or spatulate, 1–4 mm wide . . . . 15. *Brickellia frutescens*
  28. Leaf blades cordate, elliptic, lance-linear, lanceolate, lance-ovate, oblong, obovate, ovate, rhombic-ovate, subdeltate, or suborbiculate, (3–)10–100 mm wide.
    29. Florets 25–50; leaf blades elliptic, oblong, or lance-linear, margins entire . . . 26. *Brickellia oblongifolia*
    29. Florets 8–25(–45); leaf blades cordate, lanceolate, lance-ovate, oblong, obovate, ovate, rhombic-ovate, or suborbiculate, margins entire or dentate.
      30. Leaf blades elliptic, lanceolate, lance-ovate, obovate, ovate, or subdeltate.
        31. Shrubs; cypselae usually glabrous, rarely sparsely pubescent . . . . 11. *Brickellia dentata*
        31. Perennials; cypselae sparsely to densely pubescent.
          32. Phyllaries 18–22 in 5–7 series (lengths of outers to ¹/₂ inners) . . . . 10. *Brickellia cylindracea* (in part)
          32. Phyllaries 18–20 in 3–4(–5) series (lengths of outers evenly grading to inners) . . . . 22. *Brickellia lemmonii* (in part)
      30. Leaf blades mostly cordate or ovate to suborbiculate (if oblong, peduncles viscid or glandular-villous).
        33. Stems and leaves tomentose or densely puberulent, often gland-dotted as well.
          34. Involucres 8–10 mm; phyllaries 20–24; florets 8–12 . . . . . 12. *Brickellia desertorum* (in part)
          34. Involucres mostly 10–12 mm; phyllaries 30–38; florets 20–24 . . . . 25. *Brickellia nevinii*
        33. Stems and leaves sparsely pubescent to hispidulous, often gland-dotted or glandular-villous as well.
          35. Peduncles 0.5–2.5 mm, glandular-viscid; phyllaries 15–25 . . . . . . 21. *Brickellia laciniata*
          35. Peduncles (bracteate) 2–10 mm, viscid to glandular-villous; phyllaries 30–48 . . . . 24. *Brickellia microphylla*

1. Brickellia amplexicaulis B. L. Robinson, Proc. Amer. Acad. Arts 47: 199. 1911 [E]

Perennials, 100–200 cm. **Stems** branched from bases, pubescent and gland-dotted. **Leaves** opposite; petioles 0–1 mm; blades 3-nerved from bases (pinnately-veined distally), lanceolate to oblong, 40–160 × 10–75 mm, bases rounded to cordate (clasping), margins crenate to serrate, apices acute, faces densely to sparsely pubescent and stipitate-glandular. **Heads** in paniculiform arrays. **Peduncles** 7–25 mm, densely stipitate-glandular. **Involucres** cylindric to campanulate, 9–13 mm. **Phyllaries** 18–24 in 4–5 series, greenish, often purple-tinged, 3–8-striate, unequal, margins narrowly scarious (apices acute to acuminate); outer ovate (6–8-striate, sparsely glandular at apices), inner narrowly lanceolate (3–4-striate, glabrous). **Florets** 13–22; corollas pale yellow or cream, often purple- or red-tinged, 6–8 mm. **Cypselae** 2.8–4 mm, densely pubescent to hirtellous; **pappi** of 35–60 white, barbellulate bristles.

Flowering Sep–Dec. Rocky or grassy montane slopes, riparian habitats; 700–1900 m; Ariz., N.Mex.

Brickellia amplexicaulis is known from se Arizona and sw New Mexico.

2. Brickellia atractyloides A. Gray, Proc. Amer. Acad. Arts 8: 290. 1870

Shrubs, 20–50 cm. **Stems** densely branched, glandular-puberulent. **Leaves** opposite or alternate; petioles 0–3 mm; blades 3–4-nerved from bases (veins prominent), deltate, lanceolate, or ovate, 10–50 × 5–25 mm, bases acute to truncate or cordate, margins usually sharply dentate or dentate-serrate, rarely entire, apices acute to acuminate, faces glabrous or strigose and minutely glandular-puberulent. **Heads** in open, paniculiform arrays. **Peduncles** 10–70 mm, hispid to hispidulous and stipitate-glandular. **Involucres** cylindric to broadly campanulate, 9–15 mm. **Phyllaries** 24–33 in 3–4 series, green, 3–16-striate, subequal or unequal, margins narrowly scarious (apices acute to acuminate); outer (often bright green, 4–16-striate) lance-linear, linear-ovate, or broadly ovate, inner (pale green, 3–4-striate) linear or narrowly lanceolate (often chartaceous, scabrellous, often glandular). **Florets** 40–90; corollas pale yellow-green or cream, often purple-tinged, 6–8 mm. **Cypselae** 3–5.5 mm, scabrellous; **pappi** of 18–25 smooth or barbellulate bristles.

Varieties 3 (3 in the flora): sw United States, nw Mexico.

All three varieties of *Brickellia atractyloides* often intergrade and are not readily distinguished in some cases.

1. Phyllaries: outer broadly ovate . . . .
. . . . 2a. *Brickellia atractyloides* var. *atractyloides*
1. Phyllaries: outer lance-ovate to lance-linear.
  2. Phyllaries, outer: margins ± entire . . . . . .
. . . . 2b. *Brickellia atractyloides* var. *arguta*
  2. Phyllaries, outer: margins dentate . . . .
. . . . 2c. *Brickellia atractyloides* var. *odontolepis*

2a. Brickellia atractyloides A. Gray var. atractyloides

E

Leaf blades deltate, lanceolate, or ovate, 3- or 4-nerved from bases, 10–50 × 5–25 mm, margins sharply dentate or entire, apices acuminate, faces glabrous or minutely glandular-puberulent. **Involucres** 10–15 mm. **Phyllaries:** outer 7–16-striate, broadly ovate, margins entire, inner narrowly lanceolate. **Florets** 40–90; corollas pale yellow-green. $2n = 18$.

Flowering Mar–Sep. Rock crevices, cliff faces, talus slopes, outwash fans; 600–1600 m; Ariz., Calif., Colo., Nev., Utah.

2b. Brickellia atractyloides A. Gray var. arguta (B. L. Robinson) Jepson, Man. Fl. Pl. Calif., 1015. 1925

Brickellia arguta B. L. Robinson, Mem. Gray Herb. 1: 102, fig. 79. 1917

Leaf blades ovate, 10–20 × 5–15 mm, margins dentate-serrate, faces strigose, often glandular-puberulent as well. **Involucres** 10–15 mm. **Phyllaries:** outer 6–10-striate, lance-ovate to lance-linear (often stipitate-glandular), margins ± entire, inner linear (± chartaceous). **Florets** 40–55; corollas cream, often purple-tinged. $2n = 18$.

Flowering Apr–Jun. Desert mountain slopes, washes; 100–1500 m; Ariz., Calif., Nev.; Mexico (Baja California).

2c. Brickellia atractyloides A. Gray var. odontolepis (B. L. Robinson) Jepson, Man. Fl. Pl. Calif., 1016. 1925

Brickellia arguta B. L. Robinson var. *odontolepis* B. L. Robinson, Mem. Gray Herb. 1: 103, fig. 79ß. 1917

Leaf blades ovate, 10–20 × 5–15 mm, margins dentate-serrate, faces strigose, often glandular-puberulent as well. **Involucres** 10–15 mm. **Phyllaries:** outer 5–12-striate, lance-ovate to lance-linear, margins dentate, inner linear (± chartaceous). **Florets** 40–55; corollas cream, often purple-tinged.

Flowering Apr–Jun. Desert mountain slopes, washes; 100–1500 m; Calif.; Mexico (Baja California).

3. Brickellia baccharidea A. Gray, Smithsonian Contr. Knowl. 3(5): 87. 1852 C E

Shrubs, 100–150 cm. **Stems** (ascending) branched, pubescent, sometimes glandular-pubescent. **Leaves** alternate; petioles 4–10 mm; blades 3-nerved from bases, rhombic-ovate, 10–40 × 10–20 mm, bases cuneate, margins laciniate-dentate, apices acute to obtuse, faces glabrate or gland-dotted. **Heads** in dense, paniculiform arrays. **Peduncles** 5–40 mm, pubescent. **Involucres** narrowly campanulate, 7–11 mm. **Phyllaries** 35–40 in 5–7 series, 3–5-striate, unequal, (glabrous, ± chartaceous) margins narrowly scarious (apices acute to acuminate); outer ovate, inner lanceolate. **Florets** 16–17; corollas cream, 6–7 mm. **Cypselae** 2–3 mm, pubescent; **pappi** of 18–24 white, barbellate bristles. $2n = 18$.

Flowering Oct–Apr. Sandy to granitic soils, limestone slopes, granitic cliffs; of conservation concern; 500–1600 m; Ariz., N.Mex., Tex.

4. Brickellia betonicifolia   A. Gray, Smithsonian Contr. Knowl. 5(6): 72. 1853 (as betonicaefolia)

Perennials, 30–90 cm. **Stems** branched from bases, pubescent. **Leaves** opposite; petioles 1–5 mm; blades 3-nerved from bases, oblong, obovate, or ovate, (15–) 30–80 × 10–35 mm, bases cordate or truncate, margins crenate-serrate, apices acute to obtuse, faces sparsely pubescent and gland-dotted. **Heads** erect, usually in loose, paniculiform arrays, sometimes borne singly. **Peduncles** 2–15 mm, hispid to hirsute. **Involucres** cylindric to narrowly campanulate, 8–12 mm. **Phyllaries** 17–20 in 3–5 series, pale green, often purplish, 5–9-striate, unequal, margins narrowly scarious-ciliate (apices acute to acuminate); outer lance-ovate (sparsely pubescent), inner narrowly lanceolate (glabrous). **Florets** 10–16; corollas purplish, 7–9 mm. **Cypselae** 3–4 mm, densely pubescent to velutinous; **pappi** of 40–52 white, barbellate bristles. *2n* = 18.

Flowering Jul–Oct. Oak woodlands, open limestone hillsides; 1500–2000 m; Ariz., N.Mex.; Mexico.

5. Brickellia brachyphylla   (A. Gray) A. Gray, Smithsonian Contr. Knowl. 3(5): 84. 1852   E

Clavigera brachyphylla A. Gray, Mem. Amer. Acad. Arts, n. s. 4: 63. 1849

Perennials, 30–100 cm (caudices woody). **Stems** branched, pubescent. **Leaves** mostly alternate (sometimes subopposite); petioles 0–3 mm; blades 3-nerved from bases, lanceolate or lance-ovate, 10–50 × 4–20 mm, bases acute to obtuse, margins serrate or entire, apices acute, faces sparsely to densely pubescent, often gland-dotted or stipitate-glandular. **Heads** usually in open, racemiform or paniculiform arrays, rarely borne singly. **Peduncles** 4–20 mm, pubescent. **Involucres** cylindric to campanulate, 8–11 mm. **Phyllaries** 15–20 in 4–5 series, greenish, often purple-tinged, 5–9-striate, unequal, margins narrowly scarious (often ciliate, apices acute to acuminate or subaristate); outer lance-ovate (often puberulent), inner narrowly lanceolate (glabrous). **Florets** 9–12; corollas pale yellow-green, often purple-tinged, 4.5–6 mm. **Cypselae** 2.5–5.3 mm, mostly velutinous, sometimes pubescent; **pappi** of 27–32 white, usually plumose, rarely barbellate, bristles. *2n* = 18.

Flowering Jul–Oct. Limestone cliffs, rhyolitic rock, rocky ridges, canyon walls, hillsides; 1400–2400 m; Ariz., Colo., Kans., N.Mex., Tex.

6. Brickellia californica   (Torrey & A. Gray) A. Gray, Mem. Amer. Acad. Arts, n. s. 4: 64. 1849   E F

Bulbostylis californica Torrey & A. Gray, Fl. N. Amer. 2: 79. 1841

Shrubs, 50–200 cm. **Stems** branched from near bases, glandular-pubescent. **Leaves** alternate; petioles 5–60+ mm; blades 3-nerved from bases, ovate to deltate, 10–100 × 10–90 mm, bases cordate to truncate, margins crenate to serrate, apices acute to rounded, faces puberulent to glabrate, often gland-dotted. **Heads** borne in (leafy) paniculiform arrays. **Peduncles** 1–5 mm, glandular-pubescent. **Involucres** cylindric to obconic, 7–12 mm. **Phyllaries** 21–35 in 5–6 series, greenish, often purple-tinged, 3–4-striate, unequal, margins scarious; outer ovate to lance-ovate (glabrous or sparsely glandular-pubescent, apices acute to acuminate), inner lanceolate (glabrous, apices obtuse to acuminate). **Florets** 8–12; corollas pale yellow-green, 5.5–8 mm. **Cypselae** 2.5–3.5 mm, puberulent; **pappi** of 24–30 white, barbellate bristles. *2n* = 18.

Flowering Jul–Oct. Dry, rocky hillsides, arroyos, canyons; 800–2700 m; Ariz., Calif., Colo., Idaho, Nev., N.Mex., Okla., Tex., Utah, Wyo.

Floret number of 18 is reported for *Brickellia californica* in most floras; A. Cronquist (1994) noted that 18 is high by his count; he reported 12 as the maximum number for the specimens he sampled and that fits the specimens I have counted from Arizona.

7. Brickellia chenopodina   (Greene) B. L. Robinson, Mem. Gray Herb. 1: 99. 1917   C E

Coleosanthus chenopodinus Greene, Contr. U.S. Natl. Herb. 16: 177. 1913

Perennials, 100–150 cm (bases woody). **Stems** branched, stipitate-glandular. **Leaves** alternate; petioles 4–10 mm; blades 3-nerved from bases, rhombic-ovate to lanceolate, 20–50 × 10–40 mm, bases rounded to cuneate, margins entire or irregularly dentate, apices acute, faces glabrous or sparsely gland-dotted. **Heads** borne singly (terminating lateral branches). **Peduncles** (leafy) 5–50 mm, stipitate-glandular. **Involucres** cylindric to campanulate, 8–9 mm. **Phyllaries** 22–26 in 5–6 series, greenish, often purple-tinged, 3–4-striate, subequal or unequal, margins narrowly scarious; outer lanceolate to lance-ovate (often

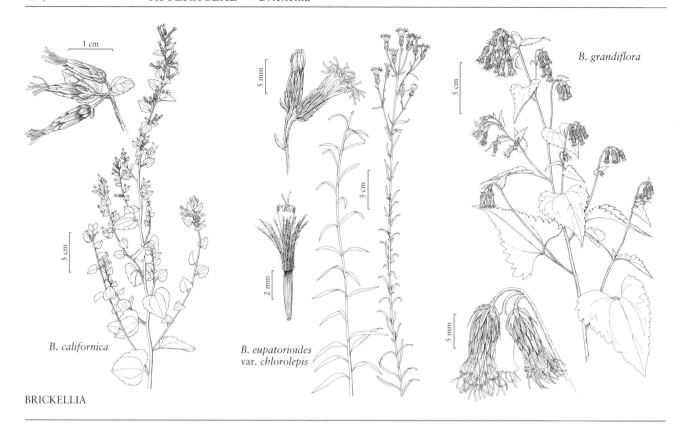

B. grandiflora

B. californica

B. eupatorioides
var. *chlorolepis*

BRICKELLIA

herbaceous, glandular-hirtellous, equaling or surpassing inner, apices acute to long-attenuate), inner lanceolate to lance-linear (glabrous, apices obtuse to acute). **Florets** 18–24; corollas pale yellow-green or greenish white, 5–6 mm. **Cypselae** 2.5–3 mm, strigose; **pappi** of 30–35 white, smooth or barbellulate bristles. $2n = 18$.

Flowering Aug–Oct. Near streams, canyon bottoms; of conservation concern; 1300–1600 m; N.Mex.

Brickellia chenopodina is known only from Grants County; it is similar in most respects to *B. floribunda* and may be little more than a shade form of the latter, with which it is sympatric.

8. Brickellia cordifolia Elliott, Sketch Bot. S. Carolina 2: 290. 1823 [C][E]

Coleosanthus cordifolius (Elliott) Kuntze

Perennials, 50–150 cm (caudices slender). **Stems** branched, pubescent. **Leaves** opposite; petioles 10–40 mm; blades 3-nerved from bases, deltate-ovate to lance-ovate, 30–150 × 20–100 mm, bases subtruncate to cordate, margins coarsely to finely serrate, apices acuminate, abaxial faces minutely pubescent and gland-dotted, adaxial faces sparsely puberulent. **Heads** (erect in flower and fruit) in corymbiform or paniculiform arrays. **Peduncles** 10–25

mm, glandular-pubescent. **Involucres** broadly campanulate, 10–12 mm. **Phyllaries** 28–35 in 4–5 series, greenish, often purple-tinged, 4–7-striate, unequal to subequal, margins scarious (ciliate); outer narrowly lanceolate to lance-ovate (glandular-pubescent, apices long-acuminate), mid lanceolate (apices acute to acuminate), inner narrowly lanceolate (apices acute to obtuse, recurved, pubescent, often gland-dotted). **Florets** 35–45; corollas pale yellow-green, often purple-tinged, 6–7.2 mm. **Cypselae** 4.5–5.5 mm, hispidulous to glabrate; **pappi** of 38–46 purple-tinged, barbellate bristles.

Flowering Aug–Oct. Moist pine and oak woodlands; of conservation concern; 10–100 m; Ala., Fla., Ga.

9. Brickellia coulteri A. Gray, Smithsonian Contr. Knowl. 3(5): 86. 1852

Shrubs, 30–150 cm. **Stems** much branched from bases, pubescent, often gland-dotted. **Leaves** opposite; petioles 4–12 mm; blades 3-nerved from bases, broadly ovate to deltate, 10–60 × 10–30 mm, bases truncate, hastate, or subcordate, margins usually toothed (1–3 sets of sharp teeth near bases), rarely subentire, apices attenuate, faces minutely pubescent. **Heads** in loose, paniculiform arrays.

Peduncles 5–40 mm, pubescent and sometimes gland-dotted. **Involucres** cylindric to campanulate, 8–12 mm. **Phyllaries** 17–22 in 4–6 series, greenish, often purple-tinged, 4–5-striate, unequal, margins scarious (apices acute to acuminate or mucronate, minutely pubescent); outer lance-ovate to narrowly lanceolate (glabrous or sparsely pubescent), inner narrowly lanceolate to linear (glabrous). **Florets** 13–25; corollas pale yellow-green, often purple-tinged, 7–8.2 mm. **Cypselae** 3–5 mm, hispidulous, strigose, or glandular-pubescent; **pappi** of 28–40 smooth or barbellulate bristles.

Varieties 3 (2 in the flora): sw United States, n Mexico.

1. Midstem leaf blades mostly 40–60 mm; peduncles mostly 15–40 mm, mostly glandular-pubescent . . . . 9a. *Brickellia coulteri* var. *coulteri*
1. Midstem leaf blades mostly 20–40(–50) mm; peduncles 5–25 mm, pubescent, mostly lacking glandular hairs . . . 9b. *Brickellia coulteri* var. *brachiata*

9a. Brickellia coulteri A. Gray var. **coulteri**

Leaf blades 20–60 × 10–30 mm, midstem blades mostly 40–60 mm. **Peduncles** mostly 15–40 mm, mostly glandular-pubescent. $2n = 18$.

Flowering Mar–Oct. Open sandy areas, arid rocky slopes, desert canyons; 10–1000 m; Ariz., Tex.; Mexico.

9b. Brickellia coulteri A. Gray var. **brachiata**
(A. Gray) B. L. Turner, Phytologia 68: 237. 1990

Brickellia brachiata A. Gray, Proc. Amer. Acad. Arts 21: 385. 1886

Leaf blades 10–40(–50) × 10–20 mm, midstem blades mostly 20–40(–50) mm. **Peduncles** 5–25 mm, pubescent, mostly lacking glandular hairs. $2n = 18$.

Flowering Mar–Oct. Open sandy areas, arid rocky slopes, desert canyons; 400–2000 m; Ariz., N. Mex., Tex.; Mexico.

Variety *brachiata* includes most specimens from Arizona, except those from lower side canyons of the Grand Canyon. Those plants are strongly gland-dotted and often have larger leaves characteristic of the type variety. Some plants from southeastern Arizona are not clearly differentiated.

10. Brickellia cylindracea A. Gray & Engelmann, Proc. Amer. Acad. Arts 1: 46. 1847 [E]

Perennials, 40–120 cm (caudices woody). **Stems** branched, pubescent, gland-dotted. **Leaves** mostly opposite, sometimes alternate or subopposite; petioles 0–3 mm; blades 3-nerved from bases, lanceolate to obovate, 20–60 × 12–30 mm, bases ± cuneate, margins crenate to serrate, apices obtuse to acute, abaxial faces tomentose, often densely gland-dotted as well (venation raised, reticulate), adaxial faces pubescent and gland-dotted. **Heads** in ± dense, racemiform or paniculiform arrays. **Peduncles** 2–15 mm, densely tomentose, often gland-dotted. **Involucres** cylindric to campanulate, 8–13 mm. **Phyllaries** 18–22 in 5–7 series, greenish, often purplish-tinged, 6–8-striate, unequal, margins scarious (apices acute to acuminate); outer ovate (sparsely villous, gland-dotted, lengths no more than ¹/₂ inner, margins long-ciliate), inner narrowly lanceolate to linear-oblong (glabrous). **Florets** 10–21; corollas greenish, yellow, or yellow-orange, often purple-tinged, 5–7 mm. **Cypselae** 3.5–5 mm, sparsely to densely pubescent; **pappi** of 30–35, white to tawny, barbellate bristles. $2n = 18$.

Flowering Jun–Nov. Dry limestone hillsides; 100–600 m; Tex.

L. D. Flyr (1970) rightly noted that *Brickellia cylindracea* is variable in leaf size, shape, and texture as well as arrangements of heads, and added that it is not easily distinguished from *B. conduplicata* (= *B. lemmonii* var. *conduplicata*), which accounts for most of the specimens in west Texas. *Brickellia cylindracea* has a more eastern distribution than *B. lemmonii* var. *conduplicata* and appears to have a strong preference for limestone soils. Some specimens from limestone outcrops in west Texas blur distinction between the two species.

11. Brickellia dentata (de Candolle) Schultz-Bipontinus in B. Seemann, Bot. Voy. Herald, 301. 1856 [E]

Clavigera dentata de Candolle in A. P. de Candolle and A. L. P. P. de Candolle, Prodr. 5: 128. 1836

Shrubs, 60–120 cm. **Stems** branched from bases, puberulent, sparsely gland-dotted. **Leaves** alternate; petioles 1–2 mm; blades obscurely 3-nerved from bases, lanceolate to lance-ovate, 5–40 × 3.5–12 mm, bases acute to cuneate, margins usually coarsely dentate, sometimes entire, apices acute, faces gland-dotted. **Heads** in (elongate) paniculiform arrays.

Peduncles 1–2 mm, puberulent, gland-dotted. **Involucres** cylindric to campanulate, 7–11 mm. **Phyllaries** 18–22 in 5–6 series, greenish or stramineous, 3–5-striate, unequal, margins scarious (apices acute to obtuse, ciliate); outer ovate (glandular-pubescent), inner lanceolate (gland-dotted). **Florets** 13–20; corollas pale green or yellow, 4.5–6.5 mm. **Cypselae** 3.5–4 mm, usually glabrous, rarely sparsely pubescent; **pappi** of 30–35 tawny, usually barbellate, sometimes subplumose, bristles. $2n = 18$.

Flowering Aug–Nov. Gravel of limestone streambeds; 100–400 m; Tex.

12. Brickellia desertorum Coville, Proc. Biol. Soc. Wash. 7: 68. 1892

Shrubs, 80–200 cm. **Stems** intricately branched, densely puberulent, often gland-dotted. **Leaves** opposite or alternate; petioles 1–2.5 mm; blades 3-nerved from bases, ovate, 3–13 × 4–14 mm, bases obtuse to truncate, margins crenate-serrate, apices obtuse, faces gland-dotted and tomentulose to densely puberulent. **Heads** in paniculiform arrays. **Peduncles** 0–3 mm, gland-dotted. **Involucres** cylindric, 8–10 mm. **Phyllaries** 20–24 in 4–7 series, greenish or brownish, 3–4-striate, unequal, margins scarious (apices acute to apiculate); outer broadly ovate (granular-puberulent), inner lanceolate to lance-linear (glabrous or tomentulose, often gland-dotted). **Florets** 8–12; corollas white, 4.5–5 mm. **Cypselae** 2–3 mm, scabrous; **pappi** of 12–15 white, smooth or barbellulate bristles.

Flowering Sep–Mar. Dry hillsides, outwash slopes, canyons; 200–1400 m; Ariz., Calif., Nev.; Mexico; West Indies; Central America (to Nicaragua).

13. Brickellia eupatorioides (Linnaeus) Shinners, Sida 4: 274. 1971

Kuhnia eupatorioides Linnaeus, Sp. Pl. ed. 2, 2: 1662. 1763

Perennials, 30–200 cm (bases woody). **Stems** branched, pubescent. **Leaves** mostly opposite (alternate in vars. *gracillima* and *texana*); petioles 0–10 mm; blades 1- or 3-nerved from bases, lanceolate, lance-linear, lance-ovate, lance-rhombic, linear, or oblong, 25–100 × 0.5–40 mm, bases acute, margins entire or ± dentate (often revolute), apices obtuse to acuminate, faces glandular-pubescent. **Heads** in paniculiform or corymbiform arrays. **Peduncles** 5–100 mm, glandular-pubescent. **Involucres** cylindric to narrowly campanulate, 7–15 mm. **Phyllaries** 22–26 in 4–6 series, green to stramineous, sometimes purple-

tinged, 3–7-striate, unequal, margins scarious (often ciliate); outer ovate to lance-ovate (puberulent, often densely gland-dotted, apices acute to acuminate), inner lanceolate (± gland-dotted, apices obtuse to aristate). **Florets** 6–35; corollas pale yellow, yellow-green, pinkish lavender, or maroon, 4.5–6 mm. **Cypselae** 2.7–5.5 mm, glabrous or strigose, sometimes hispidulous or velutinous and/or gland-dotted; **pappi** of 20–28 white or tawny, usually plumose or subplumose, sometimes barbellate, bristles.

Varieties 6 (6 in the flora): United States, Mexico.

1. Leaf blades lance-linear or linear, 0.5–3 mm wide.
   2. Florets 9–13 . . . . 13d. *Brickellia eupatorioides* var. *floridana*
2. Florets 16–30.
   3. Leaf blades 1–3 mm wide (lengths of proximal leaves greater than distals); involucres 8–13 mm . . . . 13b. *Brickellia eupatorioides* var. *chlorolepis* (in part)
   3. Leaf blades 0.5–1 mm wide (lengths of proximal leaves equaling distals); involucres 5–10 mm . . . . 13e. *Brickellia eupatorioides* var. *gracillima*
1. Leaf blades lanceolate, lance-linear, lance-ovate, or lance-rhombic, 1–40 mm wide.
   4. Phyllaries: lengths of outers and mids ½–1 times inners, apices long-acuminate, usually contorted . . . . 13f. *Brickellia eupatorioides* var. *texana*
   4. Phyllaries: lengths of outers and mids to ½ times inners, apices obtuse, acute, or acuminate, not contorted.
      5. Leaves 1-nerved from bases; petioles 0–1 mm; heads borne singly or 2–3 in loose, paniculiform arrays . . . . 13b. *Brickellia eupatorioides* var. *chlorolepis* (in part)
      5. Leaves 1- or 3-nerved from bases; petioles 0–10 mm; heads mostly 3–8 in dense, corymbiform arrays.
         6. Involucres 7–11 mm; florets 6–15 . . . . 13a. *Brickellia eupatorioides* var. *eupatorioides*
         6. Involucres 8–15 mm; florets 15–35 . . . . 13c. *Brickellia eupatorioides* var. *corymbulosa*

13a. Brickellia eupatorioides (Linnaeus) Shinners var. eupatorioides E

Kuhnia glutinosa Elliott

Plants 30–150 cm. **Leaves** opposite; petioles 1–10 mm; blades 1- or 3-nerved from bases, broadly rhombic-lanceolate to narrowly lanceolate, 25–100 × 5–40 mm, margins entire or ± dentate. **Heads** in corymbiform arrays. **Peduncles** 5–20 mm. **Involucres** 7–11 mm. **Phyllaries** green to stramineous, sometimes purple-tinged (lengths of outer to ¹/₂ inner, apices acute to acuminate, not contorted). **Florets** 6–15; corollas pale yellow to maroon, 4.5–6 mm. **Cypselae** 4–5.5 mm. **2*n*** = 18.

Flowering May–Oct. Open areas, wide range of soils; 200–2300 m; Ala., Ark., Fla., Ga., Ind., Ky., La., Md., Miss., N.J., N.C., Ohio, Pa., S.C., Tenn., Tex., Va., W.Va.

13b. Brickellia eupatorioides (Linnaeus) Shinners var. **chlorolepis** (Wooton & Standley) B. L. Turner, Phytologia 67: 129. 1989 F

Kuhnia chlorolepis Wooton & Standley, Contr. U.S. Natl. Herb. 16: 177. 1913; *Brickellia rosmarinifolia* (Ventenat) W. A. Weber subsp. *chlorolepis* (Wooton & Standley) W. A. Weber; *K. eupatorioides* Linnaeus var. *chlorolepis* (Wooton & Standley) Cronquist

Plants 30–150 cm. **Leaves** opposite; petioles 0–1 mm; blades 1-nerved from bases, lance-linear to linear, 5–90 × 1.5–9 mm, margins entire. **Heads** in paniculiform arrays. **Peduncles** 10–100 mm. **Involucres** 8–13 mm. **Phyllaries** greenish, sometimes purple-tinged (apices obtuse to acuminate). **Florets** 16–30; corollas pale yellow-green to pinkish lavender, 4.7–8 mm. **Cypselae** 4–5 mm. **2*n*** = 18.

Flowering May–Oct. Open areas, wide range of soils; 400–1400 m; Ariz., Colo., N.Mex., Tex., Utah; Mexico.

13c. Brickellia eupatorioides (Linnaeus) Shinners var. **corymbulosa** (Torrey & A. Gray) Shinners, Sida 4: 274. 1971 E

Kuhnia eupatorioides Linnaeus var. *corymbulosa* Torrey & A. Gray, Fl. N. Amer. 2: 78. 1841

Plants 30–200 cm. **Leaves** opposite; petioles 0–10 mm; blades 1- or 3-nerved from bases, lanceolate to lance-linear, 5–90 × 2–15 mm, margins entire or coarsely dentate. **Heads** in paniculiform or corymbiform arrays. **Peduncles** 5–25 mm. **Involucres** 8–15 mm. **Phyllaries** green to stramineous, sometimes purple-tinged (lengths of outer to ¹/₂ inner, apices acute to acuminate, not contorted). **Florets** 15–35; corollas pale yellow to pinkish lavender, 4.7–8 mm. **Cypselae** 2.7–5 mm. **2*n*** = 18.

Flowering May–Oct. Open areas, wide range of soils; 200–2300 m; Ark., Colo., Ill., Ind., Iowa, Kans., Mich., Minn., Mo., Mont., Nebr., N.Mex., N.Dak., S.Dak., Tex., Wis.

13d. Brickellia eupatorioides (Linnaeus) Shinners var. **floridana** (R. W. Long) B. L. Turner, Phytologia 67: 130. 1989 C E

Kuhnia eupatorioides Linnaeus var. *floridana* R. W. Long, Rhodora 72: 39. 1970; *Brickellia mosieri* (Small) Shinners; *K. eupatorioides* var. *gracilis* Torrey & A. Gray; *K. mosieri* Small

Plants 30–70 cm. **Leaves** opposite; petioles 0–2 mm; blades 1- or, obscurely, 3-nerved from bases, narrowly oblong to linear, 5–40 × 1–3.5 mm, margins entire. **Heads** in paniculiform or corymbiform arrays. **Peduncles** 5–60 mm. **Involucres** 7–11 mm. **Phyllaries** green to stramineous, sometimes purple-tinged. **Florets** 9–13; corollas pale yellow to maroon, 6–6.5 mm. **Cypselae** 3.5–5 mm.

Flowering Jun–Sep. Piney woods; of conservation concern; 10–50+ m; Fla.

Variety *floridana* is known only from southern Florida.

13e. Brickellia eupatorioides (Linnaeus) Shinners var. **gracillima** (A. Gray) B. L. Turner, Phytologia 67: 130. 1989 E

Kuhnia eupatorioides Linnaeus var. *gracillima* A. Gray, Boston J. Nat. Hist. 6: 218. 1850; *Brickellia leptophylla* (Scheele) Shinners; *K. leptophylla* Scheele

Plants 30–90 cm. **Leaves** alternate; petioles 0–1 mm; blades 1-nerved from bases, lance-linear to linear, 5–80 × 0.5–1 mm, margins entire. **Heads** in paniculiform arrays. **Peduncles** 2–20 mm. **Involucres** 5–10 mm. **Phyllaries** green to stramineous. **Florets** 15–25; corollas pale yellow to pinkish lavender, 5–7 mm. **Cypselae** 4–5.5 mm. **2*n*** = 18.

Flowering Aug–Nov., Apr. Open areas, wide range of soils; 100–800 m; Ark., Mo., Okla., Tex.

13f. Brickellia eupatorioides (Linnaeus) Shinners var. **texana** (Shinners) Shinners, Sida 4: 274. 1971 E

Kuhnia eupatorioides Linnaeus var. *texana* Shinners, Wrightia 1: 136. 1946; *Brickellia eupatorioides* var. *ozarkana* (Shinners) Shinners; *K. eupatorioides* var. *ozarkana* Shinners

Plants 30–100 cm. **Leaves** alternate; petioles 0–1 mm; blades 1- or 3-nerved from bases, lance-ovate to lance-linear, 5–70 × 4–20 mm, margins entire or coarsely dentate. **Heads** in paniculiform arrays. **Peduncles** 5–20 mm. **Involucres** 7–11 mm. **Phyllaries** green to stramineous, sometimes purple-tinged (often contorted, lengths of outer ¹/₂ or more inner, apices long-acuminate). **Florets** 12–24; corollas pale yellow to pinkish lavender, 4.7–7 mm. **Cypselae** 4–5 mm. $2n = 18$.

Flowering Aug–Nov. Open areas, wide range of soils; 50–1000 m; Ark., Kans., Mo., Okla., Tex.

14. Brickellia floribunda　A. Gray, Smithsonian Contr. Knowl. 5(6): 73. 1853

Perennials, 100–200 cm (bases woody). **Stems** branched, stipitate-glandular. **Leaves** alternate; petioles 20–35 mm; blades 3-nerved from bases, deltate-ovate or rhombic-ovate, 20–140 × 10–100 mm, bases truncate to cordate, margins irregularly dentate, apices acute, faces gland-dotted or glandular-pubescent. **Heads** in paniculiform arrays. **Peduncles** 3–18 mm, stipitate-glandular. **Involucres** cylindric to campanulate, 7.5–8.5 mm. **Phyllaries** 22–26 in 5–6 series, greenish, often purple-tinged, 3–4-striate, unequal, margins narrowly scarious (apices obtuse to acute); outer ovate to lanceolate (glandular-hirtellous, shorter than inner), inner linear-lanceolate (glabrous). **Florets** 15–24; corollas pale yellow-green or greenish white, 5–6.5 mm. **Cypselae** 2–3 mm, strigose; **pappi** of 30–35 white, barbellulate bristles. $2n = 18$.

Flowering Aug–Oct. Near streams, canyon bottoms; 1100–1800 m; Ariz., N.Mex.; Mexico.

15. Brickellia frutescens　A. Gray, Proc. Amer. Acad. Arts 17: 207. 1882

Shrubs, 30–60 cm. **Stems** intricately branched, hirtellous, sparsely gland-dotted. **Leaves** alternate; petioles 1–2 mm; blades (venation obscure) oblong to spatulate, 3–12 × 1–4 mm, bases acute to acuminate, margins entire, apices acute to obtuse, faces tomentose, gland-dotted. **Heads** in paniculiform arrays. **Peduncles** 10–30 mm, villous, gland-dotted. **Involucres** cylindric, 9–11 mm. **Phyllaries** 20–25 in 5–6 series, greenish, often purple-tinged, 4–5-striate, unequal, (sparsely villous, gland-dotted) margins narrowly scarious (apices rounded, mucronate); outer ovate to lanceolate (ciliate), inner linear-oblong. **Florets** 20–30; corollas purplish, 6.5–7.5 mm. **Cypselae** 3.5–4.8 mm, hispidulous; **pappi** of 26–30 white, barbellulate bristles.

Flowering Mar–Jun. Granitic desert slopes or sands; 600–1200 m; Calif., Nev.; Mexico (Baja California).

16. Brickellia grandiflora　(Hooker) Nuttall, Trans. Amer. Philos. Soc., n. s. 7: 287. 1840　Ｅ Ｆ

Eupatorium grandiflorum Hooker, Fl. Bor.-Amer. 2: 26. 1834

Perennials, 30–95 cm (taproots thickened). **Stems** branched, puberulent. **Leaves** opposite or alternate; petioles 10–70 mm; blades 3-nerved from bases, deltate-ovate, lance-ovate, or subcordate, 15–120 × 20–70 mm, bases acute, truncate, or subcordate, margins crenate, dentate, or serrate, apices attenuate, faces puberulent and gland-dotted. **Heads** (nodding in flower and fruit) in loose, corymbiform or paniculiform arrays. **Peduncles** 4–30 mm, pubescent. **Involucres** cylindric or obconic, 7–12 mm. **Phyllaries** 30–40 in 5–7 series, greenish, 4–5-striate, unequal, margins scarious; outer lance-ovate to lanceolate (pubescent, margins ciliate, apices long-acuminate), inner lanceolate to lance-linear (glabrous, apices acute to acuminate). **Florets** mostly 20–40(–70); corollas pale yellow-green, 6.5–7.5 mm. **Cypselae** 4–5 mm, hispidulous to hirtellous; **pappi** of 20–30 white, barbellate bristles. $2n = 18$.

Flowering Jul–Oct. Rocky hillsides, shaded forests, dry slopes, canyons; 1200–3000 m; Ariz., Ark., Calif., Colo., Idaho, Kans., Mo., Mont., Nebr., Nev., N.Mex., Oreg., Tex., Utah, Wash., Wyo.

17. Brickellia greenei　A. Gray, Proc. Amer. Acad. Arts 12: 58. 1877　Ｅ

Perennials, 20–50 cm (caudices woody). **Stems** branched, villous, stipitate-glandular. **Leaves** alternate; petioles 0–2 mm; blades 3-nerved from bases, ovate, 15–30 × 5–25 mm, bases rounded to truncate, margins evenly to irregularly serrate, apices acute, faces viscid-glandular. **Heads** borne singly or in open, cymiform arrays (each head subtended by leaflike bracts). **Peduncles** 0–2 mm. **Involucres** broadly cylindric to campanulate, 14–18 mm. **Phyllaries** (24–)36–42 in (3–)5–7 series, greenish or stramineous, often purple-tinged, 3–5-striate, unequal, margins narrowly scarious (apices acute to long-acuminate); outer lance-ovate to lanceolate (sparsely gland-dotted), inner narrowly lanceolate (glabrous). **Florets** 44–60; corollas yellow-green, often purple-tinged, 8–10 mm. **Cypselae** 5.5–7 mm, glabrate to hispidulous; **pappi** of 20–30 white, barbellate bristles. $2n = 18$.

Flowering Jul–Sep. Open rocky slopes, canyon bottoms, riparian areas, serpentine soils; 800–2200 m; Calif., Oreg.

A sweet odor is often associated with *Brickellia greenei*.

18. Brickellia hinckleyi Standley, Publ. Field Mus. Nat. Hist., Bot. Ser. 22: 61. 1940 C E

Brickellia brachyphylla (A. Gray) A. Gray var. *hinckleyi* (Standley) Flyr

Perennials, 20–90 cm. **Stems** branched, coarsely scabrous. **Leaves** alternate to subopposite; petioles 1–5 mm; blades 3-nerved from bases, lanceolate to lance-ovate, 20–40 × 7–20 mm, bases rounded to broadly cuneate, margins entire or serrate, apices acute to acuminate, faces sparsely pubescent, stipitate-glandular. **Heads** in open, racemose or paniculiform arrays. **Peduncles** 5–45 mm, pubescent, stipitate-glandular. **Involucres** cylindric to campanulate, 7–11 mm. **Phyllaries** 16–20 in 4–5 series, greenish, often purple-tinged, 4–8-striate, unequal, margins narrowly scarious, often ciliate (apices acute to long-attenuate); outer lance-ovate (pubescent), inner narrowly lanceolate (glabrous). **Florets** 11–22; corollas pale yellow or cream, 4.7–6 mm. **Cypselae** 4–5 mm, silvery-setulose. **Pappi** of 25–34 white, plumose bristles.

Varieties 2 (2 in flora): Texas.

1. Stem leaves: petioles 1–5 mm, blades relatively thin, (20–)25–45 mm, faces sparsely pubescent, stipitate-glandular . . . . 18a. *Brickellia hinckleyi* var. *hinckleyi*
1. Stem leaves: petioles 0–1 mm, blades relatively thick, (15–)17–22(–25) mm, faces scabrous, gland-dotted. . . . . 18b. *Brickellia hinckleyi* var. *terlinguensis*

18a. Brickellia hinckleyi Standley var. hinckleyi C E

Plants 20–90 cm. **Stems** coarsely scabrous. **Leaves:** petioles 1–5 mm; blades relatively thin, (20–)25–45 × 7–20 mm, margins entire or serrate, faces sparsely pubescent, stipitate-glandular.

Flowering Jul–Oct. Mixed woodlands, forests, igneous slopes, canyons; of conservation concern; 1700–2300 m; Tex.

Variety *hinckleyi* occurs in the Chisos Mountains.

18b. Brickellia hinckleyi Standley var. terlinguensis (Flyr) B. L. Turner, Comps Mexico 2: 69. 1997 C E

Brickellia brachyphylla (A. Gray) A. Gray var. *terlinguensis* Flyr, Sida 3: 254. 1968

Plants 20–90 cm. **Stems** coarsely hirsute and stipitate-glandular. **Leaves:** petioles 0–4 mm; blades relatively thick, (15–)17–22 (–25) × 4–7 mm, margins entire, crenate, or serrate (and strigillose), faces scabrous, gland-dotted.

Flowering Jul–Oct. Stream bottoms; of conservation concern; 1000–1700 m; Tex.

Variety *terlinguensis* occurs in the Chisos Mountains.

19. Brickellia incana A. Gray, Proc. Amer. Acad. Arts 7: 350. 1868 E

Shrubs, 40–130 cm. **Stems** branched from bases, densely white-tomentose. **Leaves** alternate; petioles 0–2 mm; blades 3-nerved from bases, ovate, 10–30 × 5–20 mm, bases obtuse to truncate, margins entire or serrulate, apices acute, faces densely white-tomentose, often gland-dotted. **Heads** borne singly. **Peduncles** 30–60 mm, canescent. **Involucres** campanulate, 15–20 mm. **Phyllaries** 38–44 in 6–8 series, greenish to purple, 6–9-striate, unequal, margins narrowly scarious-ciliate (faces white tomentose and gland-dotted, apices obtuse to acuminate); outer ovate, inner lanceolate. **Florets** 45–60; corollas pale yellow or cream, often purple-tinged, 8.5–14 mm. **Cypselae** 8.5–10 mm, sericeous, gland-dotted; **pappi** of 42–44 white, barbellate bristles. $2n = 18$.

Flowering May–Oct. Sandy and gravelly washes, flats; 300–1600 m; Ariz., Calif., Nev.

20. Brickellia knappiana Drew, Pittonia 1: 260. 1888 C E

Shrubs, 100–200 cm. **Stems** (arching) branched from near bases, gland-dotted. **Leaves** alternate; petioles 4–5 mm; blades 3-nerved from bases, lance-ovate to lanceolate, 10–35 × 1–16 mm, bases acute to attenuate, margins dentate to serrate, apices acute to acuminate, faces densely gland-dotted and sparsely pubescent (not shiny). **Heads** in paniculiform arrays. **Peduncles** 2–5 mm, densely glandular-pubescent. **Involucres** cylindric to narrowly campanulate, 6–9 mm. **Phyllaries** 18–22 in 5–7 series, greenish, 3–4-striate, unequal, margins scarious (apices acute to acuminate); outer lance-ovate (sparsely puberulent, gland-dotted), inner lanceolate (glabrous). **Florets** 5–7; corollas white, 5–6.5 mm. **Cypselae** 2.5–3, puberulent to sparsely pubescent; **pappi** of 28–32 white, barbellulate bristles.

Flowering Sep–Oct. Gravelly washes; of conservation concern; 700–1700 m; Calif.

Brickellia knappiana is found only in the mountains of eastern California. It is noted as a possible hybrid between *B. californica* and *B. multiflora*. Collections from mixed populations of *B. knappiana* and *B. californica* have been noted on herbarium specimens.

21. Brickellia laciniata A. Gray, Smithsonian Contr. Knowl. 3(5): 87. 1852

Shrubs, 60–120 cm. **Stems** much branched, pubescent, gland-dotted. **Leaves** alternate; petioles 2–5 mm; blades 3-nerved from bases, oblong to broadly ovate, 6–15 × 5–9 mm, bases narrowly cuneate, margins irregularly dentate, apices acute to obtuse, faces sparsely pubescent, gland-dotted. **Heads** in (leafy) paniculiform or racemiform arrays. **Peduncles** 0.5–2.5 mm, glandular-viscid. **Involucres** cylindric, 6–8 mm. **Phyllaries** 15–25 in 4–6 series, stramineous, 3–5-striate, unequal, (gland-dotted) margins scarious (apices obtuse to acute); outer ovate, inner lanceolate. **Florets** 8–14; corollas pale yellow-green, often purple-tinged, 4.8–5.7 mm. **Cypselae** 2.5–3 mm, sparsely pubescent or strigose; **pappi** of 24–28 tawny, barbellate bristles. 2*n* = 18.

Flowering Aug–Nov. Dry slopes, streambeds (arroyos); 1300–1900 m; N.Mex., Tex.; Mexico.

22. Brickellia lemmonii A. Gray, Proc. Amer. Acad. Arts 17: 206. 1882 (as lemmoni)

Coleosanthus lemmonii (A. Gray) Kuntze

Perennials, 30–50 cm (caudices woody). **Stems** branched, pubescent, gland-dotted. **Leaves** mostly alternate, sometimes opposite; petioles 0–5 mm; blades 3-nerved from bases (venation raised, reticulate), elliptic to lanceolate or ovate to subdeltate, 20–70 × 7–20 mm, bases ± cuneate, margins crenate to serrate, apices acute, faces glandular-pubescent. **Heads** in paniculiform arrays.

Peduncles 2–15 mm, tomentose, often gland-dotted. **Involucres** narrowly cylindric, 10–13 mm. **Phyllaries** 18–20 in 3–4(–5) series, green or purplish, 5–9-striate, unequal, margins scarious (apices acute, attenuate, obtuse, or nearly rounded); outer ovate (lengths evenly grading to inner, margins ciliate, faces pubescent, often gland-dotted), inner narrowly lanceolate to linear-oblong (faces glabrous). **Florets** 9–24; corollas pale yellow-green, 6–8.2 mm. **Cypselae** 2.5–4 mm, densely pubescent; **pappi** of 32–43 white, mostly barbellate, sometimes subplumose, bristles.

Varieties 4 (2 in the flora): sw United States, Mexico.

1. Leaf blades elliptic to lanceolate, lengths mostly 2–5 times widths; phyllaries: apices (inner) mostly acute . . . . 22a. *Brickellia lemmonii* var. *lemmonii*
1. Leaf blades broadly ovate to subdeltate, lengths mostly 1–2 times widths; phyllaries: apices (inner) mostly obtuse to nearly rounded . . . . 22b. *Brickellia lemmonii* var. *conduplicata*

22a. Brickellia lemmonii A. Gray var. lemmonii

Brickellia lemmonii var. *wootonii* (Greene) B. L. Robinson

Leaf blades elliptic to lanceolate, 20–70 × 7–20 mm, lengths mostly 2–5 times widths. **Peduncles** 2–15 mm. **Phyllaries:** apices (inner) mostly acute.

Flowering Aug–Nov. Igneous soils, montane and canyon slopes; 1500–2200 m; Ariz., N.Mex., Tex.; Mexico.

22b. Brickellia lemmonii A. Gray var. conduplicata (B. L. Robinson) B. L. Turner, Phytologia 68: 163. 1990

Brickellia betonicifolia A. Gray var. *conduplicata* B. L. Robinson, Proc. Amer. Acad. Arts 43: 37. 1907 (as betonicaefolia); *B. conduplicata* (B. L. Robinson) B. L. Robinson; *B. viejensis* Flyr

Leaf blades broadly ovate to subdeltate, 20–60 × 7--30 mm, lengths 1–2 times widths. **Peduncles** 2–22 mm. **Phyllaries:** apices (inner) mostly obtuse to nearly rounded.

Flowering Aug–Nov. Igneous soils, montane and canyon slopes; 1500–2200 m; N.Mex., Tex.; Mexico.

23. Brickellia longifolia S. Watson, Amer. Naturalist 7: 301. 1873 [E]

Shrubs, 20–200[–250] cm. **Stems** branched from above bases, ± glabrous. **Leaves** alternate; petioles 0–5 mm; blades 3-nerved from bases, lance-elliptic, lance-ovate, lance-linear, or linear (folded or falcate), 10–130 × 2–9 mm, bases rounded or tapering, margins entire or nearly so, apices acuminate, faces gland-dotted, often puberulent (shiny). **Heads** in (leafy) paniculiform arrays. **Peduncles** 0–3 mm, glabrous or glutinous. **Involucres** cylindric, 5–7 mm. **Phyllaries** 10–24 in 6–8 series, pale green to stramineous, 3–5-striate, unequal, (glutinous) margins scarious; outer ovate (apices acute), inner lanceolate (apices obtuse). **Florets** 3–7; corollas cream, 3.5–4.5 mm. **Cypselae** 1.8–2.5 mm, scabrous; **pappi** of 30–40 barbellulate bristles.

Varieties 2 (2 in the flora): w United States.

1. Leaf blades lance-elliptic to lance-linear or linear, 10–130 × 2–9 mm, lengths 8–25 times widths, flat (not folded or falcate), bases tapering. . . . . 23a. *Brickellia longifolia* var. *longifolia*
1. Leaf blades lance-ovate to narrowly lanceolate, 10–130 × 2–15 mm, lengths 4–8 times widths, folded along midveins and/or falcate, bases rounded. . . . . . 23b. *Brickellia longifolia* var. *multiflora*

23a. Brickellia longifolia S. Watson var. **longifolia** [E]

Leaf blades lance-elliptic to lance-linear or linear, 10–130 × 2–9 mm, lengths 8–25 times widths, flat (not folded or falcate), bases tapering.

Flowering Sep–Nov. Washes, stream margins, seeps, hanging gardens; 900–1700 m; Ariz., Colo., Nev., Utah.

23b. Brickellia longifolia S. Watson var. **multiflora** (Kellogg) Cronquist in A. Cronquist et al., Intermount. Fl. 5: 384. 1994 [E]

Brickellia multiflora Kellogg, Proc. Calif. Acad. Sci. 7: 49. 1876

Leaf blades lance-ovate to narrowly lanceolate, 10–130 × 2–15 mm, lengths 4–8 times width, folded along midveins and/or falcate, bases rounded.

Flowering Sep–Nov. Washes, stream margins, seeps, hanging gardens; 900–1700 m; Ariz., Calif., Nev.

24. Brickellia microphylla (Nuttall) A. Gray, Smithsonian Contr. Knowl. 3(5): 85. 1852

Bulbostylis microphylla Nuttall, Trans. Amer. Philos. Soc., n. s. 7: 286. 1840

Shrubs, 30–70 cm. **Stems** much branched, pubescent, gland-dotted. **Leaves** alternate; petioles 0–3 mm; blades 3-nerved from bases, ovate to suborbiculate, 3–20 × 1–15 mm, bases acute to obtuse or rounded, margins entire, coarsely dentate or serrate, apices rounded to acute, faces glandular-villous or hispidulous. **Heads** in loose, paniculiform arrays (often clustered at ends of branches). **Peduncles** (bracteate) 2–10 mm, viscid to glandular-villous. **Involucres** cylindric to narrowly campanulate, 7–12 mm. **Phyllaries** 30–48 in 5–8 series, greenish, often purple-tinged, 3–5-striate, (recurved or spreading) unequal, margins scarious (apices acute to acuminate); outer obovate to suborbiculate (glandular-pubescent; mid sometimes 3-toothed with middle tooth elongated), inner linear-oblong (glabrous or sparsely gland-dotted). **Florets** 8–28(–34); corollas pale yellow, often purple-tinged, 5.5–7 mm. **Cypselae** 3.5–4.7 mm, glabrous or hirtellous; **pappi** of 18–24 white, barbellulate bristles.

Varieties 2–3 (2 in the flora): w United States, Mexico.

1. Stems and peduncles viscid to glandular-villous; florets 15–28(–34) . . . . 24a *Brickellia microphylla* var. *microphylla*
1. Stems and peduncles glandular-viscid; florets 8–13 . . . . 24b. *Brickellia microphylla* var. *scabra*

24a. Brickellia microphylla (Nuttall) A. Gray var. **microphylla**

Brickellia microphylla var. *watsonii* (B. L. Robinson) S. L. Welsh; *B. watsonii* B. L. Robinson

Stems viscid to glandular-villous. **Peduncles** viscid to glandular-villous. **Involucres** 8–12 mm. **Florets** 15–28(–34). $2n = 18$.

Flowering Jul–Oct. Dry rocky places, canyon walls, sand dunes, washes; 1200–2400 m; Ariz., Calif., Colo., Idaho, Nev., Oreg., Utah, Wash., Wyo.; Mexico.

Brickellia watsonii, more recently recognized as *B. microphylla* var. *watsonii*, is distinguished by its tomentose-puberulent, often gland-dotted indument (contrasted with the villous to glandular-villous indument of var. *microphylla*), and heads with ca. 18 flowers (versus 15–34 in var. *microphylla*); such distinctions have proved to be weakly substantiated, at best.

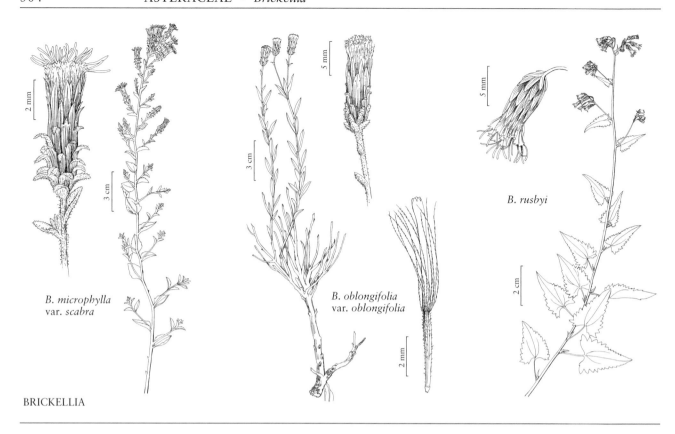

B. microphylla
var. scabra

B. oblongifolia
var. oblongifolia

B. rusbyi

BRICKELLIA

24b. Brickellia microphylla (Nuttall) A. Gray var. **scabra** A. Gray, Proc. Amer. Acad. Arts 11: 74. 1876   F

Brickellia microphylla subsp. *scabra* (A. Gray) W. A. Weber; *B. scabra* (A. Gray) A. Nelson ex B. L. Robinson

Stems glandular-viscid. **Peduncles** glandular-viscid. **Involucres** 7–11 mm. **Florets** 8–13. $2n = 18$.

Flowering Jul–Oct. Dry rocky places, canyon walls, sand dunes, washes; 1200–2400 m; Ariz., Colo., Nev., Utah, Wyo.; Mexico.

Variety *scabra* has a more eastern and southern distribution than var. *microphylla*; it grades into var. *microphylla* in southern and central Utah.

25. Brickellia nevinii A. Gray, Proc. Amer. Acad. Arts 20: 297. 1885 (as nevini)   E

Shrubs, 30–80 cm. **Stems** much branched from near bases, tomentose, gland-dotted. **Leaves** alternate; petioles 0–1 mm; blades obscurely 3–5-nerved from bases, ovate to cordate, 7–11 × 3–8 mm, bases rounded to subcordate, margins dentate-serrate, apices obtuse, faces tomentose, gland-dotted. **Heads** (terminating elongate lateral branches) in paniculiform arrays. **Peduncles** 1–5 mm, tomentose. **Involucres** cylindric to turbinate, 10–12 mm. **Phyllaries** 30–38 in 6–7 series, greenish, often red- or purplish-tinged, 3–6 striate, unequal, (recurved, tomentose, gland-dotted) margins scarious (ciliate, apices acute to acuminate); outer lance-ovate, inner lanceolate. **Florets** 20–24; corollas pale yellow or cream, often purple-tinged, 6.2–7.5 mm. **Cypselae** 3.5–5 mm, hispidulous; **pappi** of 18–24 white or tawny, barbellate bristles. $2n = 18$.

Flowering Aug–Nov. Rock crevices, rocky slopes, desert scrub; 300–1900 m; Calif., Nev.

Brickellia nevinii appears to be an incarnate form of *B. microphylla*. The variability in indument often makes distinguishing these species problematic. *Brickellia nevinii* is sometimes reported to intergrade with *B. californica*.

26. Brickellia oblongifolia Nuttall, Trans Amer. Philos. Soc., n. s. 7: 288. 1840 [E] [F]

Perennials or subshrubs, 10–60 cm (caudices woody). **Stems** branched, stipitate-glandular to glandular-pubescent. **Leaves** mostly alternate (sometimes subopposite); petioles 0; blades obscurely 3-nerved from bases, elliptic, lance-linear, or oblong, 9–40 × 1–15 mm, bases acute to attenuate, margins entire, apices acute or obtuse, faces pubescent to villous, often stipitate-glandular. **Heads** borne singly or in corymbiform arrays. **Peduncles** 2–50 mm, glandular-pubescent. **Involucres** cylindric to campanulate, 10–20 mm. **Phyllaries** 25–35 in 4–6 series, greenish, 4–5-striate, unequal, margins scarious (sometimes ciliate, apices acute to acuminate); outer ovate to lanceolate (gland-dotted to glandular-puberulent), inner linear-lanceolate to linear (glabrous). **Florets** 25–50; corollas pale yellow-green or cream, often purple-tinged, 5–10 mm. **Cypselae** 3–7 mm, stipitate-glandular or hispidulous; **pappi** of 18–25 white, barbellate to subplumose bristles.

Varieties 2 (2 in the flora): w North America.

1. Cypselae usually stipitate-glandular, sometimes hispidulous ... 26a. *Brickellia oblongifolia* var. *oblongifolia*
1. Cypselae setose-hispidulous, mostly lacking glands .... 26b. *Brickellia oblongifolia* var. *linifolia*

26a. Brickellia oblongifolia Nuttall var. **oblongifolia** [E] [F]

Cypselae usually stipitate-glandular, sometimes hispidulous. *2n* = 18.

Flowering May–Aug. Deserts, grasslands, dry rocky hillsides; 1200–2800 m; B.C.; Idaho, Mont., Nev., Oreg., Wash.

26b. Brickellia oblongifolia Nuttall var. **linifolia** (D. C. Eaton) B. L. Robinson, Mem. Gray Herb. 1: 104. 1917 [E]

Brickellia linifolia D. C. Eaton in S. Watson, Botany (Fortieth Parallel), 137, plate 15, figs. 1–6. 1871; *B. oblongifolia* subsp. *linifolia* (D. C. Eaton) Cronquist

Cypselae setose-hispidulous, mostly lacking glands. *2n* = 18.

Flowering May–Aug. Deserts, grasslands, dry rocky hillsides; 1200–2800 m; Ariz., Calif., Colo., Idaho, Mont., Nev., N.Mex., Oreg., Utah, Wash., Wyo.

27. Brickellia parvula A. Gray, Smithsonian Contr. Knowl. 3(5): 84. 1852 [E]

Perennials, 12–30 cm (caudices woody). **Stems** branched, pubescent. **Leaves** opposite; petioles 0–1 mm; blades 3-nerved from bases, deltate-ovate to lanceolate, 7–30 × 4–28 mm, bases mostly obtuse to subtruncate, sometimes acuminate, margins serrate, apices acute, faces puberulent, gland-dotted. **Heads** borne singly or in corymbiform-paniculiform arrays. **Peduncles** 10–30 mm, puberulent. **Involucres** cylindric to campanulate, 8–12 mm. **Phyllaries** 12–15 in 3–5 series, greenish, often purple-tinged, 5–7-striate, unequal, margins scarious (apices acute); outer ovate to lance-ovate (puberulent, margins ciliate), inner narrowly lanceolate (glabrous). **Florets** 8–14; corollas pale green or yellow, 5.5–6.5 mm. **Cypselae** 4.5–5 mm, densely pubescent; **pappi** of 24–30 white, usually barbellate, sometimes subplumose, bristles. *2n* = 18.

Flowering Aug–Oct. Limestone or granitic cliffs; 1500–2500 m; Ariz., N.Mex., Tex.

28. Brickellia pringlei A. Gray, Proc. Amer. Acad. Arts 17: 206. 1882

Perennials, 40–80 cm (caudices woody). **Stems** branched, pubescent, gland-dotted. **Leaves** opposite; petioles 2–4 mm; blades 3-nerved from bases (abaxial venation raised, reticulate), lanceolate to oblong, 35–60 × 10–20 mm, bases acute or rounded, margins serrate, apices acute, faces sparsely to densely pubescent, gland-dotted. **Heads** borne singly (in axils) or in paniculiform arrays. **Peduncles** (bracteate) 1–8 mm, densely tomentose. **Involucres** cylindric to turbinate, 13–15 mm. **Phyllaries** 34–40 in 6–8 series, greenish to stramineous, 5–8-striate, unequal, margins scarious (ciliate, apices acute); outer ovate to lance-ovate (often villous toward tips), inner lanceolate (glabrous). **Florets** 19–24; corollas pale yellow, 8–9.5 mm. **Cypselae** 3.5–4 mm, hirsute; **pappi** of 26–42 white, barbellate bristles. *2n* = 18.

Flowering Apr–Jun. Canyons, open rocky slopes; 1200–1700 m; Ariz.; Mexico.

29. Brickellia rusbyi   A. Gray in A. Gray et al., Syn. Fl. N. Amer. 1(2): 106. 1884   [F]

Perennials, 60–120 cm (bases woody). **Stems** branched, puberulent, gland-dotted. **Leaves** alternate; petioles 10–50 mm; blades 3-nerved from bases, deltate or rhombic-ovate, 50–100 × 20–50 mm, bases acute to truncate, margins crenate or coarsely dentate, apices long-acuminate, abaxial faces sparsely villous, gland-dotted, adaxial faces glabrous or sparsely pubescent, often gland-dotted. **Heads** in paniculiform arrays. **Peduncles** 4–6 mm, puberulent, sparsely gland-dotted. **Involucres** cylindric to campanulate, 7–10 mm. **Phyllaries** 22–26 in 4–7 series, 3–5-striate, unequal, margins scarious (apices obtuse to acute); outer ovate to lanceolate (often purple-tinged, glabrous or puberulent), inner linear to oblong (greenish, glabrous). **Florets** 13–20; corollas yellowish, 5.3–6.5 mm. **Cypselae** 2–2.5 mm, sericeous; **pappi** of 26–32 white, barbellate bristles. $2n = 18$.

Flowering Aug–Oct. Rocky slopes, shaded areas; 1500–2500 m; Ariz., N.Mex.; Mexico.

Brickellia rusbyi has been considered ill-smelling.

30. Brickellia simplex   A. Gray, Smithsonian Contr. Knowl. 5(6): 73. 1853

Perennials, 40–60 cm (caudices woody). **Stems** branched, pubescent. **Leaves** opposite or alternate; petioles 3–20 mm; blades 3-nerved from bases, deltate to lanceolate, 10–60 × 7–50 mm, bases cordate to truncate, margins crenate-dentate, apices acute, faces sparsely to densely pubescent, gland-dotted. **Heads** borne singly or in open, paniculiform arrays. **Peduncles** 20–50 mm, glandular-pubescent. **Involucres** cylindric to broadly campanulate, 12–14 mm. **Phyllaries** 40–60 in 5–6 series, greenish, 4–6-striate, unequal, margins narrowly scarious (apices acute to acuminate); outer narrowly lanceolate (glandular-pubescent), inner lanceolate (glabrous or tomentulose). **Florets** ± 60; corollas pale yellow, often purple-tinged, 9–10 mm. **Cypselae** 4–5 mm, hispidulous or glabrate; **pappi** of 32–36 white, barbellate bristles.

Flowering Aug–Oct. Oak woodlands, canyons, dry mountain slopes; 1500–2700 m; Ariz., N.Mex.; Mexico.

31. Brickellia venosa   (Wooton & Standley) B. L. Robinson, Mem. Gray Herb. 1: 50. 1917

Coleosanthus venosus Wooton & Standley, Contr. U.S. Natl. Herb. 16: 177. 1913

Perennials, 60–80 cm (caudices woody). **Stems** branched, pubescent, gland-dotted. **Leaves** opposite; petioles 0–1 mm; blades 3-nerved from bases, oblong to linear, 20–60 × 2–10 mm, bases cuneate, margins entire or dentate, apices obtuse, faces pubescent, gland-dotted. **Heads** in open, paniculiform arrays or borne singly (often terminal on lateral branches). **Peduncles** (20–)30–150 mm, puberulent, gland-dotted. **Involucres** cylindric to campanulate, 9–10 mm. **Phyllaries** 25–35 in 6–9 series, greenish, often purple-tinged, 3–9-striate, unequal, margins scarious (ciliate); outer broadly ovate to orbiculate (glandular-pubescent, apices acute to cuspidate), inner lanceolate to linear-lanceolate (glabrous, apices acute to mucronate). **Florets** 18–25; corollas pale yellow, often purple-tinged, 5.5–7 mm. **Cypselae** 3–3.5 mm, pubescent to velutinous; **pappi** of 30–40 white, barbellate bristles. $2n = 18, 18 + 1$.

Flowering Aug–Oct. Dry hills, canyon walls, mesas, limestone outcrops; 1300–1800 m; Ariz., N. Mex.; Mexico.

32. Brickellia veronicifolia   (Kunth) A. Gray, Smithsonian Contr. Knowl. 3(5): 85. 1852 (as veronicaefolia)

Eupatorium veronicifolium Kunth in A. von Humboldt et al., Nov. Gen. Sp. 4(fol.): 88, plate 341. 1818; 4(qto.): 112. 1820 (as veronicaefolium); *Brickellia veronicifolia* var. *petrophila* (B. L. Robinson) B. L. Robinson

Shrubs, 50–90 cm. **Stems** branched from bases, densely pubescent, gland-dotted. **Leaves** mostly opposite (sometimes alternate); petioles 2–5 mm; blades 3-nerved from bases, ovate to reniform, 4–15 × 7–25 mm, bases cordate to subcordate, margins crenate, apices obtuse, faces pubescent, gland-dotted. **Heads** in paniculiform arrays. **Peduncles** 0–2 mm, densely pubescent, gland-dotted. **Involucres** cylindric to broadly campanulate, 9–11 mm. **Phyllaries** 35–40 in 7–9 series, greenish, often purple-tinged, 3–6-striate, unequal, margins scarious (apices obtuse to acute); outer broadly ovate to lance-ovate (puberulent; margins ciliate), inner lanceolate (glabrous). **Florets** 18–33; corollas cream to pale yellow, often red or purple-tinged, 6–7.5 mm. **Cypselae** 3–4 mm, sparsely hispidulous; **pappi** of 24–28 white or tawny, barbellate bristles. $2n = 18$.

Flowering Sep–Oct. Moist areas, canyons, disturbed sites; 1500–2100 m; Tex.; Mexico.

Brickellia veronicifolia occurs in the Chisos Mountains.

404. BRICKELLIASTRUM   R. M. King & H. Robinson, Phytologia 24: 63. 1972

• [Generic name *Brickellia* and Latin *-astrum*, indicating inferiority or an incomplete resemblance]

Guy L. Nesom

Perennials or subshrubs, 30–80+ cm (plants not viscid). **Stems** decumbent to erect, much branched from bases. **Leaves** cauline; mostly opposite (distal sometimes alternate); petiolate; blades usually 3-nerved, deltate to ovate, margins crenate-dentate, faces pubescent or glabrous, sometimes gland-dotted. **Heads** discoid, in loose, corymbiform to paniculiform arrays. **Involucres** ± campanulate, 4–7 mm diam. **Phyllaries** persistent, 20–25 in 3–4 series, 2- or 4-nerved, lance-ovate to lanceolate, unequal (herbaceous to chartaceous). **Receptacles** convex, epaleate. **Florets** 25–35; corollas white or yellowish white, throats funnelform (lengths 2.5–4 times diams.); styles: bases not enlarged, glabrous, branches narrowly clavate. **Cypselae** prismatic, 5(–7)-ribbed, puberulent; **pappi** readily falling or fragile, of ca. 25 barbellate bristles in 1 series. $x = 10$.

Species 2 (1 in the flora): sw United States, n Mexico.

B. L. Turner (1988c, 1994c, 1996+) placed *Brickelliastrum, Asanthus,* and the Mexican genus *Dyscritogyne* R. M. King & H. Robinson within *Steviopsis* R. M. King & H. Robinson. All four are essentially segregates of *Brickellia,* distinguished especially by their lack of pubescent swellings at bases of stylar shafts. As noted by Turner, this aggregate forms a heterogeneous group; the finer division is followed here until further evidence indicates a better taxonomic disposition.

1. Brickelliastrum fendleri   (A. Gray) R. M. King & H. Robinson, Phytologia 24: 64. 1972 • Fendler's Brickellbush F

Brickellia fendleri A. Gray, Mem. Amer. Acad. Arts, n. s. 4: 63. 1849; *Eupatorium fendleri* (A. Gray) A. Gray; *Steviopsis fendleri* (A. Gray) B. L. Turner

Caudices woody, fibrous-rooted. **Stems** erect to declined, puberulent. **Leaves:** petioles 8–30 mm; blades mostly 3–9 × 2–5.5 cm, bases truncate to cordate, margins crenate-serrate to serrate, faces minutely and sparsely puberulent. **Corollas** 3.5–4.5 mm. $2n = 20$.

Flowering Jul–Sep(–Oct). Limestone boulders, crevices, ridgetops, sandstone bluffs, pine woodlands, pine and fir with Douglas fir woods; 1700–2600(–2800) m; Ariz., N.Mex., Tex.; Mexico (Coahuila).

405. FLYRIELLA   R. M. King & H. Robinson, Phytologia 24: 67. 1972 • [For Lowell David Flyr, 1937–1971, Texan, synantherologist]

John L. Strother

Perennials, 20–60[–200] cm (plants ± viscid). **Stems** erect, simple or branched from bases or ± throughout. **Leaves** cauline; mostly opposite (distal sometimes alternate); petiolate; blades

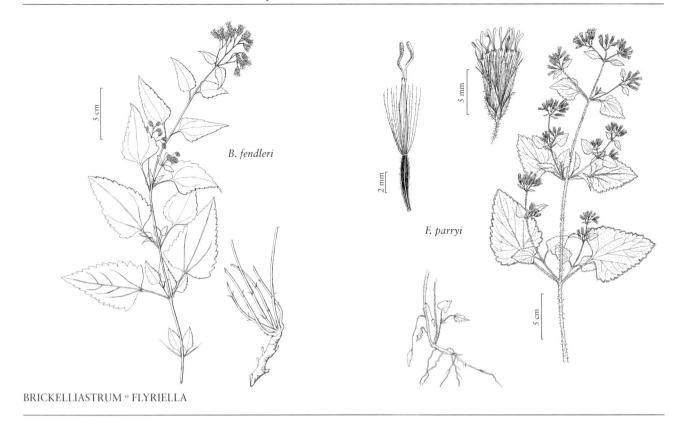

B. fendleri

F. parryi

BRICKELLIASTRUM ° FLYRIELLA

usually 3-nerved from bases, deltate to ovate, margins dentate to serrate, faces usually hirsute and/or stipitate-glandular. **Heads** discoid, in paniculiform to corymbiform arrays. **Involucres** ± cylindric [campanulate or hemispheric], 4–5+ mm diam. **Phyllaries** persistent, 24–30+ in 3–5+ series, 3–5-nerved, ovate to lance-linear, unequal (herbaceous to chartaceous). **Receptacles** flat, epaleate. **Florets** 10–30[–75]; corollas white to ochroleucous, throats ± cylindric (± contracted distally, lengths 4–6 times diams.), lobes 5, oblong-ovate; styles: bases enlarged, hirsute, branches narrowly clavate (distally dilated). **Cypselae** prismatic, 4–5-ribbed, sparsely scabrellous (at least distally), not gland-dotted; **pappi** persistent, of 20–40 smooth to barbellate bristles in 1 series. $x = 10$.

Species 6 (1 in the flora): Texas, Mexico.

SELECTED REFERENCE Baker, M. M. and B. L. Turner. 1986. Taxonomy of *Flyriella* (Asteraceae–Eupatorieae). Sida 11: 300–317.

1. **Flyriella parryi** (A. Gray) R. King & H. Robinson, Phytologia 24: 69. 1972 F

Eupatorium parryi A. Gray in W. H. Emory, Rep. U.S. Mex. Bound. 2(1): 75. 1859

Stems pilosulous and stipitate-glandular. **Leaves:** petioles 10–25(–35+) mm; blades 25–35 (–65+) × 20–35(–45+) mm. **Peduncles** 2–12(–25+) mm. **Involucres** 8–9+ mm. **Corollas** 3.5–4 mm. **Cypselae** 4.5–5 mm; **pappi** 3.5–5 mm. $2n = 20$.

Flowering Apr–Jul. Among rocks, stream banks, canyons; 400–1400 m; Tex.; Mexico (Chihuahua, Coahuila, Nuevo León).

406. ASANTHUS   R. M. King & H. Robinson, Phytologia 24: 66. 1972 • [Asa, honoring American botanist Asa Gray, 1810–1888, and Greek *anthos*, flower]

Guy L. Nesom

Perennials or subshrubs, 30–60(–100) cm. **Stems** erect, much branched. **Leaves** cauline; mostly opposite (distal sometimes alternate); ± sessile [petiolate]; blades 1-nerved (or nerves parallel, mostly linear to filiform [lanceolate] (distal usually filiform to scalelike, canescent, often with axillary, ± 4-ranked, canescent, fascicles of scale-leaves), margins entire, faces glabrous or puberulent, usually gland-dotted. **Heads** discoid, in narrow, racemiform [corymbiform to paniculiform] arrays. **Involucres** ± obconic, 4–5 mm diam. **Phyllaries** persistent, 20–25 in 4–8 series, not notably nerved, lance-ovate to oblong, unequal (herbaceous to scarious, becoming indurate). **Receptacles** flat, epaleate. **Florets** 8–14; corollas whitish, throats narrowly cylindric (lengths 6–8 times diams.); styles: bases not enlarged, glabrous, branches narrowly clavate (distally dilated, not notably papillose). **Cypselae** narrowly prismatic, (9–)10-ribbed, scabrellous on ribs; **pappi** persistent, of 20–100 barbellate (at least distally) bristles in 1–3 series.

Species 3 (1 in the flora): sw United States, w Mexico.

Asanthus has been included within *Steviopsis* (see discussion under 404. *Brickelliastrum*).

1. **Asanthus squamulosus**   (A. Gray) R. M. King &
H. Robinson, Phytologia 24: 66. 1972 • Mule
Mountain false brickellbush   F

Brickellia squamulosa A. Gray,
Proc. Amer. Acad. Arts 15: 30.
1880

Leaves: petioles 0–2 mm; blades
3–7 cm × 2–6 mm. **Involucres** 8–
10 mm. **Phyllaries** grading into
similar-sized peduncular bracts.
Cypselae 2–3 mm.

Flowering Apr–May.   Flats,
gravel banks, arroyos, pine-oak woodlands; 2000–2500
m; Ariz., N.Mex.; Mexico.

407. MALPERIA   S. Watson, Proc. Amer. Acad. Arts 24: 54. 1889 • [For Edward Palmer, 1831–1911, American field botanist who collected the type material; based on anagram of his surname]

Guy L. Nesom

Annuals, 10–40 cm. **Stems** erect, much branched. **Leaves** basal and cauline; mostly alternate (proximal opposite); mostly sessile; blades 1-nerved, linear, margins entire, faces glabrous or puberulent, sometimes gland-dotted. **Heads** discoid, in loose, corymbiform arrays or borne singly. **Involucres** ± campanulate to obconic, 5–6 mm diam. **Phyllaries** persistent, 20–30+ in 3–4+ series, not strongly nerved, lanceolate, unequal (± chartaceous, margins scarious). **Receptacles** flat, epaleate. **Florets** 20–30; corollas white, throats cylindric or narrowly urceolate (lengths 5–6 times diams.); styles: bases not enlarged, glabrous, branches filiform-clavate. **Cypselae** fusiform, 5-ribbed, scabrellous on ribs; **pappi** persistent, of 3 muticous scales plus 3,aristate scales. $x = 10$.

Species 1: w United States, nw Mexico.

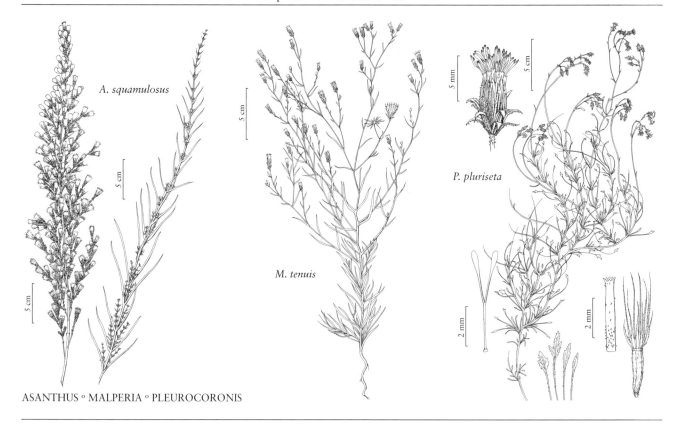

ASANTHUS ○ MALPERIA ○ PLEUROCORONIS

1. **Malperia tenuis** S. Watson, Proc. Amer. Acad. Arts 24: 54. 1889 • Brownturbans [F]

Leaves: basal often withered by flowering, cauline 2–5 cm × 0.3–1.5 mm. **Corollas** 5–6 mm. **Cypselae** ca. 3 mm; **pappi:** muticous scales ca. 0.5 mm; aristate scales ca. 5 mm. $2n = 20$.

Flowering (Jan–)Feb–Mar (–Apr). Gravelly and sandy washes and other alluvium, rocky hillsides, sandy flats, lava, desert scrub, creosote bush flats; 10–700 m; Calif.; Mexico (Baja California, Baja California Sur, Sonora).

Malperia tenuis is recognized by its annual duration, linear leaves densely arranged along stems or concentrated near bases of stems, loosely arranged heads, and pappi of two kinds of scales.

408. **PLEUROCORONIS**   R. M. King & H. Robinson, Phytologia 12: 468, fig. pp. 472, 473. 1966 • [Greek *pleura*, side, and *korone*, crown, alluding to squamellae that appear to form a crown subtending bristles]

Guy L. Nesom

Subshrubs or shrubs, (10–)30–100 cm. **Stems** erect, usually much branched. **Leaves** cauline; mostly opposite (distal often alternate); petiolate; blades 1- or 3-nerved, ovate, deltate, or rhombic to lanceolate, margins entire or toothed [lobed, 2-pinnatifid], faces usually stipitate-glandular. **Heads** discoid, borne singly or in loose, corymbiform arrays. **Involucres** ± obconic, 4–6 mm diam. **Phyllaries** persistent, 30–35 in 3–4+ series, (herbaceous to chartaceous) 1- or

3-nerved, ovate to lanceolate, unequal. **Florets** 25–30; corollas white, throats narrowly cylindric (lengths 4–6 times diams.); styles: bases not enlarged, glabrous, branches ± clavate. **Receptacles** flat or convex, epaleate. **Cypselae** prismatic, 4–5-ribbed, scabrellous; **pappi** persistent, of 3–6(–12) subulate-aristate scales plus 3–6(–12) coarse, barbellate bristles in 1 series. *x* = 9.

Species 3 (1 in the flora): w United States, nw Mexico

SELECTED REFERENCE King, R. M. 1967. Studies in the Eupatorieae, (Compositae). I–III. Rhodora 69: 35–47.

1. Pleurocoronis pluriseta (A. Gray) R. M. King & H. Robinson, Phytologia 12: 469. 1966 • Bush arrowleaf [F]

Hofmeisteria pluriseta A. Gray in War Department [U.S.], Pacif. Railr. Rep. 4(5): 96, plate 9. 1857

Plants rounded, densely branched from bases. **Stems** glandular. **Petioles** filiform, 2–6 cm, lengths 3–10 times blades. **Leaf blades** 3–10+ mm, sharply toothed, often subhastate. **Corollas** 4–5 mm. **Cypselae** 3–4 mm; **pappi**: scales 1–2+ mm, bristles 2–5 mm. *2n* = 18.

Flowering (Oct–)Mar–May(–Jun). Rocky slopes and washes, limestone outcrops, crevices of canyon walls, creosote bush scrub, acacia-mesquite; 0–200 m; Ariz., Calif., Nev., Utah; Mexico (Baja California, Baja California Sur, Sonora).

409. CARMINATIA Mociño ex de Candolle in A. P. de Candolle and A. L. P. P. de Candolle, Prodr. 7: 267. 1838 • [For Bassiani Carminati, eighteenth-century Italian author of book on hygiene, therapeutics, and materia medica]

David J. Keil

Annuals, 10–100+ cm. **Stems** erect (round to 4-angled or -ribbed), unbranched or sparingly branched (from proximal nodes, puberulent and/or villous to pilose, often in lines, sometimes glabrate). **Leaves** mostly cauline; usually opposite (proximal often withering before flowering, distal sometimes alternate); petiolate; blades (proximal) 3–5-nerved, triangular to broadly ovate (bases obtuse to cordate), margins subentire to dentate, faces glabrous or sparsely villous (at least along veins and on margins), not gland-dotted. **Heads** discoid, in spiciform or narrow, ± paniculiform arrays (subtended by lanceolate to linear, scalelike bracts). **Involucres** cylindric, [2–]3–4+ mm diam. **Phyllaries** persistent (spreading in age), (12–)17–22+ in 3–4+ series, 3-nerved, lance-deltate or lanceolate to linear, unequal (apices acute to acuminate). **Receptacles** flat, epaleate. **Florets** 8–12; corollas greenish white to cream-colored, (bases slightly enlarged, tubes and throats not markedly differentiated externally) throats cylindric to filiform, lobes 5, triangular-ovate (very short); styles: bases not enlarged, glabrous, branches ± filiform (appendages linear or weakly clavate). **Cypselae** ± prismatic (usually 5-angled) or subcylindric, usually 5-ribbed, minutely puberulent; **pappi** usually readily falling, of 8–13 plumose bristles or setiform scales in 1 series (basally coherent or weakly connate, falling together or in groups). x = 10.

Species 3 (1 in the flora): sw United States, Mexico, Central America (Guatemala).

SELECTED REFERENCE Turner, B. L. 1988b. Taxonomy of *Carminatia* (Asteraceae, Eupatorieae). Pl. Syst. Evol. 160: 169–179.

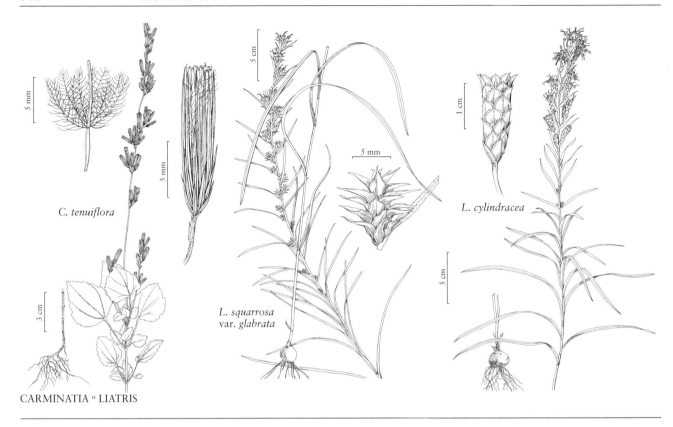

C. tenuiflora

L. squarrosa
var. glabrata

L. cylindracea

CARMINATIA ° LIATRIS

1. Carminatia tenuiflora de Candolle in A. P. de
Candolle and A. L. P. P. de Candolle, Prodr. 7: 267.
1838 F

Brickellia tenuiflora (de Candolle)
D. J. Keil & Pinkava

Stems usually puberulent and
villous (hairs crinkled, multicellular), sometimes glabrate.
Petioles 1–5 cm, villous-ciliate.
Leaf blades (proximal to mid
stem) broadly ovate to cordate, 2–
8 × 1.5–6 cm, bases obtuse to
truncate or subcordate, margins subentire to dentate,
margins villous-ciliolate (sometimes sparsely villous

along veins as well), apices obtuse to acute. **Heads** 1–
10+ per principal node, ascending to erect; sessile or on
peduncles to 7 mm. **Involucres** 9–12 mm. **Phyllaries**
glabrous (margins sometimes ciliolate). **Corollas** 5–6
mm, slightly exceeding pappi, 0.2–0.3 mm diam., apices
slightly constricted, lobes 0.15–0.2 mm; anthers
included, ca. 0.8 mm; style branches usually short-
exserted. **Cypselae** gray to nearly black, 3.5–5 mm; **pappi**
white, 5–8 mm. $2n = 20$.

Flowering Aug–Oct. Oak-juniper woodlands, cypress
woodlands, grasslands, riparian areas; 1000–2200 m;
Ariz., N.Mex., Tex.; Mexico.

---

410. LIATRIS Gaertner ex Schreber, Gen. Pl. 2: 542. 1791, name conserved • Gayfeather,
blazing star [Derivation unknown]

Guy L. Nesom

Perennials, 20–180 cm (corms globose to depressed-ovoid or napiform, sometimes elongated,
becoming rhizomes, roots all or mostly adventitious). **Stems** erect, simple or basally branched.
Leaves basal and cauline; alternate; ± petiolate (basal) or sessile (usually appressed to ascending);
blades usually 1-nerved, sometimes 3- or 5-nerved, mostly linear to ovate-lanceolate, margins

entire, faces often gland-dotted (stipitate-glandular in *L. glandulosa*). **Heads** discoid, in corymbiform, cymiform, racemiform, or spiciform arrays. **Involucres** mostly campanulate to hemispheric or turbinate-cylindric, (2.5–)3–22(–25) mm diam. **Phyllaries** persistent or tardily falling, 18–40 in (2–)3–7 series, not notably nerved, ovate to elliptic or lanceolate, usually unequal (herbaceous to petaloid, margins often hyaline, often ciliate or irregularly toothed, apices often pink-white). **Receptacles** flat, epaleate. **Florets** 3–85; corollas usually lavender to dark magenta or pinkish purple, sometimes white, throats funnelform (lengths 4–6 times diams., externally glanduliferous, glabrous inside or pilose inside near filament insertions, hairs whitish, crisped); styles: bases not enlarged, glabrous, branches linear-clavate (papillate). **Cypselae** prismatic, 8–11-ribbed, usually hirsutulous to hirtellous-pilose (glabrous in *L. oligocephala*), usually gland-dotted; **pappi** persistent, of 12–40 coarsely barbellate to plumose bristles in 1–2 series. *x* = 10.

Species 37 (37 in the flora): North America, Mexico, West Indies (Bahamas).

The globose perennating structures of *Liatris* have been described as corms and cormoid rootstocks, the elongate ones as rhizomes and penetrating rootstocks. They are here regarded as corms and rhizomes, rather than roots with adventitious buds. New stems may be produced from various lateral points (nodes) of the corms. Some taxa show various stages of transition between globose structures and elongate structures that function like horizontal rhizomes. All other Liatrinae (except perhaps the shrubby *Garberia*) apparently produce rhizomes with fibrous roots, rather than a taproot.

SELECTED REFERENCES Gaiser, L. O. 1946. The genus *Liatris*. Rhodora 48: 165–183, 216–263, 273–326, 331–382, 393–412. Gaiser, L. O. 1950. Chromosome studies in *Liatris*. I. *Spicatae* and *Pycnostachyae*. Amer. J. Bot. 37: 122–135. Gaiser, L. O. 1950b. Chromosome studies in *Liatris*. II. *Graminifoliae* and *Pauciflorae*. Amer. J. Bot. 37: 414–423. Gaiser, L. O. 1950c. Chromosome studies in *Liatris*. III. *Punctatae*. Amer. J. Bot. 37: 763–777. Gaiser, L. O. 1950d. Evidence for intersectional field hybrids in *Liatris*. Evolution 5: 52–67. Menhusen, B. R. 1963. Variation in the *Punctatae* Series of the Genus *Liatris* (Compositae). Ph.D. dissertation. University of Kansas. Nesom, G. L. 2005. Infrageneric classification of *Liatris* (Asteraceae: Eupatorieae). Sida 21: 1305–1321. Nesom, G. L. and J. M. Stucky. 2004. Taxonomy of the *Liatris pilosa* (*graminifolia*) group (Asteraceae: Eupatorieae) Sida 21: 815–826.

1. Pappi usually of plumose or subplumose bristles (barbellate to subplumose, barbels mostly 0.5–1 mm, in *L. acidota* and *L. ohlingerae*).
    2. Phyllary apices (at least inner) prolonged, ± spreading, ± dilated, petaloid (pink, purplish, white, or yellow) . . . . 10. *Liatris elegans*
    2. Phyllary apices (at least of inner) acuminate, acute, cuspidate, mucronate, obtuse, rounded, rounded-acuminate, or truncate (seldom notably spreading, not dilated or petaloid).
        3. Heads in cymiform or corymbiform to subcorymbiform arrays; peduncles (arcuate or ascending-spreading) 20–70 mm.
            4. Involucres cylindro-campanulate, 14–17 × 8–11 mm; phyllary apices usually rounded, often mucronate; Texas . . . . 7. *Liatris cymosa*
            4. Involucres hemispheric to campanulate or broadly turbinate, 17–23 × 15–20 mm; phyllary apices usually rounded to obtuse (pappus bristles barbellate to subplumose); Florida . . . . 30. *Liatris ohlingerae* (in part)
        3. Heads usually in racemiform to spiciform arrays, rarely borne singly; peduncles 0 or 1–10(–25) mm.
            5. Leaves 3–5-nerved; florets 10–60; corollas (lobes adaxially hispid).
                6. Phyllaries ± equal (outer usually as long as or longer than inner).
                    7. Heads usually borne singly (each immediately subtended by relatively broad, foliaceous bracts longer than involucres); stems glabrous . . . . 1. *Liatris compacta*
                    7. Heads in racemiform to spiciform arrays; stems glabrous or puberulent-villous . . . . 2. *Liatris squarrosa*
                6. Phyllaries ± unequal (outer shorter).

8. Stems hirsute to piloso-hirsute; phyllary apices (at least inner) acute-acuminate (all usually spreading to reflexed on distal $^1/_5$–$^1/_3$, usually without hyaline borders) .... 3. *Liatris hirsuta*

8. Stems usually glabrous; phyllary apices (at least inner) usually rounded, rounded-acuminate, or truncate, often stiffly mucronate (all essentially erect and appressed, usually with narrow hyaline borders) ...... 4. *Liatris cylindracea*

5. Leaves mostly 1-nerved, sometimes 3–5-nerved (*L. acidota*); florets (2–)3–6 (–8) or 8–14(–16); corollas (lobes glabrous).

9. Heads in loose, spiciform arrays (widely spaced, stems evident); florets 4–6(–8) or 8–14(–16).

10. Florets 4–6(–8) .... 5. *Liatris punctata* (in part)

10. Florets 8–14(–16) .... 6. *Liatris bracteata*

9. Heads in dense, spiciform arrays (closely spaced, stems usually obscured); florets (2–)3–6(–8).

11. Corms elongate or becoming rhizomes .... 5. *Liatris punctata* (in part)

11. Corms globose, subglobose, depressed-globose, or ovoid.

12. Leaves 3(–5)-nerved (bases of basal often fibrous-persistent), cauline abruptly reduced distally (pappus bristles subplumose); coastal plain, Louisiana, Texas .... 16. *Liatris acidota* (in part)

12. Leaves 1-nerved (bases not fibrous-persistent), cauline usually gradually or little reduced, sometimes abruptly reduced, distally; inland habitats.

13. Stems stipitate-glandular .... 9. *Liatris glandulosa*

13. Stems glabrous (not glandular).

14. Phyllaries (11–18) in (4–)5–6 series, mostly oblong-obovate, unequal (inner 7–9 mm), apices obtuse to truncate-rounded (each tipped by thick, often indurate apicula or mucros, lateral veins usually not evident beyond middle or at least on distal $^1/_3$); flowering mostly mid–late Aug(–Oct) .... 5. *Liatris punctata* (in part)

14. Phyllaries (6–11) in 2–3 series, oblong-lanceolate to lanceolate-triangular, subequal (inner 11–12 mm), apices acute to acute-attenuate (sometimes with narrow tips, lateral veins usually evident beyond middle, continuing nearly to tips); flowering mid Jul–Aug(–Sep) .... 8. *Liatris aestivalis*

1. Pappi usually of barbellate bristles (barbels mostly 0.1–0.4 mm).

15. Heads borne singly or (2–30) in corymbiform to subcorymbiform arrays.

16. Involucres campanulate-turbinate, 10–19(–23) × (8–)10–17(–22) mm; phyllaries in ± 3–4 series, apices mostly long-acuminate (innermost short-acuminate); Alabama .... 12. *Liatris oligocephala*

16. Involucres hemispheric to campanulate or broadly turbinate, 17–23 × 15–20 mm; phyllaries in 6–7 series, apices rounded to obtuse; Florida .... 30. *Liatris ohlingerae* (in part)

15. Heads in racemiform to spiciform arrays.

17. Leaves 3- or 5-nerved.

18. Stems glabrous (often sparsely sessile-glandular, leaf bases not fibrous-persistent); corolla tubes pilose inside .... 29. *Liatris savannensis* (in part)

18. Stems glabrous or piloso-puberulent (not glandular, leaf bases fibrous-persistent); corolla tubes glabrous inside.

19. Phyllary apices obtuse to rounded.

20. Basal leaves 6–12(–15) mm wide; Great Plains .... 13. *Liatris lancifolia*

20. Basal leaves (2–)4–10(–20) mm wide; e of Mississippi River .... 14. *Liatris spicata*

19. Phyllary apices acuminate, acute, or acute-acuminate.

21. Phyllary apices acute to acute-acuminate (reflexed to curving-spreading, tips often ± dilated, nearly petaloid); florets (4–)5–8 .... 15. *Liatris pycnostachya*

21. Phyllary apices acuminate to acute (erect and ± appressed); florets (2–)3–4(–5) .... 16. *Liatris acidota* (in part)

[17. Shifted to left margin.—Ed.]
17. Leaves usually 1-nerved, sometimes 1- (or 3-) nerved.
  22. Phyllary apices (at least mid and inner) usually acuminate to acute or rounded to rounded-retuse and minutely involute-cuspidate to acuminate or apiculate.
    23. Phyllaries narrowly triangular, margins without hyaline borders, apices (divergent), usually acuminate to acute; florets 10–12; w Louisiana, e Texas . . . . 11. *Liatris tenuis*
    23. Phyllaries elliptic-oblong, lanceolate, oblong, oblong-lanceolate, oblong-obovate, obovate, ovate, ovate-triangular, or spatulate, margins usually with narrow hyaline borders, apices (erect) acuminate, acute, obtuse-angled, rounded, or rounded-retuse, sometimes minutely involute-cuspidate to apiculate; florets 3–6(–9) or 7–10 (–12); mostly e of Mississippi River.
      24. Stems glabrous or sparsely pilose.
        25. Phyllary apices usually rounded-retuse and minutely involute-cuspidate to apiculate; corolla tubes glabrous inside.
          26. Stems glabrous or sparsely pilose; basal (and proximal cauline) leaves linear (arising from separated nodes), 1–2(–2.5) mm wide, abruptly or gradually reduced distally, margins usually ciliate proximally, faces gland-dotted . . . . 17. *Liatris tenuifolia*
          26. Stems glabrous; basal leaves (mostly arising from congested nodes) lance-linear to linear, (1–)2–6(–9) mm wide, abruptly reduced distally, margins not ciliate, faces (minutely white-dotted by stomates) weakly, if at all, gland-dotted . . . . 18. *Liatris laevigata*
        25. Phyllary apices acute, obtuse-angled, or rounded; corolla tubes pilose inside.
          27. Heads in racemiform to spiciform (often strongly secund) arrays; involucres (7–)11–15 mm; phyllaries oblong to oblong-oblanceolate; florets 3–6 . . . . 21. *Liatris pauciflora*
          27. Heads in racemiform to spiciform (sometimes secund) arrays; involucres (6–)7–9 mm; phyllaries ovate-triangular to oblong or oblong-lanceolate; florets 4–10(–12).
            28. Heads in loose arrays (not secund, internodes 6–20 mm); phyllary apices sharply acute to obtuse-angled, apiculate (apicula thickened, not markedly involute, laminae relatively thick, usually with glands in pits, without superficial glands); florets 7–10(–12); basal and proximal cauline leaves (2–)4–9(–12) mm wide (usually abruptly, sometimes gradually, reduced distally) . . . . 23. *Liatris virgata*
            28. Heads in dense arrays (often secund, internodes 1–5 mm); phyllary apices rounded, acuminate to involute-cuspidate (laminae relatively thin, glands superficial at least proximally); florets 4–7(–9); basal and proximal cauline leaves 2–5 mm wide (abruptly or gradually reduced distally) . . . . 24. *Liatris cokeri*
      24. Stems hirtellous (hairs spreading to slightly deflexed) or hirsute, piloso-puberulent, puberulent, strigoso-hirtellous, or strigoso-puberulent.
        29. Stems hirsute, puberulent, piloso-puberulent, or strigoso-puberulent.
          30. Corms elongated into rhizomes (irregularly shaped, sometimes with thickened, fibrous roots); involucres 8–10 mm . . . . 22. *Liatris garberi*
          30. Corms globose to elongate; involucres 4–6(–7) mm.
            31. Cauline leaves (proximal) usually lanceolate, linear, linear-oblanceolate, or oblanceolate, sometimes narrowly spatulate, 2–8(–10) mm wide (bracts subtending proximal heads to 2 mm wide, abruptly differentiated from distal cauline leaves); phyllary apices usually rounded or obtuse-truncate, rarely acute . . . . . . 32. *Liatris gracilis* (in part)
            31. Cauline leaves (proximal) spatulate (petioles slender, blades elliptic to lance-elliptic), (8–)11–22(–27) mm wide (bracts subtending proximal heads 2+ mm wide, grading gradually from distal cauline leaves); phyllary apices usually acuminate to acute, rarely obtuse . . . . 33. *Liatris gholsonii* (in part)

29. Stems hirtellous (hairs spreading to slightly deflexed).

    32. Heads in relatively loose, racemiform to spiciform (often secund) arrays . . . . 21. *Liatris pauciflora*

    32. Heads in dense, spiciform (cylindric) arrays.

        33. Heads (rigidly ascending, appressed to rachises and each other, overlapping); phyllary margins usually not ciliolate, apices (erect) acuminate to acute . . . . 19. *Liatris chapmanii*

        33. Heads (spreading to ascending, not strongly overlapping); phyllary margins ciliolate, apices (slightly spreading) acuminate to acute . . . . 20. *Liatris provincialis*

[22. Shifted to left margin.—Ed.]

22. Phyllary apices (at least mid and inner) usually obtuse, rounded, or truncate, rarely acute (not rounded-retuse or cuspidate to apiculate).

    34. Stems glabrous (sometimes sparsely sessile-glandular, rarely sparsely to moderately pilose in *L. pilosa*).

        35. Involucres 5–7(–9) mm; florets 4–5(–6); corolla tubes glabrous inside (pappi: lengths usually ± ¹/₂ corollas) . . . . 28. *Liatris microcephala*

        35. Involucres 6–10 mm; florets (6–)7–13(–17); corolla tubes pilose or glabrous (*L. ligulistylis*) inside (pappi: lengths usually equaling corollas, shorter in some populations of *L. helleri*).

            36. Stems glabrous (often sparsely sessile-glandular); involucres (8–)9–11(–12) mm diam.; phyllaries: hyaline borders erose to lacerate, (0.2–)0.4–1 mm wide; florets 9–17 . . . . 29. *Liatris savannensis* (in part)

            36. Stems glabrous or sparsely to moderately pilose (not glandular); involucres 5–8(–10) mm diam.; phyllaries: hyaline borders smooth or erose to lacerate, 0.2–0.4 mm wide; florets (6–)7–13(–17).

                37. Plants 15–55 cm; leaves weakly, if at all, gland-dotted; pappi: lengths ¹/₃–²/₃ or equaling corollas; montane . . . . 25. *Liatris helleri*

                37. Plants 40–120 cm; leaves ± gland-dotted; pappi: lengths ± equaling corollas; coastal plain and piedmont.

                    38. Stems glabrous or sparsely or moderately pilose; heads in loose to dense arrays (internodes 1–7 mm); peduncles 0 or 1–10(–80) mm; involucres (7–)8–10 mm; phyllaries in (3–)4–5(–6) series . . . . 26. *Liatris pilosa*

                    38. Stems glabrous; heads in loose arrays (internodes 2–14 mm); peduncles 0 or 2(–7) mm; involucres 6–8 mm; phyllaries in 3–4(–5) series . . . . 27. *Liatris elegantula*

    34. Stems hispidulo-puberulent, piloso-puberulent, puberulent, puberulent-villous, strigoso-hirtellous, or strigoso-puberulent.

        39. Involucres 2.5–7 mm diam.; florets 3–12.

            40. Stems moderately to densely strigoso-hirtellous; peduncles (divergent, arcuate-ascending) 10–25(–30) mm; involucres 5–7 mm diam.; phyllary apices rounded to subtruncate; florets 7–12 . . . . 31. *Liatris patens*

            40. Stems piloso-puberulent, puberulent, or strigoso-puberulent; peduncles 0 or (usually ascending) 2–10(–12) mm; involucres 2.5–4(–5) mm diam.; phyllary apices rounded or obtuse to acuminate or acute; florets 3–6(–9).

                41. Proximal cauline leaves usually linear to linear-oblanceolate or oblanceolate, sometimes narrowly spatulate, 2–8(–10) mm wide (proximal floral bracts to 2 mm wide, abruptly differentiated from distal cauline leaves); phyllary apices usually rounded or obtuse-truncate, rarely acute . . . . 32. *Liatris gracilis* (in part)

                41. Proximal cauline leaves spatulate (blades elliptic to lance-elliptic), (8–)11–22(–27) mm wide (proximal floral bracts 2+ mm wide, gradually differentiated from distal cauline leaves); phyllary apices usually acuminate to acute, rarely obtuse . . . . 33. *Liatris gholsonii* (in part)

        39. Involucres (6–)8–15 mm diam. (*L. squarrulosa*) or 13–22(–25) mm diam.; florets 11–80.

[42. Shifted to left margin.—Ed.]

42. Peduncles usually (5–)8–50 mm, rarely 0; phyllaries (usually erect, rarely spreading-reflexing); florets 30–80 (19–33 in *L. scariosa* var. *scariosa*); corolla tubes glabrous or pilose inside.

    43. Phyllary margins: hyaline borders erose to lacerate or irregular; corolla tubes glabrous inside; Great Plains, e slope of Rocky Mountains in United States . . . . 34. *Liatris ligulistylis*

    43. Phyllary margins: hyaline borders none or relatively smooth; corolla tubes pilose or glabrous inside; ne and nc United States, adjacent Canada . . . . 35. *Liatris scariosa*

42. Peduncles usually 0, sometimes 1–8+ (rarely to 30 in *L. squarrulosa*) mm; phyllaries (at least outer) usually reflexing, sometimes erect; florets 11–30; corolla tubes pilose inside.

    44. Phyllaries glabrous (bullate, hyaline borders relatively broad, often erose to lacerate or irregular) . . . . 36. *Liatris aspera*

    44. Phyllaries glabrous or puberulent to puberulent-hirtellous (not bullate, hyaline borders none or relatively narrow, smooth) . . . . 37. *Liatris squarrulosa*

1. Liatris compacta (Torrey & A. Gray) Rydberg, Brittonia 1: 98. 1931 • Arkansas gayfeather E

Liatris squarrosa (Linnaeus) Michaux var. *compacta* Torrey & A. Gray, Fl. N. Amer. 2: 68. 1841

Plants 22–50 cm. **Corms** globose to subglobose. **Stems** glabrous. **Leaves:** basal and proximal cauline 3–5-nerved (parallel veins evident, not persistent), elliptic-lanceolate, 80–150(–180) × 3–10(–12) mm (largest toward midstem), gradually or slightly reduced distally, glabrous, weakly gland-dotted. **Heads** borne singly or (2–5) in loose, racemiform to spiciform arrays. **Peduncles** 3–25 mm (each head subtended by broad, foliaceous bracts longer than involucres). **Involucres** cylindro-campanulate, 12–17 × 8–14 mm. **Phyllaries** (loose to spreading) in 5–7 series, narrowly lanceolate (outer, usually longer than inner, foliaceous) to broadly oblong or narrowly oblong-lanceolate, usually glabrous, sometimes sparsely puberulent, margins without hyaline borders, apices acuminate to acute or cuspidate. **Florets** 18–25; corolla tubes glabrous inside (lobes adaxially hispid). **Cypselae** 5.5–8 mm; **pappi:** lengths ± equaling corollas, bristles plumose.

Flowering (May–)Jun–Sep. Rocky ridges, bluffs, hillsides, novaculite, sandstone, open woods, openings; 100–600(–900) m; Ark.

Liatris compacta is known from west-central Arkansas. It has been treated as a variety of *L. squarrosa*, to which it appears closely related (especially var. *glabrata*). In addition to morphologic differences, *L. compacta* is distinct in habitat and is geographically disjunct from its closest relatives (*L. squarrosa* vars. *glabrata* and *squarrosa*); it is sympatric with *L. hirsuta*, which apparently is separated by habitat. The bracts subtending the heads are foliaceous; the outermost phyllaries of *L. squarrosa* are relatively elongate and usually phyllary-like, not nearly so large as in *L. compacta*.

Liatris compacta, *L. squarrosa*, *L. hirsuta*, and *L. cylindracea* are closely similar and interrelated: the basal leaves are on relatively widely spaced nodes, usually wither before flowering, and lengthen distally into the more crowded (shorter internodes) cauline leaves. At least the basal and proximal cauline leaves are basally sheathing and strongly parallel-veined. All species of this group also have at least a tendency for cylindric involucres, and corolla lobes of all are densely hirsute on the adaxial faces. Intermediates and intergrades among *L. squarrosa*, *L. hirsuta*, and *L. cylindracea* are relatively common in areas of sympatry; identifications are correspondingly arbitrary.

2. Liatris squarrosa (Linnaeus) Michaux, Fl. Bor.-Amer. 2: 92. 1803 • Loosescale gayfeather, colicroot E F

Serratula squarrosa Linnaeus, Sp. Pl. 2: 818. 1753; *Lacinaria squarrosa* (Linnaeus) Hill

Plants 30–80 cm. **Corms** globose to slightly elongate. **Stems** glabrous or puberulent-villous. **Leaves:** basal and proximal cauline 3–5-nerved, broadly linear to linear-oblanceolate, 70–220 × 2–12 mm (largest usually distal to proximalmost), gradually reduced distally, puberulent-villous or glabrous, weakly gland-dotted. **Heads** borne singly or (3–26) in loose, racemiform to spiciform arrays. **Peduncles** 0 or (spreading to ascending) 1–8(–20+) mm. **Involucres** cylindro-campanulate, 12–20(–25) × 7–10 mm. **Phyllaries** (usually erect, sometimes spreading-reflexing on the distal 1/3) in 5–7 series, narrowly ovate-triangular and foliaceous (outer) to ovate or oblong-triangular, ± equal, glabrous or pubescent, margins without hyaline borders, stiffly to softly ciliate, apices acuminate to acute. **Florets** (10–)23–45(–60); corolla tubes glabrous inside (lobes adaxially hispid). **Cypselae** 4–5.5 mm; **pappi:** lengths ± equaling corollas, bristles plumose.

Varieties 2 (2 in the flora): United States.
See discussion under 1. *Liatris compacta*.

1. Stems glabrous; leaves glabrous; phyllaries
   glabrous . . . . 2a. *Liatris squarrosa* var. *glabrata*
1. Stems usually puberulent-strigose (hairs soft,
   upturned or appressed); leaves sparsely pilose to
   glabrescent; phyllaries usually ciliate, faces glabrous
   or pubescent . . . . . . 2b. *Liatris squarrosa* var. *squarrosa*

2a. Liatris squarrosa (Linnaeus) Michaux var. **glabrata**
(Rydberg) Gaiser, Rhodora 48: 401. 1946 E F

Liatris glabrata Rydberg, Brittonia
1: 98. 1931

Stems glabrous. **Leaves** glabrous.
Phyllaries glabrous.

Flowering (Jun–)Jul–Aug.
Prairies, dunes, flood plains,
ditches, roadsides, sandy soils,
sandstone hills and outcrops,
gravelly hills, clay soils; 50–500
m; Colo., Iowa, Kans., Nebr., Okla., Tex.

Variety *glabrata* apparently is morphologically
discontinuous from var. *squarrosa* where their ranges
closely approach each other (but do not meet) in Texas,
and they might be treated as separate species, especially
in view of the treatment here of *Liatris compacta* at
specific rank. Within var. *squarrosa* as treated here,
plants along the Gulf coast tend to have much reduced
phyllary vestiture (var. *alabamensis*) and perhaps
represent the precursor to var. *glabrata*. Involucres of
var. *squarrosa* sometimes approach the morphology of
L. hirsuta; this appears to be parallel expression of
ancestral similarity, because such plants often are long-
separated from the geographic range of *L. hirsuta*. Heads
tend to be more numerous and more crowded in var.
glabrata than in var. *squarrosa*.

2b. Liatris squarrosa (Linnaeus) Michaux var.
squarrosa E

Liatris glabrata Rydberg var.
alabamensis (Alexander) Shinners;
L. squarrosa var. *alabamensis*
(Alexander) Gaiser; *L. squarrosa*
var. *gracilenta* Gaiser

Stems usually puberulent-strigose
(hairs soft, upturned or appressed.
Leaves sparsely pilose to
glabrescent. **Phyllaries** usually
ciliate, faces glabrous or pubescent.

Flowering (May–)Jul–Aug(–Sep). Oak-pine, pine,
scrub oak, roadsides, fencerows, limestone outcrops,
sand, sandy loam, chalk, and clay soils; 20–300 m; Ala.,
Del., Fla., Ga., Ill., Ind., Ky., La., Md., Mich., Mo., N.C.,
Ohio, S.C., Tenn., Va., W.Va.

3. Liatris hirsuta Rydberg, Brittonia 1: 98. 1931
   • Hairy gayfeather E

Liatris squarrosa (Linnaeus)
Michaux var. *hirsuta* (Rydberg)
Gaiser

Plants 20–70 cm. **Corms** globose
to slightly elongate. **Stems** hirsute
to piloso-hirsute. **Leaves:** basal
and proximal cauline 3–5-nerved,
linear-lanceolate, 60–180 × 2–7
mm (largest usually distal to
proximalmost), gradually reduced distally, hirsute to
piloso-hirsute. **Heads** in loose, racemiform to spiciform
arrays. **Peduncles** 0 or (peduncles spreading to
ascending) 1–10 mm. **Involucres** cylindro-campanulate,
11–17 × 6–9 mm. **Phyllaries** (spreading to reflexed) in
5–7 series, ovate-triangular (outer) to oblong-triangular,
unequal, usually sparsely hirsute, margins without
hyaline borders, coarsely hirsute-ciliate, apices acute-
acuminate. **Florets** 15–30; corolla tubes glabrous inside
(lobes adaxially hispid). **Cypselae** 5.5–6.5 mm; **pappi:**
lengths ± equaling corollas, bristles plumose.

Flowering (Jun–)Jul–Aug(–Sep). Prairies, rocky
slopes, flats, marl ridges, pine-oak woods, streamsides,
roadsides; 50–500(–900) m; Ark., Iowa, Kans., La.,
Miss., Mo., Nebr., Okla., Tex.

Liatris hirsuta occupies a geographic range separate
from and nearly contiguous with *L. squarrosa*. They
have been treated as a single species. *Liatris hirsuta* is
sympatric (without intergrades) with *L. squarrosa* var.
squarrosa in Texas, Oklahoma, Louisiana, and
Mississippi; it is sympatric with *L. compacta* in Arkansas;
it intergrades with *L. cylindracea* in Missouri. See also
discussion under 1. *L. compacta*.

4. Liatris cylindracea Michaux, Fl. Bor.-Amer. 2: 93.
   1803 • Barrelhead gayfeather E F

Lacinaria cylindracea (Michaux)
Kuntze; *Liatris intermedia* Lindley;
L. squarrosa (Linnaeus) Michaux
var. *intermedia* (Lindley) de
Candolle

Plants 20–60 cm. **Corms** usually
globose, rarely elongate. **Stems**
glabrous. **Leaves:** basal and
proximal cauline 3(–5)-nerved,
linear-oblanceolate to linear-lanceolate, 80–250 × 2–6
mm (largest usually distal to proximalmost), gradually
reduced distally, essentially glabrous (proximal margins
pilose-ciliate). **Heads** borne singly or (2–28) in loose to
dense, racemiform to spiciform arrays. **Peduncles** 0 or
(spreading-ascending) 2–10(–20) mm. **Involucres**
cylindro-campanulate, (11–)13–18 × 6–8 mm. **Phyllaries**
in 5–7 series, ovate-triangular (outer) to broadly oblong

or spatulate-oblong, strongly unequal, essentially glabrous, margins usually with narrow hyaline borders, ciliolate, apices broadly rounded, rounded-acuminate, or truncate (inner stiffly mucronate). **Florets** 10–35; corolla tubes glabrous inside (lobes adaxially hispid). **Cypselae** 5–7 mm; **pappi:** lengths ± equaling corollas, bristles plumose.

Flowering Jul–Sep. Prairies, limestone outcrops, bluffs, barrens, and glades, marl, sandstone outcrops, dunes, roadsides, sandy pine-oak, wooded northern slopes; 100–400 m; Ont.; Ala., Ark., Ga., Ill., Ind., Iowa, Kans., Ky., Mich., Minn., Mo., N.Y., Ohio, Okla., Tenn., Wis.

Stems and leaves of *Liatris cylindracea* sometimes are hairy (Kentucky, Missouri), perhaps reflecting genetic influence from *L. hirsuta*. See also discussion under 1. *L. compacta*.

5. **Liatris punctata** Hooker, Fl. Bor.-Amer. 1: 306, plate 105. 1833 • Plains or dotted gayfeather

Lacinaria punctata (Hooker) Kuntze

Plants 15–85 cm. **Corms** globose to depressed-globose or elongate, sometimes becoming simple or branched rhizomes. **Stems** glabrous. **Leaves:** basal and proximal cauline 1-nerved, linear, 50–140 × 1–7 mm, gradually or little reduced distally, essentially glabrous or sparsely piloso-hirsutulous, gland-dotted (margins sometimes ciliate). **Heads** in dense to loose, spiciform arrays. **Involucres** campanulate-cylindric, 7–15 × (4–)5–7 mm. **Peduncles** 0 or very rarely 1–2 mm. **Phyllaries** in 3–6 series, mostly oblong-obovate, unequal, essentially glabrous, margins without hyaline borders, sometimes ciliate, apices acute, acute-acuminate, obtuse, rounded, or rounded-truncate (often with indurate apicula or mucros, lateral veins usually not evident beyond middle or at least on distal 1/3). **Florets** 3–8; corolla tubes glabrous inside. **Cypselae** (5.5–)6–8.5 mm; **pappi:** lengths ± equaling corollas, bristles plumose.

Varieties 3 (3 in the flora): North America, n Mexico.

1. Corms globose to depressed-globose; involucres 7–9(–12) mm; phyllaries in (4–)5–6 series; florets (3–)4–5(–6)...... 5c. *Liatris punctata* var. *mucronata*
1. Corms elongate or becoming rhizomes; involucres 10–14 mm; phyllaries in 4(–5) series; florets 4–8.
   2. Heads in dense, spiciform arrays (closely spaced, stems usually obscured by heads) ...
   .... 5a. *Liatris punctata* var. *punctata*
   2. Heads in loose, spiciform arrays (widely spaced, stems evident)................
   .... 5b. *Liatris punctata* var. *mexicana*

5a. **Liatris punctata** Hooker var. **punctata** [E]

Lacinaria arenicola Bush; *Liatris densispicata* (Bush) Gaiser; *L. punctata* var. *coloradensis* (Gaiser) Waterfall; *L. punctata* var. *nebraskana* Gaiser

Corms elongate or becoming rhizomes. **Leaves** 10–14 cm, 1–4(–5) mm wide. **Heads** in dense, spiciform arrays (closely spaced, stems usually obscured by heads). **Involucres** 10–14 mm. **Phyllaries** in 4(–5) series. **Florets** 4–8. $2n = 20, 40, 60$.

Flowering Aug–Oct(–Nov). Grassy prairies, sagebrush prairies, rocky and gravelly ridges and slopes, roadsides, fencerows, stream banks, over granite, limestone, and sandstone, in sands, clays, gypseous clays; (50–)100–1900(–2200) m; Alta., Man., Sask.; Colo., Ill., Iowa, Kans., Mich., Minn., Mo., Mont., Nebr., N.Mex., N.Dak., Ohio, Okla., S.Dak., Tex., Wis., Wyo.

Gaiser distinguished var. *punctata* (type from Saskatchewan) from var. *nebraskana* by its thicker, often shorter stems, broader and shorter leaves, leaves and phyllaries with ciliate margins, and phyllaries with lanceolate-acuminate apices. Weak geographic trends exist in these features; variability renders var. *nebraskana* largely typological and necessitates much arbitrary identification. B. R. Menhusen (1963) also did not find justification for recognizing var. *nebraskana*. Variation in chromosome number may be associated with some of the apparent polymorphism.

Liatris densispicata was described from sand dunes in Minnesota. It was said to have "an elongate rootstock that runs horizontally in the sand, giving off clusters of aerial stems," and to differ from *L. punctata* by "the generally finer and more slender structure of the stem, spike, and leaves" (L. O. Gaiser 1946, p. 363). It was listed as a synonym of *L. punctata* by G. B. Ownbey and T. Morley (1991).

5b. **Liatris punctata** Hooker var. **mexicana** Gaiser, Rhodora 48: 354. 1946

Liatris mucronata de Candolle var. *interrupta* Gaiser

Corms elongate or becoming rhizomes. **Leaves** 50–120 × 2–7 mm. **Heads** in loose, spiciform arrays (widely spaced, stems evident). **Involucres** 10–15 mm. **Phyllaries** in 3–5 series. **Florets** 4–6. $2n = 20, 40$.

Flowering Aug–Oct(–Nov). Gravelly and rocky slopes, canyon bottoms, grassy areas, mesquite, commonly over limestone; 10–1800 m; La., N.Mex., Tex.; Mexico (Chihuahua, Coahuila, Nuevo León, San Luis Potosí, Tamaulipas).

In the southern third of Texas, and in adjacent Louisiana, New Mexico, and Mexico, heads of *Liatris punctata* tend to be more widely spaced than characteristic for the species over the rest of its range. The most distinctive plants are in trans-Pecos Texas (and Mexico), and they occur at higher elevations than those in the more eastern range. Corms usually are elongate or rhizomiform; Gaiser described *L. mucronata* var. *interrupta* from a variant with subglobose corms. Intergrades in head congestion also are encountered, even northward into the Texas panhandle, and the variety is not sharply delimited. The populations with widely separated heads probably were ancestral to *L. bracteata* and *L. cymosa*, as well as a race (yet unnamed) with piloso-hirsutulous leaves, which occurs on the Edwards Plateau.

5c. Liatris punctata Hooker var. **mucronata** (de Candolle) B. L. Turner in B. L. Turner et al., Atlas Vasc. Pl. Texas, 6. 2003 E

Liatris mucronata de Candolle in A. P. de Candolle and A. L. P. P. de Candolle, Prodr. 5: 129. 1836; *Lacinaria leptostachya* Bush; *L. ruthii* Bush; *Liatris angustifolia* (Bush) Gaiser

Corms globose to depressed-globose. **Leaves** 70–150 × 1–3 mm. **Heads** in dense, spiciform arrays (closely spaced, stems usually obscured by heads). **Involucres** 7–9(–12) mm. **Phyllaries** in (4–)5–6 series. **Florets** (3–)4–5(–6).

Flowering mid Aug–Oct. Prairies, mesquite-grasslands, roadsides, fencerows, sands, clay, sandy loams, often rocky; 50–300 m; Ark., Kans., Mo., Okla., Tex.

Varieties *mucronata* and *punctata* intergrade where their ranges meet in central Texas (G. L. Nesom and R. J. O'Kennon 2001). Plants currently identified as *Liatris punctata* or *L. mucronata* in northern Arkansas, southern Missouri, and adjacent Kansas appear to represent an undescribed race, perhaps most closely related to *L. aestivalis*.

6. Liatris bracteata Gaiser, Rhodora 48: 371. 1946 • South Texas gayfeather C E

Plants 25–75 cm. **Corms** globose. **Stems** glabrous. **Leaves:** basal and proximal cauline 1-nerved, linear, 40–100 × 1–2 mm, even-sized or gradually reduced distally, essentially glabrous, gland-dotted (proximal margins sparsely ciliate). **Heads** (widely spaced, stems evident) in loose, spiciform arrays. **Peduncles** 0. **Involucres** turbinate-cylindric, (11–)12–15 × 5–7 mm. **Phyllaries** in 5–6(–7) series, broadly oblong to lanceolate-oblong, strongly unequal, glabrous, margins without hyaline borders, finely ciliolate, apices obtuse, rounded, or truncate, sharply involute-apiculate. **Florets** 8–14(–16); corolla tubes glabrous inside. **Cypselae** 6–9 mm; **pappi:** lengths ± equaling corollas, bristles plumose. **2n = 60.**

Flowering Sep–Nov. Coastal prairies, roadsides, railroads, clays, sandy loams; of conservation concern; 0–10(–50) m; Tex.

Liatris bracteata might justifiably be treated at varietal rank within *L. punctata*; the morphologic difference appears to be primarily in head size (especially floret number). Chromosome counts (L. O. Gaiser 1950c) indicate that *L. bracteata* is hexaploid, compared to diploids and tetraploids in populations of *L. punctata*.

7. Liatris cymosa (Ness) K. Schumann, Just's Bot. Jahresber. 27(1): 528. 1901 • Aggie-land or branched gayfeather C E

Lacinaria cymosa Ness, Bull. Torrey Bot. Club 26: 21, plate 351. 1899

Plants 20–75 cm. **Corms** globose to elongate-globose. **Stems** strigoso-puberulent. **Leaves:** basal and proximal cauline 1-nerved, linear to narrowly oblanceolate, 80–120 × 2–5 mm, gradually reduced distally or abruptly reduced on distal ¹/₂ of stems, essentially glabrous (margins proximally piloso-ciliate). **Heads** (2–20+) in open, cymiform arrays. **Peduncles** 20–70 mm (bracteate). **Involucres** cylindro-campanulate, 14–17 × 8–11 mm. **Phyllaries** in 5–7 series, broadly oblong-obovate (outer) to broadly oblong, strongly unequal, sparsely short-pilose, margins without hyaline borders, ciliolate, apices rounded, sometimes mucronate. **Florets** 20–25; corolla tubes glabrous inside. **Cypselae** 6–7 mm; **pappi:** lengths ± equaling corollas, bristles plumose.

Flowering (Jul–)Aug–Oct. Post oak woodlands, fields, fencerows, openings, edges, clay soils; of conservation concern; 50–100 m; Tex.

8. Liatris aestivalis G. L. Nesom & O'Kennon, Sida 19: 768, figs. 1, 2. 2001 • Summer gayfeather E

Plants 20–65 cm. **Corms** globose or subglobose. **Stems** glabrous. **Leaves:** basal and proximal cauline 1-nerved, linear to linear-lanceolate, 70–150 × 1–2.5 mm, abruptly or gradually reduced on distal ¹/₂ of stems, essentially glabrous (margins sometimes sparsely ciliate). **Heads** (closely spaced, stems obscured) in dense, (cylindric) spiciform arrays. **Peduncles** 0. **Involucres** cylindro-turbinate,

9–12 × 3.5–5 mm. **Phyllaries** in 2–3 series, (often dark purple in anthesis) oblong-lanceolate to lanceolate-triangular, weakly unequal, essentially glabrous, margins without hyaline borders, sometimes sparsely ciliate, apices acute to acute-attenuate (sometimes with narrow tips). **Florets** 3–4(–5); corolla tubes glabrate inside. **Cypselae** 4.5–6 mm; **pappi:** lengths ± equaling corollas, bristles plumose.

Flowering mid Jul–Aug(–Sep). Shallow soils over limestone outcrops, slopes, bases of slopes, usually with yucca; 100–400 m; Okla., Tex.

9. **Liatris glandulosa** G. L. Nesom & O'Kennon, Sida 19: 778, figs. 3, 4. 2001 • Sticky gayfeather E

Plants 30–60 cm. **Corms** subglobose to depressed-globose. **Stems** stipitate-glandular. **Leaves:** basal and proximal cauline 1-nerved, linear to linear-lanceolate, 60–110 × 1.5–4(–4.5) mm, gradually or abruptly reduced on distal ¹/₂ of stems, stipitate-glandular. **Heads** in dense, (cylindric) spiciform arrays. **Peduncles** 0. **Involucres** cylindro-turbinate, 8–12 × 3–5 mm. **Phyllaries** in 3–4 series, (green, rarely purplish) oblong-lanceolate to lanceolate-triangular, unequal, densely glandular (mixture of sessile-sunken and slightly elevated glands), margins without hyaline borders, apices broadly to narrowly acute or acuminate (sometimes with indurate mucros). **Florets** 3–4(–5); corolla tubes glabrous or glabrate inside. **Cypselae** 5–6 mm; **pappi:** lengths ± equaling corollas, bristles plumose. *2n* = 20.

Flowering (Aug–)Sep–Oct. Limestone outcrops, gentle slopes or flats, shallow gravelly soils over limestone, usually with juniper and yucca; 100–300 m; Tex.

10. **Liatris elegans** (Walter) Michaux, Fl. Bor.-Amer. 2: 91. 1803 • Pinkscale or elegant gayfeather E F

Staehelina elegans Walter, Fl. Carol., 202. 1788

Plants 30–120 cm. **Corms** depressed-globose or globose to napiform. **Stems** puberulent to hirsute-puberulent. **Leaves:** (basal on relatively distant internodes usually withering before flowering) proximal cauline 1-nerved, narrowly oblanceolate, 60–200(–300) × 3–8 mm, gradually or abruptly reduced distally (becoming slightly to strongly deflexed), essentially glabrous or sparsely puberulent, gland-dotted. **Heads** in dense, spiciform arrays. **Peduncles** usually 0, sometimes 1–5(–10) mm.

Involucres turbinate-cylindric, 12–20 × 4–6 mm. **Phyllaries** in 3–4 series, narrowly lanceolate-triangular, unequal, strigose to strigoso-hispid, margins with hyaline borders, apices (at least inner) prolonged, spreading, ± dilated, petaloid (pink, purplish, white, or yellow). **Florets** 4–5; corolla tubes glabrous inside. **Cypselae** 3.5–5(–6) mm; **pappi:** lengths ± equaling corollas, bristles plumose.

Varieties 4 (4 in the flora): s United States.

Variety *elegans* extends across the geographic range of the species; the other three varieties form local enclaves essentially imbedded within var. *elegans* and sporadically intergrading with it at points of contact (see further comments under 10d. var. *kralii*).

1. Phyllary apices petaloid, blue, lavender, magenta, or pink, ± elongate and recurving, petaloid portions short relative to herbaceous bases.
  2. Corms depressed-globose . . . .
    . . . . 10a. *Liatris elegans* var. *elegans*
  2. Corms elongate, tapering . . . .
    . . . . 10b. *Liatris elegans* var. *carizzana*
1. Phyllary apices petaloid, usually yellowish to creamy white, rarely pale lavender to rose, recurving to slightly flaring or nearly straight, petaloid portions elongate or relatively short relative to bases.
  3. Heads usually crowded (peduncles 0); phyllary apices creamy white (rarely pale lavender), sometimes initially lemon yellow fading to creamy white, elongate and recurving, petaloid portions short relative to herbaceous bases . . . . 10c. *Liatris elegans* var. *bridgesii*
  3. Heads well spaced; phyllary apices light yellow or cream (rarely pale lavender), flaring-divergent slightly or not at all, petaloid portions elongate relative to herbaceous bases . . . . 10d. *Liatris elegans* var. *kralii*

10a. **Liatris elegans** (Walter) Michaux var. **elegans** E F

Lacinaria flabellata Small; *Liatris elegans* var. *flabellata* (Small) Gaiser

Corms depressed-globose. **Heads** crowded. **Phyllaries:** apices petaloid, lavender, magenta, or pink, elongate and strongly recurving, petaloid portions short relative to herbaceous bases. **Corollas** creamy white.

Flowering Aug–Oct. Dry, sandy sites, sandy clays, dunes, pine-hardwoods, pine-live oak, longleaf pine, turkey oak; 0–200 m; Ala., Ark., Fla., Ga., La., Okla., S.C., Tex.

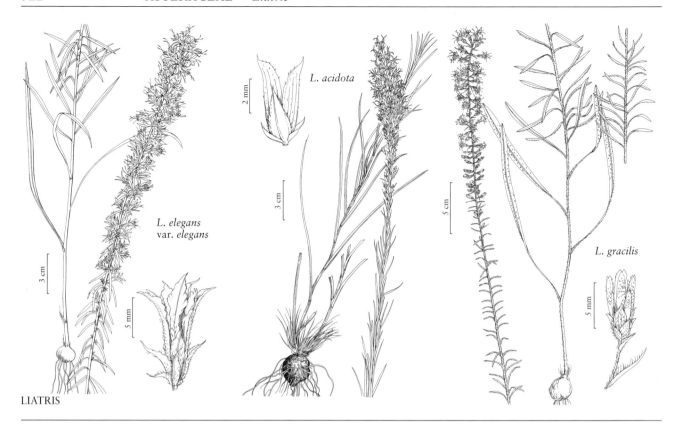

L. acidota

2 mm

L. elegans
var. *elegans*

3 cm

3 cm

5 mm

5 cm

L. gracilis

5 mm

LIATRIS

10b. Liatris elegans (Walter) Michaux var. **carizzana** Gaiser, Rhodora 48: 344. 1946 [E]

Corms elongate, tapering (± globose s of Colorado River in Bastrop and Caldwell counties). **Heads** crowded. **Phyllaries:** apices petaloid, blue, lavender, magenta, or pink, elongate and strongly recurving, petaloid portions short relative to herbaceous bases. **Corollas** creamy white.

Flowering Jul–Oct. Sandy openings, roadsides, dunes, post oak, live oak; 0–50 m; Tex.

10c. Liatris elegans (Walter) Michaux var. **bridgesii** Mayfield, Sida 20: 598, fig. 1. 2001 [E]

Corms depressed-globose. **Heads** crowded. **Phyllaries:** apices petaloid, usually creamy white, sometimes initially lemon yellow fading to creamy white, rarely pale lavender, elongate and strongly recurving, petaloid portions short relative to herbaceous bases. **Corollas** creamy white.

Flowering Jul–Oct(–Nov). Post oak savannas, open, sandy woodlands, oak and pine woodland, roadsides, fields; 100–200 m; Tex.

10d. Liatris elegans (Walter) Michaux var. **kralii** Mayfield, Sida 20: 601, fig. 4. 2001 [E]

Corms depressed-globose. **Heads** well-spaced. **Phyllaries:** apices petaloid, initially light yellow or cream, fading to cream, sometimes pale lavender to rose, slightly flaring-divergent or nearly straight, petaloid portions long relative to herbaceous bases. **Corollas** pure white.

Flowering Aug–Oct(–Nov). Sandhills, edges of wet savannas, dunes, stream banks, sandy-clay ridgetops, turkey oak, bluejack oak, longleaf pine-oak; 50–200 m; Ala., Fla., Ga., Miss.

Variety *kralii* has been collected from South Carolina (Allendale Co., *Ahles 20905,* NCU), where it appears to be disjunct from closest populations in southern Georgia. Intergrades between var. *kralii* and var. *elegans* are encountered in Alabama and Mississippi and vars. *kralii* and *elegans* appear to be sympatric in Alabama and Georgia.

11. Liatris tenuis Shinners, SouthW. Naturalist 4: 208. 1959 • Shinners's gayfeather [E]

Plants 30–55 cm. **Corms** globose to subglobose. **Stems** strigoso-puberulent. **Leaves:** basal and proximal cauline 1(–3)-nerved, linear-lanceolate to linear-oblanceolate, 120–250 × 2–3(–5) mm, abruptly reduced on distal $^1/_2$–$^2/_3$ of stems, sparsely pilose (abaxial faces), gland-dotted. **Heads** in loose, spiciform arrays (internodes 1–15 mm). **Peduncles** 0 or 1–5 mm. **Involucres** cylindro-campanulate, 10–13 × 5–6(–7) mm. **Phyllaries** in 3–4(–5) series, outermost narrowly triangular, unequal, sparsely fine-pilose to glabrate, margins without hyaline borders, ciliate, apices (loosely divergent) acute to acuminate (innermost sometimes obtuse and short-acuminate). **Florets** 10–12; corolla tubes glabrous inside. **Cypselae** 4.2–4.5 mm; **pappi:** lengths equaling corollas, bristles barbellate or proximally plumose.

Flowering (Jun–)Jul–Sep. Longleaf pine savannas, pine-hardwood edges, slopes, flats, uplands, near drainages, sands, sandy clays, fencerows, roadsides; 50–100 m; La., Tex.

Liatris tenuis is in the Center for Plant Conservation's National Collection of Endangered Plants.

12. Liatris oligocephala J. R. Allison, Castanea 66: 175, figs, 8, 9C,F. 2001 • Cahaba torch [C][E]

Plants (12–)25–48(–55) cm. **Corms** subglobose to depressed hemispheric. **Stems** glabrous. **Leaves:** basal scalelike (usually withering before flowering), cauline 1-nerved, mostly linear, 12–20(–230) × 1.2–3(–4.5) mm, gradually reduced beyond midstem, essentially glabrous (margins ciliate). **Heads** borne singly or (2–10) in subcorymbiform arrays. **Peduncles** (5–)12–40 mm. **Involucres** campanulate-turbinate, 10–19(–23) × (8–)10–17(–21) mm. **Phyllaries** in ± 3–4 series, (green or purple) lanceolate to linear-oblong, unequal, essentially glabrous or sparsely pilose, margins with or without hyaline borders, ciliate, apices long- to short-acuminate. **Florets** (11–)14–21(–25); corolla tubes glabrous inside. **Cypselae** (3.5–)5–7 mm (glabrous); **pappi:** lengths ± $^1/_2$–$^2/_3$ corollas, bristles barbellate.

Flowering Jun–Jul(–Aug). Open, grassy areas of dolomitic glades; of conservation concern; 50–200 m; Ala.

Liatris oligocephala is known only from Bibb County.

13. Liatris lancifolia (Greene) Kittell in I. Tidestrom and T. Kittell, Fl. Ariz. New Mex., 370. 1941 • Great Plains gayfeather [E]

Lacinaria lancifolia Greene, Bull. Torrey Bot. Club 25: 118. 1898; *Liatris kansana* (Britton) Rydberg

Plants (20–)40–80 cm. **Corms** globose. **Stems** glabrous. **Leaves:** basal and lower cauline 3–5-nerved, narrowly oblong-lanceolate to narrowly spatulate-oblanceolate, 60–180(–330) × 6–12(–15) mm (usually becoming more densely arranged distally), abruptly reduced near midstem (continuing densely to immediately proximal to heads), essentially glabrous, gland-dotted (bases of basal often fibrous-persistent). **Heads** in dense, spiciform arrays. **Peduncles** 0. **Involucres** turbinate-cylindric to turbinate-campanulate, 7–9 × 4–7 mm. **Phyllaries** in 3–4 series, ovate to oblong, unequal, glabrous, margins with hyaline borders, sometimes ciliolate, apices rounded to obtuse. **Florets** 5–8(–12); corolla tubes glabrous inside. **Cypselae** 4–4.5 mm; **pappi:** lengths ± equaling corollas, bristles barbellate. $2n = 20$.

Flowering Jul–Sep. Prairies (often wet or moist), banks of spring-fed streams, sandy and sandy-clay soils, saline sites; 500–1500 m; Colo., Kans., Nebr., N.Mex., S.Dak., Tex., Wyo.

The abrupt size reduction of cauline leaves in *Liatris lancifolia* is similar to that of *L. spicata* var. *resinosa*; the long, dense spikes and wider basal leaves are more like those of *L. spicata* var. *spicata*. Little differentiation exists between *L. lancifolia* and *L. spicata*, but *L. lancifolia* is maintained here at specific rank, coordinate with *L. spicata*, primarily because of its wide geographic disjunction and generally different habitat. Recognition that the two elements within *L. spicata* have a nearly analogous relationship of range and habitat might provide rationale for treating all three of these closely similar taxa at equivalent rank. *Liatris lancifolia* is expected in Oklahoma.

14. Liatris spicata (Linnaeus) Willdenow, Sp. Pl. 3: 1636. 1803 • Florist or marsh gayfeather [E]

Serratula spicata Linnaeus, Sp. Pl. 2: 819. 1753; *Lacinaria spicata* (Linnaeus) Kuntze

Plants (20–)40–110(–180) cm. **Corms** globose to slightly elongate. **Stems** glabrous. **Leaves:** basal and lower cauline 3–5-nerved, narrowly oblong-lanceolate to narrowly spatulate-oblanceolate, 120–350 × (2–)4–10(–20) mm (sometimes becoming more densely arranged distally), usually gradually reduced distally, essentially glabrous or sparsely villous, weakly gland-dotted (glandular hairs often not evident, bases of basal often fibrous-persistent). **Heads** in dense to loose, spiciform arrays. **Peduncles** usually 0, rarely 1–2 mm. **Involucres** turbinate-cylindric to turbinate-campanulate, 7–11 × 4–6 mm. **Phyllaries** in (3–)4–5 series, ovate to oblong, unequal, essentially glabrous, margins with hyaline borders, sometimes ciliolate, apices rounded to obtuse. **Florets** (4–)5–8(–14); corolla tubes glabrous inside. **Cypselae** (3.5–)4.5–6 mm; **pappi:** lengths ± equaling corollas, bristles barbellate.

Varieties 2 (2 in the flora): e North America.

Liatris spicata is sold as cut flowers. It also is commonly sold as a garden plant in various genetic permutations (probably derived from var. *spicata*, perhaps from *L. lancifolia*) and it apparently escapes cultivation. Reports from Arkansas, Connecticut, and Quebec probably reflect plants growing in or escaped from gardens.

A geographic disjunction within *Liatris spicata* occurs between the coastal plain element (var. *resinosa*) and the inland/montane element (var. *spicata*), although plants morphologically referable to var. *resinosa* occasionally are encountered in montane North Carolina and Tennessee and var. *spicata*-like plants occur in the range of var. *resinosa*. Apparent intergrades between the two taxa are common, especially in Tennessee and Alabama. The geographical gap is widest in Georgia and Alabama. Neither variety occurs naturally west of the Mississippi River, except for a historical record of var. *spicata* in Oregon County, Missouri (*Kellogg s.n.*, MO), where the population has now been genetically "swamped" by *L. pycnostachya* (G. A. Yatskievych, pers. comm.).

In both var. *spicata* and var. *resinosa*, marked variation (dimorphism) in head size occurs, the large-headed plants apparently occurring in scattered geographic enclaves without a broader geographic pattern. It seems possible that independent populational origins of polyploidy might underlie the variation.

1. Leaves: basal and proximal cauline (2–)4–8 mm wide (cauline usually abruptly reduced in size near midstem, continuing distally as linear, bractlike leaves); involucres 7–9 mm; phyllaries purplish to greenish; florets 5–6(–8); coastal plain . . . .
. . . . 14a. *Liatris spicata* var. *resinosa*
1. Leaves: basal and proximal cauline 4–10(–20) mm wide (cauline usually gradually reduced in size distally); involucres (7–)8–11 mm; phyllaries usually greenish; florets (4–)6–8(–12); montane and inland . . . . 14b. *Liatris spicata* var. *spicata*

14a. Liatris spicata (Linnaeus) Willdenow var. **resinosa** (Nuttall) Gaiser, Rhodora 48: 216. 1946

Liatris resinosa Nuttall, Gen. N. Amer. Pl. 2: 131. 1818

Leaves: basal and proximal cauline (2–)4–8 mm wide (cauline usually abruptly reduced in size near midstem, continuing distally as linear, bractlike leaves). **Heads** in dense (to loose), spiciform arrays. **Involucres** 7–9 mm. **Phyllaries** purplish to greenish. **Florets** 5–6(–8). $2n = 20$.

Flowering (Jul–)Aug–Oct(–Nov). Seeps, bogs, bog edges, wet pine and pine-palmetto flatwoods, wet ditches, canal banks, peaty sands, sandy clays; 0–30 m. Ala., Del., Fla., Ga., La., Md., Miss., N.J., N.C., S.C., Va.

Variety *resinosa* is particularly variable (especially in Florida) in number of heads (often relatively few and widely separated) and in reduction of cauline leaves.

14b. Liatris spicata (Linnaeus) Willdenow var. **spicata**

Leaves: basal and proximal cauline 4–10(–20) mm wide (cauline usually gradually reduced in size distally). **Heads** in dense to loose, spiciform arrays. **Involucres** (7–)8–11 mm. **Phyllaries** usually greenish. **Florets** (4–)6–8(–12). $2n = 20$.

Flowering Jul–Sep. Fields, road banks, fencerows, lake sides, wet to moist prairies and meadows, bogs, seepages, dunes, limestone and granite outcrops, sandy clays, sandy loams, moist woods, oak, oak-pine, and sweetgum flats, tamarack swamps; 50–1700 m; Ont., Que.; Ala., Ark., Conn., Del., Ga., Ill., Ind., Ky., Md., Mass., Mich., Miss., Mo., N.J., N.Y., N.C., Ohio, Pa., Tenn., Va., W.Va., Wis.

15. Liatris pycnostachya Michaux, Fl. Bor.-Amer. 2: 91. 1803 • Prairie or cattail gayfeather E

Lacinaria pycnostachya (Michaux) Kuntze

Plants (40–)60–120(–180) cm. **Corms** globose, sometimes becoming elongate rhizomes. **Stems** glabrous or densely piloso-puberulent. **Leaves:** basal and proximal cauline 3–5-nerved, linear to narrowly oblanceolate or oblanceolate, 110–220 × 4–10(–12) mm, gradually then abruptly reduced distally (bractlike proximal to heads), essentially glabrous or sparsely pilose to densely piloso-puberulent, weakly gland-dotted (bases of basal usually fibrous-persistent). **Heads** in dense, spiciform arrays. **Peduncles** usually 0, rarely (spreading to ascending) 1–2 mm. **Involucres** campanulate-cylindric, 7–9(–10) × 3.5–5 mm. **Phyllaries** (reflexed or curving-spreading) in 4–5 series, oblong-lanceolate, unequal, essentially glabrous or ± pilose, margins with hyaline borders, sometimes ciliate, apices usually acute to acute-acuminate (sometimes distally dilated, nearly petaloid). **Florets** (4–)5–8; corolla tubes glabrous inside. **Cypselae** 3.8–4.5(–5) mm; **pappi:** lengths ± equaling corollas, bristles barbellate.

Varieties 2 (2 in the flora): e United States.

1. Corms globose; stems usually glabrous or sparsely pilose except near heads (where sparsely piloso-puberulent); leaves glabrous. . . . .
. . . . 15a. *Liatris pycnostachya* var. *pycnostachya*
1. Corms globose to elongate; stems moderately to densely piloso-puberulent; leaves moderately to densely piloso-puberulent to nearly glabrous . . . . 15b. *Liatris pycnostachya* var. *lasiophylla*

15a. **Liatris pycnostachya** Michaux var. **pycnostachya** E

Liatris bebbiana Rydberg; *L. langloisii* (Greene) Cory

Corms globose. **Stems** usually glabrous or sparsely pilose except near heads (where sparsely piloso-puberulent). **Leaves** glabrous. $2n$ = 20, 40.

Flowering mid Jun–Aug. Moist prairies, wooded and rocky ridges, sandy woods, creek bottoms, dunes; 30–300 (–900) m; Ark., Ill., Ind., Iowa, Kans., Ky., La., Mass., Mich., Minn., Miss., Mo., Nebr., N.J., N.Y., N.Dak., Ohio, Okla., Pa., S.Dak., Tex., Wis.

Liatris pycnostachya is commonly cultivated. Collections from Michigan "doubtless represent escapes" (E. G. Voss, pers. comm.), as also do those from Massachusetts, New Jersey, and New York.

15b. **Liatris pycnostachya** Michaux var. **lasiophylla** Shinners, Field & Lab. 19: 74. 1951 E

Lacinaria serotina Greene; *Liatris serotina* (Greene) K. Schumann

Corms globose, sometimes becoming elongate rhizomes. **Stems** moderately to densely piloso-puberulent. **Leaves** moderately to densely piloso-puberulent to nearly glabrous.

Flowering mid Jul–Sep(–Nov). Sandy woods, pitcher plant and grass bogs, pine savannas, drainages, roadsides, fencerows, sands, sandy clays, clays; 0–50 m; La., Miss., Tex.

Variety *lasiophylla* occurs over most of the range of the species in Louisiana and Texas; var. *pycnostachya* occurs in the northern counties of those states and, apparently, also sporadically southward through the range of var. *lasiophylla*, at least in Texas, where it grows in drier habitats than var. *lasiophylla*. In Louisiana and Texas, var. *pycnostachya* begins flowering in mid-June and continues through July (through August more northward in its range), usually well before the main flowering period of var. *lasiophylla*. Despite these indications of reproductive isolation, apparent intermediates are commonly encountered (usually these are plants with dense cauline vesture but sparsely pubescent to glabrate leaves, compared to the densely piloso-puberulent stems and leaves of typical var. *lasiophylla*). Most of the plants with reduced vesture in the range of var. *lasiophylla* also have the later flowering period.

In Pearl River and Hancock counties, Mississippi, and St. Tammany and Washington parishes, Louisiana, plants with vesture of var. *lasiophylla* have phyllaries with rounded to slightly acute apices, different from the rest of the species. These were named *Liatris serotina*, and it seems likely that they originated as hybrids between var. *lasiophylla* and *L. spicata* var. *resinosa*, whose ranges meet in that area. Some plants of *L. spicata* in Illinois, Indiana, and Minnesota also develop cauline vesture; the tendency apparently is evolutionarily independent of that in var. *lasiophylla*.

16. **Liatris acidota** Engelmann & A. Gray, Boston J. Nat. Hist. 5: 218. 1845 • Gulf Coast or sharp gayfeather E F

Lacinaria acidota (Engelmann & A. Gray) Kuntze; *Liatris acidota* var. *vernalis* Engelmann & A. Gray

Plants 20–90(–130) cm. **Corms** usually globose to subglobose, sometimes ovoid to elongate. **Stems** glabrous. **Leaves:** basal and proximal cauline 3–5-nerved, linear to linear-oblanceolate, 100–400 × 1–3(–5) mm, gradually reduced distally or abruptly on distal 1/2 of stems, essentially glabrous (bases of basal usually fibrous-persistent). **Heads** in dense, spiciform arrays. **Peduncles** 0. **Involucres** cylindro-turbinate, 6–7(–10) × ca. 3 mm. **Phyllaries** in 3–4 series, (often purple) oblong-oblanceolate to oblong-lanceolate, unequal, essentially glabrous, margins with hyaline borders, ciliolate, apices (erect or ± appressed) acuminate to acute. **Florets** (2–)3–4(–5); corolla tubes glabrous inside. **Cypselae** 4–4.5 mm; **pappi:** lengths ± equaling corollas, bristles barbellate to subplumose. $2n$ = 20.

Flowering (Jun–)Jul–Oct(–Nov). Coastal prairies, prairie remnants, wet pine flats, savannas, oak-pine, ditches, roadsides, railroads, sand, sandy and silty loams, clays; 0–30 m; La., Tex.

17. **Liatris tenuifolia** Nuttall, Gen. N. Amer. Pl. 2: 131. 1818 • Pine-needle or shortleaf gayfeather [E]

Lacinaria tenuifolia (Nuttall) Kuntze

Plants 40–150 cm. **Corms** globose. **Stems** glabrous or sparsely pilose. **Leaves:** basal and proximal cauline (arising from separated nodes) 1-nerved, linear to linear-lanceolate, 100–300 × 1–2(–2.5) mm, gradually or abruptly reduced distally, essentially glabrous, gland-dotted (proximal margins sometimes ciliate). **Heads** in compact, racemiform arrays. **Peduncles** (ascending) 1–7 mm. **Involucres** turbinate-campanulate, 5–7 × 4–5 mm. **Phyllaries** in 2–3(–4) series, lanceolate to oblong or elliptic-oblong, unequal, essentially glabrous, margins with (pinkish purple) hyaline borders, apices usually rounded-retuse and minutely involute-cuspidate to apiculate. **Florets** 4–6; corolla tubes glabrous inside. **Cypselae** 2.5–4 mm; **pappi:** lengths ± equaling corollas, bristles barbellate.

Flowering Aug–Nov. Longleaf pine savannas, longleaf pine-scrub oak, turkey oak-bluejack oak, slash pine-sand pine-scrub, sand pine-scrub, sand ridges, hills, and flats, roadsides; 10–100 m; Ala., Fla., Ga., S.C.

18. **Liatris laevigata** Nuttall, Trans. Amer. Philos. Soc., n. s. 7: 285. 1840 • Clusterleaf gayfeather [E]

Lacinaria laevigata (Nuttall) Small; *Liatris tenuifolia* Nuttall var. *laevigata* (Nuttall) B. L. Robinson; *L. tenuifolia* Nuttall var. *quadriflora* Chapman

Plants 40–180 cm. **Corms** globose. **Stems** glabrous. **Leaves:** basal and proximal cauline (mostly arising from congested nodes) 1-nerved, lance-linear to linear, 100–380 × (1–)2–6(–9) mm, abruptly reduced distally, glabrous (minutely white-dotted by stomates), weakly, if at all, gland-dotted (glandular hairs usually not evident). **Heads** in compact, racemiform arrays. **Peduncles** (ascending) 1–7 mm. **Involucres** turbinate-campanulate, 5–7 × 4–5 mm. **Phyllaries** in 2–3(–4) series, lanceolate to oblong or elliptic-oblong, unequal, essentially glabrous, margins with hyaline borders, apices usually rounded-retuse and minutely involute-cuspidate to apiculate. **Florets** 3–5; corolla tubes glabrous inside. **Cypselae** 2.5–4 mm; **pappi:** lengths ± equaling corollas, bristles barbellate.

Flowering Aug–Oct(–Nov). Sand ridges and flats, roadsides, pine woods, sand pine-scrub, longleaf pine-scrub oak; 0–100 m; Fla., Ga.

R. P. Wunderlin (1998) treated *Liatris laevigata* and *L. tenuifolia* as varieties in one species; A. Cronquist (1980, p. 207) also treated them within one species, as "two well-marked but wholly confluent geographic vars." They have been noted to grow intermixed in Osceola County, Florida (*L. laevigata*, *Ray et al. 10472*, NCU; *L. tenuifolia*, *Ray et al. 10423*, NCU). Unequivocal intermediates have not been seen in the present study. With the observation of consistent morphologic differences in habit, vestiture, and leaf punctation, and their broad sympatry in Florida, it seems appropriate to recognize them at specific rank.

Liatris laevigata is restricted to peninsular Florida and immediately adjacent Georgia (e.g., Floyd's Island, Charlton County, *Cypert 240*, SMU). Leaf width is markedly variable, and collections of narrow-leaved forms sometimes have been identified as *L. tenuifolia*. Narrow-leaved forms of *L. laevigata* apparently are more common in southern Florida (Collier, Dade, Glades, Lee, Monroe, Okeechobee, Palm Beach, and Polk counties). Broader-leaved plants apparently tend to have larger corms, which may be related to age or habitat.

19. **Liatris chapmanii** Torrey & A. Gray, Fl. N. Amer. 2: 502. 1843 • Chapman's gayfeather [E]

Lacinaria chapmanii (Torrey & A. Gray) Kuntze

Plants 35–75(–150) cm. **Corms** globose to elongate. **Stems** hirtellous. **Leaves:** basal and proximal cauline 1-nerved, spatulate-oblanceolate to narrowly oblanceolate, 40–150(–180) × 4–8(–11) mm, abruptly reduced (linear, 1–2 mm wide), then gradually reduced distally, essentially glabrous or hirtellous (sometimes mostly along abaxial midveins), gland-dotted. **Heads** (appressed, overlapping) in dense, spiciform arrays. **Peduncles** 0. **Involucres** cylindric, 8–12 × 3.5–5 mm. **Phyllaries** in 3(–4) series, oblong to oblong-lanceolate, unequal, usually glabrous, rarely minutely puberulent, margins with hyaline borders, apices acute to acuminate. **Florets** 3–4; corolla tubes glabrous inside. **Cypselae** (3–)4–6 mm; **pappi:** lengths ± equaling corollas, bristles barbellate. $2n = 20$.

Flowering Aug–Oct. Dunes, beach strands, sand ridges, fields, roadsides, longleaf pine savannas, longleaf pine-turkey oak, turkey oak, evergreen oak-sand pine-scrub, scrub with *Ceratiola* and *Pinus clausa*; 0–50 m; Ala., Fla., Ga.

20. Liatris provincialis R. K. Godfrey, Amer. Midl. Naturalist 66: 466, fig. 1. 1961 • Godfrey's gayfeather C E

Plants 45–90 cm. **Corms** globose to elongate. **Stems** hirtellous. **Leaves:** basal and proximal cauline 1-nerved, linear-oblanceolate to narrowly oblanceolate, 60–150 × 2–6 mm, abruptly reduced, linear, 1–2 mm wide, then gradually or little reduced distally, essentially glabrous or hirtellous (sometimes mostly along abaxial midveins), gland-dotted. **Heads** in dense, spiciform arrays (spreading to ascending, not strongly overlapping). **Peduncles** 0. **Involucres** cylindric, 9–11 × 3–5 mm. **Phyllaries** in 3–4(–5) series, oblong-obovate to oblong or oblong-lanceolate, strongly unequal, glabrous, sparsely puberulent, or hirtellous, margins with hyaline borders, ciliolate, apices acute to acuminate (mid often cuspidate). **Florets** 3–4; corolla tubes glabrous inside. **Cypselae** 4–5.5 mm; **pappi:** lengths ± equaling corollas, bristles barbellate.

Flowering (Aug–)Sep–Oct. Evergreen oak-sand pine-scrub, turkey oak-longleaf pine, sand ridges, dunes; of conservation concern; 0–10 m; Fla.

Liatris provincialis is very similar to *L. chapmanii* and apparently restricted to coastal and near-coastal sites in panhandle of Florida (Franklin and Wakulla counties). Beside the difference in orientation of the heads, phyllaries of *L. provincialis* are broader (versus gradually and more narrowly lanceolate), often abruptly acute to short-acuminate or cuspidate, but there is little else to distinguish the two. *Liatris provincialis* is in the Center for Plant Conservation's National Collection of Endangered Plants.

21. Liatris pauciflora Pursh, Fl. Amer. Sept. 2: 510. 1813 • Few-flower gayfeather E

Plants 20–90 cm. **Corms** globose, sometimes depressed or elongate. **Stems** minutely puberulent-hirtellous (hairs spreading to slightly deflexed) or glabrous. **Leaves:** basal and proximal cauline 1-nerved, narrowly oblanceolate to linear-oblanceolate, 40–120 × 2–7 mm, gradually or abruptly reduced distally (continuing as linear, mostly 10–40 mm bracts), hispidulous-hirtellous or glabrous, weakly, if at all, gland-dotted (glandular hairs not evident, proximal margins sometimes ciliate). **Heads** in dense, racemiform to spiciform (strongly secund) arrays. **Peduncles** 0 or (ascending) 1–8 mm. **Involucres** cylindric, 11–15 × 4–7 mm. **Phyllaries** in 3–4 series, mostly oblong to oblong-oblanceolate, unequal, essentially glabrous, margins with hyaline borders, apices acute (sometimes with mucros or apicula). **Florets** 3–6; corolla tubes pilose inside. **Cypselae** 3–4.5 mm; **pappi:** lengths ± equaling corollas, bristles barbellate.

Varieties 2 (2 in the flora): se United States.

Varieties *pauciflora* and *secunda* are mostly allopatric, apparently overlapping in south-central Georgia and perhaps northeastern Florida. Some plants of var. *secunda* in Brunswick County, North Carolina, have nearly glabrous stems; their phyllaries have the narrower shape of more typical plants of the area and glandular punctations are strongly developed. Some plants of var. *pauciflora* in Seminole and Orange counties, Florida, have slightly hirtellous stems and lack glandular punctations. In other localities, differences in vesture and punctation are not perfectly correlated. Other differences are seen as tendencies: leaves of var. *secunda* are thicker than in var. *pauciflora* and often have strongly but narrowly thickened-revolute margins; inner phyllaries of var. *secunda* are slightly narrower than in var. *pauciflora*, and the mid and inner usually are apiculate or mucronulate.

A. Cronquist (1980) treated *Liatris pauciflora* and *L. secunda* as a single species, suggesting that they might prove to be Mendelian variants; their mostly allopatric ranges indicate otherwise. Population studies might provide insight into the evolutionary interactions.

1. Stems usually glabrous, rarely sparsely hirtellous; leaves (and phyllaries) not or weakly gland-dotted; involucres 11–15 × 4–7 mm; phyllaries: inner (9–)10–13 × 2.5–3.8 mm, apices acute (without mucros or apicula); cypselae 3–4.5 mm . . . . 21a. *Liatris pauciflora* var. *pauciflora*
1. Stems minutely puberulent-hirtellous (hairs spreading to slightly deflexed); leaves (and phyllaries) gland-dotted (each pit with a glandular hair); involucres 7–10(–14) × 3–5 mm; phyllaries: inner 10–13 × 2–3 mm, apices abruptly short-acuminate, mid and inner often apiculate to mucronulate; cypselae 4–5 mm . . . . 21b. *Liatris pauciflora* var. *secunda*

21a. Liatris pauciflora Pursh var. **pauciflora** E

Stems usually glabrous, rarely sparsely hirtellous. **Leaves** (and phyllaries) not or weakly gland-dotted. **Involucres** 11–15 × 4–7 mm. **Phyllaries:** inner (9–)10–13 × 2.5–3.8 mm, apices acute (without mucros or apicula). **Cypselae** 3–4.5 mm. $2n = 20$.

Flowering Aug–Oct. Pine flatwoods, longleaf pine-turkey oak, scrub oak, sandy peat, sand, ridges, fields, moist swales; 0–20 m; Fla., Ga.

21b. Liatris pauciflora Pursh var. **secunda** (Elliott) D. B. Ward, Novon 14: 367. 2004 [E]

Liatris secunda Elliott, Sketch Bot. S. Carolina 2: 278. 1823

Stems minutely puberulent-hirtellous (hairs spreading to slightly deflexed). **Leaves** (and phyllaries) gland-dotted (each pit with a glandular hair). **Involucres** 7–10(–14) × 3–5 mm. **Phyllaries:** inner 10–13 × 2–3 mm, apices abruptly short-acuminate, mid and inner often apiculate to mucronulate. **Cypselae** 4–5 mm. $2n = 20$.

Flowering Aug–Sep(–Oct). Longleaf pine savannas, longleaf pine-turkey oak, scrub oak, sand pine-evergreen oak scrub, pine-palmetto scrub, dunes, sand ridges, flats, woodland edges; 0–200 m; Ala., Fla., Ga., N.C., S.C.

22. Liatris garberi A. Gray, Proc. Amer. Acad. Arts 15: 48. 1880 • Garber's gayfeather

Lacinaria chlorolepis Small; *L. garberi* (A. Gray) Kuntze; *L. nashii* Small

Plants 20–80 cm. **Corms** elongated into rhizomes, irregularly shaped (sometimes with thickened, fibrous roots). **Stems** piloso-puberulent to hirsute. **Leaves:** basal and proximal cauline 1-nerved, linear to linear-oblanceolate, 120–300 × 3–8 mm, gradually reduced distally to ± midstem, abruptly reduced to ± uniform bracts distally, essentially glabrous, gland-dotted (margins sometimes sparsely ciliate). **Heads** in dense, spiciform arrays. **Peduncles** 0 or 1–5(–10) mm. **Involucres** cylindro-campanulate, 8–10 × 4–5 mm. **Phyllaries** in 3–4(–5) series, ovate to oblong-oblanceolate or oblong-lanceolate, strongly unequal, usually essentially glabrous, sometimes sparsely strigose to pilosulous, margins with hyaline borders, ciliolate, apices acute (inner with slightly subterminal, 'rolled' apicula), ciliolate. **Florets** 6–10; corolla tubes glabrous inside. **Cypselae** 3–3.5 mm; **pappi:** lengths ± equaling corollas, bristles barbellate. $2n = 20$.

Flowering Jul–Sep. Wet flats with palmetto and longleaf or slash pine, oak-palmetto, pine-live oak, grass-sedge bogs, sandy loam, sandy peat; 0–30 m; Fla.; West Indies (Bahamas).

23. Liatris virgata Nuttall, J. Acad. Nat. Sci. Philadelphia 7: 72. 1834 • King's Mountain or piedmont gayfeather [E]

Lacinaria smallii Britton; *Liatris graminifolia* Willdenow var. *racemosa* (de Candolle) Venard; *L. graminifolia* var. *smallii* (Britton) Fernald & Griscom; *L. graminifolia* var. *virgata* (Nuttall) Fernald; *L. regimontis* (Small) K. Schumann; *L. spicata* (Linnaeus) Willdenow var. *racemosa* de Candolle

Plants 35–75 cm. **Corms** mostly globose. **Stems** glabrous. **Leaves:** basal and proximal cauline 1-nerved, narrowly oblanceolate, (50–)90–150 × (2–)4–9(–12) mm wide, usually abruptly, sometimes gradually, reduced distally, glabrous or sparsely pilose (abaxially), gland-dotted (proximal margins sparsely pilose-ciliate). **Heads** in loose, racemiform to spiciform arrays (not secund, internodes 6–20 mm). **Peduncles** 0 or (ascending) mostly 2–12 mm. **Involucres** campanulate-cylindric, (6–)7–9 × 3.5–4.5(–5) mm. **Phyllaries** usually in 3–5(–6) series, in 5–6(–7) series in largest heads, ovate-triangular to oblong, unequal, glabrous, margins with narrow, hyaline borders (lacking at apices), sometimes sparsely ciliolate, apices acute to obtuse-angled (with thickened, not markedly involute, apicula). **Florets** 7–10(–12); corolla tubes pilose inside. **Cypselae** (2–)3–4 mm; **pappi:** lengths ± equaling corollas, bristles barbellate. $2n = 20$.

Flowering (Jul–)Aug–Oct(–Nov). Edges of swampy woods, stream margins, slopes, clearings, edges of upland woods, rocky woods, pine-oak woods, mixed deciduous woods, roadsides, Iredell soils, clays; 50–1000 m; Ga., N.C., S.C., Va.

Liatris virgata is characterized by its non-congested arrangement of heads, often on a slightly zigzag axis, and especially by its relatively thickened phyllaries with angled apices (acute to obtuse-apiculate). The name *Liatris regimontis* has been misapplied to the species now identified as *L. cokeri*.

24. Liatris cokeri Pyne & Stucky, Sida 14: 205. 1990 • Sandhills or Coker's gayfeather [E]

Plants 25–85 cm. **Corms** globose. **Stems** glabrous. **Leaves:** basal and proximal cauline 1-nerved, lance-linear to linear, 50–150 × 2–5 mm, gradually or abruptly reduced distally, essentially glabrous (proximal margins sparsely ciliate). **Heads** in dense, racemiform to spiciform arrays (sometimes strongly to weakly secund, especially if branches reclining, internodes 1–5 mm). **Peduncles** 0 or (ascending) 1–6(–10) mm. **Involucres** cylindro-turbinate, (7–)8–9 × 3.5–4 mm. **Phyllaries** in 3–4 series, ovate-oblong, oblong, or oblong-lanceolate (inner 7.5–10 × 1–1.8 mm), strongly unequal, essentially glabrous, margins with hyaline borders (lacking at apices), ciliolate, apices (inner and middle, sometimes outer) rounded to blunt, involute-cuspidate to short-acuminate. **Florets** 4–7(–9); corolla tubes pilose inside. **Cypselae** 3–4(–5) mm; **pappi:** lengths ± equaling corollas, bristles barbellate.

Flowering (Aug–)Sep–Oct. Sand ridges, sandy fields and roadsides, turkey-oak, longleaf pine-oak; 50–150 m; N.C., S.C.

Pyne and Stucky noted that variants of *Liatris cokeri* (apparently intermediate toward *L. virgata*) occur on the coastal plain of North Carolina and South Carolina.

25. Liatris helleri Porter, Bull. Torrey Bot. Club 18: 147. 1891 • Heller's gayfeather E

Lacinaria helleri (Porter) Porter ex A. Heller; *Liatris turgida* Gaiser

Plants 15–55 cm. **Corms** globose (sometimes knotty). **Stems** glabrous. **Leaves:** basal and proximal cauline 1-nerved, spatulate-oblanceolate to narrowly lanceolate or oblanceolate-linear, (50–)60–100(–220) × 3–8(–15) mm (usually reaching at least level of heads), quickly to gradually reduced distal to midstems, essentially glabrous or sparsely pilose (abaxially), weakly, if at all, gland-dotted. **Heads** in dense to loose, spiciform arrays. **Peduncles** 0. **Involucres** turbinate-campanulate, 7–10 × 6–8(–10) mm. **Phyllaries** in 3–4(–5) series, oblong, unequal, glabrous, margins with hyaline borders (0.2–0.4 mm wide), ciliolate, apices rounded. **Florets** 7–13 (–17); corolla tubes sparsely pilose inside. **Cypselae** (2.5–)3.5–5 mm; **pappi:** lengths $^1/_3$–$^2/_3$ or equaling corollas, bristles barbellate. $2n = 20$.

Flowering Jul–mid Sep. Rock outcrops, cliff ledges, ridges, shale, rocky openings in heath balds, roadside banks, oak, dry pine-oak, and pitch pine woods; 600–1600(–1900) m; N.C., Va. W.Va.

Liatris helleri has been regarded as a narrow endemic (of conservation concern) of high-elevation habitats in North Carolina, where it was recognized especially by a markedly shortened pappus; it has been reported (as *L. turgida*) from Alabama and Georgia (not confirmed in this study). With recognition that pappus length varies and leaves may be sparsely villous to nearly glabrous, the concept of *L. helleri* has been expanded (G. L. Nesom 2005b) to include plants identified in montane West Virginia and Virginia as *L. turgida* and *L. graminifolia* (e.g., M. J. Johnson 1971). Even in Virginia, the pappus sometimes is reduced to half the length of the corolla tube (e.g., Bedford Co., *Freer 12226*, NCU; Amherst Co., *Freer 2007*, NCU; Roanoke Co., *Uttal 10883*, NCU), and pappus length is variable in the North Carolina region of typical *L. helleri*, where the bristles characteristically are short (Avery, Burke, Caldwell, Mitchell, and Watauga counties). In exposed habitats (e.g., rock ledges, shale barrens), plants often are relatively short and the heads are relatively few and distantly spaced; in less exposed sites, plants may be taller and to 40 heads may be borne in a relatively dense array.

Liatris helleri is in the Center for Plant Conservation's National Collection of Endangered Plants.

SELECTED REFERENCE Nesom, G. L. 2005b. Broadened concept of *Liatris helleri* (Asteraceae: Eupatorieae). Sida 21: 1323–1333.

26. Liatris pilosa (Aiton) Willdenow, Sp. Pl. 3: 1636. 1803 • Grass-leaf gayfeather E

Serratula pilosa Aiton, Hort. Kew. 3: 138. 1789; *Lacinaria graminifolia* (Willdenow) Kuntze var. *pilosa* (Aiton) Britton; *Liatris graminifolia* Willdenow; *L. graminifolia* var. *dubia* (W. P. C. Barton) A. Gray; *L. graminifolia* var. *lasia* Fernald & Griscom; *L. pilosa* (Aiton) Willdenow var. *laevicaulis* de Candolle

Plants 40–120 cm. **Corms** globose. **Stems** glabrous or sparsely to moderately pilose distally or throughout. **Leaves:** basal and proximal cauline 1-nerved, narrowly oblanceolate, 60–170(–200) × 2–7(–11) mm, abruptly to gradually reduced distally (becoming linear, spreading-ascending), essentially glabrous or sparsely pilose (abaxially), gland-dotted (proximal margins piloso-ciliate). **Heads** in loose to dense, racemiform to spiciform arrays (internodes 1–7 mm). **Peduncles** 0 or (ascending) 1–10(–80) mm. **Involucres** turbinate to campanulate-cylindric, (7–)8–10 × 5–6 mm. **Phyllaries** in (3–)4–5 (–6) series, oblong, unequal, essentially glabrous, margins with hyaline borders (0.2–0.4 mm wide), erose to lacerate, ciliolate, apices usually rounded, rarely acute. **Florets** (6–)7–12(–13, mostly 9–12 in Del. and N.J.); corolla tubes pilose inside. **Cypselae** (2.5–)3–4 mm; **pappi:** lengths ± equaling corollas, bristles barbellate. $2n = 20$.

Flowering (Aug–)Sep–Oct(–Nov). Old fields, pine barrens, scrub oak-pine sandhills, openings in pine, oak, and oak-hickory woods, tidal marsh edges, sandy fields, dune hollows, wet sand near beaches, edges of tidal marshes, sand to sandy clay-loam; (0–)10–500 m; Del., Md., N.J., N.C., Pa., S.C., Va.

27. Liatris elegantula (Greene) K. Schumann, Just's Bot. Jahresber. 29(1): 569. 1903 • Greene's elegant gayfeather [E]

Lacinaria elegantula Greene, Pittonia 4: 316. 1901 (as Laciniaria); *Liatris graminifolia* Willdenow var. *elegantula* (Greene) Gaiser

Plants 60–105 cm. **Corms** mostly globose. **Stems** glabrous. **Leaves:** basal and proximal cauline 1-nerved, linear-oblanceolate to narrowly oblanceolate, 80–210 × 2–5(–10) mm, gradually or abruptly reduced distally, essentially glabrous (sparsely piloso-ciliate along proximal margins), gland-dotted. **Heads** in loose, racemiform arrays (internodes 2–14 mm). **Peduncles** 0 or (ascending) 2 (–7) mm. **Involucres** turbinate, 6–8 × 5–7 mm. **Phyllaries** in 3–4(–5) series, mostly oblong, unequal, essentially glabrous, margins with hyaline borders (0.2–0.4 mm wide), sometimes sparsely ciliolate, apices rounded. **Florets** (7–)8–11(–13); corolla tubes pilose inside. **Cypselae** 2.8–3.5(–3.8) mm; **pappi:** lengths ± equaling corollas, bristles barbellate. *2n* = 20.

Flowering Aug–Oct(–Nov). Longleaf pine-scrub oak, pine, live oak-pine, deciduous oak-pine, deciduous flatwoods, sandhills, savanna edges, edges of cypress depressions, depression meadows, live oak-pine-palmetto hammocks, sandy clay or loam, rarely clay; 0–300(–500) m; Ala., Fla., Ga., Miss.

28. Liatris microcephala (Small) K. Schumann, Just's Bot. Jahresber. 26(1): 378. 1900 • Smallhead gayfeather

Lacinaria microcephala Small, Bull. Torrey Bot. Club 25: 473. 1898; *L. polyphylla* Small

Plants 30–80 cm. **Corms** globose (fibrous roots often densely covering whole structure). **Stems** glabrous. **Leaves:** basal and proximal cauline 1-nerved, linear to linear-lanceolate, 80–200 × 1– 3(–4, –6) mm, gradually reduced or abruptly reduced on distal ¹/₂ of stems, essentially glabrous. **Heads** in loose to dense, spiciform arrays (internodes to 8 mm). **Peduncles** 0 or (ascending) 1–12 mm. **Involucres** turbinate-cylindric, 5–7(–9) × 3–4 mm. **Phyllaries** in 3–4 series, oblong to oblong-obovate, strongly unequal, essentially glabrous, margins with hyaline borders, apices rounded. **Florets** 4–5(–6); corolla tubes glabrous inside. **Cypselae** 3–3.5 mm; **pappi:** lengths ± ¹/₂ corollas, bristles barbellate. *2n* = 20, 60.

Flowering (Jul–)Aug–Oct. Outcrops of granite and sandstone, rocky slopes, sandy stream banks, openings in pine-oak, oak barrens; 200–600 m; Ala., Ga., Ky., N.C., S.C., Tenn.

29. Liatris savannensis Kral & G. L. Nesom, Sida 20: 1574, fig. 1. 2003 • Savanna gayfeather [E]

Plants (80–)90–130 cm. **Corms** globose, irregularly cylindric, narrowly ovoid, or fusiform and caudexlike. **Stems** glabrous (often sparsely sessile-glandular). **Leaves:** basal (often withering before flowering), 1- or 3–5-nerved) linear-elliptic, linear-spatulate, or linear-oblanceolate, (170–)200–400 × 3–10 mm, gradually reduced distally, essentially glabrous, gland-dotted (proximal margins pilose-ciliate). **Heads** in dense, spiciform arrays. **Peduncles** 0 or (spreading-ascending) 1–2 mm. **Involucres** turbinate-campanulate, (7–)8–10 × (8–)9–11(–12) mm. **Phyllaries** in (3–)4(–5) series, ovate to oblong, obovate, or spatulate, weakly unequal, essentially glabrous, margins with (pinkish purple) hyaline borders, erose to lacerate, ciliolate, apices broadly rounded. **Florets** 9–17; corolla tubes pilose inside. **Cypselae** ca. 4 mm; **pappi:** lengths ± equaling corollas, bristles barbellate.

Flowering (Aug–)Sep–Oct(–Nov). Sandy pinelands, moist sites dominated by slash pine, sabal palmetto, or mixture of the two; 0–30 m; Fla.

30. Liatris ohlingerae (S. F. Blake) B. L. Robinson, Contr. Gray Herb. 104: 49. 1934 • Florida gayfeather, sandtorch [E]

Lacinaria ohlingerae S. F. Blake, Bull. Torrey Bot. Club 50: 203, plate 9. 1923; *Ammopursus ohlingerae* (S. F. Blake) Small

Plants 30–100 cm. **Corms** nearly cylindric. **Stems** minutely and closely villous-puberulent, glabrescent. **Leaves:** basal and proximal cauline (usually withering before anthesis) 1-nerved, linear, 80–150 × 1–2(–2.5) mm, gradually reduced distally, essentially glabrous or sparsely puberulent and soon glabrescent, gland-dotted. **Heads** (1–30) in open, corymbiform to racemiform arrays. **Peduncles** (spreading-ascending) 20–70 mm. **Involucres** hemispheric to campanulate or broadly turbinate, 17–23 × 15–20 mm. **Phyllaries** in 6–7 series, oblong, strongly unequal, essentially glabrous, margins with (faintly purplish) hyaline borders, ciliolate, apices rounded to obtuse. **Florets** ± 20–30; corolla tubes glabrous inside. **Cypselae** 7–10 mm; **pappi:** lengths ± equaling corollas, bristles barbellate to subplumose.

Flowering (Jul–)Aug–Nov. Oak scrubs, scrubby flatwoods, rosemary scrub; 10–50 m; Fla.

Liatris ohlingerae grows in DeSoto, Highlands, and Polk counties, on Lake Wales Ridge. It is in the Center for Plant Conservation's National Collection of Endangered Plants.

31. Liatris patens G. L. Nesom & Kral, Sida 20: 1579, fig. 2. 2003 • Georgia or spreading gayfeather E

Plants 35–95 cm. **Corms** globose to depressed-globose. **Stems** moderately to densely strigoso-hirtellous. **Leaves:** basal and proximal cauline 1-nerved, narrowly oblanceolate to linear-oblanceolate, 90–180 × 2–4(–7) mm, gradually or abruptly reduced distally, essentially glabrous (proximal margins sparsely ciliate). **Heads** in loose, (columnar) racemiform arrays. **Peduncles** (divergent-ascending, initially diverging at angles of 45–90°) 10–25(–30) mm. **Involucres** turbinate-campanulate, (5–)6–7.5 × 5–7 mm. **Phyllaries** in (2–)3–4(–5) series, broadly obovate (outer) to broadly oblong-obovate, strongly unequal, sparsely strigoso-villous (outer) or essentially glabrous, margins with (pinkish purple) hyaline borders, sometimes slightly erose, usually densely ciliate, apices rounded to subtruncate. **Florets** 7–12; corolla tubes pilose inside. **Cypselae** 2.5–2.8 mm; **pappi:** lengths equaling or surpassing corollas, bristles barbellate.

Flowering Sep–Nov. Sandy pinelands, usually with longleaf pine or slash pine, pine-palmetto flats; 0–50 m; Fla., Ga., S.C.

32. Liatris gracilis Pursh, Fl. Amer. Sept. 2: 508. 1813 • Slender gayfeather E F

Lacinaria gracilis (Pursh) Kuntze; *L. laxa* Small

Plants 20–100 cm. **Corms** globose to subglobose. **Stems** puberulent to piloso- or strigoso-puberulent. **Leaves:** basal and proximal cauline 1-nerved, usually linear to linear-oblanceolate or oblanceolate, sometimes narrowly spatulate, 50–150 × 2–8(–10) mm, gradually or abruptly reduced on distal ¹⁄₂–¹⁄₃ of stems, essentially glabrous (proximal margins ciliate). **Heads** in loose, racemiform arrays. **Peduncles** 0 or (usually straight, stiffly spreading to ascending) 2–10(–12) mm. **Involucres** cylindro-campanulate, (4–)4.5–6(–7) × 3–4(–5) mm. **Phyllaries** in 3(–4) series, (green to purple) ovate to oblong, strongly to weakly unequal, sparsely puberulent, margins with hyaline borders, ciliolate, apices rounded to truncate.

Florets 3–6(–9); corolla tubes pilose inside. **Cypselae** 3–3.5 mm; **pappi:** lengths ± equaling corollas, bristles barbellate. $2n = 20$.

Flowering (Jul–)Aug–Oct(–Nov). Longleaf pine, slash pine, palmetto-scrub oak, turkey oak-bluejack oak, sand scrub, sand ridges, uplands, boggy flatwoods, sandy peat, fencerows, roadsides, woodland edges; 10–50 m; Ala., Fla., Ga., Miss., S.C.

Plants of *Liatris gracilis* from the panhandle of Florida and the southern tier of Alabama counties tend to have sharply acute phyllaries; intermediates are often encountered.

33. Liatris gholsonii L. C. Anderson, Sida 20: 98, fig. 1. 2002 • Gholson's or bluffs gayfeather C E

Plants 45–120(–180) cm. **Corms** globose. **Stems** finely puberulent. **Leaves:** basal and proximal cauline 1-nerved, spatulate (petioles slender, blades elliptic to lanceolate-elliptic), (100–)150–250 × (8–)11–22(–27) mm, gradually reduced distally, essentially glabrous, gland-dotted. **Heads** in loose, racemiform arrays. **Peduncles** (spreading-ascending) (2–)5–12(–26) mm (with bracts 2+ mm wide, grading from distal leaves). **Involucres** cylindro-turbinate, 4–5 × 2.5–3 mm. **Phyllaries** in 3(–4) series, ovate to oblong, strongly to weakly unequal, essentially glabrous, margins with hyaline borders, ciliolate, apices (at least outer) acuminate to acute. **Florets** 3–5(–6); corolla tubes pilose inside. **Cypselae** 3.5–5 mm; **pappi:** lengths ± equaling corollas, bristles barbellate. $2n = 20$.

Flowering (Jul–)Aug–Oct(–Nov). Slopes of shaded sites in deciduous woods, ecotones of mesic woods and scrub oak, sandy ridges between ravines, open sites on bluffs, loamy or gravelly sands; of conservation concern; 20–50 m; Fla.

Liatris gholsonii is closely similar to *L. gracilis*; it apparently is restricted to an area along the Apalachicola River in Liberty and Leon counties.

34. Liatris ligulistylis (A. Nelson) K. Schumann, Just's Bot. Jahresber. 29(1): 569. 1903 • Northern Plains gayfeather E

Lacinaria ligulistylis A. Nelson, Bot. Gaz. 31: 405. 1901

Plants 20–100 cm. **Corms** sub-globose (often knotty, densely fibrous-rooted). **Stems** sparsely to densely puberulent, puberulent-villous, or strigoso-puberulent. **Leaves:** basal and proximal cauline 1-nerved, narrowly oblanceolate to spatulate-lanceolate, 90–150(–220) × 4–

17(–24) mm, gradually reduced distally to near midstem, then abruptly reduced, linear or narrowly lanceolate, ascending to nearly erect, bractlike, sparsely to densely puberulent, ± gland-dotted. **Heads** (4–21) in open, racemiform arrays (terminal heads sometimes maturing first and larger). **Peduncles** usually (5–)8–15(–30) mm. **Involucres** campanulate to turbinate-campanulate, 10–15 × 13–18 mm. **Phyllaries** in (3–)4–5 series, oblong-obovate to oblong-spatulate, strongly unequal, essentially glabrous, margins with (purple) hyaline borders, erose to lacerate or irregular, apices broadly rounded to truncate. **Florets** 30–70; corolla tubes glabrous inside. **Cypselae** 5–7 mm; **pappi**: lengths ± equaling corollas, bristles barbellate. $2n = 20$.

Flowering Jul–Oct. Prairies (often wet), pine barrens, clearings in aspen and pine woods, ridges along lake shores, depressions in granite, rocky slopes, roadsides, ditches, along railroads, sand, clay; 100–2400 m; Alta., Man., Sask.; Colo., Iowa, Minn., Mont., N.Mex., N.Dak., S.Dak., Wis., Wyo.

35. Liatris scariosa (Linnaeus) Willdenow, Sp. Pl. 3: 1635. 1803 · Northern gayfeather E

Serratula scariosa Linnaeus, Sp. Pl. 2: 818. 1753; *Lacinaria scariosa* (Linnaeus) Hill

Plants 30–150 cm. **Corms** subglobose. **Stems** puberulent (at least distally, sometimes glabrous proximally). **Leaves:** basal and proximal cauline 1-nerved, elliptic- to oblanceolate-spatulate, 120–300 × 25–50(–55) mm, abruptly, gradually, or little reduced distally, essentially glabrous, weakly, if at all, gland-dotted. **Heads** in racemiform arrays. **Peduncles** usually (ascending) 10–50 mm. **Involucres** campanulate to turbinate-campanulate, 11–15 × (12–)15–22(–25) mm. **Phyllaries** in (3–)4–5 series, (erect or outer sometimes spreading-reflexing) oblong-obovate to broadly obovate (not bullate), unequal, glabrous or minutely puberulent, margins usually with relatively narrow, hyaline borders, apices broadly rounded. **Florets** 19–80; corolla tubes usually pilose inside, sometimes glabrous (Arkansas, Illinois, Missouri). **Cypselae** 4.5–6(–6.5) mm; **pappi**: lengths ± equaling corollas, bristles barbellate.

Varieties 3 (3 in the flora): United States.

1. Stems with 8–20(–25) leaves or leafy bracts proximal to heads; florets 20–40(–50). . . . . . .
. . . . 35a. *Liatris scariosa* var. *scariosa*
1. Stems with 20–85 leaves or leafy bracts proximal to heads; florets 30–80.

[2. Shifted to left margin.—Ed.]
2. Proximal cauline leaves mostly 25–50(–55) mm wide, hirtello-puberulent (and gland-dotted) . . . . 35b. *Liatris scariosa* var. *nieuwlandii*
2. Proximal cauline leaves mostly 7–20(–26) mm wide, glabrous or hirtello-puberulent (weakly, if at all, gland-dotted) . . . . 35c. *Liatris scariosa* var. *novae-angliae*

35a. Liatris scariosa (Linnaeus) Willdenow var. **scariosa** E

Lacinaria scariosa (Linnaeus) Hill var. *borealis* (Nuttall ex J. McNab) Lunell; *L. scariosa* var. *virginiana* Lunell; *Liatris borealis* Nuttall ex J. McNab; *L. scariosa* var. *virginiana* (Lunell) Gaiser

Plants 30–80(–100) cm. **Stems** with 8–20(–25) leaves or leafy bracts proximal to heads. **Leaves:** basal and proximal cauline elliptic- to oblanceolate-spatulate, mostly 100–280 × 16–40(–45) mm, glabrous (weakly, if at all, gland-dotted). **Heads** usually 19–30. **Florets** 19–33.

Flowering Aug–Sep(–Oct). Rock ledges, shale banks and barrens, limestone and sandstone outcrops, road banks, flood plains, dry woods; 600–1600 m; Md., N.C., Pa., Tenn., Va., W.Va.

Variety *scariosa* is an Appalachian entity that might justifiably be treated as distinct from vars. *novae-anglieae* and *nieuwlandii* at species level, as done by Shinners. It has shorter stems and smaller heads than vars. *nieuwlandii* and *novae-angliae*, and the transition in shape and size from basal to cauline leaves is more abrupt.

35b. Liatris scariosa (Linnaeus) Willdenow var. **nieuwlandii** (Lunell) E. G. Voss, Michigan Bot. 34: 139. 1996 E

Lacinaria scariosa (Linnaeus) Hill var. *nieuwlandii* Lunell, Amer. Midl. Naturalist 2: 176. 1912 (as Laciniaria); *Liatris ×nieuwlandii* (Lunell) Gaiser; *L. novae-angliae* (Lunell) Shinners var. *nieuwlandii* (Lunell) Shinners

Plants 30–100 cm. **Stems** with 20–85 leaves or leafy bracts proximal to heads. **Leaves:** basal and proximal cauline usually narrowly lanceolate-spatulate, sometimes broader, mostly 100–500 × 25–50(–55) mm, glabrous or hirtello-puberulent (gland-dotted). **Heads** usually 9–20. **Florets** 30–80.

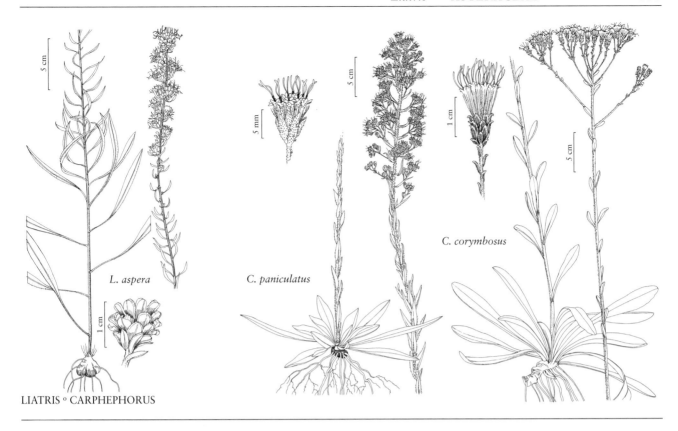

L. aspera

C. paniculatus

C. corymbosus

LIATRIS ° CARPHEPHORUS

Flowering Aug–Sep(–Oct). Prairies, glades, open woods, bluff ledges, railroads, rocky limestone soils, red clays, jack pine, pine-oak, oak-juniper, oak-hickory, aspen; 100–500 m; Ark., Ill., Ind., Mich., Mo., N.Y., Ohio, Pa., W.Va., Wis.

Plants of var. *nieuwlandii* are usually relatively tall and have relatively numerous, even-sized, densely arranged, lanceolate cauline leaves.

35c. Liatris scariosa (Linnaeus) Willdenow var. **novae-angliae** (Lunell) Gandhi, S. M. Young & P. Somers, Taxon 52: 316. 2003 [E]

Lacinaria scariosa (Linnaeus) Willdenow var. *novae-angliae* Lunell, Amer. Midl. Naturalist 2: 177. 1912 (as Laciniaria); *Liatris novae-angliae* (Lunell) Shinners

Plants 30–100 cm. **Stems** with 20–85 leaves or leafy bracts proximal to heads. **Leaves:** basal and proximal cauline lanceolate to lanceolate- or oblanceolate-spatulate, mostly 110–270 × 5–25 mm, glabrous or hirtello-puberulent, weakly, if at all, gland-dotted. **Heads** usually 5–30. **Florets** 35–60.

Flowering (Jul–)Aug–Sep. Sandy fields, woods, railroads, road banks; 20–100 m; Conn., Maine, Mass., N.H., N.J., N.Y., Pa., R.I.

36. Liatris aspera Michaux, Fl. Bor.-Amer. 2: 92. 1803
• Rough gayfeather [E] [F]

Lacinaria scariosa (Linnaeus) Willdenow var. *intermedia* Lunell; *Liatris aspera* var. *intermedia* (Lunell) Gaiser; *L. aspera* var. *salutans* (Lunell) Shinners; *L. spheroidea* var. *salutans* (Lunell) Shinners

Plants 30–180 cm. **Corms** globose. **Stems** hispidulous-puberulent. **Leaves:** basal (usually withering before flowering) and proximal cauline 1-nerved, broadly oblanceolate to elliptic- or lanceolate-spatulate to linear-lanceolate, 80–250 × (4–)6–25 mm, gradually or abruptly reduced distally, essentially glabrous or sparsely or densely puberulent-hispidulous, ± gland-dotted (distal especially). **Heads** in loose, spiciform arrays. **Peduncles** 0 or (ascending to spreading or deflexed) 1–5(–10) mm. **Involucres** campanulate-hemispheric to turbinate-campanulate, (9–)10–16 × 10–20 mm. **Phyllaries** usually in 4–5 series, (outer or outer and middle usually reflexed) broadly obovate to oblong-spatulate, unequal, essentially glabrous, margins with relatively broad, whitish-hyaline, irregular to erose or lacerate borders (especially on the distal ⅓), strongly bullate (with a sharp, low, rounded buttress delimiting the hyaline border), apices usually rounded to subtruncate. **Florets** (14–)18–24(–30);

corolla tubes pilose inside. **Cypselae** (3.5–)4.5–6 mm; **pappi:** lengths ± equaling corollas, bristles barbellate. $2n$ = 20.

Flowering Aug–Sep(–Oct). Prairies, barrens, old fields, sand dunes, ridges, fields, stream bottoms, sandstone outcrops, limestone ridges, hills, oak, oak-juniper, and oak-pine woodlands, jack pine, sand, silt loam; 50–500(–900) m; Ont.; Ala., Ark., Fla., Ga., Ill., Ind., Iowa, Kans., Ky., La., Mich., Minn., Miss., Mo., Nebr., N.C., N.Dak., Ohio, Okla., S.C., S.Dak., Tenn., Tex., Va., W.Va., Wis.

Plants of *Liatris aspera* from the Dakotas, Minnesota, Nebraska, Iowa, and Kansas have consistently sparsely to densely puberulent-hispidulous leaves (var. *aspera*); those to the east usually have glabrous leaves (var. *intermedia*); variation in vesture occurs in Missouri, Illinois, and Indiana and identities in this large area would have to be arbitrary or typological if varieties were formally recognized. Variety *salutans* was recognized by Shinners on the basis of its deflexed (versus horizontal or ascending) heads; such plants occur in Texas, western Louisiana, and adjacent areas of Arkansas and Oklahoma.

37. **Liatris squarrulosa** Michaux, Fl. Bor.-Amer. 2: 92. 1803 • Southern gayfeather E

Lacinaria shortii Alexander; *L. tracyi* Alexander; *Liatris earlei* (Greene) K. Schumann; *L. scabra* (Greene) K. Schumann; *L. scariosa* (Linnaeus) Willdenow var. *squarrulosa* (Michaux) A. Gray

Plants 25–80(–130) cm). **Corms** globose. **Stems** puberulent. **Leaves:** basal and proximal cauline 1-nerved, usually oblanceolate- to elliptic-spatulate to oblanceolate, 80–290 × 12–25(–45) mm, abruptly, then little or gradually, reduced distally, essentially glabrous or puberulent or hirtellous-hispidulous, weakly, if at all, gland-dotted. **Heads** in loose to dense, racemiform to spiciform arrays. **Peduncles** usually 0, sometimes (ascending to spreading or deflexed) 2–8(–30) mm. **Involucres** turbinate to turbinate-campanulate or campanulate-cylindric, (6–)8–12 × (6–)8–15 mm. **Phyllaries** in 4–5(–6) series, (usually spreading to distally reflexed, sometimes erect) obovate to oblong-spatulate, unequal, essentially glabrous or puberulent to puberulent-hirtellous, margins sometimes with relatively narrow, weakly differentiated hyaline borders, apices acute to obtuse (at least outer) or rounded (mid and inner). **Florets** 11–26(–28); corolla tubes pilose inside. **Cypselae** (3–)3.5–5.5(–6) mm; **pappi:** lengths ± equaling corollas, bristles barbellate.

Flowering (Jul–)Aug–Oct(–Nov). Pine, oak, and oak-hickory woods, scrub oak, prairies, rocky ridges and slopes, marl ridges, sand hills and ridges, alluvial woods (rarely), fencerows, road banks, roadsides, sand, clay, chalk; 0–600(–1000) m; Ala., Ga., Ill., Ind., Ky., La., Miss., Mo., N.C., S.C., Tenn., Tex., W.Va.

Liatris squarrulosa occurs across a range of habitats and exhibits much variation in habit, vesture, and head size (involucral dimensions, numbers of florets, lengths of cypselae); it has not been possible in herbarium study to discern discrete geographic patterns. Plants in South Carolina (including the type of *L. squarrulosa*) are shorter, leaves and phyllaries are consistently glabrous or sparsely puberulent, involucres are (6–)7–10 × 6–10 mm, and mid to inner phyllaries usually are narrowly oblong. Westward and northward, plants are taller, leaves and phyllaries characteristically have more strongly elaborated vesture (although considerable variation exists and glabrous plants are scattered nearly through the range), heads range larger, and phyllaries vary from broadly obovate to oblong-spatulate. An overview of head size is indicated by the following measurement samples: South Carolina—florets 12–16, cypselae (3–)3.5–4 mm; Georgia (including the type of *L. earlei*)—florets (11–)14–21, cypselae 4–5 mm; Alabama—florets 11–20, cypselae 3.5–5(–6) mm; Illinois (including the type of *L. scabra*) and Tennessee—florets 16–24, cypselae 3.5–5.5 mm; Louisiana—florets 16–21, cypselae 3.5–5 mm; Arkansas—florets (13–)21–26, cypselae 3.5–5 mm. Plants apparently disjunct in a cluster of north-central North Carolina counties have glabrous phyllaries and leaves, large heads, and the phyllaries are broad, more like plants to the west than those southward.

Populations in southwestern North Carolina (Cherokee, Clay, Macon, and Swain counties) and adjacent Tennessee, South Carolina, and Georgia appear to combine features of *Liatris squarrulosa* and *L. scariosa* var. *scariosa*. The plants are within the range of *L. squarrulosa* and disjunct from *L. scariosa*. The habit (large basal leaves, few cauline) is common to both taxa; heads are sessile to short-pedunculate (as in *L. squarrulosa*) and phyllaries are usually erect (the outer triangular and sometimes reflexing) and slightly broader than is characteristic of *L. squarrulosa* and often have a narrow hyaline border. These add further heterogeneity to the concept of *L. squarrulosa*, as circumscribed here; they may represent a separate evolutionary entity. The type of *L. ruthii* Alexander (an illegitimate name) from Tennessee is this form.

Hybrids:

Liatris species readily form hybrids. Named hybrids and their presumed parentals are:

Liatris ×*boykinii* Torrey & A. Gray; *Lacinaria boykinii* (Torrey & A. Gray) Kuntze—*L. elegans* × *L. tenuifolia*

Liatris ×*creditonensis* Gaiser—*L. ligulistylis* × *L. squarrosa* var. *glabrata* (known only from garden-grown plants)

Liatris ×*deamii* (Lunell) Shinners; *Lacinaria deamii* Lunell; *Liatris scariosa* (Linnaeus) Willdenow var. *deamii* (Lunell) Peattie—*L. aspera*? × *L. ligulistylis*?

Liatris ×*fallacior* (Lunell) Rydberg; *Lacinaria fallacior* Lunell—*L. ligulistylis* × *L. punctata*

Liatris ×*freemaniana* J. R. Allison—*L. cylindracea* × *L. oligocephala*

Liatris ×*frostii* Gaiser—*L. aspera* × *L. pycnostachya*

Liatris ×*gladewitzii* (Farwell) Shinners; *Lacinaria gladewitzii* Farwell—*L. aspera* × *L. cylindracea*

Liatris ×*macdanieliana* J. R. Allison—*L. cylindracea* × *L. squarrosa*

Liatris ×*ridgwayi* Standley—*L. pycnostachya* × *L. squarrosa*

Liatris ×*spheroidea* Michaux; *Lacinaria aspera* (Michaux) Greene var. *spheroidea* (Michaux) Alexander—*L. aspera* × *L. ligulistylis*? (fide L. O. Gaiser 1946)

Liatris ×*steelei* Gaiser—*L. aspera* × *L. spicata*

Liatris ×*weaveri* Shinners—*L. aspera* × *L. punctata* (known only from garden-grown plants)

411. CARPHEPHORUS Cassini, Bull. Sci. Soc. Philom. Paris 1816: 198. 1816 • [Greek *karphos*, chaff, and *phoros*, bearing, alluding to receptacular paleae] E

Guy L. Nesom

Litrisa Small; *Trilisa* (Cassini) Cassini

Perennials, 20–60+ cm (caudices relatively thick, fibrous-rooted). **Stems** erect, not branched (± scapiform). **Leaves** basal and cauline; alternate; ± petiolate (basal) or sessile; blades (usually appressed to strictly ascending) usually 1-nerved, linear to oblanceolate or spatulate, margins entire or remotely dentate (involute in *C. pseudoliatris*), faces glabrous or hairy, often gland-dotted. **Heads** discoid, in corymbiform to paniculiform arrays. **Involucres** campanulate to hemispheric, 3–12 mm diam. **Phyllaries** persistent, (5–)8–40 in (1–)2–5+ series, not notably nerved, ovate to elliptic or lanceolate, unequal (herbaceous to scarious). **Receptacles** convex, paleate or epaleate. **Florets** 12–35; corollas usually lavender to dark magenta or pinkish purple, sometimes blue, throats funnelform (sometimes externally gland-dotted, lengths 4–6 times diams.); styles: bases not enlarged, glabrous, branches linear-clavate (± papillose distally). **Cypselae** prismatic, ca. 10-ribbed, scabrellous to hispid-strigose, sometimes gland-dotted; **pappi** persistent, of 35–40, barbellulate to barbellate (subequal) bristles in 1–2 series. $x = 10$.

Species 7 (7 in the flora): se United States.

Trilisa and *Litrisa* have been maintained as separate genera (R. M. King and H. Robinson 1987), distinguished primarily by their smaller heads with fewer phyllaries and their lack of receptacular paleae. In other features, their close similarity and relationship to *Carphephorus* seem evident, and other recent students of the group at species level (e.g., H. J.-C. Hebert 1968; M. D. Correa and R. L. Wilbur 1969; A. Cronquist 1980) have treated the species within a single genus. Forthcoming molecular studies suggest that one or both segregates may be justifiably recognized.

SELECTED REFERENCES Correa, M. D. and R. L. Wilbur. 1969. A revision of the genus *Carphephorus* (Compositae–Eupatorieae). J. Elisha Mitchell Sci. Soc. 85: 79–91. Hebert, H. J.-C. 1968. Generic considerations concerning *Carphephorus* and *Trilisa* (Compositae). Rhodora 70: 474–485. James, C. W. 1958. Generic considerations concerning *Carphephorus*, *Trilisa* and *Litrisa* (Compositae). Rhodora 60: 117–122.

1. Involucres mostly 3.5–6 mm; phyllaries 5–12 in 1–2(–3) series; receptacles epaleate or partially paleate (paleae 1–2).
   2. Stems puberulous to villoso-hirsute and minutely gland-dotted; leaves gland-dotted (basal in rosettes, cauline little developed); phyllaries oblong to oblong-oblanceolate or elliptic-ovate, apices acute to obtuse-mucronate (inner pectinate-ciliate); pappus bristles in ± 2 series . . . . 1. *Carphephorus carnosus*
   2. Stems glabrous or villoso-hirsute, eglandular; leaves not gland-dotted (basal not in rosettes, cauline ± well developed); phyllaries spatulate to obovate, apices acute to obtuse or rounded (inner not ciliate); pappus bristles in ± 1 series.
      3. Stems glabrous; heads in flat-topped, corymbiform arrays; peduncles glabrous . . . . 2. *Carphephorus odoratissimus*
      3. Stems villoso-hirsute; heads in thyrsiform (often ± columnar) arrays; peduncles stipitate-glandular (each with a subtending bract) . . . . 3. *Carphephorus paniculatus*
1. Involucres mostly (6–)7–12(–15) mm; phyllaries 15–40+ in 3–5+ series; receptacles paleate (peripherally or throughout, paleae 4–33).
   4. Basal leaves narrowly linear; phyllaries triangular-lanceolate . . . . 4. *Carphephorus pseudoliatris*
   4. Basal leaves mostly oblanceolate to spatulate; phyllaries ovate-lanceolate to broadly ovate, elliptic, or elliptic-obovate.
      5. Stems, peduncles, phyllaries, and corollas gland-dotted; phyllaries ovate-lanceolate to broadly ovate, villous and gland-dotted, apices acute to obtuse . . . . 5. *Carphephorus tomentosus*
      5. Stems, peduncles, phyllaries, and corollas eglandular; phyllaries broadly elliptic to elliptic-obovate, glabrous but for ciliate margins, eglandular, apices rounded.
         6. Stems glabrous or glabrate (except peduncles); heads in open, loose, usually corymbiform, sometimes paniculiform, arrays; phyllaries broadly elliptic to elliptic-obovate (margins not hyaline, ciliate); cypselae minutely gland-dotted . . . . 6. *Carphephorus bellidifolius*
         6. Stems villoso-hirsute to short-hirsute; heads in tight, flat-topped, corymbiform arrays; phyllaries broadly elliptic (margins broad, hyaline-scarious, erose or ciliate); cypselae eglandular . . . . 7. *Carphephorus corymbosus*

1. Carphephorus carnosus (Small) C. W. James, Rhodora 60: 120. 1958 • Pineland chaffhead [E]

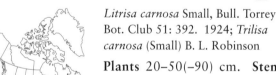

Litrisa carnosa Small, Bull. Torrey Bot. Club 51: 392. 1924; *Trilisa carnosa* (Small) B. L. Robinson

Plants 20–50(–90) cm. **Stems** densely puberulous or villoso-hirsute, densely gland-dotted. **Leaves:** basal (in rosettes) oblanceolate, mostly 3–7(–9) cm; cauline abruptly reduced, not clasping, faces gland-dotted. **Heads** in flat-topped, corymbiform arrays. **Peduncles** hirsutulous and sparsely gland-dotted. **Involucres** 4.5–6 mm. **Phyllaries** 5–12 in 1–2(–3) series, oblong to oblong-oblanceolate or elliptic-ovate (inner pectinate-ciliate), villoso-hirsute, gland-dotted, apices acute to obtuse-mucronate. **Receptacles** epaleate or partially paleate (paleae 1–2). **Corollas** glandular, lobes 0.8–1 mm. **Cypselae** eglandular; **pappus bristles** in ± 2 series. $2n = 20$.

Flowering (Jun–)Jul–Oct(–Nov). Low pinelands, pine and pine-palmetto flatwoods, grass-sedge bogs, seepage slopes, ditches, prairies; 60–100 m; Fla.

2. Carphephorus odoratissimus (J. F. Gmelin) H. J.-C. Hebert, Rhodora 70: 483. 1968 • Vanillaleaf

Chrysocoma odoratissima J. F. Gmelin, Syst. Nat. 2: 1204. 1792; *Liatris odoratissima* (J. F. Gmelin) Michaux; *Trilisa odoratissima* (J. F. Gmelin) Cassini

Plants 50–140(–180) cm. **Stems** glabrous. **Leaves:** basal oblanceolate to obovate, mostly 9–50 cm; proximal cauline well developed, gradually reduced distally, clasping, faces not gland-dotted. **Heads** in flat-topped, corymbiform arrays. **Peduncles** glabrous. **Involucres** 3.5–5 mm. **Phyllaries** 5–12 in 1–2(–3) series, oblanceolate, gland-dotted, apices obtuse. **Receptacles** epaleate or partially paleate (paleae 1–2). **Corollas** glandular, lobes 0.8–1 mm. **Cypselae** glandular; **pappus bristles** in ± 1 series.

Varieties 2 (2 in the flora): se United States.

1. Plants with strong odor of coumarin or vanilla; basal leaves usually more than 15 × 5 cm; midstem leaves broadly elliptic, margins often shallowly dentate, apices flared away from stems; primary head-bearing branches diverging from main axes at 10–20°; florets mostly 7–10 . . . .
   . . . 2a. *Carphephorus odoratissimus* var. *odoratissimus*
1. Plants with slight or no odor of coumarin or vanilla; basal leaves less than 15 × 4 cm; midstem leaves narrowly elliptic, margins entire, apices appressed to stems; primary head-bearing branches diverging from main axes at 30–45°; florets mostly 10–14 . . . .
   . . . 2b. *Carphephorus odoratissimus* var. *subtropicanus*

2a. Carphephorus odoratissimus (J. F. Gmelin) H. J.-C. Hebert var. odoratissimus [E]

Plants with strong odor of coumarin or vanilla. **Basal leaves** usually more than 15 × 5 cm; midstem leaves broadly elliptic, apices flared away from stems, margins often shallowly dentate. **Primary head-bearing branches** diverging from main axes at 10–20°. **Florets** mostly 7–10.

Flowering (Aug–)Sep–Oct(–Nov). Pine-oak borders, longleaf pine savannas, flatwoods, fields, roadsides, fencerows, stream bottoms, seepage and boggy areas, sandy sites, limestone ridges or clay hills; 0–80 m; Ala., Fla., Ga., La., Miss., N.C., S.C.

2b. Carphephorus odoratissimus (J. F. Gmelin) H. J.-C. Hebert var. subtropicanus (DeLaney, N. Bissett & Weidenhamer) Wunderlin & B. F. Hansen, Novon 11: 366. 2001 [E]

Carphephorus subtropicanus DeLaney, N. Bissett & Weidenhamer, Bot. Explor. 1: 2, figs 1, 3. 1999

Plants with slight or no odor of coumarin or vanilla. **Basal leaves** less than 15 × 4 cm; midstem leaves narrowly elliptic, margins entire, apices appressed to stems. **Primary head-bearing branches** diverging from main axes at 30–45°. **Florets** mostly 10–14.

Flowering (Sep–)Oct–Nov. Burned pine savanna-flatwoods, dry prairies; 0–50 m; Fla.

Variety *subtropicanus*, which occurs in the southern half of Florida, is mostly allopatric with var. *odoratissimus*. Intergrades occur where their ranges meet (S. L. Orzell and E. L. Bridges 2002).

3. Carphephorus paniculatus (J. F. Gmelin) H. J.-C. Hebert, Rhodora 70: 483. 1968 • Hairy chaffhead [E] [F]

Chrysocoma paniculata J. F. Gmelin, Syst. Nat. 2: 1204. 1792; *Liatris paniculata* (J. F. Gmelin) Michaux; *Trilisa paniculata* (J. F. Gmelin) Cassini

Plants (30–)50–120(–180) cm. **Stems** densely and coarsely villoso-hirsute, eglandular. **Leaves:** basal and proximal cauline oblanceolate to narrowly elliptic, mostly 5–35 cm, margins plane; cauline much reduced, faces not gland-dotted. **Heads** in thyrsiform (often ± columnar) arrays. **Peduncles** stipitate-glandular (each with a subtending bract). **Involucres** (3.5–)4–6 mm. **Phyllaries** 5–12 in 1–2(–3) series, oblanceolate, gland-dotted or stipitate-glandular (sometimes with eglandular, biseriate hairs as well), apices acute to obtuse. **Receptacles** epaleate or partially paleate (paleae 1–2). **Corollas** glandular, lobes 1–1.5 mm. **Cypselae** eglandular; **pappus bristles** in ± 1 series. **2***n* = 20.

Flowering (Aug–)Sep–Dec(–Jan). Moist or wet, low ground, wet prairies or savannas, swamp margins, pine, pine-palmetto, and palmetto flats; 0–50 m; Ala., Fla., Ga., N.C., S.C.

4. Carphephorus pseudoliatris Cassini, Bull. Sci. Soc. Philom. Paris 1816: 198. 1816 (as pseudo-liatris) • Bristleleaf chaffhead [E]

Liatris squamosa Nuttall

Plants 30–100 cm. **Stems** finely villous to villoso-hirsute, eglandular. **Leaves:** basal and proximal cauline narrowly linear, mostly 10–40 cm; cauline well developed, gradually reduced, faces gland-dotted. **Heads** in flat-topped, corymbiform arrays. **Peduncles** finely villous (or hairs somewhat appressed). **Involucres** 6–9 mm. **Phyllaries** 15–40+ in 3–5+ series, triangular-lanceolate, apices acute, abaxial faces villous, eglandular. **Receptacles** paleate (nearly throughout). **Corollas** eglandular, lobes 1–2 mm. **Cypselae** eglandular; **pappus bristles** in ± 2 series. **2***n* = 20.

Flowering (Jul–)Aug–Oct. Moist to dry sites, pine barrens, savannas, cutover pine woods, wiregrass savannas, pine-palmetto flatwoods, grass-sedge bogs, swamp edges, ditches, depressions; 10–90 m; Ala., Fla., Ga., La., Miss.

5. **Carphephorus tomentosus** (Michaux) Torrey & A. Gray, Fl. N. Amer. 2: 66. 1841 • Woolly chaffhead E

Liatris tomentosa Michaux, Fl. Bor.-Amer. 2: 93. 1803; *Carphephorus tomentosus* var. *walteri* (Elliott) Fernald

Plants 20–80 cm. **Stems** sparsely to densely hirsute to hirsute-villous, sometimes strigose distally, gland-dotted. **Leaves:** basal and proximal cauline oblanceolate, mostly 3.5–15 cm; cauline gradually reduced, faces gland-dotted. **Heads** in flat-topped, corymbiform arrays. **Peduncles** villoso-hirsute, gland-dotted. **Involucres** 7–11 mm. **Phyllaries** 15–40+ in 3–5+ series, ovate-lanceolate to broadly ovate, villous, and gland-dotted, apices acute to obtuse. **Receptacles** partially paleate (paleae often 4–5). **Corollas** glandular, lobes ca. 2 mm. **Cypselae** eglandular; **pappus bristles** in ± 1 series. $2n = 20$.

Flowering Aug–Oct. Moist to dry pine savannas, flatwoods, pine-oak woodland, wire-grass savannas, fields, sometimes more moist, peaty soils, shrub bogs, seepage; 10–50 m; Ga., N.C., S.C., Va.

6. **Carphephorus bellidifolius** (Michaux) Torrey & A. Gray, Fl. N. Amer. 2: 66. 1841 • Sandy-woods chaffhead E

Liatris bellidifolia Michaux, Fl. Bor.-Amer. 2: 93. 1803

Plants 20–60 cm. **Stems** glabrous or glabrate (except peduncles), eglandular. **Leaves:** basal and proximal cauline usually oblanceolate, sometimes nearly spatulate, mostly 4–20 cm; cauline gradually reduced, faces gland-dotted. **Heads** in open, loose, usually corymbiform, sometimes paniculiform, arrays. **Peduncles** minutely puberulent (viscid, hairs not glandular). **Involucres** 7–12(–15) mm. **Phyllaries** 15–40+ in 3–5+ series, broadly elliptic to elliptic-obovate, glabrous, eglandular (except margins ciliate), apices rounded. **Receptacles** paleate (at least peripherally). **Corollas** eglandular, lobes 1.5–2.5 mm. **Cypselae** minutely sessile-glandular; **pappus bristles** in 1(–2) series. $2n = 20$.

Flowering Jul–Oct. Dunes, sandhills, sandy rises in flatwoods, sandy fields, roadsides, weedy banks, open pine, scrub oak, and turkey oak-pine woods; 10–40 m; Ga., N.C., S.C., Va.

7. **Carphephorus corymbosus** (Nuttall) Torrey & A. Gray, Fl. N. Amer. 2: 67. 1841 • Florida paintbrush E F

Liatris corymbosa Nuttall, Gen. N. Amer. Pl. 2: 132. 1818

Plants 30–120 cm. **Stems** villoso-hirsute to hirsute, eglandular. **Leaves:** basal and proximal cauline oblanceolate, mostly 6–20 cm; cauline gradually reduced, faces gland-dotted. **Heads** in flat-topped, corymbiform arrays. **Peduncles** hirsute, eglandular. **Involucres** (6–)7–10 mm. **Phyllaries** 15–40+ in 3–5+ series, broadly elliptic, glabrous and eglandular (margins hyaline-scarious, erose or ciliate), apices rounded. **Receptacles** paleate (at least peripherally). **Corollas** eglandular, lobes ca. 1 mm. **Cypselae** eglandular; **pappus bristles** in ± 1 series. $2n = 20$.

Flowering (Jul–)Aug–Nov. Sand ridges, roadsides, flats, dry, pine and longleaf pine-turkey oak woods, pine-palmetto, palmetto flats; 0–50 m; Fla., Ga., S.C.

412. **GARBERIA** A. Gray, Proc. Acad. Nat. Sci. Philadelphia 1879: 379. 1880 • [For Abram P. Garber, 1838–1881, of Columbia, Pennsylvania, noted for his contributions to the flora of Florida] C E

Eric E. Lamont

Shrubs, 100–250 cm (± evergreen). **Stems** erect (terete, striate when dry), branched (usually gland-dotted, farinaceous to puberulent when young). **Leaves** cauline; all or mostly alternate (at flowering); petiolate or subsessile; **blades** obscurely nerved, spatulate to spatulate-obovate or orbiculate-obovate, margins entire, faces gland-dotted (viscid, farinaceous when young). **Heads** discoid, in corymbiform or paniculiform arrays. **Involucres** narrowly cylindric, 3.5–5

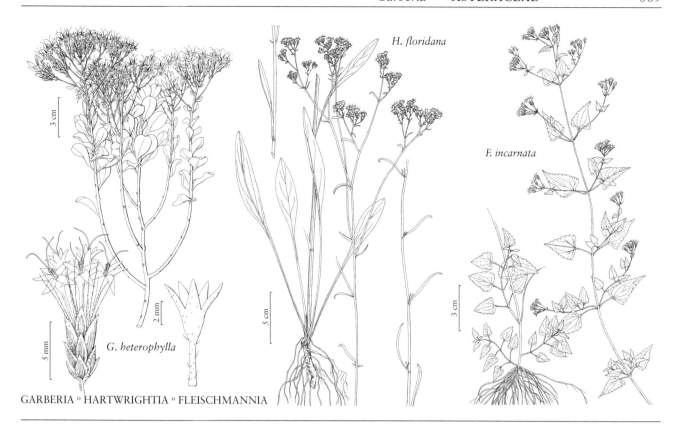

H. floridana

F. incarnata

G. heterophylla

GARBERIA ∘ HARTWRIGHTIA ∘ FLEISCHMANNIA

(–6) mm diam. **Phyllaries** persistent, (12–)15–20 in 3–5 series, ± striate, lanceolate to linear-oblong, unequal (apices acute or acuminate, abaxial faces farinaceous, usually gland-dotted). **Receptacles** weakly convex, epaleate. **Florets** usually 5 (aromatic); corollas pink to purplish, throats ± campanulate, lobes 5, triangular to lance-ovate; styles: bases not enlarged, glabrous, branches filiform to linear-clavate (distally papillose). **Cypselae** prismatic, ca. 10-ribbed, densely scabrellous; **pappi** persistent, of ca. 60–70, barbellate bristles in 2–3 series (outer shorter than inner).

Species 1: Florida.

The close relationship between *Garberia* and *Liatris* has been long recognized. T. Nuttall (1822) included *G. heterophylla* in *Liatris* sect. *Leptoclinium* (as *L. fruticosa* Nuttall). *Garberia* is distinct by its shrubby habit and karyotype (L. O. Gaiser 1954).

SELECTED REFERENCE Curtiss, A. H. 1881. *Chapmannia* and *Garberia*. Bot. Gaz. 6: 257–259.

1. Garberia heterophylla (W. Bartram) Merrill & F. Harper, Bartonia 23: 24. 1945 [C][E][F]

Cacalia heterophylla W. Bartram, Travels Carolina, 164. 1791; *Garberia fruticosa* (Nuttall) A. Gray

Leaves: (first in basal rosettes, proximal cauline sometimes opposite or subopposite, distal alternate); blades grayish green, 15–35 × 7–20 mm, bases cuneate to attenuate, apices rounded or slightly retuse, faces often viscid when fresh. **Peduncles** ± 1 cm. **Phyllaries:** outer ca. 3 mm; inner ca. 5–6 mm, more chartaceous. **Corollas** 8–10 mm, surpassing involucres. **Cypselae** (6–)7–8 mm; **pappi** often tinged with purple, 8–11(–12) mm.

Flowering mainly Oct–Dec (sometimes nearly year round). Dry, well-drained, sandy soils lacking appreciable organic materials, fire-maintained sand-pine-scrub and oak-scrub communities; of conservation concern (fide www.plantatlas.usf.edu); 0–10 m; Fla.

413. HARTWRIGHTIA   A. Gray ex S. Watson, Proc. Amer. Acad. Arts 23: 264. 1888

• [For Samuel Hart Wright, 1825–1905, collector of the specimens from which the genus was described]   C E

Guy L. Nesom

Perennials, 60–120 cm (rhizomes thickened, fibrous-rooted). **Stems** erect, branched distally. **Leaves** basal and cauline; mostly alternate; petiolate or sessile; blades 1-nerved (or pinnately nerved), elliptic to linear [spatulate to oblanceolate], margins mostly entire, faces glabrous, gland-dotted. **Heads** discoid, in loose, corymbiform arrays. **Involucres** broadly obconic, 2–3 mm diam. **Phyllaries** persistent, 12–15 in 2–3 series, not notably nerved, oblong-elliptic to lanceolate, ± equal (herbaceous). **Receptacles** convex, usually partially paleate (paleae peripheral). **Florets** 7–10; corollas white or pinkish to bluish, throats campanulate (lengths ca. 1.5 times diams.), lobes 5, ± deltate; styles: bases not enlarged, glabrous, branches filiform to weakly clavate. **Cypselae** obpyramidal, 5-angled or -grooved, gland-dotted; **pappi** usually 0, rarely 1(–5+), fragile, flexuous, ± glandular setae. $x = 10$.

Species 1: se United States.

SELECTED REFERENCE Holzinger, J. M. 1893. The systematic position of *Hartwrightia floridana*. Bull. Torrey Bot. Club 20: 287–288.

1. Hartwrightia floridana A. Gray ex S. Watson, Proc. Amer. Acad. Arts 23: 265. 1888   C E F

Leaves: basal blades mostly 5–25 × 1–8 cm, bases tapering to petiolar portions; cauline smaller, linear (bractlike) distally. **Corollas** ca. 3 mm. **Cypselae** 3–3.5 mm.

Flowering Sep–Nov. Low pastures, prairies, depressions, stream terraces, bogs, marsh edges, seepage slopes, pine flatwoods, usually in sandy peat, peat, or peat muck; of conservation concern; 0–30 m; Fla., Ga.

414. FLEISCHMANNIA   Schultz-Bipontinus, Flora 33: 417. 1850 • [For Gottfried F. Fleischmann, 1777–1850, teacher of Schultz-Bipontinus at Erlangen]

Guy L. Nesom

Annuals or perennials [subshrubs], 30–120+ cm (crowns fibrous-rooted). **Stems** erect, simple or sparingly branched (usually puberulent, hairs curled). **Leaves** cauline; opposite [alternate]; petiolate; blades 3-nerved, deltate-ovate or triangular-deltate [elliptic, rhombic], margins ± crenate to serrate, faces glabrous, sometimes gland-dotted. **Heads** discoid, in loose, corymbiform arrays. **Involucres** obconic to hemispheric, 2–4 mm diam. **Phyllaries** persistent, 20–30 in 2–4 series, 2–3-nerved, lanceolate to linear, unequal [subequal] (herbaceous to chartaceous). **Receptacles** flat or slightly convex [conic] (glabrous or with scattered hairs), epaleate. **Florets** 15–25 [10–50]; corollas bluish, pinkish, purplish, or white, throats narrowly funnelform (lengths 2.5–4 times diams.); styles: bases not enlarged, glabrous, branches linear-filiform. **Cypselae** prismatic, 5(–8)-ribbed, glabrous or sparsely hirtellous; **pappi** ± persistent or fragile, of [0 or 5–]20–40, barbellate bristles in 1 series. *x* = (4) 10.

Species ca. 80 (2 in the flora): se, s, c United States, Mexico, Central America, Andean South America.

Fleischmannia is "superficially similar to *Ageratina*, which is distinguished from it by the less well developed carpopodium (not stopper-like and sharply set off from the body as in *Fleischmannia*), usually imbricate involucral bracts, and base chromosome number of *x* = 17 (versus 4 or 10 in *Fleischmannia*)" (R. M. King and H. Robinson 1987).

1. Stems lax, sprawling or scandent; involucres 4–5 mm; outer phyllaries lanceolate, inner lanceolate to narrowly oblong-lanceolate, apices usually acute to attenuate, sometimes rounded; cypselae 1.8–2.8 mm . . . . 1. *Fleischmannia incarnata*
1. Stems erect to ascending-erect (not sprawling-scandent); involucres 3–3.5 mm; outer phyllaries ovate, inner elliptic, apices usually obtuse to rounded, rarely truncate; cypselae 1–1.2(–1.5) mm . . . . 2. *Fleischmannia sonorae*

1. **Fleischmannia incarnata** (Walter) R. M. King & H. Robinson, Phytologia 19: 203. 1970 • Pink slender-thoroughwort [F]

Eupatorium incarnatum Walter, Fl. Carol., 200. 1788

Plants 30–120(–200) cm. **Stems** lax, sprawling or scandent. **Leaves:** petioles (0.5–)1–3.5(–4) cm; blades triangular-deltate, (1–)2–5(–7) × (1–)1.5–3.5(–5) cm, bases usually truncate to cordate, sometimes obtuse, margins coarsely serrate to crenate-serrate, apices acute to acuminate. **Involucres** 4–5 mm. **Phyllaries:** outer lanceolate, inner lanceolate to narrowly oblong-lanceolate, glabrous or sparsely puberulent, apices usually acute to attenuate, sometimes rounded. **Corollas** usually pink-purple or whitish with pink to lilac lobes, rarely all white. **Cypselae** 1.8–2.8 mm, usually sparsely strigoso-hirtellous, sometimes glabrate. *2n* = 20.

Flowering Oct–Dec. Woodlands, thickets, moist soil, roads, ditches, stream banks, bottomlands, swamps, depressions, cedar glades; 10–200 m; Ala., Ark., Fla., Ga., Ill., Ind., Ky., La., Miss., Mo., N.C., Ohio, Okla., S.C., Tenn., Tex., Va., W.Va.; Mexico.

Fleischmannia incarnata was reported to occur in "s. Ariz.;" the report almost certainly was based on misidentification of *F. sonorae*.

2. **Fleischmannia sonorae** (A. Gray) R. M. King & H. Robinson, Phytologia 28: 82. 1974 • Sonoran slender-thoroughwort

Eupatorium sonorae A. Gray, Smithsonian Contr. Knowl. 5(6): 74. 1853

Plants 30–80 cm. **Stems** erect to ascending-erect (not sprawling-scandent). **Leaves:** petioles 0.5–3 cm; blades deltate-ovate, 1.5–5 × 0.6–2 cm, bases broadly cuneate to subtruncate, margins irregularly crenate-serrate, apices acute to acuminate. **Involucres** (4–)5–6 mm. **Phyllaries:** ovate (outer) or elliptic (inner), glabrous, apices usually obtuse to rounded, rarely truncate. **Corollas** usually pale purple, rarely white. **Cypselae** 1–1.2(–1.5) mm, glabrous or glabrate.

Flowering Sep–Dec. Rich soils, along streams, rocky slopes; 800–1500 m; Ariz., N.Mex.; Mexico.

Fleischmannia sonorae was reported for Arizona by T. H. Kearney and R. H. Peebles (1960) and F. Shreve and I. L. Wiggins (1964) as *Eupatorium pycnocephalum* Lessing [= *Fleischmannia pycnocephala* (Lessing) R. M. King & H. Robinson)]; revised concepts of this species group (B. L. Turner 1996+, vol. 2) place *F. pycnocephala* as a related species in Mexico and Central America, not reaching northwestward to Arizona.

415. KOANOPHYLLON Arruda in H. Koster, Trav. Brazil, 495. 1816 • Umbrella thoroughwort [Etymology unknown; possibly an oblique reference to the leaves as a source of dye like indigo]

Guy L. Nesom

Perennials, subshrubs, or shrubs [trees, vines], 40–200+ cm. **Stems** erect, branched. **Leaves** cauline; mostly opposite (distal sometimes alternate); petiolate; blades usually 3-nerved from bases, mostly deltate or ovate to triangular-lanceolate or lanceolate (lacking internal translucent lenses or streaks), margins entire or dentate, faces glabrous or sparsely hispidulous, puberulent, or strigose, sometimes sparsely to densely gland-dotted. **Heads** discoid, in corymbiform to paniculiform arrays. **Involucres** obconic to hemispheric, 2–3 mm diam. **Phyllaries** persistent (at least outermost), 7–16 in 1–2[–4] series, 3-nerved, ovate to lanceolate, ± equal or unequal. **Receptacles** flat or convex, epaleate. **Florets** 3–13[–20]; corollas whitish to pinkish [violet], throats funnelform (lengths ca. 2.5 times diams.); styles: bases sometimes enlarged, glabrous, branches filiform to obscurely clavate. **Cypselae** prismatic, 5-ribbed, sparsely hispidulous or piloso-strigose, sometimes gland-dotted; **pappi** persistent, of 30–35, barbellate bristles in 1 series. *x* = 10.

Species ca. 115 (3 in the flora): s, w United States, Mexico, West Indies, Central America, South America.

1. Leaf blades ovate to ovate-deltate or ovate-lanceolate, abaxial faces densely gland-dotted
   . . . . 1. *Koanophyllon villosum*
1. Leaf blades lanceolate to triangular-lanceolate, abaxial faces eglandular or sparsely gland-dotted.
   2. Leaves usually opposite, sometimes subopposite to alternate on distal ¹/₃ of stems, blade apices acuminate, adaxial faces glabrous; involucres 4.5–5.5 mm; phyllaries: outermost narrowly lanceolate, glabrous, margins hyaline on proximal ²/₃, eciliate
      . . . . 2. *Koanophyllon solidaginifolium*
   2. Leaves opposite, blade apices acute, adaxial faces sparsely strigose to hispidulous; involucres (3–)3.5–4 mm; phyllaries: outermost ovate-elliptic or obovate to narrowly oblong-lanceolate, puberulent, margins herbaceous, usually weakly ciliate . . . . 3. *Koanophyllon palmeri*

1. Koanophyllon villosum (Swartz) R. M. King & H. Robinson, Phytologia 32: 265. 1975 • Florida Keys umbrella thoroughwort [F]

Eupatorium villosum Swartz, Prodr., 111. 1788

Shrubs, 50–200 cm. **Stems** densely puberulent to pilose (often minutely gland-dotted). **Leaves** opposite; petioles 2–8 mm; blades ovate to ovate-deltate or ovate-lanceolate, mostly 2–7(–8) × 1–4 cm, bases truncate to cordate, margins entire or shallowly serrate to crenate, apices obtuse, abaxial faces densely gland-dotted, adaxial sparsely puberulent. **Involucres** 2.8–3 mm. **Phyllaries:** outermost elliptic-lanceolate, hispidulous-puberulent, margins herbaceous, mostly eciliate. **Corollas** white or pinkish white, 2.2–2.6 mm, lobes glandular. **Cypselae** 1.5–1.8 mm, finely hispidulous-strigose.

Flowering year round, perhaps most abundantly May–Sep. Hammocks, pinelands; 0–10 m; Fla.; West Indies (Bahamas).

2. Koanophyllon solidaginifolium (A. Gray) R. M. King & H. Robinson, Phytologia 22: 151. 1971 (as solidaginifolia) • Shrubby umbrella thoroughwort

Eupatorium solidaginifolium A. Gray, Smithsonian Contr. Knowl. 3(5): 87. 1852

Perennials or subshrubs (shrubs), 60–100 cm. **Stems** mostly sparsely puberulent (hairs upcurved or apically bent). **Leaves** mostly opposite, sometimes subopposite to alternate on distal ¹/₃ of stems; petioles 2–8(–12) mm; blades lanceolate, mostly 3–8 (–11) × 1–3 cm, bases rounded, margins entire or shallowly crenate, apices acuminate, abaxial faces eglandular,

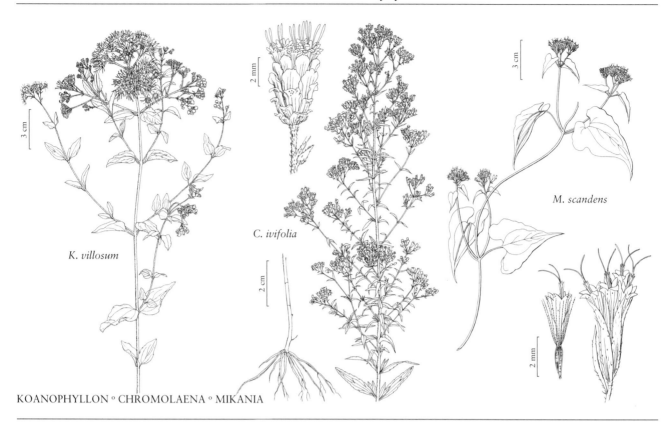

KOANOPHYLLON ∘ CHROMOLAENA ∘ MIKANIA

adaxial glabrous. **Involucres** 4.5–5.5 mm. **Phyllaries:** outermost narrowly lanceolate, glabrous, margins hyaline on proximal ²/₃, eciliate. **Corollas** usually white, sometimes purple-tinged to yellowish, 2.5–3 mm, lobes glandular. **Cypselae** 1.8–2.8 mm, sparsely hispidulous. $2n = 20$.

Flowering (Jun–)Sep–Nov(–Dec). Canyon walls, ledges, slopes, talus, limestone hills and ridges, on and among rocks, in woods along streams; 700–1600 m; Tex.; Mexico (Chihuahua, Coahuila, Durango, Zacatecas).

3. Koanophyllon palmeri (A. Gray) R. M. King & H. Robinson, Phytologia 22: 150. 1971 • Palmer's umbrella thoroughwort

Eupatorium palmeri A. Gray, Proc. Amer. Acad. Arts 21: 383. 1886

Perennials or subshrubs, 50–150 cm. **Stems** mostly sparsely puberulent (hairs upcurved or apically bent). **Leaves** opposite; petioles 0–5 mm; blades lanceolate to triangular-lanceolate, (2–)3–6 × 0.8–2 cm, bases rounded, margins shallowly serrate to subentire, apices acute (not acuminate), abaxial faces eglandular or minutely gland-dotted, adaxial sparsely strigose to hispidulous. **Involucres** (3–)3.5–4 mm. **Phyllaries:** outermost ovate-elliptic to obovate to narrowly oblong-lanceolate, puberulent, margins herbaceous, usually weakly ciliate. **Corollas** white, (2–)2.4–2.8 mm, lobes glandular. **Cypselae** 1.6–1.8 mm, sparsely hispidulous.

Flowering Sep–Nov(–Dec). Shaded rocks along streams, crevices, oak woodlands; 800–1500 m; Ariz., N.Mex.; Mexico.

Koanophyllon palmeri is native primarily to western Mexico (Sinaloa, Chihuahua, and Sonora to Durango and Jalisco; R. McVaugh 1984; B. L. Turner 1997). Similar plants (known as *Eupatorium palmeri* var. *tonsum* B. L. Robinson) occur in near-coastal localities from Michoacan to Colima, Jalisco, Nayarit, and Sinaloa, Mexico.

416. CHROMOLAENA de Candolle in A. P. de Candolle and A. L. P. P. de Candolle, Prodr. 5: 133. 1836 • [Greek *chroma*, color, and *laina*, cloak, evidently alluding to the colored phyllaries of some species, including the type]

Guy L. Nesom

Perennials or shrubs (sometimes scandent), 20–250 cm (fibrous-rooted). **Stems** erect or clambering, sparsely to densely branched. **Leaves** cauline; opposite [alternate, verticillate]; petiolate or sessile; blades usually 3-nerved from bases, mostly deltate to ovate or elliptic, sometimes linear, margins dentate or lobed, faces glabrous or puberulent to tomentose, sometimes gland-dotted. **Heads** discoid, in corymbiform to thyrsiform arrays. **Involucres** cylindric [campanulate to hemispheric], 2–7 mm diam. **Phyllaries** usually readily falling (at least in fruit), 18–65+ in 4–6+ series, 3–5-nerved, ovate to oblong or lanceolate, unequal (papery or herbaceous). **Florets** [6–]15–40[–75]; corollas white or purple to blue, lavender, or reddish, throats cylindric (lengths 3–4 times diams.); styles: bases not enlarged, glabrous, branches linear to linear-clavate. **Receptacles** flat to convex, paleate or epaleate. **Cypselae** prismatic, (3–)5-ribbed, scabrellous, usually gland-dotted as well; **pappi** persistent, of ca. 40 barbellate bristles in 1 series. *x* = 10.

Species ca. 165 (4 in the flora): se, sc United States, Mexico, Central America, South America (especially Brazil).

1. Leaves: petioles 0 or 0.5–1(–3) mm; blades linear-lanceolate to elliptic-lanceolate, bases tapered; apices of inner phyllaries ± spreading (pinkish purple, hyaline, petaloid, slightly expanded, truncate) . . . . 2. *Chromolaena ivifolia*
1. Leaves: petioles 5–20 mm; blades mostly deltate to ovate-lanceolate, bases ± abruptly contracted; apices of inner phyllaries appressed (not notably expanded or petaloid).
  2. Shrubs; heads usually in 3s, sometimes borne singly; involucres turbinate to campanulate . . . . 1. *Chromolaena bigelovii*
  2. Perennials or subshrubs; heads in clusters of 5–50+ in (terminal or lateral) corymbiform arrays; involucres cylindric.
    3. Leaf blades mostly 1.5–4 cm; involucres 5.5–7.5(–8) mm . . . . 3. *Chromolaena frustrata*
    3. Leaf blades mostly (3.5–)5–10 cm; involucres (7–)8–10 mm . . . . 4. *Chromolaena odorata*

1. **Chromolaena bigelovii** (A. Gray) R. M. King & H. Robinson, Phytologia 20: 208. 1970 • Bigelow's false thoroughwort

Eupatorium bigelovii A. Gray in W. H. Emory, Rep. U.S. Mex. Bound. 2(1): 75. 1859; *E. madrense* S. Watson

Shrubs, 50–150 cm. **Stems** erect, sparsely puberulent. **Petioles** 1–4 mm. **Leaf blades** deltate to deltate-lanceolate or ovate-lanceolate, 3–7 × 1–4 cm, margins serrate to dentate (abaxial faces reticulate-veiny with raised whitish, secondary veins, minutely gland-dotted). **Heads** usually in 3s, sometimes borne singly. **Peduncles** 0 or 2–8(–12) mm. **Involucres** turbinate to campanulate, 5–7 mm. **Phyllaries** in 4–5 series, apices of the inner appressed, acute (not petaloid or expanded). **Corollas** pale blue or white.

Flowering Oct–Dec. Dry limestone hills in oak woodlands, talus; 1000–1800 m; Tex.; Mexico (Coahuila, Nuevo León, San Luis Potosí).

2. **Chromolaena ivifolia** (Linnaeus) R. M. King & H. Robinson, Phytologia 20: 202. 1970 (as ivaefolia) • Ivy-leaf false thoroughwort F

Eupatorium ivifolium Linnaeus, Syst. Nat. ed. 10, 2: 1205. 1759 (as ivaefoliu); *Osmia ivifolia* (Linnaeus) Schultz-Bipontinus

Perennials or subshrubs, 50–150 cm. **Stems** erect, hispidulous to coarsely short-pilose. **Petioles** 0 or 0.5–1(–3) mm. **Leaf blades** linear-lanceolate to elliptic-lanceolate, 1.5–7 × 0.4–1.2 cm, margins denticulate to subentire. **Heads** usually 5–50+ in (terminal or lateral)

corymbiform arrays. **Involucres** cylindric to campanulate, 5–6 mm. **Phyllaries** in (3–)4–5 series, apices of the inner ± spreading (pinkish purple, hyaline, petaloid, slightly expanded, truncate). **Corollas** light blue to purplish or reddish. $2n = 50$.

Flowering Aug–Nov. Open woods, edges of fields and woods, roadsides, fencerows, ditches, other disturbed sites, flood plains, bottomlands, along streams; 0–50 m; Fla., La., Miss., Tex.; Mexico; West Indies; Central America; South America.

3. Chromolaena frustrata (B. L. Robinson) R. M. King & H. Robinson, Phytologia 20: 201. 1970 • Cape Sable false thoroughwort E

Eupatorium frustratum B. L. Robinson, Proc. Amer. Acad. Arts 47: 193. 1911; *Osmia frustrata* (B. L. Robinson) Small

Perennials, 15–25 cm. **Stems** erect, puberulent to hispidulous. **Petioles** 5–12 mm. **Leaf blades** (3-nerved) elliptic-lanceolate to ovate-lanceolate or deltate-ovate, mostly 1.5–4 × 0.7–2.2 cm, margins shallowly dentate to serrate. **Heads** usually in clusters of 2–6. **Involucres** cylindric, 5.5–7.5(–8) mm. **Phyllaries** in 4–6 series, apices of the inner appressed, rounded (not petaloid or expanded). **Corollas** blue to lavender.

Flowering mostly Aug–Nov, sometimes year round. Coastal rock barrens, edges of rockland hammocks, undisturbed sites; 0–10 m; Fla.

Chromolaena frustrata now grows in the Everglades National Park, Long Key State Recreation Area, Lignumvitae Key State Botanical Site, and two unprotected sites. The historical range includes more sites on the Keys.

4. Chromolaena odorata (Linnaeus) R. M. King & H. Robinson, Phytologia 20: 204. 1970 • Crucita

Eupatorium odoratum Linnaeus, Syst. Nat. ed. 10, 2: 1205. 1759; *Osmia odorata* (Linnaeus) Schultz-Bipontinus

Perennials or subshrubs, mostly 80–250 cm. **Stems** erect or sprawling to subscandent, hispidulous to coarsely short-pilose. **Petioles** 5–20 mm. **Leaf blades** (3-nerved) narrowly lanceolate to deltate-lanceolate or ovate-lanceolate, 3–10 × 1–4 cm, margins coarsely dentate to subentire. **Heads** usually 5–50+ in (terminal or lateral) corymbiform arrays. **Involucres** cylindric, (7–)8–10 mm. **Phyllaries** in 4–6(–8) series, apices of the inner appressed, rounded to truncate (sometimes slightly white-petaloid or expanded). **Corollas** purplish to light blue to nearly white or slightly pinkish. $2n = 40, 60, 70$.

Flowering mostly Oct–Dec, sometimes year round. Hammocks, thickets, pinelands, cypress-slash pine flats, canal banks, disturbed sites; 0–30 m; Fla., Tex.; Mexico; West Indies.

417. MIKANIA Willdenow, Sp. Pl. 3: 1742. 1803, name conserved • Climbing hempweed [For Josef Gottfried Mikan, 1743–1814, professor, University of Prague]

Walter C. Holmes

Vines (perennial, sometimes suffrutescent) [non-viney perennials, shrubs], to 300[–1500+] cm. **Stems** usually twining to scrambling (terete, striate, or [4-] 6-angled, sometimes winged), branched. **Leaves** cauline; opposite [whorled]; petiolate [sessile]; blades palmately 3[–7]-nerved [pinnately nerved], ± ovate or deltate-ovate to triangular [linear], margins entire or undulate to dentate or toothed to lobed, faces glabrous or puberulent to tomentose, often gland-dotted. **Heads** discoid, in corymbiform [paniculiform, racemiform, spiciform, thyrsiform] arrays. **Involucres** ± cylindric, [1–]2–3[–4] mm diam. (usually each subtended by 1 bractlet). **Phyllaries** persistent, 4 in ± 2 series (outer pair imbricate over inner pair), not notably nerved, lanceolate, linear, or oblong (bases often swollen), ± equal. **Receptacles** flat (glabrous), epaleate. **Florets** 4; corollas usually white, sometimes pink to rose or purplish, throats funnelform or campanulate, lobes 5, linear or triangular to deltate; styles: bases slightly, if at all, enlarged, glabrous, branches ± filiform [weakly clavate]. **Cypselae** ± prismatic, [4–]5[–10]-ribbed, glabrous or puberulent, sometimes gland-dotted; **pappi** persistent, of [20–]30–60 (white, buff, pinkish, or purplish) barbellulate to barbellate bristles in 1–2 series (distinct or basally connate). $x = 16$–20.

Species ca. 450 (3 in the flora): overwhelmingly neotropical (9 species in the Old World tropics), some temperate North American and South American.

All species of *Mikania* in the flora belong to *M.* sect. *Mikania* in the sense of W. C. Holmes (1996).

SELECTED REFERENCES Holmes, W. C. 1981. *Mikania* (Compositae) of the United States. Sida 9: 147–158. Holmes, W. C. 1993. The Genus *Mikania* (Compositae: Eupatorieae) in the Greater Antilles. Fort Worth. [Sida Bot. Misc. 9.] Holmes, W. C. 1996. A proposed sectional classification for *Mikania* (Eupatorieae). In: D. J. N. Hind et al., eds. 1996. Proceedings of the International Compositae Conference, Kew 1994. 2 vols. Kew. Vol. 1, pp. 621–626. Robinson, B. L. 1934. *Mikania scandens* and its near relatives. Contr. Gray Herb. 104: 55–71.

1. Stems 6-angled, gray-tomentulose to tomentose; leaf blades ovate to deltate; heads 7–10 mm; phyllaries 6–8 mm; corolla lobes linear; cypselae 3–4 mm .... 1. *Mikania cordifolia*
1. Stems terete to obscurely 6-angled, glabrate to densely pilose; leaf blades triangular to triangular-ovate; heads 4–7 mm; phyllaries 3–6 mm; corolla lobes triangular to deltate; cypselae 1.8–2.2 mm.
    2. Leaves (membranous): apices tapering; phyllaries linear-lanceolate, 5–6 mm; corollas usually pinkish to purplish, 3.5–4 mm; e United States .... 2. *Mikania scandens*
    2. Leaves (subcoriaceous to ± fleshy): apices acute to acuminate; phyllaries lanceolate to narrowly ovate, 3–4 mm; corollas white, ca. 3 mm; s Florida .... 3. *Mikania batatifolia*

1. Mikania cordifolia (Linnaeus f.) Willdenow, Sp. Pl. 3: 1746. 1803

Cacalia cordifolia Linnaeus f., Suppl. Pl., 351. 1782

Stems 6-angled, gray-tomentulose or tomentose; internodes 5–20 cm. **Petioles** 25–55 mm, densely pilose to tomentose. **Leaf blades** ovate to deltate, 5–10 × 3–8 cm, bases cordate, margins subentire to undulate-dentate, apices acute to acuminate, faces densely pilose to tomentose (abaxial paler than adaxial). **Arrays of heads** compound-corymbiform (terminal and lateral), 6 × 7+ cm. **Heads** 7–10 mm. **Phyllaries** substramineous, elliptic to narrowly ovate, 6–8 mm, apices acute to slightly rounded. **Corollas** white, 3.5–5 mm, lobes linear. **Cypselae** brown, 3–4 mm, glabrous or pubescent, sparsely gland-dotted; **pappi** of ca. 60 white, barbellate bristles 4–5 mm. 2*n* = 38.

Flowering Sep–Dec. Wet areas, woodlands, calcareous soils; 0–100 m; Ala., Fla., Ga., La., Miss., Tex.; Mexico; West Indies; Central America; South America.

Mikania cordifolia grows in all wet-tropical and subtropical America from northern Argentina to the lower Gulf Coastal Plain of the United States. It has the largest natural distribution of any species in the genus. In the tropics, *M. cordifolia* tends to be weedy, frequently occupying disturbed sites, usually in the lowlands. It is not weedy in the United States. In Louisiana, Mississippi, and Texas, *M. cordifolia* occurs in relatively open seeps and stream sides in beech (*Fagus grandiflora* Ehrhart) woods. It was collected in 1875 from the Navy Ballast Yard in Kargins Point, New Jersey (W. C. Holmes 1981); no further records for New Jersey are known.

2. Mikania scandens (Linnaeus) Willdenow, Sp. Pl. 3: 1743. 1803 [F]

Eupatorium scandens Linnaeus, Sp. Pl. 2: 836. 1753; *Mikania scandens* var. *pubescens* (Muhlenberg) Torrey & A. Gray

Stems obscurely 6-angled to terete, glabrate to densely pilose; internodes 8–15 cm. **Petioles** 20–50 mm, glabrous or puberulent. **Leaf blades** triangular to triangular-ovate, 3–15 × 2–11 cm, bases cordate to hastate, margins subentire to undulate, crenate, or dentate, apices acuminate (tips often caudate), faces puberulent. **Arrays of heads** dense, corymbiform, 12–15 × 12 cm. **Heads** 6–7 mm. **Phyllaries** green or pinkish to purplish, linear to lanceolate, 5–6 mm, apices acuminate (faces glabrous or puberulent). **Corollas** usually pinkish to purplish, sometimes white, 3–5.4 mm, sparsely gland-dotted, lobes triangular to deltate. **Cypselae** dark brown to blackish, 1.8–2.2 mm, densely gland-dotted; **pappi** of 30–37 white or pinkish to purplish bristles 4–4.5 mm. 2*n* = 38.

Flowering Jun–Dec. Wet, open areas along streams, seeps, springs, margins of lakes, swamps; 0–500 m; Ala., Ark., Conn., Del., D.C., Fla., Ga., Ill., Ind., Ky., La., Maine, Md., Mass., Mich., Miss., Mo., N.H., N.J., N.Y., N.C., Okla., Pa., R.I., S.C., Tenn., Tex., Va.; Mexico; West Indies (Bahamas).

The name *Mikania scandens* was once used to refer to most of the slender twiners with sagittate, hastate, or cordate leaf bases and corymbiform capitulescences in tropical and temperate America. As a result of work of B. L. Robinson (1934), the name is now used to refer to plants distributed primarily in eastern United States.

Mikania scandens was reported as occurring in Ontario, Canada (M. L. Fernald 1950; J. A. Steyermark 1963); it has been deleted from the flora of Canada (H. J. Scoggan 1978–1979, part 4). The Canadian reports were seemingly based upon misdeterminations and/or "too loose an application of that name with respect to present political boundaries." Records of *M. scandens* from along the Ohio River, Hamilton County, Ohio, are apparently based on non-persistent introductions; the species is apparently extirpated from Indiana, Maine, and Michigan.

3. **Mikania batatifolia** de Candolle in A. P. de Candolle and A. L. P. P. de Candolle, Prodr. 5: 197. 1836 (as batataefolia)

Stems obscurely 6-angled, sometimes winged on angles, often densely glandular, glabrous or puberulent; internodes 3.5–13 cm. **Petioles** 10–40 mm, glabrous, glandular. **Leaf blades** deltate-ovate, 1.5–6 × 1–5 cm, (subcoriaceous to somewhat fleshy) bases cordate to subcordate, margins usually hastately dentate to ± lobed (lobes divergent), apices acute to acuminate, faces glabrous (and gland-dotted). **Arrays of heads** corymbiform, 2–4 × 2–6 cm. **Heads** 4–6 mm. **Phyllaries** lanceolate to narrowly ovate, ca. 3.5 mm, apices acuminate (abaxial faces puberulent). **Corollas** white, ca. 3 mm, gland-dotted, lobes deltate. **Cypselae** brown, 1.5–2 mm, densely gland-dotted; **pappi** of ca. 40 white, barbellate bristles ca. 3 mm.

Flowering year round. Woodlands, savannas, salt marshes, swamps, usually in oölite or coral soils; Fla.; West Indies (Bahamas, Cuba).

J. K. Small (1933), B. L. Robinson (1934), R. W. Long and O. Lakela (1971), and W. C. Holmes (1981, 1993) recognized *Mikania batatifolia*. Hn. Alain (1962) referred *M. batatifolia* to *M. micrantha* Kunth, a common, polymorphic taxon of humid American tropics. A. Cronquist (1980) merged *M. batatifolia* with *M. scandens*. In reexamining the members of the *M. scandens* complex, which includes *M. batatifolia* (Robinson), it is apparent that *M. batatifolia* is distinct. Differences with *M. scandens*, including chemical evidence, were cited by Holmes (1981). For additional information, see Holmes (1981, 1993).

418. **AGERATINA** Spach, Hist. Nat. Vég. 10: 286. 1841 • Snakeroot [Generic name *Ageratum* and Latin *-ina*, diminutive]

Guy L. Nesom

Perennials, subshrubs, or shrubs [trees], mostly 20–220 cm. **Stems** usually erect, rarely scandent, sparsely to densely branched. **Leaves** cauline; mostly opposite (distal sometimes alternate); petiolate; blades 3(–5)-nerved from bases, usually deltate, lanceolate, ovate, rhombic, or triangular, sometimes orbiculate, margins entire, crenate, dentate, or serrate, faces glabrous or hispidulous, pilose, or puberulent, sometimes gland-dotted (*A. occidentalis, A. adenophora*). **Heads** discoid, usually in compact, (terminal and axillary) corymbiform arrays, sometimes borne singly. **Involucres** campanulate, 3–6 mm diam. **Phyllaries** persistent, 8–30 in 2(–3) series, 0- or 2-nerved, lanceolate to linear, ± equal (herbaceous). **Florets** 10–60; corollas white or lavender, throats obconic to campanulate (lengths 1.5–2 times diams.); styles: bases sometimes enlarged, glabrous, branches linear, seldom distally dilated. **Receptacles** convex (glabrous or hairy), epaleate. **Cypselae** prismatic or ± fusiform, usually 5-ribbed, scabrellous and/or gland-dotted; **pappi** usually persistent, sometimes fragile, rarely falling, of 5–40, barbellulate bristles in 1 series. *x* = 17.

Species ca. 250 (14 in the flora): North America, Mexico, Central America, Andean South America.

SELECTED REFERENCE Clewell, A. F. and J. W. Wooten. 1971. A revision of *Ageratina* (Compositae: Eupatorieae) from eastern North America. Brittonia 23: 123–143

1. Shrubs.
    2. Petioles (2–)3–5 mm; leaf blades 1–2 cm, margins entire or shallowly crenate, faces
        gland-dotted . . . . 10. *Ageratina wrightii*
    2. Petioles 3–10(–15) mm; leaf blades (2–)3–5(–7) cm, margins coarsely crenate, faces
        not gland-dotted . . . . 9. *Ageratina havanensis*
1. Perennials or subshrubs.
    3. Leaves alternate on at least distal ¼–½ of stems.
        4. Heads usually borne singly (rarely 2s or 3s); involucres 11–12 mm . . . . 12. *Ageratina shastensis*
        4. Heads usually 5–10 (axillary clusters usually forming elongate or broad aggregates);
            involucres 2.5–3.5(–4) mm.
            5. Leaves alternate on distal ¼–½ of stems; involucres 3–3.5(–4) mm; corollas
                pink, bluish, or white tinged with purple (not orange-veined); cypselae sessile-
                glandular . . . . 11. *Ageratina occidentalis*
            5. Leaves alternate (from bases to apices of stems); involucres 2.5–3 mm; corollas
                white (prominently orange-veined); cypselae eglandular . . . . 14. *Ageratina thyrsiflora*
    3. Leaves opposite.
        6. Peduncles densely stipitate-glandular . . . . 13. *Ageratina adenophora*
        6. Peduncles puberulent, glabrous, or glabrescent (not glandular).
            7. Flowering in spring; leaves usually narrowly lanceolate, 0.5–1.5 cm wide,
                apically long-acuminate; involucres 2.5–3 mm; cypselae glabrous . . . . . 8. *Ageratina paupercula*
            7. Flowering in fall; leaves ovate to lanceolate-ovate, triangular, or lanceolate,
                (0.5–)1.5–9 cm wide, apically acute to acuminate; involucres (in *A. jucunda*
                2.5–)3.5–7 mm; cypselae usually hairy (glabrous in *A. altissima*).
                8. Leaves mostly sessile; heads in open, loose arrays, peduncles 10–60 mm
                    . . . . 6. *Ageratina lemmonii*
                8. Leaves distinctly petiolate; heads in compact clusters, ultimate peduncles
                    1–15(–20) mm.
                    9. Petioles 1–22 mm (distal leaves greatly reduced in size well proximal
                        to heads); leaf blades 2–7(–9) × 1.5–4 cm (relatively thick).
                        10. Petioles 1–8(–12) mm; leaf margins crenate or less commonly
                            crenate-serrate to dentate or subentire . . . . 4. *Ageratina aromatica*
                        10. Petioles 7–15(–22) mm; leaf margins coarsely serrate or incised or
                            less commonly crenate to subentire . . . . 5. *Ageratina jucunda*
                    9. Petioles (5–)10–70 or 2–20 (in *A. rothrockii*) mm; leaf blades 2–11
                        (–13) × 1.5–9 cm (relatively thin).
                        11. Leaves (yellow-green or grayish yellow-green): blades triangular
                            to lanceolate-ovate or ovate, 2–5(–7) × 1.5–3.5(–4.5) cm; phyllaries
                            usually granular-puberulent . . . . 7. *Ageratina herbacea*
                        11. Leaves (green, rarely yellowish): blades lanceolate to lanceolate-
                            ovate, 4–11(–13) × 2.5–9 cm; phyllaries glabrous, villous, or
                            villous-puberulent.
                            12. Stems glabrous; leaf blades broadly deltate-ovate (thin,
                                delicate), apices obtuse; corolla lobes glabrous or sparsely
                                puberulent . . . . 3. *Ageratina luciae-brauniae*
                            12. Stems puberulent; leaf blades deltate-ovate to ovate or broadly
                                lanceolate, apices acute to acuminate; corolla lobes short-
                                villous.
                                13. Peduncles 1–5 mm; involucres 4–5 mm; cypselae glabrous;
                                    e United States and Canada . . . . 1. *Ageratina altissima*
                                13. Peduncles 5–12(–20) mm; involucres 5–7 mm; cypselae
                                    sparsely and finely strigose-hirsute; Arizona, New Mexico,
                                    sw Texas . . . . 2. *Ageratina rothrockii*

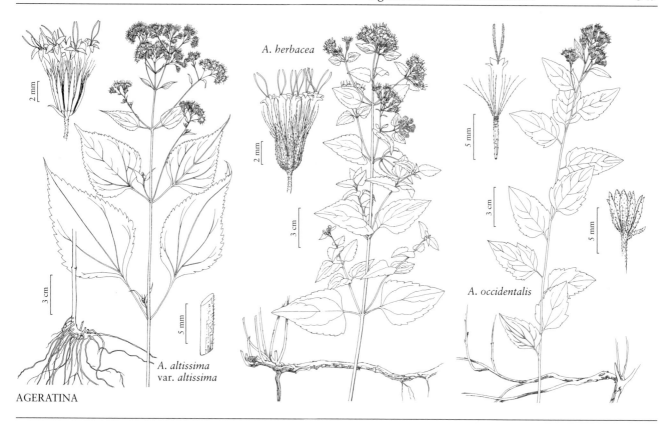

A. herbacea

A. occidentalis

A. altissima
var. altissima

AGERATINA

1. **Ageratina altissima** (Linnaeus) R. M. King & H. Robinson, Phytologia 19: 212. 1970 • White snakeroot E F

Ageratum altissimum Linnaeus, Sp. Pl. 2: 839. 1753, not *Eupatorium altissimum* Linnaeus 1753

Perennials, (30–)50–80(–120) cm (bases usually fibrous-rooted crowns, sometimes rhizomatous). **Stems** ascending to erect, sometimes semiscandent, puberulent (hairs minute, crisped). **Leaves** opposite; petioles (5–)10–30(–50) mm; blades usually deltate-ovate to ovate or broadly lanceolate, sometimes ovate-lanceolate, 4–11(–13) × 2.5–8(–9) cm, bases usually rounded to truncate or obtuse, sometimes cordate, margins coarsely and doubly incised-serrate, apices usually acuminate. **Heads** clustered. **Peduncles** 1–5 mm, puberulent. **Involucres** 4–5 mm. **Phyllaries:** apices acute, abaxial faces glabrous or sparsely and finely villous. **Corollas** white, lobes sparsely short-villous. **Cypselae** glabrous.

Varieties 2 (2 in the flora): North America.

1. Phyllaries 3–5 mm, apices not cuspidate . . . . .
. . . . 1a. *Ageratina altissima* var. *altissima*
1. Phyllaries 4–7 mm, apices cuspidate to acuminate
. . . . 1b. *Ageratina altissima* var. *roanensis*

1a. **Ageratina altissima** (Linnaeus) R. M. King & H. Robinson var. **altissima** • Common white snakeroot, eupatoire rugueuse E F

Ageratina altissima var. *angustata* (A. Gray) Clewell & Wooten; *Eupatorium rugosum* Houttuyn; *E. rugosum* var. *chlorolepis* Fernald; *E. rugosum* var. *tomentellum* (B. L. Robinson) S. F. Blake; *E. urticifolium* Reichard

Phyllaries 3–5 mm, apices not cuspidate.

Flowering Jul–Oct(–Nov). Moist forests, cove forests; 10–800 m; N.B., N.S., Ont., Que., Sask.; Ala., Ark., Conn., Del., Fla., Ill., Ind., Iowa, Kans., Ky., La., Maine, Md., Mass., Mich., Minn., Miss., Mo., Nebr., N.H., N.J., N.Y., N.C., N.Dak., Ohio, Okla., Pa., R.I., S.C., S.Dak., Tenn., Tex., Vt., W.Va., Wis.

Plants with narrow leaves, generally in the southwest part of the range of *Ageratina altissima*, have been recognized as var. *angustata* and were so mapped by A. F. Clewell and J. W. Wooten (1971), who indicated that all var. *angustata* occurs west of the Mississippi River and that this taxon was completely congruent in distribution with var. *altissima*. The present treatment confirms the westward tendency toward size reduction and observes that narrow-leaved plants occur widely through the southeast United States (including Arkansas, Kentucky, Louisiana, North Carolina, Oklahoma, Tennessee, and Texas). The transition is gradual and the region of intergradation is wide. In Texas, where the leaves mostly are narrow, plants with broad, cordate leaves are scattered through the range.

1b. Ageratina altissima (Linnaeus) R. M. King & H. Robinson var. **roanensis** (Small) Clewell & Wooten, Brittonia 23: 138. 1971 • Appalachian white snakeroot [E]

Eupatorium roanense Small, Man. S.E. Fl., 1326. 1933 (as roanensis); *E. rugosum* Houttuyn var. *roanense* (Small) Fernald

Phyllaries 4–7 mm, apices cuspidate to acuminate.

Flowering Jul–Oct. Moist forests, northern red oak, oak-chestnut, cove forests; 700–1500 m; Ala., Ga., Ky., N.C., S.C., Tenn., Va., W.Va.

2. Ageratina rothrockii (A. Gray) R. M. King & H. Robinson, Phytologia 19: 216. 1970 • Rothrock's snakeroot

Eupatorium rothrockii A. Gray in A. Gray et al., Syn. Fl. N. Amer. 1(2): 102. 1884

Perennials, (20–)40–70(–150) cm (slender, fibrous-rooted crowns, with slender rhizomes). **Stems** ascending to erect, puberulent to glabrate. **Leaves** opposite; petioles 2–20 mm; blades (3–5-nerved) lanceolate to lanceolate-ovate, mostly 3–6 × (1.5–)2–3 cm, bases obtuse to truncate, margins serrate to crenate, apices acute to acuminate, sparsely puberulent abaxially, mostly along nerves. **Heads** clustered. **Peduncles** 5–12(–20) mm, puberulent. **Involucres** 5–7 mm. **Phyllaries:** apices acute, abaxial faces glabrous or glabrescent, eglandular. **Corollas** white, lobes short-villous. **Cypselae** sparsely and finely strigose-hirsute. $2n = 85$ [ca. 100, fide A. M. Powell on label].

Flowering (Jul–)Aug–Oct. Rocky slopes and ledges, in oak-juniper, pine-oak, pine, aspen, and spruce-fir woodland; 1700–2400 m; Ariz., N.Mex., Tex.; Mexico (Chihuahua, Coahuila, Durango, Sonora).

Ageratina rothrockii is similar to *A. altissima*, probably its western vicariant, and the two perhaps would be justifiably treated as conspecific.

3. Ageratina luciae-brauniae (Fernald) R. M. King & H. Robinson, Phytologia 19: 215. 1970 • Rockhouse white snakeroot [E]

Eupatorium luciae-brauniae Fernald, Rhodora 44: 463. 1942, based on *E. deltoides* E. L. Braun, Rhodora 42: 50. 1940, not Jacquin 1798

Perennials, 30–60 cm. **Stems** erect, glabrous. **Leaves** opposite; petioles 25–70 mm; blades broadly ovate-deltate, 4–8 × 5–9 cm, (thin, delicate) bases truncate to subcordate, margins coarsely dentate, apices acute to acuminate, abaxial faces glabrous or sparsely puberulent. **Heads** clustered. **Peduncles** 1–3 mm, glabrous or sparsely puberulent. **Involucres** 3.5–4 mm. **Phyllaries:** apices acuminate, abaxial faces glabrous or sparsely puberulent. **Corollas** white, lobes glabrous or sparsely puberulent. **Cypselae** sparsely and evenly hirtellous. $2n = 34$.

Flowering Aug–Oct. Under overhanging sandstone (Pottsville formation) cliffs and ledges; 400–500 m; Ky., Tenn.

Ageratina luciae-brauniae was treated by A. F. Clewell and J. W. Wooten (1971) as a synonym of *A. altissima* and regarded by them as "bizarre plants showing extreme signs of etiolation from growing under limestone ledges" (p. 134). B. E. Wofford (1976) observed that greenhouse transplants of both species maintained distinctions that provide rationale for maintaining *A. luciae-brauniae* at specific rank.

Ageratina luciae-brauniae is in the Center for Plant Conservation's National Collection of Endangered Plants.

SELECTED REFERENCE Wofford, B. E. 1976. The taxonomic status of *Ageratina luciae-brauniae* (Fern.) King & H. Robins. Phytologia 33: 369–371.

4. Ageratina aromatica (Linnaeus) Spach, Hist. Nat. Vég. 10: 286. 1841 • Small-leaved white snakeroot [E]

Eupatorium aromaticum Linnaeus, Sp. Pl. 2: 839. 1753; *E. latidens* Small

Perennials, 30–80(–100) cm. **Stems** erect, villous-puberulent. **Leaves** opposite; petioles 1–8(–12) mm; blades narrowly to broadly deltate to nearly ovate or lanceolate, 2–7(–9) × 1.5–4 cm, (usually subcoriaceous) bases rounded or truncate to barely cuneate or subcordate, margins usually crenate, sometimes crenate-serrate to dentate or subentire, apices acute to obtuse, faces minutely pilose. **Heads** clustered. **Peduncles** 2–9 mm, densely and closely puberulent. **Involucres** 3.5–5 mm. **Phyllaries:** apices acute, abaxial faces puberulent to villous-puberulent. **Corollas** white, lobes sparsely villous. **Cypselae** usually glabrous or sparsely puberulent (near apices), rarely hirtellous on angles. $2n = 34$.

Flowering late Aug–Oct(–Nov). Sandy soils, burned pinelands, turkey oak sand ridges, pine-oak and oak-hickory upland woods, old fields, roadsides, fencerows, moist sites; 100–900 m; Ala., Conn., Del., Fla., Ky., La., Md., Mass., Miss., N.J., N.Y., N.C., Ohio, Pa., R.I., S.C., Tenn., Va., W.Va.

Intergrades (probable hybrids) between *Ageratina aromatica* and *A. altissima* were identified by A. F. Clewell and J. W. Wooten (1971) over a broad area of their sympatry. They also found intergrades between *A. aromatica* and *A. jucunda* where their ranges meet.

5. **Ageratina jucunda** (Greene) Clewell & Wooten, Brittonia 23: 142. 1971 • Hammock snakeroot E

Eupatorium jucundum Greene, Pittonia 3: 180. 1897, based on *E. aromaticum* Linnaeus var. *incisum* A. Gray in A. Gray et al., Syn. Fl. N. Amer. 1(2): 101. 1884, not *E. incisum* Richard 1792

Perennials, 40–80(–100) cm. **Stems** erect, lax, minutely pilose. **Leaves** opposite; petioles 7–15 (–22) mm; blades narrowly deltate to rhombic, 2–6(–7) × 1.5–4 cm, (usually subcoriaceous) bases usually cuneate, sometimes truncate to slightly subcordate, margins usually coarsely serrate or incised, sometimes crenate to subentire, apices acute to acuminate, abaxial faces glabrous or hairy on veins. **Heads** clustered. **Peduncles** 2–10 mm, sparsely puberulent. **Involucres** 2.5–4 mm. **Phyllaries:** apices acute, abaxial faces puberulent to villous-puberulent. **Corollas** white, lobes glabrous or sparsely short-hirtellous. **Cypselae** usually finely hirtellous-strigose on distal ¹/₃, sometimes glabrous. $2n = 34$.

Flowering (Sep–)Oct–Dec(–Jan). Sand pine scrub, longleaf pine-turkey oak sand ridges, pine-palmetto, live-oak woods, hammocks, dunes, roadsides, old fields, stream banks, dry flatwoods; 0–50 m; Fla., Ga.

6. **Ageratina lemmonii** (B. L. Robinson) R. M. King & H. Robinson, Phytologia 19: 223. 1970 • Lemmon's snakeroot

Eupatorium lemmonii B. L. Robinson, Proc. Amer. Acad. Arts 27: 171. 1892 (as lemmoni)

Perennials, 20–40(–70) cm (densely fibrous-rooted crowns). **Stems** (commonly purple) erect, puberulous (hairs usually with colored crosswalls). **Leaves** opposite; petioles usually 0, sometimes 1–2 mm; blades ovate-lanceolate, 2–4.5 × 0.5–

2.5 cm, margins shallowly serrate, abaxial faces glabrous or glabrate, gland-dotted. **Heads** in loose, open arrays. **Peduncles** 10–60 mm, puberulent. **Involucres** 4–5.5 mm. **Phyllaries:** (narrowly oblong-lanceolate) apices acute, abaxial faces sparsely hairy to glabrate. **Corollas** white, lobes densely hispid-villous. **Cypselae** sparsely and finely hispidulous. $2n = 34$.

Flowering Aug–Oct. Rocky slopes, mostly in pine-oak woodlands; 1800–2800[–3300, Mexico] m; Ariz.; Mexico.

7. **Ageratina herbacea** (A. Gray) R. M. King & H. Robinson, Phytologia 19: 222. 1970 • Fragrant snakeroot F

Eupatorium ageratifolium de Candolle var. *herbaceum* A. Gray, Smithsonian Contr. Knowl. 5(6): 74. 1853; *E. herbaceum* (A. Gray) Greene

Perennials or subshrubs, (20–)30–60(–80) cm (woody crowns and woody rhizomes). **Stems** erect (brittle), minutely puberulent. **Leaves** opposite; petioles 10–25 mm; blades triangular to lanceolate-ovate or ovate, 2–5(–7) × 1.5–3.5(–4.5) cm, bases truncate to shallowly cordate, margins dentate to serrate-dentate, abaxial faces sparsely hispidulous to glabrate, eglandular. **Heads** clustered. **Peduncles** 4–15 mm, puberulent. **Involucres** 4–5 mm. **Phyllaries:** apices acute, abaxial faces granular-puberulent. **Corollas** white, glabrous. **Cypselae** finely strigose-hispidulous. $2n = 34$.

Flowering (Jul–)Aug–Oct. Pine, pine-oak, juniper, and pinyon-juniper woodlands, rocks along streams, slopes, ridges, washes; 1400–2700(–2900) m; Ariz., Calif., Colo., Nev., N.Mex., Tex., Utah; Mexico (Baja California, Chihuahua, Coahuila, Sonora).

Ageratina herbacea is recognized by the distinctive color of its usually yellow-green, sometimes grayish, leaves, granular-puberulent involucres (with minute, thickened, eglandular hairs), and woody rhizomes.

8. **Ageratina paupercula** (A. Gray) R. M. King & H. Robinson, Phytologia 19: 215. 1970 • Santa Rita snakeroot

Eupatorium pauperculum A. Gray, Proc. Amer. Acad. Arts 17: 205. 1882

Perennials or subshrubs, to 100 cm. **Stems** erect, puberulent to glabrous. **Leaves** opposite; petioles 3–15 mm; blades narrowly to broadly lanceolate, (2–)3–7 × 0.5–1.5 cm, margins coarsely and remotely serrate, apices long-acuminate, abaxial faces sparsely pubescent. **Heads** clustered. **Peduncles** 2–6 mm, puberulent. **Involucres** 2.5–3 mm. **Phyllaries:** apices acute, abaxial faces puberulent-hispidulous. **Corollas** white, lobes sparsely hispid-villous. **Cypselae** glabrous.

Flowering Mar–May. Rocky slopes, crevices, gravelly, sandy streambeds, pine-oak woodlands; 1000–1800 m; Ariz.; Mexico.

Ageratina paupercula is recognized by its early spring flowering, relatively small and usually narrowly lanceolate and acuminate leaves, relatively small heads in clusters, and relatively small (1.2–1.4 mm versus mostly 2–3 mm in other species), glabrous cypselae.

9. **Ageratina havanensis** (Kunth) R. M. King & H. Robinson, Phytologia 19: 222. 1970 • Havana snakeroot

Eupatorium havanense Kunth in A. von Humboldt et al., Nov. Gen. Sp. 4(fol.): 100. 1818; 4(qto.): 128. 1820

Shrubs [trees], (30–)69–150 (–200) cm. **Stems** erect (brittle), puberulent to glabrous. **Leaves** persistent, opposite; petioles 3–10(–15) mm; blades deltate to broadly ovate or somewhat hastate, (2–)3–5(–8) × 2–5 cm, bases truncate to cuneate, margins dentate, apices acute, faces glabrous or nearly so, eglandular. **Heads** clustered. **Peduncles** 2–14 mm, minutely puberulent. **Involucres** 4–6 mm. **Phyllaries:** apices acute, abaxial faces glabrous or nearly so. **Corollas** white to slightly pinkish, glabrous. **Cypselae** hispid. $2n = 34$.

Flowering mainly (Sep–)Oct–Nov(–Dec), also Apr–Jul. Bluffs, limestone outcrops and slopes, ledges along streams, often in oak-juniper woodlands; 100–900 m; Tex.; Mexico; West Indies (Cuba).

Ageratina havanensis apparently is the only species of the genus in the flora area with evergreen-persistent leaves.

10. **Ageratina wrightii** (A. Gray) R. M. King & H. Robinson, Phytologia 19: 228. 1970 • Wright's snakeroot

Eupatorium wrightii A. Gray, Smithsonian Contr. Knowl. 3(5): 87. 1852

Shrubs, 50–150 cm. **Stems** erect, puberulous. **Leaves** opposite proximally, alternate on distal ¹⁄₃–¹⁄₂ of stem; petioles narrowly winged, 3–5(–8) mm; blades ovate to deltate-ovate, mostly 1–2 × 0.5–1.5(–2) cm, bases truncate to cuneate or attenuate, margins entire or shallowly crenate, apices acute to blunt or obtuse, abaxial faces gland-dotted. **Heads** clustered.

Peduncles 1–3 mm, puberulent. **Involucres** 3.5–4.5 mm. **Phyllaries:** apices acute, abaxial faces puberulent. **Corollas** pinkish white to pink, glabrous. **Cypselae** sparsely hispidulous.

Flowering (Jul–)Aug–Oct(–Nov). Limestone slopes, ledges, alluvium, in brushy vegetation; 1300–2200 m; Ariz., N.Mex., Tex.; Mexico.

11. **Ageratina occidentalis**   (Hooker) R. M. King & H. Robinson, Phytologia 19: 224. 1970 • Western snakeroot E F

Eupatorium occidentale Hooker, Fl. Bor.-Amer. 1: 305. 1833

Perennials or subshrubs, 15–70 cm (caudices woody, rhizomatous). **Stems** (green or purple) erect or ascending, puberulent. **Leaves** opposite proximally, alternate on distal ¹⁄₄–¹⁄₂ of stems; petioles 5–12 mm; blades triangular to ovate, 2.5–5 × 1.7–4 cm, bases truncate to cuneate, margins serrate, apices acute, abaxial faces gland-dotted. **Heads** clustered. **Peduncles** 2–5 mm, minutely puberulent. **Involucres** 3–3.5(–4) mm. **Phyllaries:** apices acute, abaxial faces viscid-puberulent and/or sessile-glandular. **Corollas** pink, bluish, or white tinged with purple, lobes glabrous or glabrate. **Cypselae** sessile-glandular. $2n = 34$.

Flowering (Jun–)Jul–Sep. Crevices, outcrops, rocky slopes, ridges, talus, gravelly and sandy stream bars, mixed hardwood-conifer woodlands, aspen, open and brushy vegetation; (40–50)900–2800(–3200) m; Calif., Idaho, Mont., Nev., Oreg., Utah, Wash.

Ageratina occidentalis is the only species of the genus in the flora area with sessile-glandular cypselae; peduncles also may be sessile-glandular. Its identity also can be confirmed among flora area species by its relatively long (5–6 mm) corollas.

12. **Ageratina shastensis** (D. W. Taylor & Stebbins) R. M. King & H. Robinson, Phytologia 45: 464. 1980 • Mt. Shasta snakeroot E

Eupatorium shastense D. W. Taylor & Stebbins, Madroño 25: 218, fig. 1. 1978

Perennials, 15–45 cm (bases woody, enlarged, occasionally rhizomatous). **Stems** erect (clustered from bases), puberulent to pubescent (often with some glandular hairs distally). **Leaves** opposite proximally, alternate on at least distal ¹⁄₂ of stems; petioles 4–6 mm; blades (venation raised-reticulate) orbiculate (juvenile) or deltate-ovate, 1.5–3

× 1–2 cm, (subcoriaceous) bases obtuse to truncate, margins entire or coarsely serrate to dentate, apices acute to acuminate, abaxial faces ± gland-dotted and/or stipitate-glandular. **Heads** mostly borne singly (each often subtended by a leaflike bract). **Peduncles** (0–)2–15 mm, puberulent and stipitate-glandular. **Involucres** 10–11 mm. **Phyllaries:** apices acute, abaxial faces sparsely puberulent. **Corollas** white, glabrous. **Cypselae** hispidulous. $2n = 34$.

Flowering Jun–Oct. Cracks and crevices of nearly vertical limestone cliffs; 400–1800 m; Calif.

Ageratina shastensis is recognized by relatively large solitary heads and coarsely serrate, subcoriaceous leaves alternate on distal parts of stems.

13. Ageratina adenophora (Sprengel) R. M. King & H. Robinson, Phytologia 19: 211. 1970 • Crofton weed, sticky snakeroot [I]

Eupatorium adenophorum Sprengel, Syst. Veg. 3: 420. 1826, based on *E. glandulosum* Kunth in A. von Humboldt et al., Nov. Gen. Sp. 4(fol.): 96, plate 346. 1818; 4(qto.): 122, plate 346. 1820, not Michaux 1803

Subshrubs, 50–220 cm. **Stems** (usually purplish when young) erect, stipitate-glandular. **Leaves** opposite; petioles 10–25 mm; blades (abaxially purple) ovate-lanceolate or ovate-deltate to lanceolate-ovate, (1.5–)2.5–5.5(–8) × 1.5–4(–6) cm, bases cuneate to obtuse or nearly truncate, margins serrate, apices acute to acuminate, abaxial faces stipitate- to sessile-glandular. **Heads** clustered. **Peduncles** 5–12 mm, densely stipitate-glandular and sometimes also sparsely viscid-puberulent. **Involucres** 3.5–4 mm. **Phyllaries:** apices acute, abaxial faces stipitate-glandular. **Corollas** white, pink-tinged, lobes sparsely hispidulous. **Cypselae** glabrous. $2n = 51$.

Flowering Mar–Aug(–Sep). Stream margins, ditches, road embankments, hillsides; 400–900 m; intoduced; Calif.; Mexico; also introduced in Europe, Pacific Islands (New Zealand).

14. Ageratina thyrsiflora (Greene) R. M. King & H. Robinson, Phytologia 19: 227. 1970
• Congested snakeroot

Kyrstenia thyrsiflora Greene, Leafl. Bot. Observ. Crit. 1: 9. 1903

Perennials or subshrubs, 50–100 cm. **Stems** erect (usually forming compact 'bushlets'), finely and evenly puberulent (hairs minute, bent). **Leaves** usually alternate, sometimes subopposite (densely overlapping internodes); petioles 3–20 mm; blades ovate-lanceolate to triangular, 2.5–6.5 × 1–2.5(–3) cm, bases obtuse to cuneate, margins shallowly and coarsely crenate to serrate to subentire, apices rounded-obtuse, abaxial faces hirtellous. **Heads** clustered (in dense, terminal aggregates). **Peduncles** 3–8 mm, puberulent. **Involucres** 2.5–3 mm. **Phyllaries:** apices acute to obtuse (dark orange-veined), abaxial faces. **Corollas** white (orange-veined), glabrous. **Cypselae** hispid.

Flowering Sep–Nov. Rocky sites, oak woodland; 1000–2200 m; Ariz.; Mexico (Chihuahua, Durango, Jalisco, Sinaloa, Sonora).

Ageratina thyrsiflora is known from the United States from a single collection dated 1929 from "near Nogales" in Santa Cruz County. It is recognized by its strict, unbranched or few-branched stems with alternate, densely arranged leaves, relatively small heads densely clustered in terminal aggregates, orange-veined phyllaries and corollas, and closely puberulent stems and petioles (hairs minute, sharply upwardly bent).

Literature Cited

Robert W. Kiger, Editor

This is a consolidated list of all works cited in volumes 19, 20, and 21, whether as selected references, in text, or in nomenclatural contexts. In citations of articles, both here and in the taxonomic treatments, and also in nomenclatural citations, the titles of serials are rendered in the forms recommended in G. D. R. Bridson and E. R. Smith (1991). When those forms are abbreviated, as most are, cross references to the corresponding full serial titles are interpolated here alphabetically by abbreviated form. In nomenclatural citations (only), book titles are rendered in the abbreviated forms recommended in F. A. Stafleu and R. S. Cowan (1976–1988) and F. A. Stafleu and E. A. Mennega (1992+). Here, those abbreviated forms are indicated parenthetically following the full citations of the corresponding works, and cross references to the full citations are interpolated in the list alphabetically by abbreviated form. Two or more works published in the same year by the same author or group of coauthors will be distinguished uniquely and consistently throughout all volumes of *Flora of North America* by lower-case letters (b, c, d, ...) suffixed to the date for the second and subsequent works in the set. The suffixes are assigned in order of editorial encounter and do not reflect chronological sequence of publication. The first work by any particular author or group from any given year carries the implicit date suffix "a"; thus, the sequence of explicit suffixes begins with "b". Works missing from any suffixed sequence here are ones cited elsewhere in the *Flora* that are not pertinent in these volumes.

Abh. Königl. Ges. Wiss. Göttingen = Abhandlungen der Königlichen Gesellschaft der Wissenschaften zu Göttingen.

Abrahamson, W. G. and A. E. Weis. 1997. Evolutionary Ecology Across Three Trophic Levels: Goldenrods, Gallmakers, and Natural Enemies. Princeton.

Abrams, L. and R. S. Ferris. 1923–1960. Illustrated Flora of the Pacific States: Washington, Oregon, and California. 4 vols. Stanford.

Account Exped. Pittsburgh—See: E. James 1823

Ackerman, T. L., J. Bair, and A. Tiehm. 2003. A flora of the Desert National Wildlife Refuge, Nevada. Mentzelia 7: 1–90.

Acta Bot. Fenn. = Acta Botanica Fennica.

Acta Bot. Indica = Acta Botanica Indica.

Acta Phytotax. Geobot. = Acta Phytotaxonomica et Geobotanica. [Shokubutsu Bunrui Chiri.]

Actes Soc. Hist. Nat. Paris = Actes de la Société d'Histoire Naturelle de Paris.

Adanson, M. 1763[–1764]. Familles des Plantes. 2 vols. Paris. [Vol. 1, 1764; vol. 2, 1763.] (Fam. Pl.)

Advances Cladist. = Advances in Cladistics.

Agron. J. = Agronomy Journal.

Aiton, W. 1789. Hortus Kewensis; or, a Catalogue of the Plants Cultivated in the Royal Botanic Garden at Kew. 3 vols. London. (Hort. Kew.)

Alain (Hermano). 1962. Flora de Cuba. Vol. 5. Rubiales - Valerianales - Cucurbitales - Campanulales - Asterales. Rio Piedras.

Allen, G. A. 1984. Morphological and cytological variation within the western North American *Aster occidentalis* complex (Asteraceae). Syst. Bot. 9: 175–191.

Allen, G. A. 1985. The hybrid origin of *Aster ascendens* (Asteraceae). Amer. J. Bot. 72: 268–277.

Allen, G. A. 1986. Amphiploid origin of two endemic races of *Aster* (Asteraceae) in southern California. Amer. J. Bot. 73: 330–335.

Allen, G. A. 1993. *Aster*. In: J. C. Hickman, ed. 1993. The Jepson Manual. Higher Plants of California. Berkeley, Los Angeles, and London. Pp. 205–209.

Allen, G. A., M. L. Dean, and K. L. Chambers. 1983. Hybridization studies in the *Aster occidentalis* (Asteraceae) polyploid complex of western North America. Brittonia 35: 353–361.

Allg. Gartenzeitung = Allgemeine Gartenzeitung.

Allioni, C. 1785. Flora Pedemontana sive Enumeratio Methodica Stirpium Indigenarum Pedemontii. 3 vols. Turin. (Fl. Pedem.)

Amer. Bot. Fl.—See: A. Wood 1870

Amer. J. Bot. = American Journal of Botany.

Amer. J. Sci. Arts = American Journal of Science, and Arts.

Amer. Midl. Naturalist = American Midland Naturalist; Devoted to Natural History, Primarily That of the Prairie States.

Amer. Naturalist = American Naturalist.... [Subtitle varies.]

Amer. Sci. = American Scientist....

Anais Inst. Vinho Porto = Anais do Instituto do Vinho do Porto.

Anal. Fam. Pl.—See: B. C. J. Dumortier 1829

Anales Ci. Nat. = Anales de Ciencias Naturales.

Anderberg, A. A. 1991. Taxonomy and phylogeny of the tribe Gnaphalieae (Asteraceae). Opera Bot. 104: 5–195.

Anderberg, A. A. 1991b. Taxonomy and phylogeny of the tribe Plucheeae (Asteraceae). Pl. Syst. Evol. 176: 145–177.

Anderberg, A. A. 1994. Tribe Inuleae. In: K. Bremer. 1994. Asteraceae: Cladistics & Classification. Portland. Pp. 273–291.

Anderberg, A. A. 1994b. Tribe Plucheeae. In: K. Bremer. 1994. Asteraceae: Cladistics & Classification. Portland. Pp. 292–303.

Anderson, C. E. 1972. A monograph of the Mexican and Central American species of *Trixis* (Compositae). Mem. New York Bot. Gard. 22(3): 1–68.

Anderson, J. L., J. M. Porter, and M. K. Debacon. 1996. Is *Hymenoxys helenioides* a species or a hybrid? Morphological and molecular evidence. In: J. Maschinski et al., eds. 1996. Southwestern Rare and Endangered Plants: Proceedings of the Second Conference: September 11–14, 1995, Flagstaff, Arizona. Fort Collins, Colo. Pp. 113–124.

Anderson, L. C. 1963. Studies on *Petradoria* (Compositae): Anatomy, cytology, taxonomy. Trans. Kansas Acad. Sci. 66: 632–684.

Anderson, L. C. 1970. Studies on *Bigelowia* (Astereae,

Compositae). 1. Morphology and taxonomy. Sida 3: 451–465.

Anderson, L. C. 1972. Studies on *Bigelowia* (Asteraceae), II. Xylary comparisons, woodiness and paedomorphosis. J. Arnold Arbor. 53: 499–514.

Anderson, L. C. 1977. Studies on *Bigelowia* (Asteraceae). III. Cytotaxonomy and biogeography. Syst. Bot. 2: 209–218.

Anderson, L. C. 1980. Identity of narrow-leaved *Chrysothamnus viscidiflorus* (Asteraceae). Great Basin Naturalist 40: 117–120.

Anderson, L. C. 1984. *Chrysothamnus nauseosus* ssp. *uintahensis*: A stabilized hybrid. Great Basin Naturalist 44: 416–420.

Anderson, L. C. 1986. An overview of the genus *Chrysothamnus* (Asteraceae). In: E. D. McArthur and B. L. Welch, eds. 1986. Proceedings, Symposium on the Biology of *Artemisia* and *Chrysothamnus*, Provo, Utah, July 9–13, 1984. Ogden. Pp. 29–45.

Anderson, L. C. 1986b. Sympatric subspecies in *Chrysothamnus nauseosus*. In: E. D. McArthur and B. L. Welch, eds. 1986. Proceedings, Symposium on the Biology of *Artemisia* and *Chrysothamnus*, Provo, Utah, July 9–13, 1984. Ogden. Pp. 98–103.

Anderson, L. C. 1994. A revision of *Hasteola* (Asteraceae) in the New World. Syst. Bot. 19: 211–219.

Anderson, L. C. 1995b. The *Chrysothamnus–Ericameria* connection (Asteraceae). Great Basin Naturalist 55: 84–88.

Anderson, L. C., E. L. Bridges, and S. L. Orzell. 1995. New data on distribution and morphology for the rare *Hasteola robertiorum* (Asteraceae). Phytologia 78: 246–248.

Anderson, L. C. and J. B. Creech. 1975. Comparative leaf anatomy of *Solidago* and related Asteraceae. Amer. J. Bot. 62: 486–493.

Anderson, L. C. and J. L. Reveal. 1966. *Chrysothamnus bolanderi*, an intergeneric hybrid. Madroño 18: 225–232.

Anderson, L. C. and P. S. Weberg. 1974. The anatomy and taxonomy of *Vanclevea* (Asteraceae). Great Basin Naturalist 34: 151–160.

Andreasen, M. L. and W. H. Eshbaugh. 1973. *Solidago albopilosa* Braun, a little known goldenrod from Kentucky. Castanea 38: 117–132.

Angiosperm Phylogeny Group. 2003. An update of the Angiosperm Phylogeny Group classification of the orders and families of flowering plants: APGII. Bot. J. Linn. Soc. 141: 399–436.

Ann. Bot. (Oxford) = Annals of Botany. (Oxford.)

Ann. Hist.-Nat. Mus. Natl. Hung. = Annales Historico-naturales Musei Nationalis Hungarici.

Ann. K. K. Naturhist. Hofmus. = Annalen des Kaiserlich-Königlichen Naturhistorischen Hofmuseums, Wien.

Ann. Lyceum Nat. Hist. New York = Annals of the Lyceum of Natural History of New York.

Ann. Missouri Bot. Gard. = Annals of the Missouri Botanical Garden.

Ann. Mus. Natl. Hist. Nat. = Annales du Muséum National d'Histoire Naturelle. ["National" dropped after vol. 5.]

Ann. Nat. = Annals of Nature; or, Annual Synopsis of New

Genera and Species of Animals, Plants, &c. Discovered in North America.

Ann. Sci. Nat., Bot. = Annales des Sciences Naturelles. Botanique.

Ann. Sci. Nat. (Paris) = Annales des Sciences Naturelles. (Paris.)

Ann. Soc. Linn. Lyon = Annales de la Société Linnéenne de Lyon.

Antonio, T. M. and S. Masi. 2001. The Sunflower Family in the Upper Midwest.... Indianapolis.

Applequist, W. L. 2002. A reassessment of the nomenclature of *Matricaria* L. and *Tripleurospermum* Sch. Bip. (Asteraceae). Taxon 51: 757–762.

Aquatic Bot. = Aquatic Botany; International Scientific Journal Dealing with Applied and Fundamental Research on Submerged, Floating and Emergent Plants in Marine and Freshwater Ecosystems.

Arch. Bot. (Leipzig) = Archiv für die Botanik. (Leipzig.)

Arch. Fl.—See: F. W. Schultz 1854–1869

Arctic Alpine Res. = Arctic and Alpine Research.

Arènes, J. 1950. Monographie du genre *Arctium* L. Bull. Jard. Bot. État Bruxelles 29: 67–156.

Aristeguieta, L. 1964. Compositae. In: T. Lasser. 1964+. Flora de Venezuela. 12+ vols. Caracas. Vol. 10.

Ariza E., L. 1979. Contribución al conocimiento del género *Tessaria* (Compositae), I. Consideraciónes sobre los géneros *Tessaria* y *Pluchea*. Kurtziana 12/13: 47–62.

Ark. Bot. = Arkiv för Botanik Utgivet av K. Svenska Vetenskapsakademien.

Arriagada, J. E. 1998. The genera of Inuleae (Compositae: Asteraceae) in the southeastern United States. Harvard Pap. Bot. 3: 1–48.

Arriagada, J. E. and N. G. Miller. 1997. The genera of Anthemideae (Compositae: Asteraceae) in the southeastern United States. Harvard Pap. Bot. 2: 1–46.

Ashmun, J. W., R. L. Brown, and L. F. Pitelka. 1985. Biomass allocation in *Aster acuminatus*: Variation within and among populations over 5 years. Canad. J. Bot. 63: 2035–2043.

Ashri, A. and P. F. Knowles. 1959. Further notes on *Carthamus* in California. Leafl. W. Bot. 9: 5–8.

Ashri, A. and P. F. Knowles. 1960. Cytogenetics of safflower (*Carthamus* L.) species and their hybrids. Agron. J. 52: 11–17.

Asia Life Sci. = Asia Life Sciences; the Asian International Journal of Life Sciences.

Asters Ontario ed. 2—See: J. C. Semple et al. 1996

Atlantic J. = Atlantic Journal, and Friend of Knowledge.

Atlas Vasc. Pl. Texas—See: B. L. Turner et al. 2003

Aublet, J. B. 1775. Histoire des Plantes de la Guiane Françoise.... 4 vols. Paris. [Vols. 1 and 2: text, paged consecutively; vols. 3 and 4: plates.] (Hist. Pl. Guiane)

Austral. J. Agric. Res = Australian Journal of Agricultural Research.

B. M. C. Evol. Biol. = B M C Evolutionary Biology.

Baagøe, J. 1974. The genus *Guizotia* (Compositae). A taxonomic revision. Bot. Tidsskr. 69: 1–39.

Babcock, E. B. 1947. The genus *Crepis*. Pt. 1: The taxonomy, phylogeny, distribution, and evolution of *Crepis*. Pt. 2: Systematic treatment. Univ. Calif. Publ. Bot. 21, 22.

Babcock, E. B. and H. M. Hall. 1924. *Hemizonia congesta*: A

genetic, ecologic, and taxonomic study of the hay-field tarweeds. Univ. Calif. Publ. Bot. 13: 15–100.

Babcock, E. B. and G. L. Stebbins. 1937. The genus *Youngia*. Publ. Carnegie Inst. Wash. 484.

Babcock, E. B. and G. L. Stebbins. 1938. The American species of *Crepis*. Their interrelationships and distribution as affected by polyploidy and apomixis. Publ. Carnegie Inst. Wash. 504.

Bachmann, K. 1992. Phenotypic similarity and genetic relationship among populations of *Microseris bigelovii* (Asteraceae: Lactuceae). Bot. Acta 105: 337–342.

Bachmann, K. and J. Battjes. 1994. Variability in a predominantly self-fertilizing annual with fragmented distribution, *Microseris douglasii* (Asteraceae, Lactuceae). Biol. Zentralbl. 113: 69–95.

Bachmann, K., K. L. Chambers, and H. J. Price. 1979. Genome size and phenotypic evolution in *Microseris* (Asteraceae, Cichorieae). Pl. Syst. Evol., Suppl. 2: 41–66.

Bachmann, K., K. L. Chambers, and H. J. Price. 1984. Differential geographic distribution of spatulate and pointed leaf shapes in *Microseris bigelovii* (Asteraceae–Lactuceae). Beitr. Biol. Pflanzen 59: 5–14.

Bacigalupi, R. 1931. A monograph of the genus *Perezia*, section *Acourtia*, with a provisional key to the section *Euperezia*. Contr. Gray Herb. 97: 1–81.

Baileya = Baileya; a Quarterly Journal of Horticultural Taxonomy.

Bain, J. F. 1985. Flavonoid variation in the *Senecio streptanthifolius* complex. Canad. J. Bot. 63: 1685–1690.

Bain, J. F. 1988. Taxonomy of *Senecio streptanthifolius* Greene. Rhodora 90: 277–312.

Bain, J. F. and R. K. Jansen. 1995. A phylogenetic analysis of the aureoid *Senecio* (Asteraceae) complex based on ITS sequence data. Pl. Syst. Evol. 195: 209–219.

Bain, J. F. and R. K. Jansen. 1996. Numerous chloroplast DNA polymorphisms are shared among different populations and species in the aureoid *Senecio (Packera)* complex. Canad. J. Bot. 74: 1719–1728.

Bain, J. F., B. S. Tyson, and D. F. Bray. 1997. Variation in pollen wall ultrastructure in New World Senecioneae (Asteraceae), with special reference to *Packera*. Canad. J. Bot. 75: 730–735.

Bain, J. F. and J. Whitton. 1994. Taxonomic analysis of *Senecio pauciflorus* and *S. indecorus* (Asteraceae). Nordic J. Bot. 14: 193–199.

Baird, G. I. 1996. The Systematics of *Agoseris* (Asteraceae: Lactuceae). Ph.D. dissertation. University of Texas.

Baker, M. M. and B. L. Turner. 1986. Taxonomy of *Flyriella* (Asteraceae–Eupatorieae). Sida 11: 300–317.

Baldwin, B. G. 1996. Phylogenetics of the California tarweeds and the Hawaiian silversword alliance (Madiinae; Heliantheae sensu lato). In: D. J. N. Hind et al., eds. 1996. Proceedings of the International Compositae Conference, Kew, 1994. 2 vols. Kew. Vol. 1, pp. 377–391.

Baldwin, B. G. 1999. *Constancea*, a new genus for *Eriophyllum nevinii* (Compositae–Heliantheae s. lat.). Madroño 46: 159–160.

Baldwin, B. G. 1999b. New combinations and new genera in

the North American tarweeds (Compositae–Madiinae). Novon 9: 462–471.

Baldwin, B. G. 2001. *Harmonia guggolziorum* (Compositae–Madiinae), a new tarweed from ultramafics of southern Mendocino County, California. Madroño 48: 293–297.

Baldwin, B. G. and D. W. Kyhos. 1990. A systematic and biogeographical review of *Raillardiopsis [Raillardella] muirii,* with special reference to a disjunct California Coast Range population. Madroño 37: 43–54.

Baldwin, B. G., D. W. Kyhos, J. Dvorak, and G. D. Carr. 1991. Chloroplast DNA evidence for a North American origin of the Hawaiian silversword alliance (Asteraceae). Proc. Natl. Acad. Sci. U.S.A. 88: 1840–1843.

Baldwin, B. G. and S. Markos. 1998. Phylogenetic utility of the external transcribed spacer (ETS) of 18S-26S rDNA: Congruence of ETS and ITS trees of *Calycadenia* (Compositae). Molec. Phylogen. Evol. 10: 449–463.

Baldwin, B. G., R. E. Preston, B. L. Wessa, and M. Wetherwax. 2001. A biosystematic and phylogenetic assessment of sympatric taxa in *Blepharizonia* (Compositae–Madiinae). Syst. Bot. 26: 184–194.

Baldwin, B. G. and B. L. Wessa. 2000. Origin and relationships of the tarweed–silversword lineage (Compositae–Madiinae). Amer. J. Bot. 87: 1890–1908.

Baldwin, B. G., B. L. Wessa, and J. L. Panero. 2002. Nuclear rDNA evidence for major lineages of helenioid Heliantheae (Compositae). Syst. Bot. 27: 161–198.

Ballard, R. 1986. *Bidens pilosa* complex (Asteraceae) in North and Central America. Amer. J. Bot. 73: 1452–1465.

Banckes, R. 1525. Boke of the Propreties of Herbes Called an Herball.... London.

Barber, H. N. 1941. Spontaneous hybrids between *Sonchus asper* and *S. oleraceus.* Ann. Bot. (Oxford), n. s. 5: 375–377.

Barbour, M. G. and N. L. Christensen. 1993. Vegetation. In: Flora of North America Editorial Committee, eds. 1993+. Flora of North America North of Mexico. 12+ vols. New York and Oxford. Vol. 1, pp. 97–131.

Barkley, T. M. 1962. A revision of *Senecio aureus* L. and allied species. Trans. Kansas Acad. Sci. 65: 318–408.

Barkley, T. M. 1966. A review of the origin and development of the florists' cineraria, *Senecio cruentus.* Econ. Bot. 20: 386–395.

Barkley, T. M. 1968. Taxonomy of *Senecio multilobatus* and its allies. Brittonia 20: 267–284.

Barkley, T. M. 1978. *Senecio.* In: N. L. Britton et al., eds. 1905+. North American Flora.... 47+ vols. New York. Ser. 2, part 10, pp. 50–139.

Barkley, T. M. 1980. Taxonomic notes on *Senecio tomentosus* and its allies (Asteraceae). Brittonia 32: 291–308.

Barkley, T. M. 1988. Variation among the aureoid senecios of North America: A geohistorical interpretation. Bot. Rev. (Lancaster) 54: 82–106.

Barkley, T. M. 1999. The segregates of *Senecio,* s.l., and *Cacalia,* s.l., in the flora of North America north of Mexico. Sida 18: 661–672.

Barkley, T. M., B. L. Clark, and A. M. Funston. 1996. The segregate genera of *Senecio* sensu lato and *Cacalia* sensu lato in Mexico and Central America. In: D. J. N. Hind et al., eds. 1996. Proceedings of the International Compositae Conference, Kew, 1994. 2 vols. Kew. Vol. 1, pp. 613–620.

Barkley, T. M. and A. Cronquist. 1978. *Erechtites.* In: N. L. Britton et al., eds. 1905+. North American Flora.... 47+ vols. New York. Ser. 2, part 10, pp. 139–142.

Barkley, T. M. and A. Cronquist. 1978b. *Emilia.* In: N. L. Britton et al., eds. 1905+. North American Flora.... 47+ vols. New York. Ser. 2, part 10, pp. 147–150.

Barlow-Irick, P. L. 2002. Biosystematic Analysis of the *Cirsium arizonicum* Complex of the Southwestern United States. Ph.D. dissertation. University of New Mexico.

Barratt, J. 1841. Eupatoria Verticillata.... Middletown, Ct. [Single page only, accompanied exsiccatae.] (Eupatoria Verticillata)

Barrier, M. et al. 1999. Interspecific hybrid ancestry of a plant adaptive radiation: Allopolyploidy of the Hawaiian silversword alliance (Asteraceae) inferred from floral homeotic gene duplications. Molec. Biol. Evol. 16: 1105–1113.

Bartonia = Bartonia; a Botanical Annual.

Bartram, W. 1791. Travels through North and South Carolina, Georgia, East and West Florida, the Cherokee Country, the Extensive Territories of the Muscogulges, or Creek Confederacy, and the Country of the Chactaws.... Philadelphia. (Travels Carolina)

Baskin, J. M. and C. C. Baskin. 1993. Nomenclatural history and taxonomic status of *Echinacea angustifolia, E. pallida,* and *E. tennesseensis* (Asteraceae). Sida 15: 597–604.

Bate Smith, E. C. 1980. Astringent tannins of *Cosmos bipinnatus.* Phytochemistry 19: 982.

Batra, S. W. T., J. R. Coulson, P. H. Dunn, and P. E. Boldt. 1981. Insects and Fungi Associated with *Carduus* Thistles (Compositae). Washington. [U.S.D.A. Techn. Bull. 1616.]

Battaglia, E. 1949. The alteration of meiosis during apomictic reproduction in *Chondrilla juncea* L. Caryologia 2: 23–30.

Bayer, R. J. 1984. Chromosome numbers and taxonomic notes for North American species of *Antennaria* (Asteraceae: Inuleae). Syst. Bot. 9: 74–83.

Bayer, R. J. 1985. Investigations into the evolutionary history of the polyploid complexes in *Antennaria* (Asteraceae Inuleae) I. The *A. neodioica* complex. Pl. Syst. Evol. 150: 143–163.

Bayer, R. J. 1985b. Investigations into the evolutionary history of the polyploid complexes in *Antennaria* (Asteraceae Inuleae) II. The *A. parlinii* complex. Rhodora 87: 321–329.

Bayer, R. J. 1987. Evolution and phylogenetic relationships of the *Antennaria* (Asteraceae: Inuleae) polyploid agamic complexes. Biol. Zentralbl. 106: 683–698.

Bayer, R. J. 1987b. Morphometric analysis of western North American *Antennaria* Gaertner (Asteraceae: Inuleae). I. Sexual species of sections *Alpinae, Dioicae,* and *Plantaginifoliae.* Canad. J. Bot. 65: 2389–2395.

Bayer, R. J. 1988. Patterns of isozyme variation in western North American *Antennaria* (Asteraceae: Inuleae). I. Sexual species of section *Dioicae.* Syst. Bot. 13: 525–537.

Bayer, R. J. 1989. Patterns of isozyme variation in western North American *Antennaria* (Asteraceae: Inuleae). II. Diploid and polyploid species of section *Alpinae*. Amer. J. Bot. 76: 679–691.

Bayer, R. J. 1989b. Patterns of isozyme variation in the *Antennaria rosea* (Asteraceae: Inuleae) polyploid agamic complex. Syst. Bot. 14: 389–397.

Bayer, R. J. 1989c. A systematic and phytogeographic study of *Antennaria aromatica* and *A. densifolia* (Asteraceae: Inuleae) in the western North American cordillera. Madroño 36: 248–259.

Bayer, R. J. 1989d. Nomenclatural rearrangements in *Antennaria neodioica* and *A. howellii* (Asteraceae: Inuleae: Gnaphaliinae). Brittonia 41: 396–398.

Bayer, R. J. 1989e. A taxonomic revision of the *Antennaria rosea* (Asteraceae: Inuleae: Gnaphaliinae) polyploid complex. Brittonia 41: 53–60.

Bayer, R. J. 1990. A phylogenetic reconstruction of *Antennaria* Gaertner (Asteraceae: Inuleae). Canad. J. Bot. 68: 1389–1397.

Bayer, R. J. 1990b. Investigations into the evolutionary history of the *Antennaria rosea* (Asteraceae: Inuleae) polyploid complex. Pl. Syst. Evol. 169: 97–110.

Bayer, R. J. 1990c. Patterns of clonal diversity in the *Antennaria rosea* (Asteraceae) polyploid agamic complex. Amer. J. Bot. 77: 1313–1319.

Bayer, R. J. 1990d. A systematic study of *Antennaria media*, *A. pulchella*, and *A. scabra* (Asteraceae: Inuleae) of the Sierra Nevada and White Mountains. Madroño 37: 171–183.

Bayer, R. J. 1991. Allozymic and morphologic variation in *Antennaria* (Asteraceae: Inuleae) from the low arctic of northwestern North America. Syst. Bot. 16: 492–506.

Bayer, R. J. 1992. Allozyme variation, genecology, and phytogeography of *Antennaria arcuata* (Asteraceae), a rare species from the Great Basin and Red Desert with small disjunct populations. Amer. J. Bot. 79: 872–881.

Bayer, R. J. 1993. A taxonomic revision of the genus *Antennaria* (Asteraceae: Inuleae: Gnaphaliinae) of Alaska and Yukon Territory, northwestern North America. Arctic Alpine Res. 25: 150–159.

Bayer, R. J. 1997. *Antennaria rosea* (Asteraceae)—A model group for the study of the evolution of polyploid agamic complexes. Opera Bot. 132: 53–65.

Bayer, R. J. 2000. Observations on the morphology and geographic range of *Antennaria dioica* (L.) Gaertn. (Asteraceae: Gnaphalieae). Madroño 46: 205–207.

Bayer, R. J. and D. J. Crawford. 1986. Allozyme divergence among five diploid species of *Antennaria* (Asteraceae: Inuleae) and their allopolyploid derivatives. Amer. J. Bot. 73: 287–296.

Bayer, R. J. and T. M. Minish. 1993. Isozyme variation, ecology and phytogeography of *Antennaria soliceps* (Asteraceae: Inuleae), a geographically restricted alpine apomict from the Spring Mountains, Nevada. Madroño 40: 75–89.

Bayer, R. J., B. G. Purdy, and D. G. Lebedyk. 1991. Niche differentiation among eight sexual species of *Antennaria* Gaertner (Asteraceae: Inuleae) and *A. rosea*, their allopolyploid derivative. Evol. Trends Pl. 5: 109–123.

Bayer, R. J., K. Ritland, and B. G. Purdy. 1990. Evidence of partial apomixis in *Antennaria media* (Asteraceae: Inuleae) detected by the segregation of genetic markers. Amer. J. Bot. 77: 1078–1083.

Bayer, R. J., D. E. Soltis, and P. S. Soltis. 1996. Phylogenetic inferences in *Antennaria* (Asteraceae: Inuleae: Gnaphaliinae) based on sequences from the nuclear ribosomal DNA internal transcribed spacers (ITS). Amer. J. Bot. 83: 516–527.

Bayer, R. J. and J. R. Starr. 1998. Tribal phylogeny of the Asteraceae based on two non-coding chloroplast sequences, the *trn*L intron and the *trn*L/F intergenic spacer. Ann. Missouri Bot. Gard. 85: 242–256.

Bayer, R. J. and G. L. Stebbins. 1982. A revised classification of *Antennaria* (Asteraceae: Inuleae) of the eastern United States. Syst. Bot. 7: 300–313.

Bayer, R. J. and G. L. Stebbins. 1983. Distribution of sexual and apomictic populations of *Antennaria parlinii*. Evolution 37: 555–561.

Bayer, R. J. and G. L. Stebbins. 1987. Chromosome numbers, patterns of distribution, and apomixis in *Antennaria* (Asteraceae: Inuleae). Syst. Bot. 12: 305–319.

Bayer, R. J. and G. L. Stebbins. 1993. A synopsis with keys for the genus *Antennaria* (Asteraceae: Inuleae: Gnaphaliinae) for North America. Canad. J. Bot. 71: 1589–1604.

Beaman, J. H. 1957. The systematics and evolution of *Townsendia* (Compositae). Contr. Gray Herb. 183: 1–151.

Beaman, J. H. 1990. Revision of *Hieracium* (Asteraceae) in Mexico and Central America. Syst. Bot. Monogr. 29: 1–77.

Beauchamp, R. M. 1986. A Flora of San Diego County, California. National City.

Beauchamp, R. M. and J. Henrickson. 1995. *Baccharis malibuensis* (Asteraceae): A new species from the Santa Monica Mountains, California. Aliso 14: 197–203.

Beaudry, J. R. 1959. *Solidago albopilosa* Braun and *S. flexicaulis* L. Castanea 24: 53–54.

Beaudry, J. R. 1969. Études sur les *Solidago* L. IX. Une troisième liste de nombres chromosomiques des taxons du genre *Solidago* et de certains genres voisins. Naturaliste Canad. 96: 103–122.

Beauverd, G. 1913. Le genre *Facelia* Cassini (emend. Beauverd). Bull. Soc. Bot. Genève, sér. 2, 5: 212–220.

Beauverd, G. 1913b. Contribution à l'étude des Composées, suite VIII. IV. Le genre *Micropsis* DC. (emend. Beauverd). Bull. Soc. Bot. Genève, sér. 2, 5: 221–228.

Beck, J. B. et al. 2004. Is subtribe Solidagininae (Asteraceae) monophyletic? Taxon 53: 691–698.

Becker, K. M. 1979. Monograph of the genus *Lasianthaea* (Asteraceae). Mem. New York Bot. Gard. 31(2): 1–64.

Beetle, A. A. 1960. A study of sagebrush. The section *Tridentatae* of *Artemisia*. Wyoming Agric. Exp. Sta. Bull. 368.

Beih. Bot. Centralbl. = Beihefte zum Botanischen Centralbl. Original Arbeiten.

Beitr. Biol. Pflanzen = Beiträge zur Biologie der Pflanzen.

Belcher, R. O. 1956. A revision of *Erechtites* (Compositae) with inquiries into *Senecio* and *Arrhenechthites*. Ann. Missouri Bot. Gard. 43: 1–85.

Bentham, G. 1839[–1857]. Plantas Hartwegianas Imprimis Mexicanas.... London. [Issued by gatherings with consecutive signatures and pagination.] (Pl. Hartw.)

Bentham, G. 1844[–1846]. The Botany of the Voyage of H.M.S. Sulphur, under the Command of Captain Sir Edward Belcher...during the Years 1836–1842. 6 parts. London. [Parts paged consecutively.] (Bot. Voy. Sulphur)

Bentham, G. 1873. Notes on the classification, history, and geographical distribution of Compositae. J. Linn. Soc., Bot. 13: 335–577.

Bentham, G. and J. D. Hooker. 1862–1883. Genera Plantarum ad Exemplaria Imprimis in Herbariis Kewensibus Servata Definita. 3 vols. London. (Gen. Pl.)

Ber. Geobot. Inst. E. T. H. Stiftung Rübel = Berichte des Geobotanischen Instituts der Eidg. techn. Hochschule Stiftung Rübel.

Bergius, P. J. 1767. Descriptiones Plantarum ex Capite Bonae Spei.... Stockholm. (Descr. Pl. Cap.)

Bernhardi, J. J. 1800. Systematisches Verzeichnis der Pflanzen, welche in der Gegend um Erfurt gefunden werden.... Erfurt. (Syst. Verz.)

Besser, W. S. J. G. von. 1822. Enumeratio Plantarum.... Vilna. (Enum. Pl.)

Besser, W. S. J. G. von. 1829. Lettre [on *Artemisia* including *Absinthium*]...au Directeur. Bull. Soc. Imp. Naturalistes Moscou 1: 219–265.

Biblioth. Bot. = Bibliotheca Botanica; original Abhandlungen aus dem Gesammtgebiete der Botanik.

Biblioth. Universelle Rev. Suisse = Bibliothèque Universelle: Revue Suisse (et Étrangère); Archives des Sciences Physiques et Naturelles

Biddulph, S. F. 1944. A revision of the genus *Gaillardia*. Res. Stud. State Coll. Wash. 12: 195–256.

Bierner, M. W. 1972. Taxonomy of *Helenium* sect. *Tetrodus* and a conspectus of North American *Helenium* (Compositae). Brittonia 24: 331–355.

Bierner, M. W. 1974. A systematic study of *Dugaldia* (Compositae). Brittonia 26: 385–392.

Bierner, M. W. 1989. Taxonomy of *Helenium* sect. *Amarum* (Asteraceae). Sida 13: 453–459.

Bierner, M. W. 1990. Present status of *Amblyolepis* (Asteraceae: Heliantheae). Madroño 37: 133–140.

Bierner, M. W. 1994. Submersion of *Dugaldia* and *Plummera* in *Hymenoxys* (Asteraceae: Heliantheae: Gaillardiinae). Sida 16: 1–8.

Bierner, M. W. 2001. Taxonomy of *Hymenoxys* subgenus *Picradenia* and a conspectus of the subgenera of *Hymenoxys* (Asteraceae, Helenieae, Tetraneurinae). Lundellia 4: 37–63.

Bierner, M. W. 2004. Taxonomy of *Hymenoxys* subgenus *Macdougalia* (Asteraceae, Helenieae, Tetraneurinae). Sida 21: 657–663.

Bierner, M. W. 2005. Taxonomy of *Hymenoxys* subgenus *Rydbergia* (Asteraceae, Helenieae, Tetraneurinae). Lundellia 8: 28–37.

Bierner, M. W. and R. K. Jansen. 1998. Systematic implications of DNA restriction site variation in *Hymenoxys*

and *Tetraneuris* (Asteraceae, Helenieae, Gaillardiinae). Lundellia 1: 17–26.

Bierner, M. W. and B. L. Turner. 2003. Taxonomy of *Tetraneuris* (Asteraceae: Helenieae: Tetraneurinae). Lundellia 6: 44–96.

Bigelow, J. 1824. Florula Bostoniensis. A Collection of Plants of Boston and Its Vicinity..., ed. 2. Boston. (Fl. Boston. ed. 2)

Bijdr. Fl. Ned. Ind.—See: C. L. Blume 1825–1826

Biltmore Bot. Stud. = Biltmore Botanical Studies; a Journal of Botany Embracing Papers by the Director and Associates of the Biltmore Herbarium.

Binns, S. E., B. R. Baum, and J. T. Arnason. 2002. A taxonomic revision of *Echinacea* (Asteraceae: Heliantheae). Syst. Bot. 15: 394–402.

Biochem. Syst. & Ecol. = Biochemical Systematics and Ecology.

Biol. Cent.-Amer., Bot.—See: W. B. Hemsley 1879–1888

Biol. Chem. Compositae—See: V. H. Heywood et al. 1977 [1978]

Biol. J. Linn. Soc. = Biological Journal of the Linnean Society.

Biol. Zentralbl. = Biologisches Zentralblatt.

Bittmann, M. 1990. Die Gattung *Adenocaulon* (Compositae): I. Morphologie. Candollea 45: 389–420.

Bittmann, M. 1990b. Die Gattung *Adenocaulon* (Compositae): II. Ökologie, Verbreitung und Systematik. Candollea 45: 493–518.

Blake, S. F. 1918. A revision of the genus *Viguiera*. Contr. Gray Herb. 54: 1–218.

Blake, S. F. 1921 Revision of the genus *Acanthospermum*. Contr. U.S. Natl. Herb. 20: 383–392.

Blondin, M. J., R. E. Cook, and J. C. Semple. 2005. A multivariate morphometric study of the aster genus *Sericocarpus* Nees (Asteraceae: Astereae). Sida 21: 1473–1507.

Bloom, W. L. 1977. Chromosomal differentiation between *Cirsium discolor* and C. *muticum* and the origin of supernumerary chromosomes. Syst. Bot. 2: 1–13.

Blume, C. L. 1825–1826. Bijdragen tot de Flora van Nederlandsch Indië. 10 parts. Batavia. [Parts paged consecutively.] (Bijdr. Fl. Ned. Ind.)

Böcher, T. W., B. Fredskild, K. Holmen, and K. Jakobsen. 1978. Grønlands Flora, ed. 3. Copenhagen.

Böcher, T. W., K. Holmen, and K. Jakobsen. 1968. The Flora of Greenland, ed. 2. Copenhagen.

Bogle, A. L. 1968. Evidence for the hybrid origin of *Petasites warrenii* and P. *vitifolius*. Rhodora 70: 533–551.

Bohm, B. A. and L. D. Gottlieb. 1989. Flavonoids of the annual *Stephanomeria* (Asteraceae). Biochem. Syst. & Ecol. 17: 451–453.

Boissier, P. E. 1867–1888. Flora Orientalis.... 5 vols. + suppl. Geneva. (Fl. Orient.)

Bol. Soc. Argent. Bot. = Boletín de la Sociedad Argentina de Botánica.

Bol. Soc. Bot. México = Boletín de la Sociedad Botánica de México.

Bol. Soc. Brot. = Boletim da Sociedade Broteriana.

Boldt, P. E. 1989. *Baccharis* (Asteraceae): A Review of Its Taxonomy, Phytochemistry, Ecology, Economic Status, Natural Enemies and the Potential for Its Biological Control in the United States. Temple, Tex.

Bolick, M. R. 1983. A cladistic analysis of the Ambrosiinae Lessing and Engelmanniinae Stuessy. Advances Cladist. 2: 125–141.

Bonplandia (Hanover) = Bonplandia; Zeitschrift für die gesammte Botanik.

Bossard, C. C., J. M. Randall, and M. C. Hoshovsky, eds. 2000. Invasive Plants of California's Wildlands. Berkeley.

Boston J. Nat. Hist. = Boston Journal of Natural History.

Bot. Acta = Botanica Acta; Berichte der Deutschen Botanischen Gesellschaft.

Bot. Beechey Voy.—See: W. J. Hooker and G. A. W. Arnott [1830–]1841

Bot. Beob.—See: J. C. Wendland 1798

Bot. California—See: W. H. Brewer et al. 1876–1880

Bot. Cult.—See: G. L. M. Dumont de Courset 1802–1805

Bot. Explor. = The Botanical Explorer.

Bot. Gart. Halle—See: K. Sprengel 1800–1801

Bot. Gaz. = Botanical Gazette; Paper of Botanical Notes.

Bot. Helv. = Botanica Helvetica.

Bot. J. Linn. Soc. = Botanical Journal of the Linnean Society.

Bot. Jahrb. Syst. = Botanische Jahrbücher für Systematik, Pflanzengeschichte und Pflanzengeographie.

Bot. Mag. = Botanical Magazine; or, Flower-garden Displayed.... [Edited by Wm. Curtis.] [With vol. 15, 1801, title became Curtis's Botanical Magazine; or....]

Bot. Mater. Gerb. Bot. Inst. Komarova Akad. Nauk S.S.S.R. = Botanicheskie Materialy Gerbariya Botanicheskogo Instituti Imeni V. L. Komarova, Akademii Nauk S S S R.

Bot. Misc. = Botanical Miscellany.

Bot. Not. = Botaniska Notiser.

Bot. Reg. = Botanical Register....

Bot. Rev. (Lancaster) = Botanical Review, Interpreting Botanical Progress.

Bot. Surv. Olympic Penins.—See: G. N. Jones 1936

Bot. Tidsskr. = Botanisk Tidsskrift.

Bot. Voy. Herald—See: B. Seemann 1852–1857

Bot. Voy. Sulphur—See: G. Bentham 1844[–1846]

Botany (Fortieth Parallel)—See: S. Watson 1871

Boulos, L. 1973. Révision systématique du genre Sonchus L. s.l.: IV. Sous-genre: 1. Sonchus. Bot. Not. 126: 155–196.

Boyko, E. V. 2003. Achene morphology and anatomy of Symphyllocarpus exilis Maxim. and Centipeda minima (L.) A. Br. et Aschers. (Asteraceae). Compositae Newslett. 40: 9.

Brauner, S. and L. D. Gottlieb. 1987. A self-compatible plant of Stephanomeria exigua subsp. coronaria (Asteraceae) and its relevance to the origin of its self-pollinating derivative S. malheurensis. Syst. Bot. 12: 299–304.

Brauner, S. and L. D. Gottlieb. 1989. Response to selection for time of bolting in Stephanomeria exigua subsp. coronaria and implications for the origin of Stephanomeria malheurensis. Syst. Bot. 14: 516–524.

Breitung, A. J. 1988. Distribution of the showy aster, Aster conspicuus. Canad. Field-Naturalist 102: 523–526.

Bremer, K. 1987. Tribal interrelationships of the Asteraceae. Cladistics 3: 210–253.

Bremer, K. 1994. Asteraceae: Cladistics & Classification. Portland.

Bremer, K. 1996. Major clades and grades of the Asteraceae. In: D. J. N. Hind et al., eds. 1996. Proceedings of the International Compositae Conference, Kew, 1994. 2 vols. Kew. Vol. 1, pp. 1–7.

Bremer, K., E. M. Friis, and B. Bremer. 2004. Molecular phylogenetic dating of asterid flowering plants shows early Cretaceous diversification. Syst. Biol. 53: 496–505.

Bremer, K. and M. H. G. Gustafsson. 1997. East Gondwana ancestry of the sunflower alliance of families. Proc. Natl. Acad. Sci. U.S.A. 94: 9188–9190.

Bremer, K. and C. J. Humphries. 1993. Generic monograph of the Asteraceae–Anthemideae. Bull. Nat. Hist. Mus. London, Bot. 23: 71–177.

Bremer, K. and R. K. Jansen. 1992. A new subfamily of the Asteraceae. Ann. Missouri Bot. Gard. 79: 414–415.

Brewer, W. H. et al. 1876–1880. Geological Survey of California.... Botany.... 2 vols. Cambridge, Mass. (Bot. California)

Bricker, J. S. and G. K. Brown. 1998. A molecular phylogeny for Oonopsis (Asteraceae). [Abstract.] Amer. J. Bot. 85(6, suppl.): 170.

Bridson, G. D. R. and E. R. Smith. 1991. B-P-H/S. Botanico-Periodicum-Huntianum/Supplementum. Pittsburgh.

Briquet, J. 1930. Carpologie du genre Crupina. Candollea 4: 241–278.

Briquet, J. 1930b. Les émergences et trichomes de Crupina. Candollea 4: 191–201.

Brit. Fl. Gard.—See: R. Sweet 1823–1837

Brit. J. Dermatol. = British Journal of Dermatology (and Syphilis).

Britton, N. L. 1901. Manual of the Flora of the Northern States and Canada. New York. (Man. Fl. N. States)

Britton, N. L. et al., eds. 1905+. North American Flora.... 47+ vols. New York. [Vols. 1–34, 1905–1957; ser. 2, parts 1–13+, 1954+.] (N. Amer. Fl.)

Britton, N. L. and A. Brown. 1896–1898. An Illustrated Flora of the Northern United States, Canada and the British Possessions from Newfoundland to the Parallel of the Southern Boundary of Virginia, and from the Atlantic Ocean Westward to the 102d Meridian.... 3 vols. New York. (Ill. Fl. N. U.S.)

Britton, N. L. and A. Brown. 1913. An Illustrated Flora of the Northern United States, Canada and the British Possessions from Newfoundland to the Parallel of the Southern Boundary of Virginia, and from the Atlantic Ocean Westward to the 102d Meridian..., ed. 2. 3 vols. New York. (Ill. Fl. N. U.S. ed. 2)

Britton, N. L., E. E. Sterns, J. F. Poggenburg, et al. 1888. Preliminary Catalogue of Anthophyta and Pteridophyta Reported As Growing Spontaneously within One Hundred Miles of New York City. New York. [Authorship often attributed as B.S.P. in nomenclatural contexts.] (Prelim. Cat.)

Brittonia = Brittonia; a Journal of Systematic Botany....

Brock, M. T. 2004. The potential for genetic assimilation of a native dandelion species, Taraxacum ceratophorum (Asteraceae), by the exotic congener T. officinale. Amer. J. Bot. 91: 656–663.

Brooks, R. E. 1986. Cirsium. In: Great Plains Flora Association. 1986. Flora of the Great Plains. Lawrence, Kans. Pp. 908–914.

Brotero, F. 1801. Phytographia Lusitaniae Selectior.... 1 fasc. only. Lisbon. (Phytogr. Lusitan. Select.)

Brotero, F. 1804[–1805]. Flora Lusitanica.... 2 vols. Lisbon. (Fl. Lusit.)

Brouillet, L. 1983. Nombres chromosomiques chez les *Aster* du groupe *Heterophylli* (Asteraceae–Astereae). Naturaliste Canad. 110: 171–178.

Brouillet, L. and J. Labrecque. 1987. *Aster gaspensis* Victorin: Nombre chromosomique et hybridation naturelle avec l'*A. novi-belgii* L. Naturaliste Canad. 114: 159–165.

Brouillet, L. and S. Selliah. 2005. *Symphyotrichum pygmaeum*: Transfer of *Eurybia pygmaea* from the eurybioid grade to the subtribe Symphyotrichinae (Asteraceae: Astereae). Sida 21: 1633–1635.

Brouillet, L. and J. C. Semple. 1981. A propos du status taxonomique du *Solidago ptarmicoides*. Canad. J. Bot. 59: 17–21.

Brouillet, L. and J.-P. Simon. 1979. Resource allocation and phenology of two species of *Aster* (Asteraceae) and their hybrid. Canad. J. Bot. 57: 1792–1799.

Brouillet, L. and J.-P. Simon. 1980. Adaptation and acclimation of higher plants at the enzyme level: Thermal properties of NAD malate dehydrogenase of two species of *Aster* (Asteraceae) and their hybrid adapted to contrasting habitats. Canad. J. Bot. 58: 1471–1481.

Brouillet, L. and J.-P. Simon. 1981. An ecogeographical analysis of the distribution of *Aster acuminatus* Michx. and *A. nemoralis* Aiton (Asteraceae: Astereae). Rhodora 83: 521–550.

Brouillet, L., L. E. Urbatsch, and R. P. Roberts. 2004. *Tonestus kingii* and *T. aberrans* are related to *Eurybia* and the Machaerantherinae (Asteraceae: Astereae) based on nrDNA (ITS and ETS) data: Reinstatement of *Herrickia* and a new genus, *Triniteurybia*. Sida 21: 889–900.

Brown, G. K. 1983. Chromosome numbers in *Platyschkuhria* Rydberg (Compositae) and their systematic significance. Amer. J. Bot. 70: 591–601.

Brown, G. K. 1993. Systematics of *Oonopsis* (Asteraceae). [Abstract.] Amer. J. Bot. 80(6, suppl.): 133–134.

Brown, G. K. 1993b. *Pyrrocoma*. In: J. C. Hickman, ed. 1993. The Jepson Manual. Higher Plants of California. Berkeley, Los Angeles, and London. Pp. 330–331.

Brown, G. K. and W. D. Clark. 1981. Chromosome numbers in South American *Haplopappus* Cass. (Compositae). Amer. J. Bot. 68: 1218–1221.

Brown, G. K. and W. D. Clark. 1982. Taxonomy of *Haplopappus* sect. *Gymnocoma* (Compositae). Syst. Bot. 7: 199–213.

Brown, G. K. and W. D. Clark. 1989. Flavonoids of *Haplopappus* section *Oonopsis*. [Abstract.] Amer. J. Bot. 76(6, suppl.): 190.

Brown, R. 1817. Observations on the Natural Family of Plants Called Compositae.... London. [Preprinted from Trans. Linn. Soc. London 12: 75–142. 1818.] (Observ. Compositae)

Brown, R. C. 1974. Biosystematics of *Baileya* and *Psilostrophe* (Compositae, Helenieae). Ph.D. dissertation. Arizona State University.

Brown, R. C. 1978. Biosystematics of *Psilostrophe* (Compositae: Helenieae). II. Artificial hybridization and systematic treatment. Madroño 25: 187–201.

Browne, P. 1756. The Civil and Natural History of Jamaica.... London. (Civ. Nat. Hist. Jamaica)

Brummitt, R. K. 1997. Taxonomy versus cladonomy, a fundamental controversy in biological systematics. Taxon 46: 723–734.

Brummitt, R. K. and C. E. Powell, eds. 1992. Authors of Plant Names. A List of Authors of Scientific Names of Plants, with Recommended Standard Forms of Their Names, Including Abbreviations. Kew.

Brunton, D. F. 1989. The marsh dandelion (*Taraxacum* section *Palustria;* Asteraceae) in Canada and the adjacent United States. Rhodora 91: 213–219.

Brunton, D. F. and C. W. Crompton. 1993. Marsh sow-thistle, *Sonchus palustris* L. (Asteraceae) in Ontario: An addition to the introduced flora of North America. Canad. Field-Naturalist 107: 341–344.

Budd, A. C., J. Looman, and K. F. Best. 1987. Budd's Flora of the Canadian Prairie Provinces, rev. and enl. Ottawa and Hull. [Agric. Canada, Res. Branch, Publ. 1662.]

Bull. Acad. Imp. Sci. Saint-Pétersbourg = Bulletin de l'Académie Impériale des Sciences de Saint Pétersbourg.

Bull. Bot. Lab. N. E. Forest. Inst., Harbin = Bulletin of Botanical Laboratory of North-eastern Forestry Institute. [Chih Wu Yen Chiu Shih Hui Kan.]

Bull. Brit. Mus. (Nat. Hist.), Bot. = Bulletin of the British Museum (Natural History). Botany.

Bull. Calif. Acad. Sci. = Bulletin of the California Academy of Sciences.

Bull. Illinois Nat. Hist. Surv. = Bulletin of the Illinois Natural History Survey.

Bull. Jard. Bot. État Bruxelles = Bulletin du Jardin Botanique de l'État à Bruxelles.

Bull. Nat. Hist. Mus. London, Bot. = Bulletin of the Natural History Museum. Botany Series.

Bull. Natl. Mus. Canada = Bulletin of the National Museum of Canada.

Bull. New York Bot. Gard. = Bulletin of the New York Botanical Garden.

Bull. New York State Mus. Nat. Hist. = Bulletin of the New York State Museum of Natural History.

Bull. S. Calif. Acad. Sci. = Bulletin of the Southern California Academy of Sciences.

Bull. Sci. Soc. Philom. Paris = Bulletin des Sciences, par la Société Philomatique (de Paris).

Bull. Soc. Bot. Genève = Bulletin de la Société Botanique de Genève.

Bull. Soc. Imp. Naturalistes Moscou = Bulletin de la Société Impériale des Naturalistes de Moscou.

Bull. Torrey Bot. Club = Bulletin of the Torrey Botanical Club.

Burks, K. A. 1994. The Effects of Population Size and Density on the Pollination Biology of a Threatened Thistle (*Cirsium vinaceum*). M.S. thesis. New Mexico State University.

Burrows, G. E. and R. J. Tyrl. 2001. Toxic Plants of North America. Ames.

Burry, J. N. and P. M. Kloot. 1982. The spread of composite (Compositae) weeds in Australia. Contact Dermatitis 8: 410–413.

Butterwick, M. L. 1975. A Systematic Treatment of Series *Pinnatilobatae* of *Viguiera*. M.S. thesis. University of Texas.

Cabrera, A. L. 1939. Las especies Argentinas del género "*Tessaria*." Lilloa 4: 181–189.

Cabrera, A. L. 1949. Sinopsis del género *Soliva* (Compositae). Notas Mus. La Plata, Bot. 14: 123–139.

Cabrera, A. L. 1961. Observaciónes sobre las Inuleae–Gnaphalinae (Compositae) de América del Sur. Bol. Soc. Argent. Bot. 9: 359–386.

Cabrera, A. L. 1971. Revisión del género *Gochnatia* (Compositae). Revista Mus. La Plata, Secc. Bot. 12: 1–160.

Cabrera, A. L. 1977+. Flora de la Provincia de Jujuy, República Argentina. 4+ parts. Buenos Aires.

Cabrera, A. L. and A. M. Ragonese. 1978. Revisión del género *Pterocaulon* (Compositae). Darwiniana 21: 185–257.

Cabrera R., L. 1992. Systematics of *Rzedowskiela* Gen. Nov. (Asteraceae: Mutisieae: Nassauviinae). Ph.D. dissertation. University of Texas.

Cabrera R., L. 2001. Six new species of *Acourtia* (Asteraceae) and a historical account of *Acourtia mexicana*. Brittonia 53: 416–429.

Callihan, R. H., S. L. Carson, and R. T. Dobbins. 1995. NAWEEDS, Computer-aided Weed Identification for North America. Illustrated User's Guide plus Computer Floppy Disk. Moscow, Idaho.

Campbell, J. J. N. and M. E. Medley. 1989. *Aster saxicastelli* (Asteraceae), a new species from the Rockaster River bars in southeastern Kentucky. Sida 13: 277–284.

Canad. Field-Naturalist = Canadian Field-Naturalist.

Canad. J. Bot. = Canadian Journal of Botany.

Canad. J. Genet. Cytol. = Canadian Journal of Genetics and Cytology.

Canad. J. Pl. Sci. = Canadian Journal of Plant Science.

Candolle, A. P. de. 1828–1838. Collection de Mémoires.... 10 parts. Paris and Strasbourg

Candolle, A. P. de. 1836–1838. Compositae. In: A. P. de Candolle and A. L. P. P. de Candolle, eds. 1823–1873. Prodromus Systematis Naturalis Regni Vegetabilis.... 17 vols. Paris etc. Vols. 5, 6, 7(1).

Candolle, A. P. de and A. L. P. P. de Candolle, eds. 1823–1873. Prodromus Systematis Naturalis Regni Vegetabilis.... 17 vols. Paris etc. [Vols. 1–7 edited by A. P. de Candolle, vols. 8–17 by A. L. P. P. de Candolle.] (Prodr.)

Candollea = Candollea; Organe du Conservatoire et du Jardin Botaniques de la Ville de Genève.

Canne, J. M. 1977. A revision of the genus *Galinsoga* (Compositae: Heliantheae). Rhodora 79: 319–389.

Canne, J. M. 1978. Circumscription and generic relationships of *Galinsoga* (Compositae: Heliantheae). Madroño 25: 81–93.

Carlquist, S. 1956. On the generic limits of *Eriophyllum* (Compositae) and related genera. Madroño 13: 226–239.

Carlquist, S. 1959. Studies on Madiinae: Anatomy, cytology, and evolutionary relationships. Aliso 4: 171–236.

Carlquist, S. 1966. Wood anatomy of Anthemideae, Ambrosieae, Calendulae, and Arctotideae (Compositae). Aliso 6(2): 1–23.

Carlquist, S. 1974. Island Biology. New York.

Carlquist, S. 1976. Tribal interrelationships and phylogeny of the Asteraceae. Aliso 8: 465–492.

Carlquist, S., B. G. Baldwin, and G. D. Carr, eds. 2003. Tarweeds & Silverswords: Evolution of the Madiinae (Asteraceae). St. Louis.

Carney, S. E., K. A. Gardner, and L. H. Rieseberg. 2000. Evolutionary changes over the fifty-year history of a hybrid population of sunflowers *(Helianthus)*. Evolution 54: 462–474.

Carr, G. D. 1975. Chromosome evolution and aneuploid reduction in *Calycadenia pauciflora* (Asteraceae). Evolution 29: 681–699.

Carr, G. D. 1975b. *Calycadenia hooveri* (Asteraceae), a new tarweed from California. Brittonia 27: 136–141.

Carr, G. D. 1977. A cytological conspectus of the genus *Calycadenia* (Asteraceae): An example of contrasting modes of evolution. Amer. J. Bot. 64: 694–703.

Carr, G. D. 1980. Experimental evidence for saltational chromosome evolution in *Calycadenia pauciflora* Gray (Asteraceae). Heredity 45: 107–112.

Carr, G. D., B. G. Baldwin, and D. W. Kyhos. 1996. Cytogenetic implications of artificial hybrids between the Hawaiian silversword alliance and North American tarweeds (Asteraceae: Heliantheae–Madiinae). Amer. J. Bot. 83: 653–660.

Carr, R. L. and G. D. Carr. 1983. Chromosome races and structural heterozygosity in *Calycadenia ciliosa* Greene (Asteraceae). Amer. J. Bot. 70: 744–755.

Carr, R. L. and G. D. Carr. 1993. *Calycadenia*. In: J. C. Hickman, ed. 1993. The Jepson Manual. Higher Plants of California. Berkeley, Los Angeles, and London. Pp. 218–219.

Carr, R. L. and G. D. Carr. 2000. A new chromosome race of *Calycadenia pauciflora* (Asteraceae: Heliantheae–Madiinae) from Butte County, California. Amer. J. Bot. 87: 1459–1465.

Carter, W. R. 1921. A Preliminary Catalogue of the Flora of Vancouver and Queen Charlotte Islands.... Victoria. (Prelim. Cat. Fl. Vancouver)

Caryologia = Caryologia; Giornale di Citologia, Citosistematica e Citogenetica.

Cassini, A.-H. G. de. 1826–1834. Opuscules Phytologiques.... 3 vols. Paris. (Opusc. Phytol.)

Castanea = Castanea; Journal of the Southern Appalachian Botanical Club.

Cat. Pl. Amer. Sept.—See: G. H. E. Muhlenberg 1813

Cat. Pl. Cub.—See: A. H. R. Grisebach 1866

Cat. Pl. Hort. Gott.—See: J. G. Zinn 1757

Cat. Pl. Maroc—See: É. Jahandiez et al. 1931–1941

Cat. Pl. Upper Louisiana—See: T. Nuttall 1813

Catal. Bot.—See: A. W. Roth 1797–1806

Cavanilles, A. J. 1791–1801. Icones et Descriptiones Plantarum, Quae aut Sponte in Hispania Crescunt, aut in Hortis Hospitantur. 6 vols. Madrid. (Icon.)

Cavanilles, A. J. [1801–]1802. Descripción de las Plantas.... Madrid. (Descr. Pl.)

Cent. Observ. Bot.—See: A. Zuccagni [1806]

Cent. Pl. I—See: C. Linnaeus 1755

Cent. Pl. II—See: C. Linnaeus [1756]

Chambers, K. L. 1955. A biosystematic study of the annual species of *Microseris*. Contr. Dudley Herb. 4: 207–312.

Chambers, K. L. 1957. Taxonomic notes on some Compositae of the western United States. Contr. Dudley Herb. 5: 57–68.

Chambers, K. L. 1960. *Microseris*. In: L. Abrams and R. S. Ferris. 1923–1960. Illustrated Flora of the Pacific States: Washington, Oregon, and California. 4 vols. Stanford. Vol. 4, pp. 554–562.

Chambers, K. L. 1963b. Amphitropical species pairs in *Microseris* and *Agoseris* (Compositae: Cichorieae). Quart. Rev. Biol. 38: 124–140.

Chambers, K. L. 1964. Nomenclature of *Microseris lindleyi*. Leafl. W. Bot. 10: 106–108.

Chambers, K. L. 1965. An allopolyploid *Krigia* from the southern Appalachians. [Abstract.] Amer. J. Bot. 52: 658.

Chambers, K. L. 1993c. *Microseris*. In: J. C. Hickman, ed. 1993. The Jepson Manual. Higher Plants of California. Berkeley, Los Angeles, and London. Pp. 316–319.

Chambers, K. L. 2004. Taxonomic notes on *Krigia* (Asteraceae). Sida 21: 225–236.

Chambers, K. L. 2004b. Two new subspecies of *Microseris laciniata* (Asteraceae) from the Siskiyou Mountains. Sida 21: 193–205.

Chan, R. 2000. Molecular Systematics of the Goldfield Genus *Lasthenia* (Compositae: Heliantheae Sensu Lato). Ph.D. dissertation. University of California, Berkeley.

Chan, R. 2001. Taxonomic changes and a new species in *Lasthenia* sect. *Amphiachaenia* (Compositae: Heliantheae sensu lato). Madroño 48: 205–210.

Chan, R., B. G. Baldwin, and R. Ornduff. 2001. Goldfields revisited: A molecular phylogenetic perspective on the evolution of *Lasthenia* (Compositae: Heliantheae sensu lato). Int. J. Pl. Sci. 162: 1347–1360.

Chan, R., B. G. Baldwin, and R. Ornduff. 2002. Cryptic goldfields: A molecular phylogenetic reinvestigation of *Lasthenia californica* sensu lato and close relatives (Compositae: Heliantheae sensu lato). Amer. J. Bot. 89: 1103–1112.

Channell, R. B. 1957. A revision of the genus *Marshallia* (Compositae). Contr. Gray Herb. 81: 41–132.

Chapman, A. W. 1860. Flora of the Southern United States.... New York. (Fl. South. U.S.)

Chapman, A. W. 1883. Flora of the Southern United States..., ed. 2. New York. (Fl. South. U.S. ed. 2)

Cherniawsky, D. M. and R. J. Bayer. 1998. Systematics of North American *Petasites* (Asteraceae: Senecioneae). I. Morphometric analyses. Canad. J. Bot. 76: 23–36.

Cherniawsky, D. M. and R. J. Bayer. 1998b. Systematics of North American *Petasites* (Asteraceae: Senecioneae). II. Isozyme analysis and population genetic structure. Canad. J. Bot. 76: 1476–1487.

Cherniawsky, D. M. and R. J. Bayer. 1998c. Systematics of North American *Petasites* (Asteraceae: Senecioneae). III. A taxonomic revision. Canad. J. Bot. 76: 2061–2075.

Chevallier, F. F. 1836. Flore Générale des Environs de Paris..., ed. 2. 2 vols. in 3. Paris. (Fl. Gén. Env. Paris ed. 2)

Chinnappa, C. C. 1981. Cytological studies in *Krigia* (Asteraceae). Canad. J. Genet. Cytol. 23: 671–678.

Chlor. Andina—See: H. A. Weddell 1855–1857[–1861]

Chmielewski, J. G. and J. C. Semple. 1985. The cytogeography and post-glacial migration of *Solidago flexicaulis* (Compositae) into southern Ontario. Naturaliste Canad. 112: 307–311.

Chmielewski, J. G. and J. C. Semple. 2001. The biology of Canadian weeds. 113. *Symphyotrichum lanceolatum* (Willd.) Nesom [*Aster lanceolatus* Willd.] and *S. lateriflorum* (L.) Löve & Löve [*Aster lateriflorus* (L.) Britt.]. Canad. J. Pl. Sci. 81: 829–849.

Chmielewski, J. G. and J. C. Semple. 2001b. The biology of Canadian weeds. 114. *Symphyotrichum pilosum* (Willd.) Nesom [*Aster pilosus* Willd.]. Canad. J. Pl. Sci. 81: 851–865.

Chmielewski, J. G., J. C. Semple, L. M. Burr, and W. R. Hawthorn. 1989. Comparison of achene characteristics within and among diploid and tetraploid clones of *Solidago flexicaulis* and their significance in germination and resource-allocation studies. Canad. J. Bot. 67: 1821–1832.

Civ. Nat. Hist. Jamaica—See: P. Browne 1756

Cladistics = Cladistics; the International Journal of the Willi Hennig Society.

Clapham, A. R., T. G. Tutin, and D. M. Moore. 1987. Flora of the British Isles, ed. 3. Cambridge.

Clark, C. 1998. Phylogeny and adaptation in the *Encelia* alliance (Asteraceae: Heliantheae). Aliso 17: 89–98.

Clark, W. D. 1979. The taxonomy of *Hazardia* (Compositae: Astereae). Madroño 26: 105–127.

Class-book Bot. ed. s.n.(b)—See: A. Wood 1861

Clausen, J. 1951. Stages in the Evolution of Plant Species. Ithaca, N.Y.

Clausen, J., D. D. Keck, and W. M. Hiesey. 1945. Experimental studies on the nature of species. II. Plant evolution through amphiploidy and autoploidy with examples from the Madiinae. Publ. Carnegie Inst. Wash. 564.

Clausen, J., D. D. Keck, and W. M. Hiesey. 1948. Experimental studies on the nature of species. III. Environmental responses of climatic races of *Achillea*. Publ. Carnegie Inst. Wash. 581.

Clevenger, S. and C. B. Heiser. 1963. *Helianthus laetiflorus* and *Helianthus rigidus*—hybrids or species? Rhodora 65: 121–133.

Clevinger, J. A. and J. L. Panero. 2000. Phylogenetic analysis of *Silphium* and subtribe Engelmanniinae (Asteraceae: Heliantheae) based on ITS and ETS sequence data. Amer. J. Bot. 87: 565–572.

Clewell, A. F. and J. W. Wooten. 1971. A revision of *Ageratina* (Compositae: Eupatorieae) from eastern North America. Brittonia 23: 123–143.

Clonts, J. A. 1972. A Revision of the Genus *Elephantopus* Including *Orthopappus* and *Pseudelephantopus* (Compositae). Ph.D. thesis. Mississippi State University.

Cody, W. J. 1996. Flora of the Yukon Territory. Ottawa.

Coffey, T. 1993. The History and Folklore of North American Wildflowers. New York.

Coleman, J. R. 1966. A taxonomic revision of section *Ximenesia* of the genus *Verbesina* (Compositae). Amer. Midl. Naturalist 76: 475–481.

Coleman, J. R. 1966b. A taxonomic revision of section *Sonoricola* of the genus *Verbesina* (Compositae). Madroño 19: 129–137.

Coleman, M., D. G. Forbes, and R. J. Abbott. 2001. A new subspecies of *Senecio mohavensis* (Compositae) reveals Old-New-World species disjunction. Edinburgh J. Bot. 58: 389–403.

Collectanea—See: N. J. Jacquin 1786[1787]–1796[1797]

Commentat. Soc. Phys.-Med. Univ. Lit. Caes. Mosq. = Commentationes Societatis Physico-Medicae apud Universitatem Literarum Caesaream Mosquensem Institutae.

Compan. Bot. Mag. = Companion to the Botanical Magazine....

Compositae Newslett. = Compositae Newsletter.

Comps Mexico—See: B. L. Turner 1996+

Consp. Fl. Eur.—See: C. F. Nyman 1878–1890

Constance, L. 1937. A systematic study of the genus *Eriophyllum* Lag. Univ. Calif. Publ. Bot. 18: 69–136.

Contr. Bot. India—See: R. Wight 1834

Contr. Bot. Vermont = Contributions to the Botany of Vermont.

Contr. Dudley Herb. = Contributions from the Dudley Herbarium of Stanford University.

Contr. Gray Herb. = Contributions from the Gray Herbarium of Harvard University. [Some numbers reprinted from (or in?) other periodicals, e.g. Rhodora.]

Contr. Herb. Franklin Marshall Coll. = Contributions from the Herbarium of Franklin and Marshall College.

Contr. Inst. Bot. Univ. Montréal = Contributions de l'Institut Botanique de l'Université de Montréal.

Contr. U.S. Natl. Herb. = Contributions from the United States National Herbarium.

Contr. W. Bot. = Contributions to Western Botany.

Cook, L. M., P. S. Soltis, S. J. Brunsfeld, and D. E. Soltis. 1998. Multiple independent formations of *Tragopogon* tetraploids (Asteraceae): Evidence from RAPD markers. Molec. Ecol. 7: 1293–1302.

Cook, R. E. 2002. A Biosystematic Study of *Solidago* Subsect. *Glomeruliflorae* Torr. & Gray (Asteraceae: Astereae) in Eastern North America. Ph.D. dissertation. University of Waterloo.

Correa, M. D. and R. L. Wilbur. 1969. A revision of the genus *Carphephorus* (Compositae–Eupatorieae). J. Elisha Mitchell Sci. Soc. 85: 79–91.

Correll, D. S. and M. C. Johnston. 1970. Manual of the Vascular Plants of Texas. Renner, Tex.

Cotter, D. J., L. Finkner, and D. T. Sullivan. 1982. Germination and transplanting of *Baileya multiradiata*. Res. Rep. New Mexico Agric. Exp. Sta. 489: 1–4.

Cotter, D. J., D. T. Sullivan, and R. E. Gomez. 1980. *Baileya* in Landscapes. Las Cruces. [New Mexico Coop. Ext. Serv. Guide H-704.]

Coulter, J. M. and A. Nelson. 1909. New Manual of Botany of the Central Rocky Mountains (Vascular Plants). New York. (New Man. Bot. Rocky Mt.)

Cox, P. B. and L. E. Urbatsch. 1990. A phylogenetic analysis of the coneflowers (Asteraceae: Heliantheae). Syst. Bot. 15: 394–402.

Cox, P. B. and L. E. Urbatsch. 1994. A taxonomic revision of *Rudbeckia* subg. *Macrocline* (Asteraceae: Heliantheae: Rudbeckiinae). Castanea 59: 300–318.

Crantz, H. J. N. von. 1766. Institutiones Rei Herbariae.... 2 vols. Vienna. (Inst. Rei Herb.)

Crawford, D. J. and R. Ornduff. 1989. Enzyme electrophoresis and evolutionary relationships among three species of *Lasthenia* (Asteraceae: Heliantheae). Amer. J. Bot. 76: 289–296.

Cronquist, A. 1943. The separation of *Erigeron* from *Conyza*. Bull. Torrey Bot. Club 70: 629–632.

Cronquist, A. 1945. Notes on the Compositae of the northeastern United States. III. Inuleae and Senecioneae. Rhodora 48: 116–121.

Cronquist, A. 1947. A revision of the North American species of *Erigeron*, north of Mexico. Brittonia 6: 121–302.

Cronquist, A. 1948. A revision of the *Oreastrum* group of *Aster*. Leafl. W. Bot. 5: 73–82.

Cronquist, A. 1950. A review of the genus *Psilocarphus*. Res. Stud. State Coll. Wash. 18: 71–89.

Cronquist, A. 1953. Notes on specimens of American plants in European herbaria. Leafl. W. Bot. 7: 17–31.

Cronquist, A. 1955. Compositae. In: C. L. Hitchcock et al. 1955–1969. Vascular Plants of the Pacific Northwest. 5 vols. Seattle. Vol. 5.

Cronquist, A. 1955b. Phylogeny and taxonomy of the Compositae. Amer. Midl. Naturalist 53: 478–511.

Cronquist, A. 1971. Note on *Haplopappus gracilis* and *H. ravenii*. Brittonia 23: 292.

Cronquist, A. 1977. The Compositae revisited. Brittonia 29: 137–153.

Cronquist, A. 1978c. *Gynura*. In: N. L. Britton et al., eds. 1905+. North American Flora.... 47+ vols. New York. Ser. 2, part 10, pp. 150–151.

Cronquist, A. 1980. Asteraceae. In: A. E. Radford et al., eds. 1980+. Vascular Flora of the Southeastern United States. 2+ vols. Chapel Hill. Vol. 1.

Cronquist, A. 1981. An Integrated System of Classification of Flowering Plants. New York.

Cronquist, A. 1994. Asteraceae. In: A. Cronquist et al. 1972+. Intermountain Flora. Vascular Plants of the Intermountain West, U.S.A. 5+ vols. in 6+. New York and London. Vol. 5, pp. 5–471.

Cronquist, A. et al. 1972+. Intermountain Flora. Vascular Plants of the Intermountain West, U.S.A. 5+ vols. in 6+. New York and London. (Intermount. Fl.)

Cronquist, A. and D. D. Keck. 1957. A reconstitution of the genus *Machaeranthera*. Brittonia 9: 231–239.

Crow, G. E. and C. B. Hellquist. 2000. Aquatic and Wetland Plants of Northeastern North America.... 2 vols. Madison.

Cuatrecasas, J. 1968. Notas adicionales, taxonómicas y corológicas sobre *Baccharis*. Revista Acad. Colomb. Ci. Exact. 13: 201–226.

Cufodontis, G. 1933. Kritische Revision von *Senecio* Section *Tephroseris*. Feddes Repert. Spec. Nov. Regni Veg. 70: 1–266.

Cult. Native Asters Ontario—See: J. C. Semple et al. 2002

Curr. Genet. = Current Genetics; Eukaryotes with Emphasis on Yeasts, Fungi, Protists, Cell Organelles.

Curtis, W. [1775–]1777–1798. Flora Londinensis.... 2 vols. in 6 fasc. and 72 nos. London. (Fl. Londin.)

Curtiss, A. H. 1881. *Chapmannia* and *Garberia*. Bot. Gaz. 6: 257–259.

Cuvier, F., ed. 1816–1845. Dictionnaire des Sciences Naturelles..., ed. 2. 61 vols. Strasbourg and Paris. (Dict. Sci. Nat. ed. 2)

Cycl.—See: A. Rees [1802–]1819–1820

Dabydeen, S. 1987. Natural hybridization in the genus *Cirsium*: *C. flodmanii* × *C. undulatum*. Rhodora 89: 369–373.

Dabydeen, S. 1997. Natural hybridization in the genus *Cirsium*: *C. altissimum* × *C. discolor*—cytological and morphologic evidence. Rhodora 99: 152–160.

Dahlstedt, H. 1906. Arktiska och alpina arten inom formgruppen *Taraxacum ceratophorum* (Led.) DC. Ark. Bot. 5(9): 1–44.

d'Alcontres, G. S. et al. 1973. Graveolide, a new sesquiterpene lactone of *Inula graveolens*. Gazz. Chim. Ital. 103: 239–246.

Darwiniana = Darwiniana; Carpeta del "Darwinion."

Davidson, A. and G. L. Moxley. 1923. Flora of Southern California.... Los Angeles. (Fl. S. Calif.)

Davidson, R. A. 1963. Initial biometric survey of morphologic variation in the *Cirsium altissimum–C. discolor* complex. Brittonia 15: 222–241.

Davis, F. G. 1981. The genus *Gynura* (Compositae) in Malesia and Australia. Kew Bull. 35: 711–734.

Davis, P. H., ed. 1965–1988. Flora of Turkey and the East Aegean Islands. 10 vols. Edinburgh.

Davis, W. S. 1997. The systematics of annual species of *Malacothrix* (Asteraceae: Lactuceae) endemic to the California Islands. Madroño 44: 223–244.

De Jong, D. C. D. 1965. A systematic study of the genus *Astranthium* (Compositae, Astereae). Publ. Mus. Michigan State Univ., Biol. Ser. 2: 429–528.

De Jong, D. C. D. and G. L. Nesom. 1982. *Erigeron astranthioides* (Asteraceae: Astereae), a new epappose species from western Mexico. Brittonia 34: 285–289.

De Jong, D. C. D. and G. L. Nesom. 1996. Chromosome counts in Mexican *Erigeron*. Madroño 43: 384–392.

Dean, M. L. and K. L. Chambers. 1983. Chromosome numbers and evolutionary patterns in the *Aster occidentalis* (Asteraceae) polyploid complex of western North America. Brittonia 35: 189–196.

den Nijs, J. C. M. and S. B. J. Menken. 1994. Breeding systems and evolution in *Taraxacum*. Evol. Trends Pl. 8: 11–20.

Denkschr. Österr. Akad. Wiss., Math.-Naturwiss. Kl. = Denkschriften, Österreichische Akademie der Wissenschaften. Mathematisch-naturwissenschaftliche Klasse.

Descr. Pl.—See: A. J. Cavanilles [1801–]1802

Descr. Pl. Cap.—See: P. J. Bergius 1767

Descr. Pl. Nouv.—See: É. P. Ventenat [1800–1803]

Desert Pl. = Desert Plants; a Quarterly Journal Devoted to Broadening Our Knowledge of Plants Indigenous or Adaptable to Arid and Sub-arid Regions.

Desfontaines, R. L. 1804. Tableau de l'École Botanique du Muséum d'Histoire Naturelle. Paris. (Tabl. École Bot.)

Desrochers, A. M., J. F. Bain, and S. I. Warwick. 1988. A biosystematic study of the *Carduus nutans* complex in Canada. Canad. J. Bot. 66: 1621–1631.

Desrochers, A. M., J. F. Bain, and S. I. Warwick. 1988b. The biology of Canadian weeds. 89. *Carduus nutans* L. and *Carduus acanthoides* L. Canad. J. Pl. Sci. 68: 1053–1068.

Deut. Bot. Herb.-Buch—See: H. G. L. Reichenbach 1841

DeVore, M. L. and T. F. Stuessy. 1995. The place and time of origin of the Asteraceae, with additional comments on the Calyceraceae and Goodeniaceae. In: D. J. N. Hind et al., eds. 1995. Advances in Compositae Systematics. Kew. Pp. 23–40.

Diagn. Pl. Nov. Mexic.—See: W. B. Hemsley 1878–1880

Dict. Rais. Term. Bot.—See: H. Lecoq and J. Juillet 1831

Dict. Sci. Nat. ed. 2—See: F. Cuvier 1816–1845

Dietrich, D. N. F. 1839–1852. Synopsis Plantarum.... 5 vols. Weimar. (Syn. Pl.)

Diggs, G. M., B. L. Lipscomb, and R. J. O'Kennon. 1999. Shinners' and Mahler's Illustrated Flora of North Central Texas. Fort Worth. [Sida Bot. Misc. 16.]

Dillon, M. O. 1984. A systematic study of *Flourensia* (Asteraceae, Heliantheae). Fieldiana, Bot., n. s. 16: 1–66.

Dittrich, M. 1977[1978]. Cynareae—systematic review. In: V. H. Heywood et al., eds. 1977[1978]. The Biology and Chemistry of the Compositae. 2 vols. London, New York, and San Francisco. Vol. 2, pp. 999–1015.

Dokl. Akad. Nauk S.S.S.R. = Doklady Akademii Nauk S S S R.

Doll, R. 1974. Die Gattung *Taraxacum*. Wittenberg-Lutherstadt.

Doll, R. 1977. Zur *Taraxacum*-Flora Nordamerikas. Feddes Repert. 88: 63–80.

Dominguez, X. A., M. Gutierrez, and R. Aragon. 1977. Isolation of baileyolin, a tumor inhibitory and antibiotic sesquiterpene lactone from *Baileya multiradiata*. Pl. Med. (Stuttgart) 30: 356–359.

Don, D. 1830. On the origin and nature of the ligulate rays in *Zinnia*; and on a remarkable multiplication observed in the parts of fructification of that genus. Trans. Linn. Soc. London 16: 155–158.

Dorn, R. D. 1988c. Typification of *Chaenactis alpina* (Asteraceae). Madroño 35: 161–162.

Dorn, R. D. 1992. Vascular Plants of Wyoming, ed. 2. Cheyenne. (Vasc. Pl. Wyoming ed. 2)

Dostál, J. 1976. *Centaurea*. In: T. G. Tutin et al., eds. 1964–1980. Flora Europaea. 5 vols. Cambridge. Vol. 4, pp. 254–301.

Douglas, G. W., G. B. Straley, and D. V. Meidinger. 1989–1994. The Vascular Plants of British Columbia. 4 vols. Victoria. [B.C. Minist. Forests, Special Rep. 1–4.]

Douglas, G. W., G. B. Straley, D. V. Meidinger, and J. Pojar. 1998–2002. Illustrated Flora of British Columbia. 8 vols. Victoria.

Downie, S. R. and K. E. Denford. 1988. Taxonomy of *Arnica* (Asteraceae) subgenus *Arctica*. Rhodora 90: 245–275.

Dress, W. J. 1966. Notes on the cultivated Compositae 9. *Onopordum*. Baileya 14: 74–86.

Drew, M. B. and E. C. Clebsch. 1995. Studies on the endangered *Echinacea tennesseensis* (Asteraceae): Plant community and demographic analysis. Castanea 60: 60–69.

Drury, D. G. 1970. A fresh approach to the classification of the genus *Gnaphalium* with particular reference to the species present in New Zealand (Inuleae–Compositae). New Zealand J. Bot. 9: 157–185.

Drury, D. G. 1972. The cluster and solitary-headed cudweeds native to New Zealand (*Gnaphalium* section *Euchiton*–Compositae). New Zealand J. Bot. 10: 112–179.

Dudman, A. A. and A. J. Richards. 1997. Dandelions of Great Britain and Ireland. London.

Duistermaat, H. 1996. Monograph of *Arctium* L. (Asteraceae). Gorteria, suppl. 3: 1–143.

Dumont de Courset, G. L. M. 1802–1805. Le Botaniste Cultivateur.... 5 vols. Paris. (Bot. Cult.)

Dumortier, B. C. J. 1827. Florula Belgica, Operis Majoris Prodromus.... Tournay. (Fl. Belg.)

Dumortier, B. C. J. 1829. Analyse des Familles des Plantes.... Tournay. (Anal. Fam. Pl.)

Dunford, M. P. 1964. A cytogenetic analysis of certain polyploids in *Grindelia* (Compositae). Amer. J. Bot. 51: 41–61.

Dunford, M. P. 1986. Chromosome relationships of diploid species of *Grindelia* (Compositae) from Colorado, New Mexico, and adjacent areas. Amer. J. Bot. 73: 297–303.

Durand, E. M. 1855. Plantae Prattenianae Californicae.... Philadelphia. [Preprinted from J. Acad. Nat. Sci. Philadelphia, n. s. 3: 79–104. 1855.] (Pl. Pratten. Calif.)

Durand, E. M. and T. C. Hilgard. 1854. Plantae Heermannianae.... Philadelphia. [Preprinted from J. Acad. Nat. Sci. Philadelphia, n. s. 3: 37–46. 1855.] (Pl. Heermann.)

Eaton, A. 1829. A Manual of Botany for the Northern States, ed. 5. Albany. (Man. Bot. ed. 5)

Ecology = Ecology, a Quarterly Journal Devoted to All Phases of Ecological Biology.

Econ. Bot. = Economic Botany; Devoted to Applied Botany and Plant Utilization.

Ediger, R. I. 1970. Revision of section *Suffruticosi* of the genus *Senecio* (Compositae). Sida 3: 504–524.

Edinburgh J. Bot. = Edinburgh Journal of Botany.

Edinburgh New Philos. J. = Edinburgh New Philosophical Journal.

Edwards's Bot. Reg. = Edwards's Botanical Register....

Einck, J. J., C. L. Herald, G. R. Pettit, and R. B. Von Dreele. 1978. Antineoplastic agents, part 53, the crystal structure of radiatin. J. Amer. Chem. Soc. 100: 3544–3548.

Eldenäs, P. K., M. Källersjö, and A. A. Anderberg. 1999. Phylogenetic placement and circumscription of tribes Inuleae s. str. and Plucheeae (Asteraceae): Evidence from sequences of chloroplast gene *ndh*F. Molec. Phylogen. Evol. 13: 50–58.

Elisens, W. J., R. D. Boyd, and A. D. Wolfe. 1992. Genetic and morphological divergence among varieties of *Aphanostephus skirrhobasis* (Asteraceae–Astereae) and related species with different chromosome numbers. Syst. Bot. 17: 380–394.

Elliott, S. [1816–]1821–1824. A Sketch of the Botany of South-Carolina and Georgia. 2 vols. in 13 parts. Charleston. (Sketch Bot. S. Carolina)

Ellison, W. L. 1964. A systematic study of the genus *Bahia* (Compositae). Rhodora 66: 67–86, 177–215, 281–311.

Ellison, W. L. 1971. Taxonomy of *Platyschkuhria* (Compositae). Brittonia 23: 269–279.

Elven, R. 1989. Sibirstjerna-arten som overlevde til tross for fredning. (*Aster sibiricus* L. in Norway.) Blyttia 47: 51–53.

Emory, W. H. 1848. Notes of a Military Reconnoissance, from Fort Leavenworth, in Missouri, to San Diego, in California, Including Part of the Arkansas, Del Norte, and Gila Rivers.... Made in 1846–7, with the Advanced Guard of the "Army of the West." Washington. (Not. Milit. Reconn.)

Emory, W. H. 1857–1859. Report on the United States and Mexican Boundary Survey, Made under the Direction of the Secretary of the Interior. 2 vols. in parts. Washington. (Rep. U.S. Mex. Bound.)

Encycl.—See: J. Lamarck et al. 1783–1817

Engler, H. G. A., ed. 1900–1953. Das Pflanzenreich.... 107 vols. Berlin. [Sequence of volume (Heft) numbers (order of publication) is independent of the sequence of series and family (Roman and Arabic) numbers (taxonomic order).] (Pflanzenr.)

Engler, H. G. A. and K. Prantl, eds. 1887–1915. Die natürlichen Pflanzenfamilien.... 254 fasc. Leipzig. [Sequence of fascicle (Lieferung) numbers (order of publication) is independent of the sequence of division (Teil) and subdivision (Abteilung) numbers (taxonomic order).] (Nat. Pflanzenfam.)

Englera = Englera; Veröffentlichungen aus dem Botanischen Garten und Botanischen Museum, Berlin-Dahlem.

Enum.—See: P. C. Fabricius 1759

Enum. Pl.—See: W. S. J. G. von Besser 1822; C. L. Willdenow 1809–1813[–1814]

Enum. Syst. Pl.—See: N. J. Jacquin 1760

Environm. Entomol. = Environmental Entomology.

Erythea = Erythea; a Journal of Botany, West American and General.

Estes, J. R. 1969. Evidence for autoploid evolution in the *Artemisia ludoviciana* complex of the Pacific Northwest. Brittonia 21: 29–43.

Eupatoria Verticillata—See: J. Barratt 1841

Evans, T. M. and G. K. Brown. 1991. Chloroplast DNA variation in *Haplopappus* section *Oonopsis* (Asteraceae). [Abstract.] Amer. J. Bot. 78(6, suppl.): 185.

Evol. Trends Pl. = Evolutionary Trends in Plants.

Evolution = Evolution, International Journal of Organic Evolution.

Exsicc. Genav. Conserv. Bot. Distrib. Fasc. = Exsiccatorum Genavensium e Conservatorio Botanico Distributorum Fasciculus.

Fabricius, P. C. 1759. Enumeratio Methodica Plantarum Horti Medici Helmstadiensis.... Helmstedt. (Enum.)

Fam. Pl.—See: M. Adanson 1763[–1764]

Fasc. Pl. Cantabr.—See: I. Lyons 1763

Feddema, C. 1966. Systematic Studies in the Genus *Sclerocarpus* and the Genus *Aldama* (Compositae). Ph.D. dissertation. University of Michigan.

Feddes Repert. = Feddes Repertorium.

Feddes Repert. Spec. Nov. Regni Veg. = Feddes Repertorium Specierum Novarum Regni Vegetabilis.

Fedorov, An. A., ed. 1999+. Flora of Russia: The European Part and Bordering Regions...Translated from Russian. 7+ vols. Rotterdam and Brookfield, Vt.

Fernald, M. L. 1933. Recent discoveries in the Newfoundland flora. Rhodora 35: 1–16 etc.

Fernald, M. L. 1943c. Notes on *Hieracium*. Rhodora 45: 317–325.

Fernald, M. L. 1945c. Key to *Antennaria* of the "Manual range." Rhodora 47: 221–239.

Fernald, M. L. 1946d. Technical studies on North American plants. IV. Novelties in our flora. Rhodora 48: 54–60, 65–81.

Fernald, M. L. 1950. Gray's Manual of Botany, ed. 8. New York.

Fernandes, R. 1976. *Anthemis*. In: T. G. Tutin et al., eds. 1964–1980. Flora Europaea. 5 vols. Cambridge. Vol. 4, pp. 145–159.

Field & Lab. = Field & Laboratory.

Fieldiana, Bot. = Fieldiana: Botany.

Fisher, T. R. 1957. Taxonomy of the genus *Heliopsis* (Compositae). Ohio J. Sci. 57: 171–191.

Fiz, O., V. Valcárcel, and P. Vargas. 2002. Phylogenetic position of Mediterranean Astereae and character evolution of daisies (*Bellis*, Asteraceae) inferred from nrDNA ITS sequences. Molec. Phylogen. Evol. 25: 157–171.

Fl. Alaska Yukon—See: E. Hultén 1941–1950

Fl. Amer. Sept.—See: F. Pursh [1813]1814

Fl. Arct. URSS—See: A. I. Tolmatchew 1960–1987

Fl. Ariz. New Mex.—See: I. Tidestrom and T. Kittell 1941

Fl. Belg.—See: B. C. J. Dumortier 1827

Fl. Bor.-Amer.—See: W. J. Hooker [1829–]1833–1840; A. Michaux 1803

Fl. Boston. ed. 2—See: J. Bigelow 1824

Fl. Brit.—See: J. E. Smith 1800–1804

Fl. Canada—See: H. J. Scoggan 1978–1979

Fl. Carniol. ed. 2—See: J. A. Scopoli 1771–1772

Fl. Carol.—See: T. Walter 1788

Fl. Colorado—See: P. A. Rydberg 1906

Fl. Dan.—See: G. C. Oeder et al. [1761–]1764–1883

Fl. Erlang.—See: A. F. Schweigger and H. F. F. Körte 1811

Fl. Franç.—See: J. Lamarck 1778[1779]

Fl. Francisc.—See: E. L. Greene 1891–1897

Fl. Gén. Env. Paris ed. 2—See: F. F. Chevallier 1836

Fl. Graec. Prodr.—See: J. Sibthorp and J. E. Smith 1806–1813 [–1816]

Fl. Hall.—See: E. M. Fries 1817[–1819]

Fl. Ins. Austr.—See: G. Forster 1786

Fl. Londin.—See: W. Curtis [1775–]1777–1798

Fl. Ludov.—See: C. S. Rafinesque 1817

Fl. Lusit.—See: F. Brotero 1804[–1805]

Fl. Miami—See: J. K. Small 1913b

Fl. N. Amer.—See: J. Torrey and A. Gray 1838–1843

Fl. Napol.—See: M. Tenore 1811–1836[–1838]

Fl. Norveg.—See: J. E. Gunnerus 1766–1772

Fl. Nov.-Zel.—See: J. D. Hooker 1852–1855

Fl. N.W. Amer.—See: T. J. Howell 1897–1903

Fl. Orient.—See: P. E. Boissier 1867–1888

Fl. Palouse Reg.—See: C. V. Piper and R. K. Beattie 1901

Fl. Pedem.—See: C. Allioni 1785

Fl. Peruv. Prodr.—See: H. Ruiz López and J. A. Pavón 1794

Fl. Pis.—See: G. Savi 1798

Fl. Plains N. Amer.—See: P. A. Rydberg 1932

Fl. Rocky Mts.—See: P. A. Rydberg 1917

Fl. S. Calif.—See: A. Davidson and G. L. Moxley 1923

Fl. Samojed. Cisural.—See: F. Ruprecht 1845b

Fl. Saxon.—See: H. G. L. Reichenbach 1842

Fl. S.E. U.S.—See: J. K. Small 1903

Fl. Sicul. Syn.—See: G. Gussone 1842–1844[–1845]

Fl. South. U.S.—See: A. W. Chapman 1860

Fl. South. U.S. ed. 2—See: A. W. Chapman 1883

Fl. Suec. ed. 2—See: C. Linnaeus 1755b

Fl. Taur.-Caucas.—See: F. A. Marschall von Bieberstein 1808–1819

Fl. Tellur.—See: C. S. Rafinesque 1836[1837–1838]

Fl. URSS—See: V. L. Komarov et al. 1934–1964

Fl. W. Calif.—See: W. L. Jepson 1901

Flexner, S. B. and L. C. Hauck, eds. 1987. The Random House Dictionary of the English Language, ed. 2 unabridged. New York.

Flora = Flora; oder (allgemeine) botanische Zeitung. [Vols. 1–16, 1818–33, include "Beilage" and "Ergänzungsblätter"; vols. 17–25, 1834–42, include "Beiblatt" and "Intelligenzblatt."]

Flora of North America Editorial Committee, eds. 1993+. Flora of North America North of Mexico. 12+ vols. New York and Oxford.

Flyr, L. D. 1970. *Brickellia*. In: D. S. Correll and M. C. Johnston. 1970. Manual of the Vascular Plants of Texas. Renner, Tex. Pp. 1545–1547.

Folia Geobot. Phytotax. = Folia Geobotanica et Phytotaxonomica.

Forbes, J. 1833. Hortus Woburnensis, a Descriptive Catalogue of Upwards of Six Thousand Ornamental Plants Cultivated at Woburn Abbey.... London. (Hort. Woburn.)

Forster, G. 1786. Florulae Insularum Australium Prodromus. Göttingen. (Fl. Ins. Austr.)

Foster, S. 1991. *Echinacea*: The Purple Coneflowers. Austin.

Franco, J. do A. 1976. *Carduus*. In: T. G. Tutin et al., eds. 1964–1980. Flora Europaea. 5 vols. Cambridge. Vol. 4, pp. 220–232.

Franklin, J. et al. 1823. Narrative of a Journey to the Shores of the Polar Sea, in the Years 1819, 20, 21 and 22. London. [Richardson: Appendix VII. Botanical appendix, pp. [729]–778, incl. bryophytes by Schwägrichen, algae and lichens by Hooker. Brown: Addenda [to Appendix VII], pp. 779–784.] (Narr. Journey Polar Sea)

Freire, S. E. and L. Iharlegui. 1997. Sinopsis preliminar del género *Gamochaeta* (Asteraceae, Gnaphalieae). Bol. Soc. Argent. Bot. 33: 23–35.

Frémont, J. C. 1843–1845. Report of the Exploring Expedition to the Rocky Mountains in the Year 1842, and to Oregon and North California in the Year 1843–44. 2 parts. Washington. [Parts paged consecutively.] (Rep. Exped. Rocky Mts.)

Frey, D., M. Baltisberger, and P. J. Edwards. 2003. Cytology of *Erigeron annuus* s.l. and its consequences in Europe. Bot. Helv. 113: 1–14.

Fries, E. M. 1814–1824. Novitiae Florae Svecicae.... 7 parts. Lund. [Parts paged consecutively.] (Novit. Fl. Svec.)

Fries, E. M. 1817[–1819]. Flora Hallandica.... 10 sects. Lund. [Sections paged consecutively.] (Fl. Hall.)

Fries, E. M. 1845–1849. Summa Vegetabilium Scandinaviae.... 2 sects. Uppsala, Stockholm, and Leipzig. [Sections paged consecutively.] (Summa Veg. Scand.)

Fruct. Sem. Pl.—See: J. Gaertner 1788–1791[–1792]

Funk, V. A., R. Chan, and S. C. Keeley. 2004. Insights into the evolution of the tribe Arctoteae (Compositae: subfamily Cichorioideae s.s.) using *trn*L-F, *ndh*F, and ITS. Taxon 53: 637–655.

Gabrielian, E. and C. E. Jarvis. 1996. *Amberboa moschata*, *A. glauca,* and *A. amberboi* (Asteraceae: Cardueae). A note on their taxonomy and typification of their names. Taxon 45: 213–215.

Gaertner, J. 1788–1791[–1792]. De Fructibus et Seminibus Plantarum.... 2 vols. Stuttgart and Tübingen. [Vol. 1 in 1 part only, 1788. Vol. 2 in 4 parts paged consecutively: pp. 1–184, 1790; pp. 185–352, 353–504, 1791; pp. 505–520, 1792.] (Fruct. Sem. Pl.)

Gaiser, L. O. 1946. The genus *Liatris*. Rhodora 48: 165–183, 216–263, 273–326, 331–382, 393–412.

Gaiser, L. O. 1950. Chromosome studies in *Liatris*. I. *Spicatae* and *Pycnostachyae*. Amer. J. Bot. 37: 122–135.

Gaiser, L. O. 1950b. Chromosome studies in *Liatris*. II. *Graminifoliae* and *Pauciflorae*. Amer. J. Bot. 37: 414–423.

Gaiser, L. O. 1950c. Chromosome studies in *Liatris*. III. *Punctatae*. Amer. J. Bot. 37: 763–777.

Gaiser, L. O. 1950d. Evidence for intersectional field hybrids in *Liatris*. Evolution 5: 52–67.

Gaiser, L. O. 1954. Studies in the Kuhniinae (Eupatorieae). II. J. Arnold Arbor. 35: 87–133.

Gallez, G. J. and L. D. Gottlieb. 1982. Genetic evidence for the hybrid origin of the diploid plant *Stephanomeria diegensis*. Evolution 36: 1158–1167.

Ganders, F. R., B. Klinkenberg, and R. Klinkenberg. 2003. Taxonomy in conservation: The enigmatic Vancouver Island beggarticks. Davidsonia 14: 63–70.

Gandhi, K. N. 1999. Nomenclatural novelties for the Western Hemisphere plants—II. Harvard Pap. Bot. 4: 295–299.

Gandhi, K. N. and R. D. Thomas. 1989. Asteraceae of Louisiana. Dallas. [Sida Bot. Misc. 4.]

Gandhi, K. N. and R. D. Thomas. 1991. Additional notes on the Asteraceae of Louisiana. Sida 14: 514–517.

Garcia-Jacas, N., A. Susanna, T. Garnatje, and R. Vilatersana. 2001. Generic delimitation and phylogeny of the subtribe Centaureinae (Asteraceae): A combined nuclear and chloroplast DNA analysis. Ann. Bot. (Oxford) 87: 503–515.

Garcia-Jacas, N., A. Susanna, and R. Ilarslan. 1996. Aneuploidy in the Centaureinae (Compositae): Is *n* = 7 the end of the series? Taxon 45: 39–42.

Garcia-Jacas, N., A. Susanna, V. Mozaffarian, and R. Ilarslan. 2000. The natural delimitation of *Centaurea* (Asteraceae: Cardueae): ITS sequence analysis of the *Centaurea jacea* group. Pl. Syst. Evol. 223: 185–199.

Gard. & Forest = Garden and Forest; a Journal of Horticulture, Landscape Art and Forestry.

Gard. Dict. ed. 8—See: P. Miller 1768

Gard. Dict. Abr. ed. 4—See: P. Miller 1754

Gardeau, C. 1972. Recherches biosystématiques sur la section *Jacea* Cass. et quelques sections voisines du genre *Centaurea* L. en France et dans les régiones limitrophes. Feddes Repert. 83: 311–472.

Gardner, C. A. and H. W. Bennetts. 1956. The Toxic Plants of Western Australia. Perth.

Gartenflora = Gartenflora; Monatsschrift für deutsche und schweizerische Garten- und Blumenkunde.

Gazz. Chim. Ital. = Gazzetta Chimica Italiana.

Gen. N. Amer. Pl.—See: T. Nuttall 1818

Gen. Pl.—See: G. Bentham and J. D. Hooker 1862–1883; A. L. de Jussieu 1789; J. C. Schreber 1789–1791

Gen. Pl. ed. 5—See: C. Linnaeus 1754

Gen. Sp. Aster.—See: C. G. D. Nees 1832

Gen. Sp. Pl.—See: M. Lagasca y Segura 1816b

Gentes Herb. = Gentes Herbarum; Occasional Papers on the Kinds of Plants.

Gerdes, L. B. 1998. *Aster nemoralis* and the apparent hybrid *Aster nemoralis* × *Doellingeria umbellata* in Houghton County. Michigan Bot. 37: 17–20.

Ges. Naturf. Freunde Berlin Mag. Neuesten Entdeck. Gesammten Naturk. = Der Gesellschaft naturforschender Freunde zu Berlin Magazin für die neuesten Entdeckungen in der gesammten Naturkunde.

Gillis, W. T. 1977. *Pluchea* revisited. Taxon 26: 587–591.

Giuliano, D. A. 2001. Clasificación infragenérica de las especies Argentinas de *Baccharis* (Asteraceae, Astereae). Darwiniana 39: 131–154.

Giuliano, D. A. and G. L. Nesom. 2003. A new section of *Baccharis* (Asteraceae: Astereae), and notes on allied taxa. Sida 20: 1481–1484.

Gleason, H. A. 1922. *Centratherum*. In: N. L. Britton et al., eds. 1905+. North American Flora.... 47+ vols. New York. Vol. 33, pp. 49–50.

Gleason, H. A. 1922b. *Vernonia*. In: N. L. Britton et al., eds. 1905+. North American Flora.... 47+ vols. New York. Vol. 33, pp. 52–95.

Gleason, H. A. and A. Cronquist. 1963. Manual of Vascular Plants of Northeastern United States and Adjacent Canada. Princeton.

Gleason, H. A. and A. Cronquist. 1991. Manual of Vascular Plants of Northeastern United States and Adjacent Canada, ed. 2. Bronx.

Gmelin, J. F. 1791[–1792]. Caroli à Linné...Systema Naturae per Regna Tria Naturae.... Tomus II. Editio Decima Tertia, Aucta, Reformata. 2 parts. Leipzig. (Syst. Nat.)

Godfrey, R. K. 1952. *Pluchea,* section *Stylimnus,* in North America. J. Elisha Mitchell Sci. Soc. 68: 238–271.

Godfrey, R. K. 1958. A synopsis of *Gnaphalium* (Compositae) in the southeastern United States. Quart. J. Florida Acad. Sci. 21: 177–184.

Godfrey, R. K. and J. W. Wooten. 1981. Aquatic and Wetland Plants of the Southeastern United States: Dicotyledons. Athens, Ga.

Goodrich, S. and S. L. Welsh. 1983. New variety of

Stephanomeria tenuifolia (Compositae) from Utah. Great Basin Naturalist 43: 375.

Gorteria = Gorteria; Mededelingenblad ten Dienste van de Floristiek en het Vegetatie-onderzoek van Nederland.

Gottlieb, L. D. 1971. Evolutionary relationships in the outcrossing diploid annual species of *Stephanomeria* (Compositae). Evolution 25: 312–329.

Gottlieb, L. D. 1972. A proposal for classification of the annual species of *Stephanomeria* (Compositae). Madroño 21: 463–481.

Gottlieb, L. D. 1973b. Genetic differentiation, sympatric speciation and the origin of a diploid species of *Stephanomeria*. Amer. J. Bot. 60: 545–553.

Gottlieb, L. D. 1977. Phenotypic similarity of *Stephanomeria exigua* subsp. *coronaria* and its recent derivative species "*malheurensis.*" Amer. J. Bot. 64: 873–880.

Gottlieb, L. D. 1978. *Stephanomeria malheurensis*, a new species from Oregon. Madroño 25: 44–46.

Gottlieb, L. D. 1978b. Allocation, growth rates and gas exchange in seedlings of *Stephanomeria exigua* subsp. *coronaria* and its recent derivative *S. malheurensis*. Amer. J. Bot. 65: 970–977.

Gottlieb, L. D. 1979. The origin of phenotype in a recently evolved species. In: O. T. Solbrig et al., eds. 1979. Topics in Plant Population Biology. New York. Pp. 264–286.

Gottlieb, L. D. 1991. The Malheur wire-lettuce: A rare, recently evolved Oregon species. Kalmiopsis 1: 9–13.

Gottlieb, L. D. and J. P. Bennett. 1983. Interference between individuals in pure and mixed cultures of *Stephanomeria malheurensis* and its progenitor. Amer. J. Bot. 70: 276-284.

Graf, M. 1999. Plants of the Tahoe Basin.... Sacramento and Berkeley.

Graham, A. 1996. A contribution to the geologic history of the Compositae. In: D. J. N. Hind et al., eds. 1996. Proceedings of the International Compositae Conference, Kew, 1994. 2 vols. Kew. Vol. 1, pp. 123–140.

Grana = Grana; an International Journal of Palynology Including "World Pollen and Spore Flora."

Grashoff, J. L. 1972. A Systematic Study of the North and Central American Species of *Stevia*. Ph.D. dissertation. University of Texas.

Grau, J. 1976. Chromosomenzahlen von sudamerikanischen *Haplopappus*-arten. Mitt. Bot. Staatssamml. München 12: 403–410.

Gray, A. 1848. A Manual of the Botany of the Northern United States.... Boston, Cambridge, and London. (Manual)

Gray, A. 1854. Plantae Novae Thurberianae.... Cambridge, Mass. [Preprinted from Mem. Amer. Acad. Arts, n. s. 5: 297–328. 1855.] (Pl. Nov. Thurb.)

Gray, A. 1856. A Manual of the Botany of the Northern United States..., ed. 2. New York. (Manual ed. 2)

Gray, A. 1867. A Manual of the Botany of the Northern United States..., ed. 5. New York and Chicago. [Pteridophytes by D. C. Eaton.] (Manual ed. 5)

Gray, A. 1880. Botanical Contributions. 1880. 1. Notes on Some Compositae. [N. p.] [Preprinted with pagination unchanged from Proc. Amer. Acad. Arts 16: 78–102. 1881.] (Notes Compositae)

Gray, A. 1884. *Stephanomeria* Nuttall. Proc. Amer. Acad. Arts 19: 59–63.

Gray, A. 1886. *Stephanomeria*. In: A. Gray et al. 1886. Synoptical Flora of North America: The Gamopetalae, Being a Second Edition of Vol. i Part ii, and Vol. ii Part i, Collected. 2 vols. New York, London, and Leipzig. Vol. 2, part 1, pp. 412–415.

Gray, A., et al. 1878–1897. Synoptical Flora of North America. 2 vols. in parts and fasc. New York etc. [Vol. 1(1,1), 1895; vol. 1(1,2), 1897; vol. 1(2), 1884; vol. 2(1), 1878.] (Syn. Fl. N. Amer.)

Gray, A. et al. 1886. Synoptical Flora of North America: The Gamopetalae, Being a Second Edition of Vol. i Part ii, and Vol. ii Part i, Collected. 2 vols. New York, London, and Leipzig. [Reissued 1888 as Smithsonian Misc. Collect. 591.] (Syn. Fl. N. Amer. ed. 2)

Great Basin Naturalist Mem. = Great Basin Naturalist Memoirs.

Great Plains Flora Association. 1986. Flora of the Great Plains. Lawrence, Kans.

Greene, E. L. 1891–1897. Flora Franciscana. An Attempt to Classify and Describe the Vascular Plants of Middle California. 4 parts. San Francisco. [Parts paged consecutively.] (Fl. Francisc.)

Greene, E. L. 1894. Manual of the Botany of the Region of San Francisco Bay.... San Francisco. (Man. Bot. San Francisco)

Greene, E. L. 1894b. Observations on the Compositae. Erythea 2: 105–112.

Greene, E. L. 1898. Studies in the Compositae—VII. 1. Some helenioid genera. *Tetraneuris*. Pittonia 3: 264–270.

Greene, E. L. [1901.] Plantae Bakerianae. 3 vols. [Washington.] (Pl. Baker.)

Greene, E. L. 1901b. *Taraxacum* in North America. Pittonia 4: 227–242.

Greene, E. L. 1902. A study of *Euthamia*. Pittonia 5: 72–80.

Greenman, J. M. 1917. Two exotic Compositae in North America. Ann. Missouri Bot. Gard. 4: 289–292.

Greer, L. F. 1997. *Thelesperma curvicarpum* (Asteraceae), an achene form in populations of *T. simplicifolium* var. *simplicifolium* and *T. filifolium* var. *filifolium*. SouthW. Naturalist 42: 242–244.

Gregory, S. D., E. K. Espeland, T. M. Carlsen, and E. K. Bissell. 2001. Demography and population biology of a rare tarplant, *Blepharizonia plumosa* (Asteraceae), a California summer annual forb. Madroño 48: 271–285.

Greuter, W. 2003. The Euro+Med treatment of Cichorieae (Compositae)—generic concepts and required new names. Willdenowia 33: 229–238.

Greuter, W., M. V. Aghababian, and G. Wagenitz. 2005. Vaillant on Compositae—systematic concepts and nomenclatural impact. Taxon 54: 149–174.

Greuter, W., G. Wagenitz, M. V. Agababjan, and F. H. Hellwig. 2001. Proposal to conserve the name *Centaurea* (Compositae) with a conserved type. Taxon 50: 1201–1205.

Grierson, A. J. C. 1975. *Gnaphalium*. In: P. H. Davis, ed. 1965–1988. Flora of Turkey and the East Aegean Islands. 10 vols. Edinburgh. Vol. 5, pp. 97–100.

Grisebach, A. H. R. 1866. Catalogus Plantarum Cubensium Exhibens Collectionem Wrightianam Aliasque Minores ex Insula Cuba Missas. Leipzig. (Cat. Pl. Cub.)

Grisebach, A. H. R. 1879. Symbolae ad Floram Argentinam. Zweite Bearbeitung argentinischer Pflanzen. Göttingen. (Symb. Fl. Argent.)

Groh, H. 1943. Canadian Weed Survey. Second Annual Report. Ottawa.

Gruezo, W. S. and K. E. Denford. 1994. Taxonomy of *Arnica* L. subgenus *Chamissonis* Maguire (Asteraceae). Asia Life Sci. 3: 89–212.

Gunnerus, J. E. 1766–1772. Flora Norvegica.... 2 vols. Trondhiem and Copenhagen. (Fl. Norveg.)

Guppy, G. A. 1978. Species relationships of *Hieracium* (Asteraceae) in British Columbia. Canad. J. Bot. 56: 3008–3019.

Gussone, G. 1842–1844[–1845]. Florae Siculae Synopsis.... 2 vols. Naples. (Fl. Sicul. Syn.)

Gustafsson, M. H. G. and K. Bremer. 1995. Morphology and phylogenetic interrelationships of the Asteraceae, Calyceraceae, Campanulaceae, Goodeniaceae, and related families (Asterales). Amer. J. Bot. 82: 250–265.

Haglund, G. 1943. *Taraxacum* in arctic Canada (east of 100° W.). Rhodora 45: 337–343.

Haglund, G. 1946. Contributions to the knowledge of the *Taraxacum*-flora of Alaska and Yukon. Svensk Bot. Tidskr. 40: 325–361.

Haglund, G. 1948. Further contributions to the knowledge of the *Taraxacum*-flora of Alaska and Yukon. Svensk Bot. Tidskr. 42: 297–336.

Haglund, G. 1949. Supplementary notes on the *Taraxacum*-flora of Alaska and Yukon. Svensk Bot. Tidskr. 43: 107–116.

Haines, A. 2000. Rediscovery of *Symphyotrichum anticostense* in the United States. Rhodora 102: 198–201.

Hall, E. 1873. Plantae Texanae.... Salem, Mass. (Pl. Tex.)

Hall, G. W. 1967. A Biosystematic Study of the North American Complex of the Genus *Bidens* (Compositae). Ph.D. dissertation. Indiana University.

Hall, H. M. 1928. The genus *Haplopappus*: A phylogenetic study in the Compositae. Publ. Carnegie Inst. Wash. 389.

Hall, H. M. and F. E. Clements. 1923. The phylogenetic method in taxonomy: The North American species of *Artemisia, Chrysothamnus,* and *Atriplex*. Publ. Carnegie Inst. Wash. 326.

Handb. Gewächsk. ed. 2—See: J. C. Mössler and H. G. L. Reichenbach 1827–1829[–1830]

Handel-Mazzetti, H. 1907. Monographie der Gattung *Taraxacum*. Leipzig and Vienna. (Monogr. Taraxacum)

Hanelt, P. 1963. Monographisch Übersicht der Gattung *Carthamus* L. (Compositae). Feddes Repert. Spec. Nov. Regni Veg. 67: 41–180.

Hanelt, P. 1976. *Carthamus*. In: T. G. Tutin et al., eds. 1964–1980. Flora Europaea. 5 vols. Cambridge. Vol. 4, pp. 302–303.

Hansen, C. J., L. Allphin, and M. D. Windham. 2002. Biosystematic analysis of the *Thelesperma subnudum* complex (Asteraceae). Sida 20: 71–96.

Hansen, H. V. 1990. Phylogenetic studies in the *Gerbera*-complex (Compositae, tribe Mutisieae, subtribe Mutisiinae). Nordic J. Bot. 9: 469–485.

Harms, V. L. 1965. Biosystematic studies in the *Heterotheca subaxillaris* complex (Compositae–Astereae). Trans. Kansas Acad. Sci. 68: 122–124.

Harms, V. L. 1970. *Heterotheca*. In: D. S. Correll and M. C. Johnston. 1970. Manual of the Vascular Plants of Texas. Renner, Tex. Pp. 1563–1569.

Harms, V. L. 1999. Status Report on the Large-headed Woolly Yarrow *(Achillea millefolium* var. *megacephala),* in Canada. [Ottawa.]

Harms, V. L. 2003. Checklist of the Vascular Plants of Saskatchewan.... Saskatoon.

Harrington, H. D. 1954. Manual of the Plants of Colorado. Denver. (Man. Pl. Colorado)

Harris, E. M. 1994. Developmental evidence for the derivation of syncephalia in *Lagascea* (Heliantheae; Asteraceae). Amer. J. Bot. 81: 1139–1148.

Hartman, R. L. 1976. A Conspectus of *Machaeranthera* (Compositae: Astereae) and a Biosystematic Study of the Section *Blepharodon*. Ph.D. dissertation. University of Texas.

Hartman, R. L. 1990. A conspectus of *Machaeranthera* (Asteraceae: Astereae). Phytologia 68: 439–465.

Hartman, R. L. and M. A. Lane. 1991. A natural intergeneric hybrid in the $x = 6$ group of the Astereae (Asteraceae). Sida 14: 321–329.

Harvard Pap. Bot. = Harvard Papers in Botany.

Hauber, D. P. 1986. Autotetraploidy in *Haplopappus spinulosus* hybrids: Evidence from natural and synthetic tetraploids. Amer. J. Bot. 73: 1595–1608.

Hayden, F. V. 1872. Preliminary Report of the United States Geological Survey of Montana and Portions of Adjacent Territories; Being a Fifth Annual Report of Progress. Washington. (Prelim. Rep. U.S. Geol. Surv. Montana)

Heap, J. W. 1993. Control of rush skeletonweed *(Chondrilla juncea)* with herbicides. Weed Technol. 7: 954–959.

Hebert, H. J.-C. 1968. Generic considerations concerning *Carphephorus* and *Trilisa* (Compositae). Rhodora 70: 474–485.

Hegi, G. et al. 1936–1987. Illustrierte Flora von Mitteleuropa, ed. 2. 6 vols. in 14. Munich, Berlin, and Hamburg.

Heiser, C. B. 1945. A revision of the genus *Schkuhria*. Ann. Missouri Bot. Gard. 32: 265–278.

Heiser, C. B. 1949. Study in the evolution of the sunflower species *Helianthus annuus* and *H. bolanderi*. Univ. Calif. Publ. Bot. 23: 157–208.

Heiser, C. B. 1956. Biosystematics of *Helianthus debilis*. Madroño 13: 145–176.

Heiser, C. B. 1976. The Sunflower. Norman.

Heiser, C. B., D. M. Smith, S. Clevenger, and W. C. Martin. 1969. The North American sunflowers *(Helianthus)*. Mem. Torrey Bot. Club 22(3).

Helton, N., D. Wiens, and B. A. Barlow. 1972. High polyploidy and the origin of *Balsamorhiza macrophylla*. Madroño 21: 526–535.

Hemsley, W. B. 1878–1880. Diagnoses Plantarum Novarum...Mexicanarum et Centrali-americanarum. 3 parts. London. [Parts paged consecutively.] (Diagn. Pl. Nov. Mexic.)

Hemsley, W. B. 1879–1888. Biologia Centrali-Americana.... Botany.... 5 vols. London. (Biol. Cent.-Amer., Bot.)

Herb. Brit.—See: J. Hill 1769–1770

Herb. Raf.—See: C. S. Rafinesque 1833

Heredity = Heredity; an International Journal of Genetics.

Heyn, C. C., O. Dagan, and B. Nachman. 1974. The annual *Calendula* species: Taxonomy and relationships. Israel J. Bot. 23: 169–201.

Heywood, V. H., ed. 1978. Flowering Plants of the World. Oxford.

Heywood, V. H., J. B. Harborne, and B. L. Turner, eds. 1977[1978]. The Biology and Chemistry of the Compositae. 2 vols. London, New York, and San Francisco. (Biol. Chem. Compositae)

Hickler, M. G. 1999. Notes on the habits and life-history of *Bidens discoidea*: An epiphyte in Massachusetts floodplain ponds. Rhodora 101: 298–299.

Hickman, J. C., ed. 1993. The Jepson Manual. Higher Plants of California. Berkeley, Los Angeles, and London.

Hill, D. W., E. M. Bailey, and B. J. Camp. 1980. Tissue distribution and disposition of hymenoxon (*Helenium hoopesii* and *Baileya multiradiata*, poisonous to sheep, goats, and rabbits). J. Agric. Food Chem. 28: 1269–1273.

Hill, D. W., H. L. Kim, and B. J. Camp. 1979. Quantitative analysis of hymenoxon in plant tissue of *Hymenoxys odorata*, *Helenium hoopesii* and *Baileya multiradiata*, poisonous range plants. J. Agric. Food Chem. 27: 885–887.

Hill, J. 1759–1775. The Vegetable System.... 26 vols. London. (Veg. Syst.)

Hill, J. 1768. Hortus Kewensis. London. (Hort. Kew.)

Hill, J. 1769–1770. Herbarium Britannicum.... 2 vols. London. [Volumes paged consecutively.] (Herb. Brit.)

Hill, L. M. 1976. Morphological and cytological evidence for introgression in *Aster acuminatus* Michx. in the southern Appalachians. Castanea 41: 148–155.

Hill, L. M. and O. M. Rogers. 1970. Chromosome numbers of *Aster blakei* and *A. nemoralis*. Rhodora 72: 437–438.

Hill, L. M. and O. M. Rogers. 1973. Chemical, cytological and genetic evidence for the hybrid origin of *Aster blakei* (Porter) House. Rhodora 75: 1–25.

Hilliard, O. M. and B. L. Burtt. 1981. Some generic concepts in Compositae–Gnaphaliinae. Bot. J. Linn. Soc. 82: 181–232.

Hind, D. J. N., H. J. Beentje, P. D. S. Caligari, and S. A. L. Smith, eds. 1996. Proceedings of the International Compositae Conference, Kew, 1994. 2 vols. Kew.

Hind, D. J. N., C. Jeffrey, and G. V. Pope, eds. 1995. Advances in Compositae Systematics. Kew.

Hinds, H. R. 2000. Flora of New Brunswick..., ed. 2. Fredericton.

Hist. Acad. Roy. Sci. Mém. Math. Phys. (Paris, 4to) = Histoire de l'Académie Royale des Sciences; Avec les Mémoires de Mathématique & de Phisique (Physique). [In quarto.]

Hist. Fis. Cuba—See: R. de la Sagra 1840–1855

Hist. Nat. Îles Canaries—See: P. B. Webb and S. Berthelot [1835–]1836–1850

Hist. Nat. Vég.—See: E. Spach 1834–1848

Hist. Pl. Guiane—See: J. B. Aublet 1775

Hitchcock, C. L., A. Cronquist, M. Ownbey, and J. W. Thompson. 1955–1969. Vascular Plants of the Pacific Northwest. 5 vols. Seattle. [Univ. Wash. Publ. Biol. 17.] (Vasc. Pl. Pacif. N.W.)

Hoffmann, J. J. et al. 1978. Odoratin and paucin cytotoxic sesquiterpene lactones from *Baileya pauciradiata* (Compositae). J. Pharm. Sci. 67: 1633–1634.

Hoffmann, O. 1890–1894. Compositae. In: H. G. A. Engler and K. Prantl, eds. 1887–1915. Die natürlichen Pflanzenfamilien.... 254 fasc. Leipzig. Fasc. 39, 43, 48, 54, 74, 89, 105 [IV,5], pp. 87–387.

Holm, T. 1891. Notes upon *Ovularia*, *Oakesia*, *Diclytra* and *Krigia*. Bull. Torrey Bot. Club 18: 1–11.

Holmes, W. C. 1981. *Mikania* (Compositae) of the United States. Sida 9: 147–158.

Holmes, W. C. 1993. The Genus *Mikania* (Compositae: Eupatorieae) in the Greater Antilles. Fort Worth. [Sida Bot. Misc. 9.]

Holmes, W. C. 1996. A proposed sectional classification for *Mikania* (Eupatorieae). In: D. J. N. Hind et al., eds. 1996. Proceedings of the International Compositae Conference, Kew, 1994. 2 vols. Kew. Vol. 1, pp. 621–626.

Holmgren, A. H., L. M. Shultz, and T. K. Lowrey. 1976. *Sphaeromeria*, a genus closer to *Artemisia* than to *Tanacetum* (Asteraceae: Anthemideae). Brittonia 28: 255–262.

Holub, J. 1972. On correct generic names of *Acrocentron* Cass. and *Acrolophus* Cass. (*Centaurea* L. s.l.). Preslia 44: 215–218.

Holub, J. 1976. *Filago*, *Ifloga*, *Logfia*, *Evax*, *Bombycilaena*. In: T. G. Tutin et al., eds. 1964–1980. Flora Europaea. 5 vols. Cambridge. Vol. 4, pp. 121–125.

Holub, J. 1998. Reclassifications and new names in vascular plants 1. Preslia 70: 97–122.

Holzinger, J. M. 1893. The systematic position of *Hartwrightia floridana*. Bull. Torrey Bot. Club 20: 287–288.

Hood, J. L. A. and J. C. Semple. 2003. Pappus variation in *Solidago* (Asteraceae: Astereae). Sida 20: 1617–1630.

Hooker, J. D. 1852–1855. Flora Novae-zelandiae. 2 parts. London. [Parts paged independently. Constitutes vol. 2 of: Hooker, J. D. 1844–1860. The Botany of the Antarctic Voyage of H.M. Discovery Ships Erebus and Terror in the Years 1839–1843.... 3 vols. in 6 parts. London.] (Fl. Nov.-Zel.)

Hooker, W. J. [1829–]1833–1840. Flora Boreali-Americana; or, the Botany of the Northern Parts of British America.... 2 vols. in 12 parts. London, Paris, and Strasbourg. (Fl. Bor.-Amer.)

Hooker, W. J. and G. A. W. Arnott. [1830–]1841. The Botany of Captain Beechey's Voyage; Comprising an Account of the Plants Collected by Messrs Lay and Collie, and Other Officers of the Expedition, during the Voyage to the Pacific and Bering's Strait, Performed in His Majesty's Ship Blossom, under the Command of Captain F. W. Beechey...in the Years 1825, 26, 27, and 28. 10 parts. London. [Parts paged and plates numbered consecutively.] (Bot. Beechey Voy.)

Hooker's Icon. Pl. = Hooker's Icones Plantarum....

Hooker's J. Bot. Kew Gard. Misc. = Hooker's Journal of Botany and Kew Garden Miscellany.

Hoover, R. F. 1970. The Vascular Plants of San Luis Obispo County, California. Berkeley. (Vasc. Pl. San Luis Obispo Co.)

Hort. Bot. Vindob.—See: N. J. Jacquin 1770–1776

Hort. Brit.—See: J. C. Loudon 1830; R. Sweet 1826

Hort. Brit. ed. 3—See: R. Sweet 1839

Hort. Gott.—See: H. A. Schrader 1809–1811

Hort. Herrenhus.—See: J. C. Wendland 1798–1801

Hort. Kew.—See: W. Aiton 1789; J. Hill 1768

Hort. Woburn.—See: J. Forbes 1833

Houle, F. 1988. Étude Biosystématique de la Section *Conyzopsis* du Genre *Aster* (Asteraceae). Ph.D. thesis. Université de Montréal.

Howell, J. T. 1929. A systematic study of the genus *Lessingia* Cham. Univ. Calif. Publ. Bot. 16: 1–44.

Howell, J. T. 1943. Sertulum greeneanum III. Studies in *Cirsium*. Amer. Midl. Naturalist 30: 29–39.

Howell, J. T. 1948. Concerning a California cudweed. Leafl. W. Bot. 5: 90–91.

Howell, J. T. 1949. Marin Flora: Manual of the Flowering Plants and Ferns of Marin County, California. Berkeley.

Howell, J. T. 1959. Distributional data on weedy thistles in western North America. Leafl. W. Bot. 9: 17–32.

Howell, J. T. 1959b. Studies in *Cirsium*—II. Leafl. W. Bot. 9: 9–15.

Howell, J. T. 1960b. Cynareae. In: L. Abrams and R. S. Ferris. 1923–1960. Illustrated Flora of the Pacific States: Washington, Oregon, and California. 4 vols. Stanford. Vol. 5, pp. 506–548.

Howell, J. T. 1970. Marin Flora: Manual of the Flowering Plants and Ferns of Marin County, California, ed. 2 with suppl. Berkeley.

Howell, T. J. 1897–1903. A Flora of Northwest America. 1 vol. in 8 fasc. Portland. [Fasc. 1–7 (text) paged consecutively, fasc. 8 (index) independently.] (Fl. N.W. Amer.)

Hrusa, F. et al. 2002. Catalogue of non-native vascular plants occurring spontaneously in California beyond those addressed in The Jepson Manual—Part I. Madroño 49: 61–98.

Hsi, Y.-T. 1960. Taxonomy, Distribution and Relationships of the Species of *Cirsium* Belonging to the Series *Undulata*. Ph.D. dissertation. University of Minnesota.

Huber, W. 1993. Biosystematisch-okologische Untersuchungen an den *Erigeron*-Arten (Asteraceae) der Alpen. Veröff. Geobot. Inst. E. T. H. Stiftung Rübel Zürich 114: 1–143.

Huber, W. and Ö. Nilsson. 1995. Close genetic affinity of northern and other Eurasiatic *Erigeron* species. In: D. J. N. Hind et al., eds. 1995. Advances in Compositae Systematics. Kew. Pp. 197–222.

Hughes, J. and A. J. Richards. 1988. The genetic structure of populations of sexual and asexual *Taraxacum* (dendelions). Heredity 60: 161–171.

Hughes, J. and A. J. Richards. 1989. Isozymes and the status of *Taraxacum* (Asteraceae) agamospecies. Bot. J. Linn. Soc. 99: 365–376.

Hughes, J. F. and G. K. Brown. 1994. Putative hybridization between two varieties of goldenweed (*Oonopsis foliosa*: Asteraceae). [Abstract.] Amer. J. Bot. 81(6, suppl.): 162–163.

Hultén, E. 1941–1950. Flora of Alaska and Yukon. 10 vols. Lund and Leipzig. [Vols. paged consecutively and designated as simultaneous numbers of Lunds Univ. Årsskr. (= Acta Univ. Lund.) and Kungl. Fysiogr. Sällsk. Handl.] (Fl. Alaska Yukon)

Hultén, E. 1955. Flora of Alaska and Yukon. Ark. Bot. 7: 1634–1689.

Hultén, E. 1968. Flora of Alaska and Neighboring Territories: A Manual of the Vascular Plants. Stanford.

Humboldt, A. von and A. J. Bonpland. [1805–]1808–1809 [–1817]. Plantae Aequinoctiales.... 2 vols. in 17 parts. Paris and Tübingen. [Parts numbered consecutively.] (Pl. Aequinoct.)

Humboldt, A. von, A. J. Bonpland, and C. S. Kunth. 1815[1816]–1825. Nova Genera et Species Plantarum Quas in Peregrinatione Orbis Novi Collegerunt, Descripserunt.... 7 vols. in 36 parts. Paris. (Nov. Gen. Sp.)

Humphries, C. J. 1976. A revision of the Macaronesian genus *Argyranthemum* Webb ex Schultz Bip. (Compositae–Anthemideae). Bull. Brit. Mus. (Nat. Hist.), Bot. 5: 147–240.

Hymenopappus—See: C.-L. L'Héritier de Brutelle [1788]

Iagttag. Vextrig. Marokko—See: P. K. A. Schousboe 1800

Icon.—See: A. J. Cavanilles 1791–1801

Icon. Pl. = Icones Plantarum....

Icon. Pl.—See: C. F. von Ledebour 1829–1834

Icon. Pl. Rar.—See: N. J. Jacquin 1781–1793[–1795]

Iljin, M. M. 1932. A critical survey of the genus *Amberboa* Less. Izv. Bot. Sada Akad. Nauk S.S.S.R. 30: 101–116.

Ill. Fl. N. U.S.—See: N. L. Britton and A. Brown 1896–1898

Ill. Fl. N. U.S. ed. 2—See: N. L. Britton and A. Brown 1913

Illingworth, J. M. and G. W. Douglas. 1994. Status Report on the Tall Woolly-heads *Psilocarphus elatior* (A. Gray) A. Gray. Victoria.

Illingworth, J. M. and G. W. Douglas. 1994b. Status Report on the Slender Woolly-heads *Psilocarphus tenellus* var. *tenellus* in Canada. Victoria.

Index Kew.—See: B. D. Jackson et al. [1893–]1895+

Index Seminum (Bratislava) = Delectus Seminum Horto Vratislava....

Index Seminum (Göttingen) = Index Seminum Horti Academici Goettingensis Anno...Collecta.

Index Seminum (Louvain) = Selectus Seminum Horti Botanici Lovaniensis.

Index Seminum (St. Petersburg) = Index Seminum, Quae Hortus Botanicus Imperialis Petropolitanus pro Mutua Commutatione Offert.

Inst. Rei Herb.—See: H. J. N. von Crantz 1766

Int. J. Pl. Sci. = International Journal of Plant Sciences.

Intermount. Fl.—See: A. Cronquist et al. 1972+

International Botanical Congress. [1999.] Abstracts. XVI International Botanical Congress, St. Louis, USA, August 1–7, 1999. [St. Louis.]

Intr. Hist. Nat.—See: J. A. Scopoli 1777

Irmler, C. et al. 1982. Enzymes and quantitative morphological characters compared between the allotetraploid

Microseris decipiens and its diploid parental species. Beitr. Biol. Pflanzen 57: 269–289.

Isis (Oken) = Isis, oder encyclopädische Zeitung (vorzüglich für Naturgeschichte, vergleichende Anatomie und Physiologie). [Oken ed.]

Israel J. Bot. = Israel Journal of Botany.

Iter Hispan.—See: P. Loefling 1758

Izv. Bot. Sada Akad. Nauk S.S.S.R. = Izvestiya Botanicheskogo Sada Akademii Nauk S S S R.

J. Acad. Nat. Sci. Philadelphia = Journal of the Academy of Natural Sciences of Philadelphia.

J. Agric. Food Chem. = Journal of Agricultural and Food Chemistry.

J. Amer. Chem. Soc. = Journal of the American Chemical Society.

J. Arizona Acad. Sci. = Journal of the Arizona Academy of Science.

J. Arizona-Nevada Acad. Sci. = Journal of the Arizona-Nevada Academy of Science.

J. Arnold Arbor. = Journal of the Arnold Arboretum.

J. Austral. Inst. Agric. Sci. = Journal of the Australian Institute of Agricultural Science.

J. Bot. (Hooker) = Journal of Botany, (Being a Second Series of the Botanical Miscellany), Containing Figures and Descriptions....

J. Ecol. = Journal of Ecology.

J. Elisha Mitchell Sci. Soc. = Journal of the Elisha Mitchell Scientific Society.

J. Heredity = Journal of Heredity.

J. Hist. Nat. = Journal d'Histoire Naturelle.

J. Indian Bot. Soc. = Journal of the Indian Botanical Society.

J. Jap. Bot. = Journal of Japanese Botany.

J. Linn. Soc., Bot. = Journal of the Linnean Society. Botany.

J. Pharm. Sci. = Journal of Pharmaceutical Sciences.

J. Phys. Chim. Hist. Nat. Arts = Journal de Physique, de Chimie, d'Histoire Naturelle et des Arts.

J. S. African Bot. = Journal of South African Botany.

J. S. African Veterin. Assoc. = Journal of the South African Veterinary Association.

J. Torrey Bot. Soc. = Journal of the Torrey Botanical Society.

J. Wash. Acad. Sci. = Journal of the Washington Academy of Sciences.

Jackson, B. D. et al., comps. [1893–]1895+. Index Kewensis Plantarum Phanerogamarum.... 2 vols. + 21+ suppls. Oxford. (Index Kew.)

Jackson, J. D. 1972. The evolution of functional dioecism in the genus *Baccharis* (Compositae). [Abstract.] Brittonia 24: 121.

Jackson, R. C. 1956. The hybrid origin of *Helianthus doronicoides*. Rhodora 58: 97–101.

Jackson, R. C. 1960 A revision of the genus *Iva*. Univ. Kansas Sci. Bull. 41: 793–876.

Jackson, R. C. 1962. Interspecific hybridization in *Haplopappus* and its bearing on chromosome evolution in the *Blepharodon* section. Amer. J. Bot. 49: 119–135.

Jackson, R. C. 1971. Note on *Haplopappus gracilis* and *H. ravenii*—a reply. Brittonia 23: 446.

Jackson, R. C. 1979. Intersectional hybridization in *Haplopappus: Blepharodon × Hazardia*. Syst. Bot. 4: 157–162.

Jackson, R. C. and C. T. Dimas. 1981. Experimental evidence for systematic placement of the *Haplopappus phyllocephalus* complex (Compositae). Syst. Bot. 6: 8–14.

Jackson, R. C. and A. T. Guard. 1957. Natural and artificial hybridization between *Helianthus mollis* and *H. occidentalis*. Amer. Midl. Naturalist 58: 422–433.

Jackson, R. C. and A. T. Guard. 1957b. Analysis of some natural and artificial interspecific hybrids in *Helianthus*. Proc. Indiana Acad. Sci. 66: 306–317.

Jackson, S. W. 1963. Hybridization among three species of *Ratibida*. Univ. Kansas Sci. Bull. 44: 3–27.

Jacquin, N. J. 1760. Enumeratio Systematica Plantarum, Quas in Insulis Caribaeis Vicinaque Americes Continente Detexit Novas.... Leiden. (Enum. Syst. Pl.)

Jacquin, N. J. 1770–1776. Hortus Botanicus Vindobonensis.... 3 vols. Vienna. (Hort. Bot. Vindob.)

Jacquin, N. J. 1781–1793[–1795]. Icones Plantarum Rariorum. 3 vols. in fasc. Vienna etc. [Vols. paged independently, plates numbered consecutively.] (Icon. Pl. Rar.)

Jacquin, N. J. 1786[1787]–1796[1797]. Collectanea ad Botanicam, Chemiam, et Historiam Naturalem Spectantia.... 5 vols. Vienna. (Collectanea)

Jacquin, N. J. 1797–1804. Plantarum Rariorum Horti Caesarei Schoenbrunnensis Descriptiones et Icones. 4 vols. Vienna, London, and Leiden. (Pl. Hort. Schoenbr.)

Jaeger, E. C. 1941. Desert Wild Flowers, ed. 2. Stanford and London.

Jahandiez, É., R. C. J. E. Maire, and M. L. Emberger. 1931–1941. Catalogue des Plantes du Maroc.... 4 vols. Algiers. (Cat. Pl. Maroc)

Jahresber. Pollichia = Jahresbericht der Pollichia, eines naturwissenschaftlichen Vereins der Bayerischen Pfalz.

James, C. W. 1958. Generic considerations concerning *Carphephorus, Trilisa* and *Litrisa* (Compositae). Rhodora 60: 117–122.

James, E. 1823. Account of an Expedition from Pittsburgh to the Rocky Mountains, Performed in the Years 1819 and '20...under the Command of Major Stephen H. Long. 2 vols. + atlas. Philadelphia. (Account Exped. Pittsburgh)

James, L. F. et al., eds. 1991. Noxious Range Weeds. Boulder, San Francisco, and Oxford.

Jansen, R. K. 1985. The systematics of *Acmella* (Asteraceae–Heliantheae). Syst. Bot. Monogr. 8: 1–115.

Jansen, R. K. et al. 1991. Phylogeny and character evolution in the Asteraceae based on chloroplast DNA restriction site mapping. Syst. Bot. 16: 98–115.

Jansen, R. K. et al. 1991b. Systematic implications of chloroplast DNA variation in the subtribe Microseridinae (Asteraceae: Lactuceae). Amer. J. Bot. 78: 1015–1027.

Jansen, R. K. et al. 1992. Chloroplast DNA variation in the Asteraceae: Phylogenetic and evolutionary implications. In: D. E. Soltis et al., eds. 1992. Molecular Systematics of Plants. New York. Pp. 252–294.

Jansen, R. K., K. E. Holsinger, H. J. Michaels, and J. D. Palmer. 1990. Phylogenetic analysis of chloroplast DNA restriction site data at the higher taxonomic levels: An example from the Asteraceae. Evolution 44: 2089–2105.

Jansen, R. K. and J. D. Palmer. 1987. Chloroplast DNA from lettuce and *Barnadesia* (Asteraceae): Structure, gene-localization, and characterization of a large inversion. Curr. Genet. 11: 553–564.

Jansen, R. K., E. B. Smith, and D. J. Crawford. 1987. A cladistic study of North American *Coreopsis* (Asteraceae: Heliantheae). Pl. Syst. Evol. 157: 73–84.

Jard. Malmaison—See: É. P. Ventenat 1803–1804[–1805]

Jeffrey, C. 1978. Compositae. In: V. H. Heywood, ed. 1978. Flowering Plants of the World. Oxford. Pp. 263–268.

Jeffrey, C. 1979. Note on the lectotypification of the names *Cacalia* L., *Matricaria* L. and *Gnaphalium* L. Taxon 28: 349–351.

Jeffrey, C. 1980. Generic and sectional limits in *Senecio* (Compositae): II. Evaluation of some recent studies. Kew Bull. 34: 49–58.

Jeffrey, C. 1995. Compositae systematics 1975–1993. Developments and desiderata. In: D. J. N. Hind et al., eds. 1995. Advances in Compositae Systematics. Kew. Pp. 3–22.

Jeffrey, C. and Chen Y. L. 1984. Taxonomic studies on the tribe Senecioneae (Compositae) of eastern Asia. Kew Bull. 39: 204–466.

Jepson, W. L. 1901. A Flora of Western Middle California.... Berkeley. (Fl. W. Calif.)

Jepson, W. L. [1923–1925.] A Manual of the Flowering Plants of California.... Berkeley. (Man. Fl. Pl. Calif.)

Johnson, A. W. and J. G. Packer. 1968. Chromosome numbers in the flora of Ogoturuk Creek, N.W. Alaska. Bot. Not. 121: 403–456.

Johnson, D. E. 1991. Nomenclatural conspectus of annual Eriophyllinae (Asteraceae). Novon 1: 119–124.

Johnson, M. F. 1971. A monograph of the genus *Ageratum* L. (Compositae–Eupatorieae). Ann. Missouri Bot. Gard. 58: 6–88.

Johnson, M. F. 1979. The genus *Prenanthes* L. (Cichorieae–Asteraceae) in Virginia. Castanea 45: 24–30.

Johnson, M. J. 1971. The genus *Liatris* in Virginia. Castanea 36: 137–147.

Johnson, R. R. 1969. Monograph of the plant genus *Porophyllum* (Compositae: Helenieae). Univ. Kansas Sci. Bull. 48: 225–267.

Johnston, I. M. 1938. New or noteworthy plants from temperate South America. J. Arnold Arbor. 19: 248–263.

Jones, A. G. 1978. The taxonomy of *Aster* section *Multiflori* (Asteraceae). I. Nomenclatural review and formal presentation of taxa. Rhodora 80: 319–357.

Jones, A. G. 1980. A classification of New World species of *Aster* (Asteraceae). Brittonia 32: 230–239.

Jones, A. G. 1989. *Aster* and *Brachyactis* in Illinois. Bull. Illinois Nat. Hist. Surv. 34: 139–194.

Jones, A. G. 1992. *Aster* and *Brachyactis* (Asteraceae) in Oklahoma. Fort Worth. [Sida Bot. Misc. 8.]

Jones, A. G. and D. A. Young. 1983. Generic concepts of *Aster* (Asteraceae): A comparison of cladistic, phenetic and cytological approaches. Syst. Bot. 8: 71–84.

Jones, G. N. 1936. A Botanical Survey of the Olympic Peninsula. Seattle. [Univ. Wash. Publ. Biol. 5.] (Bot. Surv. Olympic Penins.)

Jones, Q. 1954. Monograph of *Agoseris*, Tribe Cichorieae. Ph.D. dissertation. Harvard University.

Jones, R. L. 1983. A systematic study of *Aster* section *Patentes* (Asteraceae). Sida 10: 41–81.

Jones, R. L. 1994. The status of *Helianthus eggertii* Small in the southeastern U.S. Castanea 59: 319–330.

Jones, S. B. 1964. Taxonomy of the narrow-leaved *Vernonia* of the southeastern United States. Rhodora 66: 382–401.

Judd, W. S. and R. G. Olmstead. 2004. A survey of tricolpate (eudicot) phylogeny. Amer. J. Bot. 91: 1627–1644.

Jussieu, A. L. de. 1789. Genera Plantarum.... Paris. (Gen. Pl.)

Just's Bot. Jahresber. = Just's botanischer Jahresbericht; systematisch geordnetes Repertorium der botanischen Literatur aller Länder.

Källersjö, M. 1988. A generic re-classification of *Pentzia* Thunb. (Compositae: Anthemideae) from southern Africa. Bot. J. Linn. Soc. 96: 299–322.

Kalmiopsis = Kalmiopsis; Journal of the Native Plant Society of Oregon.

Kapoor, B. M. and J. R. Beaudry. 1966. Studies on *Solidago*. VII. The taxonomic status of the taxa *Brachychaeta*, *Brintonia*, *Chrysoma*, *Euthamia*, *Oligoneuron*, and *Petradoria* in relation to *Solidago*. Canad. J. Genet. Cytol. 8: 422–443.

Karis, P. O. 1995. Cladistics of the subtribe Ambrosiinae (Asteraceae: Heliantheae). Syst. Bot. 20: 40–54.

Karis, P. O. and O. Ryding. 1994. Tribe Heliantheae. In: K. Bremer. 1994. Asteraceae: Cladistics & Classification. Portland. Pp. 559–624.

Karis, P. O. and O. Ryding. 1994b. Tribe Helenieae. In: K. Bremer. 1994. Asteraceae: Cladistics & Classification. Portland. Pp. 521–558.

Kartesz, J. T. and C. A. Meacham. 1999. Synthesis of the North American Flora, ver. 1.0. Chapel Hill. [CD-ROM.] (Synth. N. Amer. Fl.)

Kawaguchi, S. T. 1996. Systematics of *Rayjacksonia* (Asteraceae: Astereae). M.S. thesis. University of Kansas.

Kay, Q. O. N. 1976. *Chamomilla*. In: T. G. Tutin et al., eds. 1964–1980. Flora Europaea. 5 vols. Cambridge. Vol. 4, p. 167.

Kay, Q. O. N. 1976b. *Matricaria*. In: T. G. Tutin et al., eds. 1964–1980. Flora Europaea. 5 vols. Cambridge. Vol. 4, pp. 165–167.

Kay, Q. O. N. 1994. Biological flora of the British Isles: *Tripleurospermum inodorum* (L.) Schultz Bip. (*Matricaria inodora* L., *Matricaria maritima* auct. p.p. non L., *Matricaria perforata* Merat, *Tripleurospermum perforatum* (Merat) Wagenitz, *Tripleurospermum maritimum* (L.) Schultz Bip. p.p. non L.). J. Ecol. 82: 681–698.

Kazmi, S. M. A. 1964. Revision der Gattung *Carduus* (Compositae). Teil II. Mitt. Bot. Staatssamml. München 5: 279–550.

Kearney, T. H. and R. H. Peebles. 1960. Arizona Flora, ed. 2. Berkeley.

Keck, D. D. 1940. New subspecies in *Haplopappus*. Madroño 5: 166–169.

Keck, D. D. 1946. A revision of the *Artemisia vulgaris*

complex in North America. Proc. Calif. Acad. Sci., ser. 4, 25: 421–468.

Keck, D. D. 1949. *Hemizonella* becomes a *Madia*. Madroño 10: 22.

Keck, D. D. 1958. Taxonomic notes on the California flora. Aliso 4: 102–103.

Keck, D. D. 1959. *Madia*. In: P. A. Munz. 1959. A California Flora. Berkeley and Los Angeles. Pp. 1113–1117.

Keck, D. D. 1959b. *Hemizonia*. In: P. A. Munz. 1959. A California Flora. Berkeley and Los Angeles. Pp. 1117–1124.

Keck, D. D. 1959c. *Baeria, Lasthenia, Crockeria*. In: P. A. Munz. 1959. A California Flora. Berkeley and Los Angeles. Pp. 1140–1144.

Keck, D. D. 1960. *Gutierrezia*. In: L. Abrams and R. S. Ferris. 1923–1960. Illustrated Flora of the Pacific States: Washington, Oregon, and California. 4 vols. Stanford. Vol. 4, pp. 260–261.

Keck, D. D. 1960b. *Calycadenia*. In: L. Abrams and R. S. Ferris. 1923–1960. Illustrated Flora of the Pacific States: Washington, Oregon, and California. 4 vols. Stanford. Vol. 4, pp. 186–192.

Keeley, J. E. and C. J. Fotheringham. 1998. Smoke-induced seed germination in California chaparral. Ecology 79: 2320–2336.

Keeley, S. C. and R. K. Jansen. 1991. Evidence from chloroplast DNA for the recognition of a new tribe, the Tarchonantheae, and the tribal placement of *Pluchea* (Asteraceae). Syst. Bot. 16: 173–181.

Keil, D. J. 1975. Revision of *Pectis* sect. *Heteropectis* (Compositae: Tageteae). Madroño 23: 181–191.

Keil, D. J. 1975b. *Pectis cylindrica* (Compositae) established as a member of the Texas flora and confirmed as a distinct species. SouthW. Naturalist 20: 286–287.

Keil, D. J. 1975c. *Pectis humifusa* new to the flora of the United States. Rhodora 77: 145–146.

Keil, D. J. 1977. A revision of *Pectis* section *Pectothrix* (Compositae: Tageteae). Rhodora 79: 32–78.

Keil, D. J. 1977b. Chromosome studies in North and Central American species of *Pectis* L. (Compositae: Tageteae). Rhodora 79: 79–94.

Keil, D. J. 1978. Revision of *Pectis* section *Pectidium* (Compositae: Tageteae). Rhodora 80: 135–146.

Keil, D. J. 1986. Synopsis of the Florida species of *Pectis* (Asteraceae). Sida 11: 385–395.

Keil, D. J. 1996. *Pectis*. In: B. L. Turner, ed. 1996+. The Comps of Mexico: A Systematic Account of the Family Asteraceae. 2+ vols. Huntsville, Tex. Vol. 6, pp. 22–43.

Keil, D. J., M. A. Luckow, and D. J. Pinkava. 1988. Chromosome studies in Asteraceae from the United States, Mexico, the West Indies, and South America. Amer. J. Bot. 75: 652–668.

Keil, D. J. and D. J. Pinkava. 1976. Chromosome counts and taxonomic notes for Compositae from the United States and Mexico. Amer. J. Bot. 63: 1393–1403.

Keil, D. J. and T. F. Stuessy. 1981. Systematics of *Isocarpha* (Compositae: Eupatorieae). Syst. Bot. 6: 258–287.

Keil, D. J. and C. E. Turner. 1992. Taxonomic notes on

California species of *Cirsium* (Asteraceae: Cardueae). Phytologia 73: 312–317.

Keil, D. J. and C. E. Turner. 1993. *Cirsium*. In: J. C. Hickman, ed. 1993. The Jepson Manual. Higher Plants of California. Berkeley, Los Angeles, and London. Pp. 232–239.

Kelch, D. G. and B. G. Baldwin. 2003. Phylogeny and ecological radiation of New World thistles (*Cirsium*, Cardueae–Compositae) based on ITS and ETS rDNA sequence data. Molec. Ecol. 12: 141–151.

Kew Bull. = Kew Bulletin.

Khan, R. and C. E. Jarvis. 1989. The correct name for the plant known as *Pluchea symphytifolia* (Miller) Gillis (Asteraceae). Taxon 38: 659–662.

Kharkevich, S. S., ed. 1985+. Plantae Vasculares Orientis Extremi Sovietici. 7+ vols. Leningrad.

Khidir, M. O. and P. F. Knowles. 1970. Cytogenetic studies of *Carthamus* species (Compositae) with 32 pairs of chromosomes. I. Intrasectional hybridization. Amer. J. Bot. 57: 123–129.

Khidir, M. O. and P. F. Knowles. 1970b. Cytogenetic studies of *Carthamus* species (Compositae) with 32 pairs of chromosomes. II. Intersectional hybridization. Canad. J. Genet. Cytol. 12: 90–99.

Kiger, R. W. and D. M. Porter. 2001. Categorical Glossary for the Flora of North America Project. Pittsburgh.

Kiku—See: S. Kitamura 1948

Kilian, N. 1997. Revision of *Launaea* Cass. (Compositae, Lactuceae, Sonchinae). Englera 17.

Kim, H. G., D. J. Loockerman, and R. K. Jansen. 2002. Systematic implications of *ndh*F sequence variation in the Mutisieae (Asteraceae). Syst. Bot. 27: 598–609.

Kim, K. J. et al. 1992. Phylogenetic implications of *rbc*L sequence variation in the Asteraceae. Ann. Missouri Bot. Gard. 79: 428–445.

Kim, K. J. et al. 1992b. Evolutionary implications of intraspecific chloroplast DNA variation in dwarf dandelions (*Krigia*–Asteraceae). Amer. J. Bot. 79: 708–715.

Kim, K. J. et al. 1992c. Phylogenetic and evolutionary implications of interspecific chloroplast DNA variation in *Krigia* (Asteraceae–Lactuceae). Syst. Bot. 17: 449–469.

Kim, K. J. and R. K. Jansen. 1994. Comparisons of phylogenetic hypotheses among differing data sets in dwarf dandelions (*Krigia*, Asteraceae): Additional information from internal transcribed spacer sequences of nuclear ribosomal DNA. Pl. Syst. Evol. 190: 157–185.

Kim, K. J. and R. K. Jansen. 1995. *ndh*F sequence evolution and the major clades in the sunflower family. Proc. Natl. Acad. Sci. U.S.A. 92: 10379–10383.

Kim, K. J. and T. J. Mabry. 1991. Phylogenetic and evolutionary implications of nuclear ribosomal DNA variation in dwarf dandelions (*Krigia*–Lactuceae–Asteraceae). Pl. Syst. Evol. 177: 53–69.

Kim, K. J. and B. L. Turner. 1992. Systematic overview of *Krigia* (Asteraceae–Lactuceae). Brittonia 44: 173–198.

Kim, S. C., D. J. Crawford, and R. K. Jansen. 1996. Phylogenetic relationships among the genera of the subtribe Sonchinae (Asteraceae): Evidence from ITS sequences. Syst. Bot. 21: 417–432.

King, L. M. 1993. Origins of genotypic variation in North

American dandelions inferred from ribosomal DNA and chloroplast DNA restriction enzyme analysis. Evolution 47: 136–151.

King, L. M. and B. A. Schaal. 1990. Genotypic variation within asexual lineages of *Taraxacum officinale*. Proc. Natl. Acad. Sci. U.S.A. 87: 998–1002.

King, R. M. 1967. Studies in the Eupatorieae (Compositae). I–III. Rhodora 69: 35–47.

King, R. M. and H. W. Dawson, eds. 1975. Cassini on Compositae.... 3 vols. New York.

King, R. M. and H. Robinson. 1970b. A monograph of the genus *Trichocoronis*. Phytologia 19: 497–500.

King, R. M. and H. Robinson. 1987. The genera of the Eupatorieae (Asteraceae). Monogr. Syst. Bot. Missouri Bot. Gard. 22.

King-Jones, S. 2001. Revision of *Pluchea* Cass. (Compositae, Plucheeae) in the Old World. Englera 23: 3–136.

Kingsbury, J. M. 1964. Poisonous Plants of the United States and Canada. Englewood Cliffs.

Kirkia = Kirkia; Journal of the Federal Herbarium [Rhodesia].

Kirkman, L. K. 1981. Taxonomic revision of *Centratherum* and *Phyllocephalum* (Compositae: Vernoniaea). Rhodora 83: 1–24.

Kirschner, J. et al. 2003. Principal features of the cpDNA evolution in *Taraxacum* (Asteraceae, Lactuceae): A conflict with taxonomy. Pl. Syst. Evol. 239: 231–255.

Kirschner, J. and J. Štěpánek. 1987. Again on the sections in *Taraxacum* (Cichoriaceae) (Studies in *Taraxacum* 6). Taxon 36: 608–617.

Kirschner, J. and J. Štěpánek. 1997. A nomenclatural checklist of supraspecific names in *Taraxacum*. Taxon 46: 87–98.

Kitamura, S. 1948. Kiku. Osaka. (Kiku)

Knowlton, C. H. and W. Deane. 1924. Reports on the flora of the Boston district. Rhodora 26: 82–88.

Koch, W. D. J. 1843–1845. Synopsis Florae Germanicae et Helveticae..., ed. 2. 3 parts. Frankfurt am Main and Leipzig. [Parts paged consecutively.] (Syn. Fl. Germ. Helv. ed. 2)

Komarov, V. L. et al., eds. 1963+. Flora of the U.S.S.R. (Flora SSSR). Translated from Russian. 28+ vols. Jerusalem etc.

Komarov, V. L., B. K. Schischkin, and E. Bobrov, eds. 1934–1964. Flora URSS.... 30 vols. Leningrad. (Fl. URSS)

Koster, H. 1816. Travels in Brazil. London. (Trav. Brazil)

Kowal, R. R. 1975. Systematics of *Senecio aureus* and allied species on the Gaspé Peninsula, Quebec. Mem. Torrey Bot. Club 23: 1–113.

Koyama, H. 1969. Taxonomic studies on the tribe Senecioneae of eastern Asia, II. Enumeration of the species of eastern Asia. Mem. Fac. Sci. Kyoto Univ., Ser. Biol. 2: 137–183.

Kral, R., ed. 1983. A Report on Some Rare, Threatened, or Endangered Forest-related Vascular Plants of the South. 2 vols. Washington. [U.S.D.A. Forest Serv., Techn. Publ. R8-TP 2.]

Kral, R. and R. K. Godfrey. 1958. Synopsis of the Florida species of *Cacalia*. Quart. J. Florida Acad. Sci. 21: 193–206.

Krasnoborov, I. M. et al., eds. 1987+. Flora Sibiri. 13+ vols. Novosibirsk.

Kubitzki, K. et al., eds. 1990+. The Families and Genera of Vascular Plants. 7+ vols. Berlin etc.

Kuntze, O. 1891–1898. Revisio Generum Plantarum Vascularium Omnium atque Cellularium Multarum.... 3 vols. Leipzig etc. [Vol. 3 in 3 parts paged independently; parts 1 & 3 unnumbered.] (Revis. Gen. Pl.)

Kupicha, F. K. 1975. *Cnicus*. In: P. H. Davis, ed. 1965–1988. Flora of Turkey and the East Aegean Islands. 10 vols. Edinburgh. Vol. 5, pp. 588–590.

Kyhos, D. W. 1965. The independent aneuploid origin of two species of *Chaenactis* (Compositae) from a common ancestor. Evolution 19: 26–43.

Kyhos, D. W., G. D. Carr, and B. G. Baldwin. 1990. Biodiversity and cytogenetics of the tarweeds (Asteraceae: Heliantheae–Madiinae). Ann. Missouri Bot. Gard. 77: 84–95.

Kyhos, D. W. and P. H. Raven. 1982. Miscellaneous chromosome counts in Asteraceae. Madroño 29: 62.

La Duke, J. C. 1982. Revision of *Tithonia*. Rhodora 84: 453–522.

La Llave, P. de and J. M. de Lexarza. 1824–1825. Novorum Vegetabilium Descriptiones. 2 fasc. Mexico City. [Fasc. 2 includes Orchidianum Opusculum, paged separately.] (Nov. Veg. Descr.)

Labrecque, J. and L. Brouillet. 1990. *Aster anticostensis*, an endemic of northeastern North America: Biology and conservation. Rhodora 92: 129–141.

Labrecque, J. and L. Brouillet. 1996. Biosystématique du complexe de l'*Aster novi-belgii* (Asteraceae: Astereae) au Québec. Canad. J. Bot. 74: 162–188.

Lack, H. W. 1975. A note on *Helminthotheca* Zinn (Compositae). Taxon 24: 111–112.

Lack, H. W. and B. E. Leuenberger. 1979. Pollen and taxonomy of *Urospermum* (Asteraceae, Lactuceae). Pollen & Spores 21: 415–425.

Lagasca y Segura, M. 1816b. Genera et Species Plantarum.... Madrid. (Gen. Sp. Pl.)

Lamarck, J. 1778[1779]. Flore Françoise ou Description Succincte de Toutes les Plantes Qui Croissent Naturellement en France.... 3 vols. Paris. (Fl. Franç.)

Lamarck, J. et al. 1783–1817. Encyclopédie Méthodique. Botanique.... 13 vols. Paris and Liège. [Vols. 1–8, suppls. 1–5.] (Encycl.)

Lamarck, J. and A. P. de Candolle. 1806. Synopsis Plantarum in Florâ Gallicâ Descriptarum. Paris. (Syn. Pl. Fl. Gall.)

Lamarck, J. and J. Poiret. 1791–1823. Tableau Encyclopédique et Méthodique des Trois Règnes de la Nature. Botanique.... 6 vols. Paris. [Vols. 1–2 = tome 1; vols. 3–5 = tome 2; vol. [6] = tome 3. Vols. paged consecutively within tomes.] (Tabl. Encycl.)

Lamboy, W. F. 1988. The status of *Aster conmixtus* and a new species of *Aster* from the southeastern United States. Syst. Bot. 13: 187–195.

Lamboy, W. F. 1992. The taxonomic status and probable origin of *Aster chlorolepis*, a southern Appalachian endemic. Castanea 57: 52–65.

Lamont, E. E. 1995. Taxonomy of *Eupatorium* section

Verticillata (Asteraceae). Mem. New York Bot. Gard. 72: 1–66.

Lamp, C. and F. Collet. 1979. A Field Guide to Weeds in Australia. Melbourne.

Lane, M. A. 1979. Taxonomy of the genus *Amphiachyris* (Asteraceae: Astereae). Syst. Bot. 4: 178–189.

Lane, M. A. 1982. Generic limits of *Xanthocephalum, Gutierrezia, Amphiachyris, Gymnosperma, Greenella,* and *Thurovia* (Compositae: Astereae). Syst. Bot. 7: 405–416.

Lane, M. A. 1983. Taxonomy of *Xanthocephalum* (Compositae: Astereae). Syst. Bot. 8: 305–316.

Lane, M. A. 1985. Taxonomy of *Gutierrezia* (Compositae: Astereae) in North America. Syst. Bot. 10: 7–28.

Lane, M. A. 1988. Generic relationships and taxonomy of *Acamptopappus* (Compositae: Astereae). Madroño 35: 247–265.

Lane, M. A. 1992. New combinations in Californian *Lessingia* (Compositae: Astereae). Novon 2: 213–214.

Lane, M. A. 1992b. New combinations in Californian *Grindelia* (Compositae: Astereae). Novon 2: 215–217.

Lane, M. A. 1993. *Gutierrezia.* In: J. C. Hickman, ed. 1993. The Jepson Manual. Higher Plants of California. Berkeley, Los Angeles, and London. Pp. 274–275.

Lane, M. A. 1993b. *Grindelia.* In: J. C. Hickman, ed. 1993. The Jepson Manual. Higher Plants of California. Berkeley, Los Angeles, and London. Pp. 271–274.

Lane, M. A. 1996. Taxonomy of *Gundlachia* (Compositae: Astereae). Brittonia 48: 532–541.

Lane, M. A. et al. 1996. Relationships of North American genera of Astereae, based on chloroplast DNA restriction site data. In: D. J. N. Hind et al., eds. 1996. Proceedings of the International Compositae Conference, Kew, 1994. 2 vols. Kew. Vol. 1, pp. 49–77.

Lane, M. A. and R. L. Hartman. 1996. Reclassification of North American *Haplopappus* (Compositae: Astereae) completed: *Rayjacksonia* gen. nov. Amer. J. Bot. 83: 356–370.

Lanzetta, R. et al. 1991. Icthytoxic sesquiterpenes and xanthoanolides from *Dittrichia graveolens.* Phytochemistry 30: 1121–1124.

Lasser, T. 1964+. Flora de Venezuela. 12+ vols. Caracas.

Leafl. Bot. Observ. Crit. = Leaflets of Botanical Observation and Criticism.

Leafl. W. Bot. = Leaflets of Western Botany.

Lecoq, H. and J. Juillet. 1831. Dictionnaire Raisonné des Termes de Botanique et des Familles Naturelles.... Paris, London, and Clermont-Ferrand. (Dict. Rais. Term. Bot.)

Ledebour, C. F. von. 1829–1834. Icones Plantarum...Floram Rossicam.... 5 vols. Riga etc. [Vols. paged independently, plates numbered consecutively.] (Icon. Pl.)

Lee, J. and B. G. Baldwin. 2004. Subtribes of principally North American genera of Cichorieae (Compositae). Novon 14: 309–313.

Lee, J., B. G. Baldwin, and L. D. Gottlieb. 2002. Phylogeny of *Stephanomeria* and related genera (Compositae–Lactuceae) based on 18S-26S nuclear rDNA ITS and ETS sequences. Amer. J. Bot. 89: 160–168.

Lee, J., B. G. Baldwin, and L. D. Gottlieb. 2003. Phylogenetic relationships among the primarily North American

genera of Cichorieae (Compositae) based on analysis of 18S–26S rDNA ITS and ETS sequences. Syst. Bot. 28: 616–626.

Legault, A. 1986. Cytogéographie et Taxonomie Infraspécifique de l'*Aster cordifolius* L. (Asteraceae) au Québec. Université de Montréal.

Legault, A. and L. Brouillet. 1989. Cytogéographie de l'*Aster cordifolius* (Asteraceae: Astereae) au Québec. Canad. J. Bot. 67: 2114–2119.

Lejeunia = Lejeunia; Bulletin des Botanistes Liégeois. Organe du Cercle de Botanique Liégeois et sa Section de Mycologue [later: ...; Revue de Botanique].

Lepage, E. 1960. *Hieracium canadense* Michx. et ses alliées en Amérique du Nord. Naturaliste Canad. 87: 59–107.

Lepage, E. 1971. Les épervières de Quebec. Naturaliste Canad. 98: 657–674.

Les, D. H. et al. 1991b. Genetic consequences of rarity in *Aster furcatus* (Asteraceae), a threatened, self-incompatible plant. Evolution 45: 1641–1650.

Les, D. H., J. A. Reinartz, and L. A. Leitner. 1992. Distribution and habitats of the forked aster (*Aster furcatus;* Asteraceae), a threatened Wisconsin plant. Michigan Bot. 31: 143–152.

Lesica, P. 2005. The resurrection of *Erigeron parryi* (Asteraceae). Brittonia 57: 47–54.

Lessing, C. F. 1832. Synopsis Generum Compositarum.... Berlin. (Syn. Gen. Compos.)

L'Héritier de Brutelle, C.-L. [1788.] *Hymenopappus.* [Paris.] (Hymenopappus)

L'Héritier de Brutelle, C.-L. 1788[1789–1792]. Sertum Anglicum.... 4 fasc. Paris. [All text in fasc. 1; plates numbered consecutively.] (Sert. Angl.)

Lilloa = Lilloa; Revista de Botánica.

Ling, Y. R. 1982. On the system of the genus *Artemisia* L. and the relationship with its allies. Bull. Bot. Lab. N. E. Forest. Inst., Harbin 2: 1–60.

Ling, Y. R. 1995. The New World *Artemisia* L. In: D. J. N. Hind et al., eds. 1995. Advances in Compositae Systematics. Kew. Pp. 225–281.

Ling, Y. R. 1995b. The New World *Seriphidium* (Besser) Fourr. In: D. J. N. Hind et al., eds. 1995. Advances in Compositae Systematics. Kew. Pp. 283–291.

Linnaea = Linnaea. Ein Journal für die Botanik in ihrem ganzen Umfange.

Linnaeus, C. 1753. Species Plantarum.... 2 vols. Stockholm. (Sp. Pl.)

Linnaeus, C. 1754. Genera Plantarum, ed. 5. Stockholm. (Gen. Pl. ed. 5)

Linnaeus, C. 1755. Centuria I. Plantarum.... Uppsala. (Cent. Pl. I)

Linnaeus, C. 1755b. Flora Suecica..., ed. 2. Stockholm. (Fl. Suec. ed. 2)

Linnaeus, C. [1756.] Centuria II. Plantarum.... Uppsala. (Cent. Pl. II)

Linnaeus, C. 1758[–1759]. Systema Naturae per Regna Tria Naturae..., ed. 10. 2 vols. Stockholm. (Syst. Nat. ed. 10)

Linnaeus, C. 1762–1763. Species Plantarum..., ed. 2. 2 vols. Stockholm. (Sp. Pl. ed. 2)

Linnaeus, C. 1766–1768. Systema Naturae per Regna Tria

Naturae..., ed. 12. 3 vols. Stockholm. (Syst. Nat. ed. 12)

Linnaeus, C. 1767[–1771]. Mantissa Plantarum. 2 parts. Stockholm. [Mantissa [1] and Mantissa [2] Altera paged consecutively.] (Mant. Pl.)

Linnaeus, C. f. 1781[1782]. Supplementum Plantarum Systematis Vegetabilium Editionis Decimae Tertiae, Generum Plantarum Editionis Sextae, et Specierum Plantarum Editionis Secundae. Braunschweig. (Suppl. Pl.)

Lipschitz, S. J. 1979. Rod *Saussurea* DC. (Asteraceae). Leningrad.

List Pl. Nevada Utah—See: S. Watson [1871]b

Lista Espécies Herb. Português—See: G. Sampaio 1913

Liston, A. and J. W. Kadereit. 1995. Chloroplast DNA evidence for introgression and long distance dispersal in the desert annual *Senecio flavus* (Asteraceae). Pl. Syst. Evol. 197: 33–41.

Liston, A., L. H. Rieseberg, and T. S. Elias. 1989. Genetic similarity is high between intercontinental disjunct species of *Senecio* (Compositae). Amer. J. Bot. 76: 383–388.

Lloyd, F. E. 1901. Some points in the anatomy of *Chrysoma pauciflosculosa*. Bull. Torrey Bot. Club 28: 445–450.

Lockwood, S. C. and R. R. Weedon. 2004. The antimicrobial effects of *Bidens comosa* (A. Gray) Wieg., *Bidens cernua* L., *Bidens frondosa* L., and *Bidens vulgata* Greene. [Abstract.] Proc. Annual Meeting Nebraska Acad. Sci. 124: 40.

Loefling, P. 1758. Iter Hispanicum, Eller Resa til Spanska Ländern uti Europa och America, Förrättad Iffrån År 1751 til År 1756 ... Utgifven Efter Dess Frånfälle af Carl Linnaeus. Stockholm. (Iter Hispan.)

Lohwasser, U. and F. R. Blattner. 2004. Phylogenetic analysis of *Microseris* (Asteraceae), including a newly discovered Andean taxon from Peru. Syst. Bot. 29: 774–780.

London J. Bot. = London Journal of Botany.

Long, R. W. 1954. Synopsis of *Helianthus giganteus* L. and related species. Rhodora 56: 198–203.

Long, R. W. 1955. Hybridization between the perennial sunflowers *Helianthus salicifolius* A. Dietr. and *H. grosseserratus* Martens. Amer. Midl. Naturalist 54: 61–64.

Long, R. W. 1966. Biosystematics of the *Helianthus nuttallii* complex (Compositae). Brittonia 18: 64–79.

Long, R. W. and O. Lakela. 1971. A Flora of Tropical Florida: A Manual of the Seed Plants and Ferns of Southern Peninsular Florida. Coral Gables. [Reprinted 1976, Miami.]

Louda, S. M. 1998. Population growth of *Rhinocyllus conicus* (Coleoptera: Curculionidae) on two species of native thistles in prairie. Environm. Entomol. 27: 834–841.

Louda, S. M., D. Kendall, J. Connor, and D. Simberloff. 1997. Ecological effects of an insect introduced for the biological control of weeds. Science 277: 1088–1090.

Loudon, J. C. 1830. Hortus Brittanicus. A Catalogue of All the Plants Indigenous, Cultivated in, or Introduced to Britain. London. (Hort. Brit.)

Löve, Á. and D. Löve. 1982b. In: IOPB chromosome number reports LXXV. Taxon 31: 342–368.

Luckow, M. A. 1983. A morphometric and cytological study of *Pectis longipes* (Compositae). [Abstract.] Amer. J. Bot. 70(suppl.): 122.

Lundellia = Lundellia; Journal of the Plant Resources Center of the University of Texas at Austin.

Lyman, J. C. and N. C. Ellstrand. 1998. Relative contribution of breeding system and endemism to genotypic diversity: The outcrossing endemic *Taraxacum californicum* vs. the widespread apomict *T. officinale* (sensu lato). Madroño 45: 283–289.

Lyons, I. 1763. Fasciculus Plantarum Circa Cantabrigiam Nascentium.... London. (Fasc. Pl. Cantabr.)

Mabry, T. J. and G. Wagenitz, eds. 1990. Research advances in the Compositae. Pl. Syst. Evol., Suppl. 4.

Mackenzie, K. K. 1920. Scientific names applicable to our purple-flowered eupatoriums. Rhodora 22: 157–165.

Madroño = Madroño; Journal of the California Botanical Society [from vol. 3: a West American Journal of Botany].

Mag. Aesth. Bot.—See: H. G. L. Reichenbach [1821–]1822–1824[–1826]

Magee, D. W. and H. E. Ahles. 1999. Flora of the Northeast: A Manual of the Vascular Flora of New England and Adjacent New York. Amherst.

Maguire, B. 1943. A monograph of the genus *Arnica*. Brittonia 4: 386–510.

Mahler, W. F. and U. T. Waterfall. 1964. *Baccharis* (Compositae) in Oklahoma, Texas, and New Mexico. SouthW. Naturalist 9: 189–202.

Malte, M. O. 1934. *Antennaria* of arctic America. Rhodora 36: 101–117.

Man. Bot. ed. 5—See: A. Eaton 1829

Man. Bot. San Francisco—See: E. L. Greene 1894

Man. Fl. N. States—See: N. L. Britton 1901

Man. Fl. Pl. Calif.—See: W. L. Jepson [1923–1925]

Man. Pl. Colorado—See: H. D. Harrington 1954

Man. S. Calif. Bot.—See: P. A. Munz 1935

Man. S.E. Fl.—See: J. K. Small 1933

Manners, G. D. and D. S. Galitz. 1985. Allelopathy of small everlasting *(Antennaria microphylla)*: Identification of constituents phytotoxic to leafy spurge *(Euphorbia esula)*. Weed Sci. 34: 8–12.

Mant. Pl.—See: C. Linnaeus 1767[–1771]

Manual—See: A. Gray 1848

Manual ed. 2—See: A. Gray 1856

Manual ed. 5—See: A. Gray 1867

Marie-Victorin, Frère. 1925. Sur quelques Composées nouvelles, rares ou critiques du Québec oriental. Proc. & Trans. Roy. Soc. Canada, ser. 3, 19: 79–96.

Marie-Victorin, Frère. 1938. Phytogeographical problems of eastern Canada. Amer. Midl. Naturalist 19: 489–558.

Marie-Victorin, Frère. 1995. Flore Laurentienne, ed. 3, updated and annotated by L. Brouillet.... Montréal.

Markos, S. 2005. Taxonomic changes in *Lessingia* (Compositae: Astereae). Madroño 52: 60–61.

Markos, S. and B. G. Baldwin. 2001. Higher-level relationships and major lineages of *Lessingia* (Compositae, Astereae) based on nuclear rDNA internal and external transcribed spacer (ITS and ETS) sequences. Syst. Bot. 26: 168–183.

Marschall von Bieberstein, F. A. 1808–1819. Flora Taurico-

Caucasica.... 3 vols. Charkow. (Fl. Taur.-Caucas.)

Marsden-Jones, E. M. and W. B. Turrill. 1954. British Knapweeds; a Study in Synthetic Taxonomy. London.

Marshall, J. B. 1974. A note on *Conyza sumatrensis* (Rezt.) E. Walker *(C. floribunda)*. Watsonia 10: 166–167.

Martin, W. C. and C. R. Hutchins. 1980. A Flora of New Mexico. 2 vols. Vaduz.

Maschinski, J., H. D. Hammond, and L. Holter, eds. 1996. Southwestern Rare and Endangered Plants: Proceedings of the Second Conference: September 11–14, 1995, Flagstaff, Arizona. Fort Collins, Colo. [U.S.D.A. Forest Serv., Gen. Techn. Rep. RM-283.]

Matthews, J. F., L. S. Barden, and C. R. Matthews. 1997. Corrections of the chromosome number, distribution and misidentifications of the federally endagered sunflower, *Helianthus schweinitzii* T. & G. J. Torrey Bot. Soc. 124: 198–209.

Mauthe, S., K. Bachmann, K. L. Chambers, and H. J. Price. 1981. Variability of the inflorescence among populations of *Microseris laciniata* (Asteraceae, Lactuceae). Beitr. Biol. Pflanzen 56: 25–52.

Mavrodiev, E. V. et al. 2005. Phylogeny of *Tragopogon* L. (Asteraceae) based on ITS and ETS sequence data. Int. J. Pl. Sci. 166: 117–133.

Mayes, R. A. 1976. A Cytotaxonomic and Chemosystematic Study of the Genus *Pyrrocoma* (Asteraceae: Astereae). Ph.D. dissertation. University of Texas.

McArthur, E. D. 1984. Natural and artificial hybridization among *Artemisia tridentata* populations. [Abstract.] Amer. J. Bot. 71(suppl.): 105.

McArthur, E. D. et al. 1998. Randomly amplified polymorphic DNA analysis (RAPD) of *Artemisia* subgenus *Tridentatae* species and hybrids. Great Basin Naturalist 58: 12–27.

McArthur, E. D. and D. J. Fairbanks, comps. 2001. Shrubland Ecosystem Genetics and Biodiversity: Proceedings: Provo, UT, June 13–15, 2000. Ogden. [U.S.D.A. Forest Serv., RMRS P-21.]

McArthur, E. D., C. L. Pope, and D. C. Freeman. 1981. Chromosomal studies of subgenus *Tridentatae* of *Artemisia*: Evidence for autopolyploidy. Amer. J. Bot. 68: 589–605.

McArthur, E. D. and S. C. Sanderson. 1999. Cytogeography and chromosome evolution of subgenus *Tridentatae* of *Artemisia*. Amer. J. Bot. 86: 1754–1775.

McArthur, E. D. and B. L. Welch. 1986. Proceedings, Symposium on the Biology of *Artemisia* and *Chrysothamnus*, Provo, Utah, July 9–13, 1984. Ogden. [U.S.D.A. Forest Serv., Gen. Techn. Rep. INT-200.]

McArthur, E. D., B. L. Welch, and S. C. Sanderson. 1988. Natural and artificial hybridization between big sagebrush *(Artemisia tridentata)* subspecies. J. Heredity 79: 268–276.

McClintock, E. 1993b. *Chamomilla*. In: J. C. Hickman, ed. 1993. The Jepson Manual. Higher Plants of California. Berkeley, Los Angeles, and London. Pp. 226–228.

McGregor, R. L. 1967. A new species and two new varieties of *Echinacea* (Compositae). Trans. Kansas Acad. Sci. 70: 366–370.

McGregor, R. L. 1968. In: IOPB chromosome number reports XVII. Taxon 17: 285–288.

McGregor, R. L. 1986. *Centaurea*. In: Great Plains Flora Association. 1986. Flora of the Great Plains. Lawrence, Kans. Pp. 898–901.

McNeill, J., E. A. Odell, L. L. Consaul, and D. S. Katz(-Downie). 1987. American Code and later lectotypifications of Linnaean generic names dating from 1753: A case study of discrepancies. Taxon 36: 350–401.

McVaugh, R. 1984. Compositae. In: R. McVaugh and W. R. Anderson, eds. 1974+. Flora Novo-Galiciana: A Descriptive Account of the Vascular Plants of Western Mexico. 8+ vols. Ann Arbor. Vol. 12.

McVaugh, R. and W. R. Anderson, eds. 1974+. Flora Novo-Galiciana: A Descriptive Account of the Vascular Plants of Western Mexico. 8+ vols. Ann Arbor.

Mears, J. A. 1975. The taxonomy of *Parthenium* section *Partheniastrum* DC. (Asteraceae–Ambrosiinae). Phytologia 31: 463–482.

Meddel. Grønland = Meddelelser om Grønland, af Kommissionen for Ledelsen af de Geologiske og Geografiske Undersølgeser i Grønland.

Mehrhoff, L. J. 1983. *Sclerolepis uniflora* (Compositae) in New England. Rhodora 85: 433–438.

Melchert, T. E. 1990. *Cosmos caudatus* (Asteraceae: Coreopsideae) in Mexico: A cytotaxonomic reappraisal. Phytologia 69: 200–215.

Mém. Acad. Imp. Sci. St. Pétersbourg Hist. Acad. = Mémoires de l'Académie Impériale des Sciences de St. Pétersbourg. Avec l'Histoire de l'Académie.

Mém. Acad. Imp. Sci. St.-Pétersbourg, Sér. 6, Sci. Math. = Mémoires de l'Académie Impériale des Sciences de St.-Pétersbourg. Sixième Série. Sciences Mathématiques, Physiques et Naturelles.

Mem. Amer. Acad. Arts = Memoirs of the American Academy of Arts and Science.

Mem. Fac. Sci. Kyoto Univ., Ser. Biol. = Memoirs of the Faculty of Science, Kyoto University. Series of Biology.

Mem. Gray Herb. = Memoirs of the Gray Herbarium of Harvard University.

Mém. Mus. Hist. Nat. = Mémoires du Muséum d'Histoire Naturelle.

Mem. New York Bot. Gard. = Memoirs of the New York Botanical Garden.

Mem. Reale Accad. Sci. Ist. Bologna = Memorie della R[eale]. Accademia delle Scienze dell' Istituto di Bologna.

Mém. Soc. Phys. Genève = Mémoires de la Société de Physique et d'Histoire Naturelle de Genève.

Mem. Torrey Bot. Club = Memoirs of the Torrey Botanical Club.

Menhusen, B. R. 1963. Variation in the *Punctatae* Series of the Genus *Liatris* (Compositae). Ph.D. dissertation. University of Kansas.

Menken, S. B. J., E. Smit, and J. C. M. den Nijs. 1995. Genetical population structure in plants: Gene flow between diploid sexual and triploid asexual dandelions *(Taraxacum* section *Ruderalia)*. Evolution 49: 1108–1118.

Mentzelia = Mentzelia; Journal of the Northern Nevada Native Plant Society.

Merriam-Webster. 1988. Webster's New Geographical Dictionary. Springfield, Mass.

Methodus—See: C. Moench 1794

Meusel, H. and A. Kästner. 1990. Lebensgeschichte der Gold- und Silberdisteln. Monographie der Mediterran-Mitteleuropaischen Compositen-Gattung *Carlina* Band I. Merkmalsspektren und Lebensraume der Gattung. Denkschr. Österr. Akad. Wiss., Math.-Naturwiss. Kl. 127: 1–294.

Meusel, H. and A. Kästner. 1994. Lebensgeschichte der Gold- und Silberdisteln. Monographie der Mediterran-Mitteleuropaischen Compositen-Gattung *Carlina* Band II. Artenvielfalt und Stammesgeschichte der Gattung. Denkschr. Österr. Akad. Wiss., Math.-Naturwiss. Kl. 128: 1–657.

Meusel, H., A. Kästner, and E. Vitek. 1996. The evolution of *Carlina*—A hypothesis based on ecogeography. In: D. J. N. Hind et al., eds. 1996. Proceedings of the International Compositae Conference, Kew, 1994. 2 vols. Kew. Vol. 1, pp. 723–737.

Miao, B. M., B. L. Turner, and T. J. Mabry. 1995. Chloroplast DNA variations in sect. *Cyclachaena* of *Iva* (Asteraceae). Amer. J. Bot. 82: 919–923.

Miao, B. M., B. L. Turner, and T. J. Mabry. 1995b. Systematic implications of chloroplast DNA variation in the subtribe Ambrosiinae (Asteraceae: Heliantheae). Amer. J. Bot. 82: 924–932.

Miao, B. M., B. L. Turner, and T. J. Mabry. 1995c. Molecular phylogeny of *Iva* (Asteraceae: Heliantheae) based on chloroplast DNA restriction site variation. Pl. Syst. Evol. 195: 1–12.

Michaux, A. 1803. Flora Boreali-Americana.... 2 vols. Paris and Strasbourg. (Fl. Bor.-Amer.)

Michigan Bot. = Michigan Botanist.

Mickelson, C. J. and H. H. Iltis. 1966. Preliminary reports on the flora of Wisconsin no. 55. Compositae IV—composite family IV (tribes Helenieae and Anthemideae). Trans. Wisconsin Acad. Sci. 55: 187–222.

Middendorff, A. T. von. 1847–1867. Reise in den äussersten Norden und Osten Siberiens während der Jahre 1843 und 1844.... 4 vols. in parts and fasc. St. Petersburg. (Reise Siber.)

Miller, P. 1754. The Gardeners Dictionary.... Abridged..., ed. 4. 3 vols. London. (Gard. Dict. Abr. ed. 4)

Miller, P. 1768. The Gardeners Dictionary..., ed. 8. London. (Gard. Dict. ed. 8)

Milstead, W. L. 1964. A Revision of the North American Species of *Prenanthes*. Ph.D. dissertation. Purdue University.

Mirbel, C. de. 1815. Éléments de Physiologie Végétale et de Botanique. 2 vols. + atlas. Paris.

Mitchell, J. C. and G. Dupuis. 1971. Allergic contact dermatitis from sesquiterpenoids of the composite family plants. Brit. J. Dermatol. 84: 139–150.

Mitt. Bot. Staatssamml. München = Mitteilungen (aus) der Botanischen Staatssammlung München.

Moench, C. 1794. Methodus Plantas Horti Botanici et Agri Marburgensis.... Marburg. (Methodus)

Moerman, D. E. 1998. Native American Ethnobotany. Portland.

Mogie, M. and H. Ford. 1988. Sexual and asexual *Taraxacum* species. Biol. J. Linn. Soc. 35: 155–168.

Mogie, M. and A. J. Richards. 1983. Satellited chromosomes, systematics and phylogeny in *Taraxacum* (Asteraceae). Pl. Syst. Evol. 141: 219–229.

Mohr, C. T. 1901. Plant life of Alabama. Contr. U.S. Natl. Herb. 6.

Molec. Biol. Evol. = Molecular Biology and Evolution.

Molec. Ecol. = Molecular Ecology.

Molec. Phylogen. Evol. = Molecular Phylogenetics and Evolution.

Molina, G. I. 1782. Saggio sulla Storia Naturale del Chili.... Bologna. (Sag. Stor. Nat. Chili)

Monogr. Syst. Bot. Missouri Bot. Gard. = Monographs in Systematic Botany from the Missouri Botanical Garden.

Monogr. Taraxacum—See: H. Handel-Mazzetti 1907

Moore, A. J. and L. Bohs. 2003. An ITS phylogeny of *Balsamorhiza* and *Wyethia* (Asteraceae: Heliantheae). Amer. J. Bot. 90: 1653–1660.

Moore, M. 1993. Medicinal Plants of the Pacific West. Santa Fe.

Moore, R. J. 1972. Distribution of native and introduced knapweeds *(Centaurea)* in Canada and the United States. Rhodora 74: 331–346.

Moore, R. J. 1975. The biology of Canadian weeds. 13. *Cirsium arvense* (L.) Scop. Canad. J. Bot. 55: 1033–1048.

Moore, R. J. and C. Frankton. 1962. Cytotaxonomy and Canadian distribution of *Cirsium edule* and C. *brevistylum*. Canad. J. Bot. 40: 1187–1196.

Moore, R. J. and C. Frankton. 1963. Cytotaxonomic notes on some *Cirsium* species of the western United States. Canad. J. Bot. 41: 1553–1567.

Moore, R. J. and C. Frankton. 1964. A clarification of *Cirsium foliosum* and *Cirsium drummondii*. Canad. J. Bot. 42: 451–461.

Moore, R. J. and C. Frankton. 1965. Cytotaxonomy of *Cirsium hookerianum* and related species. Canad. J. Bot. 43: 597–613.

Moore, R. J. and C. Frankton. 1966. An evaluation of the status of *Cirsium pumilum* and *Cirsium hillii*. Canad. J. Bot. 44: 581–595.

Moore, R. J. and C. Frankton. 1967. Cytotaxonomy of foliose thistles *(Cirsium* spp. aff. *C. foliosum)* of western North America. Canad. J. Bot. 45: 1733–1749.

Moore, R. J. and C. Frankton. 1969. Cytotaxonomy of some *Cirsium* species of the eastern United States, with a key to eastern species. Canad. J. Bot. 52: 543–551.

Moore, R. J. and C. Frankton. 1974. The Thistles of Canada. Ottawa. [Canada Dept. Agric., Res. Branch, Monogr. 10.]

Moore, R. J. and C. Frankton. 1974b. The *Cirsium arizonicum* complex of the southwestern United States. Canad. J. Bot. 52: 543–551.

Mooring, J. S. 1965. Chromosome studies in *Chaenactis* and *Chamaechaenactis* (Compositae, Helenieae). Brittonia 17: 17–25.

Mooring, J. S. 1975. A cytogeographic study of *Eriophyllum lanatum* (Compositae, Helenieae). Amer. J. Bot. 81: 919–926.

Mooring, J. S. 1980. A cytogeographic study of *Chaenactis douglasii* (Compositae, Helenieae). Amer. J. Bot. 67: 1304–1319.

Mooring, J. S. 1992. Chromosome numbers and geographic distribution in *Chaenactis douglasii* (Compositae, Helenieae). Madroño 39: 263–270.

Mooring, J. S. 1994. A cytogenetic study of *Eriophyllum confertiflorum* (Compositae, Helenieae). Amer. J. Bot. 81: 919–926.

Mooring, J. S. 1997. A new base chromosome number and phylogeny for *Eriophyllum* (Asteraceae, Helenieae). Madroño 44: 364–373.

Mooring, J. S. 2001. Barriers to interbreeding in the *Eriophyllum lanatum* (Asteraceae, Helenieae) species complex. Amer. J. Bot. 88: 285–312.

Mooring, J. S. 2002. Experimental hybridizations of *Eriophyllum* annuals (Asteraceae, Helenieae). Amer. J. Bot. 89: 1973–1983.

Moran, R. V. 1996. The Flora of Guadalupe Island, Mexico. San Francisco.

Morefield, J. D. 1992. Evolution and Systematics of *Stylocline* (Asteraceae: Inuleae). Ph.D. dissertation. Claremont Graduate School.

Morefield, J. D. 1992b. Three new species of *Stylocline* (Asteraceae: Inuleae) from California and the Mojave Desert. Madroño 39: 114–130.

Morefield, J. D. 1992c. Resurrection and revision of *Hesperevax* (Asteraceae: Inuleae). Syst. Bot. 17: 293–310.

Morefield, J. D. 1992d. Notes on the status of *Psilocarphus berteri* (Asteraceae: Inuleae). Madroño 39: 155–157.

Morefield, J. D. 2004. New taxa and names in North American *Ancistrocarphus, Diaperia,* and *Logfia* (Asteraceae: Gnaphalieae: Filagininae) and related taxa. Novon 14: 463–475.

Morgan, D. R. 1990. A Systematic Study of *Machaeranthera* (Asteraceae) and Related Groups Using Restriction Site Analysis of Chloroplast DNA and a Taxonomic Revision of *Machaeranthera* Section *Psilactis*. Ph.D. dissertation. University of Texas.

Morgan, D. R. 1993. A molecular systematic study and taxonomic revision of *Psilactis* (Asteraceae: Astereae). Syst. Bot. 18: 290–308.

Morgan, D. R. 1997. Reticulate evolution in *Machaeranthera* (Asteraceae). Syst. Bot. 22: 599–615.

Morgan, D. R. 2003. nrDNA external transcribed spacer (ETS) sequence data, reticulate evolution, and the systematics of *Machaeranthera* (Asteraceae). Syst. Bot. 28: 179–190.

Morgan, D. R. and R. L. Hartman. 2003. A synopsis of *Machaeranthera* (Asteraceae: Astereae), with recognition of segregate genera. Sida 20: 1387–1416.

Morgan, D. R. and B. B. Simpson. 1992. A systematic study of *Machaeranthera* (Asteraceae) and related groups using restriction site analysis of chloroplast DNA. Syst. Bot. 17: 511–531.

Morgan, J. T. 1966. A Taxonomic Study of the Genus *Boltonia* (Asteraceae). Ph.D. dissertation. University of North Carolina.

Morse, C. A. 1994. Systematics and Taxonomy of *Stenotus* Nutt. (Asteraceae: Astereae). M.S. thesis. University of Kansas.

Morton, G. H. 1973. The Taxonomy of the *Solidago arguta–boottii* Complex. Ph.D. dissertation. University of Tennessee.

Morton, G. H. 1975. A new subspecies and other nomenclatural changes in the *Solidago arguta* complex. Phytologia 28: 1–3.

Morton, G. H. 1978. *Tussilago.* In: N. L. Britton et al., eds. 1905+. North American Flora.... 47+ vols. New York. Ser. 2, part 10, p. 174.

Morton, G. H. 1984. A practical treatment of the *Solidago gigantea* complex. Canad. J. Bot. 62: 1279–1282.

Morton, J. K. 1979. Observations on Houghton's goldenrod *(Solidago houghtonii)*. Michigan Bot. 18: 31–35.

Morton, J. K. and J. M. Venn. 1995. The marsh sow-thistle *(Sonchus palustris)* in North America. Rhodora 97: 93–95.

Moseley, R. K. 1994. The Status and Distribution of Cusick's False Yarrow *(Chaenactis cusickii)* in Idaho. Boise.

Mosquin, T. and D. E. Hayley. 1966. Chromosome numbers and taxonomy of some Canadian arctic plants. Canad. J. Bot. 44: 1209–1218.

Moss, E. H. 1959. Flora of Alberta: A Manual of Flowering Plants, Conifers, Ferns, and Fern Allies Found Growing without Cultivation in the Province of Alberta, Canada. Toronto.

Moss, E. H. and J. G. Packer. 1983. Flora of Alberta: A Manual of Flowering Plants, Conifers, Ferns, and Fern Allies Found Growing without Cultivation in the Province of Alberta, Canada, ed. 2. Toronto.

Mössler, J. C. and H. G. L. Reichenbach. 1827–1829[–1830]. Handbuch der Gewächskunde..., ed. 2. 3 vols. in 4. Altona. (Handb. Gewächsk. ed. 2)

Muhlenberg, G. H. E. 1813. Catalogus Plantarum Americae Septentrionalis.... Lancaster, Pa. (Cat. Pl. Amer. Sept.)

Muhlenbergia = Muhlenbergia; a Journal of Botany.

Mulligan, G. A. 1984. Chromosome numbers of some plants native and naturalized in Canada. Naturaliste Canad. 111: 447–449.

Munz, P. A. 1935. A Manual of Southern California Botany.... San Francisco. (Man. S. Calif. Bot.)

Munz, P. A. 1959. A California Flora. Berkeley and Los Angeles.

Munz, P. A. 1968. Supplement to A California Flora. Berkeley and Los Angeles.

Munz, P. A. 1974. A Flora of Southern California. Berkeley.

Murray, J. A. 1784. Caroli à Linné Equitis Systema Vegetabilium.... Editio Decima Quarta.... Göttingen. (Syst. Veg. ed. 14)

Myles, T. G. and B. F. Binder. 1990. The desert marigold moth. Desert Pl. 10: 75–78.

N. Amer. Fl.—See: N. L. Britton et al. 1905+

Narr. Journey Polar Sea—See: J. Franklin et al. 1823

Nat. Pflanzenfam.—See: H. G. A. Engler and K. Prantl 1887–1915

Naturaliste Canad. = Naturaliste Canadien. Bulletin de Recherches, Observations et Découvertes se Rapportant à l'Histoire Naturelle du Canada.

Nees, C. G. D. 1818. Synopsis Specierum Generis Asterum Herbacearum.... Erlangen. (Syn. Aster. Herb.)

Nees, C. G. D. 1832. Genera et Species Asterearum. Breslau. (Gen. Sp. Aster.)

Neher, R. T. 1966. Monograph of the Genus *Tagetes* (Compositae). Ph.D. thesis. Indiana University.

Nelson, C. H. and G. S. Van Horn. 1975. A new simplified method for constructing Wagner networks and the cladistics of *Pentachaeta* (Compositae: Astereae). Brittonia 27: 362–372.

Nelson, E. E. 1901. A revision of certain species of plants of the genus *Antennaria*. Proc. U.S. Natl. Mus. 23: 697–713.

Nesom, G. L. 1983. Biology and taxonomy of American *Leibnitzia* (Asteraceae). Brittonia 35: 126–139.

Nesom, G. L. 1983b. Taxonomy of *Erigeron concinnus* (Asteraceae) and its separation from *E. pumilus*. Sida 10: 159–166.

Nesom, G. L. 1984. Taxonomy and distribution of *Chaptalia dentata* and *C. albicans* (Asteraceae: Mutisieae). Brittonia 36: 396–401.

Nesom, G. L. 1988. Synopsis of *Chaetopappa* (Compositae: Astereae) with a new species and the inclusion of *Leucelene*. Phytologia 64: 448–456.

Nesom, G. L. 1989. New species, new sections, and a taxonomic overview of American *Pluchea* (Compositae: Inuleae). Phytologia 67: 158–167.

Nesom, G. L. 1989b. Comments on the annual species of *Xanthocephalum* (Compositae: Astereae) with a new combination. Phytologia 66: 482–487.

Nesom, G. L. 1989c. Infrageneric taxonomy of New World *Erigeron* (Compositae: Astereae). Phytologia 67: 67–93.

Nesom, G. L. 1989d. The separation of *Trimorpha* (Compositae: Astereae) from *Erigeron*. Phytologia 67: 61–66.

Nesom, G. L. 1989e. The *Solidago canadensis* (Asteraceae: Astereae) complex in Texas with a new species from Texas and Mexico. Phytologia 67: 441–450.

Nesom, G. L. 1990. Taxonomy of the genus *Laënnecia* (Asteraceae: Astereae). Phytologia 68: 205–228.

Nesom, G. L. 1990b. Taxonomic summary of *Omalotheca* (Asteraceae: Inuleae). Phytologia 68: 241–246.

Nesom, G. L. 1990c. Further definition of *Conyza* (Asteraceae: Astereae). Phytologia 68: 229–233.

Nesom, G. L. 1990d. Taxonomic summary of *Ericameria* (Asteraceae: Astereae), with the inclusion of *Haplopappus* sects. *Macronema* and *Asiris*. Phytologia 68: 144–155.

Nesom, G. L. 1990e. Taxonomy of *Heterotheca* sect. *Heterotheca* (Asteraceae: Astereae) in Mexico, with comments on the taxa in the United States. Phytologia 69: 282–294.

Nesom, G. L. 1990f. The taxonomic status of *Gamochaeta* (Asteraceae: Inuleae) and the species of the United States. Phytologia 68: 186–198.

Nesom, G. L. 1990g. Taxonomy of the *Erigeron coronarius* group of *Erigeron* sect. *Geniculactis* (Asteraceae: Astereae). Phytologia 69: 237–253.

Nesom, G. L. 1990h. Infrageneric taxonomy of North and Central American *Baccharis* (Asteraceae: Astereae). Phytologia 68: 40–46.

Nesom, G. L. 1990i. Studies in the systematics of Mexican and Texan *Grindelia* (Asteraceae: Astereae). Phytologia 68: 303–332.

Nesom, G. L. 1990j. Taxonomy of *Solidago petiolaris* (Astereae: Asteraceae) and related Mexican species. Phytologia 69: 445–458.

Nesom, G. L. 1991. Union of *Bradburia* with *Chrysopsis* (Asteraceae: Astereae), with a phylogenetic hypothesis for *Chrysopsis*. Phytologia 71: 109–121.

Nesom, G. L. 1991b. A phylogenetic hypothesis for the goldenasters (Asteraceae: Astereae). Phytologia 71: 136–151.

Nesom, G. L. 1991c. Taxonomy of *Isocoma* (Asteraceae: Astereae). Phytologia 70: 69–114.

Nesom, G. L. 1991d. Morphological definition of the *Gutierrezia* group (Asteraceae: Astereae). Phytologia 71: 252–262.

Nesom, G. L. 1991e. Transfer of *Aster kingii* to *Tonestus* (Asteraceae: Astereae). Phytologia 71: 122–127.

Nesom, G. L. 1991f. Redefinition of *Hesperodoria* (Asteraceae: Astereae) and the segregation of *Columbiadoria*, a new monotypic genus from the western United States. Phytologia 71: 244–251.

Nesom, G. L. 1992. Transfer of *Chaetopappa elegans* to *Ionactis* (Asteraceae: Astereae). Phytologia 73: 416–424.

Nesom, G. L. 1992b. Revision of *Erigeron* sect. *Linearifolii* (Asteraceae: Astereae). Phytologia 72: 157–208.

Nesom, G. L. 1993b. Taxonomic infrastructure of *Solidago* and *Oligoneuron* (Asteraceae: Astereae) and observations on their phylogenetic position. Phytologia 75: 1–44.

Nesom, G. L. 1993c. Taxonomy of *Doellingeria* (Asteraceae: Astereae). Phytologia 75: 452–462.

Nesom, G. L. 1993d. The genus *Oreostemma* (Asteraceae: Astereae). Phytologia 74: 305–316.

Nesom, G. L. 1993e. Taxonomy of *Sericocarpus* (Asteraceae: Astereae). Phytologia 75: 45–54.

Nesom, G. L. 1993f. *Erigeron*. In: J. C. Hickman, ed. 1993. The Jepson Manual. Higher Plants of California. Berkeley, Los Angeles, and London. Pp. 253–261.

Nesom, G. L. 1993g. Three species of *Aster* (Asteraceae: Astereae) disjunct in northern Coahuila, Mexico. Phytologia 74: 296–304.

Nesom, G. L. 1994. Subtribal classification of the Astereae (Asteraceae). Phytologia 76: 193–274.

Nesom, G. L. 1994b. Review of the taxonomy of *Aster* sensu lato (Asteraceae: Astereae), emphasizing the New World species. Phytologia 77: 141–297.

Nesom, G. L. 1994c. Hybridization in the tribe Astereae (Asteraceae). Phytologia 77: 298–307.

Nesom, G. L. 1995. Revision of *Chaptalia* (Asteraceae: Mutisieae) from North America and continental Central America. Phytologia 78: 153–188.

Nesom, G. L. 1997. Taxonomic adjustments in North American *Aster* sensu latissimo (Asteraceae: Astereae). Phytologia 82: 281–288.

Nesom, G. L. 1997b. The status of *Aster scabricaulis* (Asteraceae: Astereae), an endemic of the Gulf coastal plain. Phytologia 82: 300–315.

Nesom, G. L. 1998. Two newly recognized species of *Baccharis* (Asteraceae: Astereae) from Mexico. Phytologia 84: 43–49.

Nesom, G. L. 1999. Review of early nomenclature in *Euthamia* (Asteraceae: Astereae). Sida 18: 1009–1018.

Nesom, G. L. 1999b. *Gamochaeta simplicicaulis* (Asteraceae: Gnaphalieae) in four southeastern states and new for North America. Sida 18: 1259–1264.

Nesom, G. L. 2000. Generic Conspectus of the Tribe Astereae (Asteraceae) in North America, Central America, the Antilles, and Hawaii. Fort Worth. [Sida Bot. Misc. 20.]

Nesom, G. L. 2000b. *Gamochaeta simplicicaulis* (Asteraceae: Gnaphalieae) in Georgia. Sida 19: 413.

Nesom, G. L. 2001. Taxonomic review of *Chrysogonum* (Asteraceae: Heliantheae). Sida 19: 811–820.

Nesom, G. L. 2001b. An anomalous population of *Aster* (Asteraceae: Astereae) in Michigan. Sida 19: 625–632.

Nesom, G. L. 2001c. Notes on variation in *Pseudognaphalium obtusifolium* (Asteraceae: Gnaphalieae). Sida 19: 615–619.

Nesom, G. L. 2002. *Euchiton* (Asteraceae: Gnaphalieae) in North America and Hawaii. Sida 20: 515–521.

Nesom, G. L. 2003. New combinations in *Xanthisma* (Asteraceae: Astereae). Sida 20: 1585–1588.

Nesom, G. L. 2004. New species of *Gamochaeta* (Asteraceae: Gnaphalieae) from the eastern United States and comments on similar species. Sida 21: 717–741.

Nesom, G. L. 2004b. New distribution records for *Gamochaeta* (Asteraceae: Gnaphalieae) in the United States. Sida 21: 1175–1185.

Nesom, G. L. 2004c. Asteraceae from wool mill sites in South Carolina, including new records for North America. Sida 21: 1215–1223.

Nesom, G. L. 2004d. *Pseudognaphalium canescens* (Asteraceae: Gnaphalieae) and putative relatives in western North America. Sida 21: 781–790.

Nesom, G. L. 2004e. Taxonomic reevaluations in North American *Erigeron* (Asteraceae: Astereae). Sida 21: 19–39.

Nesom, G. L. 2005. Infrageneric classification of *Liatris* (Asteraceae: Eupatorieae). Sida 21: 1305–1321.

Nesom, G. L. 2005b. Broadened concept of *Liatris helleri* (Asteraceae: Eupatorieae). Sida 21: 1323–1333.

Nesom, G. L. 2005c. Taxonomic review of *Astranthium integrifolium* (Asteraceae: Astereae). Sida 21: 2015–2021.

Nesom, G. L. 2005d. Taxonomy of the *Symphyotrichum (Aster) subulatum* group and *Symphyotrichum (Aster) tenuifolium* (Asteraceae: Astereae). Sida 21: 2125–2140.

Nesom, G. L. and G. I. Baird. 1993. Completion of *Ericameria* (Asteraceae: Astereae), diminution of *Chrysothamnus*. Phytologia 75: 74–93.

Nesom, G. L. and G. I. Baird. 1995. Comments on "The *Chrysothamnus–Ericameria* connection." Phytologia 78: 61–65.

Nesom, G. L. and T. J. Leary. 1992. A new species of *Ionactis* (Asteraceae: Astereae) and an overview of the genus. Brittonia 44: 247–252.

Nesom, G. L. and D. R. Morgan. 1990. Reinstatement of *Tonestus* (Asteraceae: Astereae). Phytologia 68: 174–180.

Nesom, G. L. and D. F. Murray. 2004. Notes on North American arctic and boreal species of *Erigeron* (Asteraceae: Astereae). Sida 21: 41–58.

Nesom, G. L. and R. D. Noyes. 1999. Notes on sectional delimitations in *Erigeron* (Asteraceae: Astereae). Sida 18: 1161–1165.

Nesom, G. L. and R. J. O'Kennon. 2001. Two new species of *Liatris* series *Punctatae* (Asteraceae: Eupatorieae) centered in north-central Texas. Sida 19: 767–787.

Nesom, G. L. and J. M. Stucky. 2004. Taxonomy of the *Liatris pilosa (graminifolia)* group (Asteraceae: Eupatorieae). Sida 21: 815–826.

Nesom, G. L. and B. L. Turner. 1998. Variation in the *Berlandiera pumila* (Asteraceae) complex. Sida 18: 493–502.

Nesom, G. L., L. A. Vorobik, and R. L. Hartman. 1990. The identity of *Aster blepharophyllus* (Asteraceae: Astereae). Syst. Bot. 15: 638–642.

Neue Entd.—See: K. Sprengel 1820–1822

Neues J. Bot. = Neues Journal für die Botanik.

New Fl.—See: C. S. Rafinesque 1836[–1838]

New Man. Bot. Rocky Mt.—See: J. M. Coulter and A. Nelson 1909

New Phytol. = New Phytologist; a British Botanical Journal.

New Zealand J. Bot. = New Zealand Journal of Botany.

Nicolson, D. H. 1980. A summary of cytological information on *Emilia* and the taxonomy of four Pacific taxa of *Emilia* (Asteraceae: Senecioneae). Syst. Bot. 5: 391–407.

Nomencl. Bot. ed. 3—See: E. A. Raeuschel 1797

Nordenskiöld, A. E. 1882–1887. Vega-expeditionens Vetenskapliga Iakttagelser Bearbetade af Deltagare i Resan och Andra Forskare.... 5 vols. Stockholm. (Vega Exp. Vetensk. Iakttag.)

Nordenstam, B. 1977[1978]. Senecioneae and Liabeae—systematic review. In: V. H. Heywood et al., eds. 1977[1978]. The Biology and Chemistry of the Compositae. 2 vols. London, New York, and San Francisco. Vol. 2, pp. 800–830.

Nordenstam, B. 1978. Taxonomic studies in the tribe Senecioneae (Compositae). Opera Bot. 44: 1–83.

Nordic J. Bot. = Nordic Journal of Botany.

Norlindh, N. T. 1943. Studies in the Calenduleae. 1. Monograph of the Genera *Dimorphotheca, Castalis, Osteospermum, Gibbaria*, and *Chrysanthemoides*. Lund. (Stud. Calenduleae)

Norlindh, N. T. 1964. On the identity of *Arctotis stoechadifolia* Berg. Svensk Bot. Tidskr. 58: 193–203.

Norlindh, N. T. 1965. *Arctotis venusta* T. Norl. spec. nova, an ornamental plant from southern Africa. Bot. Not. 118: 403–411.

Norlindh, N. T. 1977[1978]. Arctoteae—systematic review. In: V. H. Heywood et al., eds. 1977[1978]. The Biology and Chemistry of the Compositae. 2 vols. London, New York, and San Francisco. Vol. 2, pp. 943–959.

Northington, D. K. 1974. Systematic studies of the genus *Pyrrhopappus* (Compositae, Cichorieae). Special Publ. Mus. Texas Tech Univ. 6: 1–38.

Not. Milit. Reconn.—See: W. H. Emory 1848

Notas Mus. La Plata, Bot. = Notas del Museo de La Plata. Botânica.

Notes Compositae—See: A. Gray 1880

Notes Roy. Bot. Gard. Edinburgh = Notes from the Royal Botanic Garden, Edinburgh.

Notizbl. Königl. Bot. Gart. Berlin = Notizblatt des Königlichen botanischen Gartens und Museums zu Berlin.

Nouv. Mém. Soc. Imp. Naturalistes Moscou = Nouveaux Mémoires de la Société Impériale des Naturalistes de Moscou.

Nov. Gen. Sp.—See: A. von Humboldt et al. 1815[1816]–1825

Nov. Pl. Descr. Dec.—See: C. G. Ortega 1797–1800

Nov. Veg. Descr.—See: P. de La Llave and J. M. de Lexarza 1824–1825

Nova Acta Regiae Soc. Sci. Upsal. = Nova Acta Regiae Societatis Scientiarum Upsaliensis.

Novak, S. J., D. E. Soltis, and P. S. Soltis. 1991. Ownbey's tragopogons: Forty years later. Amer. J. Bot. 78: 1586–1600.

Novi Provent.—See: K. Sprengel [1818]

Novit. Fl. Svec.—See: E. M. Fries 1814–1824

Novon = Novon; a Journal for Botanical Nomenclature.

Novosti Sist. Vyssh. Rast. = Novosti Sistematiki Vysshikh Rastenii.

Noyes, R. D. 2000. Biogeographical and evolutionary insights on *Erigeron* and allies (Asteraceae) from ITS sequence data. Pl. Syst. Evol. 220: 93–114.

Noyes, R. D. and L. H. Rieseberg. 1999. ITS sequence data support a single origin for North American Astereae (Asteraceae) and reflect deep geographic divisions in *Aster* s.l. Amer. J. Bot. 86: 398–412.

Noyes, R. D. and D. E. Soltis. 1996. Genotypic variation in agamospermous *Erigeron compositus* (Asteraceae). Amer. J. Bot. 83: 1292–1303.

Noyes, R. D., D. E. Soltis, and P. S. Soltis. 1995. Genetic and cytological investigations in sexual *Erigeron compositus* (Asteraceae). Syst. Bot. 20: 132–146.

Nuttall, T. 1813. A Catalogue of New and Interesting Plants Collected in Upper Louisiana.... London. (Cat. Pl. Upper Louisiana)

Nuttall, T. 1818. The Genera of North American Plants, and Catalogue of the Species, to the Year 1817.... 2 vols. Philadelphia. (Gen. N. Amer. Pl.)

Nuttall, T. 1822. A catalogue of a collection of plants made in East-Florida, during the months of October and November, 1821, by A. Ware. Amer. J. Sci. Arts 5: 286–304.

Nyman, C. F. 1878–1890. Conspectus Florae Europaeae.... 4 parts + 2 suppls. Örebro. [Parts 1–4 and suppl. 1 paged consecutively; suppl. 2 paged independently.] (Consp. Fl. Eur.)

Oberprieler, C. 2001. Phylogenetic relationships in *Anthemis* L. (Compositae, Anthemideae) based on nrDNA ITS sequence variation. Taxon 50: 745–762.

Oberprieler, C. 2002. A phylogenetic analysis of *Chamaemelum* Mill. (Compositae: Anthemideae) and related genera based upon nrDNA ITS and cpDNA *trnL/trnF* IGS sequence variation. Bot. J. Linn. Soc. 138: 255–273.

Oberprieler, C. and R. Vogt. 2000. The position of *Castrilanthemum* Vogt & Oberprieler and the phylogeny of Mediterranean Anthemideae (Compositae) as inferred from nrDNA ITS and cpDNA *trnL/trnF* IGS sequence variation. Pl. Syst. Evol. 225: 145–170.

Oberprieler, C. and R. Vogt. 2002. *Cladanthus*. In: Med-Checklist notulae, 21. Willdenowia 32: 195–208.

Observ. Compositae—See: R. Brown 1817

Observ. Phys. = Observations sur la Physique, sur l'Histoire Naturelle et sur les Arts.

Occas. Pap. Rancho Santa Ana Bot. Gard. = Occasional Papers of the Rancho Santa Ana Botanical Garden.

Ochsmann, J. 2001. On the taxonomy of spotted knapweed (*Centaurea stoebe* L.). In: L. Smith, ed. 2001. Proceedings of the First International Knapweed Symposium of the Twenty-first Century, March 15–16, 2001, Coeur d'Alene, Idaho. Albany, Calif. Pp. 33–41.

Ockendon, D. L., S. M. Walters, and T. P. Whiffen. 1969. Variation within *Centaurea nigra*. Proc. Bot. Soc. Brit. Isles 7: 549–552.

Oeder, G. C. et al., eds. [1761–]1764–1883. Icones Plantarum...Florae Danicae Nomine Inscriptum. 17 vols. in 51 fasc. Copenhagen. [Fascicles paged independently and numbered consecutively throughout volumes.] (Fl. Dan.)

Ohio J. Sci. = Ohio Journal of Science.

Olsen, J. S. 1979. Taxonomy of the *Verbesina virginica* complex (Asteraceae). Sida 8: 128–134.

Opera Bot. = Opera Botanica a Societate Botanice Lundensi.

Opusc. Phytol.—See: A.-H. G. de Cassini 1826–1834

Oregon Natural Heritage Program. 2004. Rare, Threatened and Endangered Plants and Animals of Oregon. Portland.

Ornduff, R. 1964. Biosystematics of *Blennosperma* (Compositae). Brittonia 16: 289–295.

Ornduff, R. 1966b. A biosystematic survey of the goldfield genus *Lasthenia* (Compositae: Helenieae). Univ. Calif. Publ. Bot. 40: 1–92.

Ornduff, R. 1969. The origin and relationship of *Lasthenia burkei* (Compositae). Amer. J. Bot. 56: 1042–1047.

Ornduff, R. 1993. *Lasthenia*. In: J. C. Hickman, ed. 1993. The Jepson Manual. Higher Plants of California. Berkeley, Los Angeles, and London. Pp. 298–300.

Ornduff, R. and B. A. Bohm. 1975. Relationships of *Tracyina* and *Rigiopappus* (Compositae). Madroño 23: 53–55.

Ortega, C. G. 1797–1800. Novarum, aut Rariorum Plantarum Horti Reg. Botan. Matrit. Descriptionum Decades.... 10 decades in 4 parts. Madrid. [Parts paged consecutively.] (Nov. Pl. Descr. Dec.)

Orzell, S. L. and E. L. Bridges. 2002. Notes on *Carphephorus odoratissimus* (Asteraceae) in peninsular Florida, U.S.A. Sida 20: 559–569.

Ostler, W. K., C. M. McKell, and S. White. 1986. *Chrysothamnus nauseosus*: A potential source of natural rubber. In: E. D. McArthur and B. L. Welch, eds. 1986. Proceedings, Symposium on the Biology of *Artemisia* and *Chrysothamnus*, Provo, Utah, July 9–13, 1984. Ogden. Pp. 389–394.

Osvald, H. and E. Åberg, eds. 1953. Proceedings of the Seventh International Botanical Congress, Stockholm July 12–20 1950. Stockholm and Waltham, Mass.

Ottawa Naturalist = Ottawa Naturalist; Transactions of the Ottawa Field-Naturalists' Club.

Ownbey, G. B. 1951b. Natural hybridization in the genus *Cirsium*—I. *C. discolor* (Muhl. ex Willd.) Spreng. × *C. muticum* Michx. Bull. Torrey Bot. Club 37: 541–547.

Ownbey, G. B. 1964. Natural hybridization in the genus *Cirsium*—II. *C. altissimum* × *C. discolor*. Michigan Bot. 3: 87–97.

Ownbey, G. B. and Hsi Y.-T. 1963. Chromosome numbers in

some North American species of the genus *Cirsium*. Rhodora 65: 339–354.

Ownbey, G. B. and T. Morley. 1991. Vascular Plants of Minnesota: A Checklist and Atlas. Minneapolis.

Ownbey, M. 1950c. Natural hybridization and amphiploidy in the genus *Tragopogon*. Amer. J. Bot. 37: 487–499.

Ownbey, M. and W. A. Weber. 1943. Natural hybridization in the genus *Balsamorhiza*. Amer. J. Bot. 30: 179–187.

Pacif. Railr. Rep.—See: War Department 1855–1860

Packer, J. G. 1972. A taxonomic and phytogeographical review of some arctic and alpine *Senecio* species. Canad. J. Bot. 50: 507–518.

Packer, J. G. 1983. Flora of Alberta: *Sparganium angustifolium* and *Erigeron trifidus*. Canad. J. Bot. 61: 359–366.

Packer, J. G. and G. D. McPherson. 1974. Chromosome numbers in some vascular plants from northern Alaska. Canad. J. Bot. 52: 1096–1099.

Pak, J. H. and K. Bremer. 1995. Phylogeny and reclassification of the genus *Lapsana* (Asteraceae: Lactuceae). Taxon 44: 13–21.

Pallas, P. S. 1771–1776. Reise durch verschiedene Provinzen des russischen Reichs.... 3 vols. St. Petersburg. (Reise Russ. Reich.)

Palmer, R. E. 1982. Ecological and Evolutionary Patterns in *Holocarpha* (Compositae, Madiinae). Ph.D. dissertation. University of California, Davis.

Pandey, A. K., S. Chopra, and R. P. Singh. 1986. Development and structure of seeds and fruits in Compositae: *Cosmos* species. J. Indian Bot. Soc. 65: 362–368.

Panero, J. L. 2005. New combinations and infrafamilial taxa in the Asteraceae. Phytologia 87: 1–14.

Panero, J. L. and V. A. Funk. 2002. Toward a phylogenetic subfamilial classification for the Compositae (Asteraceae). Proc. Biol. Soc. Wash. 115: 909–922.

Panero, J. L., R. K. Jansen, and J. A. Clevinger. 1999. Phylogenetic relationships of subtribe Ecliptinae (Asteraceae: Heliantheae) based on chloroplast DNA restriction site data. Amer. J. Bot. 413–427.

Panetta, F. D. and J. Dodd. 1987. The biology of Australian weeds. 16. *Chondrilla juncea* L. J. Austral. Inst. Agric. Sci. 53: 83–95.

Pap. Michigan Acad. Sci. = Papers of the Michigan Academy of Sciences, Arts and Letters.

Parker, E. S. and S. B. Jones. 1975. A systematic study of the genus *Balduina* (Compositae, Heliantheae). Brittonia 27: 355–361.

Parker, K. L. 1950. New combinations in *Hymenoxys*. Madroño 10: 159.

Parks, J. C. 1973. A revision of North American and Caribbean *Melanthera* (Compositae). Rhodora 75: 169–210.

Patterson, D. T. et al. 1989. Composite List of Weeds. Champaign.

Patterson, T. F. 1994. A Taxonomic Revision of *Conoclinium* DC. (Asteraceae, Eupatorieae). M.S. thesis. University of Texas.

Payne, W. W. 1964. A re-evaluation of the genus *Ambrosia* (Compositae). J. Arnold Arbor. 45: 401–430.

Pemberton, R. W. 2000. Predictable risk to native plants in weed biological control. Oecologia 125: 489–494.

Perdue, R. E. Jr. 1957[1958]. Synopsis of *Rudbeckia* subgenus *Rudbeckia*. Rhodora 59: 293–299.

Perry, L. M. 1937. Notes on *Silphium*. Rhodora 39: 281–297.

Persoon, C. H. 1805–1807. Synopsis Plantarum.... 2 vols. Paris and Tubingen. (Syn. Pl.)

Peterson, K. M. and W. W. Payne. 1973. The genus *Hymenoclea* (Compositae: Ambrosieae). Brittonia 25: 243–256.

Petrak, F. 1917. Die nordamerikanischen Arten der Gattung *Cirsium*. Beih. Bot. Centralbl. 35(2): 223–567.

Pflanzenr.—See: H. G. A. Engler 1900–1953

Phillips, K. 1963. A Taxonomic and Morphological Study of *Chrysoma paucifosculosa* (Michx.) Greene. M.S. thesis. University of South Carolina.

Philos. Mag. Ann. Chem. = The Philosophical Magazine, or Annals of Chemistry, Mathematics, Astronomy, Natural History, and General Science; New and United Series of the Philosophical Magazine and Annals of Philosophy.

Phytogr. Lusitan. Select.—See: F. Brotero 1801

Phytographia—See: C. L. Willdenow 1794

Phytologia = Phytologia; Designed to Expedite Botanical Publication.

Pike, R. B. 1970. Evidence for the hybrid status of *Aster blakei* (Porter) House. Rhodora 72: 401–436.

Pillai, A., S. K. Pillai, and O. Jacob. 1975. Embryogeny, histogenesis and apical meristems of *Cosmos bipinnatus* Cav. Acta Bot. Indica 3: 68–78.

Pinkava, D. J. 1964. Biosystematic Study of Genus *Berlandiera* DC. (Compositae). Ph.D. dissertation. Ohio State University.

Pinkava, D. J. 1967. Biosystematic study of *Berlandiera* (Compositae). Brittonia 19: 285–298.

Piper, C. V. and R. K. Beattie. 1901. The Flora of the Palouse Region. Pullman. (Fl. Palouse Reg.)

Pippen, R. W. 1968. Mexican "cacalioid" genera allied to *Senecio* (Compositae). Contr. U.S. Natl. Herb. 34: 365–447.

Pippen, R. W. 1978. *Cacalia*. In: N. L. Britton et al., eds. 1905+. North American Flora.... 47+ vols. New York. Ser. 2, part 10, pp. 151–159.

Pitelka, L. F., J. W. Ashmun, and R. L. Brown. 1985. The relationships between seasonal variation in light intensity, ramet size, and sexual reproduction in natural and experimental populations of *Aster acuminatus* (Compositae). Amer. J. Bot. 72: 311–319.

Pitelka, L. F. and W. F. Curtis. 1986. Photosynthetic responses to light in an understory herb, *Aster acuminatus*. Amer. J. Bot. 73: 535–540.

Pitelka, L. F., D. S. Stanton, and M. O. Pekenham. 1980. Effects of light and density on resource allocation in a forest herb, *Aster acuminatus*. Amer. J. Bot. 67: 942–948.

Pitelka, L. F., M. E. Thayer, and S. B. Hansen. 1983. Variation in achene weight in *Aster acuminatus*. Canad. J. Bot. 61: 1415–1420.

Pl. Aequinoct.—See: A. von Humboldt and A. J. Bonpland [1805–]1808–1809[–1817]

Pl. Baker.—See: E. L. Greene 1901

Pl. Hartw.—See: G. Bentham 1839[–1857]

Pl. Heermann.—See: E. M. Durand and T. C. Hilgard 1854

Pl. Hort. Schoenbr.—See: N. J. Jacquin 1797–1804

Pl. Med. (Stuttgart) = Planta Medica. Zeitschrift für Arzneipflanzenanwendung und Arzneipflanzenforschung [subtitle varies]. [Supplement to: Hippokrates.]

Pl. Nov. Thurb.—See: A. Gray 1854

Pl. Pratten. Calif.—See: E. M. Durand 1855

Pl. Rar. Hort. Monac.—See: F. Schrank [1817–]1819[–1822]

Pl. Syst. Evol. = Plant Systematics and Evolution.

Pl. Syst. Evol., Suppl. = Plant Systematics and Evolution. Supplementum.

Pl. Tex.—See: E. Hall 1873

Poiret, J. 1789. Voyage en Barbarie.... 2 vols. Paris. (Voy. Barbarie)

Poljakov, P. P. 1995. *Artemisia*. In: V. L. Komarov et al., eds. 1963+. Flora of the U.S.S.R. (Flora SSSR). Translated from Russian. 28+ vols. Jerusalem etc. Vol. 26, pp. 488–723.

Pollard, C. L. 1899. The genus *Achillea* in North America. Bull. Torrey Bot. Club 26: 365–372.

Pollen & Spores = Pollen et Spores.

Poole, J. M. and B. L. Heidel. 1993. Sensitive Plant Surveys in the Big Belt and Elkhorn Mountains...Montana. Helena.

Porsild, A. E. 1950. The genus *Antennaria* in northwestern Canada. Canad. Field-Naturalist 64: 1–25.

Porsild, A. E. 1950b. Five new Compositae from Yukon-Alaska. Canad. Field-Naturalist 64: 43–45.

Porsild, A. E. 1957. Illustrated flora of the Canadian Arctic Archipelago. Bull. Natl. Mus. Canada 146: 1–209.

Porsild, A. E. 1964. Illustrated flora of the Canadian Arctic Archipelago, ed. 2. Bull. Natl. Mus. Canada 146: 1–218.

Porsild, A. E. 1965. The genus *Antennaria* in eastern arctic and subarctic America. Bot. Tidsskr. 61: 22–55.

Porsild, A. E. 1975. Materials for a flora of central Yukon Territory. Publ. Bot. (Ottawa) 4. [Date on cover 1974.]

Porsild, A. E. and W. J. Cody. 1980. Vascular Plants of Continental Northwest Territories, Canada. Ottawa.

Porsild, M. P. 1930. Stray contributions to the flora of Greenland. I–V. Meddel. Grønland 77: 1–44.

Porter, Ced. L. 1943. The genus *Amphipappus* Torr. and Gray. Amer. J. Bot. 30: 481–483.

Porter, T. C. and J. M. Coulter. 1874. Synopsis of the Flora of Colorado.... Washington. (Syn. Fl. Colorado)

Powell, A. M. 1965. Taxonomy of *Tridax* (Compositae). Brittonia 17: 47–96.

Powell, A. M. 1969. Taxonomy of *Perityle,* section *Pappothrix* (Compositae–Peritylanae). Rhodora 71: 58–93.

Powell, A. M. 1973. Taxonomy of *Pericome* (Compositae–Peritylinae). SouthW. Naturalist 18: 335–339.

Powell, A. M. 1973b. Taxonomy of *Perityle,* section *Laphamia* (Compositae–Peritylanae). Sida 5: 61–128.

Powell, A. M. 1974. Taxonomy of *Perityle,* section *Perityle* (Compositae–Peritylanae). Rhodora 76: 229–306.

Powell, A. M. 1978. Systematics of *Flaveria* (Flaveriinae-Asteraceae). Ann. Missouri Bot. Gard. 65: 590–636.

Powell, A. M., D. W. Kyhos, and P. H. Raven. 1974. Chromosome numbers in Compositae. X. Amer. J. Bot. 61: 909–913.

Preece, S. J. and B. L. Turner. 1953. A taxonomic study of the genus *Chamaechaenactis* Rydberg (Compositae). Madroño 12: 97–103.

Prelim. Cat.—See: N. L. Britton et al. 1888

Prelim. Cat. Fl. Vancouver—See: W. R. Carter 1921

Prelim. Rep. U.S. Geol. Surv. Montana—See: F. V. Hayden 1872

Preslia = Preslia. Věstník (Časopis) Československé Botanické Společnosti.

Preston, R. E. 1997. *Dittrichia graveolens* (Asteraceae), new to the California weed flora. Madroño 44: 200–203.

Price, H. J. and K. Bachmann. 1975. DNA content and evolution in the Microseridinae. Amer. J. Bot. 62: 262–267.

Price, H. J., K. L. Chambers, and K. Bachmann. 1981. Geographic and ecological distribution of genomic DNA content variation in *Microseris douglasii* (Asteraceae). Bot. Gaz. 142: 415–426.

Price, H. J., K. L. Chambers, K. Bachmann, and J. Riggs. 1983. Inheritance of nuclear 2C DNA content variation in intraspecific and interspecific hybrids of *Microseris* (Asteraceae). Amer. J. Bot. 70: 1133–1138.

Prim. Fl. Holsat.—See: F. H. Wiggers [1780]

Proc. & Trans. Roy. Soc. Canada = Proceedings and Transactions of the Royal Society of Canada.

Proc. Acad. Nat. Sci. Philadelphia = Proceedings of the Academy of Natural Sciences of Philadelphia.

Proc. Amer. Acad. Arts = Proceedings of the American Academy of Arts and Sciences.

Proc. Annual Meeting Nebraska Acad. Sci. = Proceedings of Annual Meeting. Nebraska Academy of Sciences.

Proc. Biol. Soc. Wash. = Proceedings of the Biological Society of Washington.

Proc. Boston Soc. Nat. Hist. = Proceedings of the Boston Society of Natural History.

Proc. Bot. Soc. Brit. Isles = Proceedings of the Botanical Society of the British Isles.

Proc. Calif. Acad. Sci. = Proceedings of the California Academy of Sciences.

Proc. Indiana Acad. Sci. = Proceedings of the Indiana Academy of Science.

Proc. Natl. Acad. Sci. U.S.A. = Proceedings of the National Academy of Sciences of the United States of America.

Proc. U.S. Natl. Mus. = Proceedings of the United States National Museum.

Proc. Utah Acad. Sci. = Proceedings of the Utah Academy of Sciences.

Prodr.—See: A. P. de Candolle and A. L. P. P. de Candolle 1823–1873; O. P. Swartz 1788

Prodr. Pl. Cap.—See: C. P. Thunberg 1794–1800

Prosp. Hist. Pl. Dauphiné—See: D. Villars 1779

Pruski, J. F. 2004. *Panhalea heterophylla* (Compositae: Mutisioideae: Nassauvieae), a genus and species new for the flora of North America. Sida 21: 1225–1227.

Publ. Bot. (Ottawa) = Publications in Botany, National Museum of Natural Sciences, Canada.

Publ. Carnegie Inst. Wash. = Publications of the Carnegie Institution of Washington.

Publ. Field Mus. Nat. Hist., Bot. Ser. = Publications of the Field Museum of Natural History. Botanical Series.

Publ. Mus. Michigan State Univ., Biol. Ser. = Publications of the Museum. Michigan State University. Biological Series.

Pursh, F. [1813]1814. Flora Americae Septentrionalis; or, a Systematic Arrangement and Description of the Plants of North America. 2 vols. London. (Fl. Amer. Sept.)

Quart. J. Florida Acad. Sci. = Quarterly Journal of the Florida Academy of Sciences.

Quart. Rev. Biol. = Quarterly Review of Biology.

Radford, A. E. et al., eds. 1980+. Vascular Flora of the Southeastern United States. 2+ vols. Chapel Hill. (Vasc. Fl. S.E. U.S.)

Raeuschel, E. A. 1797. Nomenclator Botanicus..., ed. 3. Leipzig. (Nomencl. Bot. ed. 3)

Rafinesque, C. S. 1817. Florula Ludoviciana; or, a Flora of the State of Louisiana. Translated, Revised, and Improved, from the French of C. C. Robin.... New York. (Fl. Ludov.)

Rafinesque, C. S. 1833. Herbarium Rafinesquianum. .... Herbals; or Botanical Collections of C. S. Rafinesque...the Labor of a Whole Life! 3 parts. Philadelphia. [Parts designated as extras of Atlantic J., paged consecutively.] (Herb. Raf.)

Rafinesque, C. S. 1836[–1838]. New Flora and Botany of North America.... 4 parts. Philadelphia. [Parts paged independently.] (New Fl.)

Rafinesque, C. S. 1836[1837–1838]. Flora Telluriana.... 4 vols. Philadelphia. (Fl. Tellur.)

Rauschert, S. 1974. Nomenklatorische Probleme in der Gattung Matricaria L. Folia Geobot. Phytotax. 9: 249–260.

Raven, P. H. 1963. A flora of San Clemente Island, California. Aliso 5: 289–347.

Raven, P. H. 1963b. Pulicaria hispanica (Compositae: Inuleae), a weed new to California. Aliso 5: 251–253.

Raven, P. H. 1963c. Amphitropical relationships in the floras of North and South America. Quart. Rev. Biol. 38: 151–177.

Raven, P. H. and D. W. Kyhos. 1961. Chromosome numbers in Compositae. II. Helenieae. Amer. J. Bot. 48: 842–850.

Rees, A. [1802–]1819–1820. The Cyclopaedia; or, Universal Dictionary of Arts, Sciences, and Literature.... 39 vols. in 79 parts. London. [Pages unnumbered.] (Cycl.)

Reichenbach, H. G. L. [1821–]1822–1824[–1826]. Magazin der aesthetischen Botanik.... 2 vols. in parts. Leipzig. [Pages unnumbered, parts and plates numbered consecutively.] (Mag. Aesth. Bot.)

Reichenbach, H. G. L. 1841. Der deutsche Botaniker.... Erster Band. Das Herbarienbuch. Dresden and Leipzig. [Alt. title: Repertorium Herbarii....] (Deut. Bot. Herb.-Buch)

Reichenbach, H. G. L. 1842. Flora Saxonica. Die Flora von Sachsen.... Dresden and Leipzig. (Fl. Saxon.)

Reinartz, J. A. and D. H. Les. 1994. Bottleneck-induced dissolution of self-incompatibility and breeding system consequences in Aster furcatus (Asteraceae). Amer. J. Bot. 81: 446–455.

Reise Nord-Amer.—See: M. P. zu Wied 1839–1841

Reise Russ. Reich.—See: P. S. Pallas 1771–1776

Reise Siber.—See: A. T. von Middendorff 1847–1867

Rep. (Annual) Arkansas Geol. Surv. = Report (Annual) of the Arkansas Geological Survey.

Rep. (Annual) Michigan Acad. Sci. = Report (Annual) of the Michigan Academy of Science, (Arts, and Letters).

Rep. (Annual) Missouri Bot. Gard. = Report (Annual) of the Missouri Botanical Garden.

Rep. Bot. Soc. Exch. Club Brit. Isles = (Report,) Botanical Society and Exchange Club of the British Isles.

Rep. Exped. Rocky Mts.—See: J. C. Frémont 1843–1845

Rep. U.S. Geogr. Surv., Wheeler—See: J. T. Rothrock 1878[1879]

Rep. U.S. Mex. Bound.—See: W. H. Emory 1857–1859

Repert. Spec. Nov. Regni Veg. = Repertorium Specierum Novarum Regni Vegetabilis.

Repert. Spec. Nov. Regni Veg. Beih. = Repertorium specierum novarum regni vegetabilis. Beihefte.

Res. Rep. New Mexico Agric. Exp. Sta. = Research Report, New Mexico Agricultural Experiment Station.

Res. Stud. State Coll. Wash. = Research Studies of the State College of Washington.

Rev. Heterotheca Phyllotheca—See: J. C. Semple 1996

Rev. Pityopsis—See: J. C. Semple and F. D. Bowers 1985

Reveal, J. L. 1968. On the names in Fraser's 1813 catalogue. Rhodora 70: 25–54.

Reveal, J. L. 1970. A revision of the Utah species of Townsendia (Compositae). Great Basin Naturalist 30: 23–52.

Reveal, J. L. 1991c. Erigeron carolinianus L., an earlier name for Euthamia (Solidago) tenuifolia (Pursh) E. Greene (Asteraceae). Taxon 40: 505–508.

Reveal, J. L. and C. S. Keener. 1981. Virgulus Raf. (1837), an earlier name for Lasallea Greene (1903) (Asteraceae). Taxon 30: 648–651.

Reveal, J. L. and R. M. King. 1973. Re-establishment of Acourtia D. Don (Asteraceae). Phytologia 27: 228–232.

Reveal, J. L., G. E. Moulton, and A. E. Schuyler. 1999. The Lewis and Clark collections of vascular plants: Names, types, and comments. Proc. Acad. Nat. Sci. Philadelphia 149: 1–64.

Revis. Gen. Pl.—See: O. Kuntze 1891–1898

Revista Acad. Colomb. Ci. Exact. = Revista de la Academia Colombiana de Ciencias Exactas, Físicas y Naturales.

Revista Mus. La Plata, Secc. Bot. = Revista del Museo de La Plata. Sección Botánica.

Rhodora = Rhodora; Journal of the New England Botanical Club.

Richards, A. J. 1970. Eutriploid facultative agamospermy in Taraxacum. New Phytol. 69: 761–774.

Richards, A. J. 1970b. Hybridization in Taraxacum. New Phytol. 69: 1103–1121.

Richards, A. J. 1973. The origin of Taraxacum agamospecies. Bot. J. Linn. Soc. 66: 189–211.

Richards, A. J. 1985. Sectional nomenclature in Taraxacum (Asteraceae). Taxon 34: 633–644.

Richards, A. J. 1989. A comparison of within plant karyological heterogeneity between agamospermous and sexual Taraxacum (Compositae) as assessed by the nucleolar organiser chromosome. Pl. Syst. Evol. 163: 177–185.

Richards, A. J. 1996. Genetic variability of obligate apomicts of the genus Taraxacum. Folia Geobot. Phytotax. 31: 405–414.

Richards, A. J. and P. D. Sell. 1973. Taraxacum. In: T. G.

Tutin et al., eds. 1964–1980. Flora Europaea. 5 vols. Cambridge. Vol. 4, pp. 332–343.

Richards, E. L. 1968. A monograph of the genus *Ratibida*. Rhodora 70: 348–393.

Rieseberg, L. H. 1991. Homoploid reticulate evolution in *Helianthus*: Evidence from ribosomal genes. Amer. J. Bot. 78: 1218–1237.

Rieseberg, L. H., R. Carter, and S. Zona. 1990. Molecular tests of the hypothesized hybrid origin of two diploid *Helianthus* species (Asteraceae). Evolution 44: 1498–1511.

Rieseberg, L. H. and M. F. Doyle. 1989. Genetic variation in a rare sunflower from southern Texas, *Helianthus praecox* ssp. *hirtus* (Asteraceae). Aliso 12: 379–386.

Rieseberg, L. H., D. E. Soltis, and J. D. Palmer. 1988. A molecular reexamination of introgression between *Helianthus annuus* and *H. bolanderi*. Evolution 42: 227–238.

Ringius, G. S. 1985. A Biosystematic Study of the *Solidago spathulata* DC.–*S. glutinosa* Nutt. Complex (Compositae: Astereae). Ph.D. dissertation. University of Waterloo.

Ringius, G. S. and J. C. Semple. 1987. Cytogeography of the *Solidago spathulata*–*S. glutinosa* complex (Compositae: Astereae). Canad. J. Bot. 65: 2458–2462.

Roalson, E. H. and K. W. Allred. 1998. A clarification of *Centaurea americana* and *Centaurea rothrockii* (Compositae: Cardueae). New Mexico Botanist 7: 3–5.

Roberts, M. L. 1982. Systematic Studies of North American *Bidens* Section *Bidens* (Compositae). Ph.D. dissertation. Ohio State University.

Roberts, M. L. 1983. Allozyme variation in *Bidens discoidea* (Compositae). Brittonia 35: 239–247.

Roberts, M. L. 1985. The cytology, biology and systematics of *Megalodonta beckii* (Compositae). Aquatic Bot. 21: 99-110.

Roberts, R. P. 2002. Phylogeny of *Ericameria, Chrysothamnus* and Related Genera (Asteraceae: Astereae) Based on Nuclear Ribosomal DNA Sequence Data. Ph.D. dissertation. Louisiana State University.

Roberts, R. P. and L. E. Urbatsch. 2003. Molecular phylogeny of *Ericameria* (Asteraceae, Astereae) based on nuclear ribosomal 3' ETS and ITS sequence data. Taxon 52: 209–228.

Roberts, R. P. and L. E. Urbatsch. 2004. Molecular phylogeny of *Chrysothamnus* (Asteraceae, Astereae) based on nuclear ribosomal 3' ETS and ITS sequence data. Syst. Bot. 29: 199–215.

Roberts, R. P., L. E. Urbatsch, and K. M. Neubig. 2005. *Nestotus* and *Toiyabea*, two new genera of Asteraceae: Astereae from the western United States and Canada. Sida 21: 1647–1655.

Robinson, B. L. 1917. A monograph of the genus *Brickellia*. Mem. Gray Herb. 1: 3–151.

Robinson, B. L. 1934. *Mikania scandens* and its near relatives. Contr. Gray Herb. 104: 55–71.

Robinson, H. 1974b. Studies in the Senecioneae (Asteraceae). VI. The genus *Arnoglossum*. Phytologia 28: 294–295.

Robinson, H. 1978. Studies in the Heliantheae (Asteraceae). XII. Re-establishment of the genus *Smallanthus*. Phytologia 39: 47–53.

Robinson, H. 1981. A revision of the tribal and subtribal limits of the Heliantheae (Asteraceae). Smithsonian Contr. Bot. 51.

Robinson, H. 1996. Recent studies in the Heliantheae and Eupatorieae. In: D. J. N. Hind et al., eds. 1996. Proceedings of the International Compositae Conference, Kew, 1994. 2 vols. Kew. Vol. 1, pp. 627–653.

Robinson, H. 1999. Generic and subtribal classification of the American Vernonieae. Smithsonian Contr. Bot. 89.

Robinson, H. and R. D. Brettell. 1973. Studies in the Senecioneae (Asteraeae). IV. The genera *Mesadenia, Syneilesis, Miricacalia, Koyamacalia,* and *Sinacalia*. Phytologia 27: 256–276.

Robinson, H. and R. D. Brettell. 1973b. Studies in the Senecioneae (Asteraceae). III. The genus *Psacalium*. Phytologia 27: 254–264.

Robinson, H. and J. Cuatrecasas. 1973. The generic limits of *Pluchea* and *Tessaria*. Phytologia 27: 277–285.

Robinson, H. and J. Cuatrecasas. 1977. Notes on the genus and species limits of *Pseudogynoxys* (Greenm.) Cabrera (Senecioneae, Asteraceae). Phytologia 36: 177–192.

Roché, B. F. and C. T. Roché. 1991. Identification, introduction, distribution, ecology, and economics of *Centaurea* species. In: L. F. James et al., eds. 1991. Noxious Range Weeds. Boulder, San Francisco, and Oxford. Pp. 274–291.

Rock, H. F. L. 1957. A revision of the vernal species of *Helenium* (Compositae). Rhodora 59: 101–116, 128–158, 168–178, 203–216.

Roelofs, D. and K. Bachmann. 1997. Comparison of chloroplast and nuclear phylogeny in the autogamous annual *Microseris douglasii* (Asteraceae: Lactuceae). Pl. Syst. Evol. 204: 49–63.

Roelofs, D., J. van Velzen, P. Kuperus, and K. Bachmann. 1997. Molecular evidence for an extinct parent of the tetraploid species *Microseris acuminata* and *M. campestris* (Asteraceae, Lactuceae). Molec. Ecol. 6: 641–649.

Rogers, C. E., T. E. Thompson, and G. J. Seiler. 1982. Sunflower Species of the United States. Fargo.

Roland, A. E. and M. Zinck. 1998. Roland's Flora of Nova Scotia…, ed. 3. 2 vols. Halifax.

Rollins, R. C. 1950. The guayule rubber plant and its relatives. Contr. Gray Herb. 172: 1–73.

Roque, N. and D. J. N. Hind. 2001. *Ianthopappus,* a new genus of the tribe Mutisieae (Compositae). Novon 11: 97–101.

Roque, N. and M. S. F. Silvestre Capelato. 2001. Generic delimitation of *Gochnatia, Richterago,* and *Ianthopappus* (Compositae–Mutisieae) based on pollen morphology. Grana 40: 197–204.

Ross, T. S. and S. Boyd. 1996. Noteworthy collections. California. Madroño 43: 432–436.

Roth, A. W. 1797–1806. Catalecta Botanica…. 3 parts. Leipzig. [Parts paged independently.] (Catal. Bot.)

Rothrock, J. T. 1878[1879]. Report upon United States Geographical Surveys West of the One Hundredth Meridian, in Charge of First Lieut. Geo. M. Wheeler…. Vol. 6—Botany. Washington. (Rep. U.S. Geogr. Surv., Wheeler)

Ruffin, J. 1974. A taxonomic evaluation of the genera *Amphiachyris, Amphipappus, Greenella, Gutierrezia, Gymnosperma, Thurovia,* and *Xanthocephalum* (Compositae). Sida 5: 301–333.

Ruffin, J. 1977. Palynological survey of the genera *Amphiachyris, Amphipappus, Greenella, Gutierrezia, Gymnosperma,* and *Xanthocephalum* (Compositae). Contr. Gray Herb. 207: 117–131.

Ruiz López, H. and J. A. Pavón. 1794. Flora Peruvianae, et Chilensis Prodromus.... Madrid. (Fl. Peruv. Prodr.)

Ruiz López, H. and J. A. Pavón. 1798. Systema Vegetabilium Florae Peruvianae et Chilensis.... [Madrid.] (Syst. Veg. Fl. Peruv. Chil.)

Ruizia = Ruizia; Monografías del Jardín Botánico.

Ruprecht, F. 1845b. Flores Samejedorum Cisuralensium. St. Petersburg. [Alt. title: Beiträge zur Pflanzenkunde des Russischen Reiches.... Zweite Lieferung.] (Fl. Samojed. Cisural.)

Rustaiyan, A. et al. 1987. Further sesquiterpene lactones from the genus *Dittrichia.* Phytochemistry 26: 2603–2606.

Rydberg, P. A. 1900c. Catalogue of the flora of Montana and the Yellowstone National Park. Mem. New York Bot. Gard. 1: 1–492.

Rydberg, P. A. 1901. Studies on the Rocky Mountain flora VI. Bull. Torrey Bot. Club 28: 499–513.

Rydberg, P. A. 1906. Flora of Colorado.... Fort Collins, Colo. (Fl. Colorado)

Rydberg, P. A. 1915. *Tagetes.* In: N. L. Britton et al., eds. 1905+. North American Flora.... 47+ vols. New York. Vol. 34, pp. 148–159.

Rydberg, P. A. 1915b. *Tetraneuris.* In: N. L. Britton et al., eds. 1905+. North American Flora.... 47+ vols. New York. Vol. 34, pp. 100–108.

Rydberg, P. A. 1917. Flora of the Rocky Mountains and Adjacent Plains. New York. (Fl. Rocky Mts.)

Rydberg, P. A. 1932. Flora of the Prairies and Plains of Central North America.... New York. (Fl. Plains N. Amer.)

Rzedowski, J. 1983. Dos especies nuevas de Compositae–Astereae del Valle de México. Bol. Soc. Bot. México 45: 111–116.

Sag. Stor. Nat. Chili—See: G. I. Molina 1782

Sagra, R. de la. 1840–1855. Historia Fisica Politica y Natural de la Isla de Cuba.... 12 vols. Paris and Madrid. (Hist. Fis. Cuba)

Samml. Phys.-Oekon. Aufsätze = Sammlung physikalisch-oekonomischer Aufsätze; zur Aufnahme der Naturkunde und deren damit verwandten Wissenschaften in Böhmen.

Sampaio, G. 1913. Lista das Espécies Representadas no Herbário Português. Oporto. (Lista Espécies Herb. Português)

Samuel, R. et al. 2003. Phylogenetic relationships among species of *Hypochaeris* (Asteraceae, Cichorieae) based on ITS, plastid *trn*L intron, *trn*L-F spacer, and *mat*K sequences. Amer. J. Bot. 90: 496–507.

Sanders, A. C., D. L. Banks, and S. Boyd. 1997. Rediscovery of *Hemizonia mohavensis* (Asteraceae) and addition of two new localities. Madroño 44: 197–200.

Sanders, D. L. and C. Clark. 1987. Comparative morphology of the capitulum of *Enceliopsis.* Amer. J. Bot. 74: 1072–1086.

Sargentia = Sargentia; Continuation of the Contributions from the Arnold Arboretum of Harvard University.

Saroyan, J. P., D. R. Parnell, and J. L. Strother. 2000. Revision of *Corethrogyne* (Compositae: Astereae). Madroño 47: 89–96.

Sauck, J. R. 1975. Distribution, chromosomes, and taxonomy of *Parthenice mollis* (Compositae). Madroño 23: 227–234.

Savi, G. 1798. Flora Pisana.... 2 vols. Pisa. (Fl. Pis.)

Sched. Crit.—See: C. F. W. Wallroth 1822

Schilling, E. E. 1990. Taxonomic revision of *Viguiera* subg. *Bahiopsis* (Asteraceae: Heliantheae). Madroño 37: 149–170.

Schilling, E. E. 1997. Phylogenetic analysis of *Helianthus* (Asteraceae) based on chloroplast DNA restriction site data. Theor. Appl. Genet. 94: 925–933.

Schilling, E. E., C. R. Lander, R. D. Noyes, and L. H. Rieseberg. 1998. Phylogenetic relationships in *Helianthus* (Asteraceae) based on nuclear ribosomal DNA internal transcribed spacer region sequence data. Syst. Bot. 23: 177–187.

Schilling, E. E. and J. L. Panero. 1996. Relationships in Heliantheae subtribe Helianthinae based on chloroplast DNA restriction site analysis. In: D. J. N. Hind et al., eds. 1996. Proceedings of the International Compositae Conference, Kew, 1994. 2 vols. Kew. Vol. 1, pp. 361–376.

Schilling, E. E. and J. L. Panero. 2002. A revised classification of subtribe Helianthinae (Asteraceae: Heliantheae). I. Basal lineages. Bot. J. Linn. Soc. 140: 65–76.

Schilling, E. E., J. L. Panero, and P. B. Cox. 1999. Chloroplast DNA restriction site data support a narrowed interpretation of *Eupatorium* (Asteraceae). Pl. Syst. Evol. 219: 209–223.

Schmidt, G. J. and E. E. Schilling. 2000. Phylogeny and biogeography of *Eupatorium* (Asteraceae: Eupatorieae) based on nuclear ITS sequence data. Amer. J. Bot. 87: 716–726.

Schneider, D. J. and J. L. Du Plessis. 1980. Enteritis in sheep due to the ingestion of *Inula graveolens* Desf. (Cape khakiweed). J. S. African Veterin. Assoc. 51: 159–161.

Schousboe, P. K. A. 1800. Iagttagelser over Vextriget i Marokko. Copenhagen. (Iagttag. Vextrig. Marokko)

Schrader, H. A. 1809–1811. Hortus Gottingensis.... 2 parts. Göttingen. [Parts paged consecutively.] (Hort. Gott.)

Schrank, F. [1817–]1819[–1822]. Plantae Rariores Horti Academici Monacensis.... 2 vols in 10 parts. Munich. [Parts and plates numbered consecutively.] (Pl. Rar. Hort. Monac.)

Schreber, J. C. 1789–1791. Caroli a Linné...Genera Plantarum.... 2 vols. Frankfurt am Main. (Gen. Pl.)

Schulthess, B. H., G. E. Giger, and T. W. Baumann. 1991. *Echinacea*—Anatomy, phytochemical pattern, and germination of the achene. Pl. Med. (Stuttgart) 57: 384–388.

Schultz, F. W. 1854–1869. Archives de Flore Journal Botanique.... 2 vols. Wissembourg and Deidesheim. [Vols. paged consecutively.] (Arch. Fl.)

Schultz-Bipontinus, C. H. 1844. Ueber die Tanaceteen.... Neustadt an der Haardt. (Tanaceteen)

Schultz-Bipontinus, C. H. 1856. Enumeration of the Compositae collected by B. Seemann and J. Potts in North-western Mexico. In: B. Seemann. 1852–1857. The Botany of the Voyage of H.M.S. Herald...during the Years 1845–51. 10 parts. London. Pp. 297–315.

Schwegman, J. E. and R. W. Nyboer. 1985. The taxonomic and population status of *Boltonia decurrens* (Torrey & Gray) Wood. Castanea 50: 112–115.

Schweigger, A. F. and H. F. F. Körte. 1811. Flora Erlangensis.... 2 parts. Erlangen. (Fl. Erlang.)

Science = Science; an Illustrated Journal [later: a Weekly Journal Devoted to the Advancement of Science]. [American Association for the Advancement of Science.]

Scoggan, H. J. 1978–1979. The Flora of Canada. 4 parts. Ottawa. [Natl. Mus. Nat. Sci. Publ. Bot. 7.] (Fl. Canada)

Scopoli, J. A. 1771–1772. Flora Carniolica..., ed. 2. 2 vols. Vienna. (Fl. Carniol. ed. 2)

Scopoli, J. A. 1777. Introductio ad Historiam Naturalem.... Prague. (Intr. Hist. Nat.)

Seemann, B. 1852–1857. The Botany of the Voyage of H.M.S. Herald...during the Years 1845–51. 10 parts. London. [Parts paged consecutively.] (Bot. Voy. Herald)

Sell, P. D. and C. West. 1976. *Hieracium.* In: T. G. Tutin et al., eds. 1964–1980. Flora Europaea. 5 vols. Cambridge. Vol. 4, pp. 358–410.

Semple, J. C. 1974. The phytogeography and systematics of *Xanthisma texanum* de Candolle (Asteraceae): Proper usage of infraspecific categories. Rhodora 76: 1–19.

Semple, J. C. 1977. Cytotaxonomy of *Chrysopsis* and *Heterotheca* (Compositae–Astereae): A new interpretation of phylogeny. Canad. J. Bot. 55: 2503–2513.

Semple, J. C. 1978. A revision of the genus *Borrichia* Adans. (Compositae). Ann. Missouri Bot. Gard. 65: 681–693.

Semple, J. C. 1981. A revision of the goldenaster genus *Chrysopsis* (Nutt.) Ell. nom. cons. (Compositae–Astereae). Rhodora 83: 323–384.

Semple, J. C. 1982. Observations on morphology and cytology of *Aster hemisphaericus, A. paludosus* and *A. chapmanii* (Asteraceae) with comments on chromosomal base number and phylogeny of *Aster* subg. *Aster* sect. *Heleastrum.* Syst. Bot. 7: 60–70.

Semple, J. C. 1983. Range expansion of *Heterotheca camporum* (Compositae: Astereae) in the southeastern United States. Brittonia 35: 140–146.

Semple, J. C. 1984. Cytogeographic studies on North American asters. I. Range surveys of *Virgulus adnatus, V. concolor, V. georgianus, V. grandiflorus, V. novae-angliae, V. oblongifolius, V. patens,* and *V. walteri.* Amer. J. Bot. 71: 522–531.

Semple, J. C. 1995. A review of hypotheses on ancestral chromosomal base-numbers in the tribe Astereae and the genus *Aster.* In: D. J. N. Hind et al., eds. 1995. Advances in Compositae Systematics. Kew. Pp. 153–165.

Semple, J. C. 1996. A Revision of *Heterotheca* Sect. *Phyllotheca* (Nutt.) Harms (Compositae: Astereae): The Prairie and Montane Goldenasters of North America. Waterloo.

[Univ. Waterloo Biol. Ser. 37.] (Rev. Heterotheca Phyllotheca)

Semple, J. C. 2003. New names and combinations in goldenrods, *Solidago* (Asteraceae: Astereae). Sida 20: 1605–1616.

Semple, J. C. 2004. Miscellaneous nomenclatural changes in Astereae (Asteraceae). Sida 21: 759–765.

Semple, J. C., V. Blok, and P. Heiman. 1980. Morphological, anatomical, habit and habitat differences among the goldenaster genera *Chrysopsis, Heterotheca* and *Pityopsis* (Compositae–Astereae). Canad. J. Bot. 58: 164–171.

Semple, J. C. and F. D. Bowers. 1985. A Revision of the Goldenaster Genus *Pityopsis* Nuttall (Compositae: Astereae). Waterloo. [Univ. Waterloo Biol. Ser. 29.] (Rev. Pityopsis)

Semple, J. C. and F. D. Bowers. 1987. Cytogeography of *Pityopsis* Nutt., the grass-leaved goldenasters (Compositae: Astereae). Rhodora 89: 381–389.

Semple, J. C. and R. A. Brammall. 1982. Wild *Aster lanceolatus* × *lateriflorus* hybrids in Ontario and comments on the origin of *A. ontarionis* (Compositae–Astereae). Canad. J. Bot. 60: 1895–1906.

Semple, J. C. and L. Brouillet. 1980. A synopsis of North American asters: The subgenera, sections and subsections of *Aster* and *Lasallea.* Amer. J. Bot. 67: 1010–1026.

Semple, J. C. and C. C. Chinnappa. 1980. Phylogenetic implications of meiosis in wild and cultivated interspecific hybrids in *Chrysopsis* (Compositae–Astereae): *C. godfreyi* ($n = 5$) × *gossypina* ssp. *cruiseana* ($n = 9$) and *C. godfreyi* ($n = 5$) × *C. linearifolia* ($n = 5$). Canad. J. Bot. 58: 172–181.

Semple, J. C. and C. C. Chinnappa. 1980b. Karyotype evolution and chromosome numbers in *Chrysopsis* (Nutt.) Ell. sensu Semple (Compositae–Astereae). Canad. J. Bot. 58: 164–171.

Semple, J. C. and C. C. Chinnappa. 1984. Observations on the cytology, morphology, and ecology of *Bradburia hirtella* (Compositae–Astereae). Syst. Bot. 9: 95–101.

Semple, J. C. and C. C. Chinnappa. 1986. The cytogeography of *Chrysopsis mariana* (Compositae: Astereae): Survey over the range of the species. Rhodora 88: 261–266.

Semple, J. C. and J. G. Chmielewski. 1987. Revision of the *Aster lanceolatus* complex, including *A. simplex* and *A. hesperius* (Compositae: Astereae): A multivariate morphometric study. Canad. J. Bot. 65: 1047–1062.

Semple, J. C., J. G. Chmielewski, and R. A. Brammall. 1990. A multivariate morphometric study of *Solidago nemoralis* (Compositae: Astereae) and comparison with *S. californica* and *S. sparsiflora.* Canad. J. Bot. 68: 2070–2082.

Semple, J. C., J. G. Chmielewski, and C. Leeder. 1991. A multivariate morphometric study and revision of *Aster* subg. *Doellingeria* sect. *Triplopappus* (Compositae: Astereae): The *Aster umbellatus* complex. Canad. J. Bot. 69: 256–276.

Semple, J. C., J. G. Chmielewski, S. Rao, and G. A. Allen. 1983. The cytogeography of *Aster lanceolatus.* II. A preliminary survey of the range including *A. hesperius.* Canad. J. Bot. 61: 434–441.

Semple, J. C. and K. N. Gandhi. 2004. *Solidago* sect.

Ptarmicoidei, a new combination to replace a "rankless" name used by Torrey and A. Gray (Asteraceae: Astereae). Sida 21: 755–757.

Semple, J. C., S. B. Heard, and L. Brouillet. 2002. Cultivated and Native Asters of Ontario (Compositae: Astereae). *Aster* L. (Including *Asteromoea* Blume, *Diplactis* Raf. and *Kalimeris* (Cass.) Cass.), *Callistephus* Cass., *Galatella* Cass., *Doellingeria* Nees, *Oclemena* E. L. Greene, *Eurybia* (Cass.) S. F. Gray, *Canadanthus* Nesom, and *Symphyotrichum* Nees (Including *Virgulus* Raf.). Waterloo. [Univ. Waterloo Biol. Ser. 41.] (Cult. Native Asters Ontario)

Semple, J. C., S. B. Heard, and Xiang C. 1996. The Asters of Ontario (Compositae: Astereae): *Diplactis* Raf., *Oclemena* Greene, *Doellingeria* Nees and *Aster* L. (Including *Canadanthus* Nesom, *Symphyotrichum* Nees and *Virgulus* Raf.), revised ed. Waterloo. [Univ. Waterloo Biol. Ser. 38.] (Asters Ontario ed. 2)

Semple, J. C. and J. L. A. Hood. 2005. Pappus variation in North American Asters. I. Double, triple and quadruple pappus in *Symphiotrichum* and related aster genera (Asteraceae: Astereae). Sida 21: 2141–2159.

Semple, J. C., C. Leeder, C. Leuty, and L. Gray. 1988. *Heterotheca* sect. *Ammodia* (Compositae: Astereae): A multivariate study of *H. oregona* and specimens of Brewer's (golden)aster. Syst. Bot. 13: 547–558.

Semple, J. C., G. S. Ringius, and J. J. Zhang. 1999. The Goldenrods of Ontario: *Solidago* L. and *Euthamia* Nutt., ed. 3. Waterloo. [Univ. Waterloo Biol. Ser. 39.]

Semple, J. C. and K. S. Semple. 1977. *Borrichia* ×*cubana* (*B. frutescens* × *arborescens*): Interspecific hybridization in the Florida Keys. Syst. Bot. 2: 292–301.

Semple, J. C. and L. Tebby. 1999. A cladistic analysis of subtribe Chrysopsidinae (Asteraceae: Astereae). In: International Botanical Congress. [1999.] Abstracts. XVI International Botanical Congress, St. Louis, USA, August 1–7, 1999. [St. Louis.] Abstr. 2852, poster 401.

Semple, J. C., J. J. Zhang, and Xiang C. 1993. Chromosome number determinations in fam. Compositae, tribe Astereae. V. Eastern North American taxa. Rhodora 95: 234–253.

Sert. Angl.—See: C.-L. L'Héritier de Brutelle 1788[1789–1792]

Sharp, W. M. 1935. A critical study of certain epappose genera of the Heliantheae–Verbesininae of the natural family Compositae. Ann. Missouri Bot. Gard. 22: 51–152.

Sherff, E. E. 1937. The genus *Bidens*. Publ. Field Mus. Nat. Hist., Bot. Ser. 16.

Sherff, E. E. 1955. *Cosmos.* In: N. L. Britton et al., eds. 1905+. North American Flora.... 47+ vols. New York. Ser. 2, part 2, pp. 130–146.

Shinners, L. H. 1946. Revision of the genus *Aphanostephus* DC. Wrightia 1: 95–121.

Shinners, L. H. 1946b. Revision of the genus *Chaetopappa* DC. Wrightia 1: 63–81.

Shinners, L. H. 1946c. Revision of the genus *Leucelene* Greene. Wrightia 1: 82–89.

Shinners, L. H. 1946d. The genus *Dichaetophora* A. Gray and its relationships. Wrightia 1: 90–94.

Shinners, L. H. 1947. Revision of the genus *Krigia* Schreb. Wrightia 1: 187–206.

Shinners, L. H. 1949. Nomenclature of species of dandelion and goats-beard (*Taraxacum* and *Tragopogon*) introduced into Texas. Field & Lab. 17: 13–19.

Shinners, L. H. 1950. Notes on Texas Compositae. IV. Field & Lab. 18: 25–32.

Shinners, L. H. 1950b. Notes on Texas Compositae. V. Field & Lab. 18: 32–42.

Shinners, L. H. 1951c. Notes on Texas Compositae. VIII. Field & Lab. 19: 134.

Shinners, L. H. 1951d. The Texas species of *Evax* (Compositae). Field & Lab. 19: 125–126.

Shinners, L. H. 1951e. Revision of the north Texas species of *Heterotheca* including *Chrysopsis* (Compositae). Field & Lab. 19: 66–71.

Shreve, F. and I. L. Wiggins. 1964. Vegetation and Flora of the Sonoran Desert. 2 vols. Stanford.

Shultz, L. M. 1983. Systematics and Anatomical Studies of *Artemisia* Subgenus *Tridentatae.* Ph.D. dissertation. Claremont Graduate School.

Shultz, L. M. 1986. Taxonomic and geographic limits of *Artemisia* subgenus *Tridentatae* (Beetle) McArthur. In: E. D. McArthur and B. L. Welch, eds. 1986. Proceedings, Symposium on the Biology of *Artemisia* and *Chrysothamnus,* Provo, Utah, July 9–13, 1984. Ogden. Pp. 20–28.

Shultz, L. M. 1986b. Comparative leaf anatomy of sagebrush. In: E. D. McArthur and B. L. Welch, eds. 1986. Proceedings, Symposium on the Biology of *Artemisia* and *Chrysothamnus,* Provo, Utah, July 9–13, 1984. Ogden. Pp. 253–264.

Sibthorp, J. and J. E. Smith. 1806–1813[–1816]. Florae Graecae Prodromus: Sive Plantarum Omnium Enumeratio, Quas in Provinciis aut Insulis Graeciae Invenit.... 2 vols. London. (Fl. Graec. Prodr.)

Sida = Sida; Contributions to Botany.

Sieren, D. J. 1981. The taxonomy of the genus *Euthamia.* Rhodora 83: 551–579.

Sieren, D. J. and J. F. Merritt. 1980. In: IOPB chromosome number reports LXIX. Taxon 29: 703–730.

Simpson, B. B. and C. E. Anderson. 1978. Mutisieae. In: N. L. Britton et al., eds. 1905+. North American Flora.... 47+ vols. New York. Ser. 2, part 10, pp. 1–13.

Simurda, M. C. and J. S. Knox. 2000. ITS sequence evidence for the disjunct distribution between Virginia and Missouri of the narrow endemic *Helenium virginicum.* J. Torrey Bot. Soc. 127: 316–323.

Singhurst, J. R., R. J. O'Kennon, and W. C. Holmes. 2004. The genus *Prenanthes* (Asteraceae: Lactuceae) in Texas. Sida 21: 181–191.

Sketch Bot. S. Carolina—See: S. Elliott [1816–]1821–1824

Skinner, M. W. and B. M. Pavlik, eds. 1994. Inventory of Rare and Endangered Vascular Plants of California, ed. 5. Sacramento.

Skr. Naturhist.-Selsk. = Skrivter af Naturhistorie-Selskabet.

Skr. Vidensk.-Selsk. Christiana, Math.-Naturvidensk. Kl. = Skrifter Udgivne af Videnskabs-Selskabet i Christiana. Mathematisk-Naturvidenskabelig Klasse.

Sledge, W. A. 1975. *Cirsium.* In: C. A. Stace, ed. 1975. Hybridization and the Flora of the British Isles. London and New York. Pp. 421–426.

Small, E. and P. M. Catling. 1999. Canadian Medicinal Crops. Ottawa.

Small, J. 1919. The Origin and Development of the Compositae. London. [New Phytol. Repr. 11.]

Small, J. K. 1903. Flora of the Southeastern United States.... New York. (Fl. S.E. U.S.)

Small, J. K. 1913b. Flora of Miami.... New York. (Fl. Miami)

Small, J. K. 1933. Manual of the Southeastern Flora, Being Descriptions of the Seed Plants Growing Naturally in Florida, Alabama, Mississippi, Eastern Louisiana, Tennessee, North Carolina, South Carolina and Georgia. New York. (Man. S.E. Fl.)

Smith, B. N. and B. L. Turner. 1975. Distribution of Kranz syndrome among Asteraceae. Amer. J. Bot. 62: 541–545.

Smith, D. M. and A. T. Guard. 1958. Hybridization between *Helianthus divaricatus* and *H. microcephalus*. Brittonia 10: 137–145.

Smith, E. B. 1965. Taxonomy of *Haplopappus* section *Isopappus* (Compositae). Rhodora 67: 217–238.

Smith, E. B. 1966. Cytogenetics and phylogeny of *Haplopappus* section *Isopappus* (Compositae). Canad. J. Genet. Cytol. 8: 14–36.

Smith, E. B. 1976. A biosystematic survey of *Coreopsis* in eastern United States and Canada. Sida 6: 123–215.

Smith, E. B. 1981. New combinations in *Croptilon* (Compositae–Astereae). Sida 9: 59–63.

Smith, E. B. 1984. Biosystematic study and typification of the Californian *Coreopsis* (Compositae) sections *Tuckermannia, Pugiopappus*, and *Euleptosyne*. Sida 10: 276–289.

Smith, E. B. 1989. A biosystematic study and revision of the genus *Coreocarpus* (Compositae). Syst. Bot. 14: 448–472.

Smith, J. E. 1800–1804. Flora Britannica. 3 vols. London. [Vols. paged consecutively. Vols. 1 and 2, 1800; vol. 3, 1804.] (Fl. Brit.)

Smith, L., ed. 2001. Proceedings of the First International Knapweed Symposium of the Twenty-first Century, March 15–16, 2001, Coeur d'Alene, Idaho. Albany, Calif.

Smithsonian Contr. Bot. = Smithsonian Contributions to Botany.

Smithsonian Contr. Knowl. = Smithsonian Contributions to Knowledge.

Solbrig, O. T. 1960. The status of the genera *Amphiachyris, Amphipappus, Greenella, Gutierrezia, Gymnosperma*, and *Xanthocephalum* (Compositae). Rhodora 62: 43–54.

Solbrig, O. T. 1960b. Cytotaxonomic and evolutionary studies in the North American species of *Gutierrezia*. Contr. Gray Herb. 188: 1–63.

Solbrig, O. T. 1961. Note on *Gymnosperma glutinosum*. Leafl. W. Bot. 9: 147–150.

Solbrig, O. T. 1961b. Synopsis of the genus *Xanthocephalum* (Compositae). Rhodora 63: 151–164.

Solbrig, O. T. 1964. Infraspecific variation in the *Gutierrezia sarothrae* complex (Compositae–Astereae). Contr. Gray Herb. 193: 67–115.

Solbrig, O. T. 1965. The California species of *Gutierrezia*. Madroño 18: 75–84.

Solbrig, O. T. 1970. The phylogeny of *Gutierrezia*: An eclectic approach. Brittonia 22: 217–229.

Solbrig, O. T. 1971. The population biology of dandelions. Amer. Sci. 59: 686–694.

Solbrig, O. T. et al. 1979. Topics in Plant Population Biology. New York.

Soltis, D. E. et al. 2004. Recent and recurrent polyploidy in *Tragopogon* (Asteraceae): Cytogenetic, genomic, and genetic comparisons. Biol. J. Linn. Soc. 82: 485–501.

Soltis, D. E., P. S. Soltis, and J. J. Doyle, eds. 1992. Molecular Systematics of Plants. New York.

Soltis, P. S., G. M. Plunkett, S. J. Novak, and D. E. Soltis. 1995. Genetic variation in *Tragopogon* species: Additional origins of the allotetraploids *T. mirus* and *T. miscellus* (Compositae). Amer. J. Bot. 82: 1329–1341.

Sorrie, B. A. and P. Somers. 1999. The Vascular Plants of Massachusetts: A County Checklist. Westborough.

Soule, J. A. 1993. Systematics of *Tagetes* (Asteraceae–Tageteae) (Mexico, Argentina). Ph.D. thesis. University of Texas.

SouthW. Naturalist = Southwestern Naturalist.

Sp. Pl.—See: C. Linnaeus 1753; C. L. Willdenow et al. 1797–1830

Sp. Pl. ed. 2—See: C. Linnaeus 1762–1763

Spach, E. 1834–1848. Histoire Naturelle des Végétaux. Phanérogames.... 14 vols., atlas. Paris. (Hist. Nat. Vég.)

Spahr, R., L. Armstrong, N. D. Atwood, and M. Rath. 1991. Threatened, Endangered, and Sensitive Species of the Intermountain Region. Ogden.

Special Publ. Mus. Texas Tech Univ. = Special Publications, Museum, Texas Tech University.

Spence, W. L. 1963. A Biosystematic Study of the Genus *Lessingia* Cham. (Compositae). Ph.D. thesis. University of California, Berkeley.

Spongberg, S. A. 1971. A Systematic and Evolutionary Study of North American Arctic and Alpine Monocephalous Species of *Erigeron* (Compositae). Ph.D. dissertation. University of North Carolina.

Spongberg, S. A. 1973. A new Alaskan species of *Erigeron* L. (Compositae). Rhodora 75: 116–119.

Spooner, D. M. 1990. Systematics of *Simsia* (Compositae–Heliantheae). Syst. Bot. Monogr. 30: 1–90.

Sprengel, K. 1800–1801. Der botanische Garten der Universität zu Halle.... 1 vol. + Nachtr. Halle. (Bot. Gart. Halle)

Sprengel, K. [1818.] Novi Proventus Hortorum Academicorum Halensis et Berolinensis. Halle. (Novi Provent.)

Sprengel, K. 1820–1822. Neue Entdeckungen im ganzen Umfang der Pflanzenkunde. 3 vols. Leipzig. (Neue Entd.)

Sprengel, K. [1824–]1825–1828. Caroli Linnaei...Systema Vegetabilium. Editio Decima Sexta.... 5 vols. Göttingen. [Vol. 4 in 2 parts paged independently; vol. 5 by A. Sprengel.] (Syst. Veg.)

Spring, O., B. Zitterell-Haid, M. W. Bierner, and T. J. Mabry. 1994. Chemistry of glandular trichomes in *Hymenoxys* and related genera. Biochem. Syst. & Ecol. 22: 171–195.

Stace, C. A., ed. 1975. Hybridization and the Flora of the British Isles. London and New York.

Stace, C. A. 1991. New Flora of the British Isles. Cambridge and New York.

Stafleu, F. A. and R. S. Cowan. 1976–1988. Taxonomic Literature: A Selective Guide to Botanical Publications and Collections with Dates, Commentaries and Types, ed. 2. 7 vols. Utrecht, Antwerp, The Hague, and Boston.

Stafleu, F. A. and E. A. Mennega. 1992+. Taxonomic Literature. A Selective Guide to Botanical Publications and Collections with Dates, Commentaries and Types. Supplement. 6+ vols. Königstein.

Stebbins, G. L. 1932. Cytology of *Antennaria*. I. Normal species. Bot. Gaz. 94: 134–151.

Stebbins, G. L. 1932b. Cytology of *Antennaria*. II. Parthenogenetic species. Bot. Gaz. 94: 322–345.

Stebbins, G. L. 1935. A new species of *Antennaria* from the Appalachian region. Rhodora 37: 229–237.

Stebbins, G. L. 1936. A note on species differentiation in *Antennaria*. Rhodora 38: 367–369.

Stebbins, G. L. 1953. A new classification of the tribe Cichorieae, family Compositae. Madroño 12: 33–64.

Stebbins, G. L. 1993b. *Glyptopleura*. In: J. C. Hickman, ed. 1993. The Jepson Manual. Higher Plants of California. Berkeley, Los Angeles, and London. P. 270.

Stebbins, G. L., J. A. Jenkins, and M. S. Walters. 1953. Chromosomes and phylogeny in the Compositae, tribe Cichorieae. Univ. Calif. Publ. Bot. 26: 401–430.

Steyermark, J. A. 1934b. Studies in *Grindelia*. II. A monograph of the North American species of the genus *Grindelia*. Ann. Missouri Bot. Gard. 21: 433–608.

Steyermark, J. A. 1937. Studies in *Grindelia*. III. Ann. Missouri Bot. Gard. 24: 225–262.

Steyermark, J. A. 1963. Flora of Missouri. Ames.

Stockwell, P. 1940. A revision of the genus *Chaenactis*. Contr. Dudley Herb. 3: 89–167.

Storbeck, T. A. 1984. A Taxonomic Study of *Helianthus* Series *Microcephali* and Allied Species. Ph.D. dissertation. Indiana University.

Strother, J. L. 1969. Systematics of *Dyssodia* Cavanilles (Compositae: Tageteae). Univ. Calif. Publ. Bot. 48: 1–88.

Strother, J. L. 1974. Taxonomy of *Tetradymia* (Compositae: Senecioneae). Brittonia 26: 177–202.

Strother, J. L. 1977. Taxonomy of *Chrysactinia*, *Harnackia*, and *Lescaillea* (Compositae: Tageteae). Madroño 24: 129–139.

Strother, J. L. 1978. Taxonomy and geography of *Nicolletia* (Compositae: Tageteae). Sida 7: 369–374.

Strother, J. L. 1979. Extradition of *Sanvitalia tenuis* to *Zinnia* (Comositae–Heliantheae). Madroño 26: 173–179.

Strother, J. L. 1986. Renovation of *Dyssodia* (Compositae: Tageteae). Sida 11: 371–378.

Strother, J. L. 1989. Chromosome numbers in *Thymophylla* (Compositae: Tageteae). Sida 13: 351–358.

Strother, J. L. 2000. Who named *Euthamia* (Compositae: Astereae) and when? Sida 19: 217–218.

Strother, J. L. and B. G. Baldwin. 2002. Hymenocleas are ambrosias (Compositae). Madroño 49: 143–144.

Strother, J. L. and L. E. Brown. 1988. Dysploidy in *Hymenoxys texana* (Compositae). Amer. J. Bot. 75: 1097–1098.

Strother, J. L. and W. J. Ferlatte. 1988. Review of *Erigeron eatonii* and allied taxa (Compositae: Astereae). Madroño 35: 77–91.

Strother, J. L. and G. E. Pilz. 1975. Taxonomy of *Psathyrotes* (Compositae: Senecioneae). Madroño 23: 24–40.

Stud. Calenduleae—See: N. T. Norlindh 1943

Stuessy, T. F. 1972. Revision of the genus *Melampodium* (Compositae: Heliantheae). Rhodora 74: 1–70, 161–219.

Stuessy, T. F. 1973. A systematic review of the subtribe Melampodiinae (Compositae, Heliantheae). Contr. Gray Herb. 203: 65–80.

Stuessy, T. F. 1976. A systematic review of the subtribe Lagasceinae (Compositae, Heliantheae). Amer. J. Bot. 63: 1289–1294.

Stuessy, T. F. 1977. Revision of *Chrysogonum* (Compositae, Heliantheae). Rhodora 79: 190–202.

Stuessy, T. F. 1977[1978]. Heliantheae—systematic review. In: V. H. Heywood et al., eds. 1977[1978]. The Biology and Chemistry of the Compositae. 2 vols. London, New York, and San Francisco. Vol. 2, pp. 621–671.

Stuessy, T. F. 1978. Revision of *Lagascea* (Compositae, Heliantheae). Fieldiana, Bot. 38: 75–133.

Stuessy, T. F., R. S. Irving, and W. L. Ellison. 1973. Hybridization and evolution in *Picradeniopsis* (Compositae). Brittonia 25: 40–56.

Stuessy, T. F., T. Sang, and M. L. DeVore. 1996. Phylogeny and biogeography of subfamily Barnadesioideae with implications for early evolution of the Compositae. In: D. J. N. Hind et al., eds. 1996. Proceedings of the International Compositae Conference, Kew, 1994. 2 vols. Kew. Vol. 1, pp. 463–490.

Suh, Y. and B. B. Simpson. 1990. Phylogenetic analysis of chloroplast DNA in North American *Gutierrezia* and related genera (Asteraceae: Astereae). Syst. Bot. 15: 660–670.

Sullivan, V. I. 1972. Investigations of the Breeding Systems, Formation of Auto- and Alloploids and the Reticulate Pattern of Hybridization of North American *Eupatorium* (Compositae). Ph.D. dissertation. Florida State University.

Summa Veg. Scand.—See: E. M. Fries 1845–1849

Sundberg, S. D. 1986. The Systematics of *Aster* Subgenus *Oxytripolium* (Compositae) and Historically Allied Species. Ph.D. dissertation. University of Texas.

Sundberg, S. D. 1991. Infraspecific classification of *Chloracantha spinosa* (Benth.) Nesom (Asteraceae) Astereae. Phytologia 70: 382–391.

Sundberg, S. D. 2004. New combinations in North American *Symphyotrichum* subgenus *Astropolium* (Asteraceae: Astereae). Sida 21: 903–910.

Suppl. Pl.—See: C. Linnaeus f. 1781[1782]

Susanna, A., N. Garcia-Jacas, D. E. Soltis, and P. S. Soltis. 1995. Phylogenetic relationships in tribe Cardueae (Asteraceae) based on ITS sequences. Amer. J. Bot. 82: 1056–1068.

Svensk Bot. Tidskr. = Svensk Botanisk Tidskrift Utgifven af Svenska Botaniska Föreningen.

Swartz, O. P. 1788. Nova Genera & Species Plantarum seu Prodromus.... Stockholm, Uppsala, and Åbo. (Prodr.)

Sweet, R. 1823–1837. The British Flower Garden.... 7 vols. London. [Vols. 4–7 also issued as vols. 1–4 of series 2. (Brit. Fl. Gard.)

Sweet, R. 1826. Hortus Britannicus.... 2 parts. London. [Parts paged consecutively.] (Hort. Brit.)

Sweet, R. 1839. Hortus Britannicus..., ed. 3, edited by G. Don. London. (Hort. Brit. ed. 3)

Swenson, U. 1995. Systematics of the Blennospermatinae (Asteraceae, Senecioneae). Ph.D. dissertation. Uppsala University.

Swenson, U. and U. Manns. 2003. Phylogeny of *Pericallis* (Asteraceae): A total evidence approach reappraising the double origin of woodiness. Taxon 52: 533–546.

Symb. Antill.—See: I. Urban 1898–1928

Symb. Bot.—See: M. Vahl 1790–1794

Symb. Fl. Argent.—See: A. H. R. Grisebach 1879

Symons, J. 1798. Synopsis Plantarum Insulis Britannicis Indigenarum.... London. (Syn. Pl. Ins. Brit.)

Syn. Aster. Herb.—See: C. G. D. Nees 1818

Syn. Fl. Colorado—See: T. C. Porter and J. M. Coulter 1874

Syn. Fl. Germ. Helv. ed. 2—See: W. D. J. Koch 1843–1845

Syn. Fl. N. Amer.—See: A. Gray et al. 1878–1897

Syn. Fl. N. Amer. ed. 2—See: A. Gray et al. 1886

Syn. Gen. Compos.—See: C. F. Lessing 1832

Syn. Pl.—See: D. N. F. Dietrich 1839–1852; C. H. Persoon 1805–1807

Syn. Pl. Fl. Gall.—See: J. Lamarck and A. P. de Candolle 1806

Syn. Pl. Ins. Brit.—See: J. Symons 1798

Synth. N. Amer. Fl.—See: J. T. Kartesz and C. A. Meacham 1999

Syst. Biol. = Systematic Biology.

Syst. Bot. = Systematic Botany; Quarterly Journal of the American Society of Plant Taxonomists.

Syst. Bot. Monogr. = Systematic Botany Monographs; Monographic Series of the American Society of Plant Taxonomists.

Syst. Nat.—See: J. F. Gmelin 1791[–1792]

Syst. Nat. ed. 10—See: C. Linnaeus 1758[–1759]

Syst. Nat. ed. 12—See: C. Linnaeus 1766[–1768]

Syst. Veg.—See: K. Sprengel [1824–]1825–1828

Syst. Veg. ed. 14—See: J. A. Murray 1784

Syst. Veg. Fl. Peruv. Chil.—See: H. Ruiz López and J. A. Pavón 1798

Syst. Verz.—See: J. J. Bernhardi 1800

Tabl. École Bot.—See: R. L. Desfontaines 1804

Tabl. Encycl.—See: J. Lamarck and J. Poiret 1791–1823

Tanaceteen—See: C. H. Schultz-Bipontinus 1844

Tanowitz, B. D. 1977. An intersectional hybrid in *Hemizonia* (Compositae: Madiinae). Madroño 24: 55–61.

Tanowitz, B. D. 1982. Taxonomy of *Hemizonia* sect. *Madiomeris* (Asteraceae: Madiinae). Syst. Bot. 7: 314–339.

Tanowitz, B. D. and J. W. Adams. 1986. Natural hybridization between *Layia glandulosa* and *L. paniculata* (Asteraceae: Madiinae). Madroño 33: 244–252.

Taxon = Taxon; Journal of the International Association for Plant Taxonomy.

Taylor, C. E. 1975. *Euthamia gymnospermoides*. Ph.D. thesis. University of Oklahoma.

Taylor, C. E. and R. J. Taylor. 1984. *Solidago* (Asteraceae) in Oklahoma and Texas. Sida 10: 223–251.

Taylor, R. J. 1987. Populational variation and biosystematic interpretations in weedy dandelions. Bull. Torrey Bot. Club 114: 109–120.

Tenore, M. 1811–1836[–1838]. Flora Napolitana.... 5 vols. Naples. (Fl. Napol.)

Theor. Appl. Genet. = Theoretical and Applied Genetics; International Journal of Breeding Research and Cell Genetics.

Thompson, W. C. 1983. A Biosystematic Study of *Lagophylla* (Compositae: Heliantheae) and Related Taxa. Ph.D. dissertation. University of California, Davis.

Thunberg, C. P. 1794–1800. Prodromus Plantarum Capensium, Quas in Promontorio Bonae Spei Africes, Annis 1772–1775, Collegit.... 2 parts. Uppsala. [Parts paged consecutively.] (Prodr. Pl. Cap.)

Tibor, D. P., ed. 2001. Inventory of Rare and Endangered Plants of California..., ed. 6. Sacramento.

Tidestrom, I. and T. Kittell. 1941. A Flora of Arizona and New Mexico.... Washington. (Fl. Ariz. New Mex.)

Tolmatchew, A. I., ed. 1960–1987. Flora Arctica URSS. 10 vols. Moscow and Leningrad. (Fl. Arct. URSS)

Toman, J. 1972. A taxonomic survey of the genera *Petasites* and *Endocellion*. Folia Geobot. Phytotax. 7: 381–406.

Tomb, A. S. 1970. Novelties in *Lygodesmia* and *Stephanomeria* (Compositae–Cichorieae). Sida 3: 530–532.

Tomb, A. S. 1972. Taxonomy of *Chaetadelpha* (Compositae: Cichorieae). Madroño 21: 459–462.

Tomb, A. S. 1972b. Re-establishment of the genus *Prenanthella* Rydb. (Compositae: Cichorieae). Brittonia 24: 223–228.

Tomb, A. S. 1974. Chromosome numbers and generic relationships in subtribe Stephanomeriinae (Compositae: Cichorieae). Brittonia 26: 203–216.

Tomb, A. S. 1980. Taxonomy of *Lygodesmia* (Asteraceae). Syst. Bot. Monogr. 1: 1–51.

Tomb, A. S. et al. 1978. Chromosome numbers in the Compositae: 14. Lactuceae. Amer. J. Bot. 65: 717–721.

Tomb, A. S., D. A. Larson, and J. J. Skvarla. 1974. Pollen morphology and detailed structure of family Compositae, tribe Cichorieae. I. Subtribe Stephanomeriinae. Amer. J. Bot. 61: 486–498.

Torrell, M., N. Garcia-Jacas, A. Susanna, and J. Valles. 1999. Phylogeny in *Artemisia* (Asteraceae, Anthemideae) inferred from nuclear ribosomal DNA (ITS) sequences. Taxon 48: 721–736.

Torres, A. M. 1963. Taxonomy of *Zinnia*. Brittonia 15: 1–25.

Torres, A. M. 1964. Revision of *Sanvitalia* (Compositae–Heliantheae). Brittonia 16: 417–433.

Torrey, J. and A. Gray. 1838–1843. A Flora of North America.... 2 vols. in 7 parts. New York, London, and Paris. (Fl. N. Amer.)

Torreya = Torreya; a Monthly Journal of Botanical Notes and News.

Trans. & Proc. Bot. Soc. Edinburgh = Transactions and Proceedings of the Botanical Society Edinburgh.

Trans. Acad. Sci. St. Louis = Transactions of the Academy of Science of St. Louis.

Trans. Amer. Philos. Soc. = Transactions of the American Philosophical Society Held at Philadelphia for Promoting Useful Knowledge.

Trans. Kansas Acad. Sci. = Transactions of the Kansas Academy of Science.

Trans. Linn. Soc. London = Transactions of the Linnean Society of London.

Trans. New York Acad. Sci. = Transactions of the New York Academy of Sciences.

Trans. Wisconsin Acad. Sci. = Transactions of the Wisconsin Academy of Sciences, Arts and Letters.

Trav. Brazil—See: H. Koster 1816

Travels Carolina—See: W. Bartram 1791

Trudy Bot. Inst. Akad. Nauk S.S.S.R, Ser. 1, Fl. Sist. Vyssh. Rast. = Trudy Botanicheskogo Instituta Akademii Nauk S S S R. Ser. 1, Flora i Sistematika Vysshikh Rastenii.

Trudy Bot. Muz. = Trudy Botanicheskogo Muzeya.

Trudy Imp. S.-Peterburgsk. Bot. Sada = Trudy Imperatorskago S.-Peterburgskago Botanicheskago Sada.

Turner, B. L. 1956. A cytotaxonomic study of the genus *Hymenopappus* (Compositae). Rhodora 58: 163–186, 208–243, 259–269, 295–308.

Turner, B. L. 1962. Taxonomy of *Hymenothrix* (Helenieae, Compositae). Brittonia 14: 101–120.

Turner, B. L. 1963. Taxonomy of *Florestina* (Helenieae, Compositae). Brittonia 15: 27–46.

Turner, B. L. 1971. Taxonomy of *Sartwellia* (Compositae Helenieae). Sida 4: 265–273.

Turner, B. L. 1975. Taxonomy of *Haploësthes* (Asteraceae–Senecioneae). Wrightia 5: 108–115.

Turner, B. L. 1978. Taxonomic study of the scapiform species of *Acourtia* (Asteraceae–Mexico). Phytologia 38: 456–468.

Turner, B. L. 1979. *Gaillardia aestivalis* var. *winkleri* (Asteraceae), a white-flowered tetraploid taxon endemic to southeastern Texas. SouthW. Naturalist 24: 621–624.

Turner, B. L. 1984. Taxonomy of the genus *Aphanostephus* (Asteraceae–Astereae). Phytologia 56: 81–101.

Turner, B. L. 1987. Taxonomy of *Carphochaete* (Asteraceae–Eupatorieae). Phytologia 64: 145–162.

Turner, B. L. 1987b. Taxonomic study of *Machaeranthera*, sections *Machaeranthera* and *Hesperastrum* (Asteraceae). Phytologia 62: 207–266.

Turner, B. L. 1988. A new species of, and observations on, the genus *Smallanthus* (Asteraceae–Heliantheae). Phytologia 64: 405–409.

Turner, B. L. 1988b. Taxonomy of *Carminatia* (Asteraceae, Eupatorieae). Pl. Syst. Evol. 160: 169–179.

Turner, B. L. 1988c. Submergence of the genera *Asanthus* and *Dyscritogyne* within *Steviopsis* (Asteraceae: Eupatorieae), including new combinations. Phytologia 64: 259–262.

Turner, B. L. 1988d. A new variety of *Berlandiera lyrata* from northwestern Mexico. Phytologia 64: 205–209.

Turner, B. L. 1993c. New taxa, new combinations, and nomenclatural comments on the genus *Acourtia* (Asteraceae: Mutisieae). Phytologia 74: 385–412.

Turner, B. L. 1994c. Taxonomic status of *Brickelliastrum villarrealii* R. M. King & H. Robins. (Asteraceae: Eupatorieae). Phytologia 76: 389–390.

Turner, B. L. 1996. Taxonomy and nomenclature of *Schkuhria pinnata* (Asteraceae, Helenieae). Phytologia 79: 364–368.

Turner, B. L., ed. 1996+. The Comps of Mexico: A Systematic Account of the Family Asteraceae. 2+ vols. Huntsville, Tex. [Phytologia Mem. 10, 11.] (Comps Mexico)

Turner, B. L. 2000. *Plateilema* (Asteraceae: Helenieae) a new generic report for the United States. Sida 19: 185–187.

Turner, B. L. and D. Dawson. 1980. Taxonomy of *Tetragonotheca* (Asteraceae–Heliantheae). Sida 8: 296–303.

Turner, B. L. and R. L. Hartman. 1976. Infraspecific categories of *Machaeranthera pinnatifida* (Compositae). Wrightia 5: 308–315.

Turner, B. L. and D. B. Horne. 1964. Taxonomy of *Machaeranthera* sect. *Psilactis* (Compositae–Astereae). Brittonia 16: 316–331.

Turner, B. L. and M. C. Johnston. 1957. Chromosome numbers and geographic distribution of *Lindheimera, Engelmannia,* and *Berlandiera* (Compositae–Heliantheae–Melampodinae). SouthW. Naturalist 1: 125–132.

Turner, B. L. and K. J. Kim. 1990. An overview of the genus *Pyrrhopappus* (Asteraceae: Lactuceae) with emphasis on chloroplast DNA restriction site data. Amer. J. Bot. 77: 845–850.

Turner, B. L. and M. I. Morris. 1976. Systematics of *Palafoxia* (Asteraceae: Helenieae). Rhodora 78: 567–628.

Turner, B. L., H. Nichols, G. Denny, and O. Doron. 2003. Atlas of the Vascular Plants of Texas. 2 vols. Fort Worth. [Sida Bot. Misc. 24.] (Atlas Vasc. Pl. Texas)

Turner, B. L. and A. M. Powell. 1977[1978]. Helenieae—systematic review. In: V. H. Heywood et al., eds. 1977[1978]. The Biology and Chemistry of the Compositae. 2 vols. London, New York, and San Francisco. Vol. 2, pp. 699–737.

Turner, B. L., M. W. Turner, and J. C. Crutchfield. 1988. Populational analyses and new combinations in *Psilostrophe tagetina* and *P. gnaphalodes* (Asteraceae, Heliantheae). Phytologia 65: 231–240.

Turner, B. L. and M. Whalen. 1975. Taxonomic study of *Gaillardia pulchella* (Asteraceae–Heliantheae). Wrightia 5: 189–192.

Turner, B. L. and D. Zippin. 1992. Taxonomic study of *Venegasia* (Asteraceae: Helenieae). Sida 15: 223–229.

Turner, M. W. 1993. Systematic study of the genus *Baileya* (Asteraceae: Helenieae). Sida 15: 491–508.

Tutin, T. G. et al., eds. 1964–1980. Flora Europaea. 5 vols. Cambridge.

Tyrl, R. J. 1975. Origin and distribution of polyploid *Achillea* (Compositae) in western North America. Brittonia 27: 187–196.

Tzvelev, N. N. 1987. *Arctanthemum*. In: A. I. Tolmatchew, ed. 1960–1987. Flora Arctica URSS. 10 vols. Moscow and Leningrad. Vol. 10, pp. 114–117.

Tzvelev, N. N. 2002b. *Tripleurospermum*. In: An. A. Fedorov, ed. 1999+. Flora of Russia: The European Part and Bordering Regions...Translated from Russian. 7+ vols. Rotterdam and Brookfield, Vt. Vol. 7, pp. 179–184.

U.S. Expl. Exped.—See: C. Wilkes et al. 1854–1876

Univ. Calif. Publ. Bot. = University of California Publications in Botany.

Univ. Kansas Sci. Bull. = University of Kansas Science Bulletin.

Univ. Wyoming Publ. Sci., Bot. = University of Wyoming Publications in Science. Botany.

University of Chicago Press. 1993. The Chicago Manual of Style, ed. 14. Chicago.

Uppsala Univ. Årsskr. = Uppsala Universitets Årsskrift.

Urban, I., ed. 1898–1928. Symbolae Antillanae seu Fundamenta Florae Indiae Occidentalis.... 9 vols. Berlin etc. (Symb. Antill.)

Urbanska, K. M. 1983. *Antennaria carpatica* (Wahlb.) Bl. et Fing. s.l. in North America. I. Chromosome numbers, geographical distribution and ecology. Ber. Geobot. Inst. E. T. H. Stiftung Rübel 50: 33–66.

Urbatsch, L. E. 1976. Systematics of the *Ericameria cuneata* complex (Compositae, Astereae). Madroño 23: 338–345.

Urbatsch, L. E. 1978. The Chihuahuan Desert species of *Ericameria* (Asteraceae). Sida 7: 298–303.

Urbatsch, L. E., B. G. Baldwin, and M. J. Donoghue. 2000. Phylogeny of the coneflowers and relatives (Heliantheae: Asteraceae) based on nuclear rDNA internal transcribed spacer (ITS) sequences and chloroplast DNA restriction site data. Syst. Bot. 25: 539–565.

Urbatsch, L. E. and R. K. Jansen. 1995. Phylogenetic affinities among and within the coneflower genera (Asteraceae: Heliantheae), a chloroplast DNA analysis. Syst. Bot. 20: 28–39.

Urbatsch, L. E., R. P. Roberts, and V. Karaman. 2003. Phylogenetic evaluation of *Xylothamia, Gundlachia,* and related genera (Asteraceae, Astereae) based on ETS and ITS nrDNA sequence data. Amer. J. Bot. 90: 634–649.

Urbatsch, L. E., R. P. Roberts, and K. M. Neubig. 2005. *Cuniculotinus* and *Lorandersonia,* two new genera of Asteraceae: Astereae and new combinations in *Chrysothamnus.* Sida 21: 1615–1632.

Utah Fl. ed. 3—See: S. L. Welsh et al. 2003

Uttal, L. J. 1962. Synthesis of *Aster herveyi.* Rhodora 64: 113–117.

Uttal, L. J. 1982. *Senecio cannabinaefolius* Hook. et Arn. (Asteraceae) in Alabama. Castanea 47: 116–117.

Uttal, L. J. and D. M. Porter. 1988. The correct name for Elliott's goldenrod. Rhodora 90: 157–168.

Vaarama, A. 1953. Cytotaxonomic studies on northern *Tripleurospermum* forms. In: H. Osvald and E. Åberg, eds. 1953. Proceedings of the Seventh International Botanical Congress, Stockholm July 12–20 1950. Stockholm and Waltham, Mass. Pp. 279–280.

Vahl, M. 1790–1794. Symbolae Botanicae.... 3 vols. Copenhagen. (Symb. Bot.)

Valles, J. and E. D. McArthur. 2001. *Artemisia* systematics and phylogeny: Cytogenetic and molecular insights. In: E. D. McArthur and D. J. Fairbanks, comps. 2001. Shrubland Ecosystem Genetics and Biodiversity: Proceedings: Provo, UT, June 13–15, 2000. Ogden. Pp. 67–74.

Van Horn, G. S. 1973. The taxonomic status of *Pentachaeta* and *Chaetopappa* with a revision of *Pentachaeta.* Univ. Calif. Publ. Bot. 65: 1–41.

Van Vliet, D. J. 1951. Phenotypic variations of *Erigeron strigosus* Muhl. (Compositae) in eastern Texas. Field & Lab. 19: 161–163.

Vasc. Fl. S.E. U.S.—See: A. E. Radford et al. 1980+

Vasc. Pl. Pacif. N.W.—See: C. L. Hitchcock et al. 1955–1969

Vasc. Pl. San Luis Obispo Co.—See: R. F. Hoover 1970

Vasc. Pl. Wyoming ed. 2—See: R. D. Dorn 1992

Veg. Syst.—See: J. Hill 1759–1775

Vega Exp. Vetensk. Iakttag.—See: A. E. Nordenskiöld 1882–1887

Venkatesh, C. S. 1958. A cyto-genetic and evolutionary study of *Hemizonia* section *Centromadia* (Compositae). Evolution 39: 1236–1241.

Ventenat, É. P. [1800–1803.] Description des Plantes Nouvelles et Peu Connues Cultivés dans le Jardin de J. M. Cels.... 10 parts. Paris. [Plates numbered consecutively.] (Descr. Pl. Nouv.)

Ventenat, É. P. 1803–1804[–1805]. Jardin de la Malmaison.... 2 vols. Paris. [Plates numbered consecutively.] (Jard. Malmaison)

Veröff. Geobot. Inst. E. T. H. Stiftung Rübel Zürich = Veröffentlichungen des Geobotanischen Institutes der Eidg. Techn. Hochschule, Stiftung Rübel, in Zürich.

Villars, D. 1779. Prospectus de l'Histoire des Plantes de Dauphiné.... Grenoble. (Prosp. Hist. Pl. Dauphiné)

Vilm. Blumengärtn. ed. 3—See: A. Voss [1894–]1896

Vogt, R. 1991. Die Gattung *Leucanthemum* (Compositae–Anthemideae) auf der Iberischen Halbinsel. Ruizia 10: 1–261.

Voss, A. [1894–]1896. Vilmorin's Blumengärtnerei..., ed. 3. 2 vols. Berlin. (Vilm. Blumengärtn. ed. 3)

Voss, E. G. 1972–1996. Michigan Flora.... 3 vols. Bloomfield Hills and Ann Arbor.

Voy. Barbarie—See: J. Poiret 1789

W. Amer. Sci. = West American Scientist.

W. J. Med. Phys. Sci. = The Western Journal of the Medical and Physical Sciences.

Wagenitz, G. 1955. Pollenmorphologie und Systematik in der Gattung *Centaurea* L. s. l. Flora 142: 213–279.

Wagenitz, G. 1965. Zur Systematik und Nomenklatur einiger Arten von *Filago* L. emend. Gaertn. Subgen. *Filago* ("*Filago germanica*"-Gruppe). Willdenowia 4: 37–59.

Wagenitz, G. 1969. Abgrenzung und Gliederung der Gattung *Filago* L. s.l. (Compositae–Inuleae). Willdenowia 5: 395–444.

Wagenitz, G. 1976. Two species of the "*Filago germanica*" group (Compositae–Inuleae) in the United States. Sida 6: 221–223.

Wagenitz, G. 1976b. Systematics and phylogeny of the Compositae (Asteraceae). Pl. Syst. Evol. 125: 29–46.

Wagenitz, G. 1987. *Centaurea.* In: G. Hegi et al. 1936–1987. Illustrierte Flora von Mitteleuropa, ed. 2. 6 vols. in 14. Munich, Berlin, and Hamburg. Vol. 6(4), pp. 1405–1413.

Wagenitz, G. and F. H. Hellwig. 1996. Evolution of characters and phylogeny of the Centaureae. In: D. J. N. Hind et al., eds. 1996. Proceedings of the International Compositae Conference, Kew, 1994. 2 vols. Kew. Vol. 1, pp. 491–510.

Wagenknecht, B. L. 1960. Revision of *Heterotheca* sect. *Heterotheca* (Compositae). Rhodora 62: 61–76, 97–107.

Wagner, W. L., R. Fletcher, and R. K. Shannon. 1999. Another look at *Hymenoxys* subgenus *Plummera* (Asteraceae: Heliantheae: Gaillardiinae) from Arizona and New Mexico. Brittonia 51: 79–86.

Wain, R. P. 1982. Genetic differentiation in the Florida subspecies of *Helianthus debilis* (Asteraceae). Amer. J. Bot. 69: 1573–1578.

Wain, R. P. 1983. Genetic differentiation during speciation in the *Helianthus debilis* complex. Evolution 37: 1119–1127.

Wallace, R. S. and R. K. Jansen. 1990. Systematic implications of chloroplast DNA variation in the genus *Microseris* (Asteraceae: Lactuceae). Syst. Bot. 15: 606–616.

Wallace, R. S. and R. K. Jansen. 1995. DNA evidence for multiple origins of intergeneric allopolyploids in annual *Microseris* (Asteraceae). Pl. Syst. Evol. 198: 253–265.

Wallroth, C. F. W. 1822. Schedulae Criticae de Plantis Florae Halensis Selectis.... Halle. (Sched. Crit.)

Walter, T. 1788. Flora Caroliniana, Secundum Systema Vegetabilium Perillustris Linnaei Digesta.... London. (Fl. Carol.)

War Department [U.S.]. 1855–1860. Reports of Explorations and Surveys, to Ascertain the Most Practicable and Economical Route for a Railroad from the Mississippi River to the Pacific Ocean. Made under the Direction of the Secretary of War, in 1853[–1856].... 12 vols. in 13. Washington. (Pacif. Railr. Rep.)

Ward, D. E. and R. Spellenberg. 1986. Chromosome counts of angiosperms of western North America. Phytologia 61: 119–125.

Ward, G. H. 1953. *Artemisia* section *Seriphidium* in North America, a cytotaxonomical study. Contr. Dudley Herb. 4: 155–206.

Warners, D. P. and D. C. Laughlin. 1999. Evidence for a species-level distinction of two co-occurring asters: *Aster puniceus* L. and *Aster firmus* Nees. Michigan Bot. 38: 19–31.

Watson, A. K. 1980. The biology of Canadian weeds. 43. *Acroptilon (Centaurea) repens* (L.) DC. Canad. J. Pl. Sci. 60: 993–1004.

Watson, L. E. et al. 2002. Molecular phylogeny of subtribe Artemisiinae (Asteraceae), including *Artemisia* and its allied and segregate genera. B. M. C. Evol. Biol. 2: 17 (12 pp.).

Watson, L. E., W. J. Elisens, and J. R. Estes. 1991. Cytogenetic and electrophoretic evidence for the allotetraploid origin of *Marshallia mohrii* (Asteraceae). Amer. J. Bot. 78: 408–416.

Watson, L. E., W. J. Elisens, and J. R. Estes. 1994. Genetic differentiation in populations of *Marshallia graminifolia* s.lat. (Asteraceae). Biochem. Syst. & Ecol. 22: 577–582.

Watson, L. E. and J. R. Estes. 1990. Biosystematic and phenetic analysis of *Marshallia* (Asteraceae). Syst. Bot. 15: 403–414.

Watson, L. E., T. M. Evans, and T. Boluarte. 2000. Molecular phylogeny of tribe Anthemideae (Asteraceae), based on chloroplast gene *ndh*F. Molec. Phylogen. Evol. 15: 59–69.

Watson, S. 1871. United States Geological Expolration [sic] of the Fortieth Parallel. Clarence King, Geologist-in-charge. [Vol. 5] Botany. By Sereno Watson.... Washington. [Botanical portion of larger work by C. King.] [Botany (Fortieth Parallel)]

Watson, S. [1871]b. United States Geological Exploration of the 40th Parallel. Clarence King, U.S. Geologist, in Charge. List of Plants Collected in Nevada and Utah 1867–'69; Numbered As Distributed. Sereno Watson, Collector. [Washington.] (List Pl. Nevada Utah)

Watson, T. J. 1977. The taxonomy of *Xylorhiza* (Asteraceae–Astereae). Brittonia 29: 199–216.

Watsonia = Watsonia; Journal of the Botanical Society of the British Isles.

Webb, C. J. 1986. Variation in achene morphology and its implications for taxonomy in *Soliva* subgenus *Soliva* (Anthemideae, Asteraceae). New Zealand J. Bot. 24: 665–669.

Webb, P. B. and S. Berthelot. [1835–]1836–1850. Histoire Naturelle des Îles Canaries.... 3 vols. in 9. Paris. [Tome troisième, Botanique: Première partie, 1 vol; deuxième partie, 4 vols.] (Hist. Nat. Îles Canaries)

Weber, W. A. 1946. A taxonomic and cytological study of the genus *Wyethia*, family Compositae, with notes on the related genus *Balsamorhiza*. Amer. Midl. Naturalist 35: 400–452.

Weber, W. A. 1948. The genus *Helianthella* in Oregon. Madroño 9: 186–189.

Weber, W. A. 1952. The genus *Helianthella* (Compositae). Amer. Midl. Naturalist 48: 1–35.

Weber, W. A. 1984b. New names in *Artemisia*. Phytologia 55: 7–9.

Weber, W. A. 1987. Colorado Flora: Western Slope. Boulder.

Weber, W. A. 1990. Colorado Flora: Eastern Slope. Niwot, Colo.

Weber, W. A. and R. C. Wittmann. 1992. Catalog of the Colorado Flora: A Biodiversity Baseline. Niwot, Colo.

Weddell, H. A. 1855–1857[–1861]. Chloris Andina. Essai d'une Flore de la Région Alpine des Cordillères de l'Amerique du Sud.... 2 vols. in 16 parts. Paris. [Volumes paged independently, parts numbered consecutively.] (Chlor. Andina)

Weed Sci. = Weed Science; Journal of the Weed Science Society of America.

Weed Technol. = Weed Technology; a Journal of the Weed Science Society of America.

Weedon, R. R. 1973. Taxonomy and Distribution of the Genus *Bidens* (Compositae) in the North-central Plains States. Ph.D. dissertation. University of Kansas.

Wells, H. 1983. Hybridization and genetic recombination of *Cirsium californicum* and *C. occidentale* (Asteraceae: Carduceae [sic]). Madroño 30: 12–30.

Wells, J. R. 1965. A taxonomic study of *Polymnia* (Compositae). Brittonia 17: 144–159.

Welsh, S. L. 1974. Anderson's Flora of Alaska and Adjacent Parts of Canada. Provo.

Welsh, S. L. 1982. New taxa of thistles (*Cirsium;* Asteraceae) in Utah. Great Basin Naturalist 42: 199–202.

Welsh, S. L. 1983. Utah flora: Compositae (Asteraceae). Great Basin Naturalist 43: 179–357.

Welsh, S. L., N. D. Atwood, S. Goodrich, and L. C. Higgins, eds. 1987. A Utah flora. Great Basin Naturalist Mem. 9.

Welsh, S. L., N. D. Atwood, S. Goodrich, and L. C. Higgins, eds. 1993. A Utah Flora, ed. 2. Provo.

Welsh, S. L., N. D. Atwood, S. Goodrich, and L. C. Higgins, eds. 2003. A Utah Flora, ed. 3. Provo. (Utah Fl. ed. 3)

Welzen, P. C. van. 1997. Paraphyletic groups or what should a classification entail. Taxon 46: 99–103.

Wendland, J. C. 1798. Botanische Beobachtungen.... Hannover. (Bot. Beob.)

Wendland, J. C. 1798–1801. Hortus Herrenhusanus.... 4 parts. Hannover. [Parts 1 and 2 paged consecutively, parts 3 and 4 paged independently; plates numbered consecutively throughout.] (Hort. Herrenhus.)

Westman, W. E., F. D. Panetta, and T. D. Stanley. 1975. Ecological studies on reproduction and establishment of the woody weed, groundsel bush (Baccharis halimifolia L.: Asteraceae). Austral. J. Agric. Res 26: 855–870.

Whalen, M. 1977. Taxonomy of Bebbia (Compositae: Heliantheae). Madroño 24: 112–123.

Whitton, J., R. S. Wallace, and R. K. Jansen. 1995. Phylogenetic relationships and patterns of character change in the tribe Lactuceae (Asteraceae) based on chloroplast DNA restriction site variation. Canad. J. Bot. 73: 1058–1073.

Wiebe, E. 2000. The annotated check-list of the tribe Senecioneae Cass. (Asteraceae) in Siberia. Turczaninowia 3(4): 58–63.

Wied, M. P. zu. 1839–1841. Reise in das innere Nord-America in den Jahren 1832 bis 1834. 2 vols. Coblenz. (Reise Nord-Amer.)

Wiegand, K. M. and C. A. Weatherby. 1937. The nomenclature of the verticillate eupatoria. Rhodora 39: 297–306.

Wiggers, F. H. [1780.] Primatiae Florae Holsaticae. Kiel. (Prim. Fl. Holsat.)

Wight, R. 1834. Contributions to the Botany of India.... London. (Contr. Bot. India)

Wiklund, A. 1992. The genus Cynara L. (Asteraceae–Cardueae). Bot. J. Linn. Soc. 109: 75–123.

Wild, H. 1965. The African species of the genus Melanthera Rohr. Kirkia 5: 1–17.

Wilken, D. H. 1972. Seasonal dimorphism in Baccharis glutinosa (Compositae). Madroño 21: 113–119.

Wilken, D. H. 1975. A systematic study of Hulsea (Asteraceae). Brittonia 27: 228–244.

Wilkes, C. et al. 1854–1876. United States Exploring Expedition. During the years 1838, 1839, 1840, 1841, 1842. Under the Command of Charles Wilkes, U.S.N..... 18 vols. (1–17, 19). Philadelphia. [Vol. 15: Botany, Phanerogamia (A. Gray), 1854; Atlas, 1856. Vol. 16: Botany, Cryptogamia, Filices (W. D. Brackenridge), 1854; Atlas, 1855. Vol. 17: incl. Phanerogamia of Pacific North America (J. Torrey), 1874. Vol. 19 (2 parts): Geographical Distribution of Animals and Plants (C. Pickering), 1854. Vol. 18: Botany, Phanerogamia, part 2 (A. Gray) not published.] (U.S. Expl. Exped.)

Willdenow, C. L. 1794. Phytographia seu Descriptio Rariorum Minus Cognitarum Plantarum. 1 fasc. only. Erlangen. (Phytographia)

Willdenow, C. L. 1809–1813[–1814]. Enumeratio Plantarum Horti Regii Botanici Berolinensis.... 2 parts + suppl. Berlin. [Parts paged consecutively.] (Enum. Pl.)

Willdenow, C. L., C. F. Schwägrichen, and J. H. F. Link. 1797–1830. Caroli a Linné Species Plantarum.... Editio Quarta.... 6 vols. Berlin. [Vols. 1–5(1), 1797–1810, by Willdenow; vol. 5(2), 1830, by Schwägrichen; vol. 6, 1824–1825, by Link.] (Sp. Pl.)

Williams, D. F. et al. 1998. Recovery Plan for Upland Species of the San Joaquin Valley, California. Portland.

Williams, E. W. 1957. The genus Malacothrix (Compositae). Amer. Midl. Naturalist 58: 494–512.

Williams, M. J. 1977. Hecastocleis shockleyi A. Gray. Mentzelia 3: 18.

Winn, A. A. and L. F. Pitelka. 1981. Some effects of density on the reproductive patterns and patch dynamics of Aster acuminatus. Bull. Torrey Bot. Club 108: 438–445.

Winward, A. H. 1970. Taxonomic and Ecological Relationships of the Big Sagebrush Complex in Idaho. Ph.D. dissertation. University of Idaho.

Wodehouse, R. P. 1935. Pollen Grains: Their Structure, Identification, and Significance in Science and Medicine. New York.

Wofford, B. E. 1976. The taxonomic status of Ageratina luciae-brauniae (Fern.) King & H. Robins. Phytologia 33: 369–371.

Wolf, C. B. 1935. Observations on Baccharis pilularis DC. Occas. Pap. Rancho Santa Ana Bot. Gard. 1: 17–29.

Wolf, S. J. and K. E. Denford. 1984. Taxonomy of Arnica (Compositae) subgenus Austromontana. Rhodora 86: 239–309.

Wood, A. 1861. A Class-book of Botany.... New York. (Class-book Bot. ed. s.n.(b))

Wood, A. 1870. The American Botanist and Florist; Including Lessons in the Structure, Life, and Growth of Plants; Together with a Simple Analytical Flora, Descriptive of the Native and Cultivated Plants Growing in the Atlantic Division of the American Union.... New York and Chicago. (Amer. Bot. Fl.)

Wrightia = Wrightia; a Botanical Journal.

Wunderlin, R. P. 1998. Guide to the Vascular Plants of Florida. Gainesville.

Wunderlin, R. P., B. F. Hansen, and E. L. Bridges. 1996. Atlas of Florida Vascular Plants. Tampa. [CD-ROM.]

Wyoming Agric. Exp. Sta. Bull. = Wyoming Agricultural Experiment Station Bulletin.

Xiang, C. 1994. Molecular Systematics Study of Aster Sensu Lato and Related Genera (Asteraceae: Astereae) Based on Chloroplast DNA Restriction Site Analysis. Ph.D. dissertation. University of Waterloo.

Xiang, C. and J. C. Semple. 1996. Molecular systematic study of Aster sensu lato and related genera (Asteraceae: Astereae) based on chloroplast DNA restriction site analyses and mainly North American taxa. In: D. J. N. Hind et al., eds. 1996. Proceedings of the International Compositae Conference, Kew, 1994. 2 vols. Kew. Vol. 1, pp. 393–423.

Yates, W. F. and C. B. Heiser. 1979. Synopsis of Heliomeris. Proc. Indiana Acad. Sci. 88: 364–372.

Zanowiak, D. J. 1991. An Analysis of Systematic and Phyletic Relationships within Baccharidinae (Asteraceae: Astereae). Ph.D. dissertation. Texas A&M University.

Zardini, E. 1992. Madia sativa Mol. (Asteraceae–Heliantheae–Madiinae): An ethnobotanical and geographical disjunct. Econ. Bot. 46: 34–44.

Zavada, M. S. and S. E. de Villiers. 2000. Pollen of the Asteraceae from the Paleocene-Eocene of South Africa. Grana 39: 39–45.

Zhang, J. J. 1996. A Molecular Biosystematic Study on North American *Solidago* and Related Genera (Asteraceae: Astereae) Based on Chloroplast DNA RFLP Analysis (Phylogenetics). Ph.D. dissertation. University of Waterloo.

Zinn, J. G. 1757. Catalogus Plantarum Horti Academici et Agri Gottingensis.... [Göttingen.] (Cat. Pl. Hort. Gott.)

Zoë = Zoë; a Biological Journal.

Zuccagni, A. [1806.] Centuria I. Observationum Botanicarum.... [Zürich.] [Preprinted from J. J. Roemer, ed. 1809. Collectanea.... Zürich. Pp. 113–162.] (Cent. Observ. Bot.)

Index

Names in *italics* are synonyms, casually mentioned hybrids, or plants not established in the flora. Page numbers in **boldface** indicate the primary entry for a taxon. Page numbers in *italics* indicate an illustration. Roman type is used for all other entries, including author names, vernacular names, and accepted scientific names for plants treated as established members of the flora.